Lecture Notes in Computer Science   3358

Commenced Publication in 1973
Founding and Former Series Editors:
Gerhard Goos, Juris Hartmanis, and Jan van Leeuwen

Editorial Board

David Hutchison
    Lancaster University, UK
Takeo Kanade
    Carnegie Mellon University, Pittsburgh, PA, USA
Josef Kittler
    University of Surrey, Guildford, UK
Jon M. Kleinberg
    Cornell University, Ithaca, NY, USA
Friedemann Mattern
    ETH Zurich, Switzerland
John C. Mitchell
    Stanford University, CA, USA
Moni Naor
    Weizmann Institute of Science, Rehovot, Israel
Oscar Nierstrasz
    University of Bern, Switzerland
C. Pandu Rangan
    Indian Institute of Technology, Madras, India
Bernhard Steffen
    University of Dortmund, Germany
Madhu Sudan
    Massachusetts Institute of Technology, MA, USA
Demetri Terzopoulos
    New York University, NY, USA
Doug Tygar
    University of California, Berkeley, CA, USA
Moshe Y. Vardi
    Rice University, Houston, TX, USA
Gerhard Weikum
    Max-Planck Institute of Computer Science, Saarbruecken, Germany

Jiannong Cao   Laurence T. Yang
Minyi Guo   Francis Lau (Eds.)

Parallel and Distributed Processing and Applications

Second International Symposium, ISPA 2004
Hong Kong, China, December 13-15, 2004
Proceedings

Volume Editors

Jiannong Cao
Hong Kong Polytechnic University, Department of Computing
Hung Hom, Kowloon, Hong Kong, China
E-mail: csjcao@comp.polyu.edu.hk

Laurence T. Yang
St. Francis Xavier University, Department of Computer Science
Antigonish, B2G 2W5, NS, Canada
E-mail: lyang@stfx.ca

Minyi Guo
The University of Aizu, School of Computer Science and Engineering
Tsuruga, Ikki-machi, Aizu-Wakamatsu City, Fukushima 965-8580, Japan
E-mail: minyi@u-aizu.ac.jp

Francis Lau
The University of Hong Kong, Department of Computer Science
Pokfulam Road, Hong Kong, China
E-mail: fcmlau@cs.hku.hk

Library of Congress Control Number: 2004116720

CR Subject Classification (1998): F.2, G.2, C.2, H.4, D.2, D.4

ISSN 0302-9743
ISBN 3-540-24128-0 Springer Berlin Heidelberg New York

This work is subject to copyright. All rights are reserved, whether the whole or part of the material is concerned, specifically the rights of translation, reprinting, re-use of illustrations, recitation, broadcasting, reproduction on microfilms or in any other way, and storage in data banks. Duplication of this publication or parts thereof is permitted only under the provisions of the German Copyright Law of September 9, 1965, in its current version, and permission for use must always be obtained from Springer. Violations are liable to prosecution under the German Copyright Law.

Springer is a part of Springer Science+Business Media

springeronline.com

© Springer-Verlag Berlin Heidelberg 2004
Printed in Germany

Typesetting: Camera-ready by author, data conversion by Scientific Publishing Services, Chennai, India
Printed on acid-free paper   SPIN: 11369110   06/3142   5 4 3 2 1 0

Preface

Welcome to the proceedings of the 2nd International Symposium on Parallel and Distributed Processing and Applications (ISPA 2004) which was held in Hong Kong, China, 13–15 December, 2004.

With the advance of computer networks and hardware technology, parallel and distributed processing has become a key technology which plays an important part in determining future research and development activities in many academic and industrial branches. It provides a means to solve computationally intensive problems by improving processing speed. It is also the only viable approach to building highly reliable and inherently distributed applications. ISPA 2004 provided a forum for scientists and engineers in academia and industry to exchange and discuss their experiences, new ideas, research results, and applications about all aspects of parallel and distributed computing.

There was a very large number of paper submissions (361) from 26 countries and regions, including not only Asia and the Pacific, but also Europe and North America. All submissions were reviewed by at least three program or technical committee members or external reviewers. It was extremely difficult to select the presentations for the conference because there were so many excellent and interesting submissions. In order to allocate as many papers as possible and keep the high quality of the conference, we finally decided to accept 78 regular papers and 38 short papers for oral technical presentations. We believe that all of these papers and topics not only provide novel ideas, new results, work in progress and state-of-the-art techniques in this field, but also stimulate the future research activities in the area of parallel and distributed computing with applications.

The exciting program for this conference was the result of the hard and excellent work of many others, such as program vice-chairs, external reviewers, and program and technical committee members. We would like to express our sincere appreciation to all authors for their valuable contributions and to all program and technical committee members and external reviewers for their cooperation in completing the program under a very tight schedule.

October 2004

Jiannong Cao, Laurence T. Yang
Minyi Guo, Francis C.M. Lau

Organization

ISPA 2004 was organized mainly by the Department of Computing, Hong Kong Polytechnic University, China.

Executive Committee

General Chairs	Minyi Guo, University of Aizu, Japan
	Francis Lau, University of Hong Kong, China
Program Chairs	Jiannong Cao, Hong Kong Polytechnic University, China
	Laurence T. Yang, St. Francis Xavier University, Canada
Program Vice-Chairs	Rajkumar Buyya, University of Melbourne, Australia
	Weijia Jia, City University of Hong Kong, China
	Beniamino Di Martino, Second University of Naples, Italy
Steering Committee	Minyi Guo, University of Aziu, Japan
	Jiannong Cao, Hong Kong Polytechnic University, China
	Laurence T. Yang, St. Francis Xavier University, Canada
	Yi Pan, Georgia State University, USA
	Jie Wu, Florida Atlantic University, USA
	Li Xie, Nanjing University, China
	Hans P. Zima, California Institute of Technology, USA
Publicity Chair	Cho-Li Wang, University of Hong Kong, China
Workshop Chair	Hong-Va Leong, Hong Kong Polytechnic University, China
Local Chair	Allan K.Y. Wong, Hong Kong Polytechnic University, China
Publication Chair	Alvin T.S. Chan, Hong Kong Polytechnic University, China
Registration Chair	Joseph K.Y. Ng, Hong Kong Baptist University, China

Sponsoring Institutions

IEEE HK Chapter
Association for Computing Machinery, HK Chapter
The Information Processing Society of Japan
Springer

Program/Technical Committee

David Abramson	Monash University, Australia
Selim G. Akl	Queen's University, Canada
Giuseppe Anastasi	University of Pisa, Italy
Hamid R. Arabnia	University of Georgia, USA
Amy Apon	University of Arkansas, USA
Eric Aubanel	University of New Brunswick, Canada
David A. Bader	University of New Mexico, USA
Mark Baker	University of Portsmouth, UK
Ioana Banicescu	Mississippi State University, USA
Virendra C. Bhavsar	University of New Brunswick, Canada
Rupak Biswas	NASA Ames Research Center, USA
Anu Bourgeois	Georgia State University, USA
Martin Büecker	Aachen University of Technology, Germany
Wentong Cai	Nanyang Technological University, Singapore
Xing Cai	University of Oslo, Norway
Jesus Carretero	Universidad Carlos III de Madrid, Spain
Vipin Chaudhary	Wayne State University, USA
Weng-Long Chang	Southern Taiwan University of Technology, Taiwan
Daoxu Chen	Nanjing University, China
Ling Chen	Yangzhou University, China
Kessler Christoph	Linköping University, Sweden
Kranzlmueller Dieter	Linz University, Austria
Ramón Doallo	Universidade da Coruña, Spain
Andrei Doncescu	LAS, France
Patricia González	Universidade da Coruña, Spain
Andzrej Goscinski	Deakin University, Australia
George A. Gravvanis	Hellenic Open University, Greece
Yanxiang He	Wuhan University, China
Bruce Hendrickson	Sandia National Laboratory, USA
Dora Blanco Heras	Universidade de Santiago de Compostela, Spain
Annika Hinze	University of Waikato, New Zealand
Hung-Chang Hsiao	National TsingHua University, Taiwan
Ching-Hsien Hsu	Chung Hua University, Taiwan
Chun-Hsi Huang	University of Connecticut, USA
Constantinos Ierotheou	University of Greenwich, UK
Xiaohua Jia	City University of Hong Kong, China
Beihong Jin	Institute of Software, CAS, China
Hai Jin	Huazhong University of Science & Technology, China
Ajay Katangur	Texas A&M University at Corpus Christi, USA
Hatsuhiko Kato	Shonan Institute of Technology, Japan
Daniel S. Katz	JPL, California Institute of Technology, USA
Raj Kettimuthu	Argonne National Laboratory, USA
Sy-Yen Kuo	National Taiwan University, Taiwan

Program/Technical Committee (continued)

Tau Leng	Supermicro Computer Inc., USA
Jie Li	University of Tsukuba, Japan
Lei Li	Hosei University, Japan
Minglu Li	Shanghai Jiaotong University, China
Wenjun Li	UT Southwestern Medical Center, USA
Xiaoming Li	Peking University, China
Wekeng Liao	Northwestern University, USA
Man Lin	St. Francis Xavier University, Canada
Jiangchuan Liu	Chinese University of Hong Kong, Hong Kong, China
Jian Lu	Nanjing University, China
Paul Lu	University of Alberta, Canada
Jianhua Ma	Hosei University, Japan
Rodrigo Mello	University of São Paulo, Brazil
Michel Schellekens	National University of Ireland, Cork, Ireland
Michael Ng	University of Hong Kong, China
Jun Ni	University of Iowa, USA
Enrique Quintana-Orti	University of Jaime I, Spain
Yi Pan	Georgia State University, USA
Manish Parashar	Rutgers University, USA
Xiangzhen Qiao	Chinese Academy of Sciences, China
Fethi Rabhi	University of New South Wales, Australia
Thomas Rauber	University of Bayreuth, Germany
Gudula Rünger	Chemnitz University of Technology, Germany
Biplab K. Sarker	University of Tsukuba, Japan
Erich Schikuta	University of Vienna, Austria
Gurdip Singh	Kansas State University, USA
Peter Strazdins	Australian National University, Australia
Yuzhong Sun	Institute of Computing Technology, CAS, China
Eric de Sturler	University of Illinois at Urbana-Champaign, USA
Sabin Tabirca	University College Cork, Ireland
David Taniar	Monash University, Australia
Ruppa K. Thulasiram	University of Manitoba, Canada
Parimala Thulasiram	University of Manitoba, Canada
Xinmin Tian	Intel, USA
Dhaene Tom	University of Antwerp, Belgium
Juan Touriño	Universidade da Coruña, Spain
Sudharshan Vazhkudai	Oak Ridge National Laboratory, USA
Lorenzo Verdoscia	ICAR, Italian National Research Council (CNR), Italy
Hui Wang	University of Aizu, Japan
Guojung Wang	Central South University, China
Andrew L. Wendelborn	University of Adelaide, Australia
Jie Wu	Florida Atlantic University, USA
Bing Xiao	Hong Kong Polytechnic University, China

Program/Technical Committee (continued)

Cheng-Zhong Xu	Wayne State University, USA
Ming Xu	National University of Defense Technology, China
Zhiwei Xu	Institute of Computing Technology, CAS, China
Jingling Xue	University of New South Wales, Australia
Jun Zhang	University of Kentucky, USA
Xiaodong Zhang	William & Mary College, USA
Wei Zhao	Texas A&M University, USA
Weimin Zheng	Tsinghua University, China
Yao Zheng	Zhejiang University, China
Bingbing Zhou	University of Sydney, Australia
Wanlei Zhou	Deakin University, Australia
Xiaobao Zhou	University of Colorado at Colorado Springs, USA
Jianping Zhu	University of Akron, USA
Ming Zhu	Drexel University, USA
Albert Y. Zomaya	University of Sydney, Australia

Additional Referees

Somasheker Akkaladevi	Hui Cheng	Rafael Mayo Gual
Peter Aronsson	Benny W.L. Cheung	David Hickey
Rocco Aversa	Eunjung Cho	Judith Hippold
Shahaan Ayyub	Chi-yin Chow	Edward K.S. Ho
Stefano Basagni	Paul Coddington	Roy S.C. Ho
Andrzej Bednarski	Yafei Dai	Tim Ho
Cinzia Bernardeschi	Franca DelMastro	Sasche Hunold
Rich Boakes	Li Du	Mauro Iacono
Luciano Bononi	Colin Enticott	Cruz Izu
Eleonora Borgia	Robert Esser	Young-Sik Jeong
Donggang Cao	Nickolas Falkner	Ziling Jhong
Ning Cao	Huifeng Fan	Hao Ji
Ricolindo L. Carino	Maria Fazio	Nanyan Jiang
John Casey	Michael Frumkin	Tian Jing
Valentina Casola	Haohuan Fu	Ning Kang
Mikhail Chalabine	Boon Ping Gan	Peter Kelly
Philip Chan	Slavisa Garic	Manabu Kobayashi
Sumir Chandra	Wojtek Goscinski	Laurel C.Y. Kong
Wei Chen	Christophe Gosset	Donny Kurniawan
Elaine L. Chen	Mat Grove	Matthias Kühnemann

Charles Lakos
Wilfred Lin
Lidong Lin
Qiu Ling
Hui Liu
Ying Liu
Hua Liu
Xiapu Luo
Praveen Madiraju
Laurent Manyri
Yang Mao
Stefano Marrone
Maria J. Martin
Hakan Mattsson
Piyush Mehrotra
Srinivasan Mullai
Mingzhong Xiao
Giacomo Morabito
Francesco Moscato
Raik Nagel
Harsha Narravula
Leanne Ngo
Maria O'Keeffe
Leonid Oliker
Hong Ong
Andrew Over
Andrea Passarella
Fang Qi
Wenny Rahayu
Massimiliano Rak
Carsten Scholtes
Aamir Shafi
Haifeng Shen
Wensheng Shen
Ji Shen
Wei Shi
Warren Smith
Guanghua Song
Makoto Suzuki
Guillermo L. Taboada
Ming Tang
Toshinori Tkabatake
Roberto Torella
Sven Trautmann
Ken C.K. Tsang
Mark C.M. Tsang
Wanqing Tu
Thierry Vallee
Rob Vander
Wijngaart
Salvatore Venticinque
Murali N. Vilayannur
Shuchao Wan
Helmut Wanek
Habin Wang
Gaocai Wang
T.Q. Wang
Chen Wang
Yin Wang
Yue Wang
Habin Wang
Richard Wu
Sui Lun Wu
Weigang Wu
Percival Xavier
Yang Xiang
Helen Xiang
Yi Xie
Shuting Xu
Jin Yang
Zhonghua Yang
Shui Yu
Zoe C.H. Yu
Connie Yuen
Tianyi Zeng
Yi Zeng
Jun Zhang
Zili Zhang
Nadia X.L. Zhang
Guangzen Zhang
Qiankun Zhao
Bill Zhong
Suiping Zhou

Table of Contents

Keynote Speech

Present and Future Supercomputer Architectures
  Jack Dongarra 1

Challenges in P2P Computing
  Linoel M. Ni 2

Multihop Wireless Ad Hoc Networking:
Current Challenges and Future Opportunities
  David B. Johnson 3

Session 1A: Parallel Algorithms and Systems I

An Inspector-Executor Algorithm for Irregular Assignment
Parallelization
  Manuel Arenaz, Juan Touriño, Ramón Doallo 4

Multi-grain Parallel Processing of Data-Clustering
on Programmable Graphics Hardware
  Hiroyki Takizawa, Hiroaki Kobayashi 16

A Parallel Reed-Solomon Decoder on the Imagine Stream Processor
  Mei Wen, Chunyuan Zhang, Nan Wu, Haiyan Li, Li Li 28

Effective Nonblocking MPI-I/O in Remote I/O Operations Using a
Multithreaded Mechanism
  Yuichi Tsujita 34

Session 1B: Data Mining and Management

Asynchronous Document Dissemination in Dynamic Ad Hoc Networks
  Frédéric Guidec, Hervé Roussain 44

Location-Dependent Query Results Retrieval in a Multi-cell Wireless
Environment
  James Jayaputera, David Taniar 49

An Efficient Mobile Data Mining Model
  Jen Ye Goh, David Taniar 54

An Integration Approach of Data Mining with Web Cache Pre-fetching
  Yingjie Fu, Haohuan Fu, Puion Au 59

Session 1C: Distributed Algorithms and Systems

Towards Correct Distributed Simulation of High-Level Petri Nets with Fine-Grained Partitioning
  *Michael Knoke, Felix Kühling, Armin Zimmermann,
  Günter Hommel* 64

M-Guard: A New Distributed Deadlock Detection Algorithm Based on Mobile Agent Technology
  *Jingyang Zhou, Xiaolin Chen, Han Dai, Jiannong Cao,
  Daoxu Chen* 75

Meta-based Distributed Computing Framework
  Andy S.Y. Lai, A.J. Beaumont 85

Locality Optimizations for Jacobi Iteration on Distributed Parallel Systems
  Yonggang Che, Zhenghua Wang, Xiaomei Li, Laurence T. Yang 91

Session 2A: Fault Tolerance Protocols and Systems

Fault-Tolerant Cycle Embedding in the WK-Recursive Network
  Jung-Sheng Fu 105

RAIDb: Redundant Array of Inexpensive Databases
  Emmanuel Cecchet 115

A Fault-Tolerant Multi-agent Development Framework
  Lin Wang, Hon F. Li, Dhrubajyoti Goswami, Zunce Wei 126

A Fault Tolerance Protocol for Uploads: Design and Evaluation
  L. Cheung, C.-F. Chou, L. Golubchik, Y. Yang 136

Topological Adaptability for the Distributed Token Circulation Paradigm in Faulty Environment
  Thibault Bernard, Alain Bui, Olivier Flauzac 146

Session 2B: Sensor Networks and Protocols

Adaptive Data Dissemination in Wireless Sensor Networks
  Jian Xu, Jianliang Xu, Shanping Li, Qing Gao, Gang Peng 156

Continuous Residual Energy Monitoring in Wireless Sensor Networks
  Song Han, Edward Chan 169

Design and Analysis of a k-Connected Topology Control Algorithm for Ad Hoc Networks
  Lei Zhang, Xuehui Wang, Wenhua Dou 178

On Using Temporal Consistency for Parallel Execution of Real-Time Queries in Wireless Sensor Systems
  Kam-Yiu Lam, Henry C.W. Pang, Sang H. Son, BiYu Liang 188

Session 2C: Cluster Systems and Applications

Cluster-Based Parallel Simulation for Large Scale Molecular Dynamics in Microscale Thermophysics
  Jiwu Shu, Bing Wang, Weimin Zheng 200

Parallel Checkpoint/Recovery on Cluster of IA-64 Computers
  Youhui Zhang, Dongsheng Wang, Weimin Zheng 212

Highly Reliable Linux HPC Clusters: Self-Awareness Approach
  Chokchai Leangsuksun, Tong Liu, Yudan Liu, Stephen L. Scott, Richard Libby, Ibrahim Haddad 217

An Enhanced Message Exchange Mechanism in Cluster-Based Mobile Ad Hoc Networks
  Wei Lou, Jie Wu 223

Session 3A: Parallel Algorithms and Systems II

Algorithmic-Parameter Optimization of a Parallelized Split-Step Fourier Transform Using a Modified BSP Cost Model
  Elankovan Sundararajan, Malin Premaratne, Shanika Karunasekera, Aaron Harwood 233

Parallel Volume Rendering with Early Ray Termination for Visualizing Large-Scale Datasets
  Manabu Matsui, Fumihiko Ino, Kenichi Hagihara 245

A Scalable Low Discrepancy Point Generator for Parallel Computing
  Kwong-Ip Liu, Fred J. Hickernell 257

Generalized Trellis Stereo Matching with Systolic Array
  Hong Jeong, Sungchan Park 263

Optimal Processor Mapping Scheme for Efficient Communication of
Data Realignment
  *Ching-Hsien Hsu, Kun-Ming Yu, Chi-Hsiu Chen, Chang Wu Yu,
  Chiu Kuo Lian* 268

Session 3B: Grid Applications and Systems

MCCF: A Distributed Grid Job Workflow Execution Framework
  Yuhong Feng, Wentong Cai 274

Gamelet: A Mobile Service Component for Building Multi-server
Distributed Virtual Environment on Grid
  Tianqi Wang, Cho-Li Wang, Francis Lau 280

The Application of Grid Computing to Real-Time Functional MRI
Analysis
  E. Bagarinao, L. Sarmenta, Y. Tanaka, K. Matsuo, T. Nakai 290

Building and Accessing Grid Services
  Xinfeng Ye 303

DRPS: A Simple Model for Locating the Tightest Link
  Dalu Zhang, Weili Huang, Chen Lin 314

Session 3C: Peer-to-Peer and Ad-Hoc Networking

A Congestion-Aware Search Protocol for Unstructured Peer-to-Peer
Networks
  Kin Wah Kwong, Danny H.K. Tsang 319

Honeycomb: A Peer-to-Peer Substrate for On-Demand Media Streaming
Service
  Dafu Deng, Hai Jin, Chao Zhang, Hao Chen, Xiaofei Liao 330

An Improved Distributed Algorithm for Connected Dominating Sets in
Wireless Ad Hoc Networks
  Hui Liu, Yi Pan, Jiannong Cao 340

A New Distributed Approximation Algorithm for Constructing
Minimum Connected Dominating Set in Wireless Ad Hoc Networks
 Bo Gao, Huiye Ma, Yuhang Yang 352

An Adaptive Routing Strategy Based on Dynamic Cache in Mobile Ad
Hoc Networks
 YueQuan Chen, XiaoFeng Guo, QingKai Zeng, Guihai Chen 357

Session 4A: Grid Scheduling and Algorithms I

On the Job Distribution in Random Brokering for Computational Grids
 Vandy Berten, Joël Goossens 367

Dividing Grid Service Discovery into 2-Stage Matchmaking
 Ye Zhu, Junzhou Luo, Teng Ma 372

Performance Evaluation of a Grid Computing Architecture Using
Realtime Network Monitoring
 Young-Sik Jeong, Cheng-Zhong Xu 382

Quartet-Based Phylogenetic Inference: A Grid Approach
 Chen Wang, Bing Bing Zhou, Albert Y. Zomaya 387

Scheduling BoT Applications in Grids Using a Slave Oriented Adaptive
Algorithm
 Tiago Ferreto, César De Rose, Caio Northfleet 392

Session 4B: Data Replication and Caching

A Clustering-Based Data Replication Algorithm in Mobile Ad Hoc
Networks for Improving Data Availability
 Jing Zheng, Jinshu Su, Xicheng Lu 399

CACHE$_{RP}$: A Novel Dynamic Cache Size Tuning Model Working with
Relative Object Popularity for Fast Web Information Retrieval
 Richard S.L. Wu, Allan K.Y. Wong, Tharam S. Dillon 410

Implementation of a New Cache and Schedule Scheme for Distributed
VOD Servers
 Han Luo, Ji-wu Shu 421

Session 4C: Software Engineering and Testing

UML Based Statistical Testing Acceleration of Distributed
Safety-Critical Software
   Jiong Yan, Ji Wang, Huo-wang Chen 433

A Metamodel for the CMM Software Process
   Juan Li, Mingshu Li, Zhanchun Wu, Qing Wang 446

Performance Tuning for Application Server OnceAS
   Wenbo Zhang, Bo Yang, Beihong Jin, Ningjing Chen, Tao Huang 451

Systematic Robustness-Testing RI-Pro of BGP
   Lechun Wang, Peidong Zhu, Zhenghu Gong 463

Session 5A: Grid Protocols

MPICH-GP: A Private-IP-Enabled MPI Over
Grid Environments
   Kumrye Park, Sungyong Park, Ohyoung Kwon, Hyoungwoo Park 469

Paradigm of Multiparty Joint Authentication: Evolving Towards Trust
Aware Grid Computing
   Hui Liu, Minglu Li 474

Design and Implementation of a 3A Accessing Paradigm Supported
Grid Application and Programming Environment
   He Ge, Liu Donghua, Sun Yuzhong, Xu Zhiwei 484

VAST: A Service Based Resource Integration System for Grid Society
   Jiulong Shan, Huaping Chen, Guangzhong Sun, Xin Chen 489

Petri-Net-Based Coordination Algorithms for Grid Transactions
   Feilong Tang, Minglu Li, Joshua Zhexue Huang, Cho-Li Wang,
   Zongwei Luo 499

Session 5B: Context-Aware and Mobile Computing

Building Infrastructure Support for Ubiquitous Context-Aware Systems
   Wei Li, Martin Jonsson, Fredrik Kilander, Carl Gustaf Jansson 509

Context-Awareness in Mobile Web Services
   Bo Han, Weijia Jia, Ji Shen, Man-Ching Yuen 519

CRL: A Context-Aware Request Language for Mobile Computing
  Alvin T.S. Chan, Peter Y.H. Wong, Siu-Nam Chuang 529

A Resource Reservation Protocol for Mobile Cellular Networks
  Ming Xu, Zhijiao Zhang, Yingwen Chen 534

Session 5C: Distributed Routing and Switching Protocols I

Using the Linking Model to Understand the Performance of DHT Routing Algorithms
  Futai Zou, Shudong Cheng, Fanyuan Ma, Liang Zhang, Junjun Tang 544

Packet-Mode Priority Scheduling for Terabit Core Routers
  Wenjie Li, Bin Liu 550

Node-to-Set Disjoint Paths Problem in Bi-rotator Graphs
  Keiichi Kaneko 556

QoSRHMM: A QoS-Aware Ring-Based Hierarchical Multi-path Multicast Routing Protocol
  Guojun Wang, Jun Luo, Jiannong Cao, Keith C.C. Chan 568

Session 6A : Grid Scheduling and Algorithms II

A Dynamic Task Scheduling Algorithm for Grid Computing System
  Yuanyuan Zhang, Yasushi Inoguchi, Hong Shen 578

Replica Selection on Co-allocation Data Grids
  Ruay-Shiung Chang, Chih-Min Wang, Po-Hung Chen 584

A Novel Checkpoint Mechanism Based on Job Progress Description for Computational Grid
  Chunjiang Li, Xuejun Yang, Nong Xiao 594

A Peer-to-Peer Mechanism for Resource Location and Allocation over the Grid
  Hung-Chang Hsiao, Mark Baker, Chung-Ta King 604

The Model, Architecture and Mechanism Behind Realcourse
  Jinyu Zhang, Xiaoming Li 615

Session 6B: Cluster Resource Scheduling and Algorithms

Managing Irregular Workloads of Cooperatively Shared Computing Clusters
    Percival Xavier, Wentong Cai, Bu-Sung Lee 625

Performance-Aware Load Balancing for Multiclusters
    Ligang He, Stephen A. Jarvis, David Bacigalupo, Daniel P. Spooner, Graham R. Nudd 635

Scheduling of a Parallel Computation-Bound Application and Sequential Applications Executing Concurrently on a Cluster – A Case Study
    Adam K.L. Wong, Andrzej M. Goscinski 648

Sequential and Parallel Ant Colony Strategies for Cluster Scheduling in Spatial Databases
    Jitian Xiao, Huaizhong Li 656

Session 6C: Distributed Routing and Switching Protocols I

Cost-Effective Buffered Wormhole Routing
    Jinming Ge 666

Efficient Routing and Broadcasting Algorithms in de Bruijn Networks
    Ngoc Chi Nguyen, Nhat Minh Dinh Vo, Sungyoung Lee 677

Fault-Tolerant Wormhole Routing Algorithm in 2D Meshes Without Virtual Channels
    Jipeng Zhou, Francis C.M. Lau 688

Fault Tolerant Routing Algorithm in Hypercube Networks with Load Balancing Support
    Xiaolin Xiao, Guojun Wang, Jianer Chen 698

Session 7A: Security I

Proxy Structured Multisignature Scheme from Bilinear Pairings
    Xiangxue Li, Kefei Chen, Longjun Zhang, Shiqun Li 705

A Threshold Proxy Signature Scheme Using Self-Certified Public Keys
    Qingshui Xue, Zhenfu Cao 715

The Authentication and Processing Performance of Session Initiation
Protocol (SIP) Based Multi-party Secure Closed Conference System
  Jongkyung Kim, Hyuncheol Kim, Seongjin Ahn, Jinwook Chung 725

Session 7B: High Performance Processing and Applications

A Method for Authenticating Based on ZKp in Distributed Environment
  Dalu Zhang, Min Liu, Zhe Yang 730

A Load-Balanced Parallel Algorithm for 2D Image Warping
  Yan-huang Jiang, Zhi-ming Chang, Xue-jun Yang 735

A Parallel Algorithm for Helix Mapping Between 3D and 1D Protein
Structure Using the Length Constraints
  Jing He, Yonggang Lu, Enrico Pontelli 746

A New Scalable Parallel Method for Molecular Dynamics Based on
Cell-Block Data Structure
  Xiaolin Cao, Zeyao Mo 757

Parallel Transient Stability Simulation for National Power Grid of China
  Wei Xue, Jiwu Shu, Weimin Zheng 765

HPL Performance Prevision to Intending System Improvement
  Wenli Zhang, Mingyu Chen, Jianping Fan 777

Session 7C: Networking and Protocols I

A Novel Fuzzy-PID Dynamic Buffer Tuning Model to Eliminate
Overflow and Shorten the End-to-End Roundtrip Time for TCP
Channels
  Wilfred W.K. Lin, Allan K.Y. Wong, Tharam S. Dillon 783

Communication Using a Reconfigurable and Reliable Transport Layer
Protocol
  Tan Wang, Ajit Singh 788

Minicast: A Multicast-Anycast Protocol for Message Delivery
  Shui Yu, Wanlei Zhou, Justin Rough 798

Dependable WDM Networks with Edge-Disjoint P-Cycles
  Chuan-Ching Sue, Yung-Chiao Chen, Min-Shao Shieh, Sy-Yen Kuo .. 804

An Efficient Fault-Tolerant Approach for MPLS Network Systems
    Jenn-Wei Lin, Hung-Yu Liu 815

Session 8A: Security II

A Novel Technique for Detecting DDoS Attacks at Its Early Stage
    Bin Xiao, Wei Chen, Yanxiang He 825

Probabilistic Inference Strategy in Distributed Intrusion Detection Systems
    *Jianguo Ding, Shihao Xu, Bernd Krämer, Yingcai Bai,
    Hansheng Chen, Jun Zhang* 835

An Authorization Framework Based on Constrained Delegation
    Gang Yin, Meng Teng, Huai-min Wang, Yan Jia, Dian-xi Shi 845

A Novel Hierarchical Key Management Scheme Based on Quadratic Residues
    Jue-Sam Chou, Chu-Hsing Lin, Ting-Ying Lee 858

Session 8B: Artificial Intelligence Systems and Applications

Soft-Computing-Based Intelligent Multi-constrained Wavelength Assignment Algorithms in IP/DWDM Optical Internet
    Xingwei Wang, Cong Liu, Min Huang 866

Data Transmission Rate Control in Computer Networks Using Neural Predictive Networks
    Yanxiang He, Naixue Xiong, Yan Yang 875

Optimal Genetic Query Algorithm for Information Retrieval
    Ziqiang Wang, Boqin Feng 888

A Genetic Algorithm for Dynamic Routing and Wavelength Assignment in WDM Networks
    *Vinh Trong Le, Son Hong Ngo, Xiaohong Jiang, Susumu Horiguchi,
    Minyi Guo* 893

Session 8C: Networking and Protocols II

Ensuring E-Transaction Through a Lightweight Protocol for Centralized Back-End Database
    Paolo Romano, Francesco Quaglia, Bruno Ciciani 903

Cayley DHTs — A Group-Theoretic Framework for Analyzing DHTs
Based on Cayley Graphs
  Changtao Qu, Wolfgang Nejdl, Matthias Kriesell 914

BR-WRR Scheduling Algorithm in PFTS
  Dengyuan Xu, Huaxin Zeng, Chao Xu 926

VIOLIN: Virtual Internetworking on Overlay Infrastructure
  Xuxian Jiang, Dongyan Xu 937

Session 9A: Hardware Architectures and Implementations

Increasing Software-Pipelined Loops in the Itanium-Like Architecture
  Wenlong Li, Haibo Lin, Yu Chen, Zhizhong Tang 947

A Space-Efficient On-Chip Compressed Cache Organization for High
Performance Computing
  *Keun Soo Yim, Jang-Soo Lee, Jihong Kim, Shin-Dug Kim,
  Kern Koh* 952

A Real Time MPEG-4 Parallel Encoder on Software Distributed Shared
Memory Systems
  *Yung-Chang Chiu, Ce-Kuen Shieh, Jing-Xin Wang,
  Alvin Wen-Yu Su, Tyng-Yeu Liang* 965

A Case of SCMP with TLS
  *Jianzhuang Lu, Chunyuan Zhang, Zhiying Wang, Yun Cheng,
  Dan Wu* 975

Session 9B: High Performance Computing and Architecture

SuperPAS: A *Parallel Architectural Skeleton* Model Supporting
Extensibility and Skeleton Composition
  Mohammad Mursalin Akon, Dhrubajyoti Goswami, Hon Fung Li 985

Optimizing I/O Server Placement for Parallel I/O on Switch-Based
Irregular Networks
  Yih-Fang Lin, Chien-Min Wang, Jan-Jan Wu 997

Designing a High Performance and Fault Tolerant Multistage
Interconnection Network with Easy Dynamic Rerouting
  Ching-Wen Chen, Phui-Si Gan, Chih-Hung Chang 1007

Evaluating Performance of BLAST on Intel Xeon and Itanium2
Processors
   *Ramesh Radhakrishnan, Rizwan Ali, Garima Kochhar,
   Kalyana Chadalavada, Ramesh Rajagopalan, Jenwei Hsieh,
   Onur Celebioglu* 1017

Session 9C: Distributed Processing and Architecture

PEZW-ID: An Algorithm for Distributed Parallel Embedded Zerotree
Wavelet Encoder
   Zhi-ming Chang, Yan-huang Jiang, Xue-jun Yang, Xiang-li Qu 1024

Enhanced-Star: A New Topology Based on the Star Graph
   Hamid Reza Tajozzakerin, Hamid Sarbazi-Azad 1030

An RFID-Based Distributed Control System for Mass Customization
Manufacturing
   Michael R. Liu, Q.L. Zhang, Lionel M. Ni, Mitchell M. Tseng 1039

Event Chain Clocks for Performance Debugging in Parallel and
Distributed Systems
   Hongliang Yu, Jian Liu, Weimin Zheng, Meiming Shen 1050

Author Index 1055

Present and Future Supercomputer Architectures

Jack Dongarra[1,2]

[1] Computer Science Department, University of Tennessee
Knoxville Tennessee 37996, USA
dongarra@cs.utk.edu
[2] Oak Ridge National Laboratory
Oak Ridge, TN 37831, USA

Abstract. In last 25 years, the field of scientific computing has undergone rapid change -- we have experienced a remarkable turnover of technologies, architectures, vendors, and the usage of systems. Despite all these changes, the long-term evolution of performance seems to be steady and continuous. The acceptance of parallel systems not only for engineering applications but also for new commercial applications especially for database applications emphasized different criteria for market success such as stability of system, continuity of the manufacturer and price/performance. Due to these factors and the consolidation in the number of vendors in the market hierarchical systems build with components designed for the broader commercial market are currently replacing homogeneous systems at the very high end of performance. Clusters build with components of the shelf also gain more and more attention and today have a dominant position in the Top500. In this talk we will look at the some of the existing and planned high performance computer architectures and look at the interconnections schemes they are using. This talk will look at a number of different high performance computing architectures.

Challenges in P2P Computing

Linoel M. Ni

Department of Computer Science, Hong Kong University of Science and Technology,
Clear Water Bay, Kowloon, Hong Kong
ni@cs.ust.hk

Abstract. Peer-to-peer (P2P) is an emerging model aiming to further utilize Internet information and resources, complementing the available client-server services. P2P has emerged as a promising paradigm for developing large-scale distributed systems due to its many unique features and its potential in future applications. P2P systems are popular because of their adaptation, self-organization, load-balancing, and highly availability. However, P2P systems also present many challenges that are currently obstacles to their widespread acceptance and usage, such as efficiency, security, and performance guarantees. For example, studies have shown that P2P traffic contributes the largest portion of the Internet traffic based on the measurements on some popular P2P systems, such as FastTrack (including KaZaA and Grokster), Gnutella, and Direct Connect. Even given that 95% of any two nodes are less than 7 hops away and the message time-to-live (TTL=7) is preponderantly used, the flooding-based routing algorithm generates 330 TB/month in a Gnutella network with only 50,000 nodes. In reality, there are millions of active P2P users at any given time. Our study has shown that the mechanism of a peer randomly choosing logical neighbors without any knowledge about the underlying physical topology causes topology mismatch between the P2P logical overlay network and the physical underlying network. A large portion of the heavy P2P traffic is caused by inefficient overlay topology and the blind flooding. Security and anonymity are other concerns in P2P systems. This talk will address the above issues as well as other potential applications of P2P computing and mobile P2P systems.

Multihop Wireless Ad Hoc Networking: Current Challenges and Future Opportunities

David B. Johnson

Department of Computer Science, Rice University
6100 Main Houston, Texas 77005, USA
dbj@cs.rice.edu

Abstract. An ad hoc network is a collection of wireless mobile nodes that form a network without existing infrastructure or centralized administration. Nodes in the network cooperate to forward packets for each other, to allow mobile nodes not within direct wireless transmission range of each other to communicate. Ad hoc networks were initially studied for military applications more than 20 years ago, and they are currently a very active area of research within academia, government, and industry. However, few real applications of ad hoc networking have yet been deployed in common usage, and the commercial potentials of this technology have yet to be realized. In this talk, I will describe some of the current research challenges in ad hoc networking, and I will present what I believe are some the future real-world applications for this promising technology.

An Inspector-Executor Algorithm for Irregular Assignment Parallelization

Manuel Arenaz, Juan Touriño, and Ramón Doallo

Computer Architecture Group,
Dep. Electronics and Systems, University of A Coruña, Spain
{arenaz, juan, doallo}@udc.es

Abstract. A loop with irregular assignment computations contains loop-carried output data dependences that can only be detected at run-time. In this paper, a load-balanced method based on the inspector-executor model is proposed to parallelize this loop pattern. The basic idea lies in splitting the iteration space of the sequential loop into sets of conflict-free iterations that can be executed concurrently on different processors. As will be demonstrated, this method outperforms existing techniques. Irregular access patterns with different load-balancing and reusability properties are considered in the experiments.

1 Introduction

Research on run-time techniques for the efficient parallelization of irregular computations has been frequently referenced in the literature in recent years [4, 5, 7, 8, 10, 14, 15]. An *irregular assignment* pattern consists of a loop with f_{size} iterations, f_{size} being the size of the subscript array f (see Figure 1). At each iteration h, value $rhs(h)$ is assigned to the array element $A(f(h))$. Neither the right-hand side expression $rhs(h)$ nor any function call make within it contain occurrences of A, thus the code is free of loop-carried true data dependences. Nevertheless, as the subscript expression $f(h)$ is loop-variant, loop-carried output data dependences may be present at run-time (unless f is a permutation array). This loop pattern can be found in different application fields such as computer graphics algorithms [3], finite elements applications [12], or routines for sparse matrix computations [11].

Knobe and Sarkar [6] describe a program representation that uses *array expansion* [13] to enable the parallel execution of irregular assignment computations. Each processor executes a set of iterations preserving the same relative order of the sequential loop. Array A is expanded in order to allow different processors to store partial results in separate memory locations. For each array entry $A(j)$, with $j = 1, ..., A_{size}$, the global result is computed by means of a reduction operation that obtains the partial result that corresponds with the highest iteration number. Each processor computes this reduction operation for a subset of array elements.

An optimization to perform element-level dead code elimination at run-time is also presented in [6]. In irregular assignments, the same array element may be

$$\boxed{\begin{array}{l} A(...) = ... \\ \textbf{DO } h = 1, f_{size} \\ \quad A(f(h)) = rhs(h) \\ \textbf{END DO} \\ ... = ...A(...)... \end{array}}$$

Fig. 1. Irregular assignment pattern

computed several times, though only the last value is used after the loop ends. Consequently, intermediate values need not be computed. Classical dead code elimination typically removes assignment statements from the source code. This technique eliminates unnecessary array element definitions at run-time.

In this paper we use the inspector-executor model to parallelize irregular assignments on scalable shared memory multiprocessors. We show that this model can be efficiently applied to the parallelization of static/adaptive irregular applications, preserving load-balancing and exploiting uniprocessor data write locality. A preliminary work [1] did not include a theoretical performance analysis based on a formal characterization of static/adaptive irregular applications, and presented a quite limited performance evaluation. The technique described in this paper is embedded in our compiler framework [2] for automatic kernel recognition and its application to automatic parallelization of irregular codes.

The rest of the paper is organized as follows. Our parallelization method is presented in Section 2. The performance of our technique is compared with the array expansion approach in Section 3. Experimental results conducted on a SGI Origin 2000 using a rasterization algorithm as case study are shown in Section 4. Finally, conclusions are discussed in Section 5.

2 Parallel Irregular Assignment

In this section, we propose a run-time technique that uses the inspector-executor model to parallelize irregular assignments. The basic idea lies in reordering loop iterations so that data write locality is exploited on each processor. Furthermore, the amount of computations assigned to each processor is adjusted so that load-balancing is preserved.

The method is as follows. In the inspector code shown in Figure 2, array A is divided into subarrays of consecutive locations, A_p ($p = 1, ..., P$ where P is the number of processors), and the computations associated with each block are assigned to different processors. Thus, the loop iteration space $(1, ..., f_{size})$ is partitioned into sets f_p that perform write operations on different blocks A_p. The sets f_p are implemented as linked lists of iteration numbers using two arrays, $count(1 : P)$ and $next(1 : f_{size} + P)$. Each processor p has an entry in both arrays, $count(p)$ and $next(f_{size} + p)$. The entry $next(f_{size} + p)$ stores the first iteration number h_1^p assigned to processor p. The next iteration number, h_2^p, is stored in array entry $next(h_1^p)$. This process is repeated $count(p)$ times, i.e. the number of elements in the list. In the executor code of Figure 3, each processor

```
! Accumulative frequency distribution
his(1 : A_size) = 0
DO h = 1, f_size
    his(f(h)) = his(f(h)) + 1
END DO
DO h = 2, A_size
    his(h) = his(h) + his(h − 1)
END DO

! Computation of the linked lists
Refs = (his(A_size)/P) + 1
count(1 : P) = 0
DO h = 1, f_size
    thread = (his(f(h))/Refs) + 1
    IF (count(thread).eq.0) THEN
        next(f_size + thread) = h
    ELSE
        next(prev(thread)) = h
    END IF
    prev(thread) = h
    count(thread) = count(thread) + 1
END DO
```

Fig. 2. Inspector code

```
A(...) = ...
DOALL p = 1, P
    h = next(f_size + p)
    DO k = 1, count(p)
        A(f(h)) = ...
        h = next(h)
    END DO
END DOALL
... = ...A(...)...
```

Fig. 3. Executor code

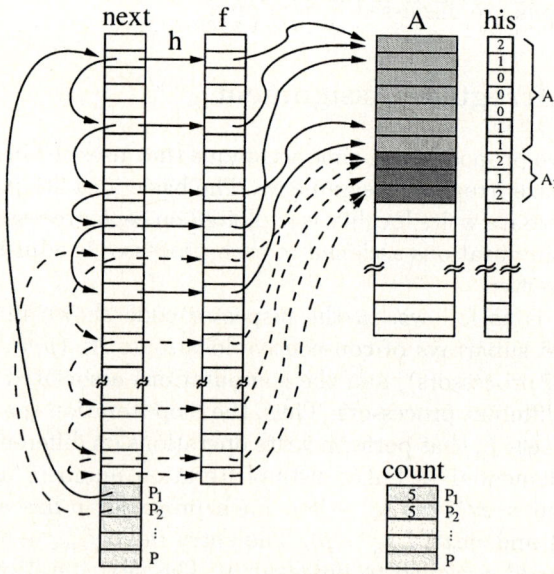

Fig. 4. Inspector-executor approach

p executes the conflict-free computations associated with the loop iterations contained in a set f_p. Figure 4 shows a graphical description of the method. The figure represents the linked-lists f_1 and f_2 of processors p_1 and p_2 as solid and dashed lines, respectively. The corresponding subarrays A_1 and A_2 are depicted as shaded regions within array A.

Load-balancing is preserved by splitting array A into subarrays A_p of different size in the inspector stage. As shown in the code of Figure 2, the inspector first computates the accumulative frequency distribution $his(1 : A_{size})$. For each array entry $A(j)$ with $j = 1, ..., A_{size}$, $his(j)$ stores the sum of the number of write references to $A(1), A(2), ..., A(j)$. The second step consists of building the linked lists f_p by determining the list corresponding to each entry of the subscript array f (see variable $thread$ in Figure 2). The appropiate list is easily computed as $his(f(h))/Refs + 1$, where $Refs$ is the mean number of iterations of the sequential loop per processor. As illustrated in Figure 4, load-balancing is preserved because, as A_1 and A_2 have different sizes (7 and 3, respectively), processors P_1 and P_2 are both assigned 5 iterations of the sequential loop.

Element-level dead code elimination can be implemented in the inspector-executor model, too. In this case, the linked lists only contain the last iteration at which array elements, $A(j)$, are modified. This difference is highlighted in Figure 5 where, unlike Figure 4, there are dotted arrows representing the loop iterations that are not computed. The code of the optimized inspector (the executor does not change) is shown in Figure 6. The accumulative frequency distribution array, $his(1 : A_{size})$, contains the number of array entries in the range $A(1), A(2), ..., A(j)$ that are modified during the execution of the irregular assignment. Note that an additional array, $iter$, is needed to store the last iteration number at which the elements of array A are modified. Finally, the phase that computes the linked lists is rewritten accordingly.

3 Performance Analysis

Memory overhead complexity of the array expansion technique proposed in [6] is $\mathcal{O}(A_{size} \times P)$ which, in practice, prevents the application of this method for large array sizes and a high number of processors. In contrast, memory overhead of our inspector-executor method is $\mathcal{O}(max(f_{size} + P, A_{size}))$. Note that the extra memory is not directly proportional to the number of processors. In practice, the complexity is usually $\mathcal{O}(f_{size})$, as $f_{size} \gg P$, or $\mathcal{O}(A_{size})$.

The efficiency of the parallelization techniques for irregular assignments is determined by the properties of the irregular access pattern. In our analysis, we have considered the following parameters proposed in [15] for the parallelization of irregular reductions: *degree of contention (C)*, number of loop iterations referencing an array element; *sparsity (SP)*, ratio of different elements referenced in the loop ($A_{updated}$) and the array size; *connectivity (CON)*, ratio of the number of loop iterations and the number of distinct array elements referenced in the loop; and *adaptivity* or *reusability (R)*, the number of times that an access pattern is reused before being updated.

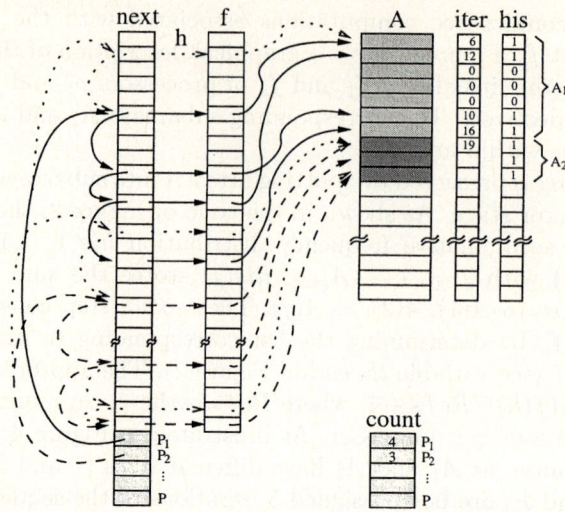

Fig. 5. Inspector-executor approach when dead code elimination is applied

```
! Accumulative frequency distribution
iter(1 : A_size) = 0
his(1 : A_size) = 0
DO h = 1, f_size
    iter(f(h)) = h
    his(f(h)) = 1
END DO
DO h = 2, A_size
    his(h) = his(h) + his(h − 1)
END DO

! Computation of the linked lists
Refs = (his(A_size)/P) + 1
count(1 : P) = 0
DO h = 1, A_size
    IF (iter(h).gt.0) THEN
        thread = (his(h)/Refs) + 1
        IF (count(thread).eq.0) THEN
            next(f_size + thread) = iter(h)
        ELSE
            next(prev(thread)) = iter(h)
        END IF
        prev(thread) = iter(h)
        count(thread) = count(thread) + 1
    END IF
END DO
```

Fig. 6. Inspector when dead code elimination is applied

Unlike the array expansion approach, the inspector-executor technique takes advantage of the adaptive nature of irregular applications. The computational overhead is associated with the inspector stage because the executor is fully parallel (it performs conflict-free computations). In static codes, the inspector overhead is negligible because it is computed only once and then reused during the program execution ($R \rightarrow \infty$). Thus, as the parallel execution time can be accurately approximated by the time of the executor, the efficiency $E \rightarrow 1$ as reusability R increases. In dynamic codes, the inspector is recomputed periodically. Supposing that the access pattern changes every time the executor is run ($R = 0$), a lower bound of the efficiency is

$$E = \frac{\#iters \; t_s}{P(T_s^{INSP} + \frac{\#iters}{P} t_s)} = \frac{\#iters \; t_s}{PT_s^{INSP} + \#iters \; t_s} \quad (1)$$

where t_s is the execution time of one iteration of the sequential irregular loop, T_s^{INSP} represents the execution time of the sequential inspector, and $\#iters$ is the number of loop iterations actually executed: f_{size} when dead-code elimination is not applied, and $A_{updated}$ when dead-code is applied. The execution time of the parallel irregular assignment is given by $T_p = T_s^{INSP} + \frac{\#iters}{P} t_s$.

As a result, the efficiency of the inspector-executor approach for any R is bounded as follows:

$$\frac{f_{size} \; t_s}{PT_s^{INSP} + \#iters \; t_s} \leq E \leq 1 \quad (2)$$

Lower efficiencies are obtained as R decreases because the irregular access pattern changes more frequently. From now on we will assume a fixed array size A_{size}. When dead code is not applied, T_s^{INSP} increases as f_{size} raises (if SP is constant, CON and f_{size} raise at the same rate). Thus, a higher lower bound is achieved if the time devoted to useful computations ($f_{size} t_s$) grows faster than the computational overhead (PT_s^{INSP}). Supposing that SP is constant, when dead code elimination is applied, the lower bound does not change because both useful computations ($A_{updated} t_s$) and overhead T_s^{INSP} remain constant as f_{size} raises.

The inspector-executor method presented in this paper preserves load-balancing, the exception being the case in which dead code elimination is not applied and the access pattern contains hot spots, i.e. array entries where most of the computation is concentrated ($SP \rightarrow 0$ and $C \rightarrow \infty$). On the other hand, the array expansion approach may unbalance workload if dead code elimination is applied. This is because as $rhs(h)$ (see Figure 1) is computed during the reduction operation that finds the partial result corresponding to the highest iteration number, it is only computed for $A_{updated}$ array elements. As a result, workload will be unbalanced if computations associated with modified elements are not uniformly distributed among processors. In other words, load-balancing is achieved if $SP \rightarrow 1$. Otherwise, the array expansion approach does not assure load-balancing because the contention distribution C of the irregular access pattern is not considered in the the mapping of computations to processors.

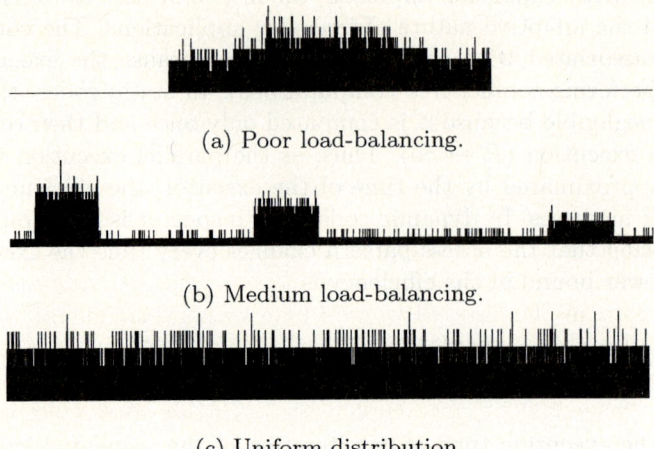

Fig. 7. Irregular access patterns

4 Performance Evaluation

In this section we present experimental results to compare the performances of our technique and the array expansion method; different parameter combinations that characterize irregular assignments are considered. The target machine was a SGI Origin2000 cc-NUMA multiprocessor. OpenMP [9] shared memory directives have been used in the parallel implementation.

4.1 Experimental Conditions

In our experiments, we have considered the parameters degree of contention (C), sparsity (SP), connectivity (CON) and reusability (R), defined in Section 3. As case study, we use the generic convex polygon scan conversion [3], a well-known rasterization algorithm from computer graphics. This algorithm presents output dependences that arise from the depiction of a set of polygons, which compose an image/scene, on a display buffer, A. A typical size for the display buffer is $A_{size} = 512 \times 512 = 262,144$ pixels. We have also considered three access patterns that represent typical cases in which the scan conversion is used (see Figure 7): a pattern with poor load-balancing that represents an scene where all the objects are displayed on a region of the buffer ($SP = 0.36$, array elements with $C > 0$ are confined in a specific region); a second pattern presents medium load-balancing that is associated with an image where most objects are concentrated on several regions of the display ($SP = 0.30$, array elements with $C > 0$ are uniformly distributed along the array, but there exist several regions with a higher C); and a third pattern that is characterized by uniformly distributed objects ($SP = 0.32$). We have considered $5,000$, $10,000$ and $20,000$ polygons to cover a reasonable range typically found in rasterization. Assuming a fixed mean number of 20 pixels per polygon, the total number of references (i.e. loop iterations, f_{size}) to the array A is $100,000$ ($CON \approx 1.20$), $200,000$ ($CON \approx 2.41$) and $400,000$ ($CON \approx 4.81$), respectively. The experimental results presented in the following sections were obtained by fixing $A_{size} = 262,144$ and $SP \approx 0.33$, on average. As a result, conclusions can be stated in terms of CON and f_{size}.

4.2 Experimental Results

When element-level dead code elimination is not applied, computational load is measured as the maximum number of loop iterations that is assigned to the processors. Both methods preserve load-balancing by assigning approximately f_{size}/P iterations to each processor, P being the number of processors. Figures 8 and 9 present execution times and speed-ups for different CON and R values. The access pattern, which is defined in terms of SP and C, is not relevant in this case. Execution times increase as CON raises because CON is related to the amount of computational load assigned to processors; it does not affect workload distribution. Note that memory overhead $\mathcal{O}(A_{size} \times P)$ prevents the execution of the array expansion approach on more than 15 processors, which is a drawback if a high number of processors is needed.

The speed-ups of the array expansion approach (dotted lines) increase as CON raises (this method does not take advantage of reusability) because the computational overhead of this method mainly depends on the reduction operation that determines the value of each array element $A(j)$, $j = 1, ..., A_{size}$, by combining the partial results computed by the processors (A_{size} and SP are constants). In the figure, speed-ups increase approximately 35% on 15 processors when CON is doubled for $A_{size} = 262,144$ and $SP \approx 0.33$. In contrast, the speed-ups of our inspector-executor technique (shaded region) depend on CON and R. In static codes ($R \rightarrow \infty$), efficiency is approximately 1 in any case (solid-star line). However, in dynamic applications, the sequential inspector imposes an upper limit on the maximum achievable speed-up (see Section 3). The curve of speed-ups for totally dynamic codes is a lower bound of the speed-up of the inspector-executor approach (see Eq. (2)). The lower bound raises when CON is increased (solid lines with $R = 0$) because the time devoted to useful computations grows faster than the computational overhead. In particular, the increment is approximately 6% on 32 processors when CON is doubled. Lower speed-ups are obtained as R decreases because the access pattern has to be rescanned a higher number of times during the execution of the program.

4.3 Results with Dead Code Elimination

The generic scan conversion algorithm depicts all the polygons that represent an image on the display buffer although, at the end, only the visible regions of the polygons remain on the display. As a result, computational resources are consumed in the depiction of invisible polygons. When element-level dead code elimination is applied, only the visible regions of the polygons are printed

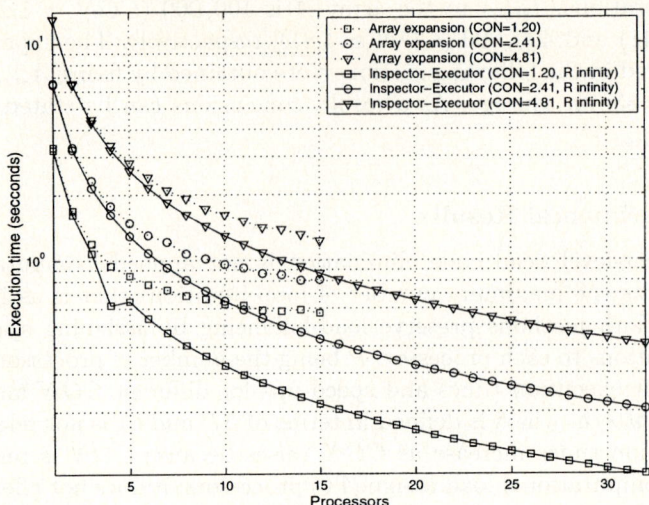

Fig. 8. Execution times

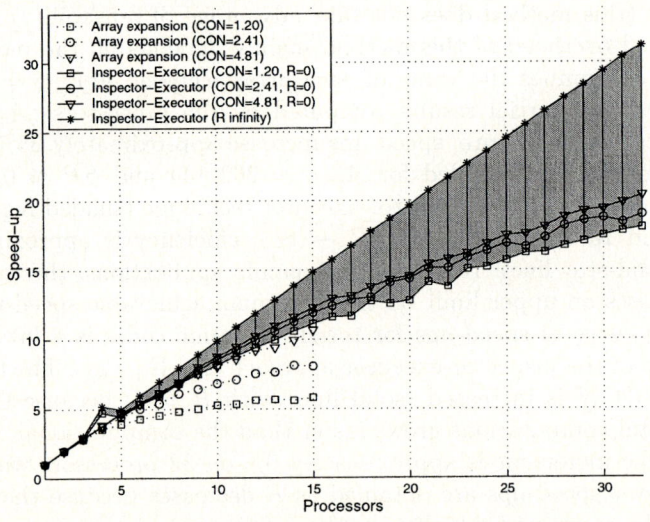

Fig. 9. Speed-ups

on the display buffer, with the corresponding saving of resources. In this case, computational load is measured as the maximum number of array elements that are computed by the processors. Figure 10 represents the computational load corresponding to the access pattern with poor load-balancing ($SP = 0.36$ and $C = 0$ for large subarrays of A) when dead code elimination is applied. Unlike our inspector-executor technique (black bars), the array expansion method (gray

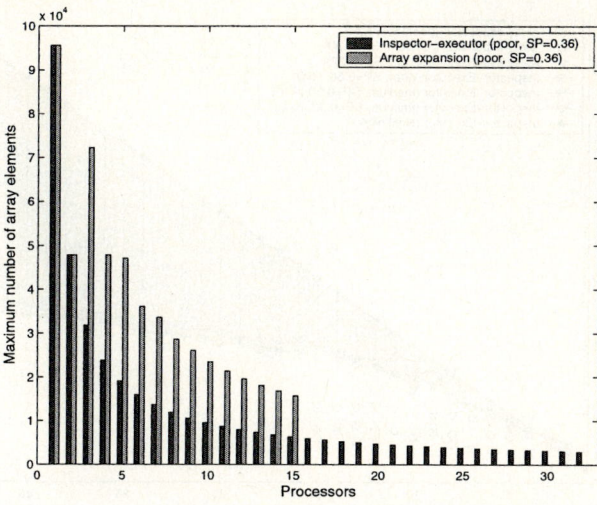

Fig. 10. Computational load when dead code elimination is applied

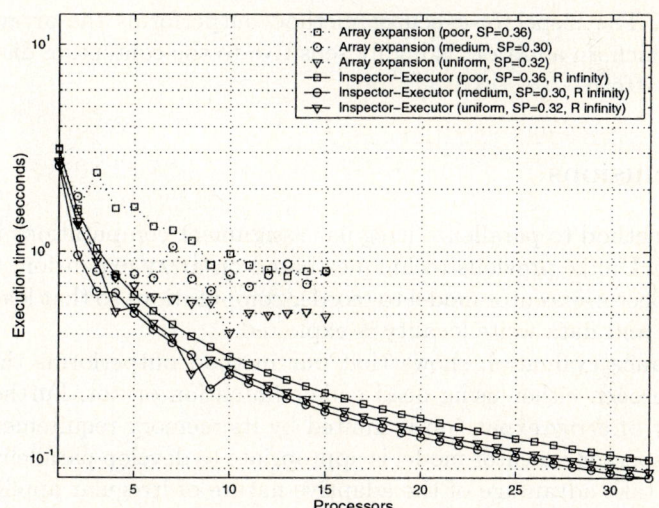

Fig. 11. Execution times when dead code elimination is applied

bars) presents load-unbalancing because array elements $A(j)$ are assigned to processors independently of the contention distribution C.

Note that workload depends on the distribution of modified array elements (SP and C), while it depends on CON if dead code elimination is not applied. Figures 11 and 12 show execution times and speed-ups when dead code elimination is applied. The parameter SP is ≈ 0.33 for all the access patterns described in Section 4.1 because load-balancing increases in the array expansion approach

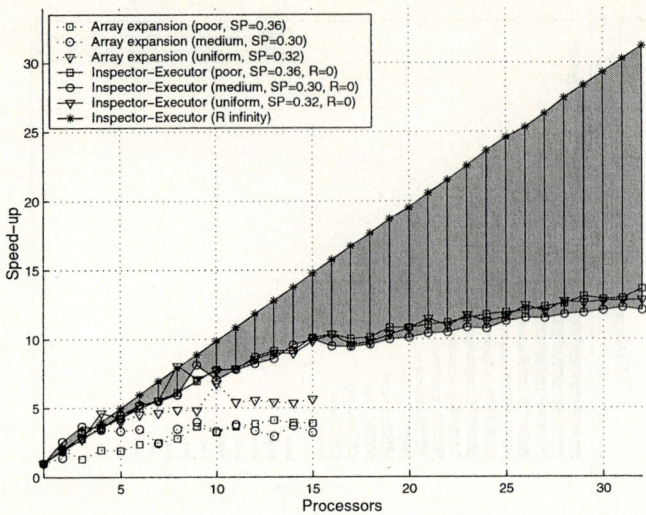

Fig. 12. Speed-ups when dead code elimination is applied

as $SP \to 1$. The inspector-executor method outperforms the array expansion technique which, in addition, is highly sensitive to the contention distribution of the access pattern.

5 Conclusions

A scalable method to parallelize irregular assignment computations is described in this work. Unlike previous techniques based on array expansion, the method uses the inspector-executor model to reorder computations so that load-balancing is preserved and data write locality is exploited.

Performance evaluation shows that our method outperforms the array expansion approach either using dead code elimination or not. Furthermore, the applicability of array expansion is limited by its memory requirements in practice. The inspector-executor model is appropriate to develop parallelization techniques that take advantage of the adaptive nature of irregular applications.

Acknowledgements

We gratefully thank Complutense Supercomputing Center in Madrid for providing access to the SGI Origin 2000 multiprocessor. This work was supported by the Ministry of Science and Technology of Spain and FEDER funds under contract TIC2001-3694-C02-02.

References

1. Arenaz, M., Touriño, J., Doallo, R.: Irregular Assignment Computations on cc-NUMA Multiprocessors. In Proceedings of 4th International Symposium on High Performance Computing, ISHPC-IV, Kansai Science City, Japan, Lecture Notes in Computer Science, Vol. 2327 (2002) 361–369
2. Arenaz, M., Touriño, J., Doallo, R.: A GSA-Based Compiler Infrastructure to Extract Parallelism from Complex Loops. In Proceedings of 17th ACM International Conference on Supercomputing, ICS'2003, San Francisco, CA (2003) 193–204
3. Glassner, A.: Graphics Gems. Academic Press (1993)
4. Gutiérrez, E., Plata, O., Zapata, E.L.: Balanced, Locality-Based Parallel Irregular Reductions. In Proceedings of 14th International Workshop on Languages and Compilers for Parallel Computing, LCPC'2001, Cumberland Falls, KY (2001)
5. Han, H., Tseng, C.-W.: Efficient Compiler and Run-Time Support for Parallel Irregular Reductions. Parallel Computing **26**(13-14) (2000) 1861–1887
6. Knobe, K., Sarkar, V.: Array SSA Form and Its Use in Parallelization. In Proceedings ACM SIGACT-SIGPLAN Symposium on the Principles of Programming Languages (1998) 107–120
7. Lin, Y., Padua, D.A.: On the Automatic Parallelization of Sparse and Irregular Fortran Programs. In Proceedings of 4th Workshop on Languages, Compilers, and Run-Time Systems for Scalable Computers, LCR'98, Pittsburgh, PA, Lecture Notes in Computer Science, Vol. 1511 (1998) 41–56
8. Martín, M.J., Singh, D.E., Touriño, J., Rivera, F.F.: Exploiting Locality in the Run-time Parallelization of Irregular Loops. In Proceedings of 31st International Conference on Parallel Processing, ICPP 2002, Vancouver, Canada (2002) 27–34
9. OpenMP Architecture Review Board: OpenMP: A Proposed Industry Standard API for Shared Memory Programming (1997)
10. Rauchwerger, L., Padua, D.A.: The LRPD Test: Speculative Run-Time Parallelization of Loops with Privatization and Reduction Parallelization. IEEE Transactions on Parallel and Distributed Systems **10**(2) (1999) 160–180
11. Saad, Y.: SPARSKIT: A Basic Tool Kit for Sparse Matrix Computations. http://www.cs.umn.edu/Research/darpa/SPARSKIT/sparskit.html (1994)
12. Turek, S., Becker, C.: Featflow: Finite Element Software for the Incompressible Navier-Stokes Equations. User Manual. http://www.featflow.de (1998)
13. Wolfe, M.J.: Optimizing Supercompilers for Supercomputers. Pitman, London and The MIT Press, Cambridge, Massachussets (1989)
14. Xu, C.-Z., Chaudhary, V.: Time Stamp Algorithms for Runtime Parallelization of DOACROSS Loops with Dynamic Dependences. IEEE Transactions on Parallel and Distributed Systems **12**(5) (2001) 433–450
15. Yu, H., Rauchwerger, L.: Adaptive Reduction Parallelization Techniques. In Proceedings of the 14th ACM International Conference on Supercomputing, Santa Fe, NM (2000) 66–77

Multi-grain Parallel Processing of Data-Clustering on Programmable Graphics Hardware

Hiroyki Takizawa[1] and Hiroaki Kobayashi[2]

[1] Graduate School of Infortmation Sciences, Tohoku University,
Aoba, Aramaki-aza, Aoba-ku, Sendai 980-8578 Japan
[2] Information Synergy Center, Tohoku University,
Aoba, Aramaki-aza, Aoba-ku, Sendai 980-8578 Japan
{tacky, koba}@isc.tohoku.ac.jp

Abstract. This paper presents an effective scheme for clustering a huge data set using a commodity programmable graphics processing unit (GPU). Due to GPU's application-specific architecture, one of the current research issues is how to bind the rendering pipeline with the data-clustering process. By taking advantage of GPU's parallel processing capability, our implementation scheme is devised to exploit the multi-grain single-instruction multiple-data (SIMD) parallelism of the nearest neighbor search, which is the most computationally-intensive part of the data-clustering process. The performance of our scheme is discussed in comparison with that of the implementation entirely running on CPU. Experimental results clearly show that the parallelism of the nearest neighbor search allows our scheme to efficiently execute the data-clustering process. Although data-transfer from GPU to CPU is generally costly, acceleration by GPU is significant to save the total execution time of data-clustering.

1 Introduction

Data clustering [1] is to group similar data units in a database, and is essential across a wide variety of research fields and their applications [2, 3, 4]. Iterative refinement clustering algorithms [5] are generally applied to finding appropriate clusters of a given data set. However, one severe problem is that the computational cost of such algorithms increases with the data set size and the dimension. For the purpose of reducing the cost for massive data clustering, therefore, many approaches to parallel data clustering have been proposed mainly using multi-processors and/or special hardware [6, 7, 8, 9].

We present an effective implementation of k-means clustering [10] on a commodity programmable graphics processing unit (GPU), which is used as a powerful SIMD-parallel coprocessor. GPU's parallel processing elements and dedicated high-bandwidth memory can drastically accelerate SIMD-parallel and streaming tasks even for non-graphics applications(e.g. [11, 12]). In addition, GPU has already become commonplace even in a low-end PC. Consequently, our scheme can significantly improve the computational efficiency of data-clustering, not requiring neither an expensive multiprocessor system nor inflexible dedicated hardware.

The GPU programming model is not universally applicable to any algorithm due to the lack of random access writes and conditionals. Our scheme hence divides the k-means algorithm into two parts. One part, called the *nearest neighbor search*, that involves huge amounts of data-parallelism without complicated control flows is executed on GPU. The other that needs random access memory writes and conditional branches is executed on CPU for effective implementation. Since the majority of the total execution time is spent for searching for the nearest cluster centroid from a data unit, accelerating the search by GPU becomes remarkable, resulting in an effective reduction in the computational cost of the k-means algorithm.

Bohn has applied graphics hardware to accelerate the self-organizing map, which is one of artificial neural networks applicable to data clustering [13]. However, since only non-programmable GPUs were available at that time, his implementation could not fully utilize the GPU performance. With a simple but effective GPU implementation, therefore, this paper examines the potential of GPU programming for data clustering.

The outline of this paper is as follows. In Section 2, we begin by briefly describing the programmable rendering pipeline of recent GPUs. Section 3 reviews k-means clustering algorithms. Section 4 presents our scheme to accelerate a k-means algorithm by dint of GPUs. In Section 5, the performance of our implementation is compared with that without using GPU. We will also discuss some implementation issues for further performance enhancement. Section 6 gives concluding remarks and our future work.

2   Graphics Processing Unit

Modern graphics hardware, such as NVIDIA's GeForce and ATI's RADEON, has two kinds of programmable processors, *vertex shader* and *fragment shader*, on the graphics pipeline to render an image. Figure 1 illustrates a block diagram of the programmable rendering pipeline of these graphics processors. The vertex shader manipulates transformation and lighting of vertices of polygons to transform them into the viewing coordinate system. Polygons projected into the viewing coordinate system are then decomposed into fragments each corresponding to a pixel on the screen. Subsequently, color and depth of a fragment is computed by the fragment shader. Finally, composition operations such as tests using depth, alpha and stencil buffers are applied to the outputs of the fragment shader to determine the final pixel colors to be written to the frame buffer.

It is emphasized here that vertex and fragment shaders are developed to utilize multi-grain parallelism in the rendering processes: the coarse-grain vertex/fragment level parallelism and the fine-grain vector component level parallelism. To exploit the coarse-grain parallelism, GPUs can operate individual vertices and fragments in parallel. That is, fragment shaders (vertex shaders) of

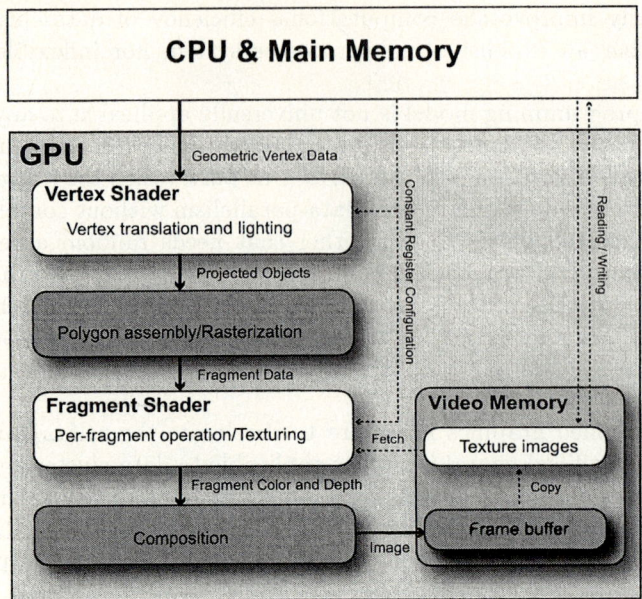

Fig. 1. Overview of a programmable rendering pipeline

recent GPUs have several processing units for parallel-processing multiple fragments (vertices). For example, NVIDIA's latest GPU has 16 processing units in the fragment shader and therefore can compute colors and depths of 16 fragments at the same time[14]. On the other hand, to exploit the fine-grain parallelism involved in all vector operations, they have SIMD instructions that can simultaneously operate on four 32-bit floating-point values within a 128-bit register. For example, one of the powerful SIMD instructions, the "multiply and add" (MAD) instruction, performs a component-wise multiply of two registers each storing four floating-point components and then does a component-wise add of the product to another register. Thus, the MAD instruction performs eight floating-point operations in one cycle.

GPU's amazing evolution on both computational capability and functionality extends application of GPUs to the field of non-graphics computations. Due to its application-specific architecture, however, it does not work well universally. To exhibit high performance for a non-graphics application, hence, we ought to consider how to bind it to GPU's programmable pipeline.

The most critical restriction in GPU programming for non-graphics applications is due to the restricted data flows in and between the vertex shader and the fragment shader. Arrows in Figure 1 show typically-permitted data flows. Both vertex and fragment shader programs have to write their outputs to write-only dedicated registers; random access writes are not provided. This is severe impediment to effective implementation of many data structures and algorithms. In addition, the lack of loop-controls, conditionals, and branching[3] is also crucial for most of practical applications. Therefore, it is not a good idea to implement an application entirely on GPUs; if the application imposes the restriction violation on the GPU programming model, the CPU should be used despite time-consuming data exchange between CPU and GPU.

From the viewpoint of data accessibility, the fragment shader is superior to the vertex shader because the fragment shader can randomly access the video memory and fetch data as texture colors. Furthermore, the fragment shader usually has more processing units than the vertex shader, and thereby the fragment shader is expected to exploit data-parallelism more effectively. Consequently, this paper presents an implementation of data clustering accelerated effectively using multi-grain parallel processing on the the fragment shader.

3 The k-Means Algorithm

In data clustering, multivariate data units are grouped according to their similarity or dissimilarity. MacQueen used the term *k-means* to denote the process of assigning each data unit to that cluster (of k clusters) with the nearest centroid[1, 10]. That is, k-means clustering employs the Euclidean distance between data units as the dissimilarity measure; a partition of data units is assessed by the squared error:

$$E[D] = \sum_{i=1}^{m} (\min_{j=1}^{k} \|\boldsymbol{x}_i - \boldsymbol{y}_j\|^2), \qquad (1)$$

where $\boldsymbol{x}_i \in R^d, i = 1, 2, \ldots, m$ is a data unit and $\boldsymbol{y}_j \in R^d, j = 1, 2, \ldots, k$ denotes the cluster centroid.

Although there are a vast variety of k-means algorithms [5], for the sake of explanation simplicity, this paper focuses on a simple and standard k-means algorithm summarized as follows:

1. Begin with any desirable initial states, e.g. initial cluster centroids may be drawn randomly from a given data set.
2. Allocate each data unit to the cluster with the nearest centroid. The centroids remain fixed through the entire data set.
3. Calculate centroids of new clusters.
4. Repeat Steps 2 and 3 until a convergence condition is met, e.g. no data units change their membership at Step 2, or the number of repetitions exceeds a predefined threshold.

The procedure above is referred to as *Forgy's method* [15] in [1]. This is essentially equivalent to the well-known LBG algorithm in vector quantization literature, named after its authors, Linde, Buzo, and Gray[16].

---

[3] The next generation GPUs, e.g. NVIDIA GeForce6800, support Shader Model 3.0 that offers dynamic controls flows.

At each repetition the assignment of m data units to k clusters in Step 2 requires km distance computations (and $(k-1)m$ distance comparisons) for finding the nearest cluster centroids, the so-called *nearest neighbor search*. The cost of each distance computation increases in proportion to the dimension of data, i.e. the number of vector elements in a data unit, d. The nearest neighbor search consists of approximately $3dkm$ floating-point operations, and thus the computational cost of the nearest neighbor search increases at the rate of $O(dkm)$. In practical applications, the nearest neighbor search consumes most of the execution time for k-means clustering because m and/or d often become tremendous. However, the nearest neighbor search involves massive SIMD parallelism; the distance between every pair of a data unit and a cluster centroid can be computed in parallel, and the distance computation can further be parallelized according to their vector components. This motivates us to implement the distance computation on recent programmable GPUs as multi-grain SIMD-parallel coprocessors.

On the other hand, there is no necessity to consider the acceleration of Steps 1 and 4 using GPU programming, because they require little execution time and further include almost no parallelism. In Step 3, cluster centroid recalculation consists of dm additions and dk divisions of floating-point values. Although most of these calculations can be performed in parallel, conditionals and random access writes are required for effective implementation of individually summing up vectors within each cluster. In addition, the divisions also require conditional branching to prevent divide-by-zero errors. Since the execution time for Step 3 is much less than that of Step 2, there is no room for performance improvement that outweighs the overheads derived from the lack of random access writes and conditionals in GPU programming. We think the CPU is suited for implementation of Steps 1, 3 and 4.

4 Mapping Data Clustering onto GPU and CPU

This section presents a novel implementation of the k-means algorithm. The overview of our implementation is shown in Figure 2. The nearest neighbor search involves multi-grain SIMD parallelism, while it is considerably time-consuming in practical uses. In our implementation scheme, therefore, the nearest neighbor search only is performed on GPU, and the remaining parts are on CPU. This division of labor leads not only to better computational efficiency, but also to easy extension of our implementation scheme to more advanced and complicated k-means clustering(e.g. [17]) thanks to the CPU's programming flexibility.

In our scheme, Step 1 first initializes the graphics library, and then copies all data units and initial cluster centroids onto GPU-side video memory. In Step 2, the nearest neighbor search is performed on GPU and the computing results, i.e. the nearest neighbor indices, are read back to CPU. According to the indices, all cluster centroids are recalculated on CPU and then passed to GPU in Step 3. Steps 2 and 3 are alternated until the termination condition is met.

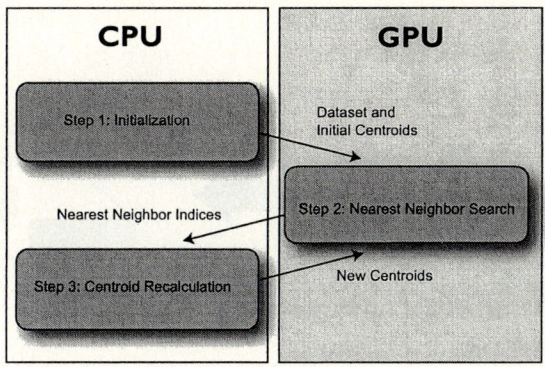

Fig. 2. Overview of the proposed implementation scheme

Our implementation scheme is aimed to effectively perform the nearest neighbor search on GPUs used as multi-grain SIMD-parallel coprocessors. To exhibit the high-performance of GPUs, we take advantage of two types of parallelism in the nearest neighbor search as mentioned in Section 3. One is the coarse-grain parallelism that all distances required for the nearest neighbor search can independently be calculated. The other is the fine-grain parallelism that all scalar operations in a vector operation can independently be done. The former parallelism is bound to GPU's fragment level parallel processing; each distance computation is bound to the per-fragment operation. The fragment shader can compute multiple distances in parallel, because it can simultaneously run several per-fragment operations using its parallel processing units. The latter parallelism is bound to GPU's vector component level parallelism. Vector operations required for the distance computation are effectively performed, because GPU's SIMD instructions perform component-wise operations between two registers each storing four floating-point components in one cycle.

In our implementation scheme, the nearest neighbor search is mapped to multi-pass rendering of a polygon of m fragments, where the polygon is rendered k times with different configuration. Figure 3 illustrates the parallel distance computation for the nearest neighbor search. Each fragment of the polygon indicates a data unit (a circle in Figure 3(a)), and each of k rendering passes corresponds to the distance computation from one cluster centroid (a star in Figure 3(a)). The fragment shader program launched per fragment calculates the distance between its own data unit and a cluster centroid in the rendering pass. Since GPUs can simultaneously operate multiple fragments, several distances of the same color in Figure 3(a) are computed in parallel.

Figure 4 shows a sample assembly code of our fragment shader program and Figure 5 illustrates how it works. A cluster centroid and its index corresponding to the rendering pass are sent to all fragment shader programs via their input registers, f[TEX1], f[TEX4], f[TEX5], f[TEX6], and f[TEX7]. Since these registers store four floating-point values, m fragments are processed using a unique

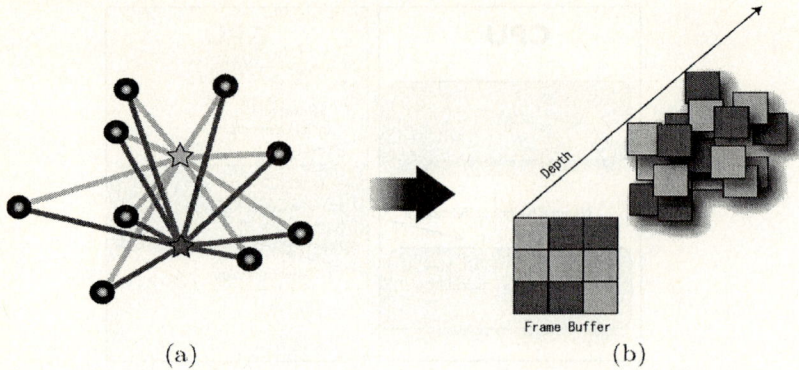

(a)   (b)

Fig. 3. Parallel distance computation. (a) distances from a cluster centroid to data units are calculated in parallel, and the centroid index and the computed distance are converted into the fragment color and depth. (b) the fragment color with the minimum depth, corresponding to the index of the nearest cluster centroid, is written to the frame buffer

```
!!FP1.0
TEX R0, f[TEX0].xyxx, TEX0, RECT;
ADDR R1, R0, -f[TEX4];
MADR R2, R1, R1, R2;
TEX R0, f[TEX0].xyxx, TEX1, RECT;
ADDR R1, R0, -f[TEX5];
MADR R2, R1, R1, R2;
TEX R0, f[TEX0].xyxx, TEX2, RECT;
ADDR R1, R0, -f[TEX6];
MADR R2, R1, R1, R2;
TEX R0, f[TEX0].xyxx, TEX3, RECT;
ADDR R1, R0, -f[TEX7];
MADR R2, R1, R1, R2;
DP4R o[DEPR], R2, {1,1,1,1};
MOVR o[COLR], f[TEX1].xyxx;
```

Fig. 4. Assembly code of fragment shader program for distance computation

16-dimensional cluster centroid in every rendering pass[4]. Meanwhile, texture-mapping is used to assign a unique data unit to each fragment. By mapping a texture image of m texels to a polygon of m fragments, each fragment is colored by a unique texel. This means that a unique data unit consisting of four floating-point values, i.e. RGBA, is assigned to each fragment. Multi-texturing is used to the assign a data unit, consisting of more than four data elements, to a fragment. The fragment shader program outputs the computed distance as the fragment depth, and the cluster centroid index as the fragment color. They are written to

---

[4] In this work, clustering of up to 16-dimensional vector data is considered, because 16-dimensional data clustering is often used for vector quantization of images.

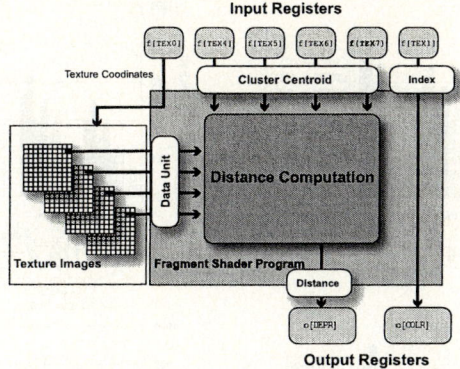

Fig. 5. The fragment shader program for distance computation

output registers, o[DEPR] and o[COLR], respectively. Then, the depth buffer test compares the fragment depth with the "so far" minimum depth, and writes the fragment color to the frame buffer only if its depth is smaller. After k rendering passes, the color corresponding to the nearest cluster centroid remains on each frame buffer element, as shown in Figure 3(b). Finally, the colors on the frame buffer are copied onto CPU-side main memory for centroid recalculation.

5 Performance Evaluation

This section evaluates the performance of our implementation scheme. All of the following results are obtained using NVIDIA GeForce5900Ultra running at 450MHz and Intel Pentium 4 3.2GHz with 3GB main memory. The operating system is Linux whose kernel version is 2.6.6 with NVIDIA's kernel module 1.0-6106. Our program code written in C++ uses OpenGL and NVIDIA's extensions, and is compiled with GNU C++ compiler 3.0.4 with "-O3" options.

In this work, the performance of our implementation scheme is investigated changing k, m, and d. The performance is assessed regarding the elapsed time for each stage of the k-means algorithm, and is compared to that of the implementation entirely on CPU without GPU co-processing.

Figure 6 shows the total execution time required for ten repetitions of the k-means algorithm in which no other termination condition is employed for fair comparison. Figure 7 shows the breakdown of the total execution time of our GPU implementation. Though the total execution times of both implementations have $O(dkm)$ growth rates, the execution time of our scheme grows more slowly. This means that our scheme can effectively perform the nearest neighbor search, which dominates the total execution time as k, m, and/or d increase. The GPU used here can perform color calculation and depth buffer test of up to four

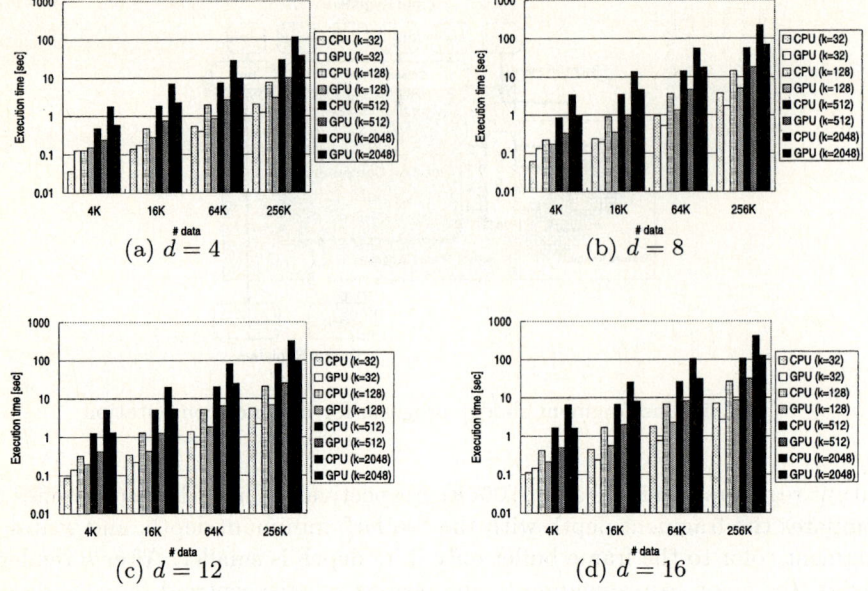

Fig. 6. Total execution time for k-means clustering

fragments per cycle; it can compute up to four distances at the same time. Moreover, each pair of `ADDR` and `MADR` instructions in Figure 4 performs 12 floating-point operations in two cycles. As a result, the GPU implementation can achieve better performance than the CPU implementation, even though the CPU used here is more than seven times faster than the GPU from the viewpoint of the clock frequency.

If the times for the stages other than the nearest neighbor search, such as initialization, are dominant in the total execution time, the CPU implementation is superior to our implementation. However, their ratios to the total execution time rapidly reduce with an increase in k, m, and/or d. Therefore, their execution times would be non-dominant in practical large-scale applications. Remember that centroid recalculation needs $d(k+m)$ floating-point operations, while the nearest neighbor search requires km distance computations consisting of approximately $3dkm$ floating-point operations. In addition, it is experimentally shown that the execution time for initialization increases with m very slowly. Accordingly, the superiority of our implementation scheme becomes more remarkable for massive data clustering.

We also discuss the performance of the data transfer between GPU and CPU. One difficulty in using GPU as a coprocessor is that the data transfer between GPU and CPU is very slow. Since our implementation needs the data transfer, its cost is one of the most important issues.

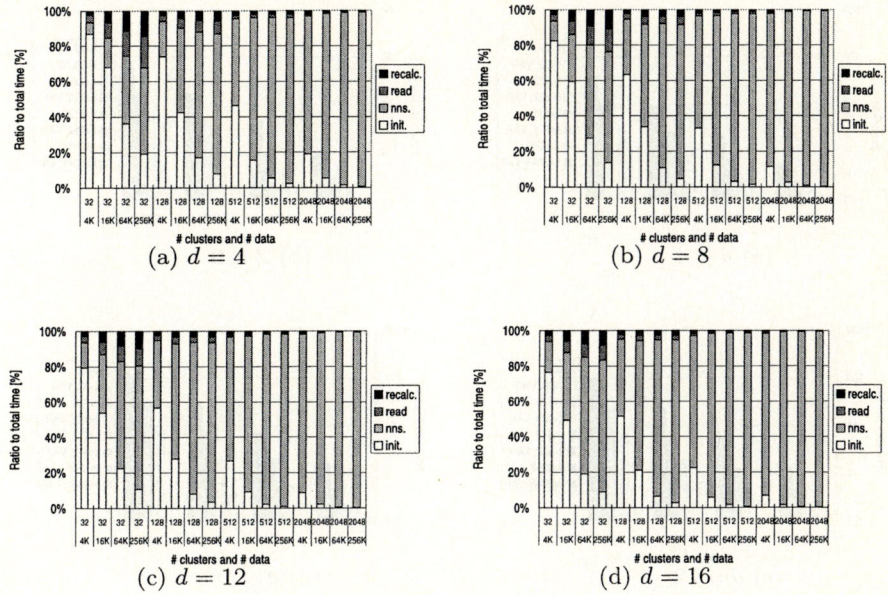

Fig. 7. Breakdown of the execution time

In our implementation, the data transfer from CPU to GPU at each rendering pass is not a bottleneck of the nearest neighbor search. This is because the large dataset has been stored in the GPU-side video memory in advance; only the geometry data of a polygon, including texture coordinates as a cluster centroid, are transferred at each rendering pass. Figure 8 depicts the experimental results with/without NVIDIA's vertex array range (VAR) extension that allows our implementation to store all geometry data on GPU-side video memory in advance. This figure clearly shows that enabling the VAR extension does not lead to the performance improvement and results in the same performance as shown in Figure 6. Accordingly, it is experimentally validated that our implementation scheme can perform the nearest neighbor search without being interrupted by the data transfer.

On the other hand, retrieving the nearest neighbor search results from the GPU-side video memory consumes a certain part of the total execution time especially for small data. The data retrieval from GPU-side video memory to the main memory is still slow; this forecasts that the entire implementation on GPUs may lead to further acceleration of the k-means clustering if GPUs support dynamic control flows necessary for effective implementation of centroid recalculation. Since the next generation GPUs with shader model 3.0 provide dynamic control flows, it would become easy to implement the k-means algorithm entirely on such GPUs. The entire implementation will be investigated in our future work.

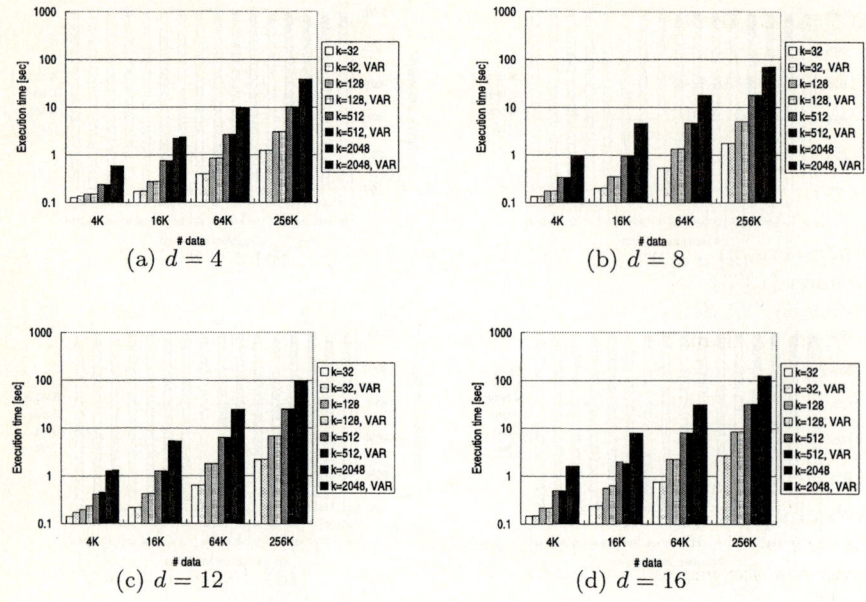

Fig. 8. Total execution time with/without the VAR extension

6 Concluding Remarks

In this paper, we have proposed an effective implementation scheme of the k-means algorithm using modern programmable GPUs. In the proposed scheme, a GPU is used as a multi-grain SIMD-parallel coprocessor to accelerate the nearest neighbor search, which consumes a considerable part of the execution time in the k-means algorithm. The distances from one cluster centroid to several data units are computed in parallel. Each distance computation is further parallelized by component-wise SIMD instructions. As a result, the implementation using GPU significantly improve the computational efficiency of the nearest neighbor search.

Experimental results clearly show the superiority of the proposed implementation scheme especially for practical large-scale data clustering. Acceleration of the nearest neighbor search by GPU is significant to save the total execution time in spite of the overhead of the data transfer between GPU and CPU.

Although the computational cost of data clustering is efficiently reduced using the GPU programming model proposed in this paper, it is still too expensive for giga- and tera-scale datasets. We are now planning to develop the implementation of data clustering entirely on the next generation GPUs that provide dynamic control flows. Furthermore, hierarchically-parallel data clustering on a PC cluster system in which each node is equipped with a programmable GPU is also an interesting research issue. These will be addressed in our future work.

Aknowledgments

This research was partially supported by Grants-in-Aid for Scientific Research(B) #14380132 and Young Scientists(B) #15700124.

References

1. Anderberg, M.: Cluster Analysis for Applications. Academic Press Inc., NY and London (1973)
2. Kohonen, T.: Self-Organizing Maps. Springer-Verlag, New York (1995)
3. Fayyad, U., Haussler, D., Stolorz, P.: KDD for science data analysis: Issues and examples. In: the Second International Conference on Knowledge Discovery and Data mining (KDD-96), AAAI Press (1996)
4. Gersho, A., Gray, R.: Vector Quantization and Signal Compression. Kluwer Academic Publishers, Norwell, MA (1992)
5. Everitt, B., Landau, S., Leese, M.: Cluster Analysis. 4th edn. Oxford University Press Inc., NY (2001)
6. Kobayashi, K., Kiyoshita, M., Onodera, H., Tamaru, K.: A memory-based parallel processor for vectror quantization: FMPP-VQ. IEICE Trans. Electron. **E80-C** (1997) 970–975
7. Abbas, H.M., Bayoumi, M.M.: Parallel codebook design for vector quantization on a message passing MI MD architecture. Parallel Computing **28** (2002) 1079–1093
8. Parhi, K., Wu, F., Genesan, K.: Sequential and parallel neural network vector quantizers. IEEE trans. Computers **43** (1994) 104–109
9. Manohar, M., Tilton, J.: Progressive vector quantization on a massively parallel SIMD machine with application to multispectral image data. IEEE transactions on Image Processing **5** (1996) 142–147
10. MacQueen, J.: Some methods for classification and analysis of multivariate observations. In: the fifth Berkley Symposium on Mathematical Statistics and Probability. Volume 1., Berkley, the University of California Press (1967) 281–297
11. Thompson, C.J., Hahn, S., Oskin, M.: Using modern graphics architectures for general-purpose computing: A fram ework and analysis. International Symposium on Microarchitecture(MICRO), Turkey (2002)
12. Moreland, K., Angel, E.: The FFT on a GPU. In SIGGRAPH/Eurographics Workshop on Graphics Hardware 2003 Proceedings (2003) 112–119
13. Bohn, C.A.: Kohonen feature mapping through graphics hardware. Computational Intelligence and Neuroscience (1998)
14. NVIDIA Corporation: GeForce 6800 product web site (2004) http://www.nvidia.com/page/geforce_6800.html.
15. Forgy, E.: Cluster analysis of multivariate data: Efficiency vs. interpretability of classification. Biometrics **21** (1965) 768–769 (Abstract)
16. Linde, Y., Buzo, A., Gray, R.: An algorithm for vector quantizer design. IEEE Transactions on Communications **COM-28** (1980) 84–95
17. Patané, G., Russo, M.: The enhanced LBG algorithm. Neural Networks **14** (2001) 1219–1237

A Parallel Reed-Solomon Decoder on the Imagine Stream Processor[1]

Mei Wen, Chunyuan Zhang, Nan Wu, Haiyan Li, and Li Li

Computer School, National University of Defense Technology,
Chang Sha, Hu Nan, P. R. of China 410073
wenmei8086@163.com

Abstract. The increasing gap between processor and memory speeds is a well-known problem in modern computer architecture. Imagine stream architecture can solve bandwidth bottleneck by its particular memory hierarchy and stream processing for computationally intensive applications. Good performance has been demonstrated on media processing and partial scientific computing domains. Reed-Solomon (RS) codes are powerful block codes widely used as an error correction method. RS decoding demands a high memory bandwidth and intensive ALUs because of complex and special processing (galois field arithmetic), and real time requirement. People usually use specialized processor or DSP to solve it that gains high performance but lacks flexibility. This paper presents a software implementation of a parallel Reed-Solomon decoder on the Imagine platform. The implementation requires complex stream programming since the memory hierarchy and cluster organization of the underlying architecture are exposed to the Imagine programmer. Results demonstrate that Imagine has comparable performance to TI C64x. This work is an ongoing effort to validate the stream architecture is efficient and makes contribution to extend the application domain.

1 Introduction

RS codes[1] are powerful block codes widely used as an error correction method in the areas such as digital communication, digital disc error correction, digital storage, wireless data communication systems etc. A RS encoder takes a block of digital data and adds extra "redundant" bits. And the RS decoder processes received code block and attempts to correct burst errors that occur in transmission or storage. RS(n,k) is defined over the finite field $GF(2^m)$. It means that the length of each block is n where $n \leq 2^m-1$ including k data symbols and (n-k) parity symbols. The decoding process uses this parity information to identify and correct up to t errors, where $t=(n-k)/2$.

Imagine[4] is a prototype processor of stream architecture[2] developed by Stanford University in 2002, which is designed to be a stream coprocessor for a general pur-

---

[1] This work was supported by the 973 Project and the 863 Project(2001AA111050) of China.
[2] There are several kinds of stream architectures, the common feature is to take stream as architectural primitives in hardware. In this paper, stream architecture is Imagine stream architecture.

pose processor that acts as the host. It contains host interface, stream controller, streaming memory system, microcontroller, 128k stream register file (SRF), eight arithmetic clusters, local register file (LRF) and network interface. Each cluster consists of eight functional units: 3 adders, 2 multipliers, 1 divide/square root unit, 1 communication unit and 1 local scratch-pad memory. The input of each functional unit is provided by LRF in a cluster. The microcontroller issues VLIW instructions to all the arithmetic clusters in a SIMD manner. The main idea of stream processing is organizing the related data words into a record. The streams are ordered finite-length sequences of data records of an arbitrary type (records in one stream are of the same type). The stream model decomposes applications into a series of computation kernels that operate on data streams. A kernel is a small program executed in arithmetic clusters that is repeated for each successive element of its input streams to produce output stream for the next kernel in the application. Imagine can be programmed at two levels: stream-level (using StreamC) and kernel-level (using KernelC) [2, 3].

2 Imagine Implementation

The Peterson-Gorenstein-Zierler (PGZ) algorithm[5] is a popular method for RS decoding. We parallelize and optimize the PGZ algorithm so that it can be adaptive to Imagine's memory hierarchy and parallel processing.

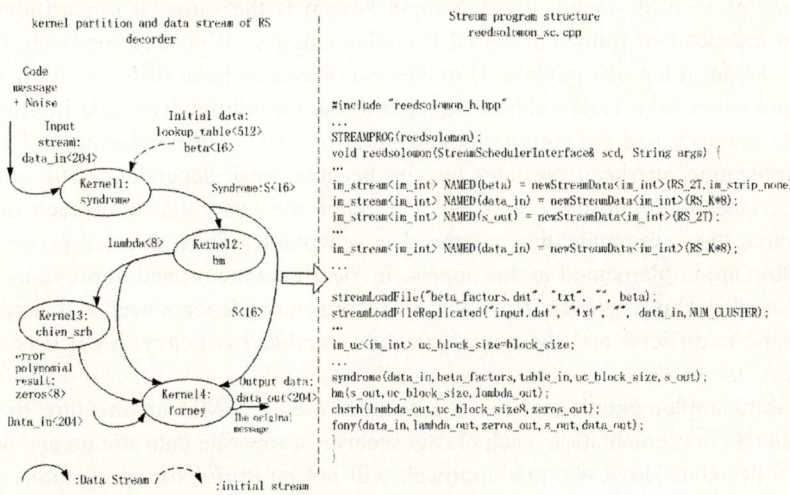

Fig. 1. Stream/kernel diagram and stream program structure for RS decoding

RS decoding application has natural stream features, and there is no dependence between RS code blocks. According to the PGZ algorithm, the whole RS decoding process can be decomposed into four kernels: *syndrome*, *bm*, *chsrh* and *forney*, respectively corresponding to syndrome computation, BM algorithm, Chien search and Forney algorithm. The data flow diagram is shown in Figure 1.Figure 1 shows that the relationship between kernels of RS decoder is a complex producer-consumer model where streams are produced and consumed, and four kernels are organized into a four-stage pipeline (not including initialization).

For RS decoding algorithm, the parallelism exploited on Imagine architecture includes instruction-level parallelism (ILP), data-level parallelism (DLP) and thread-level parallelism (TLP).

An ILP approach to partition the RS algorithm for Imagine sends elements from a single stream to all eight clusters. Obviously, the data between clusters is redundant. If necessary, records are computed redundantly on all clusters. But it is the simplest parallel approach. It minimizes serial communication blocks, and makes best use of large computing capability of Imagine architecture. It brings waste to bandwidth, LRF and computing capability. For those records with weak dependence, DLP is a better parallel approach.

DLP can exploit parallelism very well and decrease the redundancy at the same time. Here we introduce a new conception-frame. Frame is a field that consists of one or more related records. The records in a frame are not always continuous, but the interval is best to be integral times of cluster number. The dependence between frames is very weak, while the data dependence in a frame is complicated so it can be regarded as a child stream. Though input stream is the same, it is partitioned into several independent frames in the DLP implementation. With this approach, the particular stream reference method[3] makes every cluster have different frames every time and keeps high bandwidth throughput at the same time. The data in cluster has little redundancy and the computing capability of ALU is utilized enough. However, communication overhead becomes heavier because weak dependence still exists between frames of input and output stream. When the computation of each record is very large, the communication overhead is acceptable. Thus, *bm* and *forney* of RS algorithm are implemented in this approach. *Syndrome* mentioned in previous section can also adopt this approach. However, its records have dependence, so that partitioning frame is difficult and the communication overhead is heavy. So ILP is a better choice.

A third implementation of the RS decoder uses a SIMD architecture to exploit TLP. In this implementation, each cluster receives a separate data stream and acts as a full RS decoder. However, this approach will not be useful for applications that require only a fast real-time RS decoder because of its long latency.

The approaches above can be mixed to exploit parallelism efficiently. During the practical design process, it is necessary to consider the parallelism of each stream and kernel, and choose a reasonable parallel approach. More details refer to [9].

3 Performance Evaluation[3]

A statistic of bandwidth requirement of memory units in each level on Imagine and general purpose processor is in [10]. The conclusion is that the memory hierarchy of stream processor and its stream processing make bandwidth requirement distribute according to memory hierarchy (shown in figure 3(a)). The main data access is centered in LRF. It reduces the off-chip memory reference and solves the bandwidth bottleneck, so that it can increase performance greatly.

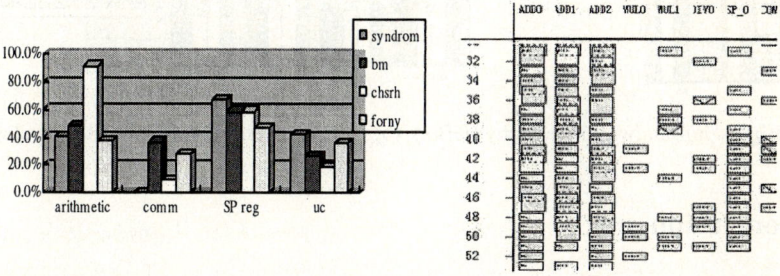

Fig. 2. (a)The utilization factor of main functional units (b) *chsrh* kernel schedule

The utilization factor of functional units of Imagine can achieve a very high level (more than 90%) by exploiting proper ILP or DLP, as shown in Figure 2(a). At the same time, Figure 2(a) expresses that the program features of each kernel. Then the bottlenecks of kernels are: the reference of scratch-pad register in the *syndrome* kernel, communication in the *bm* kernel and computation in the *chsrh* kernel, that are accordant with the theory results. It's helpful to extend the hardware of Imagine. Taking the *chrsh* kernel as an example, the main computing feature of this kernel is addition. Figure 2(b) shows a partial visualization of the inner loop of *chsrh* after scheduling and software pipelining (unrolling loop twice, and This schedule was created using iscd[2]). Each functional unit in an Imagine is shown across the top, and the cycles in the kernel are down the side. Rectangles indicate an operation on the functional unit and all the envelop-style operations are added during the scheduling process. The times of looking up tables are 16, which are accordant with the operations of SP unit in scratch-pad (SP) register shown in the figure. Addition units are almost filled and the multiplier and divide unit almost not used are spare. High local bandwidth supports the computation units to run with full loads.

For comparison, we take TMS320C67x as a reference of general purpose processor and TMS320C64x as a reference of special purpose processor because of instruction GMPY4 for RS decoding. The TI DSP algorithm is similar to the Imagine

---

[3] All the data of TI DSP in this paper are obtained by CCS2 in the -o3 flag, fast simulator. The RS decoding codes are provided by TI Corporation[5] and the C code is not optimized. All the data of Imagine in this paper are obtained by Imagine simulator ISIM.

version. Figure 3(b) presents that the running time (not including initial time) of each module on different chip simulators for RS (204,188) which is widely used for ADSL modem. Because Galois field multiply is the majority of the whole computation of RS decoding, the execution difference of Galois field multiply in different processors is the key of performance gap. As a result, the performance gap between general processor and special processor are very clear. The Imagine has comparable performance to C64x.

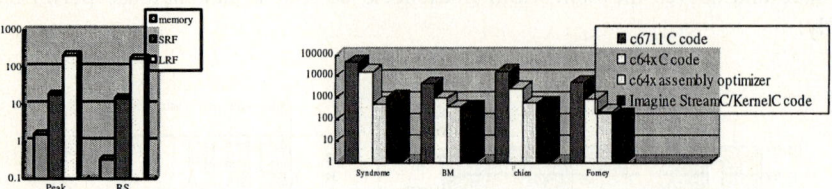

Fig. 3. (a)Bandwidth hierarchy (GB/s) (b) Performances for RS (204,188) (cycles)

4 Conclusion

This paper discusses how to develop an efficient implementation of a complete RS decoder solution on the Imagine platform and compare the experimental results with several TI DSPs. We can find that the benefits that memory hierarchy of stream architecture and stream processing bring to stream applications are significant. It is a bandwidth-effect architecture which supports a large number of ALUs. This work has shown that stream processing is applicable to RS decoding.

Researches show if application could be expressed in streams (data stream couldn't be reused once flowing, there is perfect producer and consumer model) is very important to make use of stream architecture's advantage. Typical stream applications including media processing, RS decoding, network processing and software defined radios have native stream feature, so they are best suited for stream architecture. Some classes of scientific problems are well-suited for stream processor[6]. However, Imagine processor doesn't achieve high performance for application not well-suited for its architecture, like transitive closure[7]. Its complex programming is another shortcoming. Programmers need to organize data into stream, and write program at two levels. They need to, and pay more attention because of the visible memory hierarchy. The following work is going on extending the domains of stream application, and researching on stream architecture[8] and stream scheduling deeply.

References

1. Shu Lin, D.J.Costello, Error Control Coding Fundamentals and applications, 1983
2. Peter Mattson et al, Imagine Programming System Developer's Guide, http://cva.stanford.edu, 2002.
3. Beginner's guide to Imagine Application Programming, http://cva.stanford.edu, March 2002.

4. Imagine project, http://cva.stanford.edu/Imagine/project/.
5. TI, Reed Solomon Decorder: TMS320C64x Implementation, 2000.
6. Jung ho Ahn, W.J.Dally et al, Evaluating the Imagine Stream Architecture, ISCA2004.
7. Gorden griem, Leonid oliker, Transitive Closure on the Imagine Stream Processor, 5th workshop on media and streaming processors, San Diego, CA, December 2003.
8. Mei Wen, Nan Wu, Chunyuan Zhang et al, Multiple-dimension Scalable Adaptive Stream Architecture, In: Proc of Ninth Asia-pacific Computer System Architecture Conference, Springer's LNCS 3189, 2004. 199~211
9. Nan Wu, Mei Wen, et al, Programming design patterns for the Imagine stream architecture, 13th National Conference on Information Storage Technology, Xi'an, China, 2004
10. Mei Wen, Nan Wu et al, Research of Stream Memory Hierarchy, 13th National Conference on Information Storage Technology, Xi'an, China, 2004

Effective Nonblocking MPI-I/O in Remote I/O Operations Using a Multithreaded Mechanism

Yuichi Tsujita

Department of Electronic Engineering and Computer Science,
Faculty of Engineering, Kinki University,
Umenobe, Takaya, Higashi-Hiroshima, Hiroshima 739-2116, Japan
tsujita@hiro.kindai.ac.jp

Abstract. A flexible intermediate library named Stampi realizes seamless MPI operations on interconnected parallel computers. Dynamic process creation and MPI-I/O operations both inside a computer and among computers are available with it. MPI-I/O operations to a remote computer are realized by MPI-I/O processes of the Stampi library which are invoked on a remote computer using a vendor-supplied MPI-I/O library. If the vendor-supplied one is not available, a single MPI-I/O process is invoked on a remote computer, and it uses UNIX I/O functions instead of the vendor-supplied one. In nonblocking MPI-I/O functions with multiple user processes, the single MPI-I/O process carries out I/O operations required by the processes sequentially. This results in small overlap of computation by the user processes with I/O operations by the MPI-I/O process. Therefore performance of the nonblocking functions is poor with multiple user processes. To realize effective I/O operations, a Pthreads library has been implemented in the MPI-I/O mechanism, and multi-threaded I/O operations have been realized. The newly implemented MPI-I/O mechanism has been evaluated on inter-connected PC clusters, and higher overlap of the computation with the I/O operations has been achieved.

1 Introduction

MPI [1, 2] is the de facto standard in parallel computation, and almost all computer vendors have provided their own MPI libraries. But they do not support MPI communications among different computers. To realize such mechanism, Stampi [3] was developed.

Recently, data-intensive scientific applications require a parallel I/O system, and a parallel I/O interface named MPI-I/O was proposed in the MPI-2 standard [2]. Although it has been implemented in several kinds of MPI libraries for I/O operations inside a computer (local MPI-I/O), MPI-I/O operations to a remote computer (remote MPI-I/O) have not been supported. Stampi-I/O [4] was developed as a part of the Stampi library to realize this mechanism. Users can execute remote MPI-I/O operations using a vendor-supplied MPI-I/O library with the help of its MPI-I/O processes which are invoked on a remote computer. When the vendor-supplied one is not available, a single MPI-I/O process is invoked, and it uses UNIX I/O functions instead of the vendor-supplied one (pseudo MPI-I/O method) [5].

Nonblocking MPI-I/O operation has advantage in performance compared with blocking one because overlap of computation and I/O operations is available in the nonblocking one. Unfortunately, there is not significant performance advantage in Stampi's nonblocking MPI-I/O functions compared with performance of its blocking ones due to a single task of the MPI-I/O process when UNIX I/O functions are used. To improve the performance, a Pthreads library [6] has been introduced in the MPI-I/O mechanism.

In the following sections, outline, architecture, and preliminary performance results of the MPI-I/O mechanism are described.

2 Implementation of a Pthreads Library in Stampi

A single MPI-I/O process of the Stampi library is invoked on a remote computer by user processes, and it uses UNIX I/O functions when a vendor-supplied MPI-I/O library is not available on the computer. Besides, an I/O request of a collective MPI-I/O function is translated into a combination of non-collective I/O requests, and those requests are operated sequentially by the MPI-I/O process. These architectural constraints result in poor overlap of computation by user processes in the foreground with I/O operations by the MPI-I/O process in the background. Thus, performance of its nonblocking functions is poor, typically in the collective case. To improve the performance, a LinuxThreads library [7], which is one of the Pthreads libraries, has been introduced in the MPI-I/O mechanism on a Linux PC cluster. With this implementation, the overlap ratio has been improved, and an execution time which is required to issue nonblocking I/O function has been minimized. Rest of this section describes the details of the mechanism.

2.1 Architecture of an MPI-I/O Mechanism

Architectural view of the MPI-I/O mechanism in Stampi is depicted in Figure 1. In an interface layer to user processes, intermediate interfaces which have MPI APIs (a part of a Stampi library) were implemented to relay messages between user processes and underlying communication and I/O systems.

Stampi supports both local and remote MPI-I/O operations with the same MPI-I/O APIs. In local MPI-I/O operations, a vendor-supplied MPI-I/O library is used. If the library is not available, UNIX I/O functions are used. While remote MPI-I/O operations are carried out with the help of MPI-I/O processes on a remote computer. I/O requests from the user processes are translated into message data, and they are transfered to the MPI-I/O processes. Bulk data are also transfered via the same communication path. The MPI-I/O processes play remote MPI-I/O operations using a vendor-supplied MPI-I/O library. If the vendor-supplied one is not available, the pseudo MPI-I/O method is used.

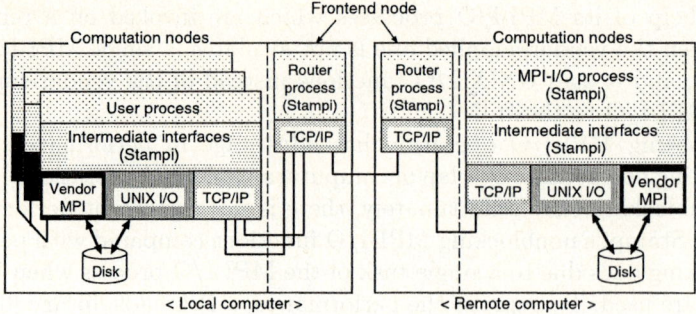

Fig. 1. Architecture of an MPI-I/O mechanism in Stampi

2.2 Execution Mechanism

Stampi supports both interactive and batch modes in executing an MPI program. Here, execution method of remote MPI-I/O operations with an interactive system which is illustrated in Figure 2 is explained. Firstly, an MPI start-up

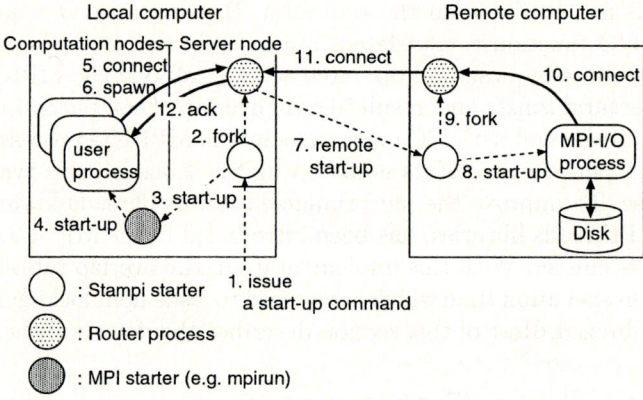

Fig. 2. Execution mechanism of remote MPI-I/O operations from a PC cluster to a remote computer

process (MPI starter) and a router process are initiated by a Stampi start-up command (Stampi starter). Then the MPI starter initiates user processes. When they call MPI_File_open(), the router process kicks off another Stampi starter process on a remote computer with the help of a remote shell command (rsh or ssh). Secondly, the starter kicks off an MPI-I/O process, and it opens a specified file. Besides, a router process is invoked on an IP-reachable node if computation nodes are not able to communicate outside directly. Remote MPI-I/O operations are available via the communication path established in this strategy. After the I/O operations, the file is closed and the MPI-I/O process is terminated when MPI_File_close() is called by the user processes.

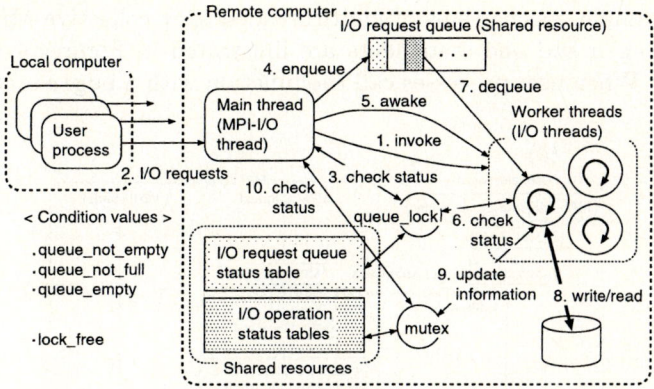

Fig. 3. Mechanism of a multithreaded MPI-I/O process

2.3 Mechanism of a Multithreaded MPI-I/O Process

With the help of the LinuxThreads library, multithreaded I/O operations have been realized in an MPI-I/O process on a Linux PC cluster. Figure 3 depicts an architecture of the mechanism. Once an MPI-I/O process is created, worker threads (I/O threads) are invoked by a main thread (MPI-I/O thread) of it using pthread_create(), and an I/O request queue is prepared. As this system is a prototype, the number of the threads is specified in a program code of it. Capability to specify the number in a user program is planed as a future work. Later, an information table for the I/O request queue (I/O request queue status table) and I/O operation status tables are created, and parameters associated with the I/O operations (I/O request, message data size, I/O status values, and so on) are stored in those tables. As the both tables are shared resources, mutual exclusion using pthread_mutex_lock()/pthread_mutex_unlock() is done during the I/O operations. Besides, condition values, queue_not_empty, queue_not_full, and queue_empty, are prepared to manage the I/O request queue. When I/O requests are sent from user processes to the MPI-I/O process, those requests and related parameters are enqueued at first by the MPI-I/O thread. Secondly, the MPI-I/O thread awakes all the I/O threads using pthread_cond_broadcast(), and each I/O thread receives a signal by pthread_cond_wait(). One of the I/O threads dequeues an I/O request and carries out a requested I/O operation and other I/O threads go to sleep state. After the I/O operation, it updates the values in the corresponding I/O status table and goes to sleep state. Completion of the I/O operation is detected by the MPI-I/O thread with checking the associated I/O operation status table.

When MPI_File_close() is issued by the user processes, a signal which terminates all the I/O threads and the tables is sent from the MPI-I/O thread to every I/O threads. Then the I/O request queue and the tables are deleted, and the MPI-I/O process which consists of all the threads is terminated.

As an example, mechanisms of multithreaded split collective MPI-I/O functions with **begin** and **end** statements are illustrated in Figures 4 (a) and (b), respectively. When user processes call the function with a **begin** statement, sev-

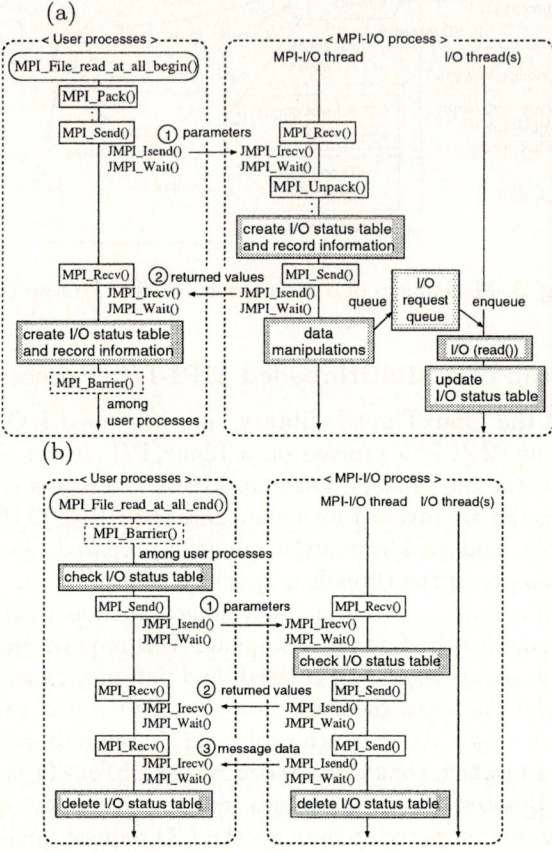

Fig. 4. Mechanisms of split collective read MPI-I/O functions with (a) **begin** and (b) **end** statements in remote MPI-I/O operations. MPI functions in solid line rectangles are MPI interfaces of Stampi. Internally, Stampi-supplied functions such as JMPI_Isend() are called by them

eral parameters which are associated with the I/O operation including an I/O request (message data size, rank of a user process which issues I/O request, and so on) are packed in a user buffer using MPI_Pack(). Then the buffer is transferred to an MPI-I/O thread using MPI_Send() and MPI_Recv() of the Stampi library. Inside the functions, Stampi-supplied underlying communication functions such as JMPI_Isend() are used for non-blocking TCP socket communications. After the message transfer, the received buffer is unpacked by MPI_Unpack(), and the I/O request and other parameters are retrieved from it. Then an I/O status table

Table 1. Specifications of PC clusters

	PC cluster (I) (DELL PowerEdge 600SC × 4)	PC cluster (II) (DELL PowerEdge 1600SC × 5)
CPU	Intel Pentium-4 2.4 GHz	Intel Xeon 2.4 GHz (dual)
Chipset	ServerWorks GC-SL	ServerWorks GC-SL
Memory	1 GByte DDR SDRAM	2 GByte DDR SDRAM
Local disk	40 GByte (ATA-100 IDE)	73 GByte (Ultra320 SCSI)
Ethernet interface	Intel PRO/1000 (on-board)	Intel PRO/1000-XT (PCI-X board)
Linux kernel	2.4.19-1SCORE (all nodes)	2.4.20-20.7smp (server node) 2.4.19-1SCOREsmp (computation nodes)
Network driver	Intel e1000 version 5.2.52	
MPI library	MPICH-SCore based on MPICH version 1.2.4	
Ethernet switch	NETGEAR GS108	3Com SuperStack4900

is created on the MPI-I/O process and they are stored in it. In addition, a ticket number, which is issued to identify each I/O operation on the MPI-I/O process, is also stored. After this operation, the ticket number and related parameters are sent to the user processes. Then, the user processes create own I/O status table and store them in it. On the MPI-I/O process, the queueing and de-queueing of the I/O request and related parameters which are previously described are carried out among the MPI-I/O thread and I/O threads.

To detect completion of the I/O operation, a split collective read function with an end statement is called. The stored information values in the I/O status table of each user process are retrieved, and the I/O request, ticket number, and related parameters are sent to the MPI-I/O thread. It finds the corresponding table which has the same ticket number, and information values associated with the I/O operation are retrieved from the table. Finally, several parameters and read data are sent to the user processes.

3 Performance Measurement

Performance of the MPI-I/O mechanism was measured on interconnected PC clusters using an SCore cluster system [8]. Specifications of the clusters are summarized in Table 1. As a server node of a PC cluster I acted as a computation node, the total number of computation nodes was four. While a PC cluster II consisted of one server node and four computation nodes. Network connections among PC nodes of the clusters I and II were established on 1 Gbps bandwidth network with full duplex mode via Gigabit Ethernet switches, NETGEAR GS108 and 3Com SuperStack4900, respectively. Interconnection between those switches was also made using 1 Gbps bandwidth network with full duplex mode.

In the both clusters, an MPICH-SCore library [9] which is based on an MPICH [10] version 1.2.4 was available, and it was used in an MPI program which was executed on the cluster I.

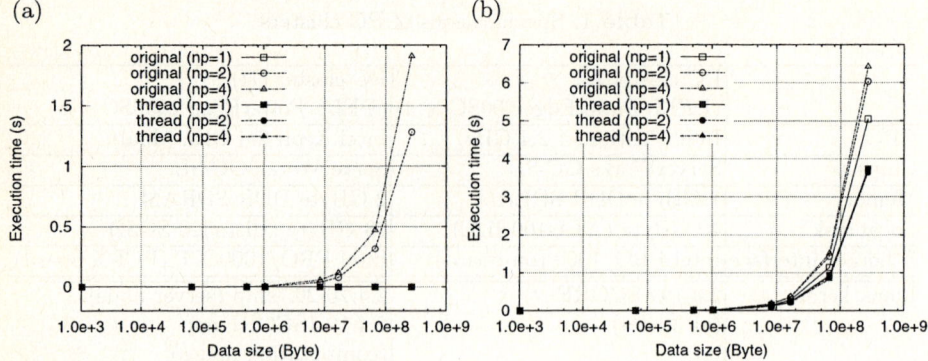

Fig. 5. Execution times of (a) read and (b) write remote MPI-I/O operations using original and multithreaded split collective MPI-I/O functions from a PC cluster I to a server node of a PC cluster II, where *original* and *thread* denote I/O operations by the original and multithreaded functions, respectively. *np* in the parentheses denotes the number of user processes

A router process was not invoked in this test because each computation node was able to communicate outside directly. Message data size was denoted as the size of whole message data to be transfered. A message data was split evenly among the user processes.

In performance measurement of remote I/O operations using nonblocking MPI-I/O functions from the cluster I to the cluster II, two kinds of nonblocking MPI-I/O functions;

– split collective MPI-I/O functions with an explicit offset value
  (MPI_File_read_at_all_begin()/ MPI_File_write_at_all_begin()) and
– nonblocking MPI-I/O functions with a shared file pointer
  (MPI_File_read_ordered_begin()/ MPI_File_write_ordered_begin()),

were evaluated. The two kinds of functions were selected for performance comparison between collective and non-collective cases. In each case, performance with and without a LinuxThreads library (multithreaded and original mechanisms, respectively) was measured. The multithreaded mechanism had two I/O threads and four segments in an I/O request queue. In this test, TCP_NODELAY option was activated by the Stampi start-up command to optimize data transfer among the two clusters.

Execution times of read and write remote split collective MPI-I/O operations are shown in Figures 5 (a) and (b), respectively. In Figure 5 (a), execution times of the read operations with a single user process are quite small in both the original and the multithreaded cases because only parameters for the I/O operation (totally 10 ∼ 30 KByte) were transfered in issuing this function. Read operation was carried out after completion of calling the function. The read operations with two and four user processes required much long time in the original method compared with the multithreaded method because I/O requests from the user

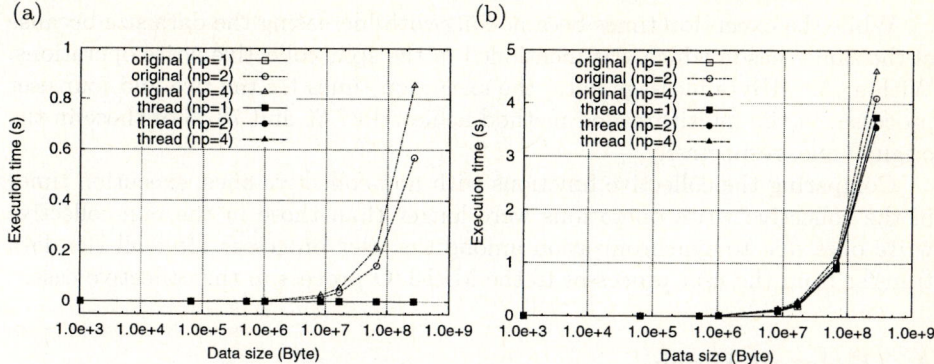

Fig. 6. Execution times of (a) read and (b) write remote MPI-I/O operations using original and multithreaded nonblocking MPI-I/O functions with a shared file pointer from a PC cluster I to a server node of a PC cluster II, where *original* and *thread* denote I/O operations by the original and multithreaded functions, respectively. *np* in the parentheses denotes the number of user processes

processes were blocked until the MPI-I/O process finished the current I/O operation due to a single MPI-I/O task. The times of the multithreaded read method achieved less than 12 % and 13 % of the times of the original method with more than 512 KByte message data in the cases of two and four user processes, respectively. The times for the two and four user processes in the multithreaded case achieved less than 0.1 % of those in in the original case with a 256 MByte data.

In Figure 5 (b), an execution time in the multithreaded case with a single user process was 74 % of that in the original case with 256 MByte message data. While the execution times of the multithreaded write operations were 60 % and 57 % of the times of the original ones with a 256 MByte data in the cases of two and four user processes, respectively. In the write operations, transfer of bulk data was also operated in addition to transfer of the parameters during calling the nonblocking function. Besides the required time for the bulk data transfer was dominant in the whole operation. Therefore the multithreaded method was not able to make the execution times short as it did in the read operations.

Execution times of read and write remote nonblocking MPI-I/O operations using a shared file pointer are shown in Figures 6 (a) and (b), respectively. In Figure 6 (a), execution times of the read function with a single user process are quite small in both the original and multithreaded cases because of the same reason as denoted in the split collective read operations. Due to a single MPI-I/O task, the original case for the two and four user processes required much longer times than the required times in the multithreaded case. The times of the multithreaded read ones were almost the same with respect to data sizes. In the two and four user processes cases, the execution times in the multithreaded case achieved less than 1 % of the times in the original case with a 256 MByte message data.

While the execution times became long with increasing the data size because of the same reason which was mentioned in the split collective write operations. With a 256 MByte message data, the execution times for the two and four user processes in the multithreaded method achieved 87 % and 78 % of those in the original one, respectively.

Comparing the collective functions with non-collective ones, execution times in the collective write operations were longer than those in the non-collective write ones due to synchronization among the user processes after all the data transfer from the user processes to the MPI-I/O process in the collective case.

4 Related Work

ROMIO [11], which is an MPI-I/O implementation in MPICH, provides seamless MPI-I/O interfaces to many kinds of file systems. Nonblocking I/O using a multithreaded library has been proposed for higher overlap of computation with I/O operations [12]. With the help of an I/O thread which is invoked in the beginning of nonblocking I/O operation, main thread is able to do next computation without waiting completion of the I/O operation. On the other hand, Stampi realizes seamless MPI-I/O operations among different computers. Besides, it realizes MPI-I/O operations using UNIX I/O functions even if a vendor-supplied MPI-I/O library is not available. A multithreaded MPI-I/O mechanism reported in this paper realizes higher overlap of computation on a local computer with I/O operations on a remote computer.

5 Summary

A multithreaded MPI-I/O mechanism using a LinuxThreads library has been realized in the Stampi library to support effective nonblocking remote MPI-I/O operations. In performance measurement which was carried out on interconnected PC clusters, execution times for both write and read operations with multiple user processes were shortened in the multithreaded method compared with the times in the original one. Typically, the effect was big in the read operations. Thus, the multithreaded method is effective in the nonblocking remote MPI-I/O operations with multiple user processes. Besides, the times for the collective write operations were longer than those for the non-collective write ones. This was due to additional time required for synchronization among user processes after all the data transfer from the user processes to an MPI-I/O process in the collective one.

Acknowledgments

The author would like to thank Genki Yagawa, director of Center for Promotion of Computational Science and Engineering (CCSE), Japan Atomic Energy Research Institute (JAERI), for his continuous encouragement. The author would like to thank the staff at CCSE, JAERI, especially Toshio Hirayama, Norihiro Nakajima, Kenji Higuchi, and Nobuhiro Yamagishi for providing a Stampi library and giving useful information.

This research was partially supported by the Ministry of Education, Culture, Sports, Science and Technology (MEXT), Grant-in-Aid for Young Scientists (B), 15700079 and Kinki University under grant number GS14.

References

1. Message Passing Interface Forum: MPI: A message-passing interface standard. (1995)
2. Message Passing Interface Forum: MPI-2: Extensions to the message-passing interface standard. (1997)
3. Imamura, T., Tsujita, Y., Koide, H., Takemiya, H.: An architecture of Stampi: MPI library on a cluster of parallel computers. In *Recent Advances in Parallel Virtual Machine and Message Passing Interface*, Volume 1908 of Lecture Notes in Computer Science., Springer (2000) 200–207
4. Tsujita, Y., Imamura, T., Takemiya, H., Yamagishi, N.: Stampi-I/O: A flexible parallel-I/O library for heterogeneous computing environment. In *Recent Advances in Parallel Virtual Machine and Message Passing Interface*, Volume 2474 of Lecture Notes in Computer Science., Springer (2002) 288–295
5. Tsujita, Y.: Flexible intermediate library for MPI-2 support on an SCore cluster system. In *Grid and Cooperative Computing*, Volume 3033 of Lecture Notes in Computer Science., Springer (2004) 129–136
6. Institute of Electrical, Electronic Engineers: Information Technology – Portable Operating Systems Interface – Part 1: System Application Program Interface (API) – Amendment 2: Threads Extensions [C Languages]. (1995)
7. LinuxThreads: (http://pauillac.inria.fr/~xleroy/linuxthreads/)
8. PC Cluster Consortium: (http://www.pccluster.org/)
9. Matsuda, M., Kudoh, T., Ishikawa, Y.: Evaluation of MPI implementations on grid-connected clusters using an emulated WAN environment. In: Proceedings of the 3rd IEEE/ACM International Symposium on Cluster Computing and the Grid (CCGrid 2003), 12-15 May 2003, Tokyo, Japan, IEEE Computer Society (2003) 10–17
10. Gropp, W., Lusk, E., Doss, N., Skjellum, A.: A high-performance, portable implementation of the MPI Message-Passing Interface standard. Parallel Computing **22** (1996) 789–828
11. Thakur, R., Gropp, W., Lusk, E.: On implementing MPI-IO portably and with high performance. In: Proceedings of the Sixth Workshop on Input/Output in Parallel and Distributed Systems. (1999) 23–32
12. Dickens, P., Thakur, R.: Improving collective I/O performance using threads. In: Proceedings of the Joint International Parallel Processing Symposium and IEEE Symposium on Parallel and Distributed Processing. (1999) 38–45

Asynchronous Document Dissemination in Dynamic Ad Hoc Networks

Frédéric Guidec and Hervé Roussain

VALORIA Laboratory – University of South Brittany – France
{Frederic.Guidec|Herve.Roussain}@univ-ubs.fr

Abstract This paper presents a document-oriented model for information dissemination in dynamic ad hoc networks, such as those composed of highly mobile and volatile communicating devices (e.g. laptops and PDAs). This model relies on an asynchronous, peer-to-peer propagation scheme where documents can be cached on intermediate devices, and be later sent again –either spontaneously or on demand– in the network.

1 Introduction

Today most laptops and personal digital assistants (PDAs) feature wireless interfaces, many of which are capable of ad hoc communication. Our work aims at fostering the design, the implementation, and the deployment of application services capable of running specifically on devices participating in a dynamic ad hoc network, that is, a network in which nodes are highly mobile and volatile. Node mobility in a dynamic network is the consequence of the fact that devices are carried by users, which are themselves mobile. Node volatility results from the fact that, since mobile devices have a low power-budget, they are frequently switched off and on by their owners. An additional problem with dynamic ad hoc networks is that in many realistic scenarios such networks present themselves as disconnected networks. As a consequence, direct transmissions between any pair of devices is not always feasible, as such transmissions require that both devices are active simultaneously in the network, and that a connected-path can be established between these devices at transmission time.

The problem of delivering messages in disconnected ad hoc networks has been approached several times and following different lines in the past few years. For example, a new network architecture relying on the general principle of message switching in store-and-forward mode has been proposed in [1]. With this approach pieces of information are transported as so-called *bundles* between *bundle forwarders*, which are capable of storing messages (or bundles) before they can be sent again in the network.

With *Epidemic Routing* [6, 4, 3], messages are buffered in mobile hosts, and random pair-wise exchanges of messages among these hosts are expected to allow eventual message delivery in partially-connected networks.

The service we present in this paper compares with the models proposed in the above-mentioned papers. However it can be observed that these papers mostly address the problem of message delivery in disconnected networks from a theoretical viewpoint: they propose new algorithms and heuristics for delivering messages in such networks, and they report the results of simulations that are meant to demonstrate how these algorithms should perform in realistic conditions. In contrast, our approach is more practical, since it consists in actually implementing a service for document dissemination in ad hoc networks, and then using this service as a building block with which application-level services can be developed and tested in realistic experimental conditions.

2 Service Overview

The general architecture of the service we propose is shown in Figure 1. This service is not meant to be used directly by end-users. Instead it is meant to serve as one of the basic building blocks with which higher-level services can later be developed. Moreover this service is document-oriented. Basically, we propose that any document sent in the network be maintained as long as possible in a local cache by as many devices as possible, so it can remain available for those devices that could not receive it at the time it was sent originally. The underlying idea is that the dissemination of multiple copies of the same document may help do with the volatility of devices, while the mobility of these devices can itself help transport information between islands in a fragmented network. Besides providing a caching system where documents can be maintained in mobile devices, our service also provides facilities for document advertisement, document discovery, and document transport between neighboring devices. For example, a device can sporadically or periodically notify its neighbors about all or part of the documents stored in its cache. It can also look for specific documents in its neighborhood, and either push documents toward –or pull documents from– its neighbors.

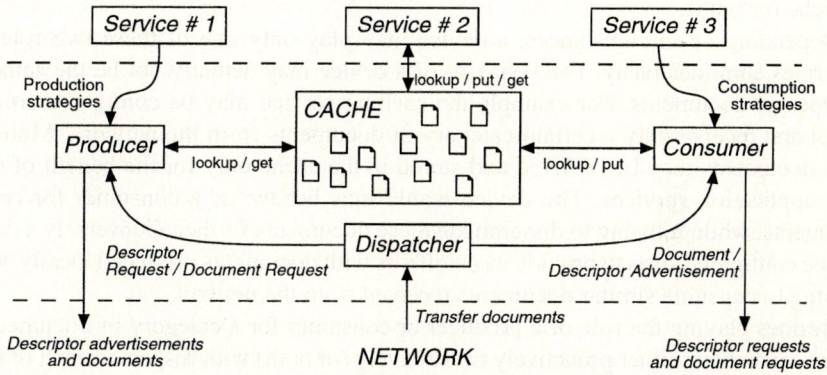

Fig. 1. Illustration of the document caching service

Structure of Documents. In the model our service relies on, each document can be associated a document descriptor, which provides information about its type, author, keywords, content, etc. A document may encapsulate its own descriptor, but the descriptor can also be handled separately (which means it can for example be transmitted, edited, stored, and displayed separately). When an application-level document must be sent in the network, it must itself be encapsulated in a *transfer document*, whose descriptor specifies transmission parameters for this document, such as its type, origin, and destination, as well as indications about how long and how far this document should propagate in the network. Examples of document descriptors can be found in [5].

Caching Documents. Once a document has been received by a device, it is expected that this document be stored for some time on this device, and possibly sent again later in the network. Each device thus maintains a cache, whose capacity can of course be adjusted depending on the resources available locally. Local strategies can additionally be defined on each device in order to specify caching modalities for documents. Possible criteria for defining such strategies are document size, type, origin, destination, lifetime, etc. The caching service is not itself responsible for deciding how it should behave with respect to documents. Instead, it provides interfaces (not detailed in this paper) with which higher-level services can specify strategies regarding how one or another category of document should be managed locally. Moreover, attributes found in a document's descriptor can help determine how this document should be managed by the caching service. For example, attributes may indicate how long a document should be considered as being valid in the network, and how often the availability of this document should be announced in the network.

Document Producers and Consumers. Each device that participates in the dissemination of documents can play several distinct roles with respect to these documents. A device is considered as the *provider* of a document if this document is stored locally (in its cache), and if it can send this document in the network. Symmetrically, a device is considered as being a potential *consumer* for a document if it can receive this document from the network, and either use this document immediately or store this document in its cache (or both).

Depending on circumstances, a device may play only one of these two roles, or both roles simultaneously. The behavior of a device may actually not be the same for all types of documents. For example the caching service may be configured so as to accept and receive only a certain category of documents from the network. Moreover these documents may be received and stored in the cache only for the benefit of other local application services. The device would thus behave as a consumer for certain documents, while refusing to disseminate these documents further. Conversely a device may be configured so as to provide its neighbors with documents produced locally, while refusing to consume similar documents received from the network.

Besides playing the role of a producer or consumer for a category of documents, a device can behave either proactively or reactively (or both) with respect to each of these roles. A device that plays the role of a document provider can behave *proactively* by sending spontaneously this document in the network. It may also behave *reactively* by sending a document in the network after this document has been explicitly requested. It can of course show a mixed behavior, sending for example one document periodically (with a rather long period so as not to load the network too much), and replying immediately to explicit requests for this document.

Similarly, a device that plays the role of a document consumer can behave either proactively or reactively, or show both kinds of behavior simultaneously. A document consumer can behave proactively by sending requests for this document in the network (thus soliciting a reactive behavior from devices that possess a copy of this document). It can also behave reactively by receiving a document from the network, and consuming this document even if it has not been explicitly requested before.

Advertisement and Request Documents. Specific kinds of documents have been defined in order to allow the advertisement, discovery, and transmission of documents between neighboring devices. For example, an "advertisement document" can be sent by a device to announce that it owns one or several documents in its cache, and that it can provide any of these documents on demand. An advertisement document is thus a special kind of transfer document whose payload is composed of one or several document descriptors, corresponding to the descriptors of the documents whose availability is being announced.

Another special kind of document, called a "request document", can likewise be sent by a device to ask for the transmission of a document, or that of several documents. A request can be addressed specifically to a given device (for example after an advertisement has been received from this device), or it can be sent to all or part of the devices in the neighborhood. The payload of a request document is composed of one or several descriptor patterns. The structure of a descriptor pattern compares with that of a descriptor, but for all or part of the attributes that can appear in a document descriptor, it specifies a regular expression to be applied to the corresponding attribute. A device receiving a descriptor pattern can thus use this pattern to examine the descriptors of the documents it maintains in its cache, and to decide which of these descriptors match the pattern. Selected documents can then be sent in the network.

3   Implementation Details and Ongoing Work

The service for asynchronous document dissemination presented in the former section has been implemented in Java. Documents and document descriptors are also reified as standard Java objects. They can be transported in the network either as serialized Java objects, or as XML-formatted documents (examples can be found in [5]). It is worth mentioning that the code we developed can be deployed equally on a single-hop network, or on a multi-hop network relying on algorithms for dynamic routing and flooding.

The development of several application-level services is also under way in our laboratory. These services all rely on the facilities offered by the document dissemination service, but each of them defines its own strategy regarding what documents must be disseminated, and in what conditions. Among these application-level services are a peer-to-peer messaging service, a presence announcement service, and a service for the distribution and the deployment of software packages. Details about the latter service can be found in [2].

4 Conclusion

The service presented in this paper permits the asynchronous dissemination of documents in dynamic ad hoc networks, such as those composed of highly mobile and volatile communicating devices. It proposes an asynchronous, peer-to-peer, document-oriented propagation model, where each document received by a device can be maintained in a local cache in this device, so it can later be sent again in the network, either spontaneously, or after a request for this document has been received from another device. This approach is expected to help do with the volatility of devices, since it permits that documents reach devices that are only active sporadically in the network. It is also expected to permit information dissemination in a fragmented network, taking advantage of the mobility of devices which can serve as carriers between disconnected parts of the network.

Acknowledgements

This work is supported by the French "Conseil Régional de Bretagne" under contract B/1042/2002/012/MASC.

References

1. Kevin Fall. A Delay-Tolerant Architecture for Challenged Internets. Technical Report IRB-TR-03-003, Intel Research, Berkeley, February 2003.
2. Nicolas Le Sommer and Hervé Roussain. JASON: an Open Platform for Discovering, Delivering and Hosting Applications in Mobile Ad Hoc Networks. In *International Conference on Pervasive Computing and Communications (PCC'04)*, pages 714–720, Las Vegas, Nevada, USA, June 2004.
3. Q. Li and D. Rus. Sending Messages to Mobile Users in Disconnected Ad Hoc Wireless Networks. In *Proceedings of the Sixth ACM/IEEE International Conference on Mobile Computing and Networking (Mobicom 2000)*, pages 44–55, August 2000.
4. Anders Lindgren, Avri Doria, and Olov Schelén. Probabilistic Routing in Intermittently Connected Networks. In *Proceedings of the Fourth ACM International Symposium on Mobile Ad Hoc Networking and Computing (MobiHoc 2003)*, June 2003.
5. Hervé Roussain and Frédéric Guidec. A Peer-to-Peer Approach to Asynchronous Data Dissemination in Ad Hoc Networks. In *International Conference on Pervasive Computing and Communications (PCC'04)*, pages 799–805, Las Vegas, Nevada, USA, June 2004.
6. Amin Vahdat and David Becker. Epidemic Routing for Partially-Connected Ad Hoc Networks. Technical Report CS-2000-06, UCSD, July 2000.

Location-Dependent Query Results Retrieval in a Multi-cell Wireless Environment

James Jayaputera and David Taniar

School of Business Systems Monash University,
Clayton Vic 3800 Australia
{James.Jayaputera, David.Taniar}@infotech.monash.edu.au

Abstract. The demand of information services is popular in recent years. However, the requested of correct answer in a mobile environment needs to have more attentions. This is due to the scope of query depends to the user location. In this paper, we propose an extension approach to handle the situation where a mobile user misses a query result at current time and expects to receive a next query result in the next interval time. The aim of this extension approach is to avoid redundant process in order to get a new query result. We show the efficiency of our proposed algorithm by giving some different examples and evaluations.

1 Introduction

Location-Dependent Information Service (LDIS) is one type of applications to generate query results based on the location of users issuing queries (requesters) [1, 2, 3]. It implies whenever users change their locations while they are sending queries, the query results have to be relied on the receiving location of the users receiving queries. Location-Dependent Query (LDQ) is one type of queries based on the data found on that particular location [4, 3]. Hence, the expected results of LDQ must accurate and depend on the new location of user.

In our past papers [5, 6], we proposed an approach to retrieve query results for LDIS applications. The query result retrieval approach allowed that method to retrieve the results produced based on the locations users requesting queries. However, that approach deals if users freely move within one cell, where a cell is an area covered by one base station (BS). [1]

In this paper, we propose an extension algorithm from our previous works. This extension algorithm is to retrieve query results in multi cells. The aim of this paper is to retrieve query results from multi cells accurately. For example, users send queries from current cell and travel with constant velocities and directions. Since a BS covers only a certain area, the users can move from one cell into another. However, delays might occur during this period, which results of

---

[1] A Base Station is a static host that does an address translation and message forwarding from a static network to wireless devices and vice-versa [7].

handover or others delays (such as transmission or processing). Handover is a process to transfer an ongoing call from one cell to another as a user changes the coverage area of a cellular system [8].

To simplify our discussion, it is assumed as follow: a geometric location is represented as two-dimensional coordinates, users travel on steady velocities and directions, every BS has knowledge about its neighbours and the expected time to leave current BS, the predicted locations are known before receiving query results and there are no errors in all partials data retrieved. Furthermore, the handover time is ignored since it does not make any changes towards the prediction of users' location. The delay occurred is static instead of variable.

The rest of this paper is organized as follows. In next section, some related works of this paper are presented. In section 3, our proposed algorithm will be discussed and later, examples will be shown. In section 4, we show the performance of our proposed algorithm. Finally, the last section will summarize the contents of this paper.

2 Related Work

In this section, we review the existing studies on those two aspects. Related works to this paper is works have been done in query results retrieval for LDIS applications, including how to retrieve query results from one Base Station and multiple Base Stations while a user is moving from one to another area [5, 6, 9, 10]

Efficiency query result retrieval in one BS for LDIS application has been discussed in our past papers [5, 6]. Figure 1 shows an illustration of our proposed algorithm to retrieve query results within a single BS. Let us considers, two locations: A and B. A user travels to east from A to B at speed 2 is sending a query, "retrieve all vending machines within 1 km from my current location". When the user accepts the query results, the query results must reflect to all vending machines which are located 1 km away from B since the user is not on A anymore. However, the user is only interested with the query results that

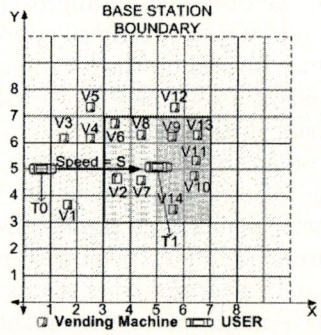

Fig. 1. Query Result Retrieval in a single BS

have not been passed (shaded area). Therefore, the valid query results are the vending machines (V9, V10, V11, V13, V14).

In work done by Sistla et al [10], the location of a moving object is conducted as dynamic attributes which is divided into three sub-attributes: function, updatetime and value. The advantage of their work is a new predication location can be found by using function of time. Therefore, we adopt their method to calculate the prediction location in our approach.

Our work is similar to their works; however, the focus of our work is to retrieve query results from multiple cells. The purpose of our work is to get query results accurately and fast.

3   Query Results Retrieval: Propose Algorithm

In this section, we propose an algorithm for query results retrieval in multi-cell for location-dependent. In retrieving query results, users can either stay in same location or move to other locations. We are only interested in moving users, however, users travel with variable velocity and directions are not discussed here. The aim of our propose algorithm is to retrieve correct query results from servers where the query scope is crossing the area of BSs.

Our proposed algorithm for query results retrieval from multiple BS is shown in figure 2. *Current_BSID* represents the current BS identification number. BS_{scope} is referred to endpoint coordinates that performed a boundary of BS. In our case,

```
Algorithm: Query_Processing_for_Multi_BS
Input: Query Scope, Number of Neighbor Base Stations, Available Base Stations
Output: result
Begin
    Query_scope ← Scope of query
    BS_scope ← current BS station scope
    BS[1..n] ← online Base Stations
    current_BSID ← ID of current base station
    result ← Get_Result(current_BSID);
    While intersection (Query_scope, BS_scope) is true
        current_BSID ← next neighbour BS ID of the current BS
        BS_scope ← get_scope(current_BSID)
        Query_scope ← Query_scope.MAX - BS_scope.MIN
        /* Append results retrieved to end of records */
        result ← result + Get_Result(current_BSID);
    End loop
    Return result
End Query_Processing_for_Multi_BS
```

Fig. 2. The proposed approach

the number of endpoints used is four since we assume that the scope of BS is a square. All online BSs are stored into a collection, called $BS[1..n]$ where n is a number of online BSs. $Query_{scope}$ is four endpoint coordinates that represents a scope of user query.

After the parameters initialization, the current BS checks whether the scope of query is intersect the scope of current BS. It generates query results in the current BS and stores the query results into *result* parameter. If there is no intersection, only the query results in the current BS are returned.

If there is any neighbour BS straight to the current BS, the neighbour BS becomes the current BS. The scope of the current BS is generated and then, the query scope is deducted against the minimum endpoint coordinates of parameter BS_{scope}. The query results within the query scope are generated. These processes keep repeating until there is no any intersection between query scope and BS scope. Then, the query results are forwarded to the user.

4 Performance Evaluation

After we discussed our proposed algorithm, evaluations on our proposed algorithms are given in this section. The objective of our evaluations are to examine situations whether our algorithms can handle situations to retrieve query results from multiple cells efficient and accurately. First, we give examples to simulate our evaluations. Then, we evaluate our examples given. The evaluation results are given at the end of this section show efficiencies of our propose approach.

Figure 3 shows processing time to process one query. The graph does not show straight line graph, because processing time to generate an answer for every query is different from one to another. The processing time to generate query results for query number 45 is the longest since there are a number of users entering cells. It can also be caused by more common data in the server. In contrast, the processing time for query number 10 is the shortest since the data and users entering the cell are rare.

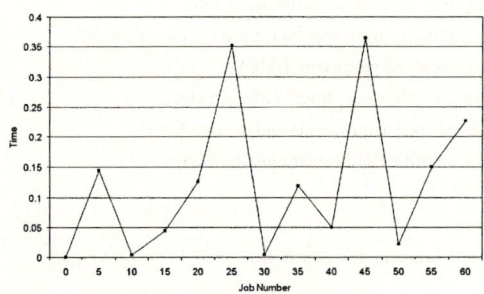

Fig. 3. Processing time for one query

5   Conclusion

In this paper, we have shown how to retrieve query results in multi cell. In early section, we give our motivation on query results retrieval in multi-cell. Afterwards, we propose our algorithm followed by analyses. We assume that the user travels on steady speed and direction. Whenever there is any intersection between query scope and BS scope, it implies the query results are within multiple cells. Therefore, the query scope must be deducted against the minimum boundary of next BS in order to process the remaining query scope inside the next BS. We also deal with delay time if any. However, we do not consider variable value of delay time. When there is any fixed delay time, it is added to the value of retrieval time. Our experiments results show accurate value. In addition, our experiments results show reasonable time to answer the user query.

References

1. Baihua, Z., Lee, D., Xu, J.: Data management in location-dependent information services. IEEE Pervasive Computing **1** (2002) 65–72
2. Tang, X., Xu, J., Lee, D.: Performance analysis of location dependent cache invalidation schemes for mobile environments. IEEE trans. on Knowledge and Data Eng. **15** (2003) 474–488
3. Zheng, B., Xu, J., Lee, D.: Cache invalidation and replacement strategies for location-dependent data in mobile environments. IEEE Trans. on Computers **51** (2002) 1141–1153
4. Dunham, M., Kumar, V.: Using semantic caching to manage location dependent data in mobile computing. Proc. of the sixth annual Int'l Conf. on Mobile Computing and Networking (2000) 210–221
5. Jayaputera, J., Taniar, D.: Defining scope of query for location-dependent information services. Int'l Conf. on Embedded and Ubiquitous Computing (2004) 366–376
6. Jayaputera, J., Taniar, D.: Query processing strategies for location-dependent information services (to appear jan 2005). Int'l Journal of Business Data Comm. and Networking (2004)
7. Goodman, D.J.: Wireless Personal Communications Systems. Addison-Wesley Wireless Communications Series (1998)
8. Markopoulos, A., Pissaris, P., Kyriazakos, S., Sykas, E.: Efficient location-based hard handoff algorithms for cellular systems. NETWORKING 2004, Third Int'l IFIP-TC6 Networking Conference (2004) 476–489
9. Kahol, A., Khurana, S., Gupta, S.K.S., Srimani, P.K.: An efficient cache management scheme for mobile environment. Proc. of the 20th Int'l Conf. on Distributed Comp. Systems (2000) 530–537
10. Sistla, P., Wolfson, O., Huang, Y.: Minimization of communication cost through caching in mobile environments. IEEE trans. on Parallel and Distributed Systems **9** (1998) 378–389

An Efficient Mobile Data Mining Model

Jen Ye Goh and David Taniar

Monash University, School of Business Systems, Clayton, Vic 3800, Australia
{Jen.Ye.Goh, David.Taniar}@infotech.monash.edu.au

Abstract. Mobile Data Mining involves the generation of interesting patterns out from datasets collected from mobile devices. Previous work are frequency pattern [3], group pattern [9] and parallel pattern [5]. As mobile applications usage increases, the volume of dataset increases dramatically leading to lag time for processing. This paper presents an efficient model that uses the principle to attack the problem early in the process. The proposed model performs minor data analysis and summary early before the source data arrives to the data mining machine. By the time the source data arrives to the data mining machine, it will be in the form of summary transactions, which reduces the amount of further processing required in order to perform data mining. Performance and evaluation shows that this proposed model is significantly more efficient than traditional model to perform mobile data mining.

Keywords: Mobile Data Mining, Mobile Applications, Mobile Users.

1 Introduction

Mobile data mining [3-5] is a research area that aims to extract interesting patterns out from mobile users. These patterns found are then filtered by means of the relevancy and accuracy of each pattern, and the end result is provided to the decision maker. The decision maker can therefore use the interesting pattern as a competitive advantage into making more accurate decision with lesser uncertainty. Mobile data mining is an important field of research because it provides support for decision makers in order to make decisions relating to mobile users. Some of the techniques in classical data mining includes time series analysis [6, 7] and spatial data mining [8]. In a mobile environment, it consists of a set of static nodes and mobile nodes. Static nodes are devices that remain static over the time. Mobile nodes are devices that remain both static and mobile over the time. An example of static node is the wireless access point and an example of mobile node is the personal digital assistant.

2 Related Work

Our related work, location dependent mobile data mining proposed a way to find out useful knowledge from mobile users. The visiting behaviour of each mobile user is first recorded down when the mobile user is near to a static node. The collective volume of the visiting records is later sent to a central server and these visiting records are aggregated based on each mobile user. The outcome of this process is a list of user profiles. Each user profile is a list of themes that relates to some static node previously visited. Each theme has a confidence in percentage assigned to it to show how confident the mobile user is interested in the particular theme.

Classical data mining methods used are association rules [1] and sequential patterns [2]. The outcome of the existing work is providing visiting behaviour mobile users and the location theme. One common weakness of all previous related work, in which this paper aims to identify and solve, is that all the transactions in the mobile data mining are gathered from the source, sending to the destination, without being analysed or modified. Many of these transactions may be irrelevant, repetitive or even contains corrupted data. Our proposed efficient model for mobile data mining aims to do minor analysis and summarizing of data, to tackle the problem early, leading to a cumulative effect where by the time the source transactions reaches the data mining machine, transactions are summarised enough and they are more readily to be processed by the data mining machine.

3 Proposed Method

Traditionally, each static node is defined by only the identification code. In the efficient model, each static node is defined the same way. but each mobile node is defined with the identification code of the mobile node itself and extra buffer for recording temporary summary transactions. In the efficient model, Static Node = {Identification}. Mobile Node = {Identification, Summary Transaction 1, Summary Transaction 2, ..., Summary Transaction n}. Summary transaction involves optimising the transaction so that each transaction is shortened to the extent that time in and time out range in the time series are defined rather than identifying each time point for the whole time series and finding the range out at a later stage using the costly high-end machines. The mobile nodes are in the best position to identify and generate the time range in the time series as *time_from* is the time point when the mobile node is in contact with other static node or mobile node, and *time_to* is the time point when the mobile node terminates or lost contact with a static node or mobile node.

Summary transaction is defined as: Summary Transaction n {Node Identification, *time_from*, *time_to*}. The node identification can be either a static node or a mobile node. If the connection was lost somewhere in the time series and re-established again, it is counted as two summary transactions as a new set of *time_from* and *time_to* will be recorded. The result of the cost efficient model is to use the strategy to first simplify the mobile data-mining problem by summarising the transactions in the mobile environment by the mobile node itself. This is because mobile node is in the best position to provide the summary transaction with minimal amount of cost. If the summary is done at a later stage, huge amount of processing power will be incurred.

3.1 Definition of Nodes

In the efficient model, the model of each static node is unchanged. Static Node = {Identification}. However, the definition of mobile node is redefined. Mobile Node =

{Identification, Summary Transaction 1, Summary Transaction 2, ..., Summary Transaction n}. Each summary transaction is defined as: Summary Transaction {Identification, *time_in*, *time_out*}. The *time_in* and *time_out* represents the time point where the mobile node established a communication status with a station. It provides a range of time by using two values instead of time series using multiple transaction and multiple values. Figure 1 shows the definition of the static node, mobile node and summary transaction.

```
Function Summary Transactions Definition
   Summary Transactions = Size Desired By Mobile User
   Transaction Refresh Frequency = Duration by Mobile User // max_time
   Transaction Refresh Size = No of Transactions by Mobile User // max_transaction
   Generate [Summary Transactions] Amount of Summary Transactions Each Holding {
      Destination Node Identification = Full Logical Name // can be either static or mobile
      time_in = Time Point Mobile Node Established Contact with Destination Node
      time_out = Time Point Mobile Node Lost Contact with Destination Node
   } End Function
```

Fig. 1. Definition of Cost Efficient Model

3.2 Gathering Summary Transactions

The mobile nodes starts to collect the summary transactions as it traverse throughout the mobile environment. It encounters different static nodes and different mobile nodes. Each established contact is recorded as summary transaction with the identification of the node and the *time_in* and *time_out* information. Due to the different amount of storage capacity available in the mobile nodes, each mobile node is configured with a *max_transaction* and *max_time*. The *max_transaction* is the number of maximum transactions that a static node is configured to store. The *max_time* is the number of time units that the mobile node must sent out the summary transactions to a mobile data mining enabled static node. Either one of the two conditions will lead to a sending of summary transactions to the mobile data mining enabled static node. Figure 2 shows the codes for refreshing transactions.

```
Function Refresh Transactions
   If (Number of Summary Transactions >= max_transaction)
      Or (Elapsed Time >= max_time) Then
         Refresh Transaction by Sending List of Summary Transactions to a Static Node
   End If
End Function
```

Fig. 2. Gathering of Summary Transactions

3.3 Perform Mobile Data Mining

By the time that all summary transactions reached to the central server for mobile data mining, all the summary transactions have made simplified by eliminating all the sequences in the time series that the mobile node have established contact, and all summary transactions that are not within the scope of the problem for the decision maker will have already be filtered out. The final list can then be fed into any kind of mobile data mining systems or classical data mining systems in order to find out useful knowledge from mobile users.

4 Performance Evaluation

Performance evaluation is done on Pentium IV 384MB machine, over 10GB of hard disk storage space. The classical model represents the gathering of all the transactions from mobile users at all times, storing them into a central repository and finally, perform data mining by first preparing the data, removing irrelevant data and feed the data into the relevant algorithms. Performance parameters: Mobile Users: 0 – 15, No of Transactions: 0 – 35,000, Level of Activity: 0% - 100%.

Figure 3 shows the performance chart between the numbers of transactions found compared to different number of mobile users required. The case scenario is based on frequency pattern mobile data mining method. The number of transactions required increases significantly for classical model while the cost efficient model increases gradually. It shows that, the classical model are not able to accommodate scalability when the number of mobile users is in the range of greater than 10,000 as the amount of processing power and memory required is huge. On the other hand, the cost efficient model significantly reduces the amount of transactions needed by using summary transactions generated by the mobile nodes themselves.

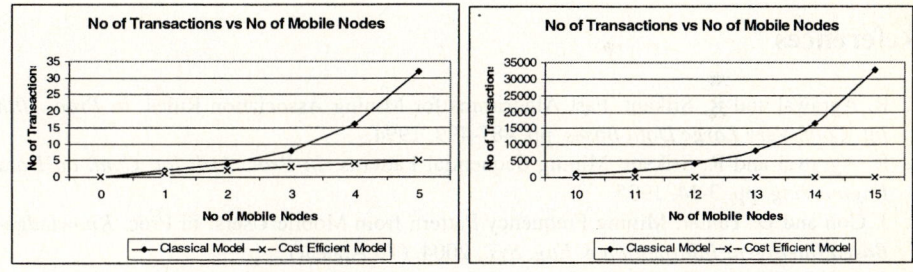

Fig. 3. Verify Summary Transactions

Figure 4 shows the performance chart between the classical model and cost efficient model by comparison the number of transactions required from each model at 50% level of activity. It can be observed that the performance for cost efficient model is similar to classical model from 0 to 2 number of mobile users and the two lines quickly deviates each other and cost efficient model maintains a better performance with lesser amount of transactions while the number of transactions required out of classical model increases significantly.

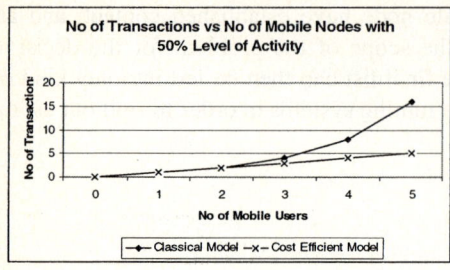 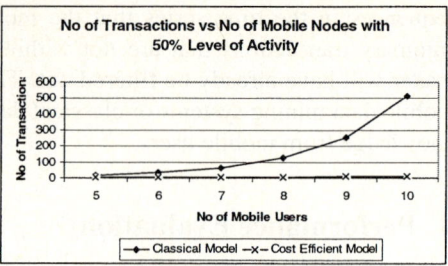

Fig. 4. Verify Summary Transactions

5 Conclusion and Future Work

The conclusion is that cost efficient model for mobile data mining, the number of transactions required to be fed into the data mining system can be reduced significantly by using summary transactions and server filtering method. The difference between the two models is that classical model delivers all transactions to the server for data mining while the cost efficient model lets the mobile nodes to summary their transactions and only then sent to the server for data mining. A filtering mechanism is placed just before the summary transactions reach the server which made it more efficient. Future work is develop novel ways to accurately determine the size of the summary transactions and the time interval before the summary transactions are uploaded to server.

References

1. R. Agrawal and R. Srikant. Fast Algorithms for Mining Association Rules. *In Proc. 20th Int. Conf. Very Large Data Bases,* pp. 487-499, 1994.
2. R. Agrawal and R. Srikant. Mining Sequential Patterns. *In Proc. 11th Int. Conf. on Data Engineering,* pp. 3-14, 1995.
3. J. Goh and D. Taniar. Mining Frequency Pattern from Mobile Users. In Proc. *Knowledge-Based Intelligent Information & Eng. Sys.,* 2004. (To Appear)
4. J. Goh and D. Taniar. Mobile Data Mining by Location Dependencies. In Proc. *5th Int. Conf. on Intelligent Data Engineering and Automated Learning,* 2004. (To Appear)
5. J. Goh and D. Taniar. Mining Physical Parallel Pattern from Mobile Users. In Proc. *Int. Conf. on Embedded and Ubiquitous Computing,* 2004. (To Appear)
6. J. Han, G. Dong, and Y. Yin. Efficient Mining of Partial Periodic Patterns in Time Series Database. *In Proc. of Int. Conf. on Data Engineering,* pp. 106-115, 1999.
7. J. Han, W. Gong, and Y. Yin. Mining Segment-Wise Periodic Patterns in Time Related Databases. *In Proc. 4th Int. Conf. on Knowledge Discovery and Data Mining,* vol. no. pp. 214-218, 1998.
8. K. Koperski and J. Han. Discovery of Spatial Association Rules in Geographical Information Databases. *4th Int. Symp. on Advances in Spatial Databases,* pp. 47-66, 1995.
9. Y. Wang, E.-P. Lim, and S.-Y. Hwang. On Mining Group Patterns of Mobile Users. *In Proc. of DEXA,* pp. 287-296, 2003.

An Integration Approach of Data Mining with Web Cache Pre-fetching

Yingjie Fu[1], Haohuan Fu[2], and Puion Au[2]

[1] Department of Computer Science,
City University of Hong Kong, Hong Kong SAR
`fuyingjie@tsinghua.org.cn`
[2] Department of Computer Engineering and Information Technology,
City University of Hong Kong, Hong Kong SAR
`fu.haohuan@student.cityu.edu.hk, poau@it.cityu.edu.hk`

Abstract. Web caching plays a very important role for improving the performance of many Web-Based systems. As web cache capacity is limited, most web cache systems are using replacement algorithm to wash out the outdated data. Our web cache prediction method is based on the fact that, many clients usually have some kinds of regular procedures to access web files, such that the regular-procedure knowledge can be mined or learned by web cache system and files can be pre-fetched accordingly.

1 Introduction

As networks become the basic infrastructure for data sharing and communication, web server response time becomes a very important measurement factor of the network and server performance. It is known that a server side web cache is placed between a web server and the clients; and the web cache buffers the copies of files requested by the clients. The usage of web cache is to improve the server responding time, decrease server load, and reduce latency to minimize "World Wide Waiting" problem.

In traditional web cache systems, file replacement policies are used to handle the limited cache size. There exist numerous replacement policies, such as LRU (Least Recently Used), LFU (Least Frequently Used), SLRU [1], LRU-MIN [2] and SIZE [3], etc. However, these pure replacement-policy strategies do not associate with any predication approach, which can enable the servers to find out what the client's access-profiles are, and the use this knowledge to predict the files to be accessed. In order to enhance web caching performance, not only the replacement policies can be applied, web cache pre-fetching technologies are also frequently used. Web cache pre-fetching needs the server-side cache to predict what kind of information would be used in the near future. In this paper, we present a new approach for web cache prediction to achieve better performance of file hit rate and byte hit rate.

2 Algorithm

At the beginning of data mining process, the web log records are loaded into a user-request pool p, which includes user record lists $l_1, l_2...l_n$. Users are identified by IP address. In each user record list l, it contains the user-accessed file URLs in a time ascending order.

As shown in Fig.1, history request pool is a table of user-request lists ls, and each l is a list of accessed file URLs. For better performance, user-request pool is designed to be a hash table with user IP as primary-key. When a request comes, according to the request source IP address, new record will be added to the user-request pool p.

User-req pool p	User request list l_1	Accessed file URLs of l_1
	User request list l_2	Accessed file URLs of l_2
	...	...
	User request list l_n	Accessed file URLs of l_n

Fig. 1. User-request Pool

There are numbers of accessed file URLs in each user-request list. Let T_{min} be the minimum threshold of URL number. In data mining process, the user-request lists with more than T_{min} URLs will be marked. Since minor URLs do not reflect user-access profile, we believe only these marked user-request lists will contribute to the data mining process. On the other hand, the URL number in each l should not grow unboundedly. When it grows to an upper threshold T_{max}, old URLs can be treated as outdated and will be washed out (deleted) from the list. It is obvious that, very old data source will not be useful for new data mining because both the user-accessing profile and web server files may have been changed already.

A new data structure is used to record user access orientation for each file URL. As shown in Fig.2, access-orientation is an integer to measure user accessing orientation for a file. It is also a hash table with file URL as primary-key.

URL_1	URL_2	...	URL_k
access-orientation$_1$	access-orientation$_2$	...	access-orientation$_k$

Fig. 2. Access orientation hash table

At the beginning in the data mining process, the value of *access-orientation*s are set to zero. While the data mining processing carrying on, they will be increased respectively. Our data mining process is not complicated. Suppose the last request from the users is URL_{last}. For each marked user-request list l, find out all of the URLs are that equal to URL_{last}. If there are N_1 URL_1s behind URL_{last}, add N_1 to *access-orientation*$_1$; if there are N_2 URL_2s behind URL_{last}, add N_2 to *access-orientation*$_2$...until add N_k to *access-orientation*$_k$. Finally, the web cache prediction can be done according to the calculation of *access-orientation*s. Our data mining procedure and web cache pre-fetching procedure can be presented into the following pseudo code:

```
Load web log into user-request pool p;
Mark the user-request lists ls containing more than T_min records
FOR each l having more than T_max records
    Wash out outdated records;
END FOR
FOR each latest URL request
    FOR each URL_j in URL_1...URL_k
        Set access-orientation_j to 0;
    END FOR
    Get the latest requested URL URL_last;
    FOR each l_i in l_1... l_n
        Find all URL==URL_last;
        FOR each URL_j in URL_1...URL_k
            Set integer N_j=0;
            Count the number (N_j) of URL_j s that follow URL_last;
            access-orientation_j= access-orientation_j+ N_j;
        END FOR
    END FOR
END FOR
Find out M URLs having the largest access-orientations from
access orientation hash table, denoted as URL_p1, URL_p2...URL_pM;
FOR each URL_j in URL_p1, URL_p2...URL_pM
    IF URL_j NOT in web cache
        copy URL_j into web cache;
    END IF
END FOR
END FOR
```

Fig. 3. Algorithm Pseudo Code

In line 20, M can be assigned according to web cache size and whole system performance. Larger the web cache size is and better the system performance is, larger value is assigned to M. As for URL_{p1}, URL_{p2}...URL_{pM} in line 22, there are two possible results. Either web cache will be full and cannot cache all of them, or web cache is large enough to hold all them. In the first case, the *URLs* that can not be cached will be discarded and web cache prediction process continues whereas in the second case, web cache prediction process will continue directly. In previous researches, there are a lot of contributions on replacement algorithm, e.g., FIFO, LRU, LFU, LRU-MIN, and SIZE etc. Web caching hit-rate and byte hit-rate can be better under the cooperation of web cache prediction method and web cache replacement algorithm. In our experiment, FIFO and LRU are used for the web cache replacement mechanism.

3 Simulation

We did the trace-driven simulation to compare the performance of our data mining approach with some well-known cache replacement algorithms. We choose three basic replacement algorithms, LFU, LRU and SIZE, which respectively consider three most basic factors, reference number, last access time and data size. We have done the trace-driven experiment for our data mining approach with 10 different cache sizes, compared with other five different cache replacement algorithms.

Fig. 4 shows the different hit rates for different algorithms. And among the different algorithms, the DM 600 means the data mining approach uses the latest 600 request-history records to do prediction, and DM 1200 means the approach uses 1200 the latest request-history records. In the web cache, LRU is used as wash-out approach. We could see that, from pure LRU to LRU plus web cache prediction, a lot achievement has been gotten. And from this figure we could see that the data mining approach has much better performance over any other pure replacement algorithms.

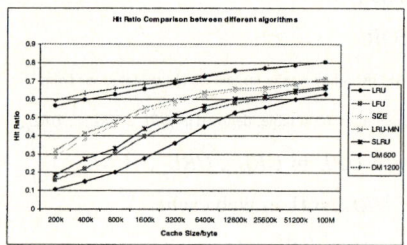

Fig. 4. Hit-rate comparison

4 Conclusion

There is a response time necessity in web accessing. In order to reduce to response time, web cache is used. Web cache prediction and pre-fetching is a very good method to improve the web response time. In this paper, we introduced a new data mining method to do prediction. The specification of the method approach and practical approach were discussed. Under the practical approach, we did experiment, and achieved satisfactory experiment results demonstrated the efficiency of integration which uses web cache prediction method and web cache replacement algorithm together in order to get better performance.

References

1. C. Aggarwal, Joel L. Wolf and P. S. Yu, *Caching on the World Wide Web*, IEEE Transactions on Knowledge and Data Engineering, vol. 11, no. 1, January/February 1999.
2. M. Abrams, C. R. Standridge, G. Abdulla, S. Williams and E. A. Fox, *Caching proxies: Limitations and potentials*, 4th International World-wide Web Conference, pages 119-133, Dec, 1995.
3. S. Williams, M. Abrams, C. R. Standridge, G. Abdulla and E. A. Fox, *Removal Policies in Network Caches for World-Wide Web Documents*, Proceedings of ACM SIGCOMM, pp. 293-305, 1996.
4. *Web_Log_2003_04.data*, City University of Hong Kong, April 2003.
5. *Web Trace, uc.sanitized-access*, ftp. ircache.org, 18th, November 2003.

Towards Correct Distributed Simulation of High-Level Petri Nets with Fine-Grained Partitioning

Michael Knoke, Felix Kühling, Armin Zimmermann, and Günter Hommel

Technische Universität Berlin,
Real-Time Systems and Robotics,
Einsteinufer 17, 10587 Berlin, Germany
knoke@cs.tu-berlin.de

Abstract. Powerful grid and cluster computers allow efficient distributed simulation. Optimistic simulation techniques have been developed which allow for more parallelism in the local simulations than conservative methods. However, they may require costly rollbacks in simulation time due to dependencies between model parts that cause violations of global causality. Different notions of time have been proposed to detect and remedy these situations. Logical time (or Lamport time) is used in many present-day distributed simulation algorithms. However, high-level colored Petri nets may contain global activity priorities, vanishing states, and global state dependencies. Thus virtual time is not sufficient to maintain the global chronological order of events for the optimistic simulation of this model class. The paper presents a new approach that guarantees a correct ordering of global states in a distributed Petri net simulation. A priority-enhanced vector time algorithm is used to detect causal dependencies.

1 Introduction

Stochastic Petri nets (PN) have been widely used for modeling the behavior of systems where synchronization of processes is crucial [1]. They provide a graphical representation and are able to represent discrete events as well as (stochastic) timing. Our simulation framework uses a variant of *colored* Petri nets (CPN) [2].

Real world systems consist of parts widely showing autonomous behavior but cooperating or communicating occasionally. This inherent concurrency and required synchronization can be modeled adequately using PNs. Distributed Petri net simulation (DPNS) can exploit this inherent parallelism efficiently using grid- and cluster computers. Hence, a partitioning algorithm is required that decomposes the model such that heavily communicating elements are not split. Each decomposed PN submodel is assigned to a *logical process* (LP) that is performing the simulation on a physical processor. A logical clock that denotes how far the simulation has progressed is assigned to a LP as well. LPs communicate using timestamped messages [3].

There has been significant work in the area of distributed simulation of PNs in the past few years. Almost all proposed algorithms assume a virtual time with an arbitrary high resolution to eliminate isochronous events. Some model specific activities can also cause events with the same virtual time, even for an assumed infinite resolution of time. Some of the activities in high-level PNs are:

- immediate transitions resulting in state changes without simulation time progress
- deterministic transitions that have a deterministic delay for state changes
- time guard functions which trigger state changes at a certain point in time

These properties of high-level PNs are either not allowed or adequately distributed to LPs, so that they are sequentially processed. Nicol and Mao [4] have contributed one of the most complete publications on distributed simulation of PNs, showing this limitation in each presented algorithm. It is obvious that in these cases the event ordering is simple and most research is focused on preferably good partitioning algorithms and early rollback detection. PN models for real world systems, such as detailed workflow modeling, may contain more than 50 percent timeless or deterministic activities.

A basic problem of distributed simulation is to avoid causality errors. Correctness of simulation can only be ensured if the (total) event ordering as produced by a sequential simulation is consistent with the (partial) event ordering due to distributed execution. Indeed, Jefferson [5] recognized this problem to be the inverse of Lamport's logical clock problem [6], i.e. providing clock values for events occurring in a distributed system such that all events appear ordered in logical time.

Lamport's algorithm allows to maintain time ordering among events [7]. However, a mapping from Lamport time to real time is not possible. Furthermore it is not sufficient to characterize causal relationships between events. But the detection of causal relationships between events is indispensable for transition priorities. Otherwise it is not possible to sort concurrent and independently fired events whose occurrence is based on a different priority. The Lamport time would impose an artificial order independent of their priority.

A logical time that characterizes causality and can be used to remedy last named problems is the *vector time* (VT) proposed by Mattern [8] and Fidge [9]. The causal relationships between events can be determined from their corresponding VT values. VT allows to detect indirect dependencies, that means comparing two VTs of different events provides information whether these events are causally dependent and if so, which event depends on which one. This has the following advantages in the context of DPNS:

- concurrent events can be identified and sorted by their priorities
- a very fine-grained model partitioning allowing deterministic and zero-firing times for output transitions of LPs is possible
- precise recovery of local LP states based on external events
- no need to solve equal Lamport time stamps

Many different high-level colored PN model classes, our class as well, allow different priorities for immediate transitions. That means if two events could be

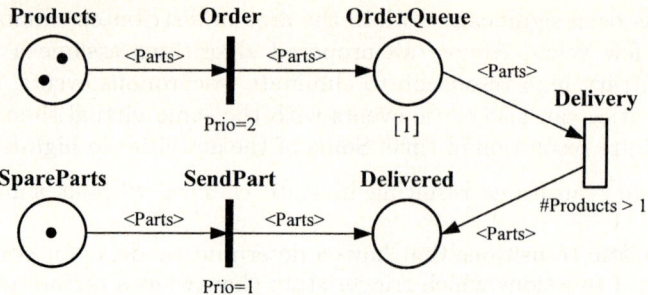

Fig. 1. Example of a high-level colored Petri net model

created at the same simulation time, the higher prioritized event is permitted first and may disable the second event through its occurrence. An example of a simple PN model is presented in Fig. 1. Transitions *Order* and *SendPart* are concurrently enabled and have different priorities, so that *Order* is processed first because of its higher priority. A sequential simulation is simple but a distributed simulation where both transitions fire optimistically, requires an order of execution.

To the best of the authors knowledge this paper presents the first time a new logical time scheme for high-level PNs which has significant advantages for partitioning without any structural limitations. It offers correctness for isochronous events and is applicable for all types of PNs, even for timeless PNs. Our extensions to the logical time fulfil today's requirements for flexibility and maximum scalability for typical real world PN models. It is not the intention of this paper to compare performance measures with any of the numerous Time Warp variations for distributed simulation of PNs. Optimistic simulation of high-level PNs is, in contrast to PDES, heavily dependent on the abilities of the underlying net class. It's always possible to design PN models perfectly fitting to a given distributed simulation algorithm. Our objective in this paper is to show new algorithms for partitioning and distributed event processing based on a new logical time scheme that opens new possibilities for DPNS performance optimization.

The paper first presents our new partitioning approach in Sect. 2. The subsequent Sect. 3 introduces a logical time scheme for prioritized globally ordered states. Some information about successfully completed test scenarios are shortly presented in Sect. 4 and finally concluding remarks are given in Sect. 5.

2 A New Partitioning Approach

Based on the correct implementation of causal dependencies that is described later in Sect. 3, the following scheme of an event-driven distributed simulation for high-level colored PNs was developed. Rollbacks can be performed more precisely and the flexibility of the partitioning is higher in particular if prioritized transitions and isochronous states are used in the model.

The simulation is composed of N sequential event driven LPs that do not share memory and operate asynchronously in parallel. Unlike in other optimistic DPNS algorithms (e.g. introduced in [10]), an *atomic unit* (AU) is defined as the smallest indivisible part of the model, whereas a LP consists of one or more of these AUs. The basic architecture and formalism of the LPs and AUs used in this paper is:

- The smallest indivisible part of the model is an atomic unit AU.
- A transition T_i is inseparably linked with all of its input places $^\bullet T_i$ and constitutes an atomic unit AU. This can lead to situations where more than one transition will be assigned to one AU, namely if a place has several output transitions.
- At least one AU is assigned to every LP_i which is running as a process on one physical node N_i.
- A communication interface attached to the LPs is responsible for the propagation of messages to the remote LPs and to dispatch incoming messages to local AUs. AUs on the same LP are communicating directly to avoid additional message overhead.
- Each LP_i, AU_j has access only to a partitioned subset of the state variables $S_{P,i} \subset S$ and $S_{U,j} \subset S_{P,i}$, disjoint to state variables assigned to other LPs, AUs. State variables of LP_i are the set of state variables of all local AUs $S_{P,i} = \bigcup S_{U,j} (\forall j)$.
- The simulation of local AUs scheduled within each LP in a way that avoids local rollbacks.

The three basic items for event-driven DPNS are state variables which denote the state of the simulation model, an event list that contains pending events, and a simulation clock which keeps track of the simulation's progress. All of these parts have been integrated into the AUs. Only two basic messages are required for simulation progress of AUs: *positive event messages* for token transfers and *negative event messages* to perform a rollback to an earlier simulation time.

A fine-grained partitioning and a discrete storage of processed states have a bunch of advantages for DPNS. First of all, in contrast to existing DPNS algorithms, e.g. described by Chiola and Ferscha [11], a rollback of complete LPs will not happen. Each AU has it's own virtual simulation time and stores its state for each local event independently from other AUs. This characteristic is essential for migration to other LPs at runtime. AUs can restore their state accurately for a given simulation time and send rollback messages to other AUs if they are affected by this rollback. Thus, rollbacks are much more precise and unnecessary rollbacks are prevented if independent AUs are simulated by a single LP. Memory consumption is lower than the classical LP approach because rarely executing AUs don't need to save their states until their own net activity.

Very important for collecting the result measures is the discrete storage of processed states. This storage mechanism allows to revert exactly to a given logical time without needing to resimulate already simulated sequences. In case of a rollback the last valid state is found with absolut precision. The disadvantage of a higher memory consumption is compensated by the much smaller size of AUs.

3 A Logical Time Scheme for Prioritized High-Level Distributed PN Simulation

In this section a logical time scheme for DPNS is presented and studied in detail. As per description in Sect. 1 it is essential for a correct ordering of states if model characteristics allows prioritized transitions and isochronous concurrent states. A distributed simulation is correct if its simulation results match the results of a traditional, single process simulator. Such sequential simulations are processing events in the order that takes the simulation time and the event priority into account. As a consequence we can conclude that a DPNS is correct if each AU is processing events in the traditional sequential manner and if incoming events are sorted exactly as they would be generated by a single process simulator. The following section presents an expanded logical time to fulfill these demands.

3.1 Event Priorities

For PN simulations on a single processor it is sufficient to have one global simulation time with an arbitrarily low resolution. All activities are running in succession and are responsible for the time increment. The simulated order of events is identical to the order in which they are simulated. Conflicts of concurrent activities are resolved by priorities or by random selection.

Immediate transitions have a priority greater than 0 and timed transitions have an implicit priority of 0. These priority values must be valid across AU borders, that means if transitions on different AUs are concurrently firing isochronously, the corresponding events must be ordered by their priority. Among identically enabled transitions one is chosen to fire first non-deterministically. For distributed simulation this approach is nonapplicable because of consistency reasons. Independent random generators on the AUs cannot guarantee the same ordering. Therefore we have decided to define a new *global event priority* (GEP) that includes the AU number into the priority value to determine an explicit relation for two equal event priorities. GEP is calculated as follows:

$$GEP = P_E * N_{AU} + i_{AU} \quad \begin{aligned} &P_E : \text{ event priority} \\ &N_{AU} : \text{ AU count} \\ &i_{AU} : \text{ current AU no.} \end{aligned} \qquad (1)$$

GEP forces the same global event ordering for concurrent events with different event priorities as a sequential simulation, but events with the same priority are ordered by the AU number in which they are created. A random selection of equal prioritized concurrently enabled transitions is non-applicable for DPNS. It forces a synchronization of model parts which acting completely autonomous. We have decided to accept this limitation because some people identify this problem as a modeling mistake.

Calculating the event priority P_E from the transition priority is nontrivial. The following order would be achieved by a sequential simulation of the model in Fig. 2: $T2 \rightarrow T4 \rightarrow T1 \rightarrow T3$. $T0$ is firing first and afterwards $T1$ and $T2$ are simultaneously enabled but $T2$ fires because of its higher priority. Now, without any simulation time elapsed, $T1$ and $T4$ are in conflict and $T4$ fires. Subsequently $T1$ and $T3$ fire in succession without taking the priority values into account. This example looks simple but it is observable that in case of a distributed simulation the firing order requires global knowledge.

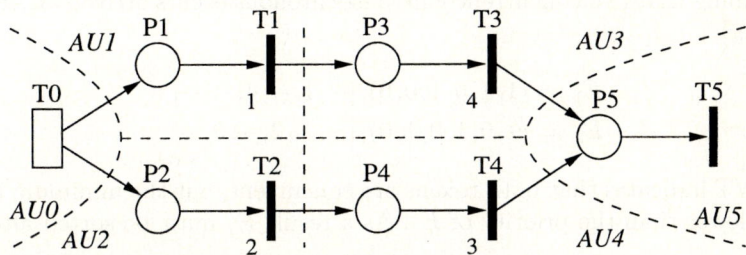

Fig. 2. An example for transition priorities

An optimistic distributed simulation doesn't need to resolve this priority problem when it appears but at the time when affected tokens are inserted into a place. This happens if at least two concurrent isochronous tokens must be ordered according to their priority. If the priority of the last fired transition was directly used to calculate the GEP it would give the token from path $T1 \to T3$ a higher order of precedence in the event queue because it was last fired from $T3$ which has a higher priority than $T4$.

To get the correct result it is important to create a priority path (herein after called *critical path*) from the last common transition or from the last timed transition. All priorities on each path must be considered for later event ordering. It can be shown that the minimum priority P_{min} of each path is decisive because the transition with the lowest priority delays the propagation of an event until no other transition with a higher priority on other paths can fire. Using the minimum priority on both paths would deliver the correct result ($P_{min_{T1,T3}} = 1, P_{min_{T2,T4}} = 2$).

An AU-sized vector of the last firing priority of each AU would be needed for calculating the minimum priority on the critical path. Events within an AU are always sequentially ordered, so it is not required to store the priorities of all transitions. This *priority vector* $p(e)$ has to be assigned to each event e. It is defined as follows:

$$p(e)_i = \begin{cases} \infty & \text{in case that } AU_i \text{ is not on the critical path} \\ & \text{otherwise the minimum priority of all preceding events on the critical path of } e \text{ in } AU_i \end{cases} \quad (2)$$

To follow the path of AUs that a token has entered and to compute the minimum priority of this path, it is just required to compute the minimum value of the priority vector. It is a precondition that all components of this vector are set to the infinite value on initialization and if a timed transition fires. On equality of two calculated minimum priorities it is obvious that a specific AU which is on both paths has randomly defined the order. For n AUs it must been AU_i with $i = p_{min} \bmod n$, as derivable through (1). The order is then explicitly observable by the corresponding VT component.

Assuming that two concurrently fired isochronous events arrived at $AU5$ with $(VT), [P_E]$:

$$E_1 := (1,1,0,1,0,0), [-,1,-,4,-,-]$$
$$E_2 := (1,0,1,0,1,0), [-,-,2,-,3,-]$$

The VT indicates that both tokens are concurrent, but the minimum priority of E_1 is lower than the priority of E_2. As a result E_1 must be sorted after E_2.

3.2 Compound Simulation Time

Distributed PN simulations running on several processors in parallel, require a logical time to detect causal dependencies and to achieve a global order of events. Certainly, the simulation progress of the distributed simulation is further on driven by the *simulation clock time* which progresses independently on each AU. PN model specific characteristics and a limited resolution of this time permit the occurrence of isochronous events. To operate with these events this time is extended by a sufficient logical time, namely the VT and the GEP introduced in Sect. 3.1. The compound time is capable of processing these isochronous events and can detect all causal dependencies. The new logical time is the *simulation time* (ST) as defined in (3), with the corresponding ordering relation (4).

$$ST = (T, V, G) \qquad (3)$$
$$T : \text{Simulation clock time}$$
$$V : \text{Vector time}$$
$$G : \text{Global event priority}$$

$$u \leq v \Leftrightarrow (T_u < T_v) \vee ((T_u = T_v) \wedge ((V_u < V_v) \vee (V_u \parallel V_v \wedge G_u \geq G_v))) \qquad (4)$$

Fig. 3. Compound simulation time

3.3 Transitivity of the Relation

A global ordering relation requires transitivity to offer explicit sorting of events. Relation (4) proposed in the last section is not transitive if it is not using the priority path for the GEP. An example of a simple Petri net that creates non-transitive events is shown in Fig. 4. Transitions $T1$, $T2$, and $T3$ create isochronous concurrent events which have to be sorted before merging the corresponding tokens at place $P4$.

Assuming that the following three simulation times S_1, S_2 and S_3 have to be compared using (4):

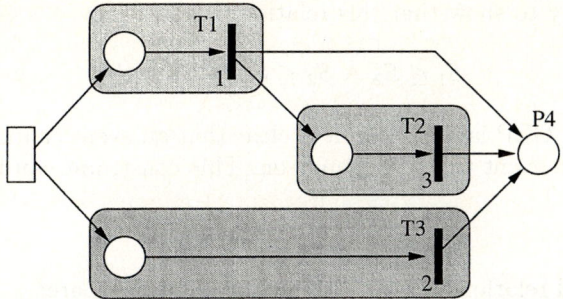

Fig. 4. Example petri net that can create non-transitive events

$$S_1 = (2004\text{-}01\text{-}01\ 00\!:\!00\!:\!00, [1,0,0], 1)$$
$$S_2 = (2004\text{-}01\text{-}01\ 00\!:\!00\!:\!00, [1,1,0], 3) \quad (5)$$
$$S_3 = (2004\text{-}01\text{-}01\ 00\!:\!00\!:\!00, [0,0,1], 2)$$

All events have the same simulation clock time which is 2004-01-01 00:00:00. By comparing the VT values it can be observed that S_2 is causally dependent on S_1 but S_3 is concurrent to S_1 and S_2. S_3 has to be sorted with its priority value which is 2. The result is $S_1 \leq S_2 \leq S_3$ and due to the transitivity theorem should follow:

$$S_1 \leq S_2 \leq S_3 \Rightarrow S_1 \leq S_3$$

In fact it is:

$$S_3 \leq S_1$$

Events corresponding to the simulation times S_1 and S_3 are concurrent and originated from simultaneous and independently activated transitions. The event with timestamp S_3 must be fired first because of the higher priority. S_2 is causally dependent on S_1 and must be sorted behind S_1 and as a result behind S_3 even though $S_2 \leq S_3$.

Theorem 1. *If the correct global event priority (as depicted in Sect. 3.1) is used then (4) is transitive.*

Proof. Consider three events e_i with $1 \leq i \leq 3$ and the corresponding priorities p_i as well as the simulation time stamps S_i. We assume that all S_i have the same simulation clock time. Then it is obvious that we have to account for the causal dependencies, namely the vector times and the priorities. The following notation is used for the causal dependency:

$$e_i \rightarrow e_j \Leftrightarrow e_j \text{ causally depends on } e_i$$
$$e_i \parallel e_j \Leftrightarrow e_i \text{ and } e_j \text{ are concurrent.}$$

Only if two events are concurrent their priorities have to be used for sorting. With this notation and on that condition (4) can be written as follows:

$$S_i \leq S_j \Leftrightarrow e_i \rightarrow e_j \vee (e_i \parallel e_j \wedge p_i \geq p_j) \quad (6)$$

It is necessary to show that this relation is transitive:

$$S_1 \leq S_2 \wedge S_2 \leq S_3 \Rightarrow S_1 \leq S_3 \qquad (7)$$

If the correct GEP is used then it is clear that an event cannot have a higher priority than the event that it depends on. This constraint can be written as:

$$e_i \rightarrow e_j \Rightarrow p_i \geq p_j \qquad (8)$$

Other helpful relationships directly deduced from (6) are:

$$e_i \rightarrow e_j \Rightarrow S_i \leq S_j \qquad (9)$$
$$e_i \parallel e_j \wedge p_i \geq p_j \Rightarrow S_i \leq S_j \qquad (10)$$
$$e_i \parallel e_j \wedge S_i \leq S_j \Rightarrow p_i \geq p_j \qquad (11)$$

Furthermore the transitivity of the causal relationship $\rightarrow$ and the sorting relation for priorities $\leq$ is assumed.

In order to prove the transitivity it is essential to consider all possibilities to combine causal relationships and priorities of the three events. Implication (7) must be valid in all cases. First of all let's focus on the causal dependencies. The implication is fulfilled if the right side of the implication is true ($e_1 \rightarrow e_3$) or the left side is false ($e_1 \leftarrow e_2 \vee e_2 \leftarrow e_3$). The following eight cases remain:

1. $e_1 \rightarrow e_2 \wedge e_2 \rightarrow e_3 \wedge e_1 \leftarrow e_3$
2. $e_1 \rightarrow e_2 \wedge e_2 \rightarrow e_3 \wedge e_1 \parallel e_3$
3. $e_1 \rightarrow e_2 \wedge e_2 \parallel e_3 \wedge e_1 \leftarrow e_3$
4. $e_1 \rightarrow e_2 \wedge e_2 \parallel e_3 \wedge e_1 \parallel e_3$
5. $e_1 \parallel e_2 \wedge e_2 \rightarrow e_3 \wedge e_1 \leftarrow e_3$
6. $e_1 \parallel e_2 \wedge e_2 \rightarrow e_3 \wedge e_1 \parallel e_3$
7. $e_1 \parallel e_2 \wedge e_2 \parallel e_3 \wedge e_1 \leftarrow e_3$
8. $e_1 \parallel e_2 \wedge e_2 \parallel e_3 \wedge e_1 \parallel e_3$

The cases 1, 2, 3, and 5 contradict the transitivity of the causality relation and need not be considered further. In case 8 the events are sorted exclusively by their priorities whose ordering relation is assumed to be transitive. Only the cases 4, 6, and 7 remain and needs to be analyzed.

Case 4. We show that the right side of (7) must be true if the left side is true.

$$\left. \begin{array}{r} e_1 \rightarrow e_2 \Rightarrow p_1 \geq p_2 \\ e_2 \parallel e_3 \wedge S_2 \leq S_3 \Rightarrow p_2 \geq p_3 \end{array} \right\} \Rightarrow p_1 \geq p_3$$
$$e_1 \parallel e_3 \wedge p1 \geq p_3 \Rightarrow S_1 \leq S_3$$

Case 6. Analogous to case 4.

$$\left. \begin{array}{r} e_1 \parallel e_2 \wedge S_1 \leq S_2 \Rightarrow p_1 \geq p_2 \\ e_2 \rightarrow e_3 \Rightarrow p_2 \geq p_3 \end{array} \right\} \Rightarrow p_1 \geq p_3$$
$$e_1 \parallel e_3 \wedge p1 \geq p_3 \Rightarrow S_1 \leq S_3$$

Case 7. The right side of the implication is false. Assuming that the left side is true the transitivity would be violated. We show that this assumption is incorrect.

$$\left.\begin{array}{r}e_1 \parallel e_2 \wedge S_1 \leq S_2 \Rightarrow p_1 \geq p_2 \\ e_2 \parallel e_3 \wedge S_2 \leq S_3 \Rightarrow p_2 \geq p_3\end{array}\right\} \Rightarrow p_1 \geq p_3$$
$$e_1 \leftarrow e_3 \Rightarrow p_1 \leq p_3$$

The outcome of this is $p_1 = p_3$. But concurrent events cannot have the same priority if unambiguous global priorities are used as depicted in Sect. 3.1. From this it follows that the left side of the implication must not be true. The transitivity is not violated.

So we can conclude that it is proven that (6) and as a result (4) are transitive.

4 Tests

In the course of our research and development we have designed a lot of models to verify our implementation of the new logical time scheme. The AU approach allows a truly distributed simulation of simple models to exploit parallelism. These simple models are not adequate for performance measurements but demonstrate the correctness of our approach. All experiments have been conducted on a 16 node, dual Intel Xeon processor, Linux cluster with 1 GB memory for each node. SCI has been used as a high-speed ring-topology-based networking.

Figure 5 shows a modified version of the model in Fig. 4 to test transitivity. Transition $T5$ may not fire if correct event ordering is used. Running this model over a long simulation time with several million events shows that $T1$ and $T6$ fire equal times, but never $T5$.

More complicated models, which have been accrued for different research projects for global operative business companies, are already successfully tested

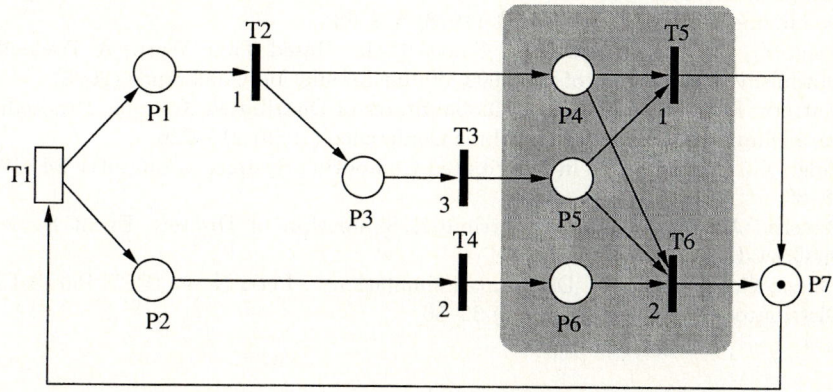

Fig. 5. Example petri net for testing transitivity

but not shown here because of lack of space. Such models are mostly not qualified for verifying substantial characteristics of the newly developed mechanisms.

5 Conclusion

This paper presented a new mechanism for distributed simulation of high-level Petri Nets. We introduced the notion of prioritized logical time which allows for a mapping between simulation clock and logical time. Applied to high-level PNs, this logical time is sufficient to allow a fine-grained partitioning not possible with Lamports logical time. It can be viewed as a total ordering scheme for high-level PN events. The Petri net model is decomposed into atomic units which have an own virtual time. Plenty of advantages for the distributed simulation arise from this approach: a better partitioning flexibility, dynamic migration with low operational expense, and efficient rollbacks.

References

1. Zimmermann, A., Freiheit, J., Huck, A.: A Petri net based design engine for manufacturing systems. Int. Journal of Production Research, special issue on Modeling, Specification and Analysis of Manufacturing Systems **39** (2001) 225–253
2. Jensen, K.: Coloured Petri Nets. Basic Concepts, Analysis Methods and Practical Use. Volume 1 : Basic Concepts. EATCS Monographs on Theoretical Computer Science, Springer-Verlag, Germany (1992)
3. Fujimoto, R.: Parallel and distributed discrete event simulation: algorithms and applications. In: Proceedings of the 1993 Winter Simulation Conference, Los Angeles, CA, Eds. ACM, New York, 1993 (1993) 106–114
4. Nicol, D.M., Mao, W.: Automated parallelization of timed petri-net simulations. Journal of Parallel and Distributed Computing **1** (1995)
5. Jefferson, D.: Virtual time. ACM Transactions on Programming Languages and Systems **7** (1985) 405–425
6. Lamport, L.: Time, Clocks, and the Ordering of Events in a Distributed System. Communications of the ACM **21** (1978) 558–565
7. Zeng, Y., Cai, W., Turner, S.: Causal Order Based Time Warp: A Tradeoff of Optimism. Proceedings of the 2003 Winter Simulation Conference (2003)
8. Mattern, F.: Virtual Time and Global States of Distributed Systems. Proceedings Parallel and Distributed Algorithms Conference (1988) 215–226
9. Fidge, C.: Logical Time in Distributed Computing Systems. Computer **24** (1991) 28–33
10. Ferscha, A.: Parallel and Distributed Simulation of Discrete Event Systems. McGraw-Hill (1995)
11. Chiola, G., Ferscha, A.: Distributed simulation of Petri Nets. IEEE Parallel and Distributed Technology **1** (1993) 33–50

M-Guard: A New Distributed Deadlock Detection Algorithm Based on Mobile Agent Technology[1]

Jingyang Zhou[1,2], Xiaolin Chen[3,1], Han Dai[1], Jiannong Cao[2], and Daoxu Chen[1]

[1] Dept. of Computer Science & Technology, Nanjing Univ., Nanjing, China
{jingyang, cdx}@nju.edu.cn; {cxl, daihan}@dislab.nju.edu.cn
[2] Dept. of Computing, Hong Kong Polytechnic Univ., Hong Kong, China
csjcao@comp.polyu.edu.hk
[3] Dept. of Computer Science, Chuxiong Normal Univ., Chuxiong, China

Abstract. Deadlock detection and resolution are of the fundamental issues in distributed systems. Although many algorithms have been proposed, these message passing based traditional solutions can hardly meet the challenges of the prevailing Internet computing and mobile computing. In this paper, we present a novel algorithm, namely the M-Guard, for deadlock detection and resolution in distributed systems based on mobile agent technology. The proposed algorithm lies in the intersection of the centralized type algorithm and the distributed type algorithm. An agent is employed in our algorithm as a guard with dual-role: when roaming in the system according to a specified itinerary algorithm, the agent collects resource request/allocation information for detecting deadlock cycles as well as propagating the collected network and resource information among the nodes. Consequently, accurate and timely detections of deadlocks can be made without any network node being the performance bottleneck. Preliminary simulation results show that, compared with several other algorithms, the M-Guard algorithm achieves both shorter deadlock persisting time and smaller phantom deadlock ratio. Moreover, the overall network communication overhead can be decreased, too.

1 Introduction

A distributed system always consists of a collection of geographically dispersed network nodes. These nodes share with each other computing resources and storage devices attached to them. System or application processes executing on any node are capable of using both local and remote shared resources, simultaneously and mutually. However, the use of blocking concurrency control is always subject to a possibility of deadlock. A deadlock occurs when a set of processes wait, in circular fashion, for exclusive access to some of the resources indefinitely hold by other processes in the same set [14]. Deadlock is not desirable because all the processes involved will be

---

[1] This work is partially supported by the National 973 Program of China under grant 2002CB312002, the National 863 Program of China under grant 2001AA113050 and the Hong Kong Polytechnic University under HK PolyU Research Grant G-YD63.

blocked indefinitely until the deadlock is resolved. Thus, the performance, such as device utilization ratio and application response time, will be degraded to an unbearable level and even the systems may crash.

Deadlock prevention, deadlock avoidance, and deadlock detection are three approaches to handle deadlocks [14]. In contrast to the former two's pessimistic and over-cautious mode, deadlock detection allows deadlocks to occur, but provides certain technique to find and resolve deadlocks. Deadlock detection is widely adopted for distributed systems due to its high flexibility and little negative influence on the system performance.

In the past two decades, many deadlock detection algorithms for distributed systems have been seen in literature, and they can be classified into three types: centralized, hierarchical and distributed [14]. In the centralized type algorithms, a certain node is assigned the sole responsibility of detecting any deadlock cycles in that system. These algorithms are easy to implement and because their knowledge about the whole system is comparatively accurate, they seldom detect deadlocks that are actually non-existing. But the coordinator may be the performance bottleneck, and even causes the problem of single-point failure. The distributed type algorithms are more flexible. Each node in the system has an equal possibility of participating in detecting deadlock cycles. They exchange information with each other and work cooperatively to solve the deadlocks. However, this type of algorithms is not accurate enough and may induce much communication overhead. A deadlock cycle is sometimes detected by multiple detectors and more than one processes involved in the same deadlock cycle might be forced to abort. The hierarchical algorithms are introduced as a compromise. They organize network nodes hierarchically as a tree, for example. Each node detects deadlocks involving only its descendant nodes. Nevertheless, the hierarchical structure is difficult to construct and maintain.

Any type of algorithms mentioned above is mostly implemented with message passing. However, the traditional message passing mechanism is not suitable for loosely coupled systems, like the Internet, and dynamically evolving systems, such as mobile systems, due to unpredictable message propagation delay and frequently changes of network topology. As a result, timely and precise detection of deadlocks cannot be guaranteed so that either an existing deadlock will not be resolved for a long time or false deadlocks are frequently detected. Furthermore, it is difficult for the message passing mechanism to accommodate dynamic changes in the system that requires adaptive and scalable solutions.

Recently, the mobile agent technology [9, 13] provides a new approach for structuring and coordinating wide-area network and distributed services that require intensive remote real-time interactions [1, 5]. In our previous research [2], we have proposed a mobile agent enabled framework for distributed deadlock detection, which is named MAEDD (Mobile Agent Enabled Deadlock Detection). In this paper, we describe a novel deadlock detection algorithm for distributed systems, i.e. the M-Guard, under the MAEDD framework, which takes advantages of the merits of both the centralized type algorithms and the distributed type algorithms. An agent is employed in the M-Guard as a dual role: when roaming in the system according to a specified itinerary algorithm, the agent collects resource request/allocation information for detecting deadlock cycles as well as propagating the collected network and resource information among the nodes. Consequently, accurate and timely detections of deadlocks can be made without any network node being the performance bottleneck. Meanwhile, because the deadlock is solely resolved by the M-Guard unique to the system, the problem that the same deadlock cycle is resolved by different detectors will not occur in our algorithm.

The remainder of this paper is organized as follows: Section 2 overviews related works. Our M-Guard algorithm is detailedly described in Section 3 together with the itinerary algorithms for the agent. Section 4 presents the preliminary simulation results and analysis. Finally, we conclude our paper in Section 5.

2 Related Works

As aforementioned, centralized, distributed, and hierarchical are three ways for deadlock detection. The representatives of the centralized type are Ho-Ramamoorthy's one-phase and two-phase algorithms [6]. In the two-phase algorithm, every node maintains a status table, which contains the status of all processes initiated at that node. The status of a process includes all resources it occupies and all resources it is waiting for. Periodically, a designated node (the coordinator) requests the status tables from all nodes, constructing a global state graph from the received information. The coordinator then tries to search for cycles in that graph. If finds, it requests every node to send the status table to it again. This time, the coordinator constructs the global state graph only using the information exists in both of the two times. If the same cycles are detected, the system is declared deadlocked.

In the distributed approach, algorithms can be classified into four categories [7]. They are *Path-Pushing (WFG-based)*, *Edge-Chasing (Probe-based)*, *Diffusing Computation* and *Global State Detection*. Instead of constructing the "state graph", the Wait-for-Graph (WFG) is used. A WFG consists of a set of processes $\{P_1, P_2, ..., P_n\}$ as the node set. An edge (P_i, P_j) exists in the graph if and only if P_i is waiting for a resource that is held by P_j [16]. Path-pushing algorithms maintain an explicit WFG. Each node periodically builds a local WFG by collecting its local wait dependencies, then searches for cycle in the WFG and tries to resolve these cycles. After that, every node sends its local WFG to its neighboring nodes. Each node updates its local WFG by inserting wait dependency received, and detects cycles in the updated WFG. The updated WFG is passed along to neighboring nodes again. This procedure will be repeated until some nodes finally detect the deadlock or announce the absence of deadlock. The most famous algorithm in this category is Obermarck's algorithm [12], which was implemented in the System R^*. Edge-chasing algorithms do not explicitly build the WFG. Instead, they send a special message called probe to detect deadlocks. A process (initiator) sends probes to each of the processes holding the locks it is waiting for. Upon receiving such a probe message, a process should forward it to all the processes it is waiting for. It is assumed that the probe message contains information to identify the initiator. If the initiator receives a probe sent by itself, it can then announce a deadlock because the probe must have traveled a cycle. This idea was originally proposed in [3] with the correctness proof presented in [8], and a revision can be seen in [15]. Similarly in the Mitchell and Merritt's Algorithm [11], a probe consists of a single number that is unique, which identifies the initiator. The probe travels along the edges in the opposite direction of global WFG, and when it returns to its initiator, a deadlock is detected.

A special case is the SET algorithm [4]. It is a combination of the edge-chasing and the path-pushing algorithms. The messages named "global candidate" are exchanged among the nodes. They carry more information than probes do, but are simpler than the local WFG information. A string of processes $(P_1, P_2, ..., P_n)$ is said to be a global candidate if for any process P_i ($1 \le i < n$), it resides on the same node and waits for P_{i+1}, and some process residing on another node is waiting for P_1 while P_n is waiting for some process on another node. When a blocked process times out, the node that process resides begins the deadlock detection process. The initiator node finds the global candidates and sends them to the nodes where resides processes blocking P_n. When a node receives a global candidate, if it can be concatenated with its own global candidates, the node checks for deadlocks and forwards the concatenation to other nodes. Otherwise, the candidate will be discarded.

3 M-Guard: Our Deadlock Detection Algorithm

3.1 System Model and Data Structure

Imaging a distributed system consists of n nodes denoted as $\{N_i | 0 \le i < n\}$. All the nodes are logically connected, i.e., for any pair of nodes, there exists a communication path between them. Each node N_i is associated with one or more resources sharable to network users. A node manager (NM) on the node is responsible for maintaining the resource allocation information. Application processes are distributed on system nodes, each has a global unique identifier, such as P_m, P_n, and etc. These processes request resources according to their own requirements. If the requested resource is mutually held by other processes, the requesting process will be blocked until the resource is released and granted to it. Some mobile agent platform (here we use the IBM Aglet) is deployed all network nodes so that the mobile agent guard (M-Guard) can migrate to and executed on each node. In this paper, we assume that the underlying communication link is free of failure and the priorities of the processes are set by their respective IDs.

We define two states for each process: *active* and *blocked*. A process can issue a request to access a resource only if it is active. If the resource P_m requests is currently occupied by another process P_n, P_m is then in the status of blocked. Process P_m may change from blocked to active if and only if it is granted the resource it requests.

NM on each node maintains a local Wait-for-Graph with data structure as follows:

Table 1. Data Structure of Local WFG

Request Process ID	Block Process ID	Remote Node ID	Flag

Table 2. Data Structure of Global WFG

Request Process ID	Request Node ID	Block Process ID	Block Node ID

As shown in Table 1, the *Request Process ID* is the request process ID; *Block Process ID* is the blocked process ID; *Remote Node ID* is the remote node ID on which block process is executing and the *Flag* marks whether the corresponding entry has been read by the agent. The data structure of the global WFG is similar to that of the local, but *Request Node ID* is the node ID on which the requesting process resides.

M-Guard also keeps a double-side queue NodeQueue (the meaning of double-side is that the delete operation is performed at both sides of the queue, while the insert operation is performed only at tail). The elements in the NodeQueue are node IDs to be visited in the future. Three operations are defined on the NodeQueue: push (insertion at the tail), popfront (deletion at the head), and popback (deletion at tail). In addition, an integer array node_visited[n] is maintained by the agent guard to record whether a node has been visited. For instance, if node N_i has been visited, then node_visited[i]=1.

When a process P_m is selected as the victim to break an existing deadlock cycle, the NM on the same node where P_m resides will receive a victim message. This message identifies the process that must abort and release all resources it occupies.

3.2 The Algorithm

The NM maintains a local WFG with regard to status of all processes and resources on the same node. Whenever a local process requests for some resource(s) locally or remotely, the corresponding NM should record the event as a new entry in the local WFG. When a local process is granted the resource or just cancel its request, the NM will delete the corresponding entry in the local WFG. The NM periodically checks the local WFG. If a local deadlock is detected, it will resolve it by informing the process with the lowest priority to abort.

In the system there is a single mobile agent named M-Guard that is responsible for detecting global deadlocks. The M-Guard visits the nodes in the distributed system according to some specified itinerary algorithm. When roaming in the system, all local WFGs are collected and combined into one global WFG. The M-Guard then analyzes the constantly updated global WFG and tries to resolve all deadlocks formed. Meanwhile, M-Guard will keep track of the network status and the resource information. For any change, for example a new resource is available on certain node, M-Guard will pick this news and inform all the NMs on the nodes it visits about that.

As we can see in our M-Guard algorithm, the M-Guard has a dual role. It can inform NMs about the status of processes, resources and the network as well as performing the task of deadlock detection. There are two key points in designing the M-Guard: one is how M-Guard collects the local WFG and how it analyzes the global WFG; the other is how to design an appropriate itinerary for the M-Guard. Apparently, a suitable travel itinerary could help M-Guard detect and resolve deadlocks more efficiently and accurately. In the following subsections, we will describe the main algorithm for deadlock detection and resolution in detail firstly. Then the three different itinerary algorithms are introduced respectively.

3.2.1 Deadlock Detection

Part A) Algorithm Executed by the Node Manager
When a local process P_m waits for process P_n on node N_j, a new entry will be appended in local WFG with the value of the field *Flag* equals to 1 meaning this entry is waiting to be read by the M-Guard.

If process P_m is granted the resource it requests or it cancels its resource request after a period of time, the NM should:

I. If the "Flag" field in the corresponding entry equals to 1, meaning that M-Guard hasn't read this entry, NM delete it directly;
II. If the "Flag" field in corresponding equals to 0, it can be inferred that the M-Guard has read the information. In order to keep consistent with information M-Guard maintains, NM should set the Flag field to -1. When the M-Guard read this entry next time, it will know the corresponding information should be deleted from its global WFG.

When NM receives a victim message sent by the M-Guard indicating that local process P_m is selected as the victim to break the deadlock cycle, it will release the resources held by P_m, and inform P_m to abort. Finally, the NM deletes the corresponding entries in the local WFG.

Part B) Algorithm Executed by the M-Guard
Every time the system initializes, the guard agent will be created on node N_0. The agent first reads the local information on N_0, and then visits other nodes in the system according to some specified itinerary algorithm depicted in Section 3.3.
On arriving at a new node N_i, the M-Guard will perform following steps:

I.   For each entry with Flag equals 1, M-Guard appends it to the global WFG. If the corresponding block process is on node N_j (i≠j and node_visited[j]=0), the agent pushes N_j into the nodeQueue. For each entry with Flag equals to -1, the agent will delete all corresponding entries in both local and global WFG;
II. When the request out-degrees of N_i in local WFG is 0, meaning that no local process waits for resources held by other processes, we know that all of the processes local to N_i are impossible to be involved in a deadlock cycle at present. If at the same time the nodeQueue is empty, the M-Guard will stay on N_i for a defined period of idle time (guard_wait_time) and then read the information again. The parameter guard_wait_time should be preset according to the deadlock frequency and network delay. Despite that none of the processes on N_i issues resource requests within that period of time, it is stipulated that M-Guard should select the next node to visit after the guard_wait_time in order to avoid sleeping on a node for too long time;
III. After reading all the local information, M-Guard checks the global WFG it maintains. If any deadlocks are detected, it will send a victim message to the NM on the same node where the victim process resides;
IV. The M-Guard collects the status information of processes, resources on N_i, which will be propagate to other NMs if necessary.

3.2.2 Deadlock Resolution
When the M-Guard detects a cycle, it must try to resolve the deadlock immediately. Traditional strategy of deadlock resolution is to choose one of the processes trapped in deadlock as the victim process. The victim process is required to release all the resources it holds and roll back to a certain status (this can be achieved by the technique of checkpoint). To minimize the loss due to the deadlock, M-Guard chooses the process with the lowest priority as the victim by sending a victim message to the NM. It is clear

that only one victim is selected because the victim message is solely sent by the M-Guard based on the global WFG. After the NM of the victim process receives the victim message sent by the M-Guard, it will release the resources held by victim and inform it to abort. Corresponding entries in the local and global WFG will be updated respectively by the NM and the agent.

3.3 The Itinerary Algorithm for M-Guard

The M-Guard should follow some specified itinerary to visit nodes in distributed systems. The itinerary algorithm has a great impact on the overall performance. A good itinerary can reduce the deadlock cycle persisting time as well as the communication cost. If an inappropriate itinerary algorithm is employed, the agent may visit too many nodes where the information maintained can contribute little to detect existing deadlock cycles. However, the performance of an itinerary algorithm may vary under different scenarios. Hence, we design three different itinerary algorithms, i.e. the random, breadth-first and depth-first itinerary. In the random itinerary, the M-Guard randomly chooses a node that hasn't been visited recently as its next visiting node. The breadth-first and depth-first itinerary refer to the different search strategy on the dependent graph. The operations of them are somewhat similar. Here, we only introduce the breadth-first itinerary algorithm as a representative.

As introduced in Sec. 3.1, the M-Guard maintains a double-side queue nodeQueue, in which elements in nodeQueue are node IDs to be visited by the agent in the future, and an integer array node_visited[n], which is used to record whether the nodes have been visited. In this itinerary algorithm for the M-Guard, when the agent arrives at a new node N_i, it should first set node_visited[i] to 1. If there is a local process P_m is waiting for a resource held by a remote processes P_n on N_j, it appends N_j into the nodeQueue at the tail. After collecting all local information, M-Guard chooses the first element in nodeQueue that hasn't been visited as the next travel target.

If all of the nodes in the nodeQueue have been visited, the M-Guard should clear the nodeQueue and set node_visited[i]=0 (0≤i<n). After that, M-Guard begins a new travel iteration from the current node; If the nodeQueue is empty and there are some nodes in the system that haven't been visited, M-Guard will randomly choose one of them to push into the nodeQueue and visit it next.

Comparatively, the random itinerary is very simple and is easy to be implemented, but sometimes is not efficient enough. For the other two itinerary algorithms, because M-Guard travels approximately along the directed edges in the WFG, it can detect deadlocks more quickly. Moreover, when the average number of nodes involved in a deadlock is relative small, the breadth-first itinerary algorithm is slightly better than depth-first itinerary algorithm.

4 Performance Evaluation

In our simulation, enough number of workstations with IBM Alget mobile agent platform deployed are organized into one network. We assume that the average transmission time for a message or an agent between a pair of nodes is T_n. N sharable resources and M processes are randomly distributed on systems nodes. C_p is defined as the average number of processes on a node. The time interval between two successful resource requests issued by a process follows the passion distributed with the expected value equals to $\lambda_1=2$ while the time a process holds a granted resource follows a passion distributed with $\lambda_2=10$. Deadlock cycles are broken by selecting the process with the largest ID as the victim. Time needed for agent execution and messages processing are omitted. The idle time of the M-Guard on a node with out-degree equals 0 is T_{gwt}. Since the scale of the system simulated is not great, we use the breadth-first itinerary algorithm for the M-Guard travel. We choose Chandy-Misra-Hass's algorithm [3] and the SET algorithm [4], which have better performance against other traditional algorithms, for comparison. Every time we run the simulation for T_{total} time slips. In the simulation, the single-resource model [7] is adopted for simple consideration, and we set $T_n=2$, $T_m=25$, $C_p=4$, $N=20$, $T_{gwt}=4$ and $T_{total}=10000$.

Three widely accepted criteria are adopted [10, 14]. They are:

I. Deadlock duration (deadlock cycle persisting time). It is the time interval between a deadlock cycle is formed and it is resolved.
II. Phantom deadlock ratio. It is the rate that the number of detected deadlocks that does not exist in the system against the number of deadlocks really exists.
III. Communication overhead. The times that the agent guard migrates and the number of messages transferred in the system.

We conduct the simulation for 10 times and take the mean value as the final result. Figure 1 shows the phantom deadlock ratio of the three algorithms as a function of the total process number. The phantom deadlock ratio of M-Guard algorithm keeps the smallest of the three at about 0.2. The phantom deadlock ratio of the Chandy et al.'s algorithm is two times bigger than that of our M-Guard algorithm while the ratio of the SET algorithm is much greater.

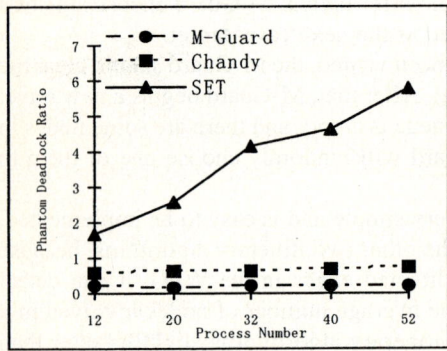

Fig. 1. Comparison of Phantom Deadlock Ratio

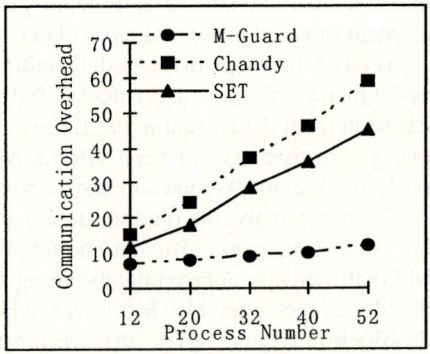

Fig. 2. Comparison of Communication Over-head

Communication overhead is another important criterion to evaluate an algorithm's performance. In M-Guard algorithm, the communication overhead refers the sum of agent migration times and the numbers of victim messages sent. In Chandy's and the SET algorithm, the overhead only refers to the messages exchanged in the system. Figure 2 depicts the average communication overhead for one successful detection of a deadlock cycle for the three algorithms respectively. As we can see, when the number of process increases, overhead of Chandy's algorithm and the SET algorithm increase much, but M-Guard algorithm has a much lower and stable communication overhead.

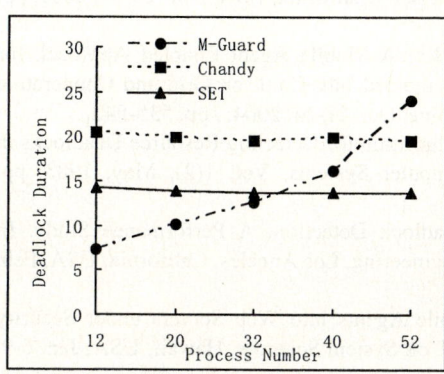

Fig. 3. Comparison of deadlock Duration

Finally, the deadlock duration is plotted in Figure 3. When the total process number is under 40, most of the deadlocks can be detected and resolved rapidly in the M-Guard algorithm. Although M-Guard has a longer deadlock persisting time when the system scale grows, we argue that this may be improved by employing a more efficient itinerary algorithm.

From the data and analysis above, we can see that the M-Guard algorithm has a better overall performance than others do. This is especially true when scale of the system is not very huge.

5 Conclusions

In this paper, we propose the M-Guard, a novel deadlock detection algorithm using mobile agent technology for distributed systems. The M-Guard employs a mobile agent, which keeps roaming in the network, to collect status information of processes and resources for detecting deadlock cycles, like a mobile guard of the system. We describe both parts of the proposed algorithm that are executed by the agent guard and the local nodes respectively. We also specify three different itinerary algorithms for the agent guard that may perform well under different scenarios. The M-Guard algorithm lies in the intersection of the centralized type and the distributed type, so that accurate and timely detections of deadlocks can be made without any network node being the performance bottleneck.

This mobile agent enabled algorithm has several advantages over the traditional message-passing based algorithms. Our simulation demonstrates that it reduces the phantom deadlock rate and the overall communication overhead, resulting in increased performance. Meanwhile, the deadlock cycle persisting time can be decreased in case that the scale of the whole system is not too big.

In the future works, we plan to design a heuristic itinerary algorithm for the M-Guard. We will conduct a further study on agent itinerary algorithms' impact on the overall performance of the M-Guard algorithm. We are also going to investigate the employment of multiple agents, each monitors a portion of the whole network, to guard a larger scale system. Finally, performance comparison and algorithm switch of our M-Guard algorithm with other deadlock detection algorithms under the MAEDD framework is desired.

References

1. J. Cao, G. H. Chan, W. Jia and T. Dillon: Checkpointing and Rollback of Wide-Area Distributed Applications Using Mobile Agents. In Proc. of 15th Intl. Parallel and Distributed Processing Symposium, San Francisco, California, USA, Apr. 23-27, 2001, pp. 1-6.
2. J. Cao, J. Zhou, W. Zhu, D. Chen and J. Lu: A Mobile Agent Enabled Approach for Distributed Deadlock Detection. In Proc. of the 3rd Intl. Conf. on Grid and Cooperative Computing (LNCS 3251), Wuhan, Hubei, China, Oct. 21-24, 2004, , pp. 535-542.
3. K. M. Chandy and J. Misra: A Distributed Algorithm for Detecting Resource Deadlocks in Distributed Systems. ACM Trans. on Computer Systems, Vol. 1(2), May, 1983, pp. 144-156.
4. A. N. Choudhary: Cost of Distributed Deadlock Detection: A Performance Study. In Proceedings of the 6th Intl. Conf. on Data Engineering, Los Angeles, California, USA, Feb. 5-9, 1990, pp. 174-181.
5. S. Funfrocken: Integrating Java-based Mobile Agents into Web Servers under Security Concerns. In Proc. of 31st Hawaii Intl. Conf. on System Sciences, Hawaii, USA, Jan. 6-9, 1998, pp. 34-43.
6. G. S. Ho and C. V. Ramamoorthy: Protocols for Deadlock Detection in Distributed Database Systems. IEEE Trans. on Software Engineering, Nov. 1982, Vol. 8(6), pp. 554-557.
7. E. Knapp: Deadlock Detection in Distributed Databases. ACM Computing Surveys, Vol. 19(4). Dec. 1987, pp.303-328.
8. A. D. Kshemkalyani and M. Singhal: Invariant-based Verification of a Distributed Deadlock Detection Algorithm. IEEE Trans. on Software Engineering, Vol. 17(8), Aug. 1991, pp. 789-799.
9. D. B. Lange and M. Oshima: Seven Good Reasons for Mobile Agents. Communications of the ACM, Vol. 42(3), 1999, pp. 88-89.
10. S. Lee and J. L. Kim: Performance Analysis of Distributed Deadlock Detection Algorithms. IEEE Trans. on Knowledge and Data Engineering, Vol. 13(4), Apr. 2001, pp. 623-636.
11. D. P. Mitchell and M. J. Merritt: A Distributed Algorithm for Deadlock Detection and Resolution. In Proc. of the 3rd ACM Symposium on Principles of Distributed Computing, New York, USA, Aug. 27-29, 1984, pp. 282-284.
12. R. Obermarck: Distributed Deadlock Detection Algorithm. ACM Trans. on Database Systems, Vol.7(2), Jun. 1982, pp.187-208.
13. V. A. Pham, A. Karmouch: Mobile Software Agents: An Overview. IEEE Communications, Vol. 36(7), Jul. 1998, pp. 26-37.
14. M. Singhal: Deadlock Detection in Distributed Systems. IEEE Computer, Vol. 22(11), Nov. 1989, pp.37-48.
15. M. K. Young, H. L. Ten and N. Soundarajan: Efficient Distributed Deadlock Detection and Resolution Using Probes, Tokens, and Barriers. In Proc. of the 1997 Intl. Conf. on Parallel and Distributed Systems, Seoul, Korea, Dec. 11-13, 1997, pp. 584-593.
16. J. Wu: Distributed System Design. CRC Press, USA, 1999.

Meta-based Distributed Computing Framework

Andy S.Y. Lai[1] and A.J. Beaumont[2]

[1] Department of Information and Communications Technology,
Institute of Vocational Education, Hong Kong
andylai@vtc.edu.hk
[2] Department of Computer Science, Aston University,
Aston Triangle, Birmingham, United Kingdom
a.j.beaumont@aston.ac.uk

Abstract. The explosive growth of distributed technologies requires frameworks to be adaptable. This paper uses design patterns as building blocks to develop an adaptive pattern-oriented framework for distributed computing applications. We describe our novel approach of combining a meta-architecture with a pattern-oriented framework, resulting in an adaptable framework which provides a mechanism to facilitate system evolution. We show how the meta-based framework can be used effectively to enable component integration and to separate system functionality from application functionality. The framework is based on modelling layers of the architecture to meet the challenges of customization, reusability, and extendibility in distributed computing technology. We also highlight how the meta-based framework will impose significant adaptability in system evolution through a simple example using a HTTP Server in conjunction with a thread pool.

1 Introduction

This work presents an approach for constructing an object-oriented (OO) design framework using distributed computing design patterns as building blocks under a meta-architecture. Such a framework can be used for constructing distributed computing applications. Design patterns and frameworks both facilitate reuse by capturing successful software development strategies. When patterns are used to structure and document frameworks in an object-oriented approach, nearly every class in the framework plays a well-defined role and collaborates effectively with other classes in the framework. In this paper, the proposed framework is not only pattern-oriented, it also employs a meta-architecture as a means of making the framework easily adaptable. The explosive growth of distributed technologies requires frameworks to be adaptable. Our meta-architecture supports dynamic adaptation of feasible design decisions in the framework design space by specifying and coordinating meta-objects that represent the building blocks within the distributed system environment. The proposed meta-based framework has the adaptability that allows the system evolution that is required in distributed computing technology. Our approach resolves the problem of the dynamic integration and adaptation within the framework, which is encountered in most distributed systems.

2 Pattern-Oriented Frameworks with Meta-architecture

Yacoub et al. [1] described pattern-oriented frameworks as containing two distinct levels: the pattern level and the class level. A simple pattern-oriented framework provides component-based design at design level but it has difficulty in addressing the framework's adaptability at the implementation level. We ensure that our framework is adaptive by using the novel combination of distributed computing design patterns within a meta-architecture. By adding a meta-architecture layer on top of the pattern-oriented framework, we will be able to provide an *adaptive level* in addition to the pattern level and class level, as shown in Figure 1.

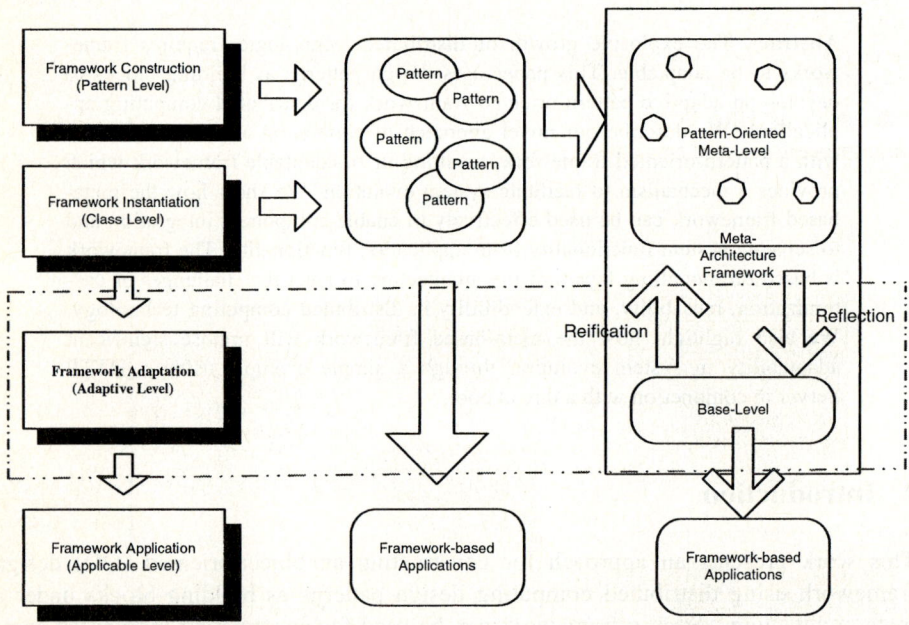

Fig. 1. Adaptive level introduced in Meta-based Pattern-Oriented Framework

The meta level provides information about selected system properties and behaviors, and the base level consists of the application logic. The implementation of system functionality is built at the meta level. Any changes in information kept at the meta level affect subsequent base-level behavior. Figure 1 shows the framework architecture where the meta level contains distributed computing patterns and the base level contains base objects (or application servers). The kernel of the framework provides the core functions for the framework adaptation between meta level and base level. It includes: *Meta objects and Meta Space Management* which handles meta level configuration, *Reflection Management* which provides dynamic reflection from the meta level to the base level, and *Reification Management* which provides dynamic reification from the base level to the meta level. In order to support the core functions for framework adaptation, we employ the Mediator, Visitor and Observer Patterns [2].

3 Patterns Instantiation and Patterns Integration in Meta Space

The kernel classes shown in Figure 2 manipulate the interaction between meta objects and are generalized for conformation from distributed computing patterns to meta objects. The class instantiation of the Thread Pool Pattern, shown in the top right of the figure, has thread pool workers defined to handle tasks assigned by the system. The management and deployment of the system-level components in the meta space is simple and uniform. On the class level, distributed computing pattern components simply extend a class called MetaObjectImpl from the kernel package to form meta components, and, on the object level, the instantiation of meta components means that they are ready to deploy to meta space. Other distributed computing patterns such as Http Server, ORB Registry and Publisher/Subscriber can follow the same procedures and be easily deployed as meta components.

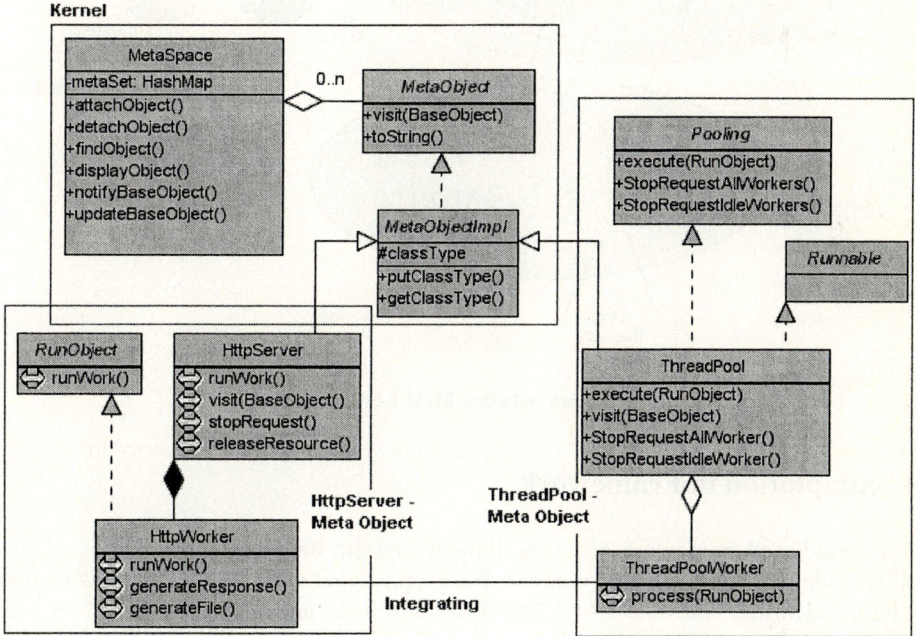

Fig. 2. Distributed Computing Patterns Integration at Meta Level

Our framework supports dynamic integration of meta objects at run-time. Thread Pooling Pattern and Http Server Pattern are meta objects and they are deployed to the meta space and that situation is illustrated in Figure 2. Every time a HTTP request comes to the base object, the meta space will handle the request by checking whether the base object is reified with the `ThreadPool` at the meta level. If the `ThreadPool` found is reified with the base object, MetaSpace will let `HTTPServer` pass its `HTTPWorker` to the thread pool workers to continue the process. Figure 2 shows that integration between meta objects, `ThreadPool` and `HttpServer` at the meta level which is used to provide a Thread Pooling HTTP Server at the base level.

Figure 3 shows the interaction between the meta level and the base level. Base object receives Http requests and passes them to `HTTPServer` at the meta level. `HTTPServer` normally processes requests using `HTTPWorker`. However, if there is a `ThreadPool` that has been reified by the base object, `HTTPWorker` will be passed to the `ThreadPool` and a `ThreadPoolWorker` handles the task for the `HTTPServer`. In this way, handling tasks will be based on the number of workers in `ThreadPool`.

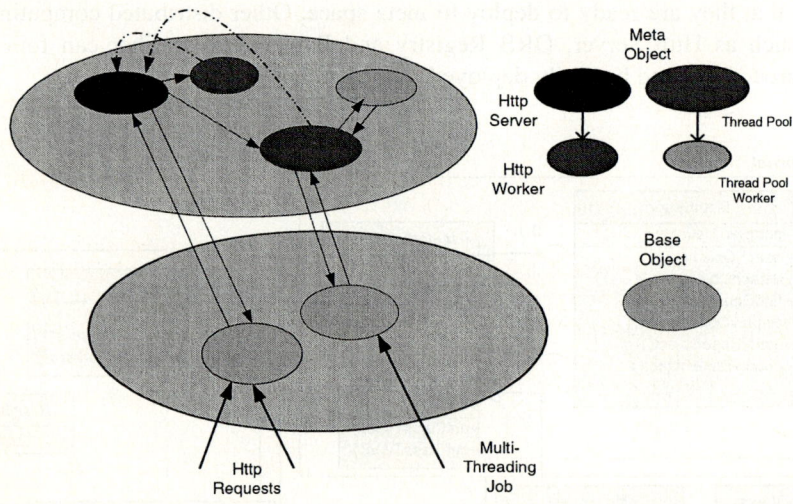

Fig. 3. Interaction between Meta Level and Base Level

4 Adaptation in Framework

Our meta-based framework provides dynamic system behavior and provides adaptability within its run-time environment. We propose that each meta object has its own internal identifier attribute `MetaID` which is private and internally stored in each meta object. Fixed Thread Pool Pattern and Growth Enabled Thread Pool Pattern are two typical examples of pooling patterns. The meta objects for these two patterns would each have its own internal identifier. However, in this case, the contents of the two internal identifiers are identical to indicate both are a kind of thread pool.

The administration utility has the ability to replace meta objects at run time by verifying their internal identifiers. One meta object can replace another if they both belong to the same type and have the same internal identifier. Therefore a fixed thread pool could be replaced with a growth enabled thread pool as described below:

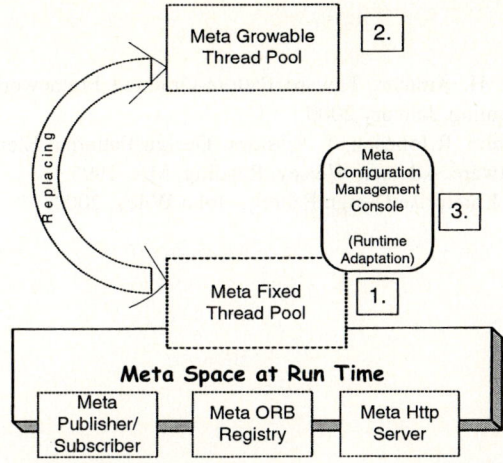

1. MetaSpace has constructed and instantiated metaFixedThreadPool which belongs to a Type called Pooling and has an internal identifier with the value *ThreadPool*. metaFixedThreadPool is then registered in the meta object repository.
2. Meta object metaGrowableThreadPool is constructed and instantiated and also belongs to a type called Pooling and so also has an internal identifier with the value *ThreadPool*. However, it has not been registered in the meta object repository. Since it has the same meta type and internal identifier as metaFixedThreadPool, they can each be replaced by the other.
3. Our configuration and management utility controls the replacements of meta objects. The changes will reflect the system behavior of the meta level and also immediately affect those base objects which have reified the meta objects which have the internal meta identifier *ThreadPool*.

5 Conclusion

We have presented a new approach to the design of a framework based on a meta-architecture employing distributed computing patterns as the building blocks for meta objects. This paper addresses how our framework can meet diverse requirements in distributed computing systems, and describes the advantage of its meta-architecture that makes the system adaptive and configurable. The proposed framework provides separation of concerns in system functionality and business functionality and makes the system's technological features open-ended for extension, and allows itself to continually evolve. The work has emerged as a promising way to meet the challenges in distributed environment currently and in the future. Nonetheless, significant work remains to be done with respect to security and retransmission, etc [3]. We view such distributed computing components as an evolutionary extension of the framework we have described.

References

1. S.M. Yacoub, H. H. Ammar: Toward Pattern-Oriented Frameworks. Journal of Object-Oriented Programming, January 2000.
2. E. Gamma, R. Helm, R.Johnson, J. Vissides: Design Patterns: Elements of Reusable Object-Oriented Software, Addison-Welsey, Reading, MA, 1995.
3. M. Grand: JAVA Enterprise Design Patterns, John Wiley, 2002.

Locality Optimizations for Jacobi Iteration on Distributed Parallel Systems

Yonggang Che, Zhenghua Wang[1], Xiaomei Li[2], and Laurence T. Yang[3]

[1] School of Computer, National University of Defense Technology,
Changsha 410073, P. R. China
[2] Institute of Equipment and Command Technology, Beijing, P.R. China
[3] Department of Computer Science, St. Francis Xavier University,
P.O. Box 5000, Antigonish, B2G 2W5, NS, Canada

Abstract. In this paper, we propose an inter-nest cache reuse optimization method for Jacobi codes. This method is easy to apply, but effective in that it enhances cache locality of the Jacobi codes while preserving their coarse grain parallelism. We compare our method to two previous locality enhancement techniques that can be used for Jacobi codes: time skewing and new tiling. We quantitatively calculate the main contributing factors to the runtime of different Jacobi codes. We also perform experiments on a PC cluster to verify our analysis. The results show that our method performs poorer than time skewing and new tiling for uniprocessor, but performs better for distributed parallel system.

1 Introduction

Jacobi iteration is widely used in Partial Differential Equations (PDE) solvers. Although more efficient schemes (e.g., multigrid) have been introduced, it remains important because it is building block of other advanced schemes, and also because it has simple computational properties. Many programs spend a large fraction of their runtime performing Jacobi-style computations.

Consider Jacobi code for Laplace equation in two dimensions. Figure 1(a) shows the original sequential Fortran code, which consists of two loop nests surrounded by an outer loop. We use the term *time-step* to refer to the iteration of loop t, and the term *spatial loop* to refer to the two inner loops (loop j and loop i). We call the time-step when t takes an odd (even) number as an *odd (even) time-step*. There is reuse between the two spatial loops in addition to the reuse carried by the time-step loop. If the sizes of array U1 and U2 exceed the size of cache, the array elements used in current time-step will be evicted from the cache before they can be used by subsequent time-step. Thereby the array elements must be fetched repeatedly to the cache in every time-step and the Jacobi code will incur a lot of cache misses and spend a good many time handling these cache misses.

Song & Li [2, 3], Wonnacott [4, 5] both proposed methods that can be used to solve this kind of problems. Song & Li used a *new tiling* technique, which utilizes a combination of loop skewing and tiling, combined with limited array expansion. They only discussed optimization for uniprocessor. Wonnacott described *time skewing*, a technique

Initiate U1 and boundary of U2
```
      do 100 t=1, mi
        do 200 j = 1,N
          do 200 i = 1, N
            U2(i,j)=(U1(i,j-1)+U1(i-1,j)
                   +U1(i+1,j)+U1(i,j+1))* 0.25
200     continue
        do 300 j = 1,N
          do 300 i = 1,N
            U1(i,j) = U2(i,j)
300     continue
100   continue
```
(a)

Initiate U2 and boundary of U1
```
      do 100 t=1, mi
        if( MOD(t, 2) .EQ. 1 ) then
          do 200 j = 1,N
            do 200 i = 1, N
              U1(i,j)=(U2(i,j-1)+U2(i-1,j)
                     +U2(i+1,j)+U2(i,j+1))* 0.25
200       continue
        else
          do 300 j = N, 1, -1
            do 300 i =N, 1, -1
              U2(i,j)=(U1(i,j-1)+U1(i-1,j)
                     +U1(i+1,j)+U1(i,j+1))* 0.25
300       continue
        end if
100   continue
```
(b)

Fig. 1. Fortran codes for serial Jacobi. (a) Original version; (b) InterNest version

that combines tiling in both the data and time domains. A time-step loop is skewed with one spatial loop over a data tile so as to preserve data dependencies across multiple time-steps. The technique involves transforming both the iteration space and array indexing functions in a loop body. Wonnacott described how to achieve scalable locality and parallelism. But the overheads due to cache miss and communication still exist and most of them can't be hidden on modern parallel systems. Other work [6, 7] discussed the data partition and task scheduling problems for Jacobi code on parallel systems. But they only considered the original Jacobi code.

In this paper, we propose a simple locality enhancement method for Jacobi code, which utilizes inter-nest reuse in the Jacobi code to improve its cache locality. In order to determine which method performs the best for a given machine and certain problem size, we quantitatively analyze the major contributing factors to overall performance of the three methods. We also perform experiments on a PC cluster to verify our analysis.

The rest of this paper is organized as follows. Section 2 presents sketch of the three locality optimization methods. Section 3 presents our quantitative analysis in detail. Section 4 presents our experimental results and discussion. In Section 5, we present a short summary.

2 Sketch of the Three Memory Locality Enhancement Methods

We briefly introduce the three memory locality enhancement methods mentioned above here. For simplicity, we don't consider Jacobi code with convergence test, which is solved by Song and Li [2]. We denote the original Jacobi codes (both serial and parallel) as *Ori*, our Jacobi codes as *InterNest*, the Jacobi codes that employ Wonnacott's time skewing as *TimeSkew*, the Jacobi codes that employ Song & Li's new tiling as *NewTile*.

2.1 InterNest

We find that the array copying operations (the second loop nest in Figure 1(a)) in the original Jacobi code can be eliminated. By odd-even duplication and using U1 or U2 to store temporary values alternately, we remove the array copying operations. There is data reuse between successive loop nests. To transform this kind of data use into cache locality, we use an Inter-nest reuse optimization method. Inter-nest reuse takes place when a set of data items is accessed in a given loop nest and then accessed again within some subsequent portion of the program. In [8], Kandemir et al presented a compiler strategy that optimizes inter-nest reuse using loop transformations. They determine the loop transformations matrix and transform the loop nest based on it. Here we do not present the transformation matrix. To make use of newly cached array elements by previous time-step, the i and j loops in even time-steps are conducted in reverse order wrt (with respect to) the i and j loops in odd time-steps. The serial code of InterNest is shown in figure 1(b). Just as Ori, the computation is performed time-step by time-step in InterNest. Figure 2 shows the parallel Jacobi codes for Ori and InterNest.

The computation order of InterNest is illustrated in figure 3, in which a point stands for the computations that update one column of array U1 or U2. Note that although each column of U1 or U2 appears many times in the figure, it actually uses the same memory location. The arrows denote the iteration order of loop j. As the 4-point stencil computation requires, for any j ($0 \leq j \leq ny + 1$), the values of column U1(*, j) depend on the values of three columns U2(*, j-1), U2(*, j) and U2(*, j+1) from previous time-step iteration. The data dependency relationship is not shown in figure 3 and other figures.

2.2 TimeSkew

Loop skewing is a program transformation proposed by Wolfe [9]. It is usually used to transform the loop and its dependences into a form for which tiling is legal. Time

Initiate U1 and boundary of U2
    do 100 t=1, mi
        Exchange boundary data of array U1
        do 200 j = 1,N
            do 200 i = 1, N
                U2(i,j)=(U1(i,j-1)+U1(i-1,j)
                +U1(i+1,j)+U1(i,j+1))* 0.25
200   continue
        do 300 j = 1,N
            do 300 i = 1,N
                U1(i,j) = U2(i,j)
300   continue
100 continue

(a)

Initiate U2 and boundary of U1
    do 100 t=1, mi
        if(MOD(t, 2) .EQ. 1) then
            Exchange boundary data of array U2
            do 200 j = 1,N
                do 200 i = 1, N
                    U1(i,j)=(U2(i,j-1)+U2(i-1,j)
                    +U2(i+1,j)+U2(i,j+1))* 0.25
200   continue
        else
            Exchange boundary data of array U1
            do 300 j = N, 1, -1
                do 300 i =N, 1, -1
                    U2(i,j)=(U1(i,j-1)+U1(i-1,j)
                    +U1(i+1,j)+U1(i,j+1))* 0.25
300   continue
        end if
100 continue

(b)

Fig. 2. Fortran codes for parallel Jacobi. (a) Original version; (b) InterNest version

skewing approach reorders the computation order of Jacobi code as figure 4 shows. In figure 4, a point stands for the computations that update one column of array U1 or U2. The time-step iterations are divided into groups (referred to as frames, separated by horizontal lines in figure 4), each contains several time-steps. The computation is conducted frame by frame. In each frame, the computations order is from upper-left to the right-below, as indicated by arrows. TimeSkew is parallelized in pipelined pattern, as illustrated in figure 5 (we only illustrate the situation for column partition to save

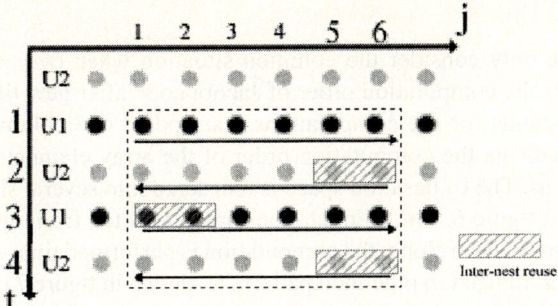

Fig. 3. Computation order of serial InterNest

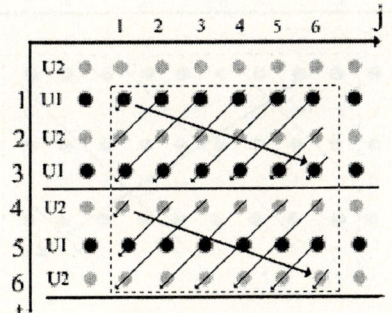

Fig. 4. Computation order of serial TimeSkew

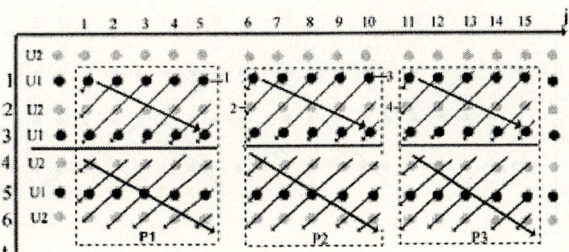

Fig. 5. Parallel version of TimeSkew (column partition)

space). Wonnacott introduced another parallelization method for time skewed stencil codes in [4]. But that method uses an inefficient memory allocation scheme, in which each processor tries to allocate the entire shared array, and then only fill and use the parts it needs. That may lose the benefit of parallel computing. We do not use that method in this paper.

2.3 NewTile

For simplicity, we only consider the common situation when ($mi < ny - ns + 1$). Figure 6 illustrates the computation order of Jacobi code after new tiling is applied. In figure 6, a point stands for the computations that update one column of array U1 or U2. The arrows indicate the computation order of the array elements. The j loops are tiled with the size ns. The t-j iteration space is reordered into several slopes as separated by the diagonals in figure 6. The computations are conducted from upper-left slopes to right-bottom slopes. In each slope, the computation is performed time-step by time-step. NewTile is also parallelized in pipelined pattern, as shown in figure 7 (we only illustrate the situation for column partition to save space).

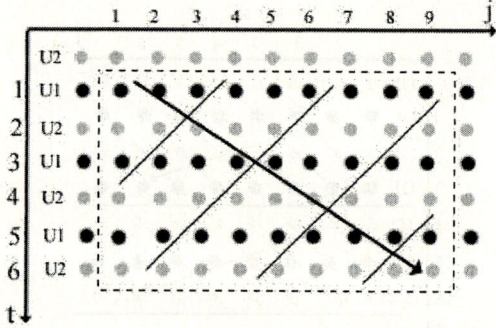

Fig. 6. Computation order of serial NewTile

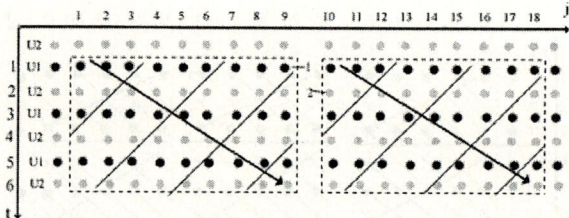

Fig. 7. Parallel version of NewTile (column partition)

3 Performance Modeling

3.1 Assumptions and Notations

We study the performance of these Jacobi codes using a quantitative method. We are only intended to roughly compare their performance here. We target our analysis to a distributed memory parallel system, in which each node has the same computation ability. The communication cost between any two nodes is the same. Each processor of

Locality Optimizations for Jacobi Iteration on Distributed Parallel Systems 97

the system has two levels of cache. We denote the cache at level i (i=1,2) as $Cache_i$. The caches use LRU (Least Recently Used) replacement algorithm. The computation domain is rectangular. Each processor is assigned a rectangular sub-domain. The program uses column-major storage order.

The architecture parameters and performance profiles of the machine are:

T_f: the average time in seconds for the processor to perform a floating-point operation;

S_1 / S_2: L1 / L2 cache size in the number of data elements;

B_1 / B_2: L1 / L2 cache line size in the number of data elements;

C_1 / C_2: L1 / L2 cache miss penalty in seconds;

W: Network bandwidth in elements/second for large message passing between two processors;

L: End to end latency or time spent sending a 0-length message between two processors.

The program parameters are:

P: the total number of processes (equal to the number of processors) used;

px / py: the number of processes in x / y dimension in the process topology;

N: total number of grid points in each dimension (boundary points not included);

nx / ny: the number of grid points in x / y dimension for one process;

mi: the number of time-step iterations, assumed to be an even number;

nf: the number of time-steps in a frame in TimeSkew;

ns: the tile size in j loop in NewTile.

For parallel codes, we have P=px*py, nx=N/px, and ny=N/py. We assume $ny \gg nf$. For column partition, py=P, nx=N, and ny=N/p. For row partition, px=P, ny=N, and nx=N/p.

3.2 Overhead Factors

The total runtime of parallel code can be divided into the following factors:

1. time spent performing useful floating-point calculations (T_{fp})
2. stalls due to processor pipeline stalls when cache misses occur (T_{miss})
3. idle time due to pipeline initiation in parallel pipeline execution pattern, or wait latency (T_{wait})
4. time spent on exchanging data with neighboring processes (T_{comm}), which can be further divided into time due to end to end latency (T_{EEL}) and time spent on actual data transfer (T_{trans}).

The serial runtime is

$$T_{ser} = T_{fp} + T_{miss}$$

The parallel runtime is

$$T_{par} = T_{fp} + T_{miss} + T_{wait} + T_{EEL} + T_{trans}$$

3.3 Cache Misses for Tree Versions of Jacobi Code

We consider only compulsory misses and capacity misses. The situation when array U1 and U2 can fit into $Cache_i$ (that is, $S_i \geq 2nx * ny$) is not considered because it rarely happens for real applications.

For all Jacobi codes, the spatial and temporal locality wrt loop i can be achieved. A whole loop i will miss $Cache_i$ no more than $4nx/B_i$ times. So for three versions of Jacobi codes, the number of misses for $Cache_i$ is no greater than $4nx * ny * mi/B_i$.

InterNest. For a whole j-i loop nest, each column U2(*,j) is accessed for three times, and each column U1(*,j) is accessed once. If $Cache_i$ can hold 4 columns of U1 and U2, this kind of data reuse will become cache locality, loop nest j-i will totally incur $(2nx * ny - S_i)/B_i$ cache misses. If $Cache_i$ can not hold 4 columns of U1 and U2, then the number of cache misses is $(4nx * ny - S_i)/B_i$. So the number of cache misses for $Cache_i$ (i=1,2) is

$$M_{Li}^{InterNest} \begin{cases} \approx (4nx * ny - S_i) * mi/B_i, (S_i < 4nx) \\ = (2nx * ny - S_i)mi/B_i, (4nx \leq S_i < 2nx * ny) \end{cases}$$

TimeSkew. As can be seen from Figure 2, the working set size of inner loop nest that computes the values of an arrow is approximately 3+2(nf-1)+1=2nf+2 columns of data. When $Cache_i$ can hold these columns of array elements, each column of array is reused for nf times in $Cache_i$. So the number of cache misses for $Cache_i$ is $(2nx * ny/B_i) * (mi/nf) = 2nx * ny * mi/B_i$. Otherwise, it will incur about $2nx * ny * mi/B_i$ cache misses. So the number of cache misses for $Cache_i$ (i=1,2) is

$$M_{Li}^{TimeSkew} \begin{cases} \approx 4nx * ny * mi/B_i, \ (S_i < 4nx) \\ = 2nx * ny * mi/B_i, \ (4nx \leq S_i < 2(nf+1)nx) \\ = 2nx * ny * mi/(B_i * nf), \\ (2(nf+1)nx \leq S_i < 2nx * ny) \end{cases}$$

NewTile. If cache can hold 2(ns+2) columns of array elements, it ensures that each column is reused for ns times. So the number of cache misses for $Cache_i$ (i=1,2) is

$$M_{Li}^{NewTile} \begin{cases} \approx 4nx * ny * mi/B_i, \ (S_i < 4nx) \\ = 2nx * ny * mi/B_i, \ (4nx \leq S_i < 2(ns+2)nx) \\ = 2nx * ny * mi/(B_i * ns), \\ (2(ns+2)nx \leq S_i < 2nx * ny) \end{cases}$$

When $2(nf+1)*nx \leq S_i$ (i=1,2), TimeSkew incurs less cache misses than InterNest at $Cache_i$. When $2(ns+2) \leq S_i$, NewTile incurs less cache misses than InterNest at $Cache_i$. For other cases, InterNest incurs slightly less cache misses than TimeSkew and NewTile. NewTile and TimeSkew incur nearly the same number of cache misses under most situations. For three versions of serial Jacobi codes, their performance is largely determined by the number of the cache misses. We can conclude that under most situations, TimeSkew and NewTile perform better than InterNest on uniprocessor, and TimeSkew and NewTile perform similarly on uniprocessor.

3.4 The Cost of Communication

For all Jacobi codes, each process must exchange boundary data with their neighbors. While the number of array elements exchanged is the same, the communication count and the waiting time due to data dependency differ greatly across four methods.

Ori and InterNest. Ori and InterNest have regular computation patterns. In each time-step, they first exchange boundary data with neighboring processes, then do stencil computations.

- **Column Partition.** Each process at the boundary of the process topology (boundary process) will communicate for 2mi times, totally send and receive 2mi*nx array elements. Each process in the interior of the process topology (interior process) will communicate for 4mi times, totally send and receive 4mi*nx array elements.
- **Row Partition.** Each boundary process will communicate for 2mi times, totally send and receive 2mi*ny array elements. Each interior process will communicate for 4mi times, totally send and receive 4mi*ny array elements.
- **2D Division.** Each process will communicate for 4mi, 6mi or 8mi times, totally send and receive 2mi*(nx+ny), 2mi*(2nx+ny), 2mi*(nx+2ny) or 4mi*(nx+ny) array elements.

TimeSkew

- **Column Partition.** As can be seen from figure 5, in the first frame, only after process P_1 finishes computing its right most column of U1 (marked by "1") at time-step 1 and send this column to process P_2, can P_2 computes its left most column of U2 (marked by "2") at time-step 2. Similarly, process P_3 must wait process P_2 because of the data dependency between array elements marked by "3"and "4". This forms a waiting chain. When P_n starts computing the left most column of U2 at time-step 2, the pipeline is "full". After that, all processes can do their work concurrently. P_n will be the last process to finish its computation. For P_n, it must wait the time for a single process to compute (P-1)((2+3+...+(nf-1))+nf(ny-nf))≈P*nf*ny columns before it can calculate its left most column of U1 at time-step 2. Performing these computations needs to exchange one column of array between any two neighboring processes. Further more, these computations also introduce cache misses. So P_n totally wait the time for a single process to perform approximately $4P * ny * nx * nf$ floating-point calculations, perform 2P communications (totally send or receive 2nx*P array elements), and handle $\frac{P*nf}{mi} M_{L1}^{TimeSkew}$ L1 data cache misses and $\frac{P*nf}{mi} M_{L2}^{TimeSkew}$ L2 cache misses.
- **Row Partition.** The number of elements exchanged is the same as InterNest. Each process can do its computation concurrently from the beginning to the end. There is no significant wait latency between processes. But the communication count is larger for it can only send or receive 1 array element at one time to maintain the computation order and the parallelism. Each boundary process will communicate

for 2mi*ny times, totally sends and receives 2mi*nx array elements. Each interior process will communicate for 4mi*ny times, totally send and receive 4mi*ny array elements.
- **2D Partition.** The communication counts and number of array elements communicated are the sum of those at column and those at row. It incurs wait latency in column direction and a large number of communications in row direction. Each process will communicate for 2mi*ny to 4mi*ny times, according to its position wrt other processes in the process topology. Its waiting pattern between processes is similar to that of the column partition. For the right most processes, the wait latency is the sum of time for a single process to perform approximately 4py*nf*nx*ny floating-point calculations, perform 2py*nf*ny communications (totally send or receive 2py*nx+2py*nf*ny array elements), and handle $\frac{py*nf}{mi}M_{L1}^{TimeSkew}$ L1 data cache misses and $\frac{py*nf}{mi}M_{L2}^{TimeSkew}$ L2 cache misses.

NewTile. The communication count is the same as TimeSkew. But as we can see from figure 7(a), P_1 must finish calculating mi(ny-ns+ny-ns-mi+1)/2≈mi(2ny-2ns-mi)/2 - columns at the upper left region in figure 7(a) before it calculates the right most column of U1 at time-step 1 (marked by "1") and send this column to P_2, then P_2 can start calculate its left most column of U2 at time-step 2 (marked by "2"). This kind of latency exists between any two neighboring processes. Totally, the right most process must wait the time for a single process to compute P*mi*(2ny-2ns-mi)/2 columns to begin calculate its left most column of U2 at time-step 2. The wait latency for P_n is the sum of time for a single process to perform $4nx*mi(P-1)(2ny-2ns-mi+1)/2 \approx 2P*nx*mi*(2ny-2ns-mi)$ floating-point calculations, perform 2P communications (totally send or receive 2nx*P array elements), and handle $\frac{P*(2ny-2ns-mi)}{2ny}M_{L1}^{NewTile}$ L1 data cache misses, $\frac{P*(2ny-2ns-mi)}{2ny}M_{L2}^{NewTile}$ L2 cache misses.

We see that the wait latency is rather long.

- **Row Partition.** The communication count is the same as TimeSkew. From figure 8, we see that each process can do its work concurrently from beginning to the end. No significant wait latency between processes.
- **2D Partition.** The communication count is the sum of those at column and those at row. It incurs wait latency in column direction and a large number of communications in row direction. Each process communicates 2mi(1+ny) to 4mi(1+ny) times, according to its position wrt other process in the process topology. The wait latency at column direction exists as column partition does. The right most processes in the process topology must wait the sum of time for a single process to perform $2py*nx*mi*(2ny-2ns-mi)$ floating-point operations, perform py*mi*(2ny-2ns-mi) communications (totally send or receive 2py*nx+py*mi*(2ny-2ns-mi) array elements), and handle $\frac{py*(2ny-2ns-mi)}{2ny}M_{L1}^{NewTile}$ L1 data cache misses and $\frac{py*(2ny-2ns-mi)}{2ny}M_{L2}^{NewTile}$ L2 cache misses.

Overall Communication Overheads. Table 1 shows the cost of communication for different Jacobi codes, where InterNest_col stands for InterNest with row partition, InterNest_row stands for InterNest with row partition, and so on. In table 1:

Table 1. The cost of communication for different versions of Jacobi

	T_{wait}	T_{EEL}	T_{trans}
InterNest_col	0	4L*mi	4mi*nx/W
InterNest_row	0	4L*mi	4mi*ny/W
InterNest_2D	0	8L*mi	4mi(nx+ny)/W
TimeSkew_col	$T_{wait}^{(1)}$	4L*mi	4mi*nx/W
TimeSkew_row	0	4L*mi*ny	4mi*ny/W
TimeSkew_2D	$T_{wait}^{(2)}$	4L*mi*ny	4mi(nx+ny)/W
NewTile_col	$T_{wait}^{(3)}$	4L*mi	4mi*nx/W
NewTile_row	0	4L*mi*ny	4mi*ny/W
NewTile_2D	$T_{wait}^{(4)}$	4L*mi*ny	4mi(nx+ny)/W

$$M^{TimeSkew} = (C_1 * M_{L1}^{TimeSkew} + C_2 * M_{L2}^{TimeSkew})$$

$$M^{NewTile} = (C_1 * M_{L1}^{NewTile} + C_2 * M_{L2}^{NewTile})$$

$$T_{wait}^{(1)} = 4P * T_f * nx * ny * nf + 2L * P + 2P * nx/W + \frac{P*nf}{mi} M^{TimeSkew}$$

$$T_{wait}^{(2)} = 4T_f * py * nf * ny * nx + 2L * py * nf * ny + 2py * (nx + nf * ny)/W + \frac{py*nf}{mi} M^{TimeSkew}$$

$$T_{wait}^{(3)} = 2T_f * P * nx * mi(2ny - mi) + 2L * P + 2P * nx/W + \frac{P(2ny-mi)}{2ny} M^{NewTile}$$

$$T_{wait}^{(4)} = 2T_f * py * nx * mi(2ny - mi) + L * py * mi(2ny - mi) + py * mi(2ny - mi)/W + \frac{py(2ny-mi)}{2ny} M^{NewTile}$$

We see that TimeSkew and NewTile incur much larger parallel overheads than InterNest.

Because 2D partition is commonly required for large parallel systems, we compare TimeSkew's and NewTile's parallel performance when 2D partition is used. Because their $T_{fp}, T_{miss}, T_{EEL}$ and T_{trans} are nearly the same, we only need to compare the difference of their T_{wait}. Her we assume $mi \ll ny$, ns=nf and $M^{TimeSkew} = M^{NewTile}$.

$$T_{par}^{NewTile} - T_{par}^{TimeSkew} = T_{wait}^{NewTile} - T_{wait}^{TimeSkew}$$
$$= py * mi(2T_f * nx(2ny - mi) + L(2ny - mi)) + py * mi(2ny - mi)/W$$
$$+ \frac{py(2ny-mi)}{2ny} M^{NewTile} - (4T_f * py * nf * ny * nx + 2py * L * nf * ny)$$
$$- \left(2py * (nx + nf * ny)/W + \frac{py*nf}{mi} M^{TimeSkew}\right)$$
$$\approx 4T_f * py * nx * ny * mi + 2py * L * ny * mi + 2py(ny * mi - nx)/W$$
$$+ py M^{TimeSkew}$$
$$> 0 \quad (ny * mi - nx) > 0$$

We see that NewTile incurs much larger wait latency than TimeSkew. It will perform poorer than TimeSkew on parallel systems.

4 Experimental Results and Discussion

The experiments are done on a PC cluster connected by 100M fast Ethernet switch. Each node is based on a 2.53GHz Intel P4 processor, which has separated L1 cache (12KB micro-ops /8KB data, 64B/line) and a unified L2 cache (512 KB, 128B/line). Each node contains 1GB 333MHz DDR RAM. Every node runs RedHat 8.0 Linux (Kernel version 2.4.18-14). We use LAM/MPI 6.5.6 as message passing library. We use G77 V7.3 compiler with the highest optimization level (-O3) turned on. Each array element is a 4-byte floating-point value. The time reported here is wall clock time. The runtime of TimeSkew depends on the parameter nf. We change nf from 2 to 64 and report the one with the shortest runtime. The runtime of NewTile depends on the parameter ns and the same approach is used to choose the best one. The performance of TimeSkew and NewTile is rather poor when row partition is used, so we do not present their results.

Table 2 shows the runtime for the four serial Jacobi codes. We see the three optimized versions outperform Ori significantly. The runtime for InterNest is slightly longer than TimeSkew and NewTile. For all problem sizes, the runtime for TimeSkew and NewTile nearly equal. These data conform to our analysis.

Table 3 shows the parallel runtime for the four methods when column partition is used. The number of time-step iteration (mi) is 100 for all problem sizes. We see that the runtime of InterNest is much shorter than that of NewTile, but is sometimes longer than that of TimeSkew.

Table 4 shows the runtime when 2D partition is used. The number of time-step iteration (mi) is 100 for all problem sizes. We see that InterNest always outperforms

Table 2. The runtime in seconds for serial Jacobi codes

N / mi	Ori	InterNest	TimeSkew	NewTile
256/10000	5.85	2.8	2.37	2.46
512/1000	11.61	9.82	9.14	9.14
1024/100	2.33	1.63	1.36	1.38
2048/100	7.48	4.79	3.82	3.87
4096/100	26.32	16.63	16.23	16.01

Table 3. The runtime in seconds for parallel Jacobi codes with column partition

N	P	Ori	InterNest	TimeSkew	NewTile
2048	2	4.112	2.627	2.266	4.226
	4	2.71	1.783	1.46	4.149
	8	1.607	1.23	1.039	4.483
4096	2	15.015	8.747	8.61	14.45
	4	8.189	5.258	4.214	13.11
	8	4.747	3.284	2.776	13.42
8192	2	48.965	30.09	39.397	73.516
	4	25.964	15.462	15.503	56.708
	8	14.622	9.339	8.912	54.608

Table 4. The runtime for parallel Jacobi codes with 2D partition

N	PX×PY	Ori	InterNest	TimeSkew	NewTile
2048	2×2	2.333	1.528	4.066	13.426
	2×4	1.318	1.025	13.324	13.676
	4×2	1.342	0.989	9.001	19.835
4096	2×2	7.650	4.748	8.693	29.967
	2×4	4.432	3.128	23.245	31.442
	4×2	4.151	2.805	14.299	40.99
8192	2×2	26.859	16.61	21.58	81.117
	2×4	13.978	9.016	45.943	81.62
	4×2	13.934	9.016	24.385	88.821

NewTile and TimeSkew significantly. The runtime for InterNest always decreases when more processors are used. For the same number of processors, the runtime of InterNest decreases when 2D partition is used instead of column partition. But for NewTile and TimeSkew, employing more processors even increases their runtime.

So we conclude that NewTile is not suitable for parallel systems. TimeSkew is only suitable for small-scale parallel systems when column partition is used. InterNest has good parallel scalability and is more suitable for large-scale parallel systems.

5 Summary

In this paper, we propose a performance enhancement method for 4-point Jacobi code that solving 2D Laplace equation. This method does not aggressively enhance the memory locality, but maintains the coarse grain parallelism in the Jacobi code. We compare our methods to Wonnacott's time skewing and Song's new tiling, both quantitatively and experimentally. We see that Wonnacott and Song's methods exhibit better memory locality and are suitable for uniprocessor. But when it comes to distributed parallel systems, their methods suffer from higher parallel overhead. Our method is more suitable for distributed memory parallel systems.

While we focus our analysis to Jacobi iteration for 2D Laplace equation, it is easy to extend this method to other scientific applications that employ Jacobi scheme over a regularly discretized domain.

References

1. Yonghong Song and Zhiyuan Li. New Tiling Techniques to Improve Cache Temporal Locality. in Proceedings of ACM SIGPLAN PLDI99, 1999: 215-228.
2. Yonghong Song and Zhiyuan Li, Effective Use of The Level-Two Cache for Skewed Tiling. Technical Report CSD-TR-01-006, Department of Computer Sciences, Purdue University, 2001.
3. Yonghong Song and Zhiyuan Li. Impact of Tile-Size Selection for Skewed Tiling. in Proceedings of the 5th Workshop on Interaction between Compliers and Architectures, 2001.

4. David Wonnacott. Using Time Skewing to Eliminate Idle Time due to Memory Bandwidth and Network Limitations. in Proceedings of International Parallel and Distributed Processing Symposim, 2000: 171-180.
5. David Wonnacott, A General Algorithm for Time Skewing. International Journal of Parallel Programming, 2002: 181-221.
6. Naraig Manjikian and Tarek S. Abdelrahman. Scheduling of wavefromnt parallelism on scalable shared memory multiprocessor. in Proceedings of International Conference on Parallel Processing. Bloomington, IL, 1996: 122-131.
7. David K. Lowenthal and Vincent W. Freeh, Architecture-independent parallelism for both shared- and distributed-memory machines using the Filaments package. Parallel Computing, 2000, 26(10): 1297-1323.
8. M. Kandemir, I. Kadayif, A. Choudhary, et al. Optimizing Inter-Nest Data Locality. in Proceedings of International Conference on Compilers, Architecture, and Synthesis for Embedded Systems. Grenoble, France, 2002: 127-135.
9. M. J. Wolfe, Loop skewing: The wavefront method revisited. International Journal of Parallel Programming, 1986, 15(4): 279-293.

Fault-Tolerant Cycle Embedding
in the WK-Recursive Network

Jung-Sheng Fu

Department of Electronics Engineering, National United University, Taiwan
jsfu@nuu.edu.tw

Abstract. [1]Recently, the WK-recursive network has received much attention due to its many favorable properties such as a high degree of scalability. By $K(d,t)$, we denote the WK-recursive network of level t, each of whose basic modules is a d-node complete graph, where $d > 1$ and $t \geq 1$. In this paper, we construct fault-free Hamiltonian cycles in $K(d,t)$ with at most $d-3$ faulty nodes, where $d \geq 4$. Since the connectivity of $K(d,t)$ is $d-1$, the result is optimal.

1 Introduction

The WK-recursive network, which was proposed in [1], is a class of recursively scalable networks. It offers a high degree scalability, which conforms very well to a modular design and implementation of distributed systems involving a large number of computing elements. A transputer implementation of a 16-node WK-recursive network has been realized at the Hybrid Computing Research Center, Naples, Italy. In this implementation, each node is implemented with the IMS T414 Transputer [2].

Previous works relating to the WK-recursive network can be found in [1, 3–12] Layout circuits in VLSI were described in [3]. A broadcasting algorithm was presented in [4]. In [9], some substructure allocation algorithms for a multiuser WK-recursive network were presented. The shortest-path routing algorithm was presented, and some topological properties, such as diameter and connectivity, were investigated in [5]. The wide diameter and fault diameter were computed in [6]. In [7], the Hamiltonian cycle embedding method with some faulty links was presented. Cycles of all possible lengths were also constructed. In [8], link-disjoint Hamiltonian cycles and link-disjoint spanning trees were constructed. In [12], a B-tree traingular coding was presented. In [10], an adaptive routing algorithm was presented. In [13], the Rabin number was derived. In [11], a communication algorithm was presented. In [14], a Hamiltonian path between arbitrary two distinct nodes was constructed.

---

[1] The author would like to thank the National Science Council of the Republic of China, Taiwan for financially supporting this research under Contract No. NSC 93-2213-E-239-010-.

Linear arrays and rings, which are two of the most fundamental networks for parallel and distributed computation, are suitable for developing simple algorithms with low communication cost. Many efficient algorithms that were designed based on linear arrays and rings for solving a variety of algebraic problems and graph problems can be found in [15, 16]. They can also be used as control/data flow structures for distributed computation in arbitrary networks. The longest path approach was applied to a practical problem that was encountered in the on-line optimization of a complex Flexible Manufacturing System in [17]. These applications motivated us to embed paths and cycles in networks.

Suppose that W is an interconnection network (network for short). A path (cycle) in W is called a Hamiltonian path (Hamiltonian cycle) if it contains every node of W exactly once. W is called Hamiltonian if there is a Hamiltonian cycle in W [18]. Since node faults and link faults may develop in a network, it is practically important to consider faulty networks. A network W is called k-node (k-link) Hamiltonian if it remains Hamiltonian after removing any k nodes (links) [19]. If W has node (link) connectivity $k + 2$ and is k-node (k-link) Hamiltonian, then it can tolerate a maximal number of node (link) faults while embedding a longest fault-free cycle. Some networks have been shown to be k-node Hamiltonian and k-link Hamiltonian. For example, the hierarchical cubic network with connectivity $n + 1$ is $(n - 1)$-link Hamiltonian [20]. The n-dimensional twisted cube [21] is $(n - 2)$-node Hamiltonian and $(n - 2)$-link Hamiltonian.

Note that an n-link Hamiltonian graph can not be guaranteed to be n-node Hamiltonian. For example, the n-cube is $(n-2)$-link Hamiltonian but not $(n-2)$-node Hamiltonian [22]. In [7], the WK-recursive network with connectivity $d-1$ was shown to be $(d - 3)$-link Hamiltonian. It is not possible to re-use their approach of replacing link failures with node failures because a faulty node will cause $d - 1$ links to fail, and because their approach can handle at most $(d - 3)$ faulty links.

In this paper, we show that the WK-recursive network with connectivity $d - 1$ is $(d - 3)$-node Hamiltonian. In addition, the WK-recursive network is Hamiltonian-connected [14] and pancyclic [7]. Hence, it can be concluded that the WK-recursive network is excellent in terms of Hamiltonicity, and that it can tolerate a maximal number of node (link) faults while embedding a longest fault-free cycle. In the next section, the WK-recursive network is formally defined.

2 Preliminaries

In this section, we introduce some important notations and concepts. The WK-recursive network can be constructed hierarchically by grouping basic modules. Any d-node complete graphs can serve as the basic modules. We use $K(d,t)$ to denote a WK-recursive network of level t, each of whose basic modules is a d-node complete graph, where $d > 1$ and $t \geq 1$. The structures of $K(4,1)$, $K(4,2)$, and $K(4,3)$ are shown in Figure 1. $K(d,t)$ is defined in terms of a graph as follows.

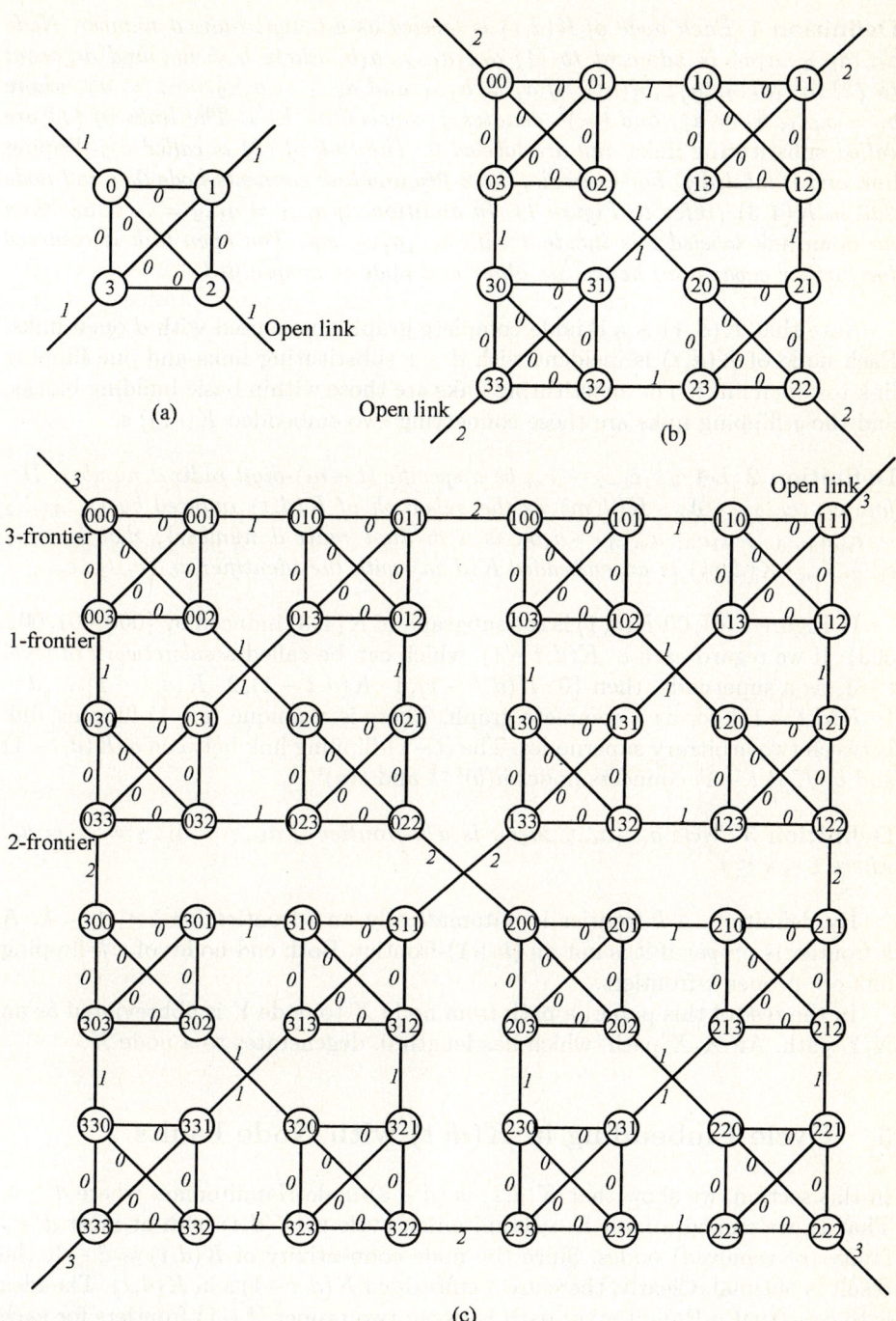

Fig. 1. The topologies of (a)$K(4,1)$, (b)$K(4,2)$, and (c)$K(4,3)$

Definition 1 *Each node of $K(d,t)$ is labeled as a t-digit radix d number. Node $a_{t-1}a_{t-2}...a_1a_0$ is adjacent to (1) $a_{t-1}a_{t-2}...a_1b$, where $b \neq a_0$, and adjacent to (2) $a_{t-1}a_{t-2}...a_{j+1}b_1(b_0)^j$ if $a_j \neq a_{j-1}$ and $a_{j-1} = a_{j-2} = ... = a_0$, where $b_1 = a_{j-1}$, $b_0 = a_j$, and $(b_0)^j$ denotes j consecutive b_0's. The links of (1) are called* substituting links *and are labeled 0. The link of (2) is called a j-*flipping link *and is labeled j. For example, the 2-flipping link connects node 033 and node 300 in $K(4,3)$ (refer to Figure 1). In addition, if $a_{t-1} = a_{t-2} = ... = a_0$, then an* open link *labeled t is incident with $a_{t-1}a_{t-2}...a_0$. The open link is reserved for further expansion; hence, its other end node is unspecified.*

Note that $K(d,1)$ is a d-node complete graph augmented with d open links. Each node of $K(d,t)$ is incident with $d-1$ substituting links and one flipping link (or open link). The substituting links are those within basic building blocks, and the j-flipping links are those connecting two embedded $K(d,j)$'s.

Definition 2 *Let $c_{t-1}c_{t-2}\cdots c_m$ be a specific $(t-m)$-digit radix d number. Define $c_{t-1}c_{t-2}\cdots c_m \cdot K(d,m)$ as the subgraph of $K(d,t)$ induced by $\{c_{t-1}c_{t-2} \cdots c_m a_{m-1} \cdots a_1 a_0 |\ a_{m-1} \cdots a_1 a_0$ is a m-digit radix d number$\}$; that is, $c_{t-1}c_{t-2}...c_m \cdot K(d,m)$ is an embedded $K(d,m)$ with the identifier $c_{t-1}c_{t-2}...c_m$.*

In Figure 1 (c), $00 \cdot K(4,1)$ is the subgraph of $K(4,3)$ induced by $\{000, 001, 002, 003\}$. If we regard each $c \cdot K(d, t-1)$, which can be called a *subnetwork* of level $t-1$, as a supernode, then $\{0 \cdot K(d,t-1), 1 \cdot K(d,t-1), 2 \cdot K(d,t-1), ..., d-1 \cdot K(d,t-1)\}$ forms a complete graph. There is a unique $(t-1)$-flipping link between two arbitrary supernodes. The $(t-1)$-flipping link between $a \cdot K(d, t-1)$ and $b \cdot K(d, t-1)$ connects nodes $a(b)^{t-1}$ and $b(a)^{t-1}$.

Definition 3 *Node $a_{t-1}a_{t-2}...a_1a_0$ is a k-*frontier *if $a_{k-1} = a_{k-2} = ... = a_0$, where $1 \leq k \leq t$.*

By definition, a k-frontier is automatically an l-frontier for $1 \leq l \leq k$. A k-frontier is *proper* if it is not an $(k+1)$-frontier. Both end nodes of a k-flipping link are proper k-frontiers.

In the rest of this paper, a path from node X to node Y is abbreviated as an X-Y path. An X-X path, which has length 0, degenerates to a node X.

3 Cycle Embedding in $K(d,t)$ with Node Faults

In this section, we show that $K(d,t)$ is $(d-3)$-node Hamiltonian, where $d \geq 4$. That is, we embed a fault free Hamiltonian cycle in $K(d,t)$ with at most $d-3$ faulty (or removed) nodes. Since the node connectivity of $K(d,t)$ is $d-1$, the result is optimal. Clearly, there are d embedded $K(d,t-1)$'s in $K(d,t)$. The idea is to construct a Hamiltonian path between two proper $(t-1)$-frontiers for each $K(d,t-1)$ and to then use d $(t-1)$-flipping links to connect these d paths to form a Hamiltonian cycle for $K(d,t)$. Suppose that a Hamiltonian path for each $K(d,t-1)$ can be constructed. The following lemma shows that the Hamiltonian cycle for $K(d,t)$ can be formed by combining these paths when at most $d-3$ nodes are removed.

Lemma 1 *If $f \leq d-3$ nodes are removed from $K(d,t)$, then we can find a sequence $v_0, v_1, \ldots, v_{d-1}$ such that nodes $v_0(v_1)^{t-1}$, $v_1(v_0)^{t-1}$, $v_1(v_2)^{t-1}$, $\ldots$, $v_{d-2}(v_{d-1})^{t-1}$, $v_{d-1}(v_{d-2})^{t-1}$, $v_{d-1}(v_0)^{t-1}$, and $v_0(v_{d-1})^{t-1}$ are not removed, where $\{v_0, v_1, \ldots, v_{d-1}\} = \{0, 1, 2, \ldots, d-1\}$.*

Proof. We can transform this problem into another problem, which is cycle embedding in a complete graph with faulty links. If we regard each $K(d,t-1)$ as a single supernode, then $K(d,t)$ can be treated as a complete graph C with d nodes. Removing a proper $(t-1)$-frontier from $K(d,t)$ is tantamount removing a link in C. Hence, one non-removed link in C can assure that two adjacent proper $(t-1)$-frontiers in $K(d,t)$ are not removed. In $K(d,t)$, at most $d-3$ proper $(t-1)$-frontiers are removed because at most $d-3$ nodes are removed. Therefore, the graph C can be treated as indicating that at most $d-3$ links are removed. In [7], a complete graph with d nodes was shown to be $(d-3)$-link Hamiltonian. That is, a Hamiltonian cycle can be found in C after at most $d-3$ links are removed. Let v_i denote a node in C corresponding to $v_i \cdot K(d, t-1)$ in $K(d,t)$, and let $\longrightarrow$ denote a link in C. Suppose that the Hamiltonian cycle in C is as follows:

$$v_0 \longrightarrow v_1 \longrightarrow v_2 \longrightarrow \cdots \longrightarrow v_{d-1} \longrightarrow v_0,$$

where $\{v_0, v_1, \ldots, v_{d-1}\} = \{0, 1, 2, \ldots, d-1\}$. These d non-removed links of the Hamiltonian cycle for C assure that nodes $v_0(v_1)^{t-1}$, $v_1(v_0)^{t-1}$, $v_1(v_2)^{t-1}$, $\ldots$, $v_{d-2}(v_{d-1})^{t-1}$, $v_{d-1}(v_{d-2})^{t-1}$, $v_{d-1}(v_0)^{t-1}$, and $v_0(v_{d-1})^{t-1}$ in $K(d,t)$ are not removed. (see Figure 2) □

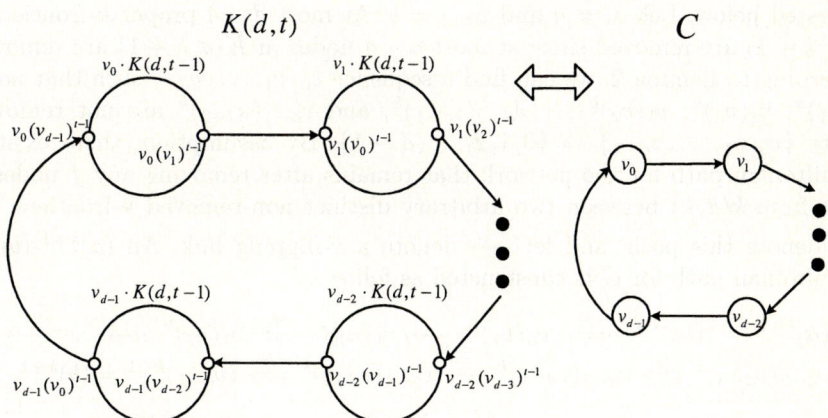

Fig. 2. The Hamiltonian cycle in C and the corresponding non-removed nodes in $K(d,t)$

Now, we want to construct Hamiltonian paths for each $K(d,t-1)$ between two proper $(t-1)$-frontiers. When at most $d-4$ nodes are removed from $K(d,t-1)$, we can construct a Hamiltonian path for the network that remains after removing nodes from $K(d, t-1)$ between two arbitrary non-removed $(t-1)$-frontiers. When $d-3$ nodes are removed from $K(d, t-1)$, we choose two non-removed proper $(t-1)$-frontiers and then construct a Hamiltonian path for the network that remains after removing nodes from $K(d, t-1)$ between the proper $(t-1)$-frontiers. We will show the former in Lemma 3 and the latter in Lemma 5. To prove Lemma 3, we need the following lemma.

Lemma 2 *Given two integers $v_0, v_{d-1} \in \{0, 1, 2, \ldots, d-1\}$, if at most $d-4$ proper $(t-1)$-frontiers are removed from $K(d, t)$, then we can find a sequence $v_0, v_1, \ldots, v_{d-1}$ such that nodes $v_0(v_1)^{t-1}$, $v_1(v_0)^{t-1}$, $v_1(v_2)^{t-1}$, $\ldots$, $v_{d-2}(v_{d-1})^{t-1}$, and $v_{d-1}(v_{d-2})^{t-1}$ are not removed, where $\{v_0, v_1, \ldots, v_{d-1}\} = \{0, 1, 2, \ldots, d-1\}$.*

Proof. There exists a Hamiltonian path between two arbitrary nodes of a complete graph with d nodes after removing $d-4$ links (see [7]). The remaining proof is similar to that of Lemma 1. □

For simplicity, we use $K(d, t)$ instead of $K(d, t-1)$. The following lemma shows that when at most $d-4$ nodes are removed from $K(d, t)$, we can construct a Hamiltonian path for the network that remains after removing nodes from $K(d, t)$ between two arbitrary non-removed t-frontiers.

Lemma 3 *Let G be the network that remains after removing any $f \leq d-4$ nodes from $K(d, t)$ and $(a)^t$ and $(b)^t$ are not removed. There exists an $(a)^t$-$(b)^t$ Hamiltonian path for G.*

Proof. We proceed by induction on t. Clearly, the lemma holds for $t = 1$. Assume that it holds for $t = k \geq 1$. The situation in the case of $t = k+1$ is discussed below. Let $v_0 = a$ and $v_{d-1} = b$. At most $d-4$ proper k-frontiers in $K(d, k+1)$ are removed since at most $d-4$ nodes in $K(d, k+1)$ are removed. According to Lemma 2, we can find a sequence $v_0, v_1, \ldots, v_{d-1}$ such that nodes $v_0(v_1)^k$, $v_1(v_0)^k$, $v_1(v_2)^k$, $\ldots$, $v_{d-2}(v_{d-1})^k$, and $v_{d-1}(v_{d-2})^k$ are not removed, where $\{v_0, v_1, \ldots, v_{d-1}\} = \{0, 1, 2, \ldots, d-1\}$. By assumption, there exists a Hamiltonian path for the network that remains after removing any f nodes at most from $k(d, k)$ between two arbitrary distinct non-removed k-frontiers. Let $\overset{H}{\Longrightarrow}$ denote this path, and let $\longrightarrow$ denote a k-flipping link. An $(a)^{k+1}$-$(b)^{k+1}$ Hamiltonian path for G is constructed as follows:

$$(a)^{k+1} = (v_0)^{k+1} \overset{H}{\Longrightarrow} v_0(v_1)^k \longrightarrow v_1(v_0)^k \overset{H}{\Longrightarrow} v_1(v_2)^k \longrightarrow \cdots \longrightarrow v_{d-2}(v_{d-3})^k \overset{H}{\Longrightarrow} v_{d-2}(v_{d-1})^k \longrightarrow v_{d-1}(v_{d-2})^k \overset{H}{\Longrightarrow} (v_{d-1})^{k+1} = (b)^{k+1}.$$
□

When $d-3$ nodes are removed from $c \cdot K(d, t-1)$ for some $c \in \{0, 1, 2, \ldots, d-1\}$, we choose two non-removed proper $(t-1)$-frontiers (i.e., $c(a)^{t-1}$ and $c(b)^{t-1}$ with $a, b \in \{0, 1, 2, \ldots, d-1\} - \{c\}$) and then construct a Hamiltonian path for the network that remains after removing $d-3$ nodes from $c \cdot K(d, t-1)$ between the proper $(t-1)$-frontiers, which is the proof of Lemma 5. In the proof of Lemma 5, there is a tricky case of $K(4, t)$. To solve this case, we need the following lemma.

Lemma 4 *Let $\{a, b, c, e\} = \{0, 1, 2, 3\}$. There exist an $(a)^t$-$(c)^t$ path and a $(b)^t$-$(e)^t$ path such that they are disjoint and contain all the nodes of $K(4, t)$, where $t \geq 1$.*

Proof. Clearly, the lemma holds for $t = 1$. The situation in the case of $t \geq 2$ is discussed below. According to Lemma 3, there exists a Hamiltonian path for $K(d, t-1)$ between two arbitrary $(t-1)$-frontiers. Let $\xRightarrow{H}$ denote this path, and let $\longrightarrow$ denote a $(t-1)$-flipping link. An $(a)^t$-$(c)^t$ path and a $(b)^t$-$(e)^t$ path in $K(4, t)$ are constructed as follows:

$$(a)^t \xRightarrow{H} a(c)^{t-1} \longrightarrow c(a)^{t-1} \xRightarrow{H} (c)^t,$$
$$(b)^t \xRightarrow{H} b(e)^{t-1} \longrightarrow e(b)^{t-1} \xRightarrow{H} (e)^t.$$
□

Lemma 5 *Let G be the network that remains after removing any $d - 3$ nodes from $K(d, t)$. Given an integer $c \in \{0, 1, 2, \ldots, d-1\}$, we can find two distinct integers $a, b \in \{0, 1, 2, \ldots, d-1\} - \{c\}$ such that nodes $(a)^t$ and $(b)^t$ are not removed and there is an $(a)^t$-$(b)^t$ Hamiltonian path for G.*

Proof. We proceed by induction on t. Clearly, the lemma holds for $t = 1$. Assume that it holds for $t = k \geq 1$. The situation in the case of $t = k + 1$ is discussed below. Let $\longrightarrow$ denote a k-flipping link. Two cases will be considered.

Case 1: The number of nodes removed from each subnetwork of level k is not greater than $d - 4$. According to Lemma 3, there exists a Hamiltonian path for the network that remains after removing at most $d - 4$ nodes from $k(d, k)$ between two arbitrary distinct non-removed proper k-frontiers. Let $\xRightarrow{H}$ denote this path. Two cases will be further considered.

Case 1.1: No $(k + 1)$-frontier in $K(d, k + 1)$ is removed. According to Lemma 1, we can find a sequence $v_0, v_1, \ldots, v_{d-1}$ such that nodes $v_0(v_1)^k$, $v_1(v_0)^k$, $v_1(v_2)^k, \ldots, v_{d-2}(v_{d-1})^k$, $v_{d-1}(v_{d-2})^k$, $v_{d-1}(v_0)^k$, and $v_0(v_{d-1})^k$ are not removed, where $\{v_0, v_1, \ldots, v_{d-1}\} = \{0, 1, 2, \ldots, d-1\}$. There exists an integer $i \in \{0, 1, 2, \ldots, d-1\}$ such that $c \notin \{v_i, v_{i+1 \bmod d}\}$. Let $a = u_0 = v_{i+1 \bmod d}$, $u_1 = v_{i+2 \bmod d}, \ldots, u_j = v_{i+1+j \bmod d}, \ldots, u_{d-2} = v_{i+(d-1) \bmod d}$, and $b = u_{d-1} = v_i$. $(a)^{k+1}$ and $(b)^{k+1}$ are not removed since no $(k + 1)$-frontier is removed. An $(a)^{k+1}$-$(b)^{k+1}$ Hamiltonian path for G is constructed as follows:

$$(a)^{k+1} = (u_0)^{k+1} \xRightarrow{H} u_0(u_1)^k \longrightarrow u_1(u_0)^k \xRightarrow{H} u_1(u_2)^k \longrightarrow \cdots \longrightarrow$$
$$u_{d-2}(u_{d-3})^k \xRightarrow{H} u_{d-2}(u_{d-1})^k \longrightarrow u_{d-1}(u_{d-2})^k \xRightarrow{H} (u_{d-1})^{k+1} = (b)^{k+1}.$$

Case 1.2: At least one $(k+1)$-frontier is removed. Clearly, at most $d - 3$ $(k+1)$-frontiers are removed since $d - 3$ nodes are removed. Hence, at least three non-removed $(k+1)$-frontiers can be found. Let $(v_0)^{k+1}$ and $(v_{d-1})^{k+1}$ be two distinct non-removed $(k + 1)$-frontiers, where $v_0, v_{d-1} \in \{0, 1, 2, \ldots, d-1\} - \{c\}$. Because at least one $(k+1)$-frontier is removed, at most $d-4$ proper k-frontiers are removed. According to Lemma 2, we can find a sequence $v_0, v_1, \ldots, v_{d-1}$ such that nodes $v_0(v_1)^k$, $v_1(v_0)^k$, $v_1(v_2)^k, \ldots, v_{d-2}(v_{d-1})^k$, and $v_{d-1}(v_{d-2})^k$ are not removed, where $\{v_0, v_1, \ldots, v_{d-1}\} = \{0, 1, 2, \ldots, d-1\}$. Let $a = v_0$ and $b = v_{d-1}$. An $(a)^{k+1}$-$(b)^{k+1}$ Hamiltonian path for G is constructed as follows:

$$(a)^{k+1} = (v_0)^{k+1} \overset{H}{\Longrightarrow} v_0(v_1)^k \longrightarrow v_1(v_0)^k \overset{H}{\Longrightarrow} v_1(v_2)^k \longrightarrow \cdots \longrightarrow$$
$$v_{d-2}(v_{d-3})^k \overset{H}{\Longrightarrow} v_{d-2}(v_{d-1})^k \longrightarrow v_{d-1}(v_{d-2})^k \overset{H}{\Longrightarrow} (v_{d-1})^{k+1} = (b)^{k+1}.$$

Case 2: There is a subnetwork of level k containing $d-3$ removed nodes. Assume that $c' \cdot K(d,k)$ contains $d-3$ removed nodes, for some $c' \in \{0,1,2,\ldots,d-1\}$. Hence, all nodes in other remaining subnetworks of level k are not removed. By assumption, we can find two distinct integers $a', b' \in \{0,1,2,\ldots,d-1\} - \{c'\}$ such that nodes $c'(a')^k$ and $c'(b')^k$ are not removed and there is a $c'(a')^k$-$c'(b')^k$ Hamiltonian path for G', which is the network that remains after removing $d-3$ nodes from $c' \cdot K(d,k)$. Let $\overset{G'}{\Longrightarrow}$ denote this path. According to Lemma 3, there exists a Hamiltonian path for $K(d,k)$ between two arbitrary k-frontiers. Let $\overset{H}{\Longrightarrow}$ denote this path. When $\{0,1,2,\ldots,d-1\} - \{a',b',c',c\} \neq \emptyset$, let $b = v_{d-1} \in \{0,1,2,\ldots,d-1\} - \{a',b',c',c\}$ and $v_1 = c'$. If $a' \neq c$, then let $a = v_0 = a'$ and $v_2 = b'$. If $a' = c$ (certainly, we have $b' \neq c$), then let $a = v_0 = b'$ and $v_2 = a'$. Let $\{v_0, v_1, \ldots, v_{d-1}\} = \{0,1,2,\ldots,d-1\}$. An $(a)^{k+1}$-$(b)^{k+1}$ Hamiltonian path for G is constructed as follows:

$$(a)^{k+1} = (v_0)^{k+1} \overset{H}{\Longrightarrow} v_0(v_1)^k \longrightarrow v_1(v_0)^k \overset{G'}{\Longrightarrow} v_1(v_2)^k \longrightarrow \cdots \longrightarrow$$
$$v_{d-2}(v_{d-3})^k \overset{H}{\Longrightarrow} v_{d-2}(v_{d-1})^k \longrightarrow v_{d-1}(v_{d-2})^k \overset{H}{\Longrightarrow} (v_{d-1})^{k+1} = (b)^{k+1}.$$

When $\{0,1,2,\ldots,d-1\} - \{a',b',c',c\} = \emptyset$, clearly, we have $d = 4$. Let $a = a'$ and $b = b'$. According to Lemma 4, there exist an $(a)^{k+1}$-$a(c')^k$ ($(b)^{k+1}$-$b(c)^k$, respectively) path and an $a(b)^k$-$a(c)^k$ ($b(a)^k$-$b(c')^k$, respectively) path containing all the nodes in $a \cdot K(4,k)$ ($b \cdot K(4,k)$, respectively). Let $\overset{P_1}{\Longrightarrow}$ denote the $(a)^{k+1}$-$a(c')^k$ path, let $\overset{P_2}{\Longrightarrow}$ denote the $a(b)^k$-$a(c)^k$ path, let $\overset{P'_1}{\Longrightarrow}$ denote the $(b)^{k+1}$-$b(c)^k$ path, and let $\overset{P'_2}{\Longrightarrow}$ denote the $b(a)^k$-$b(c')^k$ path. An $(a)^{k+1}$-$(b)^{k+1}$ Hamiltonian path for G is constructed as follows:

$$(a)^{k+1} \overset{P_1}{\Longrightarrow} a(c')^k \longrightarrow c'(a)^k \overset{G'}{\Longrightarrow} c'(b)^k \longrightarrow b(c')^k \overset{P'_2}{\Longrightarrow} b(a)^k \longrightarrow a(b)^k$$
$$\overset{P_2}{\Longrightarrow} a(c)^k \longrightarrow c(a)^k \overset{H}{\Longrightarrow} c(b)^k \longrightarrow b(c)^k \overset{P'_1}{\Longrightarrow} (b)^{k+1}. \quad \square$$

Finally, the following theorem is the main result of this paper.

Theorem 1 $K(d,t)$ is $(d-3)$-node Hamiltonian, where $d \geq 4$.

Proof. Suppose $f \leq d-3$ nodes are removed in $K(d,t)$. We proceed by induction on t. Clearly, the theorem holds for $t = 1$. Assume that it holds for $t = k \geq 1$. The situation in the case of $t = k+1$ is discussed below. Let $\longrightarrow$ denote a k-flipping link. Two cases will be considered.

Case 1: The number of nodes removed from each subnetwork of level k is not greater than $d-4$. According to Lemma 1, we can find a sequence $v_0, v_1, \ldots, v_{d-1}$ such that nodes $v_0(v_1)^k$, $v_1(v_0)^k$, $v_1(v_2)^k, \ldots, v_{d-2}(v_{d-1})^k$, $v_{d-1}(v_{d-2})^k$, $v_{d-1}(v_0)^k$, and $v_0(v_{d-1})^k$ are not removed, where $\{v_0, v_1, \ldots, v_{d-1}\} = \{0,1,2,\ldots,d-1\}$. According to Lemma 3, there exists a Hamiltonian path for the network that remains after removing any $d-4$ nodes at most from $K(d,k)$ between two arbitrary distinct non-removed k-frontiers. Let $\overset{H}{\Longrightarrow}$ denote this path. A Hamiltonian cycle is constructed as follows:

$$v_0(v_{d-1})^k \overset{H}{\Longrightarrow} v_0(v_1)^k \longrightarrow v_1(v_0)^k \overset{H}{\Longrightarrow} v_1(v_2)^k \longrightarrow \cdots \longrightarrow v_{d-2}(v_{d-3})^k$$
$$\overset{H}{\Longrightarrow} v_{d-2}(v_{d-1})^k \longrightarrow v_{d-1}(v_{d-2})^k \overset{H}{\Longrightarrow} v_0(v_{d-1})^k.$$

Case 2: There is a subnetwork of level k containing $d-3$ removed nodes. Assume that $v_0 \cdot K(d,k)$ contains $d-3$ removed nodes, for some $v_0 \in \{0,1,2,\ldots,d-1\}$. Hence, all nodes in other remaining subnetworks of level k are not removed. Let G' be the network that remains after removing $d-3$ nodes from $v_0 \cdot K(d,k)$. According to Lemma 5, we can find two distinct integers $v_1, v_{d-1} \in \{0,1,2,\ldots,d-1\} - \{v_0\}$ such that nodes $v_0(v_1)^k$ and $v_0(v_{d-1})^k$ are not removed and there is a $v_0(v_1)^k$-$v_0(v_{d-1})^k$ Hamiltonian path for G'. Let $\overset{G'}{\Longrightarrow}$ denote this path. According to Lemma 3, there exists a Hamiltonian path for $K(d,k)$ between two arbitrary k-frontiers. Let $\overset{H}{\Longrightarrow}$ denote this path. Let $\{v_0, v_1, \ldots, v_{d-1}\} = \{0,1,2,\ldots,d-1\}$. A Hamiltonian cycle is constructed as follows:

$$v_0(v_{d-1})^k \overset{G'}{\Longrightarrow} v_0(v_1)^k \longrightarrow v_1(v_0)^k \overset{H}{\Longrightarrow} v_1(v_2)^k \longrightarrow \cdots \longrightarrow v_{d-2}(v_{d-3})^k$$
$$\overset{H}{\Longrightarrow} v_{d-2}(v_{d-1})^k \longrightarrow v_{d-1}(v_{d-2})^k \overset{H}{\Longrightarrow} v_0(v_{d-1})^k. \qquad \square$$

4 Discussion and Conclusion

In this paper, using inductive proofs, we showed that $K(d,t)$ is $(d-3)$-node Hamiltonian. In addition, $K(d,t)$ was shown to be $(d-3)$-link Hamiltonian [7]. Because $K(d,t)$ has node and link connectivity $d-1$, it can tolerate a maximal number of node and link faults while embedding a longest fault-free cycle. Our results reveal that $K(d,t)$ is excellent in terms of fault-tolerant Hamiltonicity. A topic for further research is the Hamiltonian-connectedness of the WK-recursive network when there are node faults and/or link faults.

References

1. Della Vecchia, G., Sanges, C.: Recursively scalable networks for message passing architectures. In Chiricozzi, E., D'Amico, A., eds.: Parallel Processing and Applications. Elsevier North-Holland, Amsterdam (1988) 33–40
2. INMOS Limited: Transputer reference manual. Prentice-Hall, Upper Saddle River, NJ 07458, USA (1988) Includes index. Bibliography: p. 315-324.
3. Della Vecchia, G., Sanges, C.: A recursively scalable network VLSI implementation. Future Generation Computer Systems **4** (1988) 235–243
4. Della Vecchia, G., Sanges, C.: An optimized broadcasting technique for WK-recursive topologies. Future Generation Computer Systems **5** (1989/1990) 353–357
5. Chen, G.H., Duh, D.R.: Topological properties, communication, and computation on WK-recursive networks. Networks **24** (1994) 303–317

6. Duh, D.R., Chen, G.H.: Topological properties of WK-recursive networks. Journal of Parallel and Distributed Computing **23** (1994) 468–474
7. Fernandes, R., Friesen, D.K., Kanevsky, A.: Embedding rings in recursive networks. In: Proceedings of the 6th Symposium on Parallel and Distributed Processing. (1994) 273–280
8. Fernandes, R., Friesen, D.K., Kanevsky, A.: Efficient routing and broadcasting in recursive interconnection networks. In: Proceedings of the 23rd International Conference on Parallel Processing. Volume 1: Architecture. (1994) 51–58
9. Fernandes, R., Kanevsky, A.: Substructure allocation in recursive interconnection networks. In: Proceedings of the 1993 International Conference on Parallel Processing. Volume 1: Architecture. (1993) 319–323
10. Verdoscia, L., Vaccaro, R.: An adaptive routing algorithm for WK-recursive topologies. Computing **63** (1999) 171–184
11. Verdoscia, L., Scafuri, U.: CODACS project: level-node communication policies. In: Proceedings of the Eleventh Euromicro Conference on Parallel,Distributed and Network-Based Processing. (2003) 134–139
12. Della Vecchia, G., Distasi, R.: B-Tree triangular coding on WK-recursive networks. In: Parallel Computing: State-of-the-Art and Perspectives, Proceedings of the Conference ParCo'95, 19-22 September 1995, Ghent, Belgium. Volume 11 of Advances in Parallel Computing., Amsterdam, Elsevier, North-Holland (1996) 165–172
13. Liaw, S.C., Chang, G.J.: Generalized diameters and rabin numbers of networks. Journal of Combinatorial Optimization **2** (1999) 371–384
14. Fu, J.S.: Hamiltonian-connectedness of the WK-recursive network. In: Proceedings of the 7th International Symposium on Parallel Architectures, Algorithms and Networks. (2004) 569–574
15. Akl, S.G.: Parallel Computation: Models and Methods. Prentice-Hall, Upper Saddle River, NJ (1997)
16. Leighton, F.T.: Introduction to Parallel Algorithms and Architectures: arrays. trees. hypercubes. Morgan Kaufman, San Mateo (1992)
17. Ascheuer, N.: Hamiltonian path problems in the on-line optimization of flexible manufacturing systems. PhD thesis, University of Technology, Berlin, Germany (1995) Also available at ftp://ftp.zib.de/pub/zib-publications/reports/TR-96-03.ps.
18. Buckley, F., Harary, F.: Distance in Graphs. Addison-Wesley, Reading, MA, USA (1990)
19. Harary, F., Hayes, J.P.: Edge fault tolerance in graphs. Networks **23** (1993) 135–142
20. Fu, J.S., Chen, G.H.: Fault-tolerant cycle embedding in hierarchical cubic networks. Networks **43** (2004) 28–38
21. Huang, W.T., Tan, J.M., Hung, C.N., Hsu, L.H.: Fault-tolerant hamiltonicity of twisted cubes. Journal of Parallel and Distributed Computing **62** (2002) 591–604
22. Fu, J.S.: Fault-tolerant cycle embedding in the hypercube. Parallel Computing **29** (2003) 821–832

RAIDb: Redundant Array of Inexpensive Databases

Emmanuel Cecchet

INRIA, Sardes Project – 655, Avenue de l'Europe – 38330 Montbonnot – France
`Emmanuel.Cecchet@inria.fr`

Abstract. In this paper, we introduce the concept of Redundant Array of Inexpensive Databases (RAIDb). RAIDb is to databases what RAID is to disks. RAIDb aims at providing better performance and fault tolerance than a single database, at low cost, by combining multiple database instances into an array of databases. Like RAID, we define and compare different RAIDb levels that provide various cost/performance/fault tolerance tradeoffs.

We present a Java implementation of RAIDb called Clustered JDBC or C-JDBC. C-JDBC achieves both database performance scalability and high availability at the middleware level without changing existing applications nor database engines.

Keywords: database, replication clustering, middleware, scalability, dependability.

1 Introduction

Nowadays, database scalability and high availability can be achieved, but at very high expense. Existing solutions require large SMP machines or clusters with a Storage Area Network (SAN) and high-end RDBMS (Relational DataBase Management Systems). Both hardware and software licensing cost makes those solutions only available to large businesses.

In this paper, we introduce the concept of Redundant Array of Inexpensive Databases (RAIDb), in analogy to the existing RAID (Redundant Array of Inexpensive Disks) concept, that achieves scalability and high availability of disk subsystems at a low cost. RAID combines multiple inexpensive disk drives into an array of disk drives to obtain performance, capacity and reliability that exceeds that of a single large drive [5]. RAIDb is the counterpart of RAID for databases. RAIDb aims at providing better performance and fault tolerance than a single database, at a low cost, by combining multiple database instances into an array of databases.

RAIDb primarily targets low-cost commodity hardware and software such as clusters of workstations and open source databases. On such platforms, RAIDb will be mostly implemented as a software solution like the C-JDBC middleware prototype we present in this paper. However, like for RAID systems, hardware solutions could be provided to enhance RAIDb performance while still being cost effective.

Clusters of workstations are already an alternative to large parallel machines in scientific computing because of their unbeatable price/performance ratio. Clusters can also be used to provide both scalability and high availability in data server environments. Database replication has been used as a solution to improve availability and performance of distributed databases [2, 8]. Even if many protocols have been designed to provide data consistency and fault tolerance [3], few of them have been used in commercial databases [12]. Gray et al. [6] have pointed out the danger of replication and the scalability limit of this approach. However, database replication is a viable approach if an appropriate replication algorithm is used [1, 8, 14]. We propose a classification of the various distribution and fault tolerance solutions in RAIDb levels and evaluate the performance/fault tolerance tradeoff of each solution.

The outline of the rest of this paper is as follows. Section 0 gives an overview of the RAIDb architecture and its components. In section 0, we introduce a classification of the basic RAIDb levels. Then, section 0 shows how to combine those basic RAIDb levels to build larger scale RAIDb configurations. Section 0 gives an overview of C-JDBC, a Java implementation of RAIDb. Section 0 discusses related work and we conclude in section 0.

2 RAIDb Architecture

One of the goals of RAIDb is to hide the distribution complexity and provide the database clients with the view of a single database like in a centralized architecture. Fig. 1 gives an overview of the RAIDb architecture. The left side of the figure depicts the standard centralized database access by clients using the database driver to send their SQL requests. The right side of fig. 1 shows how RAIDb is used to cluster the original database.

As for RAID, a controller sits in front of the underlying resources. The clients send their requests directly to the RAIDb controller that distributes them among the set of RDBMS backends. The RAIDb controller gives the illusion of a single RDBMS to the clients by exporting a *virtual database*.

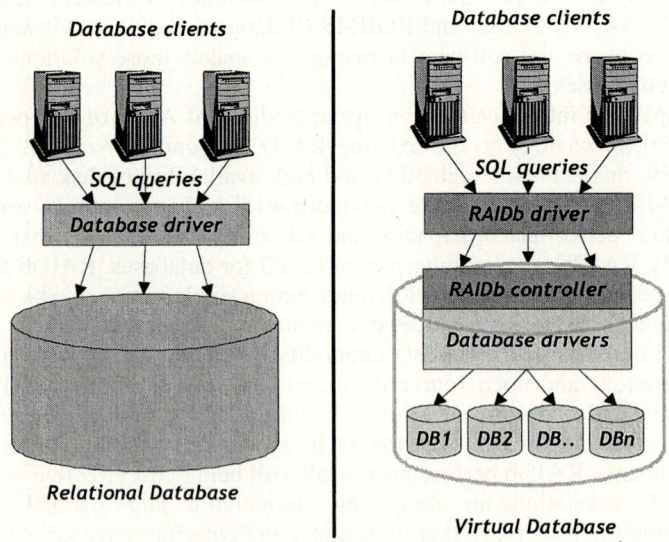

Fig. 1. RAIDb architecture overview

2.1 RAIDb Controller

RAIDb controllers may provide various degrees of services. The controller must be aware of the database tables available on each RDBMS backend so that the requests can be routed (according to a load balancing algorithm) to the right node(s) by parsing the SQL statement. This knowledge can be configured statically through configuration files or discovered dynamically by requesting the database schema directly from the RDBMS. Load balancing algorithms can range from static affinity-based or round-robin policies to dynamic decisions based on node load or other monitoring-based information.

RAIDb controllers should also provide support for dynamic backend addition and removal which is equivalent to the disks' hot swap feature. To prevent the controller from being a single point of failure, it is possible to replicate controllers and synchronize them using totally ordered reliable communication channels. This is necessary to preserve the same serializable execution order at each controller.

As RAID controllers, RAIDb controllers can offer caching to hold the replies to SQL queries. The controller is responsible for the granularity and the coherence of the cache. Additional features such as connection pooling can be provided to further enhance performance scalability. There is no restriction to the set of services implemented in the RAIDb controller. Monitoring, debugging, logging or security management services can prove to be useful for certain users.

2.2 Application and Database Requirements

In general, RAIDb does not impose any modification of the client application or the RDBMS. However, some precautions have to be taken care of, such as the fact that all requests to the databases must be sent through the RAIDb controller. It is not allowed to directly issue requests to a database backend as this might compromise the data synchronization between the backends as well as the RAIDb cache coherency.

As each RDBMS supports a different SQL subset, the application must be aware of the requests supported by the underlying databases. This problem can be easily handled if all RDBMS instances use the same version from the same vendor. For example, a cluster consisting only of MySQL 4.0 databases will behave as a single instance of MySQL 4.0. Nevertheless, heterogeneous databases can be used with RAIDb. A mix of Oracle and PostgreSQL databases is a possible RAIDb backend configuration. In such a case, the application must use an SQL subset that is common to both RDBMS. If the RAIDb controller supports user defined load balancers, the user can implement a load balancer that is aware of the respective capabilities of the underlying RDBMS. Once loaded in the RAIDb controller, the load balancer should be able to direct the queries to the appropriate database.

3 Basic RAIDb Levels

We define three basic RAIDb levels varying the degree of partitioning and replication among the databases. RAIDb-0 (database partitioning) and RAIDb-1 (database mirroring) are similar to RAID-0 (disk striping) and RAID-1 (disk mirroring), respectively. Like RAID-5, RAIDb-2 is a tradeoff between RAIDb-0 and RAIDb-1. Actually, RAIDb-2 offers partial replication of the database. We also define RAIDb-

1ec and RAIDb-2ec that adds error checking to the basic RAIDb levels 1 and 2, respectively.

Note that RAIDb is just a conceptual analogy to RAID. Data distribution in RAIDb uses a logical unit which is a database table, whereas RAID uses a physical unit defined by a disk block.

3.1 RAIDb-0: Full Partitioning

RAIDb level 0 is similar to striping provided by RAID-0. It consists in partitioning the database tables among the nodes. Each table is assigned to a unique node and every node has at least one table. RAIDb-0 uses at least 2 database backends but there is no duplication of information and therefore no fault tolerance guarantees.

RAIDb-0 allows large databases to be distributed, which could be a solution if no node has enough storage capacity to store the whole database. Also, each database engine processes a smaller working set and can possibly have better cache usage, since the requests are always hitting a reduced number of tables. As RAID-0, RAIDb-0 gives the best storage efficiency since no information is duplicated.

RAIDb-0 requires the RAIDb controller to know which tables are available on each node in order to direct the requests to the right node. This knowledge can be configured statically in configuration files or build dynamically by fetching the schema from each database.

Like for RAID systems, the Mean Time Between Failures (MTBF) of the array is equal to the MTBF of an individual database backend, divided by the number of backends in the array. Because of this, the MTBF of a RAIDb-0 system is too low for mission-critical systems.

3.2 RAIDb-1: Full Replication

RAIDb level 1 is similar to disk mirroring in RAID-1. Databases are fully replicated. RAIDb-1 requires each backend node to have enough storage capacity to hold all database data. RAIDb-1 needs at least 2 database backends, but there is (theoretically) no limit to the number of RDBMS backends.

The performance scalability is limited by the capacity of the RAIDb controller to efficiently broadcast the updates to all backends. In case of a large number of backend databases, a hierarchical structure like those discussed in section 0 would give better scalability.

Unlike RAIDb-0, the RAIDb-1 controller does not need to know the database schema, since all nodes are capable of treating any request. However, if the RAIDb controller provides a cache, it will need the database schema to maintain the cache coherence.

RAIDb-1 provides speedup for read queries because they can be balanced over the backends. Write queries are performed in parallel by all nodes, therefore they execute at the same speed as the one of a single node. However, RAIDb-1 provides good fault tolerance, since it can continue to operate with a single backend node.

3.3 RAIDb-1ec

To ensure further data integrity, we define the *RAIDb-1ec* level that adds error checking to RAIDb-1. Error checking aims at detecting Byzantine failures [9] that may occur in highly stressed clusters of PCs [7]. RAIDb-1ec detects and tolerates failures as long as a majority of nodes does not fail. RAIDb-1 requires at least 3 nodes to operate.

A read request is always sent to a majority of nodes and the replies are compared. If a consensus is reached, the reply is sent to the client. Else the request is sent to all nodes to reach a quorum. If a quorum cannot be reached, an error is returned to the client.

The RAIDb controller is responsible for choosing a set of nodes for each request. Note that the algorithm can be user defined or tuned if the controller supports it. The number of nodes always ranges from the majority (half of the nodes plus 1) to all nodes. If all nodes are chosen, it results in the most secure configuration but the performance will be the one of the slowest backend. This setting is a tradeoff between performance and data integrity.

3.4 RAIDb-2: Partial Replication

RAIDb level 2 features partial replication which is an intermediate configuration between RAIDb-0 and RAIDb-1. Unlike RAIDb-1, RAIDb-2 does not require any single node to host a full copy of the database. This is essential when the full database is too large to be hosted on a node's disks. Each database table must be replicated at least once to survive a single node failure. RAIDb-2 uses at least 3 database backends (2 nodes would be a RAIDb-1 solution). Like for RAIDb-0, RAIDb-2 requires the RAIDb controller to be aware of the underlying database schemas to route the request to the appropriate set of nodes. As RAID-5, RAIDb-2 is a good tradeoff between cost, performance and data protection.

Typically, RAIDb-2 is used in a configuration where no or few nodes host a full copy of the database and a set of nodes host partitions of the database to offload the full databases. RAIDb-2 can be useful with heterogeneous databases. An existing enterprise database using a commercial RDBMS could be too expensive to fully duplicate both in term of storage and additional licenses cost. Therefore, a RAIDb-2 configuration can add a number of smaller open-source RDBMS hosting smaller partitions of the database to offload the full database and offer better fault tolerance.

3.5 RAIDb-2ec

Like for RAIDb-1ec, RAIDb-2ec adds error checking to RAIDb-2. Three copies of each table are needed in order to achieve a quorum. RAIDb-2ec requires at least 4 RDBMS backends to operate. The choice of the nodes that will perform a read request is more complex than in RAIDb-1ec due to the data partitioning. However, nodes hosting a partition of the database may perform the request faster than nodes hosting the whole database. Therefore RAIDb-2ec might perform better than RAIDb-1ec.

3.6 RAIDb Levels Performance/Fault Tolerance Summary

Fig. 2 gives an overview of the performance/fault tolerance tradeoff offered by each RAIDb level. RAIDb-0 offers in the best case the same fault tolerance as a single database. Performance can be improved by partitioning the tables on different nodes, but scalability is limited to the number of tables and the workload distribution among the tables.

RAIDb-1 gives in the worst case the same fault tolerance as a single database, and performance scales according to the read/write distribution of the workload. On a write-only workload, performance can be lower than for a single node. At the opposite extreme, a read-only workload will scale linearly with the number of backends. RAIDb-1ec provides at least the same fault tolerance as RAIDb-1, but performance is lowered by the number of nodes used to check each read query.

RAIDb-2 offers less fault tolerance than RAIDb-1, but it scales better on write-heavy workloads by limiting the updates broadcast to a smaller set of nodes. RAIDb-2ec has better fault tolerance than RAIDb-2 but comes at the price of lower performance and a larger number of nodes.

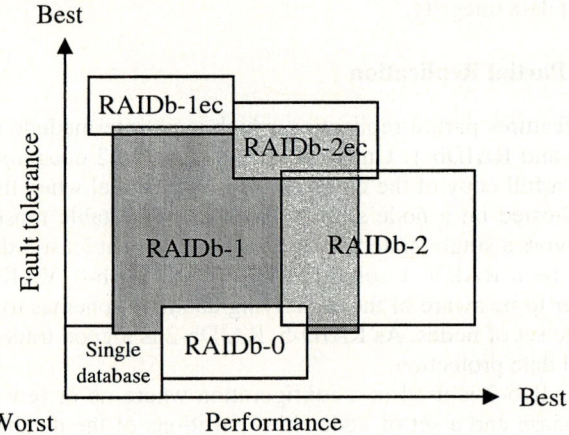

Fig. 2. RAIDb performance/fault tolerance tradeoff

3.7 RAIDb Levels Cost Effectiveness

Several parameters influence the cost and performance of every RAIDb configuration. Table 1 lists the cost and performance parameters taken into account for computing the cost/performance ratio of every RAIDb configuration.

Table 2 summarizes the performance cost tradeoff of each RAID level configuration. The optimal throughput assumes that performance scales linearly with the number of nodes regardless of the number of RAIDb controllers. Reads can occur in parallel whereas write will execute at the speed of the slowest replica. We assume that all nodes have the same performance.

RAIDb-0 performance scales up to the number of database tables if the workload is equally distributed among tables. In the best case, the disks can be fully distributed in the nodes if the tables have the same size that matches the disk size.

With RAIDb-1, each node has a full copy of the database, which makes it the most expensive solution with RAIDb-1ec. In the best case, write performance is the same as the one of a single node but reads can be parallelized on all nodes. RAIDb1-ec read performance is divided by the number of nodes involved on the error checking. This optimal case does take into account the cost of the results comparison.

Table 1. Cost/performance parameters

Name	Description
n	number of backend nodes
n$	backend node cost including software licenses but excluding disk cost
d	number of disks
d$	cost of a disk
c	number of C-JDBC controllers
c$	cost of a C-JDBC controller
t	throughput of a single database node
tab	number of tables
%read	percentage of reads in the workload
%write	percentage of writes in the workload
r	number of replicas of a table (RAIDb-2)
nec	number of nodes participating in error checking for a read request (RAIDb-*ec)

Table 2. RAIDb level cost effectiveness summary

RAIDb level	Nb of nodes	Nb of disks	Optimal throughput/cost ratio
Single DB	$n = 1$	$d \geq 1$	$\dfrac{t}{n\$ + d \times d\$}$
RAIDb0	$2 \leq n \leq tab$	$d \geq 1$	$\dfrac{n \times t}{n \times n\$ + d \times d\$ + c \times c\$}$
RAIDb1	$n \geq 2$	$n \times d$	$\dfrac{n \times t \times \%read + t \times \%write}{n(n\$ + d \times d\$) + c \times c\$}$
RAIDb1ec	$n \geq 3$	$n \times d$	$\dfrac{\dfrac{n \times t}{nec} \times \%read + t \times \%write}{n(n\$ + d \times d\$) + c \times c\$}$
RAIDb2	$n \geq 3$	$r \times d$	$\dfrac{r \times t \times \%read + \dfrac{n}{r} \times t \times \%write}{n \times n\$ + r \times d \times d\$ + c \times c\$}$
RAIDb2ec	$n \geq 4$	$r \times d$	$\dfrac{\dfrac{r \times t}{nec} \times \%read + \dfrac{n}{r} \times t \times \%write}{n \times n\$ + r \times d \times d\$ + c \times c\$}$

RAIDb-2 read throughput is a factor of the number of replicas, but only those replicas are blocked during a write allowing other writes to occur in parallel on other tables. As for RAIDb-1ec, RAIDb-2ec limits reads scalability.

4 Composing RAIDb Levels

It is possible to compose several RAIDb levels to build large-scale configurations. As a RAIDb controller may scale only to a limited number of backend databases, it is possible to cascade RAIDb controllers to support a larger number of RDBMS. As each RAIDb controller can provide its own cache, a RAIDb composition can help specialize the caches and improve the hit rate.

Fig. 3 shows an example of a 3-level RAIDb composition. The first level RAIDb-1 controllers acts as a single RAIDb-1 controller. At the second level, the database backend replica is implemented by a RAIDb-0 array with one of the partition being implemented by a RAIDb-2 configuration. Note that any controller can be replicated at any level to provide more availability.

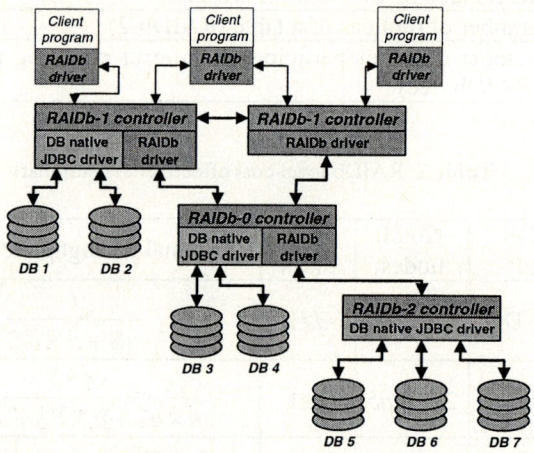

Fig. 3. RAIDb levels composition

There is potentially no limit to the depth of RAIDb compositions. It can also make sense to cascade several RAIDb controllers using the same RAIDb levels. For example, a cascade of RAIDb-1 controllers could be envisioned with a large number of mirrored databases. The tree architecture offered by RAIDb composition offers a more scalable solution for large database clusters especially if the RAIDb controller has no network support to broadcast the writes.

5 C-JDBC: A RAIDb Software Implementation

JDBC™, often referenced as Java Database Connectivity, is a Java API for accessing virtually any kind of tabular data [13]. We have implemented C-JDBC (Clustered JDBC), a Java middleware based on JDBC, that allows building all RAIDb configurations described in this paper. C-JDBC works with any RDBMS that provides a JDBC driver. The client application does not need to be modified and transparently accesses a database cluster as if it were a centralized database. The RDBMS does not need any modification either, nor does it need to provide distributed database functionalities. The distribution is handled by the C-JDBC controller that implements the logic of a RAIDb controller using a read-one write-all approach.

Fig. 4 gives an overview of the different C-JDBC components. The client application uses the generic C-JDBC driver that replaces the database specific JDBC driver. The C-JDBC controller implements a RAIDb controller logic and exposes a single database view, called virtual database, to the driver.

The authentication manager establishes the mapping between the login/password provided by the client application and the login/password to be used on each database backend. All security checks can be performed by the authentication manager. It provides a uniform and centralized resource access control.

Each virtual database has its own request manager that implement the required RAIDb level and defines the request scheduling, caching and load balancing policies. The "real" databases are defined as database backends and are accessed through their native JDBC driver. If the backend is a cascaded RAIDb controller like show on fig. 4, the C-JDBC driver is used as a regular database backend driver.

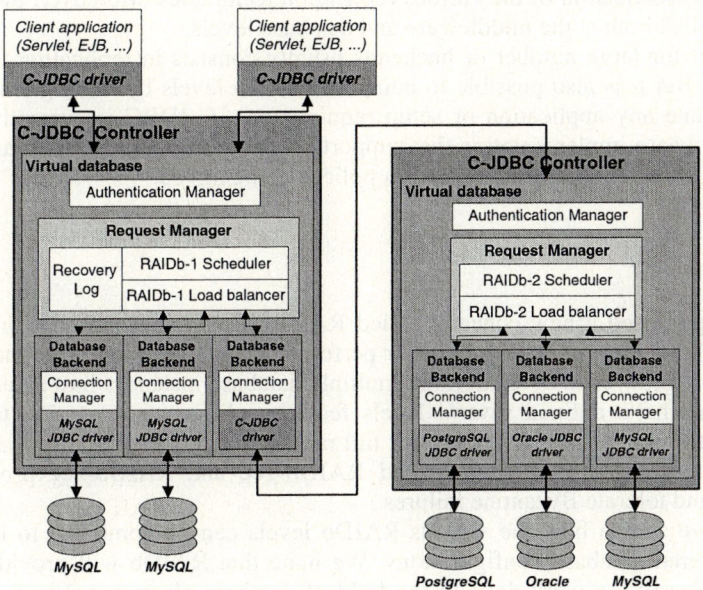

Fig. 4. C-JDBC overview

When a request comes from a C-JDBC driver, it is routed to the request manager associated to the virtual database. The *scheduler* is responsible for ordering the requests according to the desired isolation level. Once the request scheduler processing is done, the requests are sequentially ordered and sent to the *load balancer*.

Among the backends that can treat the request (all of them in RAIDb-1), one is selected according to the implemented algorithm (this applies only to reads, writes are broadcasted). Once a backend has been selected, the request is sent to its native driver through a connection manager that can perform connection pooling. The ResultSet returned by the native driver is transformed into a serializable ResultSet that is returned to the client by means of the C-JDBC driver.

C-JDBC also implements a recovery log that records all write statements between checkpoints. With each checkpoint corresponds a database dump. When a backend node is added to the cluster, a dump corresponding to a checkpoint is installed on the node. Then, all write queries since this checkpoint are replayed from the recovery log, and the backend starts accepting client queries as soon as it is synchronized with the other nodes. C-JDBC implementation and performance is further detailed in [4].

6 Related Work

Since the dangers of replication have been pointed out by Gray et al. [6], several works have investigated lazy replication techniques [10]. Ongoing efforts on eager replication have also been going on with the recent release of Postgres-R [8]. Several groups are focusing on group communications for asynchronous replication [14] or partial replication [11]. To the best of our knowledge, RAIDb is the first concept to propose a classification of the various replication techniques. Moreover, the approach can be applied both at the middleware and database levels.

Support for large number of backends usually consists in replicating the RAIDb controller. But it is also possible to compose RAIDb levels by nesting controllers to accommodate any application or setup requirements. C-JDBC is currently the only RAIDb software implementation that supports both controller replication and vertical scalability allowing different replication policies to be mixed.

7 Conclusion

We have proposed a new concept, called RAIDb (Redundant Array of Inexpensive Databases) that aims at providing better performance and fault tolerance than a single database, at a low cost, by combining multiple database instances into an array of databases. We have defined several levels featuring different replication techniques: RAIDb-0 for partitioning, RAIDb-1 for full replication and RAIDb-2 for partial replication. Additionally, two levels called RAIDb-1ec and RAIDb-2ec provide error checking and tolerate Byzantine failures.

We have shown how the various RAIDb levels can be combined to build large scale clustered database configurations. We hope that RAIDb will provide a sound basis for classifying work done in the field of database clustering. Finally, we have presented C-JDBC, a RAIDb software implementation in Java. C-JDBC is an open-source project and is available for download from http://c-jdbc.objectweb.org.

References

1. Christiana Amza, Alan L. Cox, Willy Zwaenepoel – Conflict-Aware Scheduling for Dynamic Content Applications – Proceedings of USITS 2003, March 2003.
2. Christiana Amza, Alan L. Cox, Willy Zwaenepoel – Scaling and availability for dynamic content web sites – Rice University Technical Report TR02-395, 2002.

3. P.A. Bernstein, V. Hadzilacos and N. Goodman – Concurrency Control and Recovery in Database Systems – Addison-Wesley, 1987.
4. Emmanuel Cecchet, Julie Marguerite and Willy Zwaenepoel – C-JDBC : Flexible Database Clustering Middleware – Usenix Annual Technical Conference - Freenix track, June 2004.
5. P. Chen, E. Lee, G. Gibson, R. Katz and D. Patterson – RAID: High-Performance, Reliable Secondary Storage – ACM Computing Survey, 1994.
6. Jim Gray, Pat Helland, Patrick O'Neil and Dennis Shasha – The Dangers of Replication and a Solution – Proceedings of the 1996 ACM SIGMOD International Conference on Management of Data, June 1996.
7. Monika Henziger – Google: Indexing the Web - A challenge for Supercomputers – Proceeding of the IEEE International Conference on Cluster Computing, September 2002.
8. Bettina Kemme and Gustavo Alonso – Don't be lazy, be consistent: Postgres-R, a new way to implement Database Replication –Proceedings of the 26th International Conference on Very Large Databases, September 2000.
9. L. Lamport, R. Shostak, and M. Pease – The Byzantine Generals Problem – ACM Transactions of Programming Languages and Systems, Volume 4, Number 3, July 1982.
10. E. Pacitti, P. Minet and E. Simon – Fast algorithms for maintaining replica consistency in lazy master replicated databases –Proceedings of VLDB, 1999.
11. A. Sousa, F. Pedone, R. Oliveira, and F. Moura – Partial replication in the Database State Machine – Proceeding of the IEEE International Symposium on Networking Computing and Applications (NCA'01), 2001.
12. D. Stacey – Replication: DB2, Oracle or Sybase – Database Programming & Design 7, 12.
13. S. White, M. Fisher, R. Cattel, G. Hamilton and M. Hapner – JDBC API Tutorial and Reference, Second Edition – Addison-Wesley, ISBN 0-201-43328-1, november 2001.
14. M. Wiesmann, F. Pedone, A. Schiper, B. Kemme and G. Alonso – Database replication techniques: a three parameter classification – Proceedings of the 19th IEEE Symposium on Reliable Distributed Systems (SRDS2000), October 2000.

A Fault-Tolerant Multi-agent Development Framework

Lin Wang, Hon F. Li, Dhrubajyoti Goswami, and Zunce Wei

Department of Computer Science, Concordia University,
1455 de Maisonneuve Blvd. W., Montreal, Quebec H3G 1M8, Canada
{li_wang, hfli, goswami, zunce_we}@cs.concordia.ca

Abstract. FATMAD is a fault-tolerant multi-agent development framework that is built on top of a mobile agent platform (Jade). FATMAD aims to satisfy the needs of two communities of users: Jade application developers and fault-tolerant protocol developers. Application-level fault tolerance incurs significant development-time cost. FATMAD is based on a generic fault-tolerant protocol whose refinements lead to a broad range of checkpoint and recovery protocols to be used in supporting user applications, thus significantly reducing the development time of fault-tolerant agent applications. This paper introduces the design of FATMAD and explains how fault-tolerant protocol developers can extend FATMAD with additional checkpoint and recovery protocols. The key concepts and features are illustrated through the staggered checkpoint protocol.

1 Introduction

Multi-agent systems are receiving considerable attention in many application areas [16, 20] due to the flexibility and ease of use of the agent paradigm. Fault tolerance is an important design issue in many on-line applications. Since agents are autonomous objects [2] with purposeful and also unpredictable behaviors, tolerating failures in agent applications is a non-trivial problem for agent developers and system designers.

Checkpoint and rollback-recovery [11] are popular strategies to achieve fault tolerance in distributed systems. Unlike replication-based techniques [8] that are only suitable for storage-intensive or service-oriented situations, checkpoint/recovery is applicable for all kinds of systems. In such a strategy, two inter-dependent protocols are deployed: a checkpoint/logging protocol that saves information into stable storage, and a recovery protocol that is launched upon failure to restore the execution from stable storage. In log-based recovery [1, 19], message events are logged and replayed to guarantee the deterministic recreation of the execution. Checkpoints are taken as process snapshots only in order to trim the message log. In contrast, checkpoint-based recovery [5, 11] takes coordinated checkpoints to avoid message logging/replaying. As a result, all processes that coordinate during checkpointing need to rollback together in recovery. Checkpoint/recovery strategies are usually based on the fail-stop model [18], in which processes are subject to crash failures detectable by other processes.

The distributed agent community has done a significant amount of work in agent modeling, specification, and development. Role model [9, 10] is presented to capture agent behaviors and specify the requirements. A high-level agent communication language (ACL) [7] is designed for agents to interact with each other, and agent collaborations can be modeled by AUML protocol diagram [3, 14]. On the other hand, there are agent platforms and frameworks [4, 15, 21, 23] that provide development support and runtime environment for agent applications. Common features and services are implemented as system APIs or pre-designed agent classes, with which agent developers interface to build their own agent applications.

Existing agent frameworks do not provide sufficient support for developing fault-tolerant agent applications: there is less support to tolerate agent crash and the strategies are almost fixed in the kernel. Many agent frameworks, such as Jade [4] and Aglet [21], focus on different levels of services for standard agent communication and collaboration. New version of Jade employs replication techniques to enhance the availability of system services, but it doesn't provide scheme on application level fault-tolerance for agents. Using such frameworks, in order to employ a checkpoint/recovery protocol, an application developer usually has to implement the protocol by directly programming the detailed issues like where to take a checkpoint and when to start logging messages. Some agent frameworks [15, 23] provide system services for fault-tolerant programming, but these services mainly focus on the mobility aspects of agents and are hard-coded, hence are less flexible. For example, Concordia [23] partially implements a checkpoint/recovery scheme to guarantee only the agent migration. With these frameworks, application developers are actually playing the role of checkpoint/recovery protocol designers. However, none of the existing agent frameworks provides support for protocol designers to implement and test different checkpoint/recovery strategies.

A new fault-tolerant agent framework that supports checkpoint/recovery strategies can be built to help both application developers and (checkpoint/recovery) protocol designers. Here the framework distinguishes the jobs of these two user groups. Application developers are only responsible for the application code and the selection of the appropriate checkpoint/recovery protocol from a protocol library supplied by the framework. Protocol designers build and extend the protocol library by programming their checkpoint/recovery protocols using facilities supported by the framework.

In general, the following observations apply to all checkpoint and recovery protocols: i) the protocols employ some common basic functions, such as taking a local checkpoint, logging a message, and resetting an agent to a checkpoint; ii) many protocols follow a generic behavioral pattern, e.g. upon failure, a recovery protocol has to retrieve the recorded information, decide the set of agents to be recovered, and rollback their execution to some appropriate checkpoints. When provided with a framework that has implemented most of the common functions and the generic behavioral pattern(s), protocol designers can save their efforts and focus on the protocol-specific parts. In this way, the framework can facilitate protocol implementation and application development by providing a flexible as well as sufficient agent-programming environment.

This paper presents the design of such a framework named Fault-Tolerant Multi-Agent Development framework (FATMAD). Based on checkpoint and recovery techniques, it tolerates agent failure by providing a modifiable fault tolerance kernel, which allows users to choose from a library of different strategies and to extend the library easily. The rest of the paper is organized as follows: section 2 gives an overview of the framework in terms of its objectives, structure, and its appearance to application developers. Section 3 provides more details of the design, and highlights its appearance to protocol designers. An example checkpoint and recovery protocol is used throughout to illustrate the programming interfaces. Section 4 concludes the paper.

2 Overview

FATMAD is targeted for two groups of users with different expectations and responsibilities: *protocol designers* who can extend the framework with new checkpoint/recovery protocols, and *application developers* who can use the extended or default framework to develop fault-tolerant agent applications.

From an application developer's perspective, FATMAD should act like a fault-tolerant-support layer on top of the agent application development platform. The application developer should have some basic knowledge regarding the selection of a specific checkpoint/recovery protocol that best suits the characteristics of her application, and be able to integrate with ease the selected fault-tolerant protocol into her application using the FTMAD provided API support.

A protocol designer uses FATMAD to implement a particular checkpoint/recovery protocol. Generic functionalities that are common to most protocols should be supported. FATMAD should be able to provide a meta-model that is abstract enough to cover most protocols and is also flexible enough for the protocol designer to customize her own protocols by incorporating specific "hot-spots". To be able to interface with application developers without extension, it should also come with some default checkpoint and recovery protocols that can be immediately used.

2.1 High-Level Structure

FATMAD is designed to provide all necessary supports for both user groups, at design-time and run-time. As shown in Fig. 1, FATMAD is implemented on the Jade [4] platform, and provides programming interfaces for application developers to embed fault-tolerance support. FATMAD functionalities can be abstractly divided into two parts: *FATMAD Primary* and the *Protocol Extension*. FATMAD Primary is the default framework without any extension. It provides a *kernel* with all basic services, a *protocol skeleton* that captures the generic behavioral pattern of most protocols, and a *default implementation* of the customizable parts that include two default checkpoint/recovery protocols. *Protocol Extension* is the protocol-specific part, including all customizations and configurations created by protocol designers.

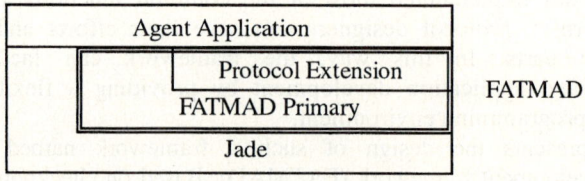

Fig. 1. Structure of FATMAD

Application developers can use either the default or the extended FATMAD for embedding agent applications with fault-tolerance support (refer to Fig. 1). In either case, they need to program with the protocol-specific classes, which are provided either by FATMAD or by protocol designers. Protocol designers can extend the framework by integrating their own protocols, which are implemented by concretizing the protocol skeleton with the protocol-specific code. The resulted protocols are provided a set of agent classes, including the protocol-specific class for application developers and other supporting classes. To ensure the correct implementation of the protocol, it is mandatory that the designers have a good understanding of the underlying design and the internal mechanisms of FATMAD. These issues are discussed in section 3.

2.2 Application Programming with FATMAD

To embed fault tolerance features into application agents, the developer has to select a checkpoint/recovery protocol that is deemed to be suitable for her application, and then adapt the application to FATMAD by applying the protocol interface(s). In order to properly use a FATMAD supported protocol, the developer needs to understand: i) the important characteristics, such as protocol scheme, suitability of application types, benefits, etc, and ii) the framework interfaces of the selected protocol. Usually, adapting an application agent into a fault-tolerant agent includes the following:

Extend Protocol Specific Agent Class. Instead of extending the Jade agent class, one needs to start by extending a protocol-specific class created by protocol designers.

Replace a Set of Agent Class Methods. Jade-provided methods need to be replaced by FATMAD-provided methods in order to perform user-defined operations.

Replace Messaging Function Calls. Jade messaging functions should be replaced by corresponding FATMAD functions, which can trigger message handling like logging.

Flag to Trigger Checkpoint/Logging. Some protocols require the application to set flags at desired points in order to trigger a checkpoint. This can be done by calling the pre-defined methods provided either by the kernel or by the protocol.

The following example shows code fragments of an agent application named *FTAppAgent*, which employs the staggered checkpoint protocol [22] via extending the protocol-specific class named *StaggeredFTAgent*. Bold words demonstrate the parts performed by application developers, e.g., the calling of *flagToCheckpoint()* method, as is required by the protocol. Details of the staggered checkpoint protocol and the implementation of *StaggeredFTAgent* class in FATMAD are discussed in Section 3.

```
public class FTAppAgent extends StaggeredFTAgent
{                                   // Class inheritance
   public void appSetup(){...} // To replace Jade setup()
   public class OEBehaviour extends SimpleBehaviour
   {          // Behaviour class conforming to Jade
      public void action(){   // Application agent action
         if(chpt_condition)   flagToCheckpoint();
            // Flag to trigger checkpoint service
         sendMessage(msg); ...} // To replace Jade send()
   ...}
...}
```

3 FATMAD for Protocol Designers

The design of FATMAD is based on the observation that different checkpoint/recovery protocols have common or similar behavioral patterns and utilize a set of common services. FATMAD provides these common parts as generic patterns/services, from which protocol designers can customize their protocol-specific parts.

3.1 The Generic Solutions

A checkpoint/recovery protocol, in general, consists of different actions that are executed or triggered under various conditions. For example, the staggered checkpoint protocol, which is a refinement of the coordinated checkpoint protocol by Chandy and Lamport [6], is triggered by a flag-to-checkpoint of an application agent (the initiator). The initiator takes a local checkpoint, sends a checkpoint request to an un-checkpointed agent, and begins to log messages from all channels. Upon receiving a checkpoint request, an agent will take similar actions. This sequence repeats until the last un-checkpointed process sends a checkpoint request back to the initiator. The initiator then sends out special marker messages to all its output channels. Once an agent receives the first marker message, it sends out marker messages to all its output channels. The message logging for a channel stops as soon as a marker message is received from that channel. The corresponding recovery protocol will rollback all agents to their most recent checkpoints, and the logged messages will be replayed accordingly.

As we can see, the above protocol involves many basic functions, e.g., taking a local checkpoint, logging/replaying of a message, rolling an agent back to a checkpoint, and so on. These basic functions are common to most checkpoint/recovery protocols, and can be supported as framework services. A protocol developer's responsibility is then to properly utilize these services while integrating her protocol into the framework.

In general, the checkpoint protocol for an agent can be viewed as a sequence of three atomic actions: checkpointing, message logging, and updating of the logging policy. Coordination actions (such as checkpoint request and marker message) induce dependencies among agents. An action is triggered when some particular conditions are satisfied. A *policy* hence can generally refer to an action and its triggering condition. These features can be supported by the framework as a generic checkpoint protocol (i.e., a generic behavioral pattern), as shown in a high level of abstraction as follows:

```
Upon checkpoint event for agent a_i:
    Take a local checkpoint;
    Update logging policy locally;
    Send checkpoint request to a subset of agents;
    Wait for feedback from a subset of agents;
    Send checkpoint commitment to a subset of agents;
    Do logging coordination with a subset of agents
    and update group logging policy;
```

A specific checkpoint protocol is a refinement of the generic protocol (by the protocol designer). In general, in a checkpoint protocol, the designer needs to specify the following for an agent: i) when a local checkpoint is taken (i.e. triggered by some specific checkpoint events like application's flag-to-checkpoint or a checkpoint request); ii) changes to the policy of logging messages that are received by that agent (e.g. start or stop logging a channel); iii) dependencies among checkpoint/logging events taken at different agents, if any; iv) defining the coordination group, etc.

The action of message logging can be implemented as a framework service that is governed by a logging policy. All checkpoints and logged messages are retrievable from some repository manager that can survive node crash. When an agent failure is detected, a recovery protocol will usually perform the following: i) gather necessary information (checkpoints and message logs) from the repository manager, ii) based on the information gathered, decide on a recovery line involving one or more agents that should roll back, and iii) enforce the rollback with an appropriate policy for replay and discard of messages. The above three steps can be modeled as a sequence of three atomic actions. Hence the framework involves a simple abstract recovery protocol. While a checkpoint protocol involves a logging policy, a recovery protocol similarly involves a message handling policy upon agent recovery. The enforcement of the policy is automatically supported by the framework.

FATMAD also supports pruning of logged data. Pruning actions are consequences of checkpoint and recovery actions. A generic pruning protocol is available for discarding logged data when they are not needed in future crashes.

The generic checkpoint/logging and recovery/pruning protocols can be refined to many checkpoint and recovery protocols rather than merely the staggered protocol. For example, during failure-free execution, checkpoint-based protocols only involve the actions of checkpointing, while log-based and hybrid protocols involve all three atomic actions but differ in their policy control. In addition, the generic protocols also fit some other protocols that are neither log-based nor globally coordinated, such as the group-based checkpoint/recovery protocol that we have recently developed [12].

FATMAD implements the previous generic solutions as follows: (a) a *kernel* that contains the essential framework services supporting all protocols; (b) a *protocol skeleton* that can be refined into specific checkpoint/recovery protocols; (c) an *application skeleton* that embodies an application agent to be interfaced properly with FATMAD (including parts of both (a) and (b)). The protocol skeleton actually implements the atomic action control and the policy triggering mechanism for the generic protocol(s). It is ready to be concretized by a protocol developer, who will incorporate protocol-specific code into the skeleton. Once extended with protocol-specific code through the use of the framework services, it forms a developer-provided protocol and can serve as an application skeleton for application developers.

3.2 Design of FATMAD

As mentioned in section 3.1, the FATMAD kernel consists of a set of basic services needed to support checkpoint and recovery protocols. These services are general purpose in nature and their functionalities are apparent from their names: (i) execution control for application agent, such as freeze, resume, abort of an agent, (ii) checkpoint service that is triggered by a checkpoint event, (iii) logging service with various logging options and management of cache/stable storage, (iv) timing service such as logical and vector clocks, (v) repository service that takes care of storage management and data access, and (vi) other message handling services such as replay/discard of messages during recovery. These services are used by the checkpoint and recovery protocols in both protocol skeleton and its extension, for creating checkpoints and message logs on the application agents as well as for deciding how to handle an agent crash at recovery time. Some services are executed by the protocol as procedure calls, while the others are triggered by the kernel as policy actions. Both procedure call and policy setting are framework- provided APIs available for protocol designers.

The protocol skeleton is provided as a set of agent classes which, when extended/concretized, implements four key functions that are associated with a typical checkpoint/recovery protocol: (i) *Agent-specific control*, including agent execution control, checkpoint service and protocol, and message handling; (ii) *Logging control* (the logging service and policy) (iii) *Recovery decision control* (the recovery protocol); (iv) *Pruning control* (the pruning protocol).

The agent-specific control part is responsible for the checkpoint/recovery behaviors of a particular agent during failure-free execution and after recovery. It is provided as an *FTBehaviour* class in FATMAD, with an embedded behavior which implements the generic checkpoint protocol by pre-setting some generic polices as outlined earlier. For example, a checkpointing action is triggered automatically upon the call of *flagToCheckpoint()* method. Protocol designers can either set the policy conditions by calling the corresponding APIs (e.g., set checkpoint interval in periodical checkpointing), or provide their protocol-specific actions as callback functions. The embedded behavior can hence be extended to accommodate different categories of checkpoint protocols (i.e., log-based, coordinated, group-based, etc.). Optionally, coordination of dependent checkpoints can be programmed in *FTBehaviour* by a protocol developer, via system messages or information piggybacking. In addition, the message handling service is integrated into the agent-specific control part, and it receives policies from the recovery decision control part. In FATMAD, a logging policy specifies the rule to be applied in logging messages on each channel and is executed by the logging control part, which is provided as a *LoggingAction* class.

To program a checkpoint/logging protocol using the *FTBehaviour* class, the designer has to deal with the issues such as setting checkpoint event, updating logging policy and checkpoint coordination. Related policies (the actions and/or the triggering conditions) have to be customized according to the needs of the particular protocol. This is done by first creating a sub-class of *FTBehaviour* and then overriding the corresponding methods as callback functions, using the kernel APIs. An example is the *StaggeredFTBehaviour* class shown in section 3.3 where the corresponding code for overridden methods is given. Often the *LoggingAction* class is not extended since in most protocols the enforcement of the logging policy can be generically shared.

The recovery decision control part, when extended by the protocol designer, is responsible for deciding on the set of agents and their rollback checkpoints, and the messages to be replayed/discarded at each agent. The decision is passed to the agent-specific control part of each involved agent. The recovery decision control part is provided as a *RecoveryManager* class which implements the generic recovery protocol, and is triggered upon detection of failure. Similar to case of checkpoint control, a protocol designer has to create a sub-class via class inheritance and override some pre-defined callback methods. The pruning control is provided as a Java interface named *PruningAction*, which in turn should be implemented if needed.

In addition to implementing the four functions in the previous discussion, a protocol designer has to put them together to provide an interface to application developers. This is done by inheriting a sub-class from *FTAgent* class, and specifying the four functions in its *ftSetup()* method that initializes all fault tolerance features for an agent. This sub-class is protocol-specific, and serves as an application skeleton in FATMAD (refer to section 2.3). Section 3.3 shows the sub-class *StaggeredFTAgent* for the staggered checkpoint protocol, with the code for its *ftSetup()* method.

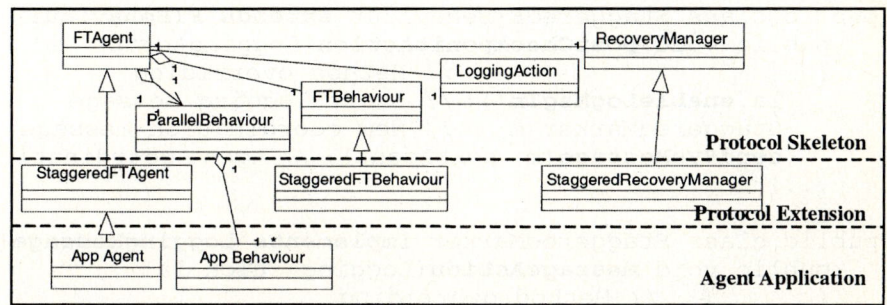

Fig. 2. Hierarchy of Key Classes in FATMAD

Fig. 2 shows the hierarchy of key classes that are relevant to a protocol developer. One should observe how fault tolerance features are added via class inheritance relations among the following classes: Jade agent → *FTAgent* (Protocol Skeleton) → *StaggeredFTAgent* (Protocol Extension) → FTAppagent (Application agent). The classes *FTBehavior*, *LoggingAction*, *RecoveryManager*, and *PruningAction* form the protocol skeleton part of FATMAD. Protocol designers need to deal with some or all of them in order to implement their protocol agents. Application developers simply inherit their application agents from the desired protocol agents.

3.3 An Example of Protocol Extension

In this section, we illustrate the use of FATMAD by showing how the staggered checkpoint protocol can be implemented using the FATMAD classes. Due to space limit, we focus on the framework support and omit the details of control flow code.

A designer of the staggered checkpoint protocol needs supports from FATMAD to do the following: i) Setting the checkpoint events as application flag and some coordination message (checkpoint request), and program the checkpointing actions; ii) Starting/stopping logging upon receipt of some special coordination message (marker); iii) Coordinating checkpointing/logging via special coordination messages.

As mentioned in Section 3.2, the designer first inherits a sub-class *StaggeredFTBehaviour* from the base agent-specific class *FTBehaviour*, which provides a method *postChecpointAction()* that is invoked by the kernel whenever the *flagToCheckpoint()* method is called by application. The designer hence needs to override *postChecpointAction()* in her sub-class to program the corresponding checkpointing action. At the same time, the designer needs to create two coordination message types via implementing a *LoggingMessage* interface provided by FATMAD to support the design of new system messages. Its *messageAction()* method should be coded subsequently for implementing the actions triggered upon receipt of such a system message, such as: checkpointing, starting/stopping logging, and coordination actions among agents. To update logging policy, the designer calls the corresponding methods of the *LoggingAction* class: *enableLogMsgIn()*, *stopLogMsgIn()*, respectively. The relevant code is shown as follows and is highlighted in bold words. The recovery protocol can be programmed similarly using corresponding FATMAD support.

```
public class StaggeredFTBehaviour extends FTBehaviour{
    public void postCheckponitAction(LoggingAction la)
    {                           // Method overriding
        la.enableLogMsgIn(); // Start logging message
        StaggeredMarker m;    // New coordination message
        sendSysMessage(m, nextAgent); // Agent coordination
    ...}
...}
public class StaggeredMarker implements LoggingMessage{
    public void messageAction(LoggingAction la)
    {           // Method overriding
        la.stopLogMsgIn();    // Stop logging message
    ...}
...}
```

In addition, the designer needs to integrate all parts of her checkpoint/recovery protocol by creating a sub-class *StaggeredFTAgent* of *FTAgent* and subsequently overriding its *ftSetup()* method. The following shows the resultant *StaggeredFTAgent*.

```
public class StaggeredFTAgent extends FTAgent{
    public void ftSetup(){ // method overriding
        setFTBehaviour(new StaggeredFTBehaviour(this));
        setLoggingAction(new LoggingAction(this));
        setEventReaction(new StaggeredRecoveryManager());
        setPrunningAction(new PrunningAction());
        ...}                       // other initialization
...}
```

4 Conclusion and Future Works

In this paper we have presented a fault-tolerant agent framework (FATMAD) based on checkpoint/recovery techniques. The framework provides supports for both application developers and protocol designers. It integrates a kernel that consists of essential framework services, a generic protocol skeleton that can be extended into specific (checkpoint/recovery) protocols, and an application skeleton that incorporates application agents with selected fault-tolerance features. The design of both protocol and application agent can be facilitated by programming with the framework services and pre-defined classes.

Ongoing extensions of FATMAD include agent monitoring and failure detection, that go beyond crash failure, by using both protocol monitoring and checking [17], and distributed predicate checking [13]. Ongoing effort also includes additional automation of more generically usable components.

References

[1] L. Alvisi, K. Marzullo, Message logging: pessimistic, optimistic, causal and optimal, IEEE Trans. Software Eng. 24(2) (1998) 149-159.
[2] Y. Aridor, D.B. Lange, Agent design patterns: elements of agent application design, in Proc. Agents'98, Minneapolis, Minnesota, May 1998, pp. 108-115.
[3] B. Bauer, J.P. Müller, J. Odell, Agent UML: a formalism for specifying multiagent interaction, in, Proc. AOSE'01, Springer-Verlag, Berlin, 2001, pp. 91-103.
[4] F. Bellifemine, A. Poggi, G. Rimassa, JADE — A FIPA-Compliant Agent Framework, in Proc. PAAM-99, London, UK, 1999. The Practical Application Company Ltd, pp. 97-108.
[5] G. Cao, M. Singhal, On coordinated checkpointing in distributed systems, IEEE Trans. Parallel and Distributed Systems, 9(12) (1998) 1213-1225.
[6] M. Chandy, L. Lamport, Distributed snapshots: determining global states of distributed systems, ACM Trans. Computing Systems, 3(1) (1985) 63-75.
[7] FIPA, FIPA'99 Specification Part 2: Agent Communication Language, http://www.fipa.org.
[8] R. Guerraoui, A. Schiper, Fault-tolerance by replication in distributed systems, in Reliable Software Technologies - Ada-Europe'96, LNCS 1088, pp. 38-57. Springer-Verlag, June 1996.
[9] E.A. Kendall, Agent roles and aspects, in: S. Demeyer, J. Bosch, (eds.), Proc. ECOOP Workshops, Springer-Verlag, LNCS 1543 (1998) 440.
[10] E.A. Kendall, Agent software engineering with role modeling, in: Proc. AOSE-2000, Springer-Verlag, Berlin, Germany, Jan. 2000, pp. 163-170.
[11] R. Koo, S. Toueg, Checkpointing and rollback recovery for distributed systems, IEEE Trans. Soft. Eng., 13(1) (1987) 23-31.
[12] H.F. Li, Z. Wei, D. Goswami, Quasi-atomic recovery for distributed agents, under revision.
[13] N. Mittal and V.K. Garg, On Detecting Global Predicates in Distributed Computations, in: Proc. IEEE ICDCS, Phoenix, May 2001, pp. 3 - 10.
[14] J. Odell, H.V.D. Paranak, B. Bauer, Extending UML for agents, in Proc. AOIS Workshop at AAAI 2000, Mar. 2000, Austin, TX, USA, pp. 3-17.
[15] H. Pals, S. Petri, C. Grewe, FANTOMAS: Fault Tolerance for Mobile Agents in Clusters, in: Proc. 15th IPDPS Workshops, Cancun, Mexico, Springer-Verlag, LNCS 1800, pp. 1236-1247.
[16] A. Pivk, M. Gams. Intelligent Agents in E-Commerce. Electrotechnical Review, 67(5)(2000) 251-260.
[17] J. Saldhana, Sol M. Shatz, UML diagrams to object petri net models: an approach for modeling and analysis, in: Proc. Intl. Conference on Software Eng. and Knowledge Eng. (SEKE), Chicago, July 2000, pp. 103-110.
[18] R.D. Schlichting, F.B. Schneider, Fail-stop processors: an approach to designing fault-tolerant computing systems, ACM Trans.Computer Systems, 1(3)(1983) 222-238.
[19] R.E. Strom, S.A. Yemini, Optimistic recovery in distributed systems, ACM Trans. Computer Systems, 3(3) (1985) 204-226.
[20] K. Sycara, K. Decker, A. Pannu, M. Williamson, D. Zeng, Distributed intelligent agents, IEEE Expert, 11(6) (1996) 36-46.
[21] H. Tai, K. Kosaka, The Aglets project, Comm. of the ACM, 42(3)(1999) 100-101.
[22] N.H. Vaidya, Staggered consistent checkpointing, IEEE Trans. Parallel and Distributed Systems, 10(7)(1999) 694-702.
[23] T. Walsh, N. Paciorek, D. Wong, Security and reliability in Concordia, in Proc. 31th Annual Hawaii International Conference on System Sciences (HICSS31), 7(1998) 44-53.

A Fault Tolerance Protocol for Uploads: Design and Evaluation*

L. Cheung[1], C.-F. Chou[2], L. Golubchik[3], and Y. Yang[1]

[1] Computer Science Department, University of Southern California, Los Angeles, CA
{lccheung, yangyan}@usc.edu
[2] Department of Computer Science and Information Engineering,
National Taiwan University
ccf@csie.ntu.edu.tw
[3] Computer Science Department, EE-Systems Department, IMSC, and ISI,
University of Southern California, Los Angeles, CA
leana@cs.usc.edu

Abstract. This paper investigates fault tolerance issues in Bistro, a wide area upload architecture. In Bistro, clients first upload their data to intermediaries, known as bistros. A destination server then pulls data from bistros as needed. However, during the server pull process, bistros can be unavailable due to failures, or they can be malicious, i.e., they might intentionally corrupt data. This degrades system performance since the destination server may need to ask for retransmissions. As a result, a fault tolerance protocol is needed within the Bistro architecture. Thus, in this paper, we develop such a protocol which employs erasure codes in order to improve the reliability of the data uploading process. We develop analytical models to study reliability and performance characteristics of this protocol, and we derive a cost function to study the tradeoff between reliability and performance in this context. We also present numerical results to illustrate this tradeoff.

1 Introduction

High demand for some services or data creates hot spots, which is a major hurdle to achieving scalability in Internet-based applications. In many cases, hot spots are associated with real life events. There are also real life deadlines associated with some events, such as submissions of papers to conferences. The demand of applications with deadlines is potentially higher when the deadlines are approaching.

---

* This work is supported in part by the NSF Digital Government Grant 0091474. It has also been funded in part by the Integrated Media Systems Center, a National Science Foundation Engineering Research Center, Cooperative Agreement No. EEC-9529152. Any opinions, findings and conclusions or recommendations expressed in this material are those of the author(s) and do not necessarily reflect those of the National Science Foundation. More information about the Bistro project can be found at http://bourbon.usc.edu/iml/bistro.

To the best of our knowledge, however, there are no research attempts to relieve hot spots in many-to-one applications, or upload applications, except for Bistro [1]. Bistro is a wide-area upload architecture built at the application layer, and previous work [2] has shown that it is scalable and secure.

In Bistro, an upload process is broken down into three steps (see Sect. 3 for details) [1]. First, in the timestamp step, clients send hashes of their files, $h(T)$, to the server, and obtain timestamps, σ. These timestamps clock clients' submission time. In the data transfer step, clients send their data, T, to intermediaries called bistros. In the last step, called the data collection step, the server coordinates bistros to transfer clients' data to itself. The server then matches the hashes of the received files against the hashes it received directly from the clients. The server accepts files that pass this test, and asks the clients to resubmit otherwise. This completes the upload procedure in the original Bistro architecture [1,2].

We are interested in developing and analyzing a fault tolerance protocol in this paper, in the context of the Bistro architecture. The original Bistro does not make any additional provisions for cases when bistros are not available during the data collection step. In addition, malicious bistros can intentionally corrupt data. Although a destination server can detect corrupted data from the hash check, it has no way of recovering the data. Hence, unavaliable bistros and malicious behavior can result in the destination server having to request client resubmissions. In this work, we are interested in using forward error correction techniques to recover corrupted or lost data in order to improve the overall system performance. The fault tolerance protocol, on the other hand, brings in additional storage and network transfer costs due to redundant data. The goal of this paper is to (a) provide better performance when intermediaries fail while reducing the amount of redundant data needed to accomplish this, and (b) to evaluate the resulting tradeoff between performance and reliability.

We propose analytical models to evaluate our fault tolerance protocol. In particular, we develop reliability models to analyze the reliability characteristics of bistros. We also derive performance models to estimate the performance penalty of employing our protocol. Moreover, we study the tradeoff between reliability and performance.

The remainder of this paper is organized as follows. Section 2 describes related work. Section 3 describes our fault tolerance protocol. We derive analytical models for this protocol in Sect. 4. Section 5 presents numerical results showing the tradeoff between performance and reliability characteristics of our protocol. Finally, we conclude in Sect. 6.

2 Related Work

This section briefly describes fault tolerance considerations in other large-scale data transfer applications, and discusses other uses of erasure codes in the context of computer networking.

One approach to achieving fault tolerance is through service replication. Replication of DNS servers is one such example. The root directory servers are replicated, so if any root server fails, DNS service is still available. Each ISP is likely to host a number of DNS servers, and most clients are configured with primary and alternate DNS servers. Therefore, even if some DNS servers fail, clients can contact an alternate DNS server to make DNS lookup requests. In Bistro, the service of intermediaries is replicated, where intermediaries provide interim storages of data until the destination server retrieves it.

In storage systems, data redundancy techniques, such as RAID techniques [3], are commonly used for providing better fault tolerance characteristics. In case of disk failures, file servers are able to reconstruct data on the failed disk once the failed disk is replaced, and data is available even before replacing the failed disks. Although data redundancy can provide better fault tolerance characteristics, the storage overhead can be high. We are interested in providing fault tolerance with small storage overhead in this work.

Erasure codes are useful in bulk data distribution, e.g., in [4] clients can reconstruct the data as long as a small fraction of erasure-encoded files are received. This scheme allows clients to choose from a large set of servers, resulting in good fault tolerance and better performance characteristics than traditional approaches.

In wireless networking, using forward error correction techniques can reduce packet loss rates by recovering parts of lost packets [5,6]. Packet loss rates in wireless networks are much higher because propagation errors occur more frequently when the data is transmitted through air. Employing forward error correction techniques can improve reliability and reduce retransmissions.

These applications of erasure codes assume that packets are either received successfully or are lost. They assume that there are other ways to detect corrupted packets. e.g., using TCP checksums. In Bistro, however, this assumption is not valid because packets can be intentionally corrupted by intermediate bistros. In Sect. 3, we describe one way to detect corrupted packets using checksums so that we can treat corrupted packets as losses.

3 Fault Tolerance Protocol

This section provides details of our fault tolerance protocol. The protocol is broken down into three parts as in the original Bistro protocol described in [2]. We provide details of each step in this section with focus on the fault tolerance aspects proposed in this paper. We also discuss related design decisions.

3.1 Timestamp Step

The timestamp step verifies clients' submissions. Clients first pass their files, T_o, to erasure code encoders to get the encoded files $T = T_1 + T_2 + \ldots + T_x$. Then, clients generate hashes of each part of their data, concatenate the hashes and send the results, H, to the destination server. The destination server replies to clients with tickets, ξ, which consist of timestamps, σ, and the hash messages clients have just sent, $h(H)$. Tickets are digitally signed by the destination server, so clients can authenticate the destination server. Fig. 1 depicts the timestamp step.

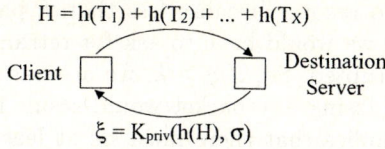

Fig. 1. Timestamp Step

In the original protocol, clients send a checksum (or a hash) of the whole file to the destination server in the timestamp step. If any packets are lost or corrupted, the checksum check would fail, and the destination server would have to discard all packets that correspond to that checksum because it does not know which packets are corrupted. This would mean that losing any packet would result in retransmissions of entire files, in the original protocol.

To solve this problem, we send multiple checksums in the fault tolerance protocol, $h(T_1)+h(T_2)+\ldots+h(T_x)$. Assume that each client has W data packets to send. The data packets are divided into Y FEC (forward error correction) groups of k packets each. For each FEC group, a client encodes k data packets into n packets (data + parity), arranges the n packets into Z checksum groups each of size g, and generates one checksum for each checksum group using a message digest algorithm such as SHA1. We assume that Z is a factor of g, because we want the size of all checksum groups to be the same, which simplifies our reliability evaluation in Sect. 4. There are altogether $X = YZ$ checksums, which are concatenated and sent in one message to the destination server. Figure 2 illustrates the relationship between FEC groups and checksum groups.

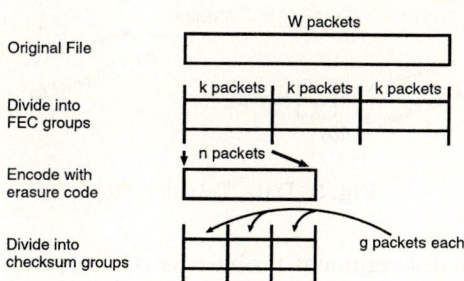

Fig. 2. FEC Groups and Checksum Groups

Note that the size of a checksum group has to be smaller than the number of data packets per FEC group ($g < k$). Recall that erasure codes do not correct corrupted packets, so we drop all packets in a checksum group if any packet within the checksum group is lost or corrupted, and then we try to recover the dropped packets using an erasure code. If $g \geq k$ and if a checksum group is dropped, then we lose more than k packets in at least one FEC group, which we would not be able to recover because less than k packets within that FEC group are received, i.e., we would have to ask for retransmissions if any packet in the file is lost or corrupted. So, if $g \geq k$, we are back to the problem of the original protocol where losing any packet would result in retransmissions. The above argument also implies that there must be at least two checksum groups per FEC group. This also explains the order in which FEC groups and checksum groups are constructed.

3.2 Data Transfer Step

In the data transfer step, clients send their files to intermediate bistros which are not trusted. Clients first choose B bistros to send their data to and then generate a session key K_{ses_i} for each chosen bistro, $1 \leq i \leq B$. After that, clients divide their files into B parts. For each part i, clients encrypt it with a session key K_{ses_i} and send that part to an intermediate bistro i. Clients also send to bistro i the session key, K_{ses_i}, and ticket, ξ, encrypted with the public key of the destination server. In addition, clients send event IDs, EID, so as to identify that the data is for a particular upload event whose event ID is EID. Each bistro i generates a receipt, ρ_i, and sends it to both an appropriate client and the destination server. The receipts contain the public key of bistro i, $K_{i,pub}$, so that both clients and the destination server can decrypt and verify the receipt[1]. Figure 3 depicts the data transfer step.

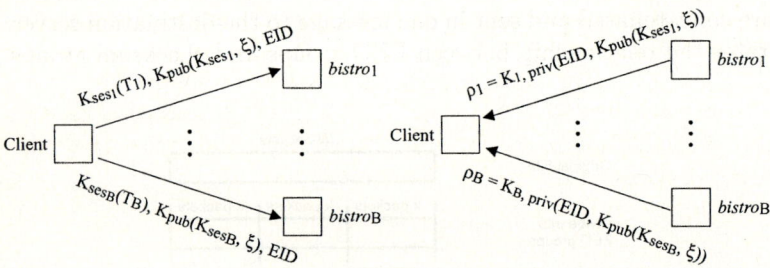

Fig. 3. Data Transfer Step

In [7], the so-called assignment problem is studied, i.e., how a client should choose a bistro to which it sends its file. However, in that case, only one bistro out of a pool of bistros is chosen. In the case of striping (our case), a client needs to choose $B \geq 1$ bistros. As shown in [7], this is a difficult problem even for $B = 1$. Hence, we leave the choice of which B bistros a client should stripe its file to, and how clients determine the value of B to future work. In the remainder of this paper, we assume that the B bistros are known.

---

[1] Note that whether the public key of an intermediate bistro is correct or not does not affect the correctness of the protocol, as in the original Bistro system, as intermediate bistros are not trusted in any case.

3.3 Data Collection Step

In the data collection step, the destination server coordinates intermediate bistros to collect data. When the destination server wants to retrieve data from bistro i, it sends a retrieval request along with the receipt ρ_i and the event ID EID. Upon receiving retrieval requests from the destination server, bistro i sends the file T_i along with the encrypted session key and ticket for decryption. Figure 4 depicts the data collection step.

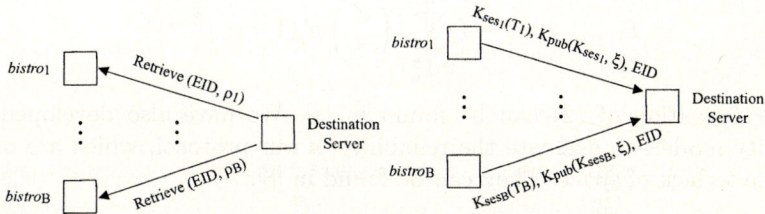

Fig. 4. Data Collection Step

When all packets within a checksum group are received, the destination server computes the checksum of the received checksum group. It then matches this checksum with what it received during the timestamp step. If these two checksums match, the destination server accepts all packets in the checksum group, and discards them otherwise.

After the destination server has retrieved data from all intermediate bistros, it passes the packets that pass the checksum check to an erasure code decoder, if it has received at least k packets from every FEC group. The erasure code decoder then reconstructs the original file T_o. If the destination server receives less than k packets from any FEC group, it contacts the appropriate clients and requests retransmissions of specific FEC group(s).

4 Analytical Models

We propose analytical models to evaluate our fault tolerance protocol in this section. We develop a reliability model to study how reliability characteristics of bistros affect system reliability. We also develop a performance model to estimate the performance penalty of employing our protocol. Lastly, we derive a cost function to study the tradeoff between reliability and performance.

4.1 Reliability Model

Let p_g be the probability that there is no loss within a checksum group. Recall that if a checksum check fails, all packets within that checksum group are discarded because we have no way of determining which of the packets are corrupted. Hence, the probability that at least one packet is lost within a checksum group is $1 - p_g$.

Let us assume that losing or corrupting one packet is independent of losing or corrupting other packets within the same checksum group. Let p be the probability that a packet is lost or corrupted. Then,

$$p_g = (1-p)^g. \tag{1}$$

Due to the lack of space, we omit the derivation of $P_{retrans}$, the probability that retransmission is needed, and simply state the result as follows:

$$P_{retrans} = 1 - (\sum_{i=\lceil \frac{k}{g} \rceil}^{Z} \binom{Z}{i} p_g^i (1-p_g)^{Z-i})^Y. \tag{2}$$

The derivation of (2) can be found in [8]. We have also developed other reliability models to evaluate the reliability of our protocol, which are omitted here due to lack of space. They can be found in [8].

4.2 Performance Model

This section describes the performance model used for evaluating our fault tolerance protocol. We limit the evaluation in this paper to the performance penalty in the timestamp step only. This is motivated by the fact that if the performance of timestamp step is poor, then we are back to the original problems where many clients are trying to send large amounts of data to a server at the same time.

We believe that it is more important to consider the potential overloading of network resources, due to sending a greater number of checksums, than the additional computational needs on the server for producing digital signatures of larger timestamp messages. This is again due to the consideration that clients sending large messages to the destination server around the deadline time would take us back to the original problem of a large number of clients trying to send large amounts of data to the server in a short period of time.

Hence, we use the number of checksum groups per data packet, $\frac{Z}{k}$, as our performance metric. This is derived by considering the total number of checksums, YZ, normalized by the file size, Yk.

4.3 Cost Function

Now that we have a reliability model and a performance model, the question is how to combine the effects of both in order to study the tradeoff between reliability and performance. This section describes a cost function which we propose to use to achieve this goal.

Let C_1 be the cost computed using the reliability model, and let C_2 be the cost computed using the performance model in the timestamp step. Thus, our cost function is

$$C = w_1 C_1 + w_2 C_2 \tag{3}$$

where w_1 and w_2 are weights of each factor.

Earlier we derived the probability that retransmission is needed, $P_{retrans}$, as a reliability metric. We can use this metric as our reliability cost. The performance cost in the timestamp step is given by the number of checksum groups per data packet, $\frac{Z}{k}$. In the next section, we study how reliability and performance metrics affect the overall cost function.

5 Numerical Results

This section provides numeric results on varying different parameters of our protocol and their effect on the cost function discussed above. Due to lack of space, we present only a subset of our experiments. Other results, which also support our findings, can be found in [8].

The parameters of interest to our system and its corresponding reliability and performance characteristics include the following.

1. Number of checksum groups per FEC group, Z. Setting Z to be large can provide better reliability because loss of a packet affects fewer other packets, as we drop the entire checksum group whenever any packet from that group is lost or corrupted. On the other hand, large values of Z result in large timestamp messages, which can have adverse effects on network resources.
2. Number of parity packets per FEC group, $n - k$. For reliability reasons, we want to send a large number of parity packets, but this increases the number of checksums we send as we are interested in adding parity checksum groups.
3. Number of data packets per FEC group, k. Given a file of W packets, we want to study the differences in dividing the file into few large FEC groups or many small FEC groups.
4. Probability of losing a packet, p. We want to see how sensitive the cost function is to p.
5. Weights w_1 and w_2. We are interested in how sensitive the cost function is to the chosen weight values. We performed a number of experiments with different weight values (please refer to [8] for the results). Due to lack of space, we omit these results here and only use representative weight values in the remainder of the paper.

The tradeoff between reliability and performance in the context of varying the number of checksum groups per FEC group, Z, is illustrated in Fig. 5(a), where $Y = 5$, $n = 20$, $k = 10$, $p = 0.01$, $w_1 = 0.9$, and $w_2 = 0.1$. In Fig. 5(a) the cost is high when Z is small because $P_{retrans}$ is high. The cost decreases when Z is between 1 and 4 because $P_{retrans}$ is improving. At $Z \geq 2$, the cost goes up again because the size of the message becomes too large.

We study the tradeoff between reliability and performance of adding parity packets, $n - k$ in Fig. 5(b), where $k = 10$, $p = 0.01$, $Z = 2$, $w_1 = 0.9$, and $w_2 = 0.1$. In Fig. 5(b), when $n - k \leq 10$, the cost decreases because $P_{retrans}$ decreases. At $n - k \geq 10$ in Fig. 5(b), the cost increases because $\frac{Z}{k}$ increases while $P_{retrans}$ approaches 0.

We study how we should choose k, the number of data packets in each FEC group, in Fig. 5(c). For this experiment we set $W = 100$, $n = 2k$, $Z = 2$, $p = 0.01$,

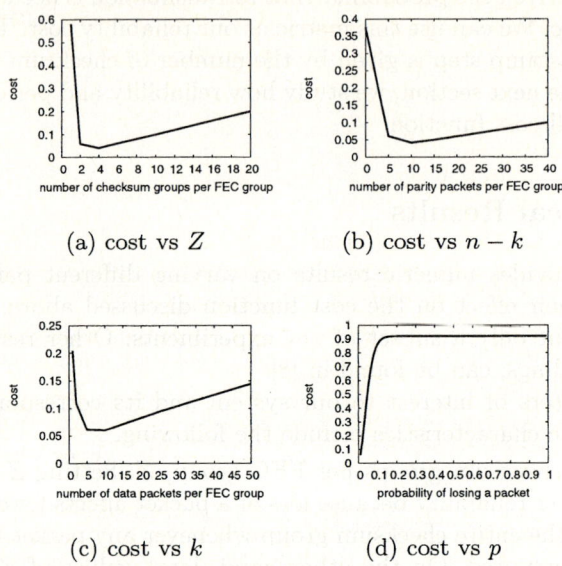

Fig. 5. Results for varying different parameters in the cost function

$w_1 = 0.9$, and $w_2 = 0.1$. Figure 5(c) shows that cost is high when k is small, because this results in a lot of checksums. The cost drops when k is between 1 and 10, as we send fewer checksums and the corresponding reliability penalty does not increase as fast. Eventually, when $k \geq 10$, cost goes up as k increases since larger FEC groups are not as fault tolerant.

We are interested in looking at how the cost function changes with the probability of losing a packet, p. We set $Y = 5$, $n = 20$, $k = 10$, $Z = 2$, $w_1 = 0.9$, and $w_2 = 0.1$ in Fig. 5(d). Since both Z and k are fixed, changes in cost reflect changes in $P_{retrans}$. Cost increases rapidly when p is between 0 and 0.1. When $p > 0.1$, since $P_{retrans}$ approaches 1, cost remains fairly constant.

6 Conclusions

Bistro is a scalable and secure wide-area upload architecture that can provide an efficient upload service. The goal of this paper was to develop a fault tolerance protocol that improves performance in the face of failures or malicious behavior of intermediaries in the context of the Bistro architecture. We developed such a protocol using a forward error correction technique. We also evaluated this protocol using proposed analytical models to study the reliability and performance characteristics. We studied the resulting cost, as a function of a number of parameters, including the number of data packets per FEC group, the number of parity packets, and the number of checksum groups per FEC group. In conclusion, we believe that fault tolerance is important in wide area data upload applications. We believe that the proposed protocol is a step in the right direction, leading to better fault tolerance characteristics with fewer retransmissions due to packet losses or corruptions, resulting in better overall system performance.

References

1. Bhattacharjee, S., Cheng, W.C., Chou, C.F., Golubchik, L., Khuller, S.: Bistro: a framework for building scalable wide-area upload applications. ACM SIGMETRICS Performance Evaluation Review **28** (2000) 29–35
2. Cheng, W.C., Chou, C.F., Golubchik, L., Khuller, S.: A secure and scalable wide-area upload service. In: Proceedings of 2nd International Conference on Internet Computing. Volume 2. (2001) 733–739
3. Patterson, D.A., Gibson, G., Katz, R.H.: A case for redundant arrays of inexpensive disks (raid). In: Proceedings of the 1988 ACM SIGMOD international conference on Management of data, ACM Press (1988) 109–116
4. Byes, J., Luby, M., Mitzenmacher, M., Rege, A.: A digital fountain approach to reliable distribution of bulk data. In: ACM SIGCOMM. (1998)
5. Ding, G., Ghafoor, H., Bhargava, B.: Resilient video transmission over wireless networks. In: 6th IEEE International Conf. on Object-oriented Real-time Distributed Computing. (2003)
6. McKinley, P., Mani, A.: An experimental study of adaptive forward error correction for wireless collaborative computing. In: IEEE Symposium on Applications and the Internet (SAINT 2001). (2001)
7. Cheng, W.C., Chou, C.F., Golubchik, L., Khuller, S.: A performance study of bistro, a scalable upload architecture. ACM SIGMETRICS Performance Evaluation Review **29** (2002) 31–39
8. Cheung, L., Chou, C.F., Golubchik, L., Yang, Y.: A fault tolerance for uploads: Design and evaluation. Technical Report 04-834, Computer Science Department, University of Southern California (2004)

Topological Adaptability for the Distributed Token Circulation Paradigm in Faulty Environment

Thibault Bernard, Alain Bui**, and Olivier Flauzac

LICA, Département de Mathématiques et Informatique,
Université de Reims Champagne-Ardenne, BP 1039,
F-51687 Reims Cedex 2, France
{thibault.bernard, alain.bui, olivier.flauzac}@univ-reims.fr

Abstract. In this paper, we combine random walks and self-stabilization to design a single token circulation algorithm. Random walks have proved their efficiency in dynamic networks and are perfectly adapted to frequent network topological changes. Self-stabilization is the most general technique to design an algorithm that tolerates transient failures. Taking account that the token circulates continually according to a random walk scheme, designing a self-stabilizing algorithm implies to solve two situations (1) no token in the system and (2) several tokens in the system. The former is generally solved by a time-out mechanism, upon timeout a new token is created. In this paper, we focus on this problem. Just state that one may choose a sufficiently long time-out period is not possible in our case: the system could never stabilize. Indeed, a random walk based token eventually cover the network but only the *expected* time to cover the network can be captured. Therefore, we introduce a mechanism "the reloaded wave propagation" to prevent unnecessary token creation and preserve self-stabilization properties.

1 Introduction

In distributed computing, the token circulation primitive is very useful for many applications. Election, spanning tree construction, mutual exclusion and several important tasks can be achieved using token circulation. This problem has been widely studied under different assumptions.

For instance, a single token which continually circulates through all the processors of a distributed system, can solved the mutual exclusion problem. The token circulation insures the liveness property - every processor enters the critical section infinitely often - and unicity of the token insures the safety property - at most one processor can be in the critical section at time. In a safe environment, these two properties are always true.

In this paper, we focus on the token circulation paradigm, for dynamic networks, including but no limited to mobile and ad-hoc networks.

** Corresponding author.

We use random walks , *i.e.* memoryless stochastic processes: the token message circulates in the system, and at each step, the processor that owns it sends it to one of its neighbors chosen uniformly at random. Random walks have proved their efficiency and are perfectly adapted to frequent network topological changes. Unlike others solutions *e.g.* [9, 3], no structure such (virtual) rings or trees have to be maintained. Random walk can be also used as an alternative method to flooding or broadcasting [6, 13]. An important result is that the token eventually visits all the processors of a system. But it is impossible to capture an upper bound on the time to visit all processors in the system, only bounds on the cover time, defined as the average time to visit all the processors are available.

In faulty environment, transient failures may occur. The concept of self-stabilization, introduced by [4], is the most general technique to design a system to tolerate arbitrary temporary faults. A self-stabilizing algorithm can be started in any global state which may occur due to failures. From any arbitrary starting state, it must be ensured that the task of the algorithm is accomplished. If during a sufficiently long period no further failures occur, the algorithm converge eventually to the intended behavior in finite time.

Designing a self-stabilizing token circulation token implies to solve two situations (1) the token lost situation and (2) the multiple tokens situation. In the latter case, tokens will merge to one in finite time [10]. Many papers with token circulation run within the state model and case (1) can be detect by reading neighbors variables. In the most general model - the message passing model - in which we are concerned, one main problem is to detect communication deadlock: processors are waiting for messages and there are no messages on communication links. A solution [7] is to use timeout.

In [14], the author proposes a message passing adaptation of Dijkstra algorithm [4]. In particular, a self-stabilizing token circulation algorithm on undirected ring is presented. Communication deadlock is solved by a timeout process in a distinguished processor called the root. Nevertheless, duplicate tokens may occur. The author introduces the counter flushing paradigm to solve this problem and thus, design a self-stabilizing token circulation algorithm. The idea of counter flushing is applied by numerous papers dealing with self-stabilization in message passing model, as [3, 8].

In [6], authors use a random walk of a mobile agent to achieve self-stabilizing communication group in ad-hoc networks. The agent (which behaves as a token) is used for broadcasting. The paper assumes the existence of a single agent in the system. A timeout mechanism is used for creation of a new agent when no agent exists (similarly to a communication deadlock). The authors suggest to choose a *"long time-out period tpi (that is function of the cover time for a graph of N nodes) to produce an agent"* and *"to avoid simultaneous creation of many agents one may choose tpi to be also a function of the identifiers [processor] pi [...] we may choose to assign $tpi = i \times C$ where i is the identifier of pi [and C is the cover time]"*. The guarantee that in finite time, eventually there exists only one agent is not clearly highlighted in [6] since C is defined as an expected time.

In this paper, we work within a general topology. As we discuss previously, we use a random walk token circulation as solution for topological changes for dynamic networks. To cover the communication deadlock problem, the self-stabilizing version of this algorithm use a decentralized timeout procedure. In a configuration where no tokens exists in the system, each processor indistinctly has the possibility of producing a new token. Simultaneous or multiple creation of token can occur in some very specific cases. But our algorithm guarantees that the system eventually stabilizes (with exactly one token). We introduce a new mechanism: the "reloading wave". Each processor maintains a timer. Upon timeout each processor may produce a token. The reloading wave prevents from creating unnecessary token: each processor that has been previously visited by the token, will not trigger the timeout and its timer will be reseted. In fact after the reloading wave procedure, the only case where a processor will produce a new token after the timeout period, corresponds to a illegitimate configuration (the processor has not been yet visited by the token). This mechanism is described in next section. There is no extra cost for self-stabilizing in terms of message complexity except the reloading wave propagation (at most $n - 1$ messages). To build the reloading wave, we use the informations collected and stored by the token through its traversal.

2 Preliminaries

Distributed Systems. A distributed system is an undirected connected graph $G = (V, E)$, where V is a set of processors with $|V| = n$ and E is the set of bidirectional communication links with $|E| = m$. (We use the terms "node", "vertex", "site" and "processor" interchangeably). A communication link (i, j) exists if and only if i and j are neighbors. Every processor i can distinguish all its links of communication. Each of them maintains a set of neighbors (denoted as N_i). The degree of i is the number of neighbors of i, i.e. $|N_i|$ (denoted as $deg(i)$). We consider a distributed system where all sites have distinct identities. We assume an upper bound $\mathcal{N}$ on the number of sites in the network, an upper bound Δ on the delay to deliver message and an upper bound Θ on processing time on each site. Moreover, we consider to have reliable channels during and after the stabilization phase.

Failures and Self-Stabilization. A transient fault is a fault that causes the state of a process (its local state, program counter, and variables) to change arbitrarily. An algorithm is called self-stabilizing if it is resilient to transient failures in the sense that, when started in an arbitrary system state, and no other transient faults occur, the processes converge to a global legal state after which they perform their task correctly (see [4, 5]).

$\mathcal{C}$ being the set of all configurations in the system, an algorithm is self-stabilizing if there is a set of legal configurations LC such as: (1) The system eventually reaches a legal configuration (convergence). (2) Starting from any legitimate configuration, the system remains in LC (closure). (3) Starting from any legitimate configuration, the execution of the algorithm verifies the specification of the problem.

Random Walks. A random walk is a sequence of vertices visited by a token that starts at i and visits other vertices according to the following transition rule: if the token is at i at time t then at time $t+1$, it will be at one of the neighbors of i, this neighbor is chosen uniformly at random among all of them [11,1]. Similarly to deterministic distributed algorithms, the time complexity of random walk based token circulation algorithms can be viewed as the number of "steps" it takes for the algorithm to achieve the network traversal. With only one walk at a time (which is the case we deal), it is also equal to the message complexity. The cover time C — the average time to visit all nodes in the system — and the hitting time denoted by h_{ij} — the average time to reach a node j for the first time starting from a given node i — are two important values that appear in the analysis of random walk-based distributed algorithms.

3 Self-Stabilizing Token Circulation Algorithm

Dealing with Communication Deadlocks. A simple single token circulation using a random walk scheme is sufficient to solve the token circulation problem under topological changes assumptions. The procedure is simple: if the token is owned by a site i at time t, at time $t+1$, it is owned by each neighbor j of i with probability $1/deg(i)$. Thus, topological changes are easily and naturally managed. Suppose the token is in a site i. If a channel (i,j) becomes unavailable (and G still connected), eventually the token will hit j for the first time after the expected time h_{ij} (h_{ij} being the hitting time defined in previous section).

The token eventually visits all the sites in the system and the expected time to achieve this task is the cover time. But a given site will receive the token in finite but unbounded time.

Consequently, a solution to a communication deadlock configuration directly inspired by [14,6] is not adapted. The choice of an "enough long timeout period" to produce new tokens, compromises the closure property of our self-stabilizing algorithm. For the same reason, the choice of a distinguished site which would take charge of the main tasks, in any case provides nothing additional. Thus, our algorithm is totally decentralized and each site maintains a timeout to send a token periodically. A timer is associated to each site. We choose to set this value at $\max_{\forall (i,j) \in V^2}(h(i,j)) \times (\Delta + \Theta)$ units of time. This value corresponds to the worst expected time for the token starting at site i to reach for the first time any site j.

During a legal execution of the algorithm (without failures), when a given site i is about to produce a new token (near timeout), it receives an information from a wave, the reloading wave. i is informed that there exists a token in the system and that i has been previously visited by the token.

The reloading wave is propagated under the following conditions: the token maintains itself a counter which is incremented at each token hop. The counter is set to 0 at token creation. This value is compared to the timeout value $Tmax = \max_{\forall (i,j) \in V^2}(h(i,j)) \times (\Delta + \Theta)$ minus the time to achieve a wave propagation (at worst case $\mathcal{N}$). When this counter value is superior or equal to the latter value, the site that holds the token launches the wave and the token counter is reseted to 0.

When a site receives the wave, it reloads its timer to $Tmax$.

Thus, a simple low cost solution to the deadlock communication problem is proposed.

In presence of multiple tokens, by [10], these tokens will merge into one.

Reloading Wave Propagation. We maintain a dynamic self-stabilizing tree [2] through the system to propagate the reloading wave. There is no additional protocol, we use the token contents. The token collects and stores identities of each site, thus its contents can be viewed as the history of the token's moves. Such a token is called a circulating word. Each time the token meets a site i, a tree rooted on i is locally computed by i using the topological informations stored in the circulating word. Although the token continually circulates through the network, the token size is bounded by $2n-1$, where n is the number of sites in the system (cf. [2]). In this paper, we use another representation of the topological information: a father-son relation table. As improvement, the token size is now bounded by n.

The following example (Fig. 1) exhibits a father-son relation table and its associated tree.

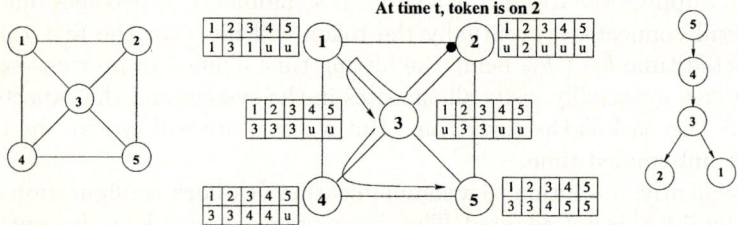

Fig. 1. Network, Father-son relation table update through random moves, and Tree (constructed on site 5)

The tree computation is achieved by updating the data structure stored in the token each time it moves. If the token moves from j to i, the update is done as follow: (i is the currently visited site, and j is the previously visited site)

1. i becomes the new root of the computed tree.
2. j becomes the son of i.

This gives the following algorithm:

Algorithm 1 Algorithm on node p

$token.table[p] \longleftarrow p$, $token.table[sender] \longleftarrow p$
sends $token$ to a site chosen randomly among N_p

$token.table$ denotes the data structure in the token.

Transient failures and topological changes can produce inconsistent token. Each time a site receives the token, it can check and locally correct token consistency by applying "Internal test procedure" Algorithm 4.

Internal Test Example. Site 5 receives the token (from the previous network). Site 5 checks the table consistency. It detects that site 1 could not be its son on the tree, since site 1 is not in its neighborhood. Locally, site 5 eliminates the subtree rooted in 1.

Node	1	2	3	4	5
Father	5	1	1	5	5

Before the test

Node	1	2	3	4	5
Father	u	u	u	5	5

After the test

As we discuss previously, several tokens can occurred in the system. In finite time, they merge into one. To improve convergence, the algorithm also merges topological informations: a single tree is built thanks to the two data structures stored in both tokens. See Algorithm 3 for details.

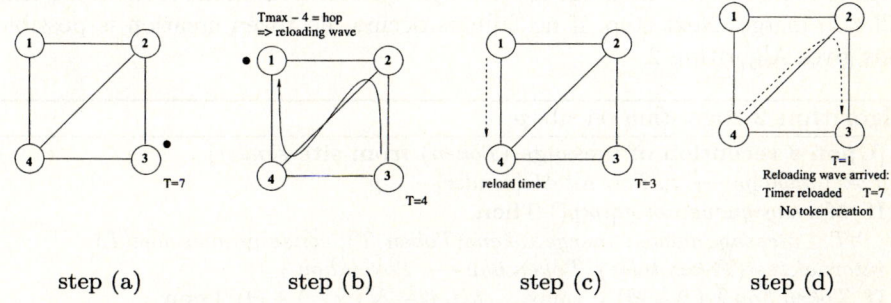

Fig. 2. Example

In Fig. 2, we give an ewample of the reloading wave in a legal configuration.

step (a) The token is in 3, the timer value is 7 and token counter is 0.
step (b) After 3 hops the token reaches site 1. The condition $Tmax - n = hop$ to launch the reloading wave is true.
step (c) In our case, the dynamic tree is the chain 1-4-2-3. Site 4 receives the wave and reload its timer.
step (d) Site 3 receives the wave. The timer of value 1 is reloaded to Tmax =7. No token is created.

In Fig. 3, we illustrate how a token creation is possible when a site is not contained on the tree built.

step (a) Site 3 has not been yet visited by the token. Token is created in site 2. Timers values are 7 for all sites
step (b) The token moves randomly through site 2, 4, 1, and back to 2. Wave launch condition is true.

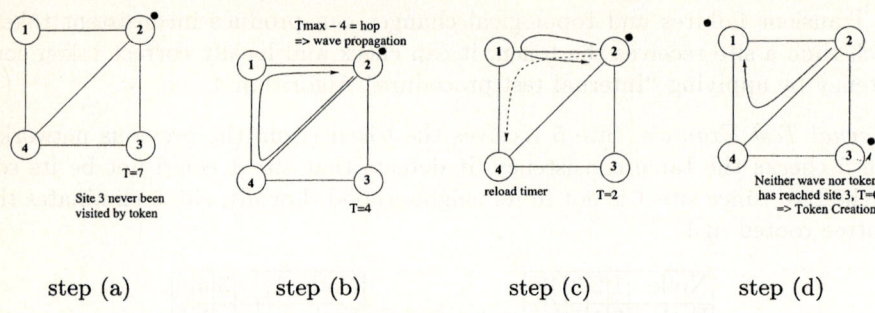

step (a) — step (b) — step (c) — step (d)

Fig. 3. Example

step (c) The wave reloads all sites timers except 3 (it is not contained in the tree)
step (d) Upon a timeout site 3 creates a new token.

In finite time, tokens will merge to one, topological informations of the two tokens will also merge. Next step, if no failures occurs, no token creation is possible. This gives Algorithm 2.

Algorithm 2 Algorithm on site p

[**Upon a reception of message** ($Token$) **from site** $sender$]
$Token.table[p] \longleftarrow p$, $Token.table[sender] \longleftarrow p$
If $Messagequeue\_not\_empty()$ **Then**
$\quad \forall T \in message\ queue : merge\_tokens(Token, T), consume\_message(T)$
$internal\_test(Token.table)$, $Token.hop \longleftarrow Token.hop + 1$
If $Token.hop \times (\Delta + \Theta) \geq (\max_{(i,j)} h(i,j) - \mathcal{N}) \times (\Delta + \Theta)$ **Then**
$\quad \forall v \in son_p$ (respecting $Token.table$): send $reload, subtree(Token.table, v)$) to v
$\quad Token.hop \longleftarrow 0$
Send $Token$ to i chosen randomly in $N(p)$, Reload the timer to $\max_{(i,j)} h(i,j)$
[**Upon a release of timer**]
For $i = 0$ to $\mathcal{N}$ **Do**
$\quad Token.table[i] \longleftarrow undefined$
$Token.table[p] \longleftarrow p$, $Token.hop \longleftarrow 0$
Send $Token$ to i chosen randomly in $N(p)$, Reload the timer to $\max_{\forall (i,j) \in V^2} h(i,j)$
[**Upon a reception of message** ($reload, table$)]
$\forall v \in son_p$ (respecting the tree $table$): send $reload, subtree(table, v)$) to v
Reload the timer to $\max_{(i,j)} h(i,j)$

This algorithm uses the following functions:

- $Messagequeue\_not\_empty$ returns true if another message is present on the site.
- $merge\_tokens$ merges the topological informations of the tokens, by adding unknown topological informations to the token. The number of hop is the maximum of the merging tokens. The result of the merger is placed in token Tok. See Algorithm 3.

- *internal_test* tests the consistency of the table, relatively to the neighborhood relation of the site p. If any node i is described as son of p in the table but is not in N_p, all the associations (identity-father) of the subtree whose root is i are deleted. See Algorithm 4.
- *subtree*(t, v) returns the subtree which root is v constructed with the table t. See Algorithm 5.

Algorithm 3 Merge token Algorithm

Procedure: merge_tokens(Tok: token T : token)
$Tok.hop = max(Tok.hop, T.hop)$, $T.table[p] \longleftarrow p$, $T.table[sender] \longleftarrow p$
For $k = 0$ to N Do
  If $(Tok.table[k] = undefined) \cap (T.table[k] \neq undefined)$ Then
    $Tok.table[k] \longleftarrow T.table[k]$

Algorithm 4 Internal Test Algorithm on site p

Procedure: internal_test(table : array)
for all $i = 0$ to $\mathcal{N}$ Do
  $list \longleftarrow \phi$
  If $(table[i] = p) \cap (i \notin \{N_p \cup \{p\}\})$ Then
    $list \longleftarrow list \cup \{i\}$
  for all $h \in list$ Do
    For $j = 0$ to $\mathcal{N}$ Do
      If $table[j] = h$ Then
        $table[j] = undefined$, $list \longleftarrow list \cup \{j\}$
    $list \longleftarrow list - \{h\}$

Algorithm 5 Subtree Algorithm

Procedure: subtree(table : array, v : integer)
$\forall i, table'[i] \longleftarrow undefined$
$table'[v] \longleftarrow v$, $list \longleftarrow \{v\}$
for all $h \in list$ Do
  For $j = 0$ to $\mathcal{N}$ Do
    If $table[j] = h$ Then
      $table'[j] = h$, $list \longleftarrow list \cup \{j\}$
  $list \longleftarrow list - \{h\}$
return $table'$

Proof Outline

We claim that algorithm 2 is self-stabilizing for the predicate $\mathcal{PL}$: *there is exactly one token with a complete and consistent table (and token counter consistent with timer), in the system*. A father-son relation (FS) table is said *consistent* if it permits the construction of a tree over the network. A FS table is said *complete*, if all sites have a defined father in the FS table.

Starting from any configuration, any execution of the algorithm satisfies the properties of the random token circulation problem. We prove that every execution starting from an arbitrary configuration, a legitimate configuration is reached - convergence - a configuration from which its satisfies $\mathcal{PL}$ forever - closure.

A site visited by the token will be reached by the reloading wave. Then it cannot trigger its timer and new token creation is impossible. Thus *when all sites have been visited by one token, no new token can be created.*

By [10, 12], and Algorithm 3, at most in expected time $8n^3/27$, all token eventually merge. All merging topological information are preserved. If there is no token in the system, then upon a timeout (cf. Algorithm 2), one or several tokens are created and next they merge to one. Then, *starting from any initial configuration, the system eventually reaches a configuration where there is exactly one token.*

Once the token has cover the network, it becomes consistent (all FS relations are updated) and since all sites have been visited, the token is complete.

We conclude that *starting from any arbitrary configuration, the system eventually satisfies* $\mathcal{PL}$.

To prove the closure property, we observe that token creation is impossible for an execution starting from a legitimate configuration. Each site is infinitely often visited by a complete and consistent token, which states the closure property and the correctness of the algorithm.

4 Conclusion

In this paper, we discuss on the token circulation paradigm in distributed computing. The token moves thanks to a random walk scheme, to provide a simple and efficient solution to network topological changes. Our protocol works on arbitrary networks, is fully distributed, and without specific structure to maintain. Our protocol is also self-stabilizing. To this aim, we focus on the communication deadlock. We use a timeout mechanism combined with the reloading wave to prevent unnecessary token creation. Finally, we prove the convergence and closure properties with the additional result of [10].

References

1. R. Aleliunas, R. Karp, R. Lipton, L. Lovasz, and C. Rackoff. Random walks, universal traversal sequences and the complexity of maze problems. In *FOCS 79*, pp 218–223, 1979.
2. T Bernard, A Bui, and O Flauzac. Random distributed self-stabilizing structures maintenance. In *ISADS'04, LNCS 3061*. pp 231 – 240. Springer, 2004.
3. Y Chen and JL Welch. Self-stabilizing mutual exclusion using tokens in mobile ad hoc networks. In *Proceedings of the 6th international workshop on Discrete algorithms and methods for mobile computing and communications*, pp 34–42. ACM Press, 2002.
4. EW Dijkstra. Self stabilizing systems in spite of distributed control. *CACM*, 17(11):643–644, 1974.
5. S Dolev. *Self-Stabilization*. MIT Press, 2000.
6. Shlomi Dolev, Elad Schiller, and Jennifer L. Welch. Random walk for self-stabilizing group communication in ad-hoc networks. In *SRDS*, 2002.

7. MG Gouda and N Multari. Stabilizing communication protocols. *IEEE Transactions on Computers*, 40(4):448–458, 1991.
8. R. Hadid and V. Villain. A new efficient tool for the design of self-stabilizing l-exclusion algorithms: the controller. In *WSS'01, LNCS 2194*. Springer, 2001.
9. L Higham and S Myers. Self-stabilizing token circulation on anonymous message passing. In *OPODIS'98*, pp 115–128, Hermes. 1998.
10. Amos Israeli and Marc Jalfon. Token management schemes and random walks yield self-stabilizing mutual exclusion. In *ACM PODC 90*, pp 119–131, 1990.
11. Laszlo Lovasz. Random walks on graphs: A survey. In , *Combinatorics: Paul Erdos is Eighty (vol. 2)*, pp 353–398. Mathematical Society, 1993.
12. P. Tetali and P. Winkler. On a random walk problem arising in self-stabilizing token management. In *10th ACM PODC 91*, pp 273–280, 1991.
13. D. Tsoumakos and N. Roussopoulos. A comparison of peer-to-peer serach methods. In *Sixth International Workshop WebDB'03*, 2003.
14. George Varghese. Self-stabilization by counter flushing. In *SIAM J. Computing (vol.30)*, pp 486–510, 2000.

Adaptive Data Dissemination in Wireless Sensor Networks[*]

Jian Xu[1,2], Jianliang Xu[1], Shanping Li[2], Qing Gao[1,2], and Gang Peng[2]

[1] Hong Kong Baptist University,
Kowloon Tong, Hong Kong
{jianxu, xujl, qgao}@comp.hkbu.edu.hk
[2] Zhejiang University,
Hangzhou, 310027, P.R. China
shan@cs.zju.edu.cn, e_pglmary@hotmail.com

Abstract. Recent years have witnessed growing research interest in wireless sensor networks. In the literature, three data storage strategies, i.e., *local*, *data-centric*, and *index-centric*, have been proposed to answer one-shot queries originated from inside the network. Through careful analysis, we show in this paper that each of the three strategies respectively achieves the best performance in a certain range of query-rate to update-rate (Q2U) ratios. We therefore propose several adaptive schemes which periodically adjust the data storage and dissemination strategy in accordance with the Q2U ratio observed in the network. Experimental results demonstrate that in dynamic environments the proposed adaptive schemes substantially outperform the three basic strategies.

1 Introduction

Recent advances in embedded systems and wireless communications have enabled the development and deployment of wireless sensor networks. A sensor network usually involves a large number of small, battery-powered sensor nodes scattered on an operational area to collect fine-grained, high-precision sensing data. With features of low cost and easy deployment, sensor networks have a wide range of applications in various domains such as habitat monitoring, surveillance, health care, and intruder detection [1], [2]. However, a wireless sensor network is constrained by limited battery power and scarce wireless bandwidth. Thus, many studies have been carried out to develop scalable and energy-efficient data dissemination schemes [3, 4].

In the literature, three data storage strategies, i.e., *local*, *data-centric*, and *index-centric*, have been proposed to answer one-shot queries originated from inside the network (see Fig. 1) [5]. With *local* data storage, the sensing data is stored on the local node that captures it. To answer a query, the query is flooded to all nodes in

---

[*] This work was supported in part by a grant from Hong Kong Baptist University (Grant FRG/04-05/I-17).

the network to find out the desired data. With *data-centric* storage, the sensing data is hashed, based on some pre-defined keys, to some geometric location and stored on the node nearest to that location (called *centric storage node,* or simply *centric node*); all data with the same hashed value will be stored on the same node. A key-based query is hashed in the same way as the sensing data, hence, it can be resolved on the corresponding centric node. With *index-centric* storage, the actual sensing data is stored on the local sensor node, and its search key value and a pointer to the local node are stored on the centric node. A query first goes to the centric node and then follows the location pointer to retrieve the actual data on the corresponding local node.

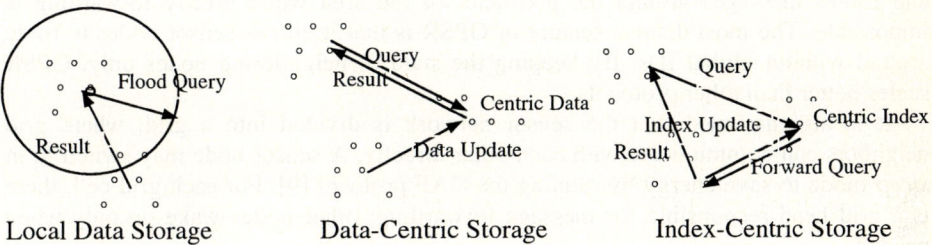

Fig. 1. Three Basic Data Storage and Dissemination Strategies

In this paper, we first analyze the performance of these storage and dissemination strategies in terms of total network traffic (which transfers to energy consumption). Through both theoretical and simulation-based studies, we find that each of these three strategies respectively achieves the best performance in a certain range of query-rate to update-rate (Q2U) ratios. Based on this observation, this paper proposes adaptive data dissemination (ADD) schemes to minimize the network traffic and, hence, to extend the network lifetime in dynamic environments. Under the adaptive schemes, the system periodically estimates the Q2U ratio and adjusts the data storage and dissemination strategy in accordance with the observed Q2U ratio. We address two practical issues in the implementation of an ADD scheme: 1) how to estimate the Q2U ratio periodically; 2) when to switch from one storage strategy to another. Extensive simulations are conducted to evaluate the performance of the proposed adaptive schemes. The results show that they substantially outperform the basic strategies when the access pattern changes over time.

The rest of the paper is organized as follows. Section 2 presents the background of this study. Section 3 describes the main design of the ADD schemes. Simulation results are presented in Section 4. Section 5 reviews the related work in this field. Finally, Section 6 concludes the paper.

2 Background

In this paper, we focus on a large-scale sensor network that covers a wide region. Some targets (e.g., animals) are moving within the region. The sensor nodes in the network detect the status of each target, which usually means an event of interest, and generate sensing data to report the status of a target. Users are also making movements within the region. From time to time, a user may want to retrieve the current status of a target via a nearby sensor node. The node where the sensing data is stored is called a *source*. The node via which the user retrieves the status is called a *sink*.

We assume that all sensor nodes are aware of their geographic locations. This can be achieved through the use of GPS [6] or some other localization techniques such as triangulation [7]. This is a reasonable assumption because in many cases the sensing data is useful only if the source location is known. We assume that the GPSR routing protocol [8] is employed to forward messages in the network. The GPSR makes greedy forwarding decisions using information about a node's immediate neighbors, and routes messages around the perimeter of the area while greedy forwarding is impossible. The most distinct feature of GPSR is that it allows sensor nodes to route around without global IDs. By keeping the state of neighboring nodes only, GPSR scales better than other protocols.

It is also assumed that the sensor network is divided into a grid, where grid neighbors can communicate with each other directly. A sensor node may switch to in *sleep* mode to save energy by running the GAF protocol [9]. For each grid cell, there is a grid head responsible for message forwarding; other nodes wake up only when needed.

3 Adaptive Data Dissemination

3.1 Analysis of Basic Strategies

As discussed in the Introduction, there are three basic data storage and dissemination strategies: local, data-centric, and index-centric storage. Under the local storage, the source stores sensing data on its local storage; a sink retrieves the sensing data by flooding the query throughout the whole network. Obviously, this strategy is the best choice when the query rate is very low. When the update rate is low, data-centric storage might be a better choice. In this case, the sensing data is pushed to some centric node using GHT [5], which hashes a search key value into a geographic location and maps it to the nearest sensor node. A query is hashed in the same way to locate the node containing desired data. Another alternative, index-centric storage, might be attractive for moderate update and query rate. As the name suggests, only the index of sensing data (rather than actual data) is pushed to some centric node. A sink always routes queries to appropriate centric index nodes, from where the queries will be forwarded to the source by following the location pointers. The source sends the result back to the sink when it receives a query.

We now quantitatively analyze the performance of each data storage strategy. Similar to the previous work [10], [11], this paper uses the metric of total network traffic to quantify the communication overhead. We consider a sensor network with $n \times n$ nodes distributed in some random grid topology. To facilitate the analysis, we introduce several notations. Denote by N the total number of sensor nodes in the network, i.e., $N = n \times n$. We use the fact that the asymptotic cost of a flooding is $O(N)$ and that of direct routing between two nodes randomly selected in the network is $O(\sqrt{N})$. Let r_q be the query rate and r_u be the data update rate. Let s_q, s_i, and s_d be the sizes of a querying message, an index updating message, and a data messages, respectively. In most cases, we have $s_d > s_q$ and $s_d > s_i$. In the following, we derive the network traffic in a unit time for each basic data storage strategy.

a) Local Data Storage (LS)
The source stores sensing data locally and hence incurs no update cost. A query is flooded to all nodes at a cost of $O(N)$. The result is sent back to the sink from the source at a cost of $O(\sqrt{N})$. Therefore, the query will produce a traffic of $O(Nr_q s_q)$ and the return of result will generate a traffic of $O(\sqrt{N} r_q s_d)$. The total traffic in a unit time is given by:

$$LS: \quad O(Nr_q s_q + \sqrt{N} r_q s_d) \tag{1}$$

b) Index-Centric Storage (IC)
The source updates the index stored on some centric node at a cost of $O(\sqrt{N})$ when an update on the search key field(s) occurs. A query is forwarded by the centric node to the source at a cost of $O(2\sqrt{N})$ and the return of result at $O(\sqrt{N})$. The total traffic is thus given by:

$$IC: \quad O(\sqrt{N} r_u s_i + 2\sqrt{N} r_q s_q + \sqrt{N} r_q s_d) \tag{2}$$

c) Data Centric Storage
The source sends the data to some centric nodes at a cost of $O(\sqrt{N})$ when an update occurs. A query is resolved also at a cost of $O(\sqrt{N})$. The same cost is incurred to return the result. The total traffic is given by:

$$DC: \quad O(\sqrt{N} r_u s_d + \sqrt{N} r_q s_q + \sqrt{N} r_q s_d) \tag{3}$$

We also observe that in a grid network, where a node can reach its vertical or horizontal neighbors only, the routing hops between two nodes under GPSR can be estimated. Specifically, in an $m \times n$ grid network, the average number of hops for routing from a node at coordinate $[p, q]$ to any other nodes is:

$$\sum_{i=1}^{m} \sum_{j=1}^{n} (2\min(|i-p|, |j-q|) + \||i-p| - |j-q|\|) \Big/ (m \times n) \tag{4}$$

Based on (4), we can estimate the average number of routing hops between two nodes randomly selected in an $n \times n$ grid network. We have conducted simulation to verify the analytical result using the simulator ns-2 [13]. In the simulation, 500 pairs of nodes are randomly selected to communicate. In Fig. 2, we show the average numbers of hops obtained from the simulation, the analytical result from (4), and the approximation of $SQRT(N)$ (i.e., $O(\sqrt{N})$), respectively. We can see that the simulation result agrees with the analytical result. It is also observed that the average number of hops is about $0.66\sqrt{N}$.

Similarly, the average number of routing hops from a fixed centric node to another randomly selected node in the network can be obtained with (4). Fig. 3 shows the comparison of routing hops from the node located at the geometrical center of the network to any other nodes. The ratio of the actual routing hops to the value of *SQRT(N)* is about 0.50. We can easily extend this result to any other fixed centric node.

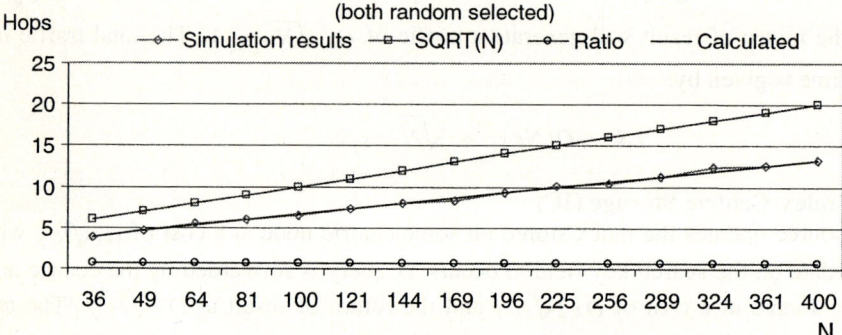

Fig. 2. Average Hops Between Two Nodes

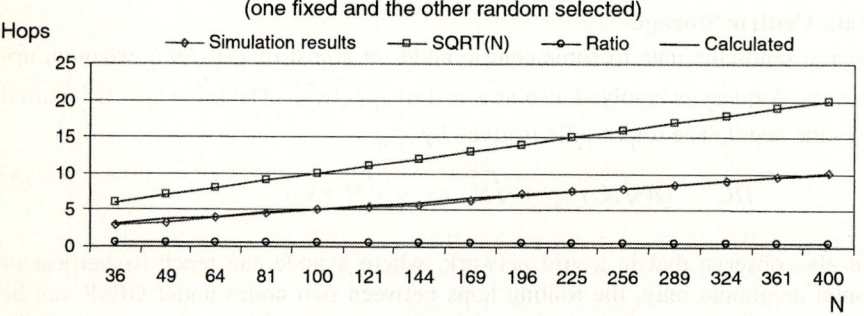

Fig. 3. Average Hops Between Two Nodes

Without loss of generality, suppose the centric node of the search key value is the geometrical center in the network. The network traffics can be re-written as follows:

$$LS: \quad Nr_q s_q + 0.66\sqrt{N} \; r_q s_d \tag{5}$$

$$IC: \quad 0.50\sqrt{N} \; r_u s_i + \sqrt{N} \; r_q s_q + 0.66\sqrt{N} r_q s_d) \tag{6}$$

$$DC: \quad 0.50\sqrt{N} \; r_u s_d + 0.50\sqrt{N} \; r_q s_q + 0.50\sqrt{N} \; r_q s_d) \tag{7}$$

In the following, we analyze the thresholds of query-rate to update-rate (Q2U) ratios for which one strategy dominates another:

$$LS\text{-}IC: \qquad r_q/r_u = 0.50\sqrt{N}s_i/(N-\sqrt{N})s_q \qquad (8)$$

$$LS\text{-}DC: \qquad r_q/r_u = 0.50\sqrt{N}s_d/(N-0.50\sqrt{N})s_q \qquad (9)$$

$$IC\text{-}DC: \qquad r_q/r_u = 0.50(s_d - s_i)/(0.50s_q + 0.16s_d) \qquad (10)$$

For a grid network of 256=16×16 nodes, if the data message size s_d is 50 bytes, and both the query and index message sizes s_q, s_i are 10 bytes. The threshold values for (8), (9) and (10) are 0.03, 0.16, and 1.54, respectively. The optimal storage strategy with respect to a certain Q2U ratio is illustrated as follows:

Q2U Ratio: 0.03   0.16   1.54
                    Low|___ ¦ __Middle_____|_High_____
Best Storage Strategy: LS | IC | DC

3.2 Proposed Adaptive Algorithms

As discussed, each of these three strategies respectively achieves the best performance in a certain range of Q2U ratios. Therefore, we propose several adaptive data storage and dissemination schemes to minimize the network traffic for dynamic access patterns. We evaluate the access pattern periodically. Denote by L the interval between two consecutive evaluations. The data storage nodes are responsible for collecting query and update rates during an evaluation interval.

To predict future access patterns, we rely on historical data accesses. Based on different prediction methods, we propose three adaptive algorithms. The first two estimate the Q2U ratio based on the history in the immediate past two intervals. We classify the ratio into three classes: L (low), M (moderate), and H (high), according to the analysis presented in the last section. The transition of storage strategy in the next interval follows the conditions shown in Fig. 4. The third predicts the Q2U ratio based on a simple exponential aging method. The final projected value is the predicted ratio plus a prediction error, which is the difference between the projected ratio and the observed ratio for the immediate past interval. The details of this algorithm are described in Algorithm 3.

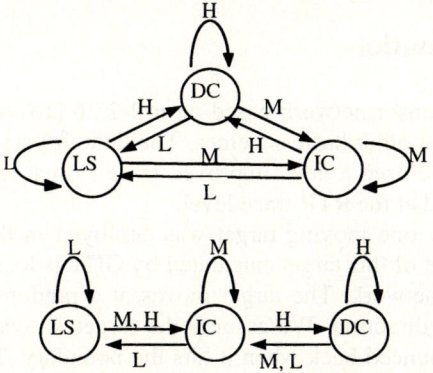

Fig. 4. Strategy Transition in **Algorithm 1** and **Algorithm 2**

Table 1. Simulation Parameters

	Parameter	Value
Topology	Random Grid	16×16
	Grid Size	19.5m
	field Size	595×595m
GPSR Routing	Beacon Interval	3.0s
	Beacon Expiration	13.5s
	Implicit Beacon	Yes
	Planarization	Yes
Radio	Range	40m
	MAC	802.11
Message Size	Index update	10 bytes
	Data update/Result	50 bytes
	Query	10 bytes

Algorithm 3: Adaptive Algorithm based on Exponential Aging Prediction

    Collect current query update ratio r_ratio_i ;
    Calculate $f\_ratio(i+1)$ using **(11)**;
    Switch $f\_ratio(i+1)$ {
        Case *L*:   Local storage in next interval; break;
        Case *M*:   Index centric storage in next interval; break;
        Case *H*:   Data centric storage in next interval; break;
    }
Exponential aging prediction:
Real query update ratio: $r\_ratio_{(i-1)}$, r_ratio_i, $r\_ratio_{(i+1)}$,
Predicted Q2U ratio: $f\_ratio_{(i-1)}$, f_ratio_i, $f\_ratio_{(i+1)}$,
Evaluation interval: __(i-1)__ |___i___|___(i+1)__

$$f\_ratio_{(i+1)} = f\_ratio_i + a\Box(f\_ratio_i - r\_ratio_i) \quad (11)$$

4 Performance Evaluation

We developed a wireless sensor network based on ns2-2.26 [13] to evaluate the proposed adaptive schemes and other basic schemes. The default parameters settings are summarized in Table 1. We used a new, improved trace format in our simulation to collect messages transferred at the RTR trace level.

    For simplification, only one moving target was deployed in the sensor network. We assume the centric node of this target calculated by GHT is located at the geometrical center of the sensor network. The target moves at a random speed lower than 50m/s and with a random direction. We assume the target is aware of the network boundary, and it will be bounced back when it hits the boundary. The target can produce stimulus from time to time. The frequency of generating stimulus is controlled in the simulation to facilitate performance comparison. At the beginning of each evaluation interval, the centric node makes storage decisions. If there is a strategy transition, the centric node will broadcast it throughout the network. Then, the storage strategy is changed on every node in the network. All sensor nodes are stationary and distributed over a 595×595 m^2 flat field, which is divided into GAF grid cells.

The simulation results to be presented in this section include two parts. The first part is the comparison of local storage, index-centric, and data-centric storage under different Q2U ratios. The second part is the comparison of different adaptive algorithms with various evaluation intervals.

4.1 Comparison of Local Storage, Index Centric and Data Centric Storage

In this subsection, we show the comparative results of total traffic produced by local storage, index-centric, and data-centric storage schemes. The data update rate is fixed throughout the experiments. We vary the query rate and obtain different Q2U ratios. The sink node which issues the query is randomly selected. All simulations last for 1000 seconds. We repeat experiments three times with different random queries for each Q2U ratio and average the total traffics of the three repetitions as the result.

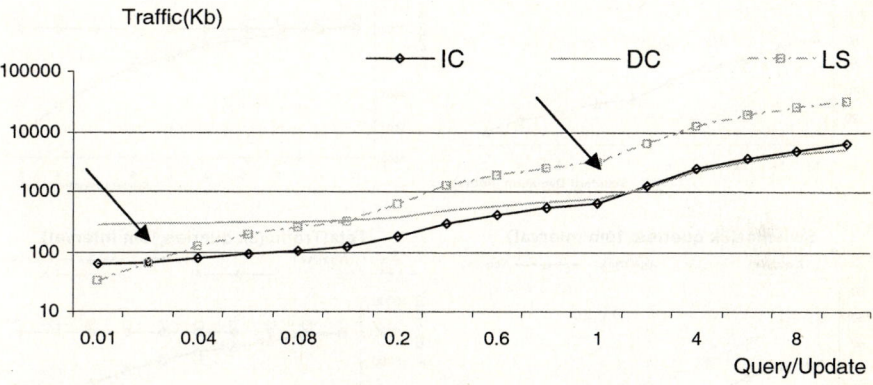

Fig. 5. Comparison of LS, IC and DC

As shown in Fig. 5, when the Q2U ratio is high, local storage is shown to have the highest total traffic. This is because the cost saving in updates brought by local storage is compromised by queries, since it needs to flood query messages to the whole network. Data-centric storage has the lowest total traffic in all three storage schemes when the Q2U ratio is higher than 1.54. In the range of the ratio between 0.03 and 1.54, index-centric storage achieves the best performance. The results are consistent with the analysis carried out in Section 3.

4.2 Comparison of Different Adaptive Algorithms and Evaluation Intervals

In this subsection, we compare different adaptive schemes with various evaluation intervals. We conduct three sets of experiments with 3K, 60K, and 120K queries in 24 hours. The evaluation intervals are set at 5, 10, and 30 minutes. The data update rate is fixed throughout the simulations. In order to simulate dynamic access patterns, we assume that queries arrive following the Gauss distribution with the peak hour at 12pm and the standard deviation of 3-12 hours.

Fig. 6 shows the results for 3K queries. The total traffic for data-centric storage in these simulations is too large (more than 36Mb) to be shown in the figure. The total traffics of local storage and index-centric storage strategies are kept constant throughout the settings of query arrival deviation. The adaptive schemes always achieve the least traffic since they adapt the storage scheme to the current Q2U ratio. The number of storage strategy transition decreases when the evaluation interval gets longer as showed in the figure. When the query arrival deviation gets smaller, i.e., the query times converge towards 12pm, the total traffic decreases since there are fewer strategy transitions. Adaptive algorithms 1 and 2 do not show obvious differences since there is no direct transition of Q2U ratio from low to high or high to low in all evaluation intervals during the 24 hours. The storage strategy transitions of adaptive algorithm 3

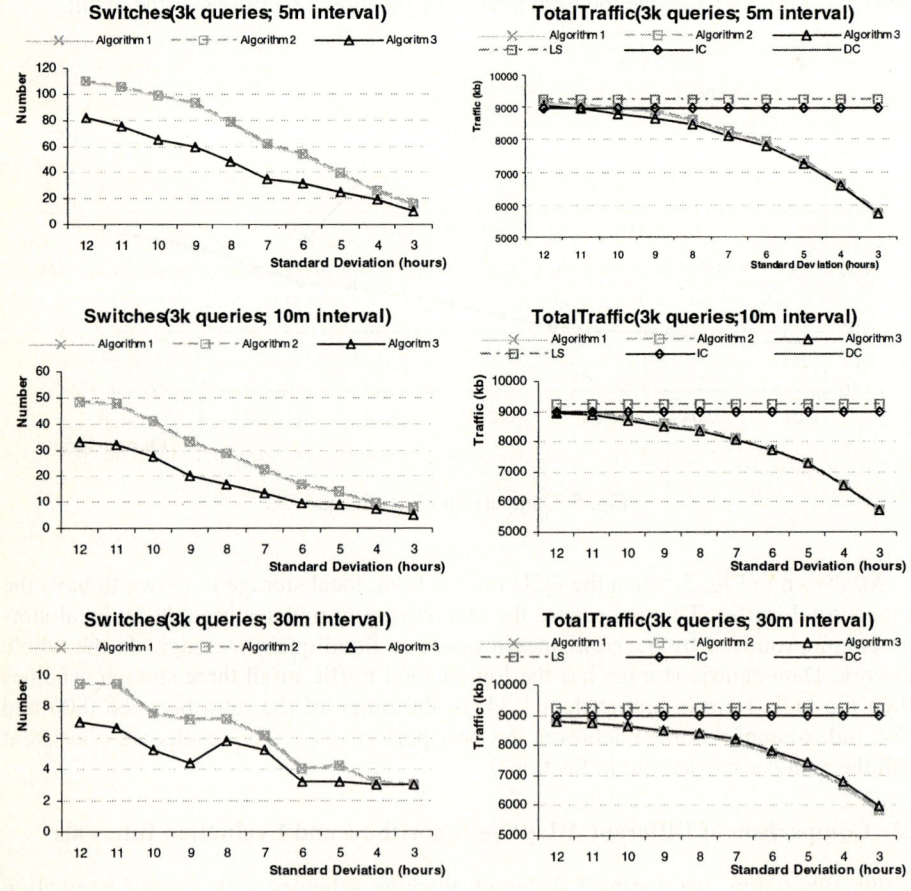

Fig. 6. 3K queries in 24 hours

are less than the other two for all three different evaluation intervals. Its total traffic is the lowest among all the storage schemes in most cases.

When we evaluate 60K queries in 24 hours, the Q2U ratio has a moderate value in most cases. Thus, less strategy transitions are expected. With different evaluation intervals, there are no obvious differences for strategy switches and total traffic for the three adaptive schemes. Fig. 7 shows the results with an evaluation interval of 5 minutes. The total traffic for local storage is too high (up to 185Mb) to be shown in the figure. Again, we can observe that the third adaptive scheme based on exponential aging prediction achieves the best performance.

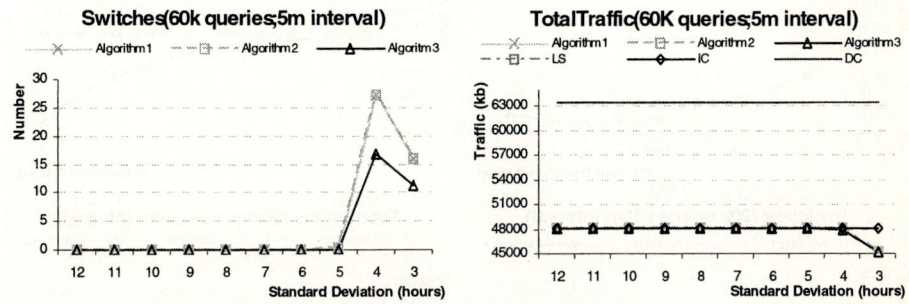

Fig. 7. 60K queries in 24 hours

In the simulations of 120K queries (see Fig. 8), the adaptive schemes also show their better performance than other basic schemes. The traffic for local storage is 369Mb and not shown in the figure. When the query arrival deviation is longer than 5 hours, the strategy transitions keep decreasing. The strategy transitions increase significantly when the deviation is within 3 hours. The reason behind this is that the Q2U ratio is fluctuated around 1.54, as shown in Fig. 9. A longer evaluation interval also results in less strategy transitions in this set of experiments. This is consistent with the previous few experiments. In addition, we can observe that the total traffic with the adaptive schemes is reduced as the query arrival deviation decreases.

5 Related Work

A number of data dissemination schemes have been proposed for wireless sensor networks in the literature [3], [14], [15]. Traditional sensor networks usually rely on a base station, which not only serves as a center for data collection and storage but also a gateway to the external clients. Centralized storing of data at the base station is convenient for analysis and management. However, if there are many queries from the sensor nodes inside the network, this is very inefficient since queries and results must swing between the base station and the sink.

Data-centric routing algorithms, such as Directed Diffusion [3] and TTDD [12], have also been proposed for *long-live queries*. In Directed Diffusion, data is named using attribute-value pairs. A query is broadcast throughout the sensor network as an interest for data. Gradients or paths are set up during this procedure and the source reinforces one of them. The results will return to the sink along this route. TTDD is a two-tier data dissemination approach. The data source that detects a target builds a grid structure to flood the advertisement of sensing data reports. A sink sends queries and receives results directly to or from the source via the grid structure. The advertisement in the grid structure, however, introduces significant traffic overhead. Therefore, these two approaches are not suitable for *one-shot queries*.

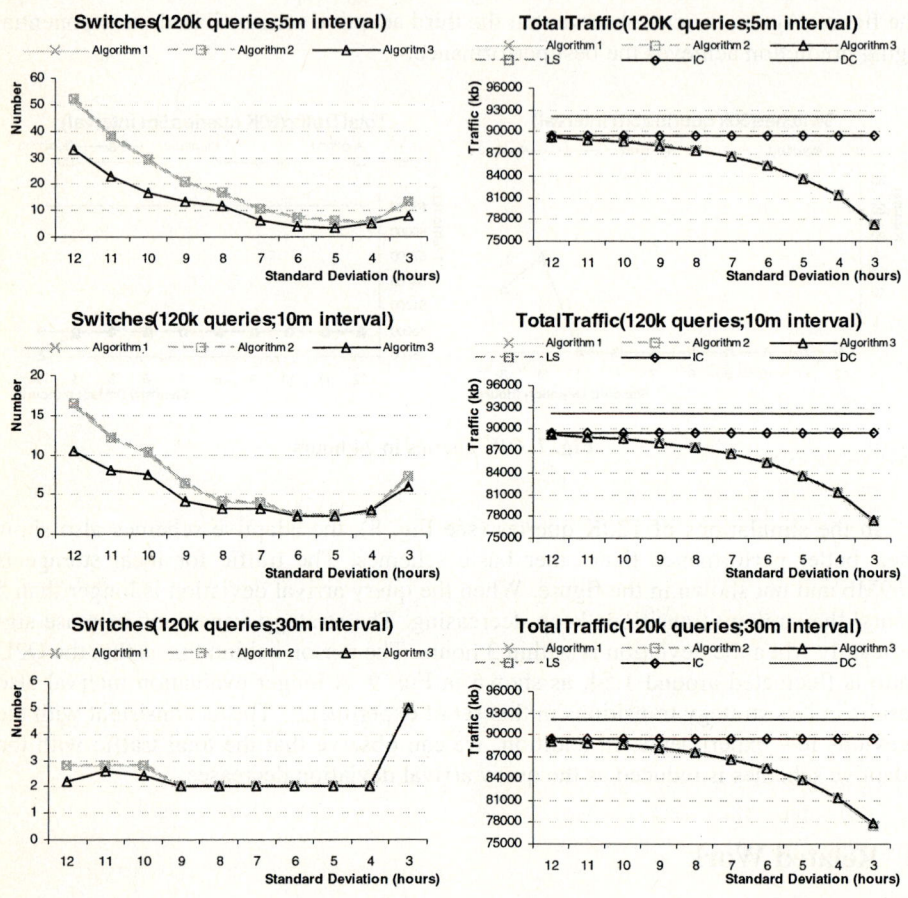

Fig. 8. 120K queries in 24 hours

6 Conclusion

This paper presents the design and evaluation of adaptive data storage and dissemination schemes for dynamic environments. The objective is to minimize the total network traffic incurred by data queries and updates. Different data storage strategies are adopted in accordance with the observed Q2U ratios. We have conducted a series of simulation experiments to evaluate the performance of the basic strategies and three proposed adaptive schemes. The results have shown the proposed adaptive schemes save a significant amount of data communications compared to the basic strategies for dynamically changing access patterns. In particular, the adaptive algorithm based on the exponential aging prediction performs better than the other two.

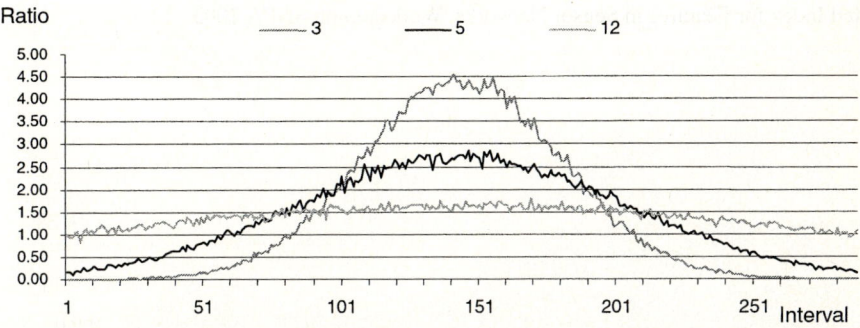

Fig. 9. Query/Update Ratio (120k queries;5m interval)

References

1. G. Pottie and W. Kaiser. Wireless Integrated Network Sensors. Communications of the ACM, 43(5):51-8, May 2000.
2. I. Akyildiz, W. Su, Y. Sankarasubramaniam, and E. Cyirci. Wireless Sensor Networks: A Survey. Computer Networks, 38(4):393-422, 2002.
3. C. Intanagonwiwat, R. Govindan, and D. Estrin. Directed Diffusion: A Scalable and Robust Communication Paradigm for Sensor Networks. Proceedings of MOBICOM2000.
4. J. Heidemann, F. Silva, and D. Estrin. Matching Data Dissemination Algorithms to Application Requirements. Proceedings of the first international conference on Embedded networked sensor systems,2003.
5. S.Ratnasamy, B. Karp, L. Yin, F. Yu, D. Estrin, R. Govindan, and S. Shenker. GHT: A Geographic Hash Table for Data-Centric Storage. Proceedings of the First ACM International Workshop on WSNA 2002.
6. US Naval Observatory GPS Operations. http://tycho.usno.navy.mil/gps.html, 2001.
7. J. Albowitz, A. Chen, and L. Zhang. Recursive Position Estimation in Sensor Networks. ICNP2001.
8. B. Karp and H.T. Kung, GPSR: Greedy Perimeter Stateless Routing for Wireless Networks, Proceedings of MOBICOM2000.
9. Y. Xu, J. Heidemann, and D. Estrin. Geography-informed Energy Conservation for Ad Hoc Routing. ACM MOBICOM2001.
10. A. Ghose, J. Grossklags, and J. Chuang. Resilient Data-Centric Storage in Wireless Ad-Hoc Sensor Networks. Proceedings of the International Conference on MDM2003.
11. W. Zhang, G. Cao and T. La. Data Dissemination with Ring-Based Index for Wireless Sensor Networks. ICNP2003.

12. F. Ye, H. Luo, J. Cheng, S. Lu and L. Zhang. A Two-Tier Data Dissemination Model for Large-scale Wireless Sensor Networks. Proceedings of Mobicom2002.
13. ns-2.26. http://www.isi.edu/nsnam/ns.
14. S. Shenker, S. Ratnasamy, B. Karp, R. Govindan, and D. Estrin. Data-Centric Storage in Sensornets. ACM SIGCOMM Computer Communication Review 33(1):137 – 142, Jan 2003.
15. B. Greenstein, De. Estrin, R. Govindan, S. Ratsanamy, and S. Schenker. DIFS: A Distributed Index for Features in Sensor Networks. Workshop on SNPA 2003.

Continuous Residual Energy Monitoring in Wireless Sensor Networks

Song Han and Edward Chan

Dept. of Computer Science, City University of Hong Kong
`csedchan@cityu.edu.hk`

Abstract. A crucial issue in the management of sensor networks is the continuous monitoring of residual energy level of the sensors in the network. In this paper, we propose a hierarchical approach to construct a continuous energy map of a sensor network. Our method consists of a topology discovery and clustering phase, followed by an aggregation phase when energy information collected are abstracted and merged into energy contours in which nodes with similar energy level are grouped into the same region. The monitoring tree is restructured periodically to distribute energy cost among all nodes fairly, which increases the lifetime of the sensor network. Simulation results indicate that our method is able to generate accurate energy maps with low energy cost.

1 Introduction

With recent advances in sensor technology, wireless sensor networks have received substantial research interest. However, sensor nodes have limited resources such as computing capability, memory and battery power, and it is particularly difficult to replenish the battery of the sensors. Hence methods to preserve energy in sensors, as well as the monitoring of the residual energy level of the nodes in a sensor network are crucial research topics.

In this paper, we focus on the issue of efficient residual energy collection. The process of monitoring energy levels in the sensors consumes energy, and since it is done on a continuous basis, it is important to make sure the process is as energy efficient as possible. Furthermore, given the large number of nodes in a sensor network, typically only an approximate instead of an exact view is required. Just like a weather map, it is sufficient if we can draw an energy map for a sensor network, in which we separate the sensor nodes into several groups according to different residual energy ranges. If such a map can be generated efficiently and reasonably accurately, we can use this energy distribution map to deploy additional nodes to the regions where the energy of the sensor will be depleted soon.

To achieve this, we propose a hierarchical approach for collecting residual energy information continuously in the sensor network in order to construct an energy map at the base station. The entire sensor network is first separated into several static clusters using the TopDisc algorithm proposed by Deb et al. [10] Each cluster is represented by a head node inside the cluster, and at the same time, a topology tree is constructed which consists of these head nodes and some bridging delivery nodes between two adjacent clusters. Based on the energy information collected, a set of polygons which represent the contours of different energy levels is produced independently for each cluster.

The topology tree is then used to collect the energy graphs from leaf nodes to the base station, and in-network aggregation is used to unify the adjacent polygons with the same energy range in order to simply the overall energy map and reduce message cost. In addition to the construction of the initial topology tree for energy collection, reorganization of the tree to evenly distribute energy consumption in the monitoring process is performed periodically to extend the battery life of the sensors in the network.

The rest of the paper is organized as follows. A summary of related work is presented in Section 2. Section 3 describes the design of the continuous hierarchical residual energy collection in detail. The performance of the algorithm is examined in Section 4 and compared with similar algorithms. The paper concludes with Section 5 where some issues and future work are discussed.

2 Related Work

Although there has been a large number of recent work on sensor networks [1,2], only a fairly small number explicitly deals with the issue of managing sensor networks and even fewer deals with the monitoring of residual energy levels. A distributed approach to sensor network monitoring is proposed in [5]. The notion of *active* neighbour monitoring is combined with low overhead *passive* monitoring by the network wide control system to provide high responsiveness without incurring the heavy energy consumption associated with a network wide active monitoring scheme based on constant updates.

The notion of an energy map for sensor networks is first proposed by Zhao at al. [7]. Called the residual energy scan (eScan), this pioneering work applies the techniques of in-network aggregation and abstracted representation of energy graphs which are also used in our algorithm. However, there is no notion of a hierarchical structure in eScan and the topology tree consists of all the sensor nodes in the network, which can lead to additional cost in message delivery. Moreover, since no topology maintenance scheme is proposed, the nodes close to the base station will consume energy at a very high rate for large-sized networks due to the large number of messages delivered, leading to quick depletion of the available energy resources for these nodes. To reduce the energy cost of collecting an eScan and to make monitoring information available to all nodes within a network, the authors propose a monitoring tool called *digest* which is an aggregate of some network properties [9].

A mechanism to predict the energy consumption by a sensor node in order to construct the energy map of a sensor network is proposed in [6,11]. With the proposed energy dissipation model, a sensor node need not transmit its energy information periodically. Instead it can just send one message with its energy information and the parameters of the model, with the major advantage of a greatly extended lifetime for the sensor.

3 Continuous Hierarchy Residual Energy Collection

In this paper we propose an energy efficient mechanism for Continuous Residual Energy Monitoring (CREM). The residual energy information is presented in the form of an energy map (Fig. 1), inspired by eScan. The energy map shows the residual energy of different regions in the sensor network. The regions are colored differently depending on the different energy ranges within that region. Using this energy map, the network manager can decide where new sensor nodes need be deployed to maintain the effectiveness of the monitoring activity of the sensor network.

3.1 System Model and Assumptions

In this section we briefly describe the system model used in our study and state the underlying assumptions used in the formulation of our model. We assume that there is a base station at the network edge. The base station, which has a continuous energy supply, sends a request to collect residual energy to all nodes in the sensor network. This collection of residual energy information occurs at regular intervals called the *monitoring cycle*. Each node is immobile and has symmetric communications to other nodes within certain range R. The nodes also know their positions, and each node is powered by battery with normalized capacity of 100. The sensor network consumes energy according to the Hotspot Dissipation model [7]. The positions of the hotspots are also generated randomly.

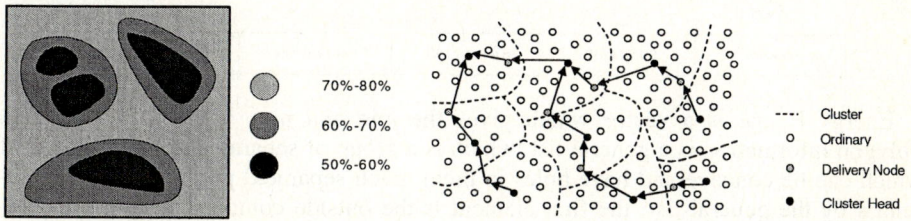

Fig. 1. Example of an Energy Map   **Fig. 2.** Topology Discovery

3.2 Topology Discovery

The initial task in the energy collection is to organize the network in a way that would reduce the required level of message transmission in the entire monitoring process. The basic approach is to divide the network into a number of clusters, each with a cluster head which acts as a representative for nodes in the neighborhood. A "Topology Discovery Request" is sent by the base station. The request is propagated through controlled flooding so that all nodes receive a request packet if they are connected. At the same time, sensors which cannot receive any message from their neighbors will be excluded. In the second step, the sensor network is divided into clusters based on the TopDisc algorithm [10]. This algorithm is based on the simple greedy log (n)-approximation algorithm for finding the set cover.

At the end of this process, the sensor network is divided into n clusters and each cluster is represented by one node, which is called the head node. The head node obtains energy information from all the nodes in the cluster because they are all within its communication range. Each head node knows its parent head node, but they can not communicate with each other directly. Instead, a *grey* (or delivery) node acts as an intermediary which delivers messages between each pair of head node. Fig. 2 is an example of the tree generated in the topology discovery phase.

This algorithm, while similar to the TopDisc algorithm, differs from the original one in one significant aspect. In constructing the topology, we reduced the radius of the communication from range R to R/2 to make sure that in each formed cluster, every node is reachable from any other node in the same cluster. This is important to our topology maintenance process, which will be described in detail in Section 3.3.4.

3.3 Periodical Residual Energy Collection

Residual energy collection is a core process that is performed on a continuous basis. It can be divided into three phases: first, determining the local energy graph; second, performing in-network aggregation of the energy and propagation of the information to the base station, and third, reconstructing the topology tree from the root to the leaves to distribute energy consumption in a more equitable manner.

3.3.1 Abstracted Representation of Energy Graph

One of the main objectives of our algorithm is to reduce the message cost. Our approach is to select an abstracted representation of the energy graph since there is no need to know the energy information for all the nodes in the sensor network. The structure of the message which is used to deliver energy information is as follows:

Header	Sender ID	Receiver ID	Energy Range	Polygon Information

Energy Range is a vector which gives the min and max values of the region. Polygon information is a general list, which is a group of separated polygons, each of which can be concave and have holes in them. Each separated polygon is also represented by the general list: the first element is the outside contour and the other elements are the outside contours of the holes.

3.3.2 Determining the Local Energy Graph

In this step, we wish to get a set of polygons which can provide the contours of the different energy regions. During the energy monitoring cycle, every node in a given cluster sends its energy information to the head node of the cluster. The energy information includes the position of the node and the energy value. After the head node received information from the nodes in its cluster, it divides all the nodes into several sets according to different energy ranges. A convex contour is generated for each node set, and only the vertexes and the sequence are stored. Boolean computing [8] is applied between each contour. The result will be general contours, it can be concave, and has holes, and also can be consisted of several parts. A general list is then used to represent the data structure of the polygons in this cluster, as explained in the previous section.

3.3.3 In-Network Aggregation of Energy Graphs

When a head node gets the local energy graph, the information will send be sent to its parent head node through the default delivery node we have mentioned above. When a head node receives all the messages from its children, it will do the in-network aggregation of the energy graphs. If two polygons are adjacent to each other physically, we can merge them into one polygon to reduce the number of the vertexes and the communication cost. For each contour, the physically adjacent nodes for each vertex of the contour are used to generate an *extended contour*. Two polygons can be joined together if their extended contours intersect. The joined contours can then be reduced in size to form the final merged contour shown in the figure.

3.3.4 Topology Maintenance

Combining the schemes proposed in the previous sections, the energy graph of the sensor network can be obtained efficiently and accurately. However, since the monitoring process is a continuous one, the head node is each cluster will deplete its energy resources at a much higher rate than other nodes due to the need to constantly transmit and receive messages. When a head node exhausts its energy, the topology tree needs to be reconstructed.

There are two methods to handle the issue of topology tree maintenance in our hierarchical clustering system: a dynamic scheme and a static scheme. In a dynamic topology maintenance scheme, the whole sensor network is re-clustered periodically to distribute the energy cost among all nodes more evenly, similar to the approach taken in [4]. The disadvantage of this approach is the large number of extra messages required in re-clustering the network. The other alternative is to use a static topology maintenance scheme (the approach used in this paper) where the hierarchical structure of the clusters is preserved, but different head nodes and delivery nodes in the cluster are used based on the energy graph stored in the former head node [3]. At the beginning of each monitoring cycle, this process is applied to the entire tree from the root to the leaves, and each parent / child cluster pair will perform the following steps:

1. In the parent cluster, the former head node (FP) selects the new head node (NP) according to the energy graph stored in its cache.
2. FP sends a RE-CONSTRCT message which includes the position of NP to the former head node (FC) of the child cluster through the former delivery node (FD).
3. In the child cluster, based on the message from FD, FC selects the new head node (NC) using the same scheme as FP.
4. FC sends a CONFIRM message in which includes the position of node NC to FP through FD.
5. As the delivery node has the same work load as the head node, it is necessary to rotate the delivery node periodically as well. Consequently, when FP receives the CONFIRM message, the FP selects another node to be the new delivery node (ND) which is different from the new head node, and sends a CONFIRM-DNODE message which includes the position of ND to FC through FD. Then it will broadcast a TOPOLOGY-CHANGE message with the position of NC to all the nodes in its cluster.
6. In the child cluster, after receiving the message CONFIRM_DNODE, FC sends a message to NC informing it of the new delivery node. Then it broadcasts the TOPOLOGY-CHANGE message (with the position of NC) to all the nodes in its cluster.

After all communicating node pairs have finished the exchange of the various messages, the new topology tree will be constructed. We will now discuss how the new head nodes and delivery nodes are selected within the cluster. Clearly, in order to reduce the size of the energy graph and also to facilitate in-network aggregation of the energy contours, we need to keep the energy graphs as regular as possible. A naïve approach is to always choose the node with the most remaining energy in the cluster as the new head node. However, by just choosing nodes with the most residual energy without regard to existing energy contours, the energy of the nodes are depleted and they drop into a lower energy range, resulting in a highly irregular energy contour.

The approach used in this paper is to select a node based not on its residual energy but on its proximity to the next lower energy range. This is shown in Fig. 8. Suppose Node X has the minimum distance to any node in the next energy range, and Y the most residual energy. If X (instead of Y) is chosen as the new head node or delivery node, after it has consumed substantial energy in its role and its residual energy drops to the next lower level, the energy graph will still be quite regular.

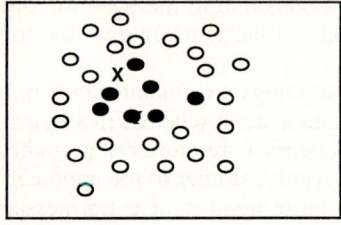 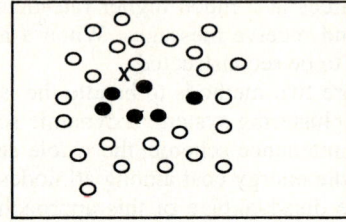

Fig. 3. Selection of new head nodes based on distance to next energy level: Fig.3a is the initial state and Fig.3b is the state after energy dropped to a lower range

4 Simulation Results

4.1 Performance Metrics and Simulation Setup

In this section we will define the following performance metrics used in the experiments (1) Residual reachable nodes (2) Fidelity and (3) Total Message Cost, and describe the simulation setup as well.

Residual Reachable Nodes measures the number of nodes which can be reached by a path from the base station. If the number of residual reachable nodes is small, then clearly many sensors have run out of energy and the sensor network is unlikely to function well. The notion of **fidelity** is used to measure how accurately the energy map represents residual energy information in the actual network. We divide the energy information in a sensor node (from 0% to 100%) into N ranges, and each is called an **energy range,** where the i th energy range stands for the range <100 %*(i-1) / N, 100%*i / N>. If a polygon consists entirely of nodes which fall into the energy range X, we call this polygon the **related polygon** to X. For any node n in a senor network S, if n is in the energy range X, and n is inside the contour or on the outside contour of its related polygon, it is a **correct node**, otherwise it is a **wrong node**. Now we define the fidelity F of the energy map as follows:

Fidelity = total no. of correct nodes / total no. of the nodes in sensor network

Since message transmission is a major source of energy consumption in a sensor network, the overhead of our algorithm is reflected in the level of message transmission, measured by the total number of bytes transmitted. The overall message cost consists of two components: message cost involved in performing the periodic energy scan and message cost that is incurred in maintaining the topology tree:

Total Message Cost (bytes) = Cost $_{local\ energy\ scan}$ + Cost $_{topology\ tree\ reconstruction}$

To study the performance of CREM, a simulation program is written using C++ and CSim-18. For each simulation run, a different set of nodes are generated and located randomly; the hotspots are generated according to the Hotspot model. The position of base station is at the origin i.e. <0, 0>, and the sensors are distributed over a field whose size increases with the number of nodes to maintain the same density of nodes for all network sizes. The energy monitoring cycle is set to 100 time units, and the sensor network is divided into 5 energy ranges from <0%, 20%> to <80%, 100%>. The energy range is assumed to be represented using a single byte, and two bytes are used to represent the position for each node. In the calculation of message cost, protocol overhead is ignored.

4.2   Simulation Results

In the first experiment, we compare CREM with centralized collection, which collects individual residual energy information directly from each node.

Fig. 4 shows the ratio of the total message cost for centralized collection and CREM respectively. It can be seen that CREM consistently outperforms centralized collection by a wide margin regardless of network size, which means that the out-performance of CREM algorithm is scalable. Moreover, Fig. 4 clearly indicates that for a given network size, the cost ratio between centralized collection and CREM increases as communication range increases. This is expected because as communication range increases, the number of nodes in the cluster will increase, which also means that many more nodes fall into the contour of a certain energy range and hence can be ignored using CREM. Fig. 5 shows the fidelity achieved using CREM, which is typically above 95%. Thus, CREM is able to significantly reduce total message cost at only a small degradation in fidelity when compared to Centralized collection (which has a fidelity of 100%).

In the second group of experiments, we examine the impact of several algorithms on the lifetime of the sensors in the sensor network. The first method, centralized collection, gathers energy information from every node in the network directly as described previously. The second method, called Static Clustering, also uses a hierarchical tree topology for residual energy monitoring, but *without* topology restructuring. By comparing the performance of CREM to Static Clustering we can isolate the impact of topology restructuring on overall performance.

The results of the experiments, with a fixed communication range of 5 for the sensor, are shown in Fig. 6-9. Fig. 6 shows the effect of network size on fidelity. The fidelity of the energy map is over 90% initially, and stays in that range for a considerable number of monitoring cycles. However, as the number of nodes which has exhausted its energy climbs, it is more and more difficult to maintain an accurate representation of the energy levels. This effect is more pronounced when networks increase in size. As the number of nodes increases, the number of levels in the hierarchy increases and the total number of messages handled by the head nodes and delivery nodes increases as well, so that their energy is consumed at a higher rate. As a result of this decrease in the number of functional nodes, accurate representation becomes more difficult.

Fig. 7 shows clearly the lifetime of the network decreases with the increasing network size. The reason is that for larger networks, there are more levels in the hierarchy and the head nodes and the delivery nodes have more energy information to deliver and hence depletes their energy more rapidly. This is consistent with results seen in other experiments discussed in this section.

The fidelity of the three algorithms over the lifetime of the sensor network is shown in Fig. 8. If we assume that a fidelity of about 90% is acceptable, then it can be seen that CREM results in an 20-25 fold increase in the lifetime of the network compared to Centralized collection, and an 7-8 fold increase compared to Static Clustering. The latter two methods do have a marginally higher fidelity than CREM (by about 3%) during the first few monitoring cycles, but even then CREM still supports an average fidelity of over 95%, which should be adequate for most energy monitoring applications. It is clear that CREM greatly extends the lifetime of the network at the expense of just a minor decrease in fidelity.

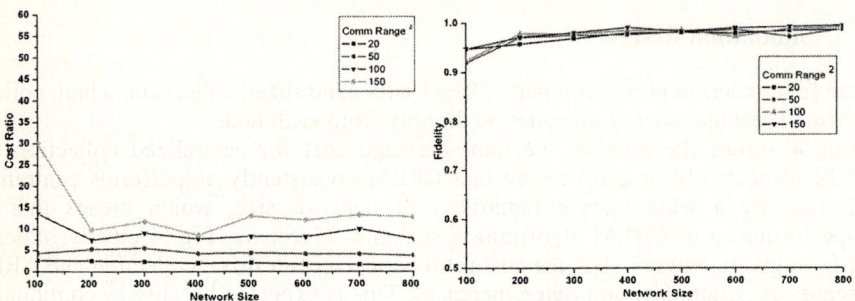

Fig. 4. Cost ratio between CREM and centralized collection

Fig. 5. Fidelity vs network size for CREM

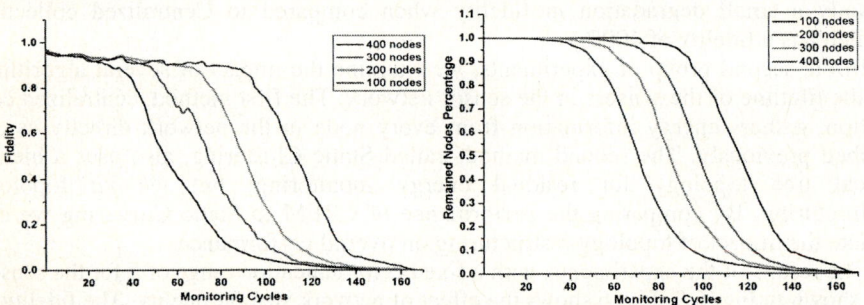

Fig. 6. Fidelity vs Monitoring cycles for CREM

Fig. 7. Residual reachable nodes versus Monitoring cycles for CREM

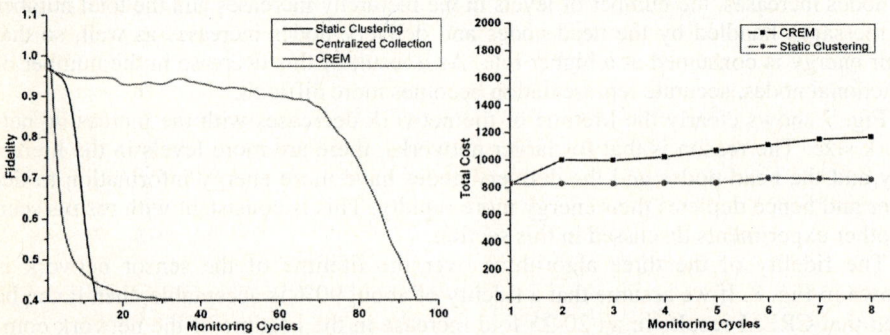

Fig. 8. Fidelity vs monitoring cycles for the three methods

Fig. 9. Total message cost: CREM versus static clustering

If we compare the total message cost between Static Clustering and CREM in Fig. 9, it can be seen that our algorithm introduces an additional overhead of around 25%. Although this is a bit high, by suitably sharing the additional energy consumption among different nodes, the lifetime of the sensor network is extended by an even greater amount as noted above, making this additional cost very worthwhile.

5 Conclusion

Energy is one of the most critical resources in a sensor network, and it is important to be able to monitor the availability of this resource with low overhead. In this paper, we propose a hierarchical approach to continuously collect and aggregate residual energy information for presentation in the form of an energy map. By using techniques such as in-network aggregation, construction of a hierarchical monitoring structure as well as the continuous rotation of the cluster heads, energy cost due to message transmission is reduced. Extensive simulation results show that our approach is energy efficient and is able to extend the lifetime of the sensor network substantially with only a small decrease in accuracy of the energy map generated.

References

1. D. Estrin, R. Govindan, J. Heidemann and S. Kumar, "Next Century Challenges: Scalable Coordination in Sensor Networks", *Proc. ACM/IEEE Int'l. Conf. on Mobile Computing and Network*, August 1999.
2. Lakshminarayanan Subramanian and Randy H.Katz, "An Architecture for Building Self-Configurable systems", *Proc. IEEE/ACM Workshop on Mobile Ad Hoc Networking and Computing*, August 2000.
3. S. Han and E. Chan, "Hierarchical Continuous Energy Maps for Sensor Networks", Technical Report, Dept. of Computer Science, City University of Hong Kong, 2004.
4. W. Heinzelman, A. Chandrakasan and H. Balakrishnan, "Energy-Efficient Communication Protocol for Wireless Microsensor Networks", *Proc. of the 33rd International Conference on System Sciences*, January 2000.
5. Chih-fan Hsin and Mingyan Liu, " A Distributed Monitoring Mechanism for Wireless Sensor Networks", *Proc. Workshop on Wireless Security* 2002.
6. A. F. Mini, Badri Nath and Antonio A. F. Loureiro, "Prediction-based Approaches to Construct the Energy Map for Wireless Sensor Networks", *Proc. 21st Brasilian Symposium on Computer Networks*, Natal, RN, Brazil, May 19-23, 2003.
7. Y. J. Zhao, R. Govindan and D. Estrin. Residual Energy Scan for Monitoring Sensor Networks", *Proc. IEEE Wireless Communications and Networking Conference*, March 2002.
8. Michael V. Leonov, Alexey G. Nikitin, "An Efficient Algorithm for a Closed Set of Boolean Operations on Polygonal Regions in the Plane", Preprint 46, Novosibirsk, A. P. Ershov Institute of Informatics Systems, 1997.
9. J. Zhao, R. Govindan, and D. Estrin, "Computing aggregates for monitoring wireless sensor networks", Technical Report 02-773, USC, September 2003.
10. B. Deb, S. Bhatangar, and B. Nath, "A Topology Discovery Algorithm for Sensor Networks with Applications to Network Management", *Proc. IEEE CAS Workshop on Wireless Communications and Networking*, Pasadena, USA, Sept. 2002.
11. A. F. Mini, Badri Nath and Antonio A. F. Loureiro, "A Probabilistic Approach to Predict the Energy Consumption in Wireless Sensor Networks", *Proc. IV Workshop de Comunicação sem Fio e Computação Móvel*, São Paulo, Brazil, October 23-25 2002.

Design and Analysis of a k-Connected Topology Control Algorithm for Ad Hoc Networks

Lei Zhang[1], Xuehui Wang[2], and Wenhua Dou[1]

[1] School of Computer,
[2] School of Mechatronics Engineering and Automation,
National University of Defense Technology, Changsha 410073, China
findzhanglei@hotmail.com

Abstract. Topology control is an effective approach to reduce the energy consumption in wireless ad hoc networks. In this paper, we propose a k-connected (KC) energy saving topology control algorithm, which is fully distributed, asynchronous and scalable with little overhead. We prove KC algorithm has several valuable properties. First, it is optimal for the local energy-saving topology control; second it preserves the network connectivity (even k-connectivity if the neighbor topology is k-connected) and dramatically reduces the energy consumption; third, each node degree obtained by the KC algorithm cannot exceed $6*k$. Performance simulation shows the effectiveness of our proposed algorithm.

1 Introduction

Minimizing energy consumption is an important challenge in wireless ad hoc networks. Topology control via per-node transmission power adjustment has been shown to be effective in extending network lifetime and increasing network capacity (due to better spatial reuse of spectrum).

Several energy-saving topology control algorithms [1]-[6] have been proposed to create a power-efficient network topology in wireless ad hoc networks with limited mobility. Ramanathan et al. [1] proposed the CONNECT and BICONN-AUGMENT algorithms to solve the 1-connected and 2-connected energy-saving topology control problems, but both CONNECT and BICONN-AUGMENT are centralized algorithms with poor scalability. Roger Wattenhofer al. [2] introduced a cone-based distributed topology control algorithm (CBTC) with the support of directional antenna, but directional antenna is usually unusable for ad hoc networks. Douglas M. Blough [3] proposed a k-Neigh approach for topology control based on the principle of maintaining the number of neighbors of every node equal to or slightly below a specific value k. The approach enforces symmetry on the resulting communication graph and does not require the knowledge of the exact number n of nodes in the network to work, as k is only loosely dependent on n(e.g., k=9 for n in the range 50-500). They estimate the value of k that guarantees connectivity of the communication graph with high probability. Ning Li [5] devised another distributed topology control algorithm LMST basing on the local minimum spanning tree theory, but LMST can only maintain 1-connectivity of the network and need the position system to identify the mutual distance between neighbor nodes, which may be inapplicable.

In this paper, we propose a k-connected (KC) energy saving topology control algorithm for wireless ad hoc networks, which need not the directional antenna or position system support, it only requires each node is able to measure the received signal power level during communication. Furthermore, CBTC has no bound on the number of messages nor on the energy expended in determining the proper transmit power, whereas in our algorithm each node need only transmit three messages at the maximum power.

2 Design of the KC Algorithm

We consider a wireless ad hoc network as a network of homogenous nodes. All nodes are arbitrarily deployed in a two-dimensional plane. Each node is equipped with an omnidirectional antenna with adjustable transmission power in the range of 0 to P_{smax}. V denotes the node set in the network, $\forall u \in V$, we define its neighbor and neighbor set as follows.

Definition 1. *Neighbor and Neighbor Set.* $\forall u \in V$, *if node u can communicate with node $v(v \in V)$ using the maximum transmission power, node v is called a neighbor of node u, all the neighbors of node u constitute its neighbor set V_N^u (including node u).*

Network connectivity can be measured by k-edge connectivity or k-vertex connectivity, the latter is stronger than the former, so we use k-vertex connectivity in this paper. The k-connected energy-saving topology control algorithm is composed of the following four phases: topology information collection, local topology construction, transmission power adjustment and mobility manipulation.

2.1 Topology Information Collection

In this phase each node collects information from neighbors and constructs a neighbor topology information graph $G_N^u = (V_N^u, E_N^u)$, where V_N^u is the neighbor set, E_N^u is the edge set among all the nodes in V_N^u. G_N^u can be obtained as follows.

Each node broadcasts a HELLO message using the maximum transmission power P_{smax} to its neighbors, by measuring the receiving power of HELLO messages, node u can determine the minimum power $P_{s\min}^{(u,v)}$ required to reach its neighbor node v as in [5]. Assume the remaining battery energy of node u is W_u, we define the lifetime of link (u,v) as

$$T_{\max}^{(u,v)} = \frac{W_u}{P_{s\min}^{(u,v)}} \tag{1}$$

so link (u,v) can be regarded as a directed edge between node u and v with two weight values: $P_{s\min}^{(u,v)}$ and $T_{\max}^{(u,v)}$, all these directed edges construct edge set E, thus $G = (V, E)$ is a directed graph representing the global network topology.

Each node collects its neighbor link information from the received HELLO messages and broadcasts these information in a Neighbor Link Information Message (NLI) using the maximum transmission power. After receiving the NLI messages from all the neighbors, node u can build its neighbor topology information graph $G_N^u = (V_N^u, E_N^u)$.

2.2 Local K-Connected Topology Construction

On obtaining the neighbor topology information graph, each node builds a k-connected subgraph $G_N^{u\,\prime} = (V_N^u, E_N^{u\,\prime})$ using the following local k-connected energy-saving topology control(LKC) algorithm.

Step 1. Construct a subgraph $G_N^{u\,\prime} = (V_N^u, E_N^{u\,\prime})$ without any edges, i.e. $E_N^{u\,\prime} = \phi$. $\forall x \in V_N^u$, set its minimum transmission power $P_{\min}^x$ to 0, $P_{\min}^x = 0$.

Step 2. $\forall x, y \in V_N^u$, merge the two directed edges $e_{P,T}^{(y,x)}$ and $e_{P,T}^{(x,y)}$ between them into a undirected edge $e_{P,T}^{xy}$, the weight values of the new undirected edge are:

$$P_{s\min}^{xy} = \max(P_{s\min}^{(x,y)}, P_{s\min}^{(y,x)}) \quad T_{\max}^{xy} = \min(T_{\max}^{(x,y)}, T_{\max}^{(y,x)})$$

The purpose of this step is to avoid the unidirectional link in the network, because MAC protocols usually require bidirectional links for proper operation.

Step 3. Sort the edges in $G_N^u = (V_N^u, E_N^u)$ according to its lifetime $T_{\max}^{xy}$ in a non-increasing order, the sort result is denoted as S.

Step 4. If S is empty, terminate the algorithm, else retrieve the first edge $e_{P,T}^{xy}$ from S.

Step 5. If node x and y are in the same k-connected subgraph of $G_N^{u\,\prime} = (V_N^u, E_N^{u\,\prime})$, go to step 4; else add edge $e_{P,T}^{xy}$ to $E_N^{u\,\prime}$ and update the transmission power by performing the following two steps.

$$if \ \ P_{\min}^x < P_{s\min}^{xy}, \ \ set \ \ P_{\min}^x = P_{s\min}^{xy}$$

$$if \ \ P_{\min}^y < P_{s\min}^{xy}, \ \ set \ \ P_{\min}^y = P_{s\min}^{xy}$$

Step 6. If $G_N^{u\,\prime}$ is k-connected, terminate the algorithm, else go to step 4.

LKC is a greedy algorithm, it iteratively adds edges to $E_N^{u\,\prime}$ to build a k-connected sub-network. LKC is the core of KC algorithm and it has the following properties.

Lemma 1. *Given a k-connected graph $G = (V, E)$, $x \in V, y \in V$, if x, y are directly connected by an edge e^{xy}, and there exist $j(j \geq k+1)$ disjoint paths between node x and y in $G = (V, E)$, then delete e^{xy} from $G = (V, E)$, the remaining graph $G_d = (V, E_d)$ is still k-connected.*

Proof. $\forall u \in V, v \in V$, there are at least k disjoint paths between u and v, if these k disjoint paths do not pass through e^{xy}, deleting e^{xy} from $G = (V, E)$ does not affect the connectivity between node u and v. Suppose one of these k disjoint paths pass through e^{xy}, we denote the k disjoint paths as

$$p_1, p_2 ... p_{k-1}, p(u, a1, a2...x, y...b1, b2, v)$$

after deleting e^{xy}, $p(u, a1, a2...x, y...b1, b2, v)$ does not exist in $G_d = (V, E_d)$, there are only $j - 1 \geq k$ disjoint paths between node x and y, and at most k-1 of them will intersect with $p_1, p_2...p_{k-1}$, assume one of the remaining paths between x and y is $p(x, c1, c2...y)$, so we can find another new disjoint path $p(u, a1, a2...x, c1, c2...y...b1, b2, v)$ between node u and v, the number of disjoint paths between node u and v is still k, therefore after deleting e^{xy} from $G = (V, E)$, the remaining graph $G_d = (V, E_d)$ is k-connected.

Theorem 1. *If the neighbor topology information graph $G_N^u = (V_N^u, E_N^u)$ is k-connected, then $G_N^{u'} = (V_N^u, E_N^{u'})$ obtained by LKC algorithm is also k-connected.*

Proof. Step 6 and step 4 in LKC algorithm guarantee that if $G_N^{u'} = (V_N^u, E_N^{u'})$ is not k-connected, it will search every edge in E_N^u. According to step 5, if node x and y are in the same k-connected subgraph, i.e. if there have been k disjoint paths between x and y in $G_N^{u'} = (V_N^u, E_N^{u'})$, edge $e_{P,T}^{xy}$ will not be added to $G_N^{u'} = (V_N^u, E_N^{u'})$, so $G_N^{u'} = (V_N^u, E_N^{u'})$ is equal to a subgraph of $G_N^u = (V_N^u, E_N^u)$ which reduces some redundant edges, By Lemma 1, reducing these kind of edges does not affect the network connectivity, therefore if $G_N^u = (V_N^u, E_N^u)$ is k-connected, $G_N^{u'} = (V_N^u, E_N^{u'})$ obtained by LKC algorithm is also k-connected.

Theorem 2. *LKC algorithm is an optimal solution to obtain $G_N^{u'} = (V_N^u, E_N^{u'})$, i.e. the lifetime $T_0 = \min\{T_{\max}^{xy} | T_{\max}^{xy} \in G_N^{u'}, x, y \in V_N^u\}$ of $G_N^{u'}$ obtained by LKC algorithm is maximized.*

Proof. We denote the last edge added by LKC algorithm as $e_{P,T}^{uv}$, since every edge is retrieved in a non-increasing lifetime order, which implies $T_{\max}^{uv} = T_0$. There are k-1 disjoint paths between node u and v before adding $e_{P,T}^{uv}$ to $E_N^{u'}$, denote them as $p_1, p_2...p_{k-1}$.

Suppose LKC algorithm is not optimal, T_0 is not the maximum lifetime of $G_N^{u'}$. Then there must exist another optimal algorithm, which can find a path $p_{uv} \neq e_{P,T}^{uv}$, p_{uv} does not intersect with $p_1, p_2...p_{k-1}$ and its lifetime is longer than T_0. This can only be achieved by adding other edges to $E_N^{u'}$ and these edges' lifetime must be longer than T_0. In fact, all the edges whose lifetime is longer than T_0 have been retrieved in step 4. Assume the last added edge to construct p_{uv} is $e_{P,T}^{xy}$, $e_{P,T}^{xy}$ was retrieved by LKC algorithm before but not added to $E_N^{u'}$ only if there had been k disjoint paths between node x and y, now to construct p_{uv}, $e_{P,T}^{xy}$ must be added, which implies the k disjoint paths between node x and y are all intersected with the k-1 disjoint paths $p_1, p_2...p_{k-1}$, that is impossible and leads to a contradiction. Therefore LKC algorithm is an optimal solution to obtain $G_N^{u'} = (V_N^u, E_N^{u'})$.

2.3 Transmission Power Adjustment

On termination of the LKC algorithm, node u obtains the transmission power $P_{\min}^x$ of each node in V_N^u, it will broadcast these information to its neighbors in a Transmission Power Control (TPC) message. After receiving the TPC message, each node adjusts its transmission power using a Max-Min method. Assume node u receives a TPC message from neighbor x, first it extracts its transmission power P^u_{TPCx} from the message (calculated by neighbor x) and compares it with $P^u_{\min}$ (calculated by itself), the less one is recorded in P_{ux}, i.e.

$$P_{ux} = \min(P^u_{\min}, P^u_{TPCx})$$

After receiving the TPC messages from all neighbors, node u set its transmission power to the maximum of P_{ux}:

$$P_u = \max(P_{ux}, \forall x \in V^u_N \text{ and } x \neq u)$$

The final network topology after transmission power adjustment is denoted as $G_0 = (V, E_0)$.

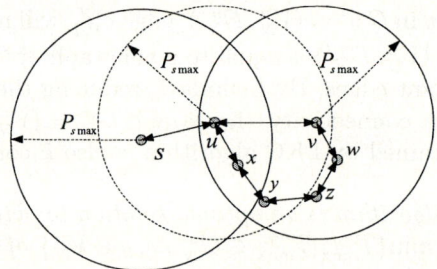

Fig. 1. An illustration of Max-Min transmission power adjustment

The Max-Min transmission power adjustment removes some redundant paths while preserving the network connectivity. As illustrated in Fig. 1 ($k=1$), because node z and w are v's neighbors but not u's, node u has to communicate with node v directly, its transmission power obtained by LKC algorithm is $P^u_{\min} = P^{uv}_{s\min}$, while node v can find a more energy efficient path $p(v, w, z, y, x, u)$, on which the transmission power of node u is $P^u_{TPCv} = P^{ux}_{s\min} < P^{uv}_{s\min}$. So the minimum transmission power between node u and v is $P_{uv} = P^{ux}_{s\min}$. Similarly, the minimum transmission power of node u to reach other neighbors can be calculated:

$$P_{us} = P^{us}_{s\min}, \quad P_{ux} = P^{ux}_{s\min}, P_{uy} = P^{uy}_{s\min}$$

By Max-Min transmission power adjustment, the final transmission power of node u is:

$$P_u = \max(P_{uv}, P_{us}, P_{ux}, P_{uy}) = P^{us}_{s\min} < P^u_{\min}$$

which implies the edge $e^{uv}_{P,T}$ will be removed from $E^{u'}_N$, while the final network is still connected. Fig. 1 only illustrates the case of $k = 1$, for the more general case ($k \geq 1$), we prove the conclusion is still correct in Lemma 2.

Lemma 2. $\forall u \in V$, if $G^{u'}_N = (G^u_N, E^{u'}_N)$ obtained by LKC algorithm is k-connected, after the Max-Min transmission power adjustment, $\forall v \in V^u_N$ and $v \neq u$, there are at least k disjoint paths between node u and v in $G_0 = (V, E_0)$.

Proof. $\forall u \in V$, $v \in V_N^u$, $v \neq u$, because $G_N^{u\prime} = (G_N^u, E_N^{u\prime})$ and $G_N^{v\prime} = (G_N^v, E_N^{v\prime})$ are k-connected graph, we denote the k disjoint paths between u and v in $G_N^{u\prime} = (G_N^u, E_N^{u\prime})$ as $p_u^1, p_u^2, ...p_u^k$, and those in $G_N^{v\prime} = (G_N^v, E_N^{v\prime})$ as $p_v^1, p_v^2, ...p_v^k$. After the Max-Min transmission power adjustment, the transmission power of node u is

$$P_u = \max\left(\min(P_{\min}^u, P_{TPCv}^u), \forall v \in V_N^u \text{ and } v \neq u\right)$$

which implies $P_u \geq \min(P_{\min}^u, P_{TPCv}^u)$, if $P_{\min}^u > P_{TPCv}^u$, the k disjoint paths between node u and v are $p_v^1, p_v^2, ...p_v^k$, else they would be $p_u^1, p_u^2, ...p_u^k$. Therefore after the Max-Min transmission power adjustment, there are at least k disjoint paths between node u and v in $G_0 = (V, E_0)$.

2.4 Mobility Manipulation

To manipulate the mobility of wireless nodes, each node should broadcast HELLO message periodically, the interval between two broadcasts is determined by the mobility speed. When any node finds the neighbor topology is changed, it will rebroadcast the Neighbor Link Information message to notify its neighbors to update the neighbor topology information graph and readjust the transmission power from scratch.

3 Properties of KC algorithm

Theorem 3. $\forall u \in V$, *if the neighbor topology* $G_N^u = (V_N^u, E_N^u)$ *is k-connected, the final topology* $G_0 = (V, E_0)$ *obtained by KC algorithm is also k-connected.*

Proof. Let $n(u,v)$ represents the number of edges along a path p_{uv} between node u and v in $G_0 = (V, E_0)$, $n(u,v)=1$ means u and v are directly connected, $n(u,v)=2$ means there is an intermediate node between u and v along path p_{uv}. $\forall u \in V, v \in V$, if $n(u,v)=1$, node v is node u's neighbor, by Lemma 2, there are k disjoint paths between u and v in $G_0 = (V, E_0)$. Assume when $n(u,v) = m$ ($m \geq 1$), there are k disjoint paths between node u and v in $G_0 = (V, E_0)$, now we prove the assumption is still held for $n(u,v) = m+1$.

When $n(u,v) = m+1$, we denote the m intermediate nodes along path p_{uv} as $a_1, a_2...a_m$, so there exists a path $p_{a_1v} = (a_1, a_2...a_m, v)$ between node a_1 and v, $n(a_1, v) = m$, according to our assumption, there exist k disjoint paths between node a_1 and v in $G_0 = (V, E_0)$ (including path p_{a_1v}). Consider the k nodes that are directly connected with a_1 on the k disjoint paths, denote them as $a_2, s_1, s_2...s_{k-1}$, if we can find k disjoint paths from u to these k nodes: $p_{ua_2}, p_{us_1}, p_{us_2}...p_{us_{k-1}}$, then there must exist k disjoint paths between node u and v, as illustrated in Fig. 2.

Because node u and $a_2...a_m$ are a_1's neighbors, they are in the same neighbor set, by Lemma 2, there exist k disjoint paths between each pair of these nodes in $G_0 = (V, E_0)$. Now we will show how to find the k disjoint paths from u to $a_2, s_1, s_2...s_{k-1}$. For node a_2, obviously a path $p_{ua_2} = (u, a_1, a_2)$ exists between node u and a_2 in $G_0 = (V, E_0)$. For node s_1, because there are k disjoint paths between node u and s_1 in $G_0 = (V, E_0)$, at most one of them will intersect with

p_{ua_2}, we can find a disjoint path p_{us_1} from the remaining k-1 ones. Similarly, for node s_{k-1}, among its k disjoint paths to node u at most k-1 of them will intersect with path $p_{ua_2}, p_{us_1}, p_{us_2} \ldots p_{us_{k-2}}$, the last disjoint one is $p_{us_{k-1}}$. So we can find k disjoint paths $p_{ua_2}, p_{us_1}, p_{us_2} \ldots p_{us_{k-2}}, p_{us_{k-1}}$ from node u to $a_2, s_1, s_2 \ldots s_{k-1}$ respectively, which implies there also exist k disjoint paths between node u and v, thus when $n(u,v) = m+1$, there are still k disjoint paths between node u and v in $G_0 = (V, E_0)$. Therefore the final topology $G_0 = (V, E_0)$ obtained by KC algorithm is k-connected.

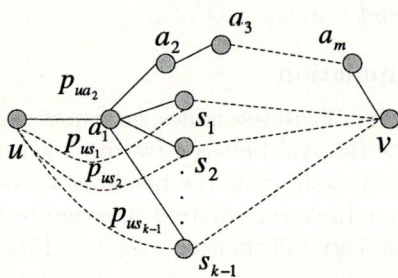

Fig. 2. The topology derived by KC algorithm is k-connected

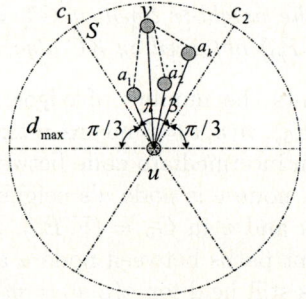

Fig. 3. The degree of any node is bounded by 6*k in the final topology

Theorem 4: If $\forall u \in V$, its remaining battery energy is W, then the degree of any node in $G_0 = (V, E_0)$ cannot exceed 6*k.

Proof: $\forall u, v \in V$, we assume the receiving threshold is $P_{r\min}$, according to the communication theory, it satisfies

$$P_{r\min} = c * \frac{P_{s\min}^{(u,v)}}{d(u,v)^r} \qquad (2)$$

where c and r are constants, $d(u,v)$ is the mutual distance between node u and v, by the lifetime definition in equation (1), we have

$$T_{\max}^{(u,v)} = \frac{c*W}{P_{r\min}*d(u,v)^r} \qquad (3)$$

which indicates the maximum lifetime of link (u,v) is determined by their mutual distance $d(u,v)$. As illustrated in Fig. 3, all the neighbors of node u are in a circle area, whose center is u and radius is equal to $d_{\max} = \sqrt[r]{c*\frac{P_{s\max}}{P_{r\min}}}$, consider a $\pi/3$ radian sector S of this circle area, we will prove there are at most k nodes directly connected to u in S. Assume in graph $G_0 = (V, E_0)$ node u has had k directly connected nodes inside S, denoted as $a_1 a_2 ... a_k$, none of them locates on the radius of S. Pick another node v in S, according to step 3 in the LKC algorithm and equation (3), we get $d(u,v) \geq d(u,a_i)$, $i = 1...k$, which implies the distance between node u and v is not less than that between u and a_i, since v is in a $\pi/3$ radian sector and $a_1 a_2 ... a_k$ are inside S, we have $d(u,v) > d(a_i, v)$, $i = 1...k$, so we can find k disjoint paths whose lifetime is not less than that of the direct connection between node u and v as follows

$$(u, a_i, v) \qquad i = 1...k$$

recall the optimality of LKC algorithm, u and v cannot be directly connect in graph $G_0 = (V, E_0)$, therefore in S the number of directly connected nodes is no more than k.

Now consider a special case, where node a_i and v locate on the two radius of sector S respectively, and satisfy $d(u,v) = d(a_i, v)$, thus u can connect with v directly or through node a_i. The latter is the same as a_i is inside S, while the former will lead to $k+1$ nodes which are directly connected with u. In this case because a_i and v are located on the radius of S, they will be shared with the neighbor sectors, so on average there are still k nodes directly connected with u.

The area covered by S is only 1/6 of node u's communication range, therefore the degree of u cannot exceed 6*k.

4 Performance Evaluation

We evaluate the performance of KC algorithm through simulations. Assume n nodes are uniformly distributed in a $l \times l$ square area, two-ray ground propagation model is used for the wireless channel, the maximum transmission power is 0.2818w, the receiving threshold is 3.652×10^{-10}w and the corresponding maximum transmission range is $250m$.

4.1 Average Node Degree

A smaller average node degree usually implies less contention/interference and better spatial reuse. Energy-saving topology control algorithm should preserve the network connectivity while reducing the average node degree. In this simulation, we set $n = 100$ and $l = 1000m$, Table 1. shows the average node degree of each topology, from which we can see CBTC, LMST and KC all dramatically reduce the average node degree. Moreover, KC outperforms both LMST and CBTC.

Table 1. Average node degree derived by different topology control algorithms

Algorithm	Max. Power	CBTC	LMST	KC
Average Degree	15.81	3.52	2.38	2.08

4.2 Scalability

To evaluate the scalability of the KC algorithm, we first fix the node density and vary the number of nodes (the network size increases) in the network from 50 to 300, the average node degrees for the topologies generated using the KC algorithm when $density = 2 \times 10^{-4}$, $density = 1 \times 10^{-4}$ and $density = 5 \times 10^{-5}$ are shown in Fig. 4(a). The average node degree increases slowly with the number of nodes under a fixed node density, and the higher the node density, the lower the average node degree. Then we fix the distribution area $l = 1000m$ and vary the number of nodes from 50 to 300, in Fig. 4(b) we compare the average node degree of the topologies generated using max transmission power, CBTC, LMST and KC algorithm. The KC algorithm outperforms the others and its average node degree (overlapped with LMST in figure 4(b)) is almost invariable with the number of nodes, which is in contrast with the observation that the average node degree of the topology generated using the maximum transmission power increases almost linearly. These two simulations show that the performance of KC algorithms is not sensitive to the node density and network size. Moreover, the KC algorithm is a localized algorithm, each node need only interact with its neighbors, the communication overhead of each node does not increase with the node density or network size.

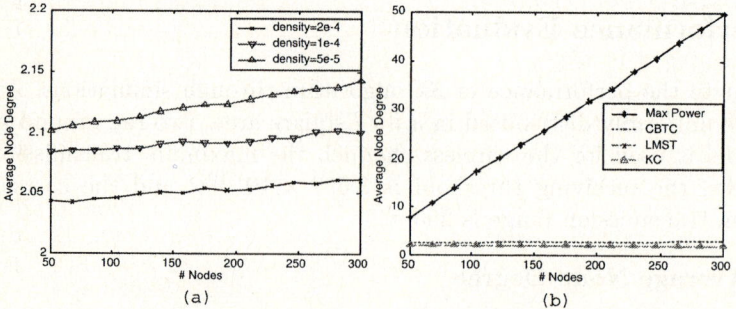

Fig. 4. Scalability of the KC algorithm

4.3 Average Transmission Power

Fig. 5 shows the average transmission power under different topology control algorithms where $l = 1000m$, from which we can see the transmission power under KC algorithm is the minimum. With the node density increasing, the average distance between each pair of neighbor nodes becomes closer. Hence, the nodes need lower transmission power to ensure the network connectivity.

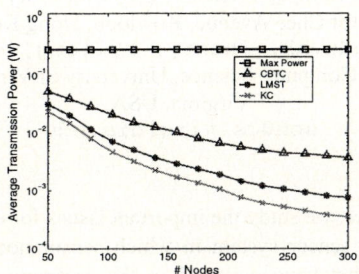

Fig. 5. Average transmission power under different topology control algorithms

References

1. R. Ramanathan and R. Rosales-Hain, "Topology control of multihop wireless networks using transmit power adjustment," in Proc. IEEE INFOCOM 2000, Tel Aviv, Israel, Mar. 2000, pp. 404–413.
2. L. Li, J. Y. Halpern, P. Bahl, Y.-M. Wang, and R. Wattenhofer, "Analysis of a cone-based distributed topology control algorithm for wireless multi-hop networks," in Proc. ACM Symposium on Principles of Distributed Computing, Newport, Rhode Island, United States, Aug. 2001, pp. 264–273.
3. D.M.Blough, M.Leoncini, G.Resta, P.Santi, "The k-Neighbors Approach to Symmetric Topology Control in Ad Hoc Networks", submitted to IEEE J. on Selected Areas in Communications.
4. Y.-C. Tseng, Y.-N. Chang, and B.-H. Tzeng, "Energy-efficient topology control for wireless ad hoc sensor networks," In Proc. Int. Conf. Parallel and Distributed Systems (ICPADS 2002).
5. N. Li, J. C. Hou, and L. Sha, "Design and analysis of an MSTbased topology control algorithm," in *Proc. IEEE INFOCOM 2003*, San Francisco, CA, USA, Apr. 2003.
6. P. Wan, A. Calinescu, X. Li, and O. Frieder, "Minimum energy broadcast routing in static ad hoc wireless networks," in Proc. Of IEEE INFOCOM. 2001.
7. S. Narayanaswamy, V. Kawadia, R. S. Sreenivas, and P. R. Kumar, "Power control in ad-hoc networks: Theory, architecture, algorithm and implementation of the compow protocol," in Proc. of European Wireless 2002, Florence, Italy, Feb. 2002, pp. 156–162.

On Using Temporal Consistency for Parallel Execution of Real-Time Queries in Wireless Sensor Systems

Kam-Yiu Lam[1], Henry C.W. Pang[1], Sang H. Son[2], and BiYu Liang[1]

[1] Department of Computer Science, City University of Hong Kong,
83 Tat Chee Avenue, Kowloon, Hong Kong
cskylam@cityu.edu.hk, {henry, byliang}@cs.cityu.edu.hk
[2] Department of Computer Science, University of Virginia, Charlotte,
Virginia, USA
son@cs.virginia.edu

Abstract. In this paper, we study the important issues for execution of real-time queries in a wireless sensor system in which sensor nodes are distributed to monitor the events that have occurred in the environment. Three important objectives in processing the real-time queries are: (1) to minimize the number of missed deadlines, (2) to minimize the processing costs, especially in data communication; and (3) to provide temporally consistent sensor data values for query execution. To reduce the data transmission cost and delay time in gathering the right versions of data items for a query, we propose the Parallel Data Shipping with Priority Transmission (PAST) scheme to determine where and how to execute a real-time query. To meet the deadlines of the queries, a deadline-driven priority policy is adopted to schedule the transmission of sensor data versions to the coordinator node.

1 Introduction

In a wireless sensor system, sensor nodes are deployed to monitor the real-time status of the environment. In this paper, we focus on processing of real-time queries which access to sensor databases. Real-time queries are usually submitted to monitor the current status of the system environment and generate a timely response if certain events are detected or emergency situations occur. Each real-time query is associated with a soft deadline. It is an important system performance objective to complete real-time queries before the deadlines. The set of data items accessed by a real-time query has to be temporally consistent [6]. Otherwise, the correctness of the results could be seriously affected. Although temporal inconsistency is an important issue to real-time query processing due to the fast changing nature of the system environment, temporal consistency of sensor data has not received much attention until very recently. In [6], they apply the concept of temporal tolerance to reduce data transmission workload in a wireless network. In [1], temporal consistency has been chosen to be the correct execution criterion for continuous aggregate queries in a wireless sensor network. As we will discuss later, the cost for meeting temporal consistency in real-time query execution could be expensive in a wireless environment. It depends heavily on how to collect the data items for processing the query.

In this paper, we adopt a *multi-version data transmission (MVDT)* scheme using relative consistency [11, 10] as the correctness criterion for execution of real-time queries. Each generated data item has an interval to identify its validity period. We provide an algorithm to ensure that all data items accessed by a query are relatively consistent. To determine where and how to execute a real-time query, we propose the *Parallel Data Shipping with Priority Transmission (PAST)* scheme for gathering the right versions of data items for a real-time query. To meet the execution deadline of a query, a *deadline-driven priority scheduling policy* is adopted to schedule the transmission of sensor data versions to the coordinator node.

2 Related Works

Wireless sensor data management for monitoring and control purposes is a new area in mobile computing research [9]. The in-network approach was proposed for processing queries in wireless sensor networks [4, 3, 8]. In [4], the Tiny Aggregation (TAG) scheme was proposed for Tiny database (TinyDB) which is a sensor data processing system with an SQL-like interface for aggregate queries. Users may specify the epoch duration for aggregating the data for a query to limit the difference in time-spans among the collected data items. It organizes the nodes into a tree structure, and partial computation and aggregation are performed on the way from the leaf nodes towards the root. To minimize the number of communication messages, a generic aggregation scheme is proposed to combine with TAG to propagate partial results of a query following the topology and connections of the sensor nodes [3]. In [3], a technique called pipeline aggregation is proposed to improve the degree of tolerance towards loss of messages and node disconnection. The tradeoff is more messages being transmitted for aggregation. In [5], the aggregation mechanism is enhanced by proposing the semantic routing tree (SRT) scheme in which several schemes for choosing the parent nodes are discussed. In [7], the ACQUIRE mechanism is proposed to obtain the required data for a query in a sensor network using a look ahead scheme together with an active query with triggered local updates.

Most of the previous works in the area have ignored the temporal consistency issue in accessing sensor data. The pioneer work on that issue is [6] which aims to reduce the aggregate workloads. In [1], we proposed a sequential aggregation scheme for execution of continuous queries using temporal consistency as the correct execution notion with the objective to minimize the aggregation cost. Most of the previous works are for aggregate queries using partial computation to minimize the data access cost and delay. On the contrary, our focus is on time-constrained queries in which the operations of a query have precedence constraints. The focus of this paper is to combine the techniques in both real-time query processing and data management in wireless sensor systems to support efficient processing of real-time queries in a wireless network. To our best knowledge, this is the first study on real-time queries on sensor data.

3 System Model and Temporal Consistency

The wireless sensor system model consists of a base station (BS) and a collection of sensor nodes distributed in the environment. The area is divided into a number of square grids with length of r as shown in Figure 1. A grid may contain multiple sensor nodes. It is assumed that the nodes within the same grid have similar functions, i.e., capturing the same signals of their surrounding environment. The length r of a grid is defined such that a node can directly communicate with all the nodes in its neighboring grids. The communication range R of a sensor node is $2\sqrt{2} \times r$. Each grid has a coordinator node elected from the set of nodes in the grid. It is responsible for reporting the operation status of the nodes within the grid to the base station periodically. The base station is responsible for the communication between the sensor nodes and the users of the system. Each sensor node maintains a tiny sensor database for storing its sensor data items (data versions). Each newly created data version is associated with a time-stamp to indicate the time when it is created. The time-stamps can be used for ensuring temporal consistency in query execution and will be discussed in details in Section 3.3.

Real-time queries are submitted for event detection in an area (or areas). Due to the responsive nature of a real-time query, it is important that the values of the data items accessed by a real-time query are representing the current information ("real-time status of the entities") in the environment. Each real-time query has a currency requirement on its accessed data items. Failing to meet the requirement implies that they are too "old" and not correctly describing the current situation of the environment. Since the action is a response to the occurred events, each real-time query is given a deadline on its completion time.

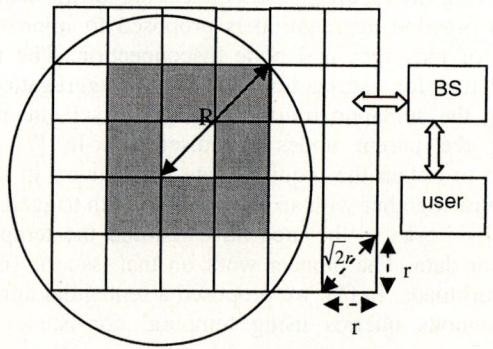

Fig. 1. System Model

A real-time query consists of a sequence of read operations on sensor data items for event detection. Unlike aggregate queries, the set of operations in a real-time query is defined with precedence constraints ($<_i$). To simplify the discussion, it is assumed that the operations are *totally* ordered and the required items of a query are defined at the grid level. Once the base station receives a real-time query, it determines the set of grids from which the set of data items need to be accessed by the query, and then it can identify the set of sensor nodes, called participating nodes, for the query. The query will then be divided into a set of sub-queries and the sub-queries are passed to the grids for retrieving the required sensor data. Since the main purpose of a real-time query is for event detection, the required sensor data are location-dependent instead of sensor dependent. For example, a query T_i accesses to sensor data items from sensor node N_j if T_i wants to access the status of grid A in the system and A is now monitored by N_j. If N_j has moved out of grid A, N_j does not need to report its sensor data for T_i. On the other hand, if another node N_k has entered grid A, N_k will start to report for T_i.

In addition to meeting the deadline and currency requirement, another important issue is the reliability of the results generated in processing a real-time query. It is quite common that the sampled sensor data items contain errors due to various types of noises in the system. Thus, the result generated from a set of data items may also contain errors. To improve the reliability and accuracy of the results, it is important to provide multiple results by accessing multiple data versions of the data items in processing a real-time query. Therefore, a real-time query is associated with a result interval requirement, which specifies the time interval of data items for generating the results. The followings formally define a real-time query T_i:

T_i = $\{D_i, Op_i, <_i, O_i, \Delta_i, R_i\}$
D_i = the deadline of T_i
Op_i = the set of operations in T_i
$<_I$ = defines the execution orders of the operation in T_i
O_i = the set of data items to be accessed by T_i
Δ_i = the currency requirement on O_i.
R_i = the time interval of the results required to be generated from T_i

3.1 Temporal Consistency of Sensor Data Items

Temporal consistency is commonly used as the correctness notion for execution of real-time queries on dynamic data items whose values change continuously with time. Data item x is *absolutely consistent* if the current data version x_i is still within its life-span at the commit time of a query which has accessed to it, i.e., Start_time (x_i) + VI_x > current time. Start_time (x_i) is the creation time of the current version x_i of data item x. VI_x is the validity interval of x. It is a pre-defined value. We use a time bound, *upper valid time (UVT)* and *lower valid time (LVT)* to label the validity interval of a data version. The sampling period equals to the validity interval such that at every time point there is a valid version of the data item. The current version will become stale once a new data version is generated. The validity interval of a data item could be small and the total number of versions generated for a data items could be large if it is defined based on the *maximum rate* of change of the data item. The set of data items for execution of a real-time query are *relatively consistent* if they are temporally correlated to each other, i.e., representing the status of entities in the environment at the *same time point* [11].

Definition of Relative Consistency: Given a set of data versions V from different data items, the versions in V are relatively consistent if $\bigcap \{VI(x_i) \mid x_i \in V\} \neq \Phi$, where $VI(x_i)$ = [$LVT(x_i)$, $UVT(x_i)$].

In addition to meeting the relative consistency requirement, the set of data items also needs to satisfy the currency requirement of the query. For query T_i, the creation time of all its accessed data versions should not be earlier than $(D_i - \Delta_i)$, where D_i and

Δ_i are the deadline and the currency requirement of T_i, respectively, i.e. $(\bigcap\{VI(x_i) | x_i \in V\}) \bigcap [D_i - \Delta_i, D_i] \neq \Phi$. The time window ($D_i$ to ($D_i - \Delta_i$)) is called the *valid time window* for the set of valid results of the query. Within the valid time window, there may have a set of multiple versions of data items meeting the query requirements. In this paper, we adopt relative consistency as the notion of correctness for execution of a real-time query. Note that the currency requirement is similar to the *epoch* used in [4] for data collection. However, due to the fast changing nature of sensor data if the epoch value is not small enough, they may not be relatively consistent. On the other hand, the use of a small epoch will make a large amount of data useless since the transmission delay may be larger than the epoch.

4 Relative Consistency Problem in Wireless Sensor Systems

Although using relative consistency can reduce the cost for processing a real-time query since it does not need to access to the latest version of a data item, how to provide the required data items for execution with minimum data transmission overhead in a wireless sensor system is still a challenge. To illustrate the problem, let us first consider the conventional *sequential* scheme for execution of a real-time query in which the sequential order is according to the operation dependencies $<_i$ defined in the query. A query is forwarded to the node where the required data item of its operation is located according to the sequence of its operations. As shown in Figure 2, it is assumed that the required data items of query T_i are managed by sensor nodes $\{N_1, N_3, N_7, N_{10}\}$ and the orders of access by T_i are: $N_1 \rightarrow N_3 \rightarrow N_7 \rightarrow N_{10}$. When T_i arrives at N_1 for data item o_1, it gets the latest version $o_{1,j}$. Then, it moves on to N_3, N_7 and finally to N_{10} to complete its computations.

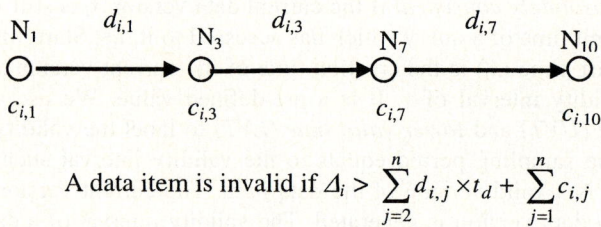

A data item is invalid if $\Delta_i > \sum_{j=2}^{n} d_{i,j} \times t_d + \sum_{j=1}^{n} c_{i,j}$

Fig. 2. Temporal consistency problem in sequential execution

In this example, two problems have to be addressed: (1) the relative consistency problem; and (2) the reliability problem. Although at the time when T_i is at N_1, the version $o_{1,j}$ is the latest one, it may become outdated and the currency requirement may not be satisfied at the time when T_i is at N_{10} due to the long propagation delay in a wireless network. Therefore, to meet the currency requirement:

$$\Delta_i > \sum_{j=2}^{n} d_{i,j} \times t_d + \sum_{j=1}^{n} c_{i,j} \quad (1)$$

where n is the number of operations in T_i and $d_{i,j}$ is the length defined in terms of number of hop counts between each successive node for execution of T_i. t_d is the communication time to send a data version through one hop. $c_{i,j}$ is the computation time to process the jth operation of T_i at a node. If the sum of the total delay in transmission plus the total computation delay is greater than the currency requirement, i.e., eqn. (1) is not satisfied, no result will be generated for the query. Even though the currency requirement can be satisfied, the set of results generated may not meet the result duration requirement R_i if the transmission delay is long.

Multiple versions of data items can be provided for execution of the query to improve the reliability of the generated results by using a *pipeline approach* in the sequential execution scheme. For example, the operations of a query will stay at the nodes until the deadline of the query. If a new version is generated from the node where the first operation of the query is residing, the operation will be evaluated and then the new result is forwarded to the next node where the next operation of the query is residing. Then, the next operation will be executed using the data version which is relatively consistent with the version from the previous node. The procedure is repeated until all the operations of the query have completed. The interval of results R_i to be generated for T_i is: $\Delta_i - \sum_{j=2}^{n} d_{i,j} \times t_d - \sum_{j=1}^{n} c_{i,j} > R_i$. The problem of this approach is heavy data transmission cost since the number of messages for data transmission is higher and the message overhead is directly proportional to the number of messages for sending data versions to the next nodes in the sequence. Further, some data transmission and computations, especially those operations in the beginning of a query, may become useless if they cannot meet the currency requirement when they reach the last operation node.

5 Parallel Data Shipping with Priority Transmission (PAST)

As discussed in the previous section, the major problem of the sequential execution scheme is the long delay in getting the required data items for the operations of a real-time query. Instead of sequentially forwarding operations to sensor nodes for execution according to their execution orders, a simple way to reduce the data access delay is to adopt a parallel scheme in which the participating nodes forward data values required by the operations to the coordinator node to execute the query. Data shipping is suitable for wireless sensor systems as the sizes of sensor data are usually small. Most sensor data values can be fitted into one to two bytes. Note that due to the precedence constraints among the operations, parallel execution of operations at multiple nodes is not easy to achieve.

Although the concept of parallel data shipping is simple, the design of an efficient data shipping scheme meeting the processing constraints of a real-time query in a wireless environment is not trivial. In PAST, the participating nodes of a query submit data versions to a carefully selected coordinator node in a parallel and synchronized fashion such as the one shown in Figure 3. We determine when and which data versions are to be sent from each participating node to the coordinator node so that the currency, result interval and deadline requirements of the query can be satisfied. The submission of data versions from the participating nodes is synchronized depending on the farthest participating node from the coordinator node. The scheduling of transmission of data versions at each node follows a priority scheme to be discussed in Section 5.2 so that the arrival times of the data versions is close to the expected time. Compared with the sequential scheme, the tradeoff of the parallel data shipping scheme is heavier workload at the coordinator node (grid). To minimize the workload at the coordinator node, the nodes within the coordinator grid may rotate to be the coordinator node. PAST consists of three phases:

(i) Analysis phase at the base station;
(ii) Collection phase at the sensor nodes; and
(iii) Processing phase at the coordinator node

In this paper, we concentrate on a two level execution model, the participating nodes and the coordinator node. If a query can be divided into hierarchical subqueries such that they can be processed atomically, the proposed scheme PAST can easily be extended to be a hierarchical scheme.

Coordinator node

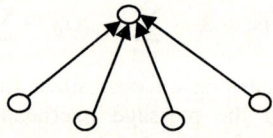

Participating nodes

Fig. 3. Parallel submission of data versions

5.1 Analysis Phase

The analysis phase is performed at the base station once it receives a real-time query. After considering the grids to be accessed by the query, it determines the set of participating nodes of the query. Then, the base station determines the answers for the following two questions in the analysis phase:

(1) Which node should be assigned as the coordinator node? Note that the coordinator node does not need to be one of the participating nodes. It is chosen such that the total transmission cost of the data versions from the participating nodes to the coordinator node is minimized. An algorithm is provided in Section 5.1.1 for choosing the coordinator node.
(2) What are the data versions from each participating node to be sent to the coordinator node? All the submitted data versions have to satisfy the relative consistency and currency requirements of the query. Therefore, data versions that are not relatively consistent with the data versions from other participating nodes or do not meet the currency requirement of the queries should not be submitted from the participating nodes in order to minimize the unnecessary data transmission workload.

5.1.1 Determination of the Coordinator Node

The coordinator node is responsible for collecting data versions from the participating nodes to execute the query. Figure 4 shows the time line for getting the data versions from a participating node j.

As shown in Figure 4, to meet the currency requirement Δ_i and the result interval requirement R_i, the participating node j has to send the data versions covering the time period of the data item from $(D_i - C_i - MT_{i,j} - R_i)$ to $(D_i - C_i - MT_{i,j})$ at time $(D_i - C_i - MT_{i,j})$. C_i is the computation time needed for node j to evaluate the query. The arrivals time of the data versions has to be earlier than $(D_i - C_i)$ and the data transmission delay $(MT_{i,j})$ of it has to be smaller than $(\Delta_i - C_i - R_i)$. Otherwise, the coordinator node may not have sufficient time to finish the query. The data transmission delay from a participating node to the coordinator node can be measured in terms of number of hops in communication between them if we assume that the data transmission delay for sending a data version through one hop is a constant t_d. (If the transmission delay is not a constant, we can use the expected value for calculation.) Let D_{max} be the maximum data transmission delay from all the participating nodes of a query. The maximum number of hops H_i of the participating nodes from the coordinator node is: $H_i = D_{max} / t_d$.

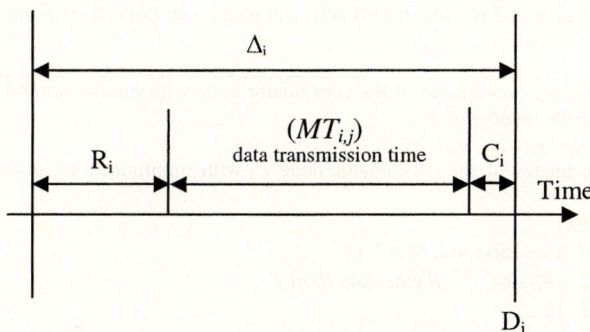

Fig. 4. Time line for sending data versions

Let $G_{all} = \{g_1, g_2, g_3, \ldots, g_n\}$ be the set of grids in the system and F_{ij} is defined as the distance (in number of hops) between grid i and grid j where $i, j = 1, \ldots, n$. Suppose T_i wants to access to u grids/nodes and its required nodes are in the grids set $G_i = \{g_{i1}, g_{i2}, g_{i3}, \ldots, g_{i,u}\}$ and $u = |G_i|$. Let F_{totalX} be the total transmission length defined in terms of hops for choosing grid X as the grid where the coordinator node is residing:

$$F_{totalX} = \sum_{j \in G_i} F_{X,j}$$

Let the coordinates of a grid k be (X_k, Y_k) and $S_{g_{ik}}(H_i)$ be the set of grids which can be reached by the data versions originated from g_{ik} with a distance of no more than H_i in hop counts, i.e.:

$$S_{g_{ik}}(H_i) = \{g_{ij} | F_{g_{ik},j} \leq H_i, j \in G_{all}\} \qquad (2)$$

Eqn. (2) defines a square region with a participating node as the central point of the square and the boundary is H_i hop counts from the grid where the participating node is residing. As shown in Figure 5, the coordinator node is resided in the area within the overlapped regions of the squares from the participating nodes. Algorithm 1 calculates the coordinates of the coordinator node which is within the intersect regions of all the participating nodes.

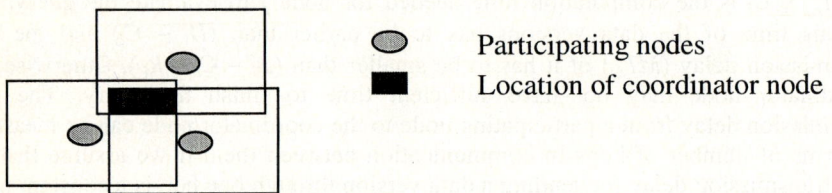

Fig. 5. Location of the coordinator node

In the above algorithm, if there is no intersection among the regions with length $(\Delta_i - C_i - R_i) / t_d$ from all the participating nodes, we will increase the region size by decreasing the value of R_i. Doing so will decrease the period of time length covered

---

Objective: Finding the coordinates of the coordinator node with minimum total hop counts from the participating grids (nodes) of T_i
input: $G_i = \{g_{i1}, g_{i2}, g_{i3}, \ldots, g_{iu}\}, R_i$
output: The coordinates of the coordinator node P_A with minimum total distance from all the participating nodes
FEASIBLE := **false**;
  while(FEASIBLE == **false** and $R_i > 1$) {
    $H_i := (\Delta_i - C_i - R_i) / t_d$;   // calculate H_i of T_i
    $S := S_{g_{i1}}(H_i)$;
    for $k := 2$ **to** u **do** {
      $S := S \cap S_{g_{ik}}(H_i)$;}
    if $(S == \Phi)$ **then** {
      FEASIBLE := **false**;   // not feasible
      $R_i := R_i - 1$;   // decrease result interval requirement}
    else {FEASIBLE := **true**;
      $F_C :=$ infinity;
      for each grid Q **in** S **do** {
        if ($F_{total\,Q,i} < F_C$) **then** {
          $P_A := (X_Q, Y_Q)$;}}
    }
  }
  If (FEASIBLE == **false**) **then**
    abort; // no feasible solution
  else
    return P_A;

Algorithm 1. Calculating the coordinates of the coordinator node

by the set of data versions submitted from the participating nodes and will reduce the result interval of a query. If the participating nodes are so apart that there is no interaction even when R_i is reduced to zero, no result will be generated for the query.

5.1.2 Start Time for Transmission

The set of data versions to be submitted from each participating node is those data versions which are valid within the interval from $(D_i - C_i - D_{max} - R_i)$ to $(D_i - C_i - D_{max})$. For the participating nodes, which transmission times are much smaller than D_{max}, the data versions may arrive at the coordinator node much earlier than $(D_i - C_i)$. They will be put into the buffer until all the required data items have arrived, and then the processing of the query will be started at the coordinator node.

5.2 Collection Phase

Once the coordinator node and the transmission time of each participating node have been determined, the information together with result interval requirement R_i will be transmitted to the participating nodes. According to the received information, the participating nodes search their databases for the required versions and then submit them to the coordinator node at the assigned start time. Each message is associated with the deadline of the query in addition to the data versions and path information.

In order to make the arrival time of a message close to the estimated arrival time $(D_i - C_i)$, a priority scheduling algorithm is adopted at the relay nodes to forward the data versions to the coordinator node. Since the processing power of a sensor node is very limited, the operating system in a sensor node may only support single thread processing and pre-emption in execution is usually not allowed. Therefore, it is usually not preferable to use a sophisticated real-time scheduling especially with pre-emption as adopted in conventional real-time systems. In a wireless sensor system, the biggest time delay in processing a real-time query is usually data communication instead of processing delay at the sensor nodes. Therefore, instead of adopting a priority-cognitive real-time scheduling algorithm in scheduling the processing of the real-time queries, we propose a priority-based message transmission algorithm, for meeting the deadlines of the queries. The priority scheduling is non-preemptive. Each sensor node maintains a data transmission queue. Once it has completed a transmission, it will examine the transmission jobs in the message queue and select the highest priority one for transmission to the next node as defined in the path associated with the data versions in the message.

In our priority assignment policy, the priority assigned to a message is based on its deadline and distance from its coordinator node such as:

Priority of a message M_i at node N_j = (D_i − Current time) / number of hops from N_j to the coordinator.

A higher priority is assigned to a message for transmission if the calculated value is smaller. Note that this is progressive scheme. If a message is far from the coordinator node, it will be assigned a higher priority. Its priority decreases when it is closer to its destination since the hop count is smaller.

5.3 Processing Phase

After the coordinator node has received all the required data versions for a query, it will process them according to the precedence orders of the operations defined in the query. It needs to ensure that the relative consistency requirement is satisfied. Algorithm 2 shows the logic in meeting the relative consistency requirement in query execution.

Objective: Query processing logic for accessing u data items
input: data streams $x_{j,k}$, $1 \leq j \leq u$, $k = 1,2,3,...$
output: query execution result stream
for $j = 1$ to u **do** {
   $k_j := 1$; // initial version index of the jth data stream}
   $LB_i := D_i - \Delta_i$; // initial lower bound of result interval
   while($LB_i < D_i - \Delta_i - R_i$) {
     $UB_i := \min_{j}\{UVT(x_{j,k_j})\}$;
     execute T_i with data $\{x_{1,k_1}, x_{2,k_2},..., x_{u,k_u}\}$;
     result validity interval $I := [LB_i, UB_i]$;
     $J := \{j | UVT(x_{j,k_j}) = UB_i\}$;
     for each j in J **do** {
       $k_j := k_j + 1$;   // use next data version of stream j}
     $LB_i := UB_i$;}

Algorithm 2. Query accessing data versions

6 Conclusions

In this paper, we have studied how to process real-time queries in a wireless sensor system. Although many queries in a wireless sensor network processes real-time properties, the problem has been greatly in previous research although query processing in sensor networks has received a lot of research interests in recent years. To our best knowledge, this is the first study on the problem. Real-time queries are associated with a deadline on their completion times. In processing a real-time query, an important requirement is the temporal consistency. This issue has also been greatly ignored in the previous work in sensor data management. In this paper, we adopt the notion of relative consistency for processing real-time queries. We propose a synchronized parallel scheme called *Parallel Data Shipping with Priority Transmission (PAST)* to collect multiple versions of data items for execution of the query with smaller data transmission cost and transmission delay and at the same time to meet the processing constraints of the queries, i.e., deadline, data currency and result requirements. To meet the transmission deadline, a deadline-based priority scheduling algorithm is designed in PAST to transmit the data versions from a participating node to the coordinator node of the query. An important feature of PAST is that multiple results are generated for a query to improve the accuracy and reliability of the results.

Acknowledgement. This work was supported, in part, by NSF grants IIS-0208758 and CCR-0329609.

References

1. Kam Yiu Lam and Henry C.W. Pang: Correct Execution of Continuous Monitoring Queries in Wireless Sensor Systems. *Proceedings of the Second International Workshop on Mobile Distributed Computing (MDC 2004)*, Tokyo,Japan, March 2004.
2. A. Mainwaring, D. Culler, J. Polastre , R. Szewczyk , J. Anderson: Wireless Sensor Networks for Habitat Monitoring. *Proceedings of the first ACM international workshop on Wireless Sensor Networks and Applications* 2002.
3. S. R. Madden, R. Szewczyk, M. J. Franklin, and D. Culler: Supporting aggregate queries over ad-hoc sensor networks. *Proceedings of Workshop on Mobile Computing and System Applications (WMCSA)*, 2002.
4. S.R. Madden, M. J. Franklin, J. M. Hellerstein, and W. Hong: Tag: A tiny aggregation service for ad-hoc sensor networks. *Proceedings of 2002 OSDI*.
5. S. Madden, M. J. Franklin and J.M. Hellerstein: The Design of an Acquisitional Query Processor For Sensor Networks. *Proceedings of SIGMOD 2003*, June 9-12, San Diego, CA.
6. Mohamed A. Sharaf, Jonathan Beaver, Alexandros Labrinidis, Panos Chrysanthis: TiNA: A Scheme for Temporal Coherency-Aware in-Network Aggregation. *Proceedings of 2003 International Workshop in Mobile Data Engineering*.
7. Narayanan Sadagopan, Bhaskar Krishnamachari, and Ahmed Helmy: Active Query Forwarding in Sensor Networks (ACQUIRE). to appear in *Ad Hoc Networks*.
8. Y. Yao and J. E. Gehrke: The Cougar Approach to In-Network Query Processing in Sensor Networks. SIGMOD Record, vol. 31, no. 3, September 2002.
9. Y. Yao and J. E. Gehrke: Query Processing in Sensor Networks. *Proceedings of the First Biennial Conference on Innovative Data Systems Research (CIDR 2003)*, Asilomar, California, January 2003.
10. Ben Kao, Kam-Yiu Lam, Brad Adelberg, Reynold Cheng and Tony Lee: Maintaining Temporal Consistency of Discrete Objects in Soft Real-Time Database Systems. IEEE Trans on Computers, vol. 52, no. 3, pp. 373-389.
11. Ramamritham, K.: Real-time Databases. International Journal of Distributed and Parallel Databases, vol. 1, no. 2, 1993.

Cluster-Based Parallel Simulation for Large Scale Molecular Dynamics in Microscale Thermophysics

Jiwu Shu, Bing Wang, and Weimin Zheng

Instiue of HPC, Dept. of Computer Science and Technology,
Tsinghua Uni., Beijing, China, 100084
shujw@tsinghua.edu.cn

Abstract. A cluster-based spatial decomposition algorithm for solving large-scale Molecular Dynamics simulation of thermophysics is proposed. Firstly, three kinds of domain division strategies are provided and their efficiency and scalability are analyzed. Secondly, a method called FLNP (Fast Location of Neighboring Particles) to accelerate the location of neighboring particles is proposed, which greatly reduces the cost of calculation and communication of interaction. Additionally, a new memory management technique called AMM (Adaptive Memory Management) is applied to meet the large memory requirement. The parallel algorithm based on these above technologies was implemented on a cluster of SMPs and tested on a system of 6,912,000 particles and achieved an efficiency of 77.0%.

1 Introduction

Recently, more and more researchers are attracted by the microscale problems in thermophysics, which include atomic beam bombardment, liquid-vapor interface and nucleation. Because those processes are all going along in microscopic time and space scale, it is so difficult to control them that up to now, there is no traditional experimental method to measure these processes directly and accurately [1]. Molecular Dynamics (MD) simulation provides a new method for the research of microscale thermophysics, by which researchers can understand these problems at the level of molecules or even atoms.

There are two important characteristics when the MD simulation method is used to study the problems of microscale thermophysics. Firstly, this kind of MD simulation must handle a great number of particles; secondly, the simulating process must go on for numerous time steps (say 1,000,000 or 10,000,000 steps). All of the above factors lead to enormous calculation especially when MD simulation is applied to a very large system with many particles. Researchers have spent much time to simplify MD model and improve MD algorithms, but those various fast methods cannot provide a satisfactory solution to the MD simulation of large-scale system. It is obvious that a complete solution of such a problem must depend on the development of efficient and scalable parallel algorithms. The simulation in PC or workstations can only be applied to a system with

10^3-10^4 particles, and it has been reported that the largest thermophysical MD simulated system in supercomputers is about of 1,000,000 particles [2]. These kinds of system scale can not meet the need of MD simulation in most of thermophysical fields, for example, nucleation. So the task of designing an efficient and scalable parallel MD simulation algorithm is much useful to the research of microscale thermophysics.

In the recent years, the technology of Cluster has been well developed. A cluster is a local computing system comprising a set of independent computers and a network interconnecting them [3]. The Cluster systems have so many advantages that they are widely used in the international areas of scientific and engineering computing [4].

This paper takes a review on current work in parallel MD simulation, and proposes a new spatial decomposition MD algorithm based on Cluster system, which can be used efficiently to simulate a large-scale multi-particle thermophysical system. Firstly, we provide three different domain division strategies and analyze their efficiency and scalability. Secondly, we propose a method called FLNP (Fast Location of Neighboring Particles) to accelerate the location of neighboring particles, which greatly reduce the cost of interaction calculation and communication. The FLNP method has the advantages of both LinkCell and NeighborList and has a good performance in practical simulation. Thirdly, a new memory management called Adaptive Memory Management is presented to reduce the tremendous memory usage, which is always the bottleneck of large-scale MD simulation. With the above algorithm and strategies, we simulate a large system with 6,912,000 particles and get satisfactory results on the PVT property calculation of this system.

2 MD Simulation of Microscale Thermophysics Problems

In MD Simulation [5], all the N simulated particles are treated as a point mass and are given initial positions and velocities. Then, the inter-particle forces are calculated and the positions and velocities are updated after a short simulating time. When repeated for hundreds of thousands of times, the above simulating process can constitute a motion picture of the whole simulated system in microscale.

In our work, we simulate a multi-particle system and calculate its PVT properties. The N particles are simulated in a 3D cubic space with periodic boundary conditions at the state point defined by the reduced density $\rho^* = 0.0685$ and the reduced temperature $T^* = 1.2$. The simulation is begun with the particles on an fcc lattice with randomized velocities. A roughly uniform spatial density persists for the duration of the simulation. The simulation is run at constant N, volume V and temperature T, a statistical sampling from the canonical ensemble.

The computational task in MD simulation is to solve the Newton's equation given by

$$\begin{cases} F_i(t) = m_i \frac{dv_i}{dt} = \sum_j F_2(r_i, r_j) + \sum_j \sum_k F_3(r_i, r_j, r_k) + \dots \\ \frac{dr_i}{dt} = v_i \end{cases} \quad (1)$$

where m_i the mass of particle i, r_i and v_i are its position and velocity vectors. F_2 is a force function describing pair-wise interactions. $F_k(k >= 3)$ describes the multi-body interaction which is ignored in our simulation.

The most time-consuming part in MD simulation is the calculation of interaction, which usually requires 90% of total simulation time. The force terms in Equation (1) are typically non-linear functions of the distance between particle i and the other particles. In our simulation, the interaction can be modeled with a Lennard-Jonse potential energy as

$$\phi(r) = 4\varepsilon[(\frac{\sigma}{r})^{12} - (\frac{\sigma}{r})^{6}] \qquad (2)$$

where r is the distance between two interacting particles, ε and σ are constants. In a long-range model, each particle interacts with all the other $(N-1)$ particles, which leads a computational complexity of $O(N^2)$. But many physical problem can be modeled with short-range interaction, that is, the summations in Equation (1) can be restricted to particles within some small region surrounding particle i. We can implement it using a cutoff distance r_c, outside of which all interactions are ignored. In this case, the interaction calculating complexity was reduced to $O(N)$. In our simulation, the cutoff distance r_c is 4.0σ. How to minimize the number of neighboring particles that must be checked for possible interactions is an important problem, which can greatly influence the speed of short-range MD simulation.

3 Parallel MD Simulation

In the past twenty years, researchers have developed three classes of parallel MD simulation algorithms. The first class is called Atom Decomposition (AD) [6]. Both the replicated-data strategy [7,8] and the systolic-loop method [9] belong to this kind of class. The main shortcoming of this algorithm is the enormous memory requirement because each processor must maintains the positions of all the particles. Only when dealing with MD system of a small number of particles on shared memory machines, the AD algorithm can gives a good performance. The second class is Force Decomposition algorithm (FD) [10]. This kind of algorithms doesn't need all of the particles' position so it requires much less memory than AD algorithm. But FD algorithm cannot maintain load balance as easily as AD algorithm can, and only when the force matrix has a uniform sparse structure, the FD algorithms can achieve a good load balance. The last class is Spatial Decomposition (SD)[11]. The main benefit of SD is that it takes full advantages of the local nature of the inter-particle forces and performs only local communication. Thus, in large-scale MD simulation, it achieves optimal $O(N/P)$ scaling and achieves better performance on Cluster than AD and FD algorithms. Therefore, we chose the SD algorithm as our simulation method and propose three kinds of domain division strategies to make the SD algorithm more efficient and more scalable.

Our work is to design an efficient and scalable parallel algorithm of MD simulation on cluster systems, in which the most popular parallel programming environment is Message Passing Interface (MPI). Under MPI programming model, an efficient parallel algorithm must meet the following needs: (1) to increase the asynchronous calculating granularity in each processor, and reduce the inter-processor synchronizing communication cost; (2) to maintain a good load balance among all processors to reduce the inter-processor waiting time; (3) a high scalability. After a elaborate comparison among the AD, FD and SD algorithms, we finally chose the SD.

3.1 Domain Division Strategies

Domain division is the most important technique in SD algorithm, which can greatly affect the load balance, communication cost and the scalability of the parallel algorithms [12, 13]. Three different kinds of domain division strategies are discussed and analyzed here, as described in figure (1.a, 1.b, 1.c). For the convenience of discussion, suppose the whole simulating domain is divided into P sub-domains $\sum_i (i = 1, 2, \ldots P)$, the number of processor is also P, $P_i(i = 1, 2, \ldots P)$. The P sub-domains are assigned to P processors separately. That is, processor P_i computes the interactions on particles in sub-domain $\sum_i$, and updates their positions and velocities. Below we discuss the differences of these three strategies in load-balance, communication and scalability.

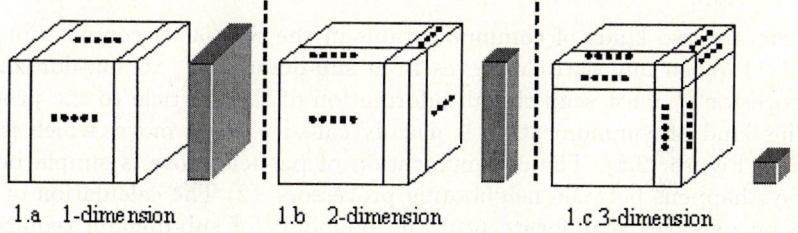

1.a 1-dimension   1.b 2-dimension   1.c 3-dimension

Fig. 1. domain division strategies

Firstly, we discuss the performance of these three strategies in load balance. We can draw a conclusion that the 1-dimension division showing in Figure (1.a) can achieve the best load balance because of two reasons. (1) The load imbalance is caused mainly because the non-uniformity of particle density, which has the least effects on 1-dimension division of the three division strategies. Suppose the simulating domain is scaled as x, y, z in three dimensions, and the 1-dimension division in Figure (1.a) is implemented in x dimension. Only the non-uniformity of particle density in x direction can influence the load balance of 1-dimension division. On the contrary, no matter which dimension the non-uniformity occurs in, the load balance of 3-dimension in Figure (1.c) can be greatly influenced. (2) The algorithm with 1-dimension division strategy can implement dynamic load balance more easily than that with 3-dimension division strategy. The communication architecture of 1-dimension division is easier than that of 3-dimension division. Thus the algorithm with 1-dimension division strategy can easily re-divide the sub-domains locally or globally when the particle density alters. On the other hand, the algorithm with 3-dimension cannot achieve an easy implementation of dynamic load balance due to complex communication. Generally speaking, the cost to maintain a good load balance of the three kinds of domain division strategies can be expressed as below:

$$B_1 < B_2 < B_3 \tag{3}$$

where $B_i (i = 1, 2, 3)$ are the total cost due to a load imbalance, in 1-dimension, 2-dimension, 3-dimension, separately.

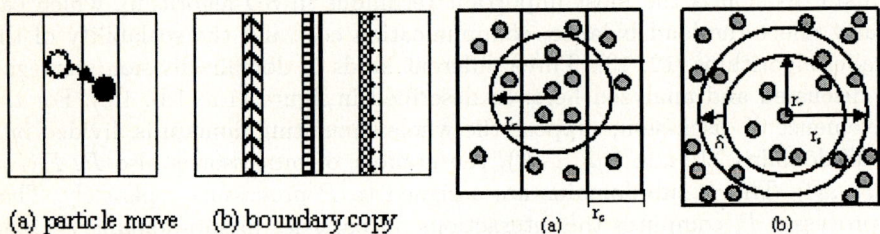

Fig. 2. communications in parallel SD algorithm

Fig. 3. NeighborList Method (a) and LinkCell Method (b)

There are two kinds of communications in the Spatial Decomposition algorithm. (1) When any particle moves from sub-domain $\sum_i$ to sub-domain $\sum_j$, the processor P_i must send the all information of this particle to the processor P_j. This kind of communication is usually called particle move, which is illustrated in Figure (2.a). The communication of particle move is simple because it always happens between neighboring processors. (2) The calculation of interaction on particles that locate near the boundary of sub-domain requires the positions of other particles that maybe belong to another processor, which lead to an exchange of particle position called boundary copy illustrated in Figure (2.b). In short-range MD simulation, the exchanges involve only those particles whose distance to boundary is within cutoff distance r_c.

We compare the communication cost of three domain division strategies in both communication data volume and communication time. (1) Under the 1-dimension division, processor P_i need only communicate with the two neighboring processors P_{i-1} and P_{i+1}, and each time step there are two communications at most. Under the 2-dimension division, processor P_i must communicate with 8 neighboring processors, which requires 4 communications even using fold [14] technology. Under the 3-dimension division, the number of neighboring processors is 26 and the times of communication are 6. (2) The main task of communication is boundary copy and so we analyze the communicate data volume of boundary copy. The data volume can be expressed in the following equations

$$\begin{cases} C_1 = (N/\rho)^{2/3} \times \rho \times 2 \\ C_2 = (N/\rho)^{2/3}/P^{1/2} \times \rho \times 4 \\ C_2 = (N/\rho)^{2/3}/P^{2/3} \times \rho \times 6 \end{cases} \tag{4}$$

where C_1, C_2, C_3 are the total communication data volumes of 1-dimension, 2-dimension and 3-dimension division strategies, respectively. N is the particle number of the simulated system, P is the processor number on Cluster system and ρ is the particle number in each box. We have

$$C_1 : C_2 : C_3 = 1 : 2/P^{1/2} : 3/P^{2/3} \tag{5}$$

Specially, when $P > 6$ (this condition can be easily achieved), we have

$$C_1 > C_2 > C_3 \tag{6}$$

Equation (6) shows that, the total communication data volume of 3-dimension is the least and that of 1-dimension the greatest. Further experimental result proves that the communication data volume is the dominant factor in total communication cost of large-scale parallel MD simulation.

At last, we will compare the scalabilities of the three strategies. Generally speaking, the 3-dimension division strategy performs better than the 1-dimension division. Two facts make us reach this conclusion. (1) Equation (4) shows that when P becomes larger, the communication data volume reduces rapidly with 3-dimension division but remains constant with 1-dimension. Thus when more and more processors are used, the algorithm with 1-dimension becomes more and more inefficient. The 1-dimension division strategy limits the scalability of parallel algorithms. (2) When N is fixed, the number of sub-domains in SD algorithms is limited by the boundary copy. Speaking in detail, the sub-domain must be longer than an individual box in the dividing direction; otherwise the communication must become more complex. The length of a box is r_c or r_s when using FLNP method, which is discussed bellow. Form the discussion in (1) and (2), we can conclude that the algorithm with 3-dimension division strategy is the most scalable MD simulation algorithm on Cluster.

From the above discussion, we can draw the conclusion that when the load balance is well maintained, the 3-dimension division is the best domain decomposition strategy for the parallel MD simulation on Cluster.

3.2 The FLNP Method

In short-range MD simulation of a system with N particles, in order to calculate the force on particle i, we need not check all of the other $(N-1)$ particles because only those particles who are within the cutoff distance r_c can contribute to the force on particle i. There are two basic techniques used to accomplish this.

In the first idea, the LinkCell [15] method, the simulating domain is divided into many 3D cells of side length d, where d equal to r_c or slightly larger, as illustrated in Figure (3.a) and each particle is mapped to some cell. This reduces the task of finding neighbors of a given particle to checking in 27 cells, that is, the cell which this particle is in and the 26 surrounding ones. Since mapping the particles to cells only requires $O(N)$ work, the original $O(N^2)$ work required by force calculation is greatly reduced.

The other technique used for speeding up MD calculation is known as NeighborList [16] as illustrated in Figure (3.b). When the list is built, all of the nearby particles within an extended cutoff distance $r_s = r_c + \delta$ are stored. The list is used to calculate interactions for a few time steps. Then before any particles could have moved from a distance $r > r_s$ to $r < r_c$, the list is rebuilt. The NeighborList method has the advantages that when the list built, checking all of the possible neighboring particles in list is much faster than checking all particles in simulated system. However, the process of list building and rebuilding still requires checking all of the simulated particles.

Based on the analysis and comparisons between them, we can see that Link-Cell and NeighborList are two speedup techniques coming in different ways. The former is an optimizing strategy in the spatial aspect. On the other hand, the latter, the NeighborList technique concerns more about optimizations in time aspect. When NeighborList applied, the neighboring particle lists can be kept unchanged in several continuous time steps, in which it is not necessary to rebuild the list and the neighboring particle searching is not time-consuming at all. Generally speaking, from two different aspects, the space and the time, two different speedup techniques are developed. If a strategy concerns not only spatial but also time aspect, it must benefit from both and a better speedup performance is expected. The FLNP (Fast Location of Neighboring Particles) strategy proposed in this paper is just the combination of both LinkCell and NeighborList. Firstly, with this method, the whole simulated domain is divided into many cells with the side length r_s. To a particular particle A, any particles which are contributed to the interactions on A are within the neighboring cells of A (including the cell where A is). This is similar to LinkCell method. Secondly, like the NeighborList technique, for each particle A, a neighbor list is maintained, recording information of all particles with a distance of r_s. It is not necessary to rebuild the list in each time step. Only when some particles outside r_s go into the cutoff distance r_c, the list is rebuilt. This new method has obvious advantages relative to basic LinkCell and NeighborList techniques. Firstly, compared to LinkCell method, it reduces the number of particles that should be checked because the there are far fewer particles to check in a sphere of volume $\frac{4}{3}\pi r_s^3$ than in a cube of volume $27r_s^3$. On the other hand, compared to basic NeighborList method, there is a significant time saving when list is rebuilt because the searching volume has been reduced from the whole simulated domain to $27r_s^3$. In general, FLNP technique combines the benefits of both LinkCell and NeighborList, and optimizes the particle searching task with reducing both the searching volume and the searching times.

Moreover, the FLNP method can reduce the calculation and communication cost due to particle move and boundary copy. This is caused mainly by two reasons. (1)The FLNP maintains a neighbor list for each particle, which stores all of the particles that can possibly contribute to the force calculation. In algorithms not using FLNP, when some particles near the boundary move from one sub-domain to another, the neighbor list has to be rebuilt. But in algorithms using FLNP, if these moving particles don't enter other particles' extended cutoff distance, the neighbor list can be rebuilt late. So the communication cost of particle move can be reduced to some extent. (2)When boundary copy (as illustrated in Figure (2)) occurs, processors must check which particles are near the boundary and must be sent to the neighboring processors. This work must be done at each time step in algorithms not using FLNP; but in algorithms using FLNP, it can be done once every few time steps, when the neighbor lists are rebuilt. During all the other time steps, we can easily send the latest position information of particles that have been checked as boundary particles in the previous time step. It must be pointed out that the basic NeighborList strategy can also bring out the two benefits.

Furthermore, the calculation cost can be half reduced, when applying Newton's Third Law, which shows that,

$$F_{ij} = -F_{ji} \tag{7}$$

In NeighborList method, we can implement this speedup by simply ignoring the particle j in the neighboring list of particle i.

3.3 Adaptive Memory Management

The large-scale MD simulation involves a great number of particles, which require a lot of memory resources. In fact, the usual workstations can not simulate a large system with about 1,000,000-100,000,000 particles due to shortage of local memory. Though a cluster may be rich in distributed memory resource, it is necessary to use the memory efficiently because memory is the bottleneck to increase the scale of simulated system.

In our algorithm, an Adaptive Memory Management (AMM) is put into practice, with which the usage of memory is optimized. The AMM technique can help to manage the memory usage efficiently by dynamically allocating and releasing memory according to the need of simulation. In parallel SD algorithms, each processor is responsible to particles in a fixed sub domain. Because of the motion of these particles, the number of particles belonging to some processor is changing from time to time. So the memory need of each processor is also unfixed. If each processor simply allocates superfluous memory for all of N particles, the memory is not used efficiently and the simulated system is restricted to a smaller scale. With AMM technique, each processor is given a fixed number of memories when parallel program is initialed. In simulating process, each processor inspects the particle number of its own sub domain. When allocated memory is not enough to hold the particles, the processor allocates double number of memory. On the other hand, when particles decrease in number to such a degree that 2/3 of the allocated memory is useless, the processor releases half of its memory. This technique brings little additional cost, because the change of density of simulated particles is not abrupt and the number of memory allocation and release is usually small. Practically, one user-level function ReallocateMem(double ratio) is provided to dynamically manage the memory usage within the SMP node. In fact, when NeighborList method is applied, the largest system our cluster can simulate is of about 1,000,000 particles. And with AMM technique, we successfully simulated a very large system with 6,912,000 particles.

4 Results and Analysis

The parallel MD algorithm of Section 3 was tested on our Cluster system. This Cluster is made up of 36 SMP nodes. Each node has 4 CPUs of Intel Xeon PIII700, 36Gbytes of hard disk, and 1Gbytes of memory. The communication medium between SMP nodes is Myrinet Switch with bandwidth of 2.56Gb/s. The software environments are Redhat Linux 7.2(kernel version 2.4.7-10smp), MPICH-1.2.7 and gm-1.5pre4 which is network protocol running on Myrinet. The simulated physical system is a canonical ensemble with 6,912,000 Argon atoms and our simulation gives the accurate phase figures.

4.1 Comparison of Domain Division Strategies

Figure (4) shows the performance curves of three kinds of domain division strategies separately. Generally speaking, the algorithm with 3-dimension division gets the highest performance and the one with 1-dimension division the lowest.

Figure (4) also shows that the three kinds of domain division strategies have similar parallel efficiency when P is small (say $P \leq 9$ processors). When more and more processors are used, the efficiency of 1-dimension division drops down quickly. On the contrary, the declination of efficiency of 2-dimension and 3-dimension division is slight. We can draw the conclusion that the algorithm with 3-dimension division is the most efficient and most scalable for MD parallel simulation on Cluster system. The algorithm with 2-dimension division also has a fine scalability but it is less efficient than that with 3-dimension division. The algorithm with 1-dimension is the worst one because of its awful efficiency and scalability, and it can provide an acceptable performance only when P is small.

Figure (5) illustrates the communication cost of three domain division strategies. With processors of similar numbers, the communication cost of 1-dimension strategy is must higher than those of 2-dimension and 3-dimension strategies and the cost of 3-dimension is a little lower than that of 2-dimension. It is maybe noted that the experimental data in Figure (5) seems not consistent with Equation (5). The reason is that it is the whole communication costs which are illustrated in Figure (5) but in Equation (5) only the communication data

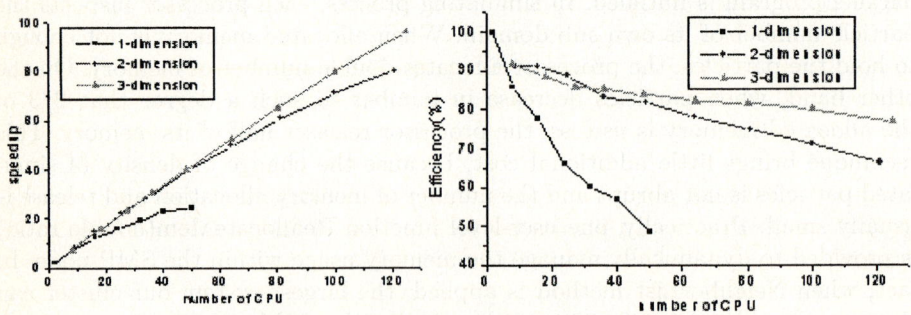

Fig. 4. speedup and efficiency of domain division strategies

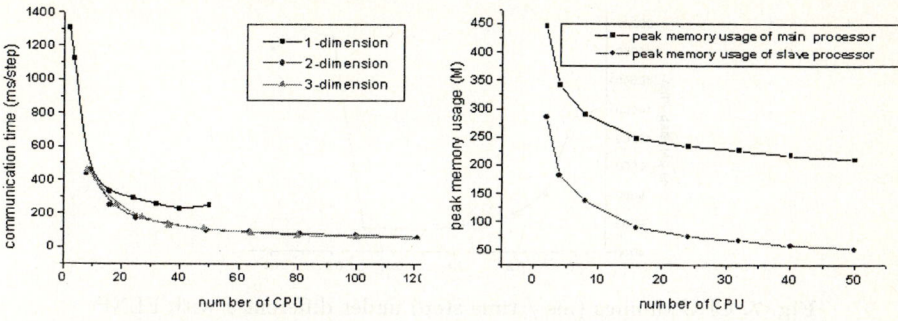

Fig. 5. the communication cost of domain division strategies

Fig. 6. peak memory usage when AMM is applied

volumes are analyzed. Obviously, the whole communication cost is not only related to the communication data volume, but also depends on other factors: (1) the communication times and communication mode. Even if equal data volumes are communicated, complex global communication mode is much more time-consuming than some simple modes. (2) The synchronization cost. With SD algorithm, global synchronization among all processes is necessary in every time step. All of the three decomposition strategies, 1-dimension, 2-dimension and 3-dimension, have similar communication mode and synchronization operation. Therefore, though there are great gaps among communication data volumes of the three strategies, the differences of the whole communication costs of them are not so apparently.

4.2 Influence of FLNP to Parallel Efficiency

In Figure (7), we plot the 3-dimension algorithm's computing time per step under different δ. The processor number is 8 and the particle number is 6,912,000.

The experimental result shows that, the FLNP method can bring much greater improvement to parallel algorithm's speed than the two basic technologies: LinkCell and NeighborList. Firstly, LinkCell is described with the result obtained when δ equals to zero in Figure (7), which shows that the speed with FLNP is about double to the speed with LinkCell. Secondly, the basic NeighborList technology cannot be use separately on MD simulation of such a large-scale system that has 6,912,000 particles. In fact, when 8 processors used, each processor must handle 864,000 particles averagely. If basic NeighborList technology would be used, it should have taken dozens of hours to build the neighbor list once.

The result also shows that, the value of δ can influence the speed of parallel algorithm, which requires a precise value of δ. The optimal value of δ for our simulation is in the scope of $[0.5\sigma, 0.6\sigma]$.

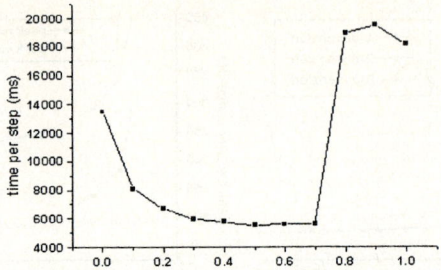

Fig. 7. CPU timings (ms / time step) under different δ with FLNP

4.3 Memory Usage with AMM

When AMM is not used, the system with 6,912,000 particles can not be simulated in our cluster because of limitation of memory. In fact, each processor must allocate more than 6GB memory to hold possible particles, which exceeds the greatest number of memory a 32-bit processor can directly access, 4GB. When AMM is applied, the memory need is greatly reduced and the 6,912,000 particle system is successfully simulated. Figure (6) shows the memory usage of parallel SD algorithm with AMM and 1-dimension strategy. If only concerning memory, the largest system our cluster can simulate may have 100,000,000 particles.

5 Conclusion

In this paper, we design and implement a Cluster-based spatial decomposition algorithm, which is suitable to the large-scale MD simulation of microscale thermophysical problems. Another important optimizing strategy in short-range MD simulation is to minimize the number of neighboring particles that must be checked for possible force calculation. This paper proposes and implements a new method called *fast location of neighboring particles*, which combines the benefits of both *link-cell* and *neighborlist* and can greatly accelerate the calculation of interaction. δ is the most important parameter in this new method, which can greatly influence the efficiency of parallel algorithm. The memory is the bottleneck in large-scale MD simulation. An adaptive memory management is provided to our algorithm, with which the memory is used efficiently and the simulated systems are increased to a larger scale.

Acknowledgements

The work in this paper was partly supported by a 985 Basic Research Foundation of Tsinghua University, P.R. China (Grant No.JC2002027, No. JC2001024).

References

1. Chou F C, Lukes J R, Liang X G et al , Molecular Dynamics in Microscale Thermophysical Engineering. Heat Transfer, 10(1999), 141-176

2. BFeng XiaoLi, Li ZhiXin, Guo ZengYuan, Molecular dynamics study on thermal conductivity and discussion on some related topics. Journal of engineering thermophysics, 2(22),(2001),195-198
3. Mark Baker, Cluster Computing White Paper - Final Release (Version 2.0) 28,December,2000
4. Kengo Nakajima, and Hiroshi Okuda, Parallel iterative solvers for unstructured grids using a directive/MPI hybrid programming model for the GeoFEM platform on SMP cluster architectures, Concurrency Computat.: Pract. Exper. 14:411,2002
5. J. M. HAILE, Molecular Dynamics Simulation Elementary Methods. (Wiley Professional Paperback Edition Published 1997)
6. D.L. Greenwell, R.K. Kalia, J.C. Patterson, P.Vashishta, Molecular Dynamics Algorithm on the connection machine, Int. J. High Speed Computing 1(2),(1989),321-328
7. W. Smith, A replicated data molecular dynamics strategy for the parallel Ewald sum, Comp. Phys. Comm. 67(3),(1992),392-406
8. W. Smith, T.R. Forester, Parallel Macromolecular simulations and the replicated data strategy, Comp. Phys. Comm. 79(1),(1994),52-62
9. D. Okunbor, Integration methods for N-body problems, Proceedings of the Second International Conference On Dynamics Systems (1996)
10. Ravi Murty, Daniel Okunbor , Efficient Parallel Algorithms For Molecular Dynamics Simulations, Parallel Computing 25(3),(1999), 217-230
11. S.Plimpton, Fast parallel algorithms for short-range molecular dynamics, J.Comput.Phys. 117(1),(1995),1-19
12. Ryoko Hayashi, Susumu Horiguchi, Parallel molecular dynamics simulations of polymers (In Japanese), Transactions of Information Processing Society of Japan, 39(6),(1998),1775-1781
13. Shu Jiwu, Zheng Weimin etc, Parallel computing for lattice Monte Carlo simulation of large-scale thin film growth, Science in China(Series F) 45(2),(2002)
14. G. C. Fox, M. A. Johnson, G. A. Lyzenga, S. W. Otto, J. K. Salmon, and D. W. Walker, Solving Problems On Concurrent Processors: Volume I (Prentice Hall, Englewood Cliffs, NJ, 1988)
15. R. W. Hockney, S. P. Goel, and J. W. Eastwood, Quiet high-resolution computer models of a plasma, J. Comput. Phys, 14(48),(1974)
16. L. Verlet, Computer experiments on classical fluids. I. Thermodynamical properties of Lennard-Jones molecules. Phys. Rev. 159(98),(1967)

Parallel Checkpoint/Recovery on Cluster of IA-64 Computers[1]

Youhui Zhang, Dongsheng Wang, and Weimin Zheng

Department of Computer Science, Tsinghua Univ.,
100084, Beijing, P.R.C
zyh02@mail.tsinghua.edu.cn

Abstract. We design and implement a high availability parallel run-time system---ChaRM64, a Checkpoint- based Rollback Recovery and Migration system for parallel running programs on a cluster of IA-64 computers. At first, we discuss our solution of a user-level, single process checkpoint/recovery library running on IA-64 systems. Based on this library, ChaRM64 is realized, which implements a user-transparent, coordinated checkpointing and rollback recovery (CRR) mechanism, quasi-asynchronous migration and the dynamic reconfiguration function. Owing to the above techniques and efficient error detection, ChaRM64 can handle cluster node crashes and hardware transient faults in a IA-64 cluster. Now ChaRM64 for PVM has been implemented in Linux and the MPI version is under construction. As we know, there are few similar projects accomplished for IA-64 architecture.

1 Introduction

Cluster of computers (COCs) is a parallel system using off-the-shelf computers connected through high-speed network. Eight of the top ten machines in the November 2003 Top500 [1] list are cluster systems. While the growth in CPU count has provided great increases in computing power, it also presents significant reliability challenges to applications. Failures in the computing environment are making it more difficult to complete long-running jobs.

PVM and MPI are popular systems for message passing parallel programming. However, their implementation itself does not specify any particular kind of fault tolerant behavior. To fully increase the system availability, it is necessary to enhance these parallel running systems. Checkpointing & Rollback Recovery (CRR) and Process Migration offer a low overhead and full solution to this problem [2,3].

On the other hand, 64-bit architecture will become the main trend of high performance computing and IA-64 is one important architecture in this field. IA-64 combined numerous innovations in its design, which contain massive resources such as rich instruction set and large register files [4]. However, some characteristic features such as huge register files, the register stack, and the backing store, bring up difficulties when the current CRR software is ported to IA-64 and few works have been done in this field, as we know.

---

[1] Supported by High Technology and Development Program of China (No. 2002AA1Z2103).

We analyze the problems on IA-64, and provide our solution to implement the user-level, single process checkpointing software for IA-64 Linux systems. Based on the tool, we implement a Checkpoint-based Rollback Recovery and Migration system for parallel running programs on the IA-64 version of Linux, which is named ChaRM64 and owns the following features:

1. User-level, user-transparent implementation. So it is not necessary for users to modify their source code.
2. It is implemented on the top of the message passing system (such as PVM and MPI), and the approach can be easily adapted for different platforms.
3. A coordinated CRR mechanism and a quasi-asynchronous process migration are realized.
4. It supports dynamic reconfiguration and n-node fault tolerance.
5. As we know, it is the first parallel CRR implementation on IA-64 platform. Now ChaRM64 for PVM has been implemented and the MPI version is under construction.

The next section illustrates the key techniques employed in our single process checkpointing software for IA-64. And then the structure of ChaRM64 is described in Section 3. Section 4 gives some future works.

2 Process Checkpointing and Rollback Recovery on the IA-64 Architecture

2.1 Traditional Methods

We will discuss the user-level implementation of checkpointing and recovery in the following sections. We take libckpt [5] as our paradigm, a transparent checkpointing library that has been ported to many platforms. Upon checkpointing, the checkpointing procedure is called from signal handler or from the user program. In the simplest case, the procedure will save the process state in the following order:

1. Call setjmp() to save the current CPU sate to a memory area.
2. Save the contents of the data segment to the checkpoint file.
3. Save the stack segment to the checkpoint file.

2.2 What Is Special on IA-64: Register Stack and Backing Store

One characteristic feature of IA-64 architecture is its register stack. There are 128 general registers in total, which are split in two subnets. The first 32 general registers are static; the following 96 registers named from r32 to r127, are organized as a register stack. Each procedure frame has two regions on the register stack: the input parameters and local variables, and the output parameters. On a procedure call, the output region of the caller frame is renamed to be the input region of the procedure frame being called. This is similar to "shifting" the register stack, so that each procedure frame accesses their registers from the same base. On return, the registers are restored to the previous state. The renaming and restoring are automatically done by hardware.

There are only 96 registers allocated for the register stack. When the depth of the register stack exceeds this limit, contents of the old registers will be stored to a memory area called backing store, so that registers are freed up for reuse.

So, a process running on IA-64 has two stacks: one conventional memory stack, and one register stack. Part of the contents of the register stack lies in the CPU registers; the rest of them are in the memory. To checkpoint/restore a process on IA-64, we must find out ways to save/restore the states of the CPU and the register stack.

2.3 Solutions on the IA-64

Most of the frame work of libckpt still works but some problems must be handled with special attention.

1. CPU State. In libckpt and most user-level checkpointing software, the setjmp/longjmp function in the GNU C Library is used to save/restore the CPU state to memory. On IA-64 the implementation of setjmp/longjmp will not save/restore any registers in the register stack, but they will save/restore the position of the register stack. Even when the registers had been flushed, the longjmp() function will operate through the RSE so that the register stack will be rewinded to the original position.
2. Saving the Register Stack. Since the setjmp() function does not save the contents of the register stack, we must save them manually. On IA-64, the instruction flushrs is responsible for flushing the registers to the backing store. After flushrs is executed, only the current procedure frame has its registers in the CPU register file. Registers of previous procedure frames will be flushed to the backing store.
3. Dual-Stack Recovery. Recovery on IA-64 requires both the memory stack and the register stack be recovered before resuming the program execution. To avoid direct operation on the register file and the RSE, we may simulate a deep series of function calls, to make the size of the register stack much larger than the one in the checkpoint file, so that the recovery will not overwrite the contents of the register files.

3 ChaRM64

3.1 Overview

ChaRM64 is a distributed run-time system with high availability for the reliable execution of parallel application programs on COCs.

This system consists of a checkpoint manager module (C_manager), a checkpointing and rollback recovery module (ChaRM_CRR), a state monitor module (WatchDaemon) and a process migration and checkpoint file mirror module (Mig & Mir).

During fault-free operation, the system is invoked periodically to save system consistent states; corresponding processes are coordinated by C_manager to take a checkpoint. When a fault is detected, the rollback recovery is trigged, and system states restoring, restarting and necessary process migration could recover the fault.

In ChaRM64 system, application processes in one node save their checkpoints information and in-transit messages to local disk instead of network file system, this can avoid producing burst network traffic and reduce the checkpointing time. Checkpointing to local disk can recovery any number of node transient faults. ChaRM64 uses RAID like checkpoint mirroring technique to tolerate one or more node permanent faults. Each node uses a background process to mirror the checkpoint file and related information to other nodes besides its local disk. When some node fails, the recovery information of application processes running on the node will be available on other nodes.

Using ChaRM64, transient faults can be recovered automatically, and the node permanent fault can also be recovered through checkpoint mirroring and process migration techniques. If any node that is running the computation drops out of the COC, due to failure, load, ownership or software/hardware maintenance, the computation can be continuous. An available node can also be rejoining the system to compute dynamically.

Now ChaRM64 for PVM has been implemented in Linux. Besides checkpointing the IA-64 process, some other key technologies, including process ID mapping, functions wrapping and renaming, exit/rejoin mechanism and signal/message notification, are employed [6].

3.2 Performance

Several programs is selected to evaluate the overload of the IA-64 checkpointing software on two Intel servers equipped with one 1.30 GHz Itanium 2 CPU. The test focuses on the running time overhead with or without checkpointing and results are listed in Table 1.

Table 1. Testing Results of ChaRM64

Program	Checkpoint Number	Number of processes	Running Time without Checkpointing (second)	Running Time with ten checkpoints (second)	Running time Overhead (%)
matrix-multiplication (512x512, 100)	10	9	127.32	129.36	1.6
matrix-multiplication (512x512, 500)	10	9	634.62	677.55	6.7
matrix-multiplication (512x512, 1000)	10	9	1290.55	1352.44	4.8
matrix-multiplication (512x512, 3000)	10	9	3991.63	4439.35	11
Pi	10	12	9.58	10.35	8
Fractal	10	9	42.76	47.26	10

The first program is a matrix multiply program. In the parallel version, the data is distributed among the worker tasks that perform the actual multiplication and send back their respective results to the master task. The second, Pi, calculates π using a "dartboard" algorithm. All processes contribute to the calculation, with the master averaging the values for π. The last is a parallel fractal program.

At first we execute programs with different arguments without checkpointing at all and calculate the running time. Then, programs are recompiled to own checkpointing functions and checkpointed 10 times during the running process. We record the running time. It is obvious that the time overload introduced by our checkpointing software is small in respect that the ratio of the extra time to the normal is always less than 15%.

4 Future Work

In this paper we focused on the mechanics of checkpointing and recovery, rather than performance and optimization. The 64-bit addressing of the IA-64 architecture provided the ability to utilize a huge memory for some super-computing tasks. For processes occupying tens of giga bytes of memory, checkpointing will be a time-consuming work, so that there will be a greater demand for optimizations, such as incremental checkpointing and user-directed checkpointing. At the same time we are waiting for USA Argonne National Laboratory to release a stable mpich2 version for IA-64. And then we will implement ChaRM64 for MPI soon.

References

1. Top500 supercomputer list, November 2003. http://www.top500.org/.
2. Elnozahy E N, Johnson D B, Wang Y M. A Survey of Rollback Recovery Protocols in Message-Passing System. Technical Report. Pittsburgh, PA: CMU-CS-96-181. Carnegie Mellon University, Oct 1996.
3. Elnozahy E N. Fault tolerance for clusters of workstations. Banatre M and Lee P (Editors), chapter 8, Spring Verlag, Aug. 1994.
4. Sverre Tarp. IA-64 architecture: A detailed tutorial. CERN-IT Division. November 1999.
5. James S. Plank, Micah Beck, Gerry Kingsley and Kai Li, "Libckpt: Transparent Checkpointing under Unix", Conference Proceedings, Usenix Winter 1995 Technical Conference, New Orleans, LA, January, 1995, pp. 213-223.
6. M. Litzkow and M. Solomon. The Evolution of Condor Checkpointing, 1998.

Highly Reliable Linux HPC Clusters: Self-Awareness Approach

Chokchai Leangsuksun[1],[*], Tong Liu[2], Yudan Liu[1],[*], Stephen L. Scott[3],[**], Richard Libby[4], and Ibrahim Haddad[5]

[1] Computer Science Department, Louisiana Tech University
[2] Enterprise Platforms Group, Dell Corp.,
[3] Oak Ridge National Laboratory
[4] Intel Corporation
[5] Ericsson Research

{box, yli010}@latech.edu, Tong_Liu@dell.com, scottsl@ornl.gov,
rml@hpc.intel.com, ibrahim.haddad@ericsson.com

Abstract. Current solutions for fault-tolerance in HPC systems focus on dealing with the result of a failure. However, most are unable to handle runtime system configuration changes caused by transient failures and require a complete restart of the entire machine. The recently released HA-OSCAR software stack is one such effort making inroads here. This paper discusses detailed solutions for the high-availability and serviceability enhancement of clusters by HA-OSCAR via multi-head-node failover and a service level fault tolerance mechanism. Our solution employs self-configuration and introduces Adaptive Self Healing (ASH) techniques. HA-OSCAR availability improvement analysis was also conducted with various sensitivity factors. Finally, the paper also entails the details of the system layering strategy, dependability modeling, and analysis of an actual experimental system by a Petri net-based model, Stochastic Reword Net (SRN).

1 Introduction

One of the challenges in a clustered environment is to keep system failure to a minimum and to provide the highest possible level of system availability. If not resolved in a timely fashion, such failures can often result in service unavailability/outage that may impact businesses, productivity, national security, and our everyday lives. High-Availability (HA) computing strives to avoid the problems of unexpected failures through active redundancy and preemptive measures. Systems that have the ability to hot-swap hardware components can be kept alive by an OS runtime environment that understands the concept of dynamic system configuration. Furthermore, multiple

---

[*] Research supported by Center for Entrepreneurship and Information Technology, Louisiana Tech University.
[**] Research supported by U. S. Department of Energy grant, under contract No. DE-AC05-00OR22725 with UT-Battelle, LLC.

head-nodes (sometimes called service nodes) can be used to distribute workload while consistently replicating OS runtime and critical services, as well as configuration information for fault-tolerance. As long as one or more head-nodes survive, the system can be kept consistent, accessible and manageable.

2 HA-OSCAR Architecture

A typical Beowulf cluster consists of two node types: a head node server and multiple identical client nodes. A server or head node is responsible for serving user requests and distributing them to clients via scheduling/queuing software. Clients or compute nodes are normally dedicated to computation [4]. However, this single head-node architecture is a single-point-of-failure-prone of which the head-node outage can render the entire cluster unusable.

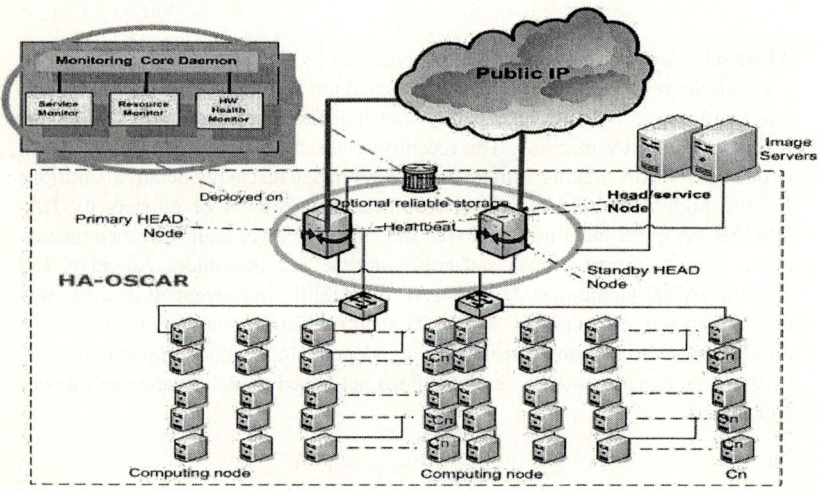

Fig. 1. HA-OSCAR architecture

There are various techniques to implement cluster architecture with high-availability. These techniques include active/active, active/hot-standby, and active/cold standby. In the active/active, both head nodes simultaneously provide services to external requests and once one head is down, the other will take over total control. Whereas, a hot standby head node monitors system health and only takes over control when there is an outage at the primary head node. The cold standby architecture is very similar to the hot standby, except that the backup head is activated from a cold start.

Our key effort focused on "simplicity" by supporting a self-cloning of cluster master node (redundancy and automatic failover). Although, the aforementioned failover concepts are not new, HA-OSCAR simple installation, combined HA and HPC architecture are unique and its 1.0 beta release is the first known field-grade HA-Beowulf cluster release.

3 HA-OSCAR Serviceability Core

Our existing HA-OSCAR and OSCAR installation and deployment mechanism employs Self-build- a self-configuration approach through an open source OS image capture and configuration tool, SystemImager, to clone and build images for both compute and standby head nodes. Cloned images can be stored in a separate image servers (see Figure 1) which facilitate upgrade and improve reliability with potential rollback and disaster recovery.

3.1 Head-Node Cloning

As stated in Section 3, our approach of removing the head node single-point-of-failure is to provide a hot-standby for the active head node. The hot-standby head node is a mirror of the active node and will process user requests when the active head node fails. To simplify the construction of this environment, we begin with a standard OSCAR cluster build on the active node and then "clone" or duplicate that to the hot-standby head node. As shown in Fig. 3, the network configuration differs between the active and hot-standby node by the public IP address, thus they are not exact duplicates of one another. Presently, hardware must be identical between the active and standby machine. Once cloned, both head nodes will contain the Linux operating system and all the necessary components for an OSCAR cluster [4] including: C3, LAM/MPI, LUI, Maui PBS Scheduler, MPICH, OpenSSH, OpenSSL, PBS, PVM, and System Installation Suite (SIS).

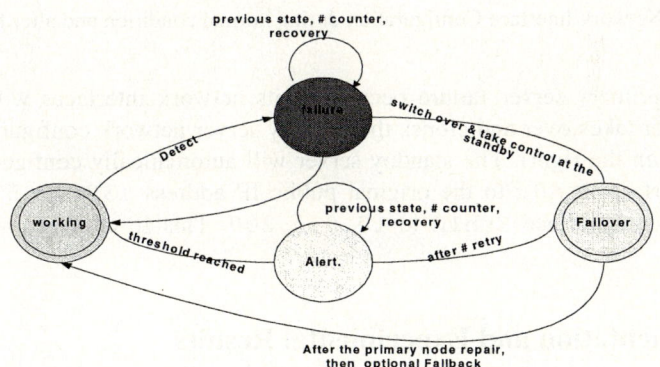

Fig. 2. Adaptive Self-Healing State Diagram

3.2 Adaptive Self-Healing (ASH) Technique

HA-OSCAR addresses service level faults via the ASH technique, whereby some service level failures are handled in a more graceful fashion. Fig. 2. illustrates an ASH state-transition diagram in order to achieve service level fault tolerance. In a perfect condition, all the services are functioning correctly. The ASH MON daemon monitors service availability and health at every tunable interval and triggers alerts upon failure detection. Our current monitoring implementation is a polling approach in which a default interval is 5 seconds. However, this polling interval is tunable for the faster detection time. Section 6 entails impacts and analysis of different polling interval times.

3.3 Server Failover

Fig. 3 shows an example of the HA-OSCAR initial network interface configuration during a normal operation. In our example, HA-OSCAR assigns the primary server public IP address "$Eth0_p$:" with 138.47.51.126 and its private IP address "$Eth1_p$:" as 10.0.0.200. We then configure an alias private IP address $Eth1:0_p$: as 10.0.0.222. For the standby server, the public IP address "$Eth0_s$:" is initially unassigned and its private IP address is $Eth1_s$: 10.0.0.150.

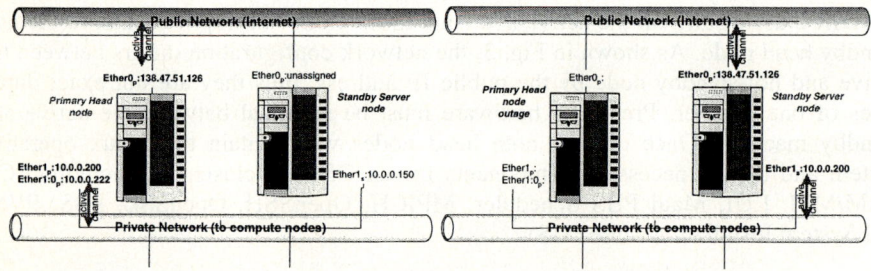

Fig. 3. Network Interface Configuration during normal condition and after failover

When a primary server failure occurs, all its network interfaces will drop. The standby server takes over and clones the primary server network configuration shown in Figure 3 (on the right). The standby server will automatically configure its public network interface $Eth0_s$: to the original public IP address 138.47.51.126 and private network interface $Eth1_s$: to 10.0.0.200. This IP cloning only takes 3-5 seconds.

4 Implementation and Experimental Results

HA-OSCAR should support any Linux Beowulf cluster. We have successfully verified a HA-OSCAR cluster system test with the OSCAR release 3.0 and RedHat 9.0. The experimental cluster consisted of two dual xeon server head nodes, each with 1 GB RAM, 40 GB HD with at least 2GB of free disk space and two network interface cards. There were 16 client nodes that were also Intel dual xeon servers with 512 MB RAM and a 40 GB hard drive. Each client node was equipped with at least one network interface card; Head and client nodes are connected to dual switches as shown in Figure 1.

5 Modeling and Availability Prediction

In order to gauge an availability improvement based on the experimental cluster, we evaluated our architecture, its system failure and recovery with a modeling approach. We also studied the overall cluster uptime and the impact of different polling interval sizes in our fault monitoring mechanism. Stochastic Reward Nets (SRN) technique has been successfully used in the availability evaluation and prediction for complicated systems, especially when the time-dependent behavior is of great interest [5]. We utilized Stochastic Petri Net Package (SPNP) [6] to build and solve our SRN model.

We calculated instantaneous availabilities of the system and its parameters. Details can be founded in [1]. We obtained a steady-state system availability of 99.993%, which was a significant improvement when compared to 99.65%, from a similar Beowulf cluster with a single head node. Furthermore, higher serviceability such as the abilities to incrementally upgrade and hot-swap cluster OS, services, applications and hardware, will further improve planned downtime which undoubtedly benefits the overall aggregate performance. Figure 4 illustrates the total availability (planned and unplanned downtime) improvement analysis of our HA-OSCAR dual-heads vs. a single service node Beowulf clusters when exploiting redundant service nodes for both fault tolerance and hot and incremental upgrade.

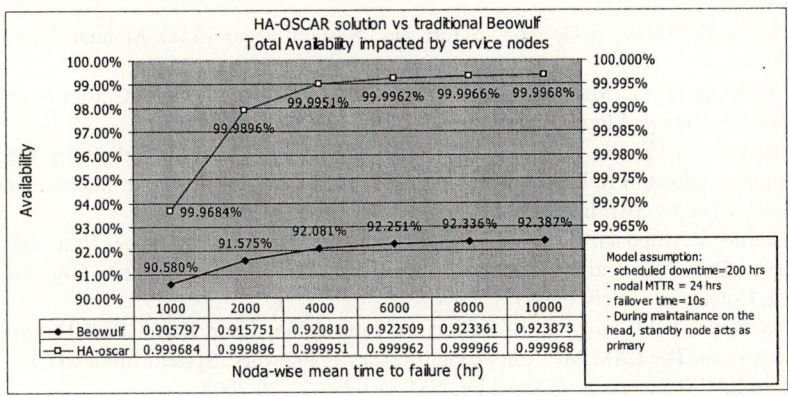

Fig. 4. HA-OSCAR and traditional Linux HPC: the total availability improvement analysis

6 Conclusions and Future Work

Our proof-of-concept implementation, experimental and analysis results [1, 2, and here] suggest that HA-OSCAR solution is a significant enhancement and promising solution to providing a high-availability Beowulf cluster class architecture. The availability of our experimental system improves substantially from 99.65% to 99.9935%. The polling interval for ASH failure detection indicates a linear behavior to the total cluster availability. The introduction of the hot-standby server is clearly cost-effective method when compared with an investment of a typical cluster since a double outage of the servers are very unlikely, therefore, clusters with HA-OSCAR are likely much more available than a typical Beowulf cluster. We have furthered our investigation from outage detection to prediction techniques [10], active-active multi-head failover and grid-enabled HA-OSCAR In addition, we have recently investigated a job queue migration mechanism based on Sun Grid Engine (SGE) and experimented with Berkley Lab Checkpoint/Restart (BLCR) and LAM/MPI [7] for an automated checkpoint and restart mechanism to enhance fault tolerance in HA-OSCAR framework. Our initial findings [9] are very promising. We will extend transparent failure recovery mechanisms for more options, including a sophisticated rule-based recovery and integration with openMPI (a unified MPI platform development effort from LA/MPI , FT/MPI [8] and LAM/MPI [7] teams).

References

1. C. Leangsuksun, L. Shen, T. Liu, H. Song, S. Scott, Availability Prediction and Modeling of High Availability OSCAR Cluster, IEEE International Conference on Cluster Computing (Cluster 2003), Hong Kong, December 2-4, 2003.
2. C. Leangsuksun, L. Shen, T. Liu, H. Song, S. Scott, Dependability Prediction of High Availability OSCAR Cluster Server, The 2003 International Conference on Parallel and Distributed Processing Techniques and Applications (PDPTA'03), Las Vegas, Nevada, USA, June 23-26, 2003.
3. B Finley, D Frazier, A Gonyou, A Jort, etc., SystemImager v3.0.x Manual, February 19, 2003.
4. M J. Brim, T G. Mattson, S L. Scott, OSCAR: Open Source Cluster Application Resources, Ottawa Linux Symposium 2001, Ottawa, Canada, 2001
5. J Muppala, G Ciardo, K. S. Trivedi, Stochastic Reward Nets for Reliability Prediction, Communications in Reliability, Maintainability and Serviceability: An International Journal published by SAE International, Vol. 1, No. 2, pp. 9-20, July 1994.
6. G Ciardo, J. Muppala, K. Trivedi, SPNP: Stochastic Petri net package. Proc. Int. Workshop on Petri Nets and Performance Models, pp 142-150, Los Alamitos, CA, Dec. 1989. IEEE Computer Society Press.
7. S. Sankaran, J. M. Squyres, B. Barrett, A. Lumsdaine, Jason Duell, Paul Hargrove, and Eric Roman. The LAM/MPI Checkpoint/Restart Framework: System-Initiated Checkpointing. In LACSI Symposium, Santa Fe, NM, October 27-29 2003.
8. G.E. Fagg, E.Gabriel, Z. Chen, T.Angskun, G. Bosilca, A.Bukovsky and J.J.Dongarra: 'Fault Tolerant Communication Library and Applications for High Performance Computing', LACSI Symposium 2003, Santa Fe, NM, October 27-29, 2003.
9. C.V Kottapalli, Intelligence based Checkpoint Placement for Parallel MPI programs on Linux Clusters, Master Thesis Report, Computer Science Program, Louisiana Tech University, August 2004 (In preparation)
10. C. Leangsuksun et al, "A Failure Predictive and Policy-Based High Availability Strategy for Linux High Performance Computing Cluster", The 5th LCI International Conference on Linux Clusters: The HPC Revolution 2004, Austin, TX, May 18-20, 2004.

An Enhanced Message Exchange Mechanism in Cluster-Based Mobile Ad Hoc Networks[*,**]

Wei Lou[1] and Jie Wu[2]

[1] Dept. of Computing, Hong Kong Polytechnic University,
Hung Hom, Kowloon, Hong Kong
[2] Dept. of Computer Science and Engineering, Florida Atlantic University,
Boca Raton, Florida, USA

Abstract. In mobile ad hoc networks (MANETs), networks can be partitioned into clusters. Clustering algorithms are localized algorithms that have the property of creating an upper bounded clusterheads in any networks even in the worst case. Generally, clusterheads and selected gateways form a connected dominating set (CDS) of the network. This CDS forms the backbone of the network and can be used for routings (broadcasting/multicasting/unicating). As clusterheads need to determine the selected gateways to connect their adjacent clusterheads within the coverage set, they periodically collect 2-hop and 3-hop clusterhead information. These information can be gathered when they hear their non-clusterhead neighbors sending out messages that contain neighboring clusterhead information. These extra maintenance cost can be reduced when some enhanced mechanism is applied. In this paper, an enhanced mechanism is proposed to reduce the total length of the messages when non-clusterhead nodes exchange their 1-hop and 2-hop neighboring clusterhead information. Simulation shows that over 50% of message overhead can be saved for dense networks.

1 Introduction

Mobile ad hoc networks (MANETs) are collections of autonomous mobile hosts without the help of center base stations. Applying such networks into practice brings many challenges to the protocol design, such as routing in highly dynamic networks, allocating shared wireless channels and saving limited bandwidth. Trade-offs are needed in the protocol design to achieve these conflicting goals. One fundamentally problem of MANETs is the scalability issue of the network. As the size of the network increases and the network becomes dense, a flat infrastructure of the network may not work properly, even for a single

---

[*] Wei Lou's work was supported in part by the Seed Project Grant of the Department of Computing, Hong Kong Polytechnic University. Contact E-mail: csweilou@comp.polyu.edu.hk
[**] Jie Wu's work was supported in part by NSF grants CCR 9900646, CCR 0329741, ANI 0073736 and EIA 0130806. Contact E-mail: jie@cse.fau.edu

broadcast operation [1]. Therefore, building some type of hieratical infrastructure for a large network is a necessity and can enhance the performance of the whole network.

The cluster structure is a two-layer hieratical network that converts a dense network to a sparse one, and therefore, relieves the communication overhead of the whole network. The clustering algorithms partition the network into a group of clusters. Each cluster has one clusterhead that dominates all other members in the cluster. Two clusterheads cannot be neighbors. Gateways are those non-clusterhead nodes that have at least one neighbor that belongs to other clusters. It is easy to see that clusterheads and gateways form a *connected dominating set* (CDS) of the original network. A *dominating set* (DS) is a subset of nodes such that every node in the graph is either in the set or has an edge linked to a node in the set. If the subgraph induced from a DS of the graph is connected, the DS is a CDS. It has been proved that finding a *minimum CDS* (MCDS) in a given graph is NP-complete; this applies to a unit disk graph as well [2,3]. In cluster networks, selecting gateways to connect clusterheads and maintaining such a CDS structure in a mobile environment is an extra cost that can be reduced.

Theoretically, we can describe a MANET as a unit disk graph $G = (V, E)$, where the node set V represents a set of wireless mobile hosts and the edge set E represents a set of bi-directional links between the neighboring hosts, assuming all hosts have the same transmission range r. Two hosts are considered neighbors if and only if their geographic distance is less than r. $N_k(v)$ is v's k-hop neighbor set, including v itself.

When building the infrastructure of the cluster-based CDS in MANETs, each clusterhead collects the clusterhead information within its coverage set so that it can determine some gateways to connect all clusterheads in its coverage set. A node v's coverage set $C(v)$ is a set of clusterheads that are within a specific coverage area of v. It can be a *3-hop coverage set*, which includes all the clusterheads in its 3-hop neighbor set $N_3(v)$, or a *2.5-hop coverage set*, which includes all the clusterheads in $N_2(v)$ and the clusterheads that have members in $N_2(v)$. These clusterhead information can be gathered by hearing the non-clusterhead neighbors sending out messages that contain neighboring clusterhead information. In this paper, we propose an enhanced mechanism that can reduce the overhead of message exchanges. Instead of sending messages that include all paths to the neighborhood clusterheads, in the enhanced mechanism, each non-clusterhead node sends the message that only provides one path per neighboring clusterhead. In Fig. 1 (a), the original mechanism requires node g to send clusterhead u a message that includes all intermediate nodes (inside the dotted circle) connected to clusterhead v. In Fig. 1 (b), the enhanced mechanism requires node g to send clusterhead u a message that only includes an intermediate nodes (grey node) connected to clusterhead v. Simulations show that a large amount of overhead can be saved.

The remaining part of the paper is organized as follows: Clustering algorithms are briefly introduced in Section 2. Section 3 describes the message exchanging mechanism. In Section 4 shows the simulation results. Conclusions are drawn in Section 5.

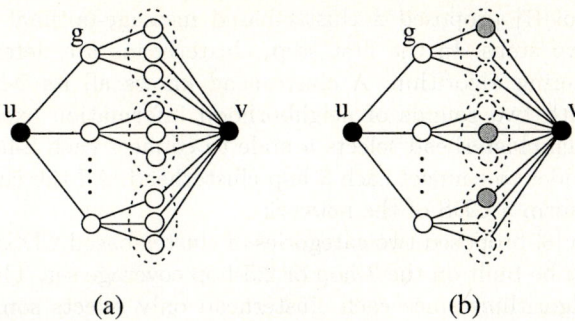

Fig. 1. Message exchange mechanism: (a) original mechanism and (b) enhanced mechanism

2 Preliminaries

The distributed clustering algorithm, lowest-ID clustering algorithm [4], is initiated by electing as a clusterhead the node whose ID is locally the smallest one among all its neighbors. At the beginning, all nodes in the network are candidates. When a candidate finds itself to be the one with the smallest ID among all its 1-hop candidate neighbors, it declares itself as the clusterhead of a new cluster and notifies all its 1-hop neighbors. When a candidate receives a clusterhead notification from a neighboring clusterhead, the candidate joins in the cluster, changes itself to a non-clusterhead member of the cluster, and announces its non-clusterhead state to all its neighbors. If it receives more than one clusterheads' declaration, it joins in the cluster whose clusterhead has the smallest ID. Non-clusterheads that have neighbors belonging to other clusters become gateways. The network will eventually be partitioned into clusters where each cluster has one clusterhead and several gateway/non-clusterhead members. Replacing the clusterhead selecting priority from node ID to effective node degree leads to the highest node degree clustering algorithm [4].

Jiang et al [5] proposed a cluster-based routing protocol (CBRP) that forms a cluster structure by first electing clusterheads and then letting each clusterhead select one or one pair of gateways to connect to each clusterhead in its adjacent clusters. Therefore, each clusterhead may build multiple paths to its adjacent clusterheads as much as it can maintain.

Kwon and Gerla [6] proposed a passive clustering scheme that constructs the cluster structure during the data propagation. A clusterhead candidate applies the "first declaration wins" rule to become a clusterhead when it successfully transmits a packet. Then, its neighbor nodes can learn the presence of this clusterhead and change their states to become gateways if they have more than one adjacent clusterhead or ordinary (non-clusterhead) nodes otherwise. The passive clustering algorithm has the advantages of no initial clustering phase, no need of the complete neighborhood information for the clusterhead election and no communication overhead for maintaining cluster structure or updating neighborhood information, but it suffers poor delivery rate and global parameter requirement.

Alzoubi et al [7] proposed a cluster-based message-optimal CDS which is formed with two steps: In the first step, clusterheads are determined by the lowest-ID clustering algorithm. A clusterhead knows all its 2-hop and 3-hop clusterheads with two rounds of neighborhood information exchanges. In the second step, each clusterhead selects a node to connect each 2-hop clusterhead and a pair of nodes to connect each 3-hop clusterhead. All the clusterheads and selected nodes form a CDS of the network.

Lou and Wu [8] proposed two categories of cluster-based CDSs. Both cluster-based CDSs can be built on the 3-hop or 2.5-hop coverage set. These algorithms are localized algorithms since each clusterhead only selects some gateways to connect the clusterheads within the coverage set. In general, the size of a clusterhead's 2.5 hop coverage set is less than that of its 3 hop coverage set. Therefore, the cost of maintaining the 2.5-hop coverage set is less than that of the 3-hop coverage set.

One common feature of all the above algorithms is that each clusterhead needs two rounds of neighborhood information exchanges and overhead of these neighborhood information is not taken into consideration when the clustering algorithm is designed.

3 An Enhanced Mechanism of Exchanging Neighborhood Information

Constructing clusters in a MANET needs several rounds of message exchanges. Also, to build the cluster-based CDS of the network, each clusterhead has to gather neighboring clusterhead information within its coverage set. These information comes from the neighboring clusterhead messages sent by those non-clusterheads.

3.1 Construction Process

The construction of the cluster-based backbone is described in detail as follows:

At the beginning, each node can learn its neighbors' IDs through HELLO messages. The network is partitioned into clusters by applying the lowest-ID clustering algorithm. A clusterhead will send out a CH message and a non-clusterhead will send out a NCH message to inform its neighbors.

After the clusters have been formed, each node knows all its 1-hop neighbors. A non-clusterhead v sends out a 1-hop neighboring clusterhead message $CH\_HOP1(v)$ which includes all v's 1-hop neighboring clusterheads. When the clusterhead u receives the CH_HOP1 messages from all its non-clusterhead neighbors, u adds an entry for each new neighboring clusterhead with its associated gateway that connects the clusterheads together.

Once another non-clusterhead w receives the message $CH\_HOP1(v)$ from v, w builds the message $CH\_HOP2(w)$ as follows: If a clusterhead u that is found in $CH\_HOP1(v)$ is also a 1-hop neighbor of w, w ignores u in the $CH\_HOP2(w)$; Otherwise, w checks if u is a new 2-hop clusterhead of w. If so, w creates a new entry that contains the 2-hop clusterhead u and the its associated node v. If u

is already included in the entry and the associated node v has a higher priority (e.g., lower ID or higher linkage quality, etc.) than the original one, the entry can be updated with the new associated node v.

When w receives the CH_HOP1 messages from all its non-clusterhead neighbors, it sends out a message CH_HOP2(w) that contains all 2-hop clusterhead entries. Unlike other algorithms that each clusterhead includes all associated gateways in the message CH_HOP2, the CH_HOP2 with enhanced mechanism builds each entry with one clusterhead and one associated gateway. Therefore, the size of the message CH_HOP2 can be much smaller. When the clusterhead u receives a CH_HOP2, u builds a new entry for each clusterhead in the CH_HOP2.

After clusterhead u receives all CH_HOP1 and CH_HOP2 messages from its non-clusterhead neighbors, u builds its coverage set $C(u) = C_2(u) \cup C_3(u)$, where $C_2(u)$ consists of all elements in CH_HOP1 and $C_3(u)$ consists of all elements in CH_HOP2. If a clusterhead appears in both $C_2(u)$ and $C_3(u)$, the one in $C_3(u)$ is removed.

Each clusterhead u selects gateways to connect all the clusterheads in $C(u)$ with the forward node set selection process. The gateways are selected by the greedy algorithm: The neighbor node or the pair of nodes, whichever has the highest "yield" is first selected as the gateway(s). A "yield" of node(s) is defined as the total number of the clusterheads in the $C(u)$ that was connected by the selected gateway(s) divided by the number of the selected gateway(s). A tie of yield is broken by selecting the smaller node ID. When a node is selected, all of the clusterheads in $C(u)$ is removed. The selection process repeats until $C(u)$ is empty.

After a clusterhead determines its gateways, it sends out a GATEWAY message that contains all selected gateways. The selected 1-hop gateways forward the GATEWAY message so that all the selected 2-hop gateways can be informed.

3.2 Example

Fig. 2 shows the construction process of a cluster-based CDS backbone. At the beginning, all the nodes are candidates (Fig. 2 (a)). With the lowest-ID clustering (LID) algorithm, nodes 1, 2, 4, 8 and 10 become clusterheads and form clusters labelled as C_1, C_2, C_3, C_4 and C_5; then nodes 7 and 9 join in cluster C_1, nodes 3 and 6 join in cluster C_2, nodes 5 joins in cluster C_3 (Fig. 2 (b)).

Fig. 2(c) illustrates message exchange when node 4 construct its 3-hop and 2.5 hop coverage sets:

(1) For 3-hop coverage set, node 6 sends CH_HOP1(6) (M1 in Fig. 2(c)) which contains its 1-hop clusterhead neighbor set {2∗}, 3 sends CH_HOP1(3) (M2 in Fig. 2(c)) contains {2∗, 8}. Here, ∗ indicates the clusterhead of the cluster that the node belongs to. Likewise, nodes 5 and 7 send CH_HOP1(5)={4∗, 10} and CH_HOP1(7)={1∗, 4} (M3 and M5 in Fig. 2(c)). After receiving M1 and M2, node 7 may form CH_HOP2(7) (M6 in Fig. 2(c)) which contains its 2-hop clusterhead neighbors and associated gateways {2[3], 8[3]}. Here, CH_HOP2(u)= {$v[w], ...$} means that clusterhead u connects to clusterhead v via w. Then node 7 sends CH_HOP2(7) out. Note that node 7 picks node 3 as the gateway instead

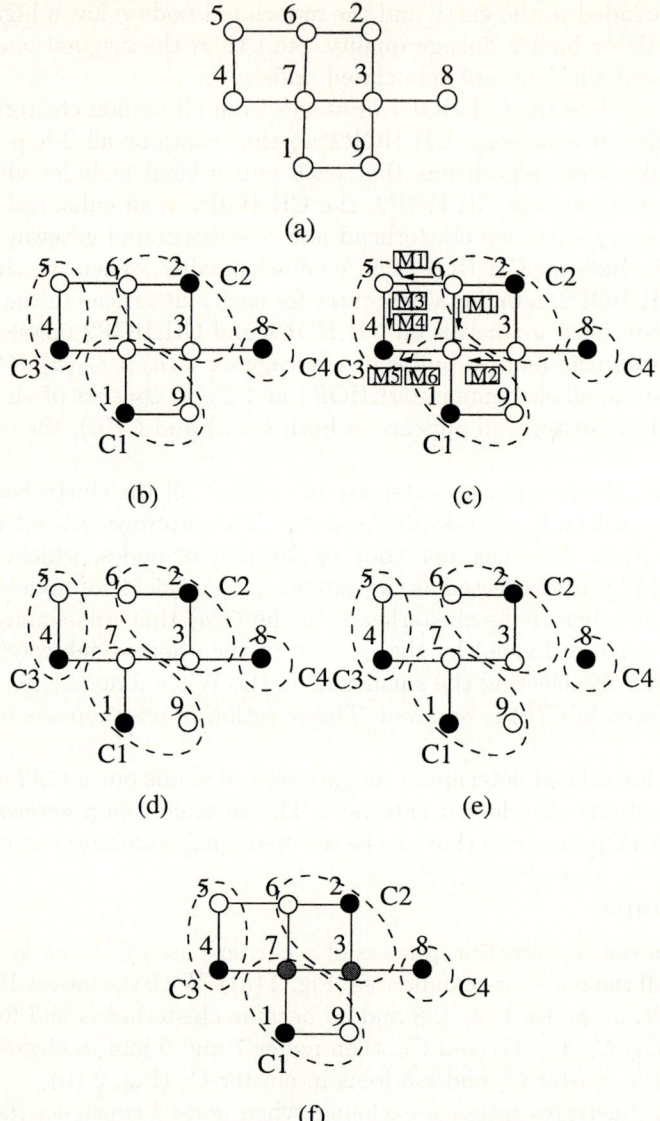

Fig. 2. Illustration of constructing a coverage set: (a) initial network, (b) clusters, (c) exchange neighboring information, (d) 3-hop coverage set, (e) 2.5-hop coverage set, and (f) cluster-based CDS

of node 6 because node 3 has two adjacent clusterheads 2 and 8. Also, the size of the CH_HOP2(7) is smaller than the one that includes all possible paths to node 2 since it only keeps the most favorable path in the message. Similarly, node 5 will send CH_HOP2(5)={2[6]} out (M4 in Fig. 2(c)). After receiving M3, M4, M5 and M6, node 4 can build its local view of its coverage set (Fig. 2(d)).

Clusterhead 4 selects node 5 as the gateway to connect to clusterhead 10, selects nodes 7 and 3 as the gateways to connect clusterheads 1, 2 and 8, and sends a message GATEWAY(1) = {3, 5, 7}. Similarly, clusterheads 1 and 8 select nodes 7 and 3 as gateways and send GATEWAY(1)=GATEWAY(8)={3, 7}; clusterhead 2 selects node 3 to connects clusterhead 8, selects nodes 3 and 7 to connect clusterheads 1 and 4, selects nodes 6 and 5 to connect clusterhead 10, and sends GATEWAY(2)={3, 5, 6, 7}; clusterhead 10 selects nodes 5 and 6 to connect clusterheads 2 and 4, and sends GATEWAY(10)={5, 6}. The final cluster-based CDS is {1,2,3,4,5,6,7,8,10} (Fig. 2 (f)).

(2) For the 2.5-hop coverage set, node 4 builds its local view as Fig. 2(e). Here, clusterhead 4 does not know clusterhead 8 since CH_HOP2(7) only includes {2[3]}. Therefore, the size of the CH_HOP2(7) is even reduced. The forward node set selection process is similar and the final cluster-based CDS is {1,2,3,4,5,6,7,8,10} (Fig. 2 (f)).

3.3 Message Overhead Complexity

The message exchange overhead for constructing a cluster-based CDS of the network is listed in Table 1. Note that the number of clusterheads within 1-hop and 2-hop neighbor set are bounded by a constant value [7], the size of CH_HOP1 and CH_HOP2 with the enhanced message exchange mechanism are $O(1)$. On contrast, the original one without this mechanism will send message CH_HOP2 that containing all possible gateways. Therefore, the size of CH_HOP2 is $O(\Delta)$, where Δ is the maximum node degree of the network.

Table 1. Message Overhead Complexity

Message type	Algorithm	
	Original	Enhanced
HELLO	$O(1)$	$O(1)$
CH(NCH)	$O(1)$	$O(1)$
CH_HOP1	$O(1)$	$O(1)$
CH_HOP2	$O(\Delta)$	$O(1)$
GATEWAY	$O(1)$	$O(1)$
Total	$O(\Delta)$	$O(1)$

4 Simulations

We measure the average sizes of the message exchange for constructing the cluster-based CDS with enhanced message exchange mechanism (referred to as Enhanced (3-hop and 2.5-hop) and without enhanced message exchange mechanism (referred to as Original (3-hop)). The size of the message is counted as the number of nodes included in the message.

The simulation runs under the following simulation environment: A number of nodes (ranging from 100 to 1000) are randomly placed in a confined working space 100×100. The nodes have the same transmission ranges, and the link between two nodes is bi-directional. The network is generated with two fixed average node degrees: $d = 6$ and 18, which are the representatives of the relative sparse and dense networks. If the generated network is not connected, it is discarded. We only consider the traffic of the packets at the network layer without any transmission collision. We repeat the simulation until the 99% confidential interval of the result is within ±5%.

Table 2. Message exchange overhead (sparse network:n=1000,d=6)

Message	Algorithm		
	Original(3-hop)	Enhanced(3-hop)	Enhanced(2.5-hop)
HELLO	1000.0	1000.0	1000.0
CH(NCH)	1000.0	1000.0	1000.0
CH_HOP1	1209.3	1209.3	1209.3
CH_HOP2	5106.3	3874.7	2973.7
GATEWAY	1134.0	1134.0	1013.3
Total	9449.6	8218.0	7197.3

Table 3. Message exchange overhead (dense network: n=1000, d=18)

Message	Algorithm		
	Original(3-hop)	Enhanced(3-hop)	Enhanced(2.5-hop)
HELLO	1000.0	1000.0	1000.0
CH(NCH)	1000.0	1000.0	1000.0
CH_HOP1	1601.8	1601.8	1601.8
CH_HOP2	15587.6	6912.5	5421.9
GATEWAY	837.7	837.7	724.4
Total	20027.1	11352.0	9748.1

The Tables 2 and 3 show the cases that the sizes of different types of messages exchanged for the construction of the clusters of the network, when the number of nodes in the network is 1000 and the average node degree is 6 and 18 respectively. We can see that the size of CH_HOP2 is the most weighted part of the total sizes (40% ~ 70%). In the sparse scenario where average node degree d is 6, the enhanced mechanism can reduce 25% of overhead of CH_HOP2 with 3-hop coverage set and 17% less with 2.5-hop coverage set compared with the original one. The overhead of the total size of the message can reduce 13% and 24% of the original one with 3-hop and 2.5-hop coverage set. In the dense network where average node degree d is 18, the reduction of the overhead of the CH_HOP2 can reach 56% with 3-hop coverage set and 66% with 2.5-hop coverage. Correspondingly, the total size of the message can reduce 43% and 51% of the original one.

The Fig. 3 shows the average size of message per node when the size of the network ranges from 100 to 1000. Fig. 3 (a) shows the scenario when the

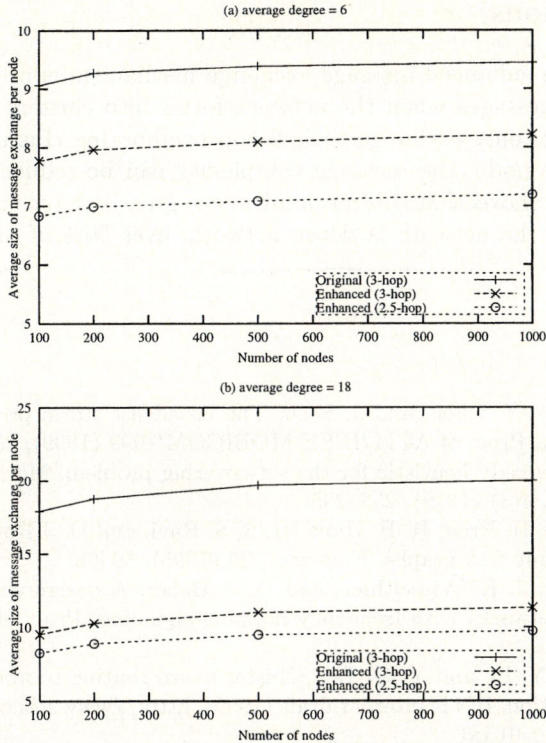

Fig. 3. Message Overhead: (a) average node degree = 6, and (b) average node degree = 18

node degree is 6. The cost decreases from over 9 for the original one to around 8 (over 10%) for the enhanced mechanism with 3-hop coverage set, and then to around 7 (over 20%) for the enhanced mechanism with 2.5-hop coverage set. Fig. 3 (b) shows the scenario when d is 18. The cost decreases more remarkably from near 20 for the original one to around 10 (50%) for the enhanced mechanism with 3-hop coverage set and even less for that with 2.5-hop coverage set. We observe that overhead total cost of message exchange per node increases remarkably (doubled) when the network becomes dense. Also, we find that the enhanced mechanism can greatly reduce the message exchange overhead, especially for the dense network. The reason is that when the network is dense and each node has more neighbors, a CH_HOP2 will include more different intermediate nodes for each 2-hop clusterhead for the original mechanism. Thus, the overhead of CH_HOP2 increases remarkably. When the CH_HOP2 includes only one intermediate node for each 2-hop clusterhead for the enhanced mechanism, the message overhead can be greatly reduced.

5 Conclusions

In this paper, an enhanced message exchange mechanism is proposed to reduce the size of the messages when the network forms into clusters. When the non-clusterhead nodes only exchange their 2-hop neighboring clusterhead with only one intermediate node, the message complexity can be reduced from $O(\Delta)$ to $O(1)$. Simulation shows that this mechanism can greatly reduce the total message overhead. When the network is dense network, over 50% of message overhead can be saved.

References

1. S. Ni, Y. Tseng, Y. Chen, and J. Sheu: The broadcast storm problem in a mobile ad hoc network. Proc. of ACM/IEEE MOBICOM'1999 (1999), 151–162.
2. V. Chvatal: A greedy heuristic for the set-covering problem. Mathematics of Operation Research, 4(3) (1979), 233–235.
3. M. V. Marathe, H. Breu, H. B. Hunt III, S. S. Ravi, and D. J Rosenkrantz: Simple heuristics for unit disk graphs. Networks, 25 (1995), 59–68.
4. A. Ephremides, J. E. Wieselthier, and D. J. Baker: A design concept for reliable mobile radio networks with frequency hopping signaling. Proc. of the IEEE, 75(1) (1987), 56–73.
5. M.L. Jiang, J. Y. Li, and Y. C. Tay: Cluster based routing protocol (CBRP) functional specification. IETF Internet draft (1999), http://www.ietf.org/ietf/draft-ietf-manet-cbrp-spec-01.txt.
6. T. J. Kwon and M. Gerla: Efficient flooding with passive clustering (PC) in ad hoc networks. ACM Computer Communication Review, 32(1) (2002), 44–56.
7. K. M. Alzoubi, P. J. Wan, and O. Frieder: Message-optimal connected dominating sets in mobile ad hoc networks. Proc. of ACM/IEEE MOBIHOC'2002 (2002), 157–164.
8. W. Lou and J. Wu: A cluster-based backbone infrastructure for broadcasting in manets. Proc. of IEEE IPDPS'2003, Workshop of WMAN (2003).

Algorithmic-Parameter Optimization of a Parallelized Split-Step Fourier Transform Using a Modified BSP Cost Model*

Elankovan Sundararajan[1], Malin Premaratne[2], Shanika Karunasekera[1], and Aaron Harwood[1]

[1] Department of Computer Science and Software Engineering,
The University of Melbourne,
ICT Building, 111 Barry Street, Carlton 3053,
Victoria, Australia
e.sundararajan@pgrad.unimelb.edu.au, {aharwood, shanika}@cs.mu.oz.au
[2] Advanced Computing and Simulation Laboratory,
Department for Electrical and Computer System Engineering,
Monash University, Clayton 3800,
Victoria, Australia
malin.premaratne@eng.monash.edu.au

Abstract. Adaptive algorithms are increasingly acknowledged in leading parallel and distributed research. In the past, algorithms were manually tuned to be executed efficiently on a particular architecture. However, interest has shifted towards algorithms that can adapt themselves to the computational resources. A cost model representing the behavior of the system (i.e. system parameters) and the algorithm (i.e algorithm parameters) plays an important role in adaptive parallel algorithms. In this paper, we contribute a computational model based on Bulk Synchronous Parallel processing that predicts performance of a parallelized split-step Fourier transform. We extracted the system parameters of a cluster (upon which our algorithm was executed) and showed the use of an algorithmic parameter in the model that exhibits optimal behavior. Our model can thus be used for the purpose of self-adaption.

1 Introduction

Optimizing software to exploit the underlying features of computational resources has been an area of research for many years. Traditionally, optimization is done by hand tuning. However, this approach is tedious and requires skilled programmers and technical knowledge in both the algorithms and complexity of target platform. Performance of highly tuned software can decline upon even slight changes in the hardware and computational load. This is further aggravated by

---

* This work was funded in part by the Australian Research Council, ARC grant number DP0452102.

the short life cycles of modern computer platforms. As a result, tuning of software to fully exploit the computational resources should be a dynamic process. Parallel algorithms should be able to take into consideration the availability of computational resources i.e. system parameters in conjunction with algorithm parameters before starting and during execution.

The bulk synchronous parallel (BSP) processing model of computation was first proposed by Valiant [1] as a bridging model between hardware and software for general purpose computation. BSP model consists of three parts: computation, communication and synchronization. Structure introduced by the BSP model for parallel programming ensures clarity and flow of complicated parallel programs. In this paper, we consider a modified BSP cost model approach as a possible way to provide performance prediction of parallel algorithms running on a cluster of PCs.

We address parallel implementation of the split-step Fourier (SSF) method for solving the nonlinear Schrödinger equation (NLSE) arising in the field of nonlinear fiber optics [2]. The selection of this problem is partially motivated by the complexity of optical network and associated processes which presents a formidable challenge for network designers because such complex systems require the modeling and simulation of basic sub-systems as well as the overall system.

In Section 2, we give the BSP details, a parallel SSF method to solve nonlinear Schrödinger equation and brief discussion on related works. Section 3 describes the modified BSP cost analysis model for parallel split-step method, Section 4 presents analysis of predicted time and real time performance. Section 5 offers some concluding remarks.

2 Background

2.1 Bulk Synchronous Parallel Processing Model

A BSP program is one which proceeds in stages, known as supersteps[1] [3, 17, 18, 19]. A superstep consists of two phases. In the first phase, processors compute using locally held data. In the second phase datasets are communicated between the processors. The amount of time required to complete a BSP program is the sum of the times required for completion of its supersteps. Thus, the required time for a superstep is the sum of: (i) the longest computation time on any of the processors, $max\{\frac{w_i}{s_i}\}$; (ii) the longest communication time on any of the processors; and (iii) the time overhead required by barrier synchronization.

The execution time for superstep i is:

$$T_{si} = max\left\{\frac{w_i}{s_i}\right\} + T_{scL}, \qquad (1)$$

where T_{si} represents the time for a superstep, w_i represents the number of floating point operations performed by processor i, s_i gives the processing speed in

---
[1] http://users.Comlab.ox.ac.uk/bill.mccoll/oparl.html

FLOPS of processor i and T_{scL} the total time taken for communication and synchronization. To minimize the execution time in Eq.(1), the programmer must: (i) balance local computation over processors in each superstep; (ii) balance communication between processors to reduce large variation; and (iii) minimize the total number of supersteps in the program.

2.2 Parallel SSF Method

A non-linear Schrödinger equation (NLSE) is a nonlinear partial differential equation that cannot be solved analytically except for a few cases. Taha and Ablowitz [4] conducted a study where the SSF method of Tappert [5] was compared with several finite difference, pseudospectral and global methods. In a majority of their experiments the split-step method turned out to be superior, thus high accuracy may be achieved at comparatively low computational cost [6].

Electromagnetic wave propagation in optical fiber is governed by the NLSE:

$$\frac{\partial u}{\partial z} - \frac{\beta}{2}\frac{\partial^2 u}{\partial t^2} + \frac{i}{2}\alpha u - g|u|^2 = 0, \qquad (2)$$

where u is the envelope of the signal wave, z is distance, t is time, β is group velocity dispersion, α is loss coefficient, g is Kerr coefficient and $i = \sqrt{-1}$.

Eq.(2) can be formally written as

$$\frac{\partial u}{\partial z} = (\hat{D} + \hat{N})u, \qquad (3)$$

where $\hat{D}$ is a differential operator that accounts for dispersion and absorption in linear medium and $\hat{N}$ is a nonlinear operator that governs the effect of fiber nonlinearities on pulse propagation[2]. In general, dispersion and nonlinearity act together along the length of the fiber. Using the SSF method, propagation of a pulse over the full length of optical fiber is simulated by dividing the full length of the fiber into relatively small segments with length h, such that changes in the envelope of optical signals can be considered sufficiently small.

More specifically, propagation from z to $z + h$ is carried out in two steps. In the first step, the nonlinearity acts alone, and $\hat{D} = 0$ in Eq.(3). In second step, dispersion acts alone, and $\hat{N} = 0$ in Eq.(3). Using Eq.(3), the following approximate expression with second order accuracy in h can be written as $u(z + h) \approx \exp(h\hat{D})\exp(h\hat{N})u(z,t)$.

The exponential operator $\exp(h\hat{D})$ is evaluated conveniently in the Fourier domain using $\exp(h\hat{D})B(z,t) = F_T^{-1}\exp[h\hat{D}(i\omega)]F_T B(z,t)$, where F_T denotes the Fourier transform operation and $\hat{D}(i\omega)$ is a number in the Fourier space.

The equation above is run over a number of iterations $\left[\frac{Z}{h}\right]$ to arrive at the desired distance. The required run time for the simulation depends on the number of datasets used for the Fourier transform and the number of iterations.

Discrete Fourier Transform. The discrete Fourier transform (DFT) plays an important role in many scientific and technical applications [7]. The discrete one dimensional forward DFT for discrete set of N points is defined in the literature [8] as

$$U(z) = \sum_{k=0}^{N-1} u(k) e^{\left(-\frac{2\pi i z k}{N}\right)} : z \in \{0, 1, ..., N-1\}. \qquad (4)$$

A direct implementation uses $O(N^2)$ computations. However a fast Fourier transform (FFT) gives $O(N \log N)$ computations. The first algorithm of Fast Fourier Transform (FFT) dates back to Gauss and was made popular by Cooley and Tukey [9]. Researchers have generated variants of this algorithm by looking at specific features [10, 11, 12].

An efficient implementation of the FFT is the FFT-Transpose Method (FFT-TM) and can be derived from the one dimensional FFT. Suppose that the number of discrete samples is N and that it has the following factorization: $N = N_0 N_1$, where $N_0, N_1 \in \mathbb{Z}^+$ then define two-dimensional arrays $u(l, m) = u(k)$ and $U(n, p) = U(z)$ where $k = l + m N_0$ and $z = n + p N_1$. Then, Eq.(4) can be transformed [8, 10, 13] to:

$$U(n, p) = \sum_{l=0}^{N_0-1} e^{-i\left(\frac{2\pi l p}{N_0}\right)} e^{-i\left(\frac{2\pi n}{N}\right)} \sum_{m=0}^{N_1-1} u(l, m) e^{-i\left(\frac{2\pi m n}{N_1}\right)}.$$

Therefore, $U(n, p)$ can be computed as two multiple FFTs. First compute N_0 FFTs of length N_1 each and multiply with $exp(\frac{-i2\pi n}{N})$:

$$U'(l, n) = e^{-i\left(\frac{2\pi n}{N}\right)} \sum_{m=0}^{N_1-1} u(l, m) e^{-i\left(\frac{2\pi m n}{N_1}\right)}.$$

Next compute N_1, FFTs of length N_0, $U(n, p) = \sum_{l=0}^{N_0-1} U'(l, n) e^{-i\left(\frac{2\pi l p}{N_0}\right)}$. The complexity of the FFT-TM algorithm is given by $O(\frac{N}{P} log_2 N)$.

Implementation Using BSP. The FFT-TM algorithm consists of computing one dimensional FFT on both dimensions using FFTW[11, 12]:

Pseudo code for the 2 Dimensional Split-step method.

```
        Calculate Initial data.
        For (Loop = 1 to NLoop)
            Superstep 1.
                If(Loop != 1)
                    Multiply conjugate phase factor.
                    Backward transform every row in each processor.
                    Nonlinear step.
                End of If
                1.  Forward transform every row in each processor.
                2.  Multiply phase factor.
                3.  Distribute and arrange the dataset.
            End of Superstep 1.
            Superstep 2.
```

```
           1. Forward transform every column in each processor.
           2. Linear step.
           3. Backward transform every column in each processor.
           4. Arrange and distribute the dataset.
        End of superstep 2.
    End of Loop.
```

This implementation can be used for any number, 2^i, $i = 0, 1, 2, \ldots$, of processors. The size of rows can be changed without changing the code by varying the product $N = N_0 N_1$, thus N_0 is an algorithmic parameter.

Related Work. A parallel implementation of the SSF method for solving the NLSE is presented in [14]. This work explored the performance of the implementation for a fixed number of rows. It is also mentioned that a perfect speedup can be achieved over sequential SSF algorithm by tuning the number of processors and problem size. Adaptive systems are currently an active area of research. These systems are expected to have the necessary 'intelligence' to adapt themselves to available resources so that efficient performance can be achieved. Many works have attempted to obtain this type of efficient systems in different fields [11, 15, 16]. In this paper, we have introduced a model which is based on empirical studies on the target cluster. This model is later used to predict the optimal behavior of algorithmic parameter, N_0 for best possible execution time.

3 Modified BSP Cost Analysis Model

In this section, we provide detailed notation to represent parts in the calculation. $T_{Lft}(m)$ and $T_{Uft}(m)$ with m being the dataset size, is the lower bound and the upper bound on time taken to transform m complex datum using the FFTW package. $T_{Lcms}(x)$ and $T_{Ucms}(x)$ with x being the message size in bytes, is the lower bound and upper bound on predicted time for all-to-all communication to complete. It also includes the barrier synchronization time for a subsequent barrier operation. We used a look up table to store information gathered from empirical experiment. The performance prediction model uses this table for prediction purposes. We have used utilities run on the target cluster to predict the time taken for arithmetic, trigonometric, and other basic mathematical functions, with results kept in a look up table.

The model described below is for the implementation using row-by-row distribution. We have used coefficients such as 221 and 479.26 in (6), (7) and (8). These are constants that represents the lower and upper bound of total normalized time taken for arithmetic, trigonometric and other mathematical functions respectively.

Total time consumed in an iteration is given by Eq.(5):

$$T_{lp} = T_C + T_{CMS}, \qquad (5)$$

where T_{lp} is total time taken to complete an iteration, T_C is total time taken for computation in a loop and T_{CMS} is total time taken for communication and synchronization in a loop. Continuing, $T_C = T_{cs1} + T_{cs2} + T_{ca}$ where T_{cs1} is time taken for computation in superstep 1, T_{cs2} is time taken for computation in superstep 2 and T_{ca} is the time taken for computation after supersteps. T_{cs1} includes the time for Fourier transform on the rows of initial data using forward FFT, multiplying phase factor to the transformed dataset and also the time in arranging the received dataset. $\frac{N_0}{P}$ is the total number of rows in a node, P represents number of processors used in the numerical simulation, N_1 represents the total number of data in a row while T_{Lpt} and T_{Upt} are the average time of the repeated runs minus the standard deviation and average time of repeated runs plus the standard deviation for the addition operation. These values are later used to normalize all the other operations. Now,

$$\frac{N_0}{P}\{T_{Lft}(N_1)\} + T_{Lpt}\left\{\frac{221N_0N_1}{P}\right\} \leq T_{cs1} \leq \frac{N_0}{P}\{T_{Uft}(N_1)\} + T_{Upt}\left\{\frac{479.26N_0N_1}{P}\right\}. \tag{6}$$

The time taken in superstep 2, T_{cs2}, includes the time for transforming the column of dataset from 1st superstep using forward FFT, operation on linear part, transforming the column of dataset using backward FFT and arranging the dataset before distribution:

$$\frac{2N_1}{P}\{T_{Lft}(N_0)\} + T_{Lpt}\left\{\frac{448.33N_0N_1}{P}\right\} \leq T_{cs2} \leq \frac{2N_1}{P}\{T_{Uft}(N_0)\} + T_{Upt}\left\{\frac{978.82N_0N_1}{P}\right\}. \tag{7}$$

Computation before supersteps involves calculating the initial value of the pulse. This Initial value is calculated only once and it is outside the iteration loop. Computation after supersteps is for multiplying phase factor, transforming the rows of datasets from 2nd superstep and computation of the nonlinear part. Let T_{cb} be the time taken for computation before supersteps:

$$T_{Lpt}\left\{\frac{389.91N_0N_1}{P}\right\} \leq T_{cb} \leq T_{Upt}\left\{\frac{895.65N_0N_1}{P}\right\}.$$

$$T_{Lpt}\left\{\frac{319.4N_0N_1}{P}\right\} + \frac{N_0 T_{Lft}(N_1)}{P} \leq T_{ca} \leq T_{Upt}\left\{\frac{693.78N_0N_1}{P}\right\} + \frac{N_0 T_{Uft}(N_1)}{P}. \tag{8}$$

Communication occurs at the end of each superstep. In both supersteps, all-to-all is used to scatter portions of the calculated data from each process to all the other processes. An experiment was conducted on a cluster of 2,4 and 8 PCs to predict the time, $T_{cms}(x)$, taken to distribute messages with different sizes using all-to-all, assuming a single user mode:

$$\frac{N_0}{P}\{T_{Lcms}(16N_1)\} \leq T_{cms1}, T_{cms2} \leq \frac{N_0}{P}\{T_{Ucms}(16N_1)\}.$$

Thus, T_{CMS} is given by

$$\frac{2N_0}{P}\{T_{Lcms}(16N_1)\} \leq T_{CMS} \leq \frac{2N_0}{P}\{T_{Ucms}(16N_1)\} \tag{9}$$

and substituting (6),(7),(8),(9) into Eq.(5) gives

$$\frac{2N_1}{P}\left\{T_{Lft}(N_0)\right\}+\frac{2N_0}{P}\left\{T_{Lft}(N_1)\right\}+T_{Lpt}\left\{\frac{988.73N_0N_1}{P}\right\}+\frac{2N_0}{P}\left\{T_{Lcms}(16N_1)\right\}\le T_{lp}$$
$$\le\frac{2N_1}{P}\left\{T_{Uft}(N_0)\right\}+\frac{2N_0}{P}\left\{T_{Uft}(N_1)\right\}+T_{Upt}\left\{\frac{2151.86N_0N_1}{P}\right\}+\frac{2N_0}{P}\left\{T_{Ucms}(16N_1)\right\}.$$
(10)

For K iterations, Eq.(10) can be written as

$$T_{Lpt}\left\{\frac{389.91N_0N_1}{P}\right\}+K\left\{\frac{2N_1}{P}\{T_{Lft}(N_0)\}+\frac{2N_0}{P}\{T_{Lft}(N_1)\}+T_{Lpt}\left\{\frac{988.73N_0N_1}{P}\right\}+\right.$$
$$\left.\frac{2N_0}{P}\{T_{Lcms}(16N_1)\}\right\}\le T_{LP}\le T_{Upt}\left\{\frac{895.65N_0N_1}{P}\right\}+K\left\{\frac{2N_1}{P}\{T_{Uft}(N_0)\}+\right.$$
$$\left.\frac{2N_0}{P}\{T_{Uft}(N_1)\}+T_{Upt}\left\{\frac{2151.86N_0N_1}{P}\right\}+\frac{2N_0}{P}\{T_{Ucms}(16N_1)\}\right\}$$

where T_{LP} is the time taken for completing the simulation.

4 Experimental and Predicted Computation Times

Experiments were conducted on a cluster which consists of PCs each having a Pentium 4 2GHz processor, 512 MB memory, 40 GB hardisk space, and a Red Hat 7.3 Linux operating system. These PCs are interconnected using 16 port Gigabit LAN Switch (Dlink DGS-1016T). We used the Message Passing Interface (MPI)[20] as the basis for parallel programming, namely the MPI all-to-all and barrier functions.

We developed two versions of the parallel SSF, one distributes datasets row-by-row and the other distributes datasets all-at-once. We find that for the former, refer Fig. 1, as the number of rows increases to more than the size of data in a row (i.e. $N_0 > N_1$), the number of calls to all-to-all collective communication primitive increases and the size of datasets that are distributed at any one time becomes less to the extend of effecting the performance of the implementation. This behavior is reflected by our model which can predict the best size of N_0 that should be used to extract the best performance from the implementation. The parallel program which distributes datasets all-at-once, Fig. 2, is slower for the best size of N_0 compared to the one sent row-by-row, Fig. 1, because the datasets must be arranged before distribution and rearranged again after distribution to obtain the transposed datasets whereas the row-by-row implementation requires arranging the datasets only once.

Fig. 1 and Fig. 2 shows actual run time and predicted (average of upper bound and lower bound) time of the implementation for different data size (i.e. 2^{20}, 2^{22} and 2^{23}) and different number of processors (i.e. $P = 2, 4, 8$) using distribution row-by-row and all-at-once respectively. We have used a maximum size of 2^{23} data, so that the speedup can be calculated (i.e. to compare speed with a single

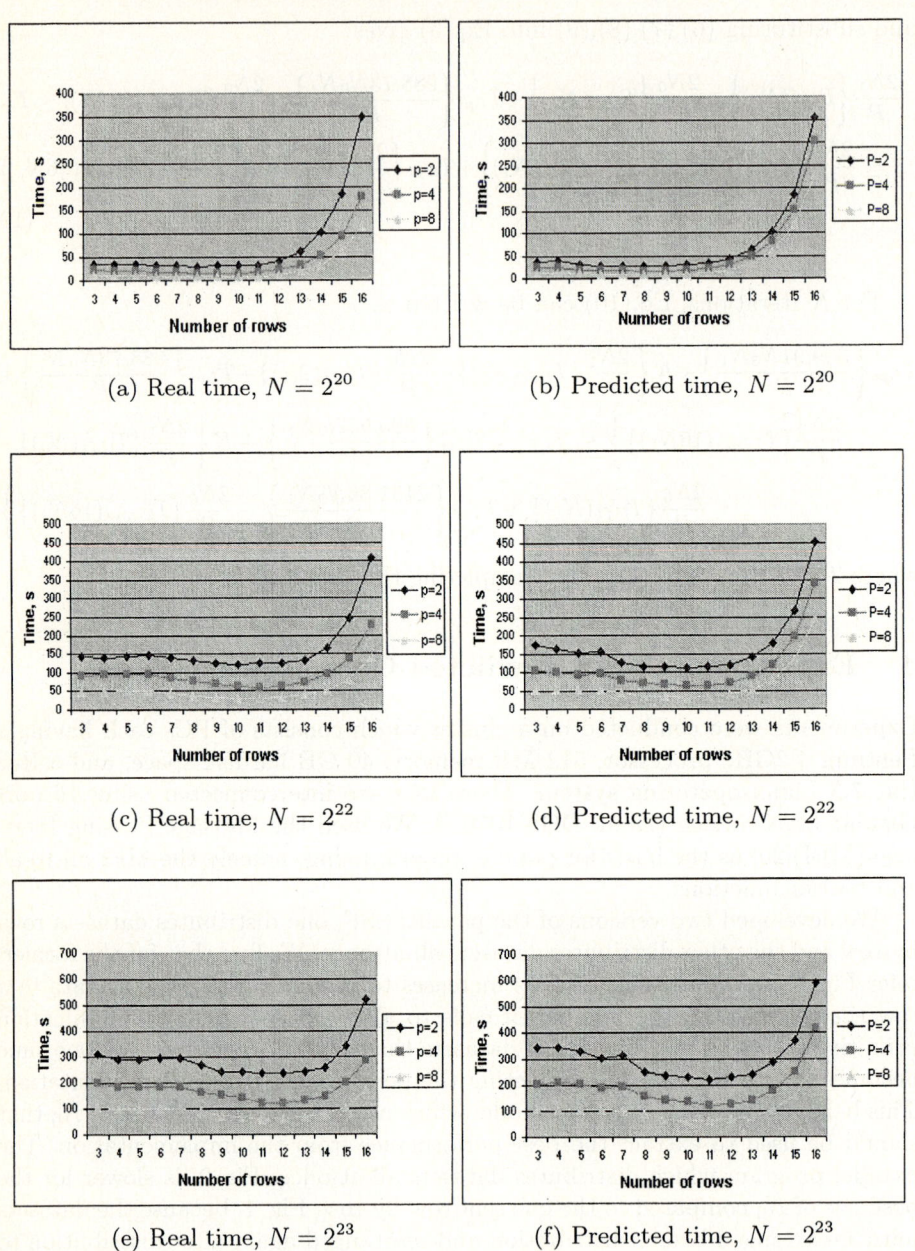

Fig. 1. Actual run time and predicted time for different data sizes, using row-by-row distribution

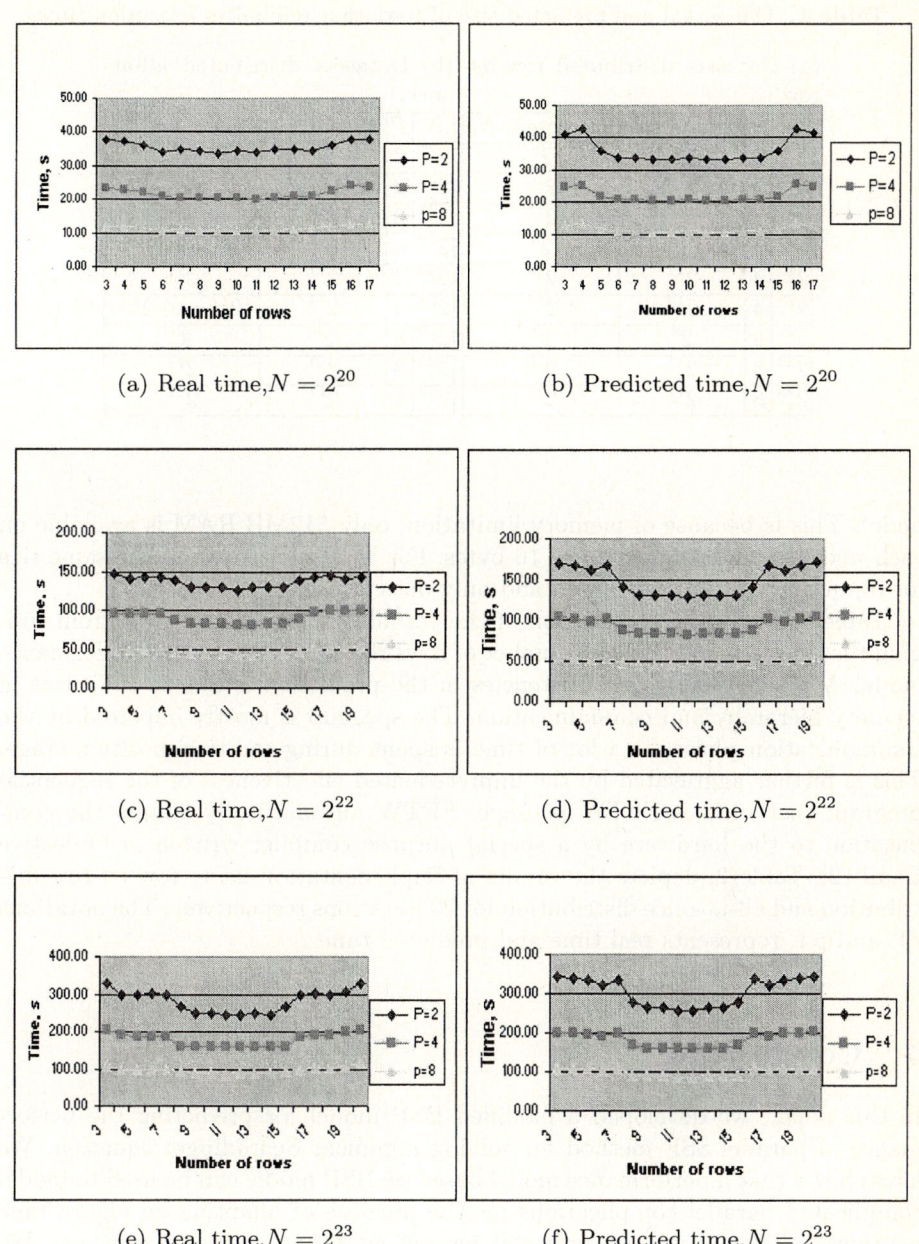

Fig. 2. Actual run time and predicted time for different data sizes, using all-at-once distribution

Table 1. The actual and predicted size of rows that minimizes execution time

(a) Datasets distributed row-by-row

N	P	Actual, N_0	Predicted, N_0
2^{20}	2	2^8	2^8
2^{22}	2	2^{10}	2^{10}
2^{23}	2	2^{11}	2^{11}
2^{20}	4	2^9	2^8
2^{22}	4	2^{11}	2^{10}
2^{23}	4	2^{12}	2^{11}
2^{20}	8	2^8	2^8
2^{22}	8	2^{11}	2^{10}
2^{23}	8	2^{11}	2^{11}

(b) Datasets distributed all-at-once

N	P	Actual, N_0	Predicted, N_0
2^{20}	2	2^9	2^9
2^{22}	2	2^{11}	2^{11}
2^{23}	2	2^{12}	2^{11}
2^{20}	4	2^{11}	2^9 and 2^{11}
2^{22}	4	2^{11}	2^{11}
2^{23}	4	2^{11}	2^{11} and 2^{12}
2^{20}	8	2^9	2^9
2^{22}	8	2^{11}	2^{11}
2^{23}	8	2^{11}	2^{11}

node). This is because of memory limitation, only 512MB RAM is available on each node. Each datum requires 16 bytes. For FFT two arrays of the same size for input and output is required and only data size of 2^n is allowed.

Table 1. shows the comparison of the actual best number of rows from running the program and the best number of rows recommended by the performance model. We believe that inconsistencies in the prediction is due to the effect of memory hierarchy and communication. The speedup is mostly impaired by the communication phase, as a lot of time is spent during the transposition stage. This is further aggravated by the unprecedented effectiveness of the sequential program that utilizes FFTW package. FFTW automatically adapts the computation to the hardware by a special purpose compiler written in Objective Caml[12]. Table 2. depicts the results of implementation using row-by-row distribution and all-at-once distribution for 20 iterations respectively. The notations r.t. and p.t. represents real time and predicted time.

5 Conclusion

In this paper, we developed a modified BSP model for predicting the performance of parallel SSF method for solving nonlinear Schrödinger equation. We have shown that a performance model based on BSP model can be used to model complicated parallel complications for the purpose of adapting an algorithmic parameter. Thus, our work is useful for self adaptive parallel algorithms. We have shown that by using the performance model we can predict an optimal value for the algorithmic parameter N_0 that will minimize the computation time of the algorithm. The predicted performance is accurate but different to the measured values and we believe it is due to the effect of the memory hierarchy and communication. Our present model does not take memory hierarchy into consideration. In future work, we would like to explore the possibility of having a performance model that could include the effect of memory hierarchy as well.

Table 2. (a-f) Speedup and timings for row-by-row distribution. (g-l) Speedup and timings for all-at-once distribution

(a) $N = 2^{20}$

P	r.t. (s)	Speedup
1	34.19	1
2	30.34	1.13
4	14.71	2.32
8	7.58	4.51

(b) $N = 2^{22}$

P	r.t. (s)	Speedup
1	203.61	1
2	122.45	1.66
4	58.62	3.47
8	30.13	6.76

(c) $N = 2^{23}$

P	r.t. (s)	Speedup
1	320.11	1
2	234.72	1.36
4	119.86	2.67
8	60.12	5.32

(d) $N = 2^{20}$

P	p.t. (s)	Speedup
1	34.19	1
2	37.10	1.26
4	15.01	2.27
8	7.58	4.51

(e) $N = 2^{22}$

P	p.t. (s)	Speedup
1	203.61	1
2	110.63	1.84
4	61.14	3.33
8	30.89	6.59

(f) $N = 2^{23}$

P	p.t. (s)	Speedup
1	320.11	1
2	220.19	1.45
4	121.75	2.63
8	61.50	5.20

(g) $N = 2^{20}$

P	r.t. (s)	Speedup
1	34.19	1
2	33.80	1.01
4	20.18	1.69
8	10.67	3.2

(h) $N = 2^{22}$

P	r.t. (s)	Speedup
1	203.61	1
2	125.14	1.63
4	80.29	2.54
8	42.17	4.83

(i) $N = 2^{23}$

P	r.t. (s)	Speedup
1	320.11	1
2	242.77	1.32
4	159.52	2.01
8	86.55	3.70

(j) $N = 2^{20}$

P	p.t. (s)	Speedup
1	34.19	1
2	33.16	1.03
4	20.55	1.66
8	10.82	3.16

(k) $N = 2^{22}$

P	p.t. (s)	Speedup
1	203.61	1
2	128.62	1.58
4	82.48	2.47
8	43.74	4.66

(l) $N = 2^{23}$

P	p.t. (s)	Speedup
1	320.11	1
2	258.68	1.24
4	158.88	2.01
8	87.99	3.64

Acknowledgement

Elankovan Sundararajan would like to thank The National University of Malaysia for providing financial assistance. We would also like to thank anonymous reviewers for their constructive comments.

References

1. Valiant, L.: A bridging model for parallel computation. Communication of the ACM **33** (1990) 103–111
2. Agrawal, G.: Nonlinear Fiber Optics. 3rd edn. Academic Press (2001)
3. Skillicorn, D., Hill, J., McColl, W.: Questions and answers about BSP. Scientific Programming **6** (1998) 249–274

4. Taha, T., Ablowitz, M.: Analytical and numerical aspects of certain nonlinear evolution equation II. numerical, nonlinear Schrödinger equation. In: J. Comp. Phys. Volume 55. (1984) 203–230
5. Tappert, F.: Numerical solutions of the Korteweg-de Vries equation and its generalizations by the split-step Fourier method. In: Lect. Appl. Math. (1974) 215–216
6. Weideman, J., Herbtz, B.: Split-step methods for the solution of the nonlinear Schrödinger equation. SIAM Journal on Numerical Analysis **23** (1986) 485–507
7. Gupta, A., Kumar, V.: The scalability of FFT on parallel computers. IEEE Transaction of Parallel and Distributed Systems **4** (1993) 922 – 932
8. Chu, E., George, A.: Inside the FFT black box: serial and parallel fast Fourier transform algorithms. Boca Raton, Fla.: CRC Press (2000)
9. Cooley, C., Tukey, J.: An algorithm for the machine calculation of complex Fourier series. Math. Comput **19** (1965) 297–301
10. Calvin, C.: Implementation of parallel FFT algorithms on distributed memory machines with a minimum overhead of communication. Parallel Computing **22** (1996) 1255–1279
11. Frigo, M., Johnson, S.: FFTW: An adaptive software architecture for the FFT. Proceedings of the IEEE International Conference on Acoustics, Speech, and Signal Processing **3** (1998) 1381–1384
12. Frigo, M.: A fast Fourier transform compiler. Proceedings of the ACM SIGPLAN'99 Conference on Programming Language Design and Implementation (PLDI) (1999) 169–180
13. Proakis, J.: Digital Communications. Boston : McGraw-Hill, McGraw-Hill Higher Education (2001)
14. Zoldi, S.M., Ruban, V., Zenchuk, A., Burtsev, S.: Parallel implementations of the split-step Fourier method for solving nonlinear Schrödinger systems. SIAM News **32** (1997)
15. Whaley, R., Petitet, A., Dongarra, J.: Automated empirical optimizations of software and the ATLAS project. Parallel Computing **27** (2001) 3–35
16. Chen, Z., Dongarra, J., Luszczek, P., Roche, K.: Self adapting software for numerical linear algebra and LAPACK for clusters. Parallel Computing **29** (2003) 1723–1743
17. McColl, W.F., Tiskin, A.: Memory-efficient matrix computations in the BSP model. Algorithmica **24** (1999) 287–297
18. Tiskin, A.: Bulk-synchronous parallel Gaussian elimination. Journal of Mathematical Sciences **108** (2002) 977–991
19. Gerbessiotis, A.V., Siniolakis, C.J., Tiskin, A.: Parallel priority queue and list contraction: The BSP approach. Computing and Informatics **21** (2002) 59–90
20. Gropp, W., Lusk, E., Skjellum, A.: Using MPI:Portable parallel programming with the Message Passing Interface. 2nd edn. The MIT Press (1999)

Parallel Volume Rendering with Early Ray Termination for Visualizing Large-Scale Datasets

Manabu Matsui, Fumihiko Ino, and Kenichi Hagihara

Graduate School of Information Science and Technology, Osaka University,
1-3 Machikaneyama, Toyonaka, Osaka 560-8531, Japan
m-matui@ist.osaka-u.ac.jp

Abstract. This paper presents an efficient parallel algorithm for volume rendering of large-scale datasets. Our algorithm focuses on an optimization technique, namely early ray termination (ERT), which aims to reduce the amount of computation by avoiding enumeration of invisible voxels in the visualizing volume. The novelty of the algorithm is that it incorporates this technique into a distributed volume rendering system with global reduction of the computational amount. The algorithm also is capable of statically balancing the processor workloads. The experimental results show that our algorithm with global ERT further achieves the maximum reduction of 33% compared to an earlier algorithm with local ERT. As a result, our load-balanced algorithm reduces the execution time to at least 66%, not only for dense objects but also for transparent objects.

1 Introduction

Direct volume rendering [1] is a technique for displaying three-dimensional (3-D) volumetric scalar data as a two-dimensional (2-D) image. Typically, the data values in the volume are made visible by mapping them to color and opacity values, which are then accumulated to determine the image pixel.

One challenging issue in volume rendering is to realize fast rendering for large-scale datasets. However, finding a solution to this issue is not easy due to the high time and space complexities of volume rendering, both represented as $O(n^3)$ for an $n \times n \times n$ voxel volume. Therefore, fast processors with large memories are necessary to carry out this compute-intensive rendering with in-core processing.

To address this issue, many acceleration techniques have been proposed in the past. Levoy [2] proposes two optimization techniques that reduce the time complexity of volume rendering. The first technique is early ray termination (ERT), which adaptively terminates accumulating color and opacity values in order to avoid useless ray casting. The second technique is a hierarchical octree data structure [3], which encodes spatial coherence in object space in order to skip empty regions of the volume. These techniques reduce the execution time by roughly a factor of between 5 and 11. Nukata et al. [4] present a cuboid-order rendering algorithm that aims to maximize the cache hit ratio by dividing the volume into cuboids, which are then rendered successively. This algorithm enables view-independent fast volume rendering on a single CPU computer.

Another promising approach is parallelization on parallel computers. Hsu [5] proposes the segmented ray casting (SRC) algorithm, which parallelizes volume rendering on a distributed memory parallel computer. This algorithm distributes the volume by using a block-block decomposition and carries out data-parallel processing to generate subimages for each decomposed portion. The subimages are then merged into a final image by using an image compositing algorithm [6,7].

Though many earlier projects propose a wide variety of acceleration schemes, data distributed parallel schemes [5–7] are essential to render large-scale datasets that cause out-of-core rendering on a single CPU computer. One problem in these schemes is that the increase of computational amount compared to sequential schemes, where the amount can easily be reduced by means of ERT. This increase is due to the low affinity between ERT and data distribution. That is, while data distribution makes processors independently render the volume data, ERT is based on the visibility of the data, determined in a front-to-back order. Therefore, earlier parallel algorithms independently apply ERT to each distributed data in order to perform data-parallel processing. This locally applied ERT, namely local ERT, increases the computational amount compared to global ERT, because the visibility is locally determined, so that processors render locally visible but globally invisible voxels. Thus, earlier data distributed schemes are lacking the capability of global ERT, so that these schemes can suffer in low efficiency especially for large-scale datasets with many transparent objects.

Gao et al. [8] address this issue by statically computing the visibility of the volume data. However, their static visibility culling approach requires pre-processing for every viewing direction, so that its pre-processing stage prevents rapid visualization.

The key contribution of this paper is the development of a parallel algorithm that dynamically realizes global ERT in a distributed volume rendering system. Our algorithm has the following two advantages.

R1: Reduction of the memory usage per processor by data distribution.
R2: Reduction of the computational amount by global ERT without pre-processing.

To realize R1, our algorithm employs a block-cyclic decomposition that is capable of statically balancing the processor workloads. To realize R2, the algorithm employs an efficient mechanism for sharing the visibility information among processors.

The remainder of the paper is organized as follows. Section 2 introduces earlier algorithms and presents the problem we tackled in this work. Section 3 describes the details of our algorithm while Section 4 presents some experimental results on a cluster of 64 PCs. Finally, Section 5 concludes the paper.

2 Volume Rendering

Figure 1(a) shows an overview of the ray casting algorithm [1]. This algorithm produces an image by casting rays from the viewpoint through the screen into the viewing volume. The image pixel $I_{s,t}$ on point (s,t) is determined by accumulating color and opacities values of penetrated voxels $V_1, V_2, \ldots, V_k$: $I_{s,t} = \sum_{i=1}^{k} \alpha(V_i) c(V_i) \prod_{j=0}^{i-1}(1 - \alpha(V_j))$, where $C(V_i)$ and $\alpha(V_i)$ are the color and opacity values of the i-th penetrated voxel V_i, respectively; $0 \leq \alpha(V_i) \leq 1$; and $\alpha(V_0) = 0$. Computing $I_{s,t}$ for all points (s,t) on the screen, where $1 \leq s \leq n_s$ and $1 \leq t \leq n_t$, generates the final image of size $n_s \times n_t$. In the following discussion, let $a_{s,t}(i)$ be the accumulated transparency for voxel V_i, where $a_{s,t}(i) = \prod_{j=0}^{i-1}(1 - \alpha(V_j))$.

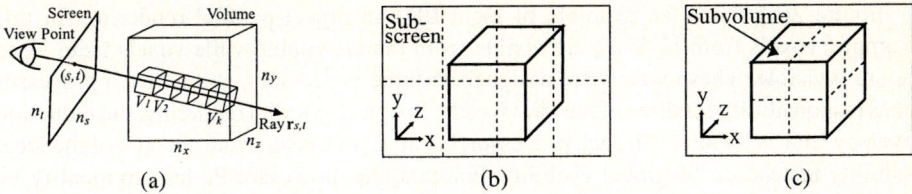

Fig. 1. Ray casting and its parallel schemes. (b) Screen-parallel rendering and (c) object-parallel rendering parallelize (a) ray casting by exploiting the parallelism in screen space and in object space, respectively

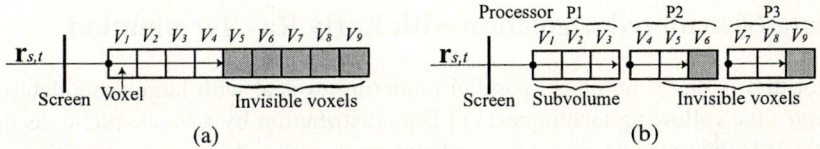

Fig. 2. Early ray termination (ERT). (a) Global ERT for sequential and screen-parallel rendering, and (b) local ERT for object-parallel rendering. While global ERT terminates the ray immediately before invisible voxel V_5, local ERT fails to avoid accumulating locally visible but globally invisible voxels: V_5, V_7, and V_8. Voxels V_6 and V_9 are invisible locally as well as globally

ERT reduces the computational amount by avoiding accumulation of color and opacity values that do not have influence on the final image. That is, ERT avoids enumerating voxels $V_l, V_{l+1}, \ldots, V_k$ if $a_{s,t}(l) = 0$.

Earlier parallel schemes can be classified into two groups: screen-parallel and object-parallel rendering as illustrated in Figure 1.

Screen-parallel rendering exploits the parallelism in screen space. In this scheme, the screen is divided into p subscreens, where p represents the number of processors, and tasks associated with each subscreen are assigned to processors. Because each processor takes responsibility for the entire of a ray as it does in sequential schemes, ERT can easily be applied to this scheme, as illustrated in Figure 2(a). Furthermore, by assigning the tasks in a cyclic manner, this scheme statically balances the processing workloads. However, it requires large main memory to provide fast rendering for any given viewpoint, because every processor need to load the entire volume into memory. Thus, though screen-parallel rendering is a good scheme for small datasets, which require no data decomposition, it does not suit for large-scale datasets.

In contrast, object-parallel rendering exploits the parallelism in object space. This scheme divides the volume into p subvolumes, and then assigns tasks associated with each subvolume to processors. Parallel rendering of each subvolume generates p distributed subimages, so that image compositing is required to merge subimages into the final image. Thus, this scheme allows us to distribute subvolumes to processors, so that is suitable for large-scale datasets. However, because accumulation tasks of a ray can be assigned to more than one processor, it is not easy to utilize global ERT in this scheme.

Figure 2(b) shows an example of local ERT in object-parallel rendering. In this example, voxels from V_1 to V_4 are visible from the viewpoint while voxels from V_5 to V_9 are invisible. These voxels are assigned to three processors, so that each processor takes responsibility for three of the nine voxels. In object-parallel rendering, the reduction given by ERT is localized in each processor, because processors take account of the local visibility instead of the global visibility. For example, processor P2 fails to identify V_5 as an invisible voxel, because it accumulates opacity values from its responsible V_4 in order to perform data-parallel processing. Furthermore, although P2 terminates the ray after V_5, its back neighborhood P3 is unaware of this termination, so that accumulates V_7 and V_8, according to the local visibility.

3 Data Distributed Algorithm with Early Ray Termination

Our algorithm is based on object-parallel rendering to deal with large-scale datasets. It integrates the following techniques: (1) Data distribution by a block-cyclic decomposition; (2) Concurrent processing of volume rendering and image compositing; (3) Visibility sharing by a master/slave paradigm; (4) Parallel image compositing.

3.1 Data Distribution

Our algorithm distributes the volume data according to a block-cyclic decomposition, aiming to maximize the parallelism that can be decreased due to global ERT. The following discussion describes why we employ this decomposition.

In order to realize global ERT in object-parallel rendering, processors have to share the visibility information, namely accumulated transparency. For example, as illustrated in Figure 2(b), in a case where processors P2 and P3 are responsible for neighborhood voxels and P2 terminates the ray, P3 can avoid accumulating all of its responsible voxels V_7, V_8, and V_9 after it obtains the value of accumulated transparency that P2 has computed for V_6. However, this indicates that casting a ray with global ERT has no parallelism in the viewing direction, because P3 has to wait for P2 to complete rendering of its responsible voxels. Thus, applying global ERT to object-parallel rendering decreases the entire parallelism in object space due to processor synchronization. Note here that the parallelism in screen space is remained.

The key idea to address this decreased parallelism is that exploiting parallelism in vertical planes perpendicular to the viewing direction. That is, we employ (A) a data decomposition that allows every processor to have equal-sized tasks on any cross sections of the volume. Such decomposition minimizes the overhead for the processor synchronization by allowing processors to overlap communication with computation. For example, processor P3 in Figure 2(b) can perform rendering for other rays during waiting for P2, because any processor has its responsible tasks on any vertical plane perpendicular to the viewing direction. Furthermore, this decomposition realizes static load balancing because tasks on any cross sections are assigned equally to processors.

As the results of the above considerations, our algorithm employs a block-cyclic decomposition. Note here that a cyclic decomposition is more appropriate than this decomposition in terms of (A). However, it possibly decreases rendering performance due to frequent communication among processors, because a task in the cyclic decomposition corresponds to a voxel, so that communication occurs for each voxel. Therefore, we use a combination of block and cyclic decompositions in order to have coarse-grained tasks without losing the nature of load balancing.

In addition to this data distribution technique, our algorithm aims to reduce the time complexity by encoding empty regions as Levoy does in [2]. We use an adaptive block decomposition rather than Levoy's hierarchical octree, because it increases traversing overheads with the degree of its hierarchy [9]. The adaptive block decomposition addresses this issue by uniformly space partitioning.

3.2 Concurrent Processing of Volume Rendering and Image Compositing

As mentioned before, ERT aims to efficiently render the volume according to the visibility. Note here that the visibility in object-parallel rendering is determined by compositing of subimages. Therefore, applying global ERT to object-parallel rendering requires concurrent processing of volume rendering and image compositing. Furthermore, to obtain better performance, (B) a rapid read/write access to the visibility information, namely the accumulated transparency for an arbitrary ray, must be provided.

In order to realize (B), we classify processors into two groups as follows: (1) Rendering Processors (RPs), defined as processors that render subvolumes in order to generate subimages; (2) Compositing Processors (CPs), defined as processors that composite subimages and manage accumulated transparency $a_{s,t}$ for all rays, where $a_{s,t}$ denotes the accumulated transparency for ray $\mathbf{r}_{s,t}$. Note here that the relation between CPs and RPs is similar to that between masters and slaves in the master/slave paradigm. The details of CPs and RPs are presented later.

3.3 Visibility Sharing Mechanism

Figure 3 shows the processing flow of our master/slave based algorithm. Let r, c, and v be the number of RPs, CPs, and subvolumes, respectively. The processing flows for RPs and CPs consist of the following phases.

Processing Flow for RPs

1. Data distribution. The volume is divided into at least r subvolumes, which are then distributed to r RPs in a round-robin manner. Data distribution phase occurs only at the beginning of the system.
2. Rendering order determination. Each RP determines the rendering order of assigned subvolumes by constructing a list of subvolumes, L, in which its responsible v/r subvolumes are sorted by the distance to the screen in an ascending order. This ascending order is essential to achieve further reduction by means of ERT. List L is updated every time the viewpoint moves.
3. Accumulated transparency acquisition. Each RP deletes a subvolume from the head of L, then obtains accumulated transparencies from CPs, for all rays that penetrate the subvolume. Let T be a set of accumulated transparencies obtained from CPs.
4. Subvolume rendering. For all rays $\mathbf{r}_{s,t}$ such that $a_{s,t} \in T$ and $a_{s,t} > 0$, each RP accumulates the voxels penetrated by $\mathbf{r}_{s,t}$ so that generates a subimage. Note here

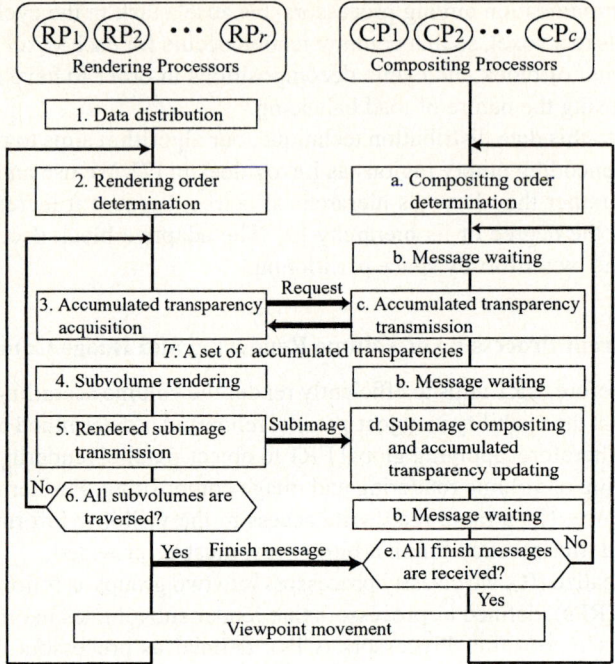

Fig. 3. Processing flow of proposed algorithm

that the algorithm avoids rendering for all rays $\mathbf{r}_{s,t}$ such that $a_{s,t} \in T$ and $a_{s,t} = 0$, according to ERT.

5. Rendered subimage transmission. Each RP transmits the rendered subimages to CPs. Because blank pixels have no influence on the final image, the algorithm transmits only pixels inside the bounding rectangle of the subimages in order to reduce the amount of communication.
6. Completion check. Phases 3., 4., and 5. are repeated until list L becomes empty. Empty L indicates that the RP completes performing all assigned tasks for the current viewpoint, so that it sends a finish message to all CPs.

Processing Flow for CPs

a. Compositing order determination. Each CP determines the compositing order of subimages by constructing a list of subvolumes, M, in which all v subvolumes are sorted by the distance to the screen in an ascending order.
b. Message waiting. Each CP waits for incoming messages from RPs. Such messages contain request messages for accumulated transparency acquisition, data messages including rendered subimages, and finish messages.
c. Accumulated transparency transmission. Each CP transmits T that RPs require.
d. Subimage compositing and accumulated transparency updating. Each CP updates T by compositing its local subimages with received subimages, according to the order

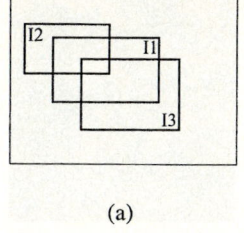
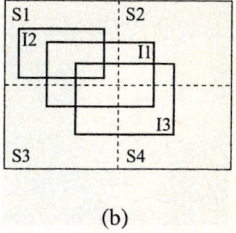

(a)   (b)

Fig. 4. Screen-parallel compositing. Dividing the screen into subscreens produces more parallelism of image compositing

of list M. If keeping this order is impossible due to the lack of the still unrendered subimages, it stores the received subimages into a local buffer for later compositing.
e. Completion check. Phases b., c. and d. are repeated until receiving finish messages from all RPs.

3.4 Parallel Processing of Image Compositing

In the master/slave paradigm, the master becomes a performance bottleneck if it is assigned many slaves beyond its capacity. Therefore, our algorithm parallelizes the master's tasks by exploiting the parallelism in screen space. That is, as screen-parallel rendering does, it divides the screen into at least c subscreens and assigns them to c CPs. Let s be the number of subscreens.

In addition to the benefits of acceleration, this screen-parallel compositing increases the parallelism of image compositing. Figure 4 gives an example of this increased parallelism. In this example, subimages I1, I2, and I3 are rendered from neighborhood subvolumes located in a front-to-back order. As shown in Figure 4(a), compositing I1 and I3 requires rendered I2 if we avoid dividing the screen. In this case, CPs have to wait for RPs to generate I2 before compositing I1 and I3. In contrast, if the screen is divided into subscreens, rendered I2 is unnecessary to perform compositing in subscreens S2, S3, and S4. Therefore, compositing in these subscreens can be carried out without waiting the rendering of I2, so that screen-parallel compositing enables compositing subimages at shorter intervals. This means that accumulated transparencies are updated at shorter intervals, which contribute to achieve further reduction by ERT.

Thus, dividing the screen allows us to exploit more parallelism of image compositing. Furthermore, it also contributes to realize (B) because it enables more frequent updating of accumulated transparencies.

4 Experimental Results

In order to evaluate the performance of our algorithm, we compare it with two earlier algorithms: SRC [5] and SRC with load balancing (SRCLB). We also present a guideline for obtaining the appropriate values for the four parameters: r, c, v, and s.

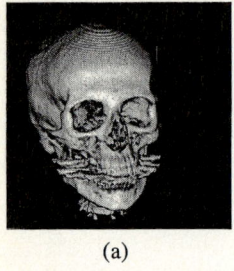 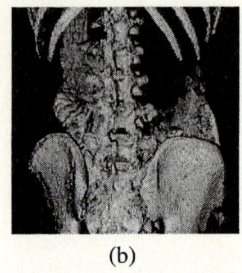 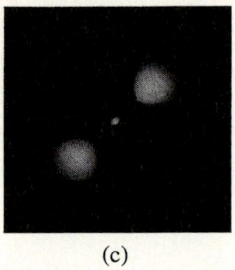

(a) (b) (c)

Fig. 5. Rendering results of volume datasets used in experiments. (a) D1: skull volume of size $512 \times 512 \times 448$, (b) D2: abdomen volume of size $512 \times 512 \times 730$, and (c) D3: hydrogen atom volume of size $512 \times 512 \times 512$. Each volume is rendered on a 512×512 pixel screen

The SRC algorithm is an object-parallel algorithm that parallelizes the ray casting algorithm with a block-block decomposition. On the other hand, the SRCLB algorithm incorporates a load balancing capability into SRC. To balance processing workloads, it divides the volume into subvolumes with marginal regions. For every viewpoint, it adaptively varies the size of responsible regions inside the subvolumes, according to the execution time measured for the last viewpoint. Therefore, SRCLB requires more physical memory compared to the remaining two algorithms, which use disjoint decompositions. However, it requires no data redistribution during volume rendering. Both the SRC and SRCLB algorithms use an improved binary-swap compositing (BSC) [7] for image compositing. Furthermore, ERT is locally applied to them. All the three algorithms use an adaptive block decomposition [9] to skip empty regions in the volume.

We have implemented the three algorithms by using the C++ language and MPICH-SCore library [10], a fast implementation of the Message Passing Interface (MPI) standard. We used a Linux cluster of 64 PCs for the experiments. Each node in this cluster has two Pentium III 1-GHz processors and 2 GB of main memory, and connects to a Myrinet switch, which provides a link bandwidth of 2 GB/s.

Figure 5 shows images rendered for three employed datasets D1, D2, and D3. In addition, we also used a large-scale dataset D4, skull-big volume of size $1024 \times 1024 \times 896$, generated by trilinear interpolation of D1. The screen sizes are 512×512 pixel for D1, D2, and D3, and 1024×1024 pixel for D4.

4.1 Performance Comparison to Earlier Algorithms

To measure rendering performance, we rotated the viewpoint around the viewing objects, so that obtained an average of 24 measured values. Table 1 shows the averaged results. In this table, T_1, T_2, and T_3 represent the averaged execution time for SRC, SRCLB, and our algorithms, respectively. N_1, N_2, and N_3 also represent the averaged number of rendered voxels. We measured them by using the best parameter values determined by the guideline presented in the next section (see Table 2). The length for marginal region of SRCLB is given by 256 voxels, which is the maximum length for performing in-core rendering on our cluster.

Table 1. Measured execution time and number of rendered voxels for proposed, SRC, and SRCLB algorithms. On less than eight processors, some execution failed due to the lack of physical memory. N_s represents the number of voxels rendered by a sequential algorithm with ERT

p	D1: skull volume $N_s = 3.1 \cdot 10^6$ voxels								D2: abdomen volume $N_s = 18.2 \cdot 10^6$ voxels							
	SRC		SRCLB		Proposed		Reduction ratio		SRC		SRCLB		Proposed		Reduction ratio	
	T_1	N_1	T_2	N_2	T_3	N_3	R_T	R_N	T_1	N_1	T_2	N_2	T_3	N_3	R_T	R_N
4	2134	6.7	—	—	2366	4.4	0.90	1.52	6155	23.7	—	—	6497	20.9	0.95	1.13
8	1471	8.3	1185	8.0	1283	6.3	1.15	1.32	3719	25.9	—	—	3316	23.8	1.12	1.09
16	1177	9.9	773	10.2	691	6.3	1.70	1.57	2684	27.9	2350	28.0	1761	23.7	1.52	1.18
32	920	11.4	480	12.3	438	6.8	2.10	1.68	1903	29.6	1432	29.9	934	24.7	2.04	1.20
64	682	12.7	302	14.0	279	8.3	2.44	1.53	1334	31.0	859	31.5	541	25.5	2.47	1.22
128	435	15.0	240	12.6	207	11.0	2.10	1.36	780	32.6	536	27.8	373	27.3	2.09	1.19
p	D3: hydrogen atom volume $N_s = 7.6 \cdot 10^6$ voxels								D4: skull-big volume N_s is unmeasurable							
	SRC		SRCLB		Proposed		Reduction ratio		SRC		SRCLB		Proposed		Reduction ratio	
	T_1	N_1	T_2	N_2	T_3	N_3	R_T	R_N	T_1	N_1	T_2	N_2	T_3	N_3	R_T	R_N
4	2168	8.1	—	—	2759	7.3	0.79	1.11	—	—	—	—	—	—	—	—
8	1201	8.3	—	—	1725	7.5	0.70	1.11	7146	40.8	—	—	5836	26.7	1.22	1.53
16	888	8.3	777	8.3	872	7.6	1.02	1.09	5386	49.2	4736	49.5	3328	31.5	1.62	1.56
32	673	8.3	479	8.4	499	7.8	1.35	1.06	4144	57.2	3172	59.4	2178	36.3	1.90	1.58
64	468	8.4	302	8.6	301	8.1	1.55	1.04	3152	64.8	2365	68.5	1311	41.0	2.40	1.58
128	325	8.5	187	7.5	226	8.2	1.44	1.04	2132	78.8	1640	86.0	927	50.2	2.30	1.57

$$R_T = T_1/T_3, R_N = N_1/N_3$$

Table 2. Parameter values employed for p processors. Parameters r, c, v, and s represent the number of RPs, that of CPs, that of volume divisions, and that of screen divisions, respectively

p	D1: skull volume of size $512 \times 512 \times 448$			D2: abdomen volume of size $512 \times 512 \times 730$			D3: hydrogen atom volume of size $512 \times 512 \times 512$			D4: skull-big volume of size $1024 \times 1024 \times 896$		
	r	v	s	r	v	s	r	v	s	r	v	s
4	3	$2 \times 2 \times 2$	23×23	3	$3 \times 3 \times 3$	24×24	3	$6 \times 6 \times 6$	24×24	—	—	—
8	7	$2 \times 2 \times 2$	22×22	7	$8 \times 8 \times 8$	24×24	7	$6 \times 6 \times 6$	18×18	7	$4 \times 4 \times 4$	24×24
16	14	$4 \times 4 \times 4$	24×24	15	$8 \times 8 \times 8$	23×23	14	$6 \times 6 \times 6$	19×19	15	$5 \times 5 \times 5$	23×23
32	27	$7 \times 7 \times 7$	22×22	29	$10 \times 10 \times 10$	20×20	27	$8 \times 8 \times 8$	16×16	27	$7 \times 7 \times 7$	23×23
64	55	$8 \times 8 \times 8$	18×18	59	$9 \times 9 \times 9$	16×16	50	$8 \times 8 \times 8$	10×10	55	$8 \times 8 \times 8$	24×24
128	109	$8 \times 8 \times 8$	14×14	111	$12 \times 12 \times 12$	11×11	106	$8 \times 8 \times 8$	14×14	109	$11 \times 11 \times 11$	20×20

This table indicates that our algorithm is generally faster than SRC, because it shows $R_T > 1.0$ for all $p > 8$, where R_T is the reduction ratio of the execution time compared to SRC. In particular, on a larger number of processors, our algorithm reduces the execution time in half for datasets D1, D2, and D4.

In contrast, the reduction ratio is relatively small on a smaller number of processors. This small improvement can be explained as follows. The first reason is that RPs in our classification based algorithm is c fewer than that in the remaining two algorithms. This indicates that processor classification is not suited for systems with smaller p, because such systems do not have computing resources enough to deal with compute-intensive rendering. In such small systems, any resource must be dedicated to the performance bottleneck, namely subvolume rendering, in order to achieve faster acceleration. Actually, as presented in Table 2, we obtain $c = 1$ for all $p \leq 8$, so that the number of RPs is insufficient to that of CPs in these situations. The second reason is that tasks associated with the same ray are assigned to a few processors in the SRC algorithm. This indicates that on a smaller number of processors, local ERT is sufficient to terminate rays in an early rendering phase, so that there is no redundant accumulation left for global ERT.

By comparing R_T and R_N, where R_N is the reduction ratio of rendered voxels compared to SRC, we can see that the reduction of the execution time is more than that of rendered voxels. In particular, although ERT achieves few reduction for transparent dataset D3, our algorithm reduces its execution time by 33%. This acceleration is provided by load balancing. As shown in Figure 5(c), the most voxel in dataset D3 is transparent and few opaque voxel is located around the center of the volume. For such datasets, the SRC algorithm constructs smaller blocks located near the center and larger blocks located far from the center, because it uses a combination of an adaptive block decomposition and block-block decomposition. Therefore, processors assigned with center blocks have relatively larger tasks than others, so that the processor workloads become imbalanced. Actually, the average and standard deviation of subvolume rendering time on 128 processors, μ and σ, respectively, are improved from $\mu = 91$ and $\sigma = 88$ ms in the SRC method to $\mu = 105$ and $\sigma = 34$ ms in our algorithm.

Finally, we compare our algorithm to SRCLB by using D1 and its larger version D4. Our algorithm shows better improvement to SRCLB for D4 rather than for D1. On 128 processors, its reduction ratio to SRCLB is $T_2/T_3 = 1.77$ for D4 but 1.16 for D1. This is due to the lack of physical memory required for marginal regions. That is, although we maximized the length for marginal regions, it was not enough to balance the workloads for D4. Thus, compared to our algorithm, SRCLB requires a larger physical memory to balance the workloads for large-scale datasets.

4.2 Parameter Setup

We now present a guideline for obtaining the appropriate values for the four parameters: $r, c, v,$ and s. Figure 6 shows the execution time averaged over RPs for different parameter values. We show the results only for skull dataset D1 because we obtained similar results for others.

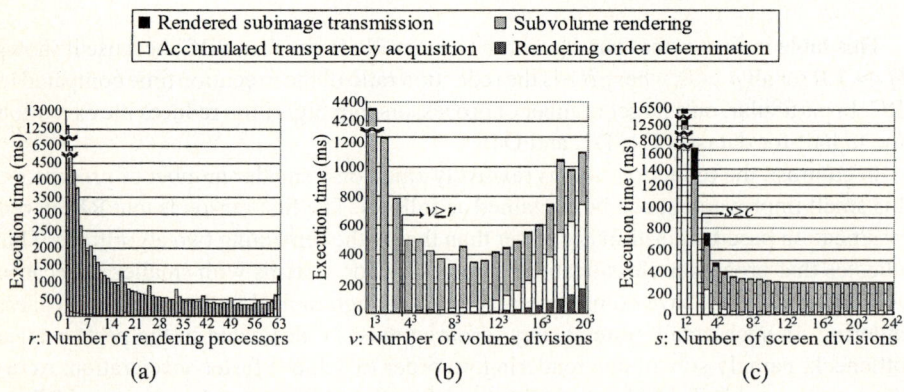

Fig. 6. Execution time measured on RPs for different parameter values using skull dataset D1. Results for (a) $v = 8 \times 8 \times 8$ and $s = 18 \times 18$, (b) $r = 55$ and $s = 18 \times 18$, and (c) $r = 55$ and $v = 8 \times 8 \times 8$

We first investigate the influence of r under fixed v and s. Figure 6(a) shows the breakdown of the execution time on $p = 64$ for different r. The time for subvolume rendering decreases as r increases, so that we obtain the shortest time when $r = 55$. This decrease is due to the reduction of computational amount per processor, because each RP is responsible for v/r subvolumes, which decreases with the increase of r. On the other hand, when $r > 55$, the execution time turns to increase because the communication time for accumulated transparency acquisition increases with r. This increase is due to the lack of CPs, which makes RPs wait in the accumulated transparency acquisition phase. Thus, there is a tradeoff between r and c.

The appropriate values for r and c are determined by finding a balancing point as follows. Given a volume of size $n \times n \times n$ and a screen of $n \times n$, the time complexities of volume rendering and image compositing are $O(n^3)$ and $O(n^2)$, respectively. Therefore, if we assume that the appropriate values for r and c balance the workloads of them, r and c are given by:

$$r/c = (w_1 + w_2)/w_2 \cdot n^3/n^2, \qquad (1)$$

$$p = r + c, \qquad (2)$$

where $(w_1 + w_2)/w_2$ is a granularity of rendering to compositing, w_1 is the time for trilinear interpolation to determine a scalar value of a voxel, and w_2 is the time for accumulating color and opacity values of a image pixel.

Next, we investigate v (Figure 6(b)). The increase of v means the downsize of task granularity. In addition, the block-cyclic decomposition becomes similar to the cyclic decomposition with the increase of v. Therefore, increasing v leads to better load balancing, which minimizes the execution time for subvolume rendering. However, fine-grained tasks cause frequent communication between RPs and CPs, because they shorten the intervals of accumulated transparency acquisition and updating. Although shorter intervals contribute to achieve further reduction by ERT, but CPs can suffer from network contention when the intervals are too short beyond the network capacity. Therefore, the execution time turns to increase in such a situation. Thus, there is a tradeoff between load balancing and communication frequency.

As same as for r and c, the appropriate value for v also is determined by finding a balancing point. In Table 2, we can see that the best value of v differs among datasets. Therefore, the appropriate value must be determined for each p and dataset by finding the saturated value of at least r. For example, by increasing v from r, we can identify the saturated value when the execution time turns from decrease to increase.

Finally, we investigate s (Figure 6(c)). When $s < c$, the parallelism of image compositing increases with s, so that increasing s reduces the execution time. However, there is no significant difference when $s \geq c$. Furthermore, Table 2 shows similar values of s for datasets D1, D2, and D3. Therefore, the appropriate value for s depends on each p and n. The value can be determined by finding the saturation point with increasing s from c.

In summary, Equations (1) and (2) determine the appropriate values for r and c, which dominate the execution time. The appropriate value for v can be determined by using the tradeoff between load balancing and communication time. The value is given by finding the saturated value of at least r for each p and dataset. The last parameter

s has relatively small significance on the execution time. The appropriate value can be determined by finding the saturated value of at least c for each p and n.

5 Conclusions

We have presented an efficient parallel volume rendering algorithm that is capable of rendering large-scale datasets on a distributed rendering system. The novelty of the algorithm is a combination of global ERT and data distribution with static load balancing. To realize this, the algorithm uses the master/slave paradigm where slave processors carry out the rendering tasks to accumulate color and opacity values of voxels while master processors perform compositing tasks and manage the accumulated values to share them among processors. The experimental results show that the reduction given by ERT increases with the size of datasets, and the improvement produced by static load balancing increases with the number of processors.

References

1. Levoy, M.: Display of surfaces from volume data. IEEE Computer Graphics and Applications **8** (1988) 29–37
2. Levoy, M.: Efficient ray tracing of volume data. ACM Trans. Graphics **9** (1990) 245–261
3. Yau, M.M., Srihari, S.N.: A hierarchical data structure for multidimensional digital images. Comm. ACM **26** (1983) 504–515
4. Nukata, M., Konishi, M., Goshima, M., Nakashima, Y., Tomita, S.: A volume rendering algorithm for maximum spatial locality of reference. IPSJ Trans. Advanced Computing Systems **44** (2003) 137–146 (In Japanese)
5. Hsu, W.M.: Segmented ray casting for data parallel volume rendering. In: Proc. 1st Parallel Rendering Symp. (PRS'93). (1993) 7–14
6. Ma, K.L., Painter, J.S., Hansen, C.D., Krogh, M.F.: Parallel volume rendering using binary-swap compositing. IEEE Computer Graphics and Applications **14** (1994) 59–68
7. Takeuchi, A., Ino, F., Hagihara, K.: An improved binary-swap compositing for sort-last parallel rendering on distributed memory multiprocessors. Parallel Computing **29** (2003) 1745–1762
8. Gao, J., Huang, J., Shen, H.W., Kohl, J.A.: Visibility culling using plenoptic opacity functions for large volume visualization. In: Proc. 14th IEEE Visualization Conf. (VIS'03). (2003) 341–348
9. Lee, C.H., Park, K.H.: Fast volume rendering using adaptive block subdivision. In: Proc. 5th Pacific Conf. Computer Graphics and Applications (PG'97). (1997) 148–158
10. O'Carroll, F., Tezuka, H., Hori, A., Ishikawa, Y.: The design and implementation of zero copy MPI using commodity hardware with a high performance network. In: Proc. 12th ACM Int'l Conf. Supercomputing (ICS'98). (1998) 243–250

A Scalable Low Discrepancy Point Generator for Parallel Computing

Kwong-Ip Liu and Fred J. Hickernell

Department of Mathematics, Hong Kong Baptist University, Kowloon, Hong Kong
kiliu@math.hkbu.edu.hk, fred@hkbu.edu.hk

Abstract. The Monte Carlo (MC) method is a simple but effective way to perform simulations involving complicated or multivariate functions. The Quasi-Monte Carlo (QMC) method is similar but replaces independent and identically distributed (i.i.d.) random points by low discrepancy points. Low discrepancy points are regularly distributed points that may be deterministic or randomized. The digital net is a kind of low discrepancy point set that is generated by number theoretical methods. A software library for low discrepancy point generation has been developed. It is thread-safe and supports MPI for parallel computation. A numerical example from physics is shown.

Keywords: Monte Carlo and Quasi-Monte Carlo methods, digital nets, parallel programming, software library.

1 Introduction

Simulations are useful for solving many problems in scientific computing. For example, since most integrals cannot be evaluated analytically, approximation by numerical methods is often the only possible solution. For an integral depending on a few variables, tensor product trapezoidal or Simpson's rule may be applied. However, for high dimensional integrals, these methods becomes less practical since the amount of work required to attain a certain error tolerance increases exponentially with the dimension. The Monte Carlo (MC) method and the Quasi-Monte Carlo (QMC) method are alteratives since the errors of these methods are less dependent on dimension. From the computational point of view, they are simple to implement and are highly parallelizable.

Consider the following approximation of an s-dimensional integral,

$$I(f) = \int_{[0,1)^s} f(\mathbf{x})\, d\mathbf{x} \approx Q(f) = \frac{1}{n} \sum_{i=1}^{n} f(\mathbf{x}_i), \qquad (1)$$

where f is a multivariate function, $\{\mathbf{x}_i\}$ is a set of n points lying in $[0,1)^s$. A simple Monte Carlo method chooses the $\{\mathbf{x}_i\}$ to be independent and identically distributed (i.i.d.) random points. By Quasi-Monte Carlo method, the $\{\mathbf{x}_i\}$ is a set of carefully selected regularly distributed points. It has been found that in many practical problems, like pricing of financial derivatives, the performance of the Quasi-Monte Carlo method is better than the Monte Carlo method, even when s is on the hundreds. This observation can be explained by the *effective dimension* [1]. This paper focuses on digital nets which is a kind of low discrepancy points.

2 Design and Implementation of the Library

A library has been implemented based on the codes in [2] and [3]. By combining different scrambling methods and using the Gray code [4], the generation of digital net can be highly simplified by the following recursive function [2],

$$\phi(x_{0,j}) = \tilde{\mathbf{e}}_j, \ j = 1, 2, \ldots, s \tag{2}$$

$$\phi(x_{i+1,j}) = \phi(x_{i,j}) - \tilde{\mathbf{C}}_{j,\hat{k}_{i+1}} \bmod 2, \quad \begin{cases} i = 0, 1, \ldots, 2^m - 1, \\ j = 1, 2, \ldots, s, \end{cases} \tag{3}$$

where $\tilde{\mathbf{C}}_{j,l}$ is the l-th column of $\tilde{\mathbf{C}}_j = \mathbf{L}_j \mathbf{C}_j \mathbf{L}^T$, $\mathbf{L}_j$ and $\mathbf{L}$ are nonsingular lower triangular random matrices, $\mathbf{C}_j$ are generating matrices of digital net, $\tilde{\mathbf{e}}_j$ are random vectors and $\hat{k}_i = \min(k : \lfloor i2^{-k} \rfloor \neq i2^{-k})$.

So far the library supports the standard C rand() function and the SPRNG library [5] for random number generation, and MPI as the parallel programming library. During initialization, a digital net may be chosen in one of four different modes:

NETGEN_SIMPLE Use rand() function in C standard library to scramble the digital net. The same sequence of digital net is generated in different processors of a parallel program, if the same seed is used.

NETGEN_SPRNG Use SPRNG to scramble the sequences. The same sequence of digital net is generated in different processors of a parallel program, if the same seed is used.

NETGEN_MPI Use SPRNG to scramble the net. Independent sequences of digital net are generated in different processors of a parallel program, even the same seed is used in all processors.

NETGEN_BLOCK Use SPRNG to scramble the digital net. A long sequence of digital net is divided into different segments for different processors of a parallel program [6].

The independent sequences in NETGEN_MPI are generated by different scrambled generator matrices. In NETGEN_BLOCK mode, the sequences in different processors are generated by the same set of matrices, but reflect different segments of a long sequence. In order to support programs with several phases, the sequences are extendable. In NETGEN_BLOCK, the point number of the beginning of extended sequence is found by,

$$MK + kn,$$

where M is the original number of points in each processor, K is the number of processors, k is the rank of processor and n is the number of extended points. Actually, MK is the last point of the original sequence, and kn is the offset of the new sequence in the k-th processor. For other modes, the point number of the beginning of extended sequence is the end of current sequences plus one.

All function prototypes and user-data structures are defined in the header file netgen.h. An abstract programming object "net handle" is defined to represent a digital net.

netHandle *nh;

After initialization, the net handle is used as a parameter to other functions in the library. The main function prototypes in the library are

- netHandle *InitNetHandle(int dim, unsigned long num, int type, int mode, unsigned seed, int scrambling); – creates a digital net. Four types of nets are supported: SOBOL [7], NIED [8], NIEDXING [9] or RANDOM. After calling this function, a pointer to the created digital net is returned.
- void FinalNetHandle(netHandle *nh); – releases all resources allocated by the net pointed by nh.
- void NextPoint(netHandle *nh); – forwards to the next point in the net pointed by nh. The coordinates of the point are stored in double array, nh->point.
- void ExtendPoint(netHandle *nh, unsigned long num); – extends num points in the net pointed by nh.

3 Numerical Example

3.1 Multivariate Integration

Many problems in physics are multivariate integration. We tested the performances of Monte Carlo and Qusai-Monte Carlo methods by a typical integration [10].

$$Q(s) = \int_{\mathbf{R}^s} \cos(\|\mathbf{x}\|) e^{-\|x\|^2} d\mathbf{x}$$

$$= \pi^{s/2} \int_{[0,1]^s} \cos\left(\sqrt{\sum_{j=1}^{s} \frac{\Phi^{-1}(\mathbf{y}_j)^2}{2}}\right) d\mathbf{y}, \qquad (4)$$

where $\|\cdot\|$ denotes the Euclidean norm in $\mathbf{R}^s$, and Φ denotes the standard multivariate Gaussian distribution function. One of the reasons for choosing this function is that the values of the integral is known.

The results of the serial (NETGEN_SPRNG mode) and the parallel (NETGEN_BLOCK and NETGEN_MPI modes) executions were the average of 100 simulations. The relative RMS errors, variances and execution times have been recorded. For serial executions, the performances by using a digital net are better than the one by using i.i.d. random points for 9-dimensional integral, for both error magnitudes

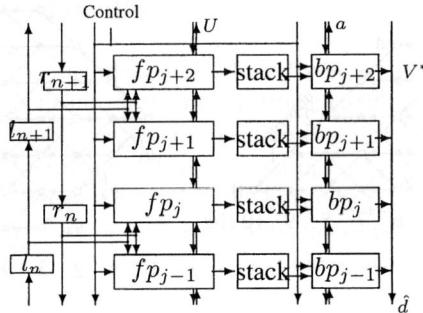

Fig. 3. Inter processor connectivity of a linear array of PEs

3. Backward Initialization:
$$\hat{d}_{2N} = 0 .$$

4. Backward Recursion: For each site $t = 2N - 1, \ldots, 0$, find the optimal disparity
$$\hat{d}_{t+1} = \hat{d}_t + V_{t,\hat{d}_t} .$$

3 Systolic Array Architecture

Based on the stereo matching algorithm given in the previous section, we wish design a scalable architecture to implement stereo matching in hardware. The overall architecture is a linear systolic array of PEs as shown in Fig. 3. Communication extends only to neighboring PEs and the array is completely regular in structure, making actual hardware design relatively simple. The area complexity is $\mathcal{O}(N)$ for N PEs for all componenents except for the best path decision registers $\mathcal{V}$ which is $\mathcal{O}(N^2)$.

4 Experimental Results

As shown in Fig. 4(a) the images received from a pair of cameras are processed by the rectification logics which use the linear interpolation method and then the generalized stereo matching part calculates disparity data from rectified images. This architecture is implemented on Xilinx Virtex - II XC2V8000 FPGA which incorporates 208 PEs. To access the FPGA, it is mounted on PCI interface board. PC then reads the computed disparity, converts it to a gray scale image and displays it. The implemented stereo matching board is shown in Fig. 4(b).

Given images of the vergent cameras the experimental results in Fig. 4 show that when the camera canting the optical axes inward, the old parallel axis system doesn't work. We can know the generalized trellis system supports the vergent case in addition to the parallel case.

where M is the original number of points in each processor, K is the number of processors, k is the rank of processor and n is the number of extended points. Actually, MK is the last point of the original sequence, and kn is the offset of the new sequence in the k-th processor. For other modes, the point number of the beginning of extended sequence is the end of current sequences plus one.

All function prototypes and user-data structures are defined in the header file netgen.h. An abstract programming object "net handle" is defined to represent a digital net.

netHandle *nh;

After initialization, the net handle is used as a parameter to other functions in the library. The main function prototypes in the library are

- netHandle *InitNetHandle(int dim, unsigned long num, int type, int mode, unsigned seed, int scrambling); – creates a digital net. Four types of nets are supported: SOBOL [7], NIED [8], NIEDXING [9] or RANDOM. After calling this function, a pointer to the created digital net is returned.
- void FinalNetHandle(netHandle *nh); – releases all resources allocated by the net pointed by nh.
- void NextPoint(netHandle *nh); – forwards to the next point in the net pointed by nh. The coordinates of the point are stored in double array, nh->point.
- void ExtendPoint(netHandle *nh, unsigned long num); – extends num points in the net pointed by nh.

3 Numerical Example

3.1 Multivariate Integration

Many problems in physics are multivariate integration. We tested the performances of Monte Carlo and Qusai-Monte Carlo methods by a typical integration [10].

$$Q(s) = \int_{\mathbf{R}^s} \cos(\|\mathbf{x}\|) e^{-\|x\|^2} d\mathbf{x}$$
$$= \pi^{s/2} \int_{[0,1]^s} \cos\left(\sqrt{\sum_{j=1}^{s} \frac{\Phi^{-1}(\mathbf{y}_j)^2}{2}}\right) d\mathbf{y}, \qquad (4)$$

where $\|\cdot\|$ denotes the Euclidean norm in $\mathbf{R}^s$, and Φ denotes the standard multivariate Gaussian distribution function. One of the reasons for choosing this function is that the values of the integral is known.

The results of the serial (NETGEN_SPRNG mode) and the parallel (NETGEN_BLOCK and NETGEN_MPI modes) executions were the average of 100 simulations. The relative RMS errors, variances and execution times have been recorded. For serial executions, the performances by using a digital net are better than the one by using i.i.d. random points for 9-dimensional integral, for both error magnitudes and the rates of error decreases. However, the rate of error decrease by using i.i.d. random points becomes compatible to the rate of error decreases by using a digital net in 100-dimensional integral. This is consistent with the error upper bounds in [11].

The relative RMSE of different net types in NETGEN_BLOCK mode parallel executions with 10 processors were plotted in Figure 1. They are almost the same as the figures of serial executions if plotted since a single sequence was distributed to different processors. Similar features have been observed for NETGEN_MPI mode.

Figure 2 shows the comparison between NETGEN_MPI and NETGEN_BLOCK on relative RMSE. The relative RMSE of NETGEN_BLOCK mode are smaller than

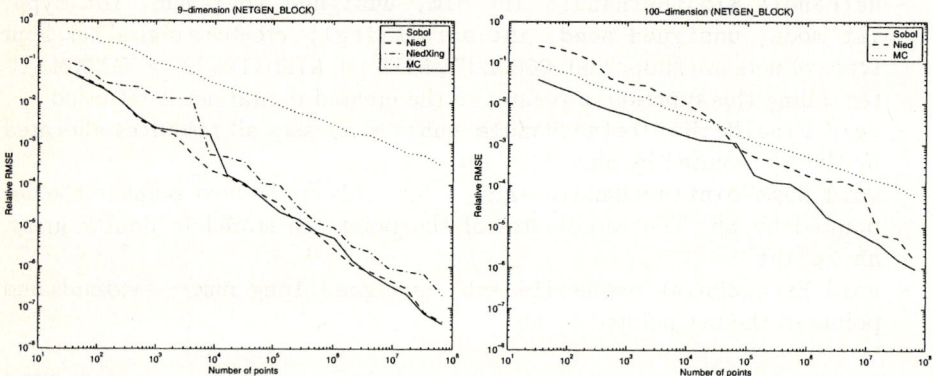

Fig. 1. Relative RMSE in NETGEN_BLOCK mode parallel executions for 9-dimensional and 100-dimensional integrals

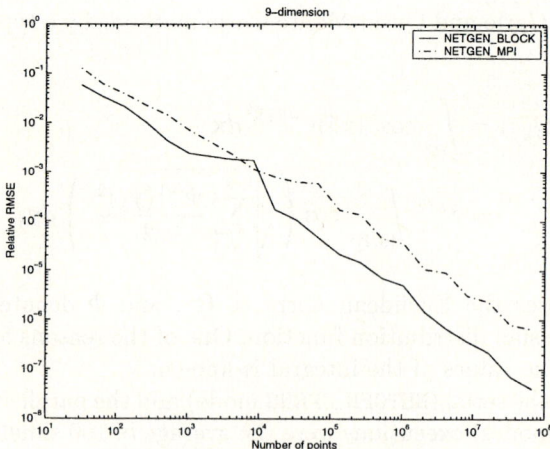

Fig. 2. Comparison between NETGEN_MPI and NETGEN_BLOCK on relative RMSE

NETGEN_MPI mode. It is because the number of effective points to estimate the integral in NETGEN_BLOCK mode are more than the number in NETGEN_MPI mode. However, it is difficult for NETGEN_BLOCK mode to estimate the error in the cases in which the exact answer is not known. It is because only one estimation is resulted in NETGEN_BLOCK mode from all processors.

The average time and the speedup of executions of 9-dimensional integrals were shown in Figure 3. The communication overheads of 10 processors is significant for the number of points less than 10^6. Moreover, the execution time using digital net is less than the time using i.i.d. random points because of the efficient generation algorithm of the nets. The random number generator is called only in the inital phase.

Assume t_p is the time for parallel, N is the number of points to be used, n is the number of processors, T_n is the communication overhead which is independent to N but is dependent on n. A simple model for the time of parallel executions of MC/QMC methods can be stated as,

$$t_p = T_n + k(N/n) \tag{5}$$

where k is another constant reflecting the processing time of one simulation. This model is particularly useful for MC/QMC methods since the simulations in different processors are basically independent. The speedup would be,

$$\frac{t_s}{t_p} = \frac{kN}{T_n + k(N/n)} = \frac{1}{T_n/kN + 1/n} \tag{6}$$

The speedup will be improved if the execution time is longer (a larger N) or the number of processors decreases (a smaller T_n). For parallel execution using Sobol net using 10 processors in Figure 3, the parameters are approximately $T_{10} = 5.195$ and $k = 4.547 \times 10^{-6}$.

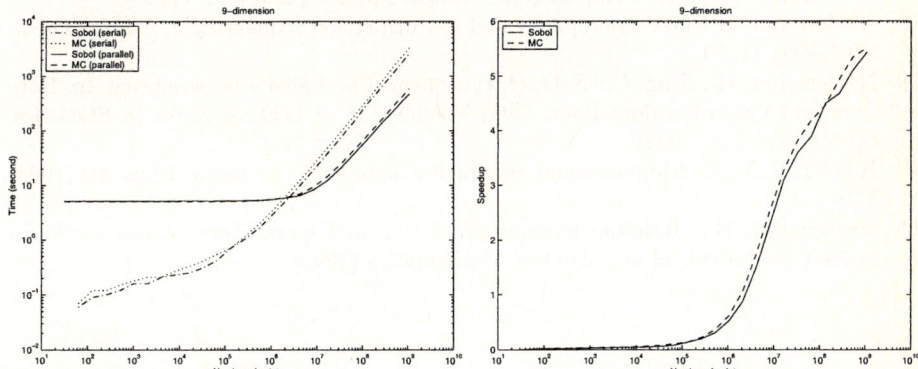

Fig. 3. The average time and speedup of the parallel executions of 9-dimensional integrals

4 Summary and Future Work

We briefly introduced our implementation of a software library to generate different digital nets. We hope the library is useful for developing the programs in both serial and parallel computers. In the future, more net types will be added to the library.

Numerical examples of the applications of digital nets have been shown. During the implementation of these programs, we observed that some applications of digital nets are quite similar and can be summaried in several standard patterns of usages. High-level functions will be added to the library for these standard patterns of usages, so that the implementation of applications will be easy and efficient.

References

1. Caflish, R.E., Morokoff, W., Owen, A.B.: Valuation of mortgage-backed securities using brownian bridges to reduce effective dimension. The Journal of Computational Finance **1** (1997) 27–46
2. Hong, H.S., Hickernell, F.J.: Algorithm 823: Implementing scrambled digital sequences. ACM Transactions on Mathematical Software **29** (2003) 95–109
3. Bratley, P., Fox, B.L., Niederreiter, H.: Implementation and tests of low-discrepancy sequences. ACM Trans. Model. Comput. Simul. **2** (1992) 195–213
4. Lichtner, J.: Iterating an α-ary gray code. SIAM Journal of Discrete Mathematics **11** (1998) 381–386
5. Mascagni, M., Srinivasan, A.: Algorithm 806: SPRNG: a scalable library for pseudorandom number generation. ACM Transactions on Mathematical Software **26** (2000) 436–461
6. Ökten, G., Srinivasan, A.: Parallel quasi-monte carlo methods on a heterogeneous cluster. In Fang, K.T., Hickernell, F.J., Niederreiter, H., eds.: Monte Carlo and Quasi-Monte Carlo Methods 2000, Springer-Verlag (2002) 406–421
7. Sobol', I.M.: The distribution of points in a cube and the approximate evaluation of integrals. U.S.S.R. Comput. Math. Math. Phys. **7** (1967) 86–112
8. Niederreiter, H.: Low-discrepancy and low dispersion sequences. J. Numb. Theor. **30** (1998) 51–70
9. Niederreiter, H., Xing, C.: Nets, (t,s)-sequences and algebraic geometry. In: Random and Quasi-Random Point Sets. Volume 138 of Lecture Notes in Statistics. Springer-Verlag (1998)
10. Keister, B.D.: Multidimensional quadrature algorithms. Comput. Phys. **10** (1996) 119–122
11. Niederreiter, H.: Random number generation and quasi-Monte Carlo methods. Society for Industrial and Applied Mathematics (1992)

Generalized Trellis Stereo Matching with Systolic Array

Hong Jeong and Sungchan Park

Pohang University of Science and Technology, Electronic amd Electrical Engineering,
Pohang, Kyungbuk, 790-784, South Korea
hjeong@postech.ac.kr
http://isp.postech.ac.kr

Abstract. We present here a real time stereo matching chip which is based on a general trellis form with vergent optical axis. The architecture can deal with general axis angle of cameras with better resolution in given space. For a pair of images with $M \times N$ pixels, only $\mathcal{O}(MN)$ time is required. The design is highly scalable and fully exploits the concurrent and configurable nature of the algorithm. We implement stereo chip on Xilix FPGA with 208 PEs(Processing Elements) that can obtain disparity range of 208 levels. It can provide the real-time stereo matching for the mega-pixel images.

1 Introduction

Stereo vision is the process of recreating depth or distance information from a pair of images of the same scene. Its methods fall into two broad categories [1]. One is the local method which uses the constraint in small window pixels like block matching or feature matching technique. And the other is the global method which uses the global constraints on scan-lines or whole image like dynamic programming and graph cuts typically. Normally many real time systems [2], [3], [4], [5] use the local methods. Although it has low complexity, there are some local problems where it fail to match, due to occlusion, uniform texture, ambiguity of low texture, and etc. Also, the popular local matching method, block matching skill makes disparity data blurred in object boundary. The global methods can solve these local problems but suffer from the huge processing time. In 2000, Jeong and Oh [6] built a stereo global matching ASIC that can deal with parallel optical axis in real time. The high speed is possible due to a parallel dynamic programming search method on a trellis solution space. It is suitable for highly parallel implementation and produces good matching results. However, a desire to improve depth resolution and depth of field has led to modifications to this algorithm to incorporate vergent cameras, that is, cameras with optical axes that intersect. Based on this algorithm, we introduce in this paper an efficient linear systolic array architecture that is appropriate for VLSI implementation. The array is highly regular, consisting of identical and simple processing elements (PEs) with only nearest-neighbor communication and external communication occurs with the end PEs. So, it makes possible to construct an inexpensive and truly portable stereo vision system.

This paper is organized as follows: A brief review of the matching algorithm based on a trellis structure is presented in Sec. 2. Sec. 3 describes the systolic array of PEs realizing this algorithm. The test results are discussed in Sec. 4 and finally conclusions are given in Sec. 5.

2 Stereo Matching Algorithm

The observed image line g is the ideal image line f where each pixel or element has been corrupted by independent and identically distributed additive white gaussian noise. A possible match is indicated wherever a projection line from one image intersects a projection line from the other image. This is shown in Fig. 1(a) for a pair of cameras with parallel optical axes. Because a pixel in the left image can only be matched to a pixel in the right image with zero or higher disparity, intersections or possible matches occur over a triangular region. However, by canting the optical axes inward, a pixel in one image can be matched to a larger range of pixels in the other image. This is known as the *vergent* camera model and mimics human vision. An example of this is shown in Fig. 1(b).

Note that the matching region now consists of two triangles and that a pixel in one image can now be matched to any pixel in the other image, effectively doubling the size of the match point set. It can have all positive and negative disparity data. Vergent cameras allow us to increase the depth range over parallel cameras, or to double the depth resolution with the same depth of field.

Given the observations g^l and g^r, we wish to obtain a maximum a priori (MAP) estimate of the disparity

$$\begin{aligned}\hat{d} &= \arg\max_{d} P(d|g^l, g^r) \\ &= \arg\max_{d} P(g^l, g^r|d) P(d) \ .\end{aligned} \quad (1)$$

The prior probability distribution $P(d)$ is modeled using a simple independent binary probability P_o of an occlusion at each pixel site. An occlusion occurs when a point is visible in one image but not in the other. We assume that each pixel is corrupted by AWGN with distribution $N(0, \sigma)$. Then, the log-likelihood of $P(d|g^l, g^r)$ is

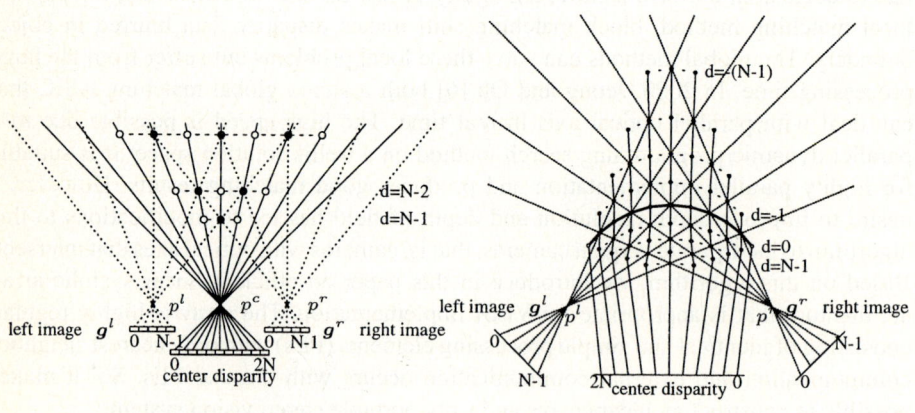

(a) Projection model for parallel cameras (b) Projection model for vergent cameras

Fig. 1. Projection model for several camera models

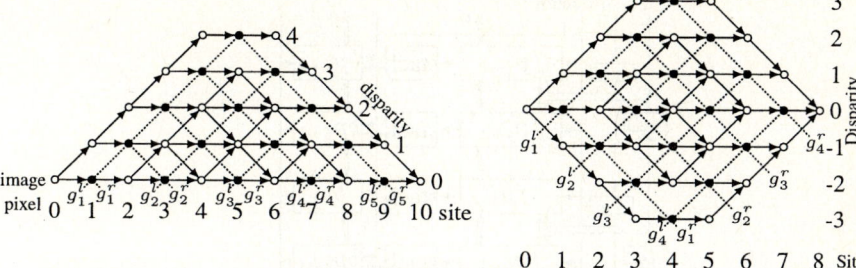

(a) Disparity trellis of parallel cameras for $N = 5$. (b) Generalized disparity trellis for $N = 4$.

Fig. 2. Disparity trellis

$$U(d) = \sum_{i=1}^{2N} [\Delta g(d_i) o(i + d_i) + \gamma |d_i - d_{i-1}|] \ , \quad (2)$$

where

$$\Delta g(d_i) = \left(g^l_{\frac{1}{2}(i-d_i+1)} - g^r_{\frac{1}{2}(i+d_i+1)} \right)^2 \quad (3)$$

is the cost of matching two pixels, $o()$ is a function that is 1 when the argument is odd and 0 otherwise,

$$\gamma = \sigma^2 \log \frac{(1 - P_o)^2}{\sqrt{2\pi}\sigma(P_o)^2} \quad (4)$$

is the cost of an occlusion.

The optimal disparity $\hat{d}$ is the disparity that minimizes (2). The structure of the trellis allows the solution $\hat{d}$ to be found efficiently using the Viterbi algorithm.

Algorithm. Given g^l and g^r, the accumulated cost $U_j(t)$, the best input path decision $V_{t,j}$ and the optimal disparity $\hat{d}$ are computed as follows:

1. Forward Initialization:

$$U_j(0) = \begin{cases} 0, & j = base, \\ \infty, & \text{otherwise.} \end{cases}$$

2. Forward Recursion: For each site $t = 1, \ldots, 2N$, find the best path into each node j: $\Delta g(a, b) = |g^l_{\frac{1}{2}(a-b+1)} - g^r_{\frac{1}{2}(a+b+1)}|$

(a) If $t + j$ is even,

$$U_j(t) = \min_{k \in [-1,1]} U(t - 1, j + k) + \gamma |k| \ ,$$

$$V_{t,j} = \arg\min_{k \in [-1,1]} U(t - 1, j + k) + \gamma |k| \ .$$

(b) If $t + j$ is odd,

$$U_j(t) = U(t - 1, j) + \Delta g(t, j - base) \ ,$$

$$V_{t,j} = 0 \ .$$

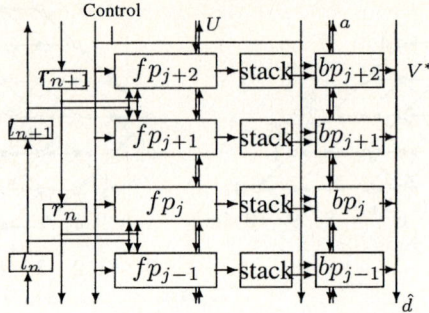

Fig. 3. Inter processor connectivity of a linear array of PEs

3. Backward Initialization:
$$\hat{d}_{2N} = 0 \ .$$

4. Backward Recursion: For each site $t = 2N - 1, \ldots, 0$, find the optimal disparity

$$\hat{d}_{t+1} = \hat{d}_t + V_{t,\hat{d}_t} \ .$$

3 Systolic Array Architecture

Based on the stereo matching algorithm given in the previous section, we wish design a scalable architecture to implement stereo matching in hardware. The overall architecture is a linear systolic array of PEs as shown in Fig. 3. Communication extends only to neighboring PEs and the array is completely regular in structure, making actual hardware design relatively simple. The area complexity is $\mathcal{O}(N)$ for N PEs for all componenents except for the best path decision registers V which is $\mathcal{O}(N^2)$.

4 Experimental Results

As shown in Fig. 4(a) the images received from a pair of cameras are processed by the rectification logics which use the linear interpolation method and then the generalized stereo matching part calculates disparity data from rectified images. This architecture is implemented on Xilinx Virtex - II XC2V8000 FPGA which incorporates 208 PEs. To access the FPGA, it is mounted on PCI interface board. PC then reads the computed disparity, converts it to a gray scale image and displays it. The implemented stereo matching board is shown in Fig. 4(b).

Given images of the vergent cameras the experimental results in Fig. 4 show that when the camera canting the optical axes inward, the old parallel axis system doesn't work. We can know the generalized trellis system supports the vergent case in addition to the parallel case.

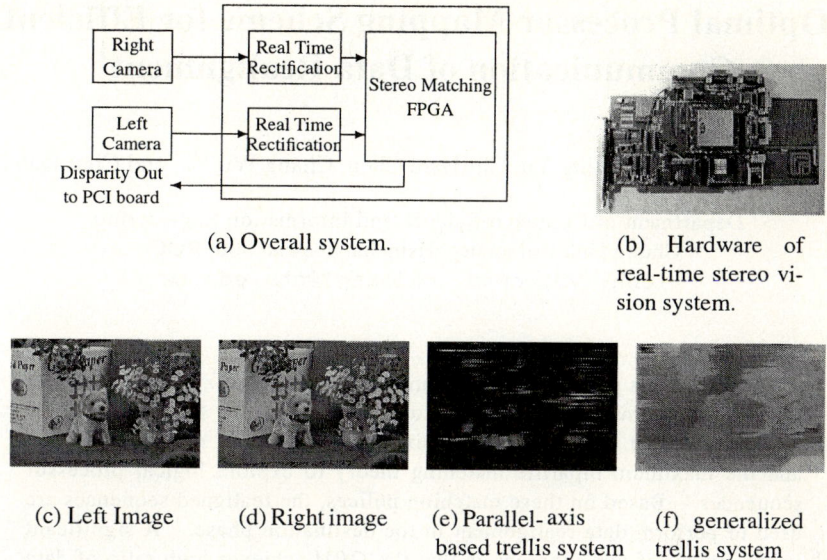

Fig. 4. Hardware system and output of images

5 Conclusions

As a step toward a real-time stereo, we have presented a fast and efficient VLSI architecture and implementation of a stereo matching algorithm in the vergent camera case. The architecture has the form of linear systolic array using simple PEs that are connected with only neighboring PEs. The design is simple to implement and scales very well. The hardware board can process 1280 by 1000 images of maga pixel cameras at 15 frames/s.

References

1. Myron Z. B. and Gregory D. H.: Advances in computational stereo. IEEE Transactions on Pattern Analysis and Machine Intelligence. vol. **25**. no. **8** (2003)993–1007
2. T. Kanade, A. Yoshida, K. Oda, H. Kano, and M. Tanaka: A stereo machine for video-rate dense depth mapping and its new applications. in Proceedings of the IEEE International Conference on Computer Vision and Pattern Recognition. IEEE Computer Society Press. (1996)
3. K. Konolige: Small vision systems: Hardware and implementation. Proc. Eighth Int Ol Symp. Robotics Research. (1997)
4. H. Yamaguchi, E. Kawamura, S. Kimura, T. Shinbo and K. Naka: A convolver-based real-time stereo machine (sazan).Proc. Computer Vision and Pattern Recognition. vol.1 (1999)457–463
5. T. Takeuchi, M. Hariyama and M. Kameyama: Vlsi processor for reliable stereo matching based on adaptive window-size selection . Proceedings of the IEEE International Conference on Robotics and Automation. (2001)1168–1173
6. H. Jeong and Y. Oh: Fast stereo matching using constraints in discrete space. IEICE Transactions on Information and Systems. (2000)

Optimal Processor Mapping Scheme for Efficient Communication of Data Realignment[1]

Ching-Hsien Hsu, Kun-Ming Yu, Chi-Hsiu Chen, Chang Wu Yu, and Chiu Kuo Lian

Department of Computer Science and Information Engineering,
Chung Hua University, Hsinchu, Taiwan 300, ROC
`{chh, yu, cwyu, ckliang}@chu.edu.tw`

Abstract. In this paper, we present an *Optimal Processor Mapping* (OPM) scheme to minimize data transmission cost for general BLOCK-CYCLIC data realignment. We examine a size oriented greedy matching method and the maximum bipartite matching theory to explore logical processor sequences. Based on these matching polices, the realigned sequences are used to perform data realignment in the destination phase. A significant improvement of our approach is that the *OPM* achieves high ratio of data remain in local space and leading minimum inter-processor communications. The *OPM* scheme could handle array realignment with arbitrary BLOCK-CYCLIC type and multidimensional arrays. Theoretical analysis and experimental results show that our technique provides considerable improvement for dynamic data realignment.

1 Introduction

In many data parallel applications, an optimal distribution of data depends on the characteristics of an algorithm, as well as on the attributes of the target architecture. Because the optimal distribution changes from one phase to another, data realignment turns out to be a critical operation during runtime. Therefore, many data parallel programming languages support run-time primitives for changing a program's data decomposition. Since data realignment is performed at run-time, there is a performance trade-off between the efficiency of the new data decomposition for a subsequent phase of an algorithm and the cost of redistributing matrix data among processors. Thus efficient methods for performing data realignment are of great importance for the development of parallelizing compilers for those languages.

Techniques for dynamic data realignment are discussed in many researches. A detailed expatiation of these techniques was described in [4]. Since the communication overheads usually dominate overall performance of a redistribution algorithm, many researches have been concentrated on the optimizations of communication. Examples are the processor mapping technique [5] for minimizing data transmission overheads, the multiphase

---

[1] This work was supported in part by NSC of Taiwan under grant number NSC92-2213-E-216-025 and in part by Chung-Hua University, under contract CHU-93-TR-010.

redistribution strategy [6] for reducing message startup cost, the communication scheduling approach [1, 2, 3, 7, 10] for avoiding node contention, the strip mining approach [8] for overlapping the communication and computation steps and the spiral mapping technique [9] for enhancing communication locality. In this paper, we present an optimal processor mapping scheme to minimize data transmission cost for general BLOCK-CYCLIC data redistribution. A major feature of the proposed technique is that it achieves the highest ratio of data remain in local space and leading minimum inter-processor communication. The ability to handle arbitrary BLOCK-CYCLIC multidimensional array is also an important extension.

2 Preliminaries and Cost Model

To simplify the presentation, we use $BC_{x \to y}$ to represent the CYCLIC(x) to CYCLIC(y) redistribution for the rest of the paper. As mentioned in [4], each processor has to compute the following four sets, Destination Processor Set (DPS[P_i]), Send Data Sets ($\bigcup_{P_j \in DPS[Pi]} SDS[P_{i \to j}]$), Source Processor Set (SPS[P_j]), and Receive Data Sets ($\bigcup_{P_i \in SPS[Pj]} RDS[P_{j \leftarrow i}]$) for a redistribution. Since a Bipartite Graph (BG) is usually used to represent the communication patterns between source and destination processors sets, the above terms, |DPS[P_i]|, the number of destination processors in DPS[P_i] and |SDS[$P_{i \to j}$]|, the number of elements in SDS[$P_{i \to j}$] could represent the out degree of node P_i and the weight (w_{ij}) of edge e_{ij} in BG, respectively.

To facilitate the analysis of next sections, we formulate the communication cost of a processor P_i in an algorithm to perform data redistribution as $T_{comm} = \alpha_i \times T_s + \delta_i \times T_d$, where T_s is the message startup cost; T_d is the data transmission costs of interconnection network of a parallel machine; $\alpha_i = |DPS(P_i)|$ is the number of destination processors in DPS(P_i) and $\delta_i = \sum_{P_j \in DPS[Pi]} |SDS[P_{i \to j}]|$ is the total number of elements in all SDS[$P_{i \to j}$], for all $j \in DPS[P_i]$.

In general, data transmission cost is directly proportional to the size of redistributing data and influence the total execution time. Recall the bipartite graph representation, for source processor P_i, δ_i is equal to the summation of edge weight for all directional edges that has source vertex P_i, i.e., $\delta_i = \sum_{\forall j \in DPS[Pi]} w_{ij}$. Because the edge e_{ij} will not incur inter-processor communication when $i = j$, we have $\delta_i = \sum_{\forall j \in DPS[Pi], i \neq j} w_{ij}$. The global volume of transition data $\delta = \sum_{i=0}^{P-1} \delta_i$.

Another structure widely used to demonstrate the inter-processor communication patterns is *Communication Table* (*CT*), which is defined as a $P \times P$ matrix in which $CT_{ij} = w_{ij}$, the edge weight of e_{ij} in BG. Since CT_{ij} will not incur inter-processor communication when $i = j$, we obtain $\delta = \sum_{i,j=0}^{P-1} CT_{ij} - \sum_{i=0}^{P-1} CT_{ii}$. Let L_d be summation of the volume of elements in the main diagonal in CT, i.e., $L_d = CT_{00} + \ldots + CT_{P-1P-1}$, we then have $\delta = \sum_{i,j=0}^{P-1} CT_{ij} - L_d$. Because $\sum_{i,j=0}^{P-1} CT_{ij}$ is a constant, increasing L_d will lead lower δ and minimize communication cost. In next section, we will describe the techniques to maximize L_d.

3 Processor Mapping Scheme

We begin by illustrating the utility of processor mapping technique for the example of $BC_{10 \to 5}$ on $A[1:100]$ over 5 processors. Figure 1 shows data alignments using two different processor sequences, the *Traditional Logical Processor Sequences* (*TLPS*) and the *Realigned Logical Processor Sequences* (*RLPS*), where *TLPS* is an ascending order of integers from 0 to $P-1$; *RLPS* is another permutation of those integers except the one of *TLPS*. The shadow blocks represent data reside in the same logical processor through the operation of data realignment. We refer these data as *Local Data Sets* (L_d). Thus we can have $L_d = CT_{00} + CT_{44} = 20$ in *TLPS* as shown in Figure 2(a) and $L_d = CT_{00} + CT_{11} + CT_{22} + CT_{33} + CT_{44} = 50$ in *RLPS* as shown in Figure 2(b).

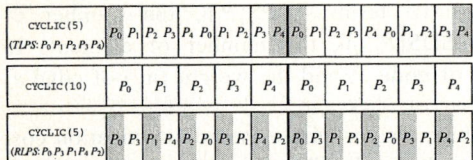

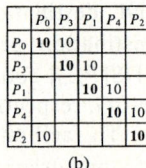

Fig. 1. Data alignments using *TLPS* and *RLPS*

Fig. 2. Communication Tables (a) CT for *TLPS* (b) CT for *RLPS*

3.1 Size Oriented Greedy Matching

According to the above example, to permute the order of logical processors for data realignment might increase the amount of elements in local data sets. A *Size-Oriented Greedy* (SOG) approach for reordering logical processors is introduced.

Let $RLPS[0:P-1]$ be the realigned sequence of logical processors and β is the largest CT_{ij} in CT, for all $0 \leq i, j \leq P - 1$. In *SOG* method, $RLPS[j]$ is set to i if $CT_{ij} = \beta$. For example, given the communication table of a $BC_{5 \to 4}$ over 12 processors as shown in Figure 3(a), we have $\beta = 4$; for source processors (the first column) P_0, P_3, P_4, P_7, P_8 and P_{11}, they have largest CT_{ij} equals to 4; therefore, the *RLPS* is set to $\{0, 3, 4, 7, 8, 11, -, -, -, -, -, -\}$. Now, processors P_1, P_2, P_5, P_6, P_9 and P_{10} have largest CT_{ij} equals to 3 (i.e., $\beta - 1$). For processor P_1, CT_{11}, CT_{14}, CT_{17} and CT_{110} are equal to 3. Since $RLPS[1]$ and $RLPS[4]$ were prior occupied by P_3 and P_8, thus, $RLPS[7]$ is set to P_1. Similarly, $RLPS[6]$ is set to P_2. Finally, we have $RLPS = \{0, 3, 4, 7, 8, 11, 2, 1, 6, 5, 10, 9\}$. The communication table for $BC_{5 \to 4}$ over 12 processors using *RLPS* is shown in Figure 3(b).

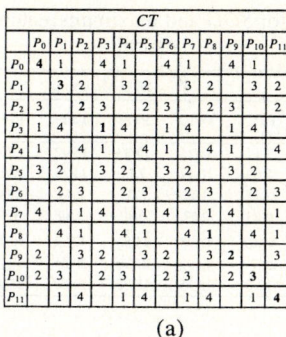
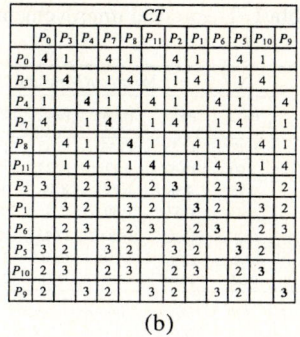

Fig. 3. Communication Tables of $BC_{5 \to 4}$ over 12 processors. (a) CT for *TLPS* (P_0 P_1 P_2 P_3 P_4 P_5 P_6 P_7 P_8 P_9 P_{10} P_{11}). (b) CT for *RLPS* (P_0 P_3 P_4 P_7 P_8 P_{11} P_2 P_1 P_6 P_5 P_{10} P_9).

3.2 Maximum Matching

It is possible that an *RLPS* derived from the *SOG* matching is not an optimal result for a given realignment request. Foe example, the *TLPS* produces $L_d = 24$ for a $BC_{8 \to 6}$ over 11 processors while the *RLPS* = {0, 5, 3, 4, 2, 7, 1, 6, 9, 8, 10} that derived from *SOG* matching yields $L_d = 58$. If we use a maximum bipartite matching algorithm for deriving new *RLPS* = {0, 5, 10, 4, 9, 3, 8, 2, 7, 1, 6}, we have $L_d = 66$. Form this results, the maximum matching conduces largest local data set. In this paper, we apply Hopcroft and Karp's algorithm for deriving *RLPS*. We will analyze the effectiveness between *SOG* matching and the maximum matching in next section.

4 Performance Analysis

To facilitate the following analysis of experiments, we have some declarations. Let L_{d_TR}, L_{d_SOG} and L_{d_MM} be the amount of elements in local data sets for *TLPS*, *RLPS* derived from *SOG* matching and maximum matching, respectively. The improvement rates of data transmission costs for *SOG* and maximum matching methods are defined as $IR_{SOG} = (L_{d\_SOG} - L_{d\_TR}) / \delta$ and $IR_{MM} = (L_{d\_MM} - L_{d\_TR}) / \delta$, respectively, where $\delta = \sum_{i,j=0}^{P-1} CT_{ij}$. Table 1 shows the theoretical improvement rate of data transmission costs for the two matching schemes. We found that the *SOG* matching and maximum matching have almost the same improvement rate for most cases, except for S4(P=48), R2 (P=16) and R2 (P=64). According to this similarity, we adopt the *RLPS* derived from maximum matching for realigning data in our tests.

To get the performance comparison, we have implemented the realignment algorithms using both *TLPS* and *RLPS*. Both programs were written in the single program multiple data (SPMD) programming paradigm with C+MPI code and executed on an SMP/Linux cluster consisted of 24 SMP nodes.

Table 1. Theoretical improvement rate of T_d for *SOG* and maximum matching

	Maximum Matching						SOG Matching					
	P=12	P=16	P=24	P=48	P=64	P=128	P=12	P=16	P=24	P=48	P=64	P=128
S1 ($BC_{5 \to 4}$)	9.17	11.25	13.33	15.42	15.94	16.72	9.17	11.25	13.33	15.42	15.94	16.72
S2 ($BC_{5 \to 6}$)	5.00	10.42	9.17	11.25	15.10	15.89	5.00	10.42	9.17	11.25	15.10	15.89
S3 ($BC_{5 \to 8}$)	3.75	4.38	6.25	8.44	9.06	9.84	3.75	4.38	6.25	8.44	9.06	9.84
S4 ($BC_{20 \to 11}$)	0.23	0.57	0.80	2.90	3.34	4.09	0.23	0.57	0.80	2.89	3.34	4.09
R1 ($BC_{10 \to 5}$)	41.67	43.75	45.83	47.92	48.44	49.22	41.67	43.75	45.83	47.92	48.44	49.22
R2 ($BC_{15 \to 5}$)	22.22	25.00	27.78	30.56	31.25	32.29	22.22	22.92	27.78	30.56	30.73	32.29
R3 ($BC_{20 \to 5}$)	12.5	18.75	18.75	21.88	23.44	24.22	12.5	18.75	18.75	21.88	23.44	24.22

Figures 4(a) and (b) show the execution time to perform S1 and R3 realignment, respectively. As the theoretical prediction, algorithm with *RLPS* scheme outperforms the *TLPS* for both cases. This is because the amount of data needs to be exchanged in *RLPS* is less than that in *TLPS*.

Figures 5(a) and (b) illustrate the improvement rate for six test samples on different number of processors. The R-type realignments has higher improvement rate than S type. This phenomenon matches the information given in Table 1.

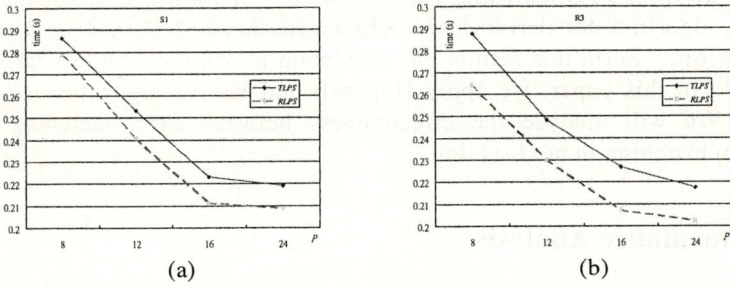

Fig. 4. Execution time of data realignment using *TLPS* and *RLPS* (a) $BC_{5 \to 4}$ (b) $BC_{20 \to 5}$ (size = 1.152×10^7 Bytes)

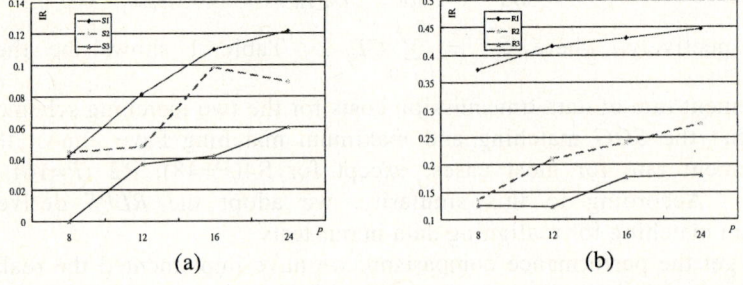

Fig. 5. Improvement rate of algorithm using RLPS on different cases. (a) S type realignment. (b) R type realignment (size = 1.152×10^7 Bytes)

5   Conclusions

We have presented an optimal processor mapping to minimize data transmission cost for general BLOCK-CYCLIC data realignment. Use data to logical processor mapping scheme, the desired destination data-layout could be accomplished over a new *RLPS*. A significant effect is the achievement of large volume of local data sets that leading minimum inter-processor communication. The maximum matching mechanism is applicable to handle arbitrary BLOCK-CYCLIC data and multidimensional arrays. The theoretical analysis and experimental tests show the efficiency of the proposed technique is superior to traditional algorithms.

References

1. Frederic Desprez, Jack Dongarra, and Antoine Petitet," Scheduling Block-Cyclic Data redistribution," *IEEE Trans. on PDS*, vol. 9, no. 2, pp. 192-205, Feb. 1998.
2. M. Guo, I. Nakata, and Y. Yamashita, "Contention-Free Communication Scheduling for Array Redistribution," *Parallel Computing*, vol. 26, no. 8, 2000.
3. S. K. S. Gupta, S. D. Kaushik, C.-H. Huang, and P. Sadayappan, "On Compiling Array Expressions for Efficient Execution on Distributed-Memory Machines," *Journal of Parallel and Distributed Computing*, Vol. 32, pp. 155-172, 1996.
4. C.-H. Hsu and Kun-Ming Yu, "Processor Mapping Technique For Communication Free Data Redistribution on Symmetrical Matrices," *Proc.* of the *7th IEEE International Symposium on Parallel Architectures, Algorithms, and Networks*, 2004.
5. E. T. Kalns, and Lionel M. Ni, "Processor Mapping Technique Toward Efficient Data Redistribution," *IEEE Trans. on PDS*, vol. 6, no. 12, December 1995.
6. S. D. Kaushik, C. H. Huang, J. Ramanujam, and P. Sadayappan, "Multiphase data redistribution: Modeling and evaluation," *Proceeding of IPPS'95*, pp. 441-445, 1995.
7. N. Park, Viktor K. Prasanna, Cauligi S. Raghavendra, "Efficient Algorithms for Block-Cyclic Data redistribution Between Processor Sets," *IEEE Transactions on Parallel and Distributed Systems*, vol. 10, No. 12, pp.1217-1240, Dec. 1999.
8. A. Wakatani and Michael Wolfe, "A New Approach to Array Redistribution: Strip Mining Redistribution," *Proc. of Parallel Architectures and Languages Europe*, 1994.
9. A. Wakatani and Michael Wolfe, "Optimization of Array Redistribution for Distributed Memory Multicomputers, "*Parallel Computing*, vol. 21, no. 9, 1995.
10. H.-G. Yook and Myung-Soon Park, "Scheduling GEN_BLOCK Array Redistribution," Proceedings of the IASTED International Conference Parallel and Distributed Computing and Systems, November, 1999.

MCCF: A Distributed Grid Job Workflow Execution Framework

Yuhong Feng* and Wentong Cai

School of Computer Engineering,
Nanyang Technological University, Singapore 639798

Abstract. With the explosion of scientific data, distributed scientific applications present great challenges to the existing job workflow execution models over the Grid. Based on the idea of having executable codes as part of Grid resources, a Mobile Code Collaboration Framework (MCCF) utilizing light-weight mobile agent and dynamic services for distributed job workflow execution is proposed in this paper. Instead of the existing light-weight mobile agents, agent core (AC), which is an XML file specifying the functional descriptions of sub-jobs, is used to adapt job workflow execution to the dynamic characteristics of the Grid and to reduce the security risks. In addition, dynamic service mechanism is also introduced to facilitate the multidisplinary scientific cooperation and application integration over the Grid. As a proof-of-concept, a prototype of MCCF is implemented.

Keywords: Grid computing, job workflow execution model, job workflow programming model, mobile agent, code mobility, dynamic service.

1 Introduction

Grid computing aims at providing scientific communities a flexible, secure, coordinated resource sharing among dynamic collections of individuals, institutions and resources [6]. However, data intensive collaborative scientific (DICS) computations, such as bioinformatics [7], often require diverse, high volume and distributed data sets. High volume data motion over the Internet makes it a bottleneck for such computations [3]. In addition, DICS computations encompass a large repository of analysis modules, each of which acts on specific kinds of data. Developing such computations as monolithic codes is a backbreaking job.

Job workflow includes the composition of a complete job from multiple sub-jobs, specification of the execution order of the sub-jobs, and the rules that define the interactions among the sub-jobs. Job workflow is a natural technology for developing DICS computations.

---

* Contact Author: Email – pg01797855@ntu.edu.sg

Pioneers have done much research on the development of effective workflow execution models. Some popular business workflow execution models, for example, Web Service Flow Language (WSFL) [9] and Business Process Execution Language for Web Services (BPEL4WS) [2], use a typical Client/Server execution model. When Client/Server model is applied, the execution depends on the workflow engine to intermediate at each step of the sub-job execution, which will produce a lot of unnecessary traffic around the engine. Grid Service Flow Language [8] allows services to deliver messages to each other directly, thus obviating the need for a centralized workflow engine to relay the data between services. However, its control flow stays with the workflow engine, which makes the workflow engine a single point of failure. Considering Grid resource dynamic membership and dynamic quality of service characteristics, mobile agent based distributed workflow enactment [10] seems a promising solution.

Service is classified into dynamic service and static service [5]. When the service deployment is location independent, meaning that it can be instantiated on any servicing host during runtime, the service is called dynamic service. On contrary, when the service can only be instantiated on certain set of hosts, it is called static service. Having executable codes as part of Grid resources, a detailed classification of job workflow execution model was introduced and a simulation study was carried out to compare the performance of various workflow execution models [5]. The simulation results show that when the sub-job executions involve large data sets from distributed data repositories, light-weight mobile agent based on dynamic service gets better performance.

In this paper, Mobile Code Collaboration Framework (MCCF) based on dynamic services together with light-weight mobile agent for distributed job workflow execution is introduced in details.

2 Mobile Code Collaboration Framework

For a given job workflow application, the sub-jobs of the job workflow are represented as a set $J = \{j_1, \ldots, j_n\}$, n is the number of sub-jobs. The sub-jobs and data dependencies among them can be viewed as a directed acyclic graph (DAG), which is represented as $G = (V, E)$, where $V = J$, i.e., sub-jobs are represented by nodes. The dependencies among the sub-jobs are represented by directed edges. A directed edge, $e_{i,k} \in E$, from j_i to j_k, represents that j_k is data dependant to j_i, i.e., j_k will take j_i's output as its input, and the execution of j_k will not be ready to start until the completion of j_i.

2.1 Grid Resources

Having executable codes as Grid resources, Grid resources include data repositories, computational resources, code repositories, and network resources. For a certain data set required by a sub-job, there may exist several data repositories having the data set, which are represented as a set $\mathcal{D} = \{d_0, \ldots, d_d\}$. Similarly, for a certain required code by a sub-job, multiple code repositories may have the codes, which are represented as a set $\mathcal{C} = \{c_0, \ldots, c_c\}$. For a certain sub-job to be executed, there will be multiple computational resources satisfying the computation requirements, which can be represented as a set $\mathcal{M} = \{m_1, \ldots, m_m\}$. Grid resources are shared among different organizations, which may involve communications over the Internet. $\mathcal{N}$ is the matrix capturing network characteristics among nodes in $\mathcal{D}$, $\mathcal{M}$, and $\mathcal{C}$. So a Grid is modelled as $\mathcal{G} = (\mathcal{D}, \mathcal{M}, \mathcal{C}, \mathcal{N})$.

2.2 Dynamic Mapping of the Job Workflows

The objective of MCCF is to map the static job workflow specification to the dynamic Grid resource on the fly for distributed job workflow execution, which is shown in Figure 1. The job workflow specification includes static specification and dynamic specification. The static specification specifies the sub-jobs and data dependencies among them, and the user policies for resource scheduling. It is generated from user input. Resources for sub-job execution, input data from predecessor sub-jobs, and locations of the required data set and codes can be specified as dynamic. This means that the information will be filled up by the workflow engine dynamically during execution. The resources for sub-job execution is scheduled on the runtime, i.e., when it is ready to run. The resource scheduling will be done according to the policies specified by users.

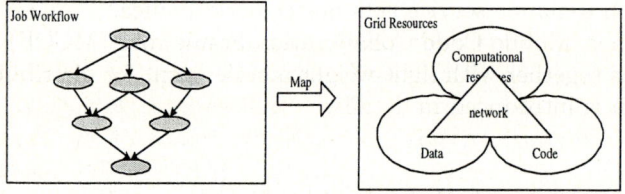

Fig. 1. Job Workflow Execution on Grid Resources

2.3 Runtime Support

Mobile agent implementations can be classified into non-functional part and functional part. The former includes codes for mobile agent communication, intelligence, migration and so on, which are common to all mobile agents. The latter is specific to application functionality. According to what codes will be transferred during the migration, current mobile agent systems can be classified into two categories: monolithic mobile agent and light-weight mobile agent [5]. When a mobile agent carries all its codes on its migration, it is a monolithic mobile agent. When a mobile agent carries only its non-functional implementation or just the description of its functional implementation, it is called light-weight mobile agent.

Code mobility is defined as the movement of the executable code over networks towards the location of needed resources for execution [4]. "Code-on-demand" (COD) is a design paradigm for code mobility. Applications developed using this paradigm can download and link on-the-fly part of their codes from remote hosts that act as code servers [11]. For COD, codes can be stored in trusted code repository, which will reduce more risks than that of any code from any machine.

The light-weight mobile agent in MCCF is different from existing light-weight mobile agents. It is defined using an agent core (AC). AC is an XML file, which contains the ID to identify it and the job workflow specification. No code is included for sub-job execution, instead, the code functionality description is provided. During runtime, only codes (including supporting packages) conforming to sub-job functionality description are selected and downloaded from code repositories for execution.

As we can see, no code is contained in AC. The functional implementations are provided in trusted code repositories as dynamic services. On migration, only AC will be transferred. COD technology is used to download executables. This greatly simplifies the existing mobile agent security problem. In addition, AC migration can be easily carried out for non-identical platforms and incur less overhead.

Common non-functional implementations, instead of being transferred over the Internet, are built into runtime supporting infrastructures. They are grouped into what is called AC agents, which work cooperatively for sub-job execution. AC agents include plan agent, task agent, and coordinator agent. They are constructed on each host when AC arrives and destructed when AC is migrated.

Plan agent is in charge of dynamic resource scheduling according to the policies specified in the AC. It locates the data repository, schedules the destination host and locates the code repository for a ready sub-job. Task agent is responsible for sub-job execution. It will download the dynamic services from selected code repository and instantiate them for sub-job execution. Coordinator agent will start and synchronize the execution of plan agent and task agent, and activate AC migration.

3 Prototype Implementation

We have built a simple prototype of MCCF to illustrate the distributed job workflow execution over the Grid. Our implementation is built on J2ME (Java 2 MicroEdition) and Java Agent Development Framework (JADE) [12]. Java is chosen as the implementation language for the following two reasons: it is platform-independent and its customerized ClassLoader and inspection mechanism make dynamic remote class loading possible. Java CoG-1.1 [13] is utilized to access Grid services. Globus2.2 needs to be installed on each selected host.

Currently only random resource selection is supported. AC agents are implemented by extending JADE Agent class. A simple code repository service is implemented as a proof-of-concept. Java byte codes are provided at specified directory on the code repository. Gridftp [1] is the common data transfer and access protocol which provides secure, efficient data movement over the Grid. It is used to download the codes, fetch required data sets and transfer AC between selected resources.

4 Conclusions & Future Work

The explosion of scientific data and dynamic nature of Grid resources pose great challenges to the existing job workflow execution models. MCCF utilizes dynamic services and light-weight mobile agent to fulfill new requirements. First, executable codes are provided as dynamic services, which eases job composition and integration, and makes COD possible. Thus, agent core can be migrated amongst resources and execution can be carried out at where the data is located. Second, AC migration simplifies the security problem of mobile agent, and also makes mobile agent platform independent. Third, in concept, MCCF provides a two-level job workflow program model: component developers develop executable code for dynamic services and application developers just need to specify the input data and code description for sub-job execution and the data dependency among the jobs. This will greatly reduce the efforts for the application developers to "gridify" their applications.

References

1. W. Allcock et al. GridFTP: Protocol Extensions to FTP for the Grid. Mar 2001. http://www-fp.mcs.anl.gov/dsl/GridFTP-Protocol-RFC-Draft.pdf
2. T. Andrews, F. Curbera, H. Dholakia, Y. Goland, J. Klein, F. Leymann, K. Liu, D. Roller, D. Smith, S. Thatte, I. Trickovic, and S. Weerawarana. Business Process Execution Language for Web Services Version 1.1, 05 May 2003. http://www-106.ibm.com/developerworks/webservices/library/ws-bpel/
3. B. Athey. Future Needs for Bioinformatics, Computational Biology, Bioengineering, and Biomedical Imaging Requiring Next Generation Supercomputing. *DARPA Biomedical Computing Needs for HPC Systems Workshop.* Arlington, VA. January 17, 2003.
4. R. Brandt and H. Reiser. Dynamic Adaptation of Mobile Agents in Heterogenous Environments. *2001 Mobile Agents: 5th International Conference*, Dec 2001.
5. Y. Feng, W. Cai, and J. Cao. A Simulation Study of Job Workflow Execution Models over the Grid. In *2003 Proceedings of the International Workshop on Grid and Cooperative Computing (GCC2003)*, GCC (2), 935-943, 2003.
6. I. Foster, C. Kesselman, and S. Tuecke. The Anatomy of the Grid. *International Journal of High Performance Computing Applications*, 15(3):200–222, 2001.
7. G. Heffelfinger, and A. Geist. Report on the ComputationalInfrastructure Workshop for theGenomes to Life Program. Jan, 2002. http://doegenomestolife.org/compbio/mtg_1_22_02/infrastructure.pdf
8. S. Krishnan, P. Wagstrom, and G. Laszewski. GSFL: A Workflow Framework for Grid Services. http://www-unix.globus.org/cog/projects/workflow/gsfl-paper.pdf
9. F. Leymann. Web services flow language. May 2001. http://www-3.ibm.com/software/solutions/webservices/pdf/WSFL.pdf

10. S. Loke, and A. Zaslavsky. Towards Distributed Towards Distributed Workflow Enactment with Itineraries and Mobile Agent Management. *E-Commerce Agents 2001*, 283-294, 2001.
11. J. Vitek and C. Tschudin. Mobile Object Systems: Towards the Programmable Internet. *Second International Workshop, MOS'96*, July 1996.
12. Java Agent Development Framework. http://sharon.cselt.it/projects/jade/
13. Java CoG Kit. http://www-unix.globus.org/cog/java/?CoGs=&

Gamelet: A Mobile Service Component for Building Multi-server Distributed Virtual Environment on Grid*

Tianqi Wang, Cho-Li Wang, and Francis Lau

Department of Computer Science, The University of Hong Kong, Pokfulam
{tqwang, clwang, fcmlau}@cs.hku.hk

Abstract. A DVE system provides a computer-generated virtual world where individuals located at different places could interact with each other. In this paper, we present the design of a grid-enabled service oriented framework for facilitating the building of DVE systems on Grid. A service component named "gamelet" is proposed. Each gamelet is characterized by its load awareness, high mobility, and embedded synchronization. Based on gamelet, we show how to re-design the existing monopolistic model of a DVE system into an open and service-oriented system that can fit into current Grid/OGSA framework. We also demonstrate an adaptive gamelet load-balancing (AGL) algorithm that helps the DVE system achieve better performance. We evaluate the performance through a multiplayer online game prototype implemented on Globus Toolkit. Results show that our approach can achieve faster response time and higher throughput.

1 Introduction

A Distributed Virtual Environment (DVE) system is a software system through which people that are geographically dispersed over the world can interact with each other by sharing a consistent environment in terms of space, presence and time [1]. These environments usually aim for a sense of realism and an immerse experience by incorporating realistic 3D graphics and providing real-time interactions for a large number of concurrent users. DVEs are now widely used in virtual shopping mall, interactive e-learning and multiplayer online games.

A typical DVE system should be able to support a life-like world and real-time interactions for a large number of users in a consistent way. This translates into intensive requirements on both the computing power and the network bandwidth. Multi-server architecture is a popular solution to tackle these problems [2, 10]. In addition, several techniques [3] such as dead reckoning, packet aggregation, and area of interest (AOI) management have been proposed to reduce the network bandwidth consumption and the servers' computing load.

---

* This research is supported in part by the China National Grid project (863 program) and the HKU Foundation Seed Grant 28506002.

In recent years, some projects have tired to build DVE systems over Grid. Cal-(IT)2 Game Grid [8] provides the first ever grid infrastructure for games research in the area of communication system and game service protocols. Butterfly Grid [9] tries to provide an easy to use commercial grid-computing environment for both the game developers and game publishers. However, a few new challenges remain to be solved. One is how to re-design the existing monopolistic model of a DVE system into an open and service-oriented system that can fit into current Grid/OGSA framework [7]. Another lies in how to ensure a certain qualities of service from the perspective of end users. In a DVE system, each user is represented as an entity called "avatar" whose state is controlled by the user's input commands. Since users can move and act freely, it may incur significant workload imbalance among servers, which in turn causes unpredictable delays in computing the global world state and delivering the state to the client machines. This makes it very difficult to achieve real-time interactions.

In this paper, we propose a flexible and scalable service-oriented framework based on a component called "gamelet" to support the building of a DVE system on Grid. Based on the gamelet framework, we further propose an adaptive gamelet load-balancing (AGL) algorithm that is able to make the DVE system more scalable and cost-effective.

The rest of the paper is organized as follows. In Section 2, we discuss the gamelet concept and system framework. In Section 3, we present the prototype design. Section 4 presents the performance evaluations. Section 5 presents some related work. Section 6 concludes the paper.

2 Gamelet-Based System Framework

2.1 Gamelet Definition

We define *gamelet* as a mobile service component that is responsible for processing the workload introduced by a partitioned virtual environment. A group of related objects, including both static and dynamic objects, form one partition and will be assigned to one gamelet for processing. For example, in a room-based virtual environment, one or several rooms can form one partition. In a session-based virtual environment, like a virtual meeting system, users joining the same session can form one partition. For system with a persistent wide-open virtual environment, overlapping partition technique [11] can be used to support smooth visual interactions among the participants across multiple partitions. A gamelet has the following unique characteristics:

- Load Awareness. A gamelet is able to detect and monitor the CPU and network load it created on its hosting server. It also provides a rich set of API for load inquiry.
- High Mobility. A gamelet can move freely and transparently to a new grid node within our framework. The mobility is achieved at the application level, and the world states are transferred using object serialization technique.

– Embedded Synchronization. In DVE, there will be some visual interactions among two partitions. Two gamelets should be able to communicate reliably to synchronize their world states whenever necessary.

With gamelets, we can accelerate the application development circles as most existing partition schemes adopted by the multi-server DVE systems can easily adapt to the gamelet framework. Moreover, a gamelet can be migrated among the server nodes to support load balancing at run-time.

2.2 System Framework

Figure 1 shows the overall picture of the system framework. It consists of three kinds of components: monitor, gamelet and communicator. A monitor is responsible to collect the workload information of each gamelet periodically and execute the load balancing algorithms. The communicator acts as the gateway for all the incoming packets and will do intelligent message routing, which means it will forward an incoming packet to all the related gamelets according to the predefined partition logic. In such a case, the gamelets are responsible to execute the embedded synchronization protocol to provide a consistent view to the clients. To join the world, a participant will first contact a well-known monitor. The monitor assigns a communicator to the client. The client will always send messages to this communicator. The communicator then forwards the messages to the corresponding gamelets for processing and the updated world states are sent back directly to the clients.

In our framework, GT3 core services provide the following supports. Grid Security Infrastructure (GSI) is used to ensure the secure communication and authentication among the gamelets over an open network. Each gamelet service instance will have a unique Grid Service Handle (GSH) managed by the Naming Service and is associated with a structured collection of information called

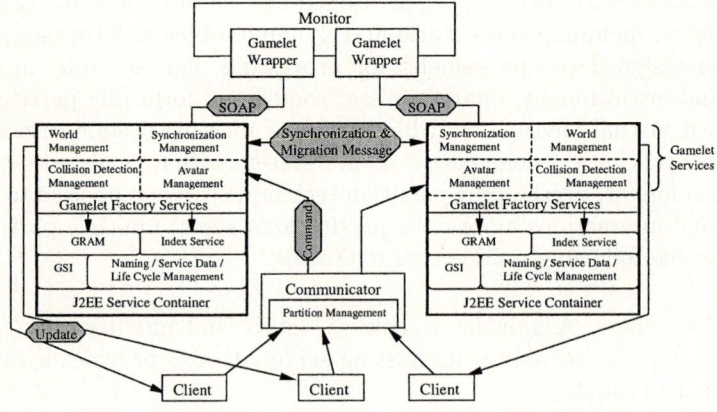

Fig. 1. Gamelet-based System Framework

Service Data that can be easily queried by requestors. Life Cycle Management service provides methods to control a gamelet throughout its life cycle from creation to destruction.

GT3 base services are based on the GT3 core services. Gamelet Factory service makes use of the Grid Resource Allocation and Management (GRAM) service to enable remote creation and management of a gamelet service. A set of service data, e.g., the cost of the gamelet service, is associated with a Gamelet Factory service through the use of Index Service. Using the Index Service, the monitor can dynamically discover the desired Gamelet Factory services that can meet certain basic requirements in terms of speed, availability, cost and etc.

To enable a flexible control over multiple gamelets, each gamelet has a corresponding wrapper in the monitor. This design encapsulates the complicated grid related stuffs and makes the monitor totally separated from the underline grid libraries. So the monitor component is more portable. The gamelet factory service enables several stateful gamelet services to be created at the same grid node concurrently. A gamelet factory will periodically register its GSH into a registry, from where a monitor can locate a set of gamelet factory services and reliably create and manage gamelet service instances. The monitor and gamelet form a client-server relationship that is in line with the OGSA client-server programming model [12].

2.3 Gamelet Migration

Gamelet migration enables load balancing. The migration protocol is as follows. When the monitor determines that a gamelet is in need of migration, it first tries to locate a GSH of another Gamelet Factory service and then creates a new gamelet to serve as the target of the migration. Once the new gamelet has been created, the monitor will notify the communicator to store the incoming packets temporarily in a resizable message queue. Then the monitor directs the old gamelet to transfer its world content to the new one. Finally, the monitor will notify the communicator the completion of the migration. Communicator will then forward the stored packets and the succeeding packets according to the updated mapping information. Therefore, the migration process won't result in notable message loss.

2.4 Load Balancing Algorithm

To support the load balancing of a DVE system in the grid environment, we propose a new adaptive gamelet load-balancing (AGL) algorithm. The word *adaptive* has two-fold meanings. Firstly, the algorithm adapts to the network latency among the grid nodes to make gamelet migration decisions. Secondly, the algorithm evaluates each gamelet based on the activities of the clients being managed and adapts to the resource heterogeneity of the grid nodes to make gamelet migration decisions.

In AGL algorithm, there is a threshold δ_m used to judge whether a new load balancing process is necessary to be performed for server m. Its value can be adjusted based on the runtime stability of grid resources. Since a grid server may become unavailable due to some reasons, it might need to be removed from the DVE system at anytime. Under such a situation, δ_m can be set to about 0 so that server m will be considered as overloaded until all its gamelets are migrated to other grid servers. Then server m can be removed from the system.

We formulate the gamelet load balancing problem as a graph repartition problem. The graph is built as follows. For the ith gamelet G_i, create a vertex V_i in the graph. For any two vertices V_i and V_j, if there are some intercommunications $C_{i,j}$ between them, create an edge between them with value $W_{i,j} = C_{i,j}$. Other notations used are: $Val(G_i)$ represents the CPU load G_i introduces to the server, defined as a weighted packet sending rate; $Syn(G_i)$ is the synchronization cost of G_i; $Cost(G_i, m)$ is the cost of migrating G_i to server m; $Percentage(G_i)$ is the percentage of CPU load that G_i introduces to the server, $Percentage(G_i) \in [0, 1]$. The following lists the main procedure of the AGL algorithm.

Adaptive Gamelet Load Balancing Algorithm:

1. Select a server s which has the highest CPU load among the overloaded servers. Server s is considered to be overloaded if its CPU load is larger than δ_s, which is loaded from a configuration file at runtime.
2. Select a server t with the least CPU load as the migration target. Calculate the migration cost of each gamelet G_i in server s:

$$Cost(G_i, t) = Syn'(G_i) - Syn(G_i)$$

   $Syn(G_i)$ is the pre-migration synchronization cost of gamelet i calculated by formula:

$$Syn(G_i) = \sum_{G_j \in \phi} W_{i,j} \times Latency_{m,n}$$

   where ϕ is the set of gamelets that have some communication traffic with gamelet i, gamelet i runs in server m and gamelet j runs in server n. $Syn'(G_i)$ is the post-migration synchronization cost calculated by assuming that gamelet i has been migrated to server t.
3. Replace server t with a less loaded server. Repeat step 2 again until all existing gamelet servers have been evaluated.
4. Assign a gamelet with the smallest value of $Cost(G_i, q)$ from server s to server q. Calculate how much percentage of the workload gamelet i contributes to the original server. Estimate how much workload it will add to server q (this can be easily done by quantifying and comparing the hardware configurations of the two servers). The algorithm estimates the percentage of workload introduced by gamelet i by this formula:

$$Percentage(G_i) = Val(G_i) / \sum_{G_s \in \psi} Val(G_s)$$

where ψ is a set of gamelets that are currently running in server s. If server q will be overloaded after migration, repeat step 4 again and try the gamelet with the second smallest value of $Cost(G_i, q)$.
5. Repeat step one to four until the estimated CPU load of the original server is under its threshold.
6. If there is still an overloaded server while all the other existing servers cannot afford to share workload, the algorithm will make a decision to discover a new grid node with certain requirements, e.g., a grid node with $> 10Mbps$ bandwidth and $< 200ms$ latency time, add it to the system as another server candidate. Wait for some time and go to step 1 again.

One strength of the algorithm is that the workload model is more accurate than other approaches that only take into account the clients number or density. Therefore, it makes the workload sharing more effective. The reason to use a weighted packet rate is that different participants will have different packet sending rate and different commands will lead to different workload too. The algorithm can also adapt to the network latency among the grid servers to make gamelet migration decisions. This is achieved through the use of a cost model that integrates both the gamelet synchronization cost and grid inter-server latency. All these factors enable the AGL algorithm work well in a grid environment.

3 Prototype Design and Implementation

There are four components in our prototype: client simulator, communicator, gamelet and monitor. The size of the 3D world is 100*100*20. The world is partitioned into 16 equal-sized cells and the overlapping length of the neighboring partitions is 5. The client simulator simulates a number of randomly distributed clients. Each will send out a packet every 100ms. Each data packet has 32 bytes, consisting of the avatar's position, command and timestamp. We define a circle with a radius of 5 to be the area that a client is interested in. The gamelet also do collision detections after each command is executed.

We define the response time (RT) to be the average time interval of each client from sending out a packet to receiving the confirmation from the server that the command has been executed and the results have been sent to all the related clients. System capacity (CP) is defined as the maximum number of participants that the system can accommodate so that the interactions in the world will have a reasonable average RT ($\leq$ 200 ms) and loss rate ($\leq 50\%$).

The gamelet and monitor are implemented on top of GT3.0 [14]. All components are implemented using J2SE 1.4.2 and run on Linux kernel 2.4.18 with P4 2.00GHz CPU, 512MB RAM and 100Mbps Ethernet [13]. The network round-trip time among the nodes is within several milliseconds if not specified explicitly.

4 Performance Results

4.1 Gamelet Creation and Migration

We study the performance of gamelet creation and migration when a monitor sequentially creates gamelets in different servers. We find that when the monitor creates the first gamelet in the first server, it usually takes about 7-8 seconds. This is because various GT3 runtime libraries need to be loaded and initialed at both the monitor and gamelet server side. But the creation of the first gamelet in a second server takes 2-3 seconds. This is because the monitor has loaded and created necessary libraries. The creation of the second gamelet in the same server will only take about 100 ms.

The gamelet migration process will introduce a short time of delay. The delay time depends on the size of the gamelet, e.g., the total number of avatars that need to be transferred. The interactions between the monitor and the communicator add about 30-40 ms to the pure gamelet content transmission time. However, the migration process won't bring obvious influence to the packet loss rate, since the communicator will store the incoming packets temporally and forward them later.

4.2 Gamelet Migration Algorithm

We compare the performance of our proposed AGL algorithm with a popular used algorithm named even-avatar load-balancing (EAL) algorithm, which tries to even the number of avatars that each server holds. Table 1 shows part of the network latency configurations among the servers in the experiments. The latency is below 1 ms within the same server.

Table 1. Network Latency Among the Servers

Latency(ms)	S1	S2	S3	S4	S5	S6	S7	S8
S2	100	1	200	150	100	100	100	50

Initially, the 16 gamelets are all in server 1. As the number of clients increases, servers are added into the system one by one according to their indices. A server is regarded as overloaded if its CPU load is larger than 90%. In the experiment, the client simulator will generate a virtual environment with 3 hotspots.

Figure 2(a) shows the average RT under the two different approaches with 2, 4, and 8 servers respectively. The graph shows that the AGL algorithm can get much better performance. Table 2 shows the performance result under 4 servers. Under EAL approach, S1, S2 and S3 are all overloaded. Therefore, the RT of the three servers are very long leading to much worse performance on average. The influence of the inter-server latency is worthy of mention. E.g., under AGL approach, S2 has similar load with S1, but the RT of S2 is about 100 ms longer than that of S1. This is because network connections with S2 is not as fast as that with S1.

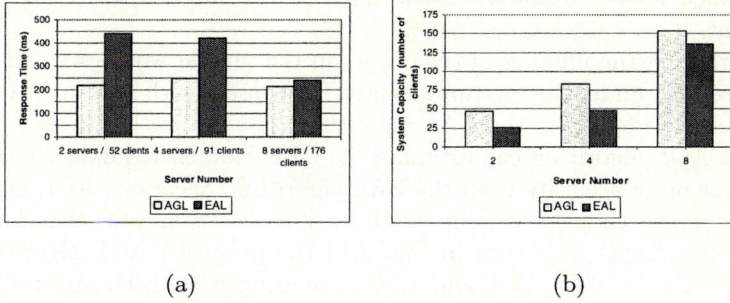

(a)   (b)

Fig. 2. Gamelet Migration Algorithms Evaluations

Table 3 and Table 4 show the CPU load and RT when 8 servers is used under the two approaches. This time the EAL approach happens to even the load among the servers (though it can not guarantee), and the inter-server traffic is even smaller than that of AGL algorithm. However, the average RT is still not as good as that of AGL algorithm. We find that the RT of S2, S3 and S4 is much worse under EAL approach. This is mainly because much of the inter-server traffic is transferred through low network, e.g., S2 and S3 under EAL approach. This situation is avoided under AGL algorithm since two gamelets with large

Table 2. Performance Comparisons with 4 Servers

91 Clients	CPU Load				Inter-server	RT (ms)				ART
4 Servers	S1	S2	S3	S4	Traffic	S1	S2	S3	S4	ms
AGL	90%	89%	79%	81%	276.8 Kbps	223	321	210	241	248.7
EAL	100%	100%	99%	42%	184.1 Kbps	503	572	512	101	422.0

Table 3. Performance Comparisons with 8 Servers (1)

176 Clients	CPU Load								Inter-server
8 Servers	S1	S2	S3	S4	S5	S6	S7	S8	Traffic
AGL	90%	72%	69%	71%	71%	90%	69%	68%	891.1 Kbps
EAL	90%	88%	89%	91%	89%	90%	35%	36%	742..6 Kbps

Table 4. Performance Comparisons with 8 Servers (2)

176 Clients	RT (ms)								ART
8 Servers	S1	S2	S3	S4	S5	S6	S7	S8	(ms)
AGL	270	323	180	180	184	247	175	158	214.6
EAL	235	402	309	278	214	235	137	110	240.0

amount of traffic will tend to be assigned to two servers with better connections provided they can not be in the same server due to the CPU balancing requirement.

By decreasing the number of the clients in the virtual world, we can get the CP of the servers under the two approaches. The system with AGL algorithm has much larger capacity than the EAL algorithm. The reason lies in two aspects. Firstly, the AGL algorithm can estimate the CPU load each gamelet introduces to the server more accurate than the EAL algorithm. Secondly, AGL algorithm tries to minimize the intercommunications among the servers that have bad network connections. As shown in Fig. 2(b) the proposed AGL algorithm can improve the CP by 80%, 72% and 13% comparing with EAL approach when using 2, 4 and 8 servers respectively, therefore making a more scalable and cost-effective system.

5 Related Work

Most of the existing multi-server approaches are based on data partition scheme and can only perform limited local load sharing strategies. Examples includes CittaTron [4], NetEffect [5] and CyberWalk [6]. CittaTron is a multi-server networked game whose world is partitioned into several regions, each of which is assigned to one server. A load adjustment scheme is introduced by transferring some users from the highest loaded region to the neighboring regions. However, this approach can hardly be effective since only the user number is considered. In NetEffect, the virtual world is partitioned into separated communities that are managed by multiple servers. A heavy loaded server can transfer some of its communities to a less loaded server dynamically. Unfortunately, in such a system the load balancing process is not transparent to the clients. A user has to log into the system again after each migration. CyberWalk is a web-based multi-server distributed virtual walkthrough environment. The region managed by each server can be adjusted by transferring the boundary areas among neighboring regions. However, if there is more than one hotspots in the concerned regions, a cascading effect may occur which will seriously affect the system performance. Therefore, all the above load-sharing approaches are not suitable for a very dynamic DVE system where the participants can move and act freely. It is more difficult for these approaches to work well in a dynamic grid environment.

6 Conclusions

In this paper, we present a novel service component called gamelet for building DVE systems over Grid. Existing multi-server DVE systems based on partition scheme can be easily mapped into the gamelet-based framework. Our gamelet migration protocol and the load balancing algorithm help enable a more scalable DVE system. We have evaluated the performance of our proposed approach through a multi-player game prototype. Results show that our approach can build a more scalable system.

In our future work, we will study how various synchronization protocols will influence the performance of gamelet-based DVE system and what are the possible improvements. Currently, we assumes a simple two-way synchronization scheme. However, more complicated synchronization protocol is needed for applications that have both high consistency and response time requirements. One possible direction is to study how the communicator can help gamelet do synchronization more efficiently.

References

1. S. Singhal and M. Zyda, *Networked Virtual Environments: Design and Implementation*, Addison Wesley, July 1999.
2. T. Funkhouser, "Network Topologies for Scalable Multi-User Virtual Environments", *Proceedings of the 1996 Virtual Reality Annual International Symposium, IEEE VRAIS 96*, San Jose, CA, USA, 1996, pp. 222-228.
3. J. Smed, T. Kaukoranta, and H. Hakonen, "A Review on Networking and Multiplayer Computer Games" Technical Report 454, Turku Centre for Computer Science, 2002.
4. M. Hori, K. Fujikawa, T. Iseri, and H. Miyahara, "CittaTron: a Multiple-server Networked Game with Load Adjustment Mechanism on the Internet", *Proceedings of the 2001 SCS Euromedia Conference*, Valencia Spain, 2001, pp.253-260.
5. Tapas K. Das, Gurminder Singh, Alex Mitchell, P. Senthil Kumar, and Kevin McGee, "NetEffect: a network architecture for large-scale multi-user virtual worlds", *Proceedings of the ACM symposium on Virtual reality software and technology*, Lausanne, Switzerland, 1997, pp.157-163.
6. N. Beatrice, S. Antonio, L. Rynson, and L. Frederick, "A Multiserver Architecture for Distributed Virtual Walkthrough", *Proceedings of ACM Symposium on Virtual Reality, Software and Technology 2002*, Hong-Kong, 2002.
7. I. Foster, C. Kesselman, J. Nick, and S. Tuecke, *The Physiology of the Grid: An Open Grid Services Architecture for Distributed Systems Integration*, Open Grid Service Infrastructure WG, Global Grid Forum, 2002.
8. California Inst. for Telecommunications & Infor. Tech. http://www.calit2.net
9. Butterfly Grid. http://www.butterfly.net/.
10. Beatrice Ng, Frederick W. B. Li, Rynson W. H. Lau, Antonio Si, and Angus M. K. Siu, "A performance study on multi-server DVE systems", *Information SciencesInformatics and Computer Science: An International Journal*, v. 154, n.1-2, 2003, pp. 85-93.
11. J.Y. Huang, Y.C. Du, and C.M. Wang, "Design of the Server Cluster to Support Avatar Migration, *IEEE Virtual Reality 2003 Conference*, LA, USA, 2003, pp.7-14.
12. S. Tuecke, K. Czajkowski, I. Foster, J. Frey, S. Graham, C. Kesselman, T. Maguire, T. Sandholm, P. Vanderbilt, and D. Snelling, *Open Grid Services Infrastructure (OGSI) Version 1.0*, Global Grid Forum Draft Recommendation, 2003.
13. HKU Gideon300 Grid. http://www.srg.csis.hku/gideon/, 2004.
14. Globus Toolkit 3.0. http://www.globus.org/.

The Application of Grid Computing to Real-Time Functional MRI Analysis

E. Bagarinao[1], L. Sarmenta[2], Y. Tanaka[3], K. Matsuo[1], and T. Nakai[1]

[1] Photonics Research Institute, National Institute of Advanced Industrial Science and Technology
(AIST) Kansai Center, Ikeda City, Osaka 563-8577, Japan
[2] Department of Information Systems and Computer Science, Ateneo de Manila University,
Quezon City 1108, Metro Manila, Philippines
[3] Grid Research Technology Center, National Institute of Advanced Industrial Science and Technology, Tsukuba City, Ibaraki 305-8566, Japan

Abstract. The analysis of brain imaging data such as functional MRI (fMRI) data often requires considerable computing resources, which in most cases are not readily available in many medical imaging facilities. This lack of computing power makes it difficult for researchers and medical practitioners alike to perform on-site analysis of the generated data. This paper proposes and demonstrates the use of Grid computing technology to provide medical imaging facilities with the capability of analyzing functional MRI data in real time with results available within seconds after data acquisition. Using PC clusters as analysis servers, and a software package that includes fMRI analysis tools, data transfer routines, and an easy-to-use graphical user interface, we are able to achieve fully real-time performance with a total processing time of 1.089 s per image volume (64 x 64 x 30 in size), much less than the per volume acquisition time set to 3.0 s. We also study the feasibility of using XML-based computational web services, and show how such web services can improve accessibility and interoperability while still making real-time analysis possible.

1 Introduction

Functional magnetic resonance imaging (fMRI) is a non-invasive technique used to investigate the functions of the human brain. In an fMRI examination or experiment, the subject's brain is repeatedly scanned while the subject is performing certain tasks. Statistical analysis is then performed on the resulting time series of brain scans in order to produce an activation map indicating regions in the brain that are active during certain tasks. Since most medical imaging facilities today do not have the computational resources to perform such an analysis in real-time, medical practitioners and researchers typically do not get the results of the examination until long after the scanning session is completed and the subject is no longer inside the MR scanner. This limits the type of examinations and analyses that can be done, since examiners are forced to run a static set of tests and cannot add or discard tests dynamically based on the results of previous tests. It can also lead to wasted work and lost opportunities in cases where the results of an experiment must be discarded due to errors encountered during the gathering that could have been avoided if the examiner had immediate feedback.

In this paper, we present a system that is capable of achieving real-time processing of functional MRI time series using remote computational resources accessed via the Grid. Several PC clusters belonging to different institutions participating in the Medical Grid Project (MGP) were utilized. A software package, *baxgrid*, was developed to support the integration of the different subsystems. We describe these subsystems and the overall system's implementation, and present the results of the evaluation of the system's performance. We also present results from a new approach involving XML-based computational web services, and show how such web services can improve accessibility and interoperability while still making real-time analysis feasible.

2 System Overview and Design

We propose the use of a computational grid composed of multiple MR imaging subsystems, analysis servers, and data servers, all interconnected via the Internet. Since the analysis of data is performed remotely by the high-performance analysis servers, real-time processing is possible even if the medical imaging facility itself has only limited computational capability. Furthermore, since the system allows sites that are geographically separated to participate in the grid, it can bring fMRI analysis capabilities to places that do not yet have it, such as developing countries and financially-constrained institutions.

2.1 The Medical Grid Project Testbed and Analysis Servers

To study and demonstrate the idea of using grid computing for medical applications such as fMRI, we have formed the Medical Grid Project (MGP) testbed composed of several sites in Japan and the Philippines, as shown in Fig. 1.

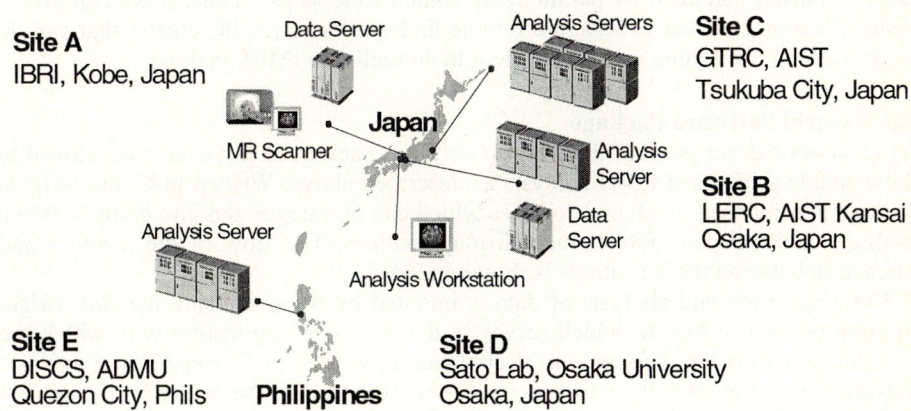

Fig. 1. The Medical Grid Project Testbed for real-time fMRI analysis

The MGP testbed provides five PC clusters serving as analysis servers (one cluster each at Sites A, B, and E, and two clusters from Site C) and two data servers (one from

Site A and another one from Site B). The MRI acquisition subsystem is located in Site A. For real-time benchmarking, the two clusters in Site C were utilized. One of them, cluster A, has 4 compute nodes and 1 server node. The other one, the cluster B, has 32 compute nodes and 1 server node. Each node (NEC Express 5800) for both clusters has two 1.4-GHz Pentium III processors. Within each cluster, the nodes are interconnected via Gigabit Ethernet link. For offline timing and code optimization, cluster C in Site A was used. It has 8 compute nodes and 1 server node. Each node has two 800-MHz PIII processors with 1-GB of physical memory. The cluster is interconnected using Myrinet (Myricom, Inc. Arcadia, CA) and Fast Ethernet links. To measure data transfer times across large geographic distances, we used the analysis workstation at Site D in Japan to send and receive data from a PC cluster at Site E in the Philippines.

2.2 Software Components

Using Grid Computing for Functional MRI Analysis
The analysis of a functional MRI time series is a computationally intensive task that includes image preprocessing (motion correction and smoothing), incremental general linear modeling (GLM) using a method such as that in [2], and statistical analysis. The analysis starts when the first image volume is received from the analysis workstation of the MRI acquisition subsystem. As more image volumes arrive, each new volume is realigned relative to the first volume in the series to correct for head movement during the scan. After realignment, spatial smoothing using a three-dimensional Gaussian filter is then applied. The estimates of the GLM coefficients, t-statistics, and t-statistics threshold are then updated, incorporating the information of the new volume. The procedure is repeated with each new volume until the last volume in the series is processed.

While a single PC doing this analysis may not be able to do it fast enough for real-time use, wherein the whole analysis process must be completed for each volume before the next volume is made available, a high-performance parallel machine, such as a PC cluster, can do it by parallelizing some of the steps. Thus, if we can give a medical imaging facility access to a remote high-performance PC cluster that can do the processing for it, then we can enable it to do real-time fMRI analysis.

The Baxgrid Software Package
Fig. 2 shows a description of the *baxgrid* software package that we have developed to allow enable grid-based fMRI analysis as described above. Written in *C*, the *baxgrid* package is composed of several routines which can be categorized into computational routines, data transfer routines, and display routines. The flow of the analysis and interaction between these routines is shown in Fig. 2.

The acquisition and analysis of data is initiated by the user using the *Baxgridgui* program (shown in **Fig. 3**), which serves as the front-end application with which the user interacts directly. This graphical application, written in C, runs on the analysis workstation and allows users to specify, among others: (1) the scanning parameters such as the number of slices and the number of volumes to process, etc., (2) the preprocessing operations that will be included in the analysis, and (3) the remote computational resources that will be used. Once the user starts the analysis, *baxgridgui* invokes several local and remote processes to facilitate the analysis and transfer of the data from the local workstation to the remote analysis server.

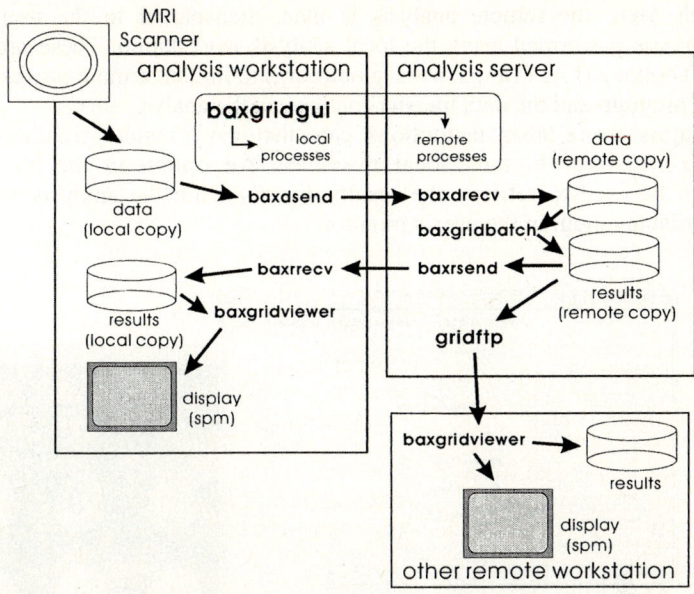

Fig. 2. Interaction of the different routines (local and remote) of the baxgrid package during real-time analysis

After the first volume is acquired from the MR image acquisition facility (image acquisition phase), it is immediately sent to the analysis server (data transfer phase) using the *baxdsend* and *baxdrecv* programs, two C programs that transfer binary data through direct TCP/IP sockets between the hosts. When the volume data is received by the analysis server, the server starts processing it (analysis phase) immediately using the *baxgridbatch* program. To be able to compute the statistics incrementally, the method proposed in [2] is used. The Message Passing Interface (MPI) library (http://www-unix.mcs.anl.gov/mpi/) is used to parallelize the analysis within a volume. The result of this analysis is an activation map file showing the areas of the brain which are active. These results are sent back (results transfer) to the analysis workstation via *baxrsend* and *baxrrecv*. On the analysis workstation, the *baxgridviewer* routine receives and displays the new activation map data (as shown in Fig. 3). The whole process is then repeated for the next volume until the last volume is processed.

Note that to achieve *fully* real-time analysis – wherein the user can see the results of the previous volume's analysis before the next volume is scanned – the total time from data transfer to the update of the activation map should be less than the repetition time (TR) of the scan, the time interval between the re-scanning of the same image slice. In the case of our experiments, TR = 3 s. However, if enough computing power is available, then lower values of TR may be used for more temporal precision.

The User Interface

The *baxgridgui* and *baxgridviewer* programs use the gtk+ library (http://www.gtk.org) for Linux to provide a user-friendly graphical user interface (GUI) as shown in Fig. 3.

Through this GUI, the remote analysis is made transparent to the user, as if all computations are performed using the local analysis workstation. These programs use the Globus Toolkit 2.0 API (http://www.globus.org) to initiate remote processes such as the analysis program and the data transfer routines on the analysis server.

Collaborators from other institutions can also view results simultaneously by running *baxgridviewer* in their local machines. An option in the *baxgridviewer* program can be used to retrieve the results directly from the analysis server using gridftp and display them in the user's monitor.

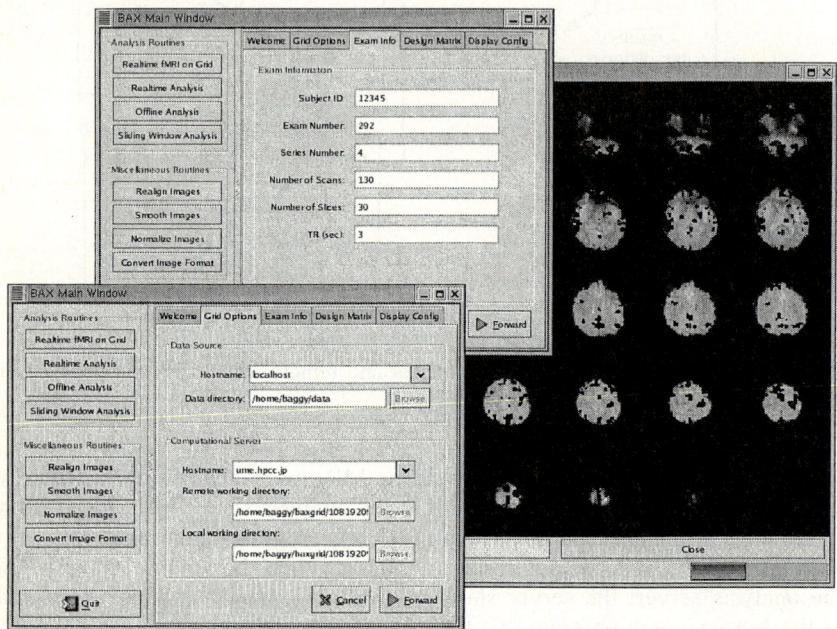

Fig. 3. Screenshots of the baxgridgui and baxgridviewer application windows

2.3 Computational Web Services

Aside from the original baxgrid software that uses C and Globus APIs for data transfer and remote execution, we also implemented a prototype system that uses *computational web services* instead.

The idea behind *computational web services* (CWS) [1] is to use XML web services to offer a simple programming interface that can make it very easy for programmers to tap the computational power of a Grid computing resource. XML web services are like older remote procedure call or remote object technologies such as RPC, CORBA, RMI, and DCOM, that allow programmers to call a function on a remote machine as easily as if the function were running on the local machine. Unlike these older technologies, however, XML Web Services use HTTP and industry-standard platform-independent XML-based protocols such as SOAP, WSDL, and UDDI. These protocols allow XML Web Services to be invoked from potentially any programming language, and to be accessed from anywhere that the Web is available, even from within a firewall. Thus, XML web services are much more interoperable and more easily accessible than these older technologies.

In the case of Medical Grid Computing and fMRI research, a CWS can provide a simple API for submitting jobs and data to the analysis server, which can then execute the job quickly on a high-performance cluster. Currently, since the baxgrid program uses Globus, it requires the complex installation procedure of Globus on both the analysis workstation and the analysis server, and also requires a complex firewall configuration (which may not even be allowed by some institutions' firewall policies). By using a CWS instead of Globus, we can greatly simplify the setup procedure, as well as eliminate the firewall problems. Furthermore, since the CWS is accessible from a variety of programming languages, it makes it much easier to write GUI-based client applications that can run on different operating systems.

Fig. 4 shows the design of a generic computational web service for MRI or fMRI applications which we are currently implementing. In this figure, the user would use a client program (Step 1), which would allow him to specify the input files as well as the analysis he wants to perform (e.g., activation map, do realignment, etc.) Then (Step 2), this program invokes the *startJob()* remote function of the MRIService computational web service. The program submits the parameters (including the common information in the image headers such as image size, etc.), and specifies the type of analysis as indicated by the user in Step 1. Given these parameters, the CWS will start the executable corresponding to the desired analysis type on the cluster, and return a job number, which the client can use to identify the job. Then (Step 3), the analysis program will loop through all the volumes and submit them to the server using the *submitVolume()* remote function of the web service. Each call to *submitVolume()* must include the *jobID* number so that the server can distinguish

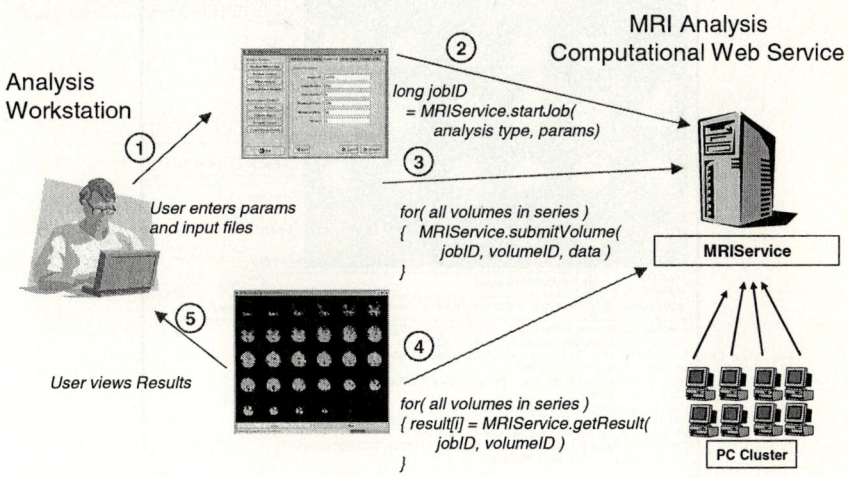

Fig. 4. Using a computational web service for MRI Analysis

which job a volume belongs to. This makes it possible for the server to serve multiple client (e.g., multiple hospitals or MRI facilities) at the same time. The call to *submitVolume()* also includes a volumeID so the server can keep the volume images in sequential order. As each volume is submitted, the MRIService puts the data in an incoming queue. Meanwhile, the process that was started in Step 2 runs as a program on the PC cluster, and periodically checks the incoming queue and processes any new volumes. Any new results produced by the cluster are put in an outgoing queue. Then (Step 4), on the client side, the viewer part of the analysis program periodically retrieves any available volume images using the *getResult()* web service. Finally, this information is displayed to the user in real time (Step 5).

Note that although initially meant to be used for fMRI analysis, this system can potentially be used to handle many different kinds of image analysis applications that process time series of medical images. All that is needed is for the user to specify a different executable (assuming that the executable binaries are already present in the analysis server and PC cluster).

At present, we have built a prototype web service using Java and Apache AXIS (http://www.apache.org/), that runs on the PC cluster at Ateneo de Manila University, and allows clients to send and retrieve data and result files, as well as to start the analysis program. To demonstrate interoperability and ease-of-programming, we have successfully written a number of client programs, including a text-based batch processing program written in Java, running on Linux and Windows, as well as an interactive graphical client similar to *baxgridviewer* and *baxgridgui*, written in Microsoft .NET C#, running on Windows, as shown in Fig. 5.

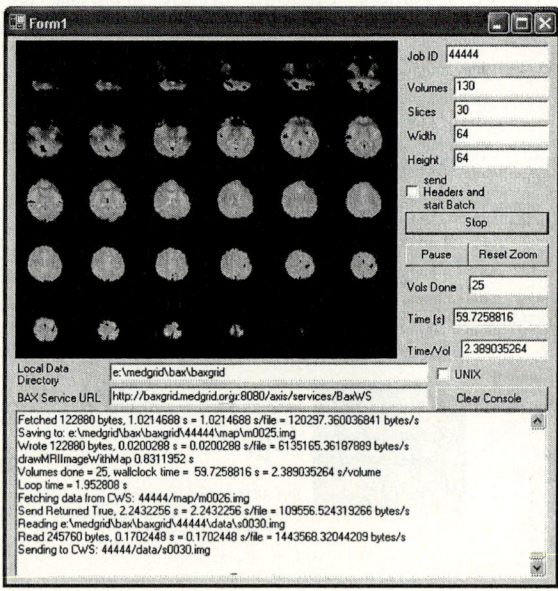

Fig. 5. A GUI-based Microsoft .NET C# client connected to a Java XML Web service

3 Results

3.1 Data Transfer Time

Two critical time factors affecting the remote real-time analysis of fMRI data are transfer times (data and results) and analysis time. Fully real-time analysis can only be achieved when the total time (roughly, data transfer time + analysis time + results transfer time + display update) is less than the used TR. Using a benchmarking data set consisting of 130 volumes of 64 x 64 x 30 voxels each, we measured the average time to transfer an image volume from one site to another repeatedly throughout the day for several days. The results are shown in Fig. 6.

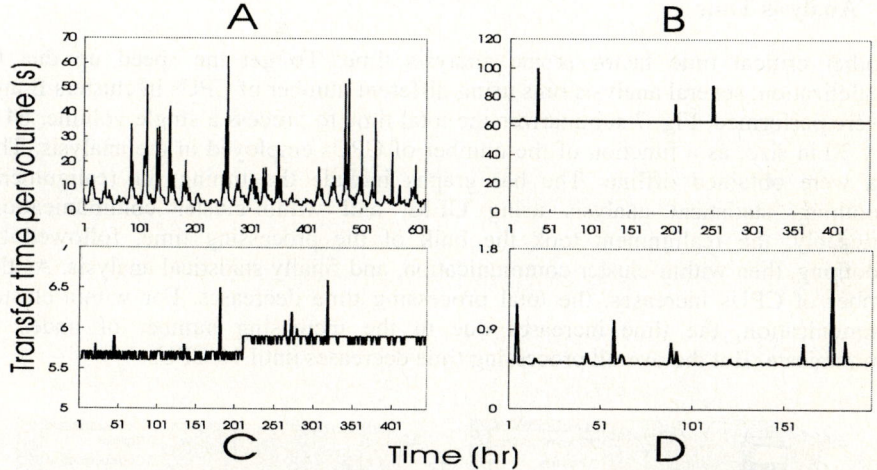

Fig. 6. Transfer times per volume from one MGP participating site to another (measured every hours for several days). Cases: (A) from Site D to Site E, (B) from Site A to Site C, (C) from Site B to Site C, and (D) from Site D to Site C

For the given data size (in this case, 480Kbytes per volume), only Case D with an average transfer time of 0.533 s has the potential for attaining fully real-time performance. Each site (D and C) connects to a Gigabit backbone. Case B is the slowest since the resources in Site A connect to the Internet via an ISDN (Integrated Services Digital Network) line. Case C is promising in spite of the fact that the resources in Site B are connected to the Internet via an ADSL (Asymmetric Digital Subscriber Line) connection. Although it is not possible to achieve fully real-time analysis with the given data size and TR (3 s), the average transfer time per volume (5.825 s) is close enough to the TR that the delay in the analysis would not be significant. It may still be possible to achieve real-time analysis by reducing the data size, or using a longer TR.

Case A of Fig. 6 illustrates the challenges faced when doing grid computing across international borders. For this case, the entire Ateneo de Manila University (Site E) shares the Internet connection to the world, consisting of multiple E1 lines. Thus, during school days (Monday to Friday), the network slows down dramatically, especially during peak hours as exhibited by the peaks of the transfer time. On weekends and at night, however, the network activity is greatly reduced and the transfer time goes down. The minimum time in this experiment is about 2.5 s, which would make real-time analysis possible if network activity is minimized. (Note: this experiment was done using volume data of 480Kbytes each. In a later experiment, described below, we were able to reduce the data to 280Kbytes by removing redundant header information, and were able to cut down the transfer time to a best case of 1.5 s or less per volume.) These results are promising, especially for researchers in developing countries, and other places with limited computational resources, since it demonstrates that it is possible to access computational resources in developed countries, and thus enhance their capabilities without too much additional cost.

3.2 Analysis Time

Another critical time factor is the analysis time. To get the speed up due to parallelization, several analysis runs using different number of CPUs in clusters B and C were performed. Fig. 7 summarizes the total time to process a single volume, 64 x 64 x 30 in size, as a function of the number of CPUs employed in the analysis. The data were obtained offline. The bar graphs include the timing for realignment, smoothing, statistical analysis using GLM, and within-cluster communication. Noticeably, the realignment took the bulk of the processing time, followed by smoothing, then within-cluster communication, and finally statistical analysis. As the number of CPUs increases, the total processing time decreases. For within-cluster communication, the time increases due to the increasing number of nodes to communicate. But the overall processing time decreases until 16 CPUs.

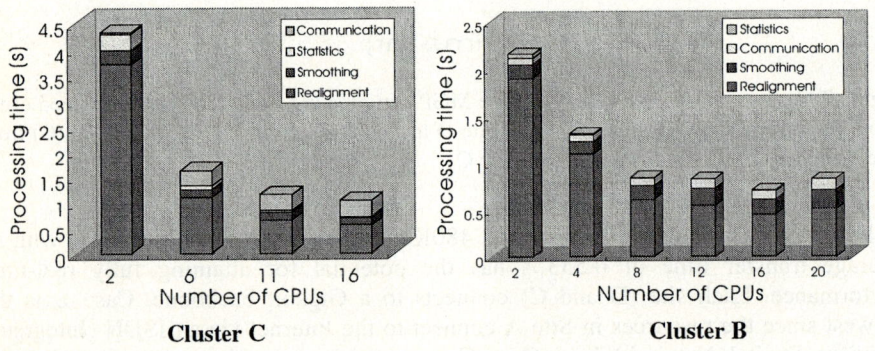

Fig. 7. Total time to process a volume (64 x 64 x 30) of fMRI data as a function of the number of CPUs, for cluster C (slower processors) and cluster B (faster processors). The processing includes realignment, smoothing, and statistical analysis. The time spent for data communication within the cluster is indicated as communication

From the above figures, we can conclude that the use of PC clusters significantly improved the performance of the system. In this case, we achieved optimal performance with 16 CPUs. Beyond this number, communication overhead causes the total processing time to start to increase.

One may note that with the newer, faster processors available in Cluster B, the analysis can already be done in less than 3 s with only 2 processors, suggesting that it would be better to just have the medical imaging facility buy an off-the-shelf dual-processor machine, and perform the real-time analysis locally. This may be true in this case for this particular experiment, since the volumes we are processing are not very large. However, as the volumes become larger (e.g., with higher resolution brain scans), then the performance of a single dual-processor machine will not be enough anymore. Furthermore, note that the analysis performed in this experiment is just one kind of fMRI analysis. In the future, we may have other, more computationally intensive and more parallelizable, algorithms that we wish to run on each volume, in which case having access to the PC cluster would be significantly beneficial.

3.3 Total Analysis Time

Finally, we simulated a real-time run by using the analysis workstation in Site D and cluster B in Site C. The running results were also displayed simultaneously in Site B. The simulated real-time run used real data from a previous MRI examination, and included everything specified in Fig. 2 except the participation of the MRI scanner itself (which is not readily accessible for computational experiments). To get the timing for the data transfer, we used the *time* function to record the total time the application was running. Dividing the resulting time by 130 would approximate the transfer time per volume. This value represents a good estimate since the data were already available locally and the program did not have to wait for the data's acquisition. Using this approach, an estimated data transfer time per volume equal to 0.215 s was obtained. (Note that this is different from the result of the transfer test in Fig. 6 (average 0.533 s) because for this run, the header information, which is redundant information that is about the same size as the image data, was excluded from the transfer, thus reducing the time by almost half.

The analysis time was also estimated similarly, that is, taking the total running time of *baxgridbatch* using 16 CPUs and dividing it by 130. The result was 0.960 s per volume. Note that this value includes waiting time, that is, before *baxgridbatch* could start the analysis for one volume, it had to wait for the data transfer to complete. Subtracting the data transfer time obtained above, we got 0.745 s. This value is close to the value obtained during the offline runs, which is 0.710 s (Fig. 7b, 16 CPUs). For the results transfer time, the waiting times for data transfer and analysis need to be subtracted from the obtained value, which was around 1.269 s. This gives 0.094 s for result transfer time. Finally, the time to update the display is recorded using the *g_timer_** routines in the gtk+ library. For local update, the time is around 0.035 s, whereas for remote update, the time is 1.6 s. These values are dependent on the size of the actual display. From these values, the total processing time per volume can be approximated and is equal to 1.089 s (for local display) and 2.654 s for remote display. Note that the reported values above varied from run-to-run, though not significantly, with the total value always within the used TR, thus demonstrating fully real-time analysis.

3.4 Computational Web Services

Aside from improving interoperability and programmability, the use of these XML Web services also greatly reduced the setup requirements compared to that of Globus.

Furthermore, even though these web services use Java and use XML, a text-based language, to encode the data during data transfer, the data transfer rate was only around 2x slower than that of the direct socket transfer program, as shown in Fig. 8. Moreover, by having the web service automatically use GZIP compression (using built-in Java routines) to compress the data before encoding it as XML, we were able to reduce the slowdown to less than 2x (around 1.5x for image files, in this case, and much less for the highly-compressible activation map files), and come close to achieving a total transfer time (including the image file and the activation map) of less than 3s between Osaka and Manila during off-peak hours, thus demonstrating at least the feasibility of real-time performance even when using XML web services. (We have actually measured total transfer times of less than 3s using a CWS but were not able to measure the time for the C socket version for the same time period, and so do not present the data here.)

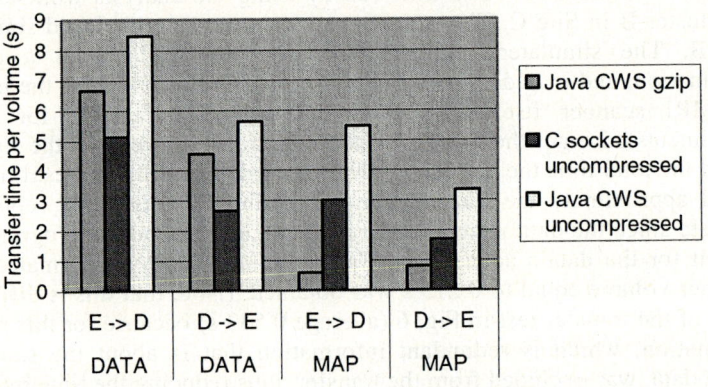

Fig. 8. A comparison of data transfer times between Sites D and E, using C sockets and a computational web service

4 Related Work

Several other groups have also studied the use of grid computing technologies for the analysis of medical imaging data. These include the MediGrid Project [4,5] and the Neurogrid Project [6]. Darthmouth College has established the fMRI Data Center [7] which uses Globus to provide researchers with access to several databases of fMRI data, as well as allows them to submit their own data and run fMRI applications on these data. Most of these projects are primarily concerned with offline processing of image data, and not so much on real-time processing as we are attempting in this project.

Our idea of computational web services [1] was originally developed independently from the idea of grid services as developed in the OGSA project [3]. Thanks to OGSA, there is now great interest in grid services in industry and many grid computing projects are now taking advantage of XML web services. As far as we know, however, most medical imaging applications using grid computing are still using the older version of Globus, which does not yet using web services.

Some grid computing projects today aimed at practical scientific applications, such as GridSphere [8] use grid portals. These grid portals typically allow users to submit jobs to the Grid by providing a web page that they can visit and upload data to using a web browser. XML-based computational web services and grid services, on the other hand, provide a *programmatic* interface that allows *programs* to communicate with a server via the HTTP protocol. Computational web services do not use web browsers and act like essentially like remote objects that can be invoked by user programs. Thus, the applications of computational web services are much more vast than those of grid portals.

5 Discussion and Conclusion

One of the critical factors in the real-time analysis of functional MRI time series is the analysis time. This in turn depends on the complexity of the analysis performed and the size of the data being considered. For analysis involving only simple comparison t-test or those involving only limited number of voxels (e.g., a slice or two), achieving real-time performance would not require considerable computing power. But the analysis results would not be as useful. It could only give a rough idea of possible regions of activation. The coverage would also be strongly limited to specific regions in the brain. To attain more reliable and complete results, more sophisticated analysis employing for instance image preprocessing to eliminate non-task related artifacts is important. Whole-brain image analysis is also necessary to have an overall perspective of the different regions activated during the experiment. For this case, the effect of analysis time in the overall real-time performance comes into play. The use of PC clusters to process the data was shown to significantly improve the analysis performance of the system.

Performing real-time analysis using remote resources offers the possibility of doing more intensive analysis even if the imaging site doesn't have enough computing resources. However, it imposes additional constraints in the overall performance of the system. The unpredictable behavior of the network, as demonstrated in Case A in Fig. 6, can strongly affect the transfer time. This in turn can affect the total analysis time, which determines whether fully real time analysis could be achieved or not. To mitigate this problem, an incremental approach was used. By incremental, we mean that the data are processed volume-by-volume in contrast to the batch approach where all the volumes are collected first before the processing starts. The advantages are apparent. Since the data are transferred volume-by-volume, network load is distributed across time during the experiment. The analysis can also be started in parallel with data acquisition, thus making the final results available within a short time after all the data are acquired. The running results can also be used as a determinant of the quality of the acquired data and as a measure on how the experiment progresses.

In this paper, we demonstrated fully real-time analysis in the simulated real-time runs for a real data set. The total analysis time per volume in this experiment was around 1.089 s, much less than the repetition time, TR, of 3 s. This implies that the TR of the scan could still be shortened to increase the data's temporal resolution. Alternatively, larger data sets with higher spatial resolution could also be considered for processing without losing fully real-time performance. Although in the current implementation it is not possible to attain real-time analysis using slow connections, the use of compression algorithms during data transfer could minimize transfer times to further improve the performance of the system. Moreover, recent advances in high-speed network connectivity may soon eliminate this problem.

The results of the experiments between Site E to Site D (Fig. 6 and Fig. 8) also illustrate the potential of grid computing as a technology that will enable researchers from different countries share their resources and collaborate with each other. For instance, medical imaging sites in developing countries will be able to access computational resources in developed countries, thus enhancing their capabilities without the additional cost associated with resource acquisition and maintenance.

Aside from demonstrating the feasibility of geographically disparate Grid computing, the results of using the XML computational web service (Fig. 8), also demonstrate the potential of web-service based Grid technologies (including our own ad hoc one, as well as standard ones such as OGSA) for use in real practical, and even data-intensive applications, such as these.

Future work on this project will include further improvement and development of the existing code (with special focus on the computational web service components), and the exploration of other medical applications of grid computing. Since the results of this paper show that data transfer times are still a significant limiting factor, we plan to consider other applications as well, where the computation-to-communication ratio is much higher than in the case studied in this paper.

References

1. Sarmenta LFG, Chua SJ, Echevarria P, Mendoza JM, Santos RR, Tan S. Bayanihan Computing .NET: Grid Computing with XML Web Services. In: 2nd IEEE International Symposium on Cluster Computing and the Grid; 2002; Berlin, Germany; 2002.
2. Bagarinao E, Matsuo K, Nakai T, Sato S. Estimation of general linear model coefficients for real-time application. Neuroimage 2003;19(2):422–429.
3. Foster I, et al. The Physiology of the Grid: An Open Grid Services Architecture for Distributed Systems Integration. http://www.globus.org/research/papers/ogsa.pdf. February 2002.
4. Montagnat J, Breton V, Magnin IE. Using grid technologies to face medical image analysis challenges. BioGrid'03, IEEE/ACM CCGrid03, Tokyo, Japan, May 2003.
5. Tweed T, Miguet S. Distributed indexation of a mammographic database using the grid. International Workshop on Grid Computing and e-Science. 17th Annual ACM International Conference on Supercomputing. San Francisco, USA, June 21st 2003.
6. Buyya R, Date S, Mizuno-Matsumoto Y, Venugopal S, Abramson D. Composition of Distributed Brain Activity Analysis and its On-Demand Deployment on Global Grids. Technical Report, Grid Computing and Distributed Systems (GRIDS) Lab, Dept. of Computer Science and Software Engineering, The University of Melbourne, Australia. 2002.
7. The fMRI Data Center Home Page. http://www.fmridc.org/
8. GridSphere Home Page. http://www.gridsphere.org/

Building and Accessing Grid Services

Xinfeng Ye

Department of Computer Science, Auckland University, New Zealand
xinfeng@cs.auckland.ac.nz

Abstract. A computation grid, which hosts services that are shared by users, is formed. The grid system is responsible for deciding which services need to be replicated for efficiency reason. A dynamic XML document is an XML document with embedded Web service calls. This paper proposes an approach in which dynamic XML documents are used to specify Grid services that users want to access.

1 Introduction

Open Grid Services Architecture (OGSA) [7,16] is built on Globus [8] and Web services. The conventional way of accessing a Web service is either by sending a SOAP message to the Web service or by using the features provided in the SDKs of programming languages to invoke the Web service. Both approaches require users have good programming skills. A dynamic XML document is an XML document with embedded Web service calls [1]. The Web service calls are made when the content of the full document is needed.

In this paper, a computation grid based on OGSA is described. Users use dynamic XML documents to specify Grid services that they want to access. A tool is developed for managing the calls to the Grid services specified in the dynamic XML documents. As the tool is responsible for making calls to services, users only need to know how to write dynamic XML documents in order to use the Grid services. The computation grid manages the service instances serving the users' requests to ensure that (a) users' requests are served within a reasonable amount of time, and (b) the resources in the computation grid are not wasted.

2 Dynamic XML Documents

2.1 Using Dynamic XML Documents

Dynamic XML documents were developed in Active XML system [1]. In a dynamic XML document, there is a special kind of elements, called *service elements*, which represent calls to Web services. In [1], a service element needs to have attributes to specify the location of the Web service. In this paper, since it is impossible to know which machine in the computation grid hosts the services specified in a dynamic XML document when the document is written, a service element only needs to specify the name of the service. The computation grid will be responsible for mapping a service name to a machine that hosts an instance of the service.

The example in Fig. 1 shows a simple dynamic XML document. The document specifies the design of a yacht's sail. In this document, the user wants to see how the sail performs when the yacht sails. The Grid service that simulates the performance of the sail is called "SailSimulation". When the service is called, two parameters, i.e. the sail's design data and the data describing the wind condition, are passed to the service. Tag **externalURL** is used in dynamic XML documents to specify the link to an external document. If the contents between the **externalURL** tags start with "http", it means that the document resides on a web site. If the contents start with "file", it means that the file resides on user's local file system.

```
<document name="SailDesign">
  <service name="SailSimulation">
    <param>
      <externalURL>file://C:/project/SailData.xml</externalURL>
    </param>
    <param>
      <externalURL>http://winddata.com/wind.xml</externalURL>
    </param>
  </service>
</document>
```

Fig. 1. A Simple Dynamic XML Document

The parameters of a Grid service can be results returned from calls to other Grid services. The example in Fig. 2 shows that the wind data is obtained by calling Grid service **HistoricalWindData**. When **HistoricalWindData** is called, the user needs to provide the parameters specifying the location and time interval for the wind data to be retrieved. Data returned from **HistoricalWindData** are passed to **SailSimulation** as a parameter. In this example, an implicit order of calling Grid services exists. That is, when service **SailSimulation** is called, a call to service **HistoricalWindData** must be made first to obtain the value of the parameter to be passed to service **SailSimulation**. The relations between Grid services can also be specified explicitly using XLink [6].

```
<document name="SailDesign">
  <service name="SailSimulation">
    <param>
      <externalURL>file://C:/project/SailData.xml</externalURL>
    </param>
    <param>
      <service name=" HistoricalWindData">
        <param> Location </param>
        <param> Time </param>
      </service>
    </param>
  </service>
</document>
```

Fig. 2. A Dynamic XML Document with Implicit Calling Order

2.2 A Tool for Handling Dynamic XML Document

A tool, dXMLTool, is developed in this project. dXMLTool is for making calls to Grid services specified in dynamic XML documents as well as for viewing and editing dynamic XML documents.

When a user wants to invoke a Grid service in a document, the user clicks the corresponding service element in the document. The dXMLTool interacts with the computation grid to locate a machine that hosts an instance of the requested Grid service. The details of the interaction between the dXMLTool and the computation grid are explained in §4.1.

Once a service instance is located, the dXMLTool calls the operation provided by the service instance on behalf of the user. If the values of the parameters to be passed to the Grid service are stored in files referred through external links, e.g. file://C:/project/SailData.xml, http://winddata.com/wind.xml, etc, the dXMLTool retrieves the contents of the files and passes the retrieved data to the Grid service. As explained in §2.1, the parameters in a Grid service call might be the results returned from other Grid service calls. Thus, before a Grid service call is made, the dXMLTool also checks whether the value of a parameter is the result returned from the call to another Grid service, say *gs*. If this is the case, the dXMLTool will call *gs* first.

Two Grid services do not depend on each other if their outputs are not used as the inputs of their counterpart. If there is not a dependency relation between two Grid services, the two services can be called simultaneously. A user might specify several Grid services that can be called concurrently in a dynamic XML document. The dXMLTool provides a button that enables a user invoke all the Grid services specified in the document. In order to determine the dependencies amongst the Grid services, the dXMLTool represents the dynamic XML document as a tree. It can be seen that, if two Grid services do not have ancestor/descendent relationship in the tree, the two services do not depend on each other. Thus, they can be called simultaneously.

3 A Computation Grid

Machines in several labs are pooled together to form a computation grid. The hosting environment provided by each lab is shown in Fig. 3. The computation grid is made up of several such environments.

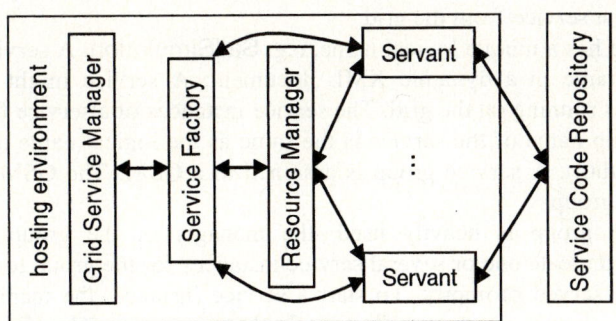

Fig. 3. Computation Grid Hosting Environment

The components managing the computation grid, e.g. Grid Service Manager, Resource Manager etc., provide their services in the form of Web services. In this paper, the services provided by these components are not regarded as Grid services. *Grid services* are defined as the services referenced by users using service elements in dynamic XML documents, e.g. SailSimulation, HistoricalWindData.

Each lab hosts a Grid Service Manager (GSM). The GSM is responsible for (a) keeping track of the Grid services available on the computation grid and the service instances running on the computation grid, (b) monitoring the utilizations of the service instances, and (c) requesting the computation grid replicate a Grid service when the utilizations of all existing instances of the service exceed a threshold.

A web server, i.e. Tomcat, is running on each of the lab machines. The web server hosts a web service, called Servant. Servant provides two operations startService and terminateService. The Servant service on a machine is responsible for starting or terminating a Grid service instance on the machine. If a machine is not hosting any Grid service, the machine is regarded as *idle*.

Each lab runs a Web service called ResourceManager which corresponds to the Resource Manager in Fig. 3. The ResourceManager service of a lab is responsible for monitoring the status of the machines in the lab, i.e. whether a machine is idle or is hosting an instance of a Grid service. The ResourceManager holds the URLs of all the Servant services in the lab. A ResourceManager service provides four operations, i.e. getServant, register, unregister and alive. The getServant operation finds the URL of a Servent (if any) running on an idle machine and returns the URL to the caller. The register operation is called by a Servant service when the machine hosting the Servant service becomes idle. unregister operation is used to inform the ResourceManager that an "idle" machine is no longer idle.

Each lab runs a ServiceFactory service which corresponds to the Service Factory component in Fig. 3. ServiceFactory service provides a createService operation. The operation creates an instance of a requested Grid service. The code implementing Grid services is stored in the Service Code Repository. The code repository is hosted by a web server and is shared by all labs.

All Grid services hosted by the computation grid are implemented as Web services. Each service provides two operations. One is for implementing the service needed by the user, e.g. simulating the design of a yacht's sail. The name of the operation is determined by the functionality of the operation, e.g. sailSimulation. The other operation is called destroy. This operation is called when the system removes an instance of the service from the grid.

Each service has a unique logical name, e.g. SailSimulation. A service is referred by its logical name in a dynamic XML document. A service might have several service instances running on the grid. The service instances of a service form a *service group*. The group name of the service is the same as the logical name of the service, e.g. SailSimulation. A service group is assigned to a GSM. The GSM is called the *manager* of the group.

If a service group is heavily used, the manager of the group requests the computation grid create one or several service instances for the group to ensure users' requests can be served promptly. To start a service instance, the manager calls the createService operation of the ServiceFactory service. The ServiceFactory service calls the getServant operation of the ResourceManager service to obtain an idle machine first. The ResourceManager passes the URL of the Servant service on the idle machine to the ServiceFactory service. The ServiceFactory service calls the startService operation of the Servant service to instruct the machine to start a service instance. When startService is called, the code of the service instance is downloaded from the Service Code Repository and is installed on the idle machine. As a result, a service instance is created on the idle machine. The Servant service that starts the service instance is called the *guardian* of the service instance. The ServiceFactory service registers the service instance to the manager of the group to which the instance belongs. When a service instance is registered to the manager, the estimated response time of the operation provided by the service instance is also logged with the manager. The ServiceFactory service also calls the unregister operation of the ResourceManager to remove the "idle" machine from the ResourceManager's list.

To better utilize the grid resources, if the utilization of a service group is below a threshold, the manager of the group will terminate some service instances in the group. To terminate a service instance, the manager calls the destroy operation of the service instance. The destroy operation calls the terminateService operation of the service instance's guardian. The guardian's terminateService operation removes the code of the service instance from the web server hosting the instance. As a result, the service instance is removed from the computation grid. The guardian also calls the register operation of the ResourceManager to indicate that the machine is idle again.

4 How the System Works

4.1 Invoking a Grid Service

When a user activates a service element in a dynamic XML document to call a Grid service, the dXMLTool interacts with the computation grid to find a service instance to serve the call. The managers of service groups maintain references to the service instances in the system. Thus, the dXMLTool must contact the manager of the service group to which the service instance belongs. However, if it is the first time for the dXMLTool to call a Grid service, the dXMLTool does not know the identity of the GSM that manages the corresponding service group. By default, the dXMLTool is set to contact the GSM hosted by the user's lab.

When a GSM is contacted by the dXMLTool, there are three possibilities:

(a) The GSM manages the service group that provides the requested service.

    In this case, the GSM finds a service instance to serve user's request and returns the service instance to the dXMLTool.

(b) The service group that provides requested service is managed by another GSM.

    In this case, the GSM informs the dXMLTool of the identity of the GSM, say *gsm'*, from which the dXMLTool can obtain the required service instance. As a consequence, the dXMLTool contacts *gsm'* to obtain the required service instance.

(c) No service instance that serves user's request has been created. That is, it is the first time that the requested Grid service is called by a user.

In this case, the GSM (i) requests the computation grid create a required service instance, (ii) forms a service group for the service and becomes the manager of the service group, and (iii) notifies the other GSMs in the system about the existence of the newly created service group.

Once the dXMLTool knows which GSM manages a requested service group, the dXMLTool caches this information. So that, future calls to the service can be sent to the correct GSM.

A service group manager uses the following rules in managing service instances:

(a) When receiving a request for a service instance:
   (i) Allocate an idle service instance that has the shortest response time. According to the response time of the service instance, the manager records the time that the service instance is expected to complete serving the request.
   (ii) If there is no idle service instance, put the request in a *wait-for-service* queue.
(b) When a service instance becomes available, if there are requests in the *wait-for-service* queue, allocate the service instance to the first request in the queue.

4.2 Replicating Grid Services

If the utilization of a service group is high, in order to ensure that users' requests for a Grid service can be responded promptly, new service instances are created for the group. The utilization of a service group is defined as below:

Let (a) δt be the time period over which the usage of the group is monitored, (b) t_i be the time that service instance i spent on serving users' request during δt, (c) SG be a set including all the service instances in the service group, and (d) t'_i be the amount of time that service instance i is alive (a service instance is alive once it is created) during δt. The utilization U of a service group SG is defined as:

$$U = \frac{\sum_{i \in SG} t_i}{\sum_{i \in SG} t'_i}$$

δt is chosen based on whether users can obtain a response from the system within a reasonable period of time. According to [14], 10 seconds is the limit for keeping users' attention focused. Some services need to run for a period longer than 10 seconds to produce their results. Thus, the goal of this project is, after a user sends a service request, the user will be notified within 10 seconds that a service instance has started working on the request. Taking into account of the time for the dXMLTool to interact with the GSMs to find a service instance to serve users' requests, in this project, δt is set to 8 seconds.

If $U = 1$ holds for a service group, it means that all service instances of the group are busy during δt. Thus, if extra requests for service instances arrive during this period, these requests cannot be served immediately. If $U < 1$, it means the service group has spare capacity to handle more users' requests within time period δt.

At the end of a δt period, the manager of a service group checks the utilization of the group. If the utilization is above a threshold ε, the service group is regarded as being heavily used. In this case, the manager creates more service instances to make the utilization below ε. In this project, ε is set to 80%. Setting ε to a value less than 100% allows a service group having spare capacities to cope with a sudden surge in request numbers.

Let SG' be the set of service instances created by the computation grid to lower the utilization of a group, say SG. The following rule is used by the group's manager to determine the number of service instances that should be in SG'.

Rule: After creating sufficient number of service instances, the following should hold:

$$U' = \frac{\sum_{i \in SG} t_i}{\sum_{i \in (SG \cup SG')} t'_i} \leq \varepsilon$$

Due to the timing of the arrival of service requests, it is possible that, even if the utilization of a group does not exceed the threshold, a user's request still cannot be responded in 10 seconds. Therefore, when a request is added to the *wait-for-service* queue, the group manager also checks the expected completion time of all existing service instances. If a request cannot be responded in 10 seconds, the manager asks the computation grid create a new service instance.

4.3 Removing Grid Service Instances

If the utilization of a service group is below a threshold for a long period, the manager of the service group will terminate one or several service instances in the service group[1]. A service group's manager uses the following principles to decide whether to terminate some service instances in the service group:

Let (i) U be service group SG's utilization, (ii) ε be the threshold indicating the service group is heavily used, and (iii) φ be the threshold indicating the service group is under utilized.

Case 1: None of the service instances in the service group have been used, i.e. $U=0$.
Terminate all service instances in the group. That is, group SG is terminated.

Case 2: $U \neq 0$

(i) Find and remove one service instance, say SI, from SG, such that one of the following conditions holds:

$$\frac{\sum_{i \in SG} t_i}{\sum_{i \in (SG-\{SI\})} t'_i} < \varphi \quad (C1)$$

$$\varphi \leq \frac{\sum_{i \in SG} t_i}{\sum_{i \in (SG-\{SI\})} t'_i} \leq \varepsilon \quad (C2)$$

If C1 holds, it means that, the utilization of SG is still low even after SI is removed from SG. Thus, more service instances can be considered to be removed from SG. Hence, (i) is applied to $SG-\{SI\}$ again.

---

[1] In this project, the threshold is set to 20% while the time interval is set to 30 minutes.

If C2 holds after applying (i), it means that, after *SI* is removed from the group, the utilization of the group is within an acceptable range. Thus, the process of terminating service instance stops after *SI* is terminated.

(ii) If no service instance can be found when applying (i), stop the process of finding and terminating service instances in *SG*.

The manager will wait for another period, e.g. 30 minutes, and starts this process again if the utilization of the service group is still below φ.

When group *SG* is terminated, i.e. Case 1, notifications should be sent to all the GSMs in the system. Thus, a future request for obtaining a service instance from *SG* will not be forwarded to *SG*'s current manager. Instead, a GSM which receives the request will be responsible for requesting the computation grid create a service instance to serve the call (as explained in §4.1).

5 Performance

Performance test has been carried out to measure the average response time to service requests under different utilization levels of a service group. The following assumptions are made for the test: (a) the size of the code that implements the Grid service is 20 Kbytes, (b) δt is set to 8 seconds, (c) data are collected from ten consecutive δt periods, (d) there are three service instances in the service group initially, (e) it takes about 550ms to start a Grid service instance after a service group manager requests the computation grid to do so, (f) the manager of a service group requests the computation grid create new service instances if the utilization of the group is above 80% or a request cannot be responded in 10 seconds, and (g) each service instance completes serving a user's request in 30 seconds.

The average response time is the average time to respond to the new requests generated during a δt period. The new requests sent to the group at a δt period spread evenly over the period. That is, the interval between the requests sent to the service group at the ith δt period is $\delta t/$(the number of new requests during i^{th} period).

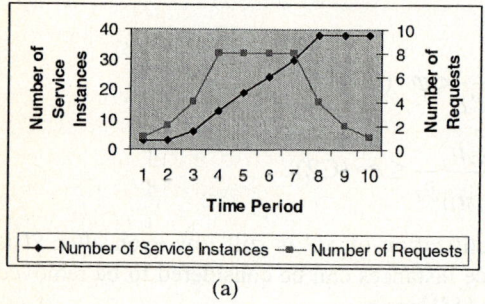
(a)

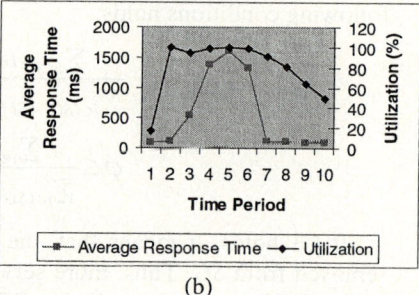
(b)

Fig. 4. Performance Test

From Fig. 4(a), as there are three service instances initially, the requests generated in the first two periods can be served immediately. Thus, the response time is about 100ms as shown in Fig. 4(b). During the 3rd period, requests are served by creating new service instances, since the existing instances cannot start serving these requests in 10 seconds. Thus, the response time is about 550ms. Since the utilization of the group is above 80% between 2nd to 7th periods, new service instances are also created at the end of these periods to lower the utilization. During the 4th to 6th periods, some requests are served by creating new service instances while others will wait to be served by the existing instances. In most cases, the time waiting for the existing instances to complete is longer than the time to create a new service instance. Thus, the response time for these three periods is above 550ms. As sufficient numbers of service instances have been created, the requests generated in the last four periods can be served immediately (i.e. in 100ms) as shown in Fig. 4(b).

6 Related Work

Numerous Grid projects are being carried out, e.g. Condor-G [10], Nimrod/G [2], ICENI [11], etc. All these systems provide a general-purpose grid computing environment while our project focuses on (a) providing the services through web services, and (b) allowing users describe their desired grid services in a coherent and structured way through the use of dynamic XML documents.

There are many projects on developing general-purpose workflow management solutions for Grid computing, e.g. Chimera [9], Pegasus [5], GridFlow [4] and Discovery Net [3]. The main difference between these systems and the approach in this paper is that our project allows the dependencies of the Grid services to be specified implicitly through nested calls to the grid services as well as explicitly using XLink while the other systems require users explicitly define the dependencies between the grid services. In addition, instead of submitting a complete workflow to the system for execution each time, the dynamic XML approach allows users only invoke the services that the users are interested by clicking on the relevant service elements. Similar to Discovery Net, a dynamic XML document not only includes dependency relations of the Grid services, it can also include information like the rationality of a design, auditing information, etc. This seems to make the document more self-contained.

A number of Grid portals have been built over the years [12,13,15]. When using a Grid portal, a user typically interacts with the portal to select a service and to provide required input data. In our approach, a user specifies all the required services in a dynamic XML file. This has the potential to describe the users' desired Grid services as well as to present the results produced by the Grid services in a more coherent and structured way. The resulting XML file can be consumed directly by domain specific application programs, e.g. visualisation tools, to present the results to the users in a more appealing format. Thus, the approach discussed in this paper complements the Grid portal approach.

7 Conclusions

In a traditional approach, users write programs to invoke grid services and synthesize the results returned from the services. This means that, in order to use the services, a user needs a lot of programming experience. This paper proposes an approach in which users use dynamic XML documents to specify the services that the users want to access. The dXMLTool handles the invocation of the services. This approach does not require the users have much programming experience. Calls to the same service at different parts of a document or in a different document can be reused easily by copying the relevant segment of the XML documents. Compared with program code, the contents of XML documents are easy to understand. Thus, reusing segments of a XML document is relatively easier than reusing code segments of a program.

The computation grid in this paper provides a management service that monitors and manages the resources in the system. The management service ensures that (a) the system resources are not wasted, and (b) users' requests can be served efficiently.

Acknowledgment

The author would like to thank the anonymous reviewers for their useful comments.

References

1. S. Abiteboul, A. Bonifati, G. Cobéna, I. Manolescu and T. Milo: Dynamic XML Documents with Distribution and Replication, Proc. of *SIGMOD* 2003, pp. 527–538
2. D. Abramson, R. Buuya and J. Giddy: A Computational Economy for Grid Computing and its Implementation in the Nimrod-G Resource Broker, Future Generation Computer Systems. Volume 18, Issue 8, Oct-2002
3. S. Al Sairafi, F. S. Emmanouil, M. Ghanem, N. Giannadakis, Y. Guo, D. Kalaitzopolous, M. Osmond, A. Rowe, J. Syed and P. Wendel: The Design of Discovery Net: Towards Open Grid Services for Knowledge Discovery, in International Journal of High Performance Computing Applications. Vol 17, Issue 3. 2003
4. J. Cao, S. A. Jarvis, S. Saini and G. R. Nudd: "WorkFlow Management for Grid Computing." In Proc. of the 3rd IEEE/ACM International Symposium on Cluster Computing and the Grid, pp. 198–205, 2003
5. E. Deelman, J. Blythe, Y. Gil, C. Kesselman, G. Mehta, K. Vahi, A. Lazzarini, A. Arbree, R. Cavanaugh, S. Koranda: Mapping Abstract Complex Workflows onto Grid Environments, Journal of Grid Computing, vol. 1, pp. 25–39, 2003
  S. DeRose, E. Maler and D. Orchard: XML Linking Language (XLink) Version 1.0, http://www.w3.org/TR/xlink
7. I. Foster, C. Kesselman, J. Nick, S. Tuecke: The Physiology of the Grid: An Open Grid Services Architecture for Distributed Systems Integration, Open Grid Service Infrastructure WG, Global Grid Forum, June 22, 2002
8. I. Foster and C. Kesselman: Globus: A Toolkit-Based Grid Architecture, in I. Foster and C. Kesselman eds, The Grid: Bluprint for a New Computing Infrustructure, Morgan Kaufmann, 1999, pp. 259–278

9. I. Foster, J. Voeckler, M. Wilde, and Y. Zhao, "Chimera: A Virtual Data System for Representing, Querying, and Automating Data Derivation," pp. 37–47, 14th International Conference on Scientific and Statistical Database Management, 2002
10. J. Frey, T. Tannenbaum, I. Foster, M. Livny, and S. Tuecke, "Condor-G: A Computation Management Agent for Multi-Institutional Grids", Proceedings of the Tenth IEEE Symposium on High Performance Distributed Computing (HPDC10), 2001
11. N. Furmento, J. Hau, W. Lee, S. Newhouse, and J. Darlington, Implementations of a Service-Oriented Architecture on top of Jini, JXTA and OGSI, Second Across Grids Conference, 2004
12. D. Gannon, M. Christie, O. Chipara, L. Fang, M. Farrellee, G. Kandaswamy, W. Lu, B. Plale, A. Slominski, A. Sarangi, Y. L. Simmhan, Building Grid Services for User Portals, http://www.extreme.indiana.edu/~gannon/GridServiceUserPortal.pdf
13. Integrated e-Science Environment for CLRC, http://esc.dl.ac.uk/IeSE
14. R.B. Miller: Response Time Man-Computer Conversational Transactions, AFIPS Fall Joint Computer Conference, pp. 267–277, Vol.33, 1968
15. NCSA Alliance Scientific Portal Project, http://www.extreme.indiana.edu/alliance
16. S. Tuecke, K. Czajkowski, I. Foster, J. Frey, S. Graham, C. Kesselman, T. Maguire, T. Sandholm, P. Vanderbilt, D. Snelling: Open Grid Services Infrastructure (OGSI) Version 1.0, Global Grid Forum Draft Recommendation, 2003

DRPS: A Simple Model for Locating the Tightest Link*

Dalu Zhang, Weili Huang, and Chen Lin

Department of Computer Science and Engineering,
Tongji University, Shanghai, P.R.China
daluz@public.sta.net.cn, gitry@163.com, spalding@sohu.com

Abstract. The tightest link of a network path is the link where the end-to-end available bandwidth is limited. We propose a new and simple probe model, called *Dual Rate Periodic Streams* (DRPS), for finding the location of the tightest link. A DRPS probe is a periodic stream with two rates. Initially, it goes through the path at a comparatively high rate. When arrived at a particular link, the probe shifts its rate to a lower level and keeps the rate. If proper rates are set to the probe, we can control whether the probe is congested or not by adjusting the shift time. When the point of rate shift is in front of the tightest link, the probe can go through the path without congestion, otherwise congestion occurs. Thus, we can find the location of the tightest link by congestion detection at the receiver.

1 Introduction

In the past few years, there are large quantities of researches on available bandwidth measurement. Besides measuring the value of available bandwidth, more and more researchers are interested in locating tight link. The purpose of this paper is to propose a new measurement model, called *Dual Rate Periodic Streams* (DRPS), for locating the tightest link of a path. DRPS based *pathload*[4] and *tailgating* technique[5]. It extends the existing measurement of available bandwidth and brings some new applications. It takes more benefits to network management and network applications. It also provides more information that is valuable to network behavior research.

Recently several tools have been developed for locating the tightest link, such as *BFind*[1], *STAB*[6], and *Pathneck*[3]. *BFind* essentially induces network congestion through continuous transmission of UDP traffic and determines the location of the tight link from *traceroute* round-trip times. *STAB* infers the location of the tightest link by measuring and evaluating the available bandwidth of the prefix path of each link. *Pathneck* combines measurement packets and load packets in a single probing packet train. It infers the position of the tightest link by estimate packet train length on the incoming link of each router.

---

* Supported by the National Natural Science Foundation of China(No. 90204010).

2 Dual Rate Periodic Streams

When a stream of rate R goes through a path, the input stream rate and the output stream rate at the ith link are denoted by IR_i^R and OR_i^R respectively. If $IR_i^R > OR_i^R$, we say that the ith link is a *tight link* at rate R. If $IR_i^R > OR_i^R$ for any $R > A$, we say that the ith link is *the tightest link* of the path. Given a path P, *the first tight link* is the tightest link of the path, denoted by tl_1^P. For any $i > 1$, *the ith tight link* of a path is the tightest link of the prefix path of the $(i-1)^{th}$ tight link, denoted by tl_i^P.

We propose a new probe model, called *Dual Rate Periodic Streams* (DRPS), for finding the location of the tightest link. It has two properties, *dual rate* and *periodic stream*.

Dual rate probe is a stream with two rates. Initially, the probe goes through the path at a high rate R_H. At the moment of *shift time* H_S, the probe shifts its rate to a low level R_L and keeps the rate until arriving at the receiver. Here *shift time* H_S can be regarded as the hops by which the probe keeps with a high rate. We denote a dual rate probe by $DRP(R_H, R_L, H_S)$ and use the following theorem to describe the property of dual rate probe.

Theorem 1. *Dual Rate Property. Given a dual rate probe $DRP(R_H, R_L, H_S)$, $A_{tl_1} < R_H < A_{tl_2}$ and $R_L < A_{tl_1}$. If $H_S < H(tl_1)$, then $OR_H = R_L$. Else, if $H_S \geq H(tl_1)$, then $OR_H < R_L$.*

If the rate shift occurs before the probe passes the tightest link of the path(i.e. $H_S < H(tl_1)$), then the probe can go through the tightest link without congestion; otherwise, it has to queue at the tightest link. So theorem 1 provides a way to find the tightest link of a path. If proper rates ($A_{tl_1} < R_H < A_{tl_2}$ and $R_L < A_{tl_1}$)[1] are set to DRP, we can find the tightest link by detecting the output rate at the last link OR_H for each H_S(from $1, \cdots H$).

A dual rate probe can be regarded as the mixture of two streams. One stream, denoted by S_1, is a stream of rate R_L, which can go through the path. Another stream, denoted by S_2, is a hop-limited stream of rate $R_H - R_L$, which expires at the hop H_S. In terms of Theorem 1, dual rate probes with too large shift time cause a difference between send rate R_L and receive rate OR_H of S_1. Fig. 1. shows that two dual rate probes with different shift times, $DRP(R_H, R_L, H(l_4))$ and $DRP(R_H, R_L, H(l_5))$, go through a 10-hop path. In each column, filled part is the cross traffic and blank part is the available bandwidth. The tightest link of the path is l_5. S_1 is the black stream, and S_2 is the gray stream. In (a), the point of rate shift is in front of the tightest link. There is no congestion in the path, so S_1 keeps its rate until arriving at the receiver. In (b), the point of rate shift is at the back of the tightest link. Congestion occurs at the tightest link and the rate of S_1 slows down.

In theorem 1, we determine whether the probe is congested in the path or not by its rate at the receiver. Generally speaking, the rate is computed by

---

[1] A_{tl_1} is the available bandwidth of the tightest link tl_1, and A_{tl_2} is the available bandwidth of the second tight link tl_2.

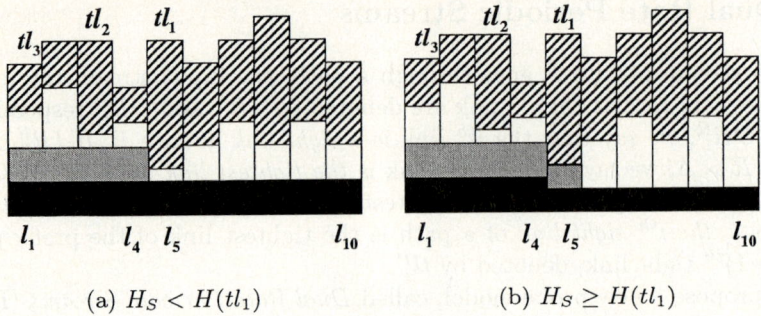

Fig. 1. (a) $DRP(R_H, R_L, H(l_4))$ shifts before arriving at the tightest link, no congestion occurs. (b) $DRP(R_H, R_L, H(l_5))$ shifts after it passed the tightest link, a congestion occurs

arriving times of the first packet and the last packet. The behavior of single packet will affect the estimation of the rate. In addition, it is difficult to make a decision when the rate at the receiver is close to the low rate set at the sender. Therefore, we use a special dual rate probe, which is called *Dual Rate Periodic Streams* (DRPS). Based on the periodic stream property given in [4], we can detect congestion by OWD difference instead of the rate at the receiver.

3 Locating the Tightest Link

DRPS is a simple and flexible model. It gives the theoretical framework for locating the tightest link. Furthermore, when we apply DRPS to practical measurements, more details need to be considered. Here we focus on three key issues, which are parameters setup, probe structure, and how to speed up the locating procedure.

Rate Selection: We set $R_H = A_{tl_1}(1 + \delta)$, and set $R_L = R_H/2$, where δ is a small proportion and set to 5% by default. Thus, we make sure that $R_H > A_{tl_1}$, $R_L < A_{tl_1}$, and the rate of probe is not too large. Although it is uncertain that $R_H < A_{tl_2}$, we believe it doesn't matter. If A_{tl_2} is close to and no 5% more than A_{tl_1}, we locate the second tight link, which is also significant to its applications and even more important than the tightest link. The selection of an approximate rate simplifies the locating process by avoiding measuring A_{tl_2}. We can only measure A_{tl_1} (i.e. end-to-end available bandwidth A) by *pathload* [4].

DRPS Probe Structure: Hop-limited packets are used to ensure back-to-back queuing at a given link in both tailgating [5] and cartouche [2] techniques. We also use hop-limited packets to shift the probe's rate. First, we construct a periodic stream of rate R_H. Next we set the TTL value for each packet to make sure those hop-limited packets and normal packets are arranged alternatively in the probe. So the low rate is half of the high rate. Fig. 2 is an illustration for probe structure and probing process. Packets in black are normal packets and packets in white are hop-limited packets. They are arranged alternatively in the stream. The egress link of the black node is the tightest link and the egress link of the gray node is the second tight link. In (a), hop-limited packets are dropped before arriving at the tightest link, so no congestion occurs. In (b), hop-limited packets are not dropped in time, so congestion occurs and the gap between two successive packets increases.

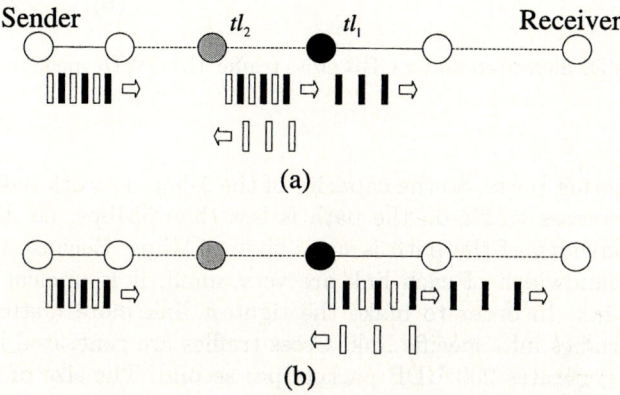

Fig. 2. DRPS probe structure and probing illustration

Quick Locating Algorithm: The problem of locating the tightest link can be regarded as a search problem for a sorted sequence in terms of Theorem 1. Thus, binary search can be used to shorten the measurement period. Binary search is the best method for searching a number in a sorted sequence. It reduces the complexity from $O(H)$ to $O(\log_2 H)$. Because H is no more than 30 hops usually, we will generate at most five probes.

4 Simulation and Experiment

We construct a path with 7 links (i.e. $H = 7$) in NS environment. At the sender, we add two CBR agents to generate two periodic streams S_1 and S_2, which make up our DRPS probe. S_1 and S_2 will send packets by turns strictly, which means one for S_1 packet and one for S_2 packet between the same interval. In our simulations, we make l_4 be the tightest link by adjusting cross traffic. We test our DRPS technique under CBR cross traffic and Pareto cross traffic respectively. Fig. 3 shows OWDs of the probes used for locating the tightest link. We find the OWDs show an increasing trend when the TTL value is set to 4 and non-increasing trend when TTL is set to 3. Then the location of the tightest link is found.

We also experiment on a practical 5-hop network path in our lab. The network path consists of four routers, which are connected with each other by their

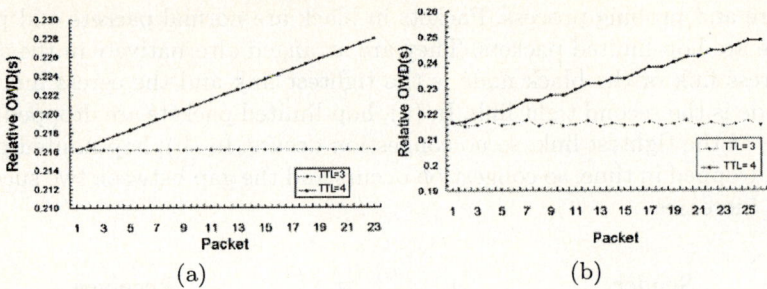

Fig. 3. (a) OWDs measured under CBR cross traffic. (b) OWDs measured under Pareto cross traffic

100Mbps Ethernet ports. So the capacity of the 5-hop network path is 100Mbps. The primitive cross traffic on the path is less than 5Mbps, i.e. the end-to-end available bandwidth of the path is more than 95Mbps. Because the differences of available bandwidth of each link are very small, it is unclear which link is the tightest link. In order to make the tightest link more distinguishable, we inject cross traffics into specific link. Cross traffics are generated by 15 threads. Each thread generates 200 UDP packets per second. The size of packets varies from 1300 bytes to 1500 bytes randomly with the average of 1400 bytes. In our experiments, cross traffics are generated on the middle links(l_2, l_3, and l_4) in turn. Thus, the location of the tightest link varies with the position where cross traffics are injected. In each case, we run DRPS simultaneously. It works well and finds the correct position of the tightest link.

References

1. A. Akella, S. Seshan, and A. Shaikh. An empirical evaluation of wide-area internet bottlenecks. In *Proceedings of ACM SIGCOMM Internet Measurement Conference*, October 2003.
2. K. Harfoush, A. Bestavros, and J. Byers. Measuring bottleneck bandwidth of targeted path segments. In *Proceedings of IEEE INFOCOM*, March 2003.
3. N. Hu, L. Li, Z. Mao, P. Steenkiste, and J. Wang. Locating internet bottlenecks: algorithms, measurements, and implications. In *Proceedings of ACM SIGCOMM*, August 2004.
4. M. Jain and C. Dovrolis. End-to-end available bandwidth: Measurement methodology, dynamics, and relation with tcp throughput. In *Proceedings of ACM SIGCOMM*, August 2002.
5. K. Lai and M. Baker. Measuring link bandwidths using a deterministic model of packet delay. In *Proceedings of ACM SIGCOMM*, August 2000.
6. V. Ribeiro, R. Riedi, and R. Baraniuk. Spatio-temporal available bandwidth estimation with stab. In *Proceedings of ACM SIGMETRICS*, June 2004.

A Congestion-Aware Search Protocol for Unstructured Peer-to-Peer Networks

Kin Wah Kwong and Danny H.K. Tsang

Department of Electrical and Electronic Engineering,
Hong Kong University of Science and Technology
{erick, eetsang}@ust.hk

Abstract. Peer-to-Peer (P2P) file sharing is the hottest, fastest growing application on the Internet. When designing Gnutella-like applications, the most important consideration is the scalability problem. Recently, different search protocols have been proposed to remedy the problems in Gnutella's flooding. However, congestion due to large query loads from users and peer heterogeneity definitely impact on the performance of search protocols, and this consideration has received little attention from the research community. In this paper, we propose a congestion-aware search protocol for unstructured P2P networks. The aim of our protocol is to integrate congestion control and object discovery functionality so that it can achieve good performance under congested networks and flash crowds. The simulation results show that our protocol can largely reduce a hit delay while maintaining a high hit rate, and the congestion problems such as query loss and system overloading can be effectively alleviated.

1 Introduction

P2P file sharing has attracted a lot of attention recently from the general public. Pioneered by Napster, users can share their files such as music and movies over the Internet, and allow other users to download them freely via a P2P approach. After that, many P2P file sharing systems such as Gnutella [1], KaZaA [2], Bit-Torrent [3] have been developed and widely used. Recently, distributed hash table (DHT) P2P systems, say CAN [4], have been advocated to organize the overlay networks into a certain structure so as to improve exact-match search performance. In this paper, we focus on unstructured, distributed P2P architecture such as Gnutella and KaZaA. We also define that congestion is caused by large query rates from users, but the physical network congestion such as router's buffer overflow is ignored.

Flooding search creates many repeated messages and hence generates massive traffic which is a very serious problem [5]. To solve this problem, different search algorithms have been recently proposed, say random walk [6], criticality-based probabilistic flooding [7] and APS [8], to replace flooding. Moreover, according to [9], among Napster users, their Internet access bandwidths are very heterogeneous, from 56Kbps dial-up connection to cable-modem connection. Therefore, we believe that a search protocol should be aware of both congestion and peer heterogeneity.

To improve the efficiency of a P2P file sharing system, many techniques have been employed. Common methods are caching (say [6]) and topology adaptation. Generally speaking, in caching, a peer caches recently-seen query hit information. After that, if a peer receives a query request for the same file, it can directly reply to the query by using the cached answer and the amount of query traffic can be reduced. However, if the cached answer is outdated or invalid, the search result is affected. Super-peer topology formation is a kind of adaptation technique. In this case, powerful users (such as high bandwidth users) form the backbone of a P2P network and most of the query traffic is processed by them. Therefore, low capacity users can be shielded from massive query traffic. The super-peer idea is used in KaZaA [10]. In [11], different methods such as biased random walk, caching, topology adaptation and flow control are used together to build a scalable P2P file sharing system called Gia.

Our Question Is: *Is it possible to design a better search protocol such that P2P systems are robust under congestion and flash crowds without using caching and topology adaptation techniques?*

To answer this question, we propose a congestion-aware search algorithm. Our protocol can be used in KaZaA-like and Gnutella-like systems and it has the following properties:

1. Maintains a low hit delay and a high hit rate[1] under congestion
2. Alleviates congestion problems such as messages loss and system overloading
3. Fully distributed algorithm and lightweight maintenance

The remainder of this paper is organized as follows. Section 2 describes our proposed congestion-aware search protocol in detail. We present the extensive simulation results in Section 3. We then conclude the paper in Section 4.

2 Proposed Protocol

Our proposed search protocol consists of three parts — **Congestion-Aware Forwarding** (CAF), **Random Early Stop** (RES) and **Emergency Signaling** (ES). We first look at the system model.

We design our search protocol based on the random walking technique. To initiate a query request, the requesting node randomly sends out w query walkers to its neighbors, same as [6]. If the number of node's neighbors is smaller than w, a node only sends one walker to each neighbor. Each walker is assigned an initial TTL value. This TTL value is decreased by one for each node visit. If TTL is

---

[1] Generally speaking, hit rate represents how good the search algorithm to locate requested files is. We will define rigorously what hit rate is in the performance evaluation section.

zero, a node discards the walker. We model each node i with a capacity constraint C_i which denotes the number of messages (such as queries and emergency signals) it can process per unit time, and its new query generation rate per unit time is represented by q_i. If the aggregate message incoming rate is higher than the node's capacity, excessive messages are queued in the node's input buffer. $Q_i(t)$ represents how many messages are backlogged in node i's input buffer at time t.

Definition 1. *Congestion level (CL) of node i at time t:*

$$CL_i(t) \triangleq \frac{1+Q_i(t)}{C_i} \qquad (1)$$

Remark 1. We use CL to denote how congested a node is. Equation (1) means how much waiting time a walker would encounter if it is forwarded to node i. For simplicity, we write CL_i only.

Each node i maintains a **forwarding table** which is built based on information carried by a **feedback message**. A forwarding table is used to assist in forwarding walkers. A feedback message is generated when a query hit occurs, and is sent back to the query originator via the query reverse path (as in Gnutella). The forwarding table is composed of **entries**. Each entry, $E_{i \to j}^x$, consists of four different data as follows:

Object x: This field stores the object x's identifier such as an object's name or hash value.
Estimated Maximum Congestion Level $p_{i \to j}^x$: The estimated CL of a bottleneck node in the direction of neighbor node j.
Estimated Hop Count $h_{i \to j}^x$: The estimated distance, in term of hop count, to locate object x in the direction of neighbor node j. This value may not be an integer, because it is updated by averaging previous estimated hop count values.
Expiry Time $t_{i \to j}^x$: The entry is expired and removed when time $t_{i \to j}^x$ has elapsed.

Definition 2. *Total Congestion (TC) in the direction of neighbor node j for object x:*

$$f_{i \to j}^x \triangleq p_{i \to j}^x \cdot h_{i \to j}^x \qquad (2)$$

Remark 2. Basically, we should forward walkers to a direction with smaller queueing delay and shorter hop count in reaching a node storing requested objects. To combine these two features, $f_{i \to j}^x$ is used for reference in forwarding walkers.

2.1 Feedback Update

To build a forwarding table, we utilize a feedback message generated when a query hit occurs. The main purpose of feedback message is to allow nodes to acquire congestion and object location information so that nodes can forward queries wisely. When a query walker successfully finds object x in node k, a

Table 1. Feedback Update Algorithm

When node i receives a feedback message M^x from its neighbor node j:
1. **IF** the entry $E^x_{i \to j}$ exists in the forwarding table **THEN**
2. $t^x_{i \to j} \leftarrow$ *new expiry time*
3. $h^x_{i \to j} \leftarrow (1 - \alpha) \times h^x_{i \to j} + \alpha \times M^x_{HC}$
4. $p^x_{i \to j} \leftarrow (1 - \beta) \times p^x_{i \to j} + \beta \times M^x_{CL}$
5. **ELSE**
6.   **IF** the forwarding table is full **THEN**
7.   Remove the soonest expiring entry
8.   **ENDIF**
9.   Initialize the new entry $E^x_{i \to j}$ as follows:
10. $t^x_{i \to j} \leftarrow$ *new expiry time*
11. $h^x_{i \to j} \leftarrow M^x_{HC}$
12. $p^x_{i \to j} \leftarrow M^x_{CL}$
13. **ENDIF**
14. $M^x_{HC} \leftarrow M^x_{HC} + 1$
15. **IF** $CL_i > M^x_{CL}$ **THEN** $M^x_{CL} \leftarrow CL_i$ **ENDIF**
16. Forward the feedback message M^x to the next node of the reverse path if node i is not the query originator. Otherwise, drop this feedback message.

---

feedback message (M^x) generated by node k is sent to the query originator via the query's reverse path. The feedback message not only contains the query hit information such as node k's IP address and object x's size, but also has Congestion Level field (M^x_{CL}) and Hop Count field (M^x_{HC}). These two fields are initialized in node k as follows:

$$M^x_{CL} \leftarrow CL_k, \quad M^x_{HC} \leftarrow 1$$

When a feedback message traverses back to the query originator, the forwarding tables of the reverse-path nodes including the query requestor as well as M^x_{CL} and M^x_{HC} of a feedback message are updated based on the algorithm shown in Table 1.

In the feedback update algorithm, $h^x_{i \to j}$ and $p^x_{i \to j}$ are updated by using the moving average approach (lines 3, 4). α and β are between 0 and 1. A feedback message *memorizes* an object distance and a congestion bottleneck along an object discovery path. M^x_{HC} is a counter to record how many hop counts from the current node to the object's hosting node. This field is increased by one in each reverse path node (line 14). M^x_{CL} is to record a maximum CL of a node along the reverse path. We only need to update M^x_{CL} if a node's CL value is larger than its current value (line 15). The reason for M^x_{CL} records a maximum CL instead of accumulating it is to prevent this field from numerical overflowing. Also, emergency signaling can help to correct the inaccuracy of an estimated CL in a node's forwarding table. Each entry is tagged with an expiry time. Expiry time is used to control the life span of an entry and prevent an outdated entry affecting the protocol decision. If the forwarding table is full, we can simply remove the soonest expiring entry in order to add another new entry (lines 9-12).

2.2 Congestion-Aware Forwarding (CAF) and Random Early Stop (RES)

Feedback update helps nodes to know the congestion and object location around them. Intuitively, queries should be forwarded in some direction so as to shorten the hit delay, increase the hit rate, prevent congestion and discover a new object's host. Moreover, we should drop a walker in preemptive manner to alleviate the congestion. However, in dropping non-zero-TTL walkers, we need to ensure that the hit rate should not be greatly degraded. To achieve the above requirement, we develop CAF and RES.

The detailed CAF and RES algorithms are shown in Tables 2 and 3 respectively ($Uniform(a, b)$ denotes a uniformly random real number between a and b). In using CAF, we firstly classify the node's neighbors into set A and set B. If a node's forwarding table contains object x's information in the direction of neighbor node j, then node j belongs to set A. Otherwise, it belongs to set B. However, a query originator (say node s_1) and the node from which a query is received (say node s_2) are excluded from these two sets, because a query should not be forwarded back to these two nodes. The mathematical expressions of set A and set B are:

$$A = \{v \in Nbr\,(i) \setminus \{s_1, s_2\} \,|\, E_{i \to v}^x \; exists\} \tag{3}$$

$$B = \{v \in Nbr\,(i) \setminus \{s_1, s_2\} \,|\, E_{i \to v}^x \; not\; exists\} \tag{4}$$

where $Nbr\,(i)$ is the set of node i's neighbors.

Therefore, there are three different cases in a forwarding decision (lines 1, 8 and 18 of Table 2). In case 1 (line 1), the forwarding table of node i contains *full information* about object x in all possible forwarding directions. As we have *full information* in case 1, the RES algorithm should be executed there (Table 2, line 2). Table 3 shows the RES algorithm, a node first calculates the difference between the shortest distance estimation $h_{smallest}$ (provided by a forwarding table) to reach a node storing object x and a walker's TTL. If the difference is larger than a RES trigger threshold ε_{min}, a walker is to be dropped with a probability proportional to this difference, but not more than a maximum RES drop probability $Drop_{max}$. If a query is not dropped by RES, node i forwards this query to one of its neighbors based on the forwarding probability $\mathrm{P}\,(j, A)$ (line 6 of Table 2) as follows:

$$\mathrm{P}\,(j, A) = \frac{\frac{1}{f_{i \to j}^x}}{\sum_{l \in A} \frac{1}{f_{i \to l}^x}} \tag{5}$$

Thus, a direction where smaller TC occurs is chosen with a higher chance.

In case 2 (line 8), a node has information for some directions only. In this case, the neighbors in set B should not be ignored in a forwarding decision, because it is possible to discover a new object host and a low congested path in the *null-information* direction. To deal with this situation, we first select one

Table 2. Congestion-Aware Forwarding Algorithm

Upon node i decides to forward walkers:
1. **IF** $(A \neq \phi, B = \phi)$ **THEN**
2. $Dropped \leftarrow$ Random Early Stop Algorithm
3.   **IF** $Dropped$ is **TRUE THEN**
4.   **RETURN**
5.   **ENDIF**
6.   Forward to node $j \in A$ with probability $P(j, A)$ and **RETURN**
7. **ENDIF**
8. **IF** $(A \neq \phi, B \neq \phi)$ **THEN**
9.   Choose a node $j \in A$ with probability $P(j, A)$
10.   **IF** $f^x_{i \to j} < \kappa$ **THEN** $temp \leftarrow Uniform(0.5, 1)$
11.   **ELSE** $temp \leftarrow Uniform(0, 0.5)$
12.   **ENDIF**
13.   **IF** $temp \geq Uniform(0, 1)$ **THEN** Forward to node j
14.   **ELSE** Forward randomly to a node $j' \in B$
15.   **ENDIF**
16.   **RETURN**
17. **ENDIF**
18. **IF** $(A = \phi, B \neq \phi)$ **THEN**
19.   Forward randomly to a node $j \in B$ and **RETURN**
20. **ENDIF**

Table 3. Random Early Stop Algorithm

Select the smallest $h^x_{i \to j}$ where $j \in A$, denoted by $h_{smallest}$
1. **IF** $\delta \triangleq h_{smallest} - TTL \geq \varepsilon_{min}$ **THEN**
2.   **IF** $\min(\frac{\delta - \varepsilon_{min}}{\varepsilon_{max} - \varepsilon_{min}}, Drop_{max}) \geq Uniform(0, 1)$ **THEN**
3.   Drop this query and **RETURN TRUE**
4.   **ENDIF**
5. **ENDIF**
6. **RETURN FALSE**

node (line 9), say j, from set A based on the probability in Eq. 5. However, we can not ensure that a chosen node j is a good path in term of TC, so we need to arbitrarily introduce a probability in the selection. The algorithm considers forwarding a query to node j depending on its TC. If the TC in the direction of node j is smaller than an Alternate Selection threshold κ (line 10), a node sends a query to node j with a probability of at least 0.5 (line 10). Otherwise, a query is forwarded to node j with, at most, the probability of 0.5 (line 11). Thus, a direction of low TC is favored. If node j is not selected, a query is forwarded to a node j' randomly chosen from set B (line 14). In case 3 (line 18), a node has no information about object x and it forwards a query randomly.

When a neighbor (say node j) of node i disconnects, node i can simply remove all the entries related to j (i.e. $\forall E^*_{i \to j}$, where $*$ means wild-card) to maintain the consistency of a forwarding table. Therefore, this maintenance procedure is very simple and lightweight. Moreover, in this paper, we assume that a walker is not to be dropped if it arrives at a node that has been visited before (i.e. looping), but dropping those walkers does not affect our protocol.

2.3 Emergency Signaling

In the flash crowd situation, the amount of a query request for a particular object increases suddenly. In this situation, a query walker takes a longer time to reach an object host, and the feedback update is delayed. In this case, a node uses outdated CL data to make decision in forwarding the query. This is a big problem because a node does not know where the current congestion bottleneck is. As a consequence of this situation, both hit delay and hit rate become worsen.

To solve this problem, each walker carries a Congestion Header (CH) field. When node i forwards a walker to node j, $p_{i \to j}^x$ value is to be stored in this walker's CH field. Then, node j calculates the difference between its current CL and a walker's CH field value. If the difference, $CL_j - p_{i \to j}^x$, is larger than an ES trigger threshold ES_{tri}, an emergency signal, tagged with value CL_j, is sent back to node i. Upon node i receiving this signal, an emergency update is triggered for all entries relating to the direction of node j as follows: $p_{i \to j}^x \leftarrow \max(CL_j, p_{i \to j}^x)$ for every object x in the forwarding table. If node i has no entry for this direction, it does nothing. When a node receives emergency signaling data from its neighbors, it processes the data at the highest priority. The rationale behind the emergency update algorithm is simple: Since $p_{i \to j}^*$ is an estimated maximum CL in the direction of node j, we update this value based on two different situations. First, if a current $p_{i \to j}^*$ is larger than CL_j, it implies that a bottleneck of a route in this direction is not in node j, and we should keep the original value. Otherwise, it means that an old estimated CL value is not suitable to be used because a current maximum CL value is CL_j and a new bottleneck is now in node j.

Furthermore, we set a rate constraint on generating emergency signals. Each node i is allowed to send at most ES_i^{rate} emergency signals per unit time to each neighbor. Then the total maximum number of emergency signals generated per unit time is $ES_i^{rate} \times |Nbr(i)|$, where $|Nbr(i)|$ denotes the number of neighbors of node i.

3 Performance Evaluation

We have developed a P2P simulator by using C++ to evaluate the performance of our search protocol (CA). We compare our protocol with random walk (RW) and Adaptive Probabilistic Search (APS) [8]. We focus on the pessimistic APS for comparison. The APS parameters are the same as [8]. Due to the limited space, we only present two simulations – flash crowd situation and dynamic change of topology. For other simulations, please refer to [12]. The total number of nodes is assumed to be 1000. Each node i is assigned an exponential service rate (capacity) of C_i in unit of messages per second. The capacity distribution among the nodes is same as [11]. The lowest capacity nodes can only process one message per second. Each node issues query requests at the Poisson arrival rate q_i (requests/second). In our simulation, we assume all nodes generate queries at the same rate.

Each node randomly connects to b neighbors where $5 \leq b \leq 20$. However, to prevent lowest capacity nodes connecting to too many neighbors, they are allowed to connect to five neighbors only. If the number of a node's neighbors is lower than five, it randomly connects to some nodes such that the node maintains b neighbors. In each simulation, we use the same topology, and thus the simulation results are not biased due to different topologies.

We randomly distribute $M = 30$ different distinct objects, with a Zipf-like distribution property, into different nodes. Object 1 is the most popular with a 60% replication factor. Object 30 is the rarest that only 1% of the nodes store it. Each node is randomly looking for an object in each request. The simulation parameters are shown in Table 4. The time unit is in second.

Table 4. Simulation Parameters

α	β	ε_{min}	ε_{max}	$Drop_{max}$	κ	ES_i^{rate}	ES_{tri}	w	TTL	Expiry time
0.6	0.8	2	8	0.8	100	0.067	50	6	15	300

To evaluate the effectiveness of our protocol, we define several performance metrics as follows.

Average hit delay of object x, D_x: the average time of each walker to find object x.

Hit rate of object x, $R_x(t)$:

$$R_x(t) \triangleq \frac{No.\ of\ walkers\ hitting\ object\ x\ in\ [0,t]}{No.\ of\ query\ requests\ for\ object\ x\ in\ [0,t]} \quad (6)$$

Furthermore, we define $R(t) \triangleq \sum_{x=1}^{M} R_x(t)$ and $D \triangleq \frac{1}{M} \sum_{x=1}^{M} D_x$ to denote **total hit rate** and **total average hit delay** respectively.

3.1 Simulation 1: Flash Crowd Situation

One of the most important considerations in designing search protocols is its robustness under flash crowds. We set the input buffer of each node to be unlimited in this simulation in order to study how good our protocol is in *releasing* the *query pressure* generated in the flash crowd situation. The query rate q_i is 0.05. At time 500s, 1300s and 2300s, we randomly select 100 nodes to increase their query rates from 0.05 to 5 for emulating a flash crowd situation. Then these 100 chosen nodes simultaneously generate queries for object 26 only,

Table 5. Summary of Hit Rate and Hit Delay in Simulation 1 & Simulation 2 at $t=3000$ (D_x and D are in unit of second)

	$R_{10}(t)$	$R_{20}(t)$	$R_{30}(t)$	$R(t)$	D_{10}	D_{20}	D_{30}	D
Simulation 1								
CA	5.4	5.2	4.5	153.8	4.3	4.4	4.9	4.7
APS	4.7	4.5	4.3	139.9	167.1	183.7	205.1	172.5
RW	1.4	0.6	0.4	40.4	331.0	352.9	399.7	355.5
Simulation 2								
CA	4.2	3.5	2.6	112.7	38.7	41.6	43.7	38.9
APS	3.1	2.5	2	85.0	63.9	70.8	81.6	67.6
RW	1.2	0.6	0.3	35.7	75.4	78.2	90.5	79

which is 1.2% replicated. Each flash crowd period lasts for 20 seconds. After that, all these nodes return to the normal setting and their query rates are back to 0.05.

This simulation shows that the hit delay is kept very low while maintaining a high hit rate by using CA. Due to the page limit, we only extract some results for presentation as shown in Table 5. The percentage of walkers dropped by RES is 1.05%. In Fig. 1(a), it is shown that the *query pressure* is released very quickly, roughly within 600 seconds, and the total amount of remaining walkers is very small compared with APS and RW. However, in APS and RW, the hit delay is very large because many walkers are backlogged in the system. In Fig. 1(b), we show the role of emergency signaling under flash crowd situation. In the normal time, no emergency signal is required because the CA's query hit feedback update is enough to gather congestion information (except earlier time 50s~250s since some nodes still randomly forward queries). In the flash crowd periods, the emergency signaling takes effect. Due to this signaling, nodes can immediately sense the new congestion situation and make the decision to forward queries to less congested neighbors. Therefore, using CA, the system overloading problem can be effectively alleviated.

3.2 Simulation 2: Dynamic Change of Topology

To simulate the dynamic behavior of P2P networks, we use the setting suggested in [11]. Each node is assigned an independent *life* which is a uniformly random number between 0 and 1000. When a node's life is up, the node disconnects from its neighbors and all queued messages in it are discarded. A leaving node also removes its forwarding table. Then, this node immediately re-joins the network again by randomly choosing some neighbors to connect to, and picks a new *life* value. The query rate q_i is 0.1. A feedback message is to be dropped if a reverse path is changed. We also assume the input buffer of each node is infinite.

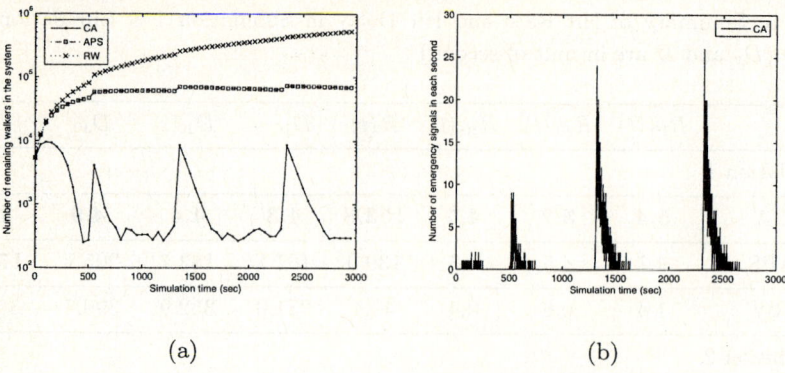

(a) (b)

Fig. 1. Flash crowd situation. (a) Number of remaining walkers in the system. (b) Number of emergency signals generated in each second

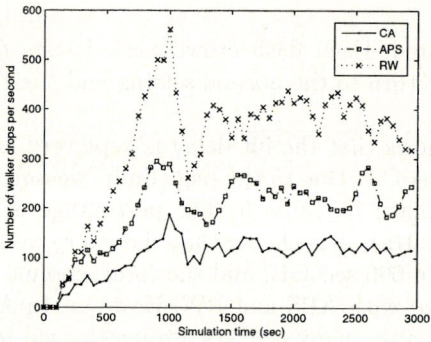

Fig. 2. Walker drops per second due to nodes leave the network (averaged over 150 seconds)

As shown in Table 5, our protocol can maintain the lowest hit delay in the dynamic environment. Moreover, CA's hit rate is the best, because CA is equipped with a load balancing function, so the queue length of each node is maintained at a low level compared with APS and RW. Therefore, the number of walker drops due to node departure can be greatly reduced, as indicated in Fig. 2. As a result, the user's experience has not suffered much.

4 Conclusion

In this paper, we proposed a congestion-aware search protocol for unstructured P2P networks. The aim of the protocol is to combine congestion control and object discovery functionality to achieve a better search performance in heterogeneous, dynamic P2P networks. By using our search protocol, no other flow control algorithm, caching and topology adaptation are required to solve the congestion problems such as query loss and system overloading. The simulation results show that it can achieve a very low hit delay while ensuring a high hit rate. Also, the number of messages lost can be kept to a minimum. Therefore, we believe that it is possible to build a larger P2P file sharing network based on our congestion-aware search protocol. We also believe that our congestion-aware idea can be applied to other fields such as agent-based resource discovery systems and sensor networks.

Acknowledgment. We thank Professor Keith Ross of Polytechnic University for many useful discussions. Also, we would like to thank anonymous reviewers for their helpful comments.

References

1. Gnutella. http://www.gnutella.com
2. KaZaA. http://www.kazaa.com/
3. BitTorrent. http://bitconjurer.org/BitTorrent/
4. S. Ratnasamy et al: A Scalable Content-addressable Network. In Proceedings of ACM SIGCOMM 2001
5. J. Ritter: Why Gnutella Can't Scale. No, Really. [Online] http://www.darkridge.com/~jpr5/doc/gnutella.html
6. Q. Lv et al.: Search and Replication in Unstructured Peer-to-Peer Networks. In Proceedings of ACM ICS 2002.
7. F. Banaei-Kashani and C. Shahabi: Criticality-based Analysis and Design of Unstructured Peer-to-Peer Networks as "Complex System". In Proceedings of GP2PC 2003.
8. D. Tsoumakos and N. Roussopoulos: Adaptive Probabilistic Search for Peer-to-Peer Networks. In Proceedings of IEEE International Conference on P2P Computing 2003
9. S. Saroui et al: Measurement Study of Peer-to-Peer File Sharing Systems. In Proceedings of Multimedia Computing and Networking 2002
10. J. Liang et al: Understanding KaZaA. [Online] http://cis.poly.edu/~ross/papers/UnderstandingKaZaA.pdf.
11. Y. Chawathe et al.: Making Gnutella-like P2P Systems Scalable. In Proceedings of ACM SIGCOMM 2003
12. Kin Wah KWONG: A Congestion-Aware Search Protocol for Unstructured Peer-to-Peer Networks. Technical Report, HKUST 2004

Honeycomb: A Peer-to-Peer Substrate for On-Demand Media Streaming Service[*]

Dafu Deng, Hai Jin, Chao Zhang, Hao Chen, and Xiaofei Liao

Cluster and Grid Computing Lab.,
Huazhong University of Science and Technology, Wuhan, 430074, China
`{dfdeng, hjin, haochen, xfliao}@hust.edu.cn`

Abstract. Peer-to-Peer media streaming service has gained tremendous momentum in recent years. However, a number of challenges in Peer-to-Peer media streaming have not been addressed. In this paper, we propose a peer-to-peer substrate for supplying on-demand media streaming service, called *Honeycomb*, which mainly addresses on the key technical issue of: constructing a locality-aware P2P overlay network with high scalability and manageability. *Honeycomb* can fully utilize the locality of underlying physical network so that the bandwidth consumption used for the overlay maintenance can be effectively saved and the QoS requirements for delivering media content can be easily satisfied.

1 Introduction

The success of P2P file-sharing and storage applications, such as Gnutella [3] and FreeNet [2], gives us a graceful reference to solve the scalability problem existed in the conventional on-demand media streaming systems with client/server model. Some recent works suggest building on-demand media streaming service on the P2P substrates. For example, GNUStream [5] proposes an application-level multicast scheme for effectively transferring video data based on fully decentralized Gnutella network. PROMISE [4] system is built on highly structured Pastry network.

However, the existed "highly structured" P2P substrates, such as Pastry [6], Chord [9], and CAN [7], are mainly designed to enhance searching performance. Those systems are poor in locality since the overlay network is constructed based on the DHT algorithm of shared objects. For on-demand media streaming services, locality of overlay network is very important because *Quality-of-Service* (QoS) requirements for media content delivery, such as latency, can be easily satisfied if nearby hosts in the underlying Internet are also nearby with each other in the overlay network.

Other fully decentralized P2P substrates, such as Gnutella, are easily to be maintained and can achieve high resilient of transient user population. However, locality issue is also not considered in those systems since they are mainly used for file-

---

[*] This paper is supported by National Hi-Tech R&D Project under grant No.2002AA1Z2102 and National Science Foundation of China under grant No.60125208 and No.60273076.

sharing systems in which clients first download the entire file before using it. There are no timing constraints on downloading the requested file object and peers often use small bandwidth to upload the requested file data. In on-demand media streaming systems, timing constraints are crucial since a packet arriving after its scheduled playback time is useless and considered lost. Furthermore, fully decentralized overlay networks make the scalability of their searching schemes, such as flooding and random walks, limited.

In papers [8] and [10], the authors proposed to construct the locality-aware topology based on landmarks, similar to our work, trying to meet the efficiency and QoS requirements. Although the efficiency can be improved, paper [8] needs extra deployment of landmarks and produces some hotspots in the underlying physical network when the overlay is heterogeneous and large. Paper [10] uses the similar method with ours. However, in [10], since that each peers needs to maintain the address information of all of classmates and neighbor classmates, as well as the searching service must use flooding scheme, the overhead for maintaining those address information list and the poor scalable flooding scheme seriously limit its scalability.

In this paper, we propose a novel peer-to-peer substrate, called *Honeycomb*, which focuses on the following aspects.

Locality: The neighbor peers in the overlay network of P2P substrate for on-demand media streaming service should be closely located in the underlying physical network. To achieve this goal, *Honeycomb* uses the concept of **closeness** to construct overlay network and clusters peers into many groups so that each group consists peers close with each other.

Scalability: The cost of the overlay construction, maintenance, and the searching scheme should be as small as possible. For this purpose, *Honeycomb* organizes the overlay network as a layered ring structure so that the maximum logical-link (or routing) information maintained on a peer is converged to a constant. The overlay construction process is implemented in a distributed fashion to reduce the number of measured hosts whenever a peer joins into the system. Furthermore, *Honeycomb* organizes the intra-group overlay as a reliable ring so that failure recovery can be done regionally with only impact on at most a constant number of peers. Since the most of peers that could not guarantee QoS are not traversed, it saves the bandwidth consumption and so improves the scalability of the P2P substrate.

The remainder of this paper is organized as follows. In section 2, we propose the mechanism for constructing and maintaining the overlay network topology of *Honeycomb*. Section 3 ends with conclusion remarks and future works.

2 Locality-Aware Overlay Network

We first introduce the concept of **closeness**. Given any two hosts i and j, the notation D used in this paper represents the maximum physical hop counts between i and j (i.e. the diameter of Internet). We uniformly divide D into H parts, denoted by a sequence $\{l_0, l_1, ..., l_{H-1}\}$, where l_i represents the distance level $[i \times [D/H](i+1) \times [D/H]-1]$ in terms of hop-counts. The closeness between i and j, denoted by $C_{i \leftrightarrow j}$, is defined as:

$$C_{i\leftrightarrow j} = \left\lfloor \frac{H \times Hopcounts_{i\leftrightarrow j}}{D} \right\rfloor \quad (1)$$

In this section, we first give out the overview of locality-aware overlay network topology based on the concept of closeness, and then propose how to construct and maintain the overlay topology.

2.1 Organization of Overlay Network

Peers in the overlay network are recursively organized as a topology of layered multi-ring hierarchy defined by the following rules (where H is the number of possible values of $C_{i\leftrightarrow j}$ for any two particular peers i and j, α is a constant). In this paper, we assume that the left direction corresponds to the anti-clockwise and the right corresponds to the clockwise on a ring.

Rule 1: Peers in the overlay network are organized in H layers. The 0-th layer contains all peers.

Rule 2: Peers at the layer k ($0 \leq k \leq H-1$) are clustered into different groups. Given a peer P, if $C_{A\leftrightarrow P} = C_{B\leftrightarrow P}$, peers A and B are clustered into the same group at the k-th layer, where $k = C_{A\leftrightarrow P} = C_{B\leftrightarrow P}$.

Rule 3: For each group at the k-th ($0 \leq k \leq H-1$) layer, a peer is selected to be the leader. This leader is automatically becomes a member of the $(k+1)$-th layer if $k<H-1$. The leader's neighbor is selected to be the vice-leader of that group. It is responsible for hotly backing up the logical link information of the leader. An exception holds for the scene that there is only one peer in a group. In this case, the survival peer is both the leader and the vice-leader.

Rule 4: Logical links among peers in the group at the k-th ($0 \leq k \leq H-1$) layer are organized as a reliable ring, called basic-ring, on which each peer maintains a hosts list, denoted by $<ID_r, I_{ld}, I_{lp}, I_{rg}, I_{ll}, I_{rr}>$, where the ID_r is the unique ID of the ring. If the peer is the leader, I_{ld} records the address of vice-leader for that group. Otherwise, I_{ld} represents the address of the leader peer. Notations I_{lp}, I_{rg}, I_{ll}, and I_{rr} represent the address of its left neighbor, right neighbor, left neighbor's left neighbor, and right neighbor's right neighbor, respectively.

Rule 5: For groups at the 0-th layer, an additional reliable ring, called inner-ring, is constructed on the basic-ring to divide it into lots of sub-rings with size in $[1, α]$ peers. For each inner-ring peer, non inner-ring peers on its left sub-ring are called children of that inner-ring peer. Each inner-ring peer is responsible for maintaining an index information list to record the information of its children peers and itself, such as the address information and keywords of shared media files.

Fig.1 (a) shows an example of the layered-groups hierarchy. In this case, we assume that $C_{i\leftrightarrow j} \in [0, 2]$. Thus, the overlay network is organized in three layers. In the 0-th layer, 20 peers are clustered into 3 groups (A, B, and C). Peers 1, 13, and 19 are leader peers of groups A, B, and C, respectively, and automatically become members of layer 1. Peers 2, 14, and 20 are vice-leader peers of these three groups, respectively. In the 1-th layer, peer 1 is clustered into group D alone since both $C_{1\leftrightarrow13}>1$ and $C_{1\leftrightarrow19}>1$. Peers 13 and 19 are clustered into the group E with the leader peer 13 and vice-leader peer 19 because $C_{13\leftrightarrow19}=1$.

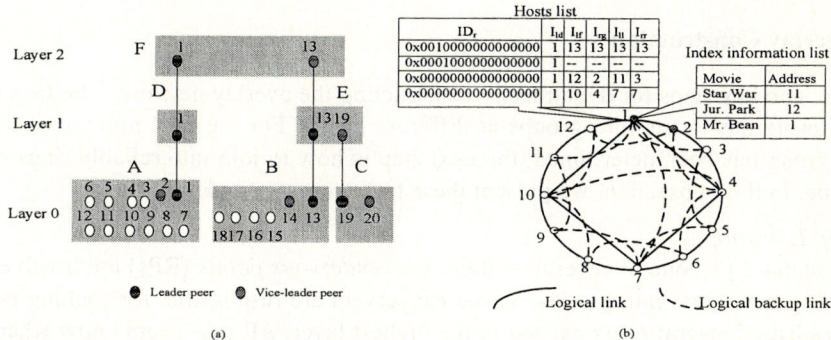

Fig. 1. An example of (a) layered-groups hierarchy and (b) the intra-group structure with $\alpha=4$ for the group A

Fig.1 (b) describes the example for intra-group topology with parameter $\alpha=4$ of the group A. In this figure, solid arcs represent logical links to the corresponding left neighbor and right neighbor, and dashed arcs illustrate logical backup links to corresponding left neighbor's left neighbor and right neighbor's right neighbor. Since the group A is located at the 0-th layer, both the basic-ring and the inner ring must be constructed. The inner-ring divides the basic ring into 4 sub-rings due to $\alpha=4$ and the total number of peers is 12. For peer 1, the left sub-ring consists of peers 1, 10, 11, and 12. Thus, it maintains the index information of media files shared by its children peer 10 and 11.

Theorem 1: No matter how the total number of peers in the system increases, the maximum logical link information maintained on a peer is a constant which is equal to $5(H+1)$ addresses.

Proof: According the definition of rule 3, peers of the group at the highest (i.e. H-1) layer maintains the maximum logical links which is related to $H+1$ reliable rings, where H is the total number of basic rings from the 0-th layer to the $(H-1)$-th layer and 1 is the number of the inner-ring in the group at the 0-th layer. Based on the definition of rule 4, there are 5 addresses must be maintained in each reliable ring. Thus, the maximum logical link information is equal to a constant $5(H+1)$ addresses.

Theorem 2: If the average number of media files shared by a peer is M, then, for each inner-peer, the average number of items in the index information list is less than $M\times(\alpha-1)+2\times(\alpha-2)$.

Proof: According to the definition of rule 5, for each inner-ring peer, the maximum number of children is α-2. Thus, the index information list must record the address information and the available bandwidth information of α-2 children. In addition, the average number of movie names is less than $M\times(\alpha-2+1)$, where $M\times(\alpha-2)$ represents keywords of movies shared by its children and $M\times 1$ is the average number of movies shared by itself. Totally, the average number of items is less than $2\times(\alpha-2)+ M\times(\alpha-2+1) = M\times(\alpha-1)+ 2\times(\alpha-2)$.

2.2 Overlay Construction

There are two key steps for dynamically constructing the overlay network. The first is how to locate new hosts into groups at different layers. For the new host, once the joining group has been determined, the next step is how to join into reliable rings of that group. In this subsection, we present these two processes in detail.

A. Group Locating

Like Gnutella-2 [3], one or several well known *rendezvous points* (RPs) are involved for locating new hosts into groups. Those RP servers are responsible for caching the addresses list of several peers existed in the highest layer. All new peers know where the RP is. Therefore, the first step for a new peer is to contact RP to randomly fetch one address of peer in the highest layer. This peer, called boot peer, will guide the new peer into groups.

The algorithm for locating a new host i into a group is presented formally in Fig.2 (a), and relies on the following notations. Let G_r be the current processing group. j and L represent the nearest peer and the layer number of group G_r, respectively. The locating algorithm starts at measuring all peers in the group at the highest layer H-1 to find out the nearest peer j. If $C_{i\leftrightarrow j} = 0$ (i.e. peer j is located at the nearest position of the host i in the physical network), the locating algorithm is finished immediately and the new host i is located into the group at the 0-th layer leaded by j. If $C_{i\leftrightarrow j} < L$ (i.e. a new nearest peer be found out in the processing group), G_r is changed to be the group at the layer $C_{i\leftrightarrow j}$ leaded by j. The procedure repeats until no new nearest peer can be found out (i.e. $C_{i\leftrightarrow j} = L$ and $L \neq 0$). In this case, a new group should be created on each layer k ($0 \leq k \leq L-1$). Meanwhile, the new host i is located into the group at the layer L leaded by j.

Fig.2 (b) illustrates an example of the group locating algorithm. In this case, the new host 21 first contacts the RP node and fetch the boot peer 13. Since F is the group on the highest layer 2, it sequentially measures the closeness value between itself and peers 1 and 13. The result shows that peer 13 is the nearest peer in F and $C_{21\leftrightarrow 13} = 1$. Then, it goes to the group E leaded by 13 and located at the layer 1. After measuring all peers in the group E, it finds that peer 19 has the same closeness value with peer 13. Thus, a new group G is created at the layer 0 and the peer 21 is located into the group E.

B. Group Joining

The group joining algorithm is presented formally in Fig.3 (a). Suppose that the located group is G_r. Based on the overlay definition, there may be two kinds of intra-group topology of the group G_r. One is the basic-ring topology for groups at the layer k ($0< k \leq H-1$). The other is the multi-rings (including a basic-ring and an inner-ring) topology for groups at the layer 0. Thus, the group joining algorithm uses two branch procedures to process these two cases.

Case 1: G_r is at the layer k $(0 < k \leq H\text{-}1)$. The group joining algorithm simply inserts the new host into the basic-ring as the leader peer's left neighbor. Note that the leader of G_r is determined by the group locating algorithm. Fig.3 (b) illustrates an example for joining into the group at the layer k $(0 < k \leq H\text{-}1)$.

Case 2: G_r is at the layer 0. In order to balance the number of children of inner-ring peers, the group joining algorithm uses a function **Find_inner_peer** to notify the leader to launch a query message at the anti-clockwise direction of the inner-ring. Once a peer at the inner-ring receives this message, it checks whether the number of children peers is less than α-2. If so, it responds a message to the leader and discards this message directly. Then the leader notifies the new host to be inserted into the basic-ring as the left neighbor of the responding peer. Otherwise, the peer forwards

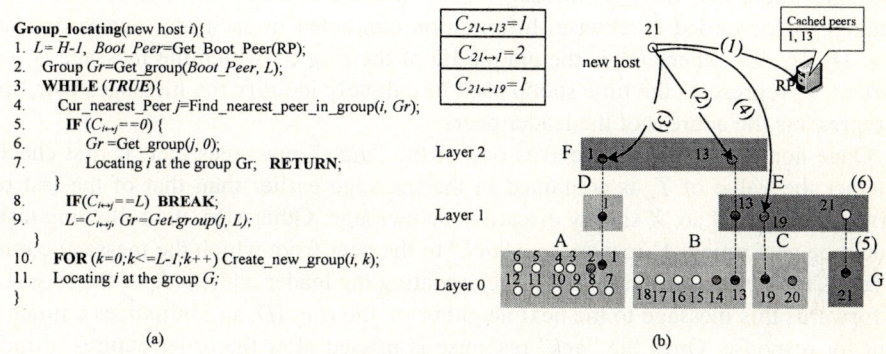

Fig. 2. (a) Pseudo-code of the algorithm to locate a new host to a group; (b) an example to locate a new host to a group

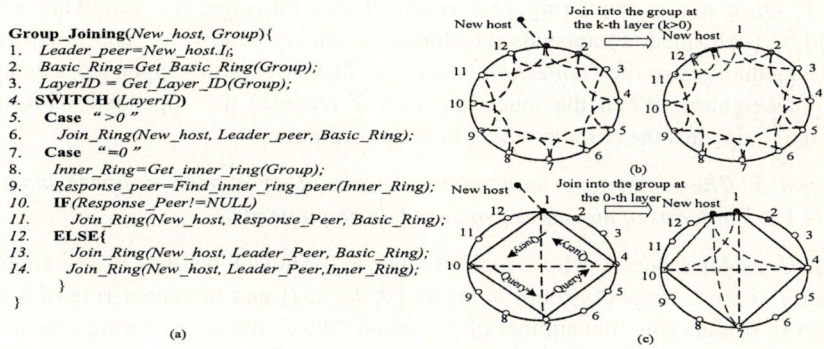

Fig. 3. (a) Pseudo-code of the group joining algorithm; (b) an example for inserting the new host into the basic-ring; and (c) an example for inserting the new host into the multi-ring

this message to its right neighbor. If all inner-ring peers have α-2 children peers (i.e. the Find_inner_peer function returns NULL), the new host is inserted into both the basic-ring and inner-ring as the left neighbor of the leader peer. Fig.3(c) shows an example for joining the multi-ring. In this figure, the logical backup links in the basic-ring are omitted just for clearness of the illustration.

2.3 General Maintenance Operations

To maintain the reliable layer ring structure (i.e. the hosts list of each peer) and the index information of inner-ring peers, two kinds of messages should be periodically launched by different peers. One is the *"alive"* message pair launched by leader peers. The other is the *"collection"* message launched by inner-ring peers.

The *"alive"* messages pair is comprised of an *"alive_l"* message and an *"alive_r"* message, where the *"alive_l"* message is forwarded anti-clockwise and the *"alive_r"* message is forwarded clockwise. Information contained in each message is denoted by <ID_r, T_m, A_l>, where ID_r is the unique ID of the ring on which the message is forwarded, T_m represents the time stamp used to uniquely identify the messages pair, and A_l represents the address of the leader peer.

Once non-leader peer X receives one of the *"alive"* messages pair, it first checks whether the value of T_m is contained in the message earlier than that of the last received message. If so, X simply discards this message. Otherwise, the following tasks should be performed: 1) responds an *"ack"* to the peer from which the message comes; 2) refers to A_l contained in the message, updating the leader address I_{ld} of the ring ID_r; 3) forwards this message to the next neighbor on the ring ID_r and initializes a timer to wait for response. Once the *"ack"* response is missed after the timer expires, it indicates that the neighbor peer fails or departs. At this moment, X forwards the received message to the backup neighbor (i.e. neighbor's neighbor) on the ring ID_r and refreshes the timer to wait for response.

The *"collection"* message is forwarded anti-clockwise along the left sub-ring of peer Y. Once a non inner-ring peer received this message, the following actions should be performed: 1) appends its address, available bandwidth, and keywords of shared media files to the *"collection"* message; 2) forwards the appended message to its right neighbor. When the inner-ring peer Y receives the appended *"collection"* message, it updates the corresponding index information.

Theorem 3: The worst-case maintenance overhead of a peer is upper bound by $2\times(H+1)+1$ in terms of messages processed during a period.

Proof: Consider a peer X whose highest layer is j. Thus, X belongs to $j+2$ rings that include $j+1$ basic-rings of groups at layers $\{0, 1, ..., j\}$ and one inner-ring of a group at layer 0. In each ring, the number of processed *"alive"* messages during a period is 2. Hence, the total number of *"alive"* messages processed by X is $2\times(j+2)$. In addition, one *"collection"* message must be processed by X. Hence, the total number of maintenance messages processed by X is $2\times(j+2)+1$. Consequently, the worst-cast maintenance overload is upper bound by $2\times(H+1)+1$ since j is upper bounded by $H-1$.

2.4 Failure Recovery

Without loss generality, we consider a peer X who fails either purposely due to a departure or accidentally due to a failure. Suppose that highest layer of X is layer j $j \in [0, H\text{-}1]$ and each item of the sequence $RS\{R_I, R_0, R_I, ..., R_j\}$ represents a ring to which X belongs, where R_I represents the inner-ring and R_i ($i \in [0, j]$) indicates the basic-ring at the layer i. As a result of the maintenance protocol, if $j=0$ and $X \in R_I$, two peers must be aware of this failure: one is a neighbor of X on the inner-ring R_I, denoted by X'_I; the other is a neighbor of X on R_0, denoted by X'_0. Otherwise, on each basic-ring R_i, a neighbor peer of X, denoted by X'_i, must be aware of this failure. Basically, the following tasks are required for recovery: 1) X should be deleted from each ring of the sequence RS; 2) peers on the ring R_k ($k \in [0, j]$) need a new leader to periodically launch maintenance messages since the leader no longer existed; 3) the basic-ring R_j needs a new vice-leader if X is the vice-leader of R_j; and 4) if X is just a inner-ring peer at the layer 0 (i.e. $j=0$ and $X \in R_I$), a new inner-ring peer should be selected to satisfy the requirement that the size of each sub-ring is in $[1, \alpha]$ peers. We propose the detail policies to remain tasks below.

We first consider the case $j=0$, that is, X belongs to only one group. If X is neither the vice-leader nor an inner-ring peer, no further work is required. If X is the vice-leader, X'_0 must be the leader since the vice-leader is the leader's right neighbor on the basic-ring. In this case, the leader X'_0 simply selects its new right neighbor on the recovered basic-ring to be the new vice-leader and backs up its hosts list to it.

If X is an inner-ring peer (i.e. $X \in R_I$), to satisfy the sub-ring size requirement, X'_I should send out a "*selection*" message including two member fields. One is a counter $C=\alpha\text{-}1$. The other is the address of X'_I. If X is X'_I's left neighbor on R_I, the message is sent to X'_I's left neighbor on R_0. Conversely, the message is sent to X'_I's right neighbor on R_0. Once a non inner-ring peer receives this "*selection*" message, it first checks whether C is equal to 1. If $C>1$, the message receiving peer updates the value of C to $C\text{-}1$ and forwards the message to its next neighbor on the basic-ring. Otherwise, it simply discards this message and joins into the inner-ring R_I.

We now consider the case $j>0$. If X is a non-vice-leader on the basic-ring R_j, X'_k ($k \in [0, j]$) must be the vice-leader of the basic-ring R_k since X is the leader of each basic-ring R_k. The following recovery actions should be performed by X'_k to work together with the task 1). It automatically a) takes place the position of X on the inner-ring R_I if $k=0$; b) becomes the new leader of R_k to launch maintenance messages; c) selects its right neighbor on R_k to be the new vice-leader and backs up its hosts list to the new vice-leader; and d) joins into the basic-ring R_{k+1} since the vice-leader of R_{k+1} will automatically become the leader and its address is backed up by X'_k. If R_{k+1} just includes one peer X, a new group should be create at the layer $k+1$ by X'_k and X'_k should join into the basic-ring R_{k+2} if $j<H\text{-}1$. If X is a vice-leader on the basic-ring R_j, a further recovery task similar with that of the case $j=0$ should be performed on the basic-ring R_j.

Fig.4 (a) and Fig.4 (b) illustrate the recovered layered-group hierarchy after peers 1 and 19 fail, respectively. The original overlay is given in Fig.1. In Fig.4 (a), peer 2 automatically becomes the leader of R_0 (i.e. the group A) and selects its right neighbor to be the vice-leader. Meanwhile, peer 2 creates a new group D' since R_I (i.e. the original group D in Fig.1 (a)) just includes one peer 1 and joins in to the basic-ring R_2 (i.e. the group F). In Fig.4 (b), peer 20 automatically becomes the leader of R_0 (i.e. the group C) and joins into the basic-ring R_1 (i.e. the group E). Since the failure peer 19 is the vice-leader of R_1 (i.e. the group E), the new joined peer 20 is selected to be the vice-leader of E. Fig.4(c) gives out the example of recovery procedure after inner-ring peer 10 fails. In this case, peer 10 is first deleted from both the inner-ring and the basic-ring of group A. Then, a *"selection"* message with initial counter value $C=3$ is initialized by the failure detecting peer 1 and forwarded anti-clockwise along the basic ring. Peer 9 receives the message with counter value $C=1$. Thus, it joins to the inner-ring as the left neighbor of peer 1.

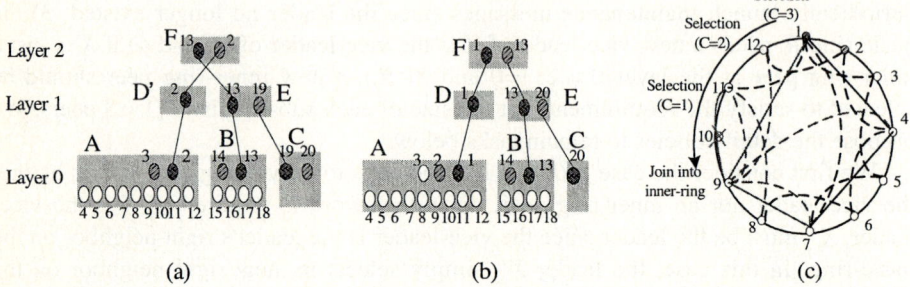

Fig. 4. The recovered layered-group hierarchy after peers (a) 1 and (b) 19 fail, as well as (c) the example recovery procedure after the inner-ring peer 10 fails. The original overlay is given in Fig.1

In overall, a peer failure requires only a few peers to be contacted for recovery. If the failure peer is just an inner-ring peer, there are at most 4 peers contacted to delete the failure peer from the inner-ring, 4 peers to delete the failure peer from the basic-ring, 2α-4 peers to select new inner-ring peer, and 4 peers to join into the inner-ring. Otherwise, there are at most 4 inner-ring peers contacted to take place the leader's position at the inner-ring, $4H$ peers to delete the failure peer from corresponding basic-rings from layer 0 to $H-1$, $4(H-1)$ peers to join into corresponding basic-rings from the layer 1 to $H-1$. Thus, we conclude the overhead of failure recovery is as follow.

Theorem 4: The number of peers needed to contact due to a failure is upper bound by $max\{8H, 2\alpha+8\}$.

Proof: The theorem has been proved above.

3 Conclusions and Future Works

In this paper, we propose a novel P2P substrate for on-demand media streaming service, called *Honeycomb*. The main contribution of *Honeycomb* is that it organizes the overlay network into layered reliable rings based on the physical location of different peers so that the underlying bandwidth consumption used for overlay maintenance can be effectively saved. Meanwhile, the locality property of *Honeycomb* makes it easily satisfy the QoS requirements for delivering the on-demand media streams. Because of space limitation, details of performance analysis are not included in this paper. Long version can be accessed in [1]. In that version, the performance of *Honeycomb* is theoretically analyzed based on modeling the underlying physical network as a random graph with power-law degree distribution. Numerical results clearly prove the significant benefits of *Honeycomb*. Our ongoing work is developing the prototype of *Honeycomb* and evaluating its performance in the real Internet environment.

References

1. D. Deng, H. Jin, C. Zhang, and H. Chen, "Honeycomb: A Peer-to-Peer Substrate for On-demand Media Streaming Service", (long version), *Technical Report*, Huazhong University of Science and Technology, June, 2004.
2. Freenet Website. [Online]. Available: http://freenet.sourceforge.net.
3. Gnutella Website. [Online]. Available: http://gnutella.wego.com.
4. M. Hefeeda, A. Habib, B. Botev, D. Xu, and B. Bhargava, "PROMISE: Peer-to-Peer Media Streaming Using CollectCast", *Proc. of ACM SIGMM'03*, Berkeley, CA, pp.45-54, Nov., 2003.
5. X. Jiang, Y. Dong, D. Xu, and B. Bhargava, "GNUSTREAM: A P2P Media Streaming system prototype", *Proc. of ICME'03*, Baltimore, MD, July, 2003.
6. A. Rowstron1 and P. Druschel, "Pastry: Scalable, distributed object location and routing for large-scale peer-to-peer systems", *Proc. of the 18th IFIP/ACM International Conference on Distributed Systems Platforms (Middleware 2001)*, Heidelberg, Germany, Nov., 2001.
7. S. Ratnasamy, P. Francis, M. Handley, R. Karp, and S. Shenker, "A scalable content-addressable network", *Proc. of ACM SIGCOMM*, San Diego, CA, Aug. 2001.
8. S. Ratnasamy, M. Handley, R. Karp, and S. Shenker, "Topologically-Aware Overlay Construction and Server Selection", *Proc. of INFOCOM*, pp.1190-1199, 2002.
9. I. Stoica, R. Morris, D. Karger, M. F. Kaashoek, and H. Balakrishnan, "Chord: A Scalable Peer-to-peer Lookup Service for Internet Applications", *Proceedings of SIGCOMM'01*, pp.149-160, San Diego, CA, Aug. 2001.
10. X. Zhang, Z. Zhang, G. Song, and W. Zhu, "A Construction of Locality-Aware Overlay Network: mOverlay and Its Performance", *IEEE Journal on selected areas in communications*, Vol.22, No.1, pp.18-28, Jan., 2004.

An Improved Distributed Algorithm for Connected Dominating Sets in Wireless Ad Hoc Networks

Hui Liu, Yi Pan[1] and Jiannong Cao[2]

[1] Department of Computer Science, Georgia State University, Atlanta, GA 30303
`hliu1@student.gsu.edu` and `pan@cs.gsu.edu`
[2] Department of Computing, Hong Kong Polytechnic University, Hong Kong
`csjcao@comp.polyu.edu.hk`

Abstract. The idea of virtual backbone routing has been proposed for efficient routing among a set of mobile nodes in wireless ad hoc networks. Virtual backbone routing can reduce communication overhead and speedup the routing process compared with many existing on-demand routing protocols for routing detection. In many studies, Minimum Connected Dominating Set (MCDS) is used to approximate virtual backbones in a unit-disk graph. However finding a MCDS is a NP-hard problem. We propose a distributed, 3-phase protocol for calculating the CDS in this paper. Our new protocol largely reduces the number of nodes in CDS compared with Wu and Li's method, while message and time complexities of our approach remain almost the same as those of Wu and Li's method. We conduct extensive simulations and show our protocol can consistently outperform Wu and Li's method. The correctness of our protocol is proved through theoretical analysis.[1]

1 Introduction

A wireless ad hoc network is a particular type of wireless networks in which an association of mobile nodes forms a temporary network, without any support of fixed infrastructure or central administration. They are widely deployed for many applications such as automated battlefield operations, wireless conferences, disaster rescues, and connection to the Internet in remote terrain, etc. Mobile nodes can control connections and disconnections by the distances between them and the willingness to collaborate during the formation of short-lived networks. That means a connection is achieved either through a single-hop radio transmission if two nodes are located within wireless transmission range of each other, or through relaying by intermediate nodes that are willing to forward packets for them.

In this paper, we assume that a wireless ad hoc network is deployed in a two-dimensional space, and each mobile node is equipped with an omni-directional antenna which has an equal maximum transmission range. Thus the topology of such a wireless ad hoc network can be modeled as a unit-disk graph (UDG). "A graph is a

---

[1] Yi Pan's research was supported in part by the National Natural Science Foundation of China (NSFC) under Grant No. 60440440451 ("two base" project).

unit graph if and only if its vertices can be put in one to one correspondence with equisized circles in a plane in such a way that two vertices are joined by an edge if and only if the corresponding circles intersect." [16]. A wireless ad hoc can be represented as a simple graph G (V, E), where V represents a set of mobile nodes and E represents a set of edges. An edge (u,v) in E indicates that nodes u and v are neighbors, and that u is within v's range of transmission, while v is within u's range.

Features of wireless ad hoc networks have posed a lot of challenges on routing protocols that are used to find a route to send a packet from a source to a destination. Mobility and lack of infrastructure cause topological changes within the network, therefore the volatility of network information is also increased. The property of limited bandwidth in wireless networks makes information collection very expensive and the power limitation factor leads to mobile nodes disconnecting frequently. Thus, an efficient and scalable routing scheme needs to be devised.

Routing protocols are classified into two main categories: topology-based and position-based. Topology-based routing protocols are based on the information concerning links [7,8,9,10,11]. In position-based routing protocols, mobile nodes know physical position information by geolocation techniques such as GPS [12,13,14]. Although a wireless ad hoc network has no fixed backbone infrastructure, many routing protocols propose the promising idea of virtual backbones such as cluster-based routing, backbone-based routing and spine-based routing [2,3,6,15]. The basic idea behind these types of algorithms is to divide a wireless ad hoc network into several small overlapping sub-networks, where each sub-network is a clique (a complete subgraph). Each sub-network has one or more virtual backbones to connect to other parts in the network. Virtual backbones are usually connected and form a dominating set of the corresponding wireless ad hoc network.

In general, a dominating set (DS) is a subset of vertices of a graph where every vertex that is not in the subset is adjacent to at least one vertex in the subset. A connected dominating set (CDS) is a dominating set that induces a connected subgraph. A virtual backbone plays a key role in routing as it simplifies the routing process to one in a smaller subgraph generated from the connected dominating set. Obviously, it is important to find a minimum connected dominating set (MCDS) of a given graph in order to reduce communication overhead, to increase the convergence speed, and to simplify the connectivity management. However, finding a MCDS is NP-complete for most graphs. Several distributed algorithms have addressed the problem of determining CDS in wireless ad hoc networks. Wu and Li [6] proposed a two-phase distributed algorithm for the construction of an approximation MCDS. The first phase is called marking process, where each node first broadcasts all IDs of its neighboring nodes to its neighbors, and after receiving two-hops information from all its neighbors it declares itself as a dominator if and only if it has two unconnected neighbors. The dominators form the initial CDS. In the second phase, the algorithm removes certain locally redundant nodes from the initial CDS. We notice that this algorithm does not mention any control messages to bridge the two consecutive stages. Further Wu's algorithm is outperformed by Das' algorithm when transmission range is very large. That means the reduction of cardinality of dominating sets is limited in dense network. The time and message complexities of this approach are $O(\Delta^2)$ and $O(\Delta n)$ respectively, where Δ is the maximum node degree and n represents the total number of vertices in graph G.

In this paper, we propose a simple and distributed heuristic algorithm for constructing the CDS based on Wu and Li's algorithm. Our protocol largely reduces the number of nodes in dominating sets, while communication and computation complexities of our algorithm remain the same polynomial complexity as those of Wu and Li's algorithm. As we know, the less the number of nodes in dominating sets, the better the performance, because our goal is to generate a small dominating set in order to speedup the routing process and decrease the communicating overhead. We implement the two algorithms and compare them. Then the simulation results show that our algorithm can consistently outperform the existing distributed algorithm of Wu and Li, and our algorithm significantly reduces the cardinality of dominating sets up to 70% compared with Wu and Li's algorithm.

The remainder of this paper is organized as follows. Section 2 gives a brief review of related work. We present our algorithm, prove the correctness of our algorithm and give performance analysis in section 3. Section 4 shows our simulation results and Section 5 concludes the paper.

2 Related Work

Various virtual backbone based routing protocols have been proposed in recent years. Distributed approximation algorithms for MCDS in mobile ad hoc networks were first developed by Das et al. [1-3]. These algorithms provided distributed implementations of the two centralized algorithms given by Guha and Khuller [4]. In Das's algorithm, a connected dominating set is found by growing a set U starting from a vertex with the maximum node degree. It then iteratively adds to U a node that is adjacent to the maximum number of nodes not yet in U until U forms a dominating set. Finally, it assigns each edge with a weight equal to the number of neighbors not in U, and then finds a minimum spanning tree T in the resulting weighted graph. All the nonleaf nodes form a CDS. This approach has two main improvements over previous protocols. Firstly, only a few nodes need to keep global information that captures the topological changes structure of the whole network, and as long as network topological changes do not affect these MCDS nodes, there is no need to recapture global information. Thus it reduces information access overhead and update overhead. Secondly, each node only needs 2-hops neighborhood information instead of information of the entire network topology. The main shortcoming of this algorithm is that the process of "constructing a spanning tree" is almost sequential, thus it needs a nonconstant number of rounds to determine a CDS. Furthermore, the algorithm suffers from high implementation complexity and message complexity.

Alzoubi et al. [5] also proposed a distributed solution with a constant approximation ratio for constructing CDS. It also consists of two phases. One phase constructs an MIS, and another constructs a dominating tree. In the first phase, a spanning tree rooted at a node v (selected through an election process) is constructed. After such construction is finished, each node is identified according to a topological sorting order of the tree. Then, nodes are marked based on their tree levels in the order starting from root v. The root v is marked as black, and other nodes are all marked as white initially. Following the order, each node is marked black if it has no black neighbor. Let U be the set of black nodes, U forms an MIS. In the second phase it constructs a tree spanning all the black nodes, and is referred to as a dominating tree. Let T be the dominating tree, where T is initially empty. The root v joins T at first. Then each black node (except v) selects a neighbor with the largest tree level but smaller than its own tree level and marks it as gray. Thus black and gray nodes form a CDS. Alzoubi et al. [5] prove that this algorithm has an approximation ratio of at most 8. Performance of this scheme is very good, however, a global infrastructure (spanning tree) is constructed before the node selection process. Also, two phases of the scheme are serialized. In addition, "locality of maintenance" is not realized in this approach, for a single change in network topology may destroy the spanning tree thus causing the dominating set to be reconstructed.

Wu and Li [6] proposed a simple and efficient distributed algorithm that can quickly find a DS in a mobile ad hoc network. Each node is marked as white initially. Let $N(v)$ be the open neighbor set of vertex v, which means $N(v)$ includes all the neighbors of vertex v. And let $N[v]$ be the closed neighbor set of vertex v, the set of all neighbors and itself. By assumption, each node has a unique ID number. This algorithm runs in two phases. In the first phase, each node broadcasts its neighbor set $N(v)$ to all its neighbors, and after collecting all adjacency information from all neighbors every node marks itself as black if there exist two unconnected neighbors. All black nodes form the initial CDS. However, considering only the first phase, there are too many nodes in dominating set. So in the second phase, the algorithm executes two extensional rules to eliminate local redundancy. Extensional rule 1 is as follows: Consider any two nodes u and v belonging to the dominating set. If $N[v] \subseteq N[u]$ and id(v) < id(u), then change v's color to white. That means if all neighbors of v and itself are covered by u, and v is connected to u and has lower id, v can be removed from the dominating set. Rule 2 is described as follows: Consider any three nodes u, v, and w belonging to a dominating set, such that u and w are two black neighbors of v. If $N(v) \subseteq N(u) \cup N(w)$ and v has the smallest id of three nodes, then v's color is changed to white. In other words, if each neighbor of v is covered by u and w together, where u and w are both connected neighbors of v, then v can be eliminated from the list of dominating nodes. Thus, the second phase removes some nodes from the original dominating set and the size of a dominating set is further reduced. However the distributed implementation lacks some type of control message to bridge the two consecutive phases. We also notice that the reduction of locally redundancy is limited and the size of dominating set is still large.

3 The Algorithm

In this section, we propose a distributed algorithm to construct CDS. This algorithm consists of three main phases: Dominators Election, Redundancy Elimination By One Neighbor and Redundancy Elimination By Two Neighbors. First, we form the initial CDS, U, which includes all the nodes once they have two unconnected neighbors. Then a node u is removed from the initial CDS, U, if there exists a neighbor of node u in U that can cover all other neighbors of u. In the third phase, we eliminate a redundant node u from U when node u has any two neighbors in U that can dominate all the neighbors of u. We will show that our algorithm is not only correct but also message and time efficient through proof.

3.1 Algorithm Description

Initially each mobile host is colored white. Dominators will be colored black and form the CDS when the algorithm terminates. We assume that each host has a unique ID, and each vertex knows its one-hop neighbors and its degree d that can be collected by periodic or event-driven hello messages. Messages are used to exchange information for computation or control. Actually messages record the information of nodes that send them. Each message contains three fields. The first field contains a unique identifier ID; the second field contains a status number, i, which is used to specify what jobs a host has finished and will do next; and the last one is a set of all one-hop neighbors represented by NEIGHBOR. The algorithm proceeds in phases. At each phase, some of the hosts are active and generate messages. Each host x first broadcasts message (x.ID, 1, x.NEIGHBOR) to the hosts at one-hop distance. In our algorithm, after a destination host y collects messages from all its neighboring hosts such as $x_1, x_2, x_3, \ldots x_d$, it does some actions according to the status number i, its own ID and IDs of all its neighbors. We set a TIMEOUT value; a neighboring node is considered passive, if y doesn't receive any message from this node after this threshold time. We also induce a new concept "Synchronization Phase" to our algorithm. Synchronization Phase lies between two continuous phases. In Synchronization phase, destination node y collects messages from its neighbors until it has received messages sent by all its neighbors who have finished jobs of previous phase and will start jobs of next phase. Notice here neighboring nodes of y can only be classified into two types. Some neighboring nodes are passive, either y has received PASSIVE messages or after the threshold time passes, y has not received anything from them, and all other nodes have the status numbers to identify of what phase jobs will begin next. The following is the details of this algorithm:

Phase 1: If y is white and i equals to 1, then it checks whether any two of its neighbors are connected. Since the information of y's two-hops neighbors can be obtained through the third field of messages, y compares the neighbor sets of its two neighbors to see whether there exist any overlaps. As we know, if such overlaps exist, then these two neighbors are connected. Otherwise, they are unconnected. Once y finds two of its neighbors are unconnected, y is colored black, declares itself as a dominator and broadcasts message (y.ID, 2.1, y.NEIGHBOR) to all y's neighbors. If y does not have unconnected neighbors, it remains white and broadcasts message (y.ID, PASSIVE). Phase 1 terminates when every host finishes this kind of judge about coloring.

Phase 2.1:
  Synchronization Phase of y, status number equals to 2.1.
  For each active neighbor x_k do:
    if (color of y is black and y.ID < x_k.ID) then
      if (x_k dominates y's neighbors and y itself) then
        y is colored white, removed from CDS and broadcasts message (y.ID,
          PASSIVE)
        break;
      end if
    end for
  y sends message (y.ID, 2.2, y. NEIGHBOR) to all y's neighbors.

Phase 2.2:
Synchronization Phase of y, status number equals to 2.2.
For each active neighbor x_k do:
  if (color of y is black and y.ID > x_k.ID) then
    if (x_k dominates y's neighbors and y itself) then
      y changes its color to white and is removed from CDS
      break;
    end if
  end if
end for
y broadcasts message (y.ID, 3.1, y. NEIGHBOR).

Phase 3.1:
Synchronization Phase of y, status number equals to 3.1.
For any two active neighbors x_k and x_j do:
  if (color of y is black and y.ID < x_k.ID and y.ID < x_j.ID) then
    if (x_k and x_j combine to dominate y's neighbors and y itself) then
      y changes its color to white and is removed from CDS
      break;
    end if
  end if
end for
y sends message (y.ID, 3.2, y. NEIGHBOR) to all y's neighbors.

Phase 3.2:
Synchronization Phase of y, status number equals to 3.2.
For any two active neighbors x_k and x_j do:
  if (color of y is black and y.ID is larger than one of x_k and x_j, but less than one of them) then
    if (x_k and x_j combine to dominate y's neighbors and y itself) then
      y changes its color to white and is removed from CDS
      break;
    end if
  end if
end for
y broadcasts message (y.ID, 3.3, y. NEIGHBOR).

Phase 3.3:
Synchronization Phase of y, status number equals to 3.3.
For any two neighbors x_k and x_j do:
  if (color of y is black and y.ID is largest one among x_k.ID, x_j.ID and y.ID) then
    if (x_k and x_j combine to dominate y's neighbors and y itself) then
      y changes its color to white and is removed from CDS
      break;
    end if
  end if
end for
y becomes passive.

When a black host finishes the jobs of phase 3.3 and becomes passive, it marks its local predicate true and propagates a token to detect termination of our algorithm as that described by Zou's algorithm [17]. Then a CDS is constructed by all the nodes remaining black, when the algorithm terminates.

3.2 Correctness and Complexity Analysis

We use a simple graph G (V, E) to represent a wireless ad hoc network, where V represents a set of mobile hosts and E denotes a set of edges. There exists an edge (u,v) in E if nodes u and v are neighbors in a wireless ad hoc network. Assume V' is the set of black nodes in V, and G' is the subgraph induced by V'. The next theorem shows that G' is a connected dominating set.

Theorem 1: If the given graph $G = (V, E)$ is not a complete graph, the graph G' induced by V', derived from our proposed approach, forms a connected dominating set.

Proof: It has been shown in 6 that nodes derived from phase 1, coloring process, form a CDS. We only need to show that whenever a node v is removed either by one neighbor or two neighbors, the remaining nodes (G'- $\{v\}$) still form a CDS. We look at the first case in which a redundant node v is removed from the dominating set by one of its neighbors. There is a requirement for removing such v that all neighbors of v and v itself must be covered by one of v' neighbors in the dominating set, and without loss of generality we assume it is node u that makes v removed. Removing v only affects the neighbor nodes of v and itself. Since u and v are neighbors and u remains in CDS, v is adjacent to CDS. Also all neighbors of v are dominated by u, so all neighbors of v are adjacent to u, a dominator. Then in the second case, if v is removed by two of its neighbors in the dominating set, u and w, u and w combine together to dominate all neighbors of node v. Obviously v is adjacent to CDS, for v is a neighbor of both u and w. While all neighbors of v are adjacent to CDS because they are covered by either u or w. And it is easy to see G'- $\{v\}$ in either case is still connected. In other word, the graph G' induced by V', derived from our proposed approach, forms a CDS.

In Wu and Li's algorithm, it mentions that the role of ID is to avoid "illegal simultaneous" removal of vertices in G'. Vertex v cannot be removed even if $N[v] \subseteq N[u]$ unless $id(v) < id(u)$. And v cannot be removed even if $N(v) \subseteq N(u) \cup N(w)$ unless v's ID is the smallest one among v, u and w. This kind of avoiding wastes a lot of chances to reduce the cardinality of CDS and is too conservative. Actually by our approach, we can remove v only if $N[v] \subseteq N[u]$ regardless of v and u's IDs. We also can remove v only if $N(v) \subseteq N(u) \cup N(w)$ without considering the order of u, v and w's IDs. At the same time, "illegal simultaneous" removal of vertices in G' is avoided by control messages in which there exist status numbers to identify the end of previous phase and beginning of next phase. In synchronization phase, a node waits for information of all neighbors. This helps avoid illegal simultaneous removal.

In phase 1, each host only broadcasts messages (ID, 1, NEIGHBOR) at most once. The message complexity is dominated by nodes' degree. Thus, the message complexity of Phase 1 is $O(\Delta n)$. Each host needs to compare the neighbor sets of its any two neighbors, and it takes $O(\Delta^2)$ time. The time complexity is $O(\Delta^2)$. In every synchronization phase, hosts only wait for receiving messages from all their neighbors, so they do not have any computation or communication jobs. In phases 2.1 and 2.2, each host checks all its black neighbors one by one, thus the time complexity is $O(\Delta)$. The message complexity is also $O(\Delta n)$ since each active host needs to broadcast message either (ID, PASSIVE) or "phase 3.1 begins". Each node checks any two of its black neighbors one pair by one pair in phases 3.1, 3.2 and 3.3, so the time complexity is $O(\Delta^2)$. The message complexity remains $O(\Delta n)$. From the above analysis we have the following theorem:

Theorem 2: The distributed algorithm has time complexity $O(\Delta^2)$ and message complexity $O(\Delta n)$.

Theorem 3: Our distributed algorithm for finding CDS is deadlock-free.

Proof: To show that our algorithm for finding CDS is deadlock-free, we need to prove that there is no mutual waiting among mobile hosts, because mutual waiting is necessary condition of the deadlock situation in message communication. Mutual waiting occurs in message communication when each of a group of hosts is waiting for a message from another member of the group, but there is no message in transit. By the definition of synchronization phase, it is easy to see that there is no mutual waiting in the network, since we set timeout value, if one host has some problem to finish jobs or send messages, other hosts who wish to exchange messages with it will regard that host as PASSIVE without infinite waiting. Thus, no mutual waiting can be created.

4 Simulation and Results

We conduct a simulation study to measure the size of the CDS derived from our algorithm and compare it with the one generated by Wu and Li's algorithm. We have simulated three algorithms: Wu and Li's algorithm without applying extensional rules, their algorithm with extensional rules, and our proposed algorithm.

In our simulation, random graphs are generated in a 600 × 600 square units of a 2-D simulation area, by randomly including a certain number of mobile nodes. We assume that each mobile node has the same transmission range r, thus the generated graph is undirected. If the distance between any two nodes is less than radius r, then there is a connection link between the two nodes. Basic, Wu's and New are three parameters used to represent the number of dominators calculated by Wu and Li's basic rule, their algorithm with the two extensional rules, and our new algorithm respectively.

Actually two groups of simulation are performed. In the first group, we set the transmission radius of mobile nodes r to 100, 125, 150, 175, 200, 225, and 250 units. In this way, we can control the density of the generated graphs because the density increases when r increases. For each transmission range r, the number of nodes n is varied from 80 to 150. For each n, the number of running times is 500 times. For each case, we calculate Basic, Wu's and New, and then record the average number ± standard deviation, minimum and maximum number of the different algorithm results in tables. We compare the results in terms of the number of dominators generated. It is well known that the lower the number of dominators, the better the result. Thus, our goal is to generate a small connected dominating set to facilitate a fast routing process and reduce access and update overhead.

Fig.1. shows the average number of dominating set nodes versus the number of nodes in the network for the increasing order of transmission radius r. We can see the performance of the basic rule without extensional rules is very poor and the ratio of nodes in the dominating set to all nodes in the network is almost 1, specifically, 0.86. Moreover, when the transmission range is very large, almost every node belongs to a dominating set. Because the basic rule is loose, when the density of generated graphs increases as the transmission range increases, it is very easy to find two neighbors of a node that are not connected. After applying the two extensional rules of Wu and Li's algorithm, the size of the dominating set is largely reduced. The ratio of the number of dominating nodes over the total number of nodes in the network changes from 60% to 22% as the transmission range increases.

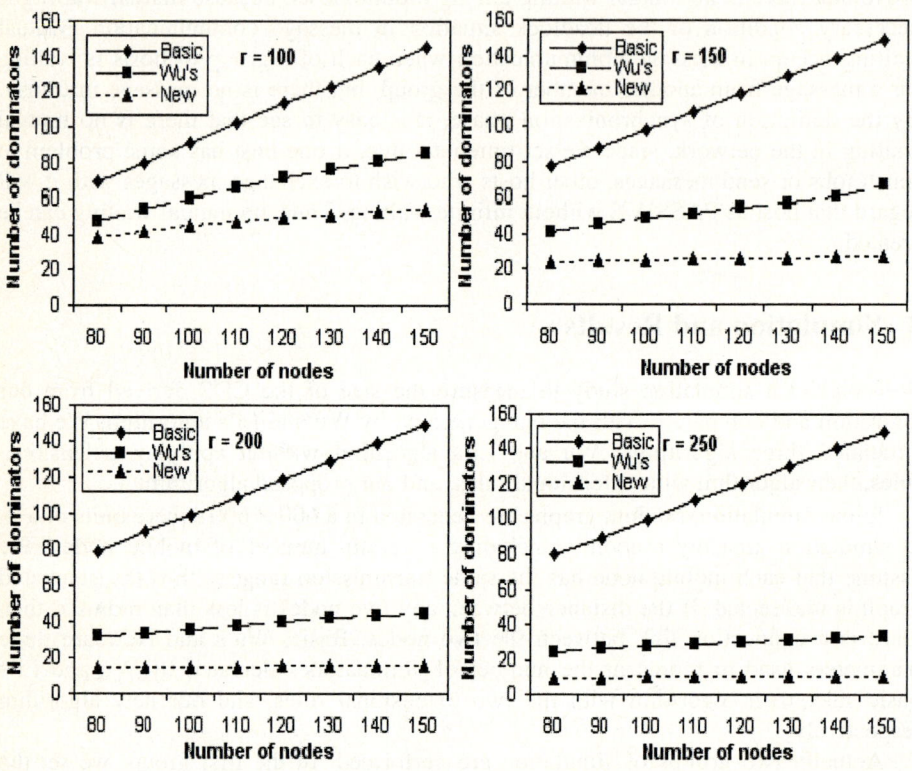

Fig. 1. Average number of dominating set nodes relative to the number of nodes n

The simulation also shows that our algorithm consistently outperforms Wu and Li's algorithm with two extensional rules. The performance of our proposed algorithm is much better than the one derived by Wu and Li's. We can see the gap between New and Wu's increases as r increases. The ratio of dominating set size induced by our algorithm over the number of nodes changes from 41% to 8% as the transmission range increases. The number of dominators produced by our algorithm is only half or one third of that derived by Wu's algorithm.

In order to fully understand this relative performance between Wu and Li's algorithm and ours, we conduct a second group of simulations. Fig.2. shows the number of dominating set nodes with respect to radius r for the increasing order of the number of nodes n. The number of dominating set nodes decreases smoothly as the transmission range increases by applying their algorithm and our new algorithm. Firstly, when the radius of mobile node's transmission range is increasing, the gap between Wu's and New becomes larger, and then when the radius of mobile node's transmission range increases more, such gap decreases a little. We find that while the network is dense, e.g. there are more mobile nodes in this ad hoc network or the transmission range of every node is very large, the reduction on the number of dominators of our approach is very large compared with that of Wu's. Since there are more chances for a node to find any one or two neighboring dominators that cover its neighbors when ad hoc networks become denser.

In a summary, our new approach consistently obtains much better performance than Wu and Li's algorithm. The cost of our approach is slightly more than Wu and Li's algorithm since more phases are induced.

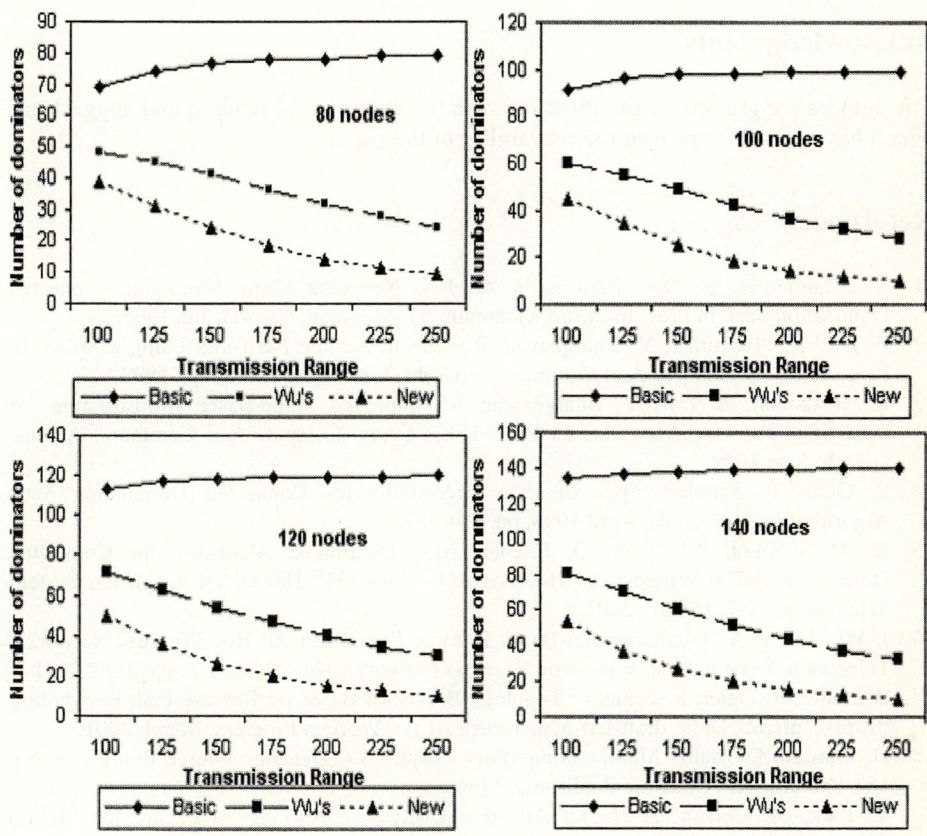

Fig. 2. Average number of dominating set nodes relative to transmission range r

5 Conclusion

In this paper, we propose a distributed algorithm for determining the connected dominating set (CDS) improved on Wu and Li's algorithm. Our approach calculates the CDS in $O(\Delta^2)$ time with 2-hops neighborhood information, where Δ is the maximum node degree of the graph. In addition, the algorithm also uses constant rounds of message exchanges, and the amount of exchanged messages is $O(\Delta n)$, where n represents the total number of vertices in graph G. We can see the time and message complexities of our approach remain the same asymptotic growth rate with Wu and Li's algorithm. While our approach significantly reduces the cardinality of CDS compared theirs. The dominator nodes selected by our new algorithm form a CDS, then a reduced graph can be generated from the CDS and the searching space for a routing process can be reduced to this graph. The effectiveness of our algorithm is confirmed through a simulation study on both sparse and dense networks. Our simulation results show that our algorithm outperforms the approach proposed by Wu and Li.

Acknowledgments

The authors are grateful to the three referees for their careful reading and suggestions which have greatly improved the readability of the paper.

References

1. V. Bharghavan, B. Das: Routing in Ad Hoc Networks Using Minimum Connected Domination Sets. In Proc. Int. Conf. Commun.'97, Montreal, Canada. Jun 1997.
2. B. Das, R. Sivakumar, V. Bharghavan: Routing in Ad-hoc Networks Using a Spine. In Proc. Int. Conf. Comput. And Commun. Networks, Las Vegas, NV., Sept. 1997.
3. R. Sivakumar, B. Das, V. Bharghavan: An Improved Spine-Based Infrastructure for Routing in Ad Hoc Networks. In Proc. IEEE Symp. Comput. And Commun., Athens, Greece, June 1998.
4. S. Guha, S. Khuller: Approximation Algorithms for Connected Dominating Sets. Algorithmica, Vol. 20(4), April 1998, pp. 374-387.
5. K. M. Alzoubi, P.J. Wan, O. Frieder: New Distributed Algorithm for Connected Dominating Set in Wireless Ad Hoc Networks. Proc. 35th Hawaii Int. Conf. On System Sciences, pp. 1-7, January 2002.
6. J. Wu, H. Li: A Dominating-Set-Based Routing Scheme in Ad Hoc Wireless Networks. Telecomm. System, Special Issue on Wireless Networks, vol. 18, no. 1-3, pp13-36, 2001
7. B. Bellur, R. Ogier, F. Templin: Topology Broadcast Based on Reverse-Path Forwarding (tbrpf). Internet Draft, draft-ietf-manet-tbrpf-01.txt, Work in Progress, March 2001.
8. D. Johnson, D. Maltz: Mobile Computing, Chapter 5 – Dynamic Source Routing, pages 153-181. Kluwer Academic Publishers, 1996.
9. V. Park, M. Corson: A Highly Adaptive Distributed Routing Algorithm for Mobile Wireless Networks. In Proc. Of INFOCOM'97, 1997.

10. C. Perkins, E. Royer: Ad-hoc On-Demand Distance Vector Routing. In Proc. Of the 2nd IEEE workshop on Mobile Computing System and Applications, pages 90-100, February 1999.
11. S. Basagni, I. Chlamatac, V. Syrotiuk, B. Woodward: A Distance Routing Effect Algorithm for Mobility (dream). In Proc. Of the 4th Annual ACM/IEEE Int. Conf. On Mobile Computing and Networking (MOBICOM)'98, pages 76-84, Dallas, TX, USA, 1998.
12. J. Li, J. Jannotti, D. S. J. De Couto, D. R. Karger, R. Morris: A Scalable Location Service for Geographic Ad Hoc Routing. In Proc. Of the 6th Annual ACM/IEEE Int. Conf. On Mobile Computing and Networking (MOBICOM) 2000, pages 120-130, Boston, MA, USA, 2000.
13. B. Karp, H. T. Kung: Greedy Perimeter Stateless Routing for Wireless Networks. In Proc. Of the 6th Annual ACM/IEEE Int. Conf. On Mobile Computing and Networking (MOBICOM) 2000, pages 243-254, Boston, MA, USA, 2000.
14. Y. B. Ko, N. H. Vaidya: Location-Aided Routing (LAR) in Mobile Ad Hoc Networks. ACM/Baltzer Wireless Networks (WINET) journal, 6(4):307-321, 2000.
15. U. C. Kozat, G. Kondylis, B. Ryu, M. K. Marina: Virtual Dynamic Backbone for Mobile Ad Hoc Networks. In IEEE International Conference on Communications (ICC), (Helsinki, Finland), June 2001.
16. M. V. Marathe, H. Breu, H. B. Hunt III, S. S. Ravi, D. J. Rosenkrantz: Simple Heuristics for Unit Disk Graphs. Networks, Vol. 25, 1995, pp. 59–68.
17. Hengming Zou: An Algorithm for Detecting Termination of Distributed Computation in Arbitrary Network Topologies within Linear Time. International Symposium on Parallel Architectures, Algorithms and Networks (ISPAN '96), 1996, pp. 168-172.

A New Distributed Approximation Algorithm for Constructing Minimum Connected Dominating Set in Wireless Ad Hoc Networks

Bo Gao[1], Huiye Ma[2], and Yuhang Yang[1]

[1] Dept. of Electronic Engineering, Shanghai Jiao Tong University, Shanghai, China
gaobo@sjtu.edu.cn
[2] Dept. of Computer Science and Engineering, Chinese University of Hong Kong
hyma@cse.cuhk.edu.hk

Abstract. In this paper, we present a new distributed approximation algorithm that constructs a minimum connected dominating set (MCDS) for wireless ad hoc networks based on a maximal independent set (MIS). Our algorithm, which is fully localized, has a constant approximation ratio, and $O(n)$ time and $O(n)$ message complexity. In this algorithm each node only requires the knowledge of its one-hop neighbors and there is only one shortest path connecting two dominators that are at most three hops away. Compared with other MCDS approximation algorithms, our algorithm shows better efficiency and performance than them.

1 Introduction

In recent years, some researchers have proposed to construct a virtual backbone by nodes in a connected dominating set (CDS) [1, 2, 3] to improve the performance in the ad hoc wireless networks. However, finding the minimum connected dominating set (MCDS) is a well-known NP-hard problem in graph theory [4]. Approximation algorithms for MCDS have been proposed in the literature. Most of these algorithms suffer from poor approximation ratio, high time complexity and message complexity.

In this paper, we present a new distributed approximation algorithm that constructs a minimum connected dominating set (MCDS) for wireless ad hoc networks. Our algorithm is based on a maximal independent set (MIS) with a constant approximation ratio, and linear time and linear message complexity. In our algorithm, there is a unique shortest path selected to connect a pair of dominators whose distance is within three hops. Our algorithm is fully localized. Each node only requires the knowledge of its single-hop neighbors.

The rest of the paper is organized as follows. In Section 2, we provide preliminaries necessary for describing our new algorithm. Section 3 presents our new distributed formation algorithm. Section 4 presents the performance results of several experiments. At last, we conclude our paper in Section 5.

2 Some Definitions and Notations

In this section, we give some definitions and notations that will be used in our paper later.

Definition 1. *A subset S of V is a Dominating Set (DS) if each node u in V is either in S or is adjacent to some node v in S. Nodes from S are called dominators, while nodes not from S are called dominatees.*

Definition 2. *A subset C of V is a Connected Dominating Set (CDS) if C is a Dominating Set and C induces a connected subgraph.*

In the CDS, the nodes in C can communicate with any other node in the same set without using nodes in $V - C$. A Dominating Set with the minimum number of nodes is called a Minimum Dominating Set, denoted by MDS. A Connected Dominating Set with minimum number is denoted by MCDS.

Definition 3. *A subset V' of vertices V in a graph G is an Independent Set (IS) if, for any pair of vertices in V', there is no edge between them.*

A MIS is a maximum cardinality subset V' of V so that there is no edge between any two vertices in V'.

3 Our New Distributed Approximation Algorithm for Constructing MCDS

Our distributed algorithm to construct approximation MCDS can be briefly described as two phases. The first phase is the construction of the MIS, where MIS nodes are referred to as dominators. In the second phase a unique shortest path is created between each pair of dominators within at most three hops distance from each other. The nodes in these shortest paths are called connectors. All the dominators and connectors form the MCDS. These two phases are described in following subsections.

3.1 Creating a MIS

In this phase, we design our method of creating a MIS inspired by the relevant work of Alzoubin [3]. Each node is in one of four states: candidate, dominator, dominatee and connector. Each node is initialized as candidate state and subsequently enters either the dominatee state or the dominator state. The connector state can only be entered from the dominatee state. There is a local variable nLower in each node. It stores the number of the current candidate neighbors with lower IDs, and is initially equal to the total number of neighbors with lower IDs.

A candidate node with nLower = 0 changes its own state to dominator (black), and then broadcasts a DOMINATOR message.

Upon receiving a DOMINATOR message, a candidate node changes its own state to dominatee (gray), and then broadcasts a DOMINATEE message.

Upon receiving a DOMINATEE message, a candidate node decreases nLower by one if the sender has a lower ID. If nLower is equal to 0 after the updating, it changes its own state to dominator, and then broadcasts a DOMINATOR message.

3.2 Creating a MCDS

In this phase, each dominator generates a REQUEST_DOMI message to find all other dominators within three hops. This message is broadcasted at most three hops before it arrives at a dominator. When a dominatee receives this message, it appends its ID into the node list included in the REQUEST_DOMI message and then broadcasts this message. In this way, when a REQUEST_DOMI message arrives at a dominator, it has already recorded the IDs of all nodes in its node list which form the path from the dominator originating this message to the dominator receiving this message. When a dominator receives a REQUEST_DOMI message for the first time from another dominator, it generates a REPLY_DOMI message including the path that this message should visit and sends this message. This path is the reverse order of the one in the REQUEST_DOMI message that it has received before. When a dominatee whose ID is included in the path of the REPLY_DOMI message receives this REPLY_DOMI message, it changes its state to connector and sends this message to the next-hop node according to

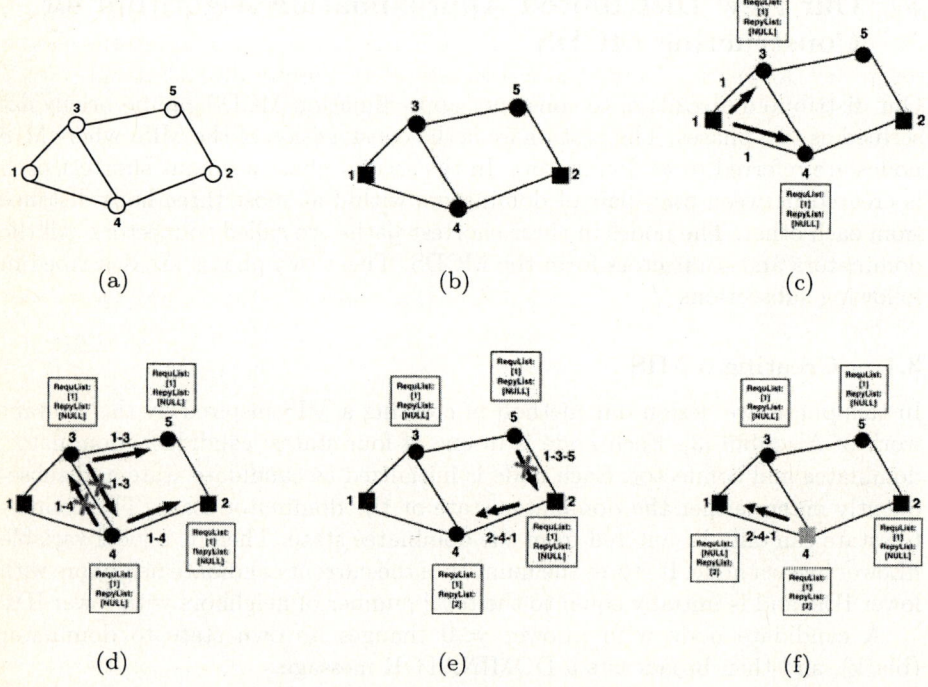

Fig. 1. Example: MCDS Constructions

the path in this message. A possible execution scenario is shown in Fig. 1(a)-Fig. 1(f). In Fig. 1, white circles represent candidates, black circles represent dominatees, black boxes represent dominators and gray boxes represent connectors.

Theorem 1. *Our distributed algorithm for constructing a MCDS has a constant approximation factor of minimum CDS in G.*

Proof. (omitted)

Theorem 2. *Our distributed algorithm for constructing a MCDS has $O(n)$ time complexity and $O(n)$ message complexity.*

Proof. (omitted)

4 Performance Evaluation

We evaluate the performance of our MCDS algorithm through simulation. We implement Wu's MCDS algorithm [2] and Alzoubin's MCDS algorithm [3], and our MCDS algorithm respectively.

Random graphs are generated in a 1000 × 1000 square units of a 2-D simulation area, by randomly throwing a certain number of nodes. For each node, its positions in the area is assumed as (x, y). The value of x coordinate is a random number distributed uniformly between 0 and 1000, and that of y coordinate distributed uniformly between 0 and 1000 too. There is a link between two nodes only if their geometric distance is less than the wireless transmission range. If the generated graph is disconnected, we simply discard it and regenerate a graph with the same parameters. In Fig. 2(a), an original topography is shown, in which the radius is 180, and connectivity is very dense in some parts of the graph, which increases the overhead of network control functions. Figure. 2(b) shows the corresponding MCDS graph of Fig. 2(a) constructed by our MCDS algorithm.

Figure. 3 shows the performance of these three algorithms in terms of the normalized MCDS size. In Fig .3(a), the number of the nodes is fixed to 300 and the transmission range of the nodes changes from 100 to 1000. In Fig .3(b), the

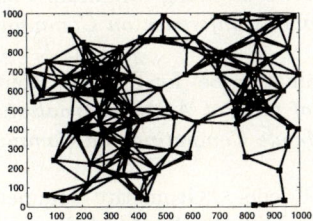
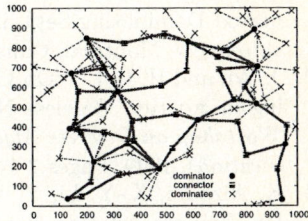

(a) A unit disk graph.   (b) Our MCDS

Fig. 2. Connected dominating set of the unit disk graph

transmission of the nodes is fixed to 300 and the number of the nodes changes from 100 to 1000. From Fig .3(b) we can see that the normalized MCDS size of our algorithm decreases with the increasing of the number of the hosts. It means that our algorithms is more efficient in the dense networks. However, this does not happen in Wu's algorithm. As shown in Fig. 3, the performance of our algorithm is obviously better than other two algorithms.

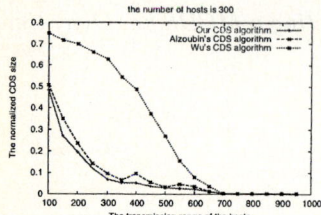

(a) The number of the nodes is 300

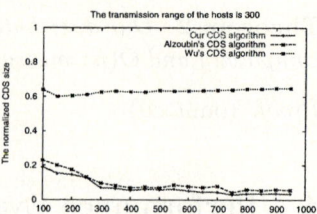

(b) The transmission range is 300

Fig. 3. The normalized MCDS size

5 Conclusions

In this paper, we have proposed a new distributed approximation algorithm based on a maximal independent set (MIS) for constructing a MCDS with a constant approximation ratio in $O(n)$ time and $O(n)$ message complexity. The algorithm is fully localized, and it does not rely on the spanning tree construction, which makes it practical for large network with dense topology. Both theory analysis and simulation show that our algorithm performs better than other two algorithms in terms of normalized MCDS size, especially, when the underlying network is large and the topology is dense.

References

1. V. Bharghavan and B. Das, "Routing in Ad Hoc Networks Using Minimum Connected Dominating Set", in *Proceedings of International Conference on Communications'97*, Montreal, Canada. June 1997.
2. J. Wu and H. L. Li, "On Calculating Connected Dominating Set for Efficient Routing in Ad Hoc Wireless Network", in *Proceedings of the 3rd ACM International Workshop on Discrete Algorithms and Methods for Mobile Computing and Communications*, 1999, Pages 7-14.
3. K. Alzoubi, X.-Y. Li, Y. Wang, P.-J. Wan, and O. Frieder, "Geometirc Spanners for Wireless Ad Hoc Network", *IEEE TRANSACTIONS ON PARALLEL AND DISTRIBUTED SYSTEMS*, VOL.14, NO.4, APRIL 2003, Pages 408-421.
4. M. R. Garey and D. S. Johnson. "Computers and Intractability. A guide to the theory of NP-completeness", Freeman, Oxford, UK, 1979.

An Adaptive Routing Strategy Based on Dynamic Cache in Mobile Ad Hoc Networks*

YueQuan Chen, XiaoFeng Guo, QingKai Zeng, and Guihai Chen

State Key Laboratory for Novel Software Technology,
Department of Computer Science and Technology, Nanjing University,
210093 Nanjing, China
chen_yuequan@yahoo.com.cn

Abstract. The dynamic changes of the topology caused by the movement of nodes makes routing become one of the key problems in the mobile Ad Hoc Networks (MANET). So how to optimize routing becomes a hot and difficult topic, among which optimizing routing cache is one of the key techniques. In this paper, we propose an adaptive dynamic cache routing (DCR) strategy based on DSR (Dynamic Source Routing), which can evaluate the link expiration time rapidly. The experimental results show that the DCR has considerable improvement in control packets, packet delivery ratio, packet drops and end-to-end average delay in the MANET.

1 Introduction

Mobile Ad Hoc Networks (MANET) are self-organized wireless networks which are multi-hops and without infrastructure [1]. Due to the absence of infrastructure, nodes can move frequently, and the topology can change dynamically. Thus, research on routing becomes a difficult and hot topic [2]. Normally, there are two main routing strategies: proactive routing and reactive routing. Proactive routing is implemented by exchanging routing tables, such as DSDV [3], WRP [4], etc. Reactive routing is on-demand routing, such as DSR [5], AODV [6], TORA [7], etc. It has been shown that reactive routing is more suitable for MANET than the proactive one [8].

In reactive routing, as the on-demand routing is based on request/reply recycle, the routing discovery cost is large, which will decrease the performance of network. If many nodes send requests at the same time, networks will congest easily. In the MANET, researchers use three main methods to decrease discovery cost: 1) Optimizing cache (such as DSR, AODV, etc). Every node has a cache to store the path from itself to destination. When it receives the route request and has a path to the specified destination, it will reply the corresponding path to source node from its cache, and if one link breaks, it can switch to an alternative path so that it can decrease the route

---

* Supported by NSF of China (No.60473053), Hi-Tech Program of china (2002AA141090), National Grand Fundamental Research 973 Program of China (No.2002CB312002) and TRAPOYT Award of China Ministry of Education.

request cost and error paths, and consequently reduces the end-to-end delay. 2) Local flooding (such as LAR [9], ZRP [10], etc). The flooding broadcast locally will reduce the discovery cost. 3) Multipath (such as SMR [11], AOMDV [12], etc]. Using multiple paths to send data parallel or concurrently and alternative path will reduce the number of requests.

Cache routing strategy, such as DSR and AODV, can reduce the discovery cost. But there are some shortcomings, for example it hasn't an efficient cache strategy and efficient automatic link expiration mechanism, so there are many researches on it. Hu. et al [13] proposed an on-demand routing protocol on cache strategy in MANET, which limited their study on expiration mechanism to a fixed level of node mobility, while a static optimal lifetime is not suitable for high mobility. Liang [14] proposed a best static optimal link expiration time based on numeric method, but it is not suitable for high mobility either. Valera et al [15] proposed a cooperative cache strategy, but it doesn't consider link expiration. Cao et al discussed how to improve performance of network using cache in application layer [16], but it is unsuitable for cache in network layer.

In this paper, we propose an adaptive dynamic cache routing strategy (DCR) based on DSR. According to DCR, the network performance can be improved by using the link-based cache organization, source or intermediate nodes caching efficient paths, and by evaluating link life-time and setting the link timeout automatically to reduce error packets and decrease end-to-end delay. Compared with DSR, DCR can reduce control packets by 10%-50%, improve packet delivery ratio by 10%-20%, decrease 20%-60% packet drops and end-to-end delay by 50%-70%.

This paper is organized as follows. Section II introduces the DSR protocol and cache problem. Section III describes the DCR link organization, cache strategy and the link expiration mechanism. Section IV analyzes the DCR Protocol. Performance evaluation by simulation is presented in Section V and conclusions and future work are given in Section VI.

2 Introduction to DSR and Cache Problem

DSR is an on-demand reactive routing protocol which is based on request/reply method. Because DSR is a classical protocol which has better performance in reactive routing [8], we take it as a reference protocol to propose our strategies and mechanisms (DCR) which are also suitable for other reactive routing protocol.

2.1 Introduction to DSR

DSR routing protocol has two main phases: routing discovery and routing maintenance.

- Routing Discovery
  Firstly, the source node broadcasts flooding route request to destination. After an intermediate node receives the request, it will check its cache to see whether it has paths to the destination. If it has a path to the destination, it will reply the corre-

sponding path to the source node; otherwise it will put its address into route request packet header, and broadcast the route request again. After the route request reaches the destination, the destination extracts the efficient information from route request packet and reply to source node through the route request path. Then intermediate node receives the route reply packet and puts the efficient routes into its cache. Afterwards, it forwards this route reply packet to upper node. When the route reply packet reaches the source node, the source node collects these paths and put them into its cache for the purpose of sending data later.

– Routing maintenance

If a node detects some broken links through MAC layer, it will judge whether its salvage bit is set or not. When the bit is not set, it will search another path leading to the destination and forward the data packet through it. Otherwise, it will drop this data packet and informs the upper node and source node to process the broken link.

2.2 DSR Cache Problem

As DSR cache is based on path organization, the intermediate node or source node will not process anything while caching the route reply packet. Meanwhile its cache has no automatic link expiration either and will delete the error link when receiving the error packets. So DSR cache is easy, but it has some problems as follows:

- Inefficient Cache Organization: In DSR, it will delete the whole path which includes error links when receiving the error packets even though there may be only one link broken in this path and other links are also existed, which will result in that the path organization can't use the link information efficiently.
- Easy Cache strategy: if intermediate nodes or source node receive route reply packets, it will cache this path without any further process. But in MANET, the longer the path is, the larger the broken probability of this path will be.
- No automatic link expiration mechanism: In DSR, it take the link for existence until receiving error link information. But the links are broken and connected dynamically in MANET. So if there is no automatic link timeout, it will increase the end-to-end delay and the number of error packets.

Due to the above mentioned reasons, we propose a Dynamic Cache Routing strategy which is based on link organization, selective cache strategy and automatic link expiration mechanism.

3 DCR Protocol Description

DCR, based on DSR, differs from DSR mainly in three aspects from DSR: link organization, cache strategy and auto link expire mechanism.

3.1 DCR Link Organization

In DSR, its cache organization is based on path (Fig. 1). This organization is easy to manage, and its routes can be directly selected when needed. The key problem is the low efficiency. In this paper, we add some features, such as the link counter and time counter, into the link organization described in [13]. The link counter indexes the numbers of one link in its cache. If a new link is added, we will initiate its link counter to one, and increase its link counter by one if the same link is added again and decrease its link counter by one when link expiration time is triggered. When receiving some link error packet or its link counter decreasing to zero, we will delete the link from the cache. The time counter indexes the link expiration which will be described in section 3.3. After these processes, we can use the link efficiently.

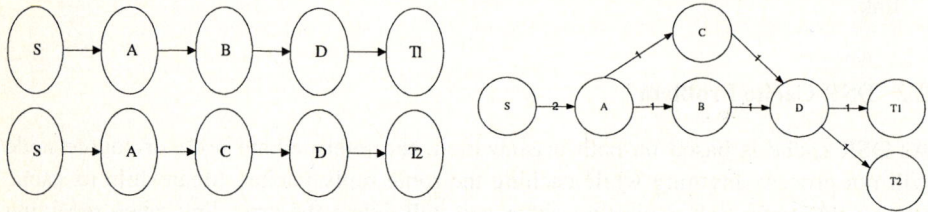

Fig. 1. Path-based Cache Organization   **Fig. 2.** Link-based Cache Organization

For example, in the path-based organization (Fig. 1), if link B→D is broken, the cache has no path to the destination T1 which results in route rediscovery and the cost increases. But in the link-based organization (Fig. 2), we can use BSF [17] (Bread First Search) to compute the shortest path from the source node to destination T1 in network topology. Thus we can achieve the path S→A→C→D→T1 and reduce the number of route discovery, decrease control packets and improve cache link efficiency greatly.

3.2 DCR Path Cache Strategy

The source node and intermediate node cache strategy have been modified as follows. When intermediate node receives route reply packet, the path with the distance from this node to destination node less than Δ_1 will be chosen to be cached based on local principle. In this way the probability for the error path can be decreased by avoiding too long path. When the intermediate node receives the route request packets, it will do the following process: if the minimum hops from the source to the intermediate node plus the minimum hops from the intermediate node to the destination already available are less than Δ_2, then it will reply the corresponding path to source node. When the source node receives route reply packets, if this route reply packet to destination is the first one, then cache it directly, otherwise we cache it if the route reply hops are less than Δ_3. The $\Delta 1_1$, Δ_2, Δ_3 is constant number.[1]

3.3 Adaptive Link Expiration Mechanism

For two nodes at one link, if one node leaves the covering region of the other node, the link will be broken. Due to the nodes' mobility, the link is broken and connected dynamically. But DSR does not consider these problems, and will not delete the broken link until it receives the error packet. This results in increasing the error packets and the end-to-end delay. Setting the link expiration time is very important: when the topology changes greatly (the pause time is shorter), if the link time is set too long, the error packets increase and the data are retransmitted, and, consequently, the delay increases; when the topology changes little (the pause time is longer), if the link lifetime is set too little, we can't use the link information efficiently, so we add an adaptive link expiration mechanism. In this paper, we propose the following method to evaluate the link time:

- When the topology changes greatly, the link is broken and connected dynamically. Suppose S_n = Link.expire$_n$ - Link.start$_n$, where the Link.start$_n$ is the starting time of link stored in the cache and the Link.expire$_n$ is the time of link whose link counter becomes zero or that receives the error packet which contains the error link information; LinkTime$_n$ is set to evaluate the link lifetime, Diff$_n$ is set to the difference of S_n and LinkTime$_n$, V_n is set to variance, TimeOut$_n$ is set to the timeout of link, α, β, γ is constant number between 0 and 1.[1]

$$\text{Diff}_n = S_n - \text{LinkTime}_{n-1}. \tag{1}$$

$$\text{LinkTime}_n = \alpha * \text{LinkTime}_{n-1} + (1-\alpha) * S_n \tag{2}$$

$$V_n = \beta * V_{n-1} + (1-\beta) * |\text{Diff}_n| \tag{3}$$

$$\text{TimeOut}_n = \begin{cases} \gamma * (\text{LinkTime}_n + V_n) & \text{if (Diff} \geq 0) \\ \text{LinkTime}_n - V_n & \text{if (Diff} < 0) \end{cases} \tag{4}$$

- When the topology changes little, that is when the pause time is larger than if we set the link expire time to related pause time, we can use the link efficiently and decrease the error packets.

$$\text{TimeOut} = \text{PauseTime} + \delta \tag{5}$$

The PauseTime is pause time of node in random WayPoint model [5], δ and δ is a constant.[1]

---

[1] Practical number will be discussed in section 5.

4 Analysis of DCR Protocol

Lemmas 1. *When the topology changes greatly, the formula (4) can converge to link time expiration rapidly.*

Proof. As the node mobility is random WayPoint model, LinkTime can be represented as an average exponent series [18], so $LinkTime_n$ can converge. We set it to L, and calculate the expectation value of the formula, so

$$E(LinkTime_n) = \alpha * E(LinkTime_{n-1}) + (1-\alpha) * E(S_n)$$

$$E(LinkTime_n) = E(LinkTime_{n-1}) = L \qquad (n \to \infty)$$

$$E(LinkTime_n) \to E(S_n) \qquad (n \to \infty)$$

Using the same treatment to formula (3), we can get:

$$E(V_n) \to |S_n - LinkTime_{n-1}|$$

Lemma 2. *When the topology changes a little, formula (5) can use the link information efficiently.*

Proof. When the topology changes little, the probability of link broken is little. As we have observed (it will be discussed in section 5.2), when the topology changes little, the link expiration time is related to pause time greatly. For we want to use the link information efficiently and avoid sending data to the error link, we add δ to pause time. The simulation shows a better result.

Lemma 3. *The time complexity of DCR is O(N) and the space complexity is O(N), N = |V|, V is the set of mobile nodes.*

Proof. By using the link-based organization and the link counter, and applying the link graph to store the link information, so the space complexity is O(N). And by using the BFS to search the path from the source node to the destination node in the cache, the time complexity becomes O(N). Compared with DSR whose space complexity is O(K× N), where K is average length of path, the space complexity of DCR is better than that of DSR. And as the time complexity of searching path in DSR is also O(N), both of them are the same.

Thus it has been shown that DCR can converge to link lifetime and set the link expiration time more efficiently than DSR.

5 Performance Evaluation of DCR

We use *GloMoSim* simulator [20] to evaluate the performance of DCR. In this simulation, the wireless bandwidth is 2Mbps, the transmit distance is 250m and the MAC layer is IEEE802.11. In 800 × 700 regions, 50 nodes can move randomly, and the model mobility is random WayPoint model. In this model, the nodes are uniformly distributed, and when one node moves to one place, it will stay there for some time and move again. In our simulation, we set the min speed as 5m/s, the max speed as 10m/s, pause time from 0s to 300s, and interval of simulation as 30s, simulation time as 300s and 30CBR, and every CBR traffic as 1kb/2s.

5.1 Performance Criteria

We evaluate the performance of DSR and DCR according to the parameters in [21]. Here, two important parameters of the packet delivery ratio and end-to-end delay are used for evaluating the performance of networks. The former one represents the capability of transmitting data, the latter represents the processing capability of packets. The control packet cost is also an important parameter. Due to the node mobility and network instability in MANET, we use the control packet to rediscover the route and maintain the route in on-demand routing algorithm. So we must decrease the control packets, including the route request packets, route reply packets and error packets. In our simulation, we use the ratio of control packets to the total received packets to evaluate the control packet cost, and the packet drops to index the case of packet dropping. In section 3.2, the Δ_1 is 5, Δ_2 is 10, Δ_3 is 10 [2]. In section 3.3, α is 0.625, β is 0.725 [3]. And as the lifetime of link complies with exponential distribution, a better performance can be got when the value of γ is 0.825. We set δ to 10s and to 210s.

5.2 Results and Analysis of Simulation

In Fig. 3, as the pause time increases, the control packets of both DCR and DSR decrease. This phenomenon can be explained as follows. As the pause time increases, the whole network topology becomes more and more stable, causing the request packets and the control packets decrease. As a whole, the control packets, in the DCR is less than 10%-50% than that in the DSR, especially when pause time is shorter (pause time 60s is an abnormal case which will be discussed at the end of this section).Because the link expiration time in DCR converges rapidly, the error packets decrease, too. But when topology changes greatly, the route requests increasing cause the control packets to increase. However, the control packet is still less than that of DSR by 10%; and when the pause time increases, the topology becomes stable and the DCR requests decrease. Therefore error packets decrease, which results in the control packets decreasing by 30%-50% than that of DSR.

In Fig. 4, as the pause time increases, the packet delivery increase rapidly in DSR, while DCR delivery ratio keeps at 86%-92% in all time. And because the link expiration time in DCR converges rapidly, when pause time is 0s, DCR can also keep on high delivery ration. Therefore, the route can refresh rapidly, which can keep the link exist in packet transmitting. But in DSR, it doesn't consider the automatic link expiration, resulting in the link broken while transmitting the data and the packet delivery

---

[2] Δ_1, Δ_2, Δ_3 is based on experience.
[3] α, β is similar of setting the parameter of RTT in TCP.

ratio decreasing. When the pause time is 60s, the packet delivery ratio decreases in both DCR and DSR. The reason is because of the network congestion; as the pause time increases, the delivery increase more rapidly, however, DCR increases 10% more than DSR in packet delivery ratio. As a whole, DCR increases the packet delivery ratio by 10%-20% than DSR.

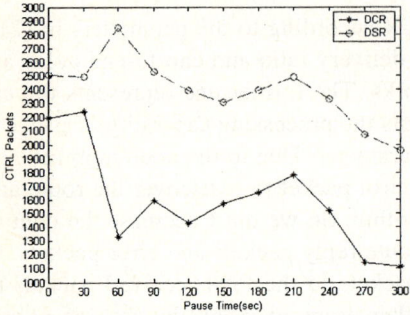 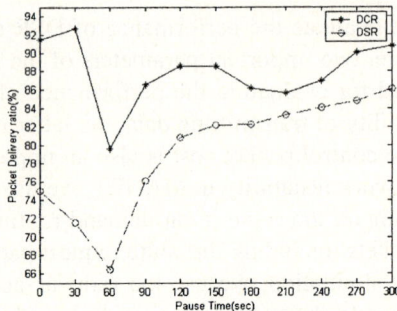

Fig. 3. Control Packets in DCR VS DSR   **Fig. 4.** Packet Delivery in DCR VS DSR

In Fig. 5, in DSR, as the pause time increases, the packet drops decrease. But in DCR, when the pause time is little, the packet drops keep small, because the link can be automatically broken, the topology can be rediscovered, thus the route is renewed. However, when the pause time is 60s, the probability of packet drops are higher caused by the network congestion. As a whole, DCR can decrease 20%-60% packet drops more than DSR.

In Fig. 6, DCR can keep a lower delay all the time. When the pause time is little, due to node mobility and topology changes, the distance from source to destination is short, so the delay is decreased. However, when the pause time is larger, the topology changes very little, making the end-to-end distance and delay longer. But the delay still keeps at low values. This is because that DCR can auto-break the error link timely to avoid transmitting data through the error path, resulting in the decrease the end-to-end delay. But DSR dose not consider this case; the packets are transmitted through the error path, which results in the end-to-end delay increasing. As a whole, DCR can decrease the delay by 50%-70%.

In general, when pause time is 60s, the network is congested, which increases DSR control packets, while that of DCR changes less. In this case, link can not be automatically broken timely. Therefore the packet delivery ratio decreases, error packets increase and end-to-end delay also increases. In other cases, DCR uses the efficient cache organization, auto link broken mechanism and efficient cache strategy to achieve less control packets, higher packet delivery, less packet drops and average end- to-end delay.

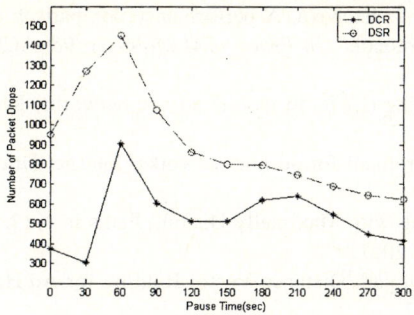 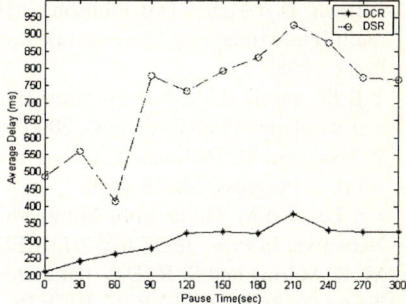

Fig. 5. Drop Packets in DCR VS DSR   **Fig. 6.** End-to-end Delay in DCR VS DSR

6 Conclusion and Future Work

Due to node mobility and topology instability in MANET, optimizing cache can improve the performance of routing. In this paper, we propose an adaptive dynamic cache strategy (DCR), which is based on link organization cache, selective cache strategy and adaptive method to evaluate the link expiration time. Comparison of DCR with DSR using simulations shows that DCR can decrease control packet by 10%-50%, increase packet delivery ratio by 10%-20%, decrease packet drops about 20%-60%, and decrease end-to-end delay by 50%. Currently we are not clear about and will further study: 1) How to use cache error informing mechanism for decreasing the error packets. 2) How to use the reactive renewing route to cache the efficient path in advance for reducing the number of requests. 3) What is the quantified influence of cache on security and QoS.

References

1. Z.J. Haas et al. Wireless Ad Hoc Networks. John Wiley, 2002.
2. R. Ramanathan and J. Redi. A Brief Overview of Ad Hoc Networks: Challenges and Directions. IEEE Commun. Magzine, 40(5), 2002.
3. C.E. Perkins and P. Bhagwat. Highly dynamic destination-sequenced distance-vector routing for mobile computers. In *Proc. ACM SIGCOMM'94*, 1994.
4. T.W. Chen and M. Gerla. Global State Routing: A New Routing Scheme for Ad-hoc Wireless Networks. In *Proc.IEEE ICC'98*, IEEE Press, 1998.
5. D. Johnson, D.A. Maltz and Y.C.Hu. Dynamic source routing in Ad hoc wireless networks. IETF Mobile Ad Hoc Networks Working Group, Internet Draft, work in progress, 2003.
6. C.E. Perkins, E.M. Royer and S. Das. Ad-hoc on demand distance vector routing. RFC3561, July 2003
7. V.D. Park and M.S. Corson. A highly adaptive distributed routing algorithm for mobile wireless networks. In *Proc. IEEE INFOCOM'97*, IEEE Press, 1997.

8. J.Broch, D.A.Maltz, D.B.Johnson, Y.Hu, and J. Jetcheva. A performance comparison of multi-hop wireless ad hoc network routing protocols. In *Proc. ACM Mobicom'98*, ACM Press, 1998.
9. Y.B.Ko and N.H.Vaidya. Location-aided routing (LAR) in mobile ad hoc networks. Kluwer Academic Publishers, 6(4), 2000.
10. Z. Haas and M. Perlman. The zone routing protocol for ad hoc networks. Internet Draft, Work in Progress, March 2000.
11. S.J. Lee and M. Gerla. Split Multipath Routing with Maximally Disjoint Paths in Ad hoc Networks. In *Proc. IEEE ICC'01*, IEEE Press, 2001.
12. M. K. Marina and S. R. Das. On-demand Multipath Distance Vector Routing for Ad Hoc Networks. In *Proc. ICNP'01*, IEEE Press, 2001
13. Y.-C.Hu and D.B. Johnson. Caching strategies in on-demand routing protocols for wireless ad hoc networks. In *Proc. ACM Mobicom'00*, ACM Press, 2000.
14. B. Liang and Z.J. Haas. Optimizing Route-Cache Lifetime in Ad Hoc Networks. In *Proc. IEEE Infocom'03*, IEEE Press, 2003.
15. A. Valera, Winston K.G. Seah and S. Rao. Cooperative Packet Caching and Shortest Multipath Routing in Mobile Ad hoc Networks. In *Proc. IEEE Infocom'03*, IEEE Press, 2003.
16. G.H. Cao, L.H. Yin and C.R. Das. Cooperative Cache-Based Data Access in Ad Hoc Networks. IEEE Computer, 37(2), IEEE Press, 2004.
17. T.H. Cormen, C.E. Leiserson, R.L. Rivest, and C. Stein. Introduction To Algorithms (second edition). MIT Press, 2001
18. G.E.P. Box and G.M. Jenkins. Time Series Analysis forecasting and Control. Holden Day publisher, 1976.
19. K.S. Trivedi. Probability and Statistics with Reliability, Queuing and Computer Science Applications. John Wiley & Sons, 2002.
20. UCLA Parallel Computing Laboratory and Wireless Adaptive Mobility Laboratory. GloMoSim: A Scalable Simulation Environment for Wireless and Wired Network Systems. http://pcl.cs.ucla.edu/projects/glomosim.html.
21. S.Corson, J.Macker. Mobile ad hoc networking (MANET):Routing protocol performance issues and evaluation considerations. RFC2501, January 1999.

On the Job Distribution in Random Brokering for Computational Grids

Vandy Berten[1,2] and Joël Goossens[2]

[1] Research Fellow for FNRS (Fond National de la Recherche Scientifique - Belgium)
[2] Université Libre de Bruxelles, Belgium
{vandy.berten, joel.goossens}@ulb.ac.be

Abstract. This paper analyses the way jobs are distributed in a computational grid environment where the brokering is done in such a way that each Computing Element has a probability to be chosen proportional to its number of CPUs. We give the asymptotic behaviour for several metrics (queue sizes, slowdown...), or, in some case, an approximation of this behaviour. We study the unsaturated case as well as the saturated case, in several stochastic distributions.

1 Introduction

Used by several popular grid systems, we shall see that "ranked brokers" can have unexpected behaviours and, to the best of our knowledge, these behaviours have not been studied so far. We focus mainly this paper – which summarises a more complete work [1] – on a particular case of ranked brokering (i.e., random brokering). Due to space limitation, we do not give proofs of our results; those proofs can be found in [1].

On a random broked grid, when a job arrives in the system, it is sent to a Computing Element (CE) with a probability proportional to that CE number of CPUs. Once a job has been dispatched towards a CE, it has to be scheduled by this CE. In this work, we consider mainly FCFS (First Come First Serve) scheduling rule.

1.1 Model of Computation

In the grids we consider, there is a central *Resource Broker* (RB), to which each CE is connected, and a client sends its jobs to that central RB. Each job j has mainly two parameters: a length (or execution time) j_ℓ, and a width (or number of parallel processes) j_w. The job j will therefore need j_w CPUs during j_ℓ units of time. We assume that, on one processor, we do not use parallelism nor preemption (and consequently migration), and that our system is greedy[1]. We also suppose that jobs are not spread across several CEs.

---

[1] A system is said to be *greedy* (sometimes called *expedient*) if it never leaves any resource idle intentionally. If a system is greedy, a resource is idle only if there is no eligible job waiting for that resource.

1.2 Mathematical Model

We will use in this paper the notations defined in [1]. Briefly, our system is composed of $\mathcal{N}$ CE, called $\mathbb{C}_i (i \in [1 \ldots \mathcal{N}])$. c_i refers to the number of CPUs of $\mathbb{C}_i$, and $C \triangleq \sum_{i=1}^{\mathcal{N}} c_i$ (the symbol $\triangleq$ means "is by definition"). $\nu(t)$ is the total amount of work received in $[0, t]$ divided by the product between the total number of CPUs (C) and the total duration (t), or in other word, the total amount of work received divided by the total amount of work that the system could provide. The interval $[0, t]$ is called the *observation period*.

$f_1(t) \underset{t}{\sim} f_2(t)$ means that $\lim_{t \to \infty} \frac{f_1(t)}{f_2(t)} = 1$.

We assume that the arrival of jobs is a random process with an average delay λ^{-1} between two successive arrivals. The average execution time $(E[j_\ell])$ has a distribution with mean μ^{-1}, and jobs are independent. $\lambda_i \triangleq \lambda \frac{c_i}{C}$, $\rho_i \triangleq \frac{\lambda_i}{\mu}$, $\rho \triangleq \frac{\lambda}{\mu}$, and the system load $\nu \triangleq \frac{\rho_i}{c_i} = \frac{\rho}{C}$.

2 Sequential Jobs

In this section, we will analyse the case where $j_\ell = 1 \; \forall j$.

2.1 Queue Size

$\nu < 1$. We focus here on a single (arbitrary) $\mathbb{C}_i$, where the arrival is a Poisson process with rate λ_i and the execution time has an Exponential distribution with mean μ^{-1}. Such a system is well known, and has been abundantly studied in the literature: this is a $M/M/c_i$ queueing system. Notice that this is a quite naïve approximation; in practice, grids are not generally as predictable as a $M/M/c$ system.

Let J_i be the number of jobs in $\mathbb{C}_i$ (running and waiting jobs) and Q_i be the number of jobs in the queue. Knowing $\mathbb{P}[J_i = n]$ (for instance by [5–page 371, section 8.5.2]) and the relationship between J_i and Q_i, we show in [1] that

Theorem 1. *If $\nu < 1$, the average queue size of $\mathbb{C}_i$ is*

$$E[Q_i] \underset{t}{\sim} b \frac{\nu^{c_i+1} c_i^{c_i}}{c_i!(\nu-1)^2} \quad \text{where} \quad b \triangleq \left[\sum_{k=0}^{c_i-1} \frac{(\nu c_i)^k}{k!} + \frac{(\nu c_i)^{c_i}}{c_i!} \frac{1}{1-\nu} \right]^{-1}.$$

$\nu > 1$. For that case, we do not longer assume that the system is $M/M/c$; we just need to know the average execution time, and the average inter-arrival delay. We will focus on the average queue size at time t, that is, if $P_n(t)$ is the probability that there are n jobs in the queue at time t, we will consider $\sum_{n=0}^{\infty} n P_n(t)$. With some subtle manipulations, we proof in [1] that:

Theorem 2. *If $\nu > 1$, we have*

$$E[Q_i(t)] \underset{t}{\sim} \lambda_i t \left(\frac{\nu - 1}{\nu} \right)$$

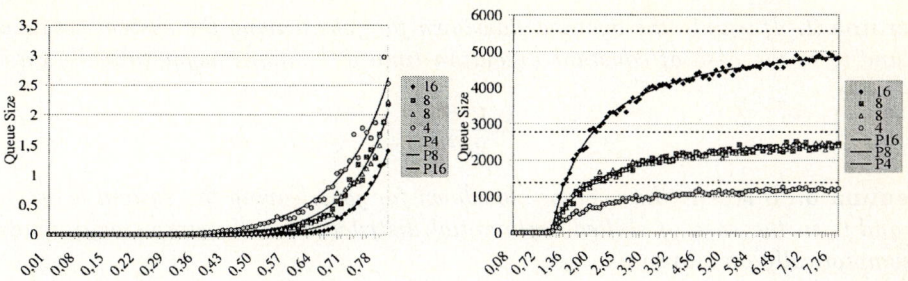

Fig. 1. Queue Size observed in simulation and theoretically expected, for non saturated (left side, with Theorem 1) and saturated systems (right side, with Theorem 2)

Experimental Results. Figure 1 (see [1] for details) highlights that our expectations (continuous lines) are really close to what we got by simulation (dots).

We observe that there is an "inversion" around $\nu = 1$: when $\nu < 1$, $E[Q_i] < E[Q_j]$ if $c_i > c_j$, and for $\nu > 1$, we have the opposite.

2.2 Used CPUs

It can be intuitively easy to see that, for $\nu > 1$, the average number of used CPUs on $\mathbb{C}_i$ is c_i, and that for $\nu < 1$, νc_i CPU's are used in average on $\mathbb{C}_i$. A formal proof is given in [1].

2.3 Slowdown

The slowdown for a particular job is classically defined as

$$\frac{\text{waiting time} + \text{execution time}}{\text{execution time}}.$$

$\nu < 1$. We have an approximation in the case where job lengths have a shifted exponential distribution with a α small (see [1]).

Lemma 1. *If $\nu < 1$, with shifted exponential distribution for job length, the average slowdown $E[\mathcal{SD}_i]$ is asymptotically close to*[2]

$$\frac{b(\nu c_i)^{c_i}}{c_i \cdot c_i!(1-\nu)^2} \frac{e^{\frac{\alpha}{1-\alpha}}}{1-\alpha} \Gamma[0, \frac{\alpha}{1-\alpha}] + 1.$$

$\nu > 1$. In the system we are studying, we measure the slowdown of completed jobs. We computed then the average for each measured job. But, at the end of our observation period, especially if $\nu \gg 1$, a lot of jobs are still in the queue and are therefore not taken into account in our average.

---

[2] $\Gamma[0, z] = \int_z^\infty \frac{e^{-\tau}}{\tau} d\tau$ is the incomplete Euler gamma function.

Lemma 2. *If $\nu > 1$, the average slowdown for jobs leaving the system between 0 and t, in the case of constant execution time μ^{-1}, tends asymptotically on t towards*
$$\lambda t \frac{\nu - 1}{2\nu^2 C}.$$

Lemma 3. *If $\nu > 1$, the average slowdown for jobs leaving the system between 0 and t, in the case of shifted exponential distribution with parameter α tends asymptotically on t towards*
$$\lambda t e^{\frac{\alpha}{1-\alpha}} \frac{\Gamma[0, \frac{\alpha}{1-\alpha}]}{1-\alpha} \frac{\nu - 1}{2\nu^2 C}.$$

(Experimental results can be found in [1].)

3 Parallel Jobs

In the previous section, we imposed the constraint that jobs required only one processor during their execution. In this section, we will be more general and relax this constraint: a job can require several CPUs, and uses them from the beginning up to the end of its execution.

We need here some more notations : w_k is the probability for a job to need for k CPUs, and $\mathcal{W}$ stands for $\sum_k k w_k$. The system load is redefined as : $\nu \triangleq \frac{\lambda \mathcal{W}}{\mu C}$.

For more details, see the full version of this paper ([1]). Notice that the case separation $\nu < 1$ and $\nu > 1$ is here changes into $\nu < \tilde{\nu}_i$ and $\nu > \tilde{\nu}_i$, where $\tilde{\nu}_i$ is the *point of saturation*.

Lemma 4. *If $\nu > \tilde{\nu}_i$, the average queue size of $\mathbb{C}_i$ is close to $t\lambda_i \frac{\nu - \tilde{\nu}_i}{\nu}$.*

Lemma 5. *Whatever the job width distribution, if the job length is fixed, the average number of used CPUs on a CE having c CPUs is*
$$\sum_{k=1}^{c} k P_k$$

Where P_k are solutions of the system
$$\begin{cases} P_k = \sum_{j=c-k+1}^{c} \left[P_j \frac{\sum_{\ell=c-j+1}^{k} w_\ell \beta(k-\ell)}{\gamma(c-j+1)} \gamma(c-k+1) \right] \\ \sum_{i=1}^{c} P_k = 1 \end{cases}$$

with $\beta(k) = \sum_{i=1}^{k} w_i \beta(k-i)$ and $\gamma(k) = \sum_{i=k}^{c} w_i$.

Lemma 6. *In the case of equidistributed job width distribution between 1 and c (the CE size), if the job length is fixed, the average number of used CPUs is*
$$\frac{3c(c+1)}{2(1+2c)}.$$

Lemma 7. *In case of equidistributed job width distribution between 1 and c_i (the CE size), if the job length is fixed, the point of saturation $\tilde{\nu}_i$ is*

$$\frac{3(c_i+1)}{2(1+2c_i)}.$$

Lemma 8. *If $\nu > \tilde{\nu}_i$, the average slowdown for jobs leaving the system between 0 and t ($\mathcal{MSD}_i(t)$), in the case of constant execution time μ^{-1}, tends (approximately) asymptotically on t towards*

$$\lambda t \frac{\nu - 1}{2(\nu - 1 + \tilde{\nu}_i)\nu C} \mathcal{W}.$$

4 Conclusion and Future Work

Our work was a first step towards a more complex analysis of general ranked based brokering. As we shown by plotting together our simulation observations and our theoretical predictions or approximations, we acquired really good knowledge of the job brokering characteristics and behaviour in the specific case we observed.

A second step in our work would be to have a look at some more complex cases; brokering based on the number of free CPUs, the queue size, or an estimation of the waiting time, for other job length and inter-arrival distributions... These new constraints will make more than probably our analysis more difficult, for instance because we introduce a feedback from Computing Elements to the Resource Broker. We believe that we now have built the tools we needed for this futher study.

Acknowledgements

The authors would like to thank Prof G. LOUCHARD and R. DEVILLERS (from ULB - Computer Science dpt) for their significant contributions.

References

1. BERTEN, V., AND GOOSSENS, J. On the job distribution in random brokering for computational grids. Tech. Rep. 518, Université Libre de Bruxelles, May 2004. http://homepages.ulb.ac.be/~vberten/Papers/RandomBrokering-Full.ps.
2. BUYYA, R. *High Performance Cluster Computing*, vol. 1, Architectures and Systems. Prentice Hall PTR, 1999.
3. ERNEMANN, C., HAMSCHER, V., SCHWIEGELSHOHN, U., STREIT, A., AND R.YAHYAPOUR. On Advantages of Grid Computing for Parallel Job Scheduling. In *Proceedings of the 2nd IEEE International Symposium on Cluster Computing and the Grid (CC-GRID 2002)* (May 2002).
4. J. KRALLMANN, U. S., AND YAHYAPOUR, R. On the design and evaluation of job scheduling algorithms. *Job Scheduling Strategies for Parallel Processing* (1999), 17-42.
5. NELSON, R. *Probability, Stochastic Processes, and Queueing Theory*. Springer-Verlag, 1995.

Dividing Grid Service Discovery into 2-Stage Matchmaking

Ye Zhu, Junzhou Luo, and Teng Ma

Department of Computer Science and Engineering, Southeast University,
210096 Nanjing, P.R. China
{tonyzhuye, jluo, mateng}@seu.edu.cn

Abstract. In the wide-area Grid environment which consists of a huge amount of stateful Grid services, the service discovery is a key and challenging issue. The user's requirements on Grid services not only include the function-related but also include the QoS or service state related requirements. Based on the analysis for the requirements of service discovery, this paper divides the service matching process into 2 stages: service type matching and instance matching, and proposes a Grid Service Discovery Model Based on 2-Stage Matching which can enable more effective service discovery. In the model VO is utilized as the managerial unit for grid services and a two-level publication architecture is adopted. The initial simulation results show that the model can effectively aggregate the service information and avoid the workload caused by frequent dynamic updating.

1 Introduction

Since the Open Grid Services Architecture (OGSA [1]) has adopted service-oriented technologies from Web Services to solve similar problems in Grid environment, the stateful Grid services and a multi-level service integration architecture is introduced. In the wide-area Grid environment which consists of a huge amount of stateful Grid services the service discovery is a key and challenging issue. The matching mechanism introduced from Web Services can be regarded as capability matching, namely focusing on whether the function offered by the service can fulfill the user's requirement. Here it emphasizes more on the satisfying of the operation logic. While in Grid services, the binding of service and resource makes the functionality no longer the only demand. Grid user will care more about the performance that service can guarantee, which can be represented by QoS parameters or service-specific state information. So the discovery process for Grid services is not merely to find that type of service which has the needed operation logic, but to find those service instances satisfying the performance requirement.

Most of the existing service discovery models didn't raise the QoS demand to the height of a query requirement, so the practical matching process is mostly the matchmaking for static service type information. Here the 'service type' emphasizes that's a type which is developed according to definite operation logic and has specific interface and behavior, while the service instance is an individual which is dynamically created and has the only GSH in OGSA. Compared with the relatively static service function and interface description, the QoS and state information are dynamically changed and should be updated in time. If we treat them in an equal way, the layer-by-layer aggregation of the large amount of dynamic information may bring quite heavy updating workload. But the real-time update for QoS and state information is absolutely necessary for improving the success rate of service instance discovery. So we propose that the static and dynamic information should be separated and adopt different policies for publication and maintenance, thus dividing the service matching process into 2 stages: service type matching and instance matching. On the basis a 2-Stage Matching Based Grid Service Discovery Model is proposed in this paper. It uses VO [2] as the managerial unit and adopts the two-level publication architecture inside VO and a Forward List based mechanism among VOs.

The rest of the paper is organized as follows. The related work is discussed in Section 2. In the next Section we introduce the 2-stage service matching process. The model architecture and key mechanisms will be presented in Section 4. Next we use the initial simulation results to show the performance of the model. Finally we conclude this paper.

2   Related Work

The performance of Grid services is usually represented by QoS parameters. The QoS-based service discovery provides a good support to resource advance reservation so as to guarantee the user's requirement on performance. In [3] the researchers classify and describe the Grid QoS properties and extend the UDDI [4] to support the QoS property based service search. However in practical Gird environments, the dynamic characteristics of the resource itself and the frequent QoS update caused by allocation can not be neglected, and there will necessarily exist large numbers of dynamic transient services, so UDDI architecture may be not suitable here or need quite more improvement. Furthermore the user's requirement may also contain some service-specific service data so the centralized aggregation for such information is unreasonable.

In OGSA each Grid service expresses its state in a standardized way as Service Data Elements (SDEs). The MDS (Monitoring and Discovery Service) [5] of Globus Project implements the OGSI-based grid information service system. MDS uses VO as the basic unit for management of Grid services and Registry for soft-state registration. It adopts decentralized maintenance mechanism and can be extended to form various types of discovery and management architectures. But MDS is more used to query the service data but not well support the service type discovery based on the service interface and activity. And MDS does not involve how to improve and optimize the service discovery through effective VO-layer organization and cooperation.

The paper [6] proposes a Grid Service Location Mechanism Based on VO and the Small-World Theory, which utilizes the similarity of the service types and properties inside the VO, and determines whether the request could be accepted by this VO through the dissemination of its service type distribution information. The mechanism is feasible but exist some problems. The neighbor relationship among VOs utilized in message dissemination is unascertainable for the numerous possibly overlapped VOs. And various types of Grid services can not be definitely represented only by type ID. Furthermore the distributed message dissemination mechanism is also unsuitable for dynamically updated service QoS.

3   Service Type Matching and Instance Matching

Based on the analysis for the main elements involved in service discovery, Grid service information can be denoted as a 4-tuple $S(P, I, S, E)$, where P is the static service property information mainly expressing the function and activity which are determined by service type; I is the interface description which usually refers to the portTypes description part in WSDL document; S is the instance-related status information mainly including QoS performance, service-specific SDE, etc; and E is the endpoint information such as GSH and GSR. Also, the user's request can be regarded as a partial description document for the service. Similarly it can be denoted as a triple $R(P', I', S')$, where these elements respectively express the corresponding aspects of the requirement to the service.

Service type matching is to discover the qualified service type that accords with the required function and interface from the numerous potential service types. Its basic input is $R(P', I')$ and $S(P, I, E)$, namely comparing the function and interface requirement contained in user's request to the service properties and portTypes interface description. The matching logic adopts the exact field-value matching and semantic similarity matching. The output is the qualified service types which could be more than one if the compatible similar service is allowed to be invoked. In practical applications the output may be the endpoint addresses of the numerous service instances corresponding to that service type, which can be used as the input for the next stage matching. It should be noticed that the service property information may be published to the Registry by the service instance, but in type matching stage it mainly involves the development-time, rather than run-time service description. The property information related to service type may be diversified or field-specific; here we just discuss some main properties:

> **Service Name:** A specific type of service will be assigned a name according to its main function or objective by the service developer. It should be standardized based on industry-specific naming criterion. But some services that have the same name may adopt various implementation logics.
> **Type ID:** It's a key element in the model which uniquely identifies the service type. It must be specified using unified mechanism, at least in a VO and similar types should have close values for their Type ID.
> **Semantic Description:** It mainly focuses on the activity of the service. Much related research has been done in the field of Semantic Web [7] and the similarity matching by ontology [8] can be adopted to enable flexible matches.
> **Categories:** The service type can be classified and construct a layered catalog.
> **Compatibility:** It can be used to describe the substitution relationship between services of the same level. If compatibility has been identified, the similar-type services provided by different developers can be replaceable under specific criteria and ensured to achieve the required functions.
> **Static Capability:** It refers to the intrinsic service capability fixed in developing stages such as the security level, lowest computing ability, etc.

Besides the above basic properties, the portTypes interface description, as a segment of WSDL document [9], is also an important factor for type matching. Though WSDL document may contain instance related information in <Bindings> element, a large part of the document can be used as the basis for service interface comparison.

Service instance matching is to find a more refined instance subset which satisfies the QoS or service data requirement from the candidate instance set. Its basic input is $R(S')$ and $S(S, E)$. One side is the QoS or specific SDE requirement contained in user's request and the other side is the QoS state and SDE values of the service instances aggregated in the Registry. The matching logic involves the field-value matching and constraint matching. The output will be the GSH of the qualified Grid service instances. However the practical course is not merely matchmaking. Grid services comprise transient services and persistent services, which have to be treated differently in instance matching stage. The transient service refers to the service whose instance is created only when it is invoked and lives for a task duration while the persistent service means that its instance has been created beforehand and may exist for quite a long time to serve many requests. For transient service the QoS requirement should be submitted to Factory interface, the latter will query the current state of its controllable resource to determine whether the potential instance to be created can satisfy the requirement. If satisfiable then the Factory Handle will be returned. For persistent service it may directly query the Registry for the QoS state and SDE information. If the request contains multi-QoS constraint it will also have to consider more factors such as the priority of diverse parameters, synthesized selecting policy, etc.

In the multi-level service integration architecture the low-level service can be dynamically organized and agglomerated to compose high-level service which has more powerful function and ability to control resource, for example a Scheduler Service may be regarded as a high-level service for on-demand resource allocation. For such kind of service which can extend its ability on demand, they should provide information about their QoS capability area in SDE. And the matching course is just to find the potential service instances whose capability covers the required QoS constraint. After that the user proxy will negotiate with the service according to its policy and establish the Service Level Agreements to ensure the QoS performance.

4 Grid Service Discovery Based on 2-Stage Matching

Based on the analysis and improvement on the existing discovery architectures [10], we realize the 2-Stage Matching Based Grid Service Discovery Model. It takes VO as the managerial unit for Grid services, adopts a two-level publication architecture which consists of one VOSR and many LSRs inside VO, and a request transfer mechanism by VO Forward List among VOs (as shown in Figure 1).

■ **LSR (Local Service Registry):** It serves as a Registry for soft-state registration and information maintenance of local Grid services, and publishes the service description document to VOSR after local publication. Here the "local" refers to a logic concept. LSR can be a Container Registry Service deployed specially for the services on the same machine or site, or it can also be a common Registry for a group of distributed but directly administered services.

■ **VOSR (VO Service Registry):** It's the centralized registry for all Grid services inside the VO. It records and maintains the type information of usable services published by LSR and the endpoint addresses of those LSR to which service instance registers. It will aggregate the distributed instances and establish the index by service type. The service type matchmaking is implemented on VOSR and its result is an address list of the LSRs that have required service instances or Factory Handles. After further filter of the result using certain policy, the request will be forwarded to these LSRs for instance matching.

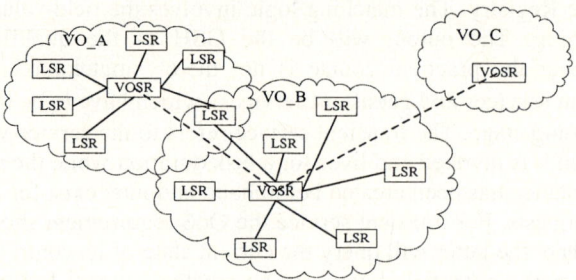

Fig. 1. The architecture of 2-Stage Matching Based Grid Service Discovery Model

On the VO layer of this model, the VOSR nodes from each VO compose a P2P network, which adopts a flexible request forward mode. It's mainly disseminated by the VO Forward List established in advance, and can also be forwarded by the history discovery record or by querying the Industry VO Directory similar to the yellow page service.

4.1 Service Publication and Update

LSR should register to the VOSR it wants to join when start-up. Then VOSR assigns it a unique ID and saves its endpoint address. The service shall register to LSR first when created. For persistent service its instance is usually created when the container startup or a long-term service subscription is accepted. LSR will record its GSH and also the GSR in HandleMap service. Meanwhile the corresponding type-related service description and SDE information will be stored in LSR. The creation for transient service should adopt the Factory Pattern, namely registering an activated Factory Handle (FH) in LSR according to the resource state, and creating its instance when necessary. Its corresponding type description and the WSDL document not containing instance-related status information will be stored to LSR. LSR needs to update them periodically or when changes reach to some extent. For dynamic QoS state or more universal SDE the update period should be relatively short.

After the local registration LSR also has to register the service to VOSR. VOSR indexes by service Type ID to aggregate the service information, as shown in Figure 2. LSR submits the corresponding service type information and WSDL document of the newly registered instance or FH to VOSR. Then VOSR queries its maintained Service Type Table. If there is no information of this type of service, it will create a new type item and store the correlative description information. If the type has been recorded before then it can locate to the LSR List pointing to those which provide this type of service currently. Similarly it may add a new LSR item or plus 1 to the <instance numbers> of the corresponding existed item and store the service data information. The service data stored in VOSR needn't to be updated in time namely they are inaccurate and just for refining the type matching result so as to reduce the query range. The information in VOSR is updated with a quite long period or when an instance or factory register or revoke in LSR. The revoking course is similar to the register. When the <instance numbers> of a item in LSR List become zero it will be deleted, but the item in Service Type Table will generally not be deleted.

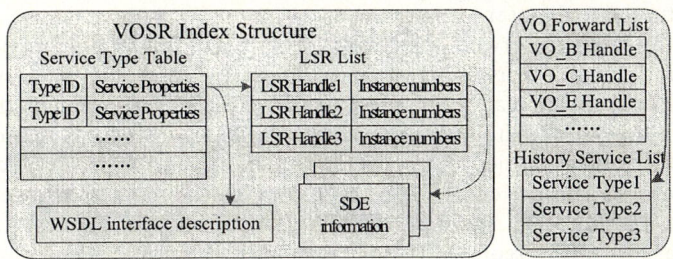

Fig. 2. VOSR Index Structure and VO Forward List

On VO Layer the service information is not aggregated. But VO needs to know the potential VO partners which it can forward the request to. Using the pre-created VO Forward List is a simple but effective dissemination approach (Figure 2). Here the VO Handle needs to register in VO Administrative Institution and get a unified ID. The creation of the Forward List has multiple modes. The VO which has related application purpose or cooperation relations should be first added to VO Forward List. The famous large-scale VO can also be added manually or by querying the Industry VO Directory similar to the Yellow Page Service. Furthermore, VO Forward List should also include a default item pointing to a VO which faces the general field.

4.2 Service Discovery

The service discovery course starts with the service request sent by the user proxy program. The request containing QoS requirement is encapsulated in SOAP message and sent to a specified VOSR. Usually it's first submitted to the VO that the user belongs to but that's not imperative:

1. When receiving service request VOSR will first authenticate the requestor and query the VO Policy Center for authorization, only authorized request can be accepted.
2. VOSR resolves the service request, searches the local maintained Service Type Table according to the service type requirement contained in request and compares them with the service properties and WSDL interface descriptions. If matched service type is found, continue with the next step; otherwise go to (5).
3. It locates to a LSR List according to Type ID of the matched item. Here further selection and refinement can be made according to certain policy to reduce the result range. It may query the service data stored on VOSR which includes QoS information as the reference basis for selection. By now we may get a LSR set and obtain their endpoint addresses by LSR handle. VOSR will add the Service Type ID into the SOAP Header of the request and forward it to these addresses.

4. After receiving the request, LSR will first find the corresponding service instances or Factory Handles by the Type ID and resolve the QoS or specific service data requirement from the request. If there is no requirement all instances of this type will be returned. Otherwise it will query the service data or submit the demand to Factory interface as described in instance matching section. Finally qualified GSH or FH are found and returned to the user proxy.

5. If not finding required service in local VO, VOSR will forward the request to other VOs according to VO Forward List and inform the user the request has been transferred. The forward can be carried out by stages, for example it may first forwarded to the VOs that have cooperation relations, which is reasonable and feasible for the project carried out cooperatively by several VOs. To accelerate the discovery course a sub-list can be used to record the most often requested outside services in a certain VO. Another VOSR will continue the above process when receiving the forwarded request. If still no matched service the request will be disseminated again through that VO List until the limited steps or time-out mechanism take effect.

4.3 Service Invoking

The user proxy receives the qualified service handles returned by LSR. The result may be quite a few service instances so the proxy has to filtrate according to certain policies such as cost, history records, or VO policy to finally get a smaller service set. For persistent service the proxy will utilize the endpoint address of LSR in return message to locate the HandleMap service and obtain the GSR which contains binding information. For transient service it can use FH to directly visit the Factory interface which may create the required service instance and return the GSR. If there is QoS requirement the proxy can adopt SLA to negotiate with the service and finally establish a BSLA specification to guarantee the performance in service invoking.

4.4 Discovery of the Large Numbers of Low-Level Services

For the VO that has common application purpose especially in scientific computing field, a possible scenario is that large numbers of low-level services of some main types are deployed on most of the nodes inside the VO, usually for fulfilling some divisible independent tasks. Such a large amount of low-level services bring a quite heavy workload to the service discovery and information maintenance. Also returning numerous low-level service handles makes it difficult for user to choose. A possible solution is to select a LSR to create a virtual composite service item which represents all the instances or factories of a certain type. In this way VOSR can only keep one item for this type of service. But certainly the capability of the composite service should be represented in a proper way in service properties by the maintainer. The user proxy may find this service through VOSR and send the invoking request to that LSR. If LSR accepts, it will disseminate the request in VO and many service instances distributed in different places actually perform the task and return the result. This solution is transparent to the users and feasible for the VO which widely deploys the basic services.

5 Simulation Results

The simulated wide-area environment (Figure 3) consists of 6 VOs and the LSR number in each VO is randomly distributed in [5, 10]. We assume there are 100 different types of services and each VO has its representative 20 service types (They may overlap). For each service, the number of the providers (persistent service or activated Factory) produced in one minute is randomly distributed in [1, 10]. It will be registered to a randomly selected LSR and its lifetime is randomly distributed in [5, 10] minutes. The popularity of each service, measured in number of requests per minute, is also randomly distributed in [5, 50] requests per minute. 50% of the requests come from the local VO and the others are randomly selected from the other VOs. The QoS capability of the service instance is dynamically changed and 50% of the grid resources are assumed to have higher QoS performance than the average QoS requirement.

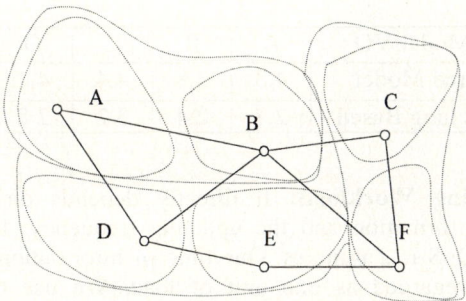

Fig. 3. Simulated Wide-Area Grid Environment

We make an initial simulation for 2000 minutes under the above scenario to evaluate the model performance from three aspects: success rate of the discovery, query responsiveness and information updating workload.

Success Rate of the Discovery: A successful service discovery includes two aspects: the potential qualified services can be discovered and the returned services can be successfully invoked and satisfy the QoS or SDE requirement. In our simulation all service information is reachable, so here we mainly focus on the latter's performance. For simplicity, we assume that the result only has two QoS levels – '1' and '0'. The service invoking results in QoS level '1', if and only if all its requirements are satisfied; otherwise it results in '0'. As a comparison, we also simulate under the basic discovery model: using the layered architecture but not aggregating QoS information, and randomly selecting the service instance. Table 1 shows the average QoS levels of the request results from every VO. The results show a significant improvement under the 2-stage matching based model, which is a necessary result for QoS-aware instance matching.

Table 1. Average QoS Level Comparison

Simulated Model/VO	A	B	C	D	E	F
Basic	0.74	0.73	0.76	0.75	0.73	0.72
2-Phase Matching Based	0.95	0.96	0.97	0.96	0.97	0.97

Query Responsiveness: It lies on the time consumed in matchmaking and the distances for request forwarding. In the wide-area environment it's mainly influenced by the second factor. Here we use the hops involved in a query request as the estimate parameter. The model for comparison is the totally layered model: AB, DE, CF respectively compose the upper-layer domains and further form the root domain. The results in Table 2 show that our model has better query responsiveness.

Table 2. Average Hop Involved in a Query Request

Simulated Model/VO	A	B	C	D	E	F
Basic Layered Model	4.3	3.8	4.4	4.1	4.6	4.5
2-Phase Matching Based	2.7	2.4	3.2	2.8	3.1	3.3

Information Updating Workload: It directly depends on the amount of the aggregated service information and the updating frequency. In our simulation the updating period for QoS is 2 minutes. One time of information update for a service type or instance is regarded as one unit of load. We use the average updating workload per-minute in LSR/VOSR as the estimate parameter and obtain the results in Table 3. As a comparison, we also simulate the scenario with no matching stage division, in which all service information is updated to VOSR (row 3). From the results we can see the division of the type and instance matching can effectively reduce the updating workload.

Table 3. Average Updating Workload Per-Minute in LSR/VOSR

Simulated Registry/VO	A	B	C	D	E	F
LSR	53	76	88	59	66	105
VOSR	246	242	243	242	244	233
VOSR(aggregating QoS)	610	605	605	610	612	607

Although it is still simple and initial, the simulation does show the effectiveness of our proposed improvement based on 2-stage matching. As a QoS-aware wide-area service discovery model, it can effectively aggregate the service information and avoid the workload brought by dynamic updating.

6 Conclusions and Future Work

Aiming at the QoS and service state requirement in Grid service discovery, the model proposed in this paper separates the relatively static and dynamic service description information and divides the service matching process into 2 stages: service type matching and instance matching. It's a compromise between the efficient service information aggregation and the workload brought by dynamic updating. The paper also discusses the differential treatment between transient service and persistent service.

Our future work is to realize a prototype system to test the feasibility of the model under the practical environment. Further analysis and improvement will be made to some details of the 2-stage service matching. Also some potential further extension on the description contents may become our next research target which may enable powerful discovery for more additive and practical functions.

Acknowledgement

This work is supported by National Natural Science Foundation of China under the Special Program "Network based Science Activity Environment" (90412014).

References

1. I. Foster, C. Kesselman, J. Nick and S. Tuecke: The Physiology of the Grid: An Open Grid Services Architecture for Distributed Systems Integration. Globus Project, 2002.
2. I. Foster, C. Kesselman and S. Tuecke: The Anatomy of the Grid: Enabling Scalable Virtual Organizations. International J. Supercomputer Applications, 15(3), 2001
3. R. Al-Ali, O. Rana, D. Walker, S. Jha, and S. Sohail: G-QoSM: Grid Service Discovery Using QoS Properties. Computing and Informatics Journal, Special Issue on Grid Computing, 21(4):363–382, 2002.
4. Universal description, discovery and integration of business of the Web. http://www.uddi.org
5. K. Czajkowski, S. Fitzgerald, I. Foster, and C. Kesselman: Grid information services for distributed resource sharing. In Proc. 10th IEEE International Symposium on High-Performance Distributed Computing (HPDC-10), pages 181–194, 7-9 August 2001.
6. Erfan Shang, Zhihui Du: Efficient Grid Service Location Mechanism Based on Virtual Organization and the Small-World Theory. Journal of Computer Research and Development, Vol.40, No.12, Dec. 2003 (Chinese)
7. Boris Motik, Andreas Abecker, "Report on Development of Web Service Discovery Framework".
8. Simone A. Ludwig, Peter van Santen, "A Grid Service Discovery Matchmaker based on Ontology Description".
9. Web Services Description Language (WSDL). http://www.w3.org/TR/wsdl
10. Feilong Tang, Minglu Li, Jian Cao, Qianni Deng: GSPD: A Middleware That Supports Publication and Discovery of Grid Services. In Proc. 2nd International Workshop on Grid and Cooperative Computing (GCC2003), pages 530-537, Dec. 2003

Performance Evaluation of a Grid Computing Architecture Using Realtime Network Monitoring[1]

Young-Sik Jeong* and Cheng-Zhong Xu

* Department of Computer Engineering, Wonkwang University,
344-2 Shinyong-Dong, Iksan, Jeonbuk 570-749, Korea
`ysjeong@wonkwang.ac.kr`
Department of Electrical and Computer Engineering, Wayne State University,
Detroit, Michigan 48202, USA
`czxu@ece.eng.wayne.edu`

Abstract. This paper integrates the concepts of realtime network monitoring and visualizations into a grid computing architecture on the Internet. We develop a Realtime Network Monitor(RNM) that performs realtime network monitoring in order to improve the performance of the grid computing framework. Through network monitoring, it is also found out that the network traffic has effects on the performance of processing for large scale applications.

1 Introduction

Recently, rapid improvements in the performance of internet service and pervasive deployment of commodity resources provide us grid computing infrastructures with tremendous potential[6],[7]. This potential is being widely tapped, and many grid middleware projects such as Globus, Condor, and NetSolve have been pursued to provide efficiently access to remote resources[4],[5],[6]. Unfortunately, there are a paucity of tools that assist a user in predicting if their application will obtain suitable performance on a particular platform. AppLes[4],[5] has been to investigated adaptive scheduling for grid application and to apply these research results to distributed applications. MicroGrid[6] is described a tool to develop and implement simulation tools that support a vehicle for the convenient scientific study of grid topologies and application performance issues.

These researches did not take into account the realtime network monitoring of sub-networks including hosts. This differentiation reduces the entire network's operation time. In height of this, this paper suggests task allocation algorithms which are based on network monitoring. This paper should use to grid computing infrastructure called Parallel Distributed Processing(PDP)[1] which is a parallel computing framework implemented with Java over the Internet. In this model, there are three kinds of participating entities; *Manager, Host, Requester*. The *Requester* is a process seeking computing resources, The *Host* is a process offering computing resources, The *Man-*

---

[1] This Research was supported by University IT Research Center Project.

ager is a process that coordinates the supply and demand for computing resources. PDP fragments the large scale application into small units, or tasks. Tasks are then processed in parallel through the key algorithms. PDP's task allocation algorithm is classified into static task allocation and dynamic task allocation, depending on whether the participating hosts are dynamically managed or not. During the static phase, the addition of new hosts and the secession of hosts do not occur. In this phase, PDP has three cockpit algorithms: the Uniform Task Allocation (UTA), the CPU Performance Task Allocation (CPTA), and the Static Adaptive Task Allocation (S_ATA). In the dynamic phase, the number of participating hosts may be changed during the execution time and PDP has one Dynamic Adaptive Task Allocation (D_ATA) algorithm based on S_ATA.

UTA is a very simple algorithm. It is executed by equally distributing a certain amount of tasks to each host regardless of the CPU capacity of the participating hosts. Each host then executes the tasks and their results are directly called back to the *requester*, using the RMI reference value. CPTA is more common than UTA because the connected hosts on this slide are distinguished by their different CPU capacities. The CPTA algorithm is executed by allocating the tasks to each host based on its CPU capacity. For evaluation of the CPU capacity, the LINPACK[2] benchmark algorithm is used. It is impossible to predict the CPU usage of each host during the execution time because the locations of hosts are geographically too far from one another. The prediction of CPU usage is further compounded by the fact that each host can be also accessed by many users, and the number of internal process of each host constantly changes. These problems are resolved by the static adaptive task allocation (S_ATA). Due to the autonomous character of the Internet, system faults can result in the secession of host connection to the manager. Specifically, network failure or internal defects of each host may unexpectedly disable the sharing of resources. In order to support that, we use to the dynamic adaptive task allocation (D_ATA).

The goal of this paper is to develop and implement the realtime network monitoring(RNM) system that is able to monitor network traffic sub-networks including hosts, is providing resources to the internet oriented grid computing architecture by task allocation depending on the network bandwidth. And also this paper provides the task allocation algorithms which are based on realtime network monitoring.

2 Realtime Network Monitoring

2.1 Monitoring Factors and Architecture

Network management, which is using SNMP methodology[3], is consisted with management agent, SNMP agent, Management Information Base(MIB) and Network Management Protocol(NMP). Total amount of network traffic and available bandwidth can be checked by simple handling. Therefore, network monitoring shall be used for data extraction method in this paper. Various network monitoring factors can be analyzed to examine network performance. In this paper, the RNM, which is going to be developed, shall be allocating tasks by knowing the capable work loading amount. Therefore, Utilization and Bandwidth shall be used for network valuation factors in network performance monitoring factors. RNM was developed by Java based toolkit and the architecture of its should be consisted as described in Fig. 1.

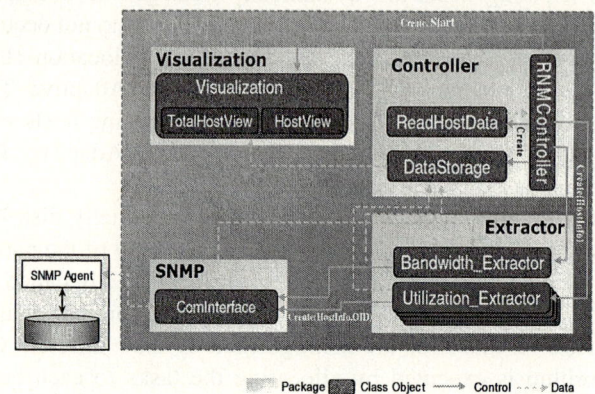

Fig. 1. Architecture of Realtime Network Monitoring System

It should be constituted with Visualization, Controller, Extractor and SNMP package. Visualization package includes TotalHostView and HostView to provide two viewing options as management interface part. Visualization class offers view to compare network usage ratio of whole host, which is able to monitor. RNMController, which is controlling RNM entire movement, is creating user interface Visualization, ReadHostData and DataStorage class. Additionally, it is creating Bandwidth Extractor and Utilization Extractor class from Extractor package by getting information from PDP. ComInterface class is connected with SNMP. Those are cooperated to extract monitoring information by generating SNMP message.

2.2 Algorithms Based on Realtime Network Monitoring

The network performance of each participating host in sub-network is different to each other. In this research, we propose the two task allocation algorithms based on realtime network monitoring; Network Performance to Task Allocation(NPTA) algorithm and Network performance Adaptive Task Allocation(NATA) algorithm. NPTA is executed by dividing the tasks based on the each host's the network performance and is assigned by allocating the appropriate tasks to each host. NATA could cope with the network traffic changing during actual execution time for application. Let's assume during the execution application at time t_i, Fig. 2 be shown that. It is based on NPTA, after pass time t_j, the available network bandwidth at Host4 is changed to the wide network bandwidth at time $t_i + t_j$, such like the sky color. This situation is resolved by the NATA algorithm.

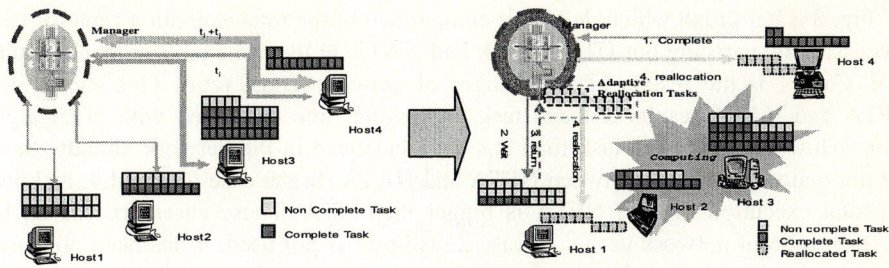

Fig. 2. Network performance Adaptive Task Allocation Algorithm

3 Performance Evaluation

Simulation environment is designed to carry out the performance evaluation without the affection by network environmental factors. For the purpose of only network performance testing, all the host's operating system has been standardized as Pentium 4.1.3GHz, 256 memory chip, same hardware and OS as Windows 2000 professional. Sub-network is consisted of Host and Traffic loader connected with switch for the testing. These 8 sub-networks are connected with main switch, which is including SNMP demon. Port between sub-network's switch and main switch is 10Mbps and sort of dual line structure. Traffic server, which is receiving package to generate traffic of manager, applier and traffic loader, is connected as 100Mbps and whole dual line communicating structure. This dual line communicating structure can prevent threshold congesting situation. Traffic loader is functioning as generating traffic charged around 20~90% in the node between sub-network switch and main switch. In this paper, we used the drawing of 3D Image as application in this experimental environment and ignored the overhead for network monitoring, which has a little network bandwidth.

NPTA and NATA based on PDP performance testing has been carried out in five ways, 20~90% of traffic can be generated by traffic loader and used to four hosts. In network environment, all the traffic is changeable. This changeable network environment can be effectively adjusted by using adaptive task allocation in network environment.

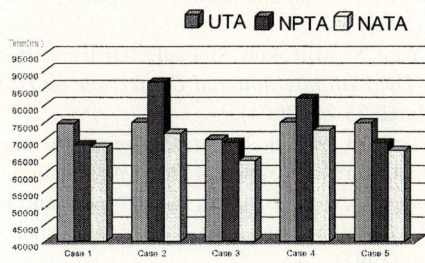

Fig. 3. Performance Evaluation of NPTA and NATA

Fig. 3 is bar graph which describes comparison of the total execution time for task processing by carrying out UTA, NPTA and NATA in the same environmental situation. Case 1 is the results of no changes of network usage ratio. This shows that NPTA and NATA results reduced task processing time compared with UTA algorithm. However, since the task load has been balanced in the network already, there are not many differences between NPTA and NATA. In the case of 2 and 4, it shows the total execution time of NPTA is bigger than UTA. These cases are caused by increase ratio of network usage. In case 2, as host1 is not used, it has been allocated with more tasks, which led increase of the total execution time of network. Same case applied with case 4. NATA allocates task to the early finished host without waiting other host's process completion. In case 3 and 5, during task processing it occurs decreasing the network usage ratio, those are similar with case 1 in differences of running time between UTA and NPTA. However, NATA reduces more the total execution time by doing other host's allocated tasks to be reallocated to other hosts, which is not charged with any works.

4 Conclusion

This paper developed the realtime network monitor system that could monitor the performance changes in each host during the actual execution time in order to improve these. By performance results, this research demonstrated that network traffic occurred in sub-network including hosts could affect the total execution time for processing application. Also it represented that effective task allocating process based on sub-network usage ratio of each host makes minimum affection not to be delayed with task processing time by monitoring network environments of each host.

References

1. Eun-Ha Song, Young-Sik Jeong, "Development of Dynamic Host Management Scheme for Parallel/Distributed Processing on the Web," KISS Vol. 8 No 3. (2002)
2. J. Dongarra, J. Bunch, D. Moler, and G. W. Stewart, "*LINPACK User`s Guide*" SIAM Philadephia PA (1979)
3. SNMP Research International, Inc., "http://www.snmp.org"
4. R. Figueiredo, P. Dinda, and J. Fortes, "A Case of Grid Computing on Virtual Machines," *The 23th International Conference on Distributed Computing Systems*, Province, Rhode Island, USA, May 19-22, (2003)
5. M. Thottan, L. Li, B. Yao, S. Mirrokni, S. Paul, "Distributed Network Monitoring for Evolving IP Networks," *The 24th International Conference on Distributed Computing Systems*, Tokyo, JAPAN, March 23-26 (2004)
6. F. Berman, et al, "The GrADS Project: Software support for high-level Grid application development" International Journal of Supercomputer Application, 15-4 (2001) 327-344
7. B. O. Christiansen, P. Cappello, M. F. Ionescu, M. O. Neary, and K. E. Schauser, "Javelin : Internet-Based Parallel Computing Using Java," ACM Workshop on Java for Science and Engineering Computation,(1997)

Quartet-Based Phylogenetic Inference: A Grid Approach

Chen Wang, Bing Bing Zhou, and Albert Y. Zomaya

School of Information Technologies, University of Sydney,
NSW 2006, Australia

Abstract. The accuracy of quartet-puzzling method, which is widely used in molecular phylogenetic analysis depends heavily on the number of intermediate trees searched and the order of molecular sequences selected to generate these trees. Because the quality of intermediate trees cannot be guaranteed, the consensus results can easily be trapped into local optima. In this paper we present a new approach to guide the intermediate tree selection. Our experimental results show that the accuracy of reconstructed trees can be improved significantly. Using our method, the task can easily be partitioned into independent subtasks of different sizes. Therefore, it can effectively be implemented and run in the heterogeneous and dynamic environment of the computational grid.

1 Introduction

Molecular hylogenetic analysis is a fundamental tool in bioinformatics and computational biology, and supports other research in diverse areas of molecular bioscience including comparative genomics, drug design, environmental biotechnology, and protection of biodiversity [1]. The task of phylogenetic analysis is to try to reconstruct the evolutionary history by inferring phylogenetic relationships from a given set of molecular sequences. The evolutionary history is organized as unrooted binary trees. The number of different unrooted binary trees is, $\prod_{i=3}^{N}(2i-5)$ where N denotes the number of given molecular sequences. As N increases, the number of possible evolutionary trees increases exponentially and it will quickly become impossible to exhaustively search the entire tree space for finding the best one that represents the true evolutionary history. Thus, heuristics are used to assist biologists with phylogenetic analysis.

Of many popular heuristic techniques, maximum-likelihood (ML) methods [2] are widely used due to their well-founded statistical basis and conceptual simplicity. Maximum likelihood methods evaluate a hypothesis about evolutionary history in terms of the probability that a proposed model of the evolutionary process and the hypothesized history would give rise to the observed data. It is conjectured that a history with a higher probability reaching the current state of affairs is a preferable hypothesis to the one with a lower probability [3]. Experimental results show that analytical results obtained by using ML methods are often more robust than those by other methods such as neighbor-joining (NJ) [4] and maximum parsimony (MP) [3, 5, 6]. However, maximum-likelihood methods are computationally more expensive. This high computational complexity often prevents the methods from being applied for large-scale analyses. As a result, there is a great effort in attempt to improve the computational speed of ML methods.

One technique used to reduce the ML computational complexity is quartet puzzling [7, 8]. The quartet puzzling is actually an approximate ML method and contains three stages. In the first, or ML stage, a set of 4-trees (four-sequences phylogenies) on all different quartets (4 sequences) is constructed using the standard maximum likelihood methods [2]. In the the second, so called puzzling stage, a number of intermediate trees are constructed based on the 4-trees generated in the ML stage. The results of this stage is dependent on the order of sequences being added and such orders are randomly generated, one for each tree, in the current TREE-PUZZLE package. In the final consensus stage, a consensus (typically majority ruling) based on the rate of occurrence of partitions is applied to construct the final tree from the intermediate trees generated in the second stage.

When the number of sequences is N, there will be a total number of $(N-4)!$ different intermediate trees. As n increases, it is impossible to exhaustively construct all different trees for the final consensus. Normally only a limited number of trees are generated in the puzzling stage. Since these intermediate trees are not evaluated, the quality of these trees is not guaranteed. Therefore, the final results can easily be trapped into local optima.

The lack of quality control in selecting intermediate trees in the quartet puzzling method is partly because there are not enough computing resources to evaluate these trees. As more and more powerful computing resources are reachable on the Internet, it is possible for us to make effective use of these resources to get the quartet puzzling out of the local optimal loops. In this paper, we present a grid-based algorithm to obtain more accurate trees.

The rest of the paper is organized as follows: Our grid quartet-based algorithm (GridQP) in Section 2. some experiment results are presented in Section 3; The conclusions are given in Section 4.

2 Grid-Based QP Algorithm

Our extensive experiments show that the accuracy of the reconstructed tree using the quartet puzzling method does not scale well with the number of intermediate trees searched and it can easily be trapped into local optima. The main reason, we believe, is there is a lack of quality control in constructing intermediate trees. We cannot expect an accurate tree to be reconstructed after the consensus if a large amount of intermediate trees are less likely, or bad trees. Thus a proper selection of the intermediate trees for consensus is crucial in obtaining a more accurate tree.

To reduce the negative impact caused by less likely trees in final consensus stage, we should evaluate the intermediate trees and delete the bad ones before the consensus. A tree with a higher likelihood value normally is the one that is closer to the correct tree. This is the base of maximum likelihood methods. Therefore, we use the likelihood value of a tree to guide the tree selection for the final consensus. When the number of different intermediate trees to be constructed are large, however, it is not feasible to compute the likelihood value of every intermediate tree. To reduce the computational complexity, we adopt a hierarchical consensus technique. It is described as follows.

For a given number of intermediate trees to be constructed, we divide them into groups of different sizes from 1 to I. For each group we apply the standard QP procedure, that is, we first construct the intermediate trees, do a consensus and then calculate the likelihood value of the consensus tree for the group. The likelihood values of all the consensus trees are compared and only K best ones are kept. Finally, we do another consensus among these K best trees. In contrast to the normal quartet puzzling method, such an arrangement will have an extra cost for evaluating the group consensus trees. How high this extra cost will depend on how the total intermediate trees are divided. When there is only one group, it is exactly the same as the normal quartet puzzling method. In our experiments we find that we are able to obtain better results using less number of intermediate trees. Therefore, the extra computational cost may not be great.

The computation for each group can be done independently and the computational costs for different groups are different. This makes our method more attractive for grid computing.

The grid environment for our modified quartet-puzzling algorithm consists of a management node, various numbers of computing nodes and a resource broker in between. The roles of the resource broker are to keep the information about available computing resources once registered there and to provide the information to the management node upon request. Once being contacted, an available computing node will receive the task from the management node, execute the task, and send the result back to the management node. The management node is responsible for contacting the resource broker for available resources, properly dividing tasks, sending tasks to and receiving partial results from computing nodes. The management node also maintains a list to record the K best trees obtained so far. After all the partial results are received, the management node will do a final consensus on the K best trees kept in the list.

To detect the actual performance of computing nodes, the management node maintains another list which records all the nodes the tasks were assigned before. When a computing node is first contacted, it will be recorded in the list and a small task is assigned to it and the start time (when the task is sent) is recorded. When the result is returned, the finish time (when the result received) is also recorded. The difference between the start and finish times will be considered in determining the size of the next task to be assigned to the node. If the computing node can finish the task quickly and immediately become available again, it will be assigned a larger task. When a computing node cannot finish the task fast enough, the size of the next task will be decreased. By detecting the actual performance of computing nodes, we are able to assign tasks of proper sizes to them and then make the computation more effective and efficient.

3 Experimental Results

We have deployed the United Devices Grid MP Platform on hundreds of PCs and the testing of our grid-based algorithm for large data sets is current under way. Here we present some results to show the accuracy of our method in comparison with the standard QP method. In our simulation, we first generate a random birth-death model tree using Phyl-O-Gen [9] with a constant per-lineage birth rate of 0.23, a constant per-lineage death rate of 0.14 and the lineage being 900. After the model tree is constructed, protein sequences are generated based on an evolution model using PSeg-Gen with commonly used software under JTT evolution model [10]. The likelihood values are compared and the Robinson-Foulds distance [11] is used to measure the differences between the model tree and reconstructed trees.

Some experimental results are depicted in Figure 1. It is clearly show that our modified quartet-puzzling algorithm can obtain much more accurate results than the standard QP.

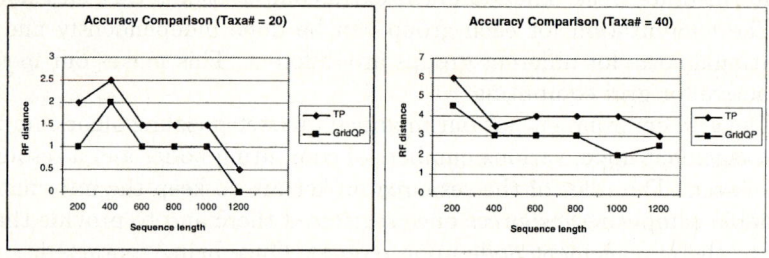

Fig. 1. RF distance comparison when the number of taxa is 20 and 40

Table 1 gives the likelihood values obtained from the same experiments. We can see that the reconstructed trees using our algorithm generally have higher maximum likelihood values than those reconstructed by the original one.

Table 1. Maximum Likelihood Comparison Between TP and GridQP

Seq. length	TP(taxa#=20)	GridQP(taxa#=20)	TP(taxa#=40)	GridQP(taxa#=40)
200	-5914.5	-5903.8	-9256.1	-9256.5
400	-11537.7	-11527.8	-19265.2	-18884.2
600	-17981.2	-17978.7	-29084.8	-29027.3
800	-23095.4	-23084.4	-37378.9	-37364.6
1000	-29228.5	-29213.7	-47634.9	-47563.5
1200	-35953.0	-35950.8	-56405.2	-56392.1

4 Conclusions

The computing complexity prevents mathematically well-founded maximum likelihood methods from being used in large problems. Quartet based approach makes the reconstruction of trees from a large number of protein sequences achievable within a reasonable amount of time. While quartet based maximum likelihood methods gain in speed, our investigation shows that the accuracy of the reconstructed trees are not increased reasonably as more computing time is devoted. In this paper we present a grid based quartet puzzling method which can overcome the loss of accuracy problem and the method is in attempt to efficiently use computing resources to obtain more accurate trees. This is the first quartet puzzling method that takes the advantage of grid environment to improve the accuracy of reconstructed phylogenetic trees. Our simulation shows the accuracies of trees reconstructed are generally better than existing quartet puzzling methods.

References

1. Swofford, D. L., Olsen, G. J., Wadell, P. J., and Hillis, D. M., Chapter 11: Phylogenetic inference. In: Systematic Biology (hillis, D. M., Moritz, C., and Mable, B. K., eds.). Sinauer Associates Sunderland, Massachusetts.
2. Tiffani L. Williams and Bernard M.E. Moret, An Investigation of Phylogenetic Likelihood Methods, Proceedings of the Third IEEE Symposium on BioInformatics and BioEngineering (BIBE'03).
3. Korbinian Strimmer and Arndt von Haeseler, Quartet puzzling: A quartet maximum-likelihood method for reconstructing tree topologies. Mol. Biol. Evol. 13(7):964-969, 1996.
4. Heiko A. Schmidt, Korbinian Strimmer, Martin Vingron and Arndt von Haeseler: TREE-PUZZLE: maximum likelihood phylogenetic analysis using quartets and parallel computing. Bioinformatics; Mar 2002; 18(3): 502-504.
5. Felsenstein, J., Evolutionary trees from DNA sequences: a maximum likelihood approach. J. Mol. Evol. 17:368-376.
6. N. Saitou and M. Nei. The neighbor-joining method: a new method for reconstructing phylogenetic trees. Mol. Biol. Evol., 4:406-425, 1987.
7. Kluge, A. G., and Farris, J. S. Quantitative phyletics and the evolution of anurans. Syst. Zool. 18: 1-32. 1969.
8. Farris, J. S. Methods for computing Wagner trees. Syst. Zool. 19: 83-92. 1970.
9. Andrew Rambaut, Phylogenetic Tree Simulator Package Version 1.1. http://evolve.zoo.ox.ac.uk/software/phylogen/manual.php. 2002.
10. Grassly NC, Adachi J, and Rambaut A., PSeq-Gen: an application for the Monte Carlo simulation of protein sequence evolution along phylogenetic trees. Comput Appl Biosci 13, 559-560. 1997.
11. Robinson, D.R., and Foulds, L.R. 1981. Comparison of phylogenetic trees. Mathematical Biosciences 53: 131-147.

Scheduling BoT Applications in Grids Using a Slave Oriented Adaptive Algorithm[*]

Tiago Ferreto[1], César De Rose[1], and Caio Northfleet[2]

[1] Faculty of Informatics - PUCRS, Brazil
`{ferreto, derose}@inf.pucrs.br`
[2] HP-Brazil
`caio.northfleet@hp.com`

Abstract. Efficient scheduling of Bag-of-Tasks (BoT) applications in a computational grid environment reveals several challenges due to its high heterogeneity, dynamic behavior, and space shared utilization. Currently, most of the scheduling algorithms proposed in the literature use a master-oriented algorithm, in which the master is the only responsible for choosing the best task size to send to each slave. We present in this paper a different approach whose main originality is to be slave-oriented, *i.e.* each slave locally determines, from a set of initial runs, which workload size is more adapted to its capacities and notifies the master of it. Finally, we show some measurements comparing our algorithm with other three well-known scheduling algorithms using the SimGrid toolkit.

1 Introduction

Computational grids as a platform to execute parallel applications is a promising research area. The possibility to allocate unprecedent amounts of resources to a parallel application and to make it with lower cost than traditional alternatives (based in parallel supercomputers) is one of the main attractives in grid computing. On the other hand, the grid characteristics, such as high heterogeneity, complexity and wide distribution (traversing multiple administrative domains), create many new technical challenges. In particular, the area of scheduling faces entirely new challenges in grid computing. Traditional schedulers (such as the operating system scheduler) control all resources of interest. In a grid, such a central control is not possible. First, the grid is just too big for a single entity to control. In a grid, a scheduler must strive for its traditional goals, improving system and application performance [1].

Bag-of-Tasks (BoT) applications are parallel master/slave applications whose tasks are independent to each other. A vast amount of work has been done in order to schedule efficiently Bag-of-Tasks applications improving the load balancing in distributed heterogeneous systems. Most of the algorithms focus on the adaptation of the workload during the execution, using either a fixed increment or decrement (*e.g.* based on an arithmetical or geometrical ratio) or a

---

[*] This research was done in cooperation with HP-Brazil.

more sophisticated function to adapt the workload. Yet the solutions presented are all based on some evaluation by the master of the slaves' capacities and of the tasks workload. This implies a significant overhead since the master has to maintain some kind of information about its slaves.

We propose in this paper the scheduling of BoT applications in Grids with a different approach whose main originality is to be slave-oriented, *i.e.* each slave locally determines, from a set of initial runs, which workload is more adapted to its capacities and informs the master of it. In turn, the master can compare the workload demanded by the slave to the network penalty paid and make the proper adjustments to adapt the workload. We have thus a workload adaptive algorithm.

2 Related Work

In this section we focus in self-scheduling algorithms [2]. These algorithms divide the total workload based on a specific distribution, providing a natural load balancing to the application during its execution. This class of algorithms is well suited for dynamic and heterogeneous environments, such as grids, and for divisible workload applications.

The Pure Self-scheduling [2] or Work Queue scheduling algorithm divides equally the workload in several chunks. A processor obtains a new chunk whenever it becomes idle. Due to the scheduling overhead and communication latency incurred in each scheduling operation, the overall finishing time may be greater than optimal [3].

The Guided Self-scheduling algorithm [4] (GSS), proposed by Polychronopoulos and Kuck, and Factoring [3], proposed by Flynn and Hummel, are based on a decreasing-size chunking scheme. GSS schedules large chunks initially, implying reduced communication/scheduling overheads in the beginning, but at the last steps too many small chunks are assigned generating more overhead [2]. Factoring was specifically designed to handle iterations with execution-time variance. Iterations are scheduled in batches of equal-sized chunks. The total size of the chunk per batch is a fixed ratio of the remaining workload.

In all algorithms shown above, the amount of workload sent to each slave is defined by the master. We propose in the following section another approach, where the evaluation of the load to be assigned to each slave is done by the slave itself.

3 Local Decision Scheduling Algorithm

The local decision scheduling algorithm (LDS) addresses BoT applications using divisible workloads, *i.e.* all independent tasks demands the same amount of computational resources. The algorithm focus on a heterogeneous, dynamic and shared environment, characterizing a typical computational grid. It is based on a distributed decision mechanism, building in each slave a performance model, which represents the application behavior based on resources utilization. Each slave computes the task received, includes this information in its performance model and, based on the analysis of its performance model, calculates the best workload size to be computed at the next iteration.

The scheduling algorithm is divided in the following phases: setup, adaptive and finalization phases. The setup phase goal is to initialize and refine the performance model of each slave. The master sends tasks to slaves using a fixed quadratic increment. This process continues until it receives a signal from the slave in order to start the adaptive phase. This signal is generated when the performance model starts presenting estimates with minimum error.

The adaptive phase goal is to adapt this performance model if any variation is observed and to generate appropriate estimates of workloads size to be computed in the next iterations. The master side of the algorithm for the adaptive phase is presented in Algorithm 1. It sends to the slave a task using the fixed quadratic increment again, but at this time, it includes information about the time slice the slave has to compute at the next iteration (*execTime* variable). This information is highly dependable on the application characteristics, workload (number of tasks), and environment conditions, and is currently static and manually defined. The master receives, after the processing of the task by the slave, the result, execution time of the task computed and an estimation of the next workload size in order to accomplish to the time slice defined at the master. At the next workload assignment for the slave, the master just changes the workload size to send to the slave accordingly to the estimate previously received. It keeps using this procedure until it reaches a specified limit (line 4). After this, it starts the finalization phase.

The slave side of the algorithm for the adaptive phase is presented in Algorithm 2. The slave starts a loop receiving tasks to be computed. Together with the task, it receives the workload size and execution time values. The workload is computed and its size with execution time inserted in a prediction table, which is used to compute the performance model. Using this prediction table and the execution time value received from the master, it computes the next workload size. After that, the slave sends to the master the result, execution time of the task received, and an estimation of the next workload size. The slave gets out from the loop when it receives a signal message from the master to initiate the finalization phase.

The finalization phase adjusts the workloads size computed in each slave in order to achieve load balancing, resulting in a better overall performance. When the master switches from the adaptive phase to the finalization phase, it stops using slaves' predictions and starts using the factoring algorithm till the end of tasks processing. After assigning the remaining tasks, the master starts a loop receiving the remaining results from the slaves.

3.1 Local Prediction of the Computational Load

In order to estimate the most suited workload, a slave needs a performance model for the execution of chunks of size $taskSize_i$. The model may include various data such as the execution time, memory utilization, cache access, etc, used to

Algorithm 1 LDS algorithm at master side

1: **while** there are tasks to schedule **do**
2:     **for** each available $slave_i$ **do**
3:        $execTime \Leftarrow maxExecTime$
4:        **if** $\sum_{i=1}^{numslaves} taskSize_i \geq$ number of tasks remaining **then**
5:           start finalization phase
6:        **else**
7:           $task \Leftarrow \mathbf{getTask}(taskSize_i)$
8:           send to $slave_i$ the $task$, $taskSize_i$ and $execTime$
9:        **end if**
10:     **end for**
11:     receive $result$, $execTime$ and $nextTaskSize$
12:     $taskSize_i \Leftarrow nextTaskSize$
13: **end while**

Algorithm 2 LDS algorithm at slave side

1: **while** there are tasks to compute **do**
2:     receive $task$, $taskSize$ and $maxExecTime$
3:     $result \Leftarrow$ compute $task$
4:     **insertPredictionTable**($taskSize$, $execTime$)
5:     $nextTaskSize \Leftarrow \mathbf{predictNextSize}(maxExecTime)$
6:     send to master the $result$, $execTime$ and $nextTaskSize$
7: **end while**

process a given task. In this preliminary version of our prototype we only take into account the execution time.

Given some N values $taskSize_1, taskSize_2, \ldots taskSize_n$ and the slaves data t (e.g. the execution time) the slave has to estimate $t(taskSize)$. In a multi-parameter model we could use algorithms such as the Singular Value Decomposition [5], one of the most robust for data modeling. It would fit the function t as a linear combination of standard base functions (e.g. $x \rightarrow e^x$, $\sqrt{}$, polynomials, ...).

Yet in the case where t only depends on the processor's speed, an affine model of the time required vs. the number of chunks to run is most realistic and used by other algorithms [6]. The modeling problem is therefore a basic linear interpolation problem of the measured running time $t_j, j = 1 \ldots n$ vs. the number of chunks $taskSize_j$. Beside the estimated coefficients a, b of the affine approximation $t = a + b \times taskSize$, the correlation coefficient is used to determine the correction of the interpolation and thus decide if more chunks should be sent in the initial phase, before entering in the adaptive phase.

The interpolation algorithm is very fast and thus does not prejudice the execution of the application. Moreover, it is trivial for a slave to determine the adapted task size, given the execution time t it has to run and the affine model (a, b). Note that in the case of a more complex, non-linear model, it would have to use a more time-consuming algorithm such as a gradient or dichotomic search to solve the $t = f(taskSize)$ equation.

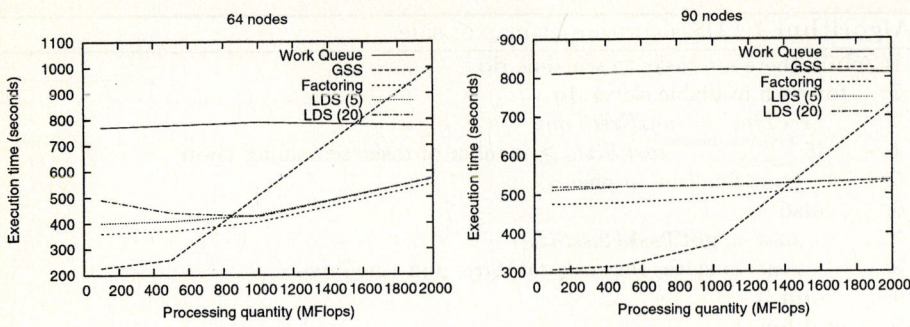

Fig. 1. Measurements scheduling 1000 tasks using 64 and 90 nodes

4 Evaluation

We used the SimGrid [7] toolkit to evaluate our scheduling algorithm. The platform used for simulation is an example of grid model included in the SimGrid package. The model is composed by 90 heterogeneous machines connected by several links with different latency and bandwidth values.

We used this platform to simulate applications with different number of tasks (1000, 10000 and 100000 tasks) and quantity of computation per task (100, 500, 1000 and 2000 MFlop/s) using deployments with 64 and 90 nodes. In our experiments we assumed that communication costs to send one task to a slave is fixed (0.001 Mbyte/s) and to receive the result is irrelevant.

Each simulated application was executed using 4 different algorithms: Work Queue, Guided Self-scheduling, Factoring (using $\alpha = 2$) and LDS (using $\beta = 5$ and 20). LDS uses the β value to compute the $maxExecTime$ parameter, which is calculated dividing the estimate of the total execution time to compute all tasks sequentially by β multiplied by the number of slaves.

Figure 1 illustrates the measurements obtained for an application containing 1000 tasks begin executed in 64 and 90 nodes of the platform, with computation amount per task varying from 100 to 2000 MFlop/s. Using 64 nodes and computation quantity ranging from 100 to approximately 800 MFlop/s, the GSS algorithm presented the best results, after 800 MFlop/s the Factoring algorithm overcomed GSS. The same behavior is presented with 90 nodes, except that this transition is observed when computation quantity is approximately 1400 MFlop/s. The Work Queue algorithm presented the worst results in both simulations. The LDS algorithm presented a behavior similar to the Factoring algorithm in both simulations.

In Figure 2 the same measurements are presented for an application with 10000 tasks. The Work Queue algorithm presented again the worst results using 64 and 90 nodes. GSS, Factoring and LDS(5) presented similar values using low computation quantity (100 MFlop/s) and 64 nodes. Using tasks with more than 400 MFlop/s the LDS(5) and LDS(20) presented the best results. Using 90 nodes and computation quantity ranging from 100 to approximately 200 MFlop/s the

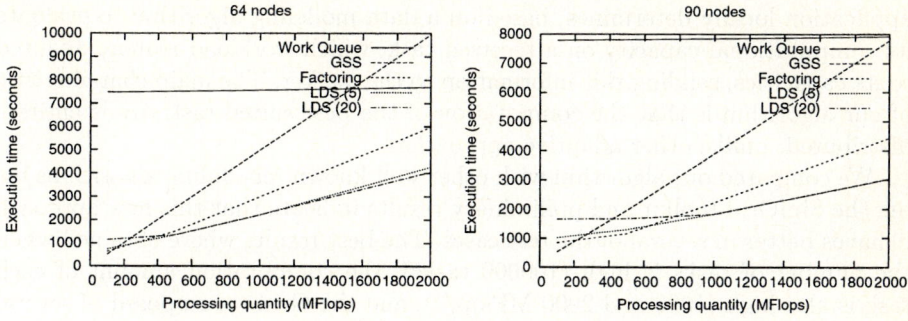

Fig. 2. Measurements scheduling 10000 tasks using 64 and 90 nodes

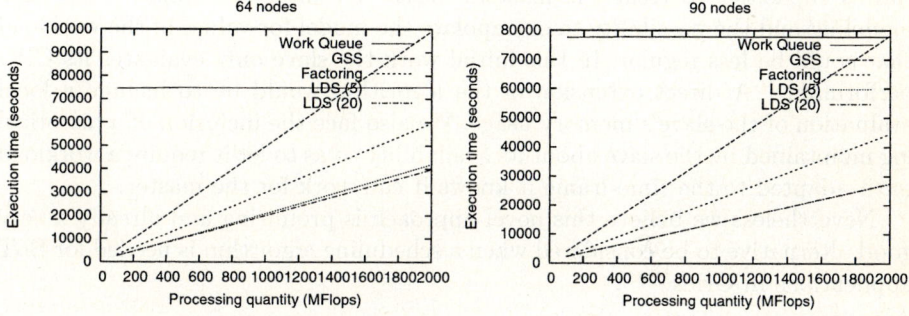

Fig. 3. Measurements scheduling 100000 tasks using 64 and 90 nodes

GSS algorithm presented the best results. Factoring behaved better from 200 to approximately 600 MFlop/s and after 600 MFlop/s, the LDS(5) and LDS(20) algorithm overcomed the other 3 algorithms.

The measurements for 64 and 90 nodes using an application with 100000 tasks (Figure 3) are very similar. The LDS algorithm presented the best results for all experiments and overcomed the Factoring algorithm in approximately 30%, *i.e.* the execution time of the application using the LDS algorithm was approximately 30% faster in comparison to the Factoring algorithm.

The measurements show that simple algorithms, such as, GSS and Factoring present good results when the total number of tasks and the computation quantity per task is low. With a higher number of tasks and computation quantity the LDS algorithm performs better, obtaining in some cases, a reduction of 30% in the execution time in comparison to the Factoring algorithm.

5 Conclusion and Future Work

In this paper we proposed a slave oriented adaptive algorithm for the scheduling of BoT applications in Grid environments. In this approach, each slave of a BoT application locally determines, based on a data modeling algorithm to evaluate its computational capacity on a received task, which workload is more adapted to its capacities, sending this information to the master. The main characteristic of our algorithm is that the computation of the best suited task size is entirely distributed, unlike other adaptive approaches.

We compared our algorithm with other well-known scheduling algorithms using the SimGrid toolkit and preliminary results indicate that this new approach behaves better in several of the test cases. The best results where obtained when the number of tasks is high (100000 tasks), the computation amount of each task is also high (1000 and 2000 MFlop/s), and the Grid is composed of several heterogeneous resources (90 nodes) resulting in a mean performance increase of approximately 30% over the Factoring algorithm.

The main limitation of our algorithm currently lies in the modeling of the slave's capacities to treat the master's tasks. We intend to improve the data modeling and the possibility to extrapolate the model for values of the workload that could be less regular. In this initial work the slave only evaluates its CPU performance. A direct extension of the algorithm would be to include a local evaluation of the slave's memory usage. We also face the inclusion of a historical log maintained by the slave about its availability so as to let it require a workload most adapted to the time-frame it knows it can work for the master.

Nevertheless we believe this novel approach is promising and already a very good alternative to be considered when a scheduling algorithm is needed for BoT applications in Grids.

References

1. da Silva, D.P., Cirne, W., Brasileiro, F.V.: Trading cycles for information: Using replication to schedule bag-of-tasks applications on computational grids. In: Euro-Par 2003. Volume 2790 of Lecture Notes in Computer Science., Springer (2003) 169–180
2. Chronopoulos, A.T., Andoine, R., Benche, M., Grosu, D.: A CLass of Loop Self-Scheduling for Heterogeneous Clusters. In: Proceedings of CLUSTER'2001. (2001)
3. Hummel, S.F., Schonberg, E., Flynn, L.E.: Factoring: A Method for Scheduling Parallel Loops. Communications of the ACM **35** (1992) 90–101
4. Polyhronopoulos, C.D., Kuck, D.: Guided Self-Scheduling: A Practical Scheduling Scheme for Parallel Supercomputers. IEEE Trans. on Computers **36** (1987) 1425–1439
5. Press, W.e.a.: Numerical Recipes in C: The Art of Scientific Computing. Number ISBN 0521431085. Cambridge University Press (1993)
6. Beaumont, O., Legrand, A., Robert, Y.: Scheduling divisible workloads on heterogeneous platforms. Parallel Computing **29** (2003) 1121–1152
7. Casanova, H.: Simgrid: A toolkit for the simulation of application scheduling. In: Proceedings of the IEEE Symposium on Cluster Computing and the Grid (CCGrid'01). (2001)

A Clustering-Based Data Replication Algorithm in Mobile Ad Hoc Networks for Improving Data Availability

Jing Zheng, Jinshu Su, and Xicheng Lu

School of Computer, National University of Defense Technology,
Changsha 410073, Hunan, China
zhengjing621@hotmail.com

Abstract. In Mobile Ad Hoc Networks (MANET), network partitioning occurs when nodes move freely and cause disconnections frequently. Network partitioning is a wide-scale topology change that can cause sudden and severe disruptions to ongoing data access, and consequently data availability is decreased. A new distributed clustering algorithm is presented in this paper for dynamically organizing mobile nodes into clusters in which the probability of path availability can be bounded. Based on this clustering algorithm, a data replication policy is proposed to improve data availability during network partitioning. Through theoretic analysis our algorithm has proper complexity. Simulation results show that the clusters created by our clustering algorithm have desirable properties and our algorithms improve the data availability effectively in MANET environment.

1 Introduction

In MANETs, mobile nodes move freely, and disconnections of links occur frequently. This may cause frequent network partitioning, which can cause sudden and severe disruptions to ongoing data access, then it poses significant challenges to data management in MANET environment.

Data availability is referred to the ratio between the number of requests to a data object and that of responses. To provide data availability during network partitioning, a clustering based data replication algorithm is proposed in this paper. The basic ideal of the algorithm is that the data object in the clusters which request this data object will be replicated to prevent deterioration of data accessibility at the point of network partitioning. The clustering mechanism is used to predict the network partitioning, and should present the following desirable properties: 1) Each cluster is connected, and there is at least one stable path between any pair of nodes in a cluster. The path stability guarantees the data availability in a cluster. 2) The number of clusters should be limited. Because there is at least one replica for every requested data in a cluster, more clusters will cause more replicas. 3) Two clusters should overlap appropriately.

This research was supported by the National Natural Science Foundation of China (No. 69933030).

Overlap between clusters can decrease the number of replicas, i.e. data replica on the common nodes will allow nodes in two clusters accessing the public replica. But all common nodes of two clusters will have to maintain cluster state for both the clusters. Hence, it is necessary to make trade-off between them. 4) Cluster should be stable across node mobility. The clustering algorithm should adapt to nodes joining in the network, disappearing from the network and moving.

Clustering based routing algorithm has been an extensive research topic. However, only a few researches have focused on clustering problem for data availability. [1] began the study of the clustering problem in ad hoc networks by proposing a *linked cluster* algorithm. Two distributed algorithms introduced in [2] choose cluster-head based lowest ID and highest degree respectively. Another method used for clustering is based on *domination set* in a graph ([3][4][5]). Above-mentioned clustering algorithms are used to support routing in MANET, forming *cluster-based hierarchical routing* or *backbone-based routing*. Because the goals of clustering are different from us, the requirements for clustering are different too. [6] proposes a method to ensure that the centralized service is available to all nodes during network partitioning for group mobility model. T.Hare' work [7] focuses on data accessibility in MANET, but does not consider topology changes and connection stability. [8] presents a stable path based clustering routing algorithm, but the method to compute path availability is complicated and the goal of clustering is different from us.

The rest of the paper is organized as follows: in section 2 we pose our problem statement in a graph theoretic framework; in section 3 a distributed algorithm to construct α-*Stable Graph* is presented in detail; in section 4, the clustering-based data replication algorithm is described; in section 5, algorithms are evaluated through simulations; and finally in section 6, the summary and some future work are presented.

2 Problem Statement

We use a widely employed model, so-called *Unit Disk Graph* (UDG) G(V, E), to study ad hoc networks: Vertices in G are nodes that are located in the Euclidean plane and assumed to have identical transmission radio. An edge between two vertices, representing that corresponding two nodes are in mutual transmission range, exists iff their Euclidean distance is not greater than the maximal transmission distance.

Definition 1. α-*Stable Neighbor*: During time slice Δt, the connectivity probability between node i and its neighbor j is denoted as $Pr(link_{i,j})$. Node j is a α-*stable neighbor* to i, if $Pr(link_{i,j})$ is larger than α (α is the threshold). Let $SN(i)=\{j| Pr(link_{i,j})\}>\alpha\}$ be the α-stable neighbors set of i.

Definition 2. α-*Stable Path*: The path $Path_{m,n}$ between node m and n is comprised of links $link_{i,j}$, $link_{i,j} \in Path_{m,n}$, and the *path availability* is denoted as $Pr(Path_{m,n})$. According to the assumption of independent link failures, the path availability is given by $Pr(path_{m,n}) = \prod_{(i,j) \in path_{m,n}} Pr(link_{i,j})$. $Path_{m,n}$ is a α-stable path, if $Pr(Path_{m,n})>\alpha$.

Definition 3. *α-stable path nodes set*: α-stable path nodes set for i, $SP(i)$, for every node j in nodes set $SP(i)$, if there is at least one α-stable path between node i and j. $SN(i) \subseteq SP(i)$.

Definition 4. *α-stable graph* G_α: The α-stable graph of graph G, $G_\alpha(V, E')$, vertices in G_α are the same as those in G, edge $(m, n) \in E'$ iff there is at least one α-stable path between node m and n, i.e. $Pr(Path_{m,n}) > \alpha$.

Fig.1. shows the UDG of a mobile ad hoc network, and Fig.2 is the G_α for Fig.1. In Fig.2, the dashed line between two nodes indicates that there is a α-stable path between them, and the solid line between two nodes indicates that they are stable neighbors each other.

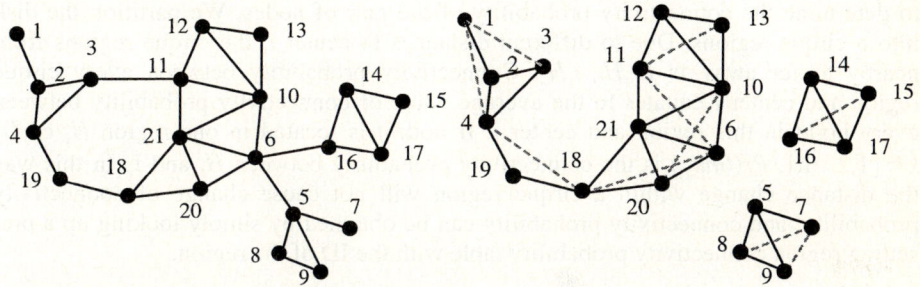

Fig. 1. UDG **Fig. 2.** *α-stable graph* G

In G_α, mobile nodes are separated into several disconnected sub-graphs, i.e. there is no any stable paths between two nodes that belong to different sub-graphs respectively; therefore, the probability of network partitioning between two sub-graphs is high. Our goal is to construct clusters in which the path availability between any two nodes is bigger than α, and the size of each cluster is as large as possible. A clique in graph G_α indicates that there is at least one α-stable path between any two nodes in the clique, therefore the clustering problem can be mapped to the *maximum clique problem* (MCP) in graph G_α. However, the cluster in our algorithm must satisfy the property: *each cluster is connected in graph G*. The graph meeting above two requirements simultaneously may not exist. We create clusters that are cliques in G_α and connected in G, so that the previous desired properties are maintained.

3 Distributed Algorithm to Construct α-Stable Graph (α-SGA)

3.1 The Connectivity Probability Between Neighbors

Assuming that link disconnection is caused by node mobility, the distance between two neighbors is used as the metric of connectivity probability between neighbors. We assume that the effective wireless communication range of node i is the disk D_i for which i is its center and R is its radius. The distance between node i and its neighbor j is d, and all nodes move in the Random Waypoint Mobility Model, i.e.

each mobile node's speed is uniformly distributed between (0, Vel_{max}) and the direction is uniformly distributed over (0, 2π) (so the maximum speed of node j relative to node i is $2Vel_{max}$.) . Therefore in time slice Δt the position of node j is uniformly distributed in disk D_j for which center is node j and radius equates to $2Vel_{max}\Delta t$. Finally, the connectivity probability between i and j in time slice Δt can be described as:

$$\Pr(link_{i,j}) = \frac{S(D_i,R) \cap S(D_j, 2Vel_{max}\Delta t)}{S(D_j, 2Vel_{max}\Delta t)} \quad (1)$$

$S(D_i)$ means the area size of disk D_i. If $Pr(link_{i,j})>\alpha$, j is a stable neighbor of i.

Δt is a systemic parameter, which presents the time interval to estimate the validity of connectivity probability. As formula (1), if Δt and Vel_{max} have been known, it was d to determine the connectivity probability of the pair of nodes. We partition the disk into n cirque regions. Due to different distances to center i, the cirque regions from nearby to far away is $H_1, H_2, ... H_n$. Connectivity probability between every cirque region and center i equates to the average value of connectivity probability between every point in this region and center i. If node j is located in one region H_k of D_i, $k \in \{1,2...n\}$, $Pr(link_{i,j})$ is the connectivity probability between H_k and i. In this way the distance change within a cirque region will not cause change of connectivity probability, and connectivity probability can be obtained by simply looking up a pre-setting region connectivity probability table with the ID of the region.

3.2 Distributed α-Stable Graph Algorithm (α-SGA)

We propose a fully distributed algorithm α-SGA to construct above α-stable graph by only exchanging information between neighbors. On each mobile node i, a global ID $id(i)$ and following local states is maintained in the distributed algorithm:

1) *Profile of Measurement of Distance to all Neighbors*, $P(i)$. Every mobile node i measures the distance to each neighbor j, and calculates the corresponding connectivity probability $Pr(link_{i,j})$, then record it in 2-tuple $< id(j), Pr(link_{i,j}) >$.

2) *The Set of Stable Neighbors*, $SN(i)$. This set is constructed by all these nodes which satisfy definition 1, using $P(i)$. In $SN(i)$, the ID of i's every stable neighbor j and its connectivity probability to i are recorded in 2-tuple $< id(j), Pr(link_{i,j}) >$.

3) *The Set of α-Stable Path Nodes*, $SP(i)$. This set records all such nodes that satisfy definition 3. For each $j \in SP(i)$, a 3-tuple $< id(j), id(N(Path_{i,j})), Pr(Path_{i,j}) >$ is maintained. If multiple paths exist between node i and j, $Pr(Path_{i,j})$ is the maximum path availability among these paths. $N(Path_{i,j})$ is the next hop node from i to j in this selected path with the maximum availability probability.

4) *The SN(j) and SP(j) of Stable Neighbor j*, $PN(j)$. For each α-stable neighbor j, i maintains $SN(j)$ and $SP(j)$ of j. Each $PN(j)$ has a 4-tuple $< id(m), Attribute(m), Pr(Path_{j,m}), Pr(Path_{i,j,m}) >$, $m \in SP(j) \cup SN(j)$. $Pr(Path_{i,j,m})$ is the availability probability of path from node i to m, passing j.

The distributed algorithm α-SGA (α-Stable Graph Algorithm) allows mobile nodes to find their stable neighbors and α-stable path nodes by exchanging information with their stable neighbors and construct their α-stable graph $G_α$. Algorithm is described as below:

1) Measurement. Every mobile node i calculates the connectivity probability to each neighbor. For each $j \in P(i)$, if definition 1 has been satisfied, i.e. $Pr(link_{i,j})>α$ then $SN(i)=SN(i) \cup \{j\}$. During initiation, $SP(i)=SN(i)$.
2) Exchange. i exchanges its $SN(i)$ and $SP(i)$ with all its stable neighbors. It sends 3-tuples $<id(m), Attribute(m), Pr(Path_{i,m})>$ to all its stable neighbors. There $m \in SP(j) \cup SN(j)$. If $Attribute(m)=0$, it means that m has been just deleted from $SP(j)$. Upon receiving information from the stable neighbors, i construct its $SP(i)$ and $SN(i)$.
3) Update. When the connectivity probability between i and its neighbor j has been changed, or i receives update information from other stable neighbors, state information $P(i)$, $SN(i)$, $SP(i)$ and $PN(j)$ should be updated. Algorithm's detail is described in Table1.

Table 1. Update Operation of α-SGA

---

if connectivity probability between i and its neighbor j has been changed **then**
  update $P(i)$;
  if $j \in SN(i)$ and $Pr(link_{i,j}) \leq α$ **then** $SN(i):= SN(i)/\{j\}$, delete all info in $PN(j)$; **endif**;
  if $j \in SN(i)$ **then** update corresponding probability value in $SN(i)$ and $PN(j)$;
      add j to update message queue;
  endif;
  if $j \notin SN(i)$ and $Pr(link_{i,j}) > α$ **then**
      $SN(i):= SN(i) \cup \{j\}$, $SP(i):= SP(i) \cup \{j\}$;
      update $SN(i)$ and $PN(j)$; add j to update message queue;
  endif;
  for all $m \in SP(i)$ & $id(N(Path_{i,m}))==id(j)$
      recalculate $Pr(Path_{i,m})$;
      if $\exists k \in SN(i)$, make $Pr(Path_{i,k,m})>α$ & $Pr(Path_{i,k,m}) == \max_{l \in SN(i)}\{Pr(Path_{i,l,m})\}$ **then**
          update $SP(i)$; $id(N(Path_{i,m})):= id(k)$; $Pr(Path_{i,m}):= Pr(Path_{i,k,m})$;
          add m to update message queue;
      else $SP(i):= SP(i)/\{m\}$; add m to update message queue; $Attribute(m):= 0$;
      endif;
  endfor;
  propagate update message to all stable neighbors except j;
endif;
if i receives update message from stable neighbor j **then**
  do similar update to $SP(i)$, $SN(i)$ and $PN(j)$ as above
  propagate update message to all stable neighbors except j;
endif;

By exchanging information with stable neighbors, α-SGA makes every node i construct a set $SP(i)$ containing all the nodes that have α-stable path to i. There is a edge between i and each node in $SP(i)$ in $G_α$. Because messages are exchanged between pairs of stable neighbors, and the change of connectivity probability caused by node's movement only is propagated among its α-stable path nodes, the communication overhead to maintain $G_α$ is dependent on the size of the α-stable path nodes set. Because usually the size of the α-stable path nodes set is much smaller than the total size of the network and only the incremental update is needed, the communication overhead of this algorithm is acceptable in MANET environment.

3.3 Heuristic Connected Clique of α-Stable Graph Algorithm (CCGA)

We propose CCGA (Connected Clique Graph Algorithm) algorithm to construct clique in G_α, based on sequential greedy heuristic method [9]. This clique contains node i and is connected in graph G.

Table 2. The CCGA algorithm

```
procedure local-search-add-move ( output: Clique (i); input: SP(i), SN(i) )
begin
 1    PA : =SP(i), Clique (i) : = i;
 2    repeat
 3        PA' : = ∅;
 4        for all j ∈ Clique (i)
              PA' : =PA' ∪ (PA∩SN(j));
          endfor;
 5        search node v ∈ PA' & SP(v)∩PA' =max_{k∈PA'}{SP(k)∩PA'}
 6        if  for all u ∈ Clique (i), v ∈ SP(u) is true then Clique (i) : = Clique (i) ∪ {v}; endif;
          PA: =PA/{v};
          if k ∈ SN(v) ∩PA and SN(k) ∩PA= ={v} then PA: =PA/{k}; endif;
 7    until PA= =∅;
 8    return Clique (i);
end;
```

In the algorithm, *Clique (i)* is the node set of clique containing node i in G_α; *PA* is the set of all candidate nodes to be added to *Clique(i)*; *PA'* is composed of all the candidate nodes which at least has a stable neighbor in *Clique(i)*. In line 6 of algorithm, if candidate node *v* at least has a stable path with every node in *Clique (i)*, then *v* is added to *Clique (i)*, so *Clique (i)* is a clique containing *i* in G_α; line 5 searches the node which have maximum number of edges with other candidate node. Algorithm complexity is $O(|SP(i)|^3)$.

4 The Clustering-Based Data Replication Algorithm

4.1 α-Stable Path Based Clustering Algorithm (α-SPCA)

The distributed α-SPCA algorithm is based on algorithms presented in the last section. The α-SPCA algorithm has two phases: Cluster Creation and Cluster Maintenance. The cluster creation is invoked when network is in the initialization phase .The cluster maintenance is an inexpensive phase of the algorithm that handles node mobility leading to the local change of the existing cluster.

■ **Cluster Creation**

1) Every node i obtains its α-stable path nodes set $SP(i)$ by exchanging information with its stable neighbors.
2) If node i does not belong to any cluster and is the least ID node among its stable neighbors which is not clustered, node i becomes a cluster head.
3) Cluster head i executes CCGA algorithm to find $Clique(i)$ in G_α. The nodes in the $Clique(i)$ compose a cluster whose cluster head is node i and *Cluster ID* (CID) equates to the ID number of i.
4) Node i which is not clustered repeats step 2) and 3) until every node belongs to at least a cluster.

■ **Cluster Maintenance**

Once cluster creation phase generates a set of clusters, the cluster maintenance phase is invoked to perform some small changes to handle node mobility as new nodes join and existing nodes leave a cluster.

1) Node Joins: When cluster head i finds any new stable path node j, i checks every cluster member in the cluster if node j is in its stable path nodes set and if node j is in some cluster member's stable neighbors set. If so, node j joins the cluster.
2) Node Leaves: If there is no longer a stable path between a pair of nodes in the same cluster, the node that has the larger ID and is not the cluster head leaves the cluster. If the leaving node does not belong to any cluster and can not join other clusters, it executes CCGA algorithm to construct a new cluster. If the node has no stable neighbor, it is called orphan node. The orphan node is a very unstable node in the network.
3) Cluster Removes: If all of the cluster members in a cluster C_i belong to multiple other clusters, the cluster head of C_i sends the apply for removing C_i to cluster heads whose cluster cover some members of C_i. If all these cluster heads agree removing C_i, the cluster head of C_i declares C_i is vanished. In case some nodes do not belong to any cluster when multiple clusters are removed simultaneously. If two clusters have the identical members, the cluster that has the larger CID is removed.

The cluster created by the α-SPCA algorithm described above has the following properties:

Property 1. The path availability between any pair of nodes in the same cluster is larger than α.

Property 2. Each cluster is connected in graph G.

Property 3. Each node i belongs to at most $|SP(i)|$ clusters.

The proofs of these properties are omitted because of limit of the space.

4.2 Clustering-Based Data Replica Allocation Algorithm (CDRA)

Based the α-SPCA algorithm, the network is clustered into several clusters. The path availability between any pair of nodes in the same cluster is bigger than α, and network partitions often present between clusters especially clusters without overlap. In the face of the challenge of network partitioning to data access, *CDRA* algorithm is proposed in this paper. The basic ideal of this algorithm is that replication of the data object in the clusters which request this data object will prevent deterioration of data accessibility at the point of network partitioning. The CDRA algorithm is described as following:

1) Every cluster head maintains states of all other cluster heads in the networks. When a node requests to access a data object, the node broadcasts the access request in the whole cluster C_i that the node belongs to. If there are some replicas of the data object in C_i, the closest replica node serves the access request.
2) If there is not replica node for the requested data object in C_i, the request is propagated from the cluster head of C_i to all other cluster heads. If there is replica in some cluster, the cluster head sends the data to the cluster head of C_i.
3) The cluster head of C_i sends the data to request nodes. A node in C_i is chosen to replicate the data object, which has request to the data object. Nodes in multiple clusters have the priority to be chosen as replica nodes. Because the path availability between any pair of nodes in a cluster is bigger than α, the data availability in a cluster is bigger than α.
4) The adaptive replica allocation algorithm (*ARAM*) [10] proposed by us is used to allocate the replica in the inter-cluster. The ARAM algorithm dynamically adjusts location and number of replicas adapting to the nodes motion and the change of read-write pattern [10].

For the several replicas of the same data object in the networks, the *ROWA (READ-ONE-WRITE- ALL)* policy is used to ensure the consistency of the replicas. 2PL is used to ensure strict consistency for intra-cluster, and optimistic concurrency control is used to ensure weak consistency for inter-cluster. The write requests for the data object are propagated to all cluster heads whose cluster has the replica of the data object, and then are forwarded to the replica nodes in the cluster. If the write request is granted, data update message forwards to all the replica nodes in the same way. Because of using the hierarchical control, replica update messages are propagated among cluster heads and among replica nodes in the same cluster. Therefore the communication overhead of replica update operations will be reduced.

5 Simulation and Analysis

To evaluate the performance of our algorithm, extensive simulations have been performed. Because our main concern is to improve data availability in the presence of frequent network partitioning events, simulations are performed in a MANET environment with sparse mobile nodes. 250 Nodes are initially randomly activated within a bounded region of 5*5 km, and transmission range R=0.5 km. All nodes move in the Random Waypoint Mobility Model. A range of node mobility with mean speeds *MVel* between 5 to 10 m/s is simulated. During each epoch the speeds of each mobile node are uniformly distributed over (0, 2*MVel*), and the direction is uniformly distributed over (0, 2π). The pause-time is 4s. Two values of the path availability threshold α are used, 0.4 and 0.6. The systemic parameter Δt is 20s. In these simulations, the read requests issued by every node are uniformly distributed from 0 to 20, the ratio between read requests and write requests is 10:1, and there is only 1 data object.

■ **Comparison Properties of the Cluster with (α, t)-Cluster Algorithm**

We compare properties of the cluster in α-SPCA with those in (α, t)-Cluster (t=1 minute) algorithm[10] in the simulation. The results show that the cluster created by α-SPCA algorithm can achieve the desirable properties described in section 1, and the α-SPCA clustering algorithm is more suitable for data replication than the (α, t)-Cluster algorithm.

Fig. 3 (a) shows the effects of mobility on mean cluster size. The results show that the α-SPCA clustering algorithm adapts cluster size to node mobility and threshold α. Another observation indicates that the mean cluster size in α-SPCA is bigger than that in (α, t)-Cluster algorithm, and the reason is that in the α-SPCA algorithm cluster head adds nodes as many as possible into its cluster and allows proper overlaps among clusters, but in (α, t)-Cluster algorithm every node only belongs to one cluster, and although there is a stable path between any pair of nodes in two clusters, these two clusters can not be combined into a single cluster. Accordingly, the number of cluster relating to the size of cluster, mean number of cluster in α-SPCA is smaller than that in the (α, t)-Cluster algorithm. According to our replication strategy, the more clusters, the more replicas may be required. Therefore, fewer replicas are required in the α-SPCA than in (α, t)-Cluster algorithm.

Fig. 3. (b) demonstrates the desirable stability property of the cluster. Cluster survival time is measured by taking the amount of elapsed time of each currently active cluster. Thus, it represents the lifetime of cluster. The chart implies that the speed affects the stability of cluster topology, and the higher link failure rates are observed at the higher speed. A cluster is removed only when all its members belong to other clusters in the α-SPCA, so the cluster is stable and the mean cluster survival times can be accepted in terms of system performance. The great jump in the point at 10m/s in (α, t)-Cluster algorithm is due to the very low probability of a node actually being clustered.

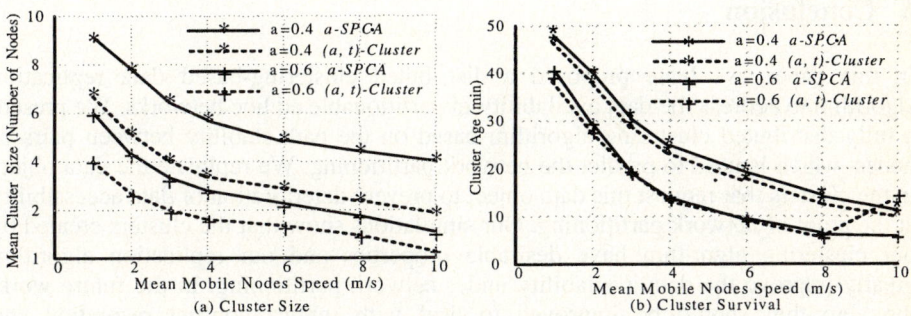

Fig. 3. Comparison Cluster Properties of the Different Algorithms

■ **Comparison Data Availability with ARAM Algorithm**

We compare *ARAM* algorithm and *Static Replica Allocation* algorithm (i.e. *SRA*, replicas are distributed on fixed nodes, and the replica allocation scheme doesn't change during the whole process of simulation) with our algorithm. The data availability of the two approaches is plotted in Fig. 4. With nodes moving, the data availability changes whenever the network partitions or merges. Because the first two approaches do not consider the effects of network partitioning on data access, Fig 4(a) shows more periodic and frequent rises and drops in data availability than Fig 4(b). Predicting that the network partitions often present between clusters, we replicate data object in every cluster that requests this data object. Thus, our algorithm greatly improves the data availability to 96.5%.

We can draw two conclusions from simulations. Firstly, in our algorithm, dynamic allocation of the replica effectively improves data availability, since the allocation is decided based on the changing network topology and the prediction of the network partitions. Secondly, our distributed algorithm only cannot ensure data availability of those clusters in which all nodes have not requested the data object before network partitioning occurs. Therefore the data availability is interrupted by chance in Fig. 4(b).

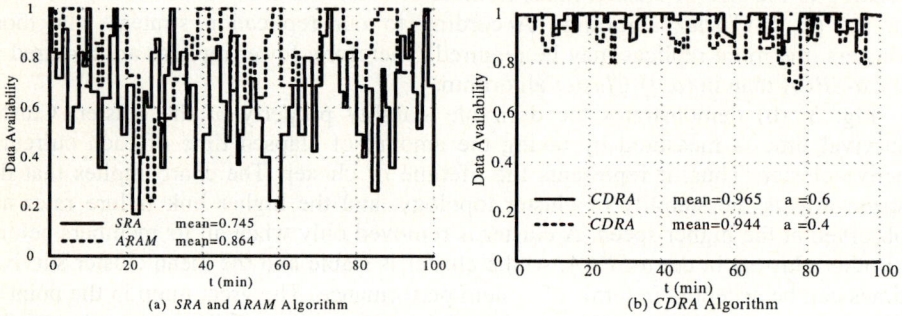

Fig. 4. Comparison Data availability of Different Algorithms

6 Conclusion

In this paper, we have proposed a distributed clustering-based data replication algorithm to address the data availability in partitionable ad hoc networks. We present a fully distributed clustering algorithm based on the path stability between pairs of nodes, which is used to predict the network partitioning. We replicate the data object in the clusters that request this data object to prevent deterioration of data accessibility at the point of network partitioning. Our simulations show that the clusters created by our clustering algorithm have desirable properties and our replication algorithm greatly improve the data availability under network partitioning. In the future work, the algorithm should be improved to deal with replica conflict resolution and reconciliation problem during network partitioning in the MANET environment.

References

[1] D. Baker, A. Ephremides, and J. A. Flyynn, "The Design and Simulation of a Mobile Radio Network with Distributed Control," *IEEE Journal on Selected Area in Communications*, SAC-2(1):226-237, 1984.

[2] M.Gerla and Jack T. Tsai, "Multicluster, Mobile, Multimedia Radio Network", *Wireless Networks*, 1:255-265, 1995.
[3] Y.P.Chen, A.L.Liestman, "Approximating Minimum Size Weakly-Connected Dominating Sets for Clustering Mobile Ad Hoc Networks", *MOBIHOC'02*, June 9-11,2002,EPFL Lausanne, Switzerland. pp. 165-172.
[4] M.Q.Rieck, S. Pai, and S. Dhar, "Distributed Routing Algorithm for Wireless Ad Hoc Networks Using d-Hop Connected d-Hop Dominating Sets", In Proceedings of *the 6th International Conference on High Performance Computing: Asia-Pacific (HPC Asia 2002)*, Tata McGraw Hill, 2002.Vol 2, pp. 443-450.
[5] J.Wu and H.Li, "A Dominating-Set-Based Routing Scheme in Ad Hoc Wireless Networks", *Special issue on wireless networks in the Telecommunication Systems Journal*. Vol. 3, 2001. pp. 63-84.
[6] K.Wang and B.Li, "Efficient and Guaranteed Service Coverage in Partitionable Mobile Ad-hoc Networks", *INFOCOM'02*, New York, June 2002. pp.1089-1098.
[7] T.Hare, "Effective Replica Allocation in Ad hoc Networks for Improving Data accessibility", *IEEE Infocom 2001*, 2001. pp. 1568-1576.
[8] A.B.McDonal and T. Znati, "A Mobility Based Framework fo Adaptive Clustering in Wireless Ad-Hoc Networks", *IEEE Journal on Selected Areas in Communication*, Vol. 17, No. 8, August 1999. pp. 1466-1487.
[9] I.M.Bomze, M.Budinich, P.M.Pardalos, and M.Pelillo, "The Maximum Clique Problem". In D.-Z. Du and P. M. Pardalos, editors, *Handbook of Combinatorial Optimization*, volume 4. Kluwer Academic Publishers, Boston, MA, 1999.
[10] J.Zheng, et al., "A Dynamic Adaptive Replica Allocation Algorithm in Mobile Ad Hoc Networks", In Proceeding of *IEEE International Conference on Pervasive Computing and Communications (PerCom'04)*. P65-70. Orlando, March 2004.

CACHE$_{RP}$: A Novel Dynamic Cache Size Tuning Model Working with Relative Object Popularity for Fast Web Information Retrieval

Richard S.L. Wu[1], Allan K.Y. Wong[1], and Tharam S. Dillon[2]

[1] Department of Computing, Hong Kong Polytechnic University, Hong Kong SAR
{csslwu, csalwong}@comp.polyu.edu.hk
[2] Faculty of Information Technology, University of Technology, Sydney Broadway,
N.S.W. 2000
{tharam}@it.uts.edu.au

Abstract. The novel CACHE$_{RP}$ model for dynamic cache size tuning leverages the relative object popularity as the sole parameter. It is capable of maintaining the given hit ratio on the fly by deriving the popularity ratio from the currently collected statistics. Being statistical the accuracy of the CACHE$_{RP}$ operation should be independent of the changes in the Internet traffic pattern, which can switch suddenly. By adaptively maintaining the given hit ratio through cache size auto-tuning in a dynamic manner the model effectively reduces the end-to-end information retrieval roundtrip time (RTT).

Keywords: dynamic cache size tuning, popularity ratio, point-estimate, statistical.

1 Introduction

Applications for running on the Internet are naturally distributed and object-based. The objects always interact in the client/server relationship, which is also called the asymmetric rendezvous. In this rendezvous it is normal for the server at one end to answer the requests from different clients at the other [6]. Since the traffic patterns of the different streams of client requests vary, the resultant pattern of the merged traffic for the server queue becomes unpredictable.

The proxy/web-server interaction is an asymmetric rendezvous and the data fetching roundtrip time (RTT) inevitably depends on the average number of trials (ANT) to get a successful transmission. If the proxy keeps a large number of hot data objects in its cache, the ANT would be reduced because of less requests made to the server. The potential and real benefits from caching have inspired different relevant areas of research, and the most-researched topic is improving hit ratio by efficient replacement algorithms [1, 5, 8]. Our literature search indicates that there is little experience, however, on how to *maintain the given hit ratio* consistently under all conditions. This maintenance ability is significant because it always guarantees a reasonable information retrieval time by reducing the average ANT value. In the process it minimizes the chance of network congestion, which could otherwise be created by the massive data transfer across the network. To achieve the goal of maintaining the given minimum hit ratio the novel dynamic/adaptive caching model, namely, $CACHE_{RP}$ (Dynamic Cache Size Tuning with Relative Object Popularity) is proposed. It leverages the relative popularity of the data objects as the sole metric for adaptive cache size tuning and maintenance. The ever changing profile of data object probability over time is depicted in Fig. 1.

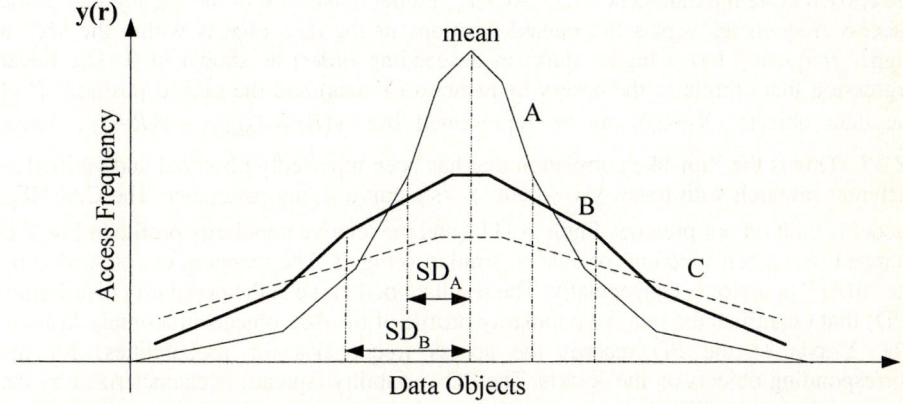

Fig. 1. Changes in the spread of the relative data object popularity profile

2 Related Work

There is a subtle difference between the effort to gain a high hit ratio and to maintain a given/prescribed hit ratio. The rationale to gain a high hit ratio is to keep as much hot data as possible in the local cache. The related issues to be addressed in order to achieve this goal include: replacement algorithms [10], caching scalability, filtration of cold data or "one-timers" [1], and adaptive web caching architectures [7]. The extant replacement algorithms, however, cannot upkeep the hit ratio consistently. The problem is that they work with a fixed-size cache, and because of this they do not accommodate the changes in the "popularity spread of the data objects". Maintaining a given hit ratio requires dynamic cache size tuning with respect to the spread of changes, best reflected by the variance or standard deviation of the relative popularity file. This profile changes with respect to the users' shift in interest toward particular objects within the data set. Fig. 1 shows the changes of the popularity spread or variance of the data objects within the same set over time. S_A and S_B are the standard deviations for the timely popularity profile A and B, which are captured at different time points. If a cache of a fixed size is able to keep enough hot data for the given confidence of one-standard deviation δ (or 68.4%) about the mean (i.e. $\delta = S_A$), then it should not work for S_B. Physically the wider S_B needs a bigger cache to hold extra hot data objects in order to satisfy the same given 68.4% confidence, and this is the essence of dynamic cache size tuning [11].

From the perspective of the changeful popularity spread in Fig. 1, any replacement algorithm designed to gain a high hit ratio with a cache of static size S works well only when the size S_δ required by the standard deviation δ is smaller than S (i.e. $S_\delta \leq S$). To maintain the given hit ratio under all conditions, for example 68.4% (i.e. one δ of the popularity profile), the cache size should be timely and adaptively adjusted by leveraging the chosen system parameters. The CACHE$_{RP}$ model makes use of the log-log plot of the "access frequencies versus the ranked positions of the data objects within the set"; a higher frequency has a higher rank (in descending order) as shown in 0. The linear regression that correlates the access frequencies (Y-axis) and the ranked positions R of the data objects (X-axis) can be represented by: $y(R) = f_{highest} - \gamma(R-1)$, where $R \geq 1$. This is the Zipf-like correlation that has been repeatedly observed and verified in different research with traces [3], where γ is a curve fitting parameter. The CACHE$_{RP}$ model is built on our previous findings [11], and the relative popularity profile in Fig. 2 is mapped into a bell function (or *bell(x)*) similar to Fig. 3. The mapping or conversion by the "*MAP*" operator is conceptually. The resultant bell curve is the popularity distribution (PD) that quantifies the relative popularity profile of the data objects in a timely fashion. The Y-axis of the PD records the access frequencies (or probabilities) for the corresponding objects on the X-axis. The PD variability (spread) is characterized by the standard deviation (SD) that measures the popularity deviations by different data objects from the "mean value": $y(1) = f_{highest}$. That is, the object with the highest access frequency in the Zipf distribution has become the "mean value" of the PD by mapping. Once the SD of the current PD is computed from the life data sampled statistically, the popularity ratio (PR) (explained in the next section) can be computed and used to adjust/tune the cache size adaptively on the fly.

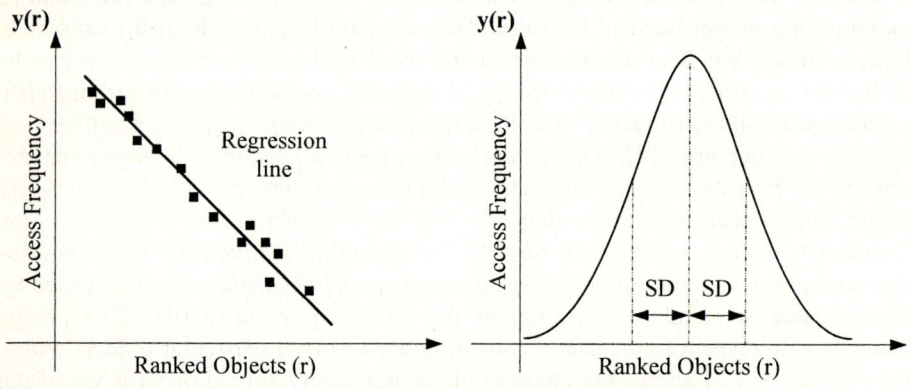

Fig. 2. Log-log regression   **Fig. 3.** Object Popularity Distribution

3 The Novel CACHE$_{RP}$ Model

The CACHE$_{RP}$ model tunes the cache size S_t adaptively to satisfy the $S_\delta \leq S_t$ condition consistently, where σ is the given hit ratio to be maintained, S_δ is the minimum cache size to achieve the goal and t of S_t indicates the time point at t. The adaptive tuning is based on the current popularity ratio (PR), which is directly computed from the sampled standard deviation or the variance of the current PD. The PR that tunes the cache size (CS) can be either the standard deviation ratio (SR) shown by equation (3.1) or the variance ratio (VR) shown by equation (3.2):

$$CS_{Tuned\_SR} = CS_{Old\_SR} * \left(SD_{ThisSample} \big/ SD_{LastSample} \right) \quad (3.1)$$

$$CS_{Tuned\_VR} = CS_{Old\_VR} * \left(SD_{ThisSample} \big/ SD_{LastSample} \right)^2 \quad (3.2)$$

where $SR = \sqrt{VR} = SD_{ThisSample} / SD_{LastSample}$.

$SD_{ThisSample}$ and $SD_{LastSample}$ are the two most current successive standard deviations for the changeful PD. Equations (3.1) and (3.2) reflect the statistical nature of the dynamic cache size tuning operation by CACHE$_{RP}$. The final cache size should be adjusted to CS_{Tuned_VR} or CS_{Tuned_SR}. The SR approach, which requires smaller amount of cache memory in the tuning process, however, is more suitable for small caching systems. In fact, the number of small caching systems, which cost less than US$1000 and have limited memory resources, is substantial in the field [9]. Therefore this paper focuses on verifying the SR approach (i.e. equation (3.1)). In the present stage of the CACHE$_{RP}$ research it is assumed that the popularity distributions are bell-shape and unimodal with a gentle mode skew.

The Internet traffic follows the power laws and changes among different patterns without warning. Therefore the accuracy of the method used to estimate the mean (μ) and the standard deviation (δ) of the timely popularity distribution (0) should be independent of the traffic patterns. For this reason the point-estimate (PE) technique, which is based on the central limit theorem [4], is chosen for the real-time μ and δ estimations. The PE has good qualities such as stability, time-proven effectiveness, and traffic insensitivity. In order to differentiate the ideal μ and δ from their estimated values with samples collected on the fly, the following are defined:

1. Estimated mean ($\bar{x}$): It is for a sample of arbitrary size n, and for normal practice the minimum value for n should be equal to or greater than 10.
2. Estimated standard deviation s_x: It is the SD or δ calculated for the n data items.

If $\delta_{\bar{x}}$ is the standard deviation for the curve plotted with many $\bar{x}$ values, then the *central limit theorem* supports the relationship of $\delta_{\bar{x}} = \delta/\sqrt{n}$, provided that n is large enough, and there is enough number of $\bar{x}$ values. The estimated value s_x from the current n data items, however, usually differs from the value $\delta_{\bar{x}}$.

If the CACHE$_{RP}$ model were to be deployed in real-life applications, then μ and δ estimations should be carried out with respect to the given tolerances. The equation (3.3), $\sqrt{N} - equation$ is derived by assuming the following tolerances:

$$E\mu = k\delta_{\bar{x}} = k(\frac{\delta}{\sqrt{N}}) \quad (3.3)$$

1. *Fractional error tolerance (E)*: It is the fractional error about true mean, μ by estimated mean, $\bar{x}$.
2. *SD tolerance (k)*: It is the number of standard deviations that $\bar{x}$ is away from the true mean μ and still be tolerated.

The $\sqrt{N}$ - *equation* dictates the minimum sample size N to yield the acceptable μ and δ predictions with respect to the specified k and E error tolerances. The $\sqrt{N}$ - *equation* can be rearranged to become the ideal equation 3.4:

$$N = \left(\frac{k\delta}{E\mu}\right)^2 \quad (3.4)$$

Previous experience shows that for real-life applications $\bar{x}$ and s_x instead of μ and δ can be used yield N, and this converts equation (3.4) to equation (3.5):

$$N = \left(\frac{ks_x}{E\bar{x}}\right)^2 \quad (3.5)$$

The following example illustrates the usefulness and application of this conversion:
1. For 50 RTT samples (i.e. sample size n is 50), the following are estimated: $\bar{x} = 25$ and $s_x = 7$.
2. The given tolerance is two standard deviations (i.e. k=2 or 95.4%), and the allowed fractional error tolerance E is therefore 4.6% (E=0.046) because k and E connote the same error.
3. The values above together yield $N = \left(\frac{2*7}{0.046*25}\right)^2 \approx 148$.

The computed N in this case indicates that for satisfying the $E\mu = k\delta_{\bar{x}}$ criterion the sample size must be at least 148. This implies that the sample size of $n=50$ used for the data collection process is not good enough. This problem should be resolved by one of the following approaches:

1. First approach: Collect at least another (126 – 50) or 76 RTT samples and then estimate $\bar{x}$ and s_x again, but this cannot guarantee the $E\mu = k\delta_{\bar{x}}$ criterion would be satisfied.
2. Second approach: Collect another $n=50$ additional RTT samples and estimate $\bar{x}$ and s_x again, and if the freshly computed N is less than $2n$ (or 100 for one additional round) then stop, otherwise repeat the process with additional n sampled data items in every new round.

The past experience in different applications shows that the second approach normally converges to satisfy $E\mu = k\delta_{\bar{x}}$ much faster. In most cases $\bar{x}$ and s_x stabilize in the second or the third trial [4].

4 CACHE$_{RP}$ Verification

4.1 Experiment Setup

The verification of the Java-based CACHE$_{RP}$ prototype is carried out by simulation over the Aglets mobile agent platform [2]. The Aglets choice is intentional because it is designed for Internet applications, and therefore it makes the verification results scalable and repeatable over the open dynamic Internet. The set up for the verification experiments is shown in Fig. 4. The basic caching architecture used is the stable "Twin-Cache System (TCS)" [1], as shown in Fig. 5. The aim of the TCS is to filter the "one-timer" cold data to make the verification result more meaningful.

In the CACHE$_{RP}$ verification experiments the dynamic cache size tuning applies only to the main cache (Main) of the TCS system. The Main holds the actual data objects while the auxiliary cache (Aux) contains only the names the data objects and their actual contents may or may not be present in Main. Their presence depends on the replacement dynamics as well as the inter-arrival times among the requests.

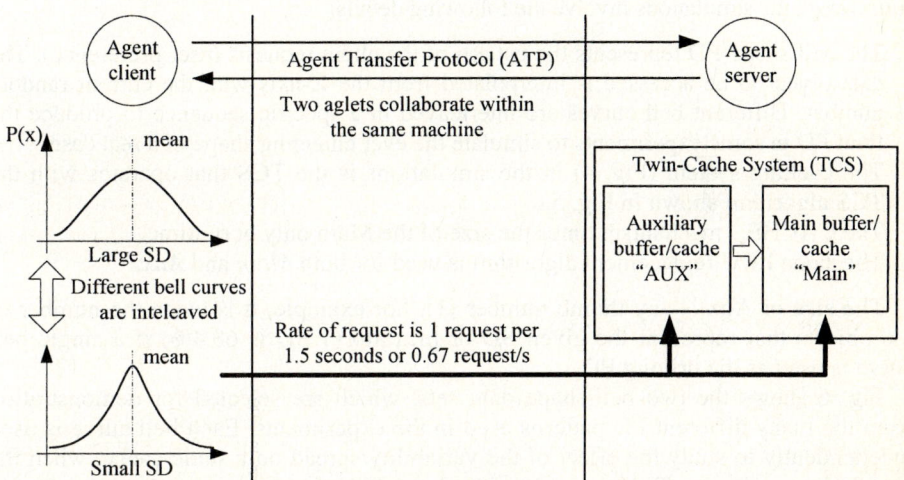

Fig. 4. Setup for the verification experiments

```
IF "Main"_has_the_request_data_object
THEN return the data
ELSE IF "Aux"_has_the name_of_the_data_object_being_requested
THEN { fetch_the_data_from_the_remote_source_and_return_it;
       cache_the_fetched_data_in_"Main"_as_well;}
ELSE {register_the_name_of_the_data_object_being_requested_in_"Aux";
      fetch_the_data_from_the_remote_source_and_return_it;}
```

Fig. 5. The "Twin-Cache System (TCS)" algorithm

In the simulations the same Least Recently Used (LRU) replacement algorithm is deployed for both Main and Aux. The replacement mechanism pushes out those aged objects that have stayed in the cache for the longest time. The simulations are carried out over the Aglets running in a Sun workstation. In each simulation two aglets (agile applets) interact in the client/server relationship over a stable designated Aglets ATP (Agent Transfer Protocol) channel. The client generates requests with respect to a bell-shape PD that represents the object popularity. A random number is generated as the access probability (Y-axis) for the distinctive object represented on the X-axis. With the access probability the corresponding object is found from the PD by interpolation. The experimental results in this paper are generated with 40,000 objects of an average size of 5k. In each simulation 1,000,000 data object requests are generated, at an average rate of 0.67 requests per second. The cache size for Main is set to be large enough for 1 standard deviation of data objects of first set of interleaving bell curves that yield the resultant PD. For example, 5k is the first curve in the interleaving sequence: $5k \rightarrow 10k$ (explained later). Such initialization has no impact on the final cache size because it is continuously tuned at runtime. The $CACHE_{RP}$ model is only verified with respect to the SR approach (Equation 3.1) in the scope of this paper, with the parameters shown in the Table 1. To summarize, the simulations involve the following details:

1. The bell-shape PD represents the pattern of the client requests (user preference). The data object to be accessed is interpolated from the X-axis with the current random number. Different bell curves are interleaved in a specific sequence to produce the final PD in some experiments to simulate the ever changing shape of a real case.
2. The caching system (Fig. 4) in the simulations is the TCS that operates with the TCS algorithm shown in Fig. 5.
3. The $CACHE_{RP}$ mechanism tunes the size of the Main only at runtime.
4. The same LRU replacement algorithm is used for both *Main* and *Aux*.

The size of Aux is any thumb number [1]. For example, it is twice the number of the objects that represent the given SD or hit ratio (1 SD is 68.4%) if a single bell curve is used as the driving PD.

Fig. 6 shows the two bell-shape data sets, which are selected for demonstration from the many different PD patterns used in the experiments. Each bell curve is used independently to study the effect of the variability/spread on a static cache, when the $CACHE_{RP}$ is absent. They are also interleaved randomly to simulate the timely changes in a PD. Other bell curves are used in the similar manner and purposes.

Table 1. The setup paramerters of the experiments

Parameter Name	Value
Number of different objects	40,000
Mean of the objects (Mean object position in PD)	20,000
Average Object Size	5k bytes
Number of requests	1,000,000
Rate of requests	0.67 request per second
Initial Main Cache Size	1 SD of first bell cuve
Initial Aux Size	2 * capacity of Main
Initial sample size of n	10
The equation used of the $CACHE_{RP}$ Model	SR (Equation 3.1)
Replacement algorithm	LRU

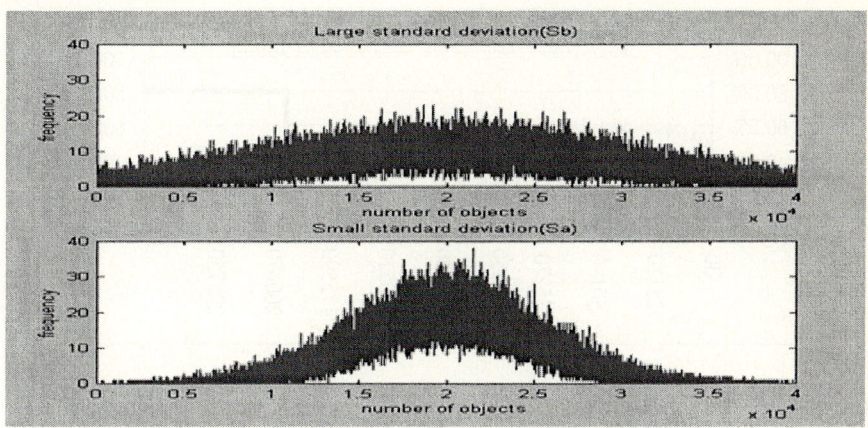

Fig. 6. The two normal distributions for the 40,000 data objects

4.2 Experimental Results

Experiments were carried with different PD patterns, and some of the results are shown in the Fig. 7. It was observed that the PE approach could generate serious hit ratio fluctuations for the $CACHE_{RP}$ dynamic tuning process. It generated high hit ratios for different PD patterns, for example, 66.5% for the $5k \rightarrow 10k$ interleaving sequence (Fig. 10). The $5k \rightarrow 10k$ sequence means that "the bell curve with the 5k standard deviation is interleaved with the 10k one" in a cyclical manner. The PE approach, however, also produced hit ratios that are lower than the expected

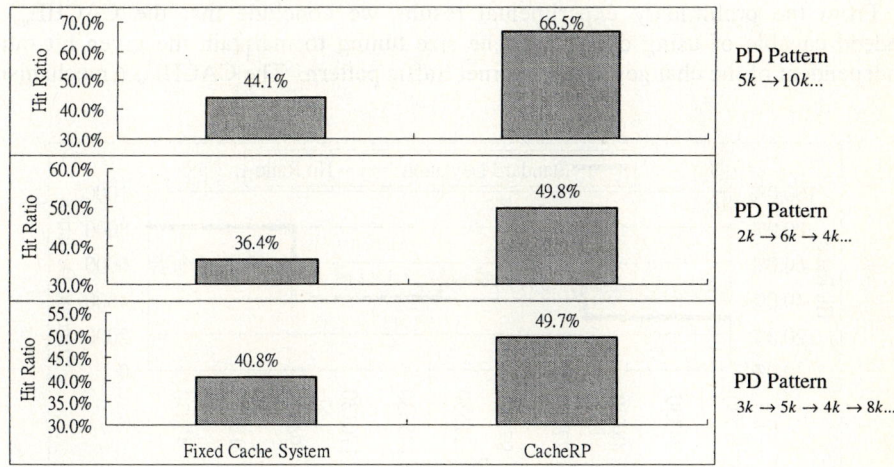

Fig. 7. Average hit ratios for Fixed Cache System and $CACHE_{RP}$ with different PD patterns

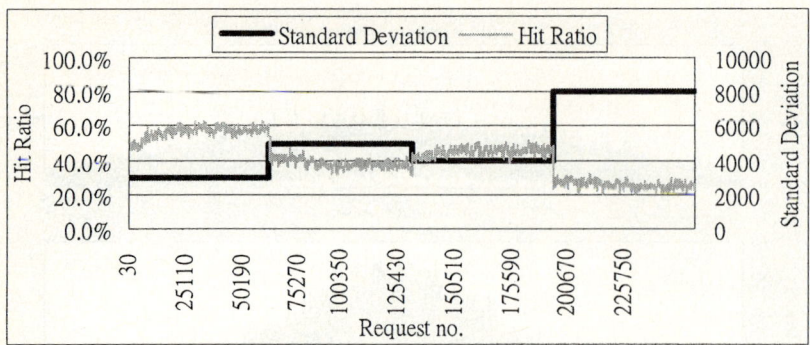

Fig. 8. The hit ratio changes for the fixed (size) cache system with the PD pattern generated by the bell curve interleaving sequence: $3k \to 5k \to 4k \to 8k$

percentage (e.g. 49.8% for the sequence: $2k \to 6k \to 4k$ in Fig.10). When compared to the fixed (size) cache system, which works with a static cache size as the control for comparison in the verification experiments, the PD sequence: $5k \to 10k$ yields an improvement of 50% [i.e. (66.5%-44.1%)/44.1%]. For the sequences: $3k \to 5k \to 4k \to 8k$ and $2k \to 6k \to 4k$, the hit ratio improvement by $CACHE_{RP}$ are only 22% [i.e. (49.7%-40.8%)/40.8%] and 37% [i.e. (49.8%-36.4%)/40.8%] respectively, with the fixed cache system as the control for comparison. Fig. 8 and Fig. 9 show the changes of the hit ratio of the fixed cache system and $CACHE_{RP}$ for the different PD sequences/patterns. Fig. 8 shows that the hit ratio of the fixed cache system drops a lot when the standard deviation of the requests increases/widens suddenly. Fig. 9 shows how the $CACHE_{RP}$ tries to tune the cache size for the current standard deviation so that the given hit ratio could be maintained.

From the preliminary experimental results we conclude that the $CACHE_{RP}$ is indeed capable of using dynamic cache size tuning to maintain the given hit ratio, independent of the changes in the Internet traffic pattern. The $CACHE_{RP}$ mechanism

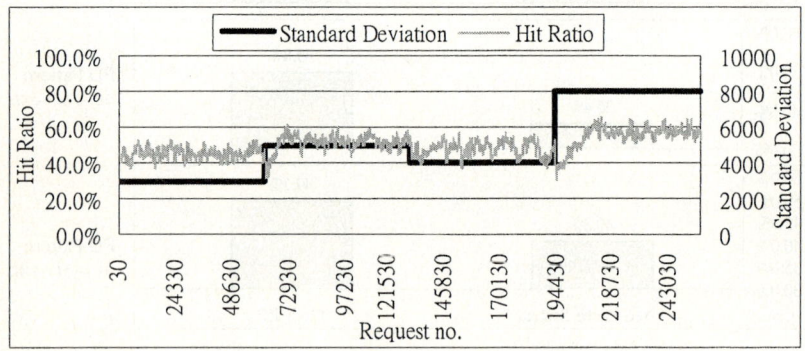

Fig. 9. The hit ratio changes for the $CACHE_{RP}$ with the PD pattern generated by the bell curve interleaving sequence $3k \to 5k \to 4k \to 8k$

working with the PE approach, however, cannot maintain the given hit ratio in a consistent manner. This implies the following: a) other alternative statistical methods based on the central limit theorem should be explored, b) accurate dynamic buffer tuning with relative object popularity as the sole parameter is viable, and c) the concept of using the popularity ratio as the tuning mechanism is technically sound because its simplicity requires short computation to minimize deleterious effects.

5 Conclusion

The novel $CACHE_{RP}$ model proposed in this paper is for dynamic cache size tuning, with the aim to maintain the given hit ratio adaptively on the fly. By doing so it reduces the chance of retransmission in the information retrieval process over the Internet and thus the ANT consistently. The $CACHE_{RP}$ tuner leverages the relative object popularity as the sole parameter for its statistical operation, which is independent of traffic pattern changes in the environment. To attain traffic insensitivity the point-estimate approach, which is based on the central limit theorem, is adopted for estimating the mean and standard deviation of the current popularity distribution. From two successively estimated standard deviations the popularity ratio, which decides the magnitude of the cache size adjustment on the fly, is computed. The results confirm that the $CACHE_{RP}$ model is indeed adaptive, but the PE approach does not upkeep the given hit ratio in a consistent manner. The next step in the near future is therefore to explore other suitable methods so that the $CACHE_{RP}$ can maintain the given hit ratio under all conditions with stability and consistency.

Acknowledgement

The authors thank the Hong Kong Polytechnic University for the grant: HZJ91.

References

1. C. Aggarwal, J.L. Wolf and P. S. Yu, Caching on the World Wide Web, IEEE Transactions on Knowledge and Data Engineering, 11(1), 1999
2. O. Mitsuru and K. Guenter, IBM Aglets Spec, http://www.trl.ibm.com/aglets/spec11.htm
3. L. Breslau et al, Web Caching and Zipf-like Distributions: Evidence and Implications, Proc. of Infocom'99, April 1999
4. J.A. Chisman, Introduction to Simulation and Modeling Using GPSS/PC, Prentice Hall, 1992
5. S. Glassman, A Caching Relay for the World Wide Web, Proc. of the 1st International Conference on the World Wide Web, Geneva, Switzerland, May 1994
6. S.M. Lewandowski, Frameworks for Component-Based Client/Server Computing, *ACM Computer Survey*, March 1998, 3–27
7. S. Michel et al., Adaptive Web Caching: Towards a New Caching Architecture, Computer Network and ISDN Systems, November 1998

8. V. Milutinovic, Caching in Distributed Systems, IEEE Concurrency, 8(3), July-September 2000
9. D. Wessels, Web Caching, O'Reilly & Associates Inc., 2001
10. Kun-lun Wu and Philip S. Yu, Latency-Sensitive Hashing for Collaborative Web Caching, IBM Research Report RC21672, February 2000
11. Richard S.L. Wu, Allan Kang Ying Wong and Tharam S. Dillon, E-MACS: A Novel Dynamic Cache Tuning Technique to Maintain the Prescribed Minimum Hit Ratio Consistently for Internet/www Application, WSEAS Transactions on Computers, 2(3), April 2004, 430–434

Implementation of a New Cache and Schedule Scheme for Distributed VOD Servers

Han Luo and Ji-wu Shu

Department of Computer Science and Technology, Tsinghua University,
Beijing 100084, China
lh98@mails.tsinghua.edu.cn

Abstract. A VOD server's caching and scheduling performance determine its service performance efficiency. This paper describes a new cache model and content replacement strategy, based on the Zipf-like Law and the characteristics of media stream service, which can reduce the disk I/O ratio by 6.22%. A performance analytical model for a disk load schedule was constructed, based on the Stochastic Process and Queuing Theory, and a new disk load strategy suitable for VOD systems was also formulated. This strategy reduces the disk block time by 3.71% on average. This paper also describes a content schedule scheme which was designed by constructing, analyzing and simplifying the SPN model deduced from the MSMQ theory. This scheme can guarantee the quality of service(QoS) and distribute program content automatically. An experiment was conducted, and the results showed that VOD servers embedded with the new cache and using the new schedule strategy could reduce the average response time to user requests by 7% to 19%.

1 Introduction

Video-On-Demand (VOD) service is an essential component of many multimedia applications. The service allows geographically distributed users to interactively access video files from remote VOD servers. Due to the real-time and synchronization of audio-video stream transport, and the large size of video files, the network transport requests and disk capacity requirements exceed those of other services, such as web and e-mail, when the server serves the same quantity of subsequent users.

2 Related Work

To support more subsequent users, VOD servers ordinarily have distributed system architecture. The multi-level distribution of service can guarantee the response time and support more subsequent users and the multi-level distribution of storage could, theoretically, increase the storage capacity to an unlimited amount of space. However, this distribution increases the complexity of the scheduling among servers. And the large scale of the video-file sizes makes traditional cache mechanisms unsuitable. In addition, the number of video files on the internet has been increasing geometrically, leading to a rapid rise in the demand for increased storage capacity on servers. At present, most VOD systems are equipped with multi-level distributed storage. When a user request arrives and the video file does not exist in local storage, there are two possible solutions: get the video file from other servers, or redirect the task to other servers. Presently, most research and implementation focuses on task redirection, not on the storage schedule. But in practical use, by reason of network and operator, task redirection may cause a load imbalance, making it impossible to guarantee the QoS. Therefore, the storage schedule must be addressed before the performance and robustness of VOD servers can be guaranteed.

Content Distribution Networks, (CDNs)[3], are commonly placed between the server and users as a cache device for network service. A CDN caches the data stream from server to user, and when the data is requested again, the CDN can provide the data directly. [4] compares the performance of a CDN device in different services. However, the kernel technology of CDNs is cache management and traditional multimedia content transport proxy, whereas the kernel technology of multimedia stream systems is the high-speed transport and distribution of stream data. And in CDN devices, any content is a Web Object, no matter what the files size or predicted characteristics are. CDN devices do not consider the application logic and data characteristics of the service; therefore low efficiency of CDN is unavoidable.

In the field of disk scheduling, most research has focused on how to provide an I/O stream more effectively on one disk. [5] described a block storage scheme to reduce disk seek time during multitasking. [6] introduced the DIOMS schedule strategy for multi-disk-request application. Howeverin real applications most servers correspond with several disk arrays, and even with multi-level storage, which is composed of flash, disk array and tape device storage. The disk service time is much longer for VOD service than for other services, which means that the probability of a disk block is much greater. Therefore, to ensure the robustness of VOD servers, how to guarantee the disk service is more important than how to provide a more effective I/O stream on one disk.

In this study, we adopted service-embedded cache architecture and designed a new cache replacement strategy, which adequately considers the application logic and service characteristics of VOD systems, such as the $Zipf-like$ [11] Law and the population of a program. The new strategy improves the cache hit rate, keeps more content of the most popular programs in the cache, and reduces the disk output. To address disk scheduling, we developed a new schedule strategy by deducting the disk-blocking probability function based on the Queue Theory, and by calculating the extremum as the reference of the schedule strategy. The new disk schedule strategy can reduce the disk-blocking probability and improve system robustness. For the content scheduling, based on the MSMQ theory, we deducted a content schedule blocking probability function, and designed a strategy by function extremum. The new content schedule scheme distributes the new video files from one server to other servers automatically and balances the load among servers.

Moreover, to implement caching and scheduling, we provided for compatibility with IA64, and utilized the high performance and large RAM capability of IA64 to improve the performance of VOD servers.

3 Multi-level Distributed VOD System Architecture

Our VOD system had a hierarchical distributed structure, consisting of several VOD server units, following the DAVIC protocol. One VOD server unit consisted of one application server (AS), one management server (MS), and one or several pump servers (PSs). The AS was responsible for the response to users' requests, the transmission of the playbill, and the load balancing of the PSs. The PSs were responsible for reading the video files, the transmission of the multimedia stream, and the response to the users' VCR requests. The MS was responsible for the management of users and video-files. Each VOD server unit could provide the self-contained VOD service for a stated amount of users, including ordering programs and managing users.

Our VOD system had a hierarchical distributed structure with n levels. The first level consisted of central server units $(0 \ldots M_1)$, where M_1, an arbitrary natural number, denotes the number of central server units. The second level consisted of region server units $(0 \ldots M_2)$, where M_2, an arbitrary natural number, denotes the number of region server units. The last level, which provides service directly to the users, consisted of end server units $(0 \ldots M_n)$, where M_n, an arbitrary natural number, denotes the number of end server units.

Any user request will be assigned to a PS, which can be located by a vector: $PS(m_1, m_2, \ldots, m_n))$, where $0 \leq m_1 \leq M_1, 0 \leq m_2 \leq M_2, \ldots, 0 \leq m_n \leq M_n$, n is level depth of PS.

The performance of VOD servers is determined by these two factors:

1. The amount of subsequent streams and network I/O bandwidth supported by one PS.
2. The response time for user requests, including ordering requests and VCR requests.

In the entire VOD system, a program has a unique identifier: P_i, for which $0 \leq i < P_{\text{count}}$, and P_{count} is the count of all programs. In order to accelerate the disk's multi-task seeking and reading, the VOD storage employs block-deposit technology. Therefore a program can be divided into a vector: $P_i = (P_{i,1}, P_{i,2}, \ldots, P_{i,\text{length}(P_i)})$, where $\text{length}(P_i)$ is the count of P_i's block. We defined $\text{tick}(P_{i,j})$ as the transmit start time of P_i's j-th block. We defined some variables for $PS(m_1, m_2, \ldots, m_n)$:

1. $C(m_1, m_2, \ldots, m_n)$: The maximal amount of subsequent streams supported by $PS(m_1, m_2, \ldots, m_n)$.
2. $\Omega_{\text{Local}}(m_1, m_2, \ldots, m_n)$: The set of all program blocks in $PS(m_1, m_2, \ldots, m_n)$'s local storage.
3. $\Omega_{\text{Cache}}(m_1, m_2, \ldots, m_n)$: The set of all program blocks in $PS(m_1, m_2, \ldots, m_n)$'s cache.

When a user request for P_i is assigned to $PS(m_1, m_2, \ldots, m_n)$, $PS(m_1, m_2, \ldots, m_n)$ starts to prepare all the blocks of P_i, copying each block to the memory, and pushing the block to the user consecutively.

There are three possibilities for the prefetching of program block $P_{i,j}$:

1. $P_{i,j} \in \Omega_{Cache}(m_1, m_2, \ldots, m_n)$.
2. $P_{i,j} \in \Omega_{Local}(m_1, m_2, \ldots, m_n)$.
3. $P_{i,j} \in [\bigcup \Omega_{Cache}(m'_1, m'_2, \ldots, m'_{n'})] \cup [\bigcup \Omega_{Local}(m'_1, m'_2, \ldots, m'_{n'})]$, where $(m'_1, m'_2, \ldots, m'_{n'}) \neq (m_1, m_2, \ldots, m_n)$.

4 Cache Scheme for Distributed VOD Servers

When $P_{i,j}$ satisfies the first condition, $P_{i,j} \in \Omega_{Cache}(m_1, m_2, \ldots, m_n)$, the block is in the local cache and prefetching has finished.

However, cache devices are much more expensive than disk storage devices. Therefore how to utilize the finite cache space is the key to the cache scheme.

By statistical analysis, we learned that the program request would allow the $Zipf - like$ Law [11]. Assuming that there are N programs in the VOD system, and sorting these program by access frequency, the access frequency can be obtained by this formula:

$$P_N(i) = \frac{\Omega}{i^\alpha}, \text{ where } \Omega = (\sum_{i=1}^{N} \frac{1}{i^\alpha})^{-1}. \qquad (1)$$

This is known as the $Zipf - like$ Law. For the $Zipf$ Law, $\alpha = 1$. But in our experiment, α was usually between 0.8 and 0.9.

We deducted the formula above, and obtained the access frequency for the top k programs:

$$\phi_N(k) = \sum_{i=1}^{k} P_N(i) = \sum_{i=1}^{k} \frac{\Omega}{i^\alpha} \approx \frac{\Omega k^{1-\alpha}}{1-\alpha}, \text{ where } \Omega = (\sum_{i=1}^{N} \frac{1}{i^\alpha})^{-1} \approx \frac{1-\alpha}{N^{1-\alpha}}. \qquad (2)$$

And we found that $\phi_N(k) = (\frac{k}{N})^{1-\alpha}$.

Making the cache hit rate equal to β, we found that

$$\phi_N(k) = (\frac{k}{N})^{1-\alpha} = \beta \Rightarrow k = N\beta^{\frac{1}{1-\alpha}}. \qquad (3)$$

For instance, if $N = 1000$ and $\alpha = 0.8$, and we want $\beta = 50\%$, then $k = 1000 \times 0.5^5 \approx 32$. This means that the cache capacity of the server unit must be large enough to contain 32 video files. If we want $\beta = 60\%$, then $k = 1000 \times 0.6^5 \approx 78$.

Among the traditional cache replacement strategies, Perfect-LFU (Least-Frequently-Used) can maintain the highest cache hit rate. We improved the Perfect-LFU strategy, adapting it for VOD service.

First the AS calculates the access frequency of every program, and sorts them. Secondly, the AS can be conscious of the cache table of all PSs in the server unit. The cache table includes the program block's ID and a count for this block. When a user request arrives, the AS queries the cache table and looks for an appropriate PS to assign the task to, matching the task to the PS's cache content. Therefore the cache content can be utilized fully.

When the AS assigns the user request to a PS, the AS tells the PS the access frequency of the program. The PS initializes or increases the count of cache page by this frequency. This ensures that the most popular programs are not deleted from the cache page.

For the program P_i, $P_i = (P_{i,1}, P_{i,2}, \ldots, P_{i,length(P_i)})$, each block corresponds to a cache page, and the initial values of the count of cache pages constitute a vector: $C_i = (C_{i,1}, C_{i,2}, \ldots, C_{i,length(P_i)})$.

Most VOD operators will establish a rule: when the program's run time does not exceed, the user does not need to pay. According to the statistical data, the access frequency of each block in one program satisfies: $Freq_i(j) = \rho(\frac{\tau u}{vj})$, where ρ is a scale factor, and v is each block's run time.

When initializing the cache page, the PS adds weight to the initial count by the access frequency, thus $C_{i,j} = H_P(\frac{\tau u}{vj})$, where H_P is the access frequency of the program.

When the AS calculates the access frequency, the AS also exchanges access frequency information with the parent AS, and thus both the AS and its parent are updated. By this means, servers can anticipate the trend of changes in the most popular programs, and maintain the cache hit rate.

5 Disk Schedule for Distributed VOD Servers

When $P_{i,j}$ satisfies the second condition, the block is in local storage and will be fetched. The stream of all the disk requests is regarded as a Poisson Process. The rate is λ and the average serving time is t, thus the traffic intensity is $A = \lambda t$.

According to the Queue Theory, each disk is an $M/G/n/n$ queue system, in which n is the count of subsequent I/O streams the disk supports. Assume disk k provides the α_k for all streams. According to the Erlang B formula, when disk k provides η_k streams, the block probability of disk k is

$$B_{\alpha_k} = \frac{(A\alpha_k)^{\eta_k}/\eta_k!}{\sum_{i=0}^{\eta_k}(A\alpha_k)^i/i!}, \text{ where } \alpha_k \geq 0 \quad (4)$$

Assume that there are D disks in a server unit. Then the block probability of the server unit is:

$$B = \sum_{k=1}^{D} \alpha_k B_{\alpha_k} = \sum_{k=1}^{D} \alpha_k \frac{(A\alpha_k)^{\eta_k}/\eta_k!}{\sum_{i=0}^{\eta_k}(A\alpha_k)^i/i!}, \text{ where } \sum_{k=1}^{D} \alpha_k = 1 \quad (5)$$

Import an arbitrary constant K, define as:

$$G = \sum_{k=1}^{D} \alpha_k \frac{(A\alpha_k)^{\eta_k}/\eta_k!}{\sum_{i=0}^{\eta_k}(A\alpha_k)^i/i!} - K(\sum_{k=1}^{D} \alpha_k - 1) \qquad (6)$$

Then let

$$\frac{\partial G}{\partial \alpha_k} = \frac{\partial}{\partial \alpha_k}\left(\alpha_k \frac{(A\alpha_k)^{\eta_k}/\eta_k!}{\sum_{i=0}^{\eta_k}(A\alpha_k)^i/i!}\right) - K = 0 \qquad (7)$$

Assume vector $\alpha^{min} = (\alpha_1^{min}, \alpha_2^{min}, \ldots, \alpha_D^{min})$ is the solution of the equation above. Then

$$\frac{\frac{(A\alpha_k^{min})^{\eta_k}}{\eta_k!} \sum_{i=0}^{\eta_k} \frac{(A\alpha_k^{min})^i(\eta_k+1-i)}{i!}}{(\sum_{i=0}^{\eta_k} \frac{(A\alpha_k^{min})^i}{i!})^2} = K \qquad (8)$$

viz.

$$B_{\alpha_k^{min}}[\eta_k + 1 - A\alpha_k^{min}(1 - B_{\alpha_k^{min}})] = K \qquad (9)$$

Using the iterative method, we can get the vector $\alpha^{min} = (\alpha_1^{min}, \alpha_2^{min}, \ldots, \alpha_D^{min})$. The disk schedule strategy is that α must satisfy $\min(|\alpha^{min} - \alpha|)$.

6 Program Schedule for Distributed VOD Servers

When $P_{i,j}$ satisfies the third condition, the block does not exist in local storage and will be fetched from another PS and pushed to the user. This operation is called as program schedule task r classified by i,j. Because there are multiple PS servers for program scheduling, the system can be considered as a multi-server-multi-queue (MSMQ) system, as illustrated in Fig. 1.

For description convenience, we make the following assumption:

1. Each system contains m servers, and accepts n class of tasks. The i-th task is named r_i, and the j-th server is named s_j.

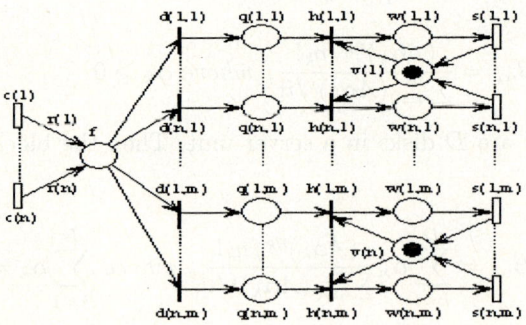

Fig. 1. MSMQ's SPN Model

2. Each server contains a cache queue. s_j's queue is marked as q_j, and q_j's cache capacity is marked as b_j. q_{ij} denotes the logistic queue of s_j to accept task r_i. b_{ij} is the capacity of q_{ij}, and $M(q_{ij})$ is the number of the token of r_i in q_{ij}.
3. The arrival of any class of task is a Poisson Process. The arrival rate for r_i is λ_i, and r_i can be assigned into any queue. When all of the queues are filled up, the task will be rejected.
4. The service time for every server to process each task can be different. s_j's service rate is defined as μ_j.

From among the RR, SQR, $SEDR$ and $OSEDR$ schedule algorithms, we selected the $SEDR$ because it can maintain high performance in most conditions. We controlled the schedule performance by setting b_{ij}.

We constructed a corresponding Markov chain of the SPN model. Because the computers currently in use can not accurately solve a $m \times n$ Markov chain, we disassembled the model to several simple sub models.

The whole SPN model is divided into $m \times n$ simple sub models. Fig. 2 illustrates the disassembled sub model A_{ij}. c_{ij} is the time transition of task r_i being assigned to q_{ij}.

We used $P[M(q_{ij})]$ to denote the steady status probability of $M(q_{ij})$, and defined some steady status probability for some conditions:

$$P[M(q_{ik}) > t] = \sum_{x=t+1} P[M(q_{ik}) = x], \qquad (10)$$

$$P[t_{ik} \times M(q_{ik}) > t] = \sum_{x>t} P[t_{ik} \times M(q_{ik}) = x] \qquad (11)$$

$$P\left[\sum_{y=1}^{n} M(q_{yk}) > \sum_{i \neq x=1}^{n} M(q_{xj}) + t\right] =$$
$$\sum \left\{ \prod_{y=1}^{n} P[M(q_{yk}) = f_{yk}] \times \prod_{i \neq x=1}^{n} P[M(q_{xj}) = h_{xj}] \right\} \qquad (12)$$

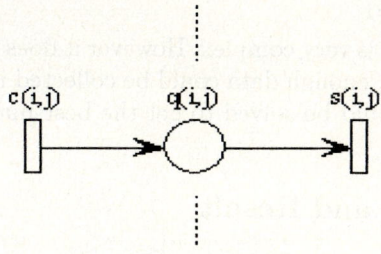

Fig. 2. Disassembled SPN model

$$P\left[\sum_{y=1}^{n} t_{yk} \times M(q_{yk}) > \sum_{i \neq x=1}^{n} t_{xj} \times M(q_{xj}) + t\right] =$$

$$\sum \left\{ \prod_{y=1}^{n} P[t_{yk} \times M(q_{yk}) = f_{yk}] \times \prod_{i \neq x=1}^{n} P[t_{xj} \times M(q_{xj}) = h_{xj}] \right\} \quad (13)$$

where

$$\sum_{y=1}^{n} f_{yk} > \sum_{i \neq x=1}^{n} h_{xj} + t$$

The schedule transition probability can be denoted in sub model A_{ij} : ($1 \leq j \leq m, 1 \leq i \leq n$). In the $SEDR$ schedule strategy, the enforceable probability of transition c_{ij} is

$$p_{ij} = \sum_{j \neq k, k=1}^{m} P[t_{ik} \times M(q_{ik}) > t] + \cdots$$

$$+ \prod_{j \neq k, k \in Q} P[t_{ik} \times M(q_{ik}) > t] \times \prod_{j \neq x, x \notin Q} P[M(q_{ix}) = b_{(ix)}] + \cdots$$

$$+ \prod_{j \neq k, k=1}^{m} P[M(q_{ik}) = b_{ik}] + \cdots + \frac{1}{\|Q\|} \prod_{k \in Q} P[t_{ik} \times M(q_{ik}) = t]$$

$$\times \prod_{x \notin Q} P[t_{ix} \times M(q_{ix}) > t] \times \prod_{y \notin Q} P[M(q_{iy}) = b_{iy}]$$

where $t_{ij} \times M(q_{ij}) = t$.

We solved the $m \times n$ submodel by the SPNP software, and obtained the response time of the system:

$$f(b_{11}, \ldots, b_{ij}, \ldots, b_{nm}) = t_{\text{ResponseTime}}. \quad (14)$$

Then

$$\frac{\partial}{\partial b_{ij}} f = 0, \text{ where } 1 \leq j \leq m, 1 \leq i \leq n. \quad (15)$$

The calculation of b is very complex. However it does not need to be calculated in real time. Therefore, enough data could be collected in the experiment period, and the SPN model could be solved to get the best matrix b.

7 Experiment and Result

We implemented the cache and schedule scheme, and conducted a series of experiments. In each experiment, we built two VOD servers, the one was the old

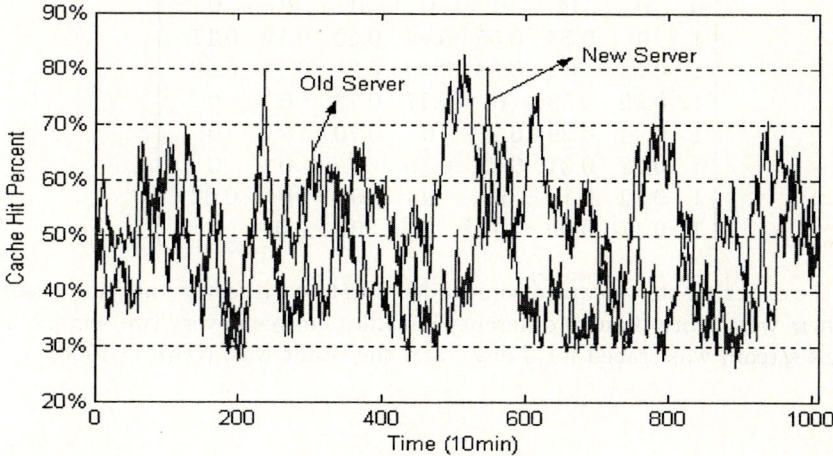

Fig. 3. Cache hit rate

server, the other was the new server embedded with the new cache and using the new schedule strategy. By this way, we could estimate the efficiency improved by the new scheme.

First, we conducted a cache hit rate experiment. The local storage contained 512 video files in each unit, and the central storage contained 4096 video files. We set the prospective cache hit rate to 50%. The data in the experiment was sampled at periods of 10 min/time.

As shown in Fig. 3, the average cache hit rate of the old server was 45.5716%, however the new server's was is 51.3697%. In addition, the network output of the new server within one week was 73.325TB, and the data read from the cache was 39.053TB, accounting for 53.26%. However the network output of the old server within one week was 68.195TB, and the data read from the cache was 32.076TB, accounting for 47.04%.

In the disk schedule experiment, we first tracked the disk block times and subsequent user counts of the new and old servers and compared the two results. When the subsequent user count was low, the two schemes were indistinguishable from each other. However when the subsequent user count exceeded 1000, the new scheme could reduce disk block times in comparison to the old one. When the subsequent user count was between 1000 and 1300, the average disk block times decreased 3.71%, as shown in Fig. 4.

We also needed to determine whether the new scheme would lead to a load imbalance. As shown in Fig. 5, the load assigned to the three disks of the eight experimental disks was very balanced.

In the program schedule experiment, we constructed 8 experimental server units. The first one was the central server unit, and the others were regional server units that provided service for users. We analyzed the network connecting the units, and simulated a multi-operator environment. Then we obtained matrix b :

$$\begin{bmatrix} 0 & 0 & 0 & 0 & 0 & 0 & 0 & 0 \\ 1 & 0 & 0.53 & 0.55 & 0.44 & 0.32 & 0.12 & 0.17 \\ 1 & 0.53 & 0 & 0.58 & 0.39 & 0.31 & 0.13 & 0.09 \\ 1 & 0.55 & 0.58 & 0 & 0.77 & 0.73 & 0 & 0 \\ 1 & 0.44 & 0.39 & 0.77 & 0 & 0.70 & 0 & 0 \\ 1 & 0.32 & 0.31 & 0.73 & 0.70 & 0 & 0 & 0 \\ 1 & 0.12 & 0.13 & 0 & 0 & 0 & 0 & 0.62 \\ 1 & 0.17 & 0.09 & 0 & 0 & 0 & 0.62 & 0 \end{bmatrix}$$

In the experiment, we copied some video files to each server unit, and ensured that there were more than 10 different video files between every two servers. The network stream was traced for 3 days, and the result was: $(Unit = 10GB)$

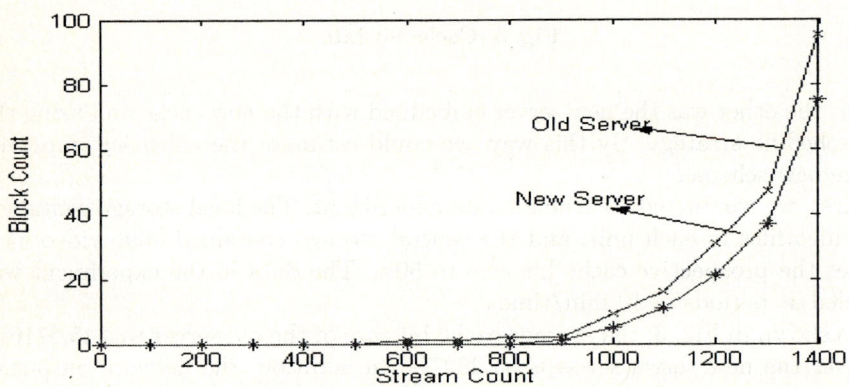

Fig. 4. Relationship of disk block times and subsequent user load

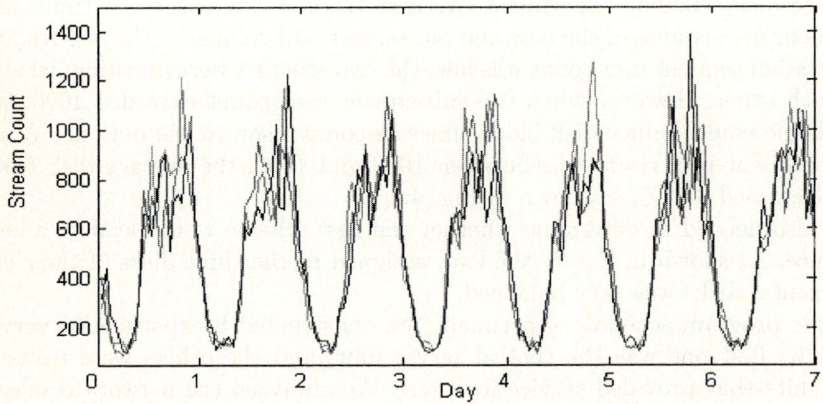

Fig. 5. Load assigned to three of eight disks used in the experiment

$$\begin{bmatrix} 0 & 0 & 0 & 0 & 0 & 0 & 0 & 0 \\ 19.3 & 0 & 22.8 & 24.9 & 23.2 & 23.9 & 9.33 & 17.3 \\ 22.1 & 24.4 & 0 & 23.4 & 39.1 & 41.3 & 12.3 & 3.43 \\ 15.3 & 33.2 & 19.2 & 0 & 32.1 & 55.2 & 0 & 0 \\ 12.3 & 25.1 & 22.8 & 32.9 & 0 & 33.2 & 0 & 0 \\ 24.3 & 17.9 & 23.9 & 22.8 & 41.2 & 0 & 0 & 0 \\ 22.1 & 10.2 & 17.3 & 0 & 0 & 0 & 0 & 32.3 \\ 18.2 & 9.78 & 7.03 & 0 & 0 & 0 & 54.2 & 0 \end{bmatrix}$$

After a week, we checked the local storage and the cache of the eight units, and found that each server unit had a copy of the top 10 video files. The program schedule worked as we designed it to.

In the last experiment, we focused on the relationship of average user response time and subsequent user count and the incremental percentage of the performance.

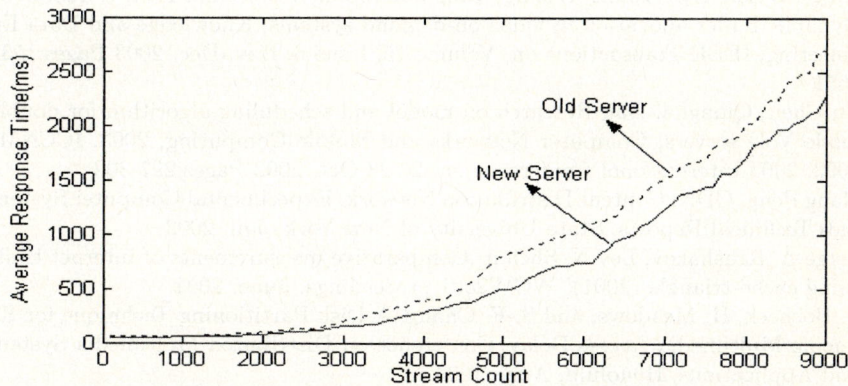

Fig. 6. Relationship of user response time and subsequent user count

As shown in Fig. 6, when the subsequent user count was the same for both servers, the new server's user response time was lower than that of the old server. And when the subsequent user count was between 500q0 and 8000, the response time was reduced by 7%-19%, which is considerable.

8 Conclusion

The new scheme described in this paper improves the VOD server's performance by implementing cache, disk schedule and program schedule. The new cache replacement strategy considers the application logic and data characteristics fully. The new disk schedule strategy considers not only the load balance but also the disk block probability. Implementing the new program schedule strategy among multiple server units can guarantee QoS and automatic program distribution. The result of the experiment proved that , by adding these three new strategies into VOD servers, the response time could be reduced effectively, and disk read and block times could also be reduced obviously. In future work, we will research how to take advantage of network bandwidth and reduce the redundancy in the network output by multicast and stream patching technology.

Acknowledgement

The work described in this paper was supported by the National High-Tech Research and Development Plan of China under Grant (No.2001AA111110 and No.2004AA111120).

References

1. Sang-Ho Lee, Kyu-Young Whang, Yang-Sae Moon, Wook-Shin Han, Il-Yeol Song, Dynamic buffer allocation in video-on-demand systems, Knowledge and Data Engineering, IEEE Transactions on, Volume 15, Issue 6, Nov.-Dec. 2003 Pages:1535-1551
2. Yu Chen, Qionghai Dai, Research on model and scheduling algorithm for double-mode VoD servers, Computer Networks and Mobile Computing, 2003. ICCNMC 2003. 2003 International Conference on, 20-23 Oct. 2003 Pages:297-302
3. Gang Peng, CDN: Content Distribution Network, Experimental Computer Systems Lab Technical Reports, State University of New York, Jan. 2003.
4. Serge A. Krashakov, Lev N. Shchur ,Comparative measurements of Internet traffic using cache-triangle (2001), WCW2001 proceedings, June, 2001
5. P. Bocheck, H. Meadows, and S.-F. Chang, A Disk Partitioning Technique for Reducing Multimedia Access Delay, Conference on Distributed Multimedia Systems and Applications, Honolulu, Aug., 1994.
6. Hai Jin, Jie Xu, Bibo Tu, Shengli Li, Disk I/O mixed scheduling strategy for VoD servers Circuits and Systems, 2003. ISCAS03. Proceedings of the 2003 International Symposium on, Volume 2, 25-28 May 2003 Pages:II-504 - II-507 vol.2
7. M. Barreiro, V. M. Gul'as, Cluster setup and its administration. In Rajkumar Buyya, editor, High Performance Cluster Computing, Vol. I. Prentice Hall, 1999.
8. S. A. Barnett and G. J. Anido, A cost comparison of distributed and centralized approaches to video-on-demand, IEEE J. Select. Areas Commun., vol. 14, pp. 1173-1183, Aug. 1996.
9. J. Armstrong, R. Virding, C. Wikstrom, M. Williams. Concurrent Programming in Erlang, Second Edition, Prentice-Hall. 1996.
10. The Digital Audio-Visual Council (DAVIC) Opening Forum
11. Lee Breslau, Pei Cao, Li Fan, Graham Phillips, and Scott Shenker. Web caching and zipf-like distributions: Evidence and implications , Technical report, University of Wieconsin-Madison, Department of Computer Science, 1210 West Dayton Street, July 1998.
12. A. Dan, er'al, Buffering and Caching in large-scale video servers, IEEE CompCon Conference, pp.217-224, 1995.

UML Based Statistical Testing Acceleration of Distributed Safety-Critical Software[1]

Jiong Yan, Ji Wang, and Huo-wang Chen

National Laboratory for Parallel and Distributed Processing, Changsha, P.R. China
yanjiong7172@vip.sina.com, jiwang@mail.edu.cn

Abstract. It is necessary to assess the reliability of distributed safety-critical systems to a high degree of confidence before they are deployed in the field. However, distributed safety-critical software systems often include some rarely executed critical functions that are often inadequately tested in statistical testing based reliability estimation. This paper presents a method that can accelerate statistical testing of distributed safety-critical software. The method starts with the derivation of scenario usage diagram model (SUD) from UML diagrams annotated with usage related attributes and reliability attributes. Then the statistical testing accelerating method based on importance sampling is presented. When both the critical scenarios and the entire software are adequately tested, the method can still compute the unbiased software reliability from the test results with much less test cases. Thus, the statistical testing cost of distributed safety-critical software can be reduced effectively.

1 Introduction

Distributed process-control systems are widely used because of the price/performance ratio, flexibility and scalability. It is important to build high dependable distributed process-control systems because these systems are often safety-critical, whose failures are often catastrophic. Thus, the validation of non-functional requirements, such as reliability, is becoming more important for distributed process-control software.

The general method to assess software reliability is to test the software according to the software Markov chain usage models, which is called statistical testing, suggested by the Cleanroom Software Engineering [11]. However, statistical testing based reliability assessment for the distributed safety-critical software has drawbacks.

Distributed safety-critical software often includes some critical functions, which are rarely executed and are often inadequately tested in statistical testing. However, any failure of these functions can result in catastrophic loss of life and property. Hence, it is essential to ensure that these functions meet their reliability requirements prior to deploying them in the field. Though we can adjust the usage distribution to

---

[1] This work is supported by National Natural Science Foundation of China under Grant No. 60233020 and No. 90104007, National Hi-Tech Program of China under Grant No. 2001AA113202 and No. 2001AA113190, and Huo Ying Dong Education Foundation under Grant No.71064.

ensure that those critical functions are tested more frequently, however, this may make the estimation from test does not yield an unbiased reliability estimate anymore.

The aim of this paper is to show that this disadvantage of statistical testing can be solved in such a way that, one may get unbiased reliability estimates, and at the same time, can effectively reduce the number of test cases needed while both the software and critical functions are tested adequately. Thus, the statistical testing can be effectively accelerated and the test cost can be reduced.

While UML has been the de-facto standard object-oriented modeling language, the UML artifacts provide a common notation ground to represent and validate software system. Thus, making the relevant technique based on UML artifacts will make the reliability estimation and statistical testing acceleration very appealing to software engineering practitioners. In this paper we use annotated UML diagrams to support statistical testing acceleration in the early stage of distributed software development.

The rest of the paper is organized as follows. Section 2 discusses the derivation of scenario usage diagram model (SUD) from UML diagrams. Section 3 proposes how to accelerate statistical testing based on SUD. Section 4 presents a case study. Section 5 reviews the related work. We conclude with a summary and future work in section 6.

2 Deriving SUD from UML Artifacts

2.1 Annotated UML Artifacts

In the UML based software development, use cases describe software high-level functionalities, and sequence diagrams describe scenarios, which depict how system components and actors interact in order to provide a system level functionality [7].

To execute a use case, specific preconditions must be satisfied. Thus, a use case UC can be defined as $UC = (pre_{uc}, SDSet_{uc})$ where $SDSet_{uc}$ is the set of sequence diagrams associated with the use case, pre_{uc} is the precondition of the use case, specified with Object Constraint Language (OCL) [7]. The environment of the software system is represented by actors.

We assume that for each use case, all the relevant scenarios have been identified and specified with sequence diagrams. Let $cmpnt_m$ be a set of identifiers and failure probabilities of the components of the software system that participate in the interaction of the sequence diagram m, that is, $cmpnt_m=\{(c_1, cpf_1), (c_2, cpf_2),..., (c_n, cpf_n)\}$, where c_i is the component identifier and cpf_i is the component failure probability.

Sequence diagrams provide the information about the order in which interactions occur. When an interaction enters a component's axis (i.e., the component receives a service request), the component is invoked. For each use case, the execution of different scenario may cause the software enter different state. Thus, we attach a postcondition for each sequence diagram. A sequence diagram m annotated with usage annotations can be defined as a 4-tuple: $m = (cmpnt_m, msg_m, post_m, pf_m)$, where

- $cmpnt_m$ is a finite set of components.
- msg_m is a finite set of messages between components. A message is labeled by a message name and corresponding parameter list. We assume that all messages in a sequence diagram are totally ordered.

- $post_m$ is the execution postcondition of the sequence diagram m, specified with OCL, which determines the next use case to be executed.
- pf_m is the execution probability of m in its associated use case, which means that for all sequence diagram m in use case U, $\sum_{m \in U} pf_m = 1$.

Use cases not only have <<extend>> and <<include>> relationships but also execution sequence relations which reflect the business process the software supports [1]. Software testing must consider the use case execution sequence relations because different execution sequences may trigger different failures. We can represent execution sequence relations between use cases by an activity diagram [2]. In such a diagram, vertices are use cases and the edges are execution sequence relations between use cases. The *join* and *fork* synchronization bars may present the synchronization among use cases. The execution sequence relations of use cases can be defined as $UCExecRelation = (UCSet, s0, sf, V, \Sigma, \delta)$, where

- *UCSet* is the set of software use cases.
- *s0* is the initial node of the software execution and *sf* is the final node.
- $V = \{s0, sf\} \cup UCSet$.
- $\Sigma = \cup(pre_{uc} \cup post_m)$, which is the state set specified in the preconditions of use cases and the postconditions of sequence diagrams.
- $\delta: V \times \Sigma \times [0, 1] \to V$ is the function that specifies the execution sequence relations between use cases, initial node and final node.

Thus, a UML model that describes the software functionalities can be defined as:

$$system = (actors, UCSet, UCExecRelation)$$

where *actors* is the set of software users, *UCSet* is the set of software use cases and *UCExecRelation* is the execution sequence relations of use cases.

It is necessary to consider the reliabilities of the communications between components in a distributed software system. The UML artifact deployment diagram shows the distributed platform configuration where the nodes represent platform sites and links represents hardware connectors. Software components are placed into the sites where they are loaded. We annotate a deployment diagram with the failure probabilities over the connectors among sites. Thus a failure probability is assigned to each interaction between two components of a sequence diagram. These failure probability annotations reflect the reliabilities of the communications between components in a distributed software system. We assume that the communications among components residing on the same site is fully reliable, i.e. the failure probability is 0.

2.2 Deriving SUD from UML Artifacts

A SUD describes the probabilistic scenario transition process that represents the software execution, which is inherently a Markov chain and is a convenient model for statistical testing and reliability estimation of a broad spectrum of applications.

A SUD is a 5-tuple: ($S, \Sigma, \delta, q0, q_{end}$), where

- S is the set of nodes of software scenarios. Each node is labeled with the corresponding scenario name.
- Σ is the set of transitions. Each transition is labeled with a transition probability.

- q_{end} is the additional node representing the end of software execution.
- q0 is the additional node representing the start of software execution.
- δ: S ∪ {q0, q_{end}} × Σ → S ∪ {q0, q_{end}}, which is the transition relation.

Figure 1 is the algorithm that derives SUD from the annotated UML artifacts.

Algorithm: Deriving SUD from UML artifacts
Input: Annotated UML artifacts.
Create a SUD and insert node q0 and q_{end};
Insert all scenario nodes into SUD, let the nodes' labels be the name of the scenarios;
FOR each pair of scenario nodes n_1 and n_2
    IF the state specified in the postcondition of the scenario n_1 ∈ the state set specified in the precondition of the use case that the scenario n_2 belongs to THEN
        Link n_1 with n_2 by a transition whose transition probability is the transition probability in *UCExecRelation* from the use case that scenario n_1 belongs to, to the use case that scenario n_2 belongs to;
    ENDIF
ENDFOR
Elicit the set *N* of scenarios nodes that are *s0*'s successive use cases in *UCExecRelation*, as well as the execution sequence relation's probabilities (*SRP*);
Link q0 with the nodes in *N* by transitions with corresponding probabilities in *SRP* ;
Elicit the set *M* of scenarios nodes that are the use cases whose successive nodes are *sf* in *UCExecRelation*, as well as the execution sequence relation's probabilities (*SRP'*);
Link the nodes in *M* with q_{end} by transitions with corresponding probabilities in *SRP'* .

Fig. 1. Algorithm for deriving SUD from annotated UML artifacts

2.3 SUD-Based Software Statistical Testing and Reliability Estimation

The software reliability *R* is the probability that no failure occurs during a particular program execution [11]. For the sake of simplicity we use failure probability *F* to represent software reliability, note that *F*=1-*R*.

The software execution can be considered as the execution of a sequence of scenarios. A particular execution of the software corresponds to a path $x=(x_1, x_2, ..., x_L)$ that traverses the SUD from node q0 to q_{end}, where x_1 is q0, and x_L is q_{end}. x_i (*i*=2, 3,...,*L-1*) are the scenarios traversed. The next node to execute is selected according to the probabilities of the outgoing transitions of current node. Each statistical test case corresponds to a particular path *x*, that is, a particular scenario execution sequence.

The simplest way of obtaining unbiased reliability estimates of software is to test the software based on SUD. We can test the software with the paths $X_1, X_2, ..., X_N$ selected according to the SUD. A function *f(X)* is defined with the parameter of path *X*: once the path fails, *f(X)*=1, otherwise, *f(X)*=0. Then, the arithmetic mean of *f(X)* is an unbiased estimate of *F*.

3 Accelerating Statistical Testing of Distributed Software

However, the method explained in Section 2.3 does not consider the consequence of potential failures. The safety-critical software often includes some critical scenarios that provide critical functions and whose execution probabilities are very low. It often requires too many test cases to test these scenarios adequately and thus makes statistical testing based safety-critical software reliability estimation infeasible [3].

3.1 Accelerating Statistical Testing with Importance Sampling

We use the Importance Sampling (IS) technique to accelerate statistical testing. IS technique can be used to speed up Monte Carlo simulations that involves rare events [12]. Assume that the SUD of software under testing is P, we can get another SUD Q by shifting the probabilities of those transitions whose transition probabilities are not 1. Select N paths from q0 to q_{end} according to Q, let X_i^j be the i-th scenario executed in the j-th path, and $P(X_i^j)$ and $Q(X_i^j)$ are the probabilities of transitions from X_i^{j-1} to X_i^j according to P and Q respectively. We assume that $P(X_i^1) = Q(X_i^1) = 1$. Let

$$W(X_j) = \frac{\prod_{t=1}^{L} P(X_t^j)}{\prod_{t=1}^{L} Q(X_t^j)} \quad (1)$$

where L is the path length. Let

$$S = f(X)W(X) \quad (2)$$

where X is a path selected according to Q and $f(X)$ is the failure probability of X, $W(X)$ is used to compensate for the transition probabilities shifting and is called likelihood ratios. The expression S is an unbiased estimator of $f(X)$, since

$$E_Q(S) = E_Q(f(X)W(X)) = \sum Q(X)f(X)W(X)$$
$$= \sum Q(X)f(x)\frac{P(X)}{Q(X)} = \sum f(X)P(X) = E_P(f(X))$$

where E_T denotes the expectation with respect to SUD T.

By increasing the execution probabilities of the transitions related to the critical scenarios, we can test the critical scenarios more frequently. At the same time, we use S with SUD Q instead of $f(X)$ with SUD P to achieve an unbiased estimate of $f(X)$. The idea behind IS based statistical testing acceleration is to increase the execution probabilities of critical scenarios, and likeliness ratio is used to compensate for the transition probabilities shifting.

3.2 Computing the Optimal SUD

For practical reasons we use the Q that produces the minimal variance while estimating the software failure probabilities, which is called the optimal SUD in this paper.

In order to compute the optimal SUD, we first calculate the failure probability of scenario m with the following steps, which is a variant of the method of [5].

We assume that the failure probability of an invoked method equals to the failure probability of the component to which the method belongs. Assume the k-th interaction in the scenario is that the component i sends a message to component j and invokes the method of component j. Thus the failure probability of the k-th interaction can be calculated with the formula:

$$pf_k = 1 - (1 - cpf_j) \times (1 - Connector_{ij}) \qquad (3)$$

where cpf_j is the failure probability of component j and $Connector_{ij}$ is the failure probability of the connector of component i and j, which represents the failure of communication. The failure probability of scenario m can be calculated as:

$$fail_m = 1 - \prod_{k \in m}(1 - pf_k) \qquad (4)$$

Algorithm: Computing optimal SUD Q from SUD P
Input: SUD P;
    $Tmax$ and $Tmin$: Temperature parameters;
    $CScnSet$: Critical scenario set;
    $FAIL$: Scenarios' failure probability vector.
Output: SUD Q.
Step 1: Add the low probabilities transitions, which lead to the execution of the scenarios in $CScnSet$, to critical transition set CTS;
Step 2: Shift the none 1 transition probabilities in SUD P and produce SUD Q;
Step 3: $k \leftarrow 0$; $T_k \leftarrow Tmax$;
Step 4: ISSimulation($P, Q, e, var, FAIL$);
Step 5: Repeat Until $T_k < Tmin$
    1. Shift the none 1 transition probabilities in SUD Q and produce SUD Q', make sure that the transition probabilities of these transitions in CTS are no less than the counterparts in P;
    2. ISSimulation($P, Q', e', var', FAIL$);
    3. If $var' < var$ Then let $Q \leftarrow Q'$, tmp$\leftarrow$ var, var$\leftarrow$ var', Goto 5;
    4. If exp[- (var – var')/ T_k] > random(0, 1) Then let $Q \leftarrow Q'$, tmp$\leftarrow$ var, var$\leftarrow$ var'; Otherwise go to 1;
    5. If |tmp-var'| $< \varepsilon$ Then go to 6, otherwise go to 1;
    6. $T_{k+1} \leftarrow$ decrease(T_k), $k \leftarrow k+1$;
Step 6: return Q.

Fig. 2. Algorithm for computing optimal SUD

The scenarios' failure probabilities form the scenarios failure probability vector, $FAIL$. We then select a number of paths with respect to SUD Q and simulate the execution of the software with these paths. Based on the simulation results, we can compute the failure probability and the variance. By adjusting the transition probabilities, we can get different Qs and corresponding variances. The Q with the minimal variance is the optimal SUD. In order to handle complex system, we use the simulated annealing algorithm [8] to compute Q, which is depicted in figure 2.

The procedure ISSimulation(P, Q, VAR e, VAR var, $FAIL$) depicted in figure 3 is used to simulate the testing process with Q and returns failure probability and variance. Note that failure probability of each path is weighted with $W(X_j)$, which is defined in formula (1).

PROCEDURE ISSimulation(P, Q, VAR e, VAR var, $FAIL$)
Step 1: Generate test paths $X_1, X_2, ..., X_N$ with respect to SUD Q;
Step 2: Repeat the following procedure for each path X_j
　　$f(X_j) \leftarrow 0$, which is the failure probability of path X_j;
　　Traverse the scenarios invoked in path X_j. During the traverse process, if scenario i fails in simulation, that is, random(0, 1) < the i-th element in $FAIL$ then let $f(X_j)$ be 1 and continue with path X_{j+1};
Step 3: $e \leftarrow \dfrac{1}{N}\sum_{j=1}^{N} f(X_j)W(X_j)$;
Step 4: $var \leftarrow \dfrac{1}{N}\sum_{j=1}^{N}(f(X_j)W(X_j)-e)$;
ENDPROCEDURE.

Fig. 3. Algorithm for IS based simulation

3.3 Assumptions and Discussions

For practical reason we have made several assumptions. First, we assume that the failures among different scenarios are independent. This assumption simplifies the estimation task. Another assumption is that the scenario failures follow the principle of regularity, i.e., that a scenario is expected to exhibit the same failure rate whenever it is invoked. The algorithms in section 3.2 are only valid under these assumptions.

The prior information, scenario failure probability vector $FAIL$, can be computed from the pre-specified upper bounds of components failure probabilities and communications failure probabilities. $FAIL$ is used to compute the optimal SUD and is not necessarily the real failure probabilities of scenarios (which is usually unknown).

Finally we adopt the simulated annealing algorithm to compute optimal SUD, which enables to handle large and complex applications, because the computation of optimal SUD with general optimization algorithm, such as Frank-Wolfe-type algorithm, is often time consuming and inapplicable to large and complex system [6].

4 Case Study

This section explains the method with a simplified distributed process-control system, which is called NPCS (Nuclear Plant Control System).

4.1 A Case of Process-Control System

The use case diagram of NPCS is shown in figure 4. The preconditions are: $pre_{Initialize}$=Idle, $pre_{Operate}$=Initialized or Operating, and $pre_{EmgcyShutdown}$=Emergency. Figure 5 depicts the architecture of the NPCS by a UML deployment diagram.

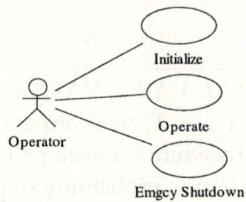

Fig. 4. The use case diagram of NPCS

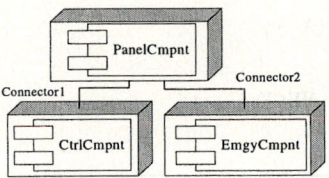

Fig. 5. The deployment diagram of NPCS

Figure 6 depicts the execution sequence relations between use cases of NPCS with a UML activity diagram (the transition probability is 1 if it is not labeled.). The transition labels are the states specified in the preconditions of use cases and postconditions of scenarios.

Figure 7 to 11 depict the scenarios.

Table 1 specifies execution probabilities of each scenario in its associated use case, as well as the states specified in the postconditions.

Figure 12(a) shows the SUD P of NPCS, which is derived with the algorithm presented in figure 1.

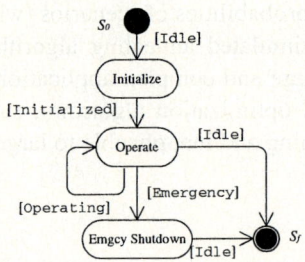

Fig. 6. NPCS use cases execution sequence relations

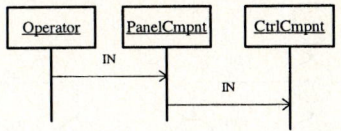

Fig. 7. Scenario 1 of use case Initialize, named as Scn_1

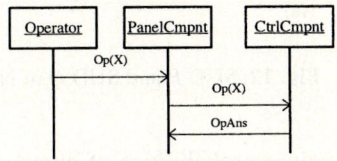

Fig. 8. Scenario 1 of use case Operate, named as Scn_2

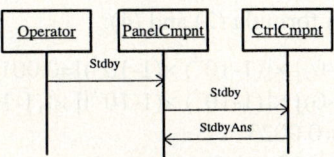

Fig. 9. Scenario 2 of use case Operate, named as Scn_3

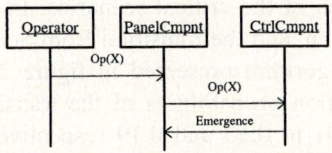

Fig. 10. Scenario 3 of use case Operate, named as Scn_4

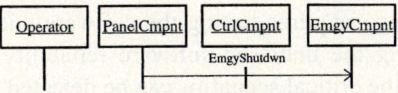

Fig. 11. Scenario 1 of use case Emgcy Shutdown, named as Scn_5

Table 1. Annotations of scenarios

	Scn_1		Scn_2		Scn_3	
	Pf_1	$post_1$	pf_2	$post_2$	pf_3	$post_3$
Operate	0.7	Operating	0.29	Idle	0.01	Emergency
Emgcy Shutdown	1	Idle	-	-	-	-
Initialize	1	Initialized	-	-	-	-

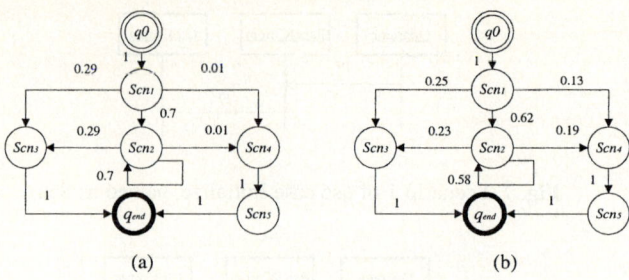

Fig. 12. SUD P and SUD Q of NPCS

We assume that the failure probabilities of components PanelCmpnt, CtrlCmpnt and EmgyCmpnt are 10^{-4}, 10^{-4} and 10^{-5}, and the failure probabilities of connector1 and connector2 are 10^{-3} and 10^{-6} respectively. We assume that the communication between operator and PanelCmpnt never fails. Thus we can compute the failure probabilities of the five scenarios with formula (3) and (4):

$fail_{Scn1} = 1-[(1-10^{-4}) \times (1-0)] \times [(1-10^{-4}) \times (1-10^{-3})] = 0.0012$
$fail_{Scn2} = 1-[(1-10^{-4}) \times (1-0)] \times [(1-10^{-4}) \times (1-10^{-3})] \times [(1-10^{-4}) \times (1-10^{-3})] = 0.0023$
$fail_{Scn3} = fail_{Scn4} = fail_{Scn2} = 0.0023$
$fail_{Scn5} = 1-[(1-10^{-5}) \times (1-10^{-6})] = 0.000011$

The critical scenario is the emergency shutdown scenario Scn_5, which is rarely executed. The critical transition set (CTS) includes all the low probability transitions that lead to the execution of the critical scenarios. In SUD P, the CTS includes the transition from Scn_1 to Scn_4 and the transition from Scn_2 to Scn_4. The optimal SUD Q is computed with the algorithm presented in figure 2 and the result is depicted in figure 12(b). The transition probabilities of the transitions in CTS in SUD P have been increased from 0.01 to 0.13 and 0.19 respectively. Thus, the critical scenario Scn_5 will acquire more tests with SUD Q.

4.2 Statistical Testing Simulation with Fixed Test Budget

Testing with fixed budge requires testing the software with a limited number of test cases. While estimating the unbiased software reliability from the test results, we hope that the faults in the critical scenarios can be detected as many as possible.

Assume that the failure probabilities of components PanelCmpnt, CtrlCmpnt and EmgyCmpnt are 10^{-5}, 10^{-5} and 10^{-7}, and that the failure probabilities of connector1 and connector2 are 10^{-6} and 10^{-7} respectively. Thus we can compute the failure probabilities of the five scenarios with formula (3) and (4), which are: 0.000021, 0.000032, 0.000032, 0.000032 and 0.0000002 respectively.

Table 2 shows the results of 10,000 test simulation. While no failure of Scn_5 is revealed in both processes, the results show that the Scn_5 acquires more tests with the acceleration method while unbiased failure probability estimate is obtained.

Table 2. Simulation results of statistical testing for 10,000 times

	Software failure probability	Variance of failure probability	Execution number of Scn_5
IS	1.281e-4	8.471e-8	3,747
Standard	1.279e-4	8.078e-9	338

4.3 Statistical Testing Simulation Under Scenarios' Reliability Constraints

In order to test the critical scenarios until their reliability requirements are met, testers often have to execute many extra test cases even the system reliability is satisfied. Since we discuss the safety-critical software testing, we assume that test reveals no failure. Thus, we can compute that we should test each critical scenario without failure no less than 10^6 times with the following formula [10], while its failure probability is no higher than 10^{-6}:

$$\theta = a/(t+a+b) \qquad (5)$$

where t is the number of testing. We assume that testers know nothing about the reliability of the program, thus a and b are both 1. The simulation results in table 3 shows that the acceleration method can effectively reduces the number of test cases required while the critical scenarios are adequately tested.

Table 3. Simulation results of testing critical scenario for 10^6 times

	Execution number of Scn_5	Software test number
IS	10^6	2,668,588
Standard	10^6	29,603,316

5 Related Work

Study on model based software reliability assessment has resulted in several techniques that can be used to estimate the application reliability. State-based models assume that the transfer of control between modules has a Markov property, and software reliability is estimated by analyzing the model that combines software control transfer with failure behavior [4]. However, it is often very difficult to analyze complex state models. The path-based models estimate system reliability based on the possible execution paths of the program[9][13]. [9] takes an experimental approach to assessing the reliability of component-based applications. However, the distributed property is not considered. [13] develops a probabilistic model to analyze component software reliability using scenarios, the model can be used to estimate the reliability of distributed applications. [5] introduces a method that assesses software reliability from annotated UML artifacts. While all these works focus on model based software reliability estimation, they are not to attempt to assess software reliability from the results of statistical testing, which is the main concern of this paper. [14] presents a method that derives Markov chain usage model from annotated UML models but gives no consideration to safety-critical scenarios' testing.

6 Conclusion and Future Work

This paper discusses a statistical testing acceleration technique for distributed safety-critical software. The method is fully integrated with annotated UML artifacts. First, SUD is derived from UML artifacts. Based on SUD and annotated UML artifacts, this paper presents a method to accelerate the statistical testing of distributed safety-critical software. The simulation results shows that the method can effectively reduce the test cost while obtaining the unbiased software reliability estimates.

We are currently working on a set of automated tools to support the method. The tools support to annotate the UML artifacts and derive the SUD P. When the critical scenarios with low execution probabilities are specified, the tool can compute the optimal SUD Q.

References

1. Binder, R.: Testing Object-Oriented Systems. Addison-Wesley, (1999)
2. Bruegge, B., Dutoit A.H.: Object-Oriented Software Engineering: Conquering Complex and Changing Systems. Prentice Hall, (2000)
3. Butler, R.W., Finelli, G.B.: The Infeasibility of Quantifying the Reliability of Life-critical Real-time Software. IEEE Transactions on Software Engineering, Vol. 19(1). (1993) 3–12
4. Cheung R.C.: A User-Oriented Software Reliability Model. IEEE Transactions on Software Engineering, Vol.6(2). (1980) 118-125
5. Cortellessa V., Singh H. Cukic B.: Early reliability assessment of UML based software models. In Proc. Of the Third International Workshop on Software and Performance (WOSP2002), Rome (2002) 302-309
6. Gutjahr W.J.: Software dependability evaluation based on Markov usage models. Performance Evaluation, Vol. 40(4). (2000) 199 – 222
7. Rumbaugh J., Jcobson I., Booch G.: The Unified Modeling Language Reference Manual. Addison-Wesley, (1999)
8. Kirkpatric S., Gelatt C.D., Vecchi M.P.: Optimization by simulated annealing. Science, Vol. 220(4598). (1983) 671 – 680
9. Krishnamurthy S., Mathur A.P.: On the Estimation of Reliability of a Software System Using Reliabilities of its Components. In Proc. Of the eighth International. Symposium of Software Reliability Engineering (ISSRE'97), (1997) 146-155
10. Miller K.W.: Estimating the Probability of Failure when Testing Reveals No Failures. IEEE Transactions on Software Engineering, Vol.18(1). (1992) 33 – 41
11. Prowell, S.J., Trammell C.J., Linger R.C., Poore J.H.: Cleanroom Software Engineering: Technology and Process. Addison-Wesley, (1999)
12. Smith P.J., Shafi H., Gao H.: Quick simulation: a review of importance sampling techniques in communication systems. IEEE Journal on Selected Areas in Communications, Vol.15(5). (1997) 597 – 613

13. Yacoub S., Cukic B., Ammar H.: Scenario-Based Reliability Analysis of Component-Based Software. In Proc. of the 10th International Symposium of Software Reliability Engineering (ISSRE'99), (1999) 22-31
14. Yan J., Wang J., Chen H.W.: Automatic Generation of Markov Chain Usage Models from Real-time Software UML Models. In: Proc of 4th International Conference On Quality Software (QSIC2004), Braunschweig, GERMANY(2004) 22-31

A Metamodel for the CMM Software Process[*]

Juan Li[1,2], Mingshu Li[1], Zhanchun Wu[1], and Qing Wang[1]

[1] Institute of Software, Chinese Academy of Sciences, Beijing, China
lijuan@itechs.iscas.ac.cn
[2] Graduate School of the Chinese Academy of Sciences, Beijing, China

Abstract. With the increasing complexity of software system, geographically distributed development has become mainstream. Managing a software process in which team members are physically distributed is challenging. How to use the Capability Maturity Model (CMM) in geographically distributed development is an area with a number of open research issues. We define a CMM Software Process (CSP) by a set of generic process elements in accordance with the requirements of the CMM. Using the Model Driven Architecture (MDA), the CSP model can be transformed into distributed CMM implementation process models. This paper presents a metamodel for the CSP model, named MM-CSP, and provides the abstract syntax and the semantic of the MM-CSP as well as a UML profile for the MM-CSP. Based on the MM-CSP, a prototype tool for CSP modeling is developed.

1 Introduction

In software engineering environment, use of the Internet as a medium for conducting distributed software engineering activities has been on the rise in recent years [1,2]. Managing a software process in which team members are physically distributed is challenging. As a widely used software process improvement model, the *Capability Maturity Model* (CMM) plays a major role in hundreds of software organizations worldwide [3,4]. However, the CMM does not specify the details of how to effectively implement the process improvement activities in the distributed development environment. In order to solve this problem, we start from the fundamental description mechanism for the CMM and define a *CMM Software process* (CSP), which is a general software process. Distributed development teams can add their characteristics into the CSP to achieve their CMM implementation processes. *Model Driven Architecture* (MDA) [5] provides a systematic framework to manage and transform models. By introducing the MDA, transforming the CSP model into distributed CMM implementation models can be performed systematically and effectively. This paper present a metamodel for the CSP model, named MM-CSP. The MM-CSP defines a language for describing the CSP model and provides the foundation for transforming the CSP model using the MDA. In addition, the MM-CSP is a CMM-based extension of the *Software Process Engineering Metamodel* (SPEM) [6]. We define the MM-CSP based on the *Meta Object Facility* (MOF) [7] and implement the MM-CSP through a CSP modeling tool.

---

[*] This research is supported by the National Natural Science Foundation of China (60273026), and the Chinese National "863" High-Tech Program (2002AA116060, 2001AA113080).

2 The Metamodel for the CSP Model

A metamodel is an explicit model of the constructs and rules needed to build specific models within a domain of interest [8]. The MM-CSP describes concepts and their relationships for the purpose of building and interpreting the CSP model. The CSP model is not a process model of a specific organization and the distributed teams can add their characteristics into the CSP model to create their CMM implementation process models. In this paper, we choose the SPEM as the basis for designing the MM-CSP, because the SPEM provides many process elements for the MM-CSP and can be used as the metamodeling backbone according to the MOF. We introduce the extension to the SPEM and represent the structural aspect of the MM-CSP in a new package, called CmmElements, which is located in the package Extension of the SPEM.

The abstract syntax of the MM-CSP is shown in Figure 1. Due to space restrictions, we present only a part of the abstract syntax.

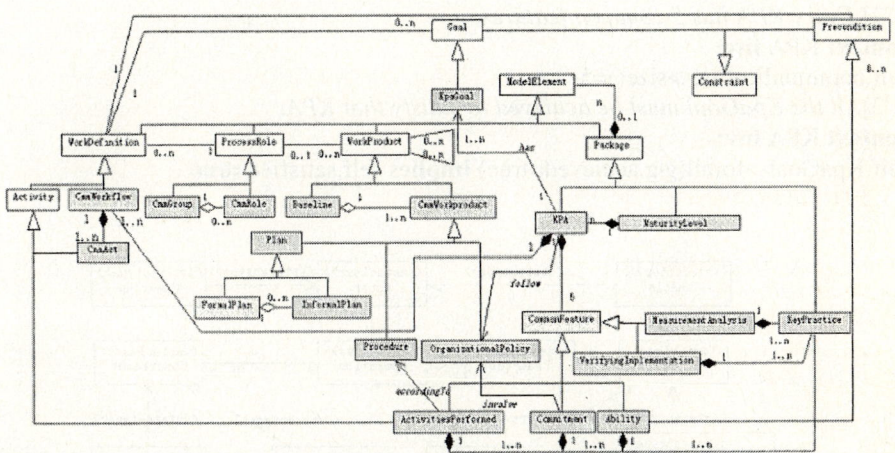

Fig. 1. Abstract syntax of the MM-CSP

- MaturityLevel
This element represents a well-defined evolutionary plateau toward achieving a mature software process. It has three attributes: name (inherited from ModelElement), isLevel and achived.
- KPA
The KPA element identifies a group of related activities that achieve a set of goals required for establishing process capability at that MaturityLevel. The KPA element has name (inherited form ModelElement) and satisified as its attributes.
- CmmAct
The CmmAct element defines actions in the CMM, such as review and audit. Its attributes contain name (inherited from ModelElement), role, precondition, postcondition, input, output, step and resource.

- CmmWorkproduct

The CmmWorkproduct element represents the work product in the CMM. It has name (inherited from ModelElement) and isDeliverable (inherited from WorkProduct) as its attributes.

- CmmRole and CmmGroup

The CmmRole element represents the roles described in the CMM. The CmmGroup element can be composed of several CmmRole elements. Both the CmmRole and CmmGroup have name (inherited from ModelElement) and responsibility (string type) as its attributes.

- Well-formed rules defined the rules and constraints on valid models. We define Well-formed rules using the *Object Constraint Language* (OCL). Due to space restrictions, we present only a portion of well-formed rules for the MM-CSP.

- MaturityLevel

[C1] *To achieve a maturity level, the KPA for that level must be satisfied.*
context MaturityLevel **inv:**
self.KPA->forall(k|k.satisfied=true) **implies** self.achieved=true

- KPA

[C2] *Every KPA has 5 common features.*
context KPA **inv:**
self.commonFeature->size()=5

[C3] *All the KpaGoal must be achieved to satisfy that KPA.*
context KPA **inv:**
self.KpaGoal->forall(g|g.achieved=true) **implies** self.satisfied=true

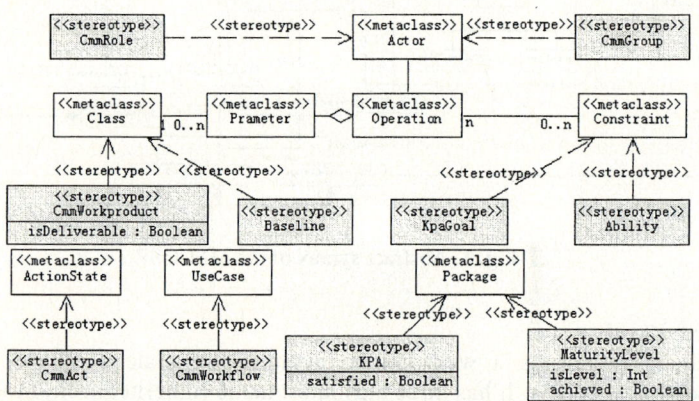

Fig. 2. Virtual metamodel

With the UML profiling mechanism [9], the MM-CSP can be implemented by integrating with UML CASE tools. The UML profile is based on stereotypes, tagged values, and constraints. The stereotype provides a way of classifying model elements as if they were instances of new virtual metamodel constructs. The formal definition of stereotypes and tags are given in the Figure 2. The isDeliverable tag of the CmmWorkproduct stereotype is true if CmmWorkproduct is defined as a formal deliverable work product of the process. The tags of the MaturityLevel stereotype are isLevel and achieved. The tag of the KPA is satisfied. The constraints are also described by the OCL. Several constraints are listed as follows:

— CmmWorkflow:
[R1] *A CmmWorkflow behavior is defined using no more than a single Activity Graph and in no other way.*
context CmmWorkflow **inv:**
self.behavior->size<=1
and
self.behavior->forall(b| b.oclIsTypeOf(ActivityGraph))
— ActionState
[R2] *An ActionState is either a CmmAct or refers to a CallAction for another CmmWorkflow.*
context ActionState **inv:**
self.stereotype.name="CmmAct" **or**
(self.entry->size=1 **and**
self.entry.oclIsKindOf(CallAction) **and**
self.entry.operation.oclIsKindOf(CmmWorkflow))

3 A Modeling Tool and Applications

Using the MM-CSP, we have developed a prototype of a modeling tool, named MDA-SPMT (*MDA-Software Process Model Transformation*), to support modeling the CSP model. The MDA-SPMT supports the MM-CSP and provides a graphical modeling environment, as shown in Figure 3. Also this tool can validate the CSP model. The CSP model consists of four parts: Roles, Work Products, KPAs and Rules. Every part is a package, which is composed of some subpackages. Besides, we use the OCL to define the rules in the CSP model. These rules prescribe the invariant to perform the activities, ensure the model's accuracy and provide the semantic support for the model transformation.

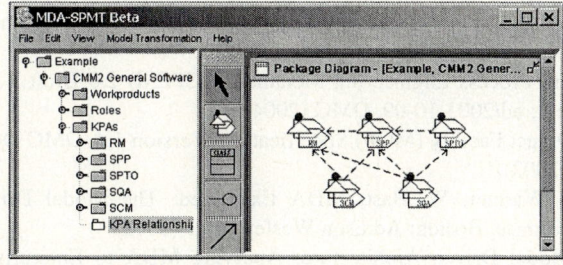

Fig. 3. MDA-SPMT

We have summarized the characteristics of the distributed development teams. Based on these characteristics, we transform the CSP model into the different CMM implementation process models manually. The distributed development teams can access their CMM implementation models via the Web. And it is important to maintain the CMM implementation model consistency in the distributed development environment.

4 Conclusions and Future Work

In recent years geographically distributed development has become mainstream. It is challenging to use the CMM in managing the software process of the distributed development environment. This paper proposes a metamodel, MM-CSP, for building the CSP model, which is the foundation for transforming the CSP model using the MDA. We choose the SPEM as the metamodeling basis for the MM-CSP and present the abstract syntax and the semantic of the MM-CSP as well as a UML profile for the MM-CSP. Based on the MM-CSP, we developed a prototype of a CSP modeling tool, named MDA-SPMT. Future work will focus on the modeling approach for the distributed organization characteristics model. Furthermore, research should be related to the study of the transformation mechanism on how to automatically or semi-automatically transform the CSP model and the distributed organization characteristics model into the distributed CMM implementation process models. The MDA-SPMT tool will be enhanced to support modeling distributed organization characteristics models and the model transformation.

References

1. J. Dominigue, P. Mulholland: Fostering debugging communities on the Web. Communications of the ACM, Vol.40, No.4 (1997) 65-71
2. L. Cai, C. K. Chang, J. Cleland-Huang: Supporting agent-based distributed software development through modeling and simulation. The Ninth IEEE Workshop on Future Trends of Distributed Computing Systems (FTDCS'03) (2003) 56-62
3. B. Pitterman: Telcordia Technologies: The Journey to High Maturity. IEEE Software, Vol.17, No.4 (2000) 89-96
4. G. Yamamura: Software Process Satisfied Employees. IEEE Software, Vol.16, No.5 (1999) 83-85
5. J. Miller, J. Mukerji, (eds.): Model Driven Architecture. OMG Document: ormsc/2001-07-01, OMG (2001)
6. OMG: Software Process Engineering Metamodel (SPEM) 2.0 Draft Request For Proposal. OMG Document: ad/2003-10-09, OMG (2004)
7. OMG: Meta Object Facility (MOF) Specification, Version 1.4. OMG Document: formal/02-04-03, OMG (2002)
8. A. Kleppe, J. Warmer, W. Bast: MDA Explained: The Model Driven Architecture™: Practice and Promise. Boston: Addison Wesley Press (2003)
9. S.F. David: Model Driven Architecture: Applying MDA to Enterprise Computing. New York: John Wiley & Sons (2003)

Performance Tuning for Application Server OnceAS[*]

Wenbo Zhang[1,2], Bo Yang[1,2], Beihong Jin[1], Ningjing Chen[1,2], and Tao Huang[1]

[1] Institute of Software, Chinese Academy of Sciences, Beijing, China
[2] Graduate School of the Chinese Academy of Sciences, Beijing, China
{wellday, yangbo, jbh, river, tao}@otcaix.iscas.ac.cn

Abstract. The J2EE application server provides a primary solution to develop enterprise-wide applications, which uses containers to hold application components. The container framework relieve developers' burden greatly because it encapsulates all the system level services and the developers are able to use these services directly without knowing underlying details. The processing capacity of application servers is becoming more and more important with the requirements of achieving higher performance and higher scalability. This paper uses ECperf, a performance benchmark tool for application servers, to studies the performance issues of the application server OnceAS, which is developed by the Institute of software, Chinese Academy of Sciences, and presents optimization approaches including bean instance pools and high speed naming service. These optimizations are implemented in OnceAS and proved to be effective through ECperf benchmark evaluation.

1 Introduction

The J2EE application server is playing a major role in developing enterprise applications. It provides a developing and runtime platform for enterprise applications with various system services. Developers need not to implement these system services by themselves. They just invoke the ready-made services in the application server and only focus their attention on actual business logics so as to achieve high development efficiency and robust application systems. Meantime, it is the application server that provides non-functional features including performance, scalability and so on for enterprise applications.

EJB (Enterprise Java Bean) is one of the most important technologies in the J2EE specification [1] to develop portable and scalable applications. EJBs are deployed in the EJB container, which acts as the runtime environment for beans. Beans are not exposed to clients directly and should be accessed by certain interfaces specified in the EJB specification. EJB container delegates all the invocations for beans. It provides the runtime services for EJB clients such as security, transaction, external resource retrieving and manages the lifecycle of beans including creation, destruction and persistence.

---

[*] This work was supported by the National Grand Fundamental Research 973 Program of China under Grant No. 2002CB312005; the National Hi-Tech Research and Development 863 Program of China under Grant No. 2001AA113010; and the National Natural Science Foundation of China under Grant No. 60173023.

Naming service [2] is one of the most basic services in a J2EE application server. It provides the function of locating the components, resources and services deployed in the system. This service is so fundamental that every application server should provide it as an indispensable component. Thus its performance usually has an impact upon the performance of application server.

OnceAS is an application server developed by Institute of Software, Chinese Academy of Sciences, following the J2EE1.3 specifications. By using ECperf [3] benchmark as performance test tool, we found there were great performance bottlenecks in the EJB container and the naming service. We tried a variety of tuning approaches for these two components, including implementing instance pool for EJBs in EJB container and localizing the invocation for the naming service.

The remainder of the paper is organized as follows. Section 2 describes background knowledge about performance tuning of application server. Section 3 analyzes the performance bottlenecks of the OnceAS by the ECperf testing. Section 4 present our optimization approaches for the OnceAS and section 5 provides a performance evaluation. Section 6 discusses some related works. Finally, section 7 provides conclusions.

2 Background

2.1 EJB Container

EJB specification [4] provides a framework to develop EJB components and EJB technology can be used to develop enterprise applications with distributed component architectures. Each EJB is a distributed and highly reusable component implementing specific business logics. EJB container provides the runtime environment for beans.

The container manages the actions and states of beans after they are deployed in the container. Additionally, the container provides the system level services for beans such as transaction management, security control, logging, etc. EJB framework defines a precise boundary between business logics and system functions. Bean developers only consider how to implement the business logics and are relieved from the burden of system programming. At the same time, each bean only provides the business function and does not depend on any specific runtime environment. This makes it convenient to manage and migrate the J2EE applications.

An EJB container should obey the following principles to provide system level services for the enterprise beans residing in it [5]:

Contract: Contracts define the boundary of different components in the EJB framework. They provide the interfaces for components to cooperate so that the components need not to know the implementation details. This ensures the compatibility of EJB technology.

Service: EJB container supplies system level services such as naming service, security service and transaction service to beans. All these services are transparent to beans.

Interference: EJB container interferes in the invocations of beans and provides the required services during the interference.

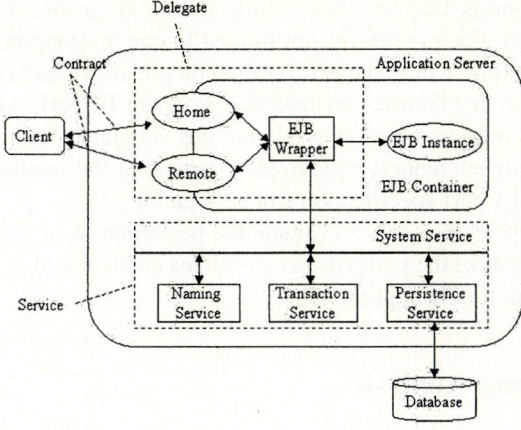

Fig. 1. The Model of EJB Container

Figure 1 describes the framework of the EJB container. The performance of EJB container is a critical issue in the overall performance of the application server because of its extremely important role. How the EJB container schedules and manages the invocation affects its throughput to process requests. Additionally, how it accesses the system service brings another potential impact for the system performance.

2.2 ECperf Benchmark Suite

ECperf is a benchmark suite proposed by Sun Corporation and some other IT companies in order to test the performance and scalability of J2EE application servers. It is complicated enough to simulate an actual electronic business system and is able to reflect the processing capability of an application server. It makes use of several kinds of beans and system services such as naming service, distributed transaction, cache mechanism, object persistence and resource management pool to fulfill the business model. So it is an ideal tool to measure the performance of J2EE application servers [6].

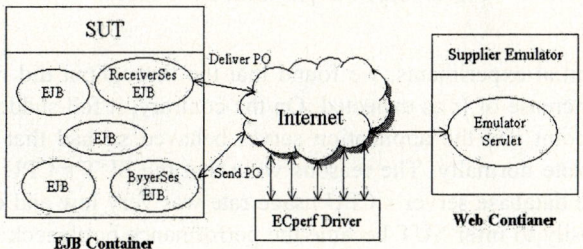

Fig. 2. Application Model of ECperf

The structure of ECperf is shown in Figure 2. The metrics for ECperf benchmark is the throughput: BBops/min (Benchmark Business Operations per minute), which is the amount of requests that are successfully processed in one minute by the SUT (Server Under Test). Each request is not limited to one transaction but could possibly be several transactions, which depend on the complexity of the request. There is an important parameter Ir (Transaction Injection Rate) in ECperf, which tightly affects the throughput. Ir reflects how many clients are simulated: larger value represents more clients and higher request concurrency arrived at the application server. More information about ECperf specification can be found in [7].

We adopt ECperf as the tool to evaluate the performance of OnceAS. The following sections will analyze the performance problems encountered in the benchmark test and then present out tuning approaches.

3 Performance Analysis

We deployed the ECperf application in a distributed system following the ECperf specification as shown in Figure 3.

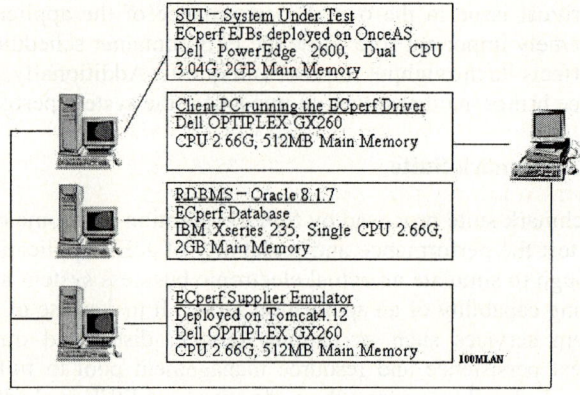

Fig. 3. ECperf Deployment Environment

During the initial experiments, we found that the throughput did not always augment with the increase of Ir as expected. On the contrary, it fell suddenly when Ir arrived at some point and the application server behaved so bad that the benchmark could not complete normally. The reasons were that the SUT's CPU usage rate was too high and the database server's CPU usage rate was very low and even idle sometimes. It was easily to infer SUT became the performance bottleneck during the tests since the database would never be idle in a normal case.

In order to further locate the bottleneck, we used the Borland's performance tuning tool Optimizeit [8] to detail the CPU usage during the benchmark. It was seen that too much CPU time was spent on accessing the naming service and the time increased more while adding Ir. This CPU time ratio is illustrated in Figure 4.

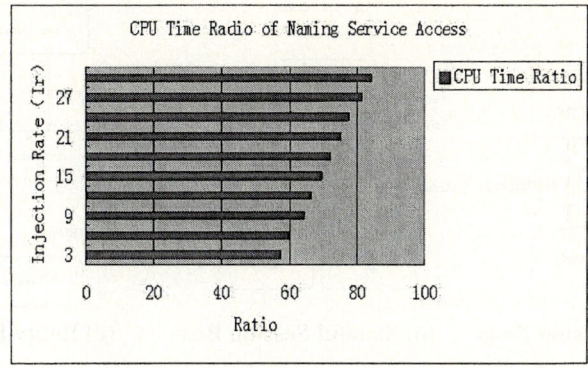

Fig. 4. CPU Time Ratio of Naming Service Access

Obviously the naming service consumes too much of CPU time. There are usually many naming service invocations in the creation of EJB instances because the beans need to query the naming service to acquire all the resources they will use. Take ECperf for an example. It will averagely create 35 EJB instances to complete a manufacturing business operation and access the naming service for 60 times. The processing pressure in ECperf is linearly proportional to Ir and the number of clients in manufacturing domain is Ir*3. When Ir reaches 30, the access to naming service will arrive at 5400 times for each manufacturing business operation. Thus, the frequent creations of EJB instances and the corresponding naming service accesses became the main performance bottlenecks for OnceAS.

There are two possible approaches to resolve this problem: one is to manage the beans' lifecycle carefully and make use of pools to cache EJB instances so as to avoid frequent creation and destruction of EJBs; the other is to optimize the naming service by minimize the overheads to access it. We will discuss the two approaches separately in the following sections and evaluate their effects.

4 Performance Tuning

4.1 EJB Instance Pool

The instance pool provides caching for managing the EJB instances effectively. Such pools cache the bean instances in some state where they are irrelevant to any particular client. For example, an instance can be placed in the pool rather than be discarded after it has been removed by the client, so that this instance can be reused for another client without creating a new instance. This technology improves the reusability of instances and reduces the performance overheads caused by frequent bean creations.

There are three types of beans used in ECperf benchmark: stateless session bean, stateful session bean and entity bean. Each of them has different lifecycles and state transitions. Figure 5 shows the state transitions of three types of bean.

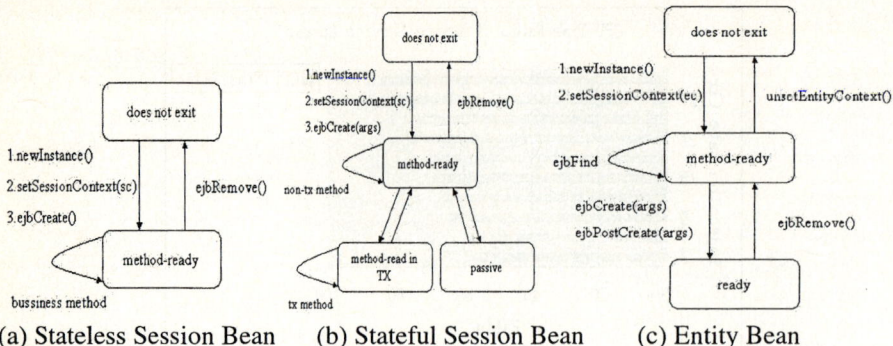

Fig. 5. LifeCycles of Enterprise Java Beans

Although different types of EJBs have different lifecycles, they all have a similar state "method-ready" where the bean instances have been created and are ready for invocations from clients. Instances in this state can be cached in a pool: an EJB instance removed by one client can be allocated to another client transparently. For different type of beans, we adopt different pooling policies.

Stateless Session Bean: Instances of a stateless session bean are identical to all clients because they do not contain any information of the clients. All such instances can be placed in the pool after the clients finish their invocations. This is called *complete pooling* policy.

Stateful Session Bean: Instances of a stateful session bean are specific for their corresponding clients so all the instances cannot be reused. Here we adopt a *non-pooling* policy.

Entity Bean: Instances of an entity bean are not specific to client in the method-ready state. They are only initialized with particular clients when ejbCreate method is invoked. In this case, instances after an ejbFind method can be pooled but instances after ejbCreate and following methods must be discarded. So we employ a *part-pooling* policy for entity bean.

We implement all the three pooling policies in OnceAS and improve the performance of EJB container because pooling the instances cuts down the access to naming service greatly.

4.2 High Efficient Naming Service

Naming service provides a convenient way to access the objects, for examples users, machines, services, components, etc., in distributed systems. It binds a unique name of each object in the system and then these objects can be located by their names. In this way, it provides a unified mechanism to store and retrieve all kinds of objects in the system. Naming service integrates all the components, services and resources together and helps them to cooperate correctly. Thus it is the most fundamental service of an application server.

OnceAS implements a naming service based on Java RMI technology. There are three roles in typical naming operations:

Client: A client acts as the invocator, who issues all the operations on naming service such as *lookup* and *bind*.

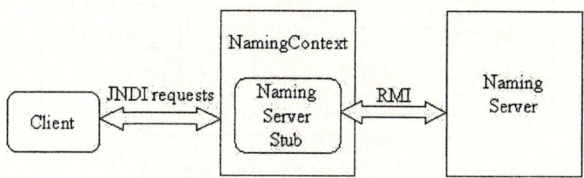

Fig. 6. Access of Naming Service

Naming Server: The naming server is a subsystem storing the objects, managing their names and responding to the clients. It has a corresponding RMI stub, which is used by client to communicate with the server remotely. All the invocations on the stub will be forwarded to the naming server by RMI mechanism.

NamingContext: NamingContext is the service provider defined by the JNDI specification [2]. It is the bridge between clients and the naming server, providing the accessing interfaces and delegating naming requests to the naming server. The NamingContext contains the naming server's stub to communicate with the server. From the view of clients, the NamingContext serves as a map from names to objects and clients can look up names and bind names to specific objects in the map.

The outline of accessing naming service in OnceAS is described in Figure 6.

A client gets an initial NamingContext as the entry if it wants to access the naming service, and then invokes the operations of the NamingContext. The naming server performs the naming operations in its storage space responding to the arrived requests transmitted by RMI and then returns the results to the client also via RMI. Examining the whole process, network communication spends much time and should be reduced as much as possible. From what is described above, the performance bottleneck may appear in the RMI communication between clients and the naming server. We present an optimized naming service solution in order to minimize the network communication as shown in Figure 7.

The tuning procedure can be divided into three parts:

1) Local Server Mechanism Once. AS initializes a naming server instance when it starts up, which called local server. The name server is running in the same JVM as OnceAS and thus can be accessed by local method invocation in OnceAS. That is to say clients in the server side can invoke the naming server directly without using RMI. Now the clients of Naming Server can be classified into two categories: local clients reside in the same JVM with OnceAS and remote clients outside the JVM of OnceAS. The former can invoke the naming server directly but the latter has to access the naming server via RMI communication. This is called the local server mechanism.

In ECperf, most requests to naming service are issued by EJBs or Servlets on the server side and they are all the local clients of the naming server. So the local server mechanism is able to convert a great number of invocations from RMI ones to local invocations and lessen the executing time seriously, which improves the performance considerably.

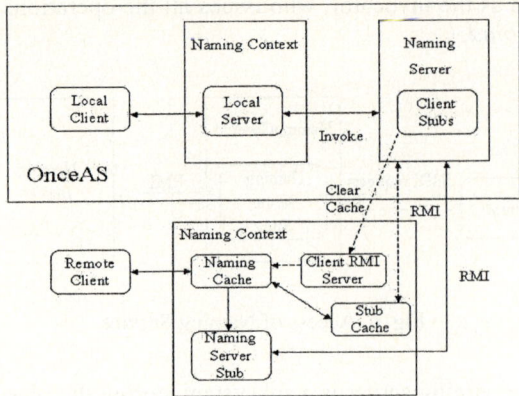

Fig. 7. Opertimized Naming Service

Any access to Naming Service needs to locate the naming server. We force all the local clients of naming server use the local server as their default one while all the remote clients need use the naming server by RMI. Because the obvious performance difference between a local invocation and an RMI invocation, the optimization will benefit OnceAS server significantly.

2) Naming Cache. As to remote clients, caching the results in NamingContext is an effective way to speed up the looking up operations and to relieve the burden of the naming server. The naming cache in Figure 7 stores the known naming bindings, which will be updated after every invocation of NamingContext. A remote client will search in the naming cache firstly every time when it tries to resolve a name. It will only connect to the remote naming server in the case it fails to get the name in the cache. The accessing pattern in ECperf is very regular and some naming entries are visited repeatedly. So the hit ratio for the cache is so high that most requests of remote clients are performed locally and the burden of the naming server is relived greatly.

There is a problem to maintain the consistence between the cache and the naming server because the server is unaware of the clients and it is hard for the server to update the content in the cache. We make the server aware of its remote clients by storing the stub of each client in the server. The naming server will communicate with all the clients to clear their cache once the naming bindings change on the server side. In actual system naming changes often happen in startup time, deployment time, redeployment time and seldom happen in regular running time. So the extra overhead for the naming service to keep consistency can be omitted when considering the overall performance of the application server.

3) Cache of Naming Server Stub. Any client to access of naming service need to locate an available naming server, and then it will ask the naming server to transfer the stub to it. Another optimization for remote clients is to cache the stub of the naming server. The stub will be stored in a cache of NamingContext once it is transferred to a remote client, which is called stub cache in Figure 7. All the following requests issued in the same JVM are able to utilize this stub instantly without transferring it again. There is no consistence problem here because the naming server does not change during the running time and thus the stub is always valid.

5 Experiments and Results

We carried out further tests on OnceAS in order to evaluate the performance improvement caused by the optimizations. There were several groups of experiments according to different optimization approaches: one is performed after implementing the instance pools in EJB container, another is performed after integrating the local server and the caches into naming service, and the last one is performed after adopting both two optimizations. We will analyze the effect of different optimizations by studying the benchmark results.

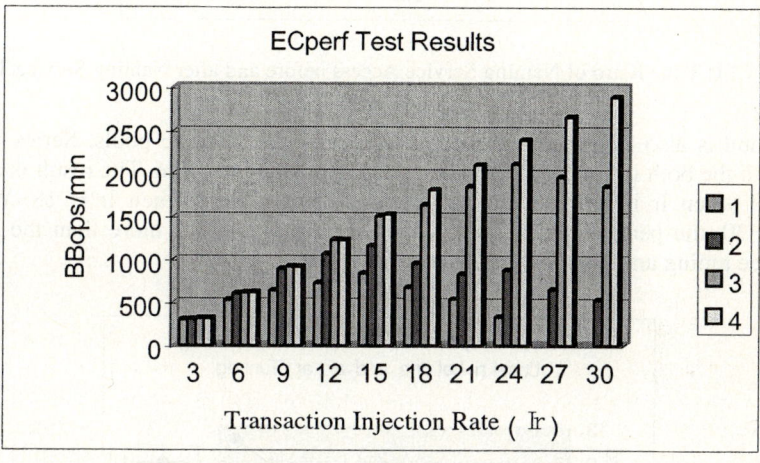

Fig. 8. Ecperf Test Results of OnceAS

The tests ran in the same environment as shown in Section 3.1 and followed several principles to avoid any impacts caused by other factors:

1) Clear and initialize the database before every test.
2) Reboot OnceAS before every test.
3) Warm up the system at least 5 minutes to utilize all the caches.
4) Specify enough pool size after employing instance pool in EJB container.

Figure 8 describes the ECperf metrics achieved in different tests. Series 1 is the case before any performance tuning, where the throughput arrives at the peak value when Ir reaches 15 and the tests fail to complete when Ir is 27 or 30 because the test pressure is too high for OnceAS. Series 2 is the case with instance pools. The maximum throughput is 1148, which is 37.9% more than the non-optimized system, and then the throughput decreases with greater Ir because the system becomes less overloaded. Series 3 is the case with the high speed naming service, which performs much better than the instance pools optimization. The throughput reaches its maximum value 2105 when Ir is 24. This result is 153% better than that without any performance

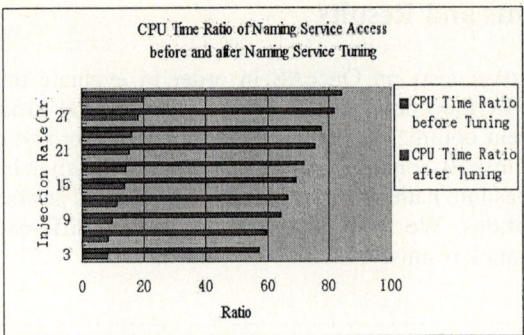

Fig. 9. CPU Time Ratio of Naming Service Access before and after Naming Service Tuning

tuning and is also 83% better than that with only the instance pools. Series 4 is the case with the both instance pools and high speed naming service. The result is close to Series 3 when Ir is low but the difference begins greater when Ir is 18. When Ir reaches 30, the peak value is 2884, which improves 246.5% more than the no performance tuning and also 37% more than Series 3.

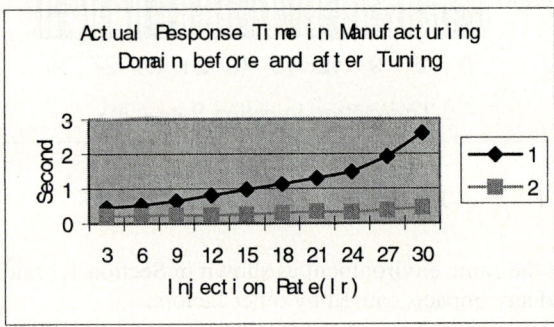

Fig. 10. Actual Response Time in Manufacturing Domain before and after Tuning

From Series 1 and 2 in Figure 8, it is shown that instance pools do help to improve the system performance but they are not able to resolve all the problems ultimately. The system does not perform well especially in heavy load situations.

After the optimization for naming service, the system performance is improved considerably. Further more, we studied the CPU usage ratio again, which is shown in Figure 9. The ratio decreases from 70.8% to 13.5% after the optimization and this reduction relieves the burden of the application server greatly, so that it helps to improve the overall performance greatly. Compared with the instance pool, the optimization for naming service is more fundamental and effective because the former only reduce the invocations of naming service partly and some other necessary invocations still hinder the improvement of system performance.

The fact that OnceAS's performance is increased furthermore after both optimization approaches are adopted shows that there is still some extra overhead after the optimization for naming service. Through proper policies for different pools, the invocations of naming service can be cut down much more. This change is easy to be observed in ECperf's another metrics – response time, which is shown in Figure 10.

These curves represent the response time in the manufacturing domain of ECperf. Series 1 is the case with only naming service optimization and Series 2 is the case with both instance pool optimization and naming service optimization. From the figures above, even though the throughput in ECperf does not improve obviously after both optimizations are adopted, the average response time is decreased by 76.3% from 1.18 seconds to 0.28 seconds because the instance pools avoid many invocations of the naming service. This kind of optimization is more evident in a heavy load system.

6   Related Work

The performance tuning for application servers have become one hot spot in the area of distributed system. ECperf, as a perfect benchmark tool in J2EE performance tests, has been noticed by many industry corporations and research institutes. Much work has been taken to the tuning for application servers based on ECperf. For example, Samuel D. Kounev [9] has discussed how to resolve the performance problems using pessimistic control policy databases and Martin Karlsson [10] has studied the behaviour of system memory in ECperf tests.

There is still much work related to the performance of application servers: Paul Brebner and Shuping Ran [11] have analyzed the performance of different submitting patterns and Emmanuel Cecchet has compared the performance and scalability of different open source J2EE application servers [12].

This paper studies more fundamental components: EJB container and especially naming service of an application server. The optimizations help to improve the performance of almost all kinds of J2EE applications and this has been validated in our ECperf benchmarks on OnceAS server.

7   Conclusions

Performance tuning for the J2EE application server is a complicated problem. We use ECperf to detect the main performance bottlenecks and present the corresponding tuning methods, including implement instance pools in EJB container and naming service optimization. Instance pools provide an effective way to manage the EJB instances and naming service optimization is mainly embodied in the local method invocation mechanism and the cache mechanism. All these optimizations are implemented in OnceAS and proved to be feasible and effective by ECperf benchmark results.

References

1. Sun Microsystems, Inc. Java 2 Platform Enterprise Edition Specification, v1.4. 2003.11
2. Java Naming and Directory Interface 1.2 Specification, Sun Microsystems Inc., July 14,1999

3. TheServerSide.com J2EE Community. The ECPerf homepage. http://ecperf.theserverside.com/ecperf/
4. Sun Microsystems, Inc. Enterprise JavaBeans 1.1 and 2.0. Specifications. http://java.sun.com/products/ejb/
5. Subrahmanyam Allamaraju. *Professional Java Server Programming J2EE 1.3 Edition*. Wrox Press, ISBN 1861005377, 2001
6. S. Deshpande, B. Martin, and S. Subramanyam. Eight Reasons ECperf is the Right Way to Evaluate J2EE Performance.*TheServerSide.com J2EE Community*, 2001. http://www.theserverside.com/
7. Sun Microsystems, Inc. The ECperf 1.0 Benchmark. Specification, June 2001. http://java.sun.com/j2ee/ecperf/
8. OptimizeIt Profiler – http://www.borland.com/optimizeit/
9. S. Kounev and A. Buchmann. Performance Issues in E-Business Systems. *In Proc. of the International Con-ference on Advances in Infrastructure for e-Business, e-Education, e-Science, and e-Medicine on the Internet -SSGRR-2002w*, 2002
10. M. Karlsson, K. Moore, E. Hagersten, and D. Wood. Memory Characterization of the ECperf Benchmark. *In Proceedings of the 2nd Annual Workshop on Memory Performance Issues (WMPI 2002)*
11. P. Brebner and S. Ran. *Entity Bean A, B, C's: Enterprise Java Beans Commit Options and Caching*. In *Proc. of IFIP/ACMInternational Conference on Distributed Systems Platforms -Middleware*, 2001
12. E. Cecchet, J. Marguerite, W. Zwaenepoel. *Performance and scalability of EJB applications*, OOPSLA 2002

Systematic Robustness-Testing RI-Pro of BGP*

Lechun Wang, Peidong Zhu, and Zhenghu Gong

School of Computer Science, National University of Defense Technology,
Changsha, 410073, P.R.China
{genjade, gzh}@nudt.edu.cn, zpd136@sina.com

Abstract. Robustness testing is a very active research area in protocol testing. This paper starts with the analysis of RI-Pro of BGP-4, and then builds Scenario Model to describe the process of route update. The new model studies the RI-Pro from the relationship of available sources instead of the function of RI-Pro. Based on this model, a novel generation method of robustness-testing suite is presented. All above compose the systematic robustness testing approach. This approach eliminates the test scaffolding of ISO9646 essentially. Robustness testing experiments of Cisco 7200 indicates that, compared with positive test suite, the error-detecting capability of negative test suite generated by this approach is enhanced 1.3 times.

1 Introduction

The *Border Gateway Protocol* version 4 [1] is the de-facto standard inter-domain routing protocol in today's Internet. Although BGP implementation has passed the conformance testing and interoperation testing, faults in BGP implementations [2] or mistakes [3] in the way it is used affect the reliability of Internet directly. Before the implementation of BGP deployed to the Internet, the robustness testing must be taken.

Routing protocols are used to update the routing table dynamically. The main functions of a routing protocol can be divided into two parts: network communicating (NC) and routing information processing (RI-Pro). The NC usually can handle low-layer network accessing, detect changes of local network connections, and establish reliable communication channels for routing information flows. In order to describe NC of BGP, [4] constructs the RFSM (Robustness Finite State Machines) and generates robustness test cases based on it. This paper is based on the assumption - NC of BGP is robustness. The RI-Pro usually consists of routing information origination and propagation as well as routing table calculation and update. Based on the insight RI-Pro of BGP, a Scenario Model is presented to describe its all the external observable behaviors. Generating negative test cases based on Scenario Model are discussed also. The result of testing practice validates its effectiveness.

The rest of this paper is organized as follows. The Scenario Model is presented in Section 2. Section 3 describes the method of generation and realization of robustness-testing suite of BGP. Finally, we draw some conclusions in Section 4.

---

* Support by National Basic Research Program of China (Grant 2003CB314802), Natural Science Foundation of China (Grant 90204005), Hi-Tech Research and Development Program (Grant 2003AA121510).

2   Analysis of RI-Pro and Scenario Model

The UPDATE message handling in [1] gives the standard of BGP RI-Pro. Figure1 shows the processing of an UPDATE. Implementations must guarantee the conformance with the specification from external observation. But the internal processing can be varied. For example, Cisco's optimum route selection modifies the specification greatly. Table1 shows the differences.

Table 1. The process of optimum route selection

RFC1771	CISCO
Check the NEXT_HOP attribute of a BGP route	Check the NEXT_HOP attribute of a BGP route
Select the only route or the one that has highest degree of preference	Select the route that has the highest value of weight
Select the route that has the lowest value of the MED (MULTI_EXIT_DISC) attribute	Select the route that has the highest value of the LOCAL_PREF attribute
Select the route that has the lowest cost	Select the route that is advertised by itself
Select the route that is advertised by the BGP speaker in a neighboring AS whose BGP Identifier has the lowest value	Select the route that has the shortest sequence of AS path segments.
Select the route that is advertised by the BGP speaker whose BGP Identifier has the lowest value	Select the route based on the ORIGIN attribute (IGP<EGP< INCOMPLETE)
	Select the route that has the lowest value of the MED attribute
	The order of selection is: EBGP > confederation > IBGP
	Select the route that can be reached by the nearest neighbor of IGP
	Select the route that is advertised by the BGP speaker whose BGP Identifier has the lowest value

Although there are great differences among implementations, different implementations can interoperate well in the Internet for having consistent external controls and behaviors. The new model of RI-Pro should keep the same external controls and behaviors, but ignore those details of internal process.

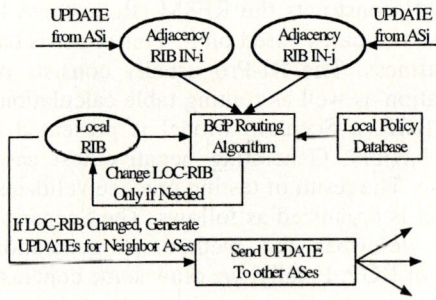

Fig. 1. RI-Pro in specification

All of the sessions, input data, output data, and control information compose the environment of RI-Pro. **Scenario Model** of RI-Pro is an abstract description of the environment for route decision process and update process, which does not concern the internal process or the implementation algorithm, but only studies the stimulations and the configuration that can reflect the behaviors of RI-Pro and trigger appropriate output data. Figure2 shows the Scenario Model of RI-Pro.

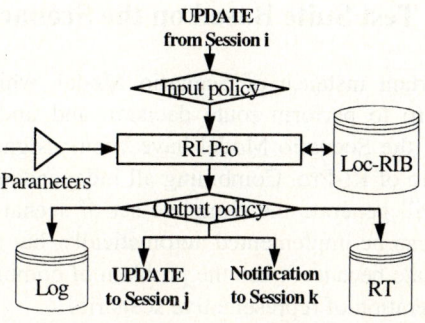

Fig. 2. Scenario Model of RI-Pro

A Scenario Model is a 5-tuple F= <S, I, K, O, Δ >, where S, I, K, O, Δ is the session set, the input set, the configuration set, the output set, the relationship set respectively.

$S = \{S_e, S_i\}$, S_e is an external session set; S_i is an internal session set.

$I = \{u_i, D, I'\}$, u_i is an UPDATE packet (in). D is a redistributed route set. I' is RIB.

$D = \{s_i, c_i, f_i, P\}$, s_i is a static route, c_i is a connect route, f_i is a default route, P is the route set generated by IGP.

$P = \{p| p$ is the route to redistribute route among routing protocols$\}$

$K = \{E, Z, H\}$, E is the input policy set, Z is the parameter set, H is the output policy set.

$E = \{\varepsilon_1, \varepsilon_2, \varepsilon_3...\}$, ε_i is an input policy.

$Z = \{\zeta_1, \zeta_2, \zeta_3...\}$, ζ_i is a parameter of BGP.

$H = \{\eta_1, \eta_2, \eta_3...\}$, η_i is an output policy.

$O = \{u_o, T, l, n\}$, u_o is an UPDATE packet (out), T is the route table, l is the Log file, n is the Notification event.

$T = \{r_1, r_2, r_3...\}$, r_i is an item in route table.

$\Delta = \{\delta| \delta$ is a relationship$\}$, δ is an abstract relationship, which expresses the dependency or restriction. For example, $k_i \delta k_j$ ($k_i, k_j \in K$) means that the two parameters must obey δ relation.

Scenario Model modifies the standard RI-Pro [1] greatly. It does not concern internal decision process or update process. But the new model reserves and classifies all the external observable behaviors. The Scenario Model adds a set of relationship to describe the relationship of available sources. Because of the adoption of relationship set, the new model is neither the duplication of the specification nor the only outside observable black box degraded from specification. The presented model provides a new view of RI-Pro from the relationship of available sources instead of the function of RI-Pro. In term of external observable behaviors, the Scenario Model is the same as the specification. However, the Scenario Model depicts the RI-Pro more comprehensively and hits the essence of RI-Pro.

3 Generate the Test Suite Based on the Scenario Model

Scenario means a certain instance of Scenario Model, which is an indispensable environment for RI-Pro to perform route decision and update process. When the elements in S, I, K of the Scenario Model have been assigned, the Scenario Model becomes a real scenario of RI-Pro. Combining all independent elements in S, I, K of the Scenario Model will generate the complete set of scenarios for the RI-Pro. The combining procedure can be implemented automatically, but this simple enumerating approach is inappropriate because it has the problem of combinational explosion, and cannot control the generation of representative scenario.

If the scenario obeys all the rules in relationship set, it can be used to conformance testing. It reflects the RI-Pro normal process with proper input data, parameters, and policies. If the scenario violates some rules in relationship set, it can be used to negative testing; it reflects the behaviors of RI-Pro in scenario with conflict relationship. The conflict relationship of Δ means that it could cause errors or failures in RI-Pro. The set of conflict relationship can be constructed from the following aspects:

- Specifications of protocol and statements of implementation
- Previous work of research
- Real-world instances
- Conflict relationship expansion

When constructing Δ, the key problem of expanding is to find out *Search space*. *Search space* gives the range to search new conflict relationships. There are two search spaces that can be used to expand BGP conflict relationships, which are about route and configuration, shown in above and below parts of figure 3 respectively.

The method of generating robustness-testing case is to reconstruct application scenario based on the conflict relationships in relationship set Δ. The principles for generating the test case are as follows:

- Constructed scenario must be as simple as possible.
- Only one conflict relationship exists in a constructed scenario.
- Constructed scenario should not use the unnecessary parameters or policies.
- Constructed scenario uses default configurations as many as possible.
- In addition, the constructed scenario should reflect the generality of conflict relationship, so it should select typical testing configuration and testing stimulation to reflect the conflict relationship.

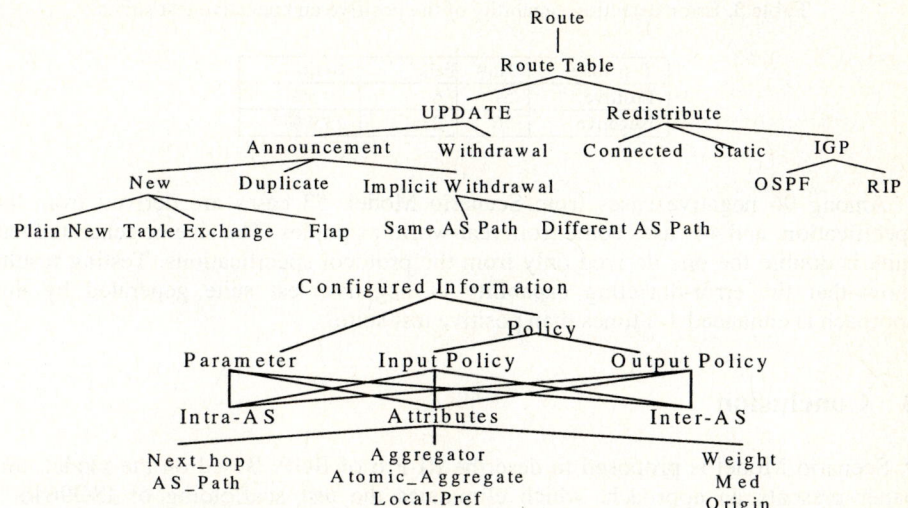

Fig. 3. Search space framework of route and configured information

3.1 Realization and Result

Based on five years practical testing experience for BGP and original instances from Cisco, NANOG, Agilent, etc, we generated the negative test suite for RI-Pro of BGP using the approach of this paper presented.

Compared with the methodology employed by Fuzz[5], our approach can generate more effective test cases than Fuzz which merely submits random input streams. Fuzz only tells whether the implementation failed or not, but it can't give the reason of failure. Our reconstructed negative test case is based on the conflict relationship. The flaws of implementation can be found out easily.

Compared with the methodology presented by Ballista[6], our approach can generate more various test cases because Ballista views protocol header fields as parameters, so it only can identify invalid test values in update message. Our test suite includes cases about route anomalies, misconfiguration, route flap, policy changes, stress testing, scalability testing, and even protocol divergence, besides the invalid update message like Ballista.

Section 6 of BGP RFC[1], i.e., "error handling", provides an important reference for the generation of negative test suite. Table2 lists the negative test cases of ANVL[7] and negative test cases based on Scenario Model. Table3 lists the error-detecting ability of the positive suite of ANVL and the negative test suite based on Scenario Model. (The target BGP platform is Cisco 7200, ISO version 11.3)

Table 2. Test cases of ANVL and Scenario Model

Test suit	No#	Overlapped	Ratio
ANVL	37	35	94.6%
Scenario Model	96	39	40.6%

Table 3. Error-detecting capability of the positive and negative test suites

Test suite	No#	Fails	Ratio
Positive	57	6	10.5%
Negative	96	23	23.9 %

Among 96 negative cases from Scenario Model, 52 cases are derived from the specification, and 44 cases come from real-world examples. The size of generated test suite is double the one derived only from the protocol specifications. Testing results show that the error-detecting capability of negative test suite generated by this approach is enhanced 1.3 times than positive test suite.

4 Conclusion

A Scenario Model is proposed to describe RI-Pro of BGP. Based on the model, this paper presents an approach, which eliminates the test scaffolding of ISO9646 [8] essentially, to generate robustness-testing suite. Based on the observation that many test cases fail because SUT cannot process unexpected events, we think the vulnerabilities come from the flaws of implementation and some inadequate definitions in specifications. The implementation should concern the environments much more; and the implementer should do more work for the user, such as doing the consistence and relativity check for configured policies and parameters. The specification should define all the objects that the RI-Pro used clearly just like the definition of update message. At last we hope to raise the level of consciousness of robustness when protocols are at the design stage. It is suggested that the RFC documents should contain a "Robustness Considerations" section, which will be of more practical use than the current "Security Considerations" section.

References

1. RFC1771: A Border Gateway Protocol 4 (BGP), March 1995
2. C. Labovitz, G. R. Malan, and F. Jahanian. Origins of Internet Routing Instability. In *IEEE INFOCOM*, June 1999
3. Ratul Mahajan, David Wetherall and Tom Anderson, Understanding BGP Misconfiguration. ACM SIGCOMM, 2002
4. Wang Lechun, Zhu Peidong, Gong Zhenghu, "Study of Robustness Testing Based on RFSM". Proceedings of INC2004, Plymouth, UK, July 2004.
5. http://www.cs.wisc.edu/~bart/fuzz.
6. http://www.ece.cmu.edu/~koopman/ballista/index.html
7. http://www.ixiacom.com/products/caa/anvl_testsuitedesc.php
8. ISO 9646(1-7): Conformance testing methodology and framework

MPICH-GP: A Private-IP-Enabled MPI Over Grid Environments

Kumrye Park[1], Sungyong Park[1,*], Ohyoung Kwon[2], and Hyoungwoo Park[3]

[1] Dept. of Computer Science, Sogang University, Seoul, Korea
{namul, parksy}@sogang.ac.kr
[2] Korea University of Technology and Education, Chonan, Korea
[3] Korea Institute of Science and Technology Information, Daejeon, Korea

Abstract. This paper presents an overview of MPICH-GP, which extends the MPICH-G2 functionality to support private IP clusters. To support the communication among private IP clusters, the MPICH-GP uses a communication relay scheme combining the NAT and a user-level proxy. The benchmarking results show that MPICH-GP outperforms the user-level two-proxy scheme used in PACX-MPI and Firewall-enabled MPICH-G, and also show comparable application performance to that of original MPICH-G2, especially in large message (or problem) size.

1 Introduction

As cluster systems become more widely available, it becomes feasible to run parallel applications across multiple private clusters at different geographic locations as a Grid environment. In the MPICH-G2 library [1], an implementation of the Message Passing Interface standard over Globus-based Grid environment, it is impossible for any two nodes located in different private clusters to communicate with each other directly across the public network.

In PACX-MPI [2], another implementation of MPI aiming to support the coupling of high performance computing systems distributed in a Grid, the communications among multiple private IP clusters are handled by *two user-level* daemons that allow the library to bundle communications and avoid having thousands of open connections between systems. However, since these daemons are implemented as proxies running in user space, the total bandwidth is only about half of the bandwidth obtained from kernel-level solutions [3]. It also suffers from higher latency due to the additional overhead of TCP/IP stack traversal and switching between kernel and user mode.

This paper presents the design and implementation of MPICH-GP, which is a private-IP-enabled MPI solution over Grid environments. To support private IP clusters, the MPICH-GP uses a communication relay scheme combining the NAT service with a user level proxy. In this approach, only incoming messages are handled by a user-level proxy to relay them into proper nodes inside the cluster, while the outgoing messages are handled by the NAT service at the front-end node of the

---

* Corresponding author

cluster. We have benchmarked our message relay scheme and compared it with the user-level two-proxy scheme used in PACX-MPI [2] and Firewall-enabled MPICH-G [4]. The performance results show that our NAT-based scheme outperforms the user-level two-proxy scheme. We have also benchmarked the performance of NAS Parallel Benchmark suite over MPICH-GP and compared it with those of MPICH-G2 and PACX-MPI. The results indicate that the overhead incurred by using a user-level proxy is minimal, especially in large message (problem) size, and the performance is at least better than that of PACX-MPI.

The rest of the paper is organized as follows. Section 2 presents an overview of the MPICH-GP architecture. The experimental results of MPICH-GP are presented in section 3. Section 4 concludes the paper.

2 Overview of MPICH-GP Architecture

In this section we present the design and implementation issues of the MPICH-GP architecture such as how the proxy process is designed including the communication relay scheme and protocol conversion, and the global rank management scheme used in MPICH-GP.

2.1 NAT-Based Communication Relay Scheme

In private IP clusters where each node within the cluster has a private IP address and thereby cannot directly communicate with public networks, a proxy that forwards incoming and outgoing messages is needed. The proxy process, in general, can be implemented either within the kernel or as a user-level process. Although the kernel-level proxy approach has the best performance, it is not widely used due to its poor portability. The user-level proxy scheme [2][4] is easy to implement but has the performance overhead such as those incurred by the TCP/IP stack traversal and the context switching between the kernel and the user mode. All packets sent from one node to the other nodes located in other clusters have to go through the user-level proxy twice, which decreases the performance further.

In MPICH-GP, only incoming messages are handled by a user-level proxy to relay them into proper nodes inside the cluster, while the outgoing messages are handled by the NAT service at the front-end node of the cluster. This brings performance improvement against the user-level two-proxy scheme since all packets pass through the user-level proxy once. By using the NAT service, which is generally provided by traditional operating systems, we could easily apply our scheme to MPICH-GP and implement a user-level proxy without modifying operating system kernel.

2.2 User-Level Proxy Daemon

In MPICH-GP, a user-level proxy daemon is running at the front-end node. This proxy daemon accepts requests from one end of the MPI process and forwards the requests to the other MPI process. Since the protocol between *MPI_Send* and *MPI_Recv* is stateful (i.e., Each MPI process maintains states such as 'await_instruction', 'await_format', 'await_header', 'await_data' and etc.), the proxy daemon is developed with a stateful server approach, where the daemon keeps the same states with the MPI processes (See Fig. 1). To maintain the source-level consistency with Globus, we designed the user-level proxy daemon with Globus I/O library [5], where we heavily used the callback mechanism for the communications among processes.

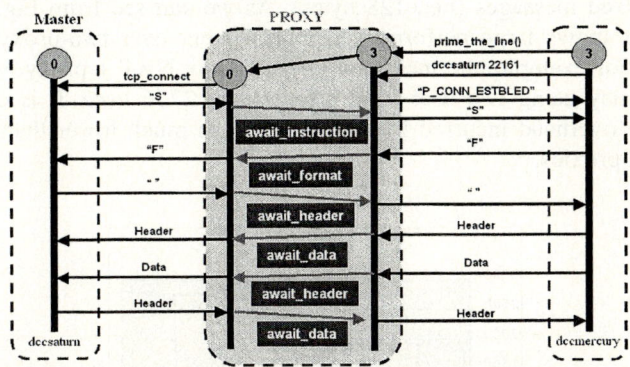

Fig. 1. User-level Proxy Daemon in MPICH-GP

2.3 Locating the Destination

When the *MPI_Send* operation is invoked in MPICH-GP, it first has to decide if the destination node is within the same cluster or outside the cluster. This allows us to determine if we need to send the data directly to the destination (if the destination is within the same cluster) or send the data via the proxy (if the destination is outside of the cluster). To decide the location of the destination, we have defined and added one proprietary field called GP_GUID into the channel data structure in Globus.

The GP_GUID structure contains the name or IP address of the front-end node where the compute node is located, the name or IP address of the compute node, etc. If the compute nodes are connected to the public network, the name of the front-end node is the same as that of the compute node. The name or IP address of the front-end node is initially set with an environment variable and is given to each MPI process. When each MPI process starts execution with *MPI_Init* function, it obtains the information from the environment variable and builds the GP_GUID structure. With the information in GP_GUID structure, any MPI process can decide whether we can directly connect to the destination node or via the proxy.

3 Experimental Results

In this section we present two benchmarking results to evaluate the performance of MPICH-GP. The first benchmarking is to check the overhead of the user-level proxy and to compare the performance with that of the user-level two-proxy scheme used in PACX-MPI. The performance of MPICH-GP is also evaluated via applications and is compared with that of PACX-MPI. For the application benchmarking, we use the NAS Parallel Benchmark suite [6]. We only report the IS benchmark in this paper.

3.1 Forwarding Performance

The goal of the measurements presented here is a comparison of two different approaches, our NAT-proxy scheme and two-proxy scheme. Fig. 2 shows the latency between two private IP clusters. The latency was measured via *ping-pong* program using small sized messages (i.e., 128 bytes). As we can see from Fig. 2, the NAT-proxy scheme shows large performance improvement over two-proxy approach by about 144%. For example, the measured latency using NAT + proxy was 1923 *usec*, while the latency using two user-level proxies was 2756 *usec*. It is clear from the result that the overhead incurred by using NAT was much lower than that of using two user-level proxies.

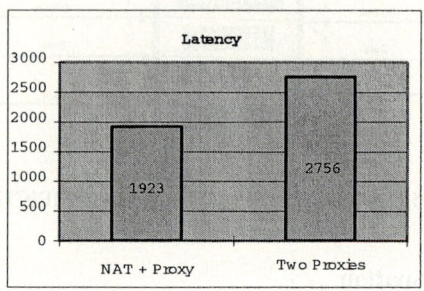

Fig. 2. Latency Between Two Private IP Clusters

3.2 Application Performance

To evaluate the latency impact on applications, the NAS Parallel Benchmark suite (NPB 3.1) [6] was run both on MPICH-GP and MPICH-G2. We have selected three benchmarks, IS, CG and LU. However, we only report the performance results of IS benchmark in this paper. By varying the problem size of the benchmarks from class S to class B (S<W<A<B), we measured the latency. The number of processors was fixed to 4.

Fig. 3 shows the experimental results of IS benchmark. The latency means the total execution time of the benchmark. The line with a tag GP/G2 indicates the ratio of the performance, MPICH-GP/MPICH-G2 (that is the overhead of MPICH-GP compared to that of MPICH-G2). As we can see from Fig. 3 (a), the latency increases as we increase the message size and the overhead of using user-level proxy keeps decreasing. This means that the overhead of using proxy gets amortized as we increase the message size. Fig. 3 (b) compares the latency between MPICH-GP, MPICH-G2 and PACX-MPI using IS benchmark. For small messages, PACX-MPI shows the worst performance, but as we increase the message size, PACX-MPI shows the best performance among them. It should be noted that PACX-MPI compresses the messages by default before sending and it can achieve better performance as we increase the message size.

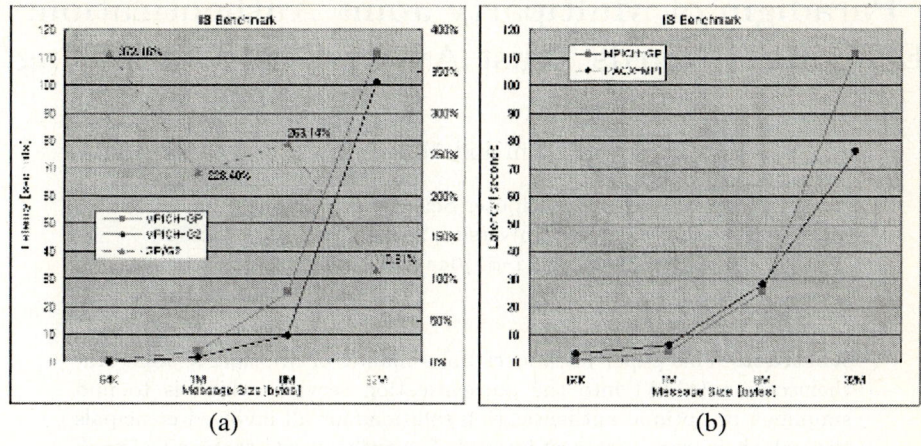

Fig. 3. IS Benchmark

4 Conclusions

In this paper we have presented the implementation issues of MPICH-GP and evaluated our implementation. We have also proposed a communication relay scheme based on the NAT and a user-level proxy, and compared our scheme with that of two user-level proxy scheme. From the forwarding bandwidth experiments, we showed that the performance of our scheme was better than that of two user-level proxy scheme. The application performance also indicated that the overhead of using proxy is minimal, especially in large message size.

References

1. Karonis, N.T., Toonen, B., Foster, I.: MPICH-G2: A Grid-Enabled Implementation of the Message Passing Interface, ANL/MCS Technical Reports P942-0402, (2002)
2. Gabriel, E., Resch, M, Beisel, T., Keller, R.: Distributed computing in a heterogeneous computing environment, LNCS 1497 (1998) 180–188
3. Müller, M., Hess, M., Gabriel, E.: Grid enabled MPI solutions for Clusters, CCGRID'03 (2003) 18–24
4. Tanaka, Y., Sata, M., Hirano, M., Nakata, H., Sekiguchi, S.: Performance Evaluation of a Firewall-compliant Globus-based Wide-area Cluster System, HPDC 2000 (2000) 121–128
5. Ian Foster, Carl Kesselman eds, Grid: Blueprint for a Future Computing Infrastructure, Morgan Kaufmann Publishers (1999)
6. NAS Parallel Benchmarks, http://www.nas.nasa.gov/Software/NPB

Paradigm of Multiparty Joint Authentication: Evolving Towards Trust Aware Grid Computing*

Hui Liu and Minglu Li

Department of Computer Science and Engineering,
Shanghai Jiaotong University, 200030 Shanghai, China
{liuhui, li-ml}@cs.sjtu.edu.cn

Abstract. This paper introduces the semantic of Multiparty Joint Authentication (MJA) into the authentication service, which is to find simplified or optimal authentication solutions for all involved principals through their transitive trust instead of to authenticate each pair of principals in turn. MJA is designed to support multiparty security contexts in grids with a specified, understood level of confidence and reduce the time cost of mutual authentications. Graph theory model is employed to define MJA, and analyze its mathematical properties. Two algorithms to find an n-principal, n-order MJA solution are also presented. MJA is indeed a trust aware mechanism and will promote trust aware grid computing eventually.

1 Introduction

Grid computing is just on the way towards popular, prosperity and progressive. However, before any desktop in the cyberworld plug into some grids to share resources, grids must prove to be security enough for both their providers and consumers. Therefore, security for grids is just like Achilles' heel, it can either boost or baffle the growth of grid computing.

According to OGSA security roadmap, future OGSA security architecture will leverage the existing and emerging WS security specifications as much as possible, fitting them into a layered grid security model [1]. The fundamental of grid security model rests with using web services to integrate different interoperability security components for the purpose of defining, managing and enforcing trust through virtual organization (VO). Obviously, authentication is an inescapable entry for any kinds of trust model.

This paper introduces the semantic of Multiparty Joint Authentication (MJA) into the authentication service, which is to find simplified or optimal authentication solutions for all involved principals as a whole instead of to authenticate each pair of principals in turn. MJA supports multiparty security contexts in

---

* This research is supported by the National Grand Fundamental Research 973 Program of China (No.2002CB312002), ChinaGrid Program of MOE, China Postdoctoral Science Foundation, and Grand Project of the Science and Technology Commission of Shanghai Municipality (No.03dz15027).

grids with a specified, understood level of confidence and reduces the total time cost of mutual authentications. Furthermore, if MJA service is designed to support OASIS SAML specification [2], authentication assertion can be used to deduce the trust emphasizing uncertain and venturesome relationship, i.e., the trust that is not processed as $true$ or $false$ binary logic. Therefore, the paradigm of MJA will facilitate developers building some trust aware grids.

The rest of this paper is organized as follows. Section 2 reviews related work about trust model and multiparty authentication. The graph theory model of multiparty authentication and the definition of MJA are proposed in section 3. In section 4, basic mathematical properties of MJA are analyzed in the form of theorems. Section 5 presents two algorithms to find a possible MJA solution. Section 6 makes some discussion about MJA and concludes this paper.

2 Related Work

Trust is a metric for principals to establish security relationships. However, trust in digital security contexts is always being processed as $true$ or $false$ binary logic, which is not consistent with the real behavior pattern in the human society. Therefore, some new models emphasize the uncertainty and venture of trust, and intend to reinstall the underlying trust mechanisms of the offline world into the online world in the same way or to the same degree [3].

We have classified this kind of trust model as mission-unaware identity trust model, mission-aware identity trust model, mission-unaware behavior trust model and mission-aware behavior trust model. If a trust model possesses only static security policies, it is an identity trust model. If some security policies evolve their values or contents according to the principals' behavior histories, it can be correspondingly termed as behavior trust model. On the other hand, if security decision is directly depending on the characteristics of the mission, such as cost, QoS, etc., it becomes mission-aware trust model. Otherwise, it falls into mission-unaware trust model [4]. All trust models can be regarded as selective projection of a mission-aware behavior trust model.

The first trust aware resource management system (TRMS) for grids suggests that the trust and reputation should decay with time [5]. The mathematical functions used in TRMS are different from that in literature [6], which suggests that recommender weights should be employed to reduce the computing cost of reputation. Trust model is also used for risk analysis in E-Commerce [7]. We have proposed a formal description of mission-aware behavior trust model based on ASM [4], and its concrete illustration can be found in literature [8].

OASIS SAML suggests a flexible and uniform architecture to define, manage and enforce trust, which is an XML based framework for exchanging security information expressed in form of assertions about subjects. Assertions convey information about authentication actions that previously performed by different SAML authorities [2]. By means of SAML architecture, authentication and authorization actions are separable, and the trust can be calculated by different authorities and be enforced at each PEP.

Authentication is an inescapable entry for any kinds of trust model. However, current mainstream security infrastructure in grid communities only supports two-party oriented authentication, and there has not any published literature dedicated to the research on multiparty authentication in grids.

Multiparty issues have been widely studied in the fields of multimedia, group communication, symmetric group encryption, security computation, etc. However, these researches concern with how to improve multicast mechanisms, how to compute, distribute, or agree on group secret key, how to protect privacy and avoid corrupted parties. One authenticated group key agreement protocol proposes implicit key authentication for multiple parties, by which the key is not directly authenticated between an arbitrary two-party of P_i and P_j $(i \neq j)$. Instead, all key authentication is performed through one fixed party P_n [9]. This protocol indicates that multiparty authentication can be simplified. However, to the best of our knowledge, none of the researches deals with using transitive trust to simplify mutual authentication for the large, dynamic principal group.

Grid security model adopts trust model, authentication services and SAML as its building blocks, the research in this paper is going to weave all these elements into a feasible and profitable solution for typical grid applications.

3 Semantic of Multiparty Joint Authentication

3.1 What Is Multiparty Authentication

The real-life network applications reveal that C/S model and two-party interaction have some inherent drawbacks. Multiparty interaction mechanism that encloses multiple participants performing a set of activities together is of great demand. In a multiparty interaction, a group of participants come together to produce some intermediate and temporary combined state, use this state to execute some activity or transfer information.

Such applications include distributed online game, distributed VR environments, distributed interactive simulation, distributed cooperative edit, distributed control and monitor, etc. When we want to improve security for such applications, the problem of multiparty authentication emerges.

Multiparty authentication is an extension of two-party authentication, which is going to confirm each participant's identity with a specified, understood level of confidence. Traditionally, multiparty authentication can be regarded as an accumulation of a series of two-party authentications.

Because a grid system intends to provide coordinated resources sharing, problem solving or services outsourcing in dynamic, multi-institutional VO [10], a typical grid application often spreads over multiple resource hosting sites and needs multiparty authentication. For example, the computational power providers of a computation job may need multiparty authentication, the idle resource providers of a cryptographic problem may need multiparty authentication, the search service providers of a parallel searching engine may need multiparty authentication, thousands of players participated in an online game may need multiparty authentication, a set of web services constituting an outsourcing workflow may need multiparty authentication, etc. In these scenarios, it seems awkward for developers to regard multiparty authentication as a series of two-party authentications, especially when thousands of or millions of participants are involved.

There are two main approaches to establish multiparty relationships: in the static manner, all the parties involved must be known and presented in advance of authentications; in the dynamic manner, old involved parties can quit the multiparty relationship while new parties involved can join the multiparty relationship. Obviously, the latter can be achieved by using the static multiparty approach together with the two-party approach repeatedly. This paper focuses on multiparty relationships formed in the static manner unless explicitly stated. On the other hand, the actions after multiparty authentication are beyond the scope of this paper.

3.2 Graph Theory Model of Multiparty Authentication

Denote each principal to be authenticated by a vertex, and let the edge connecting a pair of vertices represent that two principals have confirmed the counterparty's identity mutually. Multiparty authentication that involves n principals can be modeled as a graph of order n. We denote such a graph by MAG_n. After authenticating each pair of distinct principals, MAG_n become a complete graph K_n with $n(n-1)/2$ edges.

A straightforward simplification is to choose one principal as a trusted third party and let it to mutually authenticate with the other principals in turn. This simplification changes MAG_n into a complete bipartite graph $K_{1,n-1}$ called a star, needing only $(n-1)$ mutual authentications. A further simplification is to distribute the responsibility of the trusted third party and to establish the MJA supposition as follows:

MJA Supposition. One principal can regard another principal as a trusted third party if either of the following conditions is satisfied:

1. Two principals have authenticated each other mutually.
2. Two principals have authenticated with one common trusted third party in advance.

Based on the MJA supposition, another simplification changes MAG_n into a Hamilton chain of the complete graph K_n, which also needs $(n-1)$ mutual authentications.

3.3 Defining Multiparty Joint Authentication

Definition 1. *Multiparty Joint Authentication (MJA) is to find a simplified or optimal authentication solution in a multiparty security context that involves n principals, which is based on three conditions below:*

1. *If principal P_i, P_j have authenticated with one common trusted third party, then both P_i and P_j can confirm the counterparty's identity with a specified, understood level of confidence even without a real mutual authentication.*

2. The relationship between a principal and its trust third party must satisfy the MJA supposition (or other feasible substitute).
3. There are $m(1 \leq m \leq n)$ principals to act as the trusted third parties to serve certain subsets comprising different principals. For short, let m be the order of the n-principal MJA, denote them by $n{:}m$.

The Hamilton chain of the complete graph K_n is one possible MJA solution, however, it is not the optimal answer to MAG_n if we take some practical constraints into account:

1. Different mutual authentications have different QoS and cost.
2. Users insist on performing mutual authentication for certain pairs of principals.
3. The MJA service provider can cache some mutual authentications performed by several principals and trusted third parties for a period of time.
4. The policies for a principal to become a trusted third party are of great varieties.
5. A principal may trust a trusted third party with different policies and security level.

These constraints indicate how to find an optimal MJA solution face lots of challenges. In this paper, we focus on n-principal, n-order MJA, i.e., $n{:}n$ MJA.

4 Basic Mathematical Properties of $n{:}n$ MJA

Theorem 1. *The spanning tree of a complete graph K_n is an $n{:}n$ MJA solution.*

This theorem indicates that finding a solution of $n{:}n$ MJA can be achieved by growing a spanning tree of a complete graph K_n (The proof of different theorems are omitted in this paper).

Theorem 2. *The spanning tree of a complete graph K_n is one of the simplest $n{:}n$ MJA solution.*

In this theorem, the word "simplest" can be comprehended as "most relaxed", which means every principal can become a trusted third party and hence how to choose a trusted third party comes under no constraints.

Theorem 3. *The number of $n{:}n$ MJA solutions is n^{n-2}.*

This theorem indicates that searching the optimal solution of $n{:}n$ MJA by enumeration is an unsolvable problem. If we can design a nondeterministic Turing machine algorithm to find the optimal $n{:}n$ MJA solution with polynomial-time computation, this unsolvable problem become an NP problem.

Theorem 4. *Suppose it spends the time of T for each pair of principals to authenticate mutually, and each principal can carry out only one two-party authentication once a time. Then, the theoretical minimum time cost of an $n{:}n$ MJA solution is $2T$.*

This theorem determines the infimum of the theoretical minimum time cost of an $n{:}n$ MJA solution.

Theorem 5. *Denote a spanning tree of a completed graph K_n by T_n. Suppose it spends the time of T for each pair of principals to authenticate mutually, and each principal can carry out only one two-party authentication once a time. Then, the theoretical minimum time cost of an $n{:}n$ MJA solution is mT if the maximum value of the degree sequence of T_n is $deg(P_k) = m$.*

This theorem implies that greedy algorithm to grow a spanning tree of a completed graph K_n would not always finding the optimal $n{:}n$ MJA solution.

Theorem 6. *Suppose it spends the time of T for each pair of principals to authenticate mutually, and each principal can carry out only one two-party authentication once a time. Then, the maximum value of the theoretical minimum time cost of an $n{:}n$ MJA solution is $(n-1)T$.*

Theorem 4, 5 and 6 indicate that the bigger the number of the trusted third party is, the smaller the theoretical minimum time cost is. If all principals can act as a trusted third party, the theoretical minimum time cost takes its minimum value $2T$; If only one principal act as a trusted third party, the theoretical minimum time cost takes its maximum value $(n-1)T$.

Theorem 7. *Let T be a spanning tree of a connected graph G. Let $\alpha = \{a, b\}$ be an edge of G which is not an edge of T. Then there is an edge β of T such that the graph T' obtained from T by inserting α and deleting β is also a spanning tree of G* [11].

This theorem is very useful for us to construct some algorithms to find $n{:}n$ MJA solutions.

5 Algorithms to Find $n{:}n$ MJA Solution

If each pair of principals spends the same time to authenticate mutually (other factors can be converted into the equivalent time cost), a Hamilton chain of a completed graph K_n stands for one optimal $n{:}n$ MJA solution. Algorithm 1 is to grow this kind of spanning tree.

Algorithm 1.

1. Number n principals with one of the numbers $1, 2, \ldots, n$.
2. Generate a permutation of natural number n, denoted by $p_1, p_2, \ldots, p_i, \ldots, p_n$, where $p_i \in N \wedge 1 \leq p_i \leq n$.
3. Join $p_1, p_2, \ldots, p_i, \ldots, p_n$ in turn, the chain in the form of $p_1 - p_2 - \ldots - p_i \ldots - p_n$ is an n-principal, n-order MJA solution.

Theorem 8. *If all mutual authentication spends the same time, algorithm 1 can find one optimal n:n MJA solution.*

Generally, different pair of principals spends different time to authenticate mutually (including converted factors). Therefore, we must design some algorithms to grow a weighted spanning tree from a weighted complete graph K_n. The most straightforward algorithm is the greedy algorithm, which can be used to grow a minimum-weight spanning tree. Let K_n be a weighted complete graph with weight function c, the greedy algorithm is shown as follows.

Algorithm 2.

1. Put $F = \phi, V = \phi$.
2. Put an edge α that has minimum weight in F, at the same time, put two vertices which are incident with α in V.
3. While there exists an edge α not in F such that $F \cup \{\alpha\}$ put only one new vertex in V, determine such an edge α of minimum weight and put α in F, at the same time, put the new vertex that is incident with α in V.
4. Put $T = \{V, F\}$, which is a minimum-weight spanning tree of K_n.

Theorem 9. *If different mutual authentication spends different time, algorithm 2 can find an n:n MJA solution that has the minimum-time-cost.*

However, greedy algorithm may not find the optimal $n{:}n$ MJA solution because the real time cost of an $n{:}n$ MJA solution depends not only on the time cost for each pair of principals to authenticate mutually, but also on how many mutual authentications are to be performed by each principal (the degree sequence of the spanning tree).

This conclusion can be illustrated by one concrete example. Suppose it takes the time of T for principal P_i to authenticate mutually with all other $(n-1)$ principals, and it takes the time of $2T$ for each pair of principals, not including P_i, to authenticate mutually. Then, the minimum-weight spanning tree grown with greedy algorithm is a star whose center is P_i. The theoretical minimum time cost of such a MJA solution is $(n-1)T$. However, we can construct a Hamilton chain as follows: firstly, find a Hamilton chain for all $(n-1)$ principals, excluding P_i, with algorithm 1; secondly, join P_i with either end vertex of the previous Hamilton chain to form a new Hamilton chain. Obviously, the theoretical minimum time cost of this MJA solution is $4T$, which is better than the former.

On the other hand, different trust model can be designed to custom MJA solutions. For example, the weight of each edge may represent trust or risk between the corresponding principals, then, greedy algorithm can be used to find MJA solution with maximum level of trust and (or) minimum level of risk. This indicates that MJA will redound to trust aware grid computing.

6 Discussion and Conclusion

Grid application aims to share resource deployed within the VO. High level definitions of grid have been formalized with abstract state machine (ASM) [12]. According to such definitions, typical grid application has three important behavior rules: Resource Abstraction maps the abstract resource in the grid to the real resource on some hosting machines; User Abstraction maps the grid global user onto local users that carry their own credentials; Resource Selection is achieved by repeating Resources Abstraction and User Abstraction. Because current authentication is two-party oriented, resource selection, account and other related services are confined by this semantic, including

1. Restriction on resources selection. When abstract resource in the request is mapped onto physical resource, no authentication and authorization are performed, therefore, the selected and mapped physical resource may not be granted to the users for the reason of changing on security considerations, payment in arrears, etc. This would make an optimal resource scheduling solution abort during the execution.
2. The time cost of mutual authentication for large group users is tremendous. For example, during the start time of a popular grid game, 150,000 users may rush into the battle field for security authentication (otherwise, their time and energy will go to waste), which needs 11,249,925,000 mutual authentications without any simplification.
3. No support for the semantic of multiparty authentication. Current GSI does not directly support the semantic of multiparty authentication. Therefore, no multiparty authentication services are available now, especially for the application that needs all participants to authenticate each other simultaneously before executing their cooperative activities.
4. Restriction on dynamic resource sharing. When a grid application wants to dynamically allocate some new resources, or, resources want to migrate to some new hosting machines, the time cost of mutual authentication is tremendous. For example, new resource process would authenticate to all available resource processes, or, several different multiparty may want to participate into available grid application simultaneously. In these scenarios, the semantic of multiparty authentication seems to be very useful.
5. The Single Sign On (SSO) mechanism in the current GSI is not security enough. GSI implements SSO through proxy certificates. However, the proxy certificates may derive a vulnerable trust chains, where compromising of any proxy (child-proxy) acting on behalf of the user (parent-proxy) would destroy believes of the original user as a whole and result in trust crisis.
6. No support for the trust model that emphasizes the uncertainty and venture of trust relationships. GSI focuses on providing authentication and access control mechanism for the grid environment. It proposes and implements a security architecture based on four interoperability protocols that are used to handle U-UP, UP-RP, RP-P and P-P interactions cooperatively. In these protocols, authentication and authorization are often coupled together and expressed as *true* or *false* logic.

MJA bases itself upon multiparty authentication and has trust aware computing in mind, which can be used to eliminate restrictions above and bring some value-added points for grid applications, including

1. Support for authenticated resources selection. MJA introduces SAML assertions into the security mechanisms. Therefore, resources providers and consumers can design their own policies to establish different security level and security context. On the other hand, resource optimal selection and resource mapping can be performed by executing MJA firstly to assure their successful access control rights.
2. MJA is designed to reduce the time cost of multiparty authentication. Suppose all mutual authentications spend the same time and denote this time cost by one unit, a traditional multiparty authentication needs $n(n-1)/2$ mutual authentications and spends $n(n-1)/2$ units of time. Because an $n{:}n$ MJA needs only $(n-1)$ mutual authentications, the time cost reduces to $(n-1)$ units. That is, if there are 150,000 principals for an online game, an $n{:}n$ MJA solution spends only 149,999 units of time. On the other hand, parallel techniques can be used to speed up processing and to make response more efficiently. For example, if all principals act as a trusted party and different principals perform mutual authentications in parallel, an $n{:}n$ MJA solution for 150,000 principals can be finished just in 2 units of time.
3. MJA supports the semantic of multiparty authentication. After authentication, multiple participants can use group secret key for security group communication, or, they can use other participant's public key to establish privacy conversation. How to get the group secret key is beyond the scope of this paper.
4. Support for flexible dynamic resource sharing. By means of MJA, there are two approaches for dynamic resource sharing to assure their security. If only one mutual authentication is needed, two-party authentication would be employed; if multiparty authentications are needed, the MJA would be employed.
5. Support for new SSO mechanisms. By means of MJA, all principals have been authenticated as a whole. The group secret key together with the public key form a temporary security token for the purpose of confidentiality, integrity and nonrepudiation. MJA does not produce proxy certificates. However, some trust parties may corrupt, this makes MJA deserving further research.
6. Support for trust aware computing. MJA simplify mutual authentication through transitive trust between authenticated principals. Therefore, MJA is essentially a trust aware mechanism, which will eventually benefit from mission-aware behavior trust model. Of course, if MJA services are widely deployed within grid communities, grid applications will become trust aware correspondingly.

The essential of MJA is to increase the efficiency of multiparty authentication by distributing the responsibility of the trusted third party. It find some simplified or optimal authentication solutions for multiple principals by means of their transitive trust instead of to authenticate each pair of principals in turn. It is designed to support multiparty security scenarios in grids with a specified level of confidence and reduce the time cost of authentications.

By means of graph theory model, this paper reveals some mathematical properties of MJA and puts forward two algorithms to find an $n{:}n$ MJA solution. Future research topics on MJA includes: 1) MJA performance modeling and simulating; 2) MJA secret group key; 3) MJA assertions; 4) MJA protocols; 5) MJA services; 6) MJA based authorization; 7) MJA based Leave/Join services; 8) Policy based MJA; 8) MJA based security communication; 9) new algorithms to find MJA solutions with certain constraints; 10) MJA based trust model, etc.

References

1. Siebenlist, F., Welch, V., Tuecke, S., Foster, I., Nagaratnam, N., Janson, P., Dayka, J., Nadalin, A.: OGSA Security Roadmap. GGF OGSA Sec. Workgroup Doc. (2002)
2. Website: http://www.oasis-open.org/specs/index.php♯samlv1.1
3. Daignault, M., Shepherd, M., Marche, S., Watters, C.: Enabling Trust Online. In: Williams, A. (ed.): Proc. of IEEE 3rd Intl. Symposium on Electronic Com. IEEE Press, California (2002) 3-12
4. Li, M., Liu, H., Cao, L., Yu, J., Li, Y., Qian, Q., Jin, W.: Semantics and Formalizations of Mission-Aware Behavior Trust Model for Grids. In: Li, M., Sun, X., Deng, Q., Ni, J. (eds.): GCC 2003. Lecture Notes in Computer Science, 3032 (2004) 883-890
5. Azzedin, F., Maheswaran, M.: Evolving and Managing Trust in Grid Computing Systems. In: Kinsner, W., Sebak, A., Ferens. K., (eds.): Proc. of the IEEE Canadian Conf. on Electrical and Computer Eng. IEEE Press, Manitoba, Canada (2002) 1424-1429
6. AbdulRahman, A., Hailes, S.: Supporting Trust in Virtual Communities. In: Ralph, H., Sprague, J. (eds.): Proc. of the 33rd Ann. Hawaii Int. Conf. on Sys. Sci. IEEE Press, Hawaii (2000) 1769-1777
7. Manchala R.: E-Commerce Trust Metrics and Models. IEEE Internet Computing. 4 (2000) 36-44
8. Liu, H., Peng, Q., Shen, J., Hu, B.: A Mission-Aware Behavior Trust Model for Grid Computing Systems. In: Han, Y. (ed.): Proc. of the 2002 Int. Workshop on Grid and Cooperative Computing. Electronics Industry, Hainan China (2002) 897-909
9. Ateniese, G., Steiner, M., Tsudik, G.: New Multiparty Authentication Services and Key Agreement Protocols. IEEE J. on Selected A. in Com. 4 (2000) 628-639
10. Foster, I., Kesselman, C., Tuecke, S.: The Anatomy of the Grid, Enabling Scalable Virtual Organizations. Intl. J. of H. Performance Computing Applications 3 (2001) 200-222
11. Brualdi, R.: Introductory Combinatorics, Third Edition. Pearson Ed. Inc. (1999)
12. Zsolt, N., Vaidy, S.: Characterizing Grids: Attributes, Definitions and Formalisms. J. of Grid Computing 1 (2003) 9-23

Design and Implementation of a 3A Accessing Paradigm Supported Grid Application and Programming Environment

He Ge, Liu Donghua, Sun Yuzhong, and Xu Zhiwei

Institute of Computing Technology, Chinese Academy of Sciences, Beijing 100080
{hege, dliu, yuzhongsun}@ict.ac.cn

Abstract. The mobile grid users accessing to grid services has become a normally paradigm for getting grid resources. To improve their working productivity in dynamic and open grid environment it should provide the mobile grid users an access-point decoupled and access-time decoupled way to access grid services. In development of the VEGA Grid, corresponding to the user accessing module in the Service-Oriented Architecture, we developed a kind of we called "3A(anytime, anywhere, and on any device) accessing paradigm" supported Grid Application and Programming Environment(GAPE). To the 3A accessing paradigm, we mean a mobile grid user not only could access grid services at anytime anywhere and on any device, but also could continuously manage the executing state of his grid applications even if he changed his access point or access time. This article analyzes the 3A accessing paradigm, and introduces the implementation of the VEGA GAPE.

1 Introduction

Owing to the dynamic and open characteristics, accessing and programming grid resources have special difficulties. To make grid resources useful and accessible effectively, it requires a new software environment to improve the user's productivity, and facilitate the user's accessing and managing grid services. We define a Grid Application and Programming Environment (GAPE) as a set of tools and technologies that allow users "easy" access to Grid resources and applications.

With the development of the mobile computing technology, the mobile grid user's accessing grid services has become a normally paradigm for getting grid resources. So the need for managing the transparency of the grid services and the mobility of the grid users requires a suitable GAPE. It should provide a physics-decoupled, time-decoupled, and space-decoupled grid resources accessing paradigm, which we called "3A (anytime, anywhere, and on any device) accessing paradigm". For the 3A accessing paradigm, we mean a mobile grid user not only could access grid services at anytime anywhere and on any device, but also could continuously manage the executing state of his grid applications even if he changed his access point or access time. This interaction model could cut costs and then increase the productivity of the mobile grid users.

In GGF, there have an Applications, Programming Models and Environments Area to collect and coordinate the related works. Their works focused on grid portal such as GPDK [2], application development middleware such as NetSolve [3], or largely problem solving environments such as ECCE [4].Their works mainly focus on the implementation of the computing functions, but rarely consider the accessing models, and the 3A accessing requirement. In the mobile computing research, there have many works focus on the user mobility such as WebMC [5]. These works generally solved problems of the mobility in a computer or network architecture level. Differently, our works solve the problem in a higher application level.

In development of the VEGA Grid [6], which is the name of the grid research project at the Institute of Computing Technology, Chinese Academy of Sciences, we analysis the characteristics and requirements of the 3A accessing paradigm, develop a 3A accessing paradigm supported GAPE. By this way it adapts to the loosely-coupled characteristic of the grid environment, and provides a more powerful interaction module between the grid users and the grid resources.

2 Overview of VEGA GAPE

The system architecture of the VEGA GAPE is shown in Fig. 1.

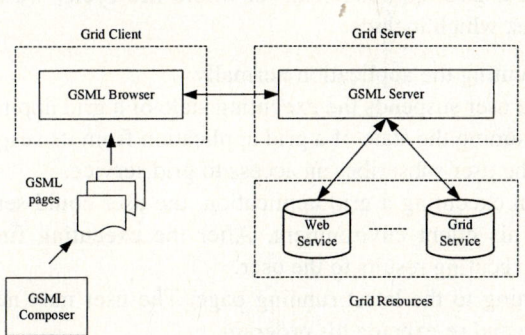

Fig. 1. VEGA grid application and programming environment overview

GSML(Grid Service Markup Language) is a Service-Oriented Programming Language implemented in VEGA Grid project, which aimed to provide the end-users with a way to express services request in a grid environment. GSML provides an easy-to-use and flexible method for the grid end-users to access and integrate the grid services. The GSML is suitable for description of the interaction between the grid users and the grid systems. It provides the grid users with a light-weight alternative to describe the requests for accessing or composing grid services. Meanwhile, it gives a declarative specification of the externally visible behaviors. We will particularly introduce the design and implementation of the GSML in another paper.

GSML Browser is designed as the interface for the end-users to access the grid resource. GSML Composer is the toolkit developed for the users to edit the ".gsml" files. The whole procedure for an end user to program and execute his grid application is: firstly, he edits ".gsml" files helped by the Composer using the GSML language at the client to express his accessing requirements to the grid services. Then he submits the file to a GSML Server, which would interpret and execute the ".gsml" file. The Server would access the physical grid services for the end users. After getting the accessing results, it would return the results to the Browser, which will display the accessing results in the client.

3 3A Accessing Paradigm Analysis

In dynamic and open grid environment, To improve the productivity of the mobile grid users, the GAPE should provide a set of decouples between the grid users and grid resources, which includes: physics decoupled, which means a grid application should be independent of the physical property of the computing node; time decoupled, which means the executing of a grid application should be decoupled of the physical time, and need not be continuous in time; and space decoupled, which means the executing space of a grid application should be independent of the physical space, especially at the access point.

From the classification of the trigger when activating a new task and the functions needed to manage a gird application in its whole life-cycle, we summarize the 3A request instructions, which include:

1. Normal. Executing the application normally.
2. Suspend. The user suspends the executing state of a grid application.
3. Resume. Resuming the state of a grid application from its suspended point.
4. Subscribe. The user subscribes an access to grid service.
5. Notify. When executing a grid application, the user could send the "notify" request and closed his client environment. After the executing finished, the system would notify the executing results to the user.
6. Back. Returning to the latest running page. The user may need go back to the latest running state and re-execute his program.

The GAPE should provide grid users with enough approaches to express his 3A accessing requirements. The user may express it in the aforehand description. Or the user may be able to press some button to express his 3A accessing request when he managing the executing state of his grid applications.

4 Architecture Design and Implementation of the GSML Server

GSML Server is a session-based 3A accessing paradigm supported server for the grid users to access grid services. It receives and interprets the user's grid service accessing requirement file, which is a ".gsml" file, and accesses to the physical grid services for the users. Meanwhile, it provides the supports for the 3A accessing paradigm. The architecture design of GSML Server is shown in Fig. 2.

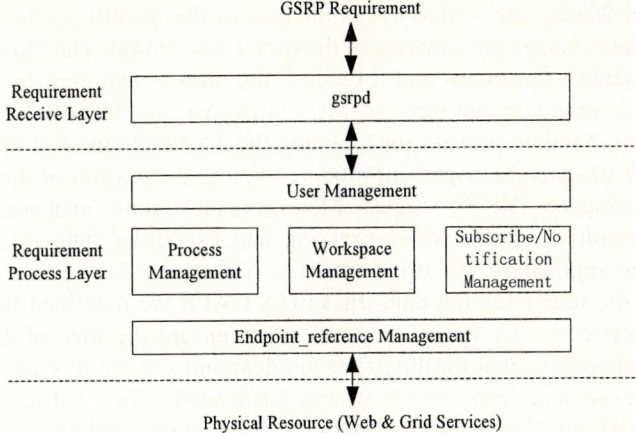

Fig. 2. Architecture design of the GSML Server

The "Requirement Receiver Layer" is a "gsrpd" daemon, which listens to the server port waiting for the user's requirement, and manages the processing handler.

The "Requirement Process Layer" interprets the user's accessing requirement file; provides the supports for the 3A accessing paradigm according to the content of the 3A requirement; and realizes the really accessing to the physical grid services. At last it returns the executing results to the upper layer. The relationship of the modules in this layer is shown in Fig. 3.

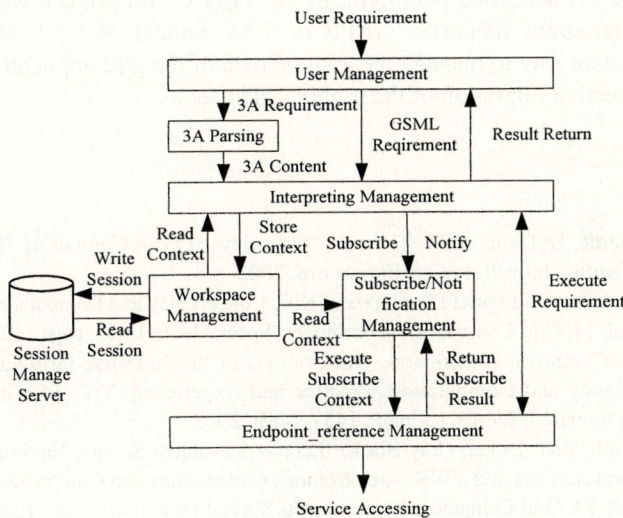

Fig. 3. Relationship of the modules in requirement process layer

The "User Management Module" authenticates the identity of the user. The "3A Parsing Module" parses the contents of the user's 3A request. The "Interpreting Management Module" interprets and executes the user's requirement according to the ".gsml" file. It listens to and receives the content and the 3A request, and then executes the corresponding process for realizing the 3A functions. It may call the interface of the Workspace Management module to store the session of the application in the user's workspace. The "Workspace Management Module" manages and maintains the middle results of the service accessing and executing state information when suspending an application. We use a database, namely session manage server to store and manage the user's session data. In VEGA GAPE we redefined the context data needed to support the 3A accessing paradigm to ensure the state of the session provide sufficient context data for the user's application to return its executing state, and continues the session from where it was suspended. The "Subscribe/Notification Management Module" realizes the function of subscribing and notifying. The "Endpoint_reference Management Module" binds the service address described in ".gsml" file by GSML language to the physical service address.

5 Conclusion

Availability is becoming one of the most important factors for the widely usage of the grid system. The users require new application paradigm and supported software platform to realize a friendly interface and environment for accessing grid services. By this way it could improve the productivity of the grid users from the whole life-cycle of the grid applications. Still the mobile grid user needs the GAPE provide the support with the 3A accessing paradigm. In the VEGA Grid project, we developed a 3A accessing paradigm supported GAPE in SOA context. We provide a flexible, robust, and efficient way to manage the context data of the gird applications to realize the virtual connection migration of the mobile grid users.

References

1. G. Fox, M. Pierce, D. Gannon, M. Thomas, "Overview of Grid Computing Environments", Global Grid Forum, http://forge.gridforum.org, 2003
2. Jason Novotny, "The Grid Portal Development Kit", Concurrency and Computation: Practice and Experience, Vol. 14, Grid Computing Environments Special Issue 13-15, page 1129-1144, 2002
3. D. Arnold, H. Casanova, J. Dongarra, "Innovations of the NetSolve Grid Computing System", Concurrency and Computation: Practice and Experience, Vol. 14, Grid Computing Environments Special Issue 13-15, page 1457-1480, 2002
4. Karen Schuchardt, Brett Didier, Gray Black, "Ecce – A Problem Solving Environment's Evolution Toward Grid Services and a Web Architecture", Concurrency and Computation: Practice and Experience, Vol. 14, Grid Computing Environments Special Issue 13-15, page 1221-1240, 2002
5. N. Li, S. Vuong, "WebMC: A Web-based Middleware for Mobile Computing", Internet Computing, 2000
6. Xu Zhiwei, Li Wei, "Research on VEGA grid architecture", Journal of Computer Research and Development, 2002, 39(8):923-929

VAST: A Service Based Resource Integration System for Grid Society*

Jiulong Shan, Huaping Chen, Guangzhong Sun, and Xin Chen

National High Performance Computing Center,
Department of Computer Science and Technology,
University of Science and Technology of China,
230027 Hefei, Anhui, China
jlshan@mail.ustc.edu.cn

Abstract. Grid and P2P are two different ways of organizing heterogeneous resources. However, the similarity in their target of resource sharing determines that they could be and should be treated together. Reference [4] proposed the concept as well as system model of 'Grid Society', which describes the Grid-P2P mixed environment and demonstrates the similarity between Grid Society and Human Society. Following this idea, we continued by an in-depth analysis of Grid Society environment from systematical aspect and designed a service-based architecture for VAST, a resource integration system. In VAST, Grid and P2P Resource Access Entry assists the Portal to co-schedule resources by providing a set of well-defined service interfaces; DCI, a P2P software system, was also developed to meet the special requirements of Portal; and Portal, coordinator of the whole system, links those two Access Entries and provides a friendly interface to the End Users for controlling and monitoring the submitted jobs. VAST has already been implemented using Java and Web Service technologies, which helps to achieve good portability. Further more, an experiment of parameter sweep application has also been conducted to test the prototype of VAST and through it VAST shows a promising potential for the collaboration of different kinds of resources.

Keywords: Resource Integration, Grid Society, Web and Grid Service.

1 Introduction

The traditional client/server model can not meet the increasing requirements for computing ability, whereas many independently deployed supercomputers, clusters and PCs do not reach an efficient usability. This unmatched situation leads to the birth of Grid and P2P.

---

* This work was supported by the National '863' High-Tech Programme of China under the grant No. 2002AA104560 and National Science Foundation of China under the grant No. 60273041.

Grid [1] is a geographically distributed computing platform that aims at resource sharing and problem solving among heterogeneous 'LARGE' resources. These are done through high speed interconnection networks and are QoS ensured. On the contrary, P2P [2] is a class of applications that accumulate available resources distributed at the Internet's edges, which are relatively 'SMALL'. And the resources provided in this way are generally non-QoS ensured.

Nevertheless, they both concerned with the coordinated sharing of distributed resources within an overlay structure. Therefore, in spite of great difference, the similarity in target determined that these two systems could be and should be treated together as a whole to get an efficient collaboration of the QoS ensured and non-ensured resources [3] [12]. The paper '*Grid Society: A System View of Grid and P2P Environment* [4]' proposed the concept and system model of 'Grid Society', which mixed access to both 'LARGE' and 'SMALL' resources.

In this paper we will discuss our further work on Grid Society, the design and implementation of the VAST resource integration system for Grid Society, which serves as a service-based architecture consisting of distributed Web Services and Grid Services.

The rest of this paper is organized as follows: The next section discusses the related works. Section 3 describes the VAST system design principle and functionality in details. Experiments of a chemical parameter sweep application is shown in section 4. Finally, some concluding remarks are made and future work is outlined in section 5.

2 Related Works

As we mentioned above, characteristic of resources, 'LARGE' or 'SMALL', distinguished Grid and P2P. On each side, there exist presently some successful projects. Besides, some other projects are making effort to collaborate them.

Globus [5] and UniCore [6] are two well-known Grid Systems that treat 'LARGE' ones, and some included technologies have already been accepted as the de facto standards for Grid Computing. Globus is an open source project, aiming to provide Grid building tools as well as environment. UniCore provides a science and engineering Grid environment combining resources of supercomputers on the Internet.

For 'SMALL' resources, projects such as SETI@Home [7], Folding@Home [8] and many others have already got a good performance. SETI@Home, with other common software such as BONIC [9], UD [10] and Entropia [11], all concern on utilizing the idle resources within a campus, a company or the whole Internet at a low cost.

Besides, more and more researchers now move their attentions to the merging of different kinds of resources. In [12], Ian Foster and Adriana Iamnitchi made a comparison between Grid and P2P on the aspects of target communities, incentives, resources and applications, and came to the conclusion that '*The Grid and P2P communities are approaching the nirvana from different directions and collaboration is the direction of future*'.

ZENTURIO [13] is an experiment management tool for cluster and Grid architectures. Based on a set of pre-defined directives, it can automatically instrument the applications and generate the corresponding set of experiments. However, it only concerns on a dedicate environment, overlooking the QoS requirements of applications and the value of 'SMALL' resources.

The Community Grids Lab of Indiana University developed a Web Service based 'Peer-to-Peer Grids system [14]' including Grids and P2P networks. In this system, all involved entities are organized in P2P mode and linked by messages. Although it can collaborate different types of resources, it loses sight of the differences in their characteristics, and has not considered the co-scheduling issue for specific applications.

3 VAST Resource Integration System

In [4], **Grid Society** is defined as a mixed system of different kinds of resources. These resources can be classified, according to the QoS they provided, into Server Entity ('LARGE ones') and Servent Entity ('SMALL' ones). It is found that Grid Society and Human Society have a strong similarity in their targets and basic elements. And such draws the conclusion that *'Grid Society and Human Society are similar system. Issues in Grid Society can be solved using corresponding solutions of similar issues in Human Society'*. However, how can a method in Human society be migrated into Grid Society and under which constraints will this migration be possible have not yet been solved.

The VAST system goes further on the above point. It facilitates the theory of Grid Society to collaborate different kinds of resources, especially to support efficient problem solving for compute intensive and data intensive applications. We performed an in-depth analysis of Grid Society environment from systematical aspect and gave the system architecture design.

3.1 System Architecture Design

In Human Society, organizations with different capabilities are made up from single persons and they are structured into a layered architecture with the help agents. All organizations will be co-scheduled by agents for the execution of complex tasks. Learned form it, we classified the involved entities of Grid Society into five basic roles according to their functionality and capability.

1. **Grid Service Provider** refers to the resources with high computing or storage capability and can provide QoS ensured resource sharing, such as the supercomputer ASCI White in LLNL and the Human Genome Database server.
2. **P2P Service Provider** includes machines who contribute their resources voluntarily. All volunteers can join and leave the system freely, which produces a volatile resource sharing.
3. **P2P Service Coordinator** plays an important role in P2P Computing system (such as in SETI@Home and Folding@Home), who is in charging of the management and coordination of the joined peers.

4. **End User** is the one who submits jobs to the system and pays for the results.
5. **Portal**, coordinator of the whole system, provides a seamless linking between Service Provider and End User.

Each of the above roles has its corresponding part in Human Society. And just the same as Agent in Human Society, Portal is the most important component of Grid Society. With the help of Portal, the involved resources are also organized into a layered architecture. First, P2P Service Providers register themselves to one or more coordinators according to their locations. After that, P2P Service Coordinators and Grid Service Providers are centralized to Portals based on their locations and the supported applications. Finally, the Portals will collaborate with each other in a self-organized way.

Based on the analysis above, a service-based architecture for VAST was designed, as shown in **Fig. 1**.

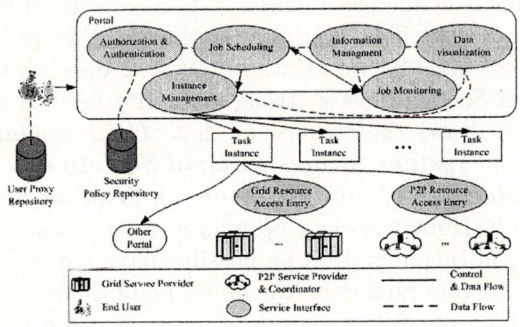

Fig. 1. VAST System Architecture

In VAST, users can submit their jobs to the system through Portal. After the authentication and authorization phase, the submitted job will be partitioned into small sub-tasks and assigned to the select resource providers through Grid and P2P Resource Access Entry. When the resources held by a Portal can not meet the user's requirements, it will redirect part or whole of the job to another friend Portal in VAST and this can achieve the sharing of resources in the whole Grid Society. Job Monitoring and Data Visualization are also provided by the Portal.

By adopting the 'Service Orientated Schema', all functions of VAST are presented as a set of well-defined Service Interfaces, which facilitate the co-scheduling of resources. The prototype of the system has already been implemented based on Java and Web Service technologies, and is currently running on our testbed which accumulates resources from PCs, clusters to supercomputers.

In the following subsections, Grid Resource Access Entry (GRAE), P2P Resource Access Entry (PRAE) and Portal, the three key pillars of VAST, will be described in details.

3.2 Access to Grid Resources

The Grid Resource Access Entry (GRAE), acting as an assistant of the Portal, works to integrate Grid Resources that are available as services. In VAST design, GRAE is defined as a set of service interfaces supporting the efficient communication between Portal and remote Grid Service Provider. Currently GRAE supports Grid Service Providers equipped with Globus Toolkit 3 (GT3), and we will make it also support UniCore in our next step work.

GT3 is the first implementation of Open Grid Service Architecture (OGSA) [15]. To integrate resources equipped on GT3, we deploy an abstract base class *AbstractGridService*, which is inherited from the *ServiceInterface* of GT3, and extend it by adding functions for access to remote Grid Services. For a specific Grid Service deployed on the remote Grid Service Provider, GRAE will first get the corresponding GSWDL through GT3's *ContainerRegistryService* and then generate the client service code by extending *AbstractGridService* automatically. Through the client service code, Instance Management Module can assign subtasks to selected remote Grid Service Providers.

3.3 Access to P2P Resources

The P2P Resource Access Entry (PRAE), another assistant component of Portal, aims at the integration of volunteer resources wide around the Internet edges. By providing a set of service interfaces, PRAE supports the communication between Portal and P2P Service Coordinator. To cooperate with the PRAE, the P2P Service Coordinator must meet some special requirements such as providing the service interface for job co-scheduling with GRAE. Because of there no existing P2P system can meet these requirements, thus we developed on our own a Distributed Computing Infrastructure (DCI). In this part, the ideas, or we can say, features of the DCI are discussed.

First, we carried out the principles of Web Service technology, both in DCI architecture design and its implementation. By inheriting a set of standard service interfaces, DCI can cooperate with PRAE seamlessly. Particularly, DCI is designed as a stand-alone component thus it can be used independently.

Second, DCI is well designed for computing intensive and data intensive applications, such as parameter sweep and coupled parallel simulation. For each kind of applications, there is a corresponding processing template inside DCI, which will binding to them automatically, just the same as the scheduler interface in Portal.

Besides, a combination of the duplication (send n copies of a sub-task simultaneously at the beginning) and timeout (resend a sub-task when the corresponding result does not come back in a time period of t) scheduling policy is also supported in DCI. We found after investigating that building a 2-dimensional scheduling metric from those two 1-dimensional metrics can achieve a better system performance [17].

3.4 Portal

Portal is the key component of the whole system [18], it provides a friendly user interface for users to submit theirs jobs, query the results as well as monitor on the execution. What's more, GRAE and PRAE can work together with the help of Portal to collaborate Grid and P2P Service Providers.

The Job Scheduling Module (JSM) will processes the submitted jobs after job owner finished their security checking. Each kind of applications has its corresponding specific processing template (scheduler) which is inherited from the interface *AbstractScheduler* (such as *BioScheduler* for Molecular Dynamic Simulation application, *ParaSweepScheduler* for all parameter sweep applications) and will binding to them automatically. Afterwards, according to the information provided by the user and the Information Management Module (InfoMM), the JSM will partition the job, select the providers, make the mapping, generate the schedule result and then pass it to the Instance Management Module (InstMM).

The InstMM receives the schedule result from JSM, starts a monitoring process for this job immediately and then assigns those sub-tasks through GRAE and PRAE to the service providers as scheduled.

During execution, the Job Monitoring Module (JMM) watches on the job until it finished. If the collected status shows a mismatch to the user's requirements, the JMM will call the JSM to adjust the initial scheduling. The JMM is also responsible for system information collection, including load of providers, job queue lists and application runtime information. All the collected information are stored by InfoMM in a repository. The monitoring functions of JMM are implemented in two ways, pull and trigger.

Data Visualization Module (DVM) is always the last step of applications. It is in charge of the visualization of the monitoring information and the result data. For the former type of information, pictures are mostly generated locally. As to the result data, the visualization work includes the following steps. First, when submitting the job, user should specify in advance the result data file name, the demanded visualization tool and the needed processing script (see **Example 4.2**). Then the DVM will choose the place to process result data. In case the needed software is not installed locally, DVM will submit this visualization work as a new job to JSM on behalf of the user and then get the result.

Finally, we use the Portlet technology [19] for arranging all these modules into Portlets. With Portlet, users or administrators can customize their own individualized user interface according to their own preferences easily. This technology also facilitates the development and management of services.

4 Experiments

In order to evaluate our prototype implementation, the experiment of a chemical parameter sweep application 'thermodynamic protein folding', which studies the protein folding mechanism from thermodynamics angle, was conducted using VAST.

Experiments were run on our testbed in USTC, consisting of one Grid Service Provider (HP Beowulf Cluster with 4 nodes, each node containing 2 Intel *Itanium 2 Madison* 1.5GHz CPU), 16 P2P Service Providers (16 single node PC, with Intel *Pentium 4* 2.4GHz CPU). All the 16 P2P Service Providers were coordinated by one P2P Service Coordinator, and the coordinator was deployed together with the Portal on the same server.

For parameter sweep applications, the submitted job will be first parsed by *ParaSweepSechduler* in Portal, then InstMM submit those partitioned sub-tasks to remote service providers through GRAE and PRAE. GRAE uses GT3's default job manager *ManagedJobFactoryService* for submitting sub-tasks and gets the result back using *GridFTP*, while PRAE communicates with DCI to handle this.

To submit a parameter sweep job, what the user only needs to do is to provide the application's source code, input parameter files, result linking program's source code, make files and two special VAST script files. These VAST script files include: (1)**source script file** (see **Example 4.1**) that describes the parameters, expected deadline and result data file name; (2)**result script file** (see **Example 4.2**) that sets the name of visualization software and corresponding manipulate scripts. The generated **schedule result file** (see **Example 4.3**) specifies the sub-task's information, including the destination provider's URL, the assigned parameter space and the arranged processor numbers of Grid Service Provider. Result data will be linked by the result linking program and visualized by DVM.

The parameter study was performed by varying the input parameter *runtimes* among 20000, 40000, 60000 and 80000. **Example 4.1** describes one scenario where the *runtime* equals to 40000 with a fixed step length 10. The expected deadline is set to 0 means no such requirement and the result data will be stored in the file *result.dat*. **Example 4.2** shows that the result data will be processed by GNUPlot with specified scripts.

Example 4.1 *vast.source*

```
    $ParaNumber      1
    $Para
        $Start   1
        $End     40000
        $Step    10
    $EndPara
    $Time            0
    $ResultDataFile  'result.dat'
```

Example 4.2 *vast.result*

```
    $VisualTool         'GNUPlot'
    $VisualResultFile   'result.png'
    $VisualScript
        set terminal png
        set output '$VisualResultFile'
```

```
            set xlabel 'Engery Bands'
            set ylabel 'Endis (1e + 05)'
            plot '$ResultDataFile' using 1:($3/100000) \
                 w lp lt 1 pt 1 t 'runtimes = 40000'
      $EndVisualScript
```

Besides, by selecting resources by *ParaSweepScheduler* automatically, users can also specify the needed service providers and the sub-tasks assignment proportion themselves. In the experiments, 2, 3 and 4 nodes of HP Cluster were selected for each value of *runtimes* separately, and for each resource selection 11 kinds of sub-tasks assignment proportions were used. **Example 4.3** is a generated schedule result file, in which the job was partitioned into equally two half and all the four nodes (8 CPUs) of HP Cluster were selected together with the P2P Service Providers.

Example 4.3 *vast.scheduling*

```
      $SubTask
            $Provider      0         /*0 representing Grid Service Provider*/
            $URL           'http://hpc.grid.ustc.edu.cn/'
            $ParaNumber    1
            $Para
                  $Start   1
                  $End     20000
                  $Step    10
            $EndPara
            $ProcessorNumber     8
            $ResultDataFile      'result.dat'
      $EndSubTask

      $SubTask
            $Provider      1         /*1 representing P2P Service Provider*/
            $URL           'http://summer.grid.ustc.edu.cn/'
            $ParaNumber    1
            $Para
                  $Start   20001
                  $End     40000
                  $Step    10
            $EndPara
            $ResultDataFile      'result.dat'
      $EndSubTask
```

Fig. 2 shows the visualized result data for those four *runtimes* values. We can see from it that with the incensement of *runtimes* user can get a more centralized result. It also results in the increasing needs for computing ability, which make collaborating different resources more necessary.

Fig. 3 shows the application's execution time under different settings while the *runtimes* is fixed at 40000, in which the X axis denotes the proportion of

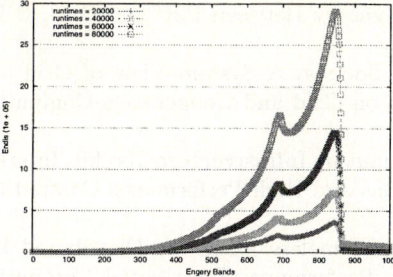 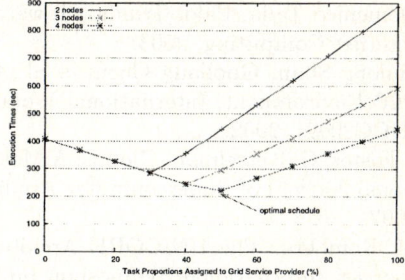

Fig. 2. Result data for diff. *runtimes*   **Fig. 3.** Execution time for diff. settings

sub-tasks assigned to HP Cluster and Y axis denotes the application's execution time (E_T). In VAST, E_T is the larger one of the Grid Service Provider's execution time (E_G) and the P2P Service Provider's execution time (E_P). From Fig. 3 we can see that with the increasing proportion of sub-tasks assigned to HP Cluster from zero to 100%, the E_T reduced first, but then increased. Moreover, the E_T also decreased when selecting more HP Cluster nodes. This means that, by selecting an optimal resource setting and sub-tasks assignment proportion we can get a minimum E_T for the application. In this experiment, it was obtained when all the 4 HP Cluster nodes were selected and the assignment proportion was 50%.

From the above experiments, we can see that VAST has a promising potential for collaboration of different kinds of resources.

5 Conclusion and Future Works

In this paper, we have presented a comprehensive view of the VAST resource integration system with description of GRAE, PRAE and Portal separately. The main contributions of VAST can be concluded in the following aspects. First, by using Human Society for reference, an efficient method for collaborating 'LARGE' and 'SMALL' resources in Grid Society was proposed. Second, the design of service-based architecture significantly increased the system scalability and facilitated the cooperation of resources. What's more, application specific processing template, employed in both Portal and DCI, helped achieving good performance.

For future works, we plan to develop necessary service interfaces in GRAE to integrate UniCore equipped resources and extend the job processing templates in Portal and DCI to support for more kinds of applications.

References

1. I. Foster, C. Kesselmann, editors, The Grid: Blueprint for a New Computing Infrastructure, Morgan Kaufman Publishers, 1998.
2. Dejans S. Milojicic, Vana Kalogeraki, et al., Peer-to-Peer Computing, HPL-2002-57R1, HP Labs 2002 Technical Reports, 2002.

3. Domenico Talia, Palol Trunfio, Toward a Synergy Between P2P and Grid, IEEE Internet Computing, 2003.
4. Jiulong Shan, Guoliang Chen, et al., Grid Society: A System View of Grid and P2P Environment, International Workshop on Grid and Cooperative Computing (GCC2002), 2002.
5. I. Foster, C. Kesselman, Globus: A Metacomputing Infrastructure Toolkit, International Journal of Supercomputer Applicaitons and High Performance Computing, 1997.
6. M. Romberg, The UNICORE Architecture: Semeless Access to Distributed Resources, International Symposium on High Performance Distributed Computing (HPDC-8), 1999.
7. David P. Anderson, Jeff Cobb, et al., SETI@home: An Experiment in Public-Resource Computing. Communications of the ACM, 2002.
8. Michael Shirts, Vijay Pande, Screen savers of the world, Unite!, Science, 2000.
9. BONIC: Berkeley Open Infrastructure for Network Computing, http://boinc.berkeley.edu/index.html
10. B. Uk, M. Taufer, et al., Implementation and characterization of protein folding on a desktop computational grid. Is CHARMM a suitable candidate for the United Devices MetaProcessor? International Parallel and Distributed Processing Symposium (IPDPS 2003), 2003.
11. Andrew Chien, Brad Calder, et al., Entropia: Architecture and Performance of an Enterprise Desktop Grid System, Journal of Parallel Distributed Computing, 2003.
12. I. Foster, A. Iamnitchi, On Death, Taxes, and the Convergence of Peer-to-Peer and Grid Computing, International Workshop on Peer-to-Peer Systems (IPTPS'03), 2003.
13. Radu Prodan, Thomas Fahringer, A Web service-based Experiment Management System for the Grid, International Parallel and Distributed Processing Symposium (IPDPS 2003), 2003.
14. Geoffrey Fox, Dennis Gannon, et al., Peer-to-Peer Grid, IBM Research Presentation, 2002.
15. I. Foster, C. Kesselman, et al., The Physiology of the Grid: An Open Grid Services Architecture for Distriubuted Systems Integration, The Globus Project, 2002.
16. Wenrui Wang, Guoliang Chen, et al., A Grid Computing Framework for Large Scale Molecular Dynamics Simulations, International Workshop on Grid and Cooperative Computing (GCC2003), 2003.
17. Guangzhong Sun, Jiulong Shan, et al., A Study on the Scheduling Policies of Global Computing System, submitted to The 3rd International Workshop on Grid and Cooperative Computing (GCC2004)
18. Jiulong Shan, Guoliang Chen, et al., Design and Implementation of Hefei Grid Portal, Journal of Mini-Micro Systems, to be published.
19. JSR-000168 Portlet Specification, http://www.jcp.org/en/jsr/detail?id=168

Petri-Net-Based Coordination Algorithms for Grid Transactions

Feilong Tang[1], Minglu Li[1], Joshua Zhexue Huang[2], Cho-Li Wang[3], and Zongwei Luo[2]

[1] Department of Computer Science and Engineering,
Shanghai Jiao Tong University, Shanghai 200030, China
{tang-fl, li-ml}@cs.sjtu.edu.cn
[2] E-Business Technology Institute, The University of Hong Kong, China
jhuang@eti.hku.hk
[3] Department of Computer Science, The University of Hong Kong, China
clwang@cs.hku.hk

Abstract. Transaction proceesing in Grid is to ensure reliable execution of inherently distributed Grid applications. This paper[1] proposes coordination algorithms for handling short-lived and long-lived Grid transactions, models and analyzes these algorithms with the Petri net. The cohesion transaction can coordinate long-lived business Grid applications by automatically generating and executing compensation transactions to semantically undo committed sub-transactions. From analysis of the reachability tree, we show that the Petri net models of above algorithms are bounded and L1-live. This demonstrates that transactional Grid applications can be realized by the proposed algorithms effectively.

1 Introduction

The goals of Grid computing are to share large-scale resources and accomplish collaborative tasks [1]. Many of inherently distributed Grid applications require high reliability, which can be achieved by transaction processing technologies.

The traditional distributed transaction has the ACID properties [2]:

- Atomicity. Participants of a transaction are either all committed or all aborted.
- Consistency. A transaction changes the system from one consistent state to another.
- Isolation. Intermediate results of a non-committed transaction are not read or written by other concurrent transactions.
- Durability. Effects of a transaction are durable once it commits.

However, the strict ACID transaction only satisfies the following conditions: (1) the transaction is short-lived, (2) the coordinator has the full control power

---

[1] This paper is supported by 973 Program of China (No.2002CB312002), ChinaGrid Program of MOE of China and grand project of the Science and Technology Commission of Shanghai Municipality (No.03dz15027).

to participants, and (3) the application systems are tightly coupled. As such, the coordination algorithms for Grid services [3] have to relax the ACID semantics because Grid transactions are mostly long-lived and lock of resources is not often allowed because of the autonomy of Grid services.

Coordination algorithms of Grid transactions consist of sets of messages and rules that regulate the interactions among the involved Grid services to achieve consistent outcomes in an orderly fashion. With the Petri net, the correctness of algorithms can be validated.

A Petri net is an abstract and formal modeling tool for representation and analysis of parallel processes. It is able to represent most coordination problems, easy to use and understand, and shares common properties, such as boundedness, liveness, deadlock-freeness, proper termination and completeness [7]. It can model systems' events, conditions and the relationships among systems. The occurrence of these events may change the state of the system, causing some of the previous conditions to cease holding and other conditions to begin to hold [12].

In this paper, we use Petri nets to explicitly model the states and transitions of Grid transactions during the coordination process, and verify the correctness of proposed coordination algorithms applicable to short-lived and long-lived transactional Grid applications.

2   Related Work

The Petri net is a powerful method for describing and analyzing the flow of information and control in systems, particularly those systems involving asynchronous and concurrent activities. Petri nets have been widely researched and applied in many areas [12,13].

Jacinto et al. [6] used Petri nets to model and analyze a distributed processing system based on the OSI-TP and OSI-CCR mechanisms for the two phase commit procedure, which takes advantage of the characteristics of the Petri nets and the abstraction concept based on the observational equivalence. [4] proposed a scheduling method using the Petri net techniques. The Petri net model incorporates the conditions that have to be met before an action can be taken. Petri nets are used to explicitly model the states and transitions of a power system during the restoration process. The time required to perform an action is also modeled in the net. A token passing and a backward search process are used to identify the sequence of restoration actions and their timing. Another Petri net model [5] of the coordination level of an intelligent mobile robot system (IMRS) can specify the integration of the individual efforts on path planning, supervisory motion control, and vision system that are necessary for autonomous operations of the mobile robot in a structured dynamic environment. The model can also be used to simulate task processing and evaluate efficiency of the operations and responsibility of the decisions in the coordination level of the intelligent mobile robot system.

3 Framework of Grid Transaction Processing

Fig. 1 shows the transaction processing framework we propose. Its core element Agent consists of following complements:

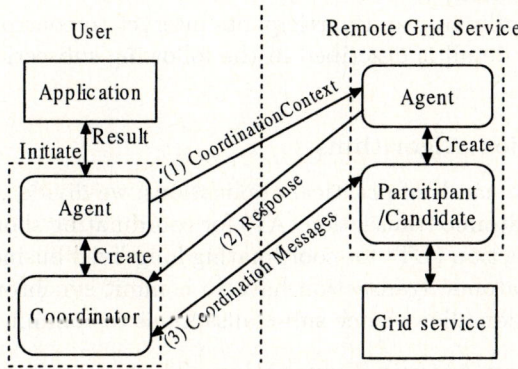

Fig. 1. Transaction processing framework for Grid services

- Time Service, which generates time-driven events to rollback failed transactions or start corresponding compensation transactions for undoing the effect of the committed sub-transactions.
- Compensation Generator, which automatically generates the compensation operations for sub-transactions of a cohesion transaction and combines them into compensation transactions when the sub-transactions commit.
- Log Service, which records coordination operations and state information.
- Interface, including application interfaces invoked by application programs, and the Rule Pre-definition interface for setting up compensation rules.
- Coordinator and Participant/Candidate, which are dynamically generated to coordinate transactions.

4 Grid Transaction Processing

4.1 Flow of Grid Transaction Processing

In the Grid service environment, transaction processing typically involves the following procedures [8,16].

- an initial agent, which initiates a global transaction for an application, discovers and selects the qualified Grid services as participants according to applications' requirements. Further information is described in [15].
- the agent negotiates with the participants. For remote Grid services to join a transaction, the agent sends the CoordinationContext (CC) messages to

them and creates a local coordinator. The CC message includes the necessary information to create a transaction, including the transaction type, the transaction identifier, the coordinator address and expire time. The agent of each participant returns a Response message to the coordinator, and locally creates a participant (for an atomic transaction) or a candidate (for a cohesion transaction).
– the created coordinator and participants interact to control the transaction execution. The detail is described in the following subsections.

4.2 Coordination Algorithms

According to the demands of practical applications, we divide Grid transactions into two types[11]: atomic transaction (AT) for coordinating short-lived activities and cohesion transaction (CT) for coordinating long-lived business activities. All participants of an atomic transaction have to commit synchronously. However, a cohesion transaction allows some sub-transactions to commit while others fail.

Coordination of an Atomic Transaction. The process includes three steps, initiation of an atomic transaction, preparation for the commit, and commit of the transaction. The coordination algorithms and the state conversion diagram are shown in Fig. 2 and Fig. 3(a) respectively.

```
ActionOfParent{
  step1: initiate an AT
  agent creates Coordinator;
  agent sends CC to all agents of P_i;
  wait for Response from agent of P_i;
  if timeout
    exit;
  step2: prepare for the transaction
  send Prepare to all Participants;
  while (t ≤ T_1) and (n1<N)
    wait for and record incoming messages;
  step3: commit the transaction
  if (n1=N)and(n1 messages are Prepared){
    record commit in log;
    send Commit to all Participants;
    while (t≤T_2) and (n2<N)
      wait for and record incoming message;
    if (n2<N)or(not N Committed messages)
    { send Rollback to all Participants;
      exit after receiving all Rollbacked; }
  } else {  send Abort to all Participants;
    exit after receiving all Aborted;
}}
```

```
ActionOfChild{
  step1: join in the transaction
  agent creates Participant after receiving
    CoordinationContext;
  agent sends Response to Coordinator;
  step2: reserve resources
  wait for Prepare from Coordinator;
  if timeout exit;
  success:=reserves resources;
  if (success){
    send Prepared to Coordinator;
  step3: commit sub-transaction
    while(t≤T_3)and(not Commit or Abort)
      wait for incoming message;
    if (message is Commit){
      allocate reserved resources;
      record commit in log;
      commit sub-transaction;
      //nested transaction,call ActionOfParent;
      send Committed to Coordinator;
  }else{ cancel reservation;
    send NotPrepared to Coordinator;
    exit; }}}
```

(a) Coordinator algorithm   (b) Participant algorithm

Fig. 2. Coordination algorithms of the atomic transaction

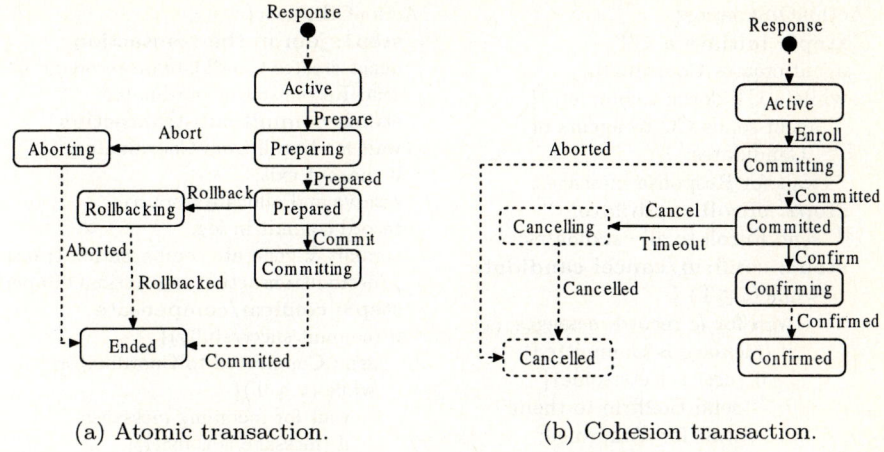

(a) Atomic transaction.   (b) Cohesion transaction.

Fig. 3. State conversion diagram

Coordination of a Cohesion Transaction. The process includes initiation of a cohesion transaction, independent commit of sub-transactions, confirmation of a user, and confirmation of successful candidates. The coordination algorithms and the state conversion diagram are shown in Fig. 4 and Fig. 3(b) respectively.

5 Petri Net Models of Grid Transactions

5.1 Modeling Coordination Algorithms with Petri Nets

The Petri net model is a powerful modeling tool. A very complex activity can often be quickly and easily translated to a Petri net representation. Petri net models have proved effective in net analysis for deadlock detection and behavior trends in asynchronous systems, thus for determining the correctness and efficiency of proposed systems [7].

A Petri net model consists of a set of places and a set of transitions. The places and transitions are connected by a set of directed arcs. A transition is said to be enabled if there are enough tokens in each of the input places as specified by the arcs connecting the input places to the transition. An enabled transition can fire if the other conditions associated with the transition are satisfied [10,12]. We model above algorithms and verify their correctness with Petri nets. In the models, the places (denoted by circles) correspond to the states and the transitions (denoted by lines) represent the actions related to the states. For the purpose of intuition, we express the coordinator states with corresponding messages.

Properties of the Petri net model with two participants are similar to those of more members. For convenience, let an atomic transaction consist of two partic-

```
ActionOfSuperior{                       ActionOfInferior{
  step1: initiate a CT                    step1: join in the transaction
  agent creates Coordinator;              agent creates Candidate on receiving CC;
  while(a CT doesn't complete){           send Response to Coordinator;
    agent sends CC to agents of           step2: commit sub-transaction
      Candidates;                         wait for Enroll from Coordinator;
    wait for Response messages;           if timeout exit;
  step2: enroll candidate                 reserve and allocate resources;
    send Enroll to all candidates;        record commit in log;
  step3: confirm/cancel candidate         commit & generate compensation transaction;
    while (t ≤T) {                        //nested transaction,call ActionOfSuperior
      wait for & record messages;         step3: confirm/compensate
      if (message is Committed)           if (commit successfully){
        if (user selects some){             send Committed to Coordinator;
          send Confirm to them;             while (t ≤ T){
          wait for Confirmed;                 wait for incoming messages;
        } else{                               if (message is Cancel){
          send Cancel to them;                  call its compensation transaction;
          wait for Cancelled;                   send Cancelled;
        }                                   } else {
                                              if (message is Confirm){
    }                                           send Confirmed;
  }
}}                                      }}}}}

     (a) Coordinator algorithm              (b) Candidate algorithm
```

Fig. 4. Coordination algorithms of the cohesion transaction

ipants and a cohesion transaction two candidates. Without losing the generality, in the cohesion transaction, we let candidate P_1 successfully commit while P_2 fail to commit. Their Petri net models are depicted in Figs. 5 and 6, where the "C" means the coordinator and the "P_i" refers to the ith participant of an atomic transaction or the ith candidate of a cohesion. The weight value above the arcs indicates the number of the changed (i.e. added or removed) tokens whenever the firing happens, and all "1"s are omitted.

5.2 Analysis of the Petri Net Models

Problems analyzed in modeling parallel systems by Petri nets are usually dealing with dynamic aspects of the control structure. Such problems are (partial or total) deadlock freeness, or liveness of the system [9].

The Petri net model can analyze the behavioral properties, which depend on the initial marking, including reachability, boundedness, liveness, coverability, reversibility, persistence and so on. For a bounded Petri net, however, all the above problems can be solved by the reachability tree [13]. Peterson pointed out that in Petri nets, many questions can often be reduced to the reachability problem [12]. Hack also showed that the liveness problem is reducible to the reachability problem and that in fact the two problems are equivalent [14].

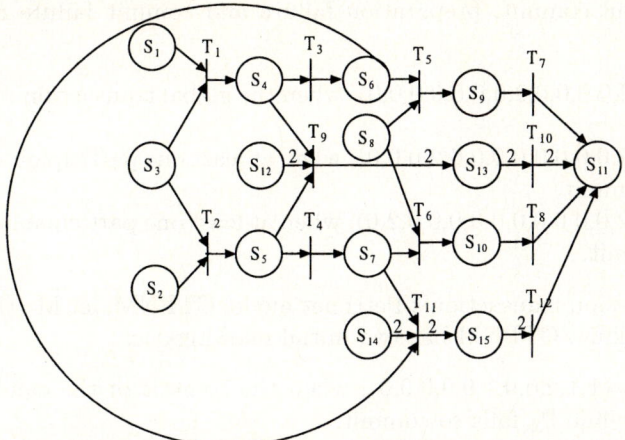

S_1: P_1-Active S_2: P_2-Active S_3: C-Prepare S_4: P_1-Preparing S_5: P_2-Preparing S_6: P_1-Prepared
S_7: P_2-Prepared S_8: C-Commit S_9: P_1-Committing S_{10}: P_2-Committing S_{11}: Ended S_{12}: C-Abort
S_{13}: P_1-Aborting and P_2-Aborting S_{14}: C-Rollback S_{15}: P_1-Rollbacking and P_2-Rollbacking
T_1: P_1 prepares for commit T_2: P_2 prepares for commit T_3: P_1 returns Prepared T_4: P_2 returns Prepared
T_5: P_1 commits T_6: P_2 commits T_7: P_1 returns Committed T_8: P_2 returns Committed T_9: P_1 and P_2 abort
T_{10}: P_1 and P_2 return Aborted T_{11}: P_1 and P_2 rollback T_{12}: P_1 and P_2 return Rollbacked

Fig. 5. Petri net model of the atomic transaction coordination algorithms(ATPNM)

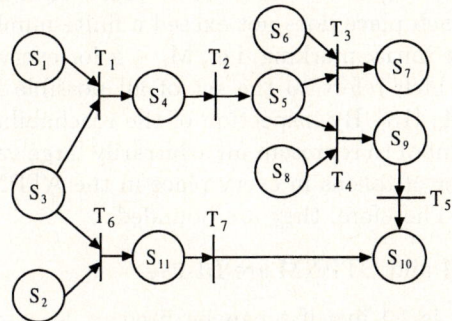

S_1: P_1-Active S_2: P_2-Active S_3: C-Enroll S_4: P_1-Committing S_5: P_1-Committed
S_6: C-Comfirm S_7: P_1-Confirmed S_8: C-Cancel S_9: P_1-Cancelling S_{10}: Cancelled
S_{11}: P_2-Committing T_1: P_1 commits T_2: P_1 returns Committed
T_3: P_1 returns Comfirmed T_4: P_1 compensates T_5: P_1 returns Cancelled
T_6: P_2 perpares for ommit T_7: P_2 fails to commit and returns Aborted

Fig. 6. Petri net model of the cohesion transaction coordination algorithms(CTPNM)

The state of a Petri net is the distribution of tokens to the places. The initial state of the Petri net is called its initial marking. For the atomic transaction's Petri net model ATPNM, let M=(M_1, M_2,..., M_{15}) be a marking, where M_i is the number of tokens in place S_i. ATPNM has three initial markings, corresponding to successful commit, preparation failure and commit failure respectively, that is:

- M_{0s}=(1,1,2,0,0,0,0,2,0,0,0,0,0,0,0), when the global transaction commits successfully.
- M_{0a}=(1,1,2,0,0,0,0,0,0,0,0,2,0,0,0), when at least one participant can not prepare for commit.
- M_{0f}=(1,1,2,0,0,0,0,0,0,0,0,0,0,2,0), when at least one participant can not correctly commit.

For the cohesion transaction's Petri net model CTPNM, let M=(M_1, M_2,..., M_{11}) be a marking. CTPNM has two initial markings as:

- $M_{0confirm}$=(1,1,2,0,0,1,0,0,0,0,0), when the commit of the candidate P_1 is confirmed while P_2 fails to commit.
- $M_{0cancel}$=(1,1,2,0,0,0,0,1,0,0,0), when the commit of the candidate P_1 is cancelled and P_2 fails to commit.

Below, we mainly analyze the boundedness and reachability, using reachability analysis which checks whether some properties can occur by considering all reachable states of the model being analyzed. By analysis of the reachability trees of ATPNM and CTPNM, we can draw the following conclusions.

Theorem 1. ATPNM and CTPNM are bounded.

Proof: A Petri net (N, M_0) is said to be k-bounded or simply bounded if the number of tokens in each place does not exceed a finite number k for any marking reachable from the initial marking, i.e., $M_i \leq k$ for every place S_i and every marking M∈R(M_0) which refers to the set of all possible markings reachable from the initial one M_0 [13]. By inspection of the reachability trees of ATPNM and CTPNM, ω (a symbol to represent an arbitrarily large value) does not occur anywhere. The number of tokens in every place in the ATPNM and CTPNM is never greater than 2. Therefore, they are bounded.

Theorem 2. ATPNM and CTPNM are L1-live.

Proof: A transition t is L1-live if t can be fired at least once in some firing sequences. A Petri net is L1-live if $\forall t \in T$, which is the set of transitions in a model, t is L1-live.

For a bounded Petri net, the reachability tree explicitly contains all possible markings. According to Theorem 1, both ATPNM and CTPNM are bounded. By inspection of the reachability trees of ATPNM and CTPNM again, any marking is reachable and every transition can be fired at least once from one of the initial markings M_{0s}, M_{0a} and M_{0f} in ATPNM, and $M_{0confirm}$ or $M_{0cancel}$ in CTPNM. Thus, ATPNM and CTPNM are L1-live.

Theorem 2 indicates that ATPNM and CTPNM can be deadlock-free as long as the firing starts with one of their initial markings, that is, the coordination process can proceed until the end of a global transaction, whether successful or not. Thus, the coordination algorithms are correct.

6 Implementation

Using Java as the main development language and SQL Server 2000 as the database server, we developed a prototype system. The action of its core component Agent depends on the request. If the Agent is requested to initiate a transaction, it creates a coordinator. If the Agent receives a CC message, it creates a participant/candidate. The coordinator and participant/candidate interact messages to coordinate the transactional activities and live until the end of a global transaction.

The Agent provides a set of interfaces called by applications.

beginTransaction(In txType, Out txHandle): starts a new transaction and gets a transaction identifier.
prepareCommit(In txHandle): reserves resources to prepare for commit.
commit (In txHandle): commits the transaction.
rollback(In txHandle): undoes operations taken previously.
enroll(In txHandle): requires a candidate of a cohesion transaction to commit a sub-transaction.
setTransactionTimeout(In txHandle): sets timeout for the specified transaction.
startCompensationTransaction(In c_i): starts the compensation transaction c_i for a cohesion transaction.
getTransactionStates(In txHandle, Out state): gets the current state of the transaction.

We encapsulate all interfaces into the TX portType of Grid service. Each interface and its input and output parameters are defined as operation and messages respectively.

7 Conclusions and Future Work

We have presented the Petri-net-based coordination algorithms for Grid applications. It has the abilities to coordinate short-lived operations and long-lived business activities. From the analysis of the reachability tree, it can be concluded that the Petri net models of these algorithms are bounded and L1-live, which theoretically prove the correctness of the algorithms. The proposed coordination algorithms can recover systems from various failures but shield the complex process from users. Next, we are going to incorporate security mechanism with the algorithms to enable them to withstand vicious attacks.

References

1. I. Foster, C. kesselman and S. Tuecke, The anatomy of the grid: enabling scalable virtual organizations. International Journal of Supercomputer Applications, 2001, 15(3): 200-222.

2. Y. Breitbart, H. Garcia-Molina and A. Silberschatz. Overview of multidatabase transaction management. The VLDB Journal-The International Journal on Very Large Data Bases, October 1992, 1(2): 181-239.
3. I. Foster, C. Kesselman, J. M. Nick and S. Tuecke, The Physiology of the Grid-An Open Grid Services Architecture for Distributed Systems Integration. June, 2002.
4. J. S. Wu, C. C. Liu, K. L. Liou et al.. A Petri Net Algorithm for Scheduling of Generic Restoration Actions. IEEE Transactions on Power Systems, Vol. 12, No. 1, February, 1997. pp. 69-76.
5. F. Y. Wang, K. J. Kyriakopoulos, A. Tsolkas et al.. A Petri-Net Coordination Model for an Intelligent Mobile Robot. IEEE Transactions on Systems, Man, and Cybernetics, Vol. 21, No. 4, July-August, 1991, pp. 777-789.
6. R. Jacinto, G. Juanole, K. Drira. On the application to OSI-TP of a structured analysis and modeling methodlogy based on Petri Net models. Proceedings of the IEEE. 1993.
7. P. B. Thomas. The Petri net as a modeling tool. Proceedings of the 14th annual Southeast regional conference. April 22 - 24, 1976. pp. 172-179.
8. F. L. Tang, M. L. Li and J. Z. X. Huang. Real-time transaction processing for autonomic Grid applications. Accepted by the Special Issue on "Autonomic Computing and Automation" at Engineering Applications of Articial Intelligence, 2004.
9. E. W. Mayr. An algorithm for the general Petri net reachability problem. Proceedings of the thirteenth annual ACM symposium on Theory of computing, May 1981. pp. 238-246.
10. A. K. Murugavel and N. Ranganathan. Petri net modeling of gate and interconnect delays for power estimation. Proceedings of the 39th conference on Design automation. June 2002. pp. 455-460.
11. F. L. Tang, M. L. Li, and J. Cao, A Transaction Model for Grid Computing. Proceedings of the 5th International Workshop on Advanced Parallel Processing Technologies (LNCS 2834), September. 2003.
12. J.L.Peterson, Petri Nets. Computing Surveys, Vol 9, No.3, September 1977, pp. 223-252.
13. T. Murata, Petri Nets: Properties, Analysis and Applications. Proceedings of the IEEE, Vol 77, No. 4, April, 1989, pp. 541-580.
14. M.Hack, The recursive equivalence of the reachability problem and the liveness problem for Petri nets and vector addition systems. Proceedings of the 15th Annual Symposium Switching and Automata, New York, 1974.
15. F. L. Tang, M. L. Li, J. Cao et al.. GSPD: A Middleware That Supports Publication and Discovery of Grid Services. Proceeding of the Second International Workshop on Grid and Cooperative Computing (LNCS 3032). December, 2003, pp.738-745.
16. F. L. Tang, M. L. Li, and J. Z. X. Huang. GridTS: A Transaction Service for Intelligent Grid Environment. To appear in "Intelligent Grid Environment: Principles and Applications", special issue of Future Generation Computer Systems, 2004.

Building Infrastructure Support for Ubiquitous Context-Aware Systems

Wei Li, Martin Jonsson, Fredrik Kilander, and Carl Gustaf Jansson

Department of Computer and Systems Sciences,
Swedish Royal Institute of Technology,
Forum 100, 164 40 KISTA, Sweden
{liwei, martinj, fk, calle}@dsv.su.se

Abstract. Many context-aware systems have been demonstrated in lab environments; however, due to some difficulties such as the scalability and privacy issues, they are not yet practical for deployment on a large scale. This paper addresses these two issues with particular interest in user's privacy protection and spontaneous system association. A person-centric service infrastructure is proposed together with a context-aware call forwarding system constructed as a proof-of-concept prototype based on the Session Initiation Protocol (SIP).

1 Introduction and Motivation

During the last decade, numerous efforts in the Ubiquitous Computing arena attempted to realize Mark Weiser's vision of the 21st century computers – to enable computers "weave themselves into the fabric of everyday life" [1]. The purpose of making computers invisible is to diminish the unnecessary distraction introduced by computers so that users can concentrate on the higher level tasks they are currently involved in. Meanwhile, computers should assist *behind the scene* [2] without the occupation of users' attention.

Today, with the rapid development in mobile and distributed computing, especially the flourish of wireless communication and resource discovery technologies, various computing resources, in terms of *services*, are becoming transparently available everywhere. In the meantime, sensor technologies have been widely adopted to provide rich information for facilitating user's interaction. In the midst of the trend that services and sensors become pervasive, the goal of ubiquitous computing has been approached by many researchers with their prototypes in different testbed lab environments [2, 3, 4].

One common issue, however, among these existing systems is scalability. Due to the fact that each of these systems uses individual data expression and communication means, they are not interoperable with each other. As none of them could be dominant in the field, a mobile user roaming from one system to another normally have to conduct a significant number of interactions, typically involving in new software installation and configuration, to be able to obtain an access to the new surroundings. These extra interaction efforts that often result in a deviation of the user's attention are in many cases unappreciated or not even *unaffordable* by mobile users who are inherently confronted with more interaction and communication constraints during the move.

Another serious issue is user's privacy in ubiquitous computing environments. With the advance of sensor technology, ever more information can be captured from both the environment and the user. This might give more possibility to simplify user's interaction with computer systems, but also introduce a strong risk towards user's privacy intrusions. The exposure of fractional personal information occasionally slowly over time may still result in a substantial privacy loss after accumulation. Therefore it is becoming a critical issue in the ubiquitous computing field how to protect the user information acquired by the systems from being misused or disclosed to other undesired parties.

By identifying these two crucial issues, along with other important concerns, we claim today's ubiquitous computing environment should have the following characteristics:

Open and Standard: A ubiquitous computing system should be based on open data formats over standard communication protocols to achieve better scalability.

Low Demands on User Devices: To become widely available, a ubiquitous computing system should not demand too much of user's terminal devices, e.g., their computation and battery power etc. It should instead accommodate the most widely used/affordable off-the-shelf devices such as today's mobile phones and Personal Digital Assistants (PDAs).

Exploiting Infrastructure Support: It is of great significance to alleviate the resource-limited user terminal devices by migrating computation into the system infrastructure, exploiting the vast resources available on the local and remote computers.

Context-Awareness Support: It is impractical or even *impossible* for a mobile user to manually select services from a big amount and maintain the interactions with them. Hence a common support is needed to enable the system make decisions on *behalf* of the users based on information regarding their current situation (or *context*).

Privacy-Protected: Due to the risk of privacy leak, the major challenge is how to ensure the users, participating in a large-scale ubiquitous context-aware system, which is likely to cross multiple organizations and administrations, with an acceptable degree of control of their personal information.

To meet the characteristics listed above, we propose a person-centric system architecture which we believe suitable for a large-scale deployment. Within this architecture, we have addressed context processing and transportation difficulties with particular interest in user's privacy protection. We have also tackled the dynamic system association (also referred to as *bootstrapping*) problem with special concerns for mobile users. We have developed a set of software components and services, for the purpose of this paper, the toolkit is called *S*ervice *T*oolkit for *A*daptive *C*ontext-*A*ware *S*ystems – *STACS*, to simplify the development of context-aware services and the construction of ubiquitous computing systems.

In section 2, we first introduce our system architecture in general, and then give more details on how to enhance security and privacy protection by enabling pseudonym-based communication. In section 3, we elaborate our service toolkit – a set of software components and services, together with a prototype context aware call forwarding system to illustrate how STACS can be used to build a ubiquitous computing system in a simple manner. Then in section 4, we discuss some related work with comparisons. And finally in section 5, we draw some conclusions based on our experience with some words of the future work.

2 System Architecture

We designed a *person*-centric software architecture in our ACAS project [5], to provide *affordable* and *adaptive* support for facilitating mobile users' interactions with services in ubiquitous computing environment. In this architecture, we assume each user has a persistent digital representative, called *Personal Server* (PS). This Personal Server is connected to the Internet, and most likely resides on the user's *Home* network. Each user is equipped with a mobile device, an off-the-shelf mobile phone or PDA. As the users move around, across *different* ubiquitous computing environments, they will be associated (and *de*-associated) with those systems spontaneously and the relevant context information (location and available services) will be reported to their individual home Personal Servers.

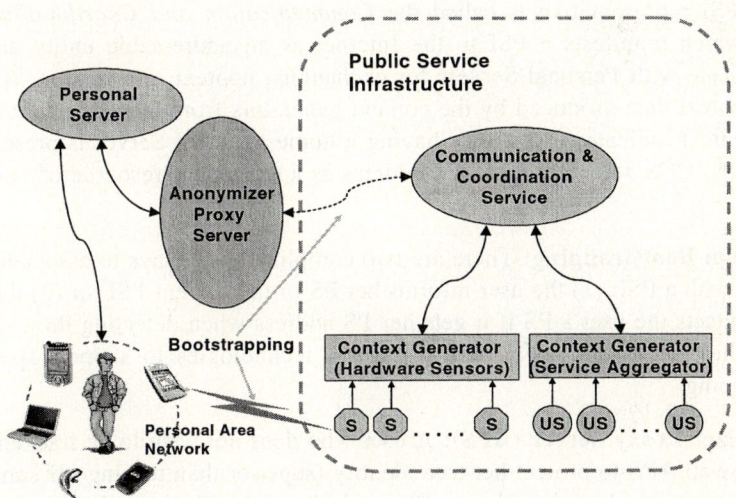

Fig. 1. System Architecture (Context Flow) US: end-User Service and S: Sensor

Two non-mutually exclusive interaction models are possible in such a setting: (1) users (with their devices) interact with the local systems directly or, (2) the PS acts as a remote intermediary between the user and local systems. We prefer the later model due to its simplicity (towards the user) that a single event to the PS is necessary for associating (bootstrapping) user's interaction with the local environment. Once the PS acknowledges the local system, it may conduct more interactions *on behalf of* the user, e.g., to employ the services in that location. Hence the computation and communication efforts required to maintain secure interaction can be reduced greatly from user's mobile devices. This complies well with what we claimed for a large-scale ubiquitous computing system: **Low demands on user devices** and **Exploiting infrastructure support**. In contrast, the former model is more subject to difficulties such as increased battery power drain and vulnerability to network-based attacks, although we do not deny the necessity and flexibility of direct interactions, especially when the PS is incapable (or unreachable) to handle the interaction with local resources. However, this method does not provide sufficient means to protect the user's privacy since the PS is exposed to the local infrastructure system, which results in the presence of our *Anonimizer Proxy Server*. As shown in Figure 1, our infrastructure consists of two sub-systems: the Public Service Infrastructure and the user's Personal Server.

Public Service Infrastructure (PSI): each PSI is an instance of a ubiquitous computing environment. A PSI may be confined to a room, a vehicle, an organization or any other natural or abstract boundary. Although PSIs are orthogonal to geo-location spaces it is likely to be conceptually convenient to create mappings between them. Within the PSI, context data is produced by *context generators* which are attached to *hardware* sensors that measure physical properties (e.g., temperature), or *software* sensors that measure computing properties (e.g., available services) of a PSI.

The PSI representative is called the *Communication and Coordination Service* (CCS), which manifests a PSI to the Internet as an addressable entity and would communicate with Personal Servers for exchanging context information. A CCS receives context data produced by the context generators from the same container PSI. When context indicates that a user having a home Personal Server is present within the PSI, the CCS adds the user's PS address as a temporary resource of context and services.

Interaction Bootstrapping: There are two complimentary ways to associate user interaction with a PSI: (1) the user informs her PS of the current PSI, or (2) the PSI actively contacts the user's PS if it gets her PS address when detecting the user. In our design, these steps are augmented by sensor technologies to support spontaneous bootstrapping.

Anonymizer Proxy Server (APS): A user who does not wish to be tracked by PSIs should use an APS to protect her true identity (superior than turning off some device capabilities). Instead of providing a PSI with the real address to the user's Personal Server (which is assumed to be persistent), the user offers a temporary token, previously negotiated with the APS. The token allows a PSI to communicate with the user's Personal Server via the APS until the token expires or is withdrawn. Disclosure of additional information is then determined by the user's Personal Server.

Personal Server (PS): As the persistent on-line representative of a user, the Personal Server contains a set of service components including the individual's context repository and a personal context manager (Figure 2). The context repository comprises different sets of information: the user device information, such as those in the Personal Area Network (PAN); personal contacts, calendar, and preferences (policies for sharing personal context); as well as the dynamic contextual data reported from the user's current PSI or personal devices.

The *Context Manager* handles the context reports and requests received through the *Personal Context Portal*, which gives an interface for external access to the user's context. It could also validate the reported context data before putting it into the *Context Repository*; or authenticate the context requester against the user's access policies. The external communication and interaction (with the owning user) requests arrive through the *Communication Portal* offering different access means, e.g., a set of Web Services over SOAP/HTTP. These communication requests are handled by different context-aware applications which can interface with the Personal Server system. Each application utilizes a relevant set of context information through the Context Manager to determine its application-specific logic and present the result to the user in different adaptive ways.

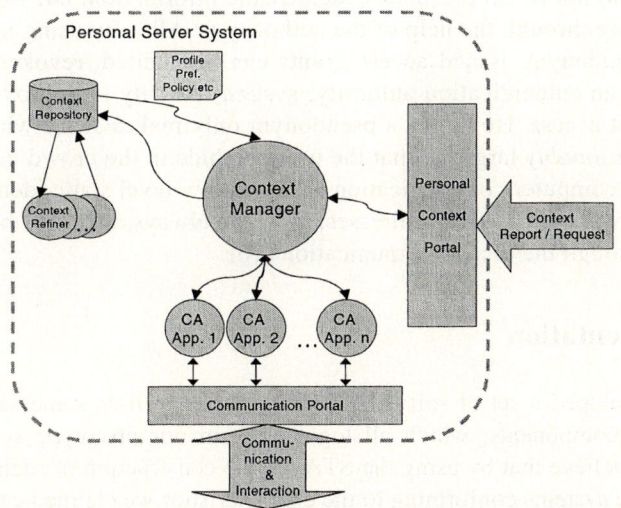

Fig. 2. Personal Server Inner Structure CA: Context-Aware

We also introduce *context refiners* which are used to produce abstract higher-level context information based on the data in the repository. A context refiner consists of a set of rules combined with a rule engine. The rules are compiled from documents, normally one set for each context-aware application. Whenever new context data arrive in the PS, the rules are applied by the reasoning engine to filter out irrelevant context information and insert new inferences. The result set derived is posted back into the context repository as available context data. For instance, a context refiner could generate an 'in-meeting' event based on the information of a user's current location and the nearby persons. This is implemented in our prototype (see Section 3.3).

2.1 Security and Privacy Protection

User preference and policy together with data encryption technologies have been commonly used to provide security and individual privacy protection in existing ubiquitous computing systems like [6, 7, 8, 9].

Different from those methods, we leverage the *pseudonym* communication mechanism (which was mentioned briefly with the introduction of Anonymizer Proxy Server (APS)), which grants access to users' context information without exposing their real identities. We thus propose a secure context exchange approach: A set of user pseudonyms (in form of some neutral tokens generated by randomization) are registered with an authentication authority in advance, and each pseudonym will be used as a subject reference to which a context requester (PSI or another user) will ask the authority for access. This request may trigger an authentication process and only the ones who pass the authentication will be granted to further contact the APS. In a simplified case, an APS can act as an authentication authority simultaneously since they are based on similar principles. By using pseudonyms, a context requester (PSI or other people) would not know and therefore could not retain personally identifiable information, but when needed can deal with abuse through the help of the authority or APS. By using multiple and replaceable pseudonym, issued access grants can be audited, revoked and blocked. Finally, with an authentication authority, system integrity is improved by preventing fraudulent access. However, a pseudonym only makes sense when the number of users is *reasonably* large, so that the user can hide in the crowd behind the APS. Also as most computer communication relies on low-level static identification such as IP and network card MAC addresses, a user is always at risk of being identified or tracked through the local communication [10].

3 Implementation

We have developed a set of software components as well as some services built on top of these components, which all together form our prototype service toolkit – STACS. We believe that by using the STACS, the construction of scalable ubiquitous context-aware systems conforming to the characteristics we claimed can be simplified and accelerated.

3.1 Communication Components

The context distributing in our architecture is based on a Subscribe/Notify/Publish mechanism. The Session Initiation Protocol (SIP) [11] is used as the underlying communication protocol, mainly because of the agile tolerance SIP defines for handling communication sessions over unreliable networks, as well as its openness and wide acceptance. Our implementation conforms to the SIP Presence Framework [12]: A SIP *Presence User Agent* (context producer) publishes sensed (low-level) context data to a SIP *Presence Agent* (context provider) which, after some processing of the data, will notify the SIP *Presence Watchers* (context consumers) who have registered interest about the generated (high-level) context results. These three SIP Presence entities have been implemented as the primary software communication components in our STACS. By making a diverse use of these SIP components (as *context sockets*), various system entities can be plugged into each other to construct a dynamic context information network.

The Context Manager in a user's home Personal Server system employs an internal Presence Watcher to subscribe and receive context information about a PSI through its Communication and Coordination Service (CCS), which is implemented based on a Presence Agent. This subscription goes through an APS which is also implemented as a Presence Agent. The Context Generators work as PUAs, publishing derived context information to the infrastructure CCS, and some of them (desired by some users' Context Manager) will be further delivered to their Personal Server systems as context notifications. The Context Manager in a Personal Server may also employ a Presence Agent for sharing personal context with PSIs or other users. There are other ways of using these components for delivering context data: e.g., a Personal Server may hire a Presence Watcher to subscribe to user's mobile devices for acquiring context data from a Presence Agent running on that device, or the user may start a PUA on the device to publish context to her Personal Server. The combination use of these components gives great flexibility for context acquisition, processing and distribution among entities across the network, or within the Personal Server and PSI systems. The superior scheme can be determined according to the requirements in concrete scenarios.

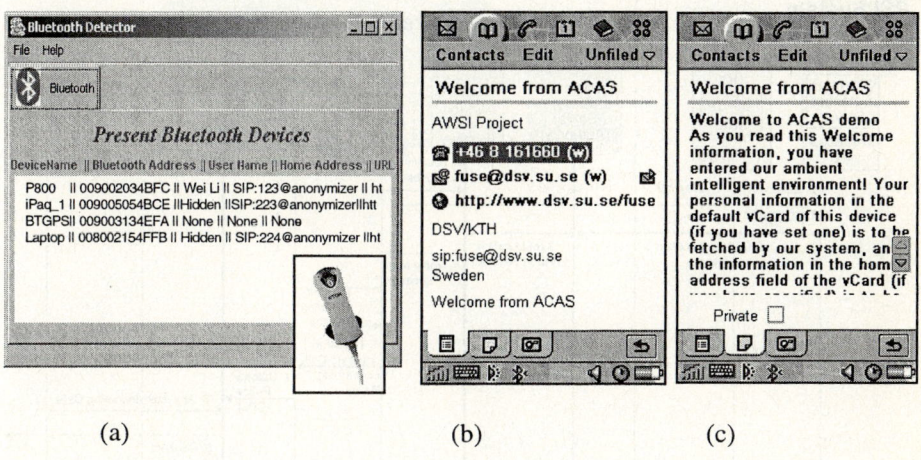

Fig. 3. (a) Bluetooth Detector using a TDK USB Bluetooth dongle. (b) The system entry addresses (sip:fuse@dsv.su.se) and (c) the Note part of a system Welcome vCard received by user's mobile phone (SonyEricsson P800).

3.3 Prototype System

A Context Aware Call Forwarding application based on our person-centric system architecture with the use of STACS has been developed as a prototype example. This application monitors a user's context changes, and will set the call forwarding when a meeting situation is determined depending on two facts: if the user is in a meeting room and not being alone. The detailed communication flow is illustrated in Figure 4.

A desktop PC in our meeting room, acting as an Infrastructure (PSI) server, runs a Bluetooth detection service to detect and retrieve a vCard file (step 2.) from a user's mobile device. It will then report the location (meeting room) to the pseudonym address (placed in the retrieved vCard) referred to as the user's Personal Server System. This context report will first arrive (3.(a)) at the corresponding Anonymous Proxy Server and then be forwarded (3.(b)) to the Context Manager (with an internal Context Repository) within the user's Personal Server system. The location report will be further delivered (step 4.) to the Meeting Monitor service (a context refiner) as a context notification, which will trigger it to infer (step 5.) the meeting status. Once approved, the Meeting Monitor will activate (step 6.) the Call Policy Generator to produce a call forwarding script (in Call Processing Language [15]) according to the user's preference, and finally upload it to the user's Personal Communication Server for enforcement (step 7.). The Personal Communication Server used is Vovida Vocal [16], an open-source SIP proxy server with support of call processing scripts. Xten X-Pro 1.0 [17] has been tested as SIP softphones on HP iPAQ 5550.

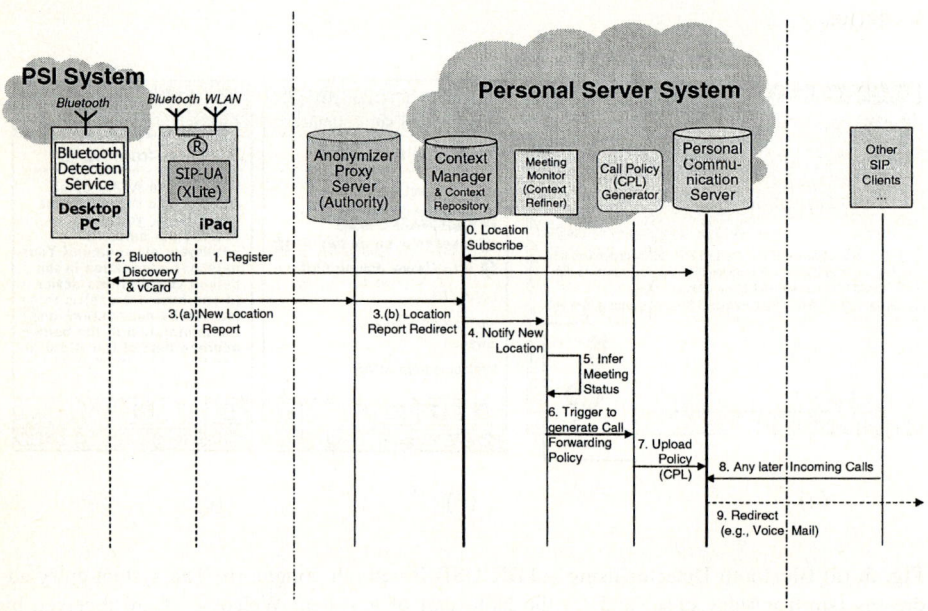

Fig. 4. Context-Aware Call Forwarding System

4 Related Work

Due to the large number of research works to date in ubiquitous and context-aware computing area, we will only address some of them which are most relevant to our work. K. El-khatib et al. [18] used Personal Agent (also implemented on SIP) to determine how to render mobile user's incoming calls in a ubiquitous computing environment with support for better performance and interaction means according to a user's profile and the available services. Stefan Beger and Henning Schulzrinne et al. have elaborated comprehensively in a recent paper [19] on how to construct ubiquitous computing system using SIP together with many other standard protocols. We agree with them that a global-scale ubiquitous computing system should be divided into different domains, and through the SIP servers in those domains, the user can utilize the rich resources in the visited domains. However, except for many similarities mainly because of the use of common technologies and standard protocols such as Bluetooth and SIP, there are clear differences to distinguish our work: firstly, we have described how the local infrastructure instead of user's mobile device, can deliver context data back to the home Personal Server system without exposing user's identifiable information; secondly we introduced a context refiner concept located at the user's home system to infer high-level context information. In additional there are many differences in design details, for example our Bluetooth detection service does not need any action by the user, while theirs needs the user's device to discover the Bluetooth access point to generate location information etc.

5 Conclusion

To achieve a practical large-scale ubiquitous computing system, we argue that lightweight and off-the-shelf mobile devices should be used by the user to interact with context-aware systems. Thus we employ the local infrastructure together with the home Personal Server system to do most of the work to support user's interaction. We have also proposed a solution for protecting the user's privacy based on an intermediary Anonymizer Proxy Server using pseudonym mechanism. We will further implement and evaluate our system to observe what and how much user will appreciate from the decreased interactions by using these intermediary proxy-based services in the infrastructure.

References

1. Mark Weiser: The Computer for the Twenty-First Century, Scientific American, pp. 94-10, September 1991
2. A. Fox, B. Johanson, P. Hanrahan, and T. Winograd: Integrating Information Appliances into an Interactive Workspace, IEEE Computer Graphics and Applications, May/June, 2000, pp. 54-65.
3. Patrik Werle, Fredrik Kilander, Martin Jonsson, Peter Lönnqvist, and Carl Gustaf Jansson: A ubiquitous service environment with active documents for teamwork support, Ubi-Comp2001, LNCS, pp. 139–155.

4. Manuel Román et al: Gaia: A Middleware Infrastructure to Enable Active Spaces, IEEE Pervasive Computing, pp. 74-83, Oct-Dec 2002.
5. http://psi.verkstad.net/ACAS/
6. Ginger Myles et al: Preserving Privacy in Environments with Location-Based Applications, IEEE Pervasive Computing, January-March 2003 (Vol. 2, No. 1) pp. 56-64.
7. W3C: A P3P Preference Exchange Language 1.0 (Appel 1.0), working draft, WorldWide Web Consortium, Apr. 2002, www.w3.org/TR/P3P-preferences
8. Xiaodong Jiang, Jason I. Hong, James A. Landay: Approximate Information Flows: Socially-Based Modeling of Privacy in Ubiquitous Computing, Proceedings of the 4th international conference on Ubiquitous Computing, p.176-193, 2002.
9. J. Cuellar, Joel Morris, and D. Mulligan: Geopriv requirements. Internet draft, Internet Engineering Task Force, March 2003. Work in progress.
10. A. Harter et al: The Anatomy of a Context-Aware Application, Proc. 5th Ann. Int'l Conf. Mobile Computing and Networking (Mobicom 99), ACM Press, New York, 1999, pp. 59-68.
11. J. Rosenberg, Henning Schulzrinne, G. Camarillo, A. R. Johnston, J. Peterson, R. Sparks, M. Handley, and E. Schooler: *SIP: session initiation protocol*. RFC 3261, Internet Engineering Task Force, June 2002.
12. Jonathan Rosenberg: A presence event package for the session initiation protocol (SIP), Internet draft, Internet Engineering Task Force, January 2003. Work in progress.
13. Internet Mail Consortium: vCard Specification, http://www.imc.org/pdi/
14. Bluetooth.org: Bluetooth Specification Volume 1, Core 1.1, February 2001
15. J. Lennox and Henning Schulzrinne: Call processing language framework and requirements. RFC 2824, Internet Engineering Task Force, May 2000.
16. http://www.vovida.org/
17. http://www.xten.com
18. K. El-Khatib, N. Hadibi, and G.v Bochmann: Support for Personal and Service Mobility in Ubiquitous Computing Environments, EuroPar 2003.
19. Stefan Berger, Henning Schulzrinne, Stylianos Sidiroglou, Xiaotao Wu: Ubiquitous computing using SIP, 13th International Workshop on Network and Operating Systems Support for Digital Audio and Video (NOSSDAV'2003)

Context-Awareness in Mobile Web Services[1]

Bo Han, Weijia Jia, Ji Shen, and Man-Ching Yuen

Department of Computer Engineering and Information Technology,
City University of Hong Kong, 83 Tat Chee Avenue, Kowloon, Hong Kong
Bo.Han@student.cityu.edu.hk

Abstract. Context-aware computing is a computing paradigm in which applications can take advantage of contextual information. Quality of network connection is a very important factor for mobile web services. However, the conditions of mobile networks may change frequently and dynamically. Thus, providing support for context-aware applications is especially important in mobile web services. Recently, a number of architectures supporting context-aware applications have been developed, but little attention is paid to the special requirements of mobile devices which particularly have many constraints. This paper discusses a client-proxy-server architecture that supports context-awareness by considering types of device, network and application characteristics. The contribution of this paper mainly lies in the division of labor between proxy and server. Application specific proxy is used to tailor the original resource based on the mobile user's context information. To prove the feasibility, a context-aware image management system is designed and realized.

1 Introduction

With the popularity of mobile devices, people can communicate easily at anytime anywhere. In current market, mobile devices are able to support different platforms. It is necessary to maintain a centralized storage with an easy access interface, so it can provide guaranteed QoS (Quality of Service) to any wired and wireless devices. To achieve this, we have proposed a web-based platform named AnyServer [10], which provides flexible services to different types of mobile (client) devices.

Context is any information that can be used to characterize the situation of an entity. An entity may be a person, place, or object that is considered relevant to the interaction between a user and an application [1]. Context-awareness means that one is able to use context information. A system is context-aware if it can extract, interpret and use context information and adapt its functionalities to the current context. The challenge for such systems lies in the complexity of collecting, representing, processing and using contextual data. The use of context information is especially important in mobile environment, where the quality of network connection provided to mobile users always changes frequently, dramatically, and without warning. Moreover, size and weight constraints on mobile devices limit the computing resources. Besides, battery life is also a nagging concern. As a consequence, mobile applications need to be capable of adapting to these changes to ensure that they could offer the best possible level of service to the user [12], while

---

[1] The work is supported by Research Grant Council (RGC) Hong Kong, SAR China, under grant nos.: CityU 1055/00E and CityU 1039/02E and CityU Strategic grant nos. 7001587.

context information is an indispensable component in these applications. In this paper, a client-proxy-server model is built based on our AnyServer platform for using contextual information. Our objectives are (1) to efficiently obtain and transmit contextual information, (2) to use context information to achieve better system performance and user experience, and (3) to implement software architecture for supporting the utilization of contextual information.

The rest of the paper is organized as follows. In Section 2, we introduce some related context-aware applications and their system models. Sections 3 and Section 4 describe our approach to integrate context-awareness in AnyServer platform and the software architecture. The implementation aspects are given in Section 5. We point out future directions and summarize major results in Section 6 and Section 7.

2 Related Work

This section introduces some context-aware mobile applications. The ParcTab system was developed at the Xerox Palo Alto Research Center [2]. The system was developed to experiment with context-awareness in an office environment. The ParcTab worked as a mobile personal digital office assistant. There were also several applications developed for context-aware experimentation. Cyberguide [3] is a computer system that provides information services to a tourist about his/her current location. For example, he/she can find directions, retrieve background information, and leave comments on the interactive map. Knowledge of the user's current location, as well as past locations, is used to provide the kind of service which comes from a real tour guide. More context-aware wireless mobile applications could be found in [6]. Most of the existing applications actually use only few context values, and the most commonly used ones are location, identity and time. The reason for this probably lies in the difficulty for computer systems to obtain and process context information.

Recently a number of architectures were proposed which provided support for context-aware applications in mobile scenarios. Thomas et al. [9] believed that applications had to adapt dynamically and transparently to the resources available at runtime. To achieve this goal, they extended the convention of the client-server model into a client-proxy-server model. Logically, the proxy hided the "mobile" client from the server, which thought that it communicated with a standard client. In [14], the authors identified requirements for a context-aware architecture including lightweight, extensibility and robustness. To meet these requirements, they proposed a three layers architecture which comprised an application layer, a management layer and an adaptor layer. However, all these three layers were located on a single device, so devices exchanged their context information in ad hoc manner and extra requirements were supposed on these mobile devices. Bruce et al. [8] presented a proxy based communication model in which filters executing on an intermediary host drop, delay or transform data moving between mobile device and fixed host. The design and capabilities of this proxy were presented, along with sample filters that addressed "real world" protocols. However, the limitation is that in order to facilitate proxy filtering, all traffic traveling to and from a particular mobile host must pass through a single gateway, the Proxy Server.

3 Context-Awareness in Any Server Platform

To make AnyServer Platform (as shown in Fig. 1) provide adaptive services based on user's context information, some additional sensors or programs are generally required to capture context information. Moreover, a common representation format for such information should be adopted, so different applications are able to use the same context information. Furthermore, AnyServer must intelligently process these information and deduce the meaning. Device-aware, network-aware and application-aware, these three factors are the most useful in AnyServer platform. The contextual data used in AnyServer include user device information (e.g. screen size, computational power and battery life), network conditions (e.g. network connectivity and transmission delay), and application type (e.g. real-time video/audio, or web browsing). Direct context information sharing is not needed in AnyServer platform.

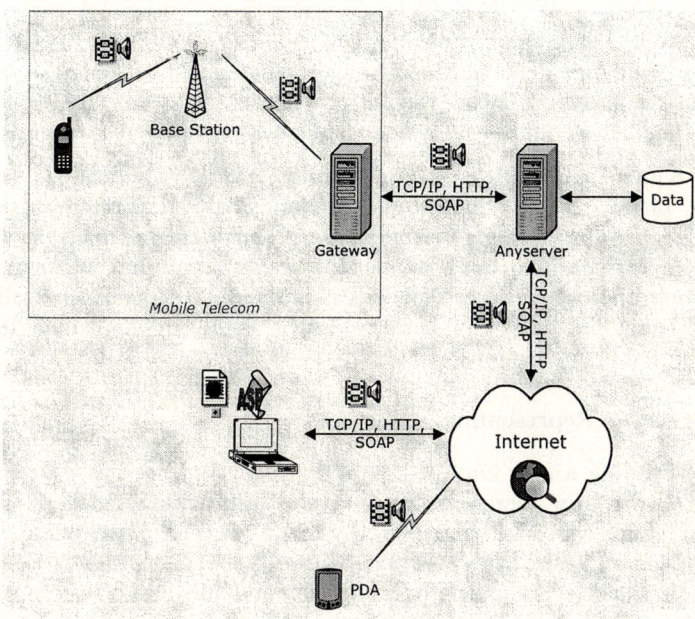

Fig. 1. Any Server Platform

3.1 Information Capturing

In this subsection we will answer the following three How-to questions:

How to Obtain Device Information? Device information retrieval can be based on Universal Plug and Play (UPnP). As depicted in [4], UPnP is designed to support zero-configuration, "invisible" networking, and automatic discovery for a breadth of device categories from a wide range of vendors. In UPnP the properties associated with the device are captured in an XML based device description document that the device would host or could be obtained from a specific URL. That is when the client sends request to AnyServer, a device description document or a specific URL accompanies with the request. The device information could be managed by AnyServer as part of user profile (to be detailed later) for future use. Based on the device information, some adaptive transmission could be carried out.

How to Capture Network Condition Information? In Any Server platform, round trip time (RTT) is chosen to measure the transmission delay. When measuring RTT several probing messages will be sent out and the following notations will be used to calculate more exact values:

RTT: the record stores the round trip time value of every probing.
EA_RTT: the record stores the exponential average of *RTT*.

Considering dynamic estimate of variability in estimating RTT, *EA_RTT* is averaged as:

$$EA\_RTT(n) = \beta * EA\_RTT(n-1) + (1-\beta) * RTT(n) \tag{1}$$

We would like to give the greater weight to more recent instance of RTT. Thus we use $\beta = 0.1$. If the mobile device has strong computational capability, it is responsible to conduct the probing. Otherwise the probing is the responsibility of the proxy. Network condition is runtime information and will be captured every now and then.

How to Know Application Related Information? Applications supported by AnyServer platform can be divided into two kinds based on timing consideration and tolerance of data loss. Timing consideration is important for many applications that are highly delay-sensitive but loss-tolerant, such as real-time interactive audio and video. Meanwhile, tolerance of data loss is important for many applications that are not loss-tolerant but delay non-sensitive, such as file downloading. In AnyServer platform, different applications will be served by different proxies.

3.2 Information Representing

Instead of creating a new kind of description language, we use eXtensible Markup Language (XML) for representing contextual information. XML is the universal format for structured data on the Web. And XML can place nearly any kind of structured data into a text file. Also the platform independent feature of XML easily tackles the obstacle that the heterogeneity of mobile devices brings to us. And using XML we can easily extend our contextual information to extra aspects. When a new family of mobile device is evolving, we can reuse the design for the old devices in an optimal way. As a result, XML is used in contextual information description. Fig. 2 is an example of device information depicted in XML.

4 Software Architecture

AnyServer must have the ability to change its behavior depending on the current context o f users. In this way, a lot of personalized applications can be built based on

```xml
<?xml version="1.0" encoding="big5"?>
  <Device>
    <Resolution>
      <Properties>
        <Property NAME = "Height"> 800 </Property>
        <Property NAME = "Width"> 600 </Property>
      </Properties>
    </Resolution>
    <Screen>
    … …
    </Screen>
    … …
  </Device>
```

Fig. 2. XML example of device information

Any Server platform. Some architectures [8][9][14] have been introduced in Section 2. Our approach is the extension of the conventional client-server model, which is also used in [9] but with quite different functionalities. In [9], the proxy shares the application logic of the "standard client" with mobile clients to adapt to dynamic wireless environment and to address the limitations of portable devices. But in our architecture the proxy relatively shares the logic of "standard server" with AnyServer. More details will be given below.

4.1 The Approach in Any Server Platform

Our client-proxy-server model naturally comes from the requirement of system scalability. To alleviate the load of AnyServer and provide the best level of service to mobile users, a proxy is introduced to the traditional client-server model. This model is primarily made up of three components listed below:

- Client: The mobile client can be PDA (Personal Digital Assistant), mobile phone or laptop. It can enjoy the attractive applications provided by AnyServer platform.
- Proxy: The user traffic would be redirected to the proxy if necessary. The proxy could also be used to probe the network condition of mobile users and tailor the original data for adaptive transmission.
- AnyServer: The server provides intelligent integrated services, such as short message services, image uploading and downloading, on-line audio and video. It is also the context information server that manages these information as user profiles.
- This communication among client, proxy and AnyServer is shown in Fig. 3. The detail is depicted as follows:

1. The client starts by transmitting the request to AnyServer.
2. Based on the application type, AnyServer responds to the client or redirects the user traffic to the application specific proxy.
3. AnyServer sends the original content and the related user profile to the proxy.
4. The client sends the same request to the proxy, maybe the network condition information gathered by the client is also sent to the proxy.
5. The proxy responds to the client with the tailored context based on the user's context information.

For the limited computational resource of mobile clients, an alternative way is that the proxy actively probes the network condition of the client. In other words, step 4 and step 5 could be replaced by unidirection traffic. However, it is also a tradeoff between the load of the client and the proxy. Moreover, the introduction of proxy can also tackle the puzzle of conflicts among adaptations of different independent applications, while this conflict is a common potential problem that can occur in a system utilizing separate adaptation mechanisms for different applications. In our approach, each proxy is designated to serve a specific application.

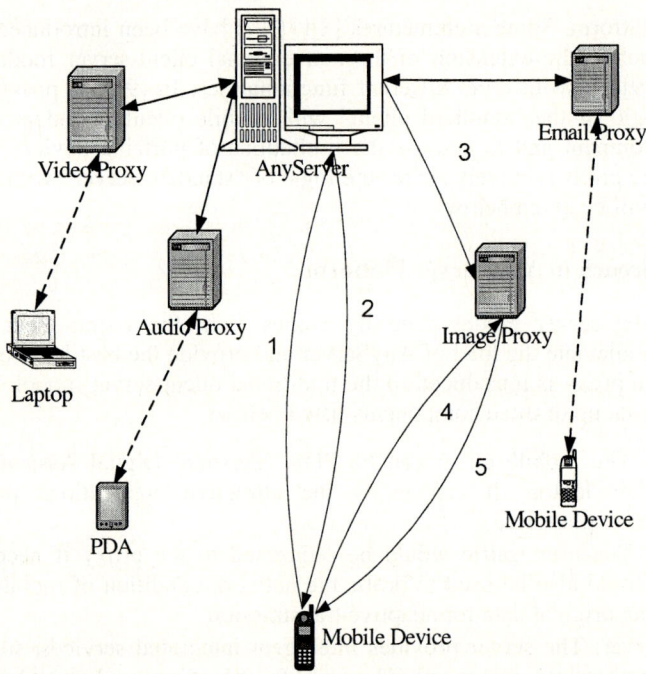

Fig. 3. Communication among client, proxy and AnyServer

4.2 User Profile Management

Customizations of mobile services to various kinds of terminals, user preferences, and varying network characteristics are attractive aspects in mobile computing. Different applications typically have different preferences for latency and integrity of data. Mobile clients need to express their preferences using voluntary profiles. The simplest way to manage the user profiles is to use a centralized context server, which provides contextual information to the applications. Schilit's mobile application customization system [5] contained dynamic environment servers, which managed a set of variable names and values representing an environment. It delivered updates to clients that had previously shown interest by subscribing to the server. The Rome system [7] developed at Stanford was based on the concept of a context trigger, which consisted of a condition and an action. This system allowed decentralized evaluation of triggers by embedding triggers in end devices. However, it did not allow context sharing.

In our architecture, user profile can be managed by AnyServer platform. Examples of user profiles are: (1) User Information Profile, such as name, identity, e-mail address; (2) Device Profile, such as device type, screen capability, memory size; (3) User Preference Profile, such as the preferred display pattern, favorite background color. User profile can also be used to save the session information. Due to the poor wireless network connection, mobile users may connect to AnyServer many times to complete one task. If the related information of each session can be properly saved, the remainder task of the previous session can be resumed in the new session when the mobile users reconnect to AnyServer platform.

5 Implementation Aspects

To prove the feasibility of our presented architecture, a prototype has been developed on AnyServer platform. Pocket PC is chosen as the first client device in the first phase. Pocket PC Client (PPCClient) is developed with Microsoft. NET Compact Framework. The reasons of this combination can be concluded as follows:

1. If a GSM (Global System for Mobile) modem is inserted into a Pocket PC, the Pocket PC will be a PDA phone that can connect to the Internet.
2. The popularity of smart phones with Microsoft .NET Compact Framework is gradually increasing, so that the programs developed for Pocket PC can be easily migrated to those smart phones.
3. More functionalities can be implemented for Pocket PC than mobile phones currently due to lack of processing power and computing capability of the mobile phones.

The PPCClient is running on Pocket PC 2003 operating system, which is the same as that for smart phones. Using Pocket PC, we can choose either IEEE 802.11 or GPRS (General Packet Radio Services) to connect to Internet. Besides, C# and embedded C++ are used as programming languages.

In Section 5.1, AnyServer, the proxy and mobile client are introduced. Uploading and downloading image are used as examples for context-awareness implementation in AnyServer platform, and they are presented in Section 5.2.

5.1 Any Server, Proxy and Mobile Client

The server side in AnyServer platform, called AnyServer, is a physical server linked with a database. For the mobile clients, the request can be sent through different network protocols in order to provide flexibility and adaptability features to those heterogeneous client devices that may only accept one specific network protocol. Database is a vital component in AnyServer platform, the user profile is managed by the database. General speaking, the proxy has two main functionalities: one is to gather the network condition information; the other is to tailor the original resource provided by AnyServer based on the mobile user's context information. In Fig. 4, the proxy may take following actions: (1) Degrade video stream, that is selectively dropping some structured data, such as frames in MPEG data stream; (2) Drop heuristically some unstructured data, such as quoted text from email body; (3) Compress data instead of dropping it. These actions are application specific. For different applications, the proxy would take different filter policies.

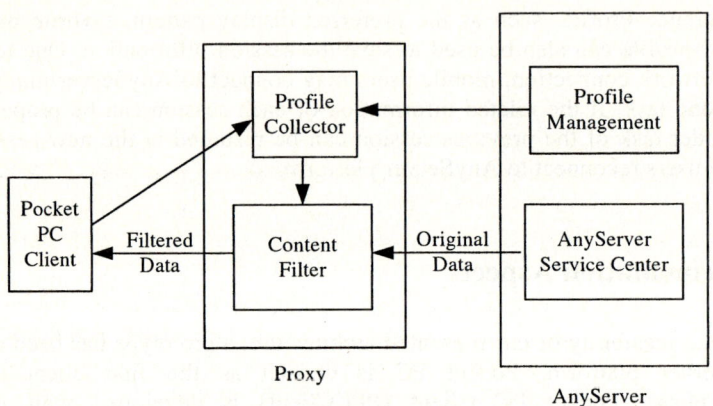

Fig. 4. Data Flow in Any Server platform

5.2 Example: Context-Aware Image Management System

In this section, an image management system is implemented based on AnyServer platform. When a user retrieves his multimedia materials on AnyServer platform, it involves two actions. Firstly, the user will get all the headers of the multimedia materials belong to him/her. As a result, the corresponding XML format headers information will be retrieved and therefore it reduces the network transmission volume. This feature will benefit those mobile users who use the service through mobile telecommunication networks which charge the user based on the volume of data sent. Then, when the user really selects a particular item to view, the actual content (maybe filtered) will be downloaded from the server side to the client device. Fig. 5 is an example of image filtering and shows both the original image and the filtered one.

Fig. 5. Image Filtering

6 Future Works

Context-aware computing has been proved to provide a reliable and flexible platform, especially for resource limited mobile devices. A natural thinking is to extend this platform to support end-to-end communications between terminals. Our future work is to build an end-to-end multimedia communication system. In this system, adaptive transmission technology will be realized and the proxy will support more functionalities.

Adaptive transmission is a hot research topic in wireless networking [11][12] where various mobile devices must be supported and the network condition fluctuates dynamically. The relative resource poverty of mobile elements as well as their lower trust and robustness argues for reliance on static servers. But the need to cope with unreliable and low-performance networks, as well as the need to be sensitive to power consumption argues for self-reliance [13]. Any viable approach to mobile computing must strike a balance between these competing concerns. As the circumstances of a mobile client change, it must react and dynamically reassign the responsibilities of the proxy. In other words, the proxy must be adaptive.

With the development of mobile devices, users may desire not only communicating with servers, but also direct end-to-end data transmission. However, it is not easy for various types of wireless devices of different capabilities to communicate with each other. For example, a mobile phone could not play the video file of some format transferred from a laptop. We are now building a system to support reliable, comfortable and compatible end-to-end communication. At the moment, the role of proxy in our AnyServer platform is just a content filter, and it tailors original content based on the context information of mobile users. In our future work, the proxy can evolve into a content server and AnyServer will take the role of an index server. Under this framework, more challenges will be faced, such as distributed content management and update, resource allocation between content servers.

7 Conclusion

Context-awareness offers many new possibilities for mobile web services. Motivated by this, we propose in this paper a client-proxy-server architecture trimmed to the special needs of mobile users. This architecture takes the special requirements of the mobile devices, such as limited capability of memory and power, into consideration. Moreover, it supports context-awareness by considering types of device, network and application. This architecture brings division of labor between proxy and server. Application specific proxy provided by AnyServer platform is used to tailor the original resource based on the mobile user's context information. We also implement our architecture and demonstrate its feasibility by a context-aware application.

References

1. Dey, A.K., Abowd, G.D.: Toward a better understanding of context and context-awareness. In Proceedings of the CHI 2000 Workshop on The What, Who, Where, When, and How of Context-Awareness (2000)
2. Want, R., Schilit, B.N., Adams, N.I., Gold, R., Pedersen, K., Goldberg, D., Ellis, J.R., Weiser, M.: An Overview of the PARCTAB Ubiquitous Computing Environment. IEEE Personal Communications 2(6) (1995) 28–43
3. Abowd, G.D., Atkeson, C.G., Hong, J., Long, S., Kooper, R., Pinkerton, M.: Cyberguide: A mobile context-aware tour guide. Baltzer/ACM Wireless Networks 3(5) (1997) 421–433
4. Universal Plug and Play Forum, Understanding Universal Plug and Play. http://www.upnp.org/download/UPNP_UnderstandingUPNP.doc (2000)
5. Schilit, B.N., Theimer, M.M., Welch, B.B.: Customizing mobile applications. In Proceedings of USENIX Mobile & Location-Independent Computing Symposium (1993) 129–138
6. Chen, G.L., Kotz, D.: A Survey of Context-Aware Mobile Computing Research. Technical Report TR2000-381, Dept. of Computer Science, Dartmouth College (2000)
7. Huang, A.C., Ling, B.C., Ponnekanti, S., Fox, A.: Pervasive computing: What is it good for? In Proceedings of the ACM International Workshop on Data Engineering for Wireless and Mobile Access (1999) 84–91
8. Zenel, B.: A general purpose proxy filtering mechanism applied to the mobile environment. Wireless Networks 5(5) (1999) 391–409
9. Kunz T., Black, J.P.: An architecture for adaptive mobile applications. In Proceedings of the 11th International Conference on Wireless Communications (1999) 27–38
10. AnyServer. http://anyserver.cityu.edu.hk
11. Nahrstedt, K., Xu, D.Y., Wichadakul, D., Li, B.C.: QoS-aware middleware for ubiquitous and heterogeneous environments. IEEE Communications Magazine 39(11) (2001) 140–148
12. Katz, R.H.: Adaptation and Mobility in Wireless Information Systems. IEEE Personal Communications 1(1) (1994) 6–17
13. Satyanarayanan, M.: Fundamental Challenges in Mobile Computing. In Proceedings of the fifteenth annual ACM symposium on Principles of distributed computing (1996)
14. Hofer, T., Schwinger, W., Pichler, M., Leonhartsberger, G., Altmann, J., Retschitzegger, W.: Context-Awareness on Mobile Devices - the Hydrogen Approach. In Proceeding of the 36th Annual Hawaii International Conference on System Sciences (2003)

CRL: A Context-Aware Request Language for Mobile Computing

Alvin T.S. Chan, Peter Y.H. Wong, and Siu-Nam Chuang

Department of Computing, The Hong Kong Polytechnic University,
Hung Hom, Kowloon, Hong Kong
{cstschan, csyhwong, cssiunam}@comp.polyu.edu.hk

Abstract. This paper introduces an XML-based generic Context Request Language (CRL), whose construction is part of a web services framework in the domain of mobile context sensing. The paper describes an implementation of the technique that is in accordance with the formal mathematical representational model, using first-order temporal language [6]. The language is an attempt to introduce intelligence into context-aware computing by defining context-sensing elements into logical entities. The use of first-order calculus in this language definition serving on web service technology allows users to utilize context aggregation and to embed user control in contextual information. By carrying out on-the-fly context inferences at the middleware level, we can achieve a complete separation of concerns between user application and context sensing. Moreover, the declaration of contextual knowledge based on situations and events within the predicate domain allows users to express changes in contextual information and to quantify these elements among times and durations.

1 Introduction

A review of current context-aware applications [1], [2], [3], [4] suggests some important areas in designing context-aware applications still need to be addressed. In particular, we believe there needs to have a methodology that allows applications to utilize and intelligently reason about contextual semantics and achieves a complete separation of concerns between applications and contextual information. Such methodology is invaluable in making intelligent context-aware applications. In this paper we describe our attempt to implement a generic rule-based language called Context Request Language (**CRL**) that provides a solution for an intelligent technique to context sensing. Our effort focuses on incorporating temporal predicate calculus, which is used in agent planning [5], with an XML-based syntax for extensibility [6]. This language allows applications to specify inferences for monitoring contextual information. The use of predicate calculus in this language definition enables users to utilize context aggregation and precise control over contextual information. CRL forms an integral part of our new Web Services architecture, CRL Framework, which supports dynamic reflective reconfiguration, asynchronous communication, and predefinition of context composition through CRL-rule definitions. In this architecture we have adopted a layered approach to relax applications participation in context sensing while still allowing applications to gain control over their context requirements. This layered approach enables the separation of mobile applications and their surrounding environment. Its key layers and components are briefly described as follows:

- **Application Layer.** This describes mobile and remote clients such as PDAs, Pocket PCs, etc. This layer is where local applications conceptually sit and they can either be dependent or independent of the context environment.
- **Web Service Requester.** This is a client agent that uses a stub that discovers context-aware web services through service registry [7]. It also hosts the Context Request Builder for constructing CRL-instances. A CRL-instance is an XML document that can either be a real-time construction of inference rules about one or more contexts or a specific definition for a particular common environment that is sometimes referred as smart space [4]. These CRL-instances are then enveloped as SOAP messages and transported to the particular web service provider that provides context inference via HTTP.
- **Web Service Provider.** This is a mediator between the context requester and the contextual environment. As it receives an SOAP call from a remote application, it uses the Context Request Parser (CRParser) that validates both the semantic and the syntax of the CRL-instances against the CRL schema, and transforms it into a set of CRL-based inference rules before passing it for context sensing.
- **Context Layer.** This layer contains a set of components that offers context-sensing ability at the hardware level and context inferences using the CRL-inference engine. There is also a CRL-rule repository that contains common CRL-rules so that applications are able to refer to pre-defined instances in the repository using key referencing.
- **CRL-Rule Repository.** It contains CRL-rules used in smart spaces [4], such as an intelligent meeting room or a smart vehicle. Since CRL-instances are written based on the CRL XML schema, these rules are intended to be organized in a tree structure. The process of updating a CRL-instance on the context layer is thus modeled as a specific algorithm in relation to the addition and deletion of nodes in a tree.
- **CRL-Management Module.** This module is part of the Context Layer and it contains a collection of components that provides administrative functionalities such as registering context sensors, providing meta-controlling, processing notification policy.

This architecture approach addresses the important issues of application participation. Using SOAP allows clients' applications to submit CRL-instances asynchronously, which means that applications are not engaged in listening to contextual feedback from the Context Layer. This is an important feature which provides the separation of concerns. The second feature is allowing user to embed predicate logic into each CRL-instance. This enhances the inference process at the context layer. The notion of reflective reconfiguration, inspired by the unique approach employed in MobiPADS [1], allows CRL-rules, either parsing through the CRL-inference engine or residing at the CRL-rule repository, to be updated at real time. Thus a highly transparent framework that does not compromise the preferred feature of separation of concerns is thus achieved.

2 An Overview of CRL

CRL plays an important role in CRL Framework, enabling it to reason about context. The complete CRL is a collection of language grammars expressed in XML for context request definition, context definition and meta-control definition. During the construction of the CRL framework we have developed a total of one core inference language (CRL-instance) and six other supplementary languages.

CRL-Instance is the core language component for constructing context request rules. Its grammar is in accordance with temporal predicate calculus [5]. We have chosen to use temporal predicate calculus as the underlying logic for the following reasons:

- It offers basic propositional logics that are common to all inference systems
- Unlike the logical systems used by other context-aware application, predicate calculus extends the premature propositional logics in that it provides a mechanism to express, and reason generically which is the key advantage of CRL.
- Furthermore, temporal predicate calculus extends predicate calculus to provide a mechanism for reasoning about contextual information and their changes at different time intervals.
- It offers a set of fundamental control structure to increase the flexibility and the complexity of CRL-instance.
- It provides a formal mathematical base in context-awareness which is novel in the field of context-aware mobile computing.

By encapsulating a proven logical system into a standard language such as XML, CRL-instance becomes a truly generic, platform independent and flexible rule-based language. The aims of CRL-instance are to provide the following:

- A language for users/applications to construct context request rules to reason about past and present contextual information.
- A set of control structures for users/applications to condition the enquiry of present and future contextual information.
- A mechanism for users or applications to specify sequential and concurrent sensing of context.
- A facility for users or applications to actively interact with the context environment using user-specific actions.
- A well-known syntax for users and applications to construct context inference rules.

Up to six supplementary languages are defined under the CRL_{sup} language group. They are system languages designed to enable meta-control, flexible user context feedback and error handling. They are briefly described as follows:

- **CRL-Rule.** This language is designed to cache common CRL rules in the CRL-rule repository used in smart spaces [4].
- **CRL-Feedback.** This is used to assist the transitioning of context information back to the Application Layer.
- **CRL-Control.** This is a meta-language that helps the monitoring and the manipulation of CRL rules that have already been defined. This enables meta-

adaptation and therefore ensures the CRL framework is working in a conflict-free environment.
- **CRL-CTree.** This defines the context entities that a CRL-inference engine monitors.
- **CRL-User.** This defines users' hierarchy, including user-specific information.
- **CRL-Error.** This language is used to assist the CRL framework in error handling during context inference process.

3 Context Layer

While constructing CRL, an implementation of the Context Layer was developed. We have chosen Java™ 2 Standard Edition (J2SE™) [8] to be the implementation language, as it is a platform-independent, object-oriented programming language that supports web service architecture and XML processing. In this implementation, Java™ API for XML Processing 1.2 specification is required to validate any CRL document against the CRL Schema. For developmental purposes, our current implementation of the Context Layer contains the CRL-Inference Engine, CRL Management Module and it also contains a fully functional CRParser which, although not part of the Context Layer, is a vital part of the framework to carry out logical inferences. Moreover, the Context Layer also contains a collection of Java objects that represent contextual sensors and a graphical user interface for purposes of emulation.

The current implementation of the Context Layer allows individual rules to be proxy-assigned. Proxy-assigned CRL-rules means that while the rules reside in the inference engine, each of them will have constructed a notification policy. These policies are a collection of triples, each containing the condition, the name of the rule that has set up the policy and the engine thread which contains the rules. Part of the implementation of the CRL Management Module is the CRL Proxy; once policies are constructed they are sent to the CRL (sensor) proxy which monitors them against any changes in the relevant contexts. There are two types of proxy-assignments – temporal notification and event notification. Two separate Java packages have been implemented to emulate our web services framework.

4 Conclusion

By introducing temporal predicate calculus into context sensing, an XML-based Context Request Language (CRL) has been defined to support the passing of a user's request to a Web Service framework for context retrieval. Importantly, CRL supports the novel concept of out-of-band separation of context co-ordination and rules from the application. The representation of CRL is intended to bridge the gap of linking the formalism in theoretical logics to the implementation of intelligent context-aware applications. A CRL definition forms an important step in bringing intelligent inferences into context sensing. We have employed the well-known temporal calculus which is an extension of predicate logics and applied this formal logic to a context-aware environment. In making CRL to be a truly flexible logical rule-based language, we have defined a collection of supplementary languages to assist in controlling context inferences and rules manipulation. We have also adopted a layered-approach to formulate the CRL Framework which leverages on web service's standard messaging protocol to achieve a complete separation of application concerns toward context environments. During the stage of designing the framework, we implemented the Context Layer, which includes a CRL-inference engine to test and demonstrate the usability of CRL.

Furthermore, the CRL, its supplementary languages and CRL-inference engine have been tested with RedPoint which is a locally developed local positioning system evolved from [9]. Their performance results have been measured and are subject to future publications.

Acknowledgement

This project is supported by the Hong Kong Polytechnic Central Research Grant GT-877.

References

1. Alvin T.S. Chan, Siu Nam Chuang, "MobiPADS: A Reflective Middleware for Context-Aware Mobile Computing", IEEE Transactions on Software Engineering, vol. 29, no. 12, Dec 2003, pp. 1072-1085.
2. A. K. Dey, "Providing Architectural Support for Building Context-Aware Applications", PhD thesis, College of Computing, Georgia Institute of Technology, December 2000.
3. H. Chen, S. Tolia, C. Sayers, T. Finin, A. Joshi, "Creating Context-Aware Software Agents", Article, First GSFC/JPL Workshop on Radical Agent Concepts, September 2001
4. H. Chen, T. Finin, A. Joshi, "An Intelligent Broker for Context-Aware Systems", InCollection, Adjunct Proceedings of Ubicomp 2003, October 2003.
5. E. Davis, "Representations of Commonsense Knowledge", Morgan Kaufmann Publishers, 1990
6. H.S. Thompson, D. Beech, M. Maloney, N. Mendelsohn, "XML Schema Part 1: Structures", May 2001, Available at http://www.w3.org/TR/xmlschema-1
7. F. Curbera, W. Nagy, S. Weerawarana, "Web Services: Why and How", OOPSLA 2001 Workshop on Object-Oriented Web Services, Florida, USA, 2001
8. Java™ 2 Platform, Standard Edition, Available at http://java.sun.com/j2se/
9. Alvin T.S. Chan, Hong Va Leong, Joseph Chan, Alan Hon, Larry Lau, Leo Li, "BluePoint: A Bluetooth-based Architecture for Location-Positioning Services", Proceedings of the ACM Symposium on Applied Computing (SAC2003), 9-12 March 2003, Florida, USA, pp. 990-995.

A Resource Reservation Protocol for Mobile Cellular Networks*

Ming Xu, Zhijiao Zhang, and Yingwen Chen

School of Computer, National University of Defense Technology,
Changsha, Hunan 410073, P.R. China
ywch_nudt@hotmail.com

Abstract. This paper proposed a protocol named RSVP-C, which aims at reserving resources for mobile cellular networks. In RSVP-C, both active and passive resource reservation routes could be established. We described the whole architecture and all the management principles. The messages format and reservation mechanism are illuminated in details. Simulation results based on a discrete-event simulation model are given as well. Compared with MRSVP, RSVP-C shows its better performance in most mobile cellular networks.

1 Introduction

With the rapid increase of mobile hosts and mobile application fields, people pay more and more attention to the Quality of Services (QoS), such as bandwidth and delay, and the methods to keep the protocol overhead to the minimum. IETF has proposed the IntServ model, which can provide QoS guarantees in data transmitting, especially multimedia real-time services in networks[1][2]. ReSource reserVation Protocol (RSVP) is a very important part of the IntServ framework, which gives real-time traffics hard QoS guarantees in network. RSVP sets up resource reservations for real-time applications and works with a routing protocol. In RSVP, the receiver initiates the reservations. From the beginning, the sender sends a *Path* message along the route determined by a routing protocol, and the receiver transmits a *Resv* message in reverse along the same path as the *Path* message was originally transmitted after it receives the *Path* message. Both the *Path* and *Resv* messages can be used to set up and maintain soft-states on the nodes along the reservation route. The soft-state mechanism maintains the group membership and route dynamically in a RSVP network. Further more, RSVP protocol supports tunneling, by which messages can be delivered through regions that do not support RSVP.

While QoS provisioning is an active topic in broadband IP networks such as the Internet, it is especially important in wireless networks, where hosts are untethered and mobile.

---

* The research is supported by the National Natural Science Foundation of China under Grand No.60073002.

2   Resource Reservation Problems in Mobile Networks

In mobile network, there is a well-known problem called handover (also called hand-off)[5], which does not occur in fixed network however. In order to get seamless handover, the system must reserve resources in advance. There are two types of resource reservations[10]. The reservation for a data flow on a link is called "active", just when packets of that flow are traveling over the link to a receiver, and called "passive" when resources are reserved for the flow on the link, but actual data packets of the flow are not being transmitted over the link.

Tazlukar et al. described a protocol named MRSVP[6], which is proposed to reserve resources for the mobile hosts. There are many problems in running this protocol. Firstly, MRSVP assumes that the multiple locations where the Mobile Host (MH) may possibly visit can be acquired before the MH starts moving, and this set of locations is defined as MSPEC. But in practice, it is very difficult to do that. If we reserve resources in any possible area where the MH will enter, there must be too many redundant resource reservation routes in the network. Secondly, the protocol depends on mobile IP[10] for routing. Mobile IP is not suitable in the condition that mobile host causes frequent handover; because route table at the home agent needs to be updated every time the handover occurs. When a mobile host moves frequently, other nodes sometimes can not get its current care-of-address because the home agent hasn't updated the route table in time, thus leads to the communication failure.

I. Mahadevan et al. proposed an architecture, which used a modified RSVP protocol to provide QoS support for mobile hosts[7]. But they validated their model only in a small network with a few base stations, and didn't put forward a complete resource reservation protocol.

In this article, we proposed a new protocol based on mobile cellular network named RSVP-C (Resource Reservation Protocol for Cellular System). We will describe the detailed structure and management principles, including network topology, process of the resource reservation, RSVP-C messages, merging different reservations and reservation switching issues, etc.

3   Resource Reservation Protocol for Cell System (RSVP-C)

Most mobile networks can be considered as the composite of two parts, the fixed stations and the mobile hosts. In cellular network systems, the whole area is divided into many regions. These small regions are called "cell", and each cell has a fixed Base Station (BS), which serves the region. BSs are always linked together and connected with upper layer nodes that can communication with the Internet. When a mobile host enters a region, the BS in the region serves it until it moves to another cell. When the mobile host enters a new cell, the BS in the new cell then supports it. Some other networks can be regarded as a cellular system, such as the satellite network, in which satellites play the role of the base stations.

Our research is based on the general cellular network. Cells represent the mobile part. The BS of the cell is linked to the fixed part by an edge router. All the BSs can provide resource reservation service for mobile hosts that enter the cell regions, and we define these region areas as Resource Reservation Domain (RRD). For example, the wireless network in a campus can be considered as a RRD. When a mobile host enters this region, the BSs in the RRD can provide resource reservation for the mobile host. When the mobile host moves out of the campus, the BSs in the campus won't provide resource reservation for it any more. Figure 1 illustrates a RRD.

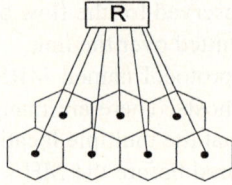

Fig. 1. A Resource Reservation Domain

In figure 1, each hexagon represents a cell, and the dot in the center of a cell represents a BS. A mobile host takes the BS of the current cell as a proxy to communicate with other fixed or mobile nodes in the network. The rectangle R represents an edge router, which can communicate with every BS in the RRD and is linked with other routers in upper network. The "node" here refers to the edge router, BS, or the mobile host. When a mobile host requests an application, the BS in its current cell reserves resources on the wireless interface.

In the cellular system, each cell has a constant maximal number of neighbors (6 neighbors), so it only needs to reserve resources for a mobile host in its neighboring cells without forecasting its movement. There are two ways to set up a reservation route: extending an existing rout, or establishing a new route from the edge router to the neighboring BS. We can choose the first one to reduce the overload of the reservation process.

Our research is based on the following hypotheses:

- All nodes in the network support RSVP-C. They can support both active and passive reservation;
- Each BS doesn't need to know the topology of the whole network. But it should know all its neighboring cells, so that they can communicate with each other directly;
- Each BS can acquire the QoS parameters of the link such as latency, bandwidth, etc.

3.1 Process of Resource Reservation When the Receiver Is a Mobile Host

As is mentioned above, a mobile network always consists of the fixed part and the mobile part. The fixed part conducts the active resource reservation process, which is the same as that in RSVP. We only need to discuss the passive reservation process in the mobile part of the network.

Process of Active Reservation. Like in RSVP, reservation is receiver-initiated in RSVP-C. After a receiver joins a multicast group by using the Internet Group Membership Protocol (IGMP), the sender starts sending a *Path* message to it. The receiver will send a *Resv* message along the reserve route as soon as it receives the *Path* message. A *Resv* message contains a FLOWSPEC object, which has two sets of parameters, i.e. RSPEC and TSPEC. They describe the desired QoS parameters and the traffic characteristics of the data flow. All the *Path* and *Resv* messages are delivered hop by hop. Each node receives a *Resv* message will perform admission control to decide whether it will reserve resources for the receiver or not.

There are two types of admission control:

1) *Admission control on wireless interface.* The BS decides whether the bandwidth on the wireless link is enough for a receiver's request. The BS will send a *ResvErr* message to the receiver if there isn't enough bandwidth.

2) *Admission control on fixed nodes.* The BS supporting the receiver acts as a proxy, and forwards a *Resv* message for active reservation upstream. Each node receiving the message will perform admission control and policy control. If the request fails at some node, the node will return a *ResvErr* message. As the condition of fixed links is much better than the wireless ones, the reservation failure probability of the fixed links is much less than that of the wireless links.

The sender periodically sends *Path* messages downstream, and the receiver periodically sends *Resv* messages upstream. All these messages maintain a soft-state on the nodes that have received them. Each RSVP message carries a SESSION object. The SESSION object contains the destination IP address of the flow, the protocol ID and the destination port number, which identifies a unique reservation from others. When the mobile host doesn't need the resources to be reserved any more, the sender or the receiver can deliver *TearDown* messages to clean up the soft-state. Figure 2 can illustrate all the process of the active reservation.

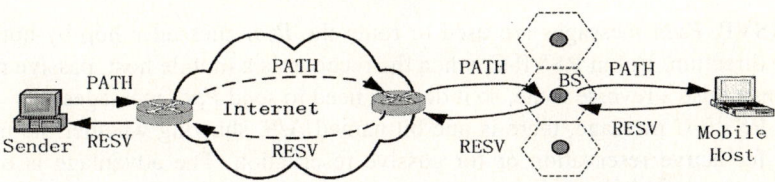

Fig. 2. Process of active reservation when the receiver is a mobile host

Process of Passive Reservation. In mobile network, the key issue of resource reservation is to make a passive reservation, which is very important for seamless handover. In RSVP-C, the passive reservation route changes while the mobile host migrates. We needn't forecast where the mobile host will move. The number of neighbors of a given cell is not more than six, so there are only a few passive reservation routes need to set up. There are two ways to extend passive reservation routes: to create passive routes between the BS supporting the mobile host and its neighboring BSs, as shown in Figure 3(a), or to create passive routes from the edge router to the neighboring BSs, as shown in Figure 3(b).

New mechanisms are needed to implement passive reservation. When the mobile receiver moves, the network chooses the current BS or the edge router as its proxy to initiate the passive reservation. The proxy sends reverse reservation request (*ReverseResv*) messages to the neighboring BS. "Reverse" here means the message is sent from the proxy to the neighbors. Its direction is the same as that of the possible dataflow in the future, and is opposite to that of the active *Resv* messages. The *ReverseResv* message also carries a FLOWSPEC object, which specifies the desired QoS parameters and the traffic characteristics. Figure 3(a) shows the BS, which supports the mobile receiver sends the *ReverseResv* message to a neighboring BS. Figure 3(b) shows the *ReverseResv* message is sent from the edge router. The neighboring BS decides whether it can provide resource reservation service for the mobile host. If admission control shows it can provide the service, the neighboring BS will send back a *Confirm* message, or it will return a *ReverseResvErr* message to reject the request.

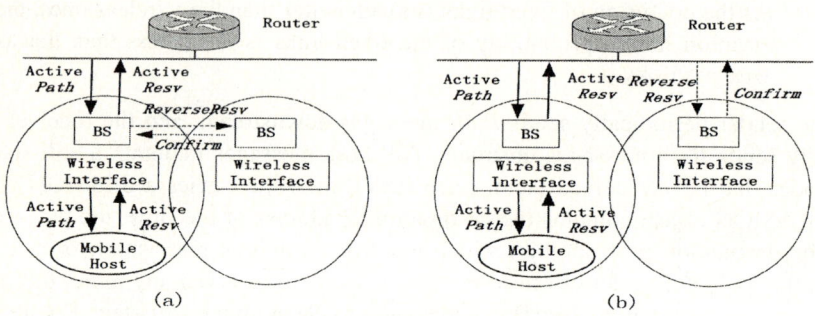

Fig. 3. Process of passive reservation when the receiver is a mobile host

In RSVP, *Path* messages are used to route the *Resv* messages hop-by-hop in the reverse direction. But in RSVP-C, when the receiver is a mobile host, passive reservation doesn't use a reverse route, so it doesn't need to send *Path* messages.

In a RSVP-C message, there is one bit named A/P, showing whether the message is sent for active reservation or for passive reservation. The advantage is obvious: resources reserved for a passive reservation can be used for some other data flows such as best-effort data flows, which prevents resources from being wasted.

Tear Down of Resource Reservation. In active reservation, if a mobile host doesn't need the resources to be reserved any more, or there are some other problems cause the reservation interrupted, *ResvTeardown* messages can be used. The sender or the receiver can initiate it to remove the reservation soft-state without waiting until lifetime expiration.

In passive reservation, when the mobile receiver migrates to a new region, the passive reservation for it in this new cell will be changed into an active one, and the passive reservations in the old neighboring cells won't be used. Then the BS in the old cell will send *ResvTeardown* messages to the neighbors to cleanup those unused passive soft-states.

If the receiver is a mobile host, the messages used for passive reservation are shown in Table 1.

Table 1. RSVP-C messages format

Message	Description
ReverseResv	The proxy of the mobile receiver sends this message to the neighboring cells to request passive reservation. The message contains a FLOWSPEC object describing the traffic characterizes and desired QoS parameters.
Confirm	The BS in the neighboring cell sends this message to the proxy if it can provide resources for passive reservation.
ReverseResvErr	The BS in the neighboring cell sends this message to reject a passive reservation request.
ResvTeardown	After the mobile receiver migrates to a new cell, the old BS will send this message to cleanup passive soft-states in the neighboring cells.

3.2 Process of Resource Reservation When the Sender Is a Mobile Host

In RSVP-C, when the sender is a mobile host the process of active resource reservation is similar with that in RSVP. The proxy of the mobile sender sends *Path* messages downstream, and the receiver sends *Resv* messages upstream. These messages maintain soft-states on the nodes along the reservation route. Besides being reserved on the fixed nodes, resources need to be reserved on the wireless interfaces as well.

When the active reservation finishes, passive reservations are still needed for the mobile sender. The BSs in neighboring cells perform admission control, and passive *Path* messages will be sent to the proxy of the sender.

For passive reservation, if a node does not have enough resources, it will send a *ResvErr* message to the proxy of the mobile sender to reject the request. The proxy receiving the passive *ResvErr* message knows that the passive reservation has failed.

3.3 Merging Reservations

Generally speaking, there are two kinds of reservation merging: (1) merging active and passive reservations of the same sender and receiver; (2) merging reservations for different applications.

There are both active and passive reservations for a mobile host, and one mobile host may request several passive reservations simultaneously. These passive and active reservations would be merged at the proxy, while the active reservations have higher prior than the passive ones. If the proxy receives a passive *Path* message from a mobile sender while it already has had an active *Path* soft-state, it won't create a new soft-state. At the same time it won't deliver the passive message downstream either.

How to merge different reservation requests is defined in the Integrated Services Specification[8]. The final FLOWSPEC after merging should be the largest of all the FLOWSPECs of the reservation requests.

There are three cases for merging reservations in mobile cellular network:

1) Merging several active reservations

This follows the definition in the Integrated Service Model.

2) Merging passive reservations of different sessions.

The final FLOWSPEC should be the largest of all the FLOWSPECs of the reservation requests, and the soft-state must be defined as a passive state. So the resources reserved can be used by other applications temporarily, especially those with best-effort requirements.

3) Merging both active and passive reservations.

In this case, a simple method is to merge reservations directly, and mark the merged reservation with an active reservation. When the resources reserved for passive reservations are larger than those for active reservations, this method may lead to resource waste. An alternative method is to calculate the FLOWSPEC of the active reservations and passive reservations separately. Hence redundant resources of passive reservations can be used by other applications before they change into an active one.

3.4 Reservation Switching

When a mobile host has an active data flow migrates to a new location, the active reservation at its previous location may be turned into a passive reservation and the passive reservation in the new location will be turned into an active reservation. When to switch between active and passive reservations must be defined in handover protocol. There are four approaches to support seamless handover[5]: redundant service, new domain service, old domain service and interrupted service. The redundant service will not interrupt the transactions performed on a mobile host, and can maintain QoS guarantees at a good level. But in nodes with redundant service will receive redundant packet, and duplicate data must be filtered out.

4 Simulation Analysis

We established a discrete-event simulation model with Markov chains to evaluate the overhead of RSVP-C protocol, which shows the blocking probability in wireless link in different conditions.

In our experiment, we suppose there are three kinds of reservation requests, i.e. Class A, Class B and Class C, which demands 10%, 20% and 50% of the whole bandwidth on a wireless link respectively, and the numbers of these reservation requests are equal. Our research focuses on a RRD, where the mobile host can move to any cell region. The MSPEC in MRSVP is the whole RRD.

Figure 4 shows the relationship between the data flow intensity and the reservation blocking probability. When the data flow intensity increases, the blocking probability increases as well. This is because when the data flow intensity becomes larger, there are more data flows to request reservations and occupy the bandwidth for more time. So the latter reservation requests are more likely to be rejected.

Figure 5 shows influence of the handover rate on the blocking probability of data flows. The blocking probability also becomes larger as the handover rate increases. This is because when a mobile host migrates to the new cell, it will request active reservation in the new cell and passive reservations in the neighboring cells.

Figure 6 and figure 7 show a comparison between RSVP-C and MRSVP when the data flow intensity is 0.15 and the handover rate is 0.5. Figure 6 indicates the difference among Class A, Class B and Class C data flows, and figure 7 shows the all the blocking probability in RSVP-C and MRSVP.

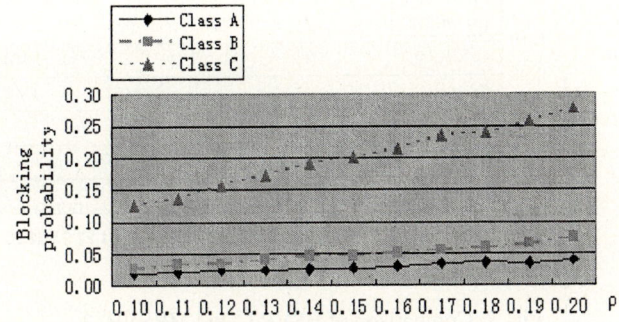

Fig. 4. Influence of Data Flow Intensity

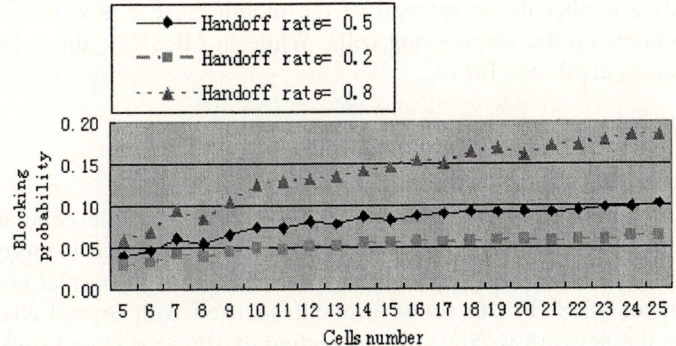

Fig. 5. Influence of handover rate

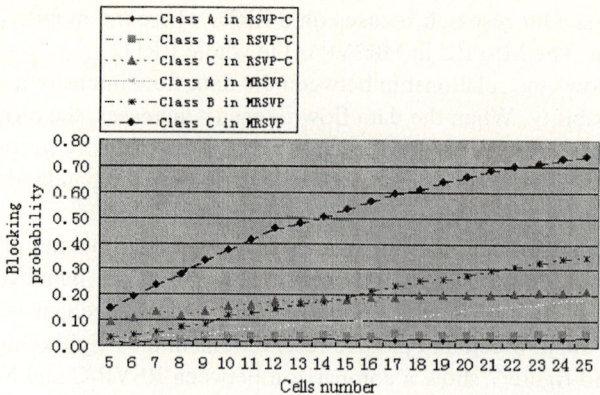

Fig. 6. Comparison of RSVP-C and MRSVP –1

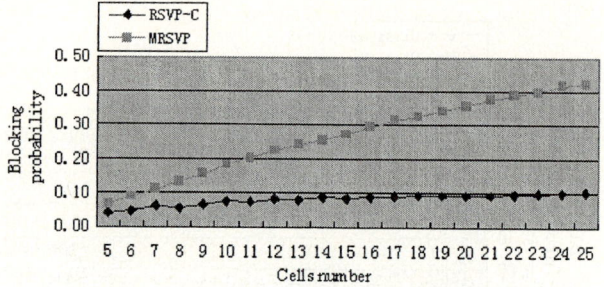

Fig. 7. Comparison of RSVP-C and MRSVP -2

We can conclude from figure 6 and figure7 that when a RRD contains a certain number of cells, RSVP-C performs much better than MRSVP does. That is because it doesn't need to predict the movement of the mobile host in RSVP-C, and it only reserves resources on the neighboring cells. While in MRSVP, the network should reserve resources in all the MSPEC.

5 Conclusion

This paper proposed a resource reservation protocol for mobile cellular networks. RSVP-C protocol supports both active and passive reservations. It is more suitable to reserve resources for real-time applications for mobile hosts in cellular networks than existing protocol MRSVP. Our protocol manifests itself with several characteristics: (1) decreases the network overhead by extending existing resource reservation route rather than setting up a new one; (2) defines reverse reservation messages, which simplify the implementation of the resource reservation; (3) results in a good performance without predicting the movement of a mobile host.

We are devoted to further studies on the following aspects:
1) Making further improvement of the protocol according to different mobile network architecture.
2) Setting up more precise mathematics model to analysis the performance of the resource reservation protocol.
3) Doing experiments in a real mobile cellular network, and measuring the metrics of the QoS guarantees in real data service.

References

1. R. Braden, D. Clark, S. Shenker. Integrated Service in the Internet Architecture: An Overview. RFC 1633, June 1994.
2. J. Wroclawski. The Use of RSVP with IETF Integrated Services. RFC 2210, September 1997.
3. L. Zhang. Resource Reservation Protocol - Version 1 Functional Specification. RFC 2205, September 1997.
4. Resource Reservation Protocol. Cisco Internetworking Technologies Overview Handbook, Chapter 43. June 1999.
5. M. Endler, V. Nagamuta. General Approaches for Implementing Seamless Handover. ACM 1-58113-511-4/02/0010, 2002.
6. A. K. Talukdar, B. R. Badrinath, A. Acharya. MRSVP: A Resource Reservation Protocol for an Integrated Services Network with Mobile Hosts. The Journal of Wireless Networks, vol. 7, no.1, 2001.
7. Mahadevan, K. M. Sivalingam. An Experimental Architecture for providing QoS guarantees in Mobile Networks using RSVP. In Proc. IEEE PIMRC '98, (Boston, MA), Sept. 1998.
8. J. Wroclawski. The Use of RSVP with IETF Integrated Services. RFC 2210, September 1997.
9. A. K. Talukdar, B. R. Badrinath, A. Acharya. On Accommodating Mobile Hosts in an Integrated Services Packet Network, In Proc. IEEE INFOCOM'97, April 1997.
10. J. D. Solomon, Mobile IP, The Internet Unplugged, Prentice Hall PTR, 1998.
11. T. Li, Y. Rekhter. A Provider Architecture for Differentiated Services and Traffic Engineering. RFC 2403, October 1998.
12. D. Awduche, J. Malcolm, J. Agogbua, M. O'Dell, J. McManus. Requirements for Traffic Engineering Over MPLS. RFC 2702, September 1999.
13. M. Seaman, A. Smith, E. Crawley, J. Wroclawski. Integrated Service Mappings on IEEE 802 Networks. RFC 2815, May 2000.
14. S. Herzog. Signaled Preemption Priority Policy Element. RFC 2751, January 2000.
15. L. Zhang. Resource Reservation Protocol - Version 1 Functional Specification. RFC 2205, September 1997.
16. D. Chalmers, M. Slornan. QoS and Context Awareness for Mobile Computing. Imperial College, London SW72BZ, U.K, the Journal of Lecture Notes in Computer Science, Vol.1707, 1999.
17. R. Yavatkar, D. Pendarakis, R. Guerin. A Framework for Policy-based Admission Control. RFC 2753, January 2000.
18. G. L. Chen, David Kotz. A Survey of Context-Aware Mobile Computing Research. Dept. of Computer Science, Dartmouth College, November 2000.
19. M. Oliveira, E. Monteiro. An Overview of QoS Service Routing Issues. In Proceedings the 5th World Multiconference on Systemics, Cybernetics and Informatics (SCI 2001), July 2001.

Using the Linking Model to Understand the Performance of DHT Routing Algorithms[1]

Futai Zou, Shudong Cheng, Fanyuan Ma, Liang Zhang, and Junjun Tang

Department of Computer Science and Engineering Shanghai Jiao Tong University,
200030 Shanghai, China
zoufutai@cs.sjtu.edu.cn

Abstract. The various proposed DHT routing algorithms embody different underlying routing geometries such as ring, hypercube etc. The routing performance depends on the geometry. In this paper, we anatomize the construction of the geometry and propose the linking model for the geometry to understand the performance of DHT routing algorithms. The effect of the different types of links on the performance is analyzed. Especially, randomized long link is considered as a new approach to optimize the performance. Our experiments show that the routing performance is greatly affected by the links, and the performance of CAN is improved with additional links.

1 Introduction

We introduce "link" to capture a dynamic geometry for the structured P2P system. The link is the relationship among nodes. In fact, it is similar to the edge among nodes in the geometry. However, it is very different the edge from the dynamic characteristic. The link can be dynamically adjusted according to the change of node's neighbors and it reflects how well the node senses the system. It may be unidirectional or bidirectional.

Though it depends on the requirements for all kinds of DHT, in essence, there exist two types of links in the geometry: *short link* and *long link*. Short links are the link from a node to its nearest nodes. Short links maintain the basic connectivity of the geometry so that a request can be routed to any node in the P2P system. Long links are the link from a node to the long distant nodes in the node space. Furthermore, short links may be divided into two types: *basic short links* and *redundant short links*. Basic short links are these linked edges in the minimal connected geometric graph, where the weight of the edge is the hops between two vertexes. Therefore the basic short links are these links between the node and its adjacent nodes with only one hop in the ID space. Similarly, redundant short links are these links sequentially following the basic short link. That means they are these links between the node and its adjacent nodes with over one hop. Redundant short links can assist the connectivity of this geometric in the face of node failure.

---

[1] Research described in this paper is supported by The Science & Technology Committee of Shanghai Municipality Key Technologies R&D Project Grant 03dz15027 and by The Science & Technology Committee of Shanghai Municipality Key Project Grant 025115032.

DHT systems can forward a request only using its basic short links; however, it is usually inefficient and unreliable. Therefore, to design a DHT system, one should add additional links to the *nude* DHT geometry so as to enhance the system performance. As mentioned above, redundant short links can enhance the connectivity of the geometry that improves the fault tolerance and long links can shorten the diameter of the geometry that reduces the average path length. Experiments have showed the linking model is an efficient approach to understand and analyze DHT routing algorithms.

2 Links Analysis and Construction

In this section, links analysis and construction are given. We use CAN [1] as a demonstrated application.

2.1 Links Analysis

We use links to capture the dynamic relationship of nodes in structured P2P systems. These links form a structured geometry. A well-organized and connected geometry in the structured P2P systems is the radical difference from the disordered geometry in the unstructured P2P systems. According to the different function in the geometry, we distinguish three types of links: the basic short link, the redundant short link, the long link. Although each kind of link has its function, the basic short link is the radical link of the geometry and is decided inherently by the geometry. Hence we emphasize on how the long links and the redundant short links impact the routing performance of P2P systems. We analyses the two metrics, average path length and resilience.

2.1.1 Average Path Length
The average path length is the average hops between every pair of nodes. It identifies how quickly a request is forward to the destination. The long link is an efficient way to improve the average path length. Chord [2] adds long links to the basic geometry to get an optimizing average path length. CAN don't have long links in that it gets a longer average path length. The methods to add long links are diversified and have still more widely space to be explored. However it needs the tradeoff between the number of links and the maintenance overheads of links. Randomizing techniques have a great help in the tradeoff [3] [4].

2.1.2 Resilience
Resilience measures the extent to which DHT can route around trouble even without the aid of recovery mechanisms that fix trouble. The basic short link is inner structure of the DHT geometry and redundant short links provide the chance to enhance its connectivity. The connectivity embodies the routing resilience to node failure. The lack of redundant short links would be less resilience, which will be frail for node failure or spend a long path to be rewound. Resilience is an important aspect of P2P systems. As for a special geometry without redundant short links, it is suggested to add redundant short links to improve the resilience.

2.2 Links Construction

The performance of CAN would be improved with additional links. In this section, we show the construction of additional links.

2.2.1 Additional Randomized Long Links

For a d-dimensional CAN, each node can establish k additional long links, where k is a constant. One node chooses a node as its long link with the probability inverse proportion to the distance between two nodes according to Kleinberg's construction [3]. Although Kleinberg's model considers all long links as being generated initially, at random, we invoke the "Principle of Deferred Decision"-a common mechanism for analyzing randomized algorithms [4] and assume that the long links of a node s are generated only when the message first reaches t. We know the probability that s chooses t as long link is $p_{s,t} = (1/c) \times \frac{1}{\| s-t \|}$, where $c = \sum_{s \neq t} \frac{1}{\| s-t \|}$ is a normalizing constant. The constructed long link is called randomized long link.

2.2.2 Additional Redundant Short Links

Redundant short links can be introduced into traditional CAN to provide improved resilience performance. Redundant short links can make routing continue to the destination even if the basic short links haven't been recovered from the failure. As a node N, its redundant short links are added by establishing links to its adjacent nodes in the hypercube with hops 2, 3, 4 etc.

Fig. 1 shows an illustration of additional links in CAN.

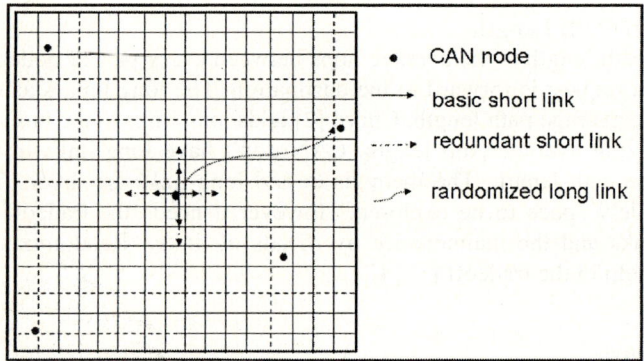

Fig. 1. An illustration of additional links in CAN with nodes distributed in a 16*16 lattice

3 Experiments

In this section, we present results from our simulation experiments. How the links impact on DHT routing performance are analyzed by adding different types of links into CAN. Our simulator is developed on the basis of [5].

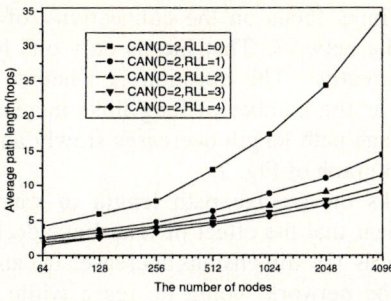 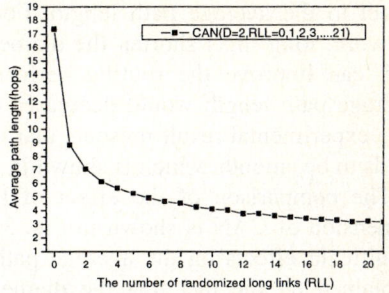

Fig. 2. The effect of randomized long links on fixed dimensional CAN. Left: Average path length for varying numbers of randomized long links (RLL) in the networks ranging from 64 to 4096 nodes in size. Right: Average path length for varying numbers of randomized long links (RLL) ranging from 0 to 21 in the network with 1024 nodes

3.1 The Effect of Randomized Long Links

Randomized long links are added to CAN according to Section 2.2.1. Due to the different dimensions for the construction of CAN, randomized long links should consider the effect of the dimension. Simply the effect is divided into two cases: fixed dimension and varying dimension. The experimental results are presented respectively in Fig. 2 and Fig. 3.

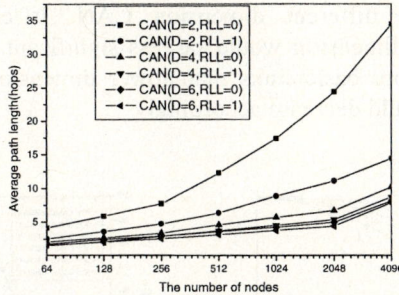

Fig. 3. The effect of randomized long links (RLL) on varying dimensional CAN. The comparison of average path length between 0 and 1 randomized long link (RLL) to different dimensional CAN in the networks ranging from 64 to 4096 nodes in size

As shown in the left graph of Fig. 2, average path length gradually decreases with the increasing long links. Obviously, with the number of nodes increases, long links decrease the average path length more significantly. It can be explained that long links have more widely space to shorten the diameter of the network as the number of nodes increases. Moreover, average path length dramatically decreases with the first long link. The reason is that the long link is very different from the short link on the effect to the average path length. For short links focus on the connectivity of the network, long links shorten the diameter of the network. That means only one long link can improve the routing performance greatly. The question is whether the average path length would decreases equally as the number of long links increases. The experimental result presents that the average path length decreases slowly and it tends to be smooth, which is shown in the right graph of Fig. 2.

The comparison of the effect of long links on average path length in varying dimension of CAN is shown in Fig. 3. It is clear that the effect of long links do less significant effects on the average path length as the dimensions increase. It can be explained by the fact that the diameter of the network would decrease while the dimension increases. As a result, the effect of long link will decrease.

3.2 The Effect of Redundant Short Links

Redundant short links are added to traditional CAN with the methods described in Section 2.2.2. As mentioned earlier, short links focus on the connectivity of the network. The basic short link is the inner structure of the DHT geometry and redundant short links provide the chance to enhance its connectivity. The connectivity embodies the routing resilience to node failure. To observe how short links impact on the resilience, we let some fixed fraction of uniformly chosen nodes fail and disable the failure recovery mechanism. Failed routing is the situation that any two alive nodes can not be connected. Fig. 4 shows the simulated results. The higher failed routing is because the failure recovery mechanism has been disabled. The left graph in Fig. 4 presents the resilience would be gradually improved with the increasing short links. It is due to the enhanced connectivity with short links. The right graph in Fig. 4 plots the resilience in different dimension CAN. It clearly shows that the improvement in higher dimension would be less significant. This is because higher dimension CAN has more basic links than lower dimension. Hence, the effect of redundant short links would decrease accordingly.

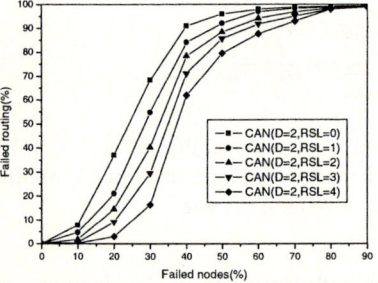 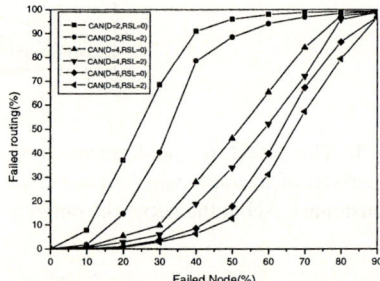

Fig. 4. The effect of redundant short links. Left: The percentage of failed routing for varying percentages of node failures considering varying numbers of redundant short links (RSL) to fixed dimensional CAN in the network of 1024 nodes. Right: The comparison of failed routing between 0 and 2 redundant short links to different dimension CAN in the network of 1024 nodes

4 Conclusions

Various DHT routing algorithms have been proposed in recent years. All these algorithms have tried to maintaining a uniform structured geometry while nodes join and leave. We use the linking model to catch the dynamic characteristics of the geometry and hence it provides a deep understanding on the performance of DHT routing algorithms. Our experiments show that the routing performance is greatly affected by the links. We expect that the linking model would provide more insight on understanding the performance of DHT routing algorithms and help to design new DHT routing algorithms or improve current DHT routing algorithms.

References

1. S. Ratnaswamy, P. Francis, M. Handley, R. Karp, and S.Shenker. A scalable content-addressable network. ACM SIGCOMM, 2001.
2. I. Stoica, R. Morris, D. Karger, F. Kaashoek, and H. Balakrishnan. Chord: A peer-to-peer lookup service for internet applications. ACM SIGCOMM, 2001.
3. J. Kleinberg. The small-world phenomenon: an algorithmic perspective. Cornell Computer Science Technical Report 99-1776, 2000.
4. R. Motwani and P. Raghavan. Randomized Algorithms. Cambridge University Press, 1995.
5. JavaSimulator. http://iris.ee.unsw.edu.au/p2p/.

Packet-Mode Priority Scheduling for Terabit Core Routers

Wenjie Li and Bin Liu

Department of Computer Science and Technology,
Tsinghua University, Beijing 100084, P.R. China
lwjie00@mails.tsinghua.edu.cn, liub@tsinghua.edu.cn

Abstract. Existing packet-mode schedulers will result in long waiting time for short control packets in IP networks. To overcome this problem, this paper proposes a packet-mode practical scheduling algorithm called short-packets-first (SPF). With uniform Poisson arrival process and low to medium offered load, we prove that SPF can reduce the average packet waiting time for overall packets by greatly lowering the average packet waiting time for short packets. Moreover, simulations under a real traffic model demonstrate that SPF performs very well compared with other packet-mode and cell-mode scheduling algorithms, especially for short packets under heavy offered load.

1 Introduction

Input queueing architecture is receiving much attention due to its high scalability and 100% throughput with virtual output queueing (VOQ) [1][2]. Based on this architecture, many practical scheduling algorithms are proposed, such as DRRM [3], iSLIP [4] and iLPF [5]. All of these algorithms operate on the fixed time slot, which is defined as the duration of a cell, or a fixed-length unit. This type of scheduling is called cell-mode scheduling. In routers, they need segmentation and reassembly (SAR) module to segment incoming packets into cells for scheduling and switching in the ingress side, and then reassemble cells after being switched for recovering original packets in the egress side.

Through modification of cell-mode scheduling, the idea of transferring cells of the same packet like a train without being interleaved with cells of other packets is first systemically studied in [6]. This type of scheduling is called packet-mode scheduling, which is the focus of this paper. Packet-mode scheduling reduces the reassembly buffer requirement at output ports and the complex control logic to reassemble packets, because cells of a packet remain contiguous in the delivery to an output port. Furthermore, it has been proven that under any admissible re-generative traffic, maximum weight packet-mode scheduling is stable [7].

This research is supported by NSFC (no. 60173009 and no. 60373007) and China 863 High-tech Plan (no. 2002AA103011-1 and no. 2003AA115110).

However, short packets (less than 64 bytes) suffer from long delay for the continuous transferring of cells of long packets in packet-mode scheduling. From the latest research results on traffic characters [8][9], we know that packets less than 64 bytes occupy approximate 50% of Internet packets, and most of these packets are TCP acknowledgment and control segments such as SYN, FIN and RST. Blocking these short packets will cause more TCP packets to be retransmitted due to timeout of acknowledgment packets, and therefore the offered load of networks will increase. So reducing the delay for short packets will improve the performance of TCP flows.

To overcome the problem of blocking short packets in packet-mode scheduling, we propose the short-packets-first (SPF) scheduling algorithm. The key idea is to buffer short packets in separate VOQs, and always schedule short packets first. The idea is simple, but there are some reasons for the idea is not introduced in the design of traditional scheduling algorithms. First of all, the first priority of short packets may cause out of sequence of packets belonging to the same flow of mixed short and long packets. Second, cell-mode scheduling is adopted widely in traditional router designs and it does not block short packets. Third, the hardware implementation complexity is increased, whatever slightly or greatly.

Through performance analysis in theory and by simulations, this paper shows that SPF achieves better average packet waiting time performance for both short packets and overall packets than general packet-mode scheduling, and also better than cell-mode scheduling under some condition. SPF is a simple practical scheduling algorithm, so it can be easily implemented in hardware.

The rest of this paper is organized as follows. Section 2 illustrates the iterative scheduling process of SPF. Section 3 presents the analysis on the performance of average packet waiting time. Section 4 performs some simulations and shows the corresponding results of SPF, cell-mode and packet-mode scheduling. Finally, the concluding remarks are given in section 5.

2 SPF Scheduling Algorithm

SPF includes three iterative steps: connection request, output grant and input accept.

Step 1: Connection Request.
An idle input port sends at most two connection requests to each output port in one time slot: one for short packets and the other for long packets. To release a connection, a disconnection request must be sent in the last cell of a packet.

Step 2: Output Grant.
Each output port j ($1 \leq j \leq N$) maintains two pointers: *OutPointer_S(j)* for short packets and *OutPointer_L(j)* for long packets.

A busy output port still grants its connected input port if it does not receive a disconnection request. A busy output port becomes idle immediately after receiving a disconnection request. An idle output port j grants an input port that sends a request for short packets to *j* and appears next in a fixed round-robin schedule from the highest priority port with the pointer *OutPointer_S(j)*. If an idle output port *j* does not receive any requests for short packets, it will grant requests for long packets with the pointer *OutPointer_L(j)* similarly. An idle output becomes busy and the corresponding pointer is updated to the one next to the granted input port, if and only if its acknowledgment is accepted by an input port in step 3.

Step 3: Input Accept.
Each input port *i* ($1 \leq i \leq N$) maintains two pointers: *InPointer_S(i)* for short packets and *InPointer_L(i)* for long packets. If input port *i* receives any grants for short packets, it will accept the one that appears next in a fixed round-robin schedule from the highest priority output port with the pointer *InPointer_S(i)* and then update *InPointer_S(i)* to the one next to the accepted output port (modulo *N*). Otherwise, if input port *i* receives any grants for long packets, it will accept one output port and update the pointer *InPointer_L(i)* as the process for short packets.

The first priority of short packets may decrease the performance of long packets, so multiple iterations of SPF can be performed to improve overall system performance. But similar to *i*SLIP, all pointers are updated only during the first iteration to avoid starvation of some input ports.

3 Performance Evaluation

Packet delay is defined as the time interval starting when the last cell of a packet arrives at an input buffer and ending when the last cell of this packet is transferred through the switch fabric. The packet service time is defined as the time a packet occupies the switch fabric, which is related to the packet length. The packet waiting time is defined as the time a packet spends in one VOQ, which equals the packet delay minus the packet service time.

If there are two or more acknowledgments to an input port in the same time slot, only one can be accepted. To estimate the average packet delay by queueing theory, we have to neglect this type of conflict. The results are only suitable for low to medium load, and not accurate for heavy load. For the convenience of the following analysis, some notations by queueing theory are defined first.

1) λ_{short}, λ_{long}, λ : the packet arrival rate of short, long and overall packets.
2) μ_{short}, μ_{long}, μ : the packet service rate of short, long and overall packets.
3) ρ_{short}, ρ_{long}, ρ : the offered load of short, long and overall packets.
4) $E(S)$: the average packet service time of overall packets.
5) C_v : the coefficient of variation of the packet service time.
6) $E(W_{cell})$, $E(W_{packet})$: the average packet waiting time for overall packets in cell-mode and packet-mode scheduling, respectively.

7) $E(W_{SPF\_S})$, $E(W_{SPF\_L})$ and $E(W_{SPF})$: the average packet waiting time for short, long and overall packets in SPF.

A cell-mode scheduling algorithm can be paralleled to processor-sharing service model [6][10]. A packet-mode scheduling algorithm is corresponding to M/G/1 FCFS queueing model [11]. Therefore

$$E(W_{cell}) = \frac{\rho E(S)}{1-\rho}, \qquad (1)$$

$$E(W_{packet}) = \frac{1+C_v^2}{2} \times E(W_{cell}). \qquad (2)$$

We use the non-preemptive priority model [11] to evaluate the average packet waiting time in SPF. There are two priorities: high priority for short packets and low priority for long packets. With the non-preemptive priority model, we have

$$E(W_{SPF\_S}) = \frac{\rho E(S)}{1-\rho_{short}} \times \frac{1+C_v^2}{2}, \qquad (3)$$

$$E(W_{SPF\_L}) = \frac{\rho E(S)}{(1-\rho_{short})(1-\rho)} \times \frac{1+C_v^2}{2}. \qquad (4)$$

Combining (3) with (4), we obtain

$$E(W_{SPF}) = \frac{\lambda_{short}}{\lambda} E(W_{SPF\_S}) + \frac{\lambda_{long}}{\lambda} E(W_{SPF\_L}) = \frac{1-\lambda_{short}/\mu}{1-\lambda_{short}/\mu_{short}} \times E(W_{packet}). \qquad (5)$$

From (5), we know that $E(W_{SPF})$ is always less than $E(W_{packet})$. This is because $\mu_{short} > \mu$ is always true when the arrival rate of long packets is greater than zero. When $C_v < 1$, the average packet waiting time for overall packets in SPF is less than both cell-mode and packet-mode scheduling.

4 Simulations

To study the performance of SPF under heavy load condition, simulations are performed in this section.

4.1 Simulation Environment

The switch size is 16 × 16. The scheduling algorithm in cell-mode is 2-*i*SLIP [4], in packet-mode is 2-*i*SLIP packet modification [6], and two iterations of SPF are simulated. 1,000,000 time slots are run and steady results between 20,000 and 980,000 are recorded. The arrival process is Poisson. Destinations of packets are uniformly distributed over all output ports. Packet length follows TRIMODEL distribution.

TRIMODEL(a, b, c, P_a, P_b): Packet lengths are chosen equal to either a cells with probability P_a, or b cells with probability P_b, or c cells with probability $1 - P_a - P_b$.

Under TRIMODEL, we set $a = 1$, $b = 9$, and $c = 24$, $P_a = 0.559$ and $P_b = 0.200$. These parameters are the same as results shown in [6], which are measured in actual backbone networks. Similar parameters are also reported in [9][12], so the model is accurate to describe real Internet traffic.

4.2 Simulation Results

Fig. 1 shows simulation results on the average packet waiting time for overall packets. When $\rho < 0.8$, the average packet waiting time in cell-mode scheduling is less than packet-mode. This is because the coefficient of variation of arrival traffic is a little larger than 1.0, cell-mode scheduling performs better under low to medium offered load. Moreover, in Fig. 1 we can also get that SPF is always better than packet-mode scheduling just as results analyzed in theory. When $\rho > 0.7$, the average packet waiting time in SPF becomes the best. With the increase of offered load, conflicts, which should be arbitrated by the scheduling algorithm, occur more and more frequently. In cell-mode scheduling, all cells from the same packet need to be scheduled separately. Compared with SPF and packet-mode, the scheduling task in cell-mode is much heavy, and thus results of scheduling algorithms have more impact on the average packet waiting time. In SPF and packet-mode scheduling, only one arbitration is needed by every packet. If a packet is granted, all its following cells can be transferred continuously. So under heavy load, SPF and packet-mode scheduling is preferable. When $\rho > 0.8$, the advantage of SPF becomes more and more obvious. E.g., the average packet waiting time in SPF, packet-mode and cell-mode scheduling is 227, 334, and 714 cells, respectively, when the offered load is 0.95.

Fig. 2 shows the average packet waiting time for short packets. Under low to medium load, the average packet waiting time of cell-mode scheduling is the best, and the performance curve of SPF is between cell-mode and packet-mode scheduling. When the offered load goes up, the average packet waiting time in cell-mode and packet-mode increases more and more sharply, and SPF becomes the best under heavy load ($\rho > 0.8$). Even when the offered load is 1.0, the average packet waiting time for short packets in SPF is still less than 160 cells.

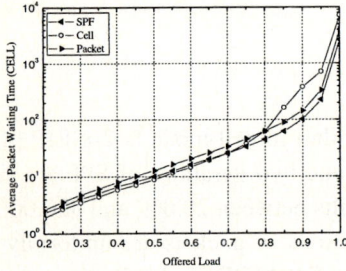
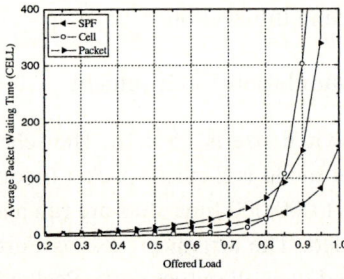

Fig. 1. Average packet waiting time for overall packets

Fig. 2. Average packet waiting time for short packets

5 Conclusions

This paper points out that short control packets in a flow should be guaranteed higher priority than long packets, and then proposes the scheduling algorithm called SPF. Under low to medium offered load, it is proven that the average packet waiting time for overall packets in SPF is less than both cell-mode and packet-mode scheduling when the coefficient of variation of the packet service time is less than 1.0. Furthermore, simulation results demonstrate that SPF can greatly reduce the average packet waiting time for short packets under heavy load. For networks full of TCP flows, such as Internet, the result is particularly valuable.

In this paper, we do not address how to solve the out-of-sequence in SPF. Whatever, the sequence of short packets and the sequence of long packets are guaranteed individually in SPF. Then we expect a low probability of out of sequence under the condition of Terabit core routers, and this will be a topic for future papers.

References

1. Anderson T.E., Owicki S.S., Saxe J.B., Thacker C.P.: High-Speed Switch Scheduling for Local Area Networks. ACM Trans. Computer Systems, Vol. 11, 4 (1993) 319–352
2. Keslassy I., McKeown N.: Analysis of Scheduling Algorithms that Provide 100% Throughput in Input-Queued Switches. Proceedings of the 39th Annual Allerton Conference on Communication, Control and Computing, Monticello, Illinois (2001)
3. Chao H.J.: Saturn: a Terabit Packet Switch Using Dual Round Robin. IEEE Commun. Mag., Vol. 38, 12 (2000) 78-84
4. McKeown N.: The iSLIP Scheduling Algorithm for Input-Queued Switches. IEEE/ACM Trans. Networking, Vol. 7, (1999) 188-201.
5. Mekkittikul A., McKeown N.: A Practical Scheduling Algorithm to Achieve 100% Throughput in Input-Queued Switches. IEEE INFOCOM 1998, Vol. 2, (1998) 792-799
6. Marsan M.A., et al.: Packet-Mode Scheduling in Input-Queued Cell-Based Switches. IEEE/ACM Trans. Networking, Vol. 10, (2002) 666-678
7. Ganjali Y., Keshavarzian A., Shah D.: Input Queued Switches: Cell Switching vs. Packet Switching. IEEE INFOCOM 2003, Vol. 3, (2003) 1651-1658
8. Thompson K., Miller G.J., Wilder R.: Wide-Area Internet Traffic Patterns and Characteristics. IEEE Network, Vol. 11, 6 (1997) 10-23
9. Fraleigh C., et al.: Packet-Level Traffic Measurements from the Sprint IP Backbone. IEEE Network, Vol. 17, 6 (2003) 6-16
10. Allen A.O.: Probability, Statistics, and Queueing Theory with Computer Science Applications, New York Academic Press (1978)
11. Wolff R.W.: Stochastic Modeling and the Theory of Queues. Prentice-Hall Inc. (1989)
12. Mellia M., Carpani A., Cigno R.L.: Measuring IP and TCP Behavior on Edge Nodes. IEEE GLOBECOM 2002, Vol. 3, (2002) 2533-2537

Node-to-Set Disjoint Paths Problem in Bi-rotator Graphs

Keiichi Kaneko

Tokyo University of Agriculture and Technology,
Koganei-shi, Tokyo 184-8588, Japan
k1kaneko@cc.tuat.ac.jp

Abstract. A rotator graph was proposed as a topology for interconnection networks of parallel computers, and it is promising because of its small diameter and small degree. However, a rotator graph is a directed graph that sometimes behaves harmful when it is applied to actual problems. A bi-rotator graph is obtained by making each edge of a rotator graph bi-directional. A bi-rotator has a Hamilton cycle and it is also pan-cyclic. In this paper, we give an algorithm for the node-to-set disjoint paths problem in bi-rotator graphs with its evaluation results. The solution achieves some fault tolerance such as file distribution based information dispersal technique. The algorithm is of polynomial order of n for an n-bi-rotator graph. It is based on recursion and divided into three cases according to the distribution of destination nodes in classes into which all the nodes in a bi-rotator graph are categorized. The sum of lengths of paths obtained and the time complexity of the algorithm are estimated. Average performance of the algorithm is also evaluated by computer experiments.

Keywords: node-to-set disjoint paths problem, bi-rotator graphs, fault tolerance, parallel computation.

1 Introduction

Currently, researches on parallel and distributed computation is getting more significant. In addition, so-called massively parallel processing systems are eagerly studied for these years. Therefore, many interconnection networks based on complicated topologies instead of simple networks such as meshes, tori, hypercubes and so on have been proposed. Most of those new topologies are variants of the Cayley graph[1, 6] and there are intensive research activities about them[2, 3, 4, 5, 7, 8, 9, 10, 11, 13, 15]. A rotator graph[6] is one of such topologies, and it is very promising because of its small diameter and its small degree. However, a rotator graph is a directed graph that sometimes behaves harmful when it is applied to actual problems.

A bi-rotator graph[12] is obtained by making each of a rotator graph bi-directional. A bi-rotator has a Hamilton cycle and it is also pan-cyclic. Among the unresolved problems exists the node-to-set disjoint paths problem: in a k-connected graph $G = (V, E)$, for a source node s and a set of k destination nodes $D = \{d_1, d_2, \cdots, d_k\}$ ($s \notin D$), find k paths from s to d_i ($1 \leq i \leq k$) that are node-disjoint except for s. This is one of important issues in parallel and distributed computing systems[5, 8, 10, 14] as well as the node-to-node disjoint paths problem[9, 15].

In general, the node-disjoint paths are obtained by using the maximum flow algorithm in polynomial order of the number of the node in the graph $|V|$. There are $n!$ nodes in an n-bi-rotator graph. Hence this approach is impractical. In this paper, we give an algorithm of polynomial order of n instead of $n!$, estimate the theoretical performance of the algorithm, and conduct the computer experiment to evaluate its average performance.

The rest of this paper is structed as follows. Section 2 gives some definitions and properties. Section 3 describes our algorithm in detail and the proof of its correctness and the estimation of its complexities are given in Section 4. In Section 5, the computer experiment is conducted. We conclude and give future works in Section 6.

2 Preliminaries

In this section, we give definitions of a bi-rotator graph and a class, then a simple routing algorithm `route` between a pair of nodes.

Definition 1. For an arbitrary permutation $\boldsymbol{u} = (u_1, u_2, \cdots, u_n)$ of n symbols $1, 2, \cdots, n$ and an integer i ($2 \leq i \leq n$), we define the positive and negative rotation operations $R_i^+(\boldsymbol{u})$ and $R_i^-(\boldsymbol{u})$ as follows:

$$R_i^+(\boldsymbol{u}) = (u_2, u_3, \cdots, u_i, u_1, u_{i+1}, u_{i+2}, \cdots, u_n),$$

$$R_i^-(\boldsymbol{u}) = (u_i, u_1, u_2, \cdots, u_{i-1}, u_{i+1}, u_{i+2}, \cdots, u_n).$$

Note that R_2^+ and R_2^- represent a same rotation operation and there are $2n - 3$ distinct rotation operations.

Definition 2. An n-bi-rotator graph, BR_n, has $n!$ nodes. Each node has a unique address that is a permutation of n symbols $1, 2, \cdots, n$. The node whose address is $\boldsymbol{u} = (u_1, u_2, \cdots, u_n)$ is adjacent to the nodes whose addresses are the elements of the set $\{R_i^+(\boldsymbol{u}), R_i^-(\boldsymbol{u}) \mid 2 \leq i \leq n\}$, and it is not adjacent to any other nodes. Let the neighbor node set of $\boldsymbol{u}$ be denoted by $N(\boldsymbol{u})$.

Table 1 shows comparisons an n-bi-rotator graph with other graphs. T_n, Q_n, $B(n,k)$, and $K(n,k)$ represent an $n \times n$ torus, an n-dimensional hypercube, an (n,k)-de Bruijn graph, and an (n,k)-Kautz graph, respectively. Currently, the average diameter of an n-bi-rotator graph is unknown. As for the integration ratio defined by Number of Nodes/(Degree × Diameter), an n-bi-rotator graph is inferior to an (n,k)-de Bruijn graph and an (n,k)-Kautz graph. However, an bi-rotator graph has recursive structure described below and it has a merit that it can easily implement algorithms such as divide-and-conquer in parallel.

Figure 1 shows examples of 2- to 4- bi-rotator graphs. Note that an address $(u_1, u_2, \cdots, u_n)$ is denoted by $u_1 u_2 \cdots u_n$ in the figure to save space.

Table 1. Comparison of a bi-rotator graph with other graphs

	Number of Nodes	Degree	Diameter	Integration
BR_n	$n!$	$2n-3$	$n-1$	$n!/(n-1)/(2n-3)$
T_n	n^2	4	n	$n/4$
Q_n	2^n	n	n	$2^n/n^2$
$B(n,k)$	n^k	n	k	n^{k-1}/k
$K(n,k)$	$n^k + n^{k-1}$	n	k	$(n^{k-1} + n^{k-2})/k$

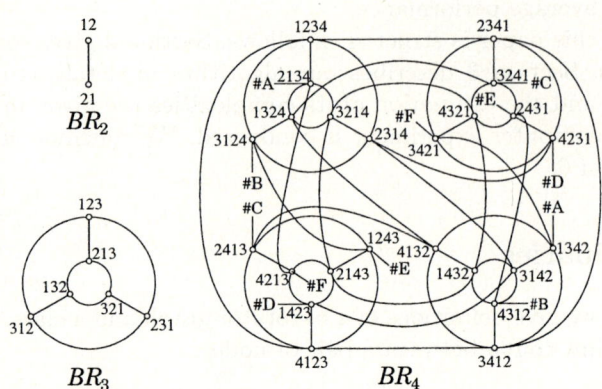

Fig. 1. Examples of 2- to 4- bi-rotator graphs

Definition 3. In an n-bi-rotator graph, a sub graph induced by the nodes that have a common symbol k at the right-most positions in their addresses comprises an $n-1$-bi-rotator graph. The sub graph is denoted by $BR_{n-1}k$ by using the common symbol k.

Definition 4. For a node u in an n-bi-rotator graph BR_n, the class of u be the set of nodes that are obtained by applying $R_n^+(\cdot)$ to u repeatedly. In the rest of this paper, let $C(u)$ denote the class of u.

For classes in an n-bi-rotator graph BR_n, following properties hold.

1. Each node belongs to exactly one class.
2. Each class consists of n nodes that comprise a ring structure.
3. In each sub bi-rotator graph, exactly one node belongs to each class.
4. Each node has two neighbor nodes that belong to the same class as that node itself. The remaining $2n-5$ neighbor nodes belong to different classes each other.

For instance, a BR_4 has 6 classes each of which consists of 4 nodes: $C_1 = \{(1,2,3,4),(2,3,4,1),(3,4,1,2),(4,1,2,3)\}$, $C_2 = \{(1,3,2,4),(3,2,4,1),(2,4,1,3),(4,1,3,2)\}$, $C_3 = \{(2,1,3,4),(1,3,4,2),(3,4,2,1),(4,2,1,3)\}$, $C_4 = \{(2,3,1,4),(3,1,4,2),(1,4,2,3),(4,2,3,1)\}$, $C_5 = \{(3,1,2,4),(1,2,4,3),(2,4,3,1),(4,3,1,2)\}$, and

$C_6 = \{(3,2,1,4), (2,1,4,3), (1,4,3,2), (4,3,2,1)\}$. Moreover, if we traverse each class from the beginning, we can find that they comprise a ring structure.

Definition 5. In a bi-rotator graph, a class path is a sub path that is a part of a ring structure formed by a class.

Following lemma holds for classes.

Lemma 1. If $n \geq 4$, for two different nodes u and v that belong to a class in BR_n, there is a node in neighbor nodes of u that belongs to the different class from those which the neighbor nodes of v belong to.

Proof. Let $u = (u_1, u_2, \cdots, u_n)$. Then from $v \in C(u)$ we can denote $v = (u_k, u_{k+1}, \cdots, u_n, u_1, \cdots, u_{k-1})$ for some k. First we assume that $k \neq n - 1$. In this case, take $w = (u_{n-1}, u_1, u_2, \cdots, u_{n-2}, u_n)$. Then w is a neighbor node of u and from the fact that $n \geq 4$ it is not possible to insert u_{n-1} between u_n and u_1 with a single rotation operation for v. Hence, there is no neighbor node of v that belongs to the same class as w. Next we assume that $k = n - 1$. In this case take $w = (u_2, u_3, \cdots, u_{n-1}, u_1, u_n)$. Then w is a neighbor node of u, and from the fact that $n \geq 4$ it is impossible to insert u_1 between u_{n-1} and u_n with a single rotation operation for v. Hence, there is no neighbor node of v that belongs to the same class as w. □

Because the shortest-path routing algorithm for an n-bi-rotator graph in polynomial-order time of n is still unknown, we use a simple unicast routing algorithm shown in [12] that is called **route** in this paper. The algorithm generates a path of which length is $O(n)$ in $O(n^2)$ time complexity. It is shown in [12] that two paths route(s, t) and route(t, s) are internally disjoint between s and t.

Moreover, in our algorithm, for four nodes s, t, u, and v in BR_n it is necessary to construct a path from s to t without including nodes u and v. If $n = 3$, then there are three internally disjoint paths between s and t if they are not adjacent, and the problem is trivial. Therefore, we give an extended routing algorithm in case of $n \geq 4$:

Case 1. If $\exists h$ s. t. $s, t, u, v \in BR_{n-1}h$, then apply the algorithm recursively inside $BR_{n-1}h$ and terminate.
Case 2. If $\exists h$ s. t. $s, t \in BR_{n-1}h$, $v \notin BR_{n-1}h$, then construct two internally disjoint paths between s and t inside $BR_{n-1}h$ and select one of them that does not include u and terminate.
Case 3. If $\exists h, l$ s. t. $h \neq l$, $s, u, v \in BR_{n-1}h$, $t \in BR_{n-1}l$, then select a class path from s to $BR_{n-1}l \cap C(s)$ and select a path from $BR_{n-1}l \cap C(s)$ to t inside $BR_{n-1}l$ and terminate.
Case 4. If $\exists h, l$ s. t. $h \neq l$, $s \in BR_{n-1}h, t \in BR_{n-1}lCu \notin BR_{n-1}h$, $v \notin BR_{n-1}l$, then select x so that $x \in \{s\} \cup N(s)$ s.t. $x \notin C(u) \cup C(v)$ holds. If $x \neq s$, then select edge (s, x). Construct a class path from x to $BR_{n-1}l \cap C(x)$. In addition, construct two internally disjoint paths between $BR_{n-1} \cap C(x)$ and t, and select one of them that does not include u and terminate.

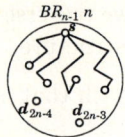
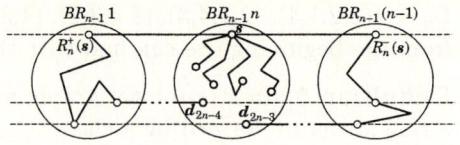

Fig. 2. Recursive application of the algorithm

Fig. 3. Construction of two disjoint paths by class paths

3 Algorithm

3.1 Classification

In a BR_3, the problem is trivial. Hence, we assume that $n \geq 4$. Taking advantage of the symmetric property of a BR_n, we fix the source node to $s = (1, 2, \cdots, n)$. Let the destination node set be $D = \{d_1, d_2, \cdots, d_{2n-3}\}$ and the set of classes to which destination nodes belong be $\mathcal{C} = \{C_1, C_2, \cdots, C_k\}$. Now let us consider the following cases:

Case 1. All destination nodes belong to the same sub bi-rotator graph as the source node ($D \subset BR_{n-1}n$)D

Case 2. Exactly one destination node is not included in the sub bi-rotator graph to which the source node belongs ($|D - BR_{n-1}n| = 1$)D

Case 3. More than one destination nodes are not included in the sub bi-rotator graph to which the source node belongs ($|D - BR_{n-1}n| \geq 2$)D

3.2 Case 1

This section gives Procedure 1 which constructs $2n - 3$ paths from the source node s to the destination node set $D = \{d_1, d_2, \cdots, d_{2n-3}\}$ that are node-disjoint except for the source node in case that $D \subset BR_{n-1}n$.

Step 1. Apply the algorithm recursively inside $BR_{n-1}n$ and obtain $2n - 5$ paths from s to $D - \{d_{2n-4}, d_{2n-3}\}$ that are node-disjoint except for s. If one of these paths, say the path from s to d_m, contains d_{2n-4}, then discard the sub path from d_{2n-4} to d_m and exchange the indices of d_{2n-4} and d_m. For d_{2n-3}, perform similar operation. After this step, we can obtain paths as shown in Figure 2.

Step 2. Select the edges $(s, R_n^+(s))$ and $(s, R_n^-(s))$. Then construct class paths from the destination nodes d_{2n-4} and d_{2n-3} to $C(d_{2n-4}) \cap BR_{n-1}1$ and $C(d_{2n-3}) \cap BR_{n-1}(n-1)$, respectively, without including the nodes $C(d_{2n-4}) \cap BR_{n-1}(n-1)$ and $C(d_{2n-3}) \cap BR_{n-1}1$. Apply route inside the sub graphs $BR_{n-1}1$ and $BR_{n-1}(n-1)$, and construct paths from $C(d_{2n-4}) \cap BR_{n-1}1$ and $C(d_{2n-3}) \cap BR_{n-1}(n-1)$ to $R_n^+(s)$ and $R_n^-(s)$, respectively. See Figure 3. The horizontal dashed lines represent class ring structures. We assume that the both edges of each dashed line is connected.

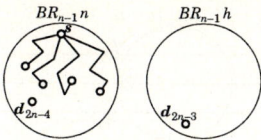
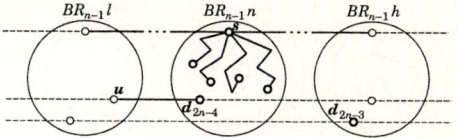

Fig. 4. Recursive application of our algorithm

Fig. 5. Construction of two class paths.

3.3 Case 2

This section gives Procedure 2 which constructs $2n - 3$ paths from the source node s to the destination node set $D = \{d_1, d_2, \cdots, d_{2n-3}\}$ that are node-disjoint except for the source node in case that $|D - BR_{n-1}n| = 1$.

Step 1. We can assume that d_{2n-3} is the destination node that is outside of BR_{n-1} without loss of generality. Inside $BR_{n-1}n$, apply our algorithm recursively to obtain $2n-5$ paths from s to $D-\{d_{2n-4}, d_{2n-3}\}$ that are node-disjoint except for s. If one of these paths, say the path from s to d_m, contains d_{2n-4}, then discard the sub path from d_{2n-4} to d_m and exchange the indices of d_{2n-4} and d_m. After this step, we can obtain paths as shown in Figure 4.

Step 2. For the destination node d_{2n-4}, apply either $R_n^+(\cdot)$ or $R_n^-(\cdot)$ to select the edge (d_{2n-3}, u) such that u belongs to the sub graph $BR_{n-1}l$ that is different from the sub graph $BR_{n-1}h$ to which d_{2n-3} belongs. Construct two class paths from s to $C(s) \cap BR_{n-1}h$ and $C(s) \cap BR_{n-1}l$ that are node-disjoint except for s. See Figure 5.

Step 3. Inside sub graphs $BR_{n-1}h$ and $BR_{n-1}l$, apply route to construct from $C(s) \cap BR_{n-1}h$ and $C(s) \cap BR_{n-1}l$ to d_{2n-3} and u, respectively, and terminate. See Figure 6.

3.4 Case 3

This section gives Procedure 3 which constructs $2n - 3$ paths from the source node s to the destination node set $D = \{d_1, d_2, \cdots, d_{2n-3}\}$ that are node-disjoint except for the source node in case that $|D - BR_{n-1}n| \geq 2$.

Step 1. Let D_1 be the set of the destination nodes that are first reachable by repeating the rotation operation $R_n^+(\cdot)$ zero or more times from each node $C_i \cap BR_{n-1}n$. Additionally, if $C(s) \in \mathcal{C}$, then let D_1 include the destination node that are first reached by repeating $R_n^-(\cdot)$ one or more times from s. Without loss of generality, we can assume that $D_1 = \{d_1, d_2, \cdots, d_p\}$Cand $D - D_1 = \{d_{p+1}, d_{p+2}, \cdots, d_{2n-3}\}$. See Figure 7.

Step 2. For each destination node d_i in $D - D_1$, find its neighbor node c_i that satisfies following conditions in a greedy manner:

- $C(c_i) \notin \mathcal{C}$,
- $C(c_i) \neq C(c_j)$, if $i \neq j$.

From the fact that $|D - D_1| \leq 2n - 4$ and Property 4 of classes, these neighbor nodes are selectable except for the final destination node, say d_{2n-3}. If all

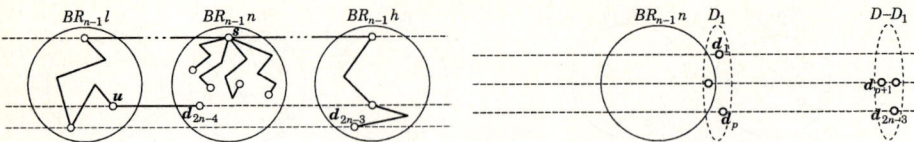

Fig. 6. Construction of paths in sub bi-rotator graphs

Fig. 7. Classification of destination nodes

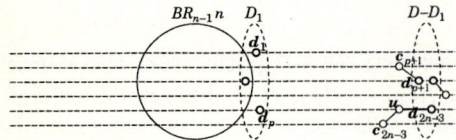
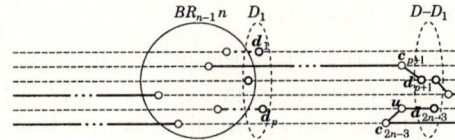

Fig. 8. Selection of neighbor nodes of $D - D_1$

Fig. 9. Selection of class paths and edges

the neighbor nodes of d_{2n-3} belong to the classes of other destination nodes or the nodes c_i, then perform the following process. If there exists a node in $\{R_n^+(d_{2n-3}), R_n^-(d_{2n-3})\} - BR_{n-1}n$ that is not a destination node, select the node u and the edge (d_{2n-3}, u). Among the neighbor nodes of u, select one node c_{2n-3} that does not belong to the class of any other destination node nor any other c_i. Lemma 1 ensures the existence of this neighbor node. Figure 8 represents this case. Otherwise, if either $R_n^+(d_{2n-3})$ or $R_n^-(d_{2n-3})$ is a destination node that does not belong to D_1, then for that destination node d_j release the node c_j that are selected previously, and among the neighbor nodes of d_j, newly select the node c_j that does not belong to the class of any other destination node nor any other c_i. Lemma 1 ensures the existence of this neighbor node. By this operation, among the neighbor nodes of d_{2n-3}, it is possible to select the neighbor node c_{2n-3} that belongs to the same class as the neighbor node of d_j that is previously released. If neither of above conditions is satisfied, it induces that $R_n^+(d_{2n-3}) \in BR_{n-1}n$ and $d_j = R_n^-(d_{2n-3}) \in D_1$. Then remove d_j from D_1 and add d_{2n-3} to D_1. Moreover, for d_j select its neighbor node c_j that does not belong to the class of any other destination nodes nor any other c_i. Lemma 1 ensures the existence of this neighbor node.

Step 3. For each destination node d_i in D_1, construct a class path from $C(d_i) \cap BR_{n-1}n$ to it so that the path does not include any other destination nodes. Next, for each c_i, construct a class path from the node in $C(c_i) \cap BR_{n-1}n$ to c_i. In these path constructions, if d_i is reachable without including other destination by repetition of both of $R_n^+(\cdot)$ and $R_n^-(\cdot)$, the shorter path is selected. Select an edge between each c_i and corresponding d_i. See Figure 9.

Step 4. If any path from s is not constructed yet, proceed to Step 5. Otherwise, if two paths from s are already constructed, proceed to Step 7. In the rest of this step, we assume that there is exactly one path from s and the path includes $R_n^-(\cdot)$ without loss of generality. Select the edge between s and $R_n^+(s)$.

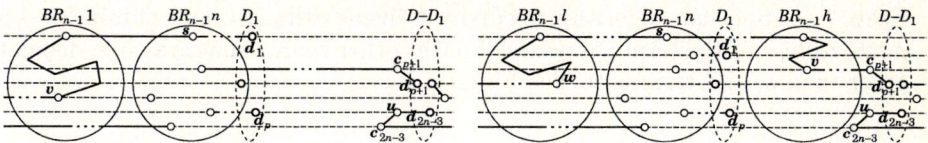

Fig. 10. Construction of a path   **Fig. 11.** Construction of two paths

If there is a node in $BR_{n-1}1$ that is included in a path already constructed, then construct a path from $R_n^+(s)$ to one of such nodes v without including other such nodes. Discard the sub path of the already constructed path from the node inside $BR_{n-1}n$ to v. See Figure 10. If there is no node in $BR_{n-1}1$ that is included in the path already constructed, for the nodes on the already constructed paths of which length is one or more, consider the set of nodes in $BR_{n-1}1$ that belong to the same class as them, but does not belong to the same class as the nodes on the path that include s as its terminal. If this set is not empty, construct a path from $R_n^+(s)$ to one node in the set v so that the path does not include any other such nodes. Let w be the first node on the already constructed path that is reachable by repeating either $R_n^+(\cdot)$ or $R_n^-(\cdot)$ from v without passing the sub graph $BR_{n-1}n$. For the path that was constructed in Step 3 and includes w, discard its sub path from the terminal node in $BR_{n-1}n$ to w. Construct a path from v to w by repeating either $R_n^+(\cdot)$ or $R_n^-(\cdot)$ without passing the sub graph $BR_{n-1}n$ and without including other c_i's and d_i's. If none of above two conditions is satisfied, then $|D - BR_{n-1}n| = 2$ holds and these two destination nodes belong to a same class. First, discard the paths to these two destination nodes. Next, construct a class path from s to the nodes in the sub graphs to which these two destination nodes belong by repeating either $R_n^+(\cdot)$ or $R_n^-(\cdot)$. Moreover, construct paths to the destination nodes inside the sub graphs. Proceed to Step 7.

Step 5. Let v and w be two nodes that are on two different paths P_1 and P_2 already constructed, respectively, and that are in different two sub graphs $BR_{n-1}h$ and $BR_{n-1}l$ that are different from $BR_{n-1}n$, respectively. If these two nodes cannot be obtained, all the nodes on the already constructed paths belong to a single sub graph except for $BR_{n-1}n$. In this case, we can obtain the above mentioned two nodes by substituting one of the paths constructed in Step 3 of which length is one or more with the longer class path.

Step 6. Obtain class paths from s to the nodes t_1 and t_2 in $BR_{n-1}h$ and $BR_{n-1}l$, respectively, by repeating $R_n^+(\cdot)$ and $R_n^-(\cdot)$. Additionally, construct paths from t_1 and t_2 to v and w avoiding at most two nodes on P_2 and P_1 by applying the extended routing algorithm in $BR_{n-1}h$ and $BR_{n-1}l$, respectively. If these paths include some nodes on the already constructed paths other than v and w, let the nearest nodes from t_1 and t_2 be v and w, respectively. However, if two different nodes on a single path are obtained, keep the nearer node to the destination and ignore the another one. Discard sub paths from sub graph $BR_{n-1}n$ to v and w. See Figure 11.

Step 7. Apply our algorithm recursively inside $BR_{n-1}n$, and obtain $2n-5$ paths from s to the terminal nodes of paths other than s that are node-disjoint other than s, and terminate. See Figure 12.

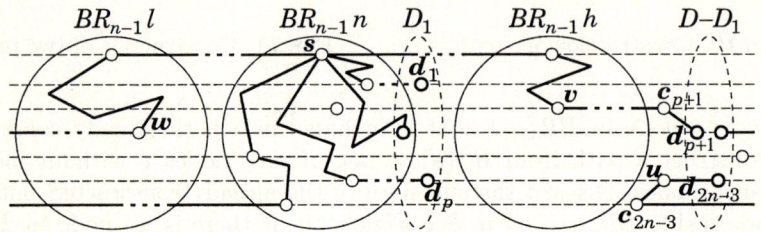

Fig. 12. Recursive application of our algorithm

4 Proof of Correctness and Estimation of Complexities

In this section, we give proof of correctness of our algorithm and the time complexity of our algorithm and the sum of path lengths obtained by our algorithm. We use the term 'disjoint' to express 'node-disjoint' in this section.

Theorem 1. The paths generated by our algorithm are disjoint except for the source node s. Let $T(n)$ and $L(n)$ be the time complexity of our algorithm and the sum of path lengths generated by our algorithm for an n-bi-rotator graph. Then $T(n) = O(n^5)$ and $L(n) = O(n^3)$.

(Proof) Based on induction on n, this theorem can be proved from the following lemmas. □

Lemma 2. The paths generated by Procedure 1 are disjoint except for the source node. The time complexity of Procedure 1 is $T(n-1) + L(n-1) \times O(n) + O(n^2)$, and the sum of path lengths is $L(n-1) + O(n)$.

(Proof) The paths obtained in Step 1 are disjoint except for s from the hypothesis of induction. One of the two paths generated in Step 2 consists of the nodes in $BR_{n-1}1$ and the sub path of the class of d_{2n-4} that does not pass $BR_{n-1}(n-1)$. Another path consists of the nodes in $BR_{n-1}(n-1)$ and the sub path of the class of d_{2n-3} that does not pass $BR_{n-1}1$. Hence these two paths are disjoint except for s each other. Additionally, these two paths contains the nodes outside of $BR_{n-1}n$ except for these terminal nodes. Therefore, these two paths are also disjoint with other paths generated in Step 1.

The time complexity of Step 1 is $T(n-1) + L(n-1) \times O(n)$. The sum of path lengths generated in Step 1 is $L(n-1)$. The time complexity of Step 2 is $O(n^2)$ and the sum of lengths of two paths obtained in Step 2 is $O(n)$. Hence, the total time complexity of Procedure 1 is $T(n-1) + L(n-1) \times O(n) + O(n^2)$ and ths sum of path length obtained by Procedure 1 is $L(n-1) + O(n)$. □

Lemma 3. The paths generated by Procedure 2 are disjoint except for the source node. The time complexity of Procedure 2 is $T(n-1) + L(n-1) \times O(n) + O(n^2)$, and the sum of path lengths is $L(n-1) + O(n)$.

(Proof) The paths obtained in Step 1 are disjoint except for s from the hypothesis of induction. The two paths generated in Steps 2 and 3 consist of nodes on different class paths that share only s, sub paths in different sub graphs, and the destination node d_{2n-3}. Hence they are disjoint each other. These paths do not have the nodes in $BR_{2n-3}n$ except for s and d_{2n-3}. Hence they are also disjoint from the paths generated in Step 1.

The time complexity of Step 1 is $T(n-1) + L(n-1) \times O(n)$. The sum of path lengths generated in Step 1 is $L(n-1)$. The time complexities of Steps 2 and 3 are both $O(n^2)$ and the sums of lengths of two paths obtained in Steps 2 and 3 are both $O(n)$. Hence, the total time complexity of Procedure 2 is $T(n-1) + L(n-1) \times O(n) + O(n^2)$ and the sum of path length obtained by Procedure 2 is $L(n-1) + O(n)$. □

Lemma 4. The paths generated by Procedure 3 are disjoint except for the source node. The time complexity of Procedure 3 is $T(n-1) + O(n^4)$, and the sum of path lengths is $L(n-1) + O(n^2)$.

(Proof) The paths constructed in Steps 2 and 3 consist of different class paths, different class paths followed by edges from their terminal nodes to different destinations, or a class path followed by two edges (c, u) and (u, d_{2n-3}). The node u, if any, is selected so as not to be shared by other paths. Hence, any pair of these paths are disjoint each other. In Step 4, if one path is generated, then it consists of a class path followed by a path in $BR_{n-1}1$. The path may be still followed by another class path, if necessary. In either case, the path is connected to the previously constructed path at the node that is first encountered. Hence, it is disjoint from other paths except for one that shares s. If two paths are generated in Step 4, we can prove that they are disjoint each other and also disjoint from other paths except for s similarly to Step 6. The two paths generated in Step 6 are disjoint except for s because they only share s. The nodes v and w are the first encountered nodes that are on other paths in different sub graphs. Hence, they are disjoint from other paths that does not include v nor w. The paths generated in Step 7 are disjoint from induction hypothesis. These paths are inside $BR_{n-1}n$ and they are connected to the previously constructed paths inside $BR_{n-1}n$. Hence the connected paths are also disjoint.

The time complexity of Step 1 is $O(n^3)$ that is necessary to obtain C. In Step 2, finding c_i's requires $O(n^4)$. The time complexity of Step 3 is $O(n^3)$. The sum of paths obtained in Step 3 is $O(n^2)$. In Step 4, finding v and w is governing and it requires $O(n^3)$ time complexity. The sum of a path or tow paths generated in Step 4 is $O(n)$. The time complexity of Step 5 is $O(n^2)$. In Step 6, the time complexity is $O(n^3)$ and the sum of path lengths is $O(n)$. Finally, the time complexity in Step 7 is $T(n-1)$ and the sum of path lengths is $L(n-1)$. □

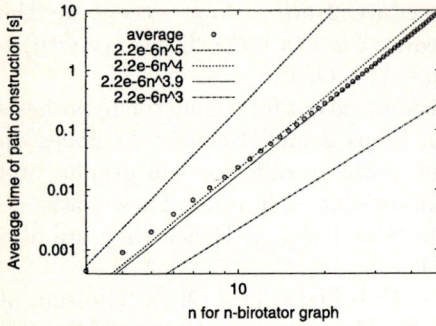 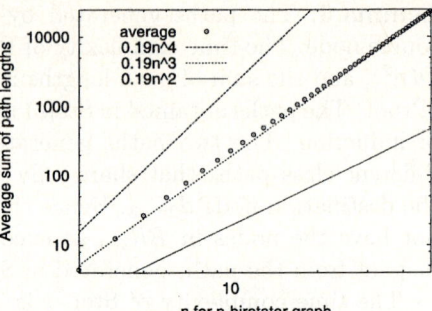

Fig. 13. The average execution time of our algorithm

Fig. 14. The average sum of path lengths obtained by our algorithm

5 Computer Experiment

To evaluate average performance of our algorithm, we conducted computer experiment by repeating following steps at least $1,000$ times for random combinations of destination nodes for each n between 3 and 50.

1. In an n-bi-rotator graph, fix the source node s to the identity permutation $(1, 2, \cdots, n)$ taking advantage of its symmetric property.
2. Set $2n - 3$ destination nodes other than s.
3. Invoke our algorithm and measure the execution time and the sum of path lengths.

Our algorithm is implemented by the functional programming language Haskell and the program is compiled by ghc (glasgow Haskell compiler) with -O and -fglasgow-exts options. The operating system of the machine is FreeBSD 3.5.1. The CPU of the machine is Celeron 400MHzCand it is equipped with a 128MB memory unit.

Figures 13 and 14 show the results of the execution time and the sum of path lengths, respectively. In each figure, the horizontal axis represents n for an n-bi-rotator graph. The vertical axis of Figure 13 represents the average execution time in second while that of Figure 14 represents the average sum of path lengths obtained by our algorithm.

From these figures, we can conclude that our algorithm obtains $2n - 3$ paths from the source node to the destination nodes that are node-disjoint except for the source node in the average execution time of $O(n^{3.9})$ for an n-bi-rotator graph, and the average sum of these path lengths is $O(n^{3.0})$.

6 Conclusion

In this paper, we have proposed an algorithm for the node-to-set disjoint paths problem in an n-bi-rotator graph. Its time complexity and the sum of path lengths are $O(n^5)$ and $O(n^3)$, respectively. Computer experiment showed that our algorithm gives disjoint paths with those complexities. The computer experiment showed that the average execution time of our algorithm is $O(n^{3.9})$ and the average sum of path lengths is $O(n^{1.9})$. Future work includes improvement of our algorithm to generate shorter paths in shorter time.

Acknowledgements

This study is partly supported by Grant-in-Aid for Scientific Research (C) of Japan Society for the Promotion of Science under the Grant No. 16500015.

References

1. Akers, S. B., Krishnamurthy, B.: A group theoretic model for symmetric interconnection networks. IEEE Trans. Comp. **38**(1989) 555–566
2. Akl, S. G., Qiu, K.: Parallel minimum spanning forest algorithms on the star and pancake interconnection networks. Proc. Joint Conference Vector and Parallel Processing (1992) 565–570
3. Akl, S. G., Qiu, K: A novel routing scheme on the star and pancake interconnection networks and its applications. Parallel Computing **19**(1993) 95–101
4. Akl, S. G., Qiu, K., Stojmenović, I.: Fundamental algorithms for the star and pancake interconnection networks with applications to computational geometry. Networks **23**(1993) 215–226
5. Berthomé, P., Ferreira, A., Perennes, S.: Optimal information dissemination in star and pancake networks. IEEE Trans. Parallel and Distributed Systems **7**(1996) 1292–1300
6. Corbett, P. F.: Rotator graphs: an efficient topology for point-to-point multiprocessor networks. IEEE Trans. Parallel and Distributed Systems **3**(1992) 622–626
7. Garfgano, L., Vaccaro, U., Vozella, A.: Fault tolerant routing in the star and pancake interconnection networks. Information Processing Letters **45**(1993) 315–320
8. Gu, Q.-P., Peng, S.: Node-to-set disjoint paths problem in star graphs. Information Processing Letters **62**(1997) 201–207
9. Hamada, Y., Bao, F., Mei, A., Igarashi, Y.: Nonadaptive fault-tolerant file transmission in rotator graphs. IEICE Trans. Fundamentals **E79-A**(1996) 477–482
10. Kaneko, K., Suzuki, Y.: An algorithm for node-to-set disjoint paths problem in rotator graphs. IEICE Trans. Inf. & Syst. **E84-D**(2001) 1155–1163
11. Kaneko, K., Suzuki, Y.: Node-to-set disjoint paths problem in pancake graphs. IEICE Trans. Inf. & Syst. **E86-D**(2001) 1628–1633
12. Lin, H.-R., Hsu, C.-C.: Topological properties of bi-rotator graphs. IEICE Trans. Inf. & Syst. **E86-D**(2003) 2172–2178
13. Qiu, K., Meijer, H., Akl, S. G.: Parallel routing and sorting on the pancake network. Proc. Int'l Conf. Computing and Information (1991) 360–371
14. Rabin, M. O.: Efficient dispersal of information for security, load balancing, and fault tolerance. JACM **36**(1989) 335–348
15. Suzuki, Y., Kaneko, K.: An algorithm for node-disjoint paths in pancake graphs. IEICE Trans. Inf. & Syst. **E86-D**(2003) 610–615

QoSRHMM: A QoS-Aware Ring-Based Hierarchical Multi-path Multicast Routing Protocol

Guojun Wang[1,2], Jun Luo[2], Jiannong Cao[1], and Keith C. C. Chan[1]

[1] Department of Computing, Hong Kong Polytechnic University,
Hung Hom, Kowloon, Hong Kong
[2] School of Information Science and Engineering, Central South University,
Changsha, Hunan Province, P. R. China 410083

Abstract. We propose a QoS-aware multicast routing protocol called QoSRHMM based on a ring-based hierarchical network structure, which establishes a *Ring-Tree* by the principle of "*ring* in the same tier of the hierarchy and *tree* in different tiers". It forms one shortest-delay path and multiple paths with delay-constrained minimal-cost between the sender and each receiver. When establishing a connection on demand, a node joins a logical ring while satisfying some QoS constraints. Mobile hosts are served in time with multicast services even in case of handoff.

1 Introduction

QoS-aware multicast shares the same goals as QoS-aware unicast: to find routing path that satisfies QoS constraints, and to utilize network resources effectively. In multimedia multicast applications, multimedia data are transmitted via a network using multicast path with delay-constrained minimal-cost. After receiving a connection request, the node establishes a multicast path as quickly as possible, and it is delay-constrained for transmitting data from a source to all destinations. The whole cost of the transmissions should be minimal. It is well known that it is an NP-complete problem to find a path satisfying the two constraint conditions.

Researchers have proposed many QoS-aware solutions such as Dijkstra, Prim, Kruskal and Bellman-Ford algorithms [2-3]. Because of the frequent change of the network topologies, these algorithms can't be directly applied to wireless networks. Wu and Hou proposed an adaptive framework to guarantee QoS in wireless IP networks [4]. Talukdar extended RSVP to MRSVP in mobile Internet [5]. Acampora and Naghshineh constructed a Virtual Connection Tree (VCT) for handoff processing in wireless ATM networks [6]. Gao and Acampora established a VCT in a micro-mobility domain to guarantee QoS [7]. We have also got some valuable information from references [8-10].

We propose a QoS-aware multicast routing protocol in mobile Internet called QoSRHMM (QoS-aware Ring-based Hierarchical Multi-path Multicast routing protocol). QoSRHMM is a protocol based on a ring-based hierarchical network structure. The main advantages of the proposed protocol are the following:

1. It inherits some advantages from the ring-based hierarchical network structure, such as high stability and scalability.

2. There are a shortest-delay path and multiple paths with delay-constrained minimal-cost between the sender and each receiver.
3. In case of handoff, mobile hosts may be served in time without reconstructing the multicast tree.
4. It is a distributed protocol that all the paths are established by local information only.

2 The Ring-Based Hierarchy

We proposed a ring-based hierarchical network structure in mobile Internet for group membership in [1]. The basic idea is that all the members in a multicast group are organized into four tiers: Border Router Tier (BRT), Access Gateway Tier (AGT), Access Proxy Tier (APT), and Mobile Host Tier (MHT). The higher three tiers are organized into logical rings. Each ring has a leader that is also responsible for interacting with upper tiers. Access Proxies (APs) are the Network Entities (NEs) that communicate directly with the Mobile Hosts (MHs). Access Gateways (AGs) are the NEs that communicate either between different wireless networks or between one wireless network and one wired network. Border Routers (BRs) are the NEs that communicate among administrative domains.

The proposed protocol builds a Ring-Tree based on the ring-based hierarchy. Network nodes are organized as rings according to some QoS requirements. The structure becomes a multicast tree if we look each ring as one node.

3 The Basic Concepts of the Ring-Tree

In this section, we define some QoS parameters and concepts. In Fig. 1, we give an example *Ring-Tree* to illustrate these concepts.

Logical Ring: There are some logical rings in which some nodes are in the same tier of the Ring-Tree to function similarly. Every node in a logical ring records information about its *leader, rear, previous,* and *next* neighbors. A logical ring can overlap with each other, i.e., a member of a ring can also be a member of another ring. A logical ring is associated with a *Ring-Strong-Path* and a *Ring-Weak-Path*. We use *ring* as *logical ring* for short. In Fig. 1, there exist 12 rings called R1, R2, ..., R12, and R12 contains only one node.

Leader: There's one leader in a ring to communicate with upper ring and maintain the ring. If upper ring exists, it is the child of the rear of its direct upper ring. For example, R1's leader is BR1 and both R8 and R3's leaders are AG1.

Rear: There's only one rear in a ring. It is responsible for the construction of a ring. When receiving a join request from a lower tier, the node sets itself the rear and sets itself the parent node of the leader of its direct lower ring. At the same time, it issues a connection request to establish a ring from itself to the leader. For example, BR2 is R1's rear, AG3 is R3's rear, and AG2 is R8's rear. R3 and R8 share the same leader AG1 and a path from AG2 to AG1. While being a member node in R3, AG4 is also the rear in R9.

Ring-Strong-Path: It is a path from rear to leader in a ring. It is the shortest-delay path and the main multicast data transmission path. A Ring-Strong-Path is shown in a solid curve, and a shared Ring-Strong-Path is shown in bold (e.g., one path is shared by R3 and R8, another by R6 and R10).

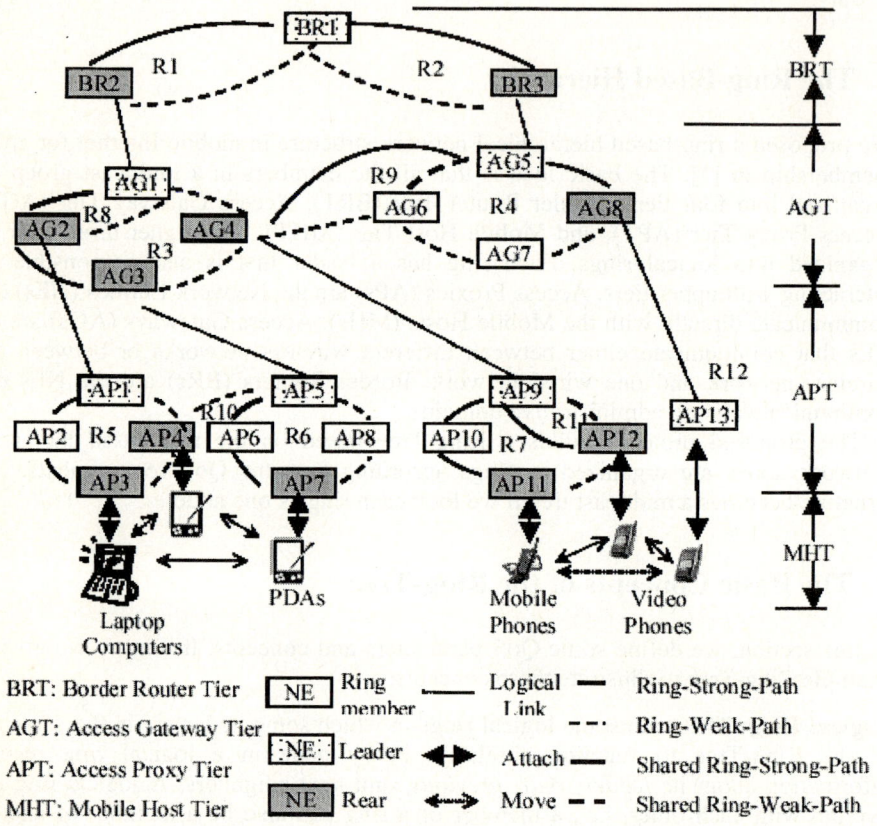

Fig. 1. The Ring-Tree

Ring-Weak-Path: It is a path with delay-constrained minimal-cost. It forms the closed logical ring together with a Ring-Strong-Path. If required, it can also function as an alternative multicast data transmission path. A Ring-Weak-Path is shown in a dashed curve, and a shared Ring-Weak-Path is shown in bold, dashed curve (e.g., one path is shared by R9 and R4, another by R7 and R11).

The network is represented by an undirected weighted graph, where $V=(v_1, v_2, \ldots, v_n)$ is the set of nodes that represent hosts or routers in the network, $E=(e_1, e_2, \ldots, e_l)$ is the set of links between node-pairs. A link from node i to node j defines two positive real weighted value ($COST_{i,j}$, $DELAY_{i,j}$). $COST_{i,j}$ is the cost of data transmission from i to j and it reflects the utilization of link resources. $DELAY_{i,j}$ is the delay of data transmission from i to j and it includes queuing-delay and transmission-delay. We assume that $COST_{i,j}=COST_{j,i}$ and $DELAY_{i,j}=DELAY_{j,i}$.

Consider a QoS-constrained multicast connection request R=(s, D, Δ), where s is the source node, D is a set of destination nodes, Δ is the delay constraint. We construct a Ring-Tree (RT) with s, D, and Δ. We define $DIST_i$ as the maximum distance between the leader and rear of ring Ri. Therefore the whole cost of Ri is $COST(DIST_i)$. If we treat each R_i as a node, then the multicast tree is $T = (E_T, V_T), \{E_T \subset E, V_{T_i} = R_i, i = 1, 2...n\}$. The whole cost of RT is:

$$COST(RT) = \sum_{i=1}^{n} COST(DISTi) + \sum_{e \in E_T(s,D)} COST(e) \quad (1)$$

Where n is the number of RT's rings, and $E_T(s, D)$ is the set of paths from source to destinations in a tree T.

Suppose delay from source s to receiver v is DELAY(s, v), then RT's delay is:

$$DELAY(RT) = \max(DELAY(s,d)), \forall d \in D \quad (2)$$

We consider the delay constraint condition:

$$DELAY(RT) \leq \Delta \quad (3)$$

In this paper, we design a distributed routing algorithm to construct a multicast Ring-Tree based on the ring-based hierarchy. It gets an optimal value of formula (1) when satisfying formula (3).

4 The QoSRHMM Protocol

4.1 The Basic Idea of QoSRHMM

The QoSRHMM protocol is a hierarchical protocol. Multicast data coming from a multicast source are forwarded by BRs along BR ring downward to its lower tiers finally to MHs. Our goal is to establish a Ring-Tree from BRs to all the member MHs. We assume all the nodes in the network have the knowledge of their neighbors through which they can attach to the Ring-Tree.

QoSRHMM establishes a Ring-Tree by the principle of "ring in the same tier of the hierarchy and tree in different tiers". If we look each ring as a node, the Ring-Tree is a tree with height of 4.

In a tier, there are some nodes that directly connect upper tier or lower tier: If one of these nodes finds a node in its direct lower tier that need to join a multicast group, it must be the rear of a ring. If one of these nodes finds a node in the same tier that need to communicate with upper tier via itself, it must be the leader of a ring. We call such nodes as *border-nodes*.

A rear first finds a leader in the same tier and establishes a path through which it can reach the source. This path from a rear to a leader called a Ring-Strong-Path. A Ring-Strong-Path has the shortest delay to the source. After that, the rear tries to find another path with delay-constrained minimal-cost using the proposed *K-BFS* algorithm, which is an extension to the Breadth-First-Search algorithm. To avoid packets flooding, at most K packets are forwarded at each node to find the path. The path is called a Ring-Weak-Path. A Ring-Strong-Path and a Ring-Weak-Path form a logical ring.

Consider that multiple rings may overlap with each other. When a node in an existing Ring-Strong-Path needs to form a new ring, there must be a shared path from itself to the leader because all the nodes in the Ring-Strong-Path have the shortest delay destined to the source. Therefore, we only need to establish a new Ring-Weak-Path that destined to the source with delay-constrained minimal-cost.

A node within an existing Ring-Weak-Path forms a new ring by re-establishing a Ring-Strong-Path and a Ring-Weak-Path. To decrease the delay of establishment, multicast data may transmit along the existing Ring-Weak-Path before new Ring-Strong-Path has been established because the existing Ring-Weak-Path is with delay-constrained minimal-cost.

We introduce three cases that build ring-strong and ring-weak paths in Fig.1:

1. Assume R6 does not exist. Consider AP7 needs to join a multicast group because of receiving the request from an attached MH. It sets itself a rear and forms a ring R6 with a Ring-Strong-Path and a Ring-Weak-Path in APT.
2. Assume R3 and R1 exist. Consider AP1 in R5 needs to attach to AG2. AG2 sets itself the rear of a new ring R8. Since R8 shared a Ring-Strong-Path with R3, it then constructs a new Ring-Weak-Path from itself to AG1 only.
3. Assume AG4 is in the Ring-Weak-Path of R3. Consider it needs to form a logical ring with AG5 as leader. It constructs a new Ring-Strong-Path from itself to AG5. After that, it tries to construct a Ring-Weak-Path to form a ring R9. It then shares a Ring-Weak-Path from AG6 to AG5 with R4.

When forming a Ring-Weak-Path, the proposed K-BFS algorithm may find multiple paths from the rear to the leader. We obey the principle of "forming a larger ring" by allowing the number of nodes to join a ring as large as possible. Therefore, we argue that it is very probable for the neighbors of an AP node where the MH attached to join a ring more easily. That is, when an MH hands-off to a new AP, it is very probable for the MH to get multicast services in time. Assuming two paths P_i and P_j, if $P_i > P_j$ (i.e., the number of nodes in path P_i is more than that in P_j), we adopt P_i as the current Ring-Weak-Path.

4.2 The Basic Data Structure and Basic Operations

CR_TAB, DR_TAB: They represent Cost-Router-Table and Delay-Router-Table, respectively. The potential output links of minimum-cost/shortest-delay are recorded in the tables. The first output link is the main link while others are candidates. These tables may be formed based on *Distance-Vector* or *Link-Status* routing algorithms.

Join_request: It is sent by a child node to inform the parent node to let the node to join the multicast group.

S_search, W_search: They are sent by a rear to initiate the Ring-Strong-Path or Ring-Weak-Path and they will be forwarded in the same tier only. When forwarding, they collect information about nodes along the path.

Strong_Path_OK, Weak_Path_OK: When the leader receives *S_search/W_search*, it assumes that the Ring-Strong-Path/Ring-Weak-Path has been correctly established.

Strong_Path_OK/Weak_Path_OK is sent back to the rear by the leader to confirm the Ring-Strong-Path/the Ring-Weak-Path.

Status(id): A set of node's current status such as its node *id*, *ring_id*, *rear(id)*, *leader(id)*, *strong-path-list(id)*, *weak-path-list(id)*. It may also record child node or parent node for a rear or leader. As mentioned before, a ring's member node can also be another ring's member. We identify these attributes of one node in different rings by *ring_id*.

Connection_OK: When a node gets to know that the path up to the sender has been successfully established, it sends a *Connection_OK* downward to the receiver along the Ring-Strong-Path to inform it to wait for data transmission.

In addition, some assisted functions such as ***search_path()*** and ***new()*** are used for searching a path and for *id* assignment.

4.3 The Description of the QoSRHMM Protocol

QoSRHMM is receiver-oriented. Assuming a mobile host MH_i needs to join a multicast group, it sends a *Join_request* message to its attached AP_j:

1. After receiving a *Join_request*, if the node is the sender, a rear, or a leader, then go to step (7); If the node is in a Ring-Strong-Path, then change itself to a new rear and the Ring-Strong-Path to be a new Ring-Strong-Path and go to step (4); Otherwise change itself to be a new rear and go to step (2).
2. The rear searches *DR_TAB* and finds a link with the shortest-delay and sends an *S_search* through the link.
3. After receiving an *S_search*, if it is a border-node that an *S_search* message first reaches, it then requests to join the upper tier by searching *DR_TAB*, changes itself to be a leader, and sends back a *Strong_Path_OK* to the rear. If not a border-node, it just forwards *S_search*.
4. After receiving a *Strong_Path_OK*, the rear sends a *W_search* to search a Ring-Weak-Path to the leader by the *K-BFS* algorithm.
5. After receiving a *W_search*, the leader sends a *Weak_Path_OK* to the rear.
6. Go to step (1).
7. Send back a *Connection_OK*.
8. The receiver receives a *Connection_OK* and then waits for data transmission.

The *K-BFS* algorithm used for searching a leader is described as follows:

1. A rear sends a *W_search* along all its adjacent links.
2. For forwarding the *W_search*, each node only selects at most *K* links with delay-constrained minimal-cost in order to reduce the flooding of messages. If more than one *W_search* message has been received, only one *W_search* is forwarded.
3. After receiving a *W_search*, the leader establishes a Ring-Weak-Path. If receiving more than one *W_search* message, the leader replaces the Ring-Weak-Path with an optimal one by the principle of "forming a larger ring".

The pseudo-code of QoSRHMM is as follows:

```
ParFor each node running QoSRHMM {
  On Receiving Join_request:
    if (node.is_sender || node.leader!=NULL || node.rear!=NULL)
```

```
        send(receiver, "Connection_OK");
    elseif (node.in_ring_strong_path){
        ring_id = new(ring_id);
        node.rear(ring_id) = node.rear(pre_ring_id);
        node.leader(ring_id) = node.leader(pre_ring_id);
        node.strong_path_list(ring_id)=
            node.strong_path_list(pre_ring_id);
        ParFor (i =1; i <= K; i++){//search a ring_weak_path
                                  //by K-BFS
          next_hop = search_path(CR_TAB, DR_TAB, QoS, i);
          send(next_hop, "W_search");}}
    else {//ready to form a new ring
        ring_id = new(ring_id);   node.rear(ring_id) = node.id;
        next_hop = search_path(DR_TAB,shortest_delay);
        send(next_hop,"S_search");}
    pre_hop = Join_request.source_id;
    node.child(ring_id) = pre_hop;
On Receiving S_search:
    ring_id = S_search.ring_id;
    next_hop = search_path(DR_TAB, shortest_delay),
"S_search");
    if (next_hop is in an upper tier){//S_search reaches
                                     //the leader
        node.parent(ring_id) = next.hop;
        node.leader(ring_id) = node.id;
        node.rear(ring_id) = S_search.source_id;
        node.strong_path_list(ring_id)=S_search.path_list;
        send(next_hop, "Join_request");
        send(node.rear(ring_id), "Strong_Path_OK");}//send back
                                                   //to confirm
    else send(next_hop,"S_search");//forward the message
On Receiving Strong_Path_OK:
    ring_id = Strong_Path_OK.ring_id;
    if (node.rear(ring_id) != NULL){ //node is the rear
       node.strong_path_list(ring_id)=Strong_Path_OK.path_list;
```

```
            for (i=1; i <= K; i++){//search a ring_weak_path
                                  //to ring leader by K-BFS
                next_hop=search_path(CR_TAB,DR_TAB,QoS,i);
                send(next_hop, "W_search");}}
         else send(Strong_Path_OK.destination,"Strong_Path_OK");
On Receiving W_search:
     ring_id = W_search.ring_id;
     if (W_search is received the first time){
       if (node. id == W_search.destination){//W_search reaches
                                             // the leader
          node.weak_path_list(ring_id)= W_search.path_list;
          send(node.rear(ring_id), "Weak_Path_OK");}
       else{ //forward W_search by K-BFS
         for (i=1; i <= K; i++){
           next_hop=search_path(CR_TAB,DR_TAB,QoS,i);
           send(node.status(next_id).leader, "W_search");}}
       elseif (node.id == W_search.destination) //multiple
                                              //W_search messages
           if(W_search.path_list > node.weak_path_list(ring_id))
                                              //replace the path
              node.status(id).weak_path_list=W_search.path_list;}
On Receiving Weak_Path_OK:
     ring_id = Weak_Path_OK.ring_id;
     if (node.rear(ring_id) != NULL){ //it is the rear
        node.weak_path_list(ring_id) = Weak_Path_OK.path_list; }
     else send(Weak_Path.destination, "Weak_Path_OK");//forward
                                                     //the message
On Receiving Connection_OK:
     send(receiver, "Connection_OK");
}
```

4.4 The Algorithmic Analysis

Theorem 4.1: A sender and all the receivers in a group are included in the Ring-Tree constructed by QoSRHMM protocol, and one shortest-delay path and multiple paths with delay-constrained minimal-cost are established between the sender and each receiver.

Proof: All the multicast paths are established on demand. Obviously, the Ring-Tree includes a sender and all the receivers in a group.

When a receiver joins a multicast group, an *S_search* is forwarded along the path that is selected with the shortest delay. When a ring is being constructed, the border-node that the *S_search* first reaches becomes the leader.

Ring-Weak-Path is established by forwarding *W_search* messages and *K* paths are selected with minimal-cost based on the delay constraint. Assuming the ring *id* is *i*, RS_i is the Ring-Strong-Path and RW_i is the Ring-Weak-Path, then $DELAY(RS_i) \le DELAY(RW_i), \forall i \in n$. Then we got:

$$DELAY(\sum_{i=1}^{n} RS_i) \le DELAY(\sum_{i=1}^{n} RW_i) \le \Delta \qquad (4)$$

which completes the proof.

Lemma 4.1: The time complexity of QoSRHMM for a member-join operation is that of ring's establishment plus that of message transmissions between different tiers.

Proof: The QoSRHMM is receiver-oriented. The relationship of different tiers is that of parent node and child node. Because firstly a ring in the same tier is formed and then the tree is constructed in different tiers, we conclude that the time complexity is that of the establishment of logical rings plus that of message transmissions between different tiers.

Theorem 4.2: The maximum running time of QoSRHMM for a member-join operation is 4Δ.

Proof: Firstly there are three transmissions of the *Join_request* message in a connection process. That is a *Join_request* transmission from MHs to APs, from APs to Ags, and from AGs to BRs. The transmissions are all from a child node to a parent node. Secondly *S_search* and *W_search* messages transmit in the same tier and the forwarding between nodes is delay-constrained. Therefore, the running time of *S_search* and *W_search* satisfies with Δ. Finally the *Strong_Path_OK* and *Connection_OK* are transmitted along Ring-Strong-Path. All these messages satisfy with Δ. Consider *Weak_Path_OK* is forwarded concurrently with *Connection_OK*, its delay is less than that of *Connection_OK*.

$$3*DELAY_{i,j}(Join\_request)+3*DELAY_{ring\_rear,ring\_leader}(S\_search) < \Delta \qquad (5)$$

$$DELAY_{ring\_rear,\ ring\_leader}(W\_search) < \Delta \qquad (6)$$

$$DELAY_{ring\_leader,\ ring\_rear}(Strong\_Path\_OK) < \Delta \qquad (7)$$

$$DELAY_{sender,\ receiver}(Connection\_OK) < \Delta \qquad (8)$$

According to (5)+(6)+(7)+(8) < 4Δ, we conclude from Lemma 1 that the maximum running time of QoSRHMM is 4Δ.

Theorem 4.3: The total communication complexity of QoSRHMM for a member-join operation is *O(n)*.

Proof: In the process of constructing a Ring-Tree, the *Join_request, S_search, Strong_Path_OK, Weak_Path_OK,* and *Connection_OK* messages are all less than $O(n)$. Since the *W_search* messages are transmitted in parallel, the complexity is $O(n)$. Therefore, the total communication complexity is $O(n)$.

Acknowledgements

This work was supported by the Hong Kong Polytechnic University Central Research Grant *G-YY41*, the University Grant Council of Hong Kong under the CERG Grant PolyU *5170/03E*, and the China Postdoctoral Science Foundation (*No. 2003033472*).

References

1. Wang G., Cao J., Chan K.C.C.: RGB: A Scalable and Reliable Group Membership Protocol in Mobile Internet, *Proceedings of the 33rd International Conference on Parallel Processing (ICPP 2004)*, Montreal, Quebec, Canada, Aug. 2004, pp. 326-333.
2. Cormen T.H., Leiserson C.E., Rivest R.L., Stein C.: *Introduction to Algorithms*, second edition, Boston: McGraw-Hill, 2001, pp. 580-607.
3. Lynch N.A.: *Distributed Algorithms*, San Francisco: Morgan Kaufmann Publishers, Inc, 1996, pp. 274-302.
4. Wu D., Hou Y.T., Zhang Y.-Q.: Scalable Video Transport over Wireless IP Networks, *Proceedings of the IEEE International Symposium on Personal, Indoor and Mobile Radio Communication (PIMRC 2000)*, London, UK, 2000, pp. 18-21.
5. Talukdar A.K., Badrinath B.R., Acharya A.: MRSVP: A Resource Reservation Protocol for an Integrated Services Network with Mobile Hosts, *Wireless Networks*, 2001, 7(1): 5-19.
6. Acampora A., Naghshineh M.: An Architecture and Methodology for Mobile-Executed Cell Hand-off in Wireless ATM Networks, *IEEE Journal on Selected Areas in Communications*, 1994, 12(8): 1365-1375.
7. Gao Q., Acampora A.: Connection Tree based Micro-mobility Management for IP-centric Mobile Networks, *Proceedings of the IEEE International Conference on Communications (ICC 2002)*, Denmark, 2002, pp. 3307-3312.
8. Du D.Z., Smith J.M., Rubinstein J.H.: *Advances in Steiner Trees*, Boston: Kluwer Academic Publishers, 2000, pp. 163-174.
9. Lee H.Y., Youn C.H.: Scalable Multicast Routing Algorithm for Delay-Variation Constrained Minimum-cost Tree, *Proceedings of the IEEE International Conference on Communications (ICC 2000)*, New Orleans, USA, 2000, pp. 1343-1347.
10. Baldi M., Ofek Y., Yenter B.: Adaptive Group Multicast with Time-driven Priority, *IEEE/ACM Transactions on Networking*, 2000, 8(1): 31-43.

A Dynamic Task Scheduling Algorithm for Grid Computing System

Yuanyuan Zhang[1], Yasushi Inoguchi[2], and Hong Shen[1]

[1] Graduate School of Information Science,
[2] Center for Information Science,
Japan Advanced Institute of Science and Technology,
1-1 Asahidai, Tatsunokuchi, Ishikawa, 923-1292, Japan
{yuanyuan, inoguchi, shen}@jaist.ac.jp

Abstract. In this paper, we propose a dynamic task scheduling algorithm which assigns tasks with precedence constraints to processors in a Grid computing system. The proposed scheduling algorithm bases on a modified static scheduling algorithm and takes into account the heterogeneous and dynamic natures of resources in Grid.

1 Introduction

Grid computing[1], the internet-based infrastructure that aggregates geographically distributed and heterogeneous resources to solve large-scale problems, is becoming increasingly popular. Heterogeneity, dynamicity, scalability and autonomy are four key characteristics of Grid.

A critical issue for the performance of Grid is that of task scheduling. Task scheduling problem is, in general, the problem of scheduling tasks to processors so that all the tasks can finish their execution in the minimal time. Since this problem is NP-complete in general, it is necessary to employ heuristics to arrive at a near-optimal solution.

Many task scheduling heuristics have been proposed for heterogeneous system [2, 3, 4], however, most of these algorithms can't work for Grid directly because resources in Grid are typically heterogeneous and the performance of the resources dynamically fluctuates over time as the resources are not dedicated to Grid.

In this paper, we propose a new scheduling algorithm which assigns precedence-constrained tasks to the processors in a Grid computing system. Our algorithm takes the dynamicity and heterogeneity of Grid into consideration so that it can better deal with the dynamics of dynamically varying resource state in Grid.

The rest of this paper is organized as follows: We formalize the problem and describe the proposed algorithm in Section 2 and Section 3 concludes the paper.

2 The Proposed Algorithm

2.1 Problem Statement

An application to be executed on the Grid system consists of many precedence-constrained tasks. The application is represented by a Directed Acyclic Graph in which every node represents a task of the application, and each directed edge represents a communication(comm. in short) link between two tasks. Edge $e_{i,j}$ represents the dependence relationship between task T_i and task T_j that T_i must finish before T_j can start.

For edge $e_{i,j}$, we call T_j a *successor* of T_i, and T_i a *predecessor* of T_j. The size of data communicated from T_i to T_j is $data_{i,j}$. A task is called an entry task if it has no predecessor, while a task is called an exit task if it has no successor. We assume that there is only one entry task and one exit task in the application.

The communication cost between task T_i and T_j, denoted by $C_{i,j}$, is the expected transfer time of the data for the communication given current network conditions. For this purpose, Network Weather Service(NWS) can be used to obtain an estimate of the current network latency and bandwidth. It is assumed that the communication cost on a processor is neglectable.

Let τ_i denote the workload of Grid task T_i, and v_m be the speed of processor P_m, then τ_i/v_m is the time to implement T_i on P_m when P_m is dedicated to execute T_i. However, the actual speed of P_m delivered to Grid is less than v_m and varies over time since the resource owner also uses it and the local jobs have higher priority over the Grid tasks. This may dramatically impact the performance of Grid resources, and makes the problem more difficult.

The objective function is to schedule the tasks of the application to the processors in Grid so that to minimize the total execution time.

2.2 The Proposed Task Scheduling Algorithm

Most static scheduling algorithms assume that task execution times and data transfer times can be estimated accurately and are commonly used with dedicated resources. However, the fact in Grid is that it is often difficult to obtain such accurate estimation before execution. Therefore, dynamic scheduling algorithms may have better performance on Grid because they are able to adapt to environmental dynamicity. In this paper we propose a new dynamic algorithm to deal with the dynamic and heterogeneous features of Grid.

First we must solve the nondedicated feature of Grid: during the implementation of a Grid task on a processor, the local jobs on the processor will arrive and interrupt the implementation of the Grid task. We consider the execution of the local jobs as non-preemptive, i.e., a local job must run until completion once it starts. The execution of the local jobs follows the rule of *first-come first-serve*. From the viewpoint of the Grid task, the state of the processor alternates between available and unavailable: when the processor is executing its own jobs, it's unavailable for the Grid task, otherwise it's available for the Grid task.

If we assume the arrival of the local jobs in processor P_m follows a Poisson distribution with arrival rate λ_m, and their execution process follows an exponential distribution with service rate μ_m, then the local job process in P_m is an M/M/1 queuing system.

The expected execution time $w^e_{i,m}$ of task T_i on P_m can be expressed as:

$$w^e_{i,m} = X^m_1 + Y^m_1 + X^m_2 + Y^m_2 ... + X^m_{N_m} + Y^m_{N_m}, \quad (1)$$

where N_m is the number of local jobs which arrive during the execution of T_i, and $X^m_j, Y^m_j (j = 1, ..., N_m)$, are respectively the computing time of a section of the Grid task and a local job. $Y^m_j (j = 1, 2, ..., N_m)$ are independent identical distribution(i.i.d.) random variables. We have:

$$X^m_1 + X^m_2 + ... + X^m_{N_m} = \tau_i/v_m. \quad (2)$$

From the knowledge of queuing theory, we have:

$$E(N_m) = \lambda_m \tau_i / v_m, \quad E(Y^m_j) = 1/(\mu_m - \lambda_m). \quad (3)$$

Since N_m and Y_j are independent$(j = 1, ..., N_m)$, we can derive:

$$\mathrm{E}(w^e_{i,m}) = E(E(w^e_{i,m}|N_m)) = \frac{\tau_i}{v_m(1-\rho_m)}. \quad (4)$$

where $\rho_m = \lambda_m/\mu_m$, is the utilization rate of P_m.

For a processor P_m with utilization rate ρ_m, we can use $\frac{\tau_i}{v_m(1-\rho_m)}$ as the expected execution time of T_i on P_m. However, ρ_m is a value that reflects the dynamicity of the Grid during a long time. It can not reflect the dynamicity during the execution of the application, therefore we introduce the concept of *processor credibility* which reflects the history of prediction accuracy for a processor during the execution of a Grid application. We denote the credibility of P_m as δ_m, which has the original value of 1 when we schedule the application. After a task T_i finishes execution on P_m, we can obtain its actual execution time $w^a_{i,m}$ and so that δ_m is modified as:

$$\delta_m = (1-\alpha)\delta_m + \alpha * w^a_{i,m}/w^e_{i,m}, \quad (5)$$

where α is a value between 0 and 1 and can be modified.

Therefore the expected execution time $w^e_{i,m}$ of T_i on P_m is modified as:

$$w^e_{i,m} = \frac{\tau_i \delta_m}{v_m(1-\rho_m)}. \quad (6)$$

The *average execution time* of T_i, which is denoted by w_i, is defined as:

$$w_i = \sum_{m=1}^{q} w^e_{i,m}/q = \sum_{m=1}^{q} \frac{\tau_i \delta_m}{v_m q (1-\rho_m)}, \quad (7)$$

where q is the number of processors in Grid.

The dynamic algorithm we propose bases on a static algorithm which is a modified Heterogeneous Earliest Finish Time(HEFT) algorithm[3]. HEFT is a traditional task scheduling algorithm for precedence-constrained tasks in heterogeneous systems. The scheduling process of this algorithm includes two phases: task selection and processor selection. In task selection phase, the tasks are queued by non-increasing ranks. For task T_i, its rank $rank_u(T_i)$, is computed as:

$$rank_u(T_i) = \omega_i + \max_{T_j \in succ(T_i)} (C_{i,j} + rank_u(T_j)). \tag{8}$$

$$\text{Here} \quad C_{i,j} = L + \frac{data_{i,j}}{B}. \tag{9}$$

In the above equation $succ(T_i)$ is the set of the successors of T_i. L and B are respectively the *average communication startup time* and *average bandwidth* among all the processors. $rank_u$ of each task is computed by traversing all the tasks upward, starting from the exit task whose $rank_u$ is defined as:

$$rank_u(T_{exit}) = \omega_{exit}. \tag{10}$$

In processor selection phase, the first task in the list is selected and allocated to the processor which gives it the minimal earliest finish time(EFT).

For task T_i, let $\text{EST}(T_i, P_m)$ and $\text{EFT}(T_i, P_m)$ denote its *earliest start time* and *earliest finish time* on processor P_m. $\text{AFT}(T_i)$ is its *actual finish time*. $\text{avail}[P_m]$ denotes the time when P_m is ready for executing new tasks. We also denote the *comm. finish time* between T_i and its successor T_j as $\text{CFT}(T_i, T_j)$.

In the HEFT algorithm, when computing $\text{EFT}(T_i, P_m)$ of T_i on P_m, the algorithm computes $\text{CFT}(T_j, T_i)$ between T_i and its predecessor T_j which has not been scheduled to P_m, and determines $\text{EST}(T_i, P_m)$ as the maximum value between $\text{CFT}(T_j, T_i)$ and $\text{avail}[P_m]$. If we have decided to schedule T_i on processor P_m, and if $\text{avail}[P_m]$ is less than $\text{CFT}(T_j, T_i)$, then P_m will remain idle when it is waiting for the finish of communication between T_j and T_i. However, if we adopt the idea of task duplication scheme in this algorithm, that is, when the communication time is long compared with computation time, we duplicate T_j which has been scheduled to other processors on P_m, thus the communication time will be removed, so that the total execution time of the tasks will be probably shortened. We name this modified HEFT algorithm as Duplication-based HEFT(DHEFT) algorithm.

The performance of DHEFT depends heavily on the granularity of communication between the tasks. For communication-intensive applications, it can achieve much better performance than HEFT algorithm, otherwise its advantage will be trivial. So we use DHEFT algorithm only for communication-intensive Grid application, otherwise we use the original HEFT algorithm. In our future work the threshold of comm.-comp. ratio for judging if an application should be scheduled using DHEFT or not will be determined.

In our dynamic scheduling algorithm, at first we relate every task with a level so that tasks with same level are independent. The level of a task is defined as the length of the longest path from the entry task to the present task. By the length of a path we mean the number of tasks on the path. The level of

T_{entry} is 1. After we schedule T_{entry} to a processor, we begin its execution immediately. Then when tasks with level i-1 start execution($i>1$) we dynamically schedule tasks with level i using DHEFT algorithm for communication-intensive tasks or HEFT algorithm. When $i>2$, since tasks with level less than i-1 have finished execution when we schedule tasks with level i , we use the estimation error of the finished tasks to modify the credibility of every processor using equation (6), so that to regulate the estimated execution time of the tasks in level i .

The following pseudocode shows the implementation process of DHEFT algorithm and the dynamic scheduling algorithm:

Duplication-based HEFT Algorithm (DHEFT)	Dynamic Task Scheduling Algorithm
Sort the tasks in list Q_1 by decreasing $rank_u$	Compute the level of every task
While there are tasks remained in Q_1 {	$EST(T_{entry}) = 0$
Select the first task T_i from Q_1	For every processor P_m
For processor P_k { t = avail[P_k]	avail[P_m] = 0
If all the predecessors of T_i are on P_k or	$Q_1 \leftarrow$ set of tasks whose level is 1
T_i is the entry task	Compute $rank_u$ of every task in Q_1
$EFT(T_i, P_k) = t + \omega_{i,k}^e$	Schedule the tasks in Q_1
Else	Start the execution of tasks in Q_1
$EST(T_i, P_k) = t$	$Q_1 \leftarrow$ set of tasks with level 2
Sort predecessors of T_i which are not	Compute $rank_u$ of tasks in Q_1
on P_k in Q_2 by decreasing $rank_u$	Schedule the tasks in Q_1
Select the first task T_j from Q_2	For i = 2; i < level(T_{exit}); i++;
If $EFT(T_j) + C_{j,i} \leq t + \omega_{j,k}^e$	Wait until any task in level i begins
$EST(T_i, P_k) = MAX(t, EFT(T_j) + C_{j,i})$	execution
Else	Refresh the credibility of every
$t = t + \omega_{j,k}^e$	processor using actual execution
$EST(T_i, P_k) = t$	times of tasks in level i-1
$EFT(T_i, P_k) = EST(T_i, P_k) + \omega_{i,k}^e$}	$Q_1 \leftarrow$ set of tasks with level i+1
Allocate T_i to P_m which gives it the minimal	Compute $rank_u$ of tasks in Q_1
EFT(T_i, P_m)	Schedule the tasks in Q_1
avail[P_m] = EFT(T_i,P_m) }	

3 Conclusion

In this paper, we propose a task scheduling algorithm for Grid computing system. The proposed scheduling algorithm considers the dynamicity and heterogeneity of Grid. It combines the list scheduling and task duplication scheme. It also uses the execution results of the finished tasks to regulate the expected execution times of the tasks to be scheduled, so that to obtain more accurate estimation and arrive at better scheduling result.

References

1. I. Foster and C. Kesselman, *The Grid: Blueprint for a New Computing Infrastructure*, Morgan Kaufmann Publishers, San Fransisco, CA, 1999.
2. C. Banino, O. Beaumont, A. Legrand, and Y. Robert, "Scheduling strategies for master-slave tasking on heterogeneous processor grids," Applied Parallel Computing: Advanced Scientific Computing: 6th Int'l Conf., pp. 423-432, Jun. 2002.
3. H. Topcuoglu, S. Hariri, and M.-Y. Wu, "Performance-Effective and Low-Complexity Task Scheduling for Heterogeneous Computing," *IEEE Trans. Parallel and Distributed Systems*, vol. 13, no. 3, pp. 260-274, Mar. 2002.
4. J.-C. Liou and M.A. Palis, "An Efficient Task Clustering Heuristic for Scheduling DAGs on Multiprocessors," Workshop on Resource Management, Symposium on Parallel and Distributed Processing. 1996.

Replica Selection on Co-allocation Data Grids

Ruay-Shiung Chang, Chih-Min Wang, and Po-Hung Chen

Department of Computer Science and Information Engineering,
National Dong Hwa University,
Shoufeng, Hualien 974, Taiwan
rschang@mail.ndhu.edu.tw

Abstract. Data Grid supports data-intensive applications in a large scale grid environment. It makes use of storage systems as distributed data stores by replicating contents. On the co-allocation architecture, the client can divide a file into k blocks of equal size and download the blocks dynamically from multiple servers by GridFTP in parallel. But the drawback is that faster servers must wait for the slowest server to deliver the final block. Therefore, designing efficient strategies for accessing a file from multiple copies is very import. In this paper, we propose two replica retrieval approaches, abort-and-retransfer and one by one co-allocation, to improve the performance of the data grids. Our schemes decrease the completion time of data transfer and reduce the workload of slower serves. Experiment results are also done to demonstrate its performances.

1 Introduction

Grid computing can be seen as one of the most important next generation network applications. In the future, people can exchange information and share knowledge using Grid technologies. Grid computing concepts can be applied to various application areas. Grid, via the broadband networks, efficiently integrates various distributed computing devices, databases, software, instruments, and even professional expertise. At the same time, Grid provides a platform that is secure, stable and simple to operate by the use of middleware. In a Grid system, each device shares their resources and coordinates the cooperation through the network.

In the future, there will be more and more data-intensive applications in some specific realms. Examples of these applications include data generated by supercomputers, experimental analyses, and simulations in scientific discoveries. These applications all have the same characteristics of accessing and handling a large data set. The ability of a single computer is difficult to meet all requirements of these scientific applications. This is where Grid can be of help. Grids seek to make use of geographically distributed computing resources around the world. For example, the Large Hardron Collider (LHC) [3] at the European physics center CERN will continuously generate several terabytes, even petabytes, of raw data per year for approximately fifteen years. The data produced are of two types: experimental data (information gathered by the experiment) and metadata (information about this experiment). Many experts in this project will access experimental physics data and metadata at many sites around the world. These users will number in the hundreds or thousands. Data Grids provide a basic architecture for such data-intensive applications to work.

Replicating [12][13][16][17] popular content in many servers is a widely used practice. In a DATAGRID environment [9][10][11], datasets are significantly large in size. Recently, this practice is being put to use in large-scale, data-sharing scientific communities where large datasets are replicated over several sites. Because these replica locations have different architecture, system load and network connectivity, clients downloading large datasets from anyone of the replica locations can leads to a very different end-user experience. There is only one stream in a typical internet download between a client and a server. This may suffer from some problem when we download large datasets in a DATAGRID environment. The bandwidth achievable is limited by several bottlenecks. For example, one is the congestion in the link connecting the server and the client. The other is the bandwidth of the server's connection to the internet. One way to improve download speeds is to download data from multiple locations in parallel. This is the replica selection problem that users or brokers may want to be able to determine the site from which particular data sets can be retrieved most efficiently, especially as data sets of interest tend to be large.

The co-allocator of data transfers handles the selection of downloading servers. The client can access the data from multiple servers around the networks. And the aggregate bandwidth of downloading is the summation of the individual transfer rates of each flow. If some servers fail, the co-allocator adapts to the situation to keep downloading the data from other servers. Therefore, it can improve the performances compared to the single server case and ease the internet congestion problem [16].

In this paper, based on the basic architecture for co-allocating Grid data transfers, we improve the performances for downloading data in parallel from multiple servers. We propose two methods, the abort-and-retransfer approach and the one by one co-allocation approach. Simulation Results demonstrate their superiority over previous methods.

The rest of this paper is organized as follows. Section 2 is the related work about replica selection. In Section 3, we introduce the co-allocation architecture. We propose our approaches in Section 4. We analyze the performances in Section 5. Section 6 provides the conclusions.

2 Related Works

There are many different sites in a Data Grids environment. These sites may have various performance characteristics, such as different storage system architectures, network connectivity, and system load. The datasets which we want to access are probably located at some of these sites. The information infrastructure is handled by MDS (Metacomputing Directory Service) within Globus [1]. We can query MDS about the information of these replica locations. The storage broker architecture [7][12] provides a replica selection mechanism to find a suitable replica for accessing using a high-speed file transfer protocol, GridFTP. But this protocol just accesses a file from the best matched server. It does not perform the operation in parallel.

In [14][15], an approach to predict future performances based on the observations of past transfers is proposed. The goal is to obtain an accurate prediction of the time required to transfer a file. The predictive framework combines three main parts. First, it needs to record the information about every data transfer. It modified GridFTP by adding mechanisms to log performance information for every file transfer. The second step is making prediction of future behavior based on past information. The third step is integrating this information with a resource provider and then allowing this information to be discovered in the context of an information service.

In [16], the author employed the prediction techniques and developed several co-allocation mechanisms to establish connections between servers and a client. The most interesting one is called Dynamic Co-Allocation. The dataset that the client wants is divided into "k" disjoint blocks of equal size. Each available server is assigned to deliver one block in parallel. When a server finishes delivering a block, another block is requested, and so on, until the entire file is downloaded. Faster servers can deliver the data quickly, thus serving larger portions of the file requested when compared to slower servers. One downside of this approach is that faster servers must wait for the slowest server to deliver the final block. [16] also adds some functionality to the load balancing scheme: (1) progressively increase the volume of data requested from faster servers; and (2) reduce the volume of data requested from slower servers. An alternative way is to stop the slowest flow and to retransfer the data from the other servers. Our work is based on the co-allocation architecture and employs the prediction technique to improve the Dynamic Co-Allocation. We propose two techniques: (1) abort and re-transfer, (2) one by one co-allocation. These techniques can increase the volume of data requested from faster servers and reduce the volume of data fetched from slower servers.

3 Co-allocation Architecture

Based on the Globus resource management architecture [1][9], the co-allocation architecture, is shown in Figure 1 [16]. It has three components [1][4][5][16]:

1. Application: Most application developers focus on the problem of achieving high efficient and simultaneous computing. Data accesses from different sources and resources sharing are also very important in Grid applications. When a user executes an application to access the data, the job of the application is to present the description of the data to broker.
2. Broker/co-allocator: The user needs a broker to identify the available and appropriate resources in the Grid system for his tasks. Then, across the co-allocator agent queries MDS about the information of these replica locations to download the data in parallel using GridFTP, an extension of ftp standard. GridFTP has features to fit the Data Grid architecture. It can also provide secure transfer through GSI (Grid Secure Infrastructure) which is also one of the Globus Toolkit components.

3. Local storage systems: The storage systems provide basic mechanisms for accessing and managing the data located such as high performance storage system (HPSS) and distributed parallel storage system (DHSS).

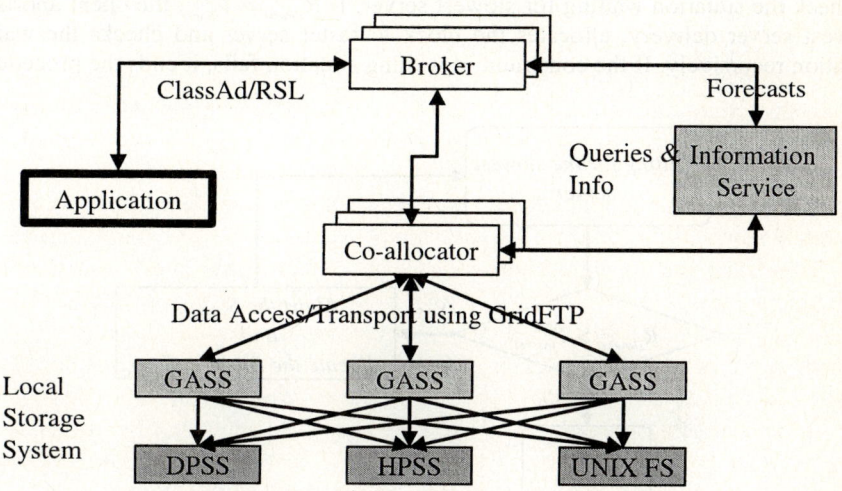

Fig. 1. The Co-allocation Architecture for Grid Data Transfer

4 Replica Selection Mechanisms

In Co-allocation architecture [16], three techniques, brute-force co-allocation, history-based co-allocation and dynamic co-allocation were proposed to allocate the data blocks and to map the replica locations for downloading. The most efficient one is the Dynamic Co-Allocation described previously. One drawback of this approach is that faster servers must wait for the slowest server to deliver the final block. In this following, we propose two techniques to improve it.

4.1 Abort and Retransfer

In consideration of the drawback that the faster servers must wait for the slowest servers, we propose a scheme to allow a client to abort the slower server delivery and retransfer data block from faster servers. First, the client can calculate two parameters, $T_{fastest}$ and $R_{slowest}$.

$$T_{fastest} = \frac{data-block-size}{fastest-available-server-bandwidth} \quad (1)$$

$$R_{slowest} = \frac{data-block-remained}{slowest-server-bandwidth} \quad (2)$$

$T_{fastest}$ is the time needed for the fastest available sever to transmit a complete data block. $R_{Slowest}$ is the remaining time needed before the slowest server completes a block delivery. The scheme is shown in Figure 2. It adapts the dynamic co-allocation to allocate the data blocks. When all data blocks are assigned, it executes our procedure to check the situation waiting for slowest server. If $R_{Slowest} > T_{fastest}$, the client aborts the slowest server delivery, allocates the block to faster server and checks the waiting situation recursively. If the constraint of waiting situation fails, it ends the procedure.

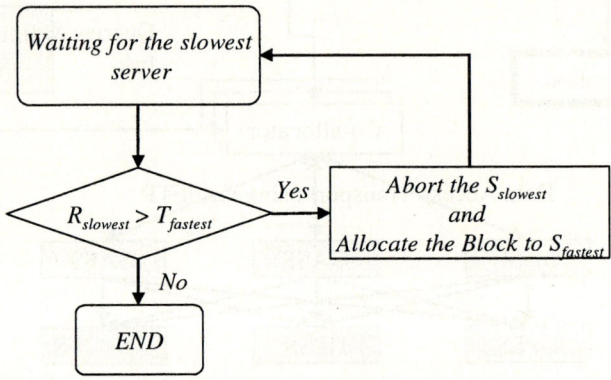

Fig. 2. Abort and Re-allocation

The approach to determining the bandwidth is to observe the historical information about data transfer rate of GridFTP [13][14][15][16]. We can calculate the average transfer bandwidths as a predictor of future transfer times. For example, NWS (Network Weather Service) and Iperf are used to measure network performance for wide area networks. We can use these techniques for predicting the bandwidth.

4.2 One by One Co-allocation

Even though the abort and retransfer approach improves the performance of data transfer, it still can not prevent the allocation to the slowest server. Assume there are n servers. For each server S_i, $1 \leq i \leq n$, we can obtain the transfer rate B_i of each server to a client by the prediction technique [14][15]. Using this information, we can calculate the time needed to transmit one data block.

$$T_i = \frac{M}{B_i} \qquad (3)$$

Where M is the data block size and B_i is the predicted transfer rate. If a file in question is divided into k blocks of equal size, the progress of the one by one co-allocation is illustrated in Figure 3. Let $C_i = T_i \times A_i$, where A_i is the number of blocks assigned to S_i. C_i is the completion time of transmitting A_i blocks from S_i.

Let $E_i = T_i \times (A_i + 1)$, where E_i is the estimated completion time assigned one more block. R is the number of blocks remained.

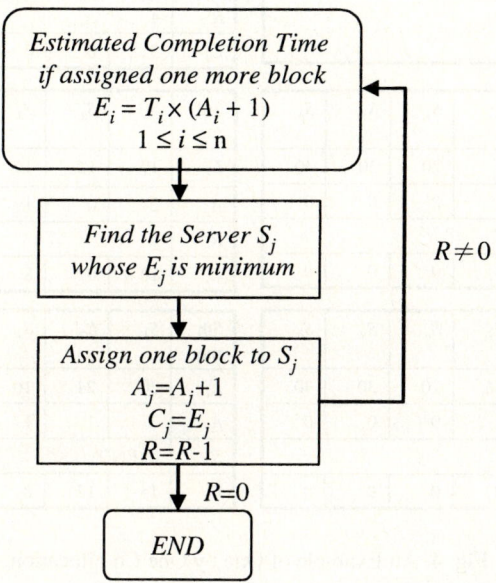

Fig. 3. One by One Co-allocation

With one by one approach, the client divides the files into k blocks and the initial value of R is equal to k. On the Co-allocation architecture, the client presents the description of k blocks to the broker by running the application. The broker identifies the file locations that can be fetched across the information services and informs the co-allocator agent. Then, the agent runs the recursive program to assign the data blocks to faster servers and downloads the data in parallel using GridFTP.

We present an example for one by one co-allocation. Figure 4 shows the example. In Figure 4, assume that there are 6 servers. The file is divided into 5 blocks. T_i is also shown. In the first run, if we assign one block to S_1, the increase in completion time will be minimized. Therefore, the co-allocator assigns one block to it (A_1 =1). In the next run, a block is assigned to S_3. The rest may be deduced by analogy. After the 5th run, we have the final allocation: We can see that S_1 gets 3 blocks, S_2 gets 1 block and S_3 gets 1 block. Set C_i= {15, 12, 8, 0, 0, 0} is the final completion time for each server and the maximum value 15 is the completion time of this downloading job. Relatively our approach avoids fetching the data blocks from slower servers in looks. It not only decreases the completion time of data transfers, but also reduces the workload of slower serves.

S_i	$i = 6$					
T_i	5	12	8	20	30	40
A_i	0	0	0	0	0	0
The number of blocks	5					

first Run	S_1	S_2	S_3	S_4	S_5	S_6
E_i	5	12	8	20	30	40
A_i	1	0	0	0	0	0
R	4					
C_i	5	0	0	0	0	0

2nd run	S_1	S_2	S_3	S_4	S_5	S_6
E_i	10	12	8	20	30	40
A_i	1	0	1	0	0	0
R	3					
C_i	5	0	8	0	0	0

3rd run	S_1	S_2	S_3	S_4	S_5	S_6
E_i	10	12	16	20	30	40
A_i	2	0	1	0	0	0
R	2					
C_i	10	0	8	0	0	0

4th run	S_1	S_2	S_3	S_4	S_5	S_6
E_i	15	12	16	20	30	40
A_i	2	1	1	0	0	0
R	1					
C_i	10	12	8	0	0	0

5th run	S_1	S_2	S_3	S_4	S_5	S_6
E_i	20	24	16	20	30	40
A_i	3	1	1	0	0	0
R	0 (END)					
C_i	15	12	8	0	0	0

Fig. 4. An Example of One by One Co-allocation

5 Result and Analysis

In this section, we analyze the performances of our approaches. The results are compared with dynamic co-allocation. We evaluate three co-allocation schemes: (1) dynamic co-allocation, (2) abort and retransfer and (3) one by one co-allocation. In

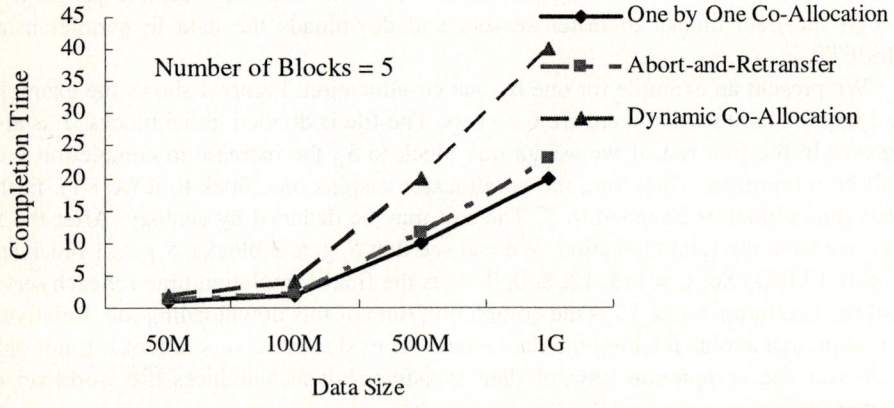

Fig. 5. Completion time of different Methods

the Data Grid environment, we assume the client can download a file from 6 servers across the broker. We calculate the ratio of the cost time of different schemes by obtaining the prediction information.

In Figure 5, we analyze the effect of the data size. The required data is divided into 5 blocks. We increase the data size from 50M to 1G. Obviously, Figure 5 shows that one by one co-allocation reduces the time efficiently. And the results of the abort and retransfer method are better than the dynamic co-allocation and worse than the one by one co-allocation method.

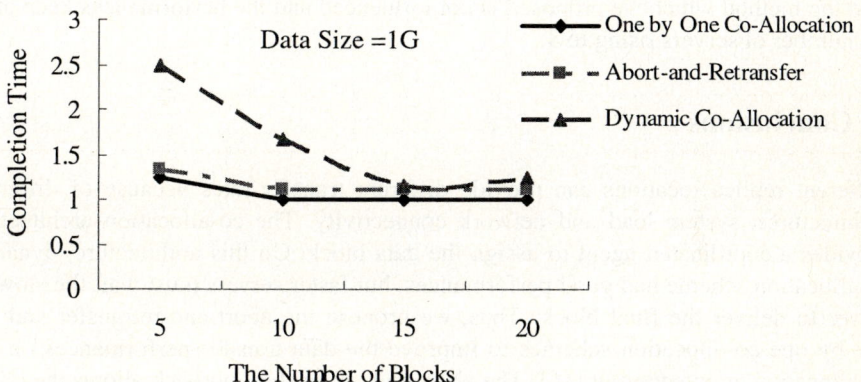

Fig. 6. Completion time using different number of blocks

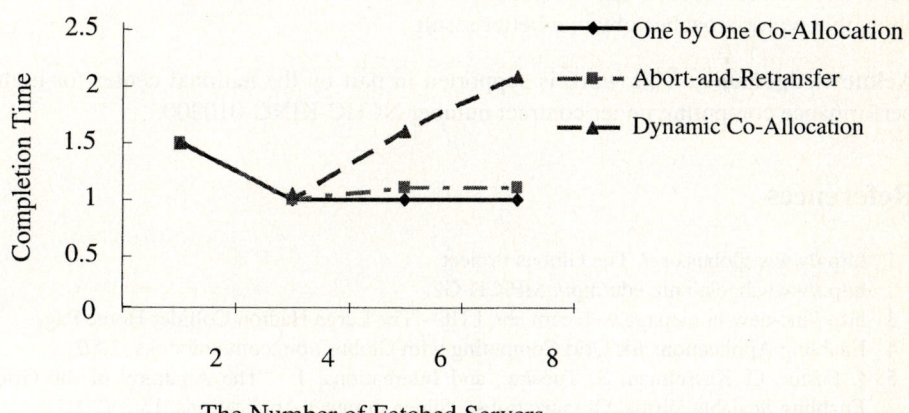

Fig. 7. Completion time using different number of fetched servers

In Figure 6, we discuss the influence of the amount of data blocks. We fix the data size to 1G and adjust the number of blocks from 5 to 20. We observe that there are no performance improvements when a file is divided into too many blocks. The completion time of the one by one co-allocation and the abort-and-retransfer do not improve when the amount of data blocks is larger than 10. And the completion of the dynamic co-allocation does not decrease when the data is divided into 20 blocks. It increases instead.

If there are a lot of servers providing to fetch a file, we can divide the file and download quickly in parallel. But if these fetched servers include some slower servers, they may influence the performances described in Figure 7. The dynamic co-allocation downs the performances when the number of fetched servers from 4 to 8. And the method which we proposed is not influenced and the performances keep after the number of servers rising to 4.

6 Conclusions

Different replica locations can provide different transfer rates because of different architectures, system load and network connectivity. The co-allocation architecture provides a coordinated agent to assign the data block. On this architecture, dynamic co-allocation scheme had good performances, but faster servers must wait the slowest server to deliver the final block. Thus, we propose the abort-and-retransfer and the one by one co-allocation schemes to improve the data transfer performances on the co-allocation architecture in [16]. The abort-and–retransfer approach allows the client to abort the slower servers if there are some faster servers that can complete the transfer fast. The one by one co-allocation was pre-scheduling to allocate the data block in advance. In this paper, we not only decrease the completed time of data transfer, but also reduce the workload of slower serves. The experimental results show that our approaches obtain a better result.

Acknowledgements. This work is supported in part by the national center for high-performance computing under contract number NCHC-KING-010200.

References

1. http://www.globus.org/, The Globus Project.
2. http://www.hpclab.niu.edu/mpi/, MPICH-G2.
3. http://lhc-new-homepage.web.cern.ch/, LHC - The Large Hadron Collider Home Page.
4. Enabling Applications for Grid Computing with Globus, ibm.com/redbooks, 2002.
5. I. Foster, C. Kesselman, S. Tuecke., and International J. "The Anatomy of the Grid: Enabling Scalable Virtual Organizations," Supercomputer Applications, 15(3)(2001).
6. G. Aloisio, M. Cafaro, E. Blasi and I. Epicoco, "The Grid Resource Broker, a Ubiquitous Grid Computing Framework," To appear in Journal of Scientific Programming, Special Issue on Grid Computing, IOS Press, Amsterdam. http://sara.unile.it/grb/grb.html
7. "Storage Resource Broker, Version 2.0," SDSC (http://www.npaci.edu/dice/srb).
8. Moore R., and A. Rajasekar, "Data and Metadata Collections for Scientific Applications," High Performance Computing and Networking, Amsterdam, NL, June 2001.

9. K. Czajkowski, l. Foster, N. Karonis, C. Kesselman, S. Martin, W. Smith, and S. Tueche, "A Resource Management Architecture for Metacomputing Systems," IPPS/SPDP '98 Workshop on Job Scheduling Strategies for Parallel Processing, 1998.
10. Bill Allcock, Joe Bester, John Bresnahan, Ann L. Chervenak, Ian foster, Carl Kesselman, Sam Meder, Veronika Nefedova, Darcy Quesnel, Steven Tuecke, "Data Management and Transfer in High-performance Computational Grid Environments," *Parallel Computing Journal*, Vol. 28 (5), May 2002, pp. 749-771.
11. Ann Chervenak, Ian Foster, Carl Kesselman, Charles Salisbury, Steven Tuecke, "The Data Grid: Towards an Architecture for the Distributed Management and Analysis of Large Scientific Datasets," *Journal of Network and Computer Applications*, 23:187-200, 2001 (based on conference publication from Proceedings of NetStore Conference 1999).
12. Heinz Stockinger, Asad Samar, Bill Allcock, Ian Foster, Koen Holtman, Brain Tierney, "File and Object Replication in Data Grids," *Journal of Cluster Computing*, 5(3)305-314, 2002.
13. S. Vazhkudai, S. Tuecke, I. Foster, "Replica Selection in the Globus Data Grid," IEEE/ACM International Symposium on Cluster Computing and the Grid, May 2001, pp. 106 – 113
14. S. Vazhkudai, J.M. Schopf, "Predicting Sporadic Grid Data Transfers," IEEE International Symposium on High Performance Distributed Computing, HPDC-11 2002.July 2002, pp. 188 – 196.
15. S. Vazhkudai, J.M. Schopf, I. Foster, "Predicting the Performance of Wide Area Data Transfers," International Parallel and Distributed Processing Symposium, IPDPS 2002, April 2002, pp. 34 – 43.
16. Sudharshan Vazhkudai, "Enabling the Co-Allocation of Grid Data Transfers," International Workshop on Grid Computing, 17 Nov. 2003, pp. 44 – 51.

A Novel Checkpoint Mechanism Based on Job Progress Description for Computational Grid*

Chunjiang Li, Xuejun Yang, and Nong Xiao

School of Computer,
National University of Defense Technology,
Changsha, 410073 China, +86 731 4575984
chunjiangli@263.net

Abstract. In this paper, we argue that application-level uncoordinated checkpointing with user-defined checkpoint data is the favorable in grid environment where heterogeneity is essentially popular. We present a novel application-level uncoordinated checkpoint protocol based on Job Progress Description (**JPD**) which is composed by a Job Progress Record Object and a group of Job Progress State Objects, these two kinds of objects act as checkpoint data for the job and the methods of them can be used as checkpoint APIs. By extending this protocol with sender-based message logging, it can be used by the message passing applications in computational grid. Emulation with a kind of master-worker message-passing applications shows that using this checkpointing protocol can dramatically reduce the wall-time of the application when failure occurs.

1 Introduction

Computational Grid [1, 2] enable the coupling and coordinated use of geographically distributed resources for such purposes as large-scale computation, distributed data analysis, and remote visualization. A computational grid that contains hundreds to thousands of machines and multiple networks has a small mean time to failure. The most common failure modes include machine faults in which hosts go down and network faults where links go down. When some resources fail, the applications using such resources have to stall, then wait for repair or migrate to other resources. In order to reduce the recovery time of the jobs, checkpoint and recovery service is absolutely necessary, which can save partial results and job states, avoids restarting the job from the very beginning.

In this paper,for simplicity, we focus on the stoping model in which a faulty process (due to resource failure) hangs and stops responding to the rest of the system, neither sending nor receiving messages [3].There are many interesting problems to be solved even in this restricted domain. Moreover, a good solution

---

* This work is supported by the National Science Foundation of China under Grant No.60203016 and the National High Technology Development 863 Program of China under Grant No.2002AA131010 and 973 project NO. 2003CB316900.

for this failure model can be a useful mechanism in addressing the other fault models.

Checkpoint and recovery (CPR) techniques have been studied for a few decades. But in computational grid which coupling a lot of independent, wide area distributed and essentially heterogeneous computing systems, how to checkpoint jobs is still a open issue that needs hard work and needs some standards to be established. In this paper, we present an application-level checkpointing protocol for computational grid based on Job Progress Description. Which requires the user of the application to make slight modifications to both source code and job scripts, and each job should be divided into independent job progresses and the job states of each progress are defined by the user. At the end of each progress, checkpoint APIs are called to save the partial results and the job states to stable storage. The partial results and the job states facilitate the recovery of the failed jobs. We argue that by extending this protocol with sender-based message logging, it is also suitable for the message passing applications in computational grid.

The rest of the paper is organized as follows. In section 2, related work for checkpoint and recovery in computational grid is discussed. The idea of our application-level checkpoint protocol is explained in section 3. The sender-based message logging mechanism used with this checkpoint protocol is discussed in section 4. Emulation results of this checkpoint mechanism with a massage-passing application are presented in section 5. Conclusion and future work are discussed in section 6.

2   Related Work

Checkpoint recovery is a service to facilitate automated recovery and continuation of interrupted computations with the aids of periodically recorded checkpoint data. A checkpoint data set represents a sufficient collection of information required to allow the checkpoint recovery services to correctly reschedule the computational job and the job can resume computations from a "known good" state. The primary objective is to avoid having to restart a job from the very beginning.

Traditionally, for a single computing system, checkpointing can be realized at three levels: kernel level, library level and application level [4]. Most operating systems do not support checkpointing, especially for parallel programs. The widely used checkpointing mechanisms is implemented at library level, such as Condor [5] and Libckpt [6], but most current checkpointing libraries are static, meaning the application source code must be available. The application level implementation of checkpointing needs the application to be modified by inserting checkpointing functions, and the transparency is lost, however, it has the advantage that the user can specify where to checkpoint, this is helpful to reduce checkpointing overhead.

A key requirement of Grid Checkpoint Recovery (GridCPR) service is recoverability of jobs among heterogeneous grid resources. In other words, resources

on which jobs are checkpointed need not be of the same type as those on which
the jobs are recovered, as long as the application code operating on the checkpointing resource can be built for and run on the recovery platforms. So, the
system level (includes kernel level and library level) checkpoint protocols which
record the process image of the running job as checkpoint data is not suitable
for the grid environment, because the process image on one computing system
can hardly be restored on a heterogeneous platform. As well as, the overhead of
checkpoint in system level checkpointing is very high and can not be controlled
by the application it self. On the contrary, the application level checkpointing protocol is suitable for the grid environment, because user can construct
portable application code and can manipulate the overhead of checkpointing.
Even if checkpointing at application level, the checkpoint data must be defined
by the user other than the process image of the running job in order to facilitate
recovery on heterogeneous platforms.

In computational grid which couples multiple heterogeneous computing systems, how to chechpoint jobs is still challenging. In several meetings of GGF held
recently, the Grid Checkpoint Recovery (GridCPR) Working Group presented
some memo [7] and drafts [8, 9] which provided information to the grid community regarding a proposed architecture for grid checkpoint recovery services and
a set of associated Application Programmer Interface (API). The consensus is
that application level checkpointing protocol is the only one suitable for the grid
environment. But their work was also very elementary, and there is still much
work has to be done.

In the work of the European DataGrid [10] project, they proposed an application level checkpoint protocol for the data processing applications [11], which
divides the job into several job steps, and records user-defined job states at the
end of each job step as the checkpoint data for job recovery.

Message-passing models are still well-suited for computational grids. Many
implementations and variants of MPI have been produced. The most prominent
for grid computing is MPICH-G2 [12, 13]. But MPICH-G2 does not support any
fault tolerant mechanism. MPICH-V [14] is an implementation of MPI standards
for grid computing, it presents fault tolerance by checkpointing and message logging. It supports fully transparency of fault tolerance. But it has two limitations.
First, its checkpointing mechanism is at the library level recording the image of
MPI processes, which makes it difficult for recovering a MPI process on another
heterogeneous computing node. Second, its message logging records all the communication context of each process and store them in a Channel Memory (CM),
which makes it less general enough.

3 Basic Idea of JPD-Based Checkpointing

In this section, we describe our application level checkpointing mechanism based
on Job Progress Description. Firstly we give following definitions:

Job Progress (JP). A group of continuous operations in a job. For example,
a group of continuous statements in the source code of the job make up a Job

Progress. So, the single threaded job can be divided into a series of continuous Job Progresses.

Independent Job Progress (IJP). If the running of one job progress does not depend on the running environment of the other job progresses, then this job progress is called independent job progress.

Job Progress States (JPS). The user defined states of the job at the end of each independent job progress. It can be expressed by a series of $<var, value>$ pairs or other directive comments on the work has been done.

3.1 Checkpoint Data Set

We use two kinds of abstract data objects as the checkpoint data of a job, the definition of these two kinds of objects is given below:

Job Progress States Object (JPSO). The object which records the job states for one **IJP**.

Job Progress Record Object (JPRO). The object which records the series of **IJP**s of a job and the latest **IJP** finished by the job.

So, during the running of a job, one **JPRO** and a series of **JPSO**s can record the job states, we call them as **Job Progress Description (JPD)**, which acts as the checkpoint data of the job. The structure of **JPD** is illustrated in Fig. 1.

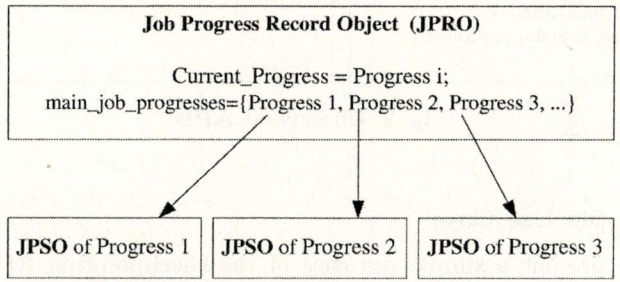

Fig. 1. Data Elements in **JPD**

This kind of organization of checkpoint data can benefit to the job control in computational grid. For in computational grid, most applications consist multiple tasks, by checking the **JPRO** of each task, the job manager can grasp the information about the states of each task and can determine the progress of the whole work.

3.2 Checkpoint APIs

The methods of **JPRO** and **JPSO** act as checkpoint APIs. The detailed definition is given in Fig. 2. Each **IJP** has a distinguished name that identifies it unambiguously. The first data member (Label_t) in **JPSO** represents such identifier. In the **JPRO** of the job, the identifiers of all the **IJP**s and the identifier of current **IJP** are recorded. So, when the job restarts, it reads the **JPRO** of it, can know the progress of the whole work. The methods of **JPRO** mainly focus on processing the **JPSO**s of the job, such as load and save **JPSO**s, get the identifier of current, former, next **IJP**s of the job, check the ending of the job, measure the size of one **JPSO**. The **JPRO** acts as the general controller of the whole job. On the other hand, the methods of **JPSO** mainly focus on recording and retrieving states values into and from **JPSO**. Using the methods in these two objects, user can write application level checkpoint jobs. Naturally, the methods of each job can be extended for particular purpose.

```
Object JPRO:
{
//Data Members
ProgressSet main_progress=
{progress1, progress2,... };
Label_t current_progress;

//Methods
Label_t load_current_progress();
Label_t load_next_progress();
int save_JPRO();
int save_JPSO(JPSO_id);
JPSO load_current_JPSO();
JPSO get_former_JPSO();
JPSO get_next_JPSO();
int set_final();
bool is_final(Label_t);
float max_P-State_size(Label_t);
}
```

```
Object JPSO :
{
//Data Members
Label_t JPSO_id=Label_t;
VarValueSet var_value_pairs[]=
{var1=value1, var2=value2,... };

//Methods
int save_value(<var, value>);
string get_string_value(string);
int get_int_value(string);
double get_double_value(string);
float get_JPSO_size();
}
```

Fig. 2. Checkpoint **APIs**

3.3 A Simple Use Case

In Fig. 3, we present a simple use case of the chechpointing APIs in a serial job. It is obvious that once restart a failed job on a computing resource, the job retrieves the **JPRO** and the nearest **JPSO** from the checkpoint data set of this job, and reconstruct the running environment of the latest **IJP**, then reenter the program segment of the latest **IJP**, continue the interrupted work of the job. In order to facilitate the control of running job, the user of the computational grid usually writes a job scripts and submit it together with the source code file of the job. In the job scripts of a checkpoint job, the user should define all the **IJP**s of the job. When initiating gridCPR service, the runtime library reads the job scripts then constructs the data structure of checkpoint data (mainly one **JPRO** and some **JPSO**s) for the job which record the states of the job in the running process.

When restarting a failed job on a heterogeneous resource, the job should be recompiled and relinked with the running time library of checkpoint service.

```
# include <cpr-api.h>

Label_t current_progress
JPRO Jobprogress;           // load JPRO;
JPSO current _P-State;
current _JPSO = Jobprogress.get_current_JPSO();    // load newest JPSO;

<Call the methods of JPSO to restore the latest states of the job >

current_progress = Jobprogress.load_next_progress ();  // get the identifier of next IJP;

switch(current_progress) {

case progress1:
    new JPSO current_JPSO;
    <do the operations in progress1>
    ......

    <call the methods of current_P-State object to save current states>
    Jobprogress.save_JPSO(current_JPSO);  // save current JPSO;
    Jobprogress.save_JPRO();  //save current JPRO;

case progress2:
    new JPSO current_JPSO;
    <do the operations in progress2>
    ......

    <call the methods of current_P-State object to save current states>
    Jobprogress.save_JPSO(current_JPSO);  // save current JPSO;
    Jobprogress.save_JPRO();  //save current JPRO;

    ......

} // end of switch;

Jobprogress.set_final();
Jobprogress.save_JPRO();
```

Fig. 3. An Sample Serial Job Using **JPD**-based Checkpoint **APIs**

During the running process of the restarted job, the job retrieves **JPRO** and **JPSO**s of the job from the global storage used for checkpoint and recovery. This global storage can be accessed from every computing resource. According to the retrieved **JPRO** and **JPSO**, the job can determine the latest states of the job before failure occurs, then reestablish the running environment and continue to do the rest of the work.

In practice, the data objects of **JPRO** and **JPSO** could be expressed in a XML file. Each checkpoint job can use one XML file to store its checkpoint data: including one **JPRO** and a series of **JPSO**s. The data objects in **JPD** (**JPRO** and **JPSO**s) themselves can be recorded as a data item in XML file.

Using the checkpoint protocol proposed in this paper, users can submit its program with a XML file as checkpoint data file which initially record the **IJP**s of the program. During the running of this program at any platform, when checkpointing, it write the mane and value of each variables,partial result (expressed as $<var, value>$ pairs) of the program into the checkpoint file. Then once the program restart at another heterogeneous platform, after rebuilding the source code of the program, the program can rerunning from the nearest **IJP** of the program as well as the XML file transmit with the program at the same time.

Each time the program start to run, it read its **JPRO** from its checkpoint file, to get the identifier of the **IJP** that has been done last time, then use it to reload the corresponding **JPRO** which records the environmental variables and partial result at last time. Then it sets up the running environment for the next **IJP**, and goes to the location of next **IJP** in the program, begins to do the residual work.

4 Sender-Base Message Logging

In order to support message-passing applications with our grid checkpointing mechanism, we record each sending message during a job progress into the corresponding **JPSO**. That is, we use sender-based message logging mechanism to expand this application level checkpoint protocol. Sender-based message logging can dramatically reduce the overhead of message logging, because when one process sends a message to another, the sender and the receiver naturally get a copy of the message, it is faster to simply save a copy in local storage on the sending machine [15].

A logging record for a message includes the following data elements:
SSN: the sequence number of the message sent by the sender process;
$SJPN$: the job progress number of the sender process;
$RJPN$: the job progress number of the receiver process;
$MCON$: the content of the message;
RSN: the sequence number of the message received by the process.

It is obvious that we record the correspondence of message to job progress in the logging record. The checkpoint architecture of application level checkpointing mechanism for message-passing applications is shown in Fig. 4

Fig. 4 depicts two processing node in computational grid for running message-passing applications. The Grid Level Checkpoint Service presents a stable global storage for checkpoint data and provides mechanisms for storing and retrieving

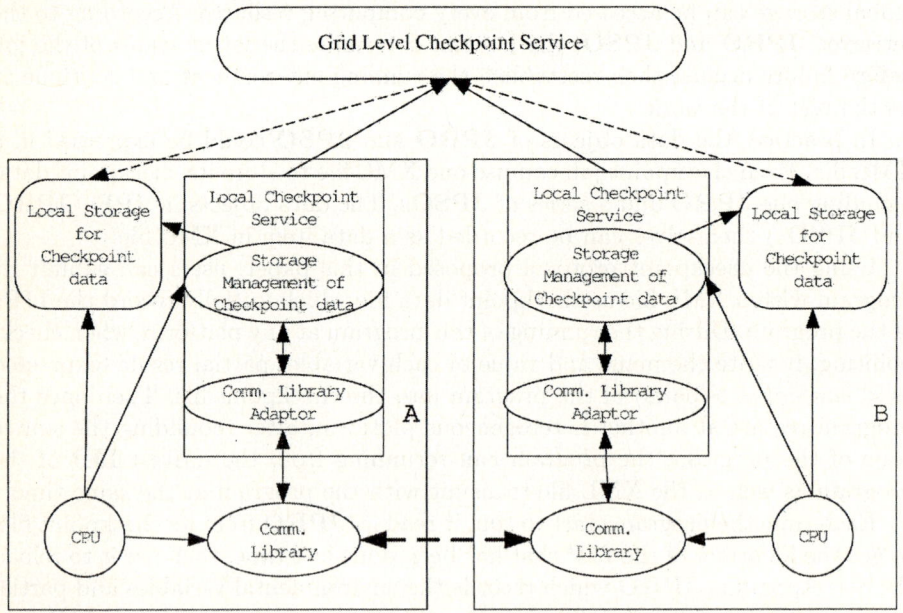

Fig. 4. The Architecture for Checkpointing Message-Passing Applications in Computational Grid

checkpoint data. At each processing node, there is a local storage for buffering checkpoint data. The storage management of checkpoint data in local checkpoint service is responsible for storing and retrieving **JPSO**s in and from local and global checkpoint data storage. In order to conduct message logging and replay messages, we design a communication library adapter over the MPI communication library. It hooks the message-passing functions presented by MPI, and expanding message logging information to each **JPSO** of the task. When replaying messages, it retrieves the logging of the messages from **JPSO**s by calling the storage management of checkpoint data, and resend-ing the messages in ascending order by RSN.

5 Emulations

We evaluate the prototype implementation of this checkpointing protocol by emulating in a master-worker style application which is running on four heterogeneous clusters that contain more than 50 computing nodes and we suppose there are 50 worker jobs and one master job, each worker job contains only one program. The master job partitions the input data into 50 segments and each data segment is sent to a worker job that runs on a computing node. Each worker job applies four algorithms to process the input data. The results generated by the worker are gathered by the master job. We divide each worker job into four **IJP**s according to the four algorithms, and insert checkpoint APIs in the source code at the end of each **IJP**. And write a simple job script for each worker job. The **JPRO** of each worker job records the progress of it. In the job scripts of the master job, all the **IJP**s of the whole application are recorded, and the **JPRO** of the master job records the progress of the whole application. During the running process, each worker job checkpoint itself un-coordinately, creating **JPSO**s and writing job states user-defined checkpoint data into it. The **JPRO** and **JPSO**s of the worker job are saved in a file. And all the checkpoint file of worker jobs can be accessed across these clusters. So, when one computing node failed, the worker job running on this failed node can be migrated to another node, recompiled and relinked with the checkpoint library for the sake of heterogeneity, then with the aids of the checkpoint file, it can continue to run the rest **IJP**s of the job. In order to evaluate the effectiveness of this checkpointing protocol, we insert failure mode manually to the worker nodes, and randomly generate failure sequences of them. We suppose there are at most 18 nodes in the failure sequence. Under the same failure sequence, we compare the wall-time of whole application under two different failure process methods, one is checkpointing, the other is restarting. The result is shown in Fig. 5, in this figure, we suppose the wall-time without any failure is 100 seconds,and the each **IJP** of one job uses 25 seconds to fulfill its work. It is obvious that with this checkpointing protocol, the application need not restart the job from the very beginning, which can dramatically reduce the wasted computation.

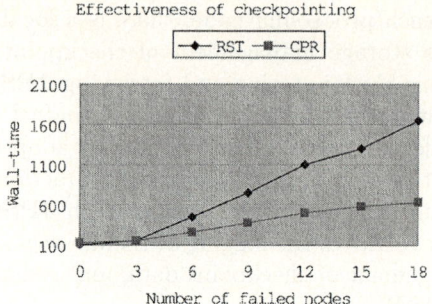

Fig. 5. Emulation Results

6 Conclusion and Future Works

In this paper, we proposed the Job Progress Description based checkpoint mechanism for checkpointing at application level in computational grid, and also described the implementation of combining this checkpoint protocol with sender-based message-logging mechanism. **JPD**-based checkpoint mechanism uses markup method to partition job into Independent Job Progresses, and defines checkpoint data by programmer at application level. So, this mechanism is suitable for checkpointing in computational grid in which resource heterogeneity is popular.

There are still much work need to be done for consummating this checkpointing mechanism, such as how to design programming model and tools for facilitating the composition of checkpoint jobs with this checkpoint protocol, how to define the job progress in the job script, how to design tools for segmenting the jobs in to subjobs containing **IJP**s, how to manage the checkpoint data generated in checkpointing, how to optimize the performance of checkpointing and recovery.

References

1. Foster, I., Kesselman, C.: The Grid: Blueprint for a New Computing Infrastructure. Morgan Kaufmann Publishers (1999)
2. Foster, I.: The grid: A new infrastructure for 21st century science. Physics Today **54(2)** (2002)
3. Lynch, N.: Distributed Algorithms. Morgan Kaufmann Publishers (1996)
4. Hwang, K., Xu, Z.: Scalable Parallel Computing, Technology, Architecture, Programming. McGraw-Hill Companies, Inc. (1997)
5. Litzkow, J.B.M., Tannenbaum, T., Livny, M.: Checkpoint and migration of unix processes in the condor distributed processing system. Technical Report 1346, University of Wisconsin-Madison (1997)
6. Plank, J.S., Beck, M., Kingsley, G., Li, K.: Libckpt: Transparent checkpointing under unix. Technical Report UT-CS-94-242 (1994)

7. GridCPR Working Group: An architecture for grid checkpoint recovery services and a gridcpr api. http://www.gridforum.org/Meetings/ggf7/drafts/GridCPR001.doc (2003)
8. GridCPR Working Group: Gwd-i: An architecture for grid checkpoint recovery services and a gridcpr api, current draft is version 1.0. (http://gridcpr.psc.edu/GGF/docs/draft-ggf-gridcpr-Architecture-1.0.pdf)
9. GridCPR Working Group: Gwd-i: Use cases for grid checkpoint and recovery, current draft is version 1.0. (http://gridcpr.psc. edu/GGF/docs/draft-ggf-gridcpr-UseCases-1.0.pdf)
10. DataGrid: European datagrid project. (http://www.eu-datagrid.org/)
11. Gianelle, A., Peluso, R., Sgaravatto, M.: Job partitioning and checkpointing. (Technical Report DataGrid-01-TED-0119-0-3)
12. Foster, I., Karonis, N.T.: A grid-enabled mpi: Message-passing in heterogeneous distributed computing systems. In Proceedings of International Conference on High Performance Networking and Computing,SC98,IEEE (1998)
13. Karonis, N.T., Toonen, B., Foster, I.: Mpich-g2: A grid-enabled implementation of the message passing interface. Journal of Parallel and Distributed Computing (JPDC) (2003) 551–563
14. Bosilca, G., Bouteiller, A., Cappello, F., Djilali, S., Fedak, G., Germain, C., Herault, T., Lemarinier, P., Lodygensky, O., Magniette, F., Neri, V., Selikhov., A.: Mpich-v: Toward a scalable fault tolerant mpi for volatile nodes. In Conference on High Performance Networking and Computing archive Proceedings of the 2002 ACM/IEEE conference on Supercomputing (2002)
15. Johnson, D., Zwaenepoel, W.: Sender-based message logging. In Digest of Papers, FTCS-17, The 17th Annual International Symposium on Fault-Tolerant Computing (1987) 14–19

A Peer-to-Peer Mechanism for Resource Location and Allocation Over the Grid

Hung-Chang Hsiao[1], Mark Baker[2], and Chung-Ta King[3,*]

[1] Computer and Communication Research Center, National Tsing-Hua University,
Taiwan 300
hchsiao@cs.nthu.edu.tw
[2] The Distributed Systems Group, University of Portsmouth, UK, PO1 2EG
Mark.Baker@computer.org
[3] Department of Computer Science, National Tsing-Hua University, Taiwan 300
king@cs.nthu.edu.tw

Abstract. Recent advances in P2P lookup overlays provide an appealing solution for distributed search without relying on a single database server. In addition to performing resource discovery, these P2P substrates also offer membership management for dynamic peers. In this paper, we propose a publicly shared architecture called VC^2A that takes advantage of a P2P lookup substrate for computational applications. VC^2A targets computational master-slave applications. An application running in VC^2A dynamically allocates resources from the system on the fly. These allocated resources then self-manage and -heal. We have implemented an architecture based on our previous efforts that include an enhanced P2P lookup overlay and a mobile agent system on top of this overlay. We show that VC^2A is not only scalable but robust, and takes advantage of heterogeneity of the resources.

1 Introduction

Peer-to-peer (P2P) computing research is extremely active. The first generation of P2P substrates, such as Gnutella, implemented the flooding-based search protocol to locate resources. Due to the potential burden of network traffic introduced by the flooding protocol, and the consequential inefficiency of resource discovery, research projects such as CAN, Chord, Pastry, Tapestry and Tornado implement *distributed hash tables* (DHT) for efficiently locating resources [1]. The DHT-based P2P overlay is thus appealing, and can serve as an ideal overlay infrastructures for managing resource information. A DHT-based P2P overlay, however, is unable to offer complex search capabilities for resource discovery. Recent proposals have shown that a DHT-based P2P overlay can be enhanced to perform keyword and/or range searches [1].

---

* This work was supported in part by National Science Council, Taiwan, under Grant NSC 93-2752-E-007-004-PAE, and National Center for High-performance Computing, under Grant NCHC–KING_010200.

One use of DHT-based P2P substrates is for computational applications, which are based on the master-slave programming paradigm. Here, the master is responsible for control the workflow of the entire task, which includes providing the input to, and collecting output from, computational tasks, while the slaves perform the computation. Examples of this type of P2P application includes SETI@home, computational biology, and parameter-sweep applications. In this study, we are interested in designing a "publicly shared" P2P substrate that allows the execution of multiple master-slave applications. Unlike a proprietary design, such as SETI@home, running multiple applications in a shared infrastructure can have several benefits. For example, (1) application developers only need to take care of the development of their applications regardless of design, deployment and maintenance the infrastructure. (2) A resource provider can have a simple, but complete, control of their resources and can delegate its control to the shared substrate without paying attention to each application that may be executed using their resources, and (3) running multiple applications can further increase the utilization of resources in the substrate, and pooling resources together increases the possibility for an application to discover those resources that can meet the application's demand.

Running these applications in a shared substrate can have the following requirements. (1) **Expressiveness.** First, the substrate should be capable of allowing an application to designate what specific characteristics allocated resources should be met. These parameters could include, for example, the current loads of the designated resources, the processor speed, or the available memory size. (2) **Code mobility.** The shared infrastructure should be capable of running any application. It could be possible that the infrastructure does not know what applications will be deployed in advance and how many resources will be allocated to a particular application. The shared infrastructure should allow an application to arbitrarily deploy its computational logic at any place within the playground formed by the resources that are allocated on-the-fly. (3) **Autonomy.** Since resources in the substrate may have their own utilisation plans and each of these plans may vary, resources allocated to a particular application should not violate their *local policies*[1]. In addition, resources allocated to an application should self-configure and -heal in order to relieve the burden on an application developer of needing to set up the dynamically formed execution environment. (4) **Robustness.** In a publicly shared environment, resources may dynamically join and leave the system. The substrate should be reliable enough for any application. A robust computing environment can also mitigate the load of an application developer to handle failure during the execution of his application. It is unlikely to rely on a central control point for managing information, and allocating and scheduling resources. This is because the central point can

---

[1] The local policy that is defined for a resource can be for example, the load that a resource can sustain, or the time over which it can be. The local policy may further depend on the policy defined by its domain administrator. Defining a local policy is out of the scope of this paper.

become the single point of failure. (5) **Scalability.** The number of resources appearing in the shared environment is unknown, but may be very large. An approach based on a central point of management will become a performance bottleneck if the number of resources is large. A substrate should thus be designed with the scalability in mind to accommodate any number of resources allocated to an application. (6) **High performance.** The resources allocated to an application should be effectively utilized. This will accelerate the execution of the application. A P2P infrastructure for supporting computational applications should bear performance criteria in mind.

In this work, we propose a publicly shared P2P infrastructure architecture called VC^2A for computational applications. Particularly, we implement the VC^2A architecture using our previous efforts, i.e., a DHT-based overlay called Tornado [2] along with an enhanced search ability [1]. The enhanced lookup overlay allows an application to efficiently and effectively perform similarity and range searches for resources in a large and dynamic environment. In addition, VC^2A is integrated with a mobile agent system called Armada [3]. In VC^2A, an application designer develops their application using Armada agents. Agents in Armada can be deployed on any resources that are dynamically discovered; they can communicate with each other regardless of their movements. VC^2A is thus programmable on the fly. Since in VC^2A a resource defines its own local policy and helps overall computation according to this local policy. When the resources are discovered for an application without violating their local policies, they self form an individual tree overlay [2]. The tree overlay self manages without involving any effort from the application. It can arbitrarily be scaled to any size while each resource only maintains a constant number of members in the associated tree. VC^2A ensures that a tree overlay is robust by taking advantage of the heterogeneity of resources. Moreover, a tree in VC^2A is nearly balanced. This allows the spread of computational tasks to the allocated resources and to collect the results, rapidly. This also provides the responsiveness needed for maintenance.

Our contributions of this work include that (1) we propose a novel architecture to leverage DHTs for grid resource discovery. (2) We design a scalable tree-based overlay which robustly organizes allocated resources, helps disseminate computational tasks, and provide an execution environment for mobile agents. (3) We integrate the DHT-based and tree-based overlays for resource discovery and organization.

2 VC^2A

Overview. Figure 1(a) shows an example that illustrates the operations of VC^2A. Each peer in VC^2A initially publishes its resource descriptor to the system using the Tornado DHT interfaces (i.e., the `publish` interface). For example, node 3

---

[2] Actually, VC^2A maintains "tree-like" overlays in the system, where redundant links in a tree are used for tolerating faults. We will not differentiate the terms–"tree" and "tree-like" overlays.

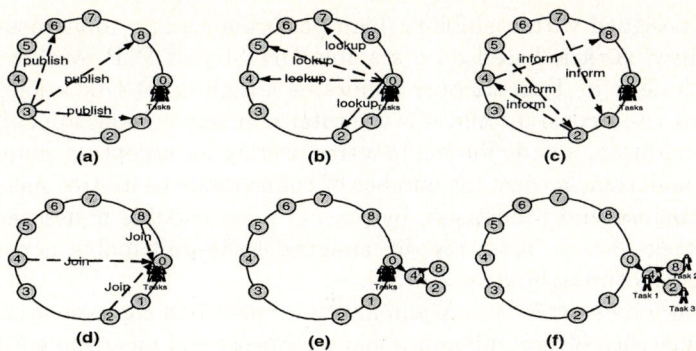

Fig. 1. An example of VC²A consisting of nine nodes, where (a) node 3 publishes its resource descriptor to the system, (b) node 0 looks up the resources for its computational task by consulting nodes 2, 4, 5 and 7, (c) nodes 2, 4, 5 and 7 respectively inform nodes 4, 8, 2, and 1 that have the designated resources, (d) nodes 2, 4 and 8 can help perform the computational task for node 0, (e) nodes 2, 4 and 8 self-form a tree overlay for receiving the computational tasks issued by node 0, and (e) node 0 submits the tasks to the tree overlay using Armada agents

publishes its resource descriptor to nodes 1, 6 and 8. A peer that has a computational task to accomplish then discovers the resources that can satisfy its requirements using the `lookup` interface. In Figure 1(b), node 0 issues the `lookup` requests to nodes 2, 4, 5 and 7. Those peers receiving the `lookup` requests inform the nodes (i.e., nodes 4, 8, 2 and 1) that can provide designated resources for the computational demands (see Figure 1(c)). If the matched peers can accept the computational tasks (i.e., nodes 2, 4 and 8 in Figure 1(d)) according to their local policies, they self-form a tree overlay, where the root of the tree is the entry point that can accept the tasks submitted by the task producer (Figures 1(e) and (f)).

We assume that a peer registers its resource descriptor into the system using enhanced DHT lookup interfaces. Enhanced DHTs include PeerSearch, Meteorograph, Neuron, Squid and XenoSearch [1]. These enhanced DHTs can efficiently perform keyword and/or range searches, for example. Figure 2 shows an example of describing demanded resources. VC²A can discover resources that have the attribute values shown in Lines 4 and 5 of Figure 2 via searching by keywords as well as the range search for Lines 1, 2, 3 and 6. A peer publishes its resources using the similar description.

1:	CPU	$\geq$ 1.5 GHz
2:	Main memory	$\geq$ 256 Mbytes
3:	Disk	$\geq$ 256 Mbytes & $\leq$ 1024 Gbytes
4:	OS	Linux kernel 2.4 version
5:	Domain	cs.nthu.edu.tw | nchc.gov.tw
6:	Idle	from GMT 10:00 PM to GMT 7:00 AM

Fig. 2. An example of a resource descriptor

VC^2A is designed with scalability, high performance and robustness in mind. For scalability, VC^2A is based on a scalable DHT-based P2P overlay (i.e., Tornado). Each node in Tornado only maintains a number of $O(k \times \log_k^n)$ connections, where k is a constant and n is the total number of nodes appearing in the system. In addition, a node joining in a tree overlay for accepting computational tasks only maintains a constant number of connections to its tree members. For high performance and robustness, by placing more capable nodes near the job producer, tasks can be more rapidly assigned to more capable nodes and the overall system's throughput is boosted.

Note that a node in VC^2A may join multiple trees if its corresponding resource descriptor matches several different lookup requests and meanwhile it is capable of performing multiple tasks without violating its local policy. We also note that the publish, lookup, inform and join messages are routed using the routing algorithm provided by the DHT-based overlay. The routing algorithm ensures that the routing is resilient.

Notations. Without loss of generality, we assume that the capability[3] of a node x is measured by the maximum number of connections (we note that connections maintained in a VC^2A overlay are duplex TCP/IP links) that it can accommodate, which is denoted $c(x)$. Based on $c(x)$, x estimates the maximum number of connections $m_i(x)$ that are used by an overlay i. If x joins p tree overlays, then $\sum_{j=1}^{p} m_j(x) \leq c(x)$. Note that the overlays may have different sizes. The larger the overlay, the larger the number of connections allocated. To simplify the discussion, we concentrate on operations that manipulate "one" overlay. We thus drop the subscript i in the following discussion, i.e., using $m(x)$ instead of $m_i(x)$.

A node x will maintain at most $m(x)$ pointers pointing to its children ($x.child[i]$, where $i = 1, 2, 3, \ldots, m(x) - 1$) and parent ($x.parent$). These nodes are the active neighbors of x. The nodes $x.parent$ and $x.child[i]$ are initialized with null pointers. Node x also maintains a *look-ahead cache* indicating potential parent nodes of x. Each entry in the look-ahead cache consists of the IP address, the port number and the maximum number of connections of a node that can become the parent of x. Finally, x maintains a value $height(x)$ that indicates its height in the tree, where $height(x) = 0$ initially. The variable $d(x)$ is used to indicate the number of connections in use.

Self-Configuration and -Healing. When a node intends to join a tree overlay, it first connects to the root node of the tree corresponding to the overlay. The two nodes then compare their heights. Since the height of the new node is always zero, it will become the child of the root, as long as the root node has not reached its capability limit, i.e., the maximum number of connections. In general, a node

---

[3] A node's capacity (in terms of number of connections) is defined by mixing its "static" machine characteristics such as the computational power and communication speed as well as the policy of how the resource provider contributes his resources.

x becomes a child of another node y if y has not reached the limit of its capacity and $height(y) > height(x)$.

On the other hand, if the root of the tree has exhausted its capacity, it designates a child that has the maximum remaining capacity as the parent of the new node. If no child node can accept that node as a child due to capacity limitation, the root node randomly picks a child to forward the joining request of the new node. That child node performs similar operations as the root node until a child can accommodate the joining request of the new node. After the new node joins the overlay, the nodes along the path to the root update their heights if necessary. Due to the space limitation, we refer readers to [4] for the details.

A VC²A tree may be fragmented due to node failure or departure. To solve this problem, each node periodically broadcasts its IP address, port number and maximum capacity to its children. The broadcast horizon is specified by a small constant s. If a node receives such a message, it pushes the associated IP address, port number and capacity value into its look-ahead cache. In this way, a node can "look ahead" s nodes (including the parent) to the tree root.

A node x first checks whether its parent is active. If the parent is not active, the node then creates a connection to the closest ascendant node u, which is found through the look-ahead cache and has a free connection. If u does not have a free connection, then it helps to discover a descendant node that can accept x's connection. Note that if the ascendant nodes maintained in the look-ahead cache are not available, x consults the root of the tree for rejoining.

Since nodes periodically monitor the "aliveness" of their parent, the monitoring message can piggyback the value $height(x)$ to the parent. The parent can then determine whether its height is maintained according to the height values from all the children. We note that there must be a sibling of x, j, such that $height(x.parent) = height(j) + 1$, where $height(j)$ is the maximum compared with other siblings. If the height value of x's parent is not equal to $height(j)+1$, $height(j) + 1$ is taken as the height value of x's parent.

Exploiting Heterogeneity of Nodes. The algorithms mentioned above do not guarantee that nodes with higher capacities are placed close to the root of a tree, because a joining node may be connected to a node whose capacity is smaller than it own. Thus, a node in VC²A needs to monitor the capacities of its children and to determine whether a child node more capable of providing connections to accommodate new nodes. If a node can be replaced, the node is moved to the position of the child node in the tree. Notably, a node is only replaced by one of its children by locking the data structures it maintains.

We note that in our algorithm we not only organize more capable nodes around the task producer in order to boost the system throughput, but ensure a robust tree overlay. This is because volatile nodes will not appear in the nearby of the task producer and will not fail the tree structure. This thus allows the task producer to submit its jobs to the peers in the tree overlay.

Tasks Multicasting and Termination. The task producer submits its computational tasks using Armada mobile agents to the root of its associated overlay tree. The root then clones the mobile agents and migrates the cloned agents to each of its children. A node receiving an agent performs a similar operation.

In addition, to submit a cloned agent, an installed agent retrieves the associated data set before performing the computation associated with that agent. In VC²A an agent will be automatically programmed by implicitly specifying its input as its parent agent or explicitly designating a location outside the system. Similarly, an agent can designate its output using the same way. We note that an application developer may manually configure the input and output for an agent. In addition, VC²A allows an agent to designate the communication protocol used, such as HTTP and FTP, in order to interact with an entity that is not in the system.

If an agent has the input data set to compute and the node hosting the agent is not overloaded according to its local policy, then the agent performs the computation. If the agent does not overload its hosting peer and its input queue is not full, it tries to obtain (or wait for) another data set from its parent agent for further computation. We note that (1) the delivery of data sets and the computation performed by the agent can be handled, simultaneously. In the current implementation, a node computes its tasks, sends tasks in its input queue, and returns the results stored in its output queue in the round-robin fashion. (2) It is possible that a number of k' input data sets can be only sent k' children nodes out of the k requesting ones, where $k' < k$. In such a case, the parent node selects top-k' more capable peers from k ones to receive the input data sets. (3) We also note that the scheduling of computation and communication in order to maximize the utilization of resources in a tree overlay is an orthogonal design space. In addition, with the application knowledge to push a task to its children or pull a task from its parent is left to the application developer.

VC²A is based on the Armada mobile agent system in which agents in different locations can communicate with each other. This allows VC²A to support an application that has complex interactions among agents. We preserve this capability for investigating applications that have complex communications among agents in the future.

For termination of an application, the task producer broadcasts a termination message to the root of the associated tree overlay. The termination message is then broadcasted down to the tree. Each of the tree nodes receiving the termination message leaves the tree overlay.

Multiple Tree Overlays. We have discussed operations that manipulate a single overlay. However, a node may join several overlays. Here, we discuss how a node manages the available connections for the participating overlays.

Initially, a VC²A node x allocates an equal number of connections to each of the overlays that it joins. Assume that x may participate in g overlays at most. Then, x will have $m_i(x) = \frac{c(x)}{g}$, where $i = 1, 2, 3, \ldots, g$. The node periodically monitors the number of connections ($d_i(x)$) used by the ith participating overlay.

It then assigns the connections to the participating overlays proportionally as $m_i(x) = \left\lfloor \frac{d_i(x)}{\sum_{j=1}^{g} d_j(x)} \times c(x) \right\rfloor$.

Consequently, if an overlay comprises of a small number of nodes, these nodes must use few connections to link with each other. The above assignment ensures that nodes use few connections to participate in a small overlay and more connections for a large overlay. This reduces the height of a tree since nodes joining in a large overlay have relatively large degrees; can allow tasks to be rapidly sent to nodes by traversing a smaller number of nodes.

To estimate the size of a tree overlay, the root node of a tree broadcasts a counting message to each of its child nodes. Upon receiving a counting message from the parent, a node recursively performs the same operation. If a node (i.e., a leaf node) does not have any children, it returns the number of children (i.e., zero) to its parent. When a node receives the number of child nodes from each of its children, it can then estimates the total number of its descendant nodes and reports such a number to its parent. In this way, the root node can eventually determine the total number of nodes in the tree. It then broadcasts such a value to each tree member. Finally, a peer can determine how many connections should be allocated for each tree overlay that it participates. We show that the depth of a tree overlay is bounded by the logarithmical size of a tree overlay (see [4]).

Clustering Nearby Nodes. It is possible that an application may restrict the resources that meet its requirements on their network-specific constrains in order to conform to the performance aim of the application. For example, an application may want to allocate resources that are geographically nearby by designating the delay or the bandwidth between the task producer and the resources having above or below a certain value. If so, in the absence of network weather services resources in VC^2A that can meet an application's demand need to measure the delay or bandwidth between the designated location and itself. Only those resources that satisfy the delay or bandwidth requirement can join the tree overlay formed for the application.

Security. First, malicious peers have several ways to potentially break down VC^2A. For example, a malicious peer may not honestly perform the protocol operations, such as, `publish`, `lookup`, `inform` and `join`. To prevent malicious peers joining the system, VC^2A includes additional "authentication nodes". It should be noted that these nodes are not used for any other VC^2A purpose, apart from authentication. A node intending to join the system needs to access an authentication node, and only after being authenticated can nodes join VC^2A and participate in a computation.

However, authenticated nodes that have joined the system may be compromised, i.e., authenticated nodes may not be trusted. Since VC^2A integrates the security mechanism in [5], the compromised nodes cannot modify the protocol codes and thus change their behavior (e.g., by stopping forwarding or maliciously forward a message to another compromised node). This ensures that a message will be correctly forwarded to its destination.

Third, it is possible that a number of compromised nodes could introduce cycles to the system by repeatedly and rapidly joining and leaving the system. Since VC^2A is implemented on top of the resilient DHT-based overlay, which can self-heal and consequently tolerate cycles introduced by compromised nodes. Unless all VC^2A nodes are compromised, it is difficult to fail the entire system. Possibly, malicious nodes could overhear messages exchanged by overlay nodes and then tailor their messages to break down the system. For such a case, VC^2A is further integrated with the secure communication mechanism [6] proposed for DHT-based overlay networks. The secure communication mechanism can set up secure channels among overlay peers.

Fourth, compromised nodes can be used to "attack" the system. This can be achieved by letting compromised nodes send a large amount of normal traffic (e.g., the join messages) to the root of a tree overlay. In this way, the root could be busy processing large amounts of normal traffic, which may prevent further participants from joining the tree. The root may also be unable to clone agents for its established child nodes, or send tasks to, and receives results back from these nodes.

For relieving such an attack, the root (denoted as r) of a tree overlay designates a few nodes in VC^2A as "secret" entry points into the overlay. r can only accept messages from its designated secret nodes. These secret nodes ensure the rate-limited communication (r informs the secret nodes its maximum communication bandwidth). To select k secret points requires k distinct hash functions ($\mathcal{H}_i$, where $i = 1, 2, \ldots, k$). The secret nodes will have the node ID closest to $\mathcal{H}_i(name)$ (this can be simply achieved by routing a message to the node, i.e., the secret node, that has the closest ID to $\mathcal{H}_i(name)$) and the $name$ is r's IP address. Thus, node x in VC^2A, which meets the resource description and intends to join r's tree, sends a join message with the destination address r. The join message will reach one of the secret node that can help x to join r's tree.

Since the message communication in VC^2A is a multi-hop end-to-end communication, compromised nodes cannot easily determine the ultimate destination (i.e., the secret nodes) of a message transfer by simply eavesdropping the source and the destination fields of one TCP/IP end-to-end connection. To determine which nodes are secret nodes requires globally monitoring all communication traffic in VC^2A and then analyzing the traffic. Consequently, compromised nodes cannot simply shut down the secret nodes in order to break down the associated tree overlay.

Finally, compromised nodes could arbitrarily send a large amount of overlay-irrelevant traffic (e.g., the TCP SYN flooding) to a tree root at the same time in order to the shut down the root. For such a case, the tree root needs to set up a few proxy nodes (the proxy node is not part of VC^2A computation) in its vicinity. The proxy node can act as a router that is responsible for forwarding network messages to the tree root. The tree root simply configures those proxy nodes to accept traffic from the secret nodes that had been previously established. Traffic that is not from a secret node will be discarded. Note that if the proxy nodes are the high-powered routers, attackers cannot easily attack the network routers since these devices can discard traffic that is not from the secret nodes at line speed.

3 Comparison with Alternatives

We compare the previously mentioned architectures with VC^2A from two aspects: *resource discovery* and *organization of allocated resources*. Both are the fundamental components for assembling a computational infrastructure. The details including references can be found in [4].

Resource Discovery. Recent projects such as SETI@home/BONIC, Entropia and XtremWeb have taken the first steps towards realizing a large-scale systems to include worldwide desktop machines. Both systems employ a central server to manage the participating resources. However, the use of a central control leads to the usual issues, such as the performance bottlenecks and a single point of failure.

In contrast, OurGrid and NaradaBrokering are based on a Gnutella-like P2P network model that adopts broadcasting to discover resources without relying on a central server. Although broadcasting does not have the issues as those introduced by SETI@home/BONIC, Entropia and XtremWeb, it may generate a significant amount of traffic and does not have the guarantees for discovering demanded resources in terms of efficiency and effectiveness. Triana is a workflow management system, which is based on JXTA. JXTA does not exactly mandate how resource discovery is done. A possible decentralized discovery mechanism in JXTA is through utilizing rendezvous peers. Resource providers publish their resource metadata information to rendezvous peers. A set of rendezvous peers acts as a search network that implements a Gnutella-like search protocol. To discover resources is to send a query message to the search network.

Butt et al. [7] propose to take advantage of network locality for resource discovery by exploiting the geographically close neighbors in a DHT-based overlay. However, nearby resources may not meet the request demand. Although their design has incorporated with an expanded ring search mechanism, this may lead to the system to degenerate into a broadcasting-based P2P network similar to Gnutella.

In contrast to previous architectures, VC^2A is based on the enhanced DHTs [1] to support complex queries. These enhanced DHTs provide the efficient performance bounds for discovering the demanded resources.

Organization of Allocated Resources. To our best knowledge, SETI@home/ BONIC, Entropia, XtremWeb, OurGrid and Triana all rely on a central control point to manage resources that satisfy the demand of the requester. The central point in SETI@home/BONIC, Entropia and XtremWeb is a standalone sever, while in OurGrid and Triana the requester handles all requested peers. Clearly, this is a non-scalable design.

Butt et al. extend Condor to the Internet using Pastry. No explicit mechanism for managing resources discovered from different Condor pools is provided. The architecture Butt *et al.* propose strongly relies on each central manager of each Condor pool. Similarly, NaradaBrokering does not offer the ability for discovered resources to self-manage. However, NaradaBrokering provides the messaging capabilities including the publishing/subscribing and the secure channel for nodes in the system.

In VC^2A, discovered resources autonomously structure into a tree overlay according to their capabilities without involving entities that not are not allocated in the publicly shared infrastructure of VC^2A. Such a design is not only scalable, but can tolerate faults. Moreover, this allows the requester to utilize relatively more capable resources in order to speed up their computation.

4 Conclusions and Project Status

We have presented the design of P2P-based system called VC^2A. VC^2A is based on a DHT-based overlay, which further incorporates a tree structured P2P overlays and takes advantage of nodes' heterogeneity. We have shown how to manage the overlay and manipulate it using mobile agents. Due to space limitation, we refer readers the performance results of VC^2A to [4]. We have prototyped VC^2A and will implement a real application to examine the system in the near future. The results will be reported when they are available, as will be our experiences using the system.

References

1. Hsiao, H.C., King, C.T.: Resource Discovery in Peer-to-Peer Infrastructure. In: High Performance Computing: Paradigm and Infrastructure. John Wiley & Sons Ltd (2004)
2. Hsiao, H.C., King, C.T.: Tornado: A Capability-Aware Peer-to-Peer Storage Overlay. Journal of Parallel and Distributed Computing **64** (2004) 747–758
3. Hsiao, H.C., Huang, P.S., Banerjee, A., King, C.T.: Taking Advantage of the Overlay Geometrical Structures for Mobile Agent Communications. In: Proceedings of the International Parallel and Distributed Processing Symposium, IEEE Computer Society (2004)
4. Hsiao, H.C., Baker, M., King, C.T.: VC^2A: Virtual Cluster Computing Architecture. Technical report, Department of Computer Science, National Tsing-Hua University, Hsinchu, Taiwan (2004) http://www.cs.nthu.edu.tw/~hchsiao/projects.htm.
5. Castro, M., Druschel, P., Ganesh, A., Rowstron, A., Wallach, D.S.: Security for Structured Peer-to-Peer Overlay Networks. In: Proceedings of the International Symposium on Operating Systems Design and Implementaion. (2002)
6. Freedman, M.J., Morris, R.: Tarzan: A Peer-to-Peer Anonymizing Network Layer. In: Proceedings of the International Conference on Computer and Communications Security, ACM Press (2002) 193–206
7. Butt, A.R., Zhang, R., Hu, Y.C.: A Self-Organizing Flock of Condors. In: Proceedings of the International Conference on High Performance Networking and Computing. (2003)

The Model, Architecture and Mechanism Behind *Realcourse**

Jinyu Zhang and Xiaoming Li

Department of Computer Science,
Peking University, Beijing, 100871, China
{zjy2,lxm}@pku.edu.cn

Abstract. *Realcourse* is a video stream service supported by a collection of physical servers distributed all over China. This article provides a comprehensive description on the model, architecture, and operation mechanism behind *realcourse*. In particular, the highly-availability feature and dynamic re-configurability is emphasized.

1 Introduction

Since its birth in April of 2003, *University Course Online*, http://realcourse.grids.cn (*realcourse* in short) has been in non-stop operation for more than one year, in spite of some of the servers up-and-down, in-and-out from time to time. As of this writing, *realcourse* is composed of 20 servers distributed on CERNET as in the following, which was grown up from initially 4 servers within Peking University campus. Table 1.

Beijing	Shanghai	Guangzhou	Harbin	Xian	Wuhan	Shenyang	Jinan	Nanjing	chengdu
9	1	2	1	2	1	1	1	1	1

Besides the servers that share the single URL as above, there are more than 1500 hours of quality course video being served, and as a result we normally observe around 100 on line users at any time. The course videos are from different universities, including Peking University, Tsinghua University, and Huazhong Unversity of Science and Technology, etc. They are uploaded to different servers concurrently with permanent backup copies created in servers other than original ones, and flow from server to server on demand.

2 Model

A few elements can be identified behind *realcourse*. $S = \{s_1, s_2, ..., s_m\}$, an open collection of servers. s_i also represents the name of the server. The word "open" indicates its dynamic nature, namely m may be different from time to time, which reflects

---

* This work is supported by ChinaGrid project and 973 grant (G1999032706).

servers down-and-up or in-and-out. $C = \{c_1, c_2, ..., c_n\}$, an open collection of video clip names. While C is of particular importance in maintaining system single image, there are two physical copies of each video, the original copy and the backup copies) respectively. The distribution of C among S is modeled by two mappings $f : C \rightarrow S$; $g : C \rightarrow S$ $\forall i, f(c) \neq g(c)$. $f(c)$ is the name of the server where the original copy of c is stored while $g(c)$ is the name of the server where the backup copy of c is stored. There is a collection of users or client, $U = \{u_1, u_2, ...\}$. An access is started by the execution of a mapping from U to S. $v : U \rightarrow S$. v(u) indicates a particular server of S is chosen to serve a user.

2.1 Functional Objectives

1. If a user u places a visit to the system, no matter which server it is connected to, a complete C should be presented to him.
2. Any download operation should be completed successfully and transparently, whether the physic content of c is initially on v(u) or not.
3. The result of any upload operation should be reflected in C, all servers in S host an identical C. The execution of f and g on every server should get the same result.

2.2 Collective and Dynamic Objectives

Realcourse is meant to be used by open public. We think failures should be allowed on servers or network. Thus, we have the following special objectives:

1. A user should not notice the failure or shutdown of a server. More specifically, the function $v : U \rightarrow S$ should always be mapped to active servers and they should always collectively host a complete C. Each server in *Realcourse* must host a complete C, which we defined as the consistency of C in this paper.
2. A new server or a recovery server should be brought to a consistent state within certain time frame.

Before proceeding to the following sections that describe our design to implement these functions, we indicate so-called system state of *realcourse* is nothing but the quaternary (S, C, f, g). The task of maintaining a consistent system state means to have all the servers have identical (S, C, f, g) as timely as possible (or as needed as possible). In one word, the technical essence of *realcourse* is maintaining a consistent system state with respect to gradual change of S and with the promise of non-stop, complete, and efficient service.

3 Architecture

The architecture of *realcourse* reflects a design and implementation of f, g and v as defined in the model above. In particular, each s_i in S is a Linux server with a unique name of s_i currently deployed on the Internet at 10 different cities in China. All servers form a loop, as shown in Fig 1.

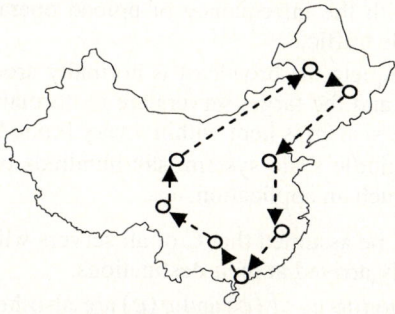

Fig. 1

On each server, a circular linked-list is used to represent the loop. Each server knows its predecessor and successor by checking the linked-list. Formally, an operator *suc* can be defined as: $suc(s_i) = s_{i+1 \bmod N}, i \in \{0,1,...,N-1\}$, N is the total number of elements in S. Note, *Realcourse* allows the servers to be inserted or deleted without stopping the service. Thus, N is actually a changing number (though not dramatically). The server-adding and server-deleting procedures are designed to keep track the live status of S and send the changed information to all active members of S. The modification of this linked-list on each server is triggered by a Linux signal send to the service process (instead of restarting the process). This signal causes the process to re-read the file that contains the linked-list distributed in advance. The consistency of this linked-list among all running servers is equal to the consistent state of S. It is also fundamental to the consistency of C because each server needs to inform any uploading event to all other servers.

In realcourse, g is then defined as: $g(c) = suc(f(c))$

In other words, the two permanent physical copies of c are stored on two successive servers. Then, it is clear that the consistency of g is based on f and S.

Finally, the consistency of C and f among all servers is the major challenge to *realcourse*. We define the consistency of C as "all servers must host an identical C" while the consistency of f is defined as "all servers know the location of c^o and the location of c^b ". Once a video clip is uploaded to one of the servers, the fact of this uploading (not the clip itself) must be known by all other active servers as quick as possible. The key technology we use to maintain the consistency is broadcast based on a reliable asynchronous messaging middleware. Asynchronous messaging middleware[1] guarantees that a message sent from place A to place B would finally arrive at B even if the network between A and B fails or B is temporally unavailable. How long the middleware keep the message can be adjusted by changing the length of the queue used to store the messages. Any event occurred in *realcourse* on server s_i that changes the state of C is transformed into a message by s_i and broadcasted to all other servers. Although the treatment is simple and effective, we understand there are times that the system is not consistent. Nevertheless, with respect the nature of our application and observing the following three points, we feel the hazard is neglectable:

- The length of this broadcast message is kept very short, usually several bytes in *realcourse*. Together with the infrequency of upload operations, the broadcast will not cause observable traffic.
- The time it takes to complete the broadcast is normally around one millisecond when both the network and the target servers are in normal status. This implies that inconsistency of the system is kept within a very limited situation.
- *Realcourse* aims at a middle scale system, say hundreds of servers, which we consider is enough for such an application.

With these in mind, it can be assumed that C of all servers will be put into consistency after all messages finally arrived at their destinations.

Upon an upload, in addition to c, $f(c)$ and $g(c)$ are also broadcasted. All servers will generate a global directory based on messages they receive in the form of $(c, f(c), g(c))$. By checking this directory, each server can know what c is in *realcourse* and where to find those two permanent copies of c. f(c) is the place from where c is first uploaded. If a server fails in the middle of an uploading of one file, it will simply delete the uploaded partial content of file and no broadcast is done. If a server fails in the middle of a downloading, the client can resume the downloading later like most other ftp servers do.

$v : U \to S$ is a more complicated function which can be explained like $v(D, L, S)$, D stands for the a vector of network distance between u and servers, L means a vector of the current loads of all servers, S is the vector of free space of all servers. The execution of v actually means that *realcourse* will choose the "fastest", "light loaded" server with enough storage space for users. The execute mechanism of v is described in the next section.

Before jumping into the description of the mechanism, some other implementation issues should be mentioned. In *realcourse*, each "c" is a video clip with a size of around 100MB. The communication applied in *realcourse* includes:

Between two servers: FTP is used for video data, MQSeires is used for message broadcast.

Between client and server: HTTP or FTP. In order to browse the video course, a video player is required. currently *realcourse* supports only RealPlayer and will be extent to support Media Player of Windows.

In terms of software, a server can be best described as in Fig 2.

4 Mechanism

With emphasis of the dynamic nature of the system, several critical events that occur in *realcourse* operations are carefully studied. They are: uploading a video-clip to *realcourse;* download (watch) a video from *realcourse*; add a server, and remove a server. ConsistencyCheck() function is run by each server every day. First of all, the implementation of v is introduced because all user-related events begin by the execution of v. 'v' is essential to make failures of some server invisible to users. *Realcourse* always choose the currently "active" and "best" server to serve the user.

4.1 Implementation of $v(D, L, S)$

Users access *realcourse* in two ways. Pointing web browser at the URL http://*realcourse*.grids.cn/ or by installing a detecting plug-in and pointing the browser at a special semi-URL "realcourse".

For the first way, an enhanced central DNS server "ds" (Fig 3) is deployed. It is the authorized server for sub domain grids.cn. Each server in *realcourse* collects the load information and storage space information of its own and sends them to **ds** in every minute. The resolution request for domain name ".realcorse.grids.cn" will finally arrived at **ds** following the protocol of Domain Name Service[2][3] in 8 steps, explained in Fig3. Note: The CACHE of Domain Name Servers are disabled by setting the TTL of DNS record to zero.

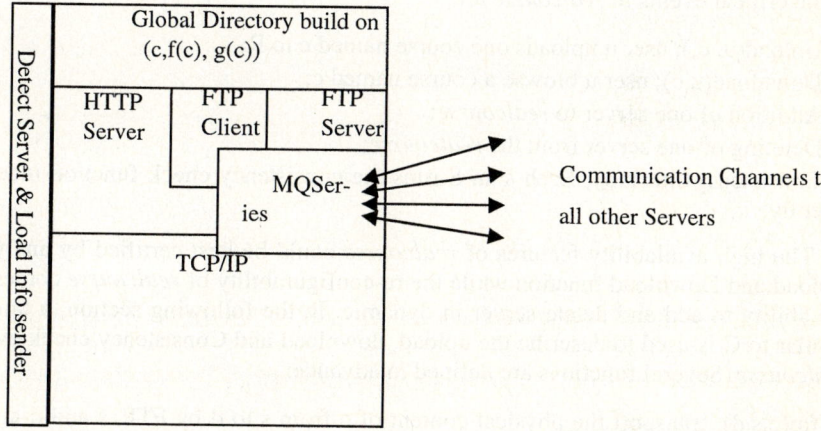

Fig. 2.

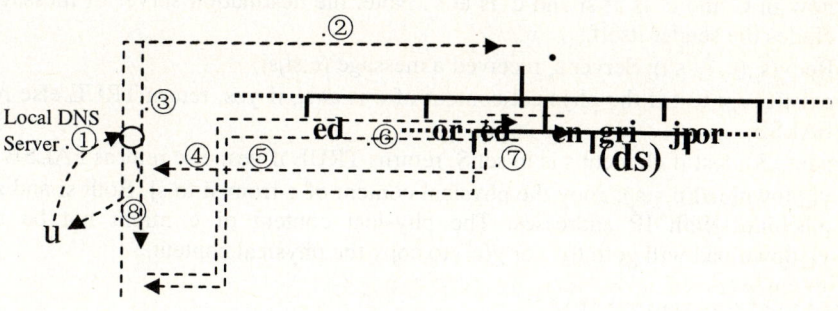

Fig. 3.

Base on the current load and storage space information about all servers and the network position of the local DNS Server, **ds** will choose a "best" server and return the IP address of this server to **u** at step 8. Here, we can assume that **u** is close to the "local" DNS server.

For the second way, user is required to install a piece of browser helper object on Windows. This piece of hook code will start a detecting procedure once user input "realcourse" in the address area of IE browser. First, it communicates with one server and retrieves current Linked-List of S, it then sends a detecting package to all servers consecutively. Each server runs a Detect Server as showed in Fig2. After receiving the detecting package, the Detect Server will decide a delay of response based on the current load of itself and free space storage. The first response that arrives at user includes the IP address of the "best" server and it will also stop the following delivery of detecting package.

4.2 Implementation of Key Events in *Realcourse*

Some critical events in *realcourse* are:

- Upload(u, c,); user u uploads one course named c to R;
- Download(u,c); user u browse a course named c;
- Addition of one server to *realcourse*;
- Deleting of one server from the *realcourse*;
- ConsistencyCheck(s_i); Each s_i in S runs the consistency check function independently.

The high availability features of *realcourse* could be best certified by analysis of Upload and Download function while the re-configurability of *realcourse* comes from the ability to add and delete server in dynamic. In the following section, a language similar to C is used to describe the upload, download and Consistency check event of *realcourse*. Several functions are defined in advance:

- ftp(c,s,d); transport the physical content of c from s to d by FTP, s and d are both computers with unique IP address. It may be a client or a server. This function returns TRUE if succeed and FALSE if fail.
- Broadcast(c, s_i, s_j); one server send a message to all servers in S to inform that c is now in C and c^o is at si and c^b is at s_j. Note: the destination server of message includes the sender itself.
- Recv(s_k,(c, s_i, s_j)); Server s_k received a message (c,si,sj)
- exist(c, s_i); test if the physical content of c is at s_i, if yes, return TRUE else return FALSE.
- isin(s,S); test if element s is in set S, returns TRUE if yes, else returns FALSE.
- yj_download(c,s_i,s_j); copy the physical content of c from si to sj. Both si and sj are machines with IP addresses. The physical content of c might not be at si, yj_download will goto f(c^o) or g(c^b) to copy the physical content.

```
/********Upload********/
Upload(u,c){
    si =v(u);/*the currently best server for u is si */
    ftp(c,u, si);
    sj=suc(si);
    broadcast(c,si,sj); /* send message (c,si,sj) to all server */
```

```
    for each sk in S{  /*include the sender itself*/
      recv(sk, (c,si,sj))
    }; /*see below*/
}

  recv(sk, (c,si,sj)){
    C=C+c;     //put the C into consistency
    f(co)=si;  //put the f into consistency
    g(cb)=sj;  //put the g into consistency
    if(sk==sj){
         ftp(c,si,sj); //copy the physical content of c to sj
    } // end of if(sk==sj)
} //end of recv
/*********Download********/
Download(u,c){
  si=v(u);
  yj_download(c,si,u); //download the file from si to u
  return;
}
  yj_download(c,si,sj){
    if(exist(c,si)==TRUE){
      ftp(c,si,sj);
      return;
    }else if(f(co)==si){ /*The content of c should be at si , but
                            it is corrupted */
      yj_download(c,g(cb),si);
    }else if(f(cb)==si){ //Server si should have the cb ,but it is
corrupted
         exit;
    }else{
      if(yj_download(c,f(co), si)==FALSE){
        if(yj_download(c,f(cb), si)==FALSE{
           exit;
        }
      }
    }
    ftp(c.si.sj);
    return;
}
```

/*Note 1: yj_download( ) is a distributed recursive function. Since the migration of data between servers is done by FTP, one endless loop could happen in case of both the original copy and backup copy of c are corrupted. The "exit" statement at Δ is necessary to avoid this endless loop.

Note 2: Execution of download will transfer file among servers as needed. We name these physical copies of c as c^c. The existence of c^c will minimize the response time of the next download for c.*/

```
  /********ConsistencyCheck********/
  ConsistencyCheck(si){
  for each c in C{
     if(!isin(f(co), S)&& !isin(g(cb),S)){
        exit(-1);
```

```
        } /*Both the server of f(co) and g(cb) are deleted. This
           situation should be avoid */
        if(!isin(f(co), S)&& isin(g(cb),S)){
            f(co)=g(cb);
            g(cb)=suc(g(cb));
        } /*The server of f(co) was deleted, the g(cb) becomes f(co)
           and suc(g(cb)) becomes g(cb)*/
        if(isin(f(co), S)&& !isin(g(cb),S)){
            g(cb)=suc(f(co);
        } //The server of g(cb) was deleted, the suc(f(co) becomes g(cb)
        if(isin(f(co), S)&& isin(g(cb),S)){
            if(g(cb)!= suc(f(co){
                g(cb)= suc(f(co);
            }
        } /*One server is added between f(co) and g(cb), it becomes
           g(cb)*/
        if((g(cb)==si)&&exist(c,si)==FALSE)){
            ftp(c,f(co),si); /*to be sure that one physical copy of c
                              is at g(cb) */
        }
    }//end of for
}
```

Addition of one server s_n to *realcourse* starts by capturing the state of (C,f,g) of one old server s_o and copy it to s_n. The major challenge comes from that during the capture of (C, f, g), events may happen and the state we put the new server in may not be in consistency with other servers. The temporal storage capacity of Asynchronous Middleware can be used to overcome this issue. What we do in *realcourse* are in four steps:

1. At time t_1, choose one light-loaded server s_o, halt the execution of recv function, inform the addition of s_n to all other servers in S including s_o. All events happen in *realcourse* before time t2(see below) will be stored as messages in the Messaging Middleware.
2. Capture the state of s_o at t1, copy it to s_n and put the s_n in state of t1. Suppose this step end at t_2,
3. Right at time t2, recover the executing of recv function on s_o and start the executing of recv function at s_n. Then, both s_n and s_o will catch up with all other servers in *realcourse*.
4. Put s_n into the candidate list of best server selection.

Note: the addition of one server into *realcourse* will put the state of f and g into an inconsistent state. But, this addition will not affect the normal download and upload function of *realcourse*. After execution of ConsistencyCheck( ) on each server, the consistency of f and g will recover.

Deleting one server from *realcourse* is somewhat easy. If a server s_i fail for a long time, it should be deleted from the system. The Linked-List on all other servers will be updated to tell the absence of s_i. *Realcourse* will finally go into consistency after the execution of ConsistencyCheck( )

Note: If two consecutive servers fail between two executions of ConsistencyCheck( ). Some c will finally be deleted by a process TaoTai( ) which is run once a month on each server. Process TaoTai( ) deletes those c^c that have not be accessed since its last execution.

5   Related Work

Realcourse is a successful application of distributed computing [4] technologies in a geographically wide area. Different from some traditional distributed fault-tolerant services like ISIS[5], *realcourse* emphasizes on giving clients access to the service-with reasonable response times-for as much of the time as possible. In fault-tolerant systems, all replica-managers(servers in case of *realcourse*)are informed of the updates operation of users in an "eager" fashion: all replica-managers receive the updates as soon as possible and they reach collective agreement before passing control back to client. Compare to other distributed highly-available services such as The Gossip architecture[6], Bayou storage system[7] and The coda file system[8], *realcourse* owns some distinguished characters:

Realcourse aims at non-critical application of video on demand and file storage in which temporal inconsistency is acceptable only if consistency of whole system will finally reached. Coda, on the contrast is designed and implemented for a general file system. By eliminating the "update" operation, *realcourse* greatly reduce the consistency requirement, while in Gossip architecture and Bayou much effort is made to maintain the casual order of "updates" operation among replica managers.

The servers in *realcourse* are not equal to each other. Only two servers are chosen to keep two permanent physical copies of files. No effort is made to keep track of those temporal copies of files but the existence of these copies greatly improve the performance of downloading operation in a way of Wide Area Network Cache. In the new version of consistency-check procedure a pervasive search is started to find those "lost" copies in case that both two permanent copies of file are corrupted.

The loop topology of all servers makes it possible for each server to run a Consistency-check procedure independently without any overhead.

By exploiting the reliable communication provided by Asynchronous Messaging Middleware, *realcourse* hides the failures of network, which is not rare by our observation in a Wide Area Network from servers. The consistency of servers is eventually kept in a "delayed" fashion.

6   Conclusions

Realcourse has been running for more than one year. To our knowledge, with its 20 servers spanned over 10 major cities, it is the first non commercial wide area distributed content delivery system deployed in China*. In our opinion, the success mainly comes from the following:

---

* We don't know if there is a comparable commercial system running in China or not.

- A conceptually clear design. For example, to avoid complexity, we did not choose to employ a more sophisticated topology among the servers, and we chose to use simple double backup strategy.
- Careful implementation. For example, to provide efficiency in across-server download, we chose to use asynchronous socket communication to effectively form a pipeline, instead of store-forward fashion.
- Using as much open source as possible. Besides the MQ (which is a great help), all our code is based on opensource, including FTP and Appache.

As for future work, we see the most meaningful next step is to introduce the concept of *cache-only servers*. Let call the servers discussed above *backbone server*. The cache-only server differs from backbone server in: it will neither be a point of upload operation, nor have the duty to hold permanent backup copies for video clips. Once this concept is implemented, a lot more servers may be added to the system from the places with less advanced network conditions, but to serve local client better. We should see this happen in next few months.

References

1. IBM, MQSeries System Administration SC33-1873-01 [DB/CD]
2. P. Mockapetris, DOMAIN NAMES - CONCEPTS AND FACILITIES, RFC1034 (1987) [S/OL]
3. P. Mockapetris, DOMAIN NAMES - IMPLEMENTATION AND SPECIFICATION, RFC1035 (1987) [S/OL]
4. George Coulouris, Jean Dollimore, Tim Kindberg, Distributed System Concepts and Design(Third version), ISBN 7-111-11749-2, China Machine Press
5. Birman, K.P. (1993). The process group approach to reliable distributed computing. Comms. ACM, Vol. 36, No. 12, pp. 36-53
6. Ladin, R., Liskov, B., Shrira, L. and Ghemawat, S.(1992). Providing Availability Using Lazy Replication. ACM Transactions on Computer System, Vol. 10, No. 4, pp. 360-91.
7. Terry, D., Theimer, M.., Petersen, K., Demers, A, Spreitzer, M. and Hauser, C. (1995). Managing update conflicts in Bayou, a weakly connected replicated storage system. Proceeding of the 15th ACM Symposium on Operation Systems Principles, pp. 172-183
8. Satyanarayanan, M., Kistler, J.J., Kumar, P., Okasaki, M.E., Siegel, E.H. and Steere, D.C. (1990). Coda: A Highly Available File System for a Distributed Workstation Environment. IEEE Transactions on Computers, Vol. 39, No. 4, pp. 447-59

Managing Irregular Workloads of Cooperatively Shared Computing Clusters

Percival Xavier, Wentong Cai, and Bu-Sung Lee

School of Computer Engineering,
Nanyang Technological University,
Nanyang Avenue, Singapore 639798
{asxpercival,aswtcai,ebslee}@ntu.edu.sg

Abstract. Cooperative resource sharing enables distinct organizations to form a federation of computing resources. A functional broker is deployed to facilitate remote resource access within the community grid. A major issue is the problem of correlations in job arrivals caused by seasonal usage and/or coincident resource usage demand patterns where high levels of burstiness in job arrivals can cause the job queue of the broker to grow to an extent such that its performance becomes severely impaired. Since job arrivals cannot be controlled, management strategies must be employed to admit jobs to sustain the resource allocation performance of the broker. In this paper, we present a theoretical analysis of the problem of job traffic burstiness on resource allocation performance in order to elicit the general job management strategies to be employed. Based on the analysis, we define and justify a job management framework for the resource broker to cope with overload conditions caused by job arrival correlations.

1 Introduction

Grid computing evolved from the need to deploy large-scale applications on a multitude of computing systems across computer networks. The ability for such applications to cross administrative domains not only contributes to the processing capabilities of the grid, but also broadens users' space for resource acquisition on the grid. Middleware for computational grids makes it very easy to connect the computer of a potential user to the grid. We look at the possibility where organizations are willing to 'pool' their computing and data resources together to form a virtual computing platform. The reason for this is that since a grid is likely to involve a diverse community of resource users who are themselves also contributors of resources, there is a likelihood that computing cycles can frequently be traded in the event that, while some organizations are experiencing resource shortage, others may not be utilizing their resources at all. Although this mode of sharing has been described in many instances especially in the domains of peer-to-peer methodologies [2, 5], little publicly known work has theoretically explored the management issues associated with a grid formed by cooperative federated computing clusters.

Following this section, we present a theoretical study of the problems and explain the necessary mechanisms to counteract them. In the analysis, we first evaluate the size of queue storms based on the contributions made by participating organizations. Next, we observe the effect of introducing slack time into deadline jobs to evaluate the improvement in resource allocation performance. In section 3, we then propose a job queue management framework based on the conclusions derived from the experiments. In section 4, we present the related work, which is followed by the conclusions and future directions in section 5.

2 Analysis of Cooperative Resource Sharing

The grid consists of a set of computing clusters belonging to a consortium of physical organizations that are located geographically within the same WAN. A Clearinghouse, which is essentially a resource broker, manages the federation of shared resources and is therefore the access point where requests for remote resources are made [14]. In this experiment, we demonstrate the physical consequence on resource management if all resource providing agents do not share out sufficient resources with respect to the community's workload. Specifically, the following experiments reflect the importance of a job admissions control strategy to deal with resource overload conditions that lead to queue storms.

2.1 Experimental Model

Variables. The parameters in the following list reflects the experimental variables that were expressed in the analysis.

- T_s - Defines the total number of simulation intervals, where each interval is assumed to be 1 day. T_s was set to 1000 for all experiments.
- N - The total number of organizations to be simulated.
- L - The peak number of jobs created for a single simulation interval for each organization (assumed to be the same for all organizations).
- η - The load-contribution ratio of all organizations.
- H - This value is proportional to the extent of job arrival correlations.

Job Submissions. The frequency of job arrivals was simulated based on the fast fractional gaussian noise generator (FGN). This mathematical workload has been used to model long-range dependencies of job inter-arrivals on supercomputers and a myriad of other applications [7]. FGN is modelled using the synthesis of a high frequency Gaussian-markov process and a low frequency Gaussian process. The parameter that dictates the extent of job arrival correlations resulting from long-range dependencies is the H. Synthetic traces were first generated from the FGN workload model. It was then modified to obtain the community workload intensity profile. This information is necessary to define the proportion of job arrivals for the entire simulation window. To generate n number of jobs at t, the simulator uses a generator function $n(t) = w_H(t) \times LN$, where $w_H(t)$

is the community workload intensity profile for a given H value for $0 < t \leq T_s$. The generator simulates jobs for all organizations by assuming that they have the same profile of traffic burstiness and load (number of submitted jobs) for any given time index t. This is done because we want to observe the 'collective' behavior of the community as a whole and then examine its impact on the overall resource allocation performance. Each job is assumed to require one CPU. The length of jobs were simulated using a random normal distribution with a mean value of 3 days with a standard deviation of 2 days.

Simulation Model. The diagram in Figure 1, is the simulation model designed for the following experiments. The community workload defined by $n(t)$ is directed to the clearinghouse. It acts as a central broker to process job submissions and relies on a job dispatcher to submit admitted jobs to an appropriate cluster resource manager. The round-robin allocation policy is used by the job dispatcher to handle the submission of jobs from the broker's queue to the respective clusters. The job dispatcher will only submit jobs to clusters that are not utilized at their full capacity. Each resource as shown in Figure 1 represents a cluster that belongs to a distinct organization. It has a constant quantity representing the number of jobs it can execute concurrently. Each cluster manages its jobs based on the EASY backfilling algorithm [11].

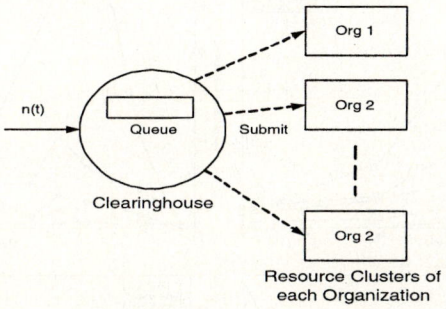

Fig. 1. Simulation Model of the Clearinghouse

Performance Measures

- Q - The measure of a queue storm is a value that is obtained from the Cumulative Distribution Function (CDF) of queue length, q, from a given simulation run. This value is expressed as $q > Q$ at $p = 5\%$. This expression gives queue size Q, so that q is greater than Q with 0.05 probability for the entire simulation run.
- $JobAllocationSuccess$ - It is the ratio of the number of jobs that meet their required execution deadlines and the total number of submitted jobs. This is a common performance measure for many scheduling algorithms involving jobs with deadlines.

- *JobAllocationFailure* - It is the ratio of the number of jobs that cannot meet their required execution deadlines and the total number of submitted jobs.

2.2 Analysis of Queue Storms Based on Load-Contribution Ratio

In order to analyze the impact that the simulation parameters have on queue storms, measurements on the rate of change of queue storms were taken for different values of N and η. Figure 2 shows the results of the analysis. MRC_N and MRC_η are measurements of the rate of change of Q at a given N and η respectively for two intensities of burstiness (i.e. $H = 0.6$ and $H = 0.8$). From Figure 2, we can observe from MRC_N that, provided if all organizations are consistently cooperating to contribute resources at a given η, then the initial rate of change of queue storms would first increase to a certain peak value before dropping to a very low value with further increases in N. This means that, once the level of membership exits the critical point (peak), then further increments in participation would not further degrade the performance of the system. Also, notice that if with improved load-contribution ratios (smaller values), this transition would become less visible.

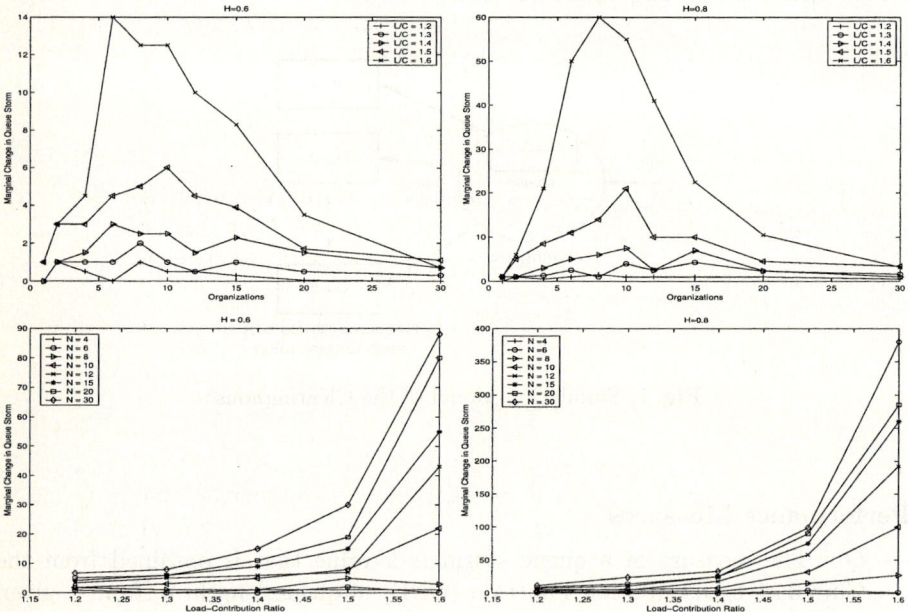

Fig. 2. Marginal Effect of N (MRC_N) for H=0.6 and H=0.8 and the Marginal Effect of η (MRC_η) for H=0.6 and H=0.8 on Queue Storms

Figure 2 also shows the change of queue storms for MRC_η by varying η. It can be seen that if the entire community (at worst case) were to increase their η, that is to say, lower their contributions or increase their average loads alone, then the rate of change of queue storms would increase considerably. It can be especially so if N is large for $H=0.8$ when compared with $H=0.6$. The MRC_N graph shows that as N increases to a large value, the size of queue storms would not increase indefinitely. The extent of a storm would be influenced more significantly by the stipulated η for the community. If unacceptable queue storms subsequently do occur due to changes in average load for some organizations, instead of limiting or reducing membership N, it is more sensible to keep η of each organization in check. This is especially true if correlations in job arrivals are likely to exist.

2.3 Effect of Introducing Slack Time for Job Management

In this experiment, we will investigate how the introduction of allowable slack time can lead to improved system management given the presence of queue storms. Slack time allowance is the additional time delay allowed for a single job to remain in the queue. Longer slack time of jobs allow a greater chance for the management system to schedule jobs more effectively, especially when demand exceeds availability. While there are many schemes suggested in the literature of supercomputing to handle the scheduling of jobs with allowable slack time [1, 3, 4], we want to investigate how the introduction of slack time can influence the successful management of jobs by a brokering system. For successful execution of an arbitrary job p, it means that the following conditions must hold:

$$p.startTime \leq p.slackTime + p.submitTime \qquad (1)$$

An issue concerned is then how users in a community would define the allowable slack time for their submitted jobs. Although, there may be a plethora of schemes that reflect the users' psychology on tolerable slack time, we assume that it is proportionally related to the execution time of the job. As such, a slack tolerance factor (STF) was used. STF influences the allowable slack time of a job as given:

$$p.slackTime = STF \times p.jobLength \qquad (2)$$

We varied STF from 0.1 to 0.5 for both $H=0.6$ and $H=0.8$ in Figure 3. The impact of introducing slack tolerance to counteract the effects of queue storms only if there is relatively lower level of burstiness ($H=0.6$). Note that when $H=0.8$, if the queue size is large, then the resource allocation performance remains poorer than $H=0.6$ regardless of the extent of relaxing STF.

In a community of cooperators where job arrivals are not that highly correlated, allowable slack time is highly effective to counteract the presence of queue storms. This fact however, cannot be applied if the correlations are high. As such, reducing the size of queues therefore becomes a greater necessity. In reality, it can be difficult to measure correlations on job arrivals of the community workload and hence apply differential treatment to the ensuing burstiness. Therefore, as a matter of convenience and assurance, queue storms must be exacerbated to

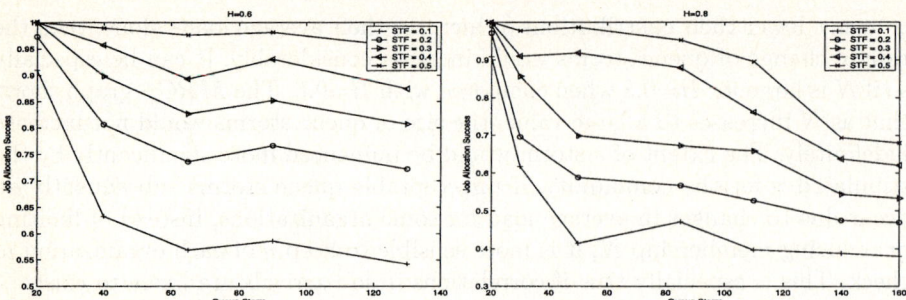

Fig. 3. Impact of Slack Tolerance on Job Allocation Success given the Presence of Queue Storms for H=0.6 and H=0.8

an extent such that its effects are acceptable before mechanisms for enforcing allowable slack time on jobs are implemented.

3   Job Queue Management Framework

We have shown that it is necessary to control the size of queue storms prior to the introduction of slack allowance in order for job management to be effective. To attack the problem of queue storms, a critical question we asked was if an optimal threshold on queue size exists such that it minimizes the number of unsuccessful job allocations given a certain level of load in the system. We expect this because due to the incidence of sudden surges in job arrivals, those with longer execution times may inadvertently slow down the servicing time of the system. Since longer jobs cannot be delayed indefinitely, admission control on such jobs may significantly improve the global performance of resource allocation. On the other extreme, it must be controlled otherwise no jobs would be executed. Therefore, if an optimal queue size threshold does exist, how and what variables would influence this value? We noted that the threshold was generally influenced by the statistical nature of the job mix. We define the statistical structure of a job mix as follows:

$$T = (t_{avg}, \sigma_{t_{avg}}, STF_{avg}, \sigma_{STF_{avg}}) \tag{3}$$

t_{avg} and STF_{avg} are the average execution time and average allowable slack factor. $\sigma_{t_{avg}}$ and $\sigma_{STF_{avg}}$ are their corresponding standard deviations, assuming a normal distribution. Also, due to the dynamism of the community, we expect fluctuations in job mix characteristics because of seasonal usage coming from different workflows. From our experiments, we found that STF_{avg} has a more significant influence on the optimal queue size threshold using the current performance measure. To cope with the resulting variations of optimal thresholds due to STF, we employed both static and adaptive thresholding for admission control. In the static-based approach, we chose the threshold that gave the average job allocation performance. In addition, since we expected the job mix fluctuations to be quite severe especially in the case that the level of membership as well as the servicing capacity of resources in the sharing community may be varying in the long run, we designed an adaptive mechanism to dynamically compute the optimal threshold according to the nature of submitted jobs at discrete time intervals. The pseudocode in Algorithm 1 gives an overview of the process of job admissions control by the clearinghouse as in Figure 1. The admissions control process is triggered only upon an arrival of a new job. When a job arrives, the queue size threshold will be obtained from $GetThreshold$. If a static scheme is employed, a constant value will be returned. But if an adaptive scheme is used, $GetThreshold$ will communicate with a $JobPropertyMonitoring$ thread and apply a mapping function to compute the actual queue threshold size. If the current queue size is greater than the threshold, the job's originating organization will be found through $Membership$. η is the instantaneous load-contribution ratio of organization j. We measure the average consumed load using $JobAvgLoad$ and the contribution by calculating the total number of executed jobs not belonging to j using $JobAvgSrv$. We employ this mode of measure to dynamically assess an organization's true contribution. Due to the fact that the shared resources are likely to be heterogeneous, the value computed by $JobAvgSrv$ is a direct measure of users' preferences over any computing clusters contributed by their respective organizations. ϕ represents a fixed time interval prior to the arrival time of the job. If the load-contribution ratio η is smaller than the stipulated ratio denoted by η_{nom}, then the job is admitted and therefore pushed into the job queue. Otherwise, the job is rejected.

Algorithm 1: Admission Control Process

On Event Job p Arrival
$K = GetThreshold$
if $Queue\_size > K$ then
    $j = Membership(p)$
    $\eta = \frac{JobAvgLoad(j,\phi)}{JobAvgSrv(j,\phi)}$
    if $\eta < \eta_{nom}$ then
        $Enqueue(p)$
    else
        $ReplyFail$
    end if
else
    $Enqueue(p)$
end if

We observed the effect of thresholding on resource allocation, for four distinct classes of hypothesized workloads with different statistical properties of submitted jobs. They are listed in Table 1.

The results for resource allocation with respect to the predefined workloads, are given in Figure 4. For job mix 1, because of the presence of thresholding, the deletion of jobs in the queue has benefitted the majority of jobs given their very

Table 1. Workloads to Examine Effect of Queue Size Thresholding for Job Management

No	Description	Specification
1	Low STF	STF_{avg}=0.3 $\sigma_{STF_{avg}}$=0.2
2	High STF	STF_{avg}=0.6 $\sigma_{STF_{avg}}$=0.2
3	Rand1 STF	Alternate 25 Days
		(1)STF_{avg}=0.2 $\sigma_{STF_{avg}}$=0.1
		(2)STF_{avg}=0.6 $\sigma_{STF_{avg}}$=0.1
4	Rand2 STF	Alternate 50 Days
		(1)STF_{avg}=0.2 $\sigma_{STF_{avg}}$=0.1
		(2)STF_{avg}=0.6 $\sigma_{STF_{avg}}$=0.1

short STF. This improvement was less observable when job mix 2 was used as more jobs were able to tolerate allocation delays. Also, note that adaptive thresholding was more successful on job mix 1 than 2 because the nominal threshold for static management was much closer to the latter job mix. Since the adaptive scheme can track the statistical nature of the workload, it was able to progressively modify the optimal threshold to improve the management of the queue size. Job mix 3 and 4, were used to demonstrate the effectiveness of adaptive management of the queue threshold under different degrees of fluctuations in the overall workload statistics. It can be seen that if fluctuations occur at a lower frequency as in job mix 4, the differences in performance is more significant than high frequency fluctuations. When fluctuations are less frequent, the threshold mechanism is able to respond more effectively to changes in job mix statistics.

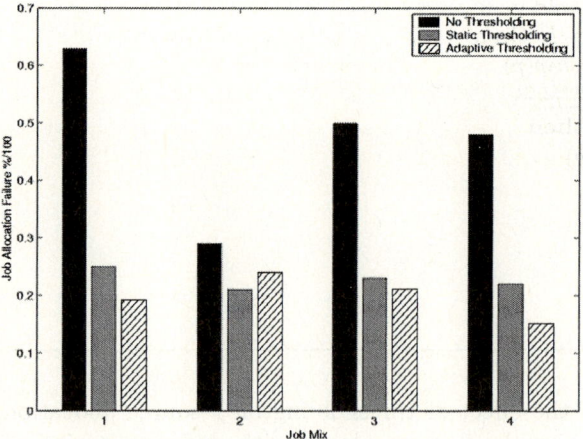

Fig. 4. Performance of Resource Allocation for Different Job Mixes given Different Queue Size Thresholding Methods

4 Related Work

In our work, jobs are handled in FCFS manner. Common FCFS batch job managers support different types of priority FCFS algorithms without admission control mechanisms. Job admissions on computing resources have largely focused on real-time, periodic applications in which the tasks have highly variable utilization requirements and that their deadlines are firm [3]. A popular method of jobs admissions is to apply slack stealing into the scheduling process that has been applied in many instances [9, 10]. With complete knowledge of the system's execution requirements, slack stealing determines whether there is adequate time in the system to admit an aperiodic task at a given priority level.

Tia et. el. employed a task classification method (using a threshold) to separate jobs guaranteed under the Rate-Monotonic Scheduled conditions from those which would require additional scheduling effort [13] . Slack-based (SB) Algorithm proposed in [12] uses a backfilling algorithm to improve system throughput and user response times. The main idea of of this algorithm is also to assign allowable slack for each job. The scheduler assigns the slack by using cost functions that take into consideration certain priorities associated with the job. The Real-Time (RT) algorithm, proposed in [8] is an approach to schedule uni-processor tasks with hard real time deadlines on multi-processor systems. This algorithm tries to meet the specified deadlines for the jobs by using heuristic functions. QoPS proposed in [6] tries to find a feasible schedule that is able to meet the deadlines of all jobs for every newly arriving job. When a violation occurs, the newly arrived job will fail to be scheduled.

Although the above works perform job admission functions, they generally re-process the current schedule of allocated jobs to 'find' an allocation for a newly arrived jobs. Although, such methodologies when translated to a grid can be performed distributively at each organization's resource manager, they would still require substantial computational effort. In our work, we focus on employing 'adaptiveness' - that is, by making use of a separate running thread to evaluate the current statistical properties of the community's job mix to control the queue size. Without the need for direct scheduling based on the slack allowable for each newly arriving job, the job management framework eliminates the need to use of algorithms to perform admission control.

5 Conclusions and Future Work

We have justified a job management framework for a computational grid to reduce the negative effects of irregular job arrivals. The critical aspect of managing community workloads is to take into account the fluctuating statistical properties of submitted jobs to compute a threshold for the job admissions process. Also, in a cooperative community, fair play is achieved by enforcing a certain load-contribution ratio for each organization to act as a throttle to prevent overloading due to unusually high number of job arrivals. The admitted jobs are further filtered by controlling the size of queues according to statistical properties of the job mix before being allocated to resources. Although the generic framework is in place, the practical competence of the system has not been fully investigated. For this to be done, realistic workloads characterizing the specific behavior of users and their application models must be derived and applied into the analysis.

References

1. P. Altenbernd and H. Hansson. The slack method: A new method for static allocation of hard real-time tasks. *Real-Time Systems*, 15(2):103–130, 1998.
2. N. Andrade, W. Cirne, and F Brasileiro. Our Grid: An approach to easily assemble grids with equitable resource sharing. *9th Workshop on Job Scheduling Strategies for Parallel Processing*, pages 53–68, Jun 2003.
3. A. Atlas and A. Bestavros. Slack stealing job admission control scheduling. Technical Report 1998-009, Boston University, 2 1998.
4. R. I. Davis, K. W. Tindell, and A. Burns. Scheduling slack time in fixed priority preemptive systems. In *IEEE Real-Time Systems Symposium, IEEE Computer Society Press*, pages 222–231, 1993.
5. D. Epema, M. Livny, R. Van Dantzig, X. Evers, and J. Pruyne. A worldwide flock of condors: Load sharing among workstation clusters. *Future Generation Computer Systems*, 12:53–65, 1996.
6. M. Islam, Balaji P., Sadayappani P., and D. K. Pandai. QoPS: A QoS based scheme for parallel job scheduling. In *Job Scheduling Strategies for Parallel Processing: 9th International Workshop*, Jun 2003.
7. S. Kleban and S. Clearwater. Quelling queue storms. In *13th International Conference High-performance and Distributed Computing*, Jul 2003.
8. K. Ramamritham, J. A. Stankovic, and P. Shiah. Efficient scheduling algorithms for real-time multiprocessor systems. *IEEE Transactions on Parallel and Distributed Systems*, 1(2), Apr 1990.
9. S. Ramos-Thuel and J. Lehoczky. On-line scheduling of hard deadline aperiodic tasks in fixed-priority systems. *Real-time Systems Symposium*, Dec 1993.
10. S. Ramos-Thuel and J. Lehoczky. Algorithms for scheduling hard aperiodic tasks in fixed-priority systems using slack stealing. *Real-time Systems Symposium*, Dec 1994.
11. J. Skovira, W. Chan, H. Zhou, and D. Lifka. The EASY-loadleveler api project. *Job Scheduling Strategies for Parallel Processing*, pages 41–47, 1996.
12. D. Talby and D. G. Feitelson. Supporting priorities and improving utilization of the ibm sp2 scheduler using slack based backfilling. In *13th Intl. Parallel Processing Symposium*, pages 513–517, Apr 1997.
13. T. Tia, Z. Deng, M. Shankar, M. Storch, J. Sun, L. Wu, and J. Liu. Probabilistic performance guarantees for real-time tasks with varying computation times. In *Real-time Technology and Applications Symposium*, Jun 1997.
14. P. Xavier, B. Lee, and W. Cai. A decentralized hierarchical scheduler for a grid-based clearinghouse. In *Workshop on Massively Parallel Processing in conjunction with IPDPS'03*, Apr 2003.

Performance-Aware Load Balancing for Multiclusters*

Ligang He, Stephen A. Jarvis, David Bacigalupo, Daniel P. Spooner,
and Graham R. Nudd

Department of Computer Science, University of Warwick,
Coventry, CV4 7AL, United Kingdom
liganghe@dcs.warwick.ac.uk

Abstract. In a multicluster architecture, where jobs can be submitted through each constituent cluster, the job arrival rates in individual clusters may be uneven and the load therefore needs to be balanced among clusters. In this paper we investigate load balancing for two types of jobs, namely *non-QoS* and *QoS-demanding* jobs and as a result, two performance-specific load balancing strategies (called ORT and OMR) are developed. The ORT strategy is used to obtain the optimised mean response time for non-QoS jobs and the OMR strategy is used to achieve the optimised mean miss rate for QoS-demanding jobs. The ORT and OMR strategies are mathematically modelled combining queuing network theory to establish sets of optimisation equations. Numerical solutions are developed to solve these optimisation equations, and a so called *fair workload level* is determined for each cluster. When the current workload in a cluster reaches this pre-calculated fair workload level, the jobs subsequently submitted to the cluster are transferred to other clusters for execution. The effectiveness of both strategies is demonstrated through theoretical analysis and experimental verification. The results show that the proposed load balancing mechanisms bring about considerable performance gains for both job types, while the job transfer frequency among clusters is considerably reduced. This has a number of advantages, in particular in the case where scheduling jobs to remote resources involves the transfer of large executable and data files.

1 Introduction

As grid technologies gain in popularity, separate clusters are increasingly being interconnected to create multicluster computing architectures for the processing of scientific and commercial applications [1][2][5]. These constituent clusters may be located within a single organization or across different geographical sites [3][6]. Load balancing across such architectures is recognised as a key research issue. If users located at different administrative domains submit jobs through domain-specific portals, there may be different submission patterns. Without intervention, this may lead to an unbalanced workload distribution among different domains; and the overall performance may be compromised.

---

* This work is sponsored in part by grants from the NASA AMES Research Center (administrated by USARDSG, contract no. N68171-01-C-9012), the EPSRC (contract no. GR/R47424/01) and the EPSRC e-Science Core Programme (contract no. GR/S03058/01).

Performance requirements are likely to vary depending on the job type. When the jobs have associated QoS demands (which we call *QoS-demanding jobs* or QDJs), performance is usually used to measure the extent of QoS-demand compliance; when jobs have no associated QoS demands (which we call *non-QoS jobs* or NQJs), a common performance criteria is to reduce the *mean response time* [9][12]. An example of the QoS is job *slack* [10][14]. The QoS of a job is satisfied if the job' waiting time (in the system) is less than its slack [15]; otherwise, the QoS is failed. Mean *miss rate* captures the aggregate slack failure and is therefore used as a performance metric to measure the proportion of jobs whose QoS demands fail.

In this paper, load balancing techniques are addressed for both *QoS-demanding jobs* (QDJs) and *non-QoS demanding jobs* (NQJs) to improve the job type-specific performance requirements in a multicluster. Two multicluster load balancing strategies, *Optimised mean Response Time* (ORT) and *Optimised mean Miss Rate* (OMR), are developed. The aim of ORT is to achieve optimised mean response time for NQJ workloads and the aim of OMR is to achieve the optimised mean miss rate for QDJ workloads. The ORT and OMR strategies are mathematically modelled combining queuing network theory to establish sets of optimisation equations. Numerical solutions are developed to solve the optimisation equation sets and determine a *fair workload level* for each cluster. When the current workload in a cluster reaches the pre-calculated fair workload level, the jobs subsequently submitted to the cluster are transferred to remote less-loaded clusters for execution.

There are a number of established workload allocation techniques in parallel and distributed systems [12][9]. A static workload allocation strategy is investigated in [12] to achieve the optimised mean response time; this strategy is specifically limited to a single cluster environment. Workload allocation techniques for multiclusters are addressed in [9] where it is assumed that the multicluster has a central entry point for the receipt of submitted jobs. In this paper, we assume that jobs can be submitted through each local cluster, and therefore the further problems brought about by the uneven submission patterns of jobs at the clusters also needs to be considered.

The problem of uneven job arrival patterns in different resources is addressed in a number of load balancing techniques [8][13]. A load balancing mechanism is presented in [8] for multi-domain environments. The mechanism identifies the least loaded computer among all domains and when a job is submitted to the system it is scheduled on that computer, whichever domain the job is actually submitted to. Hence, a job has to be transferred to a remote domain if the local domain does not contain the current least loaded computer.

In this paper however, a *fair workload level* is calculated for each cluster. Only when the current workload in a cluster exceeds its specified fair workload, does the cluster transfer the newly submitted job to a remote cluster. Although the load balancing technique presented in this paper may not achieve the best possible response time for a specific job, theoretical analysis and experimental studies show that considerable performance gains are still achieved in terms of the jobs' mean response time. Moreover, the job transfer frequency among clusters is dramatically reduced. This is desirable when jobs require the transfer of large executable and data files.

The workload allocation and load balancing techniques referenced above are applied to non-QoS jobs (NQJs). In this paper, techniques for allocating QoS-demanding jobs (QDJs) in multiclusters are also presented. The technique is similar to that for NQJs, except that a fair workload level for each cluster is otherwise calculated to obtain the optimised mean miss rate for the QDJs.

The rest of the paper is organized as follows. The system and workload model is discussed in Section 2. Section 3 presents the load balancing techniques for NQJs and QDJs in multiclusters. The performance of these techniques is evaluated in Section 4. Section 5 concludes the paper.

2 System and Workload Model

The multicluster architecture assumed in this paper is shown in Fig.1. The system consists of n different clusters, where each cluster comprises a set of homogeneous computers. Cluster i ($1 \leq i \leq n$) is modelled using an M/M/m_i queue, where m_i is the number of computers in cluster i. Jobs can be submitted through each local cluster, where they are queued for scheduling. The clusters in the multicluster are interconnected through an agent system [7], which is able to monitor the job submission in each cluster and determine the mean arrival rate of the jobs submitted to the entire multicluster. The mean arrival rate is utilized by the multicluster workload manager (which we call MUSCLE) to calculate the fair workload level for each cluster through the ORT and OMR strategies (for NQJs and QDJs, respectively).

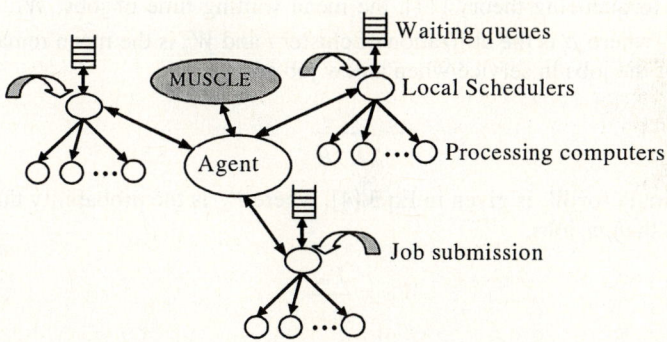

Fig. 1. The multicluster architecture

The fair workload level for each cluster is measured by the mean number of jobs in its waiting queue. The local scheduler in each cluster is informed of its fair workload level. When the current number of jobs in the waiting queue in a cluster reaches its specified fair workload level, subsequent jobs are then transferred by MUSCLE (and the supporting agent system) to other suitable clusters. Each local scheduler processes locally submitted and remotely transferred jobs based on a First-Come-First-Served basis. The scheduling itself is non-preemptive. The jobs investigated in this paper are independent and each QDJ has a slack which follows a uniform distribution in $[s_l, s_u]$.

3 Load Balancing Techniques

Suppose that the mean arrival rate of the jobs submitted to the multicluster is λ. The overall performance of the job execution depends on the workload distribution among the clusters. The fair workload level for each cluster, measured by the mean number of jobs in its waiting queue, is determined in this section. The approaches for NQJs and QDJs differ as they have different performance requirements.

3.1 ORT (Optimised Mean Response Time) Strategies for NQJs

The ORT strategy aims to optimise the mean response time of the NQJs in the multicluster. The response time of a job is defined as the time from when the job arrives at the system until it is completed. The following analysis first establishes the optimisation equations for the mean response time, then a numerical solution to the optimisation equations is developed and the fair workload level is determined.

The response time of a job is its waiting time in the queue plus its execution time. Hence, the average response time of the jobs in cluster i, denoted as R_i, can be computed using Eq.1, where W_i is the mean waiting time of the jobs in cluster i and u_i is the service rate of the computers in cluster i.

$$R_i = W_i + \frac{1}{u_i} \tag{1}$$

Cluster i, containing m_i computers, is modelled using an M/M/m_i queue ($1 \leq i \leq n$). According to queueing theory [11], the mean waiting time of jobs, W_i, is computed using Eq.2, where ρ_i is the utilization of cluster i and W_{0i} is the mean remaining execution time of the jobs in service when a new job arrives.

$$W_i = \frac{W_{0i}}{1 - \rho_i} \tag{2}$$

The formula for W_{0i} is given in Eq.3 [4], where P_{mi} is the probability that the system has no less than m_i jobs.

$$W_{0i} = \frac{P_{mi}}{m_i u_i} \tag{3}$$

Suppose that in the total workload in the entire multicluster, the fraction of workload allocated to cluster i is α_i, then,

$$\rho_i = \frac{\alpha_i \lambda}{m_i u_i} \tag{4}$$

P_{mi} in Eq.3 is given in Eq.5 [4][11].

$$P_{mi} = \frac{(m_i \rho_i)^{m_i}}{(1 - \rho_i) m_i! \left[\sum_{k=0}^{m_i - 1} \frac{(m_i \rho_i)^k}{k!} + \frac{(m_i \rho_i)^k}{(1 - \rho_i) m_i!} \right]} \tag{5}$$

Using Eqns.1-5, the formula for R_i, in terms of α_i, is derived and is shown in Eq.6.

$$R_i = \frac{m_i u_i (\frac{\alpha_i \lambda}{u_i})^{m_i}}{\left[m_i! \sum_{k=0}^{m_i-1} \frac{(\frac{\alpha_i \lambda}{u_i})^k}{k!} + \frac{(\frac{\alpha_i \lambda}{u_i})^{m_i}}{(1 - \frac{\alpha_i \lambda}{m_i u_i})} \right] (m_i u_i - \alpha_i \lambda)^2} + \frac{1}{u_i} \quad (6)$$

Thus, the mean response time of the incoming jobs over these n clusters, denoted by R, can be computed - see Eq.7.

$$R = \sum_{i=1}^{n} R_i \alpha_i \quad (7)$$

Hence, in order to achieve the optimal mean response time of the job stream in the multicluster, the aim is to find a workload allocation $\{\alpha_1, \alpha_2, ..., \alpha_n\}$ that minimizes Eq.7 subject to $\sum_{i=1}^{n} \alpha_i = 1$ and $0 \leq \alpha_i \leq \frac{m_i u_i}{\lambda}$ (the constraint $\alpha_i \leq \frac{m_i u_i}{\lambda}$ is used to ensure that cluster i does not become saturated). This is a constrained-minimum problem and according to the Lagrange multiplier theorem, solving this problem is equivalent to solving the following equation set.

$$\begin{cases} \sum_{i=1}^{n} \alpha_i = 1, \quad 0 \leq \alpha_i \leq \frac{m_i u_i}{\lambda} & (a) \\ \frac{\partial}{\partial \alpha_k} (\sum_{i=1}^{n} R_i \alpha_i) - v \frac{\partial}{\partial \alpha_k} (\sum_{i=1}^{n} \alpha_i - 1) = 0 \quad 1 \leq k \leq n & (b) \end{cases} \quad (8)$$

Since α_i is the only unknown variable in the expression of R_i, Eq.8 can be reduced to Eq.9 by solving the partial differential equations in Eq.8.b.

$$\begin{cases} \sum_{i=1}^{n} \alpha_i = 1, \quad 0 \leq \alpha_i \leq \frac{m_i u_i}{\lambda} & (a) \\ \frac{\partial}{\partial \alpha_k} (R_k \alpha_k) = v \quad 1 \leq k \leq n & (b) \end{cases} \quad (9)$$

It is impossible to find the general symbolic solution $\{\alpha_1, \alpha_2, ..., \alpha_n\}$ from Eq.9, however, we identify a property of Eq.9.b that enables us to develop a numerical solution. The property is shown in Theorem 1 (the proof is omitted). Based on this property, we develop a numerical solution to solve Eq.9 and therefore derive the optimised workload allocation $\{\alpha_1, \alpha_2, ..., \alpha_n\}$. The numerical solution is shown in Algorithm 1.

Theorem 1. $\frac{\partial}{\partial \alpha_k}(R_k \alpha_k)$ is a monotonically increasing function of α_k.

```
Algorithm 1. Computation of workload allocation among
clusters for optimised mean response time
1.    Let lower and upper limits of the mean response
time be v_lower and v_upper;
2.    while (v_lower≤v_upper)
3.        v_mid=(v_lower+ v_upper)/2;
4.        for each cluster i (1≤i≤n) do
```

5.   **if** ($v\_mid < \frac{\partial}{\partial \alpha_i}(R_i\alpha_i)|_{\alpha_i=0}$)
6. $\alpha_i = 0$;
7.   **else if** ($v\_mid > \frac{m_i u_i}{\lambda}$)
8. $v\_upper = v\_mid$;
9.   **continue**;
10.   **while** ($\alpha\_lower \leq \alpha\_upper$)
11. $\alpha\_mid = (\alpha\_lower + \alpha\_upper)/2$;
12. $v\_cur = \frac{\partial}{\partial \alpha_i}(R_i\alpha_i)|_{\alpha_i = \alpha\_mid}$;
13.   **if** (the difference between v_cur and v_mid is less than v_valve)
14. $\alpha_i = \alpha\_mid$;
15.   **if** (v_cur is less than v_mid)
16. $\alpha\_lower = \alpha\_mid$;
17.   **else**
18. $\alpha\_upper = a\_mid$;
19.   **end for**
20. $\alpha\_sum = \sum_{i=1}^{n} \alpha_i$;
21.   **if** (the difference between α_sum and 1 is less than α_valve)
22.   the current set of α_i ($1 \leq i \leq n$) is the correct workload allocation;
23.   **else if** (α_sum is less than 1)
24. $v\_lower = v\_mid$;
25.   **else**
26. $v\_upper = v\_mid$;
27. **end while**

Since a binary search technique is used to search for v and α_i in their respective search spaces [v_lower, v_upper] and [α_lower, α_upper], the time complexity of Algorithm 1 is $O(n \log k_v \log k_\alpha)$, where k_α and k_v are the number of elements in the search space of v and α_i, which equal $\frac{v\_upper - v\_lower}{\varphi}$ and $\frac{\alpha\_upper - \alpha\_lower}{\gamma}$, respectively ($\varphi$ and γ represent the precision in the calculation). Since φ and γ are predefined constants, the time complexity is linear with the number of clusters, that is n.

The feasibility and effectiveness of Algorithm 1 are shown in Theorem 2.

Theorem 2. The workload allocation strategy $\{\alpha_1, \alpha_2, ..., \alpha_n\}$ computed by Algorithm 1 minimizes the average response time of the incoming job stream in a multicluster system of n clusters.

After α_i ($1 \leq i \leq n$) is determined, the mean number of jobs in the waiting queue of cluster C_i, denoted by N_i, can be calculated using Eq.10 [11]; where W_i is the jobs' mean waiting time. W_i can be calculated using the first item to the right of Eq.6.

$$N_i = \lambda \alpha_i \times W_i \qquad (10)$$

$\lceil N_i \rceil$ is regarded as the *fair workload level* for cluster i. When the current number of jobs in the waiting queue of cluster i is less than $\lceil N_i \rceil$, the arriving jobs are scheduled locally; otherwise, the local scheduler of cluster i transfers the arriving jobs to the supporting agent system where they are further dispatched to the cluster with the least load (defined by Eq.11) among those clusters whose current number of jobs in the waiting queue is less than its fair workload level (if such a cluster does not exist, the jobs are scheduled to the cluster with the least load among all clusters).

$$load_i = \frac{\text{the job number in the waiting queue} + m_i}{m_i u_i} \qquad (11)$$

3.2 OMR (Optimised Mean Miss Rate) Strategy for QDJs

The QoS of a QDJ fails if its waiting time is greater than its slack. The performance criterion for evaluating the scheduling of QDJs differs from that for NQJs in that it typically aims to minimize the fraction of jobs that miss their QoS, termed the *miss rate*. In this subsection, a workload allocation strategy called OMR is developed, whose aim is to optimise the mean miss rate of the submitted QDJs in the multicluster. Every QDJ has some slack that follows a uniform distribution. Its probability density function $S(x)$ is given in Eq.12, where s_u and s_l are the upper and lower limits of the slack, respectively.

$$S(x) = \frac{1}{s_u - s_l} \qquad (12)$$

We continue to model cluster i (of m_i computers) as an M/M/m_i queue ($1 \leq i \leq n$). As in queuing theory [11], in an M/M/m_i queue the probability distribution function of the job waiting time, $P_w(x)$ (which means that the probability that the job waiting time is less than x), is given by Eq.13 [11], where ρ_i and P_{mi} are the same variables as those found in Eq.2 and Eq.3.

$$P_w(x) = 1 - P_{mi} e^{-m_i u_i (1 - \rho_i) x} \qquad (13)$$

Using the probability density function of slack, the miss rate of the QDJs allocated to cluster i, denoted by MR_i, and can be computed by Eq.14.

$$MR_i = \int_{s_l}^{s_u} S(x)(1 - P_w(x)) dx \qquad (14)$$

$$MR_i = \frac{m_i u_i (\frac{\alpha_i \lambda}{u_i})^{m_i} [e^{-(m_i u_i - \alpha_i \lambda) s_l} - e^{-(m_i u_i - \alpha_i \lambda) s_u}]}{\left[m_i! \sum_{k=0}^{m_i - 1} \frac{(\frac{\alpha_i \lambda}{u_i})^k}{k!} + \frac{(\frac{\alpha_i \lambda}{u_i})^k}{(1 - \frac{\alpha_i \lambda}{m_i u_i})} \right](m_i u_i - \alpha_i \lambda)^2 (s_u - s_l)} \qquad (15)$$

Applying Eq.12 and Eq.13 and solving the integral, Eq.14 becomes Eq.15, where the workload fraction α_i for cluster i is the only unknown variable.

The mean miss rate (denoted by MR) of the QDJs over these n clusters can be computed using Eq.16.

$$MR = \sum_{i=1}^{n} MR_i \times \alpha_i \qquad (16)$$

Similar to the case of minimizing the mean response time, this is a constrained-minimum problem. This requires identifying a workload allocation that minimizes MR in Eq.16 subject to $\sum_{i=1}^{n} \alpha_i = 1$ and $0 \leq \alpha_i \leq \frac{milli}{\lambda}$. This is equivalent to solving the following equation set.

$$\begin{cases} \sum_{i=1}^{n} \alpha_i = 1, \quad 0 \leq \alpha_i \leq \frac{milli}{\lambda} & (a) \\ \frac{\partial}{\partial \alpha_k}(MR_k \times \alpha_k) = v \quad 1 \leq k \leq n & (b) \end{cases} \qquad (17)$$

In the previous subsection, we state that the numerical solution to Eq.9 is based on the property that $\frac{\partial}{\partial \alpha_k}(R_k \alpha_k)$ is a monotonically increasing function of α_k. Theorem 3 is introduced to establish the case that $\frac{\partial}{\partial \alpha_k}(MR_k \times \alpha_k)$ in Eq.17 also monotonically increases over α_k. The proof of the theorem is omitted for brevity. With this property, a numerical solution is also developed to solve Eq.17. The solving algorithm is similar to that found in Algorithm 1 and the proof of the algorithms effectiveness is similar to Theorem 2. Hence, they are omitted in the paper.

Theorem 3. $\frac{\partial}{\partial \alpha_k}(MR_k \times \alpha_k)$ is a monotonically increasing function of α_k.

As in the case of NQJs, the mean number of jobs in the waiting queue of cluster i (i.e. N_i) can be obtained using Eq.10. The *fair workload level* for cluster i for QDJs can be subsequently determined. If the current number of jobs in the waiting queue of cluster i is greater than $\lceil N_i \rceil$, then the arriving jobs are transferred to the cluster with the least miss rate among those clusters whose number of jobs in the waiting queue is less than its fair workload level (if such clusters do not exist, the jobs are scheduled to the cluster with the least miss rate among all clusters).

4 Experimental Evaluation

An experimental simulator is developed to evaluate the performance of the proposed workload allocation techniques under a wide range of system settings and workload levels. Two types of job stream (NQJs and QDJs) are generated using the same parameters, with one exception, in that every QDJ has an additional slack metric which follows a uniform distribution. Each job stream includes 500,000 independent jobs. The job arrival follows a Poisson process and a job is submitted to the multicluster through a randomly selected cluster. The run of the first 100,000 jobs is considered as the initiation period, allowing the system to achieve a steady state, and the run of the last 100,000 jobs is considered the ending period. Statistical data are collected from the middle 300,000 jobs. The job size follows an exponential distribution. The mean size of the incoming jobs is set to be the inverse of the average of the speeds of all processing computers multiplied by the average of the number of computers in each cluster, that is,

$$\frac{n^2}{\sum_{i=1}^{n} u_i \sum_{i=1}^{n} m_i} \text{(sec)}$$

Based on the mean job size, the job arrival rate at which the system becomes saturated, can be computed. The incoming workload levels in the experiments are measured using the percentage of the saturated arrival rate.

An intuitive load balancing strategy, the *weighted* strategy [12][9], takes into account the heterogeneity of the clusters' performance. In this strategy, the workload fraction α_i allocated to cluster i ($1 \leq i \leq n$) is proportional to its processing capability, $m_i u_i$. Hence, α_i is computed as Eq.18.

$$\alpha_i = \frac{m_i u_i}{\sum_{i=1}^{n} m_i u_i} \tag{18}$$

Consequently, under the weighted strategy the corresponding fair workload level for each cluster can be determined using Eq.10.

The *Multi-domain Load Balancing mechanism* (MLB) of [8], which is based on a dynamic least load algorithm, can be used as the ideal bound of the mean response time obtained by the ORT strategy. Using this approach the arriving jobs are scheduled on the computer with the least load. Hence a job is transferred to a remote domain while the local domain does not contain the least loaded computer. In the load balancing mechanism presented in this paper, the job transfer frequency among clusters can be dramatically reduced so as to improve the scheduling cost. Similarly, a dynamic least miss-rate (DLM) strategy is used as the upper bound of the mean miss rate of QDJs in the multicluster. The DLM strategy schedules newly arriving QDJs to the cluster with the least miss rate.

These five load balancing strategies (ORT, OMR, Weighted, MLB and DLM) are evaluated in these experiments. The performance metrics used in the experiments include the *mean response time* (for NQJs) and the *mean miss rate* (for QDJs). Each point in the performance curves are plotted as the average result of 5 independent runs of the job streams with different initialisation random numbers.

4.1 Workload

Fig.2.a and Fig.2.b show the impact of the incoming workload levels on the mean response time and the mean miss rate of the incoming jobs under these load balancing strategies. The multicluster in this experiment consists of 4 clusters whose configurations are listed in Table 1. For QDJs, the job slacks follow a uniform distribution in the range [0, 30]. In order to gain insight into the difference of the load balancing behaviours between OMR and ORT, the ORT strategy is also used to balance QDJs, and OMR is used to balance NQJs.

Table 1. System setting in Figure 2

	Cluster 1	Cluster 2	Cluster 3	Cluster 4
m_i	3	5	7	9
u_i	20	16	12	8

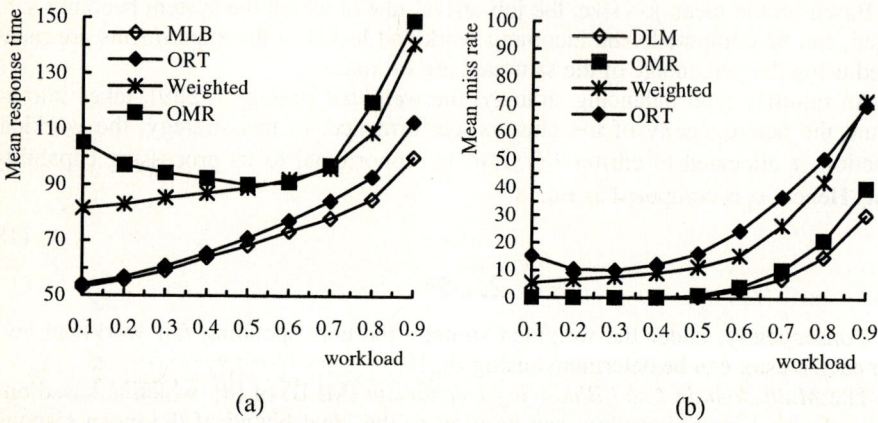

Fig. 2. Impact of the incoming workload levels on a) mean response time and b) mean miss rate

It can be observed from Fig.2.a that the ORT strategy performs significantly better than the weighted strategy in terms of the mean response time. Furthermore, the performance difference increases as the workload decreases. This trend can be explained as follows. The weighted strategy allocates the same fraction of workload to a cluster even if the workload varies. However, the waiting time accounts for a lower proportion of the response time as the workload decreases. Hence, in order to reduce the response time, a higher proportion of the incoming workload should be allocated to the cluster with the greater u_i (the number of computers m_i in each cluster has less impact). The ORT strategy is able to satisfy this allocation requirement. Fig.2.b shows the impact of the incoming workload on the mean miss rate. It can be observed that the OMR strategy outperforms the weighted strategy at all incoming workload levels.

In Fig.2.a, although the MLB outperforms ORT, the performance difference is small, especially when the workload is low. A similar pattern can be observed between the DLM and OMR strategies. This suggests that applying the ORT and OMR schemes will achieve competitive performance with relatively low cost, especially when the system load is low.

It can be observed from Fig.2.a and Fig.2.b that in most incoming workload levels, the OMR strategy obtains the worst performance in terms of mean response time while the ORT strategy obtains the worst performance in terms of mean miss rate. These results suggest that the performance-specific load balancing strategies are necessary to achieve good respective performance.

4.2 Computer Speed

Fig.3 demonstrates the impact of the difference of computer speed. Here the multicluster consists of 4 clusters and the number of computers in each cluster is set to be 4. The speed of the computers in cluster 1 varies from 21 to 6 with a decrement of 3, while the speed of all computers in the other three clusters increases from 1 to 6 with an increment of 1. Thus, the multicluster ranges from a highly heterogeneous system to a homogeneous system, while the average speed of all computers remains constant (i.e., 6). The slack of the QDJs follows a uniform distribution in [0, 10].

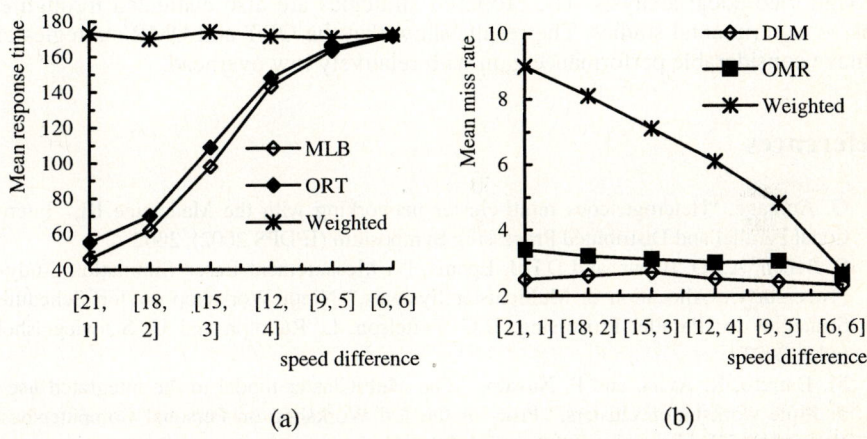

Fig. 3. The impact of speed difference on a) the mean response time and b) the mean miss rate; the arrival rate is 50% of the saturated arrival rate

Fig.3.a shows the impact of the difference of computer speed on the mean response time. It can be observed in this figure that the mean response time decreases significantly under the ORT strategy as the speed difference increases, while it remains approximately the same under the weighted strategy. This is because as the speed difference increases, despite the average computer speed remaining constant, a higher proportion of the workload is allocated to cluster 1 under the ORT strategy (higher than $m_1 u_1 / \sum_{i=1}^{n} m_i u_i$), while the weighted strategy does not make full use of the computing power of cluster 1. This suggests that under the ORT strategy, the speed difference among the clusters is a critical factor for the mean response time.

The first observation from Fig.3.b is that the OMR strategy performs better than the weighted strategy under all speed combinations, as is to be expected. A further observation is that under OMR, the mean miss rate remains approximately the same as the speed difference varies. The experimental results for other incoming workload levels show similar patterns. This suggests that under OMR, the speed difference among the clusters is not an important parameter for the mean miss rate. This differs from the characteristic of the ORT for mean response time. This divergence may originate from the difference between the expressions of the response time and the miss rate (see Eq.6 and Eq.15): we note the occurrence of $1/u_i$ in Eq.6 while this is absent in Eq.15.

5 Conclusion

Two load balancing strategies (ORT and OMR) for multicluster architectures are proposed that deal with different types of jobs. The ORT strategy can optimize the mean response time of NQJs, while the OMR strategy can optimize the mean miss rate of QDJs. The effectiveness of these proposed load balancing strategies is demonstrated through theoretical analysis. The proposed strategies are also evaluated through extensive experimental studies. The results show that the ORT and OMR strategies can achieve considerable performance gain with relatively low overhead.

References

1. O. Aumage, "Heterogeneous multi-cluster networking with the Madeleine III," International Parallel and Distributed Processing Symposium (IPDPS 2002), 2002.
2. S. Banen, A.I.D. Bucur, and D.H.J. Epema, "A Measurement-Based Simulation Study of Processor Co-Allocation in Multicluster Systems," Ninth Workshop on Job Scheduling Strategies for Parallel Processing, D.G. Feitelson, L. Rudolph and U. Schwiegelshohn (eds), 2003.
3. M. Barreto, R. Avila, and P. Navaux, "The MultiCluster model to the integrated use of multiple workstation clusters," Proc. of the 3rd Workshop on Personal Computer-based Networks of Workstations, 2000, pp. 71–80.
4. G. Bolch, Performance Modeling of Computer Systems, 2002.
5. A.I.D. Bucur and D.H.J. Epema, "The maximal utilization of processor co-allocation in multicluster Systems," Int'l Parallel and Distributed Processing Symp. (IPDPS 2003), 2003, pp. 60-69.
6. R. Buyya and M. Baker, "Emerging Technologies for Multicluster/Grid Computing," Proceedings of the 2001 IEEE International Conference on Cluster Computing, 2001.
7. J. Cao, D. J. Kerbyson, and G. R. Nudd, "Performance Evaluation of an Agent-Based Resource Management Infrastructure for Grid Computing," Proceedings of 1st IEEE/ACM International Symposium on Cluster Computing and the Grid (CCGrid'01), 2001.
8. S.T. Chanson, W. Deng, C. Hui, X. Tang, M. To, "Multidomain Load Balancing," 2000 International Conf. on Network Protocols, Japan, 2000.
9. L. He, S.A. Jarvis, D.P. Spooner, G.R. Nudd, "Optimising static workload allocation in multiclusters," Proceedings of 18th IEEE International and Distributed Processing Symposium (IPDPS'04), 2004.
10. B. Kao and H. Garcia-Molina, "Scheduling soft real-time jobs over dual non-real-time servers," IEEE Trans. on Parallel and Distributed Systems, 7(1): 56-68, 1996.
11. L. Kleinrock, Queueing system, John Wiley & Sons, 1975.
12. X.Y. Tang, S.T. Chanson, "Optimizing static job scheduling in a network of heterogeneous computers," the 29th International Conference on Parallel Processing, 2000.
13. M. Wu, "On Runtime Parallel Scheduling for Processor Load Balancing," IEEE Transaction on Parallel and Distributed Systems, 8(2): pp.173-186, Feb. 1997.

14. W. Zhu, "Scheduling soft real-time tasks on cluster," In proc. of 1999 Annual Australian Parallel and Real-Time Conference, 1999.
15. W. Zhu and B. Fleisch, "Performance evaluation of soft real-time scheduling on a multi-computer cluster," The 20th International Conference on Distributed Computing Systems (ICDCS 2000), 2000.

Scheduling of a Parallel Computation-Bound Application and Sequential Applications Executing Concurrently on a Cluster – A Case Study

Adam K. L. Wong and Andrzej M. Goscinski

School of Information Technology, Deakin University,
Geelong, Vic 3216, Australia
{aklwong, ang}@deakin.edu.au

Abstract. Studies have shown that most of the computers in a non-dedicated cluster are often idle or lightly loaded. The underutilized computers in a non-dedicated cluster can be employed to execute parallel applications. The aim of this study is to learn how concurrent execution of a computation-bound and sequential applications influence their execution performance and cluster utilization. The result of the study has demonstrated that a computation-bound parallel application benefits from load balancing, and at the same time sequential applications suffer only an insignificant slowdown of execution. Overall, the utilization of a non-dedicated cluster is improved.

1 Introduction

Although individual PCs of a non-dedicated cluster (called a cluster) are used by their owners to run sequential applications (local jobs), the cluster as a whole or its subset can also be employed to run parallel applications (cluster jobs) even during working hours. The reason is that PCs in their working environments are on average idle for much more than 50% of time [2,5,13]. Therefore, a cluster has the potential of concurrently running a mixture of parallel and sequential applications that could lead to performance improvement and better utilization of computing resources.

Computer applications can share the computational resources of a cluster in two dimensions: space and time. However, space and time sharing of computational resources in a cluster does not have to be considered separately. Consequently, an effective global scheduling scheme is needed in such a multi-user and multi-process environment to allocate resources among different applications to improve the execution performance of applications and the utilization of a cluster.

To our knowledge, only a small number of projects studied the influence of a mixture of parallel and sequential applications on their execution performance in a cluster. Also, those studies were carried out by simulation [1,3,12,15]. The only existing but often referenced paper which had addressed this scheduling problem by an experimental approach is [14]. However, the scope and depth of this study is unsatisfactory. Therefore, we are strongly convinced that a detailed experimental study of this problem will provide not only a better understanding of the problem but also form a background for the development of global scheduling facilities for computer clusters.

The aim of this paper is to report on the outcome of our experimental study into the scheduling and influence of a mixture of parallel and sequential applications on their execution performance and the cluster utilization. Here, we address the computation-bound parallel applications executing with both CPU-bound and IO-bound sequential applications concurrently on a non-dedicated cluster.

2 Dynamic Load Balancing Based Scheduling of a Computation-Bound Parallel Application and Sequential Applications

To study the behavior of a concurrently executing mixture of parallel and sequential applications on a cluster, we propose to use a two level scheduling system, where the upper level is responsible for global scheduling of processes of a parallel application and the lower level schedules local processes (parallel and sequential) running on each local computer of the cluster.

We selected dynamic load balancing to provide scheduling at the upper level for the following reasons. Dynamic load balancing is an efficient method of scheduling processes of computer applications in a cluster. By taking advantage of a process migration facility, allocation of processes to computers of a cluster can be changed dynamically according to the actual workload on each of the computers [9]. This method can also provide a unified way to utilize both space- and time-sharing.

2.1 The openMosix Dynamic Load-Balancing System and LAM/MPI

Here, we describe openMosix and an implementation of MPI [8] that can support the scheduling of a mixture of a parallel and sequential applications.

The openMosix [10] system is a Linux-based open source version of the Mosix dynamic load balancing system developed by Barak et al [6]. The openMosix/Mosix system consists of the Preemptive Process Migration mechanism and a set of algorithms for adaptive resource sharing. We have set openMosix to make load balancing decisions based on the workload of each of the computers. The version of the openMosix-enabled Linux kernel that we used is openMosix-2.4.20.

Executing parallel applications on a cluster relies on some support from a run-time environment such that the parallel applications can utilize distributed computers. One of the most important supports is provided by the IPC mechanism that allows processes of a parallel application to communicate. LAM/MPI [4] was selected for our project because it could be used with the openMosix system without too much difficulty. The version of LAM/MPI used in our project is LAM/MPI-6.5.9.

2.2 Structure of the Global Scheduling Prototype

A flexible way to execute an MPI parallel application in our openMosix cluster was to place its processes on one computer initially and allow the openMosix system to migrate them from that computer to other computers to balance the cluster load [6]. This can be achieved by providing a LAM/MPI network topology to the openMosix cluster, which specifies that only one computer is used to initiate an MPI application. Consequently, all processes are located on one computer in the MPI level but the processes are migrated to many computers in the openMosix level. The actual communication mechanism for the distributed processes is therefore relied on the IPC subsystem of openMosix rather than the LAM/MPI daemon.

3 Experimental Environment and Experiments

In this section, the parallel and sequential applications used and the experiments constructed for the execution of the selected applications are described.

3.1 Parallel Application

Our study of scheduling a mixture of parallel and sequential applications on a cluster at this stage is concentrated on how a computation-bound parallel application interacts with CPU-bound and I/O-bound sequential applications. Therefore, a computation-bound parallel application, MPI-Povray [15] (a parallelized Povray ray-tracer [11]), has been chosen. MPI-Povray distributes work among a number of processors and the communication between the processes, that is infrequent, is achieved by MPI message passing.

3.2 Sequential Benchmarks

To achieve the aim of our research, there was a need to identify and determine the influence of different sequential applications (with workloads ranging from CPU-bound to I/O-bound) executing concurrently with a parallel application on a cluster.

Ideally, our scheduling experiments should be constructed by running a parallel application together with some real user-oriented sequential applications. However, there are three major drawbacks of this approach. First, it is difficult to control the amount of workload to be executed on a computer (controllability). Second, the same amount of workload on each of the computers is also difficult to be repeated for different experiments (repeatability). Third, sequential applications should execute for a period of execution comparable to the period of execution of a parallel application, which changes with the number of computers that application runs on (durability).

The BYTE's Unix Benchmark Suite. Controllability, repeatability and durability can be achieved by using sequential benchmarks. Following our study of sequential benchmarks, we have selected the BYTE's Unix Benchmark Suite [7].

Table 1. Classification of the BYTE Unix Benchmark applications

Category	Program
CPU-bound	dhry2reg, whetstone-double, pipe, spawn, shell, syscall, arithoh, short, int, long, float, double, C, dc, Hanoi
IO-bound	execl, fstime, fsbuffer, fsdisk, context1

The BYTE Suite consists of a set of sequential applications, which were designed to test the performance of a single-processor computer system. Each application can be classified as either I/O-bound or CPU-bound as shown in Table 1.

Micro-benchmarks. A set of sequential benchmarks can be constructed by choosing different applications from the BYTE suite and packing them together into groups according to the particular workloads required.

We constructed three sets of sequential benchmarks with different workload compositions, Seq_{IO}, Seq_{IB} and Seq_{CPU}. Seq_{IO}, Seq_{IB} and Seq_{CPU} represent I/O-bound, In-Between and CPU-bound sequential benchmarks with a workload of 20%, 50% and 80% CPU utilization respectively, as shown in Table 2.

Table 2. Workload Compositions

Workloads	Components
Seq_{IO}	fstime, idle-burst, fsbuffer, idle-burst, fsdisk, idle-burst
Seq_{IB}	execl, context1, spawn, fsdisk, whetstone-double, C, fstime, syscall, hanoi, dc
Seq_{CPU}	dhry2reg, whetstone-double, pipe, spawn, shell, syscall, arithoh, int, double, C, execl, context1, hanoi, short, float

3.3 Experiments

All scheduling experiments were carried out on our openMosix cluster. The cluster contains 16 Pentium III nodes, each with 383 Mb memory, connected by a 100Mbit/s Fast Ethernet. Each computer runs the Red Hat Linux operation system with the dynamic load-balancing support using the openMosix system.

We carried out computations of the MPI-Povray parallel application with each of the Seq_{IO}, Seq_{IB} and Seq_{CPU} sequential benchmarks executed on 2, 6 or 10 computers. In each case the parallel application was run on 14, 10 and 6 computers, respectively. Depending on the total number of sequential benchmarks, N, used in each case of the scheduling experiments, the total number of computers initially available for the parallel application became varied, i.e., (16 – N). However, 16 processes were created for the parallel application to run on the cluster such that some of the parallel processes could take the advantage of dynamic load-balancing by migrating to computers executing sequential processes.

4 Results and Analysis

The execution times of the parallel and the sequential applications in each of the nine scheduling experiments described in Section 3.3 were measured. We defined the *relative speedup* to refer to the speedup achieved by the parallel application executed with sequential benchmarks on the cluster. It was calculated by dividing the execution time of the parallel application measured when there was no sharing by that when sharing was exploited. We also defined the *relative slowdown* to refer to the slowing down of a sequential benchmark executed when computers in the cluster were shared with processes of the parallel application. It was calculated by dividing the execution time of the sequential application measured when there was no sharing by that when sharing was exploited.[1] We calculated the *relative speedup* of the parallel application and the *relative slowdown* of the sequential benchmarks for each scheduling experiment. Furthermore, the summation of the execution times of the parallel and sequential applications in the two situations: sharing and non-sharing were also compared. The results are presented in Table 3.

Table 3. Speedup, Slowdown and Execution time of Applications

Number of Active Owner-Users	Relative Speedup of Parallel Application (MPI-Povray)	Relative Slowdown of Sequential Application	Execution Time (min.) of both Sequential Application and Parallel Application	
			Non-sharing	Sharing
(a) I/O-Bound				
2	+1.13	-0.09	244	240
6	+1.44	-0.06	349	307
10	+2.28	-0.04	581	429
(b) In-Between				
2	+1.08	-0.17	250	266
6	+1.28	-0.13	360	349
10	+1.87	-0.10	594	491
(c) CPU-Bound				
2	+1.04	-0.15	249	265
6	+1.16	-0.16	358	368
10	+1.57	-0.12	598	534

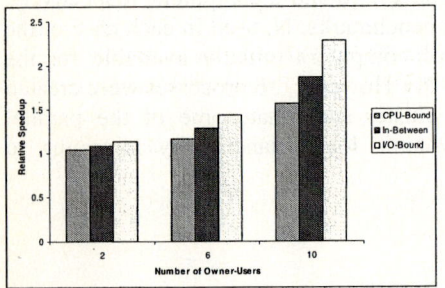
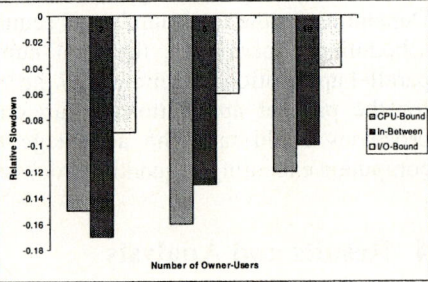

Fig. 1. (a) Relative Speedup of parallel application (b) Relative Slowdown of sequential applications

Two observations related to the execution performance of the parallel application can be made using Table 3. First, as the workload of sequential applications decreases, that is, moving from CPU-Bound towards I/O-Bound, the *relative speedup* increases. Second, as the number of computers that are used by their owner-users increases, the *relative speedup* of the parallel application executing with each of the

---

[1] We use a negative value to represent a relative slowdown.

three categories of sequential workload also increases. Thus, a computation-bound parallel application benefits from the sharing of cluster computers executing concurrently with sequential applications. These observations are shown by the plotting of the *relative speedup* of the parallel application with different sequential applications in Fig. 1(a).

On the other hand, sequential applications suffer in all cases as can be seen from the plotting of the *relative slowdown* of sequential applications as shown in Fig. 1(b).

However, the *relative slowdown* of the sequential applications in the case of I/O-Bound workload is much less significant to affect its owner-user than the other two cases: In-Between and CPU-Bound workloads. In general, although the relative slowdown generated in each case may be noticeable to an owner-user, we contend that it would be acceptable to most of the users.

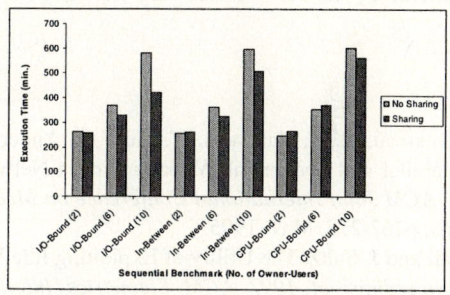

Fig. 2. Total execution turnaround time of parallel and sequential applications with and without sharing

The second part of our study aimed at cluster utilization. The summations of execution times of the parallel and sequential applications are compared in two situations: sharing and no sharing. Fig. 2 shows that the cluster is best utilized under the sharing of a computation-bound parallel application with I/O-Bound sequential applications followed by a computation-bound parallel application with In-Between sequential applications; and a computation-bound parallel application with CPU-Bound sequential applications.

5 Conclusions

The concurrent execution of a computation-bound parallel application and sequential applications of various workloads on a non-dedicated cluster demonstrated that a parallel application can benefit from migrating its processes from heavily loaded computers to lightly loaded computers executing sequential applications. Such a dynamic load-balancing based scheduling of a mixture of parallel and sequential applications offers gains particularly when the workload of the computers is low (I/O-Bound workload) and the number of such computers is large. By sharing the computers executing sequential applications, which are not accessible in a dedicated cluster, parallel applications can gain extra processing power to perform 'CPU-hungry' computations. On the other hand, users of computers executing sequential applications suffer from a slight degradation of CPU services in such sharing. The degradation of the CPU services trends to be insignificant when the workload of the computers move towards I/O-bound and the number of computers executing sequential applications is large in the cluster. Nevertheless, the overall utilization of the non-dedicated cluster can be improved. In conclusion, parallel computation-bound applications can be successfully executed on a non-dedicated cluster.

Acknowledgement

We wish to express our gratitude to Dr J. Abawajy and Dr J. Silcock for their constructive comments that helped us to generate this version of the paper. This project was partly supported by the ARC Discovery Scheme.

References

1. R H. Arpaci, A. C. Dusseau, A. M. Vahdat, L. T. Liu T. E. Anderson and D. A. Patterson. The Interaction of Parallel and Sequential Workloads on a Network of Workstations. In *Proceedings of 1995 ACM Joint International Conference on Measurement and Modeling of Computing Systems*, p267-278, May, 1995.
2. A. Acharya, G. Edjlali and J. Saltz. The Utility of Exploiting Idle Workstations for Parallel Computation. In *Proceedings of 1997 ACM Sigmetrics International Conference on Measurement and Modeling of Computer Systems*, p225-236, May, 1997.
3. C. Anglano. A Comparative Evaluation of Implicit Coscheduling Strategies for Networks of Workstations. In *Proceedings of 9th International Symposium on High Performance Distributed Computing (HPDC9)*, p221-228, August, 2000.
4. G. Burns, R. Daoud and J. Vaigl. LAM: An Open Cluster Environment for MPI. In *Proceedings of Supercomputing Symposium*, p379-386, University of Toronto, 1994.
5. W. Becker. Dynamic Balancing Complex Workload in Workstation Networks -- Challenge, Concepts and Experience. In *Proceedings High Performance Computing and Networking (HPCN) Europe Lecture Notes on Computer Science (LNCS)*, p407-412, 1995.
6. A Barak, S. Guday and R. G. Wheeler. The MOSIX Distributed Operating System, Load Balancing for UNIX. Springer-Verlag.
7. BYTE's UnixBench. The BYTE's Unix Benchmark Suite. URL: http:// www.tux.org/pub /tux/niemi/unixbench.
8. The MPI Forum. MPI: a message passing interface. In *Proceedings of the 1993 Conference on Supercomputing*, p878-883, 1993.
9. A. M. Goscinski. Distributed Operating Systems, The Logical Design. Addison-Wesley, Sydney, 1991.
10. The openMosix Homepage. URL: http://openmosix.sourceforge.net.
11. The POVRAY Homepage. Persistence of Vision Ray-tracer. URL: http:// www.povray. org.
12. K. D. Ryu and J. K. Hollingsworth. Linger Longer: Fine-Grain Cycle Stealing for Networks of Workstations. In Proceedings of the 1998 ACM/IEEE Conference on Supercomputing (CDROM), p1-12, 1998.

13. F. Tandiary, S. C. Kothari, A. Dixit and E. W. Anderson. Batrun: Utilizing Idle Workstations for Large-scale Computing. *IEEE Parallel and Distributed Technology*, 4(2):41-48, 1996.
14. F. C. Wong, A. C. Arpaci-Dusseau and D. E. Culler. Building MPI for Multi-Programming Systems using Implicit Information. In *Proceedings of the 6th European PVM/MPI User's Group Meeting*, p215-222, 1999.
15. L. Verrall. MPI-Povray: Distributed Povray Using MPI Message Passing. URL: http://www.verrall.demon.co.uk/mpipov.
16. B. B. Zhou, X. Qu and R. P. Brent. Effective Scheduling in a Mixed Parallel and Sequential Computing Environment. In *Proceedings of the 6th Euromicro Workshop of Parallel and Distributed Processing*, p32-37, January, 1998.

Sequential and Parallel Ant Colony Strategies for Cluster Scheduling in Spatial Databases

Jitian Xiao and Huaizhong Li

School of Computer and Information Science, Edith Cowan University,
2 Bradford Street, Mount Lawley, WA 6050, Australia
{j.xiao, h.li}@ecu.edu.au

Abstract. In spatial join processing, a common method to minimize the I/O cost is to partition the spatial objects into clusters and then to schedule the processing of the clusters such that the number of times the same objects to be fetched into memory can be minimized. The key issue of cluster scheduling is how to produce a better sequence of clusters to guide the scheduling. This paper describes strategies that apply the ant colony optimization (ACO) algorithm to produce cluster scheduling sequence. Since the structure of the ACO is highly suitable for parallelization, parallel algorithms are also developed to improve the performance of the algorithms. We evaluated and illustrated that that the scheduling sequence produced by the new method is much better than existing approaches.

1 Introduction

Spatial join queries in spatial databases usually access a large number of spatial data. They are the common spatial query type that requires a high processing cost due to the large volume of spatial data and the computation-intensive spatial operations. To reduce the CPU and I/O costs for spatial join processing, most spatial join processing methods are performed in two steps (i.e., *filter-and-refine* approach) [1]. The first step chooses pairs of data that are likely to satisfy the join predicate. The second step examines the predicate satisfaction for all those pairs of data passing through the filtering step. During the filtering step, a conservative approximation of each spatial object is used to eliminate objects that cannot contribute to the join result, and a *weaker* condition for the spatial predicate is applied on the approximations. This step produces a list of *candidates* that is a superset of the joinable candidates. These candidates are usually represented as pairs of object identifiers. All candidates are then checked in the refinement step by applying the spatial operation on the full descriptions of the spatial objects to eliminate the "false drops". The join cost can be reduced because the weaker condition is usually computationally less expensive to evaluate and the approximations are small in size than the full geometry of spatial objects. Generally, the refinement cost consists of two parts: one is the cost for fetching objects from the database, and the other is the cost for checking spatial relationship of pairs of objects using computational geometry algorithm. The former cost dominates the refinement cost when the spatial objects are small (e.g., tens to hundreds of vertices), and the later cost dominates for large spatial objects. In this paper, we focus on reduction of former cost in the refinement step.

To reduce the I/O cost, we proposed a method to partition candidates into clusters so that spatial objects join as many other objects as possible within their cluster and join as few objects as possible across clusters [2], thus significantly reduce the I/O cost when the join operations are processed cluster by cluster. In [3], we proposed an approach to further reduce I/O cost by carefully scheduling the clusters such that a maximum number of overlapping objects between consecutive clusters in the scheduling sequence can be reused when processing the next clusters (i.e., they do not need to be fetched into memory again). In this way, a significant reduction on disk accesses has been achieved and demonstrated through simulations.

Even so, our simulation shows that, in some cases, it may still take long time (e.g., up to hours) to complete a single spatial join operation. We propose to improve the performance of spatial join processing by employing parallel mechanism. The goal is to be achieved by three steps: (1) partition in parallel the spatial candidates into clusters such that the number of overlapping objects among clusters is minimized; (2) produce a cluster scheduling order such that the total number of overlapping objects between consecutive clusters is maximized; and (3) schedule the spatial join, cluster by cluster, in the cluster scheduling order produced in step (2) in the refinement step for spatial join processing. Preliminary experiments have shown that the parallel algorithms work well for spatial join processing, especially when the number of spatial objects in a join is large. In this paper we focus on producing the cluster scheduling order using parallel Ant Colony Optimization (ACO) approach (the parallelization of step (1) will be discussed separately). We will illustrate that the new approach produces better solutions than those produced by the former method.

The reminder of the paper is organized as follows. In Section 2, we formalize the problem by reviewing our previous work in this area. Section 3 summarizes the sequential algorithm of producing cluster scheduling order developed in [3]. Section 4 presents strategies using ant colony optimization approach to produce better orders for cluster scheduling in spatial join processing. Conclusions are presented in Section 5.

2 Problem Definition

Let S and T be the two spatial database tables for spatial join operation, denoted by $S \bowtie T$. Objects in S and T are indexed by their unique IDs. The filter operation of the spatial join produces a set of pairs of S and T objects. Let F be the set of ID pairs produced by the filter operation:

$$F = \{(sid, tid)| \ sid \text{ and } tid \text{ are IDs of objects in } S \text{ and } T,$$
$$\text{respectively, that meet the weaker join condition}\}$$

where an ID pair $(sid, tid) \in F$ is called a *candidate*. Fig. 1 (a) shows an example of F. Note that F is available in the main memory after the filter operation. F contains only IDs of the candidates, not the data objects.

The *refinement* step is to perform $S \bowtie T$ on the pairs of objects indexed by F to produce the final join results. At this step, the S and T objects need to be fetched into the main memory for the full spatial join test. Since the memory size is limited and it can not keep all objects of F in memory at the same time, the objects need to be partitioned into clusters. Objects in the same clusters are brought into the memory together and processed in a batch. For example, Fig. 1 (b) is a partitioning of the candidate set shown in Fig. 1 (a).

S id	T id
A1	B1
A2	B1
A3	B2
A3	B3
A4	B3
A5	B1
A6	B2
A6	B4
B7	B1
A8	B3
A8	B4

(a) A candidate set   (b) A candidate clustering

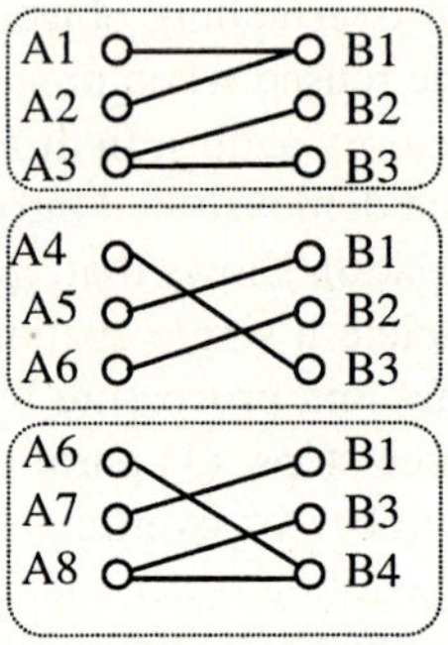

Fig. 1. An example of a candidate set and its clustering

Assume that the spatial objects referenced in F have been partitioned into clusters (e.g., by using approaches given in [2]). Our goal is to schedule the clusters in a way such that the repeatedly fetch of the overlapping objects between consecutive clusters is minimized. The I/O cost, in this paper, is measured in terms of the size of object data (e.g., number of vertices of the spatial object) that are fetched into the memory for the refinement operation.

Let $V = \{v_1, v_2, ..., v_k\}$ be the set of objects referenced in F, and $V_1, V_2, ..., V_n$ the clusters of V. For each i $(1 = i = n)$, $V_i = \{v_{i_1}, v_{i_2}, ..., v_{i_m}\}$ $(m = 1)$, $v_{i_j} \in V$ $(1 = j = m)$. That is, $\bigcup_{i=1}^{n} V_i = V$ and $V_i \neq \phi$ for each i $(1 = i = n)$. Define $size(V_i) = \sum_{v \in V_i} s(v)$ as the sum of the sizes of objects in V_i, where $s(v)$ is the size of object v. We introduce a weighted graph $G = (V, E, w)$, upon V, called *cluster overlapping* (CO) graph, to represent the overlapping relationships between data clusters. The node set $V = \{V_1, V_2, ..., V_n\}$ is a set of clusters, and the edge set E is defined as: for each node pair V_i and V_j $(i \neq j)$, there is an edge $E_{ij}=(V_i, V_j)$ if $w(V_i, V_j) = size(V_i \cap V_j) \neq 0$. Here $w(V_i, V_j)$ is the weight of edge E_{ij}.

At refinement step, if the clusters are processed in the sequence of $V_1, V_2, ..., V_n$ (i.e., no scheduling), then the total I/O cost is:

$$C_{I/O} = \sum_{i=1}^{n} size(V_i) - \sum_{i=1}^{n-1} size(V_i \cap V_{i+1}). \tag{1}$$

When processing cluster V_{i+1}, objects in $V_i \cap V_{i+1}$ are already in memory just after processing V_i. There is no need to load these objects again.

Generally, for a schedule p which determines the processing sequence of $V_1, ..., V_n$ as $V_{\pi_1}, V_{\pi_2}, ..., V_{\pi_n}$, where $V_{\pi_i} \in V$ and $V_{\pi_i} \neq V_{\pi_j}$ for $i \neq j$, the I/O cost for schedule p is

$$C_{I/O}^{\pi} = \sum_{i=1}^{n} size(V_{\pi_i}) - \sum_{i=1}^{n-1} size(V_{\pi_i} \cap V_{\pi_{i+1}}). \quad (2)$$

When the clusters are given, $\sum_{i=1}^{n} size(V_{\pi_i})$ is a constant. Let y be:

$$y = \sum_{i=1}^{n-1} size(V_{\pi_i} \cap V_{\pi_{i+1}}). \quad (3)$$

Our goal is to find a schedule such that y is maximized, which is the case that $C_{I/O}^{\pi}$ is minimized.

3 Cluster Scheduling Sequence

In our previous work [3], a *maximum overlapping (MO) order* was defined in a CO graph, and a sequential algorithm was developed to produce an approximation to the MO (AMO) order for an arbitrary CO graph. The AMO order was then used as a scheduling sequence for cluster scheduling in spatial join processing.

Given a CO graph $G = (V, E, w)$ with $V = \{V_1, V_2, ..., V_n\}$, a *maximum overlapping (MO) order* among sets $V_1, V_2, ..., V_n$ is a sequence $V_{i_1}, V_{i_2}, ..., V_{i_n}$ such that w_{MO} = $\sum_{l=1}^{n-1} size(V_{i_l} \cap V_{i_{l+1}})$ reaches the maximum among all permutations of V. w_{MO} is called the (total) overlapping weight of the MO order.

An MO order in a CO graph G is a permutation of nodes in G such that the total size of overlapping objects between adjacent nodes reaches the maximum. A simplest algorithm to find an MO order is to check all permutations of V to examine which one makes the max{ $\sum_{l}^{n-1} size(V_{i_l} \cap V_{i_{l+1}})$ }. The complexity of such a method clearly has factorial order and is certainly not practical. Although an MO order exists for each CO graph G, the problem of finding an MO order in a CO graph is *NP*-complete [3]. However, the task of finding an MO order can be reduced to the case where G is a connected graph (see Theorem 2 in [3]).

In our previous work, a maximum spanning tree (MST) based algorithm was developed in [3] to produce an AMO order of relative "high" overlapping weight in the sense that the weight of the AMO order produced by the algorithm is always greater than or equal to half the overlapping weight of an optimal MO order. The algorithm consists of three steps: The first step produces a maximum spanning tree T of the CO graph G; the second step conducts a *depth-first search* (DFS) on T and, in the third step, an AMO order is built, which is the traversal order of the DFS on T. The complexity of the algorithm is $O(m^2 \log_2 m)$, where $m = max(|V|, |E|)$.

4 The Ant Colony Optimization Based Algorithms

We now develop sequential and parallel algorithms, based on the ant colony optimization algorithm [4, 5, 6], to produce a better AMO order of higher overlapping weight. For simplicity, we limit our discussion to connected CO graphs. The algorithms and the related discussion can be easily extended to the case of unconnected CO graphs (see Theorem 2 in [3]).

4.1 The ACO Approach

The Ant Colony Optimization (ACO) metaheuristic is a population-based approach to the solution of discrete optimization problems. It has been applied to many combinatorial optimization problems [5, 6]. It imitates real ants searching for food, e.g., finding the shortest path from a food source to their nest without strength of vision. The ants use an aromatic essence, called *pheromone*, to communicate information regarding the food source. While ants move along, they lay pheromone on the ground which stimulates other ants rather to follow that trail than to use a new path. The quantity of pheromone a single ant deposits on the path depends on the total length of the path and on the quality of the food source discovered. As other ants observe the pheromone trail and are attracted to follow it, the pheromone on the path will intensified and reinforced and will therefore attract even more ants. In other words, pheromone trails leading to rich, nearby food source will be more frequented and will grow faster than trails leading to low-quality, far-away food source.

The typical application of ACO is the traveling salesman problem (TSP) [4, 6], defined as follows: A graph $G=(V, E, w)$ with node set V and edge set E is given; each edge $e \in E$ has a weight $w(e)$ associated, representing the length of it. The problem is to find a minimal-length closed tour that visits all the nodes once and only once[1]. In the ACO approach each edge of the graph has two associated measures: the heuristic desirability η_{ij}, which is defined as the inverse of the edge length and never changes for a given problem instance, and the pheromone trail τ_{ij}, which is modified at runtime by ants. Each ant has a starting node and its goal is to build a solution, that is, a complete tour. A tour is built node by node: when ant k is in node i it chooses to move to node j using a probabilistic rule that favors nodes that are close and connected by edges with a high pheromone trail value. Nodes are always chosen among those not yet visited in order to enforce the construction of feasible solutions. Then pheromone trail is updated on the edges of the solutions. The guiding principle is to increase pheromone trail on the edges that belong to short tours. Pheromone trails also evaporate so that memory of the past is gradually lost (this prevents bad initial choices from having a lasting effect on the search process). The approach, here informally described, can be implemented in many different ways and details about specific implementation choices for the TSP can be found in [4, 5] and therein.

---

[1] The algorithm is called an *ACO-TSP* algorithm in this paper, and the shortest closed tour produced by the algorithm is called a *TSP solution*.

The structure of the ACO is highly suitable for parallelization of the algorithm. A parallel version algorithm for the TSP can be found from [6].

4.2 The Sequential Algorithms for Finding Cluster Scheduling Order

To apply the parallel ACO approach for finding an AMO order for an arbitrary CO graph $G=(V, E, w)$, we make some modifications to the ACO-TSP algorithm.

The first change is to make the ACO-TSP algorithm find a longest tour instead of shortest one. For this purpose, we extend G to a complete graph $G'=(V, E', w')$: for any pair of nodes $v_i, v_j \in V$, $1 \leq i, j \leq n$, add an edge (v_i, v_j) to E'. If $(v_i, v_j) \in E$, we define $w'(v_i, v_j) = w_{max} - w(v_i, v_j)+1$; otherwise define $w'(v_i, v_j) = w_{max}+1$, where $w_{max} = \max\{w(v_i, v_j) | (v_i, v_j) \in E\}$. We take $w'(v_i, v_j)$ as the length between nodes v_i and v_j, and call G' the *converted* CO graph. It is evident that a shortest closed tour in G' corresponds to a longest closed tour in G, and an MO order of G corresponds to a longest path of G (we conceptually assume that G is a completed graph, i.e., if there exists not an edge between $v_i \in V$ and $v_j \in V$, for any pair i and j, we conceptually add an edge (v_i, v_j) with a weight $w(v_i, v_j) = 0$).

One way to obtain and AMO order for a CO graph G is to apply the ACO-TSP algorithm to its converted CO graph G' to find a TSP solution, which is a shortest closed tour on G' (or, a longest closed tour on G). And then take a longest path within the tour of G (shortest path within the tour of G'), i.e., by simply removing a shortest edge from the tour, as an AMO order of G. A simplified algorithm of such strategy is described as follows:

```
Algorithm acoBasedAMO(G)
Input: G=(V, E, w); // A CO graph with V = {V₁,V₂,...,Vₙ}
Output: Vᵢ₁,Vᵢ₂,...,Vᵢₙ; // An AMO order of G.
BEGIN
(a) Convert G into a converted CO graph G';
(b) Find a TSP solution of G' using ACO-TSP algorithm;
(c) Find an AMO order of G based on the output produced
    in step (b);
END
```

In the above step (c), the AMO order of G is formed as: Without loss generality, assume that the TSP solution of G produced in step (b) is $V_1, V_2, \ldots, V_n, V_1$. We select one edge (V_l, V_{l+1}) in the tour that makes $min\{w(V_i, V_{i+1}) | i=1,2, \ldots, n\} \cup \{w(V_n, V_1)\})$, and choose the sequence $V_{l+1}, V_{l+2}, \ldots, V_n, V_1, V_2, \ldots, V_l$ as the AMO order.

However, the above algorithm does not always produce the best solution because a longest (shortest) closed tour does not necessarily contain a longest (shortest) path of the graph. For example, in the CO graph shown in Fig. 2 (a), the longest closed tour is V1, V2, V3, V4, V1. When removing one of the minimal weighted edge (say, (V3, V4), as shown in thick dotted line) from it, a longest path within the tour can be produced as V4, V1, V2, V3, with a total edge weight 11. However, the longest path of the figure is actually V4, V1, V3, V2, with a total edge weight 12, as shown in solid lines in Fig. 2 (b).

To produce a better solution, we make further modification to the ACO-TSP algorithm on G': each ant builds a shortest path instead of a shortest closed tour. For each iteration, an ant's shortest path can be obtained from its shortest closed tour by removing one edge of maximal weight from the tour. Since the length of the shortest path is shorter than that of the shortest closed tour, this modification results in a reduction on the quantity per unit of length of pheromone laid on edge (v_i, v_j) (i.e., $\Delta\tau_{ij}$ in [4]) by each ant.

Another modification to the ACO-TSP algorithm is to set the predetermined iteration parameter, NC, of the algorithm. Although NC is independent from the number of nodes of G' and the number of ants placed at each node, its value significantly affects the quality of the TSP solution [4, 5]. To make an efficient algorithm without loss much of the convergence speed, we set NC as the same number of nodes in the graph. Our experiments show that this setting works well. The resultant algorithm based on this strategy is called Modified ACO-TSP, and named $m\_acoBasesAMO(G)$. The description of the algorithm is similar to $acoBasedAMO(G)$, and is thus omitted here.

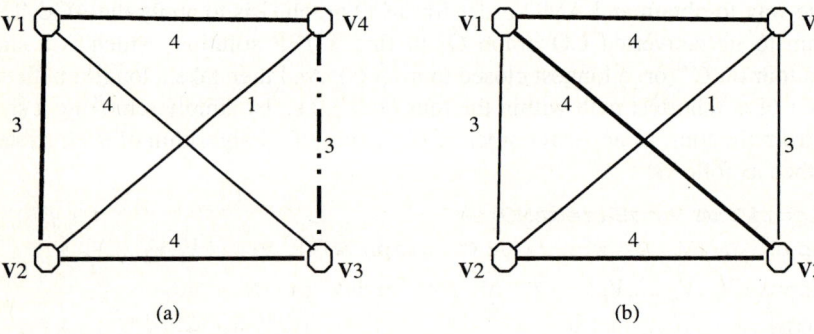

Fig. 2. (a) A CO graph, with a longest closed tour shown in thick lines. If the dotted line was removed, an AMO (but not an MO) order can be formed from the tour. (b) An MO order of the CO graph (shown in thick lines)

The complexity of both algorithms is the same as that of ACO algorithm, which is $O(NC \cdot n^2 m)$ [4], where NC is the predetermined iteration parameter of the ACO algorithm, and m the number of ants. It was proved that a reasonable setting is to set $m=n$, i.e., one ant for each node of the graph. In this case, the complexity of both algorithms can be written as $O(NC \cdot n^3)$.

Example 1. Let $V = \{a_1, a_2, ..., a_{36}\}$ be a spatial object set and $V = (V_1, V_2, ..., V_6)$ a set of clusters on V. The corresponding CO graph is given in Fig. 3. In this example, the sizes of all objects are identical, thus are not important. For simplicity, an object a_i is expressed by its index i in the figure, and the size of a_i is taken as 1 unit, for all $1 \le i \le 36$. By applying the MST based algorithm [3] to the CO graph, an AMO order was produced as shown in thick solid lines in Fig. 4 (a), which is $V_1, V_4, V_2, V_3, V_6, V_5$, with the total overlapping weight 31.

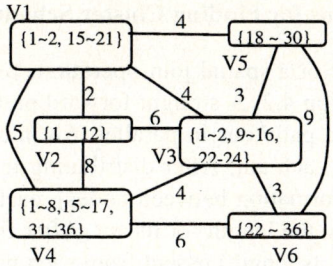

Fig. 3. A CO graph reproduced from [3]

By applying algorithm *acoBasedAMO*( ) to the CO graph G in Fig. 3, a TSP solution, which corresponds to a longest closed tour of G, was produced as shown in thick solid/dotted lines in Fig. 4 (b). In the third step of the algorithm, an edge with minimum weight in the tour is removed to produce an AMO order. In this example, both edges (V_1, V_5) and (V_1, V_3), as shown in dotted lines in the figure, have the minimum weight (i.e., 4). By removing (V_1, V_5) we get an AMO order $V_1, V_3, V_2, V_4, V_5, V_6$. If ($V_1$, V_3) was removed, we would obtain an alternative AMO order $V_3, V_2, V_4, V_5, V_6, V_1$. Similarly, when applying *m_acoBasedAMO*( ) to G, an AMO order $V_3, V_2, V_4, V_5, V_6, V_1$ is produced, which is the same as one of the AMO orders produced by *acoBasedAMO*( ).

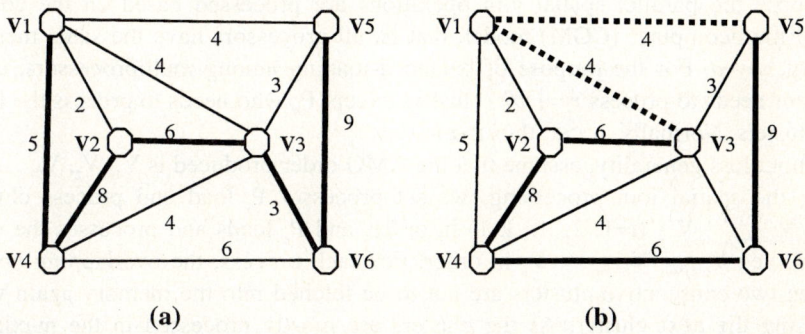

Fig. 4. (a) AMO order produced by the MST based algorithm. (b) An ACO-TSP solution

By comparing the outputs of the AMO orders produced by the MST based and the ACO-based algorithms, we find that the ACO-based algorithms produce better solutions. This can be illustrated by the fact that the total overlapping weight of the AMO order produced by the MST algorithm is 31, while the overlapping weight of the AMO orders produced by the ACO-based algorithms is 33, which is the optimal MO order for this example. Furthermore, initial experiments show that *acoBasedAMO*( ) and *m_acoBasedAMO*( ) always produce a better solution than the MST based algorithm does, and *m_acoBasedAMO*( ) always produces the best solution among the three algorithms.

4.3 The Parallel Algorithms for Finding Cluster Scheduling Order

To reduce the response time of a spatial join operation, we parallelize the sequential algorithms described in section 4.2. A straight forward parallelization strategy for the algorithms is to compute the path/tour in parallel. An initial process would spawn a set of subprocesses, one for each ant. After distributing initial information about the problem (e.g., the length information between nodes), each process can start to draw up the path/tour and compute the length for its ant. After finishing this procedure, the result (i.e., the path/tour and its length) is sent from each processor back to the master process. The master process updates the trail levels and checks for the best path/tour found so far. Then a new iteration is initiated by sending out the updated trail levels.

Assume that N processors are available for parallel computation. If ignoring any communication overhead, this approach may imply optimum (linear) speedup $T_{seq}(n)/T_{para}(n) = O(NC \cdot n^3)/ O(NC \cdot n^3/N) = O(N)$, where $T_{seq}(n)$ is the computational complexity of the sequential algorithm and $T_{para}(n)$ the computational complexity of the parallel algorithm.

4.4 Parallel Cluster Scheduling

Once the AMO order is produced, the next step is to schedule the spatial data of clusters among the parallel processors for the refinement step in spatial join processing. Assume processors $P_1, P_2, \ldots P_k$, $k \geq 2$, are available to do the join operations in parallel, i.e., clusters of spatial data can be scheduled to k sites for join processing.

Suppose the parallel spatial join operations are processed based on the coarse-grained multicomputer (CGM) model, that is, all processors have the same memory capacity, say q. For the purpose of balanced-loading among multiprocessors, every processor needs to process $r = \lceil n/k \rceil$ clusters except P_k who needs to process the last $n - r*k$ clusters. Normally, n>>k, thus r>>1.

Without loss generality, assume that the AMO order produced is $V_1, V_2, V_3, \ldots, V_n$. During the spatial join processing, we set processor P_i load and process clusters $V_{(i-1)*r+1}, V_{(i-1)*r+2}, \ldots, V_{i*r}$, (i=1, 2, …, k-1) in order, and P_k loads and processes the other clusters, i.e., $V_{(i-1)*r+1}, V_{(i-1)*r+2}, \ldots, V_n$ in order. For each process, the overlapping objects between two consecutive clusters are not to be fetched into the memory again when processing the next cluster. As the clusters are mostly processed in the maximum overlapping order, most I/O cost for fetching overlapping objects will be saved.

Note that when $k=1$, the above scheduling method degenerates to sequential scheduling over one processor. In this case the saving the I/O cost for fetching overlapping objects between consecutive clusters reaches the maximum.

5 Conclusion

In spatial join processing, spatial objects are usually partitioned into clusters and then are processed cluster by cluster. Since two clusters may have overlapping, the overlapping objects may be repeatedly loaded into memory. We proposed a method in [3] that schedules the processing of the clusters in such a sequence that two consecutive clusters in the sequence have maximal number of overlapping objects. As it is not needed to load those overlapping objects when processing the next cluster because they are already in the memory, the total I/O cost can, therefore, be minimized.

The key issue behind this method is how to produce a better AMO order to guide the cluster scheduling. This paper proposed to apply ACO approach for finding AMO orders. Both sequential and parallel version algorithms are developed. We illustrated that the new algorithms can produce a better AMO order than the existing MST based algorithm in the sense that the average overlapping weight of the AMO order produced by the new algorithms can be greater than that of the MST.

References

1. H. Samet and Walid Aref, Spatial Data Models and Query Processing. *Modern Database Systems*, Addison-Wesley Publishing Company, Inc, 1995
2. J. Xiao, Y. Zhang and X. Jia. Clustering Non-uniform- Sized Spatial Objects to Reduce I/O Cost for Spatial Join Processing, *The Computer Journal*, Vol. 44, No.5, 2001
3. J. Xiao, Y. Zhang, X. Jia and X. Zhou. A Schedule of Join Operations to Reduce I/O Cost in Spatial Database Systems, *Data & Knowledge Engineering*, Elsevier Science B.V, Vol. 35, pp. 299–317, 2000
4. L. M. Gambardella, M.Dorigo, An Ant Colony System Hybridized with a New Local Search for the Sequential Ordering Problem, INFORMS Journal on Computing, vol.12 (3), pp. 237–255, 2000
5. M. Dorigo, V. Maniezzo & A. Colorni. The Ant System: Optimization by a colony of cooperating agents. *IEEE Transactions on Systems, Man, and Cybernetics, Part B*, Vol. 26, No. 1, pp. 29–41, 1996
6. B. Bullnheimer, G. Kotsis, C. Strauss. in R. De~Leone, A. Murli, P. M. Pardalos, G. Toraldo, (eds.): Parallelization Strategies for the ANT System, *High Performance Algorithms and Software in Nonlinear Optimization*, Kluwer International Series in Applied Optimization, Vol. 24, Kluwer Academic Publishers, Dordrecht, 1998

Cost-Effective Buffered Wormhole Routing

Jinming Ge

Engineering & Computer Science,
Wilberforce University,
Wilberforce, OH 45384, USA
jge@wilberforce.edu

Abstract. The cost-effectiveness of wormhole torus-networks is systematically evaluated with emphasis on new buffered-wormhole routing algorithms. These algorithms use hardwired datelines and escape channels to alleviate bottleneck and bubble problems caused by previous algorithms and so to improve the buffer efficiency with little hardware overhead. A two-part evaluation environment is developed consisting of a cycle-driven simulator for high-level measurements such as network capacity and an ASIC design package for low-level measurements such as operating frequency and chip area. This environment is used to demonstrate that the new routers are more cost-effective than their counterparts under various workload parameters.

1 Introduction

The principal objective of this paper is to look for a cost-effective buffered wormhole router in a multicomputer torus network, especially that implemented on a single chip where cost is as important as performance. This involves several tasks. First, techniques are developed to enhance routing performance. Second, the performance, especially the buffer usage efficiency of various routing algorithms is evaluated on a common platform. Third, the router implementation cost, chip area and clock timing are derived using standard ASIC design flow with respect to current technology. Finally, combining the simulation result and the synthesis report, the cost-effectiveness of each router can be compared.

Buffered wormhole routing has gained popularity with improved VLSI technology. As more and more gates can be integrated in a single chip, it is possible to make the buffers larger and so release some congested physical links. A special case of buffered wormhole, the Virtual Cut Through (VCT), guarantees buffer space for any incoming packets, but it is less flexible than the Buffered wormhole routing because the restriction on the ratio of buffer space to maximum packet size.

A key challenge of routing design is the deadlock problem. The most common and practical way to avoid deadlock is to design deadlock-free algorithm [1][2] because of the high cost and performance degradation of recovery schemes [3][4]. There are several adaptive wormhole routing algorithms developed by using the concept of virtual channels and channel dependency graphs [2][5][6]. As presented by Aoyama and Chien, however, the virtual channels also substantially increase the router cost [2]. Duato devised the most attractive algorithm [7][8] of this type and its revised versions have been integrated into the design of the Cray T3E [9] and the MIT Reliable Router [10]. This algorithm relaxed Dally's deadlock-free condition by allowing cycles to exist in the channel dependency graph as long as there is a routing sub-function whose extended channel dependency graph is acyclic. In this cycle-tolerance scheme, only two or three VCs per direction are necessary for partial adaptive (named as *DP_like* in the following discussion) or fully adaptive (named as *DF_like* in the following discussion) routing respectively. Although it can allocate the traffic evenly among virtual channels, a bottleneck link is formed. The 'non-co-residence' assumption on all buffers [3][7] also results in bubbles along the communication link and degraded performance for small packets. To tackle the bottleneck link and the bubble problems, a routing strategy was introduced in [11]. A hardwired dateline is defined along each uni-directional ring of the torus network to alleviate the bottleneck; and flits belong to different packets can co-reside in a buffer of one of the virtual channels – the escape channel so that the buffer space can be used more efficiently. Algorithms based on this strategy, the DynBal with *DP_like* as its counterpart, and the F_DynBal with *DF_like* as its counterpart, were proven to be deadlock-free and shown to be more efficient in the buffer space usage by a flit-level performance simulator.

This paper continues the previous evaluation by considering both the performance and the buffer and control implementation cost in chip area and clock frequency. The cost of various routing schemes, DynBal, F_DynBal, DP_like, and DF_like are reported after actually modeling the routers in Verilog HDL, and synthesizing using Synopsys and targeting to a state-of-the-art ASIC (LSI G11-p) technology. By combining previous performance simulation result and the synthesis report, the cost-effectiveness of each router can be compared with others. The major results are characterized by the chip area budget. Constrained by a low chip area budget, the DynBal router is the most cost-effective among the routers evaluated. F_DynBal is the most cost-effective one if high chip area budget available and highly non-uniform workload applied.

The rest of this paper is organized as follows. The routing strategy and algorithms presented in [11] are discussed briefly in Section 2. Section 3 presents the basic performance evaluation results of the algorithms developed. The chip area and speed of the routers ASIC design derived in Section 4 are then combined with the performance evaluations to compare the cost-effectiveness in Section 5. Finally, concluding remarks are given and further directions are discussed in Section 6.

2 Routing Strategy and Algorithms

This section briefly reviews the two definitions, the routing schemes and the deadlock-free proofs introduced in [11].

2.1 Hardwired Dateline and Escape Channel

The analysis is based on two definitions. The basic dateline concept [12] is redefined to remove the bottleneck formed in the original *DP_like* routing algorithm. An escape channel distinguishes itself from a cyclic channel, uses the buffer space more efficiently, and still avoids the deadlock situation.

Hardwired Dateline. It is a dateline that cuts a ring so that it can only be crossed by traffic in the appropriate VC. The position of the dateline is recorded as the node address where the dateline crosses the switch, and is known by each node on the ring.

In the DP_like [11], since only a packet heading to its right, as illustrated in Fig. 1, can use virtual channel VC1, the virtual channel between No and N3 and buffer C30 cannot be used while there are two channels for all other links along the ring. The hardwired dateline for the ring crosses the link where the bottleneck is formed. The purpose of making the dateline hardwired is that we want to do the VC balancing dynamically and in a distributed manner rather than by using static off-line optimization as used in the T3D [12]. There will be no real connection between the input buffer of C00 and the output buffers C31 or C30. That is, a packet that arrives in input buffer C01 can use either C30 or C31 to reach its destination so that the bottleneck is removed.

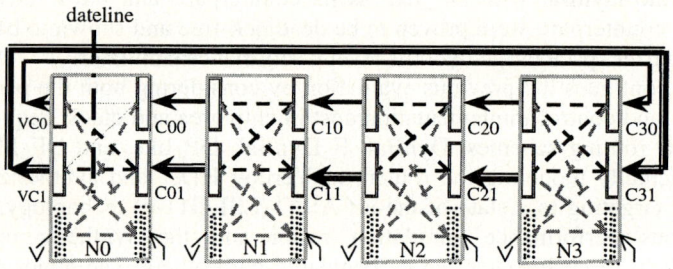

Fig. 1. The hardwired dateline crossing a unidirectional ring

Escape Channel. It is an open-loop channel that consists of a set of VC buffers without a cyclic direct-dependency among them.

The escape channel can be identified from Fig. 1 or from its corresponding channel-dependency graph [2]. The virtual channel VC0, consisting of the set of CX0, is open-looped so that it is an escape channel, while the virtual channel VC1 is a cyclic channel. The critical difference between the two kinds of channels is that routing on the former will never cross the dateline, while on the latter, it may cross if a packet is heading to the right of its current position.

The assumptions of the cycle-tolerance theory [7] are used except for one change: the queue in an escape channel can hold flits belonging to different packets, while the queue in a cyclic channel cannot. In addition, we assume that a hardwired dateline exists for each unidirectional ring of the network.

2.2 The Basic Algorithm

The ideas of hardwired dateline and the escape channel can be used to develop a new routing strategy. The algorithm, *DynBal*, for routing in a single unidirectional ring is presented first and then adapted to a bi-directional ring and finally to a multiple dimensional torus.

Routing in a single ring is depicted in Fig. 1:

For routes intended to crosses the hardwired dateline, VC1, the cyclic channel (CX1) only, must be selected.

If the route is not intended to cross the hardwired dateline, VC0, the escape channel (CX0), for a high priority, or VC1, the cyclic channel (CX1), for a low priority may be selected.

The requests to the cyclic channel can be granted if the output queue is empty and those to the escape channel can be granted if output queue space is available, so as to satisfy the assumptions.

The proof of deadlock-freedom of the basic algorithm is detailed in [11].

2.3 Algorithms Extended to Multiple Dimensions and Fully Adaptive Routing

The DynBal can be used for routing along a bi-directional ring by breaking the tie between the rings in opposite directions. A packet is routed to a specific directional ring at the source node, and keeps proceeding in the same ring until it reaches its destination. There are no dependencies between the rings in opposite directions, so that a packet in one ring never blocks a packet in the other ring. The selection of rings in the source node is based on the shortest path, and there are two different datelines for each ring.

Just as shown in the DP_like algorithm, crossing dimensions on a multi-dimensional torus should use the dimension order routing (DOR), and in each dimension routing, the packet along a particular ring should follow the rules of the above algorithms.

The DynBal algorithm adopts adaptivity within a ring. If a third virtual channel is introduced to DynBal as a fully adaptive channel used to cross dimensions in a arbitrary order, by a similar method used in the development of DF_like from DP_like [7], a fully adaptive wormhole routing algorithm, *F_DynBal*, can be developed. F_DynBal can be proven to be deadlock-free by the same approach used above: the fully adaptive channel works in a similar way to a cyclic channel in DynBal.

3 Performance Evaluation

To evaluate routing performance, the router and the network must be first modeled with reasonable assumptions. Following a description of the evaluation tool, a flit-level cycle-driven simulator, the efficiency of buffer usage by different algorithms under both dynamic and static workloads is discussed.

The following discussion is based on a bi-directional, 2-dimension 16x16 torus where each node can communicate with each other node by an attached router. Each node has an input (injector) and output (ejector) port to communicate with the router. Buffering at both input and output is modeled inside the router. Each virtual channel can be constructed with several lanes for performance reasons rather than for deadlock prevention [13]. All virtual channels (and the lanes constructed within them) in the same direction share the same physical channel. The width of the physical channel (phits) is assumed to be equal to that of flits.

The message unit of workload is a packet that consists of a header flit containing the destination node address as well as other control bits, data flits, and a tail flit. Packets are mapped to each node based on a global communication pattern, but packet flits are injected to an input buffer iff buffer space is available. A packet is consumed as soon as it reaches its destination because of an un-limited ejection buffer.

The workload can be mapped either dynamically or statically. A *dynamic workload*, in which packets are generated at every cycle of the simulation, can be used to test the performance in throughput and average latency. On the other hand, in a *static workload*, the generation is done just once and is used to measure how many cycles are needed to deliver all the packets mapped. For example, a data matrix (e.g. a picture) can be mapped to the physical torus network where each node is mapped on a certain fraction of the matrix.

A flit takes one cycle to transfer from an input buffer to an output buffer or from an output buffer to a downstream input buffer if there is no congestion. A header flit entering into an empty input buffer will have to take one extra cycle to get the routing done. The routing for the header flit following the tail of another packet (note only if the algorithm for this VC permits packet co-residence) will be done parallel with the transmission of the flit ahead of it. The routing algorithms evaluated are those discussed so far, which are DynBal, DP_like, F_DynBal, and DF_like.

3.1 Efficiency of Using Node Buffer Space

For a particular algorithm and packet size, the buffer size affects performance in several ways. A minimum of two flits per buffer is required for asynchronous Physical Channel wormhole routing; otherwise, it would introduce bubbles [1][3][11]. When

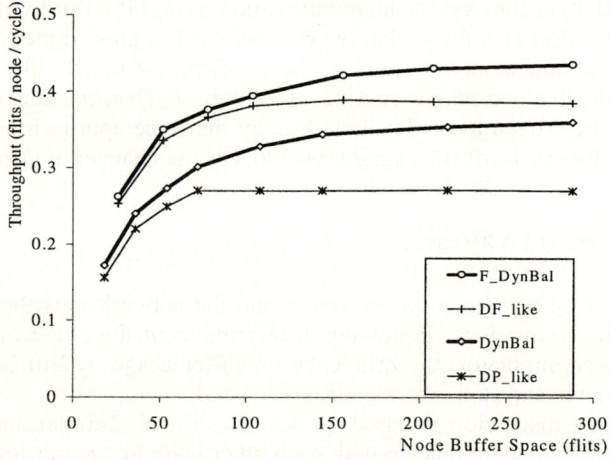

Fig. 2. The effect of buffer space to various routing algorithms (8-flits-packet, uniform traffic

the buffer size gets bigger, blocked packets hold fewer buffers as well as fewer links so that a big buffer improves the performance. However, increasing buffer size also exacerbates the Head-Of-Line problem, so there is an optimal size that should be considered. We also found that various algorithms use buffer space with different efficiencies, as shown in Fig. 2, where all of the four algorithms are evaluated based on the same buffer space per node.

Both of the two fully adaptive algorithms perform better than non-fully adaptive ones. One of the reasons is that the average buffer size per lane for these two fully adaptive ones is smaller than the two non-fully adaptive ones and so that less bubbles introduced.

If the buffer gets bigger, the bubble problem of the DP_like and the DF_like becomes significant because the bigger buffer space along all channels is just wasted due to the restriction of non-packet-co-residence. In general, the dynamic balance scheme (DynBal and F_DynBal) uses buffer space more efficiently.

3.2 Dynamic and Static Workload

Static workloads are applied to measure the number of clock cycles required to complete a task, while dynamic workload is used to measure the throughput. During the static workload delivery process different algorithms may use network capacity differently, as shown in Fig. 3, where DynBal uses fewer cycles than DP_like and F_DynBal uses fewer cycles than DF_like to deliver all packets.

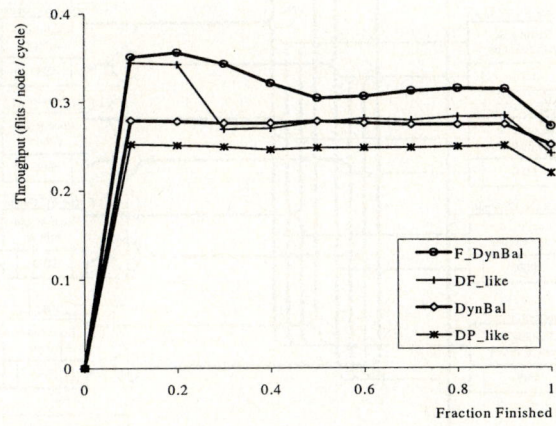

Fig. 3. Throughput during the delivery of a static load

The intensive bubbles introduced by DP_like, especially when more buffers have already resided some flits of other packets, cause inefficient use of bandwidth. This phenomenon is more typical for DF_like. Both DF_like and F_DynBal reach nearly the same delivery rate after 10% of the packets are drawn, but the rate is forced to slow down more quickly for DF_like because more "cyclic-channel" buffers need to be emptied before accepting other packets. This leads to the better performance of F_DynBal over DF_like under static workload.

4 ASIC Design of Routers

The implementation cost in the chip area and timing is extracted after modeling the routing algorithms and synthesis onto ASIC technology. All routers discussed are modeled similarly to the F_DynBal shown in Fig.4, depending on the number of VCs used for deadlock avoidance. We assume routing on a two dimensional torus with a physical link for each direction. For clarity, only part of the router, the schematic graph on the positive direction along x dimension and the local node interface (injection and ejection) is presented. The switch is used as the data path from input lanes to output lanes and is implemented as a multiplexer on the side of output channels. The number of physical link bits (phits) is assumed equal to that of flow control bits (flits). The header flit of a packet contains eight bits for the destination node address (x, y) and two bits for the header/tail tag. The buffer depth unit is a bit, and the buffer width is the same as a flit. It has a centralized output channel status monitor and an optimal path generator. These facilitate the parallelization of header decoding and path selection. There is a round-robin arbiter sitting on each output virtual channel and the physical channel, and a routing and path selection controller on each input virtual channel.

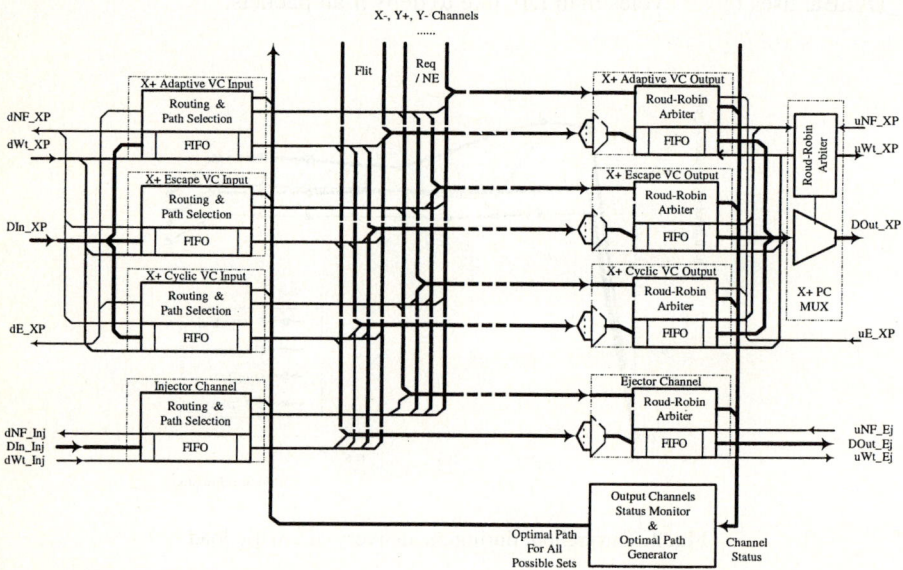

Fig. 4. The Schematic graph of the XP channel and the interface to the local injector/ejector of the F_DynBal router

The model is then described in Verilog HDL and the logic can then be verified by using simulation tools like SILOS. Finally, the design can be mapped into a specific process technology (LSI G11-p here) to be synthesized and optimized for area cost and clock frequency.

The chip area/clock timing for each of the four wormhole routers: DynBal, DP_like, F_DynaBal and DF_like is: 155K cells/1.25ns, 154K cells/1.25ns, 229K cells/1.47ns, and 223K cells/1.47ns respectively. Thirteen lanes are used for fully adaptive algorithms, nine lanes for the other two algorithms – one lane per VC and one lane for ejector/injector. No lanes used for performance enhancement, as it is too expensive as shown in [13].

5 The Cost-Effective Router

By combining the performance simulation results with the synthesis results, the cost-effectiveness of the routing algorithms under various workloads can be compared.

5.1 Uniform Traffic

The cost-effectiveness of various algorithms discussed is shown in Fig. 5. The DP_like and DF_like routing perform worse than DynBal and F_DynBal, respectively, because of both the bottleneck and the bubbles. Compared with the performance evaluation (Fig. 2), although the performance simulation shows that fully adaptive routing has much higher throughput than non-fully adaptive, it becomes much less attractive when the cost and speed are considered [9].

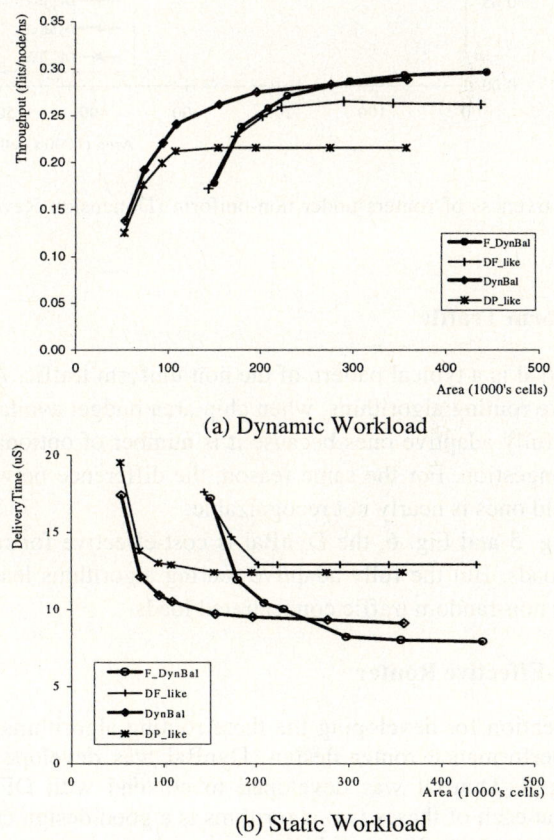

Fig. 5. Cost-effectiveness of routers under uniform workload with 8-flits maximal packets

Compared with the dynamic workload performance, the F_DynBal router gets much better performance than DF_like under static workload, as shown in Fig. 5b.

F_DynBal may use as low as 60% of the time for DF_like router to drain off the packets.

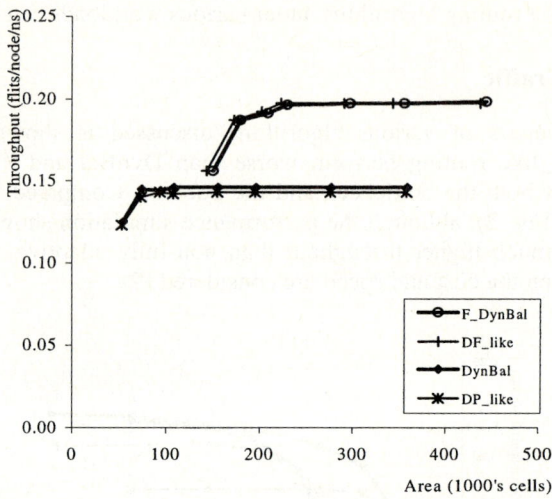

Fig. 6. Cost-effectiveness of routers under non-uniform (Dimension Reversal) traffic (8-flits-packet)

5.2 Non-uniform Traffic

Dimension reversal is a typical pattern of the non-uniform traffic. As shown in Fig. 6, the fully adaptive routing algorithms, when chip area budget available, perform much better than non-fully adaptive ones because it is number of optional paths lead to the avoidance of congestion. For the same reason, the difference between the new algorithms and the old ones is nearly not recognizable.

Compared Fig. 5 and Fig. 6, the DynBal is cost-effective for random traffic concentrated workloads. But the fully adaptive routing algorithms lead to a much better performance for non-random traffic concentrated loads.

5.3 The Cost-Effective Router

The original intention for developing the three routing algorithms was to find a low cost yet high performance router design. DynBal was developed to contend with DP_like routing; F_DynBal was developed to contend with DF_like routing. We demonstrated that each of theses two algorithms is a good design choice between two different chip area budgets: low, and high respectively.

A SoC multi-computer is plausible following the trends of process technology. If EZ4102 TinyRISC from LSI Logic is used as a local node PE, each node (PE and router) takes 6.18 mm^2 if using the same ASIC technology, promising a SoC with tens even hundred nodes.

6 Discussion

The main goal of this paper is to evaluate the cost-effectiveness of router designs for the domain of torus buffered wormhole routing. This has been accomplished in a systematic manner. First, a new scheme was presented to enhance the performance of previous algorithms. Then, performance was evaluated under various workloads by using a flit-level, cycle-driven simulator. Finally, routers based on state-of-the-art ASIC technology were constructed for detailed cost and speed evaluation. The cost-effective router is recommended based on two chip area budgets. We suggest that the DynBal be used as a low cost yet high performance router for random traffic dominated workloads and that the F_DynBal be used as a high cost yet high performance router for broader band traffic.

The routers implementing the various routing algorithms are constructed based on the same canonical model and the same state-of-the-art ASIC technology: this results in a fair comparison. As micron technology changes following Moore's Law, we can implement buffered wormhole algorithms and more complicated control as more gates are available, but there is still a challenge in making the routers faster to meet the needs of higher speed local nodes.

In the current simulator, a workload is generated according to the communication pattern specified and the mapping style – either dynamic or static. More realistic workloads are preferable. One way to accomplish this is to let a top level massively parallel processing (MPP) simulator control the execution of the cycle-driven network simulator and measure the performance under a real workload environment established by the top level.

References

1. Chien, A. A. and Kim, J. H., Planar-Adaptive Routing: Low-cost Adaptive Networks for Multiprocessors, *Proceedings of the 19th International Symposium on Computer Architecture*, IEEE Computer Society, May 1992, 268-277
2. Dally, W. J. Virtual channel flow control. *IEEE Trans. On Parallel and Distributed Systems* Vol. 3, No. 2, March, 1992, 194-205
3. Duato, J., Yalamanchili, S., and Ni, L. *Interconnection Networks: An Engineering Approach* (IEEE Computer Society Press, Los Alamitos, CA, 1997)
4. Pinkston, T.M., Flexible and Efficient Routing Based on Progressive Deadlock Recovery, *IEEE Transactions on Computers*, Vol. 48, No. 7, July 1999, 649-669.
5. Dally, W. J., and Seitz, C. L. The torus routing chip. *Distributed Computing 1* (1986), 187-196
6. Dally, W. J., Deadlock-Free Adaptive Routing in Multicomputer Networks Using Virtual Channels. *IEEE Trans. On Parallel and Distributed Systems* Vol.4, No.4, 1993, 466-475
7. Duato, J., A Necessary and Sufficient Condition for Deadlock-free Adaptive Routing in Wormhole Networks, *IEEE Trans. on Parallel and Distributed Systems*, Vol.6, No.10, 1995, 1055-1067
8. Ould-Khaoua, M., A Performance Model for Duato's Fully Adaptive Routing Algorithm in k-Ary n-Cubes, *IEEE Transactions on Computers*, Vol. 48, No. 12, Dec. 1999, 1297-1304.

9. Scott, S., and Thorson, G. The Cray T3E Network: Adaptive Routing in a High Performance 3D Torus, *Proc. of HOT Interconnects Symposium IV*, Stanford U., Aug. 1996, 147-156.
10. Dally, W.J., Dennison L.R. and etc., Architecture and Implementation of the Reliable Router, *Proceedings of Hot Interconnects II*, Palo Alto, CA, Aug. 1994, 122-133
11. Ge, J., An Efficient Buffered Wormhole Routing Scheme, Proceedings of the IASTED International Conference on Communications and Computer Networks, Cambridge (MIT), MA. November 4-6, 2002, pp281-286.
12. Scott, S., and Thorson, G. Optimized routing in the Cray T3D. In *Proc. Of the Workshop on Parallel Computer Routing and Communication* (1994), 281-294
13. Herbordt, M. C., Ge, J., Sanikp, S., Olin, K., Le, H., Design Trade-Offs of Low-Cost Multicomputer Network Switches, *Proceedings of the 7th Symposium on the Frontiers of Massively Parallel Computation*, Annapolis, MD, Feb. 1999, 25-34

Efficient Routing and Broadcasting Algorithms in de Bruijn Networks

Ngoc Chi Nguyen, Nhat Minh Dinh Vo, and Sungyoung Lee

Computer Engineering Department, Kyung Hee Univerity,
1, Seocheon, Giheung, Yongin, Gyeonggi 449-701 Korea
{ncngoc, vdmnhat, sylee}@oslab.khu.ac.kr

Abstract. Recently, routing on dBG has been investigated as shortest path and fault tolerant routing but investigation into shortest path in failure mode on dBG has been non-existent. Furthermore, dBG based broadcasting has been studied as local broadcasting and arc-disjoint spanning trees based broadcasting. However, their broadcasting algorithms can only work in dBG(2,k). In this paper, we investigate shortest path routing algorithms in the condition of existing failure, based on the Bidirectional de Bruijn graph (BdBG). And we also investigate broadcasting in BdBG for a degree greater than or equal to two[1].

1 Introduction

For routing in dBG, Z. Liu and T.Y. Sung [1] proposed eight cases shortest paths in BdBG. Nevertheless, Z. Liu's algorithms do not support fault tolerance. J.W. Mao [4] has also proposed the general cases for shortest path in BdBG (case RLR or LRL). For fault tolerance issue, he provides another node-disjoint path of length at most $k + log_2 k + 4$ (in dBG(2,k)) beside shortest path. However, his algorithm can tolerate only one failure node in binary de Bruijn networks and it cannot achieve shortest path if there is failure node on the path. Broadcasting problems on dBG have been investigated as local broadcasting[6] and arc-disjoint spanning trees[7][8]. Nonetheless, the above can only work in a binary de Bruijn network (dBG(2,k)).

Considering limitations of routing and broadcasting in dBG, we intend to investigate shortest path routing in the condition of failure existence and broadcasting in BdBG with a degree greater than or equal to two. Two Fault Free Shortest Path (FFSP) routing algorithms and one broadcasting algorithm (for one-to-all broadcasting in the all port communication model) are proposed. Time complexity of FFSP2 in the worst case is $0(2^{\frac{k}{2}+1}d)$ in comparison with $0((2d)^{\frac{k}{2}+1})$ of FFSP1 (in dBG(d,k) and k=2h). Our study shows that the maximum time steps to finish broadcast procedure is k regardless of the broadcast originator, time complexity at each node is $0(\frac{3}{2}d)$, and no overhead happens in the broadcast message.

---

[1] This research was partially supported by ITRC project of Sunmoon University.

The rest of this paper is organized as follows. Background is discussed in section 2. In section 3, FFSP routing algorithms are presented. Performance analysis for FFSP routing algorithms is carried in section 4. Section 5 discuss about broadcasting algorithm in dBG(d,k). Finally, some conclusions will be given in Section 6.

2 Background

The BdBG graph denoted as BdBG(d,k)[1] has N=d^k nodes with diameter k and degree 2d. If we represent a node by $d_0d_1...d_{k-2}d_{k-1}$, where $d_j \in 0,1,...,(d-1)$, $0 \leq j \leq (k-1)$, then its neighbor are represented by $d_1...d_{k-2}d_{k-1}p$(L neighbors, by shifting left or L path) and $pd_0d_1...d_{k-2}$(R neighbors, by shifting right or R path), where $p = 0, 1, ..., (d-1)$. We write if the path $P = R_1L_1R_2L_2$ consists of an R-path called R_1, followed by an L-path called L_1, an R-path called R_2, an L-path called L_2, and so on, where subscripts are used to distinguish different sub-paths. Subscripts of these sub-paths can be omitted if no ambiguity will occur, e.g., $P = R_1LR_2$ or P=RL. Shift string of a node A is a binary string (0 for left shift and 1 for right shift) which represents path from originator to A.

For simplest broadcasting mechanism, the originator initiates the process by making a "call" to other neighboring vertices in the graph informing them of the message. Subsequently, the informed vertices call their neighboring vertices and the process continues until all vertices in the graph are informed. Basically, this mechanism is like flooding phenomenon. Note that the interval during which a call takes place will be referred to as a time step or simply step. In flooding broadcasting (FB), level of a node A is the number of steps by which a message from originator reaches A (or shortest path length between A and originator).

The following fig. 1a shows us an example for BdBG(2,4). Fig. 1b shows us eight cases of shortest path routing on BdBG. The gray areas are the maximum substring between source (s) and destination (d). The number inside each block represents the number of bits in the block.

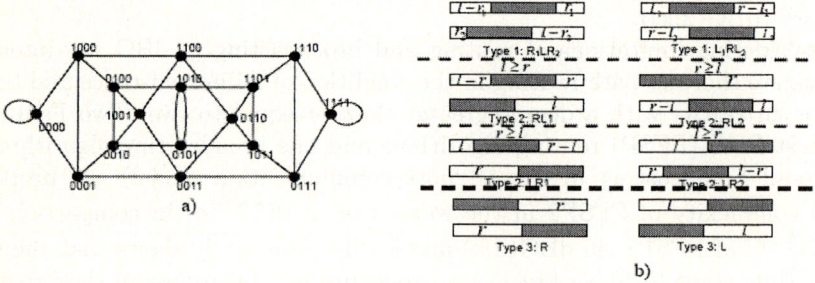

Fig. 1. a)The BdBG(2,4); b)Shortest path types[1]

3 Fault Free Shortest Path Routing Algorithms

In order to provide shortest path in the condition of failure existing, several paths of a specific source destination pair must be provided. Then FFSP is found among those paths. Therefore, the following concepts are proposed. For those, we assume that there is a separately protocol which detects failure nodes and then let other nodes know in periodically.

Definition 1: *the level m^{th} discrete set (DS_m) is a set which contains all neighbors of each element in discrete set level m-1; in the constraint that there is no existent element of discrete set level m coincides with another element of discrete set level q^{th} ($q \leq m$) or failure node set.*

Lemma 1: DS_m *is fault free.*

Lemma 2: *all the neighbors of a node belong to DS_m are in DS_{m-1}, DS_m and DS_{m+1}, except failure nodes.*

Proof: obviously we see that DS_1 and DS_2 contain all the neighbors of DS_1 except failure nodes; DS_1, DS_2 and DS_3 contain all the neighbors of DS_2 except failure nodes. So Lemma 2 is right at m=1,2. Assuming that lemma 2 is right until p, now we prove it is right at p+1. Suppose it's wrong at p+1. That means there exist a neighbor A of an element B$\in DS_{p+1}$, and A $\in DS_i, i < p$. Because lemma 2 is right until p, hence all the neighbors of A are in DS_{i-1}, DS_i and DS_{i+1} except failure nodes. Therefore, there exists an element B'$\in DS_{i-1}, DS_i$ or DS_{i+1}, and B'=B. It contradicts with definition 1. So Lemma 2 is right at p+1. Following inductive method, lemma 2 is proved.

Lemma 3: *there exists no neighbor of any element of DS_m, which is a duplicate of any element of DS_h, $\forall h \leq m\text{-}2$.*

Proof: suppose there is a neighbor A of an element B $\in DS_m$ duplicates with an element A' of DS_h (h $\leq$ m-2). Following Lemma 2, all the neighbors of A' are in DS_{h-1}, DS_h and DS_{h+1}. Therefore, there must exist a neighbor B' of A' in level h-1 or h or h+1, and B'=B. It contradicts with definition 1.

Corollary 1: *for duplicate checking at the next level of DS_q, it is not necessary to check with any element of $DS_m, \forall m \leq q\text{-}2$.*

By assigning source node S to DS_1, then expanding to the higher level, we have the following theorem.

Theorem 1: *in $BdBG(d,k)$, we can always find a FFSP from node S$\in DS_1$ to node $A_x \in DS_x$ ($\forall x \leq k$), if it exists.*

Proof: we use inductive method to prove this theorem. When x=1, 2, theorem 1 is right. Assuming that theorem 1 is right until m, m $\leq$ k. Now we prove it is right until m+1. Suppose that path from S to A_{m+1} is not the FFSP. Then we have the following cases,

- There exist $A_p \in DS_p$, $A_p = A_{m+1}$ and $p < m+1$. It contradicts definition 1.
- There exists a FFSP, $S \to B_1 \to B_2 \to ... \to B_k \to ... \to B_z \to ... \to A_{m+1}$, and B_k, $B_{k+1},..., B_z$ not belonging to any DS_i ($\forall i \le m+1$). Because $B_{k-1} \in DS_j$ ($j \le m+1$). Following Lemma 2, all the neighbors of B_{k-1} are in DS_{j-1} or DS_j or DS_{j+1}, except failure nodes. Therefore, B_k must be a failure node.

→Theorem 1 is right at m+1. Theorem 1 is proved.

Corollary 2: *path length of a path from $S \in DS_1$ to $A_x \in DS_x$ is x-1.*

Fault free shortest path algorithm 1 (FFSP1) is proposed as a result of theorem 1 (shown in fig. 2a). It can always find FFSP in all cases (fault free mode, arbitrary failure mode) if the network still remain connected.

Proof of FFSP1: suppose path $s \to ... \to a_{ip} \to b_{jk} \to ... \to d$ is not FFSP, and then we have the following cases,

- There exist a FFSP $s \to ... \to a_{i'p'} \to b_{j'k'} \to ... \to d$ (i'≤i, j'≤j). It contradicts with the above assumption that a_{ip} and b_{jk} are the first neighbors between discrete sets A and B.
- There exist a FFSP $s \to ... \to a_{i'p'} \to c_1 \to ... \to c_m \to b_{j'k'} \to ... \to d$ ($i' < i, j' < j$), and $c_1, c_2, ..., c_m$ do not belong to any discrete set A_p or B_q (p≤i, q≤j). Due to $a_{i'p'} \in A_{i'}$ and following lemma 2, all the neighbors of $a_{i'p'}$ are in $A_{i'-1}, A_{i'}$ and $A_{i'+1}$ except failure nodes. Therefore c_1 must be a failure node.

Example 1: we want to find a FFSP from source 10000 to destination 01021, failure node 00102 (dBG(3,5)).

Applying FFSP1, we have, A_1 = (10000) B_1 = (01021) A_2 = (00000, 00001, 00002, 01000, 11000, 21000) B_2 = (10210, 10211, 10212, 00102, 10102, 20102).

However, 00102 is a failure node. So B_2=(10210, 10211, 10212, 10102, 20102). A_3 = (20000, 00010, 00011, 00012, 00020, 00021, 00022, 10001, 10002, 00100, 10100, 20100, 01100, 11100, 21100, 02100, 12100, 22100).

Then we find that 02100 and 10210 in A_3 and B_2 are the first neighbors. FFSP is found by tracking back from 02100 to 10000 and 10210 to 01021. We have FFSP 10000 → 21000 → 02100 → 10210 → 01021. In this example, FFSP1 can provide 2 shortest paths (in the case of no failure node) 10000 → 21000 → 02100 → 10210 → 01021 and 10000 → 00001 → 00010 → 00102 → 01021. We pick up one FFSP 10000→21000→02100→10210→01021 (node 00102 is fail).

Furthermore, we shall see that other elements like 00000, 00002, 01000, 11000 in A_2 are useless in constructing a FFSP. So, eliminating these elements can reduce the size of A_3 (reduce the cost at extending to next level) and improve the performance of our algorithm. It shows the motivation of FFSP2. Before investigating FFSP2, we give some definition and theorem.

Definition 2: *a dominant element is an element which makes a shorter path from source to a specific destination, if the path goes through it.*

Example 2: from the above example 1 we have 2 shortest paths (in the case 00102 is not a failure node) $10000 \to 21000 \to 02100 \to 10210 \to 01021$ and $10000 \to 00001 \to 00010 \to 00102 \to 01021$. Thus 00001 and 21000 are dominant elements of A_2, because they make shorter path than others of A_2.

Therefore, by eliminating some non-dominant elements in a level, we can reduce the size of each level in FFSP1 and hence, improve the performance of FFSP1. A question raised here is how we can determine some dominant elements in a DS_k and how many dominant elements, in a level, are enough to find FFSP. The following theorem 2 is for determining dominant elements and corollary 3 answer the question, how many dominant elements are enough.

Theorem 2: *If there are some elements different in 1 bit address at leftmost or rightmost, the dominant element among them is an element which has shorter path length toward destination for cases RL2, R (shown in fig. 1b) for leftmost bit difference and LR2, L for rightmost bit difference.*

Proof: as showing in fig. 1b, there are eight cases for shortest path. Only four cases RL2, R, LR2 and L make different paths when sources are different in leftmost bit or rightmost bit.

Example 3: following example 1, we check the dominant characteristic of three nodes A 01000, B 11000 and C 21000 (in A_2) to destination D 01021. Three nodes A, B and C are leftmost bit difference. So, type RL2, R are applied.

• Apply type R: the maximum match string between A 01000 and D 01021 is 0, between B **1**1000 and D 01021 is 1, and between C **21**000 and D 0**1**0**21** is 2 → min path length is 3, in case of node C.

• Apply type RL2: the maximum match string [5] between A **0**1000 and D **0**1021 is 1 (path length: 6), between B **1**1000 and D 0**1**021 is 1 (path length: 7), and between C **21**000 and D 0**1**0**21** is 2 (same as case R) → min is 3, node C.

Therefore, minimum path length is 3 and dominant element is C.

Corollary 3: *when we apply theorem 2 to determine dominant elements, the maximum elements of DS_{m+1} are $2p$ (p is the total elements of DS_m).*

Proof: the maximum elements of DS_{m+1} by definition 1 are 2pd (dBG(d,k)). We see that in 2pd there are 2p series of d elements which are different in 1 bit at leftmost or rightmost. By applying theorem 2 to DS_{m+1}, we obtain 1 dominant element in d elements differed in 1 bit at leftmost or rightmost.

Fault Free Shortest Path Algorithm 2 (FFSP2) is proposed in fig. 2b.

The condition in line 5 and line 8 (fig. 2a, 2b) let us know whether there exists a neighbor of array A and B of discrete set, $\forall a_{ip} \in$A[i],$\forall b_{jk} \in$B[j] . The SPD(M) function, line 14 fig. 2b, finds the next level of DS M (DS N) and eliminates non-dominant elements in N followed theorem 2. Expand(M) function, line 14 fig. 2a, finds the next level of DS M. Pathlength type p function, line 19,23 fig. 2b, checks path length followed type p of each element in T toward destination. Eliminate function, line 20, 24, eliminates element in T, which has longer path length than

```
 1    DECLARE two array A[n] and B[n] of DS
        type
 2    SET A[1] to s
 3    SET B[1] to d
 4    SET i and j to 1
 5    WHILE (a_ip is not neighbor of b_jk)
 6      CALL Expand function of A[i] and return
            value to A[i+1]
 7      i = i+1
 8      IF (a_ip is neighbor of b_jk) THEN
 9        Exit while loop
10      ENDIF
11      CALL Expand function of B[j] and return
            value to B[j+1]
12      j=j+1
13    ENDWHILE

14    DEFINE Expand function of a DS M
15      DECLARE array N of DS type
16      SET N to DS of all neighbors of each
            element in M
17      CALL duplicate_check(N)
18      RETURN(N)
19    ENDDEFINE
```

a)

```
 1    DECLARE two array A[n] and B[n] of DS type
 2    SET A[1] to s
 3    SET B[1] to d
 4    SET i and j to 1
 5    WHILE (a_ip is not neighbor of b_jk)
 6      CALL SPD function with A[i] and return value to A[i+1]
 7      i = i+1
 8      IF (a_ip is neighbor of b_jk) THEN
 9        EXIT while loop
10      ENDIF
11      CALL SPD function with B[j] and return value to B[j+1]
12      j=j+1
13    ENDWHILE

14    DEFINE  SPD function of a DS M
15      DECLARE array N of DS type
16      SET N to DS of all neighbors of each element in M
17      WHILE (there exist set of elements T which differ in 1 bit
              at leftmost or rightmost in N)
18        IF (T is leftmost bit difference) THEN
19          CALL Pathlength function type RL2 and R
20          CALL Eliminate function to eliminate non-dominant
                elements
21        ENDIF
22        IF (T is rightmost bit difference) THEN
23          CALL Pathlength function type LR2 and L
24          CALL Eliminate function to eliminate non-dominant
                elements
25        ENDIF
26      ENDWHILE
27      CALL duplicate_check(N)
28      RETURN(N)
29    ENDDEFINE
```

b)

Fig. 2. a)Fault Free Shortest Path Algorithm 1 (FFSP1); b)Fault Free Shortest Path Algorithm 2 (FFSP2)

the other. The duplicate_check(N) function, line 17 fig. 2a and line 27 fig. 2b, check if there is a duplicate of any element in N with other higher level DS of N. For duplication checking, we use the result from corollary 1. Then, we get FFSP by going back from a_{ip} to s and b_{jk} to d.

Example 4: we try to find FFSP as in example 1. By applying FFSP2, we have, $A_1 = (10000)$ $B_1 = (01021)$ $A_2 = (00001, 21000)$ $B_2 = (10210, 00102)$. However, 00102 is a failure node. So B_2 becomes (10210).

$A_3 = (00010, 10000, 10001, 02100)$. However, node 10000 coincides with 10000 of A_1. So A_3 becomes (00010, 10001, 02100). Then we find that 02100 and 10210 in A_3 and B_2 are the first neighbors. FFSP is found by tracking back from 02100 to 10000 and 10210 to 01021. We have FFSP 10000 → 21000 → 02100 → 10210 → 01021.

4 Performance Analysis for FFSP1 and FFSP2

Mean path length is the significant to analyze and compare our algorithm to others. Z. Feng and Yang [2] have calculated it based on the original formula, Mean path length = $\frac{Total internal traffic}{Total external traffic}$ for their routing performance. We can use the above equation to get the mean path length in the case of failure. We assume that failure is random, and our network is uniform. That means the probability to get failure is equal at every node in the network.

Table 1 shows the results in the simulation of mean path length using six algorithms, SCP[3], RFR, NSC, PMC[2], FFSP1 and FFSP2. Our two algorithms show to be outstanding in comparison with the four algorithms. They always achieve shorter mean path length than the other algorithms.

Table 1. Mean path length of FFSP1, FFSP2 in comparison with others

d,k	no. of nodes	Mean path lengths								
		SCP (fault free mode)	RFR (fault free mode)	NSC (fault free mode)	PMC (fault free mode)	FFSP1/2 (fault free mode)	FFSP1 (1 node fail)	FFSP2 (1 node fail)	FFSP1 (2 nodes fail)	FFSP2 (2 nodes fail)
2,2	4	1.167	1.167	1.167	1.167	1.167	1.333	1.333	1	1
2,3	8	1.643	1.643	1.643	1.643	1.643	1.762	1.762	1.733	1.733
2,4	16	2.258	2.188	2.146	2.142	2.142	2.1	2.21	2.286	2.286
2,5	32	2.984	2.796	2.794	2.766	2.754	2.787	2.794	2.285	2.832
2,6	64	3.801	3.653	3.551	3.495	3.453	3.467	3.483	3.487	3.506
3,2	9	1.417	1.471	1.417	1.417	1.417	1.429	1.429	1.476	1.476
3,3	27	2.128	2.105	2.007	2.077	2.077	2.095	2.095	2.117	2.117
3,4	81	2.978	2.911	2.865	2.849	2.833	2.84	2.842	2.846	2.848
3,5	243	3.907	3.800	3.755	3.736	3.674	3.676	3.696	3.678	3.698
3,6	729	4.875	4.775	4.704	4.722	4.572	4.573	4.626	4.573	4.627
4,2	16	1.550	1.550	1.550	1.550	1.550	1.552	1.552	1.56	1.56
4,3	64	2.369	2.343	2.321	2.321	2.321	2.327	2.327	2.335	2.335
4,4	256	3.298	3.251	3.214	3.218	3.178	3.18	3.185	3.184	3.189
4,5	1024	4.273	4.207	4.172	4.209	4.1	4.101	4.13	4.101	4.13
5,2	25	1.633	1.633	1.633	1.633	1.633	1.634	1.634	1.636	1.636
5,3	125	2.510	2.491	2.471	2.471	2.471	2.473	2.473	2.476	2.476
5,4	625	3.471	3.438	3.41	3.437	3.378	3.379	3.385	3.379	3.385
6,2	36	1.690	1.690	1.690	1.690	1.690	1.691	1.691	1.693	1.693
6,3	216	2.601	2.585	2.570	2.570	2.570	2.571	2.571	2.573	2.573

This section is completed with study in time complexity of our algorithms. As A. Sengupta [9] has shown that dBG(d,k) has connectivity of d-1. Hence, our time complexity study is based on assumption that the number of failures is at most d-1 and our study is focused on large network with high degree (d>>1). Therefore, diameter of our network in this case is k. We have the following cases,

• For FFSP1, the second level DS lies in the complexity class $0(2d)$, the third level DS lies in the complexity class $0(2d(2d-1)) \approx 0(4d^2)$, the fourth lies in $0(2d(2d-1)^2) \approx 0(8d^3)$, etc... Hence, time complexity of FFSP1 lies in the complexity class $0((2d)^n)$, the value of n equals to the maximum level DS provided by FFSP1. In the worst case, time complexity of FFSP1 lies in $0((2d)^{\frac{k}{2}+1})$ (k=2h), or $0((2d)^{\frac{k+1}{2}})$ (k=2h+1), k is maximum path length from source to destination (the diameter).

• The computation time of FFSP2 can be divided into 2 parts. One is performing computation on expanding to next level, checking for duplicate and neighboring checking between DS A[m] and B[q]. This part is like FFSP1, the difference is that each DS here grows following a geometric progression with common quotient 2 and initial term 1 (as shown in corollary 3). The other part is performing computation on finding dominant elements. Hence, the second level DS lies in the complexity class $0(2+2d) \approx 0(2d)$, the third level DS lies in the complexity class $0(4+4d) \approx 0(4d)$, the fourth lies in $0(8+8d) \approx 0(8d)$, etc... Hence time complexity of FFSP2 lies in the complexity class $0(2^n d)$, the value of n equals to the maximum level DS provided by FFSP2. FFSP2 would cost us $0(2^{\frac{k}{2}+1}d)$ (k=2h), or $0(2^{\frac{k+1}{2}})$ (k=2h+1) time in the worst cases, k is maximum path length from source to destination (the diameter).

5 Broadcasting Algorithm in dBG(d,k)

By applying FB, we can easily obtain k as the maximum number of steps to finish broadcasting. However, message overhead is very high in FB. Thus, how to reduce message overhead (or letting each informed vertices call its uninformed neighbors only) in FB states the motivation for our algorithm. We assume that each packet sent to the other node must contain originator address, sender's level, a shift string of receiver and all calls take the same amount of time.

There are two cases of message overhead when an informed node A wants to inform node X. Case 1, node X has been informed already. Thus, X must have lower or equal level to A. Case 2, uninformed node X can be informed by nodes B,C,D, which have the same level as A, at the same time. For case 1, we need to compare the shortest-path length between X and A to originator. And X is informed by A if X level is higher than A's level and case 2 not happen. For case 2, we have to define some conditions, based on these conditions only A or B or C or... inform X. The following theorems are proposed for calculating path length.

Theorem 3: *given p is shortest-path length between node a and b, the minimum length of matched strings between a and b is k-p (dBG(d,k)).*

Proof: as shown in fig. 1b, there are 3 types for determining shortest path (R,L; RL,LR; $R_1 L R_2, L_1 R L_2$). The minimum matched string[5] can be obtained in type R,L among them. And length for this minimum matched string is k-p.

Theorem 4: *path length between node s and d is $min(2s_j + s_i + d_i, 2s_i + s_j + d_j)$, where s_i and d_i are the left indices, and s_j and d_j are the right indices of matched string in s and d respectively.*

Proof: path length $2s_j + s_i + d_i$, $2s_i + s_j + d_j$ are for case $R_{s_j} L_{s_j + s_i} R_{d_i}$ and $L_{s_i} R_{s_i + s_j} L_{d_j}$ respectively. These cases are the general cases for 3 types presented in fig. 1b(ex. if $s_i, s_j, d_i, d_j \neq 0$, they become type $R_1 L R_2$ and $L_1 R L_2$).

To solve the above two cases of message overhead, a Boolean valued function SPL is proposed. SPL has inputs: originator S, current node P, neighboring node X, current level n (level of P), shift string Q ($q_0 q_1 q_2 ... q_{z-1}$, length $z \leq k$) (from S to X through P). Fig. 3a shows SPL algorithm. Step 1,2,3 solve message overhead of case 1. Step 1 is a result of theorem 3. Step 4,5,6 solve case 2 message overhead. In case 2, we have several shortest paths from S to X. One shortest path must be chosen based on the following conditions:

- Shortest path corresponds with shortest matched string of S and X(step5).
- In the case, there exist 2 shortest path from the first condition. Then, shortest path which begin with shifting right is chosen. (step 6)

Step 7 compares shift string Q to the condition gotten from step 5 and 6 to determine whether X should be informed or not.

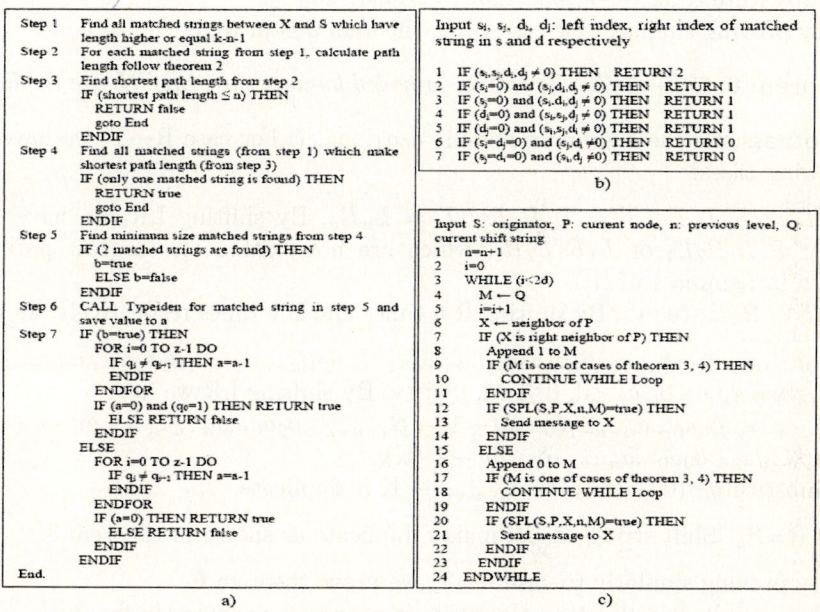

Fig. 3. a)SPL function algorithm; b)Typeiden function algorithm; c)Broadcasting algorithm for dBG(d,k)

Example 5: in dBG(3,10), given input S: 0012111001, P: 0111012110, n=7, X: 1110121100, Q=01111100. By applying SPL, we have

Step 1: find all matched strings[5] which have length higher or equal 10-7-1=2. These strings are 11, 111, 1110, 01, 012, 0121, 01211, 110, 1100.
Step 2: path lengths for strings in step 1 are 12, 10, 8, 14, 12, 10, 8, 13, 11.
Step 3: shortest path length is 8.
Step 4: matched string, which make shortest path length 8, are 1110, 01211.
Step 5: minimum size string from step 4 is 1110, b=false.
Step 6: Typeiden(input $s_i = 0, s_j = 6, d_i = 4, d_j = 2$)→returned value: 1,a=1.
Step 7: there are 2 places in Q in which two adjacent bits are different → a=-1 $\neq$0. Consequently, X is an uninformed node (step 3,8>n), but it isn't informed by P (message overhead case 2) due to our priority given in step 5 and 6.

If we apply SPL for all 2d neighbors of one node, then it cost O(2d) for running our algorithm. The following theorems reduce from O(2d) to O(1.5d). Following are some notations used, where T is the previous shifting string.

R↔T: total number of right shift in T > total number of left shift in T
L↔T: total number of left shift in T > total number of right shift in T

Theorem 5: *by shifting RLR/LRL, results are duplicate with shifting R/L.*

Proof: given a node $a_0a_1...a_{n-1}$. By shifting RLR in dBG(d,k), we have $a_0a_1...a_{n-1} \rightarrow \alpha a_0 a_1...a_{n-2} \rightarrow a_0a_1...a_{n-2}\beta \rightarrow \gamma a_0a_1...a_{n-2}$, $0 \leq \alpha, \beta, \gamma < d$.
Substitute α for $\gamma \rightarrow \gamma a_0a_1...a_{n-2} \equiv \alpha a_0a_1...a_{n-2}$.
By proving similarly for case LRL, theorem 5 is proved.

Theorem 6: *if R↔T/L↔T, results provided by next shift LR/RL are duplicate.*

Proof: assume the beginning node is $a_0a_1...a_{n-1}$. For case R↔T, we have the following cases:

• $T = R_uL_vR_w, T = L_uR_vR_w, T = L_uR_v$. By shifting LR, we have shift string $R_1L_1R_2L_2$ or $L_1R_1L_2R_2$, which are not existed for shortest path (as shown in Lemma 1 of [1]).

• $T = R_uL_v$ (u>v). By shifting R u times and L v times respectively, we have

$a_0a_1...a_{n-1} \rightarrow \beta_{u-1}...\beta_1\beta_0a_0a_1...a_{n-u-1} \rightarrow \beta_{u-v-1}...\beta_1\beta_0a_0a_1...a_{n-u-1}\delta_0\delta_1$
$...\delta_{v-1}$ where $0 \leq \beta_i, \delta_j < d$, $0 \leq i < u$, $0 \leq j < v$. By shifting LR we have,
$\beta_{u-v-1}...\beta_0a_0...a_{n-u-1}\delta_0...\delta v - 1 \rightarrow \beta_{u-v-2}...\beta_0a_0...a_{n-u-1}\delta_0...\delta v \rightarrow \gamma\beta_{u-v-2}...\beta_0a_0...a_{n-u-1}\delta_0...\delta v - 1$ (K)
Substitute $\gamma(0 \leq \gamma < d)$ for $\beta_{u-v-1} \rightarrow$ K is duplicate.

• $R=R_u$. Shift string R_uLR makes duplicate as shown in theorem 3.

By proving similarly to case L↔T, we prove theorem 6.
As a result, broadcasting algorithm is proposed as shown in fig. 3c.

Theorem 7: *in the worst case, time complexity for our broadcasting algorithm is O(1.5d).*

Proof: probability for theorem 5 happening is 25%, and for theorem 6 is less than 25%. Therefore, the probability for CONTINUE command (line 10, 18 fig. 3c) happening is 25%. So, theorem 7 is proved.

6 Conclusion

We have proposed new concepts, routing algorithms and distributed broadcasting algorithm in dBG(d,k). Our routing algorithms can provide shortest path in the case of failure existence. Our simulation result shows that FFSP2 is an appropriate candidate for the real networks with high degree and large number of nodes, while FFSP1 is a good choice for high fault tolerant network with low degree and small/medium number of nodes. In broadcasting, our algorithm requires maximum k steps to finish broadcasting process in dBG(d,k). And there is no message overhead during broadcasting. Time complexity at each node is $0(\frac{3}{2}d)$. Therefore, the algorithms can be considered feasible for routing and broadcasting in real interconnection networks.

References

1. Zhen Liu, Ting-Yi Sung, "Routing and Transmitting Problem in de Bruijn Networks" IEEE Trans. on Comp., Vol. 45, Issue 9, Sept. 1996, pp 1056 - 1062.
2. O.W.W. Yang, Z. Feng, "DBG MANs and their routing performance", Comm., IEEE Proc., Vol. 147, Issue 1, Feb. 2000 pp 32 - 40.
3. A.H. Esfahanian and S.L. Hakimi, "Fault-tolerant routing in de Bruijn communication networks", IEEE Trans. Comp. C-34 (1985), 777.788.
4. Jyh-Wen Mao and Chang-Biau Yang, "Shortest path routing and fault tolerant routing on de Bruijn networks", Networks, Vol. 35, Issue 3, Pages 207-215 2000.
5. Alfred V. Aho, Margaret J. Corasick, "Efficient String Matching: An Aid to Bibliographic Search", Comm. of the ACM, Vol. 18 Issue 6, June 1975.
6. A.H.Esfahanian, G. Zimmerman, "A Distributed Broadcast Algorithm for Binary De Bruijn networks", IEEE Conf. on Comp. and Comm., March 1988.
7. E.Ganesan, D.K.Pradhan, "Optimal Broadcasting in Binary de Bruijn Networks and Hyper-deBruijn Networks", IEEE Symposium on Parallel Processing, April 1993.
8. S.R.Ohring, D.H.Hondel, "Optimal Fault-Tolerant Communication Algorithms on Product Networks using Spanning Trees", IEEE Symp. on Parallel Processing, 1994.
9. A. Sengupta, A.Sen, and S.Bandyopadhyay, "Fault tolerant distributed system design", IEEE Trans. Circuit Syst., Vol. CAS-35, pp. 168-172, Feb. 1988

Fault-Tolerant Wormhole Routing Algorithm in 2D Meshes Without Virtual Channels*

Jipeng Zhou[1] and Francis C.M. Lau[2]

[1] Department of Computer Science, Jinan University,
Guang Zhou 510632, P.R. China
jpzhoucn@sohu.com

[2] Department of Computer Science,
The University of Hong Kong, Pokfulam, Hong Kong, P.R. China
fcmlau@cs.hku.hk

Abstract. In wormhole meshes, many routing algorithms prevent deadlocks by enclosing faulty nodes within faulty blocks. None of them however can tolerate the convex fault model without virtual channels. We propose a deterministic fault-tolerant wormhole routing algorithm for mesh networks that can handle disjoint convex faulty regions. These regions would not contain any nonfaulty nodes. The proposed algorithm does not use any virtual channels, which routes the messages using an extended X-Y routing algorithm in the fault-free regions. The algorithm is deadlock- and livelock-free.

1 Related Work

The efficiency of routing algorithms is important for achieving high performance in multiprocessor systems. There are two types of routing methods: *deterministic routing*, which uses a single path from the source to destination; and *adaptive routing*, which allows more freedom in selecting message paths. Most commercial multiprocessor computers use deterministic routing because of its deadlock freedom and ease of implementation.

Numerous deterministic fault-tolerant routing algorithms [2][7] in meshes have been proposed in recent years, most of which augment the dimension-order routing algorithm to tolerate certain faults. Boppana and Chalasani proposed a fault-tolerant routing algorithm in mesh networks [2], or in mesh and torus networks [1]. The key idea of their algorithms is that, for each fault region, a fault ring or fault chain consisting of fault-free nodes and channels can be formed around it; if a message comes in contact with the fault region, the fault ring or chain is used to route the message around the fault region. Deadlocks can be prevented by using four virtual channels per physical channel for deterministic (dimension-order) fault-tolerant routing. Their fault model is rectangle or special convex. Sui and Wang [7] proposed a fault-tolerant routing algorithm using

---

* This research is supported by GDNSF (04300769), SRF for ROCS, SEM and JNU Grant (640581).

three virtual channels per physical channel to tolerate the rectangle fault model. Zhou and Lau [10] proposed a fault-tolerant wormhole routing algorithm using three virtual channels, which can tolerate the convex fault model. Tsai [8] and Ho [5] proposed fault-tolerant routing algorithms using two virtual channels.

Wu [9] proposed a fault-tolerant extended X-Y routing protocol, which is based on dimension-order routing and the odd-even turn model and does not use any virtual channels in 2D meshes. He uses extended faulty blocks (disjoint rectangles), consisting of connected unsafe and faulty nodes. Wu's protocol can be applied in 2D meshes with orthogonal faulty blocks (convex polygons). The extended X-Y routing protocol, however, cannot reach certain locations of faults and destinations, and their fault model may include nonfaulty nodes.

Our deterministic fault-tolerant wormhole routing algorithm can be contrasted to [2][7][8][9][10]. The convex fault model used would not include any nonfaulty nodes and would not prohibit routing to any desired location. The proposed fault-tolerant algorithm does not use virtual channels; it routes the messages by X-Y routing in fault-free region, and is deadlock- and livelock-free.

2 Fault Model and Disjoint Boundaries of Fault Regions

There are two different types of faults, either the entire processing element (PE) and its associated router can fail or a physical link may fail. When a PE and its router fail, all physical links incident on the failed PE are also marked as being faulty. Only faulty PEs are considered a basic faulty element, for simplicity, and the disjoint fault-connected regions are used as fault model in this paper. In order to find fault-connected regions, each PE should know the states (faulty or safe) of its encompassing PEs after running a diagnostic program. Two PEs $p_0 = (x_0, y_0)$ and $p_1 = (x_1, y_1)$ are called 4-neighbors if $|x_0 - x_1| + |y_0 - y_1| = 1$, 8-neighbors if $max\{|x_0 - x_1|, |y_0 - y_1|\} \leq 1$. Connectivity of faulty PEs is defined in terms of adjacency relation among faulty PEs.

Definition 1. *Given that C is the connectivity relation in a mesh M. For any pair of processors $p, q \in M$ and a given faulty processor set F, we have $(p, q) \in C$ if and only if $p, q \in F$ and there exist $a_1, a_2, \ldots, a_n \in F$ such that $a_1 a_2 \ldots a_n$ is a connected path in F, where $p = a_1, q = a_n$, and the connectivity is 8-connectivity; that is, a_{i+1} is a 8-neighbor of a_i, $1 \leq i \leq n-1$.*

Definition 2. *Let $d(p_1, p_2) = |x_1 - x_2| + |y_1 - y_2|$ denote the distance between PEs $p_1 = (x_1, y_1)$ and $p_2 = (x_2, y_2)$. Two fault regions, F_1 and F_2, are said to be disjoint if for any processor $p_1 = (x_1, y_1) \in F_1$ and $p_2 = (x_2, y_2) \in F_2$, $d(p_1, p_2) \geq 3$ when $x_1 = x_2$ or $y_1 = y_2$; $d(p_1, p_2) \geq 4$ when $x_1 \neq x_2$ and $y_1 \neq y_2$.*

It is clear that the relation C is an equivalence relation that partitions the faulty processors into disjoint equivalence classes. Each equivalence class is called a *fault-connected region*. Two fault-connected regions are disjoint if the boundary ring of one fault-connected region does not intersect with the other fault-connected region. Two convex fault-connected regions are given in Fig. 7, one consisting of the set of PEs $\{((1,2),(1,3),(1,4),(2,4)\}$, and another consisting of $\{(4,2)\}$. The fault model proposed in this paper is that of the disjoint convex fault-connected region, which is different from the fault model in [6][9]. It does not demand that the distance of two fault regions in at least one dimension is not smaller than 3. For example, in Fig. 7, the distance of two fault region is smaller than 3 in two dimensions, but they are disjoint. The proposed fault model does not contain any nonfaulty nodes.

The routing algorithm presented in the paper can tolerate disjoint convex faults without using any virtual channels. In order to make the problem easy to understand, we do not consider boundary faulty nodes of the network.

2.1 Setting Up Disjoint f-Rings

To present the fault-tolerant wormhole routing algorithm, we need to connect the safe PEs and channels around the fault regions to form directional rings which are to be used as detour routes. If a fault-connected region does not touch a boundary PE of the mesh, then the safe PEs around it can form a directional ring, which is called an *f-ring*.

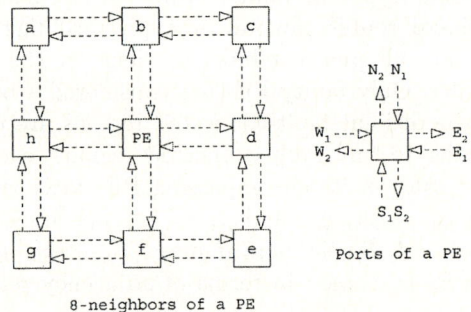

Fig. 1. The eight neighbors and ports of a PE

In this paper, every PE (except a boundary PE) of the mesh has eight neighbors which are represented by the eight Boolean variables a,b,c,d,e,f,g,h. The PE communicates with its neighbors through its input ports $\{W_1, E_1, N_1, S_1\}$ and output ports $\{W_2, E_2, N_2, S_2\}$, as shown in Fig. 1. A variable has value 1 if the corresponding neighbor is a safe PE; 0 if the corresponding neighbor is a faulty PE. The routing paths around the fault-connected regions will be set up in this section. The disjoint f-rings in clockwise or counter-clockwise direction are set in the network.

The disjoint f-rings around the fault-connected regions in counter clockwise and clockwise direction can be set up according to a simple procedure. In constant time, every PE can detect the states of its eight neighbors, which are either safe or faulty. Each then PE sets its port connections according to Table 1 and

Table 1. The port setting for disjoint f-rings in counter clockwise direction

$\{W_1, N_2\} = h\bar{a}b$	$\{E_1, S_2\} = d\bar{e}f$	$\{N_1, E_2\} = b\bar{c}d$	$\{S_1, W_2\} = f\bar{g}h$
$\{W_1, E_2\} = h\bar{b}d$	$\{E_1, W_2\} = d\bar{f}h$	$\{N_1, S_2\} = b\bar{d}f$	$\{S_1, N_2\} = f\bar{h}b$
$\{W_1, S_2\} = h\bar{b}\bar{d}f$	$\{E_1, N_2\} = d\bar{f}\bar{h}b$	$\{N_1, W_2\} = b\bar{d}\bar{f}h$	$\{S_1, E_2\} = f\bar{h}\bar{b}d$

Table 2. The port setting for disjoint f-rings in clockwise direction

$\{N_1, W_2\} = h\bar{a}b$	$\{S_1, E_2\} = d\bar{e}f$	$\{E_1, N_2\} = b\bar{c}d$	$\{W_1, S_2\} = f\bar{g}h$
$\{E_1, W_2\} = h\bar{b}d$	$\{W_1, E_2\} = d\bar{f}h$	$\{S_1, N_2\} = b\bar{d}f$	$\{N_1, S_2\} = f\bar{h}b$
$\{S_1, W_2\} = h\bar{b}\bar{d}f$	$\{N_1, E_2\} = d\bar{f}\bar{h}b$	$\{W_1, N_2\} = b\bar{d}\bar{f}h$	$\{E_1, S_2\} = f\bar{h}\bar{b}d$

Table 2. in clockwise orientation. The f-rings of different connected fault regions can all be set up in O(1) time.

2.2 Preprocessing for Fault-Tolerant Routing

When the f-rings are set up, we need to locate four special nodes, called the *boundary points*, which are used to direct the misrouting around the fault regions. Since we only consider fault regions with f-rings, the four nodes with minimal/maximal x-coordinate/y-coordinate on the contour of a fault region are selected as boundary points of the fault region.

Definition 3. *Given a fault region F in the mesh, the four special nodes are called the left-, right-, up- and down-boundary point of F, which are defined as follows:*

Left- (right-) boundary point: $left = (left_x, left_y)$ $(right = (right_x, right_y))$; *a node with minimal (maximal) x-coordinate on the contour of the fault region F.*

Up- (down-) boundary point: $up = (up_x, up_y)$ $(down = (down_x, down_y))$; *a node with minimal (maximal) y-coordinate on the contour of F.*

Note that by Definition 3, the boundary points of a fault region are not unique. The rectangle, which consists of the four edges in X and Y direction with the four boundary points, is called the boundary rectangle. Examples of boundary points are shown in Fig.7. We can run a preprocessing program, to let every node on the contour of a fault region obtain the coordinates of the four boundary points.

3 Fault-Tolerant Wormhole Routing in 2D Meshes

3.1 Extended X-Y Routing

In a 2D mesh, X-Y routing is made deadlock-free by prohibiting a turn from the Y dimension to the X dimension. Glass and Ni [4] presented the turn model method for designing partially adaptive wormhole routing algorithms that require no virtual channel. Chiu [3] proposed an odd-even turn model for designing partially adaptive and deadlock-free routing algorithms in meshes without virtual channel. The basic idea of the odd-even turn model is to restrict the locations where some of the turns can occur so that an EN turn and an NW turn would not be taken at nodes in the same column; similarly for an ES turn and an SW turn. Specifically, the odd-even turn model tries to prevent the formation of the rightmost column segment of a cycle. In a 2D mesh, a column is called an even (respectively, odd) column if the X-coordinate of the column is an even (respectively, odd) number. Chiu gave two rules for the odd-even model [3]:

Rule 1. *Any packet is not allowed to take an EN turn at any nodes located in an even column, and not allowed to take an NW turn at any nodes located in an odd column.*

Rule 2. *Any packet is not allowed to take an ES turn at any nodes located in an even column, and not allowed to take an SW turn at any nodes located in an odd column.*

Fig. 2 shows these two rules on the EN, NW, ES, and SW turns. The permissible turns are represented as small triangles, and forbidden turns are represented as ones with dashed lines. A turn in an even (odd) column is denoted by E (O). The turn model does not eliminate any type of turns for message routing; it only restricts the locations at which certain turns can be taken so that a circular wait can never occur.

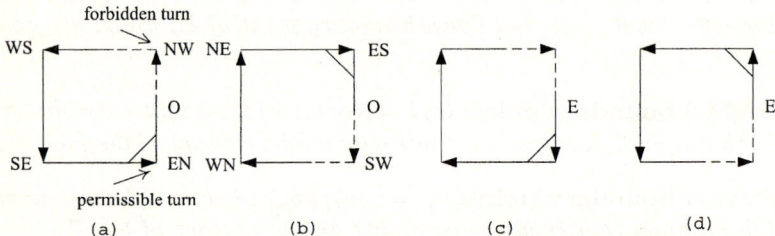

Fig. 2. Permissible and forbidden turns in the odd-even model

To propose deadlock-free wormhole routing by using the odd-even turn model, we modify the regular X-Y routing. Let $S = (x_s, y_s)$ and $D = (x_d, y_d)$ be the source and destination nodes, respectively. Let $\Delta_x = |x_s - x_d|$ and $\Delta_y = |y_s - y_d|$. The routing is divided into two situations:

– If $x_s > x_d$, then regular X-Y routing is used, which consists of two phases. In the first phase, the offset Δ_x along the X dimension is reduced to zero; in the second phase, the offset Δ_y along the Y dimension is reduced to zero, such as in Fig. 3(a) and Fig. 3(b).
– If $x_s \leq x_d$ and x_d is odd, then regular X-Y routing is used; If $x_s \leq x_d$ and x_d is even, then first the offset Δ_x along the X dimension is reduced to 1;

i.e., the routing head is on $x_d - 1$ column; second, the offset Δ_y along the Y dimension is reduced to zero; finally, the offset Δ_x along the X dimension is reduced to zero; as in Fig. 3(c) and Fig. 3(d), where if the destination is on an even column, the routing in the Y dimension is along the dashed line.

Since the above extended X-Y routing meets the two rules for the odd-even model proposed by Chiu [3], it is deadlock-free.

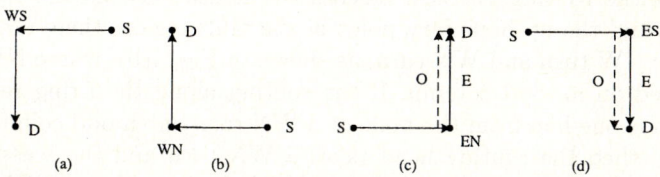

Fig. 3. Extended X-Y routing by using the turn model

3.2 Fault-Tolerant Routing Algorithm

We propose a fault-tolerant routing algorithm here. In a 2D mesh, messages are routed by the extended X-Y routing algorithm in fault-free regions. When routing becomes misrouting, it is routed around a fault-region in clockwise or counter-clockwise direction, which is based on certain rules. The goal is to route around the fault region without using any forbidden turns.

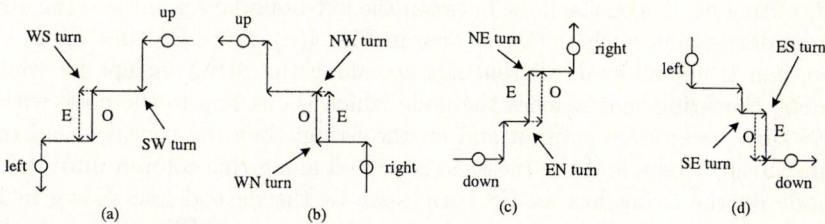

Fig. 4. Routing along the different sections around a fault region in counter clockwise direction

Since the f-ring of a fault region in clockwise or counter clockwise direction is divided into four segments by its boundary-points, all routings, (X^+, X^-, Y^+, or Y^-), in different sections around the fault region are according to certain odd-even routing rules.

When the message is routed around a fault region in counter clockwise direction, the routing in different sections along the f-ring of the fault region is carried out as follows.

1. When the routing message reaches the nodes between the up-boundary point and the right-boundary point of the fault region, there are two kinds of turns: SW turn and WS turn, as shown in Fig. 4(a), where WS turn is not allowed on odd column. If a WS turn along the f-ring of the fault region is on an odd column, the routing message takes one hop to the west, then takes the WS turn on an even column and routes the message along this column to the f-ring, such as the dashed line shown in Fig. 4(a). Since all WS turns are on even columns, so all turns in the section are permissible turns.
2. When the routing message reaches the nodes between the right-boundary point and the up-boundary point of the fault region, there are two kinds of turns: NW turn and WN turn, as shown in Fig. 4(b), where NW turn is not allowed on an odd column. If the routing along the f-ring reaches a node which is one hop from the node at a WN turn on an odd column and on the f-ring, then the routing head takes a WN turn and the message is routed along this column until its right node on the f-ring has a NW turn, such as the dashed line shown in Fig. 4(b). In this section, all routings are along the f-ring of the fault region, except the channels on the ring and on the odd column. Since all NW turns are on even columns, so all turns are permissible turns.
3. When routing message reaches the nodes between the down-boundary point and the right-boundary point of the fault region, there are two kinds of turns: NE turn and EN turn, such as those shown in Fig. 4(c), where EN turn is not allowed on an even column. So if an EN turn along the f-ring of the fault region is on an even column, the routing message goes one hop to the east, then takes the EN turn on an odd column and proceeds to the f-ring, such as the dashed line shown in Fig. 4(c). Since all EN turns are on the odd columns, so all routings in this section are permissible turns.
4. If routing head is on the node between the left-boundary point and the down-boundary point, such as that shown in Fig. 4(d). Since ES turn on an even column is not allowed, all routings are along the f-ring, except for routing along the f-ring that reaches the node which is one hop to the node with an ES turn on an even column and on the f-ring; then the routing head takes an ES turn there and the message is routed along this column until its east node on the f-ring has an ES turn, such as the dashed line shown in Fig. 4(d). Since there are two kinds of turns, SE turns and ES turns, and all ES turns are on the odd columns, so all turns are permissible turns.

When the message is routed around a fault region in clockwise direction, the routing in different sections along the f-ring of the fault region is similar and as depicted in Fig. 5. We omit the detailed descriptions here.

The fault-tolerant wormhole routing algorithm, FTRouting, is given in Fig. 6. For simplicity and because of space limitation, we assume that the destination is not a boundary node of any faulty region, as in Wu's paper [9]. In the FTRouting algorithm, although X^+ and X^- may be changed to Y^+ or Y^- routing, when X^+ routing is changed to Y^+ (Y^-) routing, Y^+ (Y^-) will leave the f-ring of the fault region; when X^- routing is changed to Y^+ (Y^-) routing, Y^+ (Y^-) will

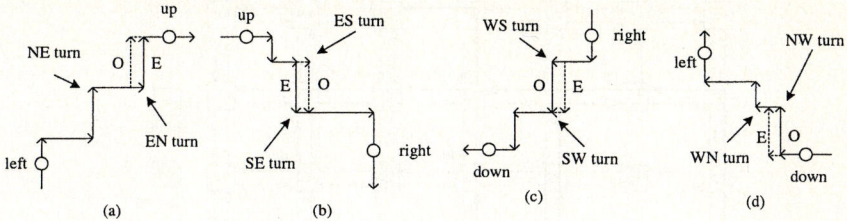

Fig. 5. Routing along the different sections around a fault region in clockwise direction

/* Let $S = (x_s, y_s)$ and $D = (x_d, y_d)$ be the addresses */
/* of source and destination. Let $C = (x_c, y_c)$ denote */
/* the current PE. Initially, $x_c = x_s, y_c = y_s$ */
1. **If** $x_c = x_d$ and $y_c = y_d$,
   Then consume routing message and return.
2. **If** routing message does not reach the boundary of a fault region
   Then use extended X-Y routing
3. **If** routing message will reach the boundary of a fault region
   Then
   (1) **If** the routing is X^+,
       Then
       (a) **If** $(d_y \geq left_y$ and $d_x \leq down_x)$ or $(d_y \geq right_y$ and $d_x \geq down_x)$
           Then message is forwarded along f-ring in counter clockwise direction
       (b) **If** $(d_y \leq left_y$ and $d_x \leq up_x)$ or $(d_y \leq right_y$ and $d_x \geq up_x)$
           Then message is forwarded along f-ring in clockwise direction
   (2) **If** the routing is X^- routing,
       Then
       (a) **If** $c_y \leq right_y$
           Then message is forwarded along f-ring in counter clockwise direction
       (b) **If** $c_y > right_y$
           Then message is forwarded along f-ring in clockwise direction
   (3) Y^+ routing is forwarded along f-ring in counter clockwise direction
   (4) Y^- routing is forwarded along f-ring in clockwise direction

Fig. 6. Fault-tolerant wormhole routing algorithm

go around the fault region in the same direction, or leave the f-ring of the fault region, so a 180° turn cannot occur. All routings ($X^+, X^-, Y^+,$ or Y^-) never pass through the right-boundary point of a fault region, since NW turn and SW turn are forbidden on an even column and EN turn and ES turn are forbidden on an odd column there, such as in Fig. 2.

An example of using algorithm FTRouting is shown in Fig. 7, where there are two routing messages, one from source $(0,2)$ to destination $(4,0)$, and another from source $(5,4)$ to destination $(0,1)$. For the routing from $(0,2)$ to $(4,0)$, when the routing header is at source $(0,2)$, it is X^+ misrouting, and the routing head is at a boundary node of faulty region, since $d_y \leq right_y$ and $d_x \geq up_x$, where

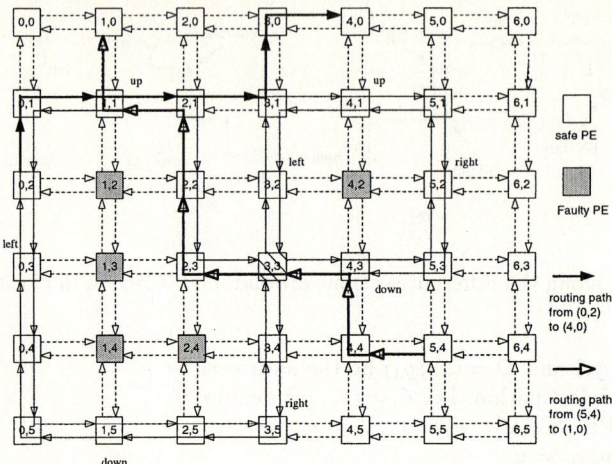

Fig. 7. Two fault-connected regions and the routing paths for the fault-regions with f-rings by using the algorithm FTRouting

$d_y = 0, right_y = 5, d_x = 4, up_x = 1$, so X^+ misrouting is in the clockwise direction. Since the destination is on an even column, X^+ routing has an EN turn at $(3, 1)$ on $x_d - 1$ column, and then takes an NE turn at $(3, 0)$ to destination.

For the routing from $(5, 4)$ to $(1, 0)$, when the routing header is at source $(5, 4)$, it is X^- normal routing; its forward node is on the boundary of a fault region, and since $c_y \leq right_y$, where $c_y = 4, right_y = 5$, the X^- routing would be in the counter-clockwise direction; and since $(3, 4)$ is on an odd column, so it takes a WN turn at $(4, 4)$ on an even column; the routing head reaches $(4, 3)$, where it has an NW turn on an even column; the message is routed to $(3, 3)$ in the X^- direction, where its forward node has a WN turn on an even column; so the message is routed along the f-ring in the clockwise direction to $(1, 1)$ on the destination column, and the message is routed to destination from there.

Theorem 1. *The fault-tolerant routing algorithm FTRouting is deadlock-free and livelock-free. (Proof is omitted)*

4 Conclusion

Deadlock- and livelock-free fault-tolerant routing is essential to guarantee the delivery of messages in real multicomputer environments that are faulty. We propose a fault-tolerant wormhole routing algorithm in 2D meshes without virtual channels in this paper. The proposed algorithm can tolerate the disjoint convex fault model, which does not contain any nonfaulty nodes and would not prohibit any routing as long as nodes outside the faulty regions are connected in the mesh network, except when the destination is on the boundary of a faulty region. We will try to derive adaptive and optimal solutions (uni- and multi-cast) for more popular fault models without virtual channels in meshes in the future.

References

1. Boppana, R.V., Chalasani, S.: Fault-tolerant communication with partitioned dimension-order routers. *IEEE Trans. on Parallel and Distributes systems*, Vol. 10, No. 10 (1999), 1026–1039.
2. Chalasani, S., Boppana, R.V.: Communication in multicomputers with nonconvex faults. *IEEE Trans. Computers*, Vol. 46, No. 5 (1997), 616–622.
3. Chiu, G.M.: The odd-even turn model for adaptive routing. *IEEE Trans. Parallel and Distributed Systems*, Vol. 11, No. 7 (2000), 729–737.
4. Glass, C.J., Ni, L.M.: The turn model for adaptive routing. *Proc. 19th Ann. Int'l Symp. Computer Architecture* (1992), 278–287.
5. Ho, C.-T., Stockmeyer, L.: A new approach to fault-tolerant wormhole routing for mesh-connected parallel computers. *IEEE Trans. Computers*, Vol. 53, No. 4 (2004), 427–438.
6. Su, C.C., Shin, K.G.: Adaptive fault-tolerant deadlock-free routing in meshes and hypercubes. *IEEE Trans. Computers*, Vol. 45, No. 6 (1996), 666–683.
7. Sui, P.H., Wang, S.D.: An improved algorithm for fault-tolerant wormhole routing in meshes. *IEEE Trans. Computers*, Vol. 46, No.9 (1997), 1040–1042.
8. Tsai, M.J.: Fault-tolerant routing in wormhole meshes. *Journal of Interconnection Networks*, Vol. 4, No. 4 (2003), 463–495.
9. Wu, J.: A fault-tolerant and deadlock-free routing in 2D meshes based on odd-even turn model. *IEEE Trans. Computers*, Vol. 52, No. 9 (2003), 1154–1169.
10. Zhou, J.P., Lau, F.C.M.: Fault-tolerant Wormhole Routing in 2D Meshes. *Proc. 2000 International Symposium on Parallel Architectures, Algorithms and Networks* (2000), Dallas/Richardson, Texas, USA.
11. Zhou, J.P., Lau, F.C.M.: Multiphase minimal Fault-tolerant Wormhole Routing in Meshes. *Parallel Computing*, Vol. 30 (2004), 423–442.

Fault Tolerant Routing Algorithm in Hypercube Networks with Load Balancing Support

Xiaolin Xiao[1], Guojun Wang[1,2], and Jianer Chen[1,3]

[1] School of Information Science and Engineering, Central South University,
Changsha, Hunan Province, P.R. China, 410083
[2] Department of Computing, Hong Kong Polytechnic University,
Hung Hom, Kowloon, Hong Kong
[3] Department of Computer Science, Texas A&M University,
College Station, TX77843-3112, USA

Abstract. We introduce load balancing support to our original fault tolerant routing algorithm in hypercube networks. We design an improved routing algorithm based on load balancing in hypercube networks. Our routing algorithm is simple and powerful, which preserves the advantages of our original algorithm. Firstly, our routing algorithm is applicable no matter whether the given hypercube network satisfies the required conditions or not. Secondly, our algorithm is distributed and local-information-based. Finally, our algorithm is effective and efficient. The contribution is that we realize load balancing in fault tolerant routing algorithm in hypercube networks in a creative way to make our algorithm perform better.

1 Introduction

Routing has been a popular topic in the study of computer networks, researchers have proposed many interconnection network topologies and related routing algorithms, among which the most typical one is routing in hypercube networks.

Because of the quick increase in network size, the self-contradiction of supply and demand on the network is increasingly revealed. Due to the fact that users provide traffic load to the network, which is usually over the capacity of network resources and above the ability that network can process, it will cause the phenomenon of dead lock and congestion. These phenomena are mainly caused by unbalanced distribution of network resources and traffic flow. However, these problems cannot be completely avoided, when relying only on the increase of network size to enhance network capacity. If we do not use some mechanisms to coordinate and control the equipments, these problems will appear consequently. Therefore, research on techniques to solve these problems is popular, especially on congestion control, flow control and load balancing. However, most of the current research on these problems focuses on Internet, ATM networks, and adaptive mesh networks [2-8].

Routing usually suffers the conflicts of network resources, regardless of concrete structure of networks. Then congestion and dead lock may occur in hypercube networks. Researchers have put forward many load balancing algorithms [2-8].

However, most of them are designed according to these network structures. To the authors' knowledge, there has been little development of efficient routing algorithms based on load balancing for the hypercube that allows faulty nodes. Some researchers have been engaged in research on a non-uniform analytical model of adaptive wormhole routing in hypercube in the presence of hot spot traffic [9-10]. But their research primarily aims at solving dead lock in network communication, which doesn't consider fault tolerance.

To overcome the above shortcomings, we introduce load balancing support to our original fault tolerant routing algorithm in hypercube networks [1]. Our improved routing algorithm is efficient and powerful. The proposed routing algorithm in hypercube networks can be extended in the Internet if combined with methods such as *Dynamic Virtual Hypercubes* and *Logical Hypercubes* in [7-8], which introduced the good properties of hypercube into the Internet.

The rest of the paper is organized as follows. Section 2 describes the improved routing algorithm in locally connected hypercube networks with load balancing support. Section 3 validates the performance of the improved algorithm through simulations. Section 4 concludes this paper.

2 Routing in Locally-Connected Hypercube Networks with Load Balancing Support

We have proposed a *locally k-subcube-connnected fault-tolerance* model and some routing algorithms in hypercube networks in [1]. However, load balancing is not involved in the original algorithms. Therefore, we take the load of flow as an important parameter to improve traditional routing algorithms and design an improved routing algorithm in hypercube networks with load balancing support in this paper. Our routing algorithm uses some load balancing strategy, such as the *least connection balancing* [6], to choose to arrange routing in some links with less load of flow for realizing load balancing. Our improved algorithm can be divided into two major parts as follows.

1. Based on the locally *k*-subcube-connnected fault-tolerance model, we set a *flow index* for each non-faulty node in every basic *k*-dimensional subcube H_k to record the number of connections flowing along each non-faulty node. Specifically, the flow index of each non-faulty node is to record the number of routing paths passed through the node. Firstly, routing is done in the hypercube network with faulty nodes according to the original algorithm. After a period of time, the value of flow index for each node will not be the same as each other, which indicates the unbalanced distribution of load of flow in the whole hypercube network. We calculate an *average flow index* for every basic *k*-dimensional subcube H_k from the number of nodes in H_k and the sum of all nodes' value of flow index in H_k. The value of average index is uniform for each node in the same basic *k*-dimensional subcube.

2. According to the above definition of average flow index, routing is done based on some load balancing strategy.

The improved algorithm with load balancing support is given in Fig. 1. Most of the algorithm's running time is spent at Step 5 and Step 6. In conclusion, if the input

n–cube H_n is locally k-subcube-connected [1], then the algorithm constructs a path of non-faulty nodes from u to v. The running time of the algorithm is bounded by $O((n-k)2^k) + O(2^k) = O(n\ 2^k)$. The proof of some theorems and corollaries that are related to the running time of the algorithm can be found in [1]. We give some remarks here.

Remark 1. The input and output conditions of the improved algorithm are the same as that of the original algorithm, except for the concrete implementation of finding the path from a non-faulty node in H_k to another non-faulty node in some k–subcube next to H_k. In the improved algorithm, we choose the routing path through a k–subcube H_k-*min* whose "average flow index" is minimum among all the non-visited H_k in H_n. While in the original algorithm, the path of non-faulty nodes is extended according to a fixed sequence [1].

The Improved Algorithm
Input: an n-cube H_n with faults and two non-faulty nodes u and v in H_n
Output: a path of non-faulty nodes in H_n that connects u and v
1.   let $u = u_1u_2...u_n$ and $v = v_1v_2...v_n$, where u_i and v_i are binary bits;
2. $w = w_1w_2...w_n$, where $w_j = u_j$ for all j;
3.   initialize the path $P = [u]$; let $i = 1$;
4.   initialize the visited flag of all the basic k–subcubes;
5.   **while** ($w_1w_2...w_{n-k} \neq v_1v_2...v_{n-k}$) **do**
   {at this point, we have, inductively, constructed a fault-free path P from $u = u_1u_2...u_n$ to $w = w_1w_2...w_n$, where $w_j = v_j$ for $j = 1,2,...,i-1$.}
   5.1   find a k–subcube H_k-*min* whose average flow index is the minimum within all of the n-k non-visited k–subcubes adjacent to w.
   5.2   find two adjacent non-faulty nodes
      $w' = w_1w_2...w_{n-k}x_{n-k+1}...x_n$ and v', and $v' \in H_k$-*min*;
   5.3   extend P from w to w' in the k–subcube $w_1w_2...w_{n-k}$**, then to v';
   5.4   reset the visited flag of k–subcube $w_1w_2...w_{n-k}$** and H_k-*min*;
   5.5   let $w = v'$; let $i = i + 1$;
6.   {at this point, we have $w_j = v_j$ for all $j = 1, 2, ..., n - k$}
   extend the path P from w to v in the k–subcube $v_1v_2...v_{n-k}$**.
{**Remark.**   if at any point the algorithm could not proceed, then
Stop: the n-cube H_n is not locally k-subcube-connected. }

Fig. 1. Routing in a Locally k-Subcube-Connected n-Cube with Load Balancing Support

Remark 2. The improved algorithm is based on load balancing. In the best case, only one shift in a non-faulty node makes the path extend into the k–subcube where destination node v resides. In the worse case, it needs the running time of $O((n-k)2^k)$ after visiting a sequence of H_k in H_n. Obviously, the improved algorithm terminates within limited steps, though the path length may be longer than that of the original algorithm. Moreover, the improved algorithm always tries to search all of the n-k non-visited adjacent k–subcubes for a k–subcube H_k-*min*.

Remark 3. In the improved algorithm, our definition of average flow index is the average value of the total number of routing paths passed for each non-faulty node in every basic k–subcube H_k in H_n. Our definition is easy to apply and meaningful, since this definition has close relation with *least connection balancing* [6], one of the typical load balancing algorithms. Of course, it is possible to seek other definitions for average flow index and contrast their performance.

Remark 4. The improved algorithm is distributed and local-information-based, in the sense that no global information about the network H_n is required by the algorithm. The only thing we assume is that from each non-faulty node, our algorithm can request the status of its neighbors.

Remark 5. If our algorithm cannot proceed at any step, then the algorithm stops and reports that the n–cube H_n is not locally k-subcube-connected. Therefore, our algorithm is applicable no matter whether the given hypercube network satisfies the required conditions or not.

3 Performance Analysis

3.1 Simulation Results

We have conducted simulations to validate the performance of the improved algorithm in hypercube networks with a variety of probability distributions of node failures. Especially, we concentrate more on whether the improved algorithm can realize load balancing or not. Our simulations are based on uniform probability distributions of node failures, i.e., we assume that each node has an equal and independent failure probability p_f. The simulation results are given in Table 1.

The explanation of each parameter and testing way in Table 1 is the same as those of the original algorithm [1]. Obviously, the parameter "*NodesExamined*" is essentially the running time of the algorithm. Considering that the main purpose of our algorithm is to realize load balancing, we add some background flow to the hypercube networks before the algorithm runs, which makes our simulations more reasonable. These background flows are generated randomly by a random function.

3.2 Discussions

In this subsection, we make a comparison of the simulation results between the improved algorithm and the original algorithm [1]. We present the difference of their performance in probability of constructing the routing paths successfully in Fig. 2 and Fig. 3. We observe that the improved algorithm preserves the advantages of the original algorithm. Moreover, probability of successfully constructing the routing paths of the improved algorithm is obviously greater than that of the original algorithm, which is very important to show that the proposed algorithm really improves the original algorithm.

However, the improved algorithm is not perfect. The time in constructing a routing path and length of the constructed routing path are longer than the original algorithm. This is mainly due to the searching procedure of H_k-*min* whose average flow index is minimal among the n-k non-visited adjacent k–subcubes. But we point out that the increase in *"PathLen"* is not significant. Since few researchers focus on studying on routing in hypercube networks with load balancing support, the improved algorithm is a breakthrough.

Table 1. Simulation Results of the Improved Algorithm

n	k	p_f (%)	k-sc (%)	PathFound (%)	PathLen (%)	Nodes Examined (%)
10	3	0.1	100.0	100.0	100.1	138.7
10	3	0.5	100.0	100.0	100.3	139.2
10	3	30.0	0.0	94.1	127.7	223.2
10	4	0.1	100.0	100.0	100.1	222.9
10	4	0.5	100.0	100.0	100.3	223.3
10	4	30.0	0.0	99.7	123.0	337.9
10	5	0.1	100.0	100.0	100.1	519.7
10	5	0.5	100.0	100.0	100.4	521.8
10	5	30.0	14.4	100.0	124.7	619.7
15	3	0.1	100.0	100.0	100.1	139.1
15	3	0.5	100.0	100.0	100.4	141.7
15	3	30.0	0.0	90.1	131.4	225.5
15	4	0.1	100.0	100.0	100.1	223.7
15	4	0.5	100.0	100.0	100.4	224.0
15	4	30.0	0.0	98.3	132.0	338.2
15	5	0.1	100.0	100.0	100.1	520.1
15	5	0.5	100.0	100.0	100.4	523.0
15	5	30.0	0.0	100.0	130.1	619.9
20	3	0.1	100.0	100.0	100.1	144.1
20	3	0.5	84.0	100.0	100.4	145.7
20	3	30.0	0.0	87.9	132.8	226.4
20	4	0.1	100.0	100.0	100.1	223.9
20	4	0.5	100.0	100.0	100.4	224.4
20	4	30.0	0.0	96.8	133.5	339.0
20	5	0.1	100.0	100.0	100.1	520.6
20	5	0.5	100.0	100.0	100.4	523.2
20	5	30.0	0.0	99.7	131.9	621.1

To overcome the above shortcomings, we give some modifications to the original definition of "average flow index". For example, we can design another definition — *"average flow threshold"*. When executing Step 5.1 in Fig. 1, if the average flow index of current adjacent k-subcube H_k is lower than average flow threshold, the routing will directly proceed in hypercube network without finding H_k-min in all n-k non-visited adjacent k–subcubes. Furthermore, we can define "average flow index" in another way that we record "average flow index" in a *representative* node in each basic k-subcube H_k, not in every non-faulty node in H_k. These definitions will bring us different results, with the former saving some running time and the later reducing some memory requirements.

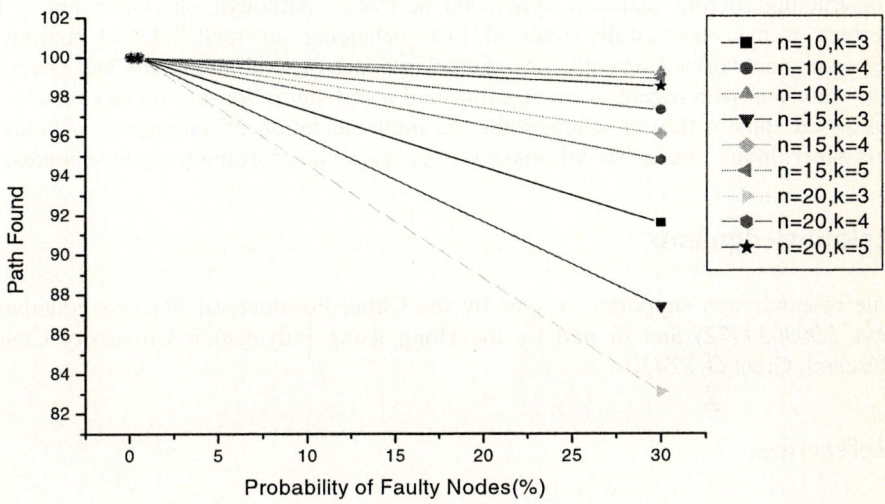

Fig. 2. Probability of Successfully Constructing the Routing Paths in the Original Algorithm

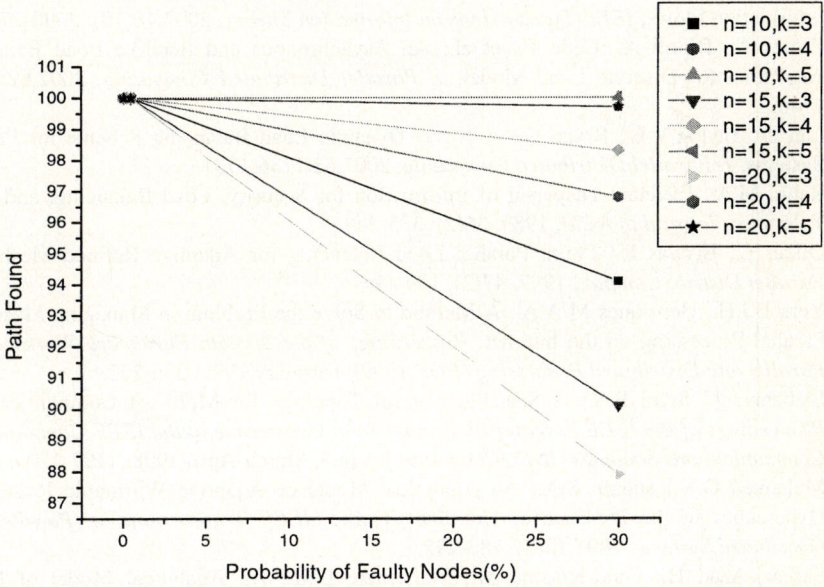

Fig. 3. Probability of Successfully Constructing the Routing Paths in the Improved Algorithm

4 Conclusions

Based on load balancing, we improved the original fault tolerant routing algorithm in locally subcube-connected hypercube networks. The improved algorithm preserves the advantages of the original algorithm and increases the probability of successfully constructing routing paths in hypercube networks. Although our algorithm is not perfect, it has successfully realized load balancing on fault tolerant routing in hypercube networks. Currently we are working on how to eliminate the bad effects of this algorithm with regard to routing time and path length. For preliminary results, we discussed some different schemes for the implementation of average flow index in this paper. In the future, we will make further research according to these schemes.

Acknowledgments

The research was supported in part by the China Postdoctoral Science Foundation (*No. 2003033472*) and in part by the Hong Kong Polytechnic University Central Research Grant *G-YY41*.

References

1. Chen J., Wang G., Chen S.: Locally Subcube-Connected Hypercube Networks: Theoretical Analysis and Experimental Results, *IEEE Transactions on Computers*, 2002,51(5):530-540.
2. Sarkar S., Tassiulas L.: A Framework for Routing and Congestion Control for Multicast Information Flows, *IEEE Transactions on Information Theory*, 2002,48(10): 2690-2708.
3. Cortes A., Ripoll A., Cedo F., et al.: An Asynchronous and Iterative Load Balancing Algorithm for Discrete Load Model, *J. Parallel Distributed Computing*, 2002,62:1730-1746.
4. Lan Z., Taylor V.E., Bryan G.: A Novel Dynamic Load Balancing Scheme for Parallel Systems, *J. Parallel Distributed Computing*, 2002,62:1736-1781.
5. Rabin M.A.: Efficient Dispersal of Information for Security, Load Balancing, and Fault Tolerance, *Journal of ACM*, 1989, 36(2): 335-348.
6. Oliker L., Biswas R.: Plum: Parallel Load Balancing for Adaptive Refined Meshes, *J. Parallel Distrib. Comput.*, 1997, 47(2), 109-124.
7. Yero E.J.H., Henriques M.A.A.: A Method to Solve the Problem in Managing Massively Parallel Processing on the Internet, *Proceedings of the Seventh Euromicro Workshop on Parallel and Distributed Processing (PDP 1999)*, February 1999: 256-262.
8. Liebeherr J., Sethi B.S.: A Scalable Control Topology for Multicast Communications, *Proceedings of the IEEE Seventeenth Annual Joint Conference of the IEEE Computer and Communications Societies (INFOCOM 1998)*, Vol.3, March-April 1998: 1197-1204.
9. Mohamed O.K., Hamid S.A.: An Analytical Model of Adaptive Wormhole Routing in Hypercubes in the Presence of Hot Spot Traffic, *IEEE Transactions on Parallel and Distributed Systems*, 2001,12(3): 283-292.
10. Sarbazi-Azad H., Ould-Khaoua M., Mackenzie L.M.: An Analytical Model of Fully-Adaptive Wormhole-Routed k-Ary n-Cubes in the Presence of Hot Spot Traffic, *IEEE Transactions on Computers*, 2000, 49(2): 1042-1049.

Proxy Structured Multisignature Scheme from Bilinear Pairings

Xiangxue Li, Kefei Chen, Longjun Zhang, and Shiqun Li

Department of Computer Science and Engineering, Shanghai Jiaotong University,
Shanghai 200030, P.R. China
xxli@sjtu.edu.cn

Abstract. In the past few years, proxy signatures have become an important research area and many excellent schemes have been proposed. Proxy signatures can combine other special signatures to obtain some new types of proxy signatures. A multisignature scheme is said to be structured if the group of signers is structured. Due to the various applications of the bilinear pairings in cryptography, many pairing-based signature schemes have been proposed. In this paper, we propose a proxy structured multisignature scheme from bilinear pairings. This scheme provides a possible solution to the problem that ordinary structured multisignature relies on the presence and cooperating of all the entities of the signing group. We support the proposed scheme with security and efficiency analysis.

1 Introduction

In 1996, Mambo, Usuda and Okamoto first introduced the concept of proxy signatures([13]). In the proxy signature scheme, an original signer is allowed to authorize a designated person as his proxy signer. Then the proxy signer is able to sign on behalf of the original signer. Since then, many proxy signature schemes have been proposed([8], [9], [14]). Proxy signature schemes have been shown to be useful in many applications, particularly in distributed computing where delegation of rights is quite common, such as e-cash systems, mobile agents for electronic commerce, mobile communications, grid computing, global distribution networks, and distributed shared object systems. Proxy signatures can combine other special signatures to obtain some new types of proxy signatures. Till now, there are various kinds of proxy signature schemes have been proposed([3], [5], [6], [16], [17]).

A multisignature is a digital signature that allows multiple signers to generate a single signature in a collaborative and simultaneous manner([7]). A multisignature scheme is said to be structured if the group of signers is structured([2]). The structure takes into account the signing order of the entities of the signing group. A structured multisignature scheme may play important role in the following scenario: When multiple entities sign a document, the signing order often reflects the role/position of each signer and signatures generated by the same group with different signing order are regarded as multisignatures with different meanings. The multisignature has to be checked against the organizational structure, or it will be considered as invalid. Till now, several structured multisignature schemes have been proposed([11], [12], [15], [18]). In [12], C. Lin et al. proposed a scheme that could resolve signing structures of serial, parallel, and the mix of them.

One problem of structured multisignatures is that they rely on the cooperating of the entities of the signing group. If one of the entities should become unavailable, such as might be expected in case of default on the signing operations, then the structured multisignature cannot be generated. Especially, when the same signing group has made agreement on a number of different signing structures, the problem becomes more serious. One solution to this case is to use proxy structured multisignature. In this paper, based on Boneh et al.'s short signature proposed at the Asiacrypt'01 conference ([1]), we present a new proxy signature scheme from bilinear pairings, which can be viewed as a special case of proxy multisignatures. The proposed scheme provides a possible solution to this problem.

The paper will proceed as follows. In section 2 we will briefly review the security requirements of ordinary proxy signatures, the definition and basic properties of bilinear pairings, some problems, and Boneh et al.'s pairing-based short signature scheme. Section 3 discusses the definition of series-parallel graph, which represents the signing structure in a structured multisignature scheme. We will present our pairing-based proxy structured multisignature scheme and its security and efficiency analysis in Section 4 and Section 5, respectively. The paper ends with some concluding remarks and acknowledgements.

2 Preliminaries

2.1 Security Requirements of Proxy Signatures

In the proxy signature scheme, an original signer is allowed to authorize a designated person as his proxy signer. Then the proxy signer is able to sign some messages on behalf of the original signer. When receiving a proxy signature, the verifier can validate its correctness and then is convinced of the original signer's agreement on the signed message. Basically, a secure proxy signature scheme should satisfy the following requirements([8]).

Strong Unforgeability: Only the legitimate proxy signer can generate a valid proxy signature; even the original signer can not.
Verifiability: Anyone can verify the signature and then is convinced of the original signer's agreement on the signed message.
Strong Identifiability: Anyone can determine the identity of the corresponding proxy signer.
Strong Undeniability: The proxy signer can not repudiate the signature which he ever generates.
Prevention of Misuse: The signed message should conform to the delegation warrant, and the proxy key pair should not be used for other purposes.

Like the general proxy signature, a secure proxy structured multisignature scheme should provide the security properties as above.

2.2 Bilinear Pairings

In this subsection, we will describe the basic definition and properties of the bilinear pairings.

Let G_1 be a cyclic additive group generated by P, whose order is a prime q, and G_2 be a cyclic multiplicative group of the same order q. Let a, b be elements of Z_q^*. We assume that the discrete logarithm problems (DLP) in both G_1 and G_2 are hard. A bilinear pairings is a map $e : G_1 \times G_1 \longrightarrow G_2$ with the following properties:

1) Bilinear: $e(aP, bQ) = e(P, Q)^{ab}$;
2) Non-degenerate: There exists P and $Q \in G_1$ such that $e(P, Q) \neq 1$;
3) Computable: There is an efficient algorithm to computa $e(P, Q)$ for all $P, Q \in G_1$.

Modified Weil Pairing and Tate Pairing are examples of cryptographic bilinear maps. Currently active research is being carried out to obtain efficient algorithms to compute pairings. Our work excludes this area.

2.3 Some Problems

Now we specify some versions of Diffie-Hellman problems. Let G_1 be a cyclic additive group generated by P, whose order is a prime q, and G_2 be a cyclic multiplicative group of the same order q, a bilinear pairing $e : G_1 \times G_1 \longrightarrow G_2$:

1) Computational Diffie-Hellman Problem (CDHP): Given P, aP, bP for $a, b \in Z_q^*$, to compute abP.
2) Decision Diffie-Hellman Problem (DDHP): Given P, aP, bP, cP for $a, b, c \in Z_q^*$, to decide whether $c = ab \mod q$.

Gap Diffie-Hellman(GDH) group: A prime order group G_1 is a GDH group if there exists an efficient polynomial-time algorithm which solves the Decision Diffie-Hellman Problem in G_1 and there is no probabilistic polynomial-time algorithm which solves the Computational Diffie-Hellman problem with non-negligible probability of success. The domains of bilinear pairings provide examples of GDH groups.

2.4 Pairing-Based Short Signature Scheme

At the Asiacrypt'01 conference, Boneh *et al.* introduced a short signature scheme, which is secure against existential forgery under a chosen message attack (in the random oracle model) assuming the Computational Diffie-Hellman assumption is hard on certain elliptic curves([1]). Their scheme works in any Gap Diffie-Hellman group where the Computational Diffie-Hellman problem (CDHP) is hard, but the Decision Diffie-Hellman problem (DDHP) is easy. Now we are ready to review Boneh *et al.*'s scheme.

Throughout this paper, we define the system's parameters as follows. Let G_1 be a cyclic additive group of prime order q, G_2 be a cyclic multiplicative group of the same order q, P be a generator of G_1. The bilinear pairing is giving by $e : G_1 \times G_1 \longrightarrow G_2$, two secure cryptographic hash functions $H_1 : \{0,1\}^* \longrightarrow Z_q$ and $H_2 : \{0,1\}^* \longrightarrow G_1$ are also required.

[**Key generation**]. Pick $s \longleftarrow_R Z_q^*$, and compute $PK = sP$. The public key is PK. The secret key is s.

[**Signing**]. Given a secret key s and a message $m \in \{0,1\}^*$, compute $P_m \longleftarrow H_2(m)$, and $S_m \longleftarrow sP_m$. The signature is $S_m \in G_1^*$.

[**Verification**]. Given a public key PK, a message m and a signature S_m, check whether the following equation holds:

$$e(S_m, P) = e(H_2(m), PK).$$

Based on Boneh et al.'s short signatures, we will propose a proxy structured multisignature scheme from bilinear pairings, and support the proposed scheme with security analysis. Before the proposal, we will first recall the graph representation of the signing group with a signing structure Λ in a structured multisignature scheme in the following section.

3 Series-Parallel Graph

Assume the signing group is $\{u_1, ..., u_n\}$. In a structured multisignature scheme, the structure Λ of the signing group can be represented as a directed group G, called series-parallel graph. A series-parallel graph is a graph which is generated by series and parallel compositions of series-parallel graphs. The simplest series-parallel graph is a base graph of two vertices and an edge. Interested readers can refer to [10] for more details. Figure 1 illustrates a typical series-parallel graph.

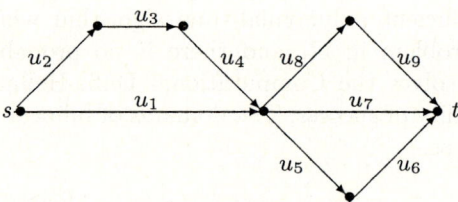

Fig. 1. A series-parallel graph G

In the signing structure shown in Figure 1, the entity u_3 in the signing group has to sign after the entity u_2 and in advance to the entity u_4, and u_8 has to sign after u_1 and u_4. In a signing structure, we denote G_{init} as the group of signers with no incoming edges in the graph representation G, $G_{prev(u_i)}$ as the group of signers whose edges are connected to u_i in G, and G_{last} as the group of signers whose edges are combined into the output in G. Take the structure represented in Figure 1 for example, $G_{init} = \{u_1, u_2\}$, $G_{prev(u_8)} = \{u_1, u_4\}$, and $G_{last} = \{u_6, u_7, u_9\}$.

4 Proxy Structured Multisignature Scheme from Bilinear Pairings

Based on Boneh et al.'s pair-based short signature, this section proposes a proxy structured multisignature scheme. Some initial settings of the scheme are assumed in Boneh et al.'s scheme.

Our proxy structured multisignature scheme consists of five phases: System setup phase, Public key of the group generation phase, Proxy generation phase, Proxy structured multisignature generation phase, and Verification phase.

[System Setup Phase]
The system's parameters are $\{G_1, G_2, e, q, P, H_1, H_2\}$.

Let $\{u_1, ..., u_n\}$ be n original signers in the signing group with signing structure Λ. For $1 \leq \forall i \leq n$, u_i has private key s_i and corresponding public key $PK_i = s_i P$. Let B be a proxy signer designated by all u_i. B has secret key s_b and corresponding public key $PK_b = s_b P$.

[Public Key of the Group Generation Phase]
This algorithm generates partial public keys for each original signer u_i and the public key of the group.

The partial public keys of the original signers in the signing structure Λ are computed as the following two steps.

- The partial public key of signer u_i in G_{init} is simply $y_i = s_i P (= PK_i)$.
- For all i such that u_i does not belong to G_{init}, u_i's partial public key is

$$y_i = s_i(P + \sum_{j: u_j \in G_{prev(u_i)}} y_j).$$

The public key y of the signing group with the structure Λ is generated by computing $y = \sum_{j: u_j \in G_{last}} y_j$.

[Proxy Generations Phase]
With signing structure Λ, the original signers $\{u_i\}$ jointly ask the proxy signer B to carry out signing a document m for them altogether. To do this, $\{u_i\}$ generate U from their private keys $\{s_i\}$, and send U to the proxy signer B. On receiving U, the proxy signer B can create the proxy key S_P from U and his own secret key s_b. From the security requirements, s_i should not be computed from S_P and U, and S_P should not be computed from U.

To delegate the signing capability to B, the original signers in the structure Λ have to make the signed warrant m_w. The warrant m_w specifies the necessary proxy details, such as the identity information of the proxy signer and all the original signers who have made agreements on the special signing structure Λ and on delegating signing capability to the designated proxy signer, the partial public key, the public key of the group, what kind of messages can be delegated, etc. If the following process is finished successfully, the proxy signer B gets a proxy key S_P.

- All u_i, which belong to G_{init},
  compute $U_i = s_i H_2(m_w)$.
- All u_i, which do not belong to G_{init},
  compute $U_i = s_i(H_2(m_w) + \sum_{j:u_j \in G_{prev(u_i)}} U_j)$.

The structured multisignature on the warrant m_w is computed as:

$$U = \sum_{j:u_j \in G_{last}} U_j$$

(m_w, U) is delivered to the proxy signer B.

- B confirms U by checking whether the following equality holds:

$$e(P, U) = e(y, H_2(m_w)).$$

- If U passes above equality, then B computes the proxy key as:

$$S_P = U + s_b H_2(m_w).$$

Using S_P, the proxy signer B can sign any message which conforms to the delegation warrant m_w on behalf of original signers $\{A_i\}$ with the signing structure Λ.

[Proxy Structured Multisignature Generation Phase]
When B signs a document m for $u_1, ..., u_n$, he performs the following process:
  chooses $x \longleftarrow_R Z_q^*$
  computes $r = e(P, P)^x$
  $c = H_1(m||r)$
  $S = xP - cS_P$
  outputs (c, S, m_w) as the proxy structured multisignature on the document m.

[Verification Phase]
After receiving the proxy structured multisignature (c, S, m_w), and the message m, the verifier operates as follows.

(1) Verify the authenticity of the public key y of the original signing group by checking whether or not the following equalities hold:

for i such that $u_i \in G_{init}, y_i = PK_i$,

for i such that u_i does not belong to G_{init},

$$e(PK_i, P + \sum_{j:u_j \in G_{prev(u_i)}} y_j) = e(P, y_i),$$

$y = \sum_{j:u_j \in G_{last}} y_j$.

If y does not pass the authenticity verification, stop. Otherwise, continue.

(2) Check whether or not the message m conforms to the warrant m_w. If not, stop. Otherwise, continue.

(3) Check whether or not the proxy signer B is authorized by these n original signers $\{u_i\}$ in the warrant m_w. If not, stop. Otherwise, continue.

(4) Compute $r = e(P, S)e(PK_b + y, H_2(m_w))^c$,

(5) Accept the proxy structured multisignature if and only if $c = H_1(m||r)$.

5 Analysis of the Proposed Proxy Structured Multi-signature Scheme

5.1 Corectness

The propertiy of correctness is satisfied. In effect, if the proxy structured multisignature is correctly generated, then:

$$r = e(P, P)^x$$
$$= e(P, S + cS_P)$$
$$= e(P, S)e(P, S_P)^c$$
$$= e(P, S)e(P, s_b H_2(m_w) + U)^c$$
$$= e(P, S)e(PK_b + y, H_2(m_w))^c$$

5.2 Efficiency

When analyzing computational costs, we assume that the pairing evaluation is the operation which takes the most running time. Under this assumption, we consider the number of pairing computations in the proposed scheme.

In the proxy structured multisignature generation phase, B needs one exponentiation in G_2, and one evaluation of the type of $aP + bQ$ in G_1 (the pairing evaluation $e(P, P)$ can be pre-computed). As for the verification phase, if the verifier often has to communicate with the original group $u_1, ..., u_n$ with the signing structure Λ and the proxy signer B, he does not have to verify the authenticity of the public key y of the group, in other words, all $e(PK_i, P + \sum_{j:u_j \in G_{prev(u_i)}} y_j)$ and $e(P, y_i)$ (for i such that u_i does not belong to G_{init}), $\sum_{j:u_j \in G_{last}} y_j$, and $e(PK_b + y, H_2(m_w))$ can be pre-computed. Thus, in every verification operation, the verifier only needs one pairing evaluation.

From above discussions, we can conclude that our proxy structured multisignature scheme not only provides a possible solution to the problem that ordinary structured multisignatures rely on the cooperating of the entities of the signing group , but also is more efficient than the ordinary structured multisignatures([12]).

5.3 Security Concerns

In this subsection, we will show that the proposed proxy structured multisignature scheme satisfies all the requirements stated in Section 2. During the proof, we assume that the warrant m_w specifies the necessary proxy details, such as the identity information of the proxy signer and all the original signers who have agreement on the signing structure Λ and on delegating signing capability to the designated proxy signer, the signing structure, the partial public key, the public key of the group, what kind of message can be delegated, etc.

Unforgeability-On the one hand, any third party, who wants to forge the proxy signature of an message m' for the proxy signer and these original signers, must have the original signers' structured multisignature on the warrant m_w. But he can not forge the signature. To see this, suppose that the adversary knows P and y, and tries to find U such that the equality $e(P, U) = e(y, H_2(m_w))$ holds. This is equivalent to forge Boneh *et al.*'s short signature. Any third party that can forge a proxy structured multisignature must be able to forge a Boneh *et al.*'s short signature. Since Boneh *et al.*'s pairing-based short signature scheme([1]) is proven to be secure against existential forgery under a chosen message attacks (in the random oracle model) assuming the Computational Diffie-Hellman Problem(CDHP) is hard on the chosen elliptic curves, the adversary cannot forge a valid proxy structured multisignature by this way.

On the other hand, since the proxy signer's private key is used in the algorithm of the proxy structured multisignature generation, and we use Hess's scheme([4]), which is secure under the hardness of CDHP and the random oracle model, to generate the proxy multisignature, even these original signers can not generate a valid proxy signature.

Verifiability-Because the warrant contains the identity information and the limit of the delegated signing capability, the verifier can verify the signature and check whether the signed message conforms to the delegation warrant or not.

Identifiability-Since there is the delegation warrant m_w in a valid proxy structured multisignature, anyone can determine the identity of the corresponding proxy signer from m_w.

Undeniability-Since the warrant is signed by the original signers in the structure Λ using Boneh *et al.*'s short signature scheme in the proxy generation phase and must be verified in the verification phase, the proxy signer can not modify the warrant. Thus, once he creates a valid proxy signature, he can not repudiate the creation of a valid proxy signature against anyone.

Prevention of Misuse-There is the warrant that has an explicit description of delegated signing capability, so the proxy signer can not sign any message that has not been authorized by the original signing group.

From above discussions, we can conclude that the proposed scheme is a secure proxy signature scheme, and actually provides all the security properties stated in Section 2.

6 Conclusions

In electronic world, proxy signature is a solution of delegation of signing capabilities. Proxy signatures can combine other special signatures to obtain some new types of proxy signatures. Various type proxy signatures are important in many applications. A multisignature scheme is said to be structured if the group of signers is structured. The structure takes into account the signing order of the entities of the signing group. In this paper, we proposed a proxy structured multisignature scheme from bilinear pairings. We analyzed the proposed scheme from security and efficiency points of view. We have shown that the scheme actually satisfied all the security properties required by proxy signatures. The proposed scheme provides a possible solution to the problem that ordinary structured multisignatures rely on the cooperating of the entities of the signing group, especially when the group has many different signing structures.

Acknowledgements

The authors thank to the anonymous reviewers for helpful comments. This work is supported by NSFC under the grants 60273049 and 90104005.

References

[1] D.Boneh, B.Lynn, H.Shacham. Short signatures from the weil pairing. In: *Advances in Cryptology-Asiacrypt* 2001, LNCS 2248, pages 514-532. Springer-Verlag, 2003.
[2] M.Burmester, Y.Desmedt, H.Doi, M.Mambo, E.Okamoto, M.Tada, Y.Yoshifuji. A structured ElGamal-type multisignature scheme. Proceedings of 3rd International Workshop on Practice and Theory in Public Key Cryptosystems(PKC 2000), Springer-Verlag, pages 466-483, 2000.
[3] X.Chen, F.Zhang, K.Kim. ID-based Multi-Proxy Signature and Blind Multisignature from Bilinear Pairings. In: *Proceeding of KIISC conference* 2003, pages 11-19, 2003, Korea.
[4] F.Hess. Efficient identity based signature schemes based on pairings. *Proceedings of 9th workshop on selected areas in cryptography-SAC*2002. Lecture Notes in Computer Science. Springer-Verlag.
[5] S.Hwang, C.Chen. New multi-proxy multi-signature schemes. Applied Mathematics and Computation, 147 (2004), pages 57-67.
[6] S.Huang. C.Shi. A simple multi-proxy signature scheme. Proceedings of the 10th National Conference on Information Security. Hualien, Taiwan, ROC, 2000, pages 134-138.

[7] K.Itakura, K.Nakamura. A public-key cryptosystem suitable for digital multisignature. NEC Research and Development, Vol 71, Oct 1983, pages 1-8.
[8] S.Kim, S.Park, D.Won. Proxy signatures,revisited. *Proc.of ICICS'97, International Conference on Information and Communications Security*, LNCS 1334, pages 223-232, 1997.
[9] B.Lee, H.Kim, K.Kim. Strong proxy signature and its applications. *Proc.of SCIS*. pages 603-608, 2001.
[10] T.Lengauer. Combinatorial algorithms for integerated circuit layout. B.G.Teubner Stuttgart, John Wiley & Sons, 1990, 468.
[11] C.Lin, T.Wu, J.Hwang. ID-based structured multisigature schemes. Advances in Network and Distributed systems Security. Kluwer Academic Publisher(IFIP Conference Proceedings 206), pages 45-59, 2001.
[12] C.Lin, T.Wu, F.Zhang. A structured multisignature scheme from the Gap Diffie-Hellman Group. Cryptology ePrint Archive, Report 2003/090. Available at http://eprint.iacr.org/2003/090.
[13] M.Mambo, K.Usuda, E.Okamoto. Proxy signature: delegation of the power to sign messages. *IEICE Trans.Fundamentals*. E79-A:9, pages 1338-1353, 1996.
[14] M.Mambo, K.Usuda, E.Okamoto. Proxy signature for delegating signing opertion. *Proc. of 3rd ACM Conference on Computer and Communications Security, ACM Press New York*, pages 48-57, 1996.
[15] C.Mitchell. An attack on an ID-based multisignature scheme. Royal Holloway, University of Landon, Mathemaitics Department Technical Report RHUL-MA-2001-9, December 2001.
[16] L.Yi, G.Bai, G.Xiao. Proxy multi-signature scheme: a new type of proxy signature scheme. Electronics Letters 36(6), 2000, pages 527-528.
[17] F.Zhang, R.Safavi-Naini, C.Lin. New proxy signature, proxy blind signature, proxy ring signature schemes from bilinear pairings. Cryptology ePrint Archive, Report 2003/104. Available at http://eprint.iacr.org/2003/104
[18] D.Zheng, K.Chen. An attack on a multisignature scheme. Cryptology ePrint Archive, Report 2003/201. Available at http://eprint.iacr.org/2003/201.

A Threshold Proxy Signature Scheme Using Self-Certified Public Keys

Qingshui Xue[1,2] and Zhenfu Cao[1]

[1] Department of Computer Science and Engineering, Shanghai Jiao Tong University,
200030, Shanghai, P.R.C.
{xue-qsh, zfcao}@cs.sjtu.edu.cn
[2] Department of Basic Theory, Shandong P.E. Institute, 250063, Jinan, P.R.C.
{ares, xue-qsh}@sjtu.edu.cn

Abstract. So far, all of proposed threshold proxy signature schemes are based on public key systems with certificates and most of them use Shamir's threshold secret share scheme. Self-certified public key system has attracted more and more attention because of its advantages. Based on Hsu et al's self-certified public key system and Li et al's proxy signature scheme, one threshold proxy signature scheme is proposed. The new scheme can provide the properties of proxy protection, verifiability, strong identifiability, strong unforgeability, strong repudiability, distinguishability, known signers and prevention of misuse. In the proxy signature verification phase, the authentication of original and proxy signers' public keys and the verification of the threshold proxy signature are executed together. In addition, the computation overhead and communication cost of the scheme are less than previous works with Shamir's secret share protocol and the public key system based on certificates.

1 Introduction

The proxy signature scheme [1], a variation of ordinary digital signature schemes, enables a proxy signer to sign messages on behalf of the original signer. Proxy signature schemes are very useful in many applications such as electronics transaction and mobile agent environment.

Mambo et al. [1] provided three levels of delegation in proxy signature: full delegation, partial delegation and delegation by warrant. In full delegation, the original signer gives its private key to the proxy signer. In partial delegation, the original signer produces a proxy signature key from its private key and gives it to the proxy signer. The proxy signer uses the proxy key to sign. As far as delegation by warrant is concerned, warrant is a certificate composed of a message part and a public signature key. The proxy signer gets the warrant from the original signer and uses the corresponding private key to sign. Since the conception of the proxy signature was brought forward, a lot of proxy signature schemes have been proposed [2-19].

Recently, many threshold proxy signature schemes were proposed [2, 6-14]. In threshold proxy signature schemes, a group of n proxy signers share the secret proxy signature key. To produce a valid proxy signature on the message m, individual proxy signers produce their partial signatures on that message, and then combine them into a full proxy signature on m. In a (t,n) threshold proxy signature scheme, the original signer authorizes a proxy group with n proxy members. Only the cooperation of t or more proxy members is allowed to generate the proxy signature. Threshold signatures are motivated both by the demand which arises in some organizations to have a group of employees agree on a given message or document before signing, and by the need to protect signature keys from attacks of internal and external adversaries.

In 1999, Sun proposed a threshold proxy signature scheme with known signers [9]. Then Hwang et al. [7] pointed out that Sun's scheme was insecure against collusion attack. By the collusion, any t-1 proxy signers among t proxy signers can cooperatively obtain the secret key of the remainder one. Meanwhile, they also proposed an improved scheme which can guard against the collusion attack. After that, [6] showed that Sun's scheme was also insecure against the conspiracy attack. In the conspiracy attack, t malicious proxy signers can impersonate some other proxy signers to generate valid proxy signatures. To resist the attack, they also proposed a scheme. Hwang et al pointed out [8] that the scheme in [7] was also insecure against the attack by the cooperation of one malicious proxy signer and the original signer. In 2002, Li et al. [2] proposed a threshold proxy signature scheme full of good properties and performance. In [14], we pointed out there were some errors in Sun's and Hwang et al.'s scheme and also proposed an improved version.

All of the proposed schemes are based on Shamir's secret share protocol and the public key systems using certificates, which have some disadvantages such as finding the certificate list when needing certificates, and high computation overheads and communication cost.

So far, there are three kinds of public key systems involving using certificates, identity-based and self-certified public keys. Currently, identity-based public systems are not pretty mature, as makes it not used in the real life.

The self-certified public key system is first introduced by Girault in 1991 [20]. In self-certified public key systems, each user's public key is produced by the CA (Certification Authority), while the corresponding private key is only known to the user. The authenticity of public keys is implicitly verified without the certificate. That is, the verification of public keys can be performed with the subsequent cryptographic applications such as key exchange protocols and signature schemes in a single step. Compared with other two public systems, the system has the following advantages [18]: (1) the storage space and the communication overheads can be reduced since the certificate is not needed; (2) the computation overhead can be reduced as it requires no public key verification.

There are four trust levels for the security of public key systems [12]. Hsu et al [18] pointed out that the self-certified public key systems might be the ideal choice for realizing cryptographic applications according to security and efficiency. Further, Hsu et al proposed a kind of self-certified public key system. In 2003, Li et al [17] proposed a generalization of proxy signature based on discrete logarithms. In the paper, based on Hsu et al's self-certified public key system and Li et al's proxy signature scheme, a threshold proxy signature scheme using self-certified public system is proposed. The main advantage of the proposed scheme is that the authenticity of original and proxy signers' public keys, and the verification of the proxy signature can be simultaneously executed in a single step. As far as we know, this threshold proxy signature scheme is the first one using self-certified public key system. The proposed scheme can provide the security properties of proxy protection, verifiability, strong identifiability, strong unforgeability, strong nonrepudiability, distinguishability, known signers and prevention of misuse of proxy signing power. That is, internal attacks, external attacks, collusion attacks and public key substitution attacks can be efficiently resisted. In addition, the new scheme is more efficient than previous works, which were implemented with Shamir's secret share protocols and public key systems based on certificates, in terms of computational overheads and communication cost.

In the paper, we organize the content as follows. In section 2, we will detail the proposed threshold proxy signature scheme. The security of the new scheme will be analyzed and discussed in section 3. In section 4, we will analyze the computational overheads and communication cost of the new scheme. Finally, the conclusion is given.

2 The Proposed Scheme

In the scheme, a system authority (SA) whose tasks are to initialize the system, the original signer U_o, certification authority (CA) whose tasks are to generate the public key for each user, the proxy group of n proxy signers $G_P = \{U_{P_1}, U_{P_2}, ..., U_{P_n}\}$, one designated clerk C whose tasks are to collect and verify the individual proxy signatures generated by the proxy signers, and construct the final threshold proxy signature, and the signature verifier are needed.

Throughout the paper, p and q are two large primes with $q|(p-1)$ and g is a generator of $GF(p)$ with order q. h is a secure one-way hash function. The parameters (p, q, g) and the function h are made public. Let ID_i be the identifier of the user U_i. Assume that x_{CA} and y_{CA} are the private and public keys of the CA, respectively, where $x_{CA} \in Z_q^*$ and

$$y_{CA} = g^{x_{CA}} \bmod p .\tag{1}$$

m_w is a warrant which records the identities of the original signer and the proxy signers of the proxy group, parameters t and n, message type to sign, the valid delegation time, etc. $ASID$ denotes the identities of the actual proxy signers.

The proposed scheme consists of four phases: registration, proxy share generation, proxy signature issuing without revealing proxy shares and proxy signature verification. We detail them below.

2.1 Registration

Step 1. Each user U_i selects an integer $t_i \in Z_q^*$ at random, computes

$$v_i = g^{h(t_i, ID_i)} \bmod p \tag{2}$$

and sends (v_i, ID_i) to the CA.

Step 2. Upon receiving (v_i, ID_i) from U_i, the CA selects $z_i \in Z_q^*$, calculates

$$y_i = v_i h(ID_i)^{-1} g^{z_i} \bmod p \tag{3}$$

$$e_i = z_i + h(y_i, ID_i) x_{CA} \bmod q \tag{4}$$

and returns (y_i, e_i) to U_i.

Step 3. U_i computes

$$x_i = e_i + h(t_i, ID_i) \bmod q \tag{5}$$

and confirms its validity by checking that

$$y_{CA}^{h(y_i, ID_i)} h(ID_i) y_i = g^{x_i} \pmod{p} \tag{6}$$

If it holds, U_i accepts (x_i, y_i) as his private and public keys. Moreover, the CA publishes U_i's public key y_i when the registration is complete. Note that the CA needn't issue extra certificate associated with y_i.

2.2 Proxy Share Generation

Step 1. The original signer chooses an integer $k_o \in Z_p^*$ randomly, computes

$$K_o = g^{k_o} \bmod p \tag{7}$$

$$\sigma_o = k_o K_o + x_o h(m_w) \bmod q \tag{8}$$

and sends (m_w, K_o, σ_o) to each of proxy signers.

Step 2. After receiving (m_w, K_o, σ_o), each of proxy signers confirms the validity of (m_w, K_o, σ_o) by

$$g^{\sigma_o} = K_o^{K_o} (y_{CA}^{h(y_o, ID_o)} h(ID_o) y_o)^{h(m_w)} \pmod{p} \tag{9}$$

If it holds, each of proxy signers regards σ_o as its proxy share.

2.3 Proxy Signature Issuing Without Revealing Proxy Shares

Without loss of generality, the proposed scheme allows any t or more proxy signers to represent the proxy group to sign a message m cooperatively on behalf of the original signer U_o.

Let $G_{P'} = \{U_{P_1}, U_{P_2}, ..., U_{P_{t'}}\}$ be the actual proxy signers for $t \leq t' \leq n$. $G_{P'}$ as a group performs the following steps to generate a threshold proxy signature.

Step 1. Each proxy signer $U_{P_i} \in G_{P'}$ chooses an integer $k_i \in Z_q^*$ at random, computes

$$K_i = g^{k_i} \bmod p \tag{10}$$

and sends it to the other $t'-1$ proxy signers in $G_{P'}$ and the designated clerk C.

Step 2. Upon receiving K_j ($j = 1,2,...,t'$, $j \neq i$), each $U_{P_i} \in G_P$ computes K and s_i as follows:

$$K = \prod_{j=1}^{t'} K_j \bmod p \tag{11}$$

$$s_i = k_i K + (\sigma_o t'^{-1} + x_{P_i}) h(m, ASID) \pmod{q} \tag{12}$$

Here, s_i is an individual proxy signature which is sent to C.

Step 3. For each received s_j ($j = 1,2,...,t'$), C checks whether the following congruence holds:

$$g^{s_i} = K_j^K \{[K_o^{K_o}(y_{CA}^{h(y_o,ID_o)} h(ID_o) y_o)^{h(m_w)}]^{t'^{-1}} y_{CA}^{h(y_i,ID_i)} h(ID_j) y_j\}^{h(m,ASID)} \bmod p \tag{13}$$

If it does, (K_i, s_i) is a valid individual proxy signature of m. If all the individual proxy signatures of m are valid, the clerk C computes

$$S = \sum_{j=1}^{t'} s_j \bmod q \tag{14}$$

Then, $(m_w, K_o, m, K, S, ASID)$ is the proxy signature of m.

2.4 Proxy Signature Verification

After receiving the proxy signature $(m_w, K_o, m, K, S, ASID)$ for m, any verifier can verify the validity of the threshold proxy signature by the steps below.

Step 1. According to m_w and $ASID$, the verifier can obtain the value of t and n, the public keys of the original signer and proxy signers from CA and knows the number t' of the actual proxy signers. Then the verifier checks whether $t' \geq t$, if it holds, he/she continues the following steps, or else, he/she will regard the threshold proxy signature $(m_w, K_o, m, K, S, ASID)$ invalid.

Step 2. The verifier confirms the validity of the proxy signature on m by checking

$$g^S = K^K [K_o^{K_o} y_{CA}^{[h(y_o,ID_o)h(m_w) + \sum_{j=1}^{t'} h(y_j,ID_j)]}$$
$$h(ID_o)^{h(m_w)} y_o^{h(m_w)} \prod_{j=1}^{t'} h(ID_j) \prod_{j=1}^{t'} y_j]^{h(m,ASID)} \pmod{p} \tag{15}$$

If it holds, the proxy signature $(m_w, K_o, m, K, S, ASID)$ is valid.

3 Security Analysis

In the section, we will analyze the security properties of the proposed scheme. Several theorems about the security will be proposed and proved below.

Theorem 1. The user can't forge his/her private key without interaction with the CA and the CA can forge the user's public key without the interaction with the user neither.

Proof. From Eq. 4, we know that although the user can select a random integer $z_i \in Z_q^*$ and compute $y_i = v_i h(ID_i)^{-1} g^{z_i} \mod p$, because of having no the knowledge of the CA's private key x_{CA}, he/she can't get a valid value of e_i to construct his self-certified private key. Obviously, the user can forge a valid private key with the probability of $1/q$. That's, the user's private key has to be set up by the interaction with the CA. Similarly, if the CA wants to forge the user's new public key which satisfies Eq. 6, he/she has to solve the difficult discrete logarithm problem, as we know it is impossible. Thus, the CA can't forge a new public key of the user. To generate the user's public key, the CA has to interact with the user.

Theorem 2. The user can't forge his public key by its private key without the interaction with the CA and the CA can't get the user's private key from the interaction with the user either.

Proof. If the user wants to forge his/her new public key which satisfies Eq. 6, he/she has to solve the difficult discrete logarithm problem, as we know it is impossible. Thus the user can't forge a new public key of the user without the interaction with the CA. From Eq. 5, we know that because the CA has no the knowledge of $t_i \in Z_q^*$ selected by the user, the CA can't obtain the user's private key x_i. In addition, from the verification Eq. 6, the CA is unable to get the user's private key x_i since he/she is faced with the difficulty of solving discrete logarithms. Therefore, we can draw the above conclusion.

Theorem 3. Any $t'' < t$ proxy signers can't generate a valid threshold proxy signature on a new message m'.

Proof. From Eqs. 12 and 14, we have

$$S = \sum_{j=1}^{t'} k_j K + \sigma_o h(m, ASID) + \sum_{j=1}^{t'} x_{P_j} h(m, ASID) (\mod q) \qquad (16)$$

Because any $t'' < t$ proxy signers have no the knowledge of k_j or $\sum k_j$, and x_{P_j} or $\sum x_{P_j}$ of other $t - t''$ proxy signers, any $t'' < t$ proxy signers are unable to cooperate to generate the valid proxy signature on a new message m'. Although any $t'' < t$ proxy signers can generate $(m_w, K_o, m', K, S, ASID)$ and it can pass the verification Eq. 15, the number of actual proxy signers is less than t, as makes it can't pass the verification step 1. Thus the forged proxy signature $(m_w, K_o, m', K, S, ASID)$ by $t'' < t$ proxy signers is invalid.

From the Eq. 16, any $t'' < t$ proxy signers are unable to get the values of k_j or $\sum k_j$, and x_{P_j} or $\sum x_{P_j}$ of other $t - t''$ proxy signers, if the message m is replaced with m', any $t'' < t$ proxy signers can't get the new value of S'. Also, from the verification Eq. 15, when m is replaced with m', given fixed some variables of the set $\{m_w, K_o, K, S, ASID\}$, the values of the other variables in the set $\{m_w, K_o, K, S, ASID\}$ will be unable to be gotten because of the difficult discrete logarithm and secure hash function. That is, from a known proxy signature $(m_w, K_o, m, K, S, ASID)$, any $t'' < t$ proxy signers can't generate valid threshold proxy signature on a new message m'

So the theorem is proved to be true.

Theorem 4. Any $t'' < t$ proxy signers can't forge another valid threshold proxy signature on the original message m from the proxy signature $(m_w, K_o, m, K, S, ASID)$.

Proof. On one hand, from Eq. 16, any $t'' < t$ proxy signers can't get the knowledge of k_j or $\sum k_j$, and x_{P_i} or $\sum x_{P_i}$ of other t t'' proxy signers. Thus, the values of some variables in set $\{K, S, ASID\}$ can't be changed by changing the values of the other variables in set $\{K, S, ASID\}$. Here note that as far as any $t'' < t$ proxy signers are concerned, the values of m_w and K_o can't be changed, as can be guaranteed by Eq. 9. On the other hand, from the verification Eq. 15, by fixing some variables of the set $\{m_w, K_o, K, S, ASID\}$, the values of the other variables in the set $\{m_w, K_o, K, S, ASID\}$ will not be able to be obtained due to the difficult discrete logarithm and secure hash function. Therefore, the theorem is proved.

Because of the similar causes, we can obtain the following theorems. The related proofs will be omitted.

Theorem 5. The original signer and any $t'' < t$ proxy signers can't cooperatively generate a valid threshold proxy signature on a new message m'.

Theorem 6. The original signer and any $t'' < t$ proxy signers can't cooperatively forge another valid threshold proxy signature on the original message m from the proxy signature $(m_w, K_o, m, K, S, ASID)$.

Theorem 7. Any third party, the original signer and any $t'' < t$ proxy signers can't cooperatively generate a valid threshold proxy signature on a new message m'.

Theorem 8. Any third party, the original signer and any $t'' < t$ proxy signers can't cooperatively forge another valid threshold proxy signature on the original message m from the proxy signature $(m_w, K_o, m, K, S, ASID)$.

Theorem 9. Anyone can be convinced of the original signer's agreement on the signed message from the proxy signature $(m_w, K_o, m, K, S, ASID)$.

Theorem 10. Anyone can identify the actual proxy signers from the proxy signature $(m_w, K_o, m, K, S, ASID)$.

Theorem 11. Anyone can distinguish proxy signatures from normal signatures.

Theorem 12. The proxy signers can't repudiate having produced the proxy signature which had ever been signed to any one.

Theorem 13. The proxy signers can't misuse the proxy signing power on other purposes.

From the above several theorems, we know that the proposed scheme can fulfill the securities of verifiability, strong identifiability, distinguishability, strong unforgeability, strong nonrepudiation, proxy protection and prevention of misuse of proxy signing power. In other words, the proposed scheme can resist equation attacks, collaboration attacks, public key substitution attacks, internal attacks and external attacks. In addition, the certificates of users are not needed in the proposed scheme. If users want to change his private key or the CA wants to change users' public key, both have to interact to finish it, or else neither of the two parties can succeed. The verifier only needs to obtain the public keys of the original signer and the proxy signers and needn't verify their validity as the verification of their public keys and the proxy signature is executed together. Thus, the self-certification of public keys can be realized.

4 Performance Analysis

We denote the following notations to facilitate the performance evaluation:

T_h: The time for performing a one-way hash function h.

T_{exp}: The time for performing a modular exponentiation computation.

T_{mul}: The time for performing a modular multiplication computation.

T_{add}: The time for performing a modular addition computation.

T_{inv}: The time for performing a modular inverse computation.

$|x|$: The bit-length of an integer x.

The computational overhead and communication cost of the proposed scheme are stated in Table 1 and 2, respectively.

Table 1. Computational overhead of the proposed scheme

Phases	Computational overhead
Registration	User: $3T_{exp} + T_{mul} + T_{add} + 4T_h$
	The CA: $T_{exp} + 3T_{mul} + T_{add} + T_{inv} + 2T_h$
Proxy share generation	The original signer: $T_{exp} + 2T_{mul} + T_h$
	Each proxy signer: $4T_{exp} + 3T_{mul} + 3T_h$
Proxy signature issuing	Each proxy signer: $T_{exp} + (t'+2)T_{mul} + 2T_{add} + T_{inv} + T_h$
	The clerk: $8t'T_{exp} + 7t'T_{mul} + (t'-1)T_{add} + t'T_{inv} + 6t'T_h$
Proxy signature verification	$7T_{exp} + (2t'+4)T_{mul} + t'T_{add} + (2t'+5)T_h$
Total [a]	$(8t'+14)T_{exp} + (10t'+11)T_{mul} + (2t'+1)T_{add} +$
	$(t'+1)T_{inv} + (8t'+10)T_h$

[a] The total computation overhead excludes the registration phase.

Table 2. Communication cost of the proposed scheme

Phases	Communication cost
Registration	$2|p|+2|q|+|ID_i|$
Proxy share generation	$|p|+|q|+|m_w|$
Proxy signature issuing	$t'|p|+t'|q|$
Proxy signature verification	$2|p|+|q|+|m_w|+|m|+|ASID|$
Total [b]	$(t'+3)|p|+(t'+2)|q|+2|m_w|+|m|+|ASID|$

[b] The total communication cost excludes the registration phase.

From the Tables 1 and 2, we know that the proposed scheme needs less computation overhead and communication cost than other threshold proxy signature schemes [2, 6-13] which were based on the certificate-based public key system and Shamir's secret share scheme. Therefore, the proposed scheme is more efficient than those schemes in terms of computational complexity and communication cost.

5 Conclusions

In the paper, based on Hsu et al's self-certified public key system and Li et al's proxy signature scheme, one threshold proxy signature scheme with self-certified public key system and non Shamir's secret share protocol has been proposed. As far as we know, it is the first scheme using self-certified public keys. The new scheme can provide the security properties of proxy protection, verifiability, strong identifiability, strong unforgeability, strong nonrepudiability, distinguishability, known signers and prevention of misuse of proxy signing power, i.e., internal attacks, external attacks, collusion attacks, equation attacks and public key substitution attacks can be resisted. In the proxy signature verification phase, the authentication of the original signer and the proxy signers' public keys and the verification of the threshold proxy signature are executed together. In addition, the computation overhead and communication cost of the proposed scheme are more efficient than previous works with Shamir's secret share protocol and the public key system with certificates.

Acknowledgment

The author would like to thank anonymous referees for their suggestions to improve the paper. This paper is supported by the National Science Fund for Distinguished Young Scholars under Grant No. 60225007 and the National Research Fund for the Doctoral Program of Higher Education of China under Grant No. 20020248024.

References

1. Mambo, M., Usuda, K. Okamoto, E.: Proxy Signature for Delegating Signing Operation. In: Proceedings of the 3.th ACM Conference on Computer and Communications Security. ACM Press, New York (1996) 48–57
2. Li, J.G., Cao, Z.F.: Improvement of a Threshold Proxy Signature Scheme. Journal of Computer Research and Development 9(11) (2002) 515-518 (in Chinese)
3. Li, J.G., Cao, Z.F., Zhang, Y.C.: Improvement of M-U-O and K-P-W Proxy Signature Schemes. Journal of Harbin Institute of Technology (New Series) 9(2) (2002) 145–148
4. Li, J.G., Cao, Z.F., Zhang, Y.C.: Nonrepudiable Proxy Multi-signature Scheme. Journal of Computer Science and Technology 18(3) (2003) 399–402
5. Li, J.G., Cao, Z.F., Zhang, Y.C., Li, J.Z.: Cryptographic Analysis and Modification of Proxy Multi-signature Scheme. High Technology Letters 13(4) (2003) 1–5 (in Chinese)
6. Hsu, C.L., Wu, T.S., Wu, T.C.: New Nonrepudiable Threshold Proxy Signature Scheme with Known Signers. The Journal of Systems and Software 58 (2001) 119~124
7. Hwang, M.S., Lin, I.C., Lu Eric, J.L.: A Secure Nonrepudiable Threshold Proxy Signature Scheme with Known Signers. INFORMATICA 11(2) (2000) 1–8
8. Hwang, S.J., Chen, C.C.: Cryptanalysis of Nonrepudiable Threshold Proxy Signature Scheme with Known Signers INFORMATICA 14(2) (2003) 205–212
9. Sun, H.M.: An Efficient Nonrepudiable Threshold Proxy Signature Scheme with Known Signers. Computer Communications 22(8) (1999) 717–722
10. Sun, H.M., Lee, N.Y., Hwang, T.: Threshold Proxy Signature. IEEE Proceedings-computers & Digital Techniques 146(5) (1999) 259–263
11. Zhang, K.: Threshold Proxy Signature Schemes. In: Information Security Workshop, Japan (1997) 191–197
12. Hsu, C.L., Wu, T.S., Wu, T.C.: Improvement of Threshold Proxy Signature Scheme. Applied Mathematics and Computation 136 (2003) 315–321
13. Tsai, C.S., Tzeng, S.F., Hwang, M.S.: Improved Non-Repudiable Threshold Proxy Signature Scheme with Known Signers. INFORMATICA 14(3) (2003) 393–402
14. Xue, Q.S., Cao, Z.F.: On Two Nonrepudiable Threshold Proxy Signature Schemes with Known Signers. INFORMATICA to appear
15. Denning, D.E.R.: Cryptography and Data Security. Addison-Wesley, Reading, MA, (1983)
16. Pedersen, T.: Distributed Provers with Applications to Undeniable Signatures. LNCS, Vol.547, Springer-Verlag New York (1991)
17. Li, L.H., Tzeng, S.F., Hwang, M.S.: Generalization of proxy signature-based on discrete logarithms. Computers & Security 22(3) (2003) 245–255
18. Hsu, C.L., Wu, T.S.: Efficient proxy signature schemes using self-certified public keys. Applied Mathematics and Computation, In Press, Corrected Proof, Available online 9 (2003)
19. Hwang, M.S., Tzeng, S.F., Tsai, C.S.: Generalization of proxy signature based on elliptic curves. Computer Standards & Interfaces 26(2) (2004) 73–84
20. Girault: Selt-certified public keys. In: Advance in Cryptology-EUROCRYPT, Springer-Verlag, (1991) 491–497

The Authentication and Processing Performance of Session Initiation Protocol (SIP) Based Multi-party Secure Closed Conference System

Jongkyung Kim[1], Hyuncheol Kim[1], Seongjin Ahn[2], and Jinwook Chung[1]

[1] Department of Computer Engineering, Sungkyunkwan University, Suwon, Korea
{jongkkim,hckim,jwchung}@songgang.skku.ac.kr
[2] Department of Computer Education, Sungkyunkwan University, Seoul, Korea
sjahn@comedu.skku.ac.kr

Abstract. This paper presents, a new SIP based multi-party secure closed conference system. In the traditional participants, except of captain, conference system doesn't have a privilege for UA (User Agent) that accepts or declines new participants. On the other hand, closed conference system supports this function. We also propose for closed conference system authentication procedure. By means of a real implementation, we provide an experimental performance analysis of SIP security mechanisms.

1 Introduction

The system configuration for multi-party SIP based conference (including 3 parties or over) can be largely classified into three models. As shown in Fig. 1(a), an end user system should cover all the signaling and media mixing during the conference, resulting in a significant overload on the system. In the case of distributed multipoint conference model, as shown in Fig. 1(b), an additional terminal leads to an increase in multicast recipient. This means that an additional INVITE message is inefficiently required whenever such process is repeated. Finally, Ad-hoc conference model adopts a simple structure where the central media server or MCU (Multipoint Control Unit) processes all the signaling messages and performs media mixing so that each end user manages only his own traffics [1][2].

The SIP message contains information that a user or server wishes to keep private. Securing SIP header and body information can be motivated by two reasons. One is that we need maintain private user and network information in order to guarantee a certain level of privacy. Another is that we have to avoid SIP sessions being set up or changed by attackers faking the identity of someone else. The SIP UAC (User Agent Client), calling side, can identity itself to a UAS (User Agent Server), called side. Therefore, SIP authentication applies only to user-to-user, user-proxy or user-registrar communication [3]

A representation of the SIP digest authentication procedure is given in Fig. 2. The function *F()* used to compute the response specifies how to combine the input parameter with some iterations of a digest algorithm [4][5]. The authentication procedure is run when the UAS, a proxy server, or the registrar server requires the calling side (UAC) to be authenticated before accepting the call, forwarding the call, or accepting the registration.

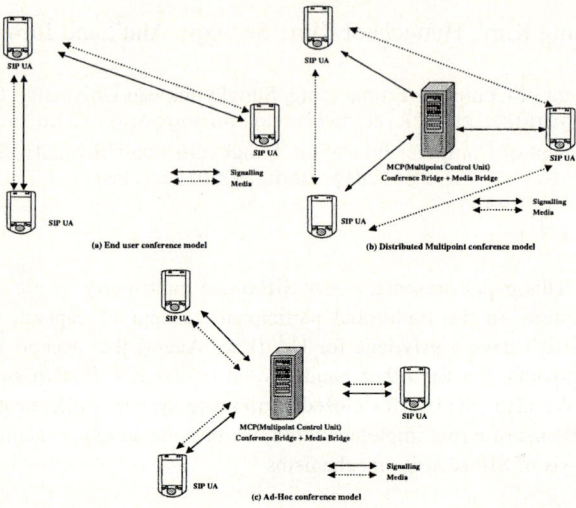

Fig. 1. Configuration for SIP-Based Multi-Party Conference System

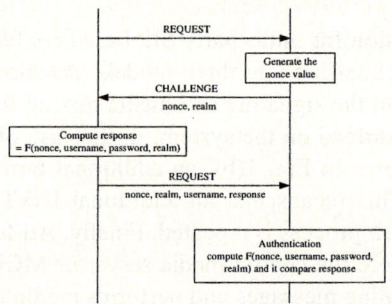

Fig. 2. The Procedure of Digest Authentication

In case of the existing conference system, as shown above, only the captain of conference has the right of invitation of additional participants. However, closed conference system needs all participants' anonymous agreement is needed to invite additional participants. In this paper, the closed conference model and the extended header will be proposed. The reminders of this paper are organized as follows. The authentication procedure for closed conference system is proposed in section 2. The section 3 describes the methodology for the evaluation of processing cost and experimental results. Finally, we make a conclusion.

2 Proposed Authentication Procedure for Closed Conference System

The simple expansion of 1:1 connection isn't able to guarantee satisfactory confidence. For instance, when a new participant joins conference session, existing participants and new one have to know the information of who has joined conference. That is why any one among them doesn't want to join conference together. Therefore, we need to know all participants of conference. We use the method that is the usage of INVITE message's SDP (Session Description Protocol) from the conference server [6].

Fig. 3 shows the procedure of the invitation of a participant in the closed conference system. To begin with Alice and Bob setup call session through authentication procedure. When the captain, Bob, invites Carol, Bob's UA sends NOTIFY message to Alice in order to agree an additional invitation. When Alice agrees to invite Carol, UA sends 200 OK message (F25). However, when he doesn't want to do that, it sends 603 declined message (F25). If Alice agrees the invitation of Carol, next time, Conference Server asks Bob to join conference. Bob checks the received information of participants. If he doesn't want to join that conference, he sends 603 messages (F43). This procedure presents with gray box in Fig. 3.

The states of the proposed UA can be divided into StateIdle (initial state), StateTrying, StateRinging, and StateInCall. The initial state is changed to StateRinging state when a phone is answered and followed by a receipt of ringing message, as shown in Fig. 4. When a ring is given someone, INVITE message is sent so that the status is set at StateTrying state. In the case of conference during the process, the REFER message is sent. If an INVITE message is received during StateIdle state, a negotiation of multimedia to be used in the session is followed by sending Ringing OK message (Real message: 180 Ringing). After that, it is changed to StateRinging state. At that time, if network resources are lacking or the phone is on the line, a receipt of PRACK message during StateRinging state results in sending 200 OK and INVITE OK(200) messages. After that, it is changed to StateInCall state.

3 Methodology for the Evaluation of Processing Cost and Experimental Results

In order to experiment with advanced feature in SIP, we have realized a test bed [7]. The goal of this test bed was twofold: firstly verification of the function behavior of the various elements and their interoperability; secondly the possibility of making some performance analysis. In particular regarding performance analysis, an interesting point is the evaluation of the cost to be paid in terms of performance for in the introduction of security mechanisms.

The results of our evaluation are reported in Table 1. The Third column reports the experiential average cost of twenty times in terms of second. Note that this throughput corresponds to 100 percent utilization of elements processing resources. The two rightmost columns are the most important ones and report the throughput value converted in a relative processing cost.

The result show that the introduction of SIP security accounts for nearly 30 percent of processing cost of no authentication procedure. This increase can be explained with the increase in number of exchanged SIP messages. Another interesting finding is that the incensement of processing time cost increase accidentally, when third attendant joins conference.

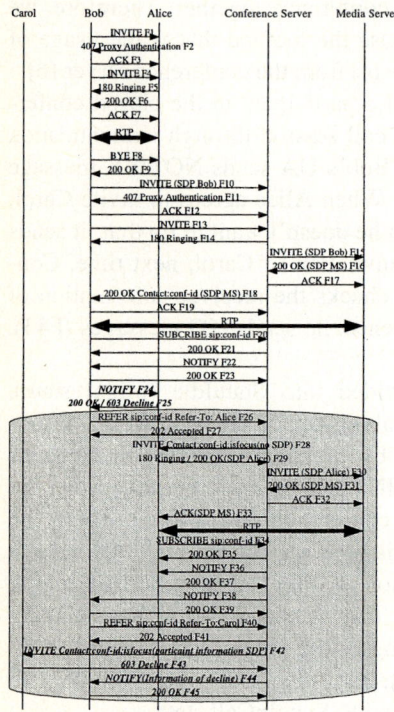

Fig. 3. The Procedure of Closed Conference System

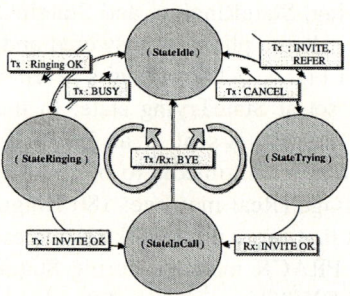

Fig. 4. Proposed SIP UA State Transition Diagram

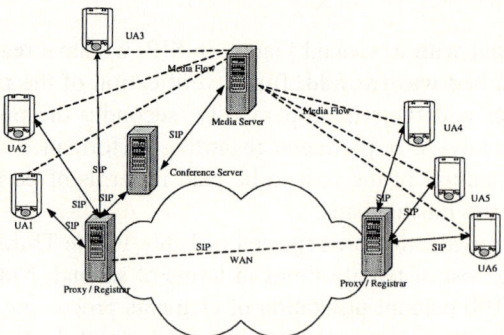

Fig. 5. The Test Bed Layout

Table 1. Experimental results

	Procedure/scenario	Processing time cost(second)	Relative cost
1	No authentication, 2 attendants	$2.873 * 10^{-2}$	100
2	No authentication, 3 attendants	$4.022 * 10^{-2}$	140
3	No authentication, 4 attendants	$4.252 * 10^{-2}$	148
4	No authentication, 5 attendants	$4.453 * 10^{-2}$	155
5	No authentication, 6 attendants	$4.683 * 10^{-2}$	163
6	Authentication, 2 attendants	$3.735 * 10^{-2}$	130
7	Authentication, 3 attendants	$5.200 * 10^{-2}$	181
8	Authentication, 4 attendants	$5.545 * 10^{-2}$	193
9	Authentication, 5 attendants	$5.746 * 10^{-2}$	200
10	Authentication, 6 attendants	$6.119 * 10^{-2}$	213

4 Conclusion

In this study, SIP based multi-party closed conference system and authentication procedure is described. The performance aspects of SIP authentication for closed conference system are considered with pure experimental approach. The processing costs of different security procedure/scenario are compared under a reference implementation. Although the performance results are obviously conditioned by the specific implementation aspects, they can be rough idea of relative processing cost of SIP security procedures.

References

[1] Jonathan Rosenberg, et al.: Models for Multi Party Conferencing in SIP, Jan. 2003
[2] A. Johnson: SIP Call Control-Conferencing for User Agents, Oct. 2003
[3] Rohan Mahy: A Call Control and Multi-party usage Framework for the Session Initiation Protocol (SIP), September 2003
[4] J, Franks et al.: HTTP Authentication: Basic and Digest Access Authentication, IETF RFC 2617, June 1999
[5] R. Rivest: The MD5 Message-digest Algorithm, IETF RFC 1321, Apr 1992
[6] Janet R. Dianda, et al.: SIP, Bell Labs Technical Journal 7(1), 3-23(2002)
[7] Guy J. Zenner, Mark H.Jones, Amit A. Patel: Emerging Uses of SIP in Service Provider Networks, Bell Labs Technical Journal 8(1), 43-63(2003)

A Method for Authenticating Based on ZKp in Distributed Environment[1]

Dalu Zhang, Min Liu, and Zhe Yang

Department of Computer Science and Technology, Tongji University,
Shanghai, P.R.C. 200092
daluz@public.sta.net.cn, {judygsf, hatasen}@hotmail.com

Abstract. A new ZKp identity protocol is proposed in this paper. It is more appropriate than the traditional identity protocol in distributed environment without an identical trusted third party. The security of this protocol relies on the discrete logarithm problem on conic over finite fields. It can be designed and implemented easier than those on elliptic curve. A simple solution is proposed to prevent a potential leak of our protocol.

1 Introduction

Zero-knowledge protocol is introduced to overcome restrict of D-H key distribution that some trust in the identity of the communicating parties has been established. In distributed environment, such as Grid and P2P, this kind of trust is not easy to establish. The Globus uses the Grid Security Infrastructure (GSI) [15] for enabling secure authentication and communication over an open network. GSI public-key security mechanisms are used to verify credentials and to achieve mutual authentication between information consumers and information providers. Public key authentication requires consideration of the trustworthiness of the certifying authorities for the purpose of public key certification. With traditional public key encryption the authentication chain must set up firstly. When a fixed trusted third party is not available, that becomes difficult. Certificates can comprise a chain, where a certificate of its issuer follows each certificate. Reliable authentication of a public key must be based on a complete chain of certification which starts at an end-entity certificate, includes zero or more CA certificates and ends at a self-signed root certificate. Zero Knowledge Proofs of Identity [1] is the first widely accepted paper that presented a sound, complete, and resource sensitive ZKp(zero knowledge proofs) algorithm for identity verification. A fixed trusted third party is not indispensable. In ZKp schemes, a trusted third party is helpful in the first phrase to produce the public information. The author of [1] thought that zero knowledge proofs might reveal one bit of knowledge. Its security depends on the difficulty of factoring. In fact, the algorithm is not satisfying because of vast information to be exchanged. Guillou proposed an algorithm for ZKp of Identity [11] in 1988. They decreased the information need to be exchanged by increasing the amount of computation. In [12]

---

[1] Supported by the National Natural Science Foundation of China under Grant No. 90204010

Brandt carried out ZK authentication scheme and secret key exchange in a single algorithm. And the identification is mutual, that is to say, both parties identify to one another concurrently. But unfortunately, it is not clear how valid and applicable this algorithm is [14]. In our paper a zero-knowledge proofs of identity protocol based on ElGamal on conic over finite field is proposed.

2   ElGamal Cryptosystem on Conic

Miller [2] applied ellipse curve to cryptography, he showed that elliptic curves had a rich enough arithmetic structure so that they would provide a fertile ground for planting the seeds of cryptography.These years, in China, some people considered that implementing cryptosystem on conic was easier and more convenient than that on ellipse curve.

F_p is a finite field (p is a prime number except 2), considering the conic C over affine plane $A^2(F_p)$:

$$C: y^2 = ax^2 - bx \mod p. \qquad a,b \in F_p, a,b \neq 0 \qquad (1)$$

Origin O (0, 0) is a point on C. if $x \neq 0$, we suppose $y = xt$, and we can infer from equation (1) that

$$x(a - t^2) = b \mod p. \qquad (2)$$

if $a \neq t^2$ then

$$x = b(a - t^2)^{-1}.$$
$$y = bt(a - t^2)^{-1}.$$

Paper [6] defined the characteristic of operator $\oplus$ for points on $C(F_p)$(where $C(F_p)$ means conic C over finite field). The origin O can be denoted by $P(\infty)$. Let H represent the set of $\{t \in F_p, t^2 \neq a\} \cup \{\infty\}$.

For every point $P(t) \in C(F_p)$, $t \in H$, defining:

$$P(t) \oplus P(\infty) = P(\infty) \oplus P(t) = P(t). \qquad (3)$$

$$\underbrace{P(t) \oplus \ldots \oplus P(t)}_{n} = nP(t).$$

For points $P(t_1), P(t_2) \in C(F_p)$, $t_1, t_2 \in H$, $t_1, t_2 \neq \infty$, $P(t_1) \oplus P(t_2) = P(t_3)$

$$t_3 = \begin{cases} (t_1 t_2 + a)(t_1 + t_2)^{-1}, & (t_1 + t_2) \neq 0 \\ \infty, & (t_1 + t_2) = 0 \end{cases}. \qquad (4)$$

For points $P(t_1), P(t_2), P(t_3) \in C(F_p)$, $(P(t_1) \oplus P(t_2)) \oplus P(t_3) = P(t_1) \oplus (P(t_2) \oplus P(t_3))$ is also easy to be proved.The operator $\oplus$ on conic over finite field satisfies commutative and associative law of composition, and $(C(F_p), \oplus, P(\infty))$ constitutes a finite Abel group.

3 ZKp of Identity on Conic

Our protocol can be implemented on ellipse curve as defined above as well as on conic as defined below:

One selects a conic over finite field F_p (p is a prime number except 2), and a point $P \in C$. Peggy chooses a random integer $a \in \{1, 2, \cdots, p-1\}$ as Peggy's private key. Peggy computes aP, and uses (C, P, aP) as her public key.

The simple version of our identification protocol can be described as follows:

(1) Peggy chooses an integer k randomly, and computes $X = a(kP)$. Then she sends X to Victor.
(2) Victor generates a bit of random binary digit b. If $b=0$, $M=0$; otherwise Victor generates a random integer i, and sets $M=iP$; Victor sends (b, M) to Peggy.
(3) If $b=0$, Peggy sends k to Victor; or else Peggy computes $Y = a(M)$, and sends Y to Victor.
(4) If $b=0$, Victor verifies if X is equal to $k(aP)$. So he believes that Peggy knows the value of k or not. If $b=1$, Victor verifies if Y is equal to $i(aP)$. So he is convinced that Peggy knows the value of a or not.

This is a single round of the protocol. Peggy and Victor can repeat it t-rounds until Victor concludes that the prover is Peggy indeed or not.

Let's suppose that an adverse person Alice wants to cheat Victor that she is Peggy. She can find out an appropriate integer k and cheats Victor when b is equal to zero though she doesn't know the value of a. But when b is not equal to zero, Alice can't find out the value of i from iP, so she can't choose an appropriate $i(aP)$ to cheat Victor. Obviously, it is impossible for Alice to deceive Victor in both cases. The probability for Alice to cheat Victor in one round is 50% even if the difficulty of extract i from iP and P is not taken into consideration. After t-rounds of protocol, the probability for Alice to cheat Victor successfully is 2^{-t}. We deem that with this protocol no one except Peggy can prove to Victor that "I am Peggy".

In order to convince Bob that he "is" Peggy, Victor can send integer k received from Peggy to Bob. There is 1/2 chance for Bob that chooses the same value of b with Victor. So after one round of the protocol, for Victor, the chance of cheating Bob is 50%. After t rounds of this protocol, the chance of succeeding for Victor is not more than 2^{-t}. So we deem that nobody can prove to others that he is the person who has proved identity to him before.

4 Chess Grandmaster Leak Analysis

The default condition of carrying out our protocol correctly is that all of the participants follow the protocol [13]. Now we assume that there is an adversary named Mallory. Mallory pretends to be a friendly party, while she doesn't follow the protocol. Mallory acts as a middle-man. It is obviously that Mallory doesn't know Peggy's private key a, but she makes Victor convinced that she knew it. Mallory acts as a middleman in this process. It is Peggy who proves to Victor that she know the value of a. But the result is that Victor believed that Mallory knew a, and believed that Mallory had the identity of "Peggy". This is similar to the problem of chess grandmaster in ZKp of identity. Y.Desmedt pointed out the same kind of attack aimed at identity proofs of Feige-Fiat-Shamir in [4]. If Mallory doesn't generate b randomly, she could send b obtained from Victor to Peggy. In this case Mallory can easily impersonate Peggy by referring Victor's questions to Peggy and answering as Peggy does.

The author of [4] introduced some solutions applied to Smart Card and proposed a new solution which makes use of a trusted active warden. The difference between active warden and former passive warden is that the warden does not only listen to catch up subliminal senders, but also interacts in the communication in a special way to better enforce the subliminal freeness. In other words, he participates actively in the communication between all participants and can modify the information that is communicated. But there are some shortcomings for this solution. It is not easy to understand and complicated to be implemented. Moreover its security depends on warden extremely.

Our main idea is to take an overtime limit $t_{timeout}$ into consideration, and enhance our protocol through simple modification.

5 Unsolved Problems

The value of round number t affects the speed and security degree of our protocol directly. If t is too large, the demand on security is satisfied. But the speed of protocol drops because the calculation is increased. If t is not large enough the speed of protocol increase, but the security is not guaranteed in this case. Further more the value of t has impact on the characteristic of ZKp. So the selection for t is basilic.

Peggy's commitment integer k in the first step can be regard as a commitment Peggy makes to Victor. Its randomicity also affects the security degree and characteristic of ZKp. The randomicity of Victor's random binary digit b in the second step is the same. We won't emphasize these problems here.

The enhanced protocol with overtime limit is simple in form, but it is difficult to decide the $t_{timeout}$. The overtime limit $t_{timeout}$ must satisfy two terms:

1) The probability that timeout is true equals to zero with negligible probability if Mallory is absent;

2) The probability that timeout is false equals to zero with negligible probability if Mallory tries to intervene in the protocol. Moreover, without the ideal condition (participants are synchronous), the timestamps is difficult to deal with.

6 State of the Art

In 1985, T.ElGamal proposed El-Gamal signature scheme, and its security relied on the difficulty of computing discrete logarithms over finite fields in [10]. It can be used to encrypt messages as well as sign a document. N.Koblitz and Miller [2] presented encryption based on ellipse curve respectively in the same year. Most of the encryption can be implemented on ellipse curve over finite field. In 1988, Beth brought a Fiat-Shamir like authentication protocol for the El-Gamal Scheme forward in [9]. Cao Zhenfu used conic instead of ellipse curve, and simulated El-Gamal, Massey-Omura and RSA in [3] and [8] apart.

7 Conclusion

We have presented a ZKp identify protocol based on ElGamal on conic and proposed the parallel version of it. Then a potential leak to our protocol is described, and a description for the enhanced version of it is given. At last we described some problems that have not been solved in this paper. Some implicit preconditions were considered ideally in this paper though the implementation in real world was more complicated.

References

1. U. Feige, A. Fiat and A. Shamir, "Zero Knowledge Proofs of Identity," Journal of Cryptology, Vol.1, No. 2, 1988, pp.77-94.
2. V. Miller. "Use of Elliptic Curves in Cryptography," Advances in Cryptology - CRYPTO'85 Proceedings, Springer- Verlag, 1986, pp.417-426.
3. Zhenfu Cao, "A Public Key Crypto -system Based on a Conic Over Finite Fields F_p," CHINACRYPT'98, 1998, pp.45-49
4. Y. Dsemedt, "Abuses in Cryptology and How to Fight Them," Advances in Cryptology - CRYPTO'88 Proceedings, Springer-Verlag, 1990,pp.375-389.
5. O. Goldreich H. Krawcyzk, "On the Composition of Zero-Knowledge Proof Systems,"17th ICALP, 1990,pp. 268-282.
6. Mingzhi Zhang, "Factoring Integers With Conic," Journal of Sichuan University(Natural Science Edition), Vol.33,No.4, 1996,pp.356-359
7. Dingyi Pei, Xueli Wang, "Encryption Authentication Code on Conic over Finite Field," Science In China (Series E), Vol.26 No.5 1996,pp.385-394
8. Zhenfu Cao "Conic analog of RSA cryptosystem and some improved RSA cryptosystems", Journal of Natural Science of Heilongjiang University, Vol.16 No.4, 1999,pp15-18
9. T.Beth, "Efficient zero-knowledge identification scheme for smart cards," Advances in Cryptology: Proceedings of Euro-crypt '88, Springer-Verlag, NY1988, pp77-84
10. ElGamal, "A public key cryptosystem and a signature scheme based on discrete logarithms." IEEE Trans. Inform. Theory, 31(4): 1985,pp469-472.
11. L.C.Guillou J-J.Quisquater, "A practical zero-knowledge protocol fitted to security microprocessor minimizing both transmission and memory," Advances in CryptologyEUROCRYPT'88 proceedings, Springer-Verlag, 1988,pp123-128
12. Jorgen Brandt, Ivan Damgard, Peter Landrock, et al. "Zero knowledge scheme with secret key exchange." In Proceedings on Advances in Cryptology, 1990,pp 583-585
13. O.Goldreich, "Foundations of crypto- graphy: basic tools," Cambridge: New York, Cambridge University Press, 2001,pp270-274
14. J. Binder and H-P. Bischof. " Zero knowledge proofs of identity for ad hoc wireless networks," 2003. http://www.cs.rit.edu/~jsb7384/zkp-survey.pdf
15. I.Foster, C. Kesselman, G. Tsudik, and S. Tuecke. A security architecture for computational grids. In ACM Conference on Computers and Security, pages 83–91. ACM Press, 1998.

A Load-Balanced Parallel Algorithm for 2D Image Warping

Yan-huang Jiang, Zhi-ming Chang, and Xue-jun Yang

School of Computer Science, National University of Defense Technology,
Changsha, 410073, HuNan Province, P.R. China
yanhuangjiang@21cn.com

Abstract. 2D image warping is a computation-intensive task and plays an important role in many fields. All existing parallel image warping algorithms are only suitable for limited geometric transformations. This paper proposes a distributed parallel image warping algorithm PIWA-LIC, which partitions the valid area of the output image in a balanced way, and each computing node does resampling for one sub output image. To guarantee data locality, a line segment approximation (LSA) method is exploited to get the corresponding input area for each sub output image. For any 2D geometric transformations with one-to-one mapping, PIWA-LIC achieves high performance because of good load balance and data locality.

1 Introduction

2D image warping [1] deals with a 2D geometric transformation between two images. It has played an important role in a variety of applications, such as medical imaging [2], remote sensing [3] and computer vision [4].

Image warping methods can be classified as forward mapping, inverse mapping, and separable mapping [5]. All the three strategies are computationally intensive and time consuming. Parallel processing is an effective way to speed up computing in many applications. Existing parallel algorithms for 2D image warping mainly include: Warpenburg [6] developed a parallel resampling method through adjusting local offsets under a SIMD environment. The performance of the algorithm will degrade with the severity of the skew between the original and resampled images. Wittenbrink [7] presented a parallel image warping algorithm under a SIMD architecture. The algorithm is only suitable for nonscaling affine transform including skewings, translations, and rotations. Sylvain [8] proposed a load-balanced parallel algorithm for forward mapping methods, yet it can't partition the input image properly for complex geometric transformations. Lee [9] implemented a parallel convolution method on a network of workstations, which can be only used to enlarge or shrink images. Jiang [10] exploited a distributed parallel image warping algorithm (We name it PIWA-LOC, which is the abbreviation for Parallel Image Warping Algorithm based on Local Output-area Computing) with good data locality. However, for some complex geometric transformations, the load distribution is unbalanced.

This paper presents a distributed parallel image warping algorithm based on local input-area computing (PIWA-LIC) for inverse mapping methods. PIWA-LIC partitions the load of resampling computation uniformly, and achieves data locality through calculating the input area for each sub output image. PIWA-LIC is suitable for any 2D geometric transformations with one-to-one mapping.

In the following sections, the discussed distributed parallel environments adopt host-node mode, where one node is the managing node engaged in global operations, and the others are computing nodes dealing with parallel computing.

2 Limitation of PIWA-LOC

PIWA-LOC [10] divides the input image into p subimages with equal sizes, where p is the number of computing nodes. Each computing node saves one of the subimages, then gets the corresponding area in the output space for the locally saved sub input image, and does resampling for the output pixels in the area. At last, all sub output images are gathered and stitched into the integrated output image. PIWA-LOC has good data locality and saves a lot of data communication time, since almost all the data needed during parallel computing are stored in local memory.

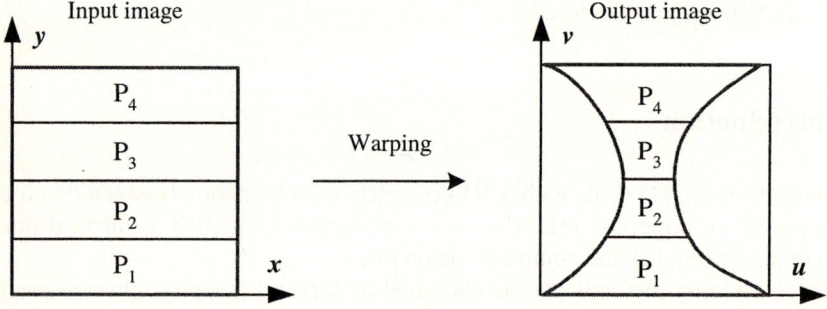

Fig. 1. Unbalanced load distribution of PIWA-LOC

However, for some complex geometric transformations, regular partition of the input image may lead to unbalanced load distribution. Fig. 1 depicts such an instance, where the regular vertical partition of the input image makes the load of computing nodes P_1 and P_4 much more than that of P_2 and P_3. To sufficiently exploit the performance of each computing node, we propose a load–balanced algorithm PIWA-LIC for 2D image warping in the next section.

3 PIWA-LIC

Different from PIWA-LOC, PIWA-LIC partitions the output image into p sub output images with almost the same computational load. Each computing node does resampling for one sub output image, and saves the corresponding input image blocks of the sub output image locally.

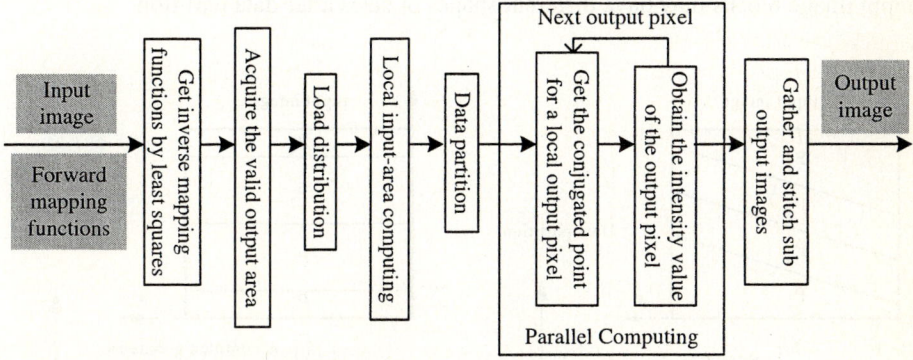

Fig. 2. Flow chart of PIWA-LIC

Fig. 2 shows the flow chart of PIWA-LIC. The main phases of the algorithm are as follows:

1. Get inverse mapping functions: This step is just the same as that of PIWA-LOC.
2. Get the valid area of the output image: We use LSA method (see section 3.3) to get the corresponding output area for the input image according to the given forward mapping functions.
3. Load partition: The computational load of the image warping task can be measured by the number of the pixels in the valid output area. We distribute the load in a balanced way, where each computing node does resampling for almost the same number of output pixels (see section 3.4).
4. Local input-area computing: For each sub output image, we get its corresponding input area. In this step, LSA method is also used to obtain the irregular-shaped input image blocks.
5. Data partition: The IIB structures (see section 3.2) of both one sub output image and the corresponding input area are sent to a related computing node.
6. Local resampling: Each computing node does resampling for all the pixels in the local sub output image.
7. Data gathering: All the resulting sub output images are gathered and stitched into the integrated output image on the managing node.

Among all the above phases, the local resampling phase is implemented parallelly on the computing nodes, while all the other phases are global operations realized on the managing node.

3.1 Load Balance and Data Locality of PIWA-PIC

Fig. 3 depicts the load balance and data locality of PIWA-LIC, where Fig. 3(a) shows the processing of a rotation, and Fig 3(b) shows the processing of a complex geometric transformation. After load distribution, each sub output image is irregularly shaped and has almost the same number of pixels in it. Partition of the input image is decided by the result of local input-area computing. From Fig. 3, we can see that, the input image blocks may have different shapes or sizes after data partition.

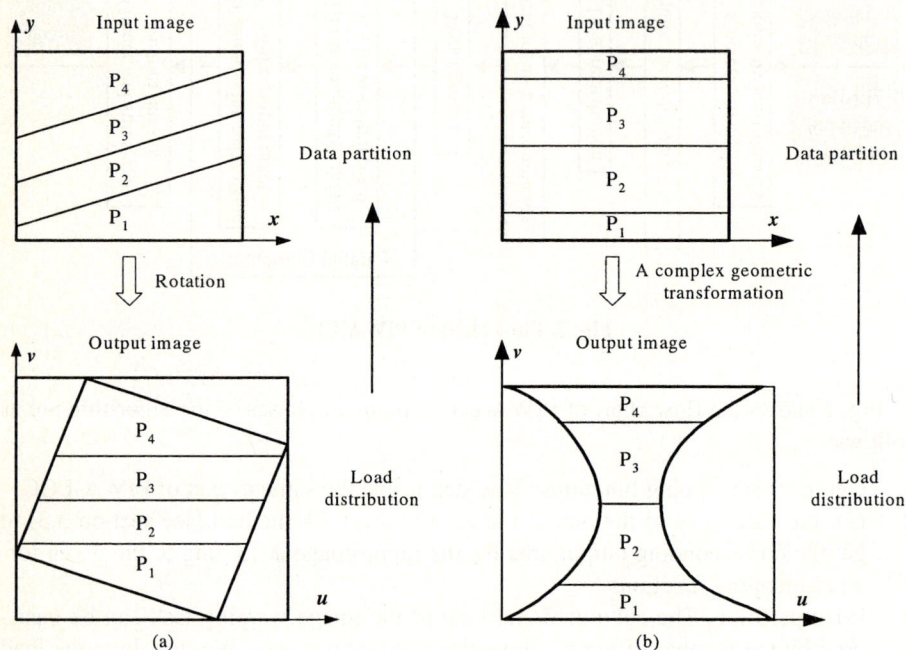

Fig. 3. Load balance and data locality of PIWA-LIC: (a) rotation; (b) a special warping

PIWA-LIC is a load-balanced algorithm, since it distributes the computational load uniformly. Furthermore, PIWA-LIC is a parallel algorithm with good data locality, since all the needed input data are stored locally during the local resampling phase. The above properties make PIWA-LIC a suitable approach for any 2D geometric transformations with one-to-one mapping.

3.2 IIB Structure

To facilitate processing, we adopt a dedicated data structure named IIB (Irregular-shaped Image Block) structure to save irregular-shaped image blocks, which is firstly provided in reference [10]. Besides the intensity values of the pixels, IIB structure has several information items used to decide the coordinates of the pixels. Let the image block be in $[u, v]$ space, these items can be described as follows:

v_1 : the minimum v-coordinate value of the pixels in the image block;
L : the number of scanlines on which there are valid pixels;

s_i: the number of pixel segments on the $i^{th}\,(1\leq i\leq L)$ scanline of the image block;

u_{ik}: the u-coordinate value of the first pixel on the $k^{th}\,(1\leq k\leq s_i)$ pixel segment;

r_{ik}: the number of pixels on the $k^{th}\,(1\leq k\leq s_i)$ pixel segment;

IIB structure ignores the background pixels, and only saves the valid pixels line by line. Figure 4(a) depicts an irregular-shaped image block *abcd* in the image ABCD. There are two pixel segments on the ith scanline of *abcd*. The starting u-coordinate values of these two pixel segments are u_{i1} and u_{i2}, and the numbers of pixels on them are r_{i1} and r_{i2} respectively.

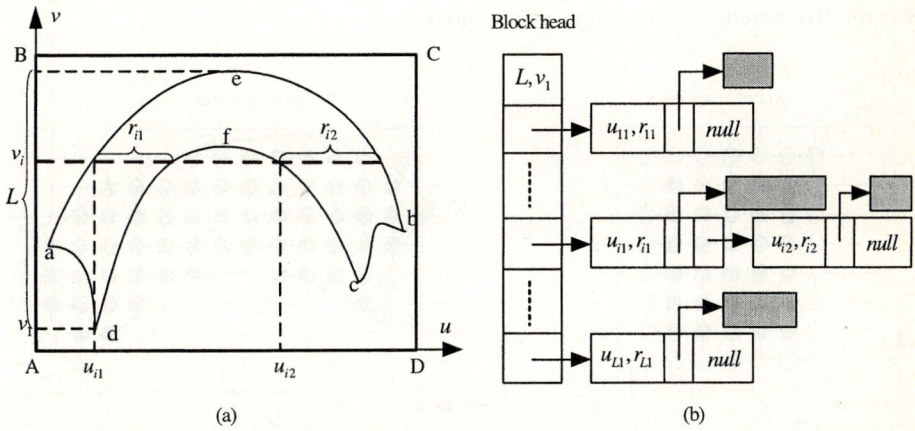

Fig. 4. Data structure for irregular-shaped image blocks: (a) information items of the irregular-shaped image block *abcd*; (b) IIB structure of *abcd*

IIB structure has two sub data structures. One is *BlockHead*, which saves L, v_1, and a pointer array with L elements. Each element points to the pixel segment information of the corresponding scanline. The other sub structure is *SegHead*, which saves the information of each pixel segment, including u_{ik}, r_{ik}, memory address of the intensity values for these r_{ik} pixels, and the address of the next pixel segment on the same scanline. The definitions of these two sub data structures are as follows:

```
struct BlockHead                    struct SegHead
{                                   {
    int    StartVertical;               int    StartHorizontal;
    int    BlockHeight;                 int    SegWidth;
    struct SegHead** SegInfo;           int*   Intensity;
}                                       struct SegHead* NextSeg;
                                    }
```

Fig. 4(b) gives the IIB structure of *abcd* in Fig. 4(a). The value of item s_i need not be saved since it can be gotten from *SegHead* structure.

To save irregular-shaped image blocks, IIB structure has two advantages: (1). The data structure ignores invalid output pixels (background pixels), so it saves memory space and data communication time for PIWA-LIC; (2). The data structure provides both the coordinate values and the intensity value for each valid pixel, which affords facilities for local resampling phase.

3.3 Line Segment Approximation Method

Both phase 2 and phase 4 of PIWA-LIC use line segment approximation (LSA) method to get the transformed area of an image block. LSA method is provided to acquire the target area of a source image block after a geometric transformation. The input of LSA method is the IIB structure of the source image block, and the output of it is the IIB structure of the target image block.

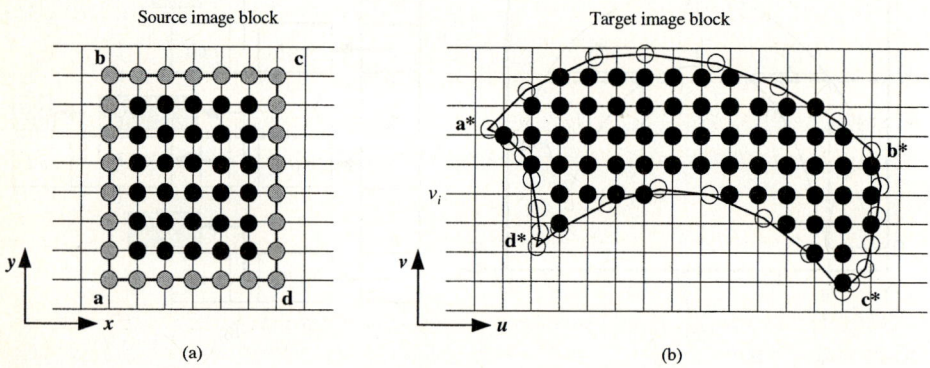

Fig. 5. The area of image blocks before and after geometric transformation: (a) boundary pixels of the source image block; (b) approximation boundary for the target image block

LSA method gets the conjugated point for each boundary pixel of the source image block, then uses line segments connected end to end with each other to depict the ideal boundary of the transformed image block, where two end points of each line segment are the conjugated points of two neighbored source boundary pixels. All line segments will shape a closed area, whose information is saved as an IIB structure. Let the source space be [x, y], and the target space is [u, v], the following is the detailed description of LSA method:

1. Get all the boundary pixels of the source image block in sequence: In Fig. 5(a), all gray dots are the boundary pixels of the source image block. We can select any of the boundary pixels as the starting one, and save them in clockwise direction.
2. Compute the conjugated points of the source boundary pixels: We access the source boundary pixels in the same sequence as step 1. For each boundary pixel (a,b), we get its conjugated point (c,d), where $c = T_1(a,b)$ and $d = T_2(a,b)$. Thereinto, T_1 and T_2 are the transform functions from the source space to the target space for the target horizontal and vertical coordinate values respectively.

3. Acquire the item values in *BlockHead* structure for the target image block: $v_1 = \lceil v_{min} \rceil$, $L = \lfloor v_{max} \rfloor - \lceil v_{min} \rceil + 1$, where v_{min} (v_{max}) is the minimum (maximum) *v*-coordinate value of the conjugated points obtained in step 2. We let $v_0 = \lceil v_{min} \rceil - 1$ and $i = 0$.

4. Obtain all the pixel segments on the $(i+1)^{th}$ scanline of the target image block:

    (1). If $i \geq L$, LSA method has been finished; else $i = i+1$, $v_i = v_{i-1} + 1$.

    (2). Compute the intersections of the ith scanline with line-segment boundary: We access all the conjugated points in sequence. Suppose there are two neighbor conjugated points *s* and *s'*, which satisfy $v_s \leq v_i \leq v_{s'}$, the *u*-coordinate value of the intersection *t* is $u_t = \dfrac{v_i - v_{s'}}{v_s - v_{s'}} \cdot (u_s - u_{s'}) + u_{s'}$. This step is repeated until all the conjugated points have been accessed.

    (3). Get the pixel segments on this scanline: After step (2), we will get even number of intersections on the scanline. Let the sequence of their *u*-coordinate values be $u_{l1}, u_{r1}, u_{l2}, u_{r2}, \cdots, u_{lm}, u_{rm}$ from the left to the right of the scanline, there are *m* pixel segments on this scanline. The starting *u*-coordinate value of the kth $(1 \leq k \leq m)$ pixel segment is $u_{ik} = \lceil u_{lk} \rceil$, and the number of the pixels on it is $r_{ik} = \lfloor u_{rk} \rfloor - \lceil u_{lk} \rceil + 1$. In Fig. 5(b), there are two pixel segments with 4 and 6 pixels on them respectively for the scanline $v = v_i$.

5. Turn to step 4, and calculate the pixel segment information for the next scanline.

After the above process, we will get all the information items of the IIB structure for the target image block.

3.4 Load Distribution

According to the information items of the IIB structure for the valid output area obtained in phase 2 of PIWA-LIC, we can distribute the computational load in a balanced way. The load distribution method can be described as follows: (1) Get the number of all the pixels in the valid output area; (2) Divide the area into *p* sub output image, each of which has almost the same number of valid output pixels.

For the output image, we let *L* be the number of scanlines that have valid pixels, s_i be the number of pixel segments on the ith $(1 \leq i \leq L)$ scanline, and r_{ij} be the number of pixels on the jth pixel segment $(1 \leq j \leq s_i)$, the number of total valid output pixels is:

$$Total = \sum_{1 \leq i \leq L} r_i, \text{ where } r_i = \sum_{1 \leq j \leq s_i} r_{ij}$$

The best load balance is that every computing node does resampling for the same number of valid output pixels, which equals to $\bar{r} = Total / p$. In PIWA-LIC, the pixels on the same scanline are not distributed to different computing nodes. Fig. 6 gives the pseudo code of our load distribution method.

$k=1$; $i=0$; $m=0$;
While ($k \leq p$)

    1. Initialize: $num_k=0$; $flag=1$; $m=i$
    2. while ($i < L$ && $flag=1$)
        1. $i=i+1$;
        2. $num_k=num_k+r_i$;
        3. If ($num_k \geq \bar{r}$ || $(\bar{r}-num_i) < r_{i+1}/2$) then $flag=0$;
    3. The valid output pixels from the mth scanline to the ith scanline are allocated to the kth computing node;
    4. $k=k+1$;

Fig. 6. Pseudo code of the load partition

By means of the above load partition method, each computing node does resampling for almost the same number of valid output pixels, so PIWA-LIC is a load-balanced parallel algorithm. The stitching phase of PIWA-LIC is simpler than that of PIWA-LOC since the valid pixels of one scanline are in the same sub output image.

3.5 Data Partition and Local Resampling

After phase 4, we will get the information items of the IIB structure of the corresponding input image block for each sub output image by LSA method. The intensity information of the IIB structure can be obtained easily since the position of each pixel in the input image blocks is known. In phase 5, the IIB structure of each sub output image and the IIB structure of the corresponding input image blocks are sent to a related computing node.

In local resampling phase, the position of each pixel in the locally saved sub output image is computed firstly. Let the position be (u,v), the inverse mapping functions are used to acquire its conjugated point (x,y) in the input space. According to the neighbor input pixels of (x,y), we can get the intensity value of (x,y) through convolution. This value is just the intensity value of the output pixel (u,v).

After local resampling, each computing node gets the intensity values for all pixels in the local sub output image. At last, all the resulting sub output images are transferred to the managing node and stitched into the integrated output image.

4 Experimental Results

We compare the performance of PIWA-LIC with that of PIWA-LOC under two distributed parallel platforms. The first platform is a cluster system (CL) which has 16 PCs connected by 100Mbps fast ethernet switch. Each PC is configured with a Pentium4-2GHz CPU and 1GB RAM, and the operating system is LINUX. The second one is a parallel machine (PM) with 16 nodes, and the network topology is a fat tree with transfer speed of 1.2Gb/s. Each node has 1GB local memory, and the operating system is UNIX. The communication of the algorithms is implemented based on MPI.

Two image warping tasks are selected to evaluate the performance. Both tasks have the same input image with the size of 10000×10000. The inverse mapping functions for either of the tasks are complete cubic polynomials obtained by least squares. Intensity interpolation is implemented by cubic convolution [11]. Fig. 8 gives the input and output images of these two tasks. For the first task, both PIWA-LOC and PIWA-LIC has good load balance and data locality. For the second one, PIWA-LIC achieves better load balance than PIWA-LOC.

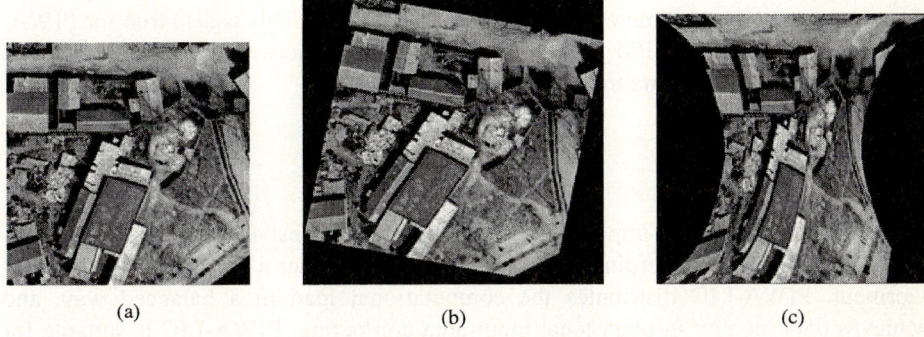

Fig. 8. Results of the image warping tasks: (a) input image; (b) output image of task1; (b) output image of task2

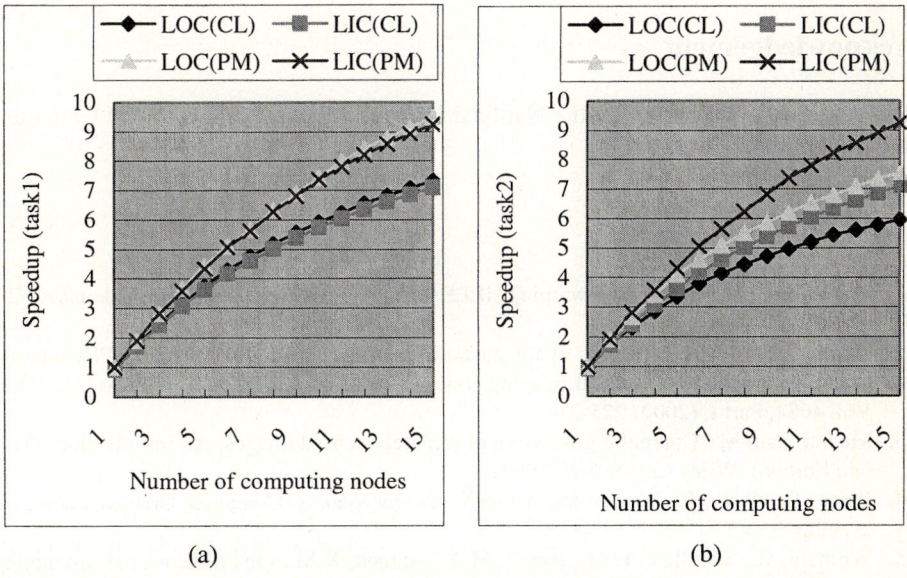

Fig. 9. Speedup results of PIWA-LOC and PIWA-LIC: (a) task1; (b) task2

The time results of the serial processing for the first task (task1) under CL and PM are 173.765s and 156.869s respectively, and those of the second task (task2) are 166.382s and 152.347s accordingly. That is, the computing capabilities of the nodes in CL and PM platforms have only a slight difference.

We abbreviate PIWA-LOC and PIWA-LIC to LOC and LIC respectively. Fig. 9(a) shows the speedup results of PIWA-LOC and PIWA-LIC for task1. We can see that the performance of these two algorithms has less difference. Fig. 9(b) illustrates the speedup results for task2, where PIWA-LIC gets much higher performance than PIWA-LOC under the same platform. This result indicates that PIWA-LIC suitable for more geometric transformations than PIW-LOC because of better load balance. All results illustrate that the performance of PIWA-LIC implemented on PM platform is much better than that implemented on the CL platform. This is also true for PIWA-LOC. The reason is that PM platform has much higher network bandwidth than CL one, which reduces the data transfer time considerably.

5 Conclusion

2D image warping is an important and computation-intensive task. This paper proposes a parallel image warping algorithm PIWA-LIC under a distributed parallel enviorment. PIWA-LIC distributes the computational load in a balanced way, and achieves data locality through local input-area computing. PIWA-LIC is suitable for more complex geometric transformations than PIWA-LOC, and achieves good parallel performance, especially under the distributed parallel systems with high network bandwidth.

Acknowledgement

This research is supported by the National Natural Science Foundation of China under grant number 69825104.

References

1. Wolberg, G.: Digital image warping. IEEE Computer Society Press, Los Alamitos, CA (1990)
2. Unser, M.: Splines: a perfect fit for medical imaging. Proceedings of the SPIE International Symposium on Medical Imaging: Image Processing (MI'02), San Diego CA, USA, Vol. 4684, Part I, (2002) 225-236
3. Mather, Paul M.: Computer processing of remotely-sensed images: An Introduction, (Second Edition). Wiley Chichester, (1999)
4. Beier, T., Neely, S.: Feature-based image metamorphosis. Computer Graphics, Vol. 26. 2(1992), 35-42
5. Wolberg, G., Sueyllam. H.M., Ismail, M.A., Ahmed, K.M.: One-dimensional resampling with inverse and forward mapping functions. Journal of Graphics Tools, Vol. 5. 3 (2001) 11–33

6. Warpenburg, M.R., Siegel, L.J.: SIMD image resampling. IEEE Transactions on Computers, Vol.31. 10(1982) 934-942
7. Wittenbrink, C.M., Somani, A.K.: 2D and 3D optimal image warping. Journal of Parallel and Distributed Computing, Vol.25. 2(1995) 197-208
8. Sylvain, C.-V., Miguet, S.: A load-balanced algorithm for parallel digital image warping. International Journal of Pattern Recognition and Artificial Intelligence, Vol.13. 4(1999) 445-463
9. C.-K., Hamdi, M.: Parallel image processing applications on a network of workstations. Parallel Computing, Vol.21. (1995) 137-160
10. Jiang, Y.-H., Yang, X.-J., Dai, H.-D., Yi, H.-Z.: A distributed parallel resampling algorithm for large images. APPT2003, Xiamen, China, September 2003. Lecture Notes in Computer Science, Vol. 2834. Springer Verlag (2003) 608-618
11. Keys, R.G.: Cubic convolution interpolation for digital image processing. IEEE Transactions on Acoustics, Speech, and Signal Processing, Vol.29. 6(1981) 1153-1160

A Parallel Algorithm for Helix Mapping Between 3D and 1D Protein Structure Using the Length Constraints

Jing He, Yonggang Lu, and Enrico Pontelli

Dept. of Computer Science, New Mexico State University,
Las Cruces, NM 88003, USA
{jinghe, ylu, epontell}@cs.nmsu.edu

Abstract. Determining 3-dimensional (3D) structures of proteins is still a challenging problem. Certain experimental techniques can produce partial information about protein structures, yet not enough to solve the structure. In this paper, we investigate the problem of relating such partial information to its protein sequence. We developed an algorithm of building a library to map helices in a 3D structure to its 1-dimensional (1D) structure using the length constraints of helices, obtained from such partial information. We present a parallel algorithm for building a mapping tree using dynamic distributed scheduling for load balancing. The algorithm shows near linear speedup for up to 20 processors tested. If the protein secondary structure prediction is good, the library contains a mapping that correctly assigns the majority of the helices in the protein.

1 Introduction

Proteins are responsible for nearly every function required for life. The sequence of a protein is considered as the primary (1D) structure and is composed of a chain of 20 types of amino acids. A functional protein can be thought of as a properly folded chain of amino acids in 3-dimensional space (3D). The 3D structure of a protein is crucial in understanding its function, because the function of a protein is determined by the 3D structure rather than the primary sequence. Due to the constraints imposed by the local environments in a folded protein and the linearity of an amino acid chain, segments of protein exist as three types of *secondary structure* elements: helices, β-sheets and coils in a folded protein. Figure 1 shows the sequence of a protein, 184 amino acids long, and its corresponding 3D structure from the Protein Data Bank (PDB). Each of the helices in 3D corresponds to a segment of the protein sequence. The length of a helix is the number of amino acids it has. The four helices with length greater than five are labeled with their corresponding amino acids on the sequence.

3D protein structure determination is a challenging problem particularly for large protein complexes and membrane proteins that are not easy to crystallize[1], [2]. Electron cryomicroscopy is a potential experimental technique to determine 3D structures of these proteins to atomic resolution [2]. Current advances of electron cryomicroscopy technique are able to determine a rough protein shape in the form of the density distribution of electrons. Such a rough protein shape is usually called the *density map* of a protein and is at resolution of 6-9 Å, the intermediate resolution [3], [4]. At this resolution, amino acid side chains that can be used to distinguish different amino acids are not generally visible. Therefore, it is a challenging problem to identify the trace of amino acid sequence in 3D space. For example, it is not clear in general where the beginning of the protein sequence is located in the 3D density map. The density map of the protein (1CC5 in Figure 1) is shown at 7 Å resolution in Figure 2 using the simulated method in the EMAN software package [5]. The orientation of the density map in Figure 2 is the same as that in the structure in Figure 1. The corresponding helices are marked using a stick in the density map, and their lengths can be computationally determined [6]. The helical nature of the helices can be visualized in the density map. However, it is not clear which segment on the protein sequence corresponds to which of the helices in the density map if only the density map is given. In this paper, we use protein secondary structure prediction results as a guide to locate the corresponding sequence segments for the helices visualized in 3D.

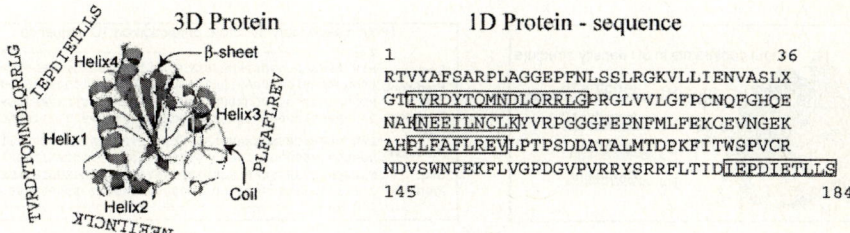

Fig. 1. The 3D and 1D structure of protein 1CC5 in the PDB. Helices in the 3D structure are labeled with their corresponding segments of amino acids (in boxes) on the sequence

For any protein sequences, computational methods such as PHD and PSI-PRED can generate a prediction on possible locations for secondary structures [7], [8], [9]. Hence, likelihood for each amino acid being involved in a helix is provided by the prediction method. Figure 2 shows an example of secondary structure prediction with the predicted likelihood, expressed as a value in the range of 0 to 9, where 9 indicates the highest likelihood of presence of a helical structure. The prediction was obtained using the PHD prediction server. As many prediction methods, protein secondary structure prediction is not 100% correct. On average, the accuracy of the secondary structure methods is about 76-78% for the best prediction methods [9]. Such an error in the protein secondary structure prediction potentially generates a lot of freedom in mapping the 3D helices to the 1D sequence using the prediction as a guide.

2 The Problem

The methodology proposed in this paper is outlined in Figure 2. Let us assume that we are given the length distribution of the helices in a 3D protein density map (illustrated in the middle box in Figure 2), obtained from the data collected from electron cryomicroscopy technique. Then our method will produce a library of possible mappings of such helices onto the protein sequence. The idea is to combine the result of protein secondary structure prediction and the helix length information to map helices in a way to maximize the likelihood for all the helices.

Because of the exponential nature of the problem and the errors in the secondary structure prediction result, the method proposed here does not attempt to generate all the possible mappings for the helices in the 3D density map. Rather, it generates a subset of all the possible mappings which is likely to contain the true solution. For example, if the secondary structure prediction is 100% correct, the goal of this method is to generate a subset of all the possible mappings in which the true solution is ranked on top. Because of the use of the secondary structure prediction as a guide, the helices in 3D are mostly placed near the high score populations (e.g. 8999999998) of the "pred-conf" that indicates the likelihood of having helices. Thus, the problem becomes how to assign 3D helices with known lengths onto the protein sequence so that the score of the overall assignment is above certain threshold.

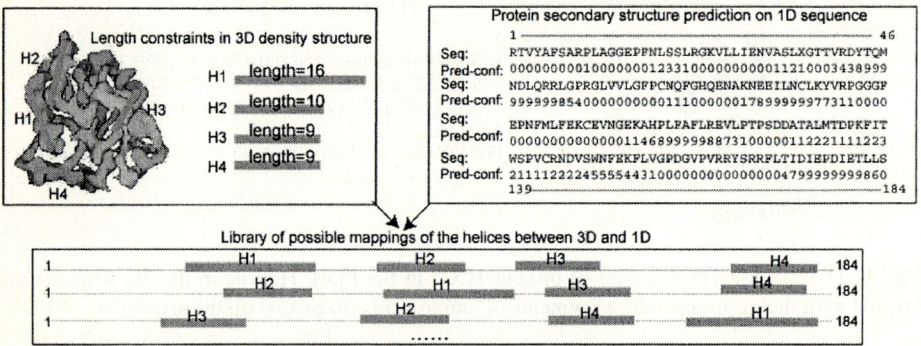

Fig. 2. The approach of building mapping library for helices between 3D and 1D structure by combining the information from protein secondary structure prediction and the protein density map at intermediate resolution

In this paper, we approached the problem of identifying the sequence segments for helices in 3D by building a library of possible mappings. Each element in the library suggests an overall arrangement of the sequence in 3D by mapping segments of sequence to the helices in 3D. The significance of building such a possible mapping library is that the library can be used as a starting point for further structural prediction. Since the problem addressed in this paper is closely related to the very recent advances in the electron cryomicroscopy technique, we are not aware of any previous approaches to the problem.

3 Building Mapping Library

The goal of our method is to map a given number of 3D helices to the protein sequence so that the score of the overall assignment is higher than a threshold provided by a user. We define the score of assigning n helices as the sum of the confidence for all the positions where the n helices are assigned. In particular, Assignment score = Σ (Pred-conf$_i$) where i $\in$ S, and S is a set of positions, where helices are assigned, on the sequence. The confidence (Pred-conf) at each amino acid position is the likelihood for that amino acid being in a helix (in our case, determined using PHD). The average score of assigning n (n>=1) helices is the score of assigning the n helices divided by the total length of the n helices. For example, the assignment of Helix 4 in Figure 1 is 77 and its average score is 7.7 (using Figure 1 and Figure 2).

The overall process of building the helix mapping library consists of two steps: building an initial library and assembling a mapping tree. The step of building an initial library is not computationally intensive, and it is not paralleled. The computationally intensive step of building a mapping tree is paralleled.

3.1 Initial Library

The first step is to build an initial library that provides a list of possible locations on the sequence for each helix in 3D. For example, for helix H_1 (from Figure 2) that is visualized in 3D, a list of possible assignment locations ($L^1_1, L^1_2, ..., L^1_{n1}$) is generated by scanning the protein sequence using a window of length 16 (Figure 2 and 3). The number of elements in each list is smaller than m-h+1 where m is the length of the protein sequence and h is the length of the helix. This is because we eliminated those elements with average score less than 3. In order to reduce the size of the mapping tree, we eliminated the locations that are less likely to be selected for an overall assignment that results in a high score. Moreover, each list in the initial library is arranged in the decreasing order of the score. For example, L^1_1 is the location with the highest score for assigning H_1 among $L^1_1, L^1_2, ..., L^1_{n1}$. Again, the initial library gives priority to the locations with high scores. This is because the locations on top of the lists will be considered first in the next step of building the mapping tree.

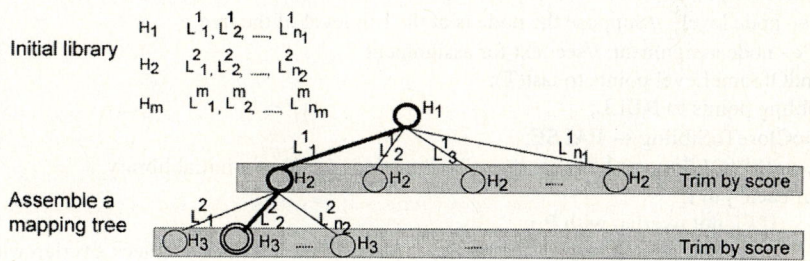

Fig. 3. Building a mapping tree in two steps: initial library and assembling the tree

3.2 Mapping Tree

3.2.1 Overall Approach

The first step provides assignment choices for each helix. During the first step, helices are dealt independently from each other. The second step is to assemble the choices to form a single assignment for all the helices on the sequence. Each node in the mapping tree corresponds to the assignment of a helix to a certain location. Each path from the root to a leaf is a possible assignment for all the helices in 3D. It is noticed that long helices are predicted more correctly than shorter helices in general (data not shown), presumably because the signal of a long helix is easier to be identified than that of a shorter helix. During the mapping tree assembly, priority is given to longer helices. Thus, the longer helices are assigned in the higher levels while the shorter helices appear in the nodes close to the leaves.

3.2.2 Adding a Node to the Mapping Tree

The computation of tree building is a process that expands a node of the tree by adding children to it (Figure 3). The nodes that are to be expanded are stored in one (in sequential algorithm) or more (in parallel algorithm) task queues. A node is composed of information about this particular node (such as the level of the tree it resides, the length of the helix) as well as the information about the assignment down to the node. A node contains a linked list of the placement for helices from the root to the current node. For example, if a node is on the 3rd level of the tree (double circled node in Figure 2), node.assignment (line 7 in Algorithm Computation) contains the information of assigning H_1 at L^1_1 and H_2 at L^2_2. For each node n, that is in charge of assigning helix H_i, to be expanded, the procedure scans the list of possible choices (L^i_1, L^i_2, ..., L^i_{ni}) provided in the initial library (line 11 in Computation). Before adding a new branch, the algorithm performs three checks: overlapping, sparseness, and average score (Computation).

1. Algorithm **Computation** (T, MinDistance, Threshold)
2. //This algorithm expands a node in the mapping tree. It takes a task queue (T) of nodes as input and output a task queue with the node replaced with its children.
3. //MinDistance and Threshold are user defined parameters.
4.   node ← last (T);
5.   delete last(T) from T;
6.   i ← node.level; //Suppose the node is at the i-th level of the tree.
7.   P ← node.assignment; //see text for assignment
8.   endOfSameLevel points to last(T);
9.   sibling points to NULL;
10.   tooCloseToSibling ← FALSE;
11.   L ← (L^{i+1}_1, L^{i+1}_2, ..., L^{i+1}_{ni+1}); //possible locations for H_{i+1} in initial library
12.   for each j in L
13.   if (j not overlap with P)
14.   while (sibling != endOfSameLevel && sibling != NULL) //check overlap with left siblings that reside in the tail portion of T
15.   k ← sibling.location;
16.   if (| j – k | < MinDistance)
17.   tooCloseToSibling ← TRUE; //j overlaps with a left sibling, ignore j
18.   sibling ← NULL;
19.   else
20.   sibling--; //points to the next node in T
21.   if (tooCloseToSibling =FALSE)
22.   score ← node.score + score of assigning H_{i+1} to j;
23.   if (score / (node.length + length of H_{i+1}) > Threshold)

24.   CREATE newNode; //update its level, score, location etc.
25.   newNode.assignment ← add j to P;
26.   push newNode to the tail of T;
27.   sibling points to last(T);
28.   return T;

The first is to make sure that the new branch will not result in overlapping placement of helices on the path from the root to the current node (line 13). The second is to ensure that the branches out from the same node are sparse enough to have good representation (lines 14-20). A user given parameter "MinDistance" is used to specify the sparseness. For example, when branch L^2_2 is created from H_2 to H_3, L^2_2 is at least 10 (MinDistance=10) amino acids away from L^2_1 who is to the left of L^2_2 (Figure 3). By introducing MinDistance, the number of branches from each node is reduced because not all the choices in the list of the initial library are used to create branches.

In order to further reduce the size of the tree, we continuously check the assignment score during the construction, and use it to trim the tree in those nodes that represent assignments with insufficient global score (Figure 3). Each node on the tree is associated with a path from the root to the node (Figure 3 and line 7 in Computation). The score of a node is the score of assigning all the helices along such a path (section 3.1). If the average score of a node is less than a given threshold (called Threshold in the algorithm) that is predetermined by the user, the last node on the path is eliminated before any further consideration (line 23 in Computation). As we mentioned earlier, only part of the mapping tree is actually generated, our approach tries to focus on those branches that are most likely to contain the actual solution. Our experiments have indicated that it is generally preferable to keep nodes in the upper levels (i.e. closer to the root) of the tree than those at the lower levels. In order to accommodate this heuristics, we implemented a variable threshold, which assumes a different value at a different level of the tree. For example, the threshold is 3 at the upper most level while it is 7 at the lowest level. This scheme allows our algorithm to explore more mappings with large likelihood.

4 Parallel Algorithm

In this project we propose a parallel scheme for the construction of the mapping library. In particular, we focus on the parallelization of the construction of the mapping tree, as this step is considerably more expensive than the step of building initial library. The general idea is to allow different processors to concurrently build different parts of the mapping tree. In order to support the parallel execution, we designed a complex data structure for the representation of each node of the mapping tree. The representation of each node stores sufficient information to summarize the mapping of helices performed from the root of the tree down to the node in question. This choice is analogous to the idea of Environment Closure, widely adopted in other applications that parallelize the construction of search trees [10]. Environment Closure allows an idle processor to start constructing a subtree of the mapping tree – thus with minimal need for communication between processors.

4.1 Scheduling

Our approach is to proceed in the parallelization by distributing the task of constructing a tree among different processors. In particular, we put a processor in charge of constructing a subtree, by initially assigning a root of the subtree it is in charge of. The parallel algorithm uses a dynamic distributed scheduling for load balancing among processors [11], [12]. The problem solved here bears many similarities to the problem of distributing a search-tree construction [10].

Each processor manages a local task queue. Initially, each processor goes through a common initialization phase, which includes activities such as reading the initial library file (line 3 in Algorithm Schedule). Processor 0 initiates the process by creating its own task queue. Each processor then begins by sending requests and processing the messages it receives. If the message requests a task, it pops a node from its task queue only if the length of its task queue is greater than a threshold. If the message is a task that was requested earlier by the processor, it is pushed to the tail of its task queue (line 13 in Schedule). A ring-termination is used in our algorithm to determine global termination of the computation [13], [14]. After processing messages, the processor performs computation. When the task queue is empty, it sends a request to other processors in the order of (0, 1, 2, ..., m) where m is the total number of processors. Only when a request fails, it sends another request to the next processor.

After processing messages, each processor performs computation by extracting a node from the local task queue. We noticed that the nodes at the upper levels of the tree are to be expanded more than those at lower levels of the tree. Therefore, the upper level nodes carry heavier work than the lower level nodes. Our algorithm implements a strategy to send heavier work to other processors to keep them busy. This strategy corresponds to what is called *top level scheduling* in the context of parallel search-tree exploration [15]. When a node is to be extracted from a task queue, if it is for its own processor, the node is removed from the tail of the queue (line 4-5 in Algorithm Computation). If a node is to be extracted for another processor, the node is removed from the head of the tail (line 18 in Schedule). This is because that the head has heavier work than the tail in general. As a result, each processor tends to do a depth-first expansion for the tree. During the depth-first expansion, the processor is busy with the work that resides in its own machine.

1. Algorithm **Schedule** (Threshold)
2. //Each processor uses this algorithm to schedules the communication and computation. A task is given out if the length of the task queue is greater than Threshold that is set by the user.
3.   Initialization;
4.   probing ← false;
5.   if (processor = 0)
6.   i ← 1;
7.   else
8.   i ← 0;
9.   While (1)
10.   message ← receive messages;
11.   if (message = termination) {terminate;}
12.   if (message = task)

```
13.                push task to the tail of my task queue; probing = false;
14.          if (message = reject)
15.                probing = false; i ← next processor;
16.          if (message = request)
17.                if (size of my task queue > Threshold)
18.                    extract a task from the head of my task queue;
19.                    send task to the requesting processor;
20.          termination process using ring termination;
21     if (task queue not empty) {do computation on the queue;} // see Computation
22     if (task queue empty && probing = false)
23          send a request message to processor i;
24          probing = true;
```

4.2 Performance and Quality of the Mapping Library

We have implemented the parallel algorithm for the step of building a mapping tree using C++ and MPI on a PC cluster running Linux. Each node of the Beowulf cluster is a dual processor with 1.7 GHz CPUs. The PC nodes are connected by Myrinet-2000 SAN switches. We tested the parallel algorithm on nine proteins, each of which has 4 to 10 helices (Table 1). The ten proteins are taken from the Protein Data Bank where 3D structures of the proteins are available (http://www.pdb.org). We extracted the length distribution information from the 3D structures for the ten proteins. Then, we sent the amino acid sequences of the ten proteins to the PHD prediction server. Since the prediction accuracy for short helices are generally low, helices with length less than 7 have not been considered.

Table 1. Run time (in seconds) of assembling a mapping tree is shown for sequential and parallel algorithm. The resulting mapping library is shown with its size and the best match, expressed as the number of matched helices over the total number of helices (in parenthesis)

Protein Id(# of Helices)	1(Sequential)	# of Processors						Best_Match matched(total)	Size of library
		1	2	4	6	8	10	20		
1CC5(4)	0.001	0.002	0.001	0.001	0.001	0.002	0.002	0.004	4(4)	9
6TMN_E(7)	0.103	0.107	0.055	0.030	0.021	0.017	0.015	0.014	0(7)	0
3TIM_A(5)	0.820	0.912	0.457	0.233	0.157	0.119	0.097	0.061	5(5)	24881
2TSC_A(7)	0.852	0.961	0.485	0.246	0.166	0.129	0.104	0.060	6(7)	3286
1ECA(7)	1.191	1.329	0.666	0.337	0.228	0.173	0.140	0.081	7(7)	3338
1GD1_O(6)	1.602	1.805	0.905	0.460	0.308	0.234	0.189	0.107	6(6)	14643
1L58(8)	7.803	8.833	4.439	2.230	1.484	1.120	0.901	0.469	6(8)	6948
2PHH(7)	58.026	64.000	32.100	16.010	10.710	8.031	6.428	3.249	7(7)	883327
2CYP(10)	547.617	600.75	305.7	153.15	102.08	76.54	61.23	30.9	7(10)	39616

Table 1 shows the running time for different number of processor used. The running time is the period of time after initialization to the time when all the processors are terminated. Barriers were put after the initialization and after all the processors are terminated. The time for initialization is not counted in the running time in Table 1, because it is quite short compared to that of computation. The time for copying the resulting library to a central location is not included in the running time in Table 1. Therefore, the running time in Table 1 reflects purely the computation and scheduling. We performed the same test three times and calculated the average of running time, although the variation is small among the three sets of running time collected. The running time for the sequential algorithm that does not include MPI message handling is also shown in Table 1. The running time of the sequential algorithm (column 2 in Table 1) is comparable to, slightly less than, the running time of the parallel algorithm using 1 processor (column 3 in Table 1).

Each protein has its PDB ID with the number of helices considered in parenthesis (e.g. 1CC5(4) in Table 1). We noticed that there is a big variation for the running time for different proteins even for a single processor. The rows in Table 1 are ordered by the running time of the sequential algorithm. For example, 1CC5 has the smallest amount of work, while 2CYP has the largest amount of work. The factors affecting the running time includes the length of the protein sequence and the length distribution of the helices. Figure 4 shows the speedup compared to a single processor. We noticed that the speedup for protein 1CC5 is not good. However, it is to do with the small amount of work 1CC5 has. Even for the sequential algorithm, the running time for 1CC5 is 0.001 second (Table 1). As the amount of work increases, the speedup becomes better (Figure 4). The speedup is very close to linear even for 20 processors for proteins such as 2PHH and 2CYP that have larger amount of work (Table 1 and Figure 4). For example, the speedup for protein 2CYP, that has the longest running time, is 19.4 when 20 processors are used.

In order to measure the communication among the processors, we measured the major time factor during the communication – the number of tasks received for 20 processors for each protein. Ranging from a small job to a big job, the total number of tasks received for protein 1ECA, 2PHH, and 2CPY are 391, 1600, and 3158 respectively. Considering the running time for the above three proteins, the job of 2PHH is about 47 times heavier than that of 1ECA, and the job of 2CYP is about 458 times heavier than that of 1ECA. However, the number of tasks received for 2PHH and 2CYP are only about 3 times and 7 times more than that of 1ECA. This means that the number of tasks circulating among the processors grows significantly slower than the amount of work. Therefore, we think that the strategies we applied to reduce the

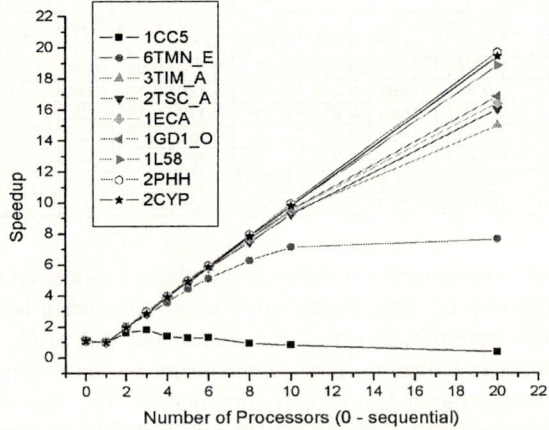

Fig. 4. Speedup. The sequential algorithm result is numbered as 0

communication have been effective. A number of techniques we implemented have contributed to the near linear speedup: the data structure of a node, depth-first expansion of the tree, and sending heavy jobs to other processors.

Among the possible mappings provided in the mapping library, Table 1 also shows the result of the best mapping in the library. For example, the best mapping in the library of 3TIM_A is able to assign all the 5 helices correctly. Since the library is not a complete library and is meant to have as sparse representative as it could be, a helix is considered assigned correctly if it is assigned within 5 amino acids distance from the true solution. Such a matching difference is expected to be corrected in further refinement steps using energy minimization. Among the nine proteins tested, our method is able to map most of the helices in the protein by the best match in the library except protein 6TMN_E. We noticed that the protein secondary structure prediction accuracy plays an important role in the quality of the library. The secondary structure prediction result of 6TMN_E completely missed two out of the seven helices so that the initial library is poorly constructed. In fact, we have done a test to generate fake predictions in which we corrected the major mistakes in the secondary structure prediction result. Our method is then able to rank all of the tested proteins at the top of the library using the fake predictions (data not shown).

5 Conclusion

In this paper, we presented a parallel algorithm for building the possible mapping library for helices between 3D and 1D protein structure. The algorithm generates a near linear speedup for up to 20 processors we tested. A number of techniques appear to contribute to the near linear speedup: the data structure of the node, the depth-first expansion of the tree and keeping light works for itself while sending heavy works to other processors. The results indicate that our method is able to work with small to medium-sized proteins. Improvement needs to be made to work with large proteins and to increase the quality of the library.

Acknowledgement. NSF Advance Program at NMSU, NSF EIA-0220590, NSF-HRD-0420407, and the SURP program at Sandia National Laboratories.

References

1. Gonen, T., Sliz, P., Kistler, J., Cheng, Y. Walz, T.: Aquaporin-0 Membrane Junctions Reveal the Structure of a Closed Water Pore. Nature.429, (2004) 193–7.
2. Chiu, W., Baker, M. L., Jiang, W. Zhou, Z. H.: Deriving Folds of Macromolecular Complexes through Electron Cryomicroscopy and Bioinformatics Approaches. Curr Opin Struct Biol.12, (2002) 263–9.
3. Zhou, Z. H., Dougherty, M., Jakana, J., He, J., Rixon, F. J. Chiu, W.: Seeing the Herpesvirus Capsid at 8.5 Å. Science.288, (2000) 877–880.
4. Zhou, Z. H., Baker, M. L., Jiang, W., Dougherty, M., Jakana, J., Dong, G., Lu, G. Chiu, W.: Electron Cryomicroscopy and Bioinformatics Suggest Protein Fold Models for Rice Dwarf Virus. Nat Struct Biol.8, (2001) 868–73.

5. Ludtke, S. J., Baldwin, P. R. Chiu, W.: Eman: Semi-Automated Software for High Resolution Single Particle Reconstructions. J Struct Biol.128, (1999) 82–97.
6. Jiang, W., Baker, M. L., Ludtke, S. J. Chiu, W.: Bridging the Information Gap: Computational Tools for Intermediate Resolution Structure Interpretation. J Mol Biol.308, (2001) 1033–44.
7. Pollastri, G., Przybylski, D., Rost, B. Baldi, P.: Improving the Prediction of Protein Secondary Structure in Three and Eight Classes Using Recurrent Neural Networks and Profiles. Proteins.47, (2002) 228–35.
8. Przybylski, D. Rost, B.: Alignments Grow, Secondary Structure Prediction Improves. Proteins.46, (2002) 197–205.
9. Jones, D. T.: Protein Secondary Structure Prediction Based on Position-Specific Scoring Matrices. J Mol Biol.292, (1999) 195–202.
10. Gupta, G., Pontelli, E., Ali, K., Carlsson, M. Hermenegildo, M.: Parallel Execution of Prolog: A Survey. ACM TOPLAS.23, (2001) 472–602.
11. Chow, K.-P. Kwok, Y.-K.: On Load Balancing for Distributed Multiagent Computing. IEEE Transactions on Parallel and Distributed Systems.13, (2002) 787–801.
12. Shivaratri, N. G., Krueger, P. Singhal, M.: Load Distributing for Locally Distributed Systems. Computer.25, (1992) 33–44.
13. Misra, J.:Detecting Termination of Distributed Computations Using Markers In: Annual Symposium on Principles of Distributed Computing. ACM Press, (1983) 290–4.
14. Wilkinson, B. Allen, M.: Parallel Programming. Prentice Hall(1998)
15. Beaumont, A. J. Warren, D. H. D.:Scheduling Speculative Work in or-Parallel Prolog Systems In: Proceedings of the International Conference on Logic Programming. MIT Press, (1993) 135–49.

A New Scalable Parallel Method for Molecular Dynamics Based on Cell-Block Data Structure[1]

Xiaolin Cao and Zeyao Mo

High Performance Computing Center, State Key Laboratory of Computational Physics, Institute of Applied Physics and Computational Mathematics, P. O. Box 8009, 100088, Bei-Jing P. R. China
{xiaolincao, zeyao_mo}@iapcm.ac.cn

Abstract. A scalable parallel algorithm especially for large-scale three dimensional simulations with seriously non-uniform particles distributions is presented. In particular, based on cell-block data structures, this algorithm uses Hilbert space filling curve to convert three-dimensional domain decomposition for load distribution across processors into one-dimensional load balancing problems for which measurement-based multilevel averaging weights(MAW) method can be applied successfully. Against inverse space-filling partitioning(ISP), MAW redistributes blocks by monitoring change of total load in each processor. Numerical experimental results have shown that MAW is superior to ISP in rendering balanced load for large-scale multi-medium MD simulation in high temperature and high pressure physics. Excellent scalability was demonstrated, with a speedup larger than 200 with 240 processors of one MPP. The largest run with 1.1×10^9 particles on 500 processors took 80 seconds per time step.

1 Introduction

Molecular dynamics(MD) simulation is an important tool in studying the properties of condensed matter and their dynamic interactions that can be difficult to obtain by other means. In order to make reasonable comparison with experiment, it is often necessary to simulate features on a micron scale. Realistic MD simulations of this size require at least $10^8 \sim 10^9$ particles, preferably more. Non-uniform distribution of various kinds of particles in space produces a highly irregular computational load. Also, as the simulations evolves, the movement of particles causes changes in the load distribution of processor used. These factors make it difficult to achieve high scalability. Therefore a good load balancing scheme is necessary to enhance scalability.

Mo[1] has presented a robust iterative 1-D DLB algorithm(*i.e.* MAW) to be suitable for 2-D parallel link-cell MD simulation. Hayashi[2] generalized the cellular automation diffusion scheme to the 3-D simulation by introducing a concept of permanent cell to

---

[1] Research supported by Chinese NSF(60273030), Chinese 863 program(2002AA104570) and CAEP Funds.

minimize inter-processor communication overheads. Against ORB and ORB-MM, Pilkington[3] have shown that of the three strategies, only Inverse Space-filling Partitioning(ISP) is able to render highly balanced workloads without incurring elaborate bookkeeping on a uniform mesh of N-body problem. NAMD[4] relies on a measurement-based DLB scheme to achieve high scalability for biomolecular systems. However, these DLB schemes are not suitable for our real application.

Based on these research described above, a measurement-based multilevel averaging weight DLB scheme based on Hilbert space-filling curve(HSFC)[5] was presented for our large-scale multi-medium MD simulation in high temperature and high pressure physics, which computational load is unpredictably and non-uniform with position and time. Then a new cell-block data structure required to describe MD simulation was constructed in order to help DLB scheme provide assistance with the movement of data. After data are moved between processors, the data structures must be rebuilt and the inter-processor communication patterns need to be updated. The DLB scheme and cell-block data structure, along with some auxiliary function, were integrated into a new parallel MD algorithm aimed at utilizing large parallel machines in a scalable manner.

The new parallel MD algorithm is described in section 2. The results of some performance evaluation are discussed in section 3. Against ISP, our DLB scheme can get better load balance with monitoring change of total load in each processor instead of monitoring change of workload in each block. Two numerical experimental results have showed that this new parallel MD algorithm can achieve high scalability for large-scale multi-medium MD simulation in high temperature and high pressure physics. Finally, we give some conclusions in section 4.

2 Parallelization Strategy

Provided that P is the number of processors, traditional link-cell domain decomposition method(DDM) partitions space into P sub-spaces (one per processor). It is highly scalable while particle densities are uniform. However, it has some disadvantages: (1) It is hard to use if the number of processors cannot be factored into three roughly equal factors; (2) Non-uniform distribution of particles can result in load imbalance; (3) Its data structure is not suitable for most DLB methods. In order to solve these problems, our method firstly partitions space into $Q(Q >> P)$ fixed-size blocks and creates cell-block data structure. Secondly, Hilbert space filling curve(HSFC) imposes a linear order(*i.e.* HSFC index) of the blocks in the high-dimensional space, which is the foundation of our DDM and DLB scheme. Thirdly, it constructs cell-block DDM based on HSFC and maps multiple blocks to each processor. Finally, a measurement- based multilevel averaging weight DLB scheme based on HSFC is used to get better load balance by redistributing blocks.

2.1 Cell-Block Data Structure

Our method firstly partitions physical space into Q blocks and then each block is subdivided into smaller volumes named cells with a side length $R_L = r_c + \delta$, where

r_c is "cut-off radius", δ is a small positive number. In Fig. 1, each block includes 4×4 computing cells (white). A layer of empty cells (shade) is padded to each block. We call these extra cells auxiliary cell, which stores temporarily some indices of particles moving to neighbor blocks. Because link-cell data structure can result in irregular memory access, we adopt compact memory management in cell-block data structure. Particles in the same cell are always stored sequentially in a list. So we designed a cell head pointer describing initial memory address of particles in the cell and an integer number describing the number of particles in the cell.

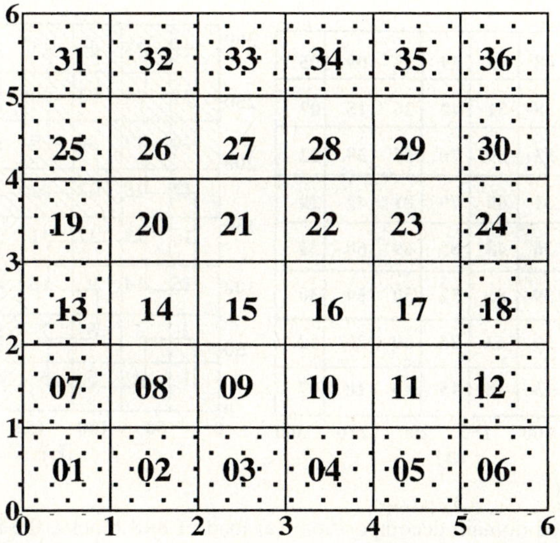

Fig. 1. Cell-Block structure include compute cell(white) and auxiliary cell(shade), where number in cell is index of cell

2.2 Cell-Block DDM Based on HSFC

Given a non-uniform model as shown in Fig. 2(a), simulation space was divided into 8×8 blocks, the total number of blocks Q is 64. Load of each block was first evaluated approximately by counting particle number in block. Then we adopt Zoltan method[6] generating HSFC, which is valid for any shape space. As shown in Fig. 2(b), HSFC visit every block of 2-D space. Meanwhile, we numbered all blocks from 1 to 64 along HSFC. The mapping of the hyperspace to the line is done once only, and is therefore a pre-processing step. Moreover, we construct a fast transform table in order to manage the mapping information. Finally, we apply 1-D recursive bisection to cut HSFC into 4 logically contiguous segments containing almost equal loads that correspond to physically irregular partitions, as shown in Fig. 2(b). Loads of 4 sub-domain is 511(dot: 1-18), 508(up diagonal line: 19-33), 525(white: 34-49), 543(down diagonal line: 50-64), respectively. Each segment includes a collection of blocks, which can be assigned to corresponding processors. So we can implicitly partition the hyperspace by partitioning simply and effectively 1-D line, which transforms hyperspace load balance problem into 1-D load balance problem. It can achieve better load balance, but arise irregular hyperspace partitions, which must manage unstructured communication. Therefore, bookkeeping information is required for each block. In real large-scale simulation, the number of blocks are generally far less than the number of cells, which reduce overheads of managing communication and the memory of bookkeeping information to $O(Q)$. Moreover, 1D recursive bisection partitioning cost is $O(Q)$. So we use block instead of cell as a allocable and mobile unit in order to reduce these overheads.

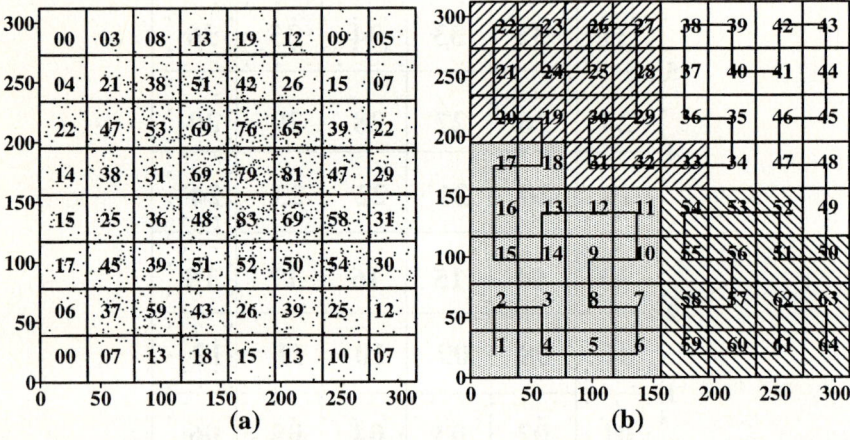

Fig. 2. HSFC-based domain decomposition. (a) load of 8×8 blocks, the number in each block represent load of owned block; (b) partitioning results, the number in each block represent the HSFC index of owned block

2.3 Measured-Based MAW DLB Based on HSFC

In our MD application, the movement of particles cause load fluctuation, and this phenomenon becomes more and more critical as the time evolves. This may cause severe load imbalance. It is often necessary to adjust the loads very quickly to be balanced. Paper[3] adopt ISP method to solve this imbalance in N-body problem. It exhibits several disadvantages in our large-scale multi-medium MD simulation in high temperature and high pressure physics. The main difficulty was determining the computational load of single block. ISP evaluates load of each block by a simple time-dependent function, such as the number of particles in the block and maybe the number of particles in neighboring blocks. It is not suitable for our MD simulation because load of each block is dependent on the number of particles, the geometric distribution of particles and the type of particles. Moreover, processor speed and cache effect both also affect this load. So a new DLB scheme needs to be designed for this MD simulation.

Once the hyperspace has been mapped to the line by HSFC, 1-D MAW[7] can be used to solve load balance problem arising from our MD application. So a measurement-based MAW DLB scheme based on HSFC and cell-block data structure was presented. It redistributes blocks sorted by HSFC index by monitoring change of total load in single processor instead of monitoring change of workload in single block. Simulation executes in the following procedure to perform DLB. First, the simulation runs for a small number of steps, typically lasting a few minutes. Actual computing time of local processor is measured during this time. After a particular simulation step, the main processor collects the load data from each processor, computes a new blocks distribution by calling MAW method. Since its partitioning time is linear with the number of blocks(Q), that is $O(Q)$, partitioning overheads on a single processor will remain modest for large problems so long as Q scaled accordingly. A table describing the complete partitioning with an $O(P)$ storage may be broadcasted to all processors, where P is the number of processors. Each processor can maintain the change in its load. Then, only the blocks near the boundary of HSFC contiguous segments need to be considered for exchanging with the neighbor processor to balance the load assuming that the particles do not move quickly. If the processors are sorted by one dimension, a processor typically communicates with only its two neighbors on the line. Communication is therefore inexpensive for adjusting load by migrating block. After migrating blocks are moved between processors, the data structures must be rebuilt and the inter-processor communication patterns need to be updated.

2.4 Force Calculation Schemes

The calculation of force on each particle is the most expensive step in our MD simulation. So it must be calculated both efficiently and in a manner which can be readily parallelized and load-balanced. We have enabled parallelization within our MD algorithm by dividing force computation into two classes of compute function: self-block and pair-block. The self-block function calculates pair interactions between particles within a particular block. The pair-block function calculates pair interactions between pairs of particles residing in neighboring blocks. If one neighbor block lies in the other processor, the system triggers a pair-block function when these data in the neighbor block are received. For managing these calculations, we create two index tables with exploitation of Newton's 3rd law. One is pair cell in single block, the other is pair cell in neighbor blocks sorted by neighbor relationship between block. These tables along with cell head pointer and the number of particles in the cell can improve executing efficiency by avoiding some jump instruction. These are much more suitable for instruction-level parallelism in advanced computer architecture.

3 Parallel Performance and Scalability

We have implemented a MD code based on these algorithm described above on the distributed memory MPP with MPI. It is suitable for our large-scale multi-medium MD simulation in high temperature and high pressure physics. For simplicity, we call this code PMD2D/3D. In this section, we examine parallel efficiency and scalability of our algorithm. These are run on a MPP including hundreds of processors. All units are given in a dimensionless form. We define the following variables: N = number of particles; Q = number of blocks; P = number of Processor used; PE is parallel efficiency; LBE is load balancing efficiency.

3.1 Parallel Performance

The first model is: N=1,560,000, Q=2000, P=1~64. Each block includes 3×3×3 cells in x, y, z direction. MD simulation lasted 1,000 time-steps. We adopt three parallel strategies: (1) regularly geometric partitioning (RGP); (2) ISP; (3) our method. RGP is often used by classical link-cell MD, which doesn't adjust load balance. So PE of RGP decreases quickly while P increases. ISP and our DLB method are much better than RGP in rendering balanced loads because they can adjust load distribution on time. Our DLB method is superior to ISP, which has improved LBE by 10%. The main reason is that load in single block change quickly and is very difficult to evaluate accurately. By comparison, our DLB relies on actual measurement of time spent by each processor to achieve a much more efficient load distribution as shown in figure 3(b). So its PE decreases slowly while P increases, while P=64, PE•60%. Part of the efficiency loss is inevitable due to communication overhead because communication and compute ratio grows when P increases. Part of this loss is idle time and partitioning time. We believe that improvements to our DLB method will allow us to decrease the idle time further.

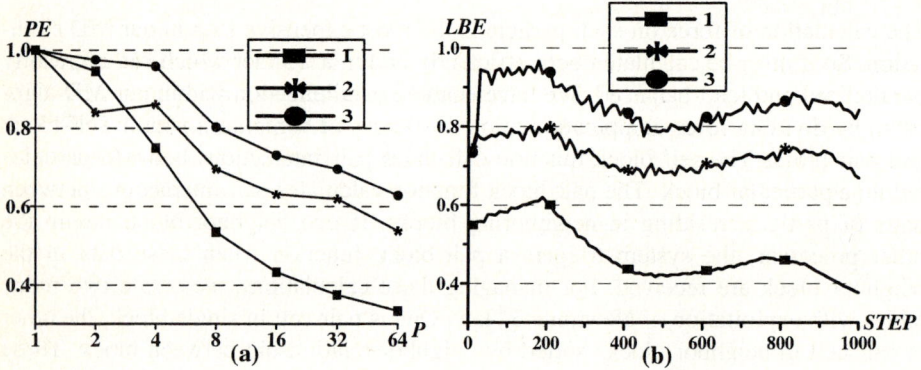

Fig. 3. N=1,560,000, parallel efficiency (a) and load balance efficiency while P=64 (b) of PMD2D/3D using RGP(line 1), ISP(line 2) and our method(line3) respectively

3.2 Scalability

In order to make reasonable comparison with experimental data, it is often necessary to be able to simulate features on a micro-scale system with at least hundreds of millions of particles, preferably more. So we simultaneously increase the system size and the number of processors, such that N/P=const. Table 1 gives the corresponding parallel efficiency of 2-D simulation while keeping N/P=1,600,000 and 3-D simulation keeping N/P=1,100,000, where t_{step} is time per integration step in seconds. Obviously, PMD2D/3D has gained very good performance for all number of processors ranging from 2 to 240. Both achieve good scalability with speedup of over 200 on 240 processors even on large numbers of processor for sufficiently large simulations. On 240 processors, it takes about 10 second every step to simulate 380,000,000 particles in 2-D and about 37 second to simulate 276,000,000 particles in 3-D. These results have showed that our algorithm is very effective in modeling hundreds of millions of particles in both 2-D and 3-D.

Table 1. Parallel efficiency with N/P=constant

P	2-D t_{step} (s)	2-D E (%)	3-D t_{step} (s)	3-D E (%)
1	8.882	100.0	32.509	100.0
4	9.127	97.3	33.331	97.5
16	9.240	96.1	35.757	90.9
64	9.340	95.1	36.802	88.3
120	9.564	92.8	37.002	87.8
240	9.991	88.9	37.328	87.1

3.3 Comparable Performances

In the last few years, many research groups[8-10] reported their record of MD simulation. For comparison, we list these results and our current record together in Table 2. However, we have not yet done experiments to compare the performance of MD with other programs for identical molecules with identical potential parameters and identical machines. From Table 2, it is shown that our MD code can simulate the same magnitude number of particles within the same magnitude time costing compared with the world record reported.

Table 2. Comparable Performance in 3-D MD simulation

Groups	Machine	P	N	t_{step} (s)
Lohmdahl [8]	CM-5	1024	100,000,000	3.5
Plimpton [9]	Paragon	3680	600,000,000	242
Stadler [9]	T3E	512	1,213,857,792	180
Roth [10]	T3E	512	5,180,116,000	387
Our team	One MPP	500	1,100,000,000	80

4 Conclusion

We have described the design of the new scalable parallel MD algorithm for large-scale multi-medium MD simulation in high temperature and high pressure physics. It uses cell-block DDM based on HSFC to attain practical scalability. It uses a Measured-based MAW DLB based on HSFC to attain high parallel efficiency even while simulating non-uniform MD systems. Our DLB scheme is superior to ISP in real application. Excellent scalability was gained, with a speedup of above 200 on 240 processors of one MPP in both 2-D and 3-D.

Although the parallel performance of our algorithm is quite good, it still leaves room for improvement. We believe that improvements to our DLB scheme, combined with the use of computation and communication overlap will allow us to decrease the idle time further. Moreover, parallel I/O and corresponding parallel visualization will be developed in order to help physicists analyze results and penetrate deeply motion of particles.

References

1. Mo Zeyao, Zhang Jinglin : Dynamic Load Balancing for Short-Range Parallel Molecular Dynamics Simulations. Intern. J. Computer Math. **79** (2002) 165-172.
2. Hayashi R., Horiguchi S.: Efficiency of Dynamic Load Balancing Based on Permanent Cells for Parallel Molecular Dynamics Simulation. Proc. Of IPDPS, Cancun, Mexio(2000) 85-92.
3. Pilkington R., Baden B.: Dynamic Partitioning of Non-Uniform Structured Workloads with Spacefilling Curves. IEEE Trans. on Parallel and Distributed Systems. **7** (1996) 288-299.
4. Kale, L., Skeel, R., Bhandarkar M.: NAMD2: Greater Scalability for Parallel Molecular Dynamics. J. Computational Physics, **151**(1999) 283-312.
5. Sagan, H.: Space-Filling Curves. Springer, New York (1994).
6. http://www.cs.sandia.gov/Zoltan/ug_html/ug_alg_hsfc.html.
7. Mo Zeyao, Zhang Baolin: Multilayer Averaged Weight Method for Dynamic Load Imbalance Problems. Intern. J. Computer Math. **76** (2001) 463-477.
8. Plimpton S.: Fast Parallel Algorithms for Short-range Molecular Dynamics. J. of Computational Physics, **117** (1995) 1-19.
9. Stadler J., Mikulla R., Trebin H. R.: IMD: A Software Package for Molecular Dynamics Studies on Parallel Computes. Intern. J. Modern Physics. **8** (1997) 1131-1140..
10. Roth J., Gahler F., Trebin H.: A Molecular Dynamics Run with 5,180,116,000 Particles. Int. J. Modern Physics C. **11** (2000) 317-322.

Parallel Transient Stability Simulation for National Power Grid of China

Wei Xue, Jiwu Shu, and Weimin Zheng

Department of Computer Science, Tsinghua Univ.,
Beijing, P.R.China, 100084
{xuewei, shujw, zwm-dcs}@tsinghua.edu.cn

Abstract. With the development of modern power system, real-time simulation and on-line control are becoming more and more critical. Transient stability analysis, where the most intensive computation locates, is the bottleneck of real-time simulation for large-scale power system. Thus, the key to achieve the real-time simulation of large scale power systems is to find the new transient stability algorithms and parallel software with high-performance computers. This paper presents a new spatial parallel transient stability algorithm including an improved parallel network algorithm and an optimal convergence checking method. The simulation software with the spatial parallel algorithm is designed and implemented on a SMP-cluster system. The test cases of national power grid of China show that the optimal computation time of the parallel software is only 38% of the actual dynamic process. It is suggested that the algorithms described in this paper can achieve the super real-time simulation of very large scale power system and make the complex on-line power applications, especially on-line supervision and control for large scale power system, feasible.

1 Introduction

Power system transient stability analysis is a powerful simulation tool in power system research. For power system operation and plan of modern power system, real-time simulation and on-line control are becoming more and more critical. However, with the development of the power system scale, computation tasks are increasingly becoming heavier and more complex. Traditional sequential computation is inadequate for real-time simulation and on-line control in power system plan and operation. Thus, the key to realize the real-time simulation of large scale power system dynamics is to find the new transient stability algorithms and parallel software with high-performance computers. Meanwhile, the new development of high performance computing technology, especially the mature use of cluster systems with a good balance of high performance and low price, makes real-time simulation for large-scale power network feasible.

To accomplish the task of power system transient stability computation, a set of DAEs (Differential Algebraic Equations) must be solved.

$$\begin{cases} \dot{\mathbf{X}} = f(\mathbf{X}, \mathbf{V}) = \mathbf{A}\mathbf{X} + \mathbf{B}u(\mathbf{X}, \mathbf{V}) \\ 0 = \mathbf{I} - Y(\mathbf{X}) * \mathbf{V} \end{cases} \quad (1)$$

In (1), the nonlinear differential equations describe the dynamic characteristics of the power system devices and the second algebraic nonlinear equations present the restriction of the power network, where $\mathbf{X}$ is the state vector of individual dynamic devices; $\mathbf{I}$ is the vector of current injection from the devices into the network; $\mathbf{V}$ is the node voltage vector; $Y(\mathbf{X})$ is the complex sparse matrix, which is not constant with time; and u is the function of $\mathbf{X}$ and $\mathbf{V}$.

The most common used sequential algorithm for transient stability analysis is the interlaced alternating implicit approach (IAI algorithm) [1]. The IAI algorithm uses the trapezoidal rule as its integration method, and solves differential equations and algebraic equations alternately and iteratively, which not only maintains the advantages of the implicit integration approach, but also has modelling and computing flexibility. According to the IAI algorithm, network equation solution is the most intensive computation part of the algorithm. With the optimal schemes for node ordering and sparse matrix computation, the computation complexity of factoring of the network equations is $O(N^{1.4})$ and the complexity of forward and backward iteration of the network equations is $O(N^{1.2})$ [2], where N is the dimension of the coefficient matrix. Thus, the time consumed for transient stability analysis increases super-linearly as the power system's size increases. The performance of sequential transient stability simulations is not adequate for real-time simulation of large-scale power grid. Therefore, it is very important to study practical parallel algorithms and software. Currently, the wide use of scalable cluster systems, which have high performance and low cost, has made parallel and distributed real-time transient stability analysis possible for large-scale power system. The research on cluster-based parallel algorithms for transient stability analysis has become a new hotspot in this field [3] [4].

This paper analyzes the known parallel algorithms of transient stability analysis, presents a cluster-based spatial parallel algorithm including an improved parallel network algorithm and an optimal convergence checking method. In the new spatial parallel algorithm, a hierarchical Block Bordered Diagonal Form power network algorithm, which uses message-passing and share-memory models simultaneously, is presented to optimize the computation of sequential part in the parallel algorithm and to improve the scalability of the algorithm on cluster system. The convergence checking method, in which the preferential local convergence scheme is introduced and the relative deviation of current injection is regarded as the new convergence checking variable, is developed to cancel the redundant computation and reduce the communication costs. Finally, the simulation software with the spatial parallel algorithm is designed and implemented on a SMP-cluster system, and the numerical results of national power grid of China show that the optimal computation time of the parallel simulation is only 38% of the actual dynamic process. It can be concluded that the algorithm described in this paper can achieve the super real-time simulation of very large scale power system and fulfill the requirements of on-line transient stability analysis and on-line control for the future nationwide power grid in China. Furthermore, the cluster computing technology is the most promising high performance computing technology for future complex power applications.

2   The New Spatial Parallel Algorithm for Transient Stability Analysis

Research on the parallel algorithm for transient stability analysis is focused on the parallel algorithm construction and task scheduling. According to the analysis of IAI algorithm, the solution of power network equations is the most intensive computation part of the computation. Thus, an effective parallel algorithm for power network equations is very critical for parallel transient stability analysis. Meanwhile, for the convergence checking scheme has to deal with the cooperation of each computing process in parallel environment and then more communication will be introduced, the smart design of convergence checking scheme is also very important for the parallel algorithm on cluster system. In this paper, our work is focused on power network algorithm and the corresponding convergence checking scheme. As to the task scheduling scheme, the latest research advance can be found in [5] [6] and the multilevel partition scheme described in [6] is used in this paper.

2.1   Research Advance on Parallel Transient Stability Algorithm

Through two decades' studies, parallel algorithms for transient stability analysis are well developed in two directions, spatial parallelism and time parallelism [7] [8] [9] [10] [11]. Spatial parallel algorithms, including *Partition method* and *Parallel Factoring algorithm*, take a time-domain integration method and decompose each time step computation into sub-tasks among processors. As a coarse granularity parallel algorithm, Partition method can be implemented easily and achieve higher efficiency and gains on distributed architecture, while Parallel Factoring scheme is better on shared memory machines. In order to get better performance on more processors, the simultaneous multiple time step solutions, such as *WR (Waveform Relaxation) method* and *parallel-in-time Newton algorithm*, are introduced into the parallel computation of transient stability analysis. These time parallel algorithms enlarge the scale of transient stability problem to be solved simultaneously. At the same time, the overall speedups of the algorithms are improved effectively. However, it is difficult to achieve good parallelism degree while maintaining a high convergence rate. In addition, another limiting factor hindering parallelism in time is that more invalid computation may be brought into simulation when random events happen in the "time window" of computing. Therefore, the coarse granularity spatial parallel algorithm is the most appropriate selection for implementation on cluster system.

2.2 Hierarchical BBDF Algorithm for Power Network Equation Solution

It is well known that large scale power system comes from the connection of regional networks. The nodes in the regional power network are more strongly connected than those between the region power networks. Therefore, the partition algorithm is well fit for solving network equations on the cluster system. For more inherent parallelism developed in network computation, a node-oriented branch splitting scheme is used to construct the network equations. In this scheme, the branches between subsystems and the fault branches are separated from the original network to form the boundary system, which suggests the relation of the subsystems. And the boundary system equations are formulated by admittance matrix of the terminals of the branches. Based on the node-oriented branch splitting scheme, the network equations can be reformed in the Block Bordered Diagonal Form [12], as shown in (2). The following network equations are for two sub-areas.

$$\begin{bmatrix} \mathbf{Y} & \mathbf{M} \\ \mathbf{M}^T & \mathbf{Z}_{CF} \end{bmatrix} \begin{bmatrix} \mathbf{U} \\ \mathbf{I}_{CF} \end{bmatrix} = \begin{bmatrix} \mathbf{I}_p \\ 0 \end{bmatrix} \quad (2)$$

In which,

$$\mathbf{Y} = \begin{bmatrix} \mathbf{Y}_{1p} & & & & & \\ & \mathbf{Y}_{2p} & & & & \\ & & \mathbf{Y}_{1n} & & & \\ & & & \mathbf{Y}_{2n} & & \\ & & & & \mathbf{Y}_{1z} & \\ & & & & & \mathbf{Y}_{2z} \end{bmatrix}, \mathbf{U} = \begin{bmatrix} \mathbf{U}_{1p} \\ \mathbf{U}_{2p} \\ \mathbf{U}_{1n} \\ \mathbf{U}_{2n} \\ \mathbf{U}_{1z} \\ \mathbf{U}_{2z} \end{bmatrix}, \mathbf{Y} = \begin{bmatrix} \mathbf{Y}_{1p} \\ \mathbf{Y}_{2p} \\ 0 \\ 0 \\ 0 \\ 0 \end{bmatrix},$$

$$\mathbf{M} = \begin{bmatrix} \mathbf{M}_{CF-1p} & & & & & \\ & \mathbf{M}_{CF-2p} & & & & \\ & & \mathbf{M}_{CF-1n} & & & \\ & & & \mathbf{M}_{CF-2n} & & \\ & & & & \mathbf{M}_{CF-1z} & \\ & & & & & \mathbf{M}_{CF-2z} \end{bmatrix},$$

$$\mathbf{Z}_{CF} = (\mathbf{Y}_C + \mathbf{Y}_F)^{-1}.$$

In (2), the subscripts 1 and 2 represent the sub-area number, and the subscripts p, n, z represent the positive, negative, and zero sequence networks respectively. $\mathbf{Y}_{1p}$, $\mathbf{Y}_{2p}$, $\mathbf{Y}_{1n}$, $\mathbf{Y}_{2n}$, $\mathbf{Y}_{1z}$ and $\mathbf{Y}_{2z}$ are the admittance matrices of three (positive, negative and zero) sequence networks respectively. $\mathbf{Z}_{CF}$ is the impedance matrix for cutting branches and fault branches. $\mathbf{Y}_C$ and $\mathbf{Y}_F$ are the admittance matrices of cutting branches and fault branches respectively. They are also the coefficient matrix of boundary equations in the BBDF computation. $\mathbf{M}$ and $\mathbf{M}^T$ are the associated matrices between $\mathbf{Y}$ and $\mathbf{Z}_{CF}$.

According to the traditional parallel scheme used in BBDF equations, the limiting factor hindering parallelism is the computation of the boundary system, which is the sequential part in the whole parallel algorithm. With the increase of sub-areas, the time required completing the boundary equations and the time spends on the communication between processors increase sharply. Because of the network equations constructed with the node-oriented branch splitting scheme, the nodes of the boundary system in different sub-areas are different and then the equivalent circuits of the boundary system are not correlative. Therefore further parallel schemes can be introduced to the computation of the boundary system.

First of all, the formulation of the admittance matrix and the equivalent current vector of the boundary system converted from each sequence network and each sub-area equations can be solved simultaneously as well as the current injection from the boundary system to the sub-areas.

Furthermore, to improve the computation performance of boundary equations, the Block Bordered Diagonal Form scheme is introduced in this paper. In the algorithm, the coefficient matrix of the boundary system are reordered by positive, negative, zero sequence parts of cutting branches in front of fault branches, as described in (3).

$$\begin{bmatrix} \mathbf{Y}_{Cp} & & & \mathbf{N}_{Cp} & \\ & \mathbf{Y}_{Cn} & & \mathbf{N}_{Cn} & \\ & & \mathbf{Y}_{Cz} & & \mathbf{N}_{Cz} \\ \mathbf{N}_{pC} & & & & \\ & \mathbf{N}_{nC} & & \mathbf{Y}_F & \\ & & \mathbf{N}_{zC} & & \end{bmatrix} \quad (3)$$

Where $\mathbf{Y}_{Cp}$, $\mathbf{Y}_{Cn}$ and $\mathbf{Y}_{Cz}$ are the three sequence node admittance matrices of cutting branches; $\mathbf{Y}_F$ is the admittance matrix of fault nodes; and $\mathbf{N}_{Cp}$, $\mathbf{N}_{Cn}$, $\mathbf{N}_{Cz}$, $\mathbf{N}_{pC}$, $\mathbf{N}_{nC}$, and $\mathbf{N}_{zC}$ are the associated matrices. Because some fault forms have invalid impedance matrices, an admittance matrix is adopted in (2) and (3).

It is noted that the boundary equations are much smaller than network equations. And most of their computation focuses on reforming the boundary equations and factoring the coefficient matrix when events happen. Therefore, a dynamic multithread scheme is used to solve the boundary equations in this paper, which is more effective for clusters that consist of shared memory machines (SMP-Cluster). When this hierarchical network equation algorithm is applied to transient stability problems, it not only dramatically improves the efficiency and gains, but also enhances the scalability of the program. At the same time, the new boundary system computation scheme provides two other benefits. One is reduction of computation time based on the symmetry of negative and zero parts in the boundary system, and the other is fewer conditions in the boundary matrix. This improves the precision and robustness of the algorithm.

Compared with the traditional boundary system equations algorithm, the algorithm proposed in this paper reduces the time consumed for solving boundary system equations largely. And the performance comparison between these two algorithms for different scale boundary system is shown in Fig. 1. For the same boundary system, the computation time with the BBDF algorithm presented in this paper is only one fifth to one sixth of the computation time with traditional algorithm. In the hierarchical BBDF power network algorithm proposed in this

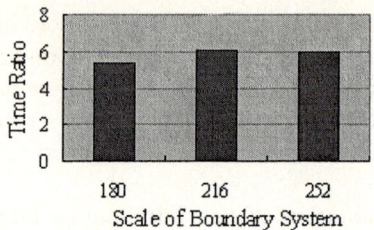

Fig. 1. Time-ratio of traditional boundary system equations algorithm to BBDF boundary system equations algorithm proposed in this paper

paper, BBDF parallel scheme is applied to both the power network solution and the boundary system solution. By this way, the computation performance of the power network equations is improved and the overall efficiency and scalability of the spatial algorithm are enhanced.

2.3 Optimal Convergence Checking Scheme

The convergence checking scheme on parallel architecture is more complex than that on sequential systems because it had to take into account the cooperation of the computing processes. In this paper, based on the analysis of different convergence rate of sub-tasks, a new convergence checking scheme is developed to improve the computation's efficiency.

As we known, for large scale power system, the electric variables near faults vary more intensively than faraway power networks. It is the famous phenomena "influence localization of fault". In the dynamic process simulation, the convergence rate of solving network equations near faults will be much slower. Therefore, the convergence met in the near fault subsystems becomes the main factor hindering the simulation. And the other subsystems have to be solved repeatedly before global convergence is reached. But these computations didn't bring any precision improvement. According to the phenomena "influence localization of fault", the computation of subsystem can be finished early before the global convergence is reached. The preferential local convergence checking scheme reduces the redundant subsystem computation and enhances parallel efficiency of the simulation.

Meanwhile, in traditional sequential algorithms, the derivation of node voltage in two successive iterations is checked to decide whether the convergence is reached. In parallel environment, the node voltage scheme introduces another global communication into each iteration of spatial parallel algorithm. In order to reduce the communication costs and get the similar convergence rate to the node voltage scheme, the relative derivation of current injection is used as the variables to be checked in convergence scheme.

The convergence checking scheme proposed in this paper can be described in detail as follows:

1. Relative derivation of currents injected is regarded as variables to check local area convergence.

2. Every subsystem checks convergence locally. If local convergence occurs, the corresponding process informs the control process (the same as the process for computing the boundary system) with a local convergence flag instead of with the new corrected vector. Then this subsystem waits for the global convergence flag or the new solutions of the boundary equations. If global convergence is reached, the simulation enters into the next time step; if not, whether or not the local computation is performed depends on the preferential convergence checking.

3. The control process collects the local convergence flags from all subsystems. After the global convergence is met, the results are sent to each subsystem. If global convergence is not reached, the boundary equations have to be solved once more.

It is noted that the solution of subsystem equations depends on not only the current injection from devices into subsystem, but also the current injection from boundary system into subsystem. So the local preferential convergence is met only when the current injection derivation from device to subsystem and from boundary system to subsystem are less than the threshold simultaneously.

2.4 Flow Chart of the Improved Spatial Parallel Algorithm

In the simulation, a large system represented by (1) is broken into N subsystems based on the partition scheme proposed in [6]. N subsystems are respectively assigned to N processor for computing; for example, the k-th subsystem is processed by the processor $P_k(k = 1, 2, , N)$. Each subsystem is calculated independently with the solution of the boundary system. Then the solution to the boundary equations comprises the computation results for each subsystem. This process is repeated until convergence is reached, as described in the following flow chart.

3 Test Results

The national power grid of China is constructed to test the transient stability algorithm proposed in this paper. The Chinese power grid test system includes six power grids of China, such as North-east power grid, North power grid, Central power grid, Chuanyu power grid, South power grid and East power grid. Shandong province power system, Fujian province power system and Yangcheng power system are also considered. The national power grid is a huge scale power system with 10188 buses, 13499 branches, 1072 generators (the traditional 5 Order dynamic model is used), 3003 loads (induction motor model is concerned) and 4 Direct Current Systems, as shown in Fig. 2.

For testing the parallel software, a cluster of Tsinghua Univ. is used in this paper. The cluster consists of multi nodes. Each node is a Symmetry Multi-Processor (SMP) computer and has four Intel Xeon PIII700 MHz CPUs, 36-gigabyte hard disks, and 1 gigabyte of memory. The communication medium between SMP nodes is Myrinet with a bandwidth of 2.56Gb/s. The software

Parallel Simulation by Processor $P_1, ..., P_N$
For TimeStep = 1, ..., MaxTimeStep (Simulation Loop)
 For Iter = 0, ..., MaxIters (Iteration Loop)
 1: Include new events;
 2: Solve the differential equations of subsystem k with the trapezoidal rule;
 $$\dot{\mathbf{X}}_k = f(\mathbf{X}_k, \mathbf{V}_k) = \mathbf{A}\mathbf{X}_k + \mathbf{B}u(\mathbf{X}_k, \mathbf{V}_k)$$
 3: Compute the current injected into subsystem k
 $\mathbf{I}_k(\mathbf{X}_k, \mathbf{V}_k)$;
 4: Check local convergence of subsystem k;
 $$\left\|\frac{\mathbf{I}_{kt} - \mathbf{I}_{k(t-1)}}{\mathbf{I}_{k(t-1)}}\right\| < \varepsilon_\mathbf{I}$$
 t is the iteration number, $\varepsilon_\mathbf{I}$ is the threshold of current convergence checking.
 5: Solve the vector and matrix corrected in subsystem k;
 $$\begin{cases} \Delta \mathbf{Z}_{km} = \mathbf{M}_{km}^T \mathbf{Y}_{km}^{-1} \mathbf{M}_{km} \\ \Delta \mathbf{U}_{kp} = \mathbf{M}_{kp}^T \mathbf{Y}_{kp}^{-1} \mathbf{I}_{kp} \\ \Delta \mathbf{Y}_{km} = \Delta \mathbf{Z}_{km}^{-1} \\ \Delta \mathbf{I}_{kp} = \Delta \mathbf{Y}_{kp} \Delta \mathbf{U}_{kp} \end{cases}, \quad m = p, n, z$$

Communication of collection between processors

 6: Global convergence checking, If convergence is reached, then break the iteration loop;
 7: Solve boundary systems;
 $$\begin{bmatrix} \mathbf{Y}_{Cp} & & & \mathbf{N}_{Cp} & & \\ & \mathbf{Y}_{Cn} & & & \mathbf{N}_{Cn} & \\ & & \mathbf{Y}_{Cz} & & & \mathbf{N}_{Cz} \\ \mathbf{N}_{pC} & & & & & \\ & \mathbf{N}_{nC} & & & \mathbf{Y}_F & \\ & & \mathbf{N}_{zC} & & & \end{bmatrix} \begin{bmatrix} \mathbf{U}_{Cp} \\ \mathbf{U}_{Cn} \\ \mathbf{U}_{Cz} \\ \mathbf{U}_{Fp} \\ \mathbf{U}_{Fn} \\ \mathbf{U}_{Fz} \end{bmatrix} = \begin{bmatrix} \mathbf{I}_{Tp} \\ 0 \\ 0 \\ \mathbf{I}_{Fp} \\ 0 \\ 0 \end{bmatrix} + \sum_{m=1}^{k} \Delta \mathbf{I}_{kp}$$

Communication of scattering the solution of boundary equations or the global convergence flag

 8: If local convergence of subsystem k is reached, Check preferential convergence in subsystem k;
 $$\left\|\frac{\mathbf{I}_{km-CF,t} - \mathbf{I}_{km-CF,t-1}}{\mathbf{I}_{km-CF,t-1}}\right\| < \varepsilon_\mathbf{I}$$
 $$\begin{cases} \mathbf{I}_{km-CF,t} = \mathbf{M}_{km} \Delta \mathbf{Y}_{km} (\mathbf{U}_{CF-km,t} - \Delta \mathbf{U}_{km,t}) \\ \Delta \mathbf{U}_{kn,t} = \Delta \mathbf{U}_{kz,t} = 0 \end{cases}$$
 $m = p, n, z$, t is the iteration number.
 9: Solve the node voltages in subsystem k;
 $$\begin{cases} \mathbf{U}_{kp} = \mathbf{Y}_{kp}^{-1}(\mathbf{I}_{kp} - \mathbf{M}_{kp} \Delta \mathbf{Y}_{kp}(\Delta \mathbf{U}_{kp} - \mathbf{U}_{CF-kp})) \\ \mathbf{U}_{kn} = \mathbf{Y}_{kn}^{-1} \mathbf{M}_{kn} \Delta \mathbf{Y}_{kn} \mathbf{U}_{CF-kn} \\ \mathbf{U}_{kz} = \mathbf{Y}_{kz}^{-1} \mathbf{M}_{kz} \Delta \mathbf{Y}_{kz} \mathbf{U}_{CF-kz} \end{cases}$$
 End For iter (End Iteration Loop)
End For TimeStep (End Simulation Loop)
End Parallel Simulation

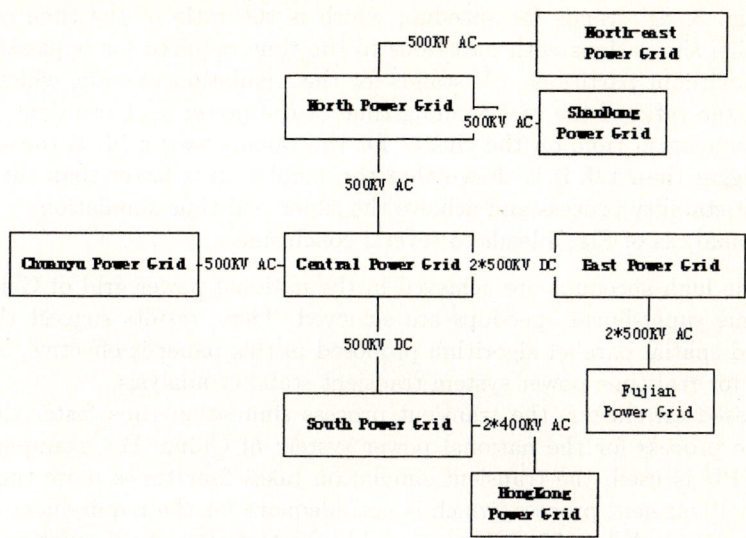

Fig. 2. The National power system of China tested in this paper

environments are Redhat Linux 7.2 (kernel version 2.4.7-10smp), Intel compiler for Linux (version 6.0), high performance math library MKL (version 6.1) and gm-1.5pre4, which is the network protocol running on Myrinet. The programming environments supported are Message-Passing Interface (MPI) and Open Multi Processing programming (OPENMP).

For all the test cases of national power system of China, a permanent A-phase fault in the middle point of the transmission line between North-east and North power grid is assumed. The A-phase line is tripped when 0.08 seconds passed, and re-closed on 1.08s. Then the lines of three phases are re-tripped on 1.16s. The time step is 0.01s, and convergence tolerance is 10^{-4}.

Fig. 3 show the speedups and simulation velocity of the parallel computing with the national power grid data described above.

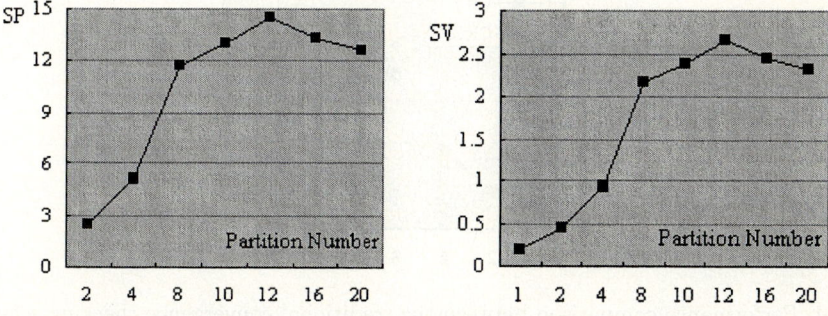

Fig. 3. The speedups (SP) and simulation velocity (SV) of test case

In Fig. 3, SP stands for speedup, which is the ratio of the time required for parallel simulations with partitions to the time required for sequential simulations without partitions. SV stands for the simulation velocity, which is the same as the ratio of the real running time of the power grid transient process to the simulation time on the cluster for the same power grid. If the value of SV is bigger than 1.0, it is shown that the simulation is faster than the actual transient stability process and achieve the super real-time simulation.

The analysis of Fig. 3 leads to several conclusions.

1. The high speedups are achieved in the national power grid of China and even some super-linear speedups are achieved. These results suggest that the improved spatial parallel algorithm proposed in this paper is effective, and can be used for real-time power system transient stability analysis.

2. Based on cluster, the transient process simulation runs faster than the real-time process for the national power system of China. For example, when single CPU is used, the transient simulation takes four times more time than the actual transient process, which is not adequate for the requirement of real-time simulation. When 12 CPUs are used in the test case, the simulation time is only 38% of the actual transient process time and a super real-time simulation is achieved. It is almost fifteen times faster than on single CPU. So the parallel processing is necessary to improve the simulation velocity. This also proves that the improved spatial algorithm for transient stability analysis proposed in this paper is very satisfying.

The performance improvement of the new convergence checking scheme is shown in Fig. 4. This convergence checking scheme uses the relative derivation of current injection as the checking variable and performs the preferential subsystem convergence checking to cancel the redundant computation and reduce the communication costs. So its advantages are emerged with the partitions increased. In Fig. 4, when 12 processors used, the simulation with the new convergence scheme in this paper is 20% faster than that with the traditional convergence scheme.

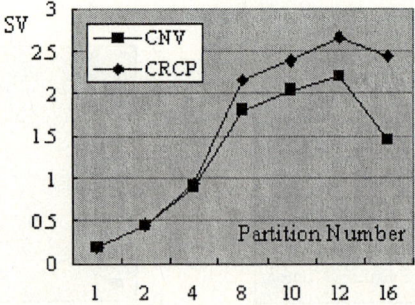

Fig. 4. Performance comparison between the traditional convergence checking scheme with the node voltage checking (CNV) and the optimal scheme in this paper (CRCP)

In addition, it is shown in Fig. 3 that the efficiency of this parallel algorithm exceeds 100% with 2 CPUs to 12 CPUs. In the sequential computation, the data size of factoring and iteration in power network computation, which is 1.5 MB, exceeds the cache size of Xeon PIII CPU (1MB). So more memory access, more time is consumed in the sequential simulation. In parallel simulations, more CPUs are used and larger cache size is available. And the overall time required by this algorithm decreases for higher cache hitting. Therefore, the well-known "cache effect" brings super-linear speedups in the transient simulation of large scale power system.

4 Conclusion

This paper proposes an improved spatial parallel algorithm for transient stability analysis, which includes a hierarchical Block Bordered Diagonal Form power network algorithm and a convergence scheme with preferential local convergence checking. In this spatial parallel algorithm, message-passing and share-memory models are used simultaneously. The spatial parallel algorithm is designed and implemented on a SMP-cluster system. Simulations are performed for national power system of China show that the optimal speedup with the parallel algorithm is 14.5 with 12 CPUs, and the corresponding parallel efficiency is 121%. The time consumed for the parallel simulation is only 38% of the actual dynamic process. The numerical results suggest that the algorithm gains much higher performance than the spatial algorithms described in [13] and [14]. This algorithm, with adequate efficiency and scalability, is a feasible selection for the real-time transient simulation of future on-line power applications. And the cluster computing technology is becoming the most promising high performance computing technology for on-line power applications of large scale power systems.

References

1. B. Stott.: Power System Dynamic Response Calculations. IEEE Proc., 1979, 67(2):219–241
2. F.L.Alvarado: Computational Complexity in Power Systems. IEEE Trans. On PAS, 1976, PAS-95(4):1028–1036
3. W. Xue, J. W. Shu, X. F. Wang, W. M. Zheng: Advance of parallel algorithm for power system transient stability simulation, Journal of system simulation, 2002, 14(2):177–182(In Chinese)
4. Y. L. Li, X. X. Zhou, Z. X. Wu: Parallel algorithms for transient stability simulation on PC cluster. PowerCon 2002, Vol.3:1592–1596
5. K.W.Chan, R.W.Dunn, A.R.Daniels: Efficient heuristic partitioning algorithm for parallel processing of large power systems network equations. IEE Proc.-Gener. Transm. Distrib..1995, 142(6):625–630
6. Shu Jiwu, Xue Wei, Zheng Weimin: An Optimal Partition Scheme of Transient Stable Parallel Computing in Power System. Automation of Electric Power Systems. 2003, 27(19):6–10(In Chinese)

7. Daniel J.Tylavsky, Anjan Bose: Parallel processing in power systems computation, IEEE Trans. On PWRS, 1992, 7(2):629–638
8. Chai J.S, Bose A.J.: Bottlenecks in Parallel Algorithms for Power System Stability Analysis. IEEE Trans. On PWRS, 1993, 8(1):9–15
9. D.M.Falcao: High Performance Computing in Power System Applications. Proc. 2nd International Meeting on Vector and Parallel Processing, Porto, Portugal, 1996, 1–23
10. Han Xiaoyan, Han Zhenxiang: The Research on Inherent Parallel Algorithm for Power System Transient Stability Analysis. Proceedings of CSEE. 1997, 17(3):145–148(In Chinese)
11. Wang Fangzong: Parallel algorithm of highly parallel relaxed Newton method for real-time simulation of transient stability. Proceedings of CSEE. 1999, 19(11):14–17(In Chinese)
12. A. Torralba: Three methods for the parallel solution of a large, sparse system of linear equations by multiprocessors, International journal of energy systems, 1992, 12(1):1–5
13. I. C. Decker, D. M. Falcao: Conjugate gradient methods for power system dynamic simulation on parallel computers, IEEE Trans. On PWRS, 1996, 11(3):1218–1227
14. M. Nagata, N. Uchida: Parallel processing of network calculations in order to speed up transient stability analysis, Electrical Engineering in Japan, 2001, 135(3):26–36

HPL Performance Prevision to Intending System Improvement[1]

Wenli Zhang[1,2], Mingyu Chen[1], and Jianping Fan[1]

[1] National Research Center for Intelligent Computing Systems,
Institute of Computing Technology, Chinese Academy of Sciences
[2] Graduate School of the Chinese Academy of Sciences,
NCIC, P.O. Box 2704, Beijing, P.R. China, 100080
zhangwl@ncic.ac.cn

Abstract. HPL is a parallel Linpack benchmark package widely adopted in massive cluster system performance test. On HPL data layout among processors, a law to determine block size NB theoretically, which breaks through dependence on trial-and-error experiments, is found based on in-depth analysis of blocked parallel solution algorithm of linear algebra equations and implementation mechanics in HPL. According to that law, an emulation model to toughly estimate HPL execution time is constructed. Verified by real system, the model is used to do some scientific prevision on the benefits to Linpack test brought by intending system improvement, such as respectively memory size increase, communication bandwidth increase and so on. It is expected to conduce to direct system improvement on optimizing HPL test in the future.

1 Introduction

Linpack[1] is a prevailing performance test benchmark at present. HPL[2] (high performance Linpack) is the first open standard parallel Linpack test package on large-scale distributed-memory parallel computing, used in Top500[3] test widely. In order to obtain optimal results, users can use any number of CPUs to any problem size and use various kinds of optimization methods based on Gaussian Elimination.

Performance test is actually to calculate the floating-point operation rate— Gflops. In LU factorization, it is

$$(2n^3/3 + 3n^2/2) / t. \tag{1}$$

Generally, to obtain HPL peak value, the problem size should be as large as close to 80% of total memory capacity. NB is influential to HPL test time, but till now, its determination mainly depends on experience, which brings about the deficiency of reliability emulation model of performance test.

---

[1] This research is supported by Chinese National High-tech Research and Development (863) Program (grants 2003AA 1Z2070) and by the foundation of Knowledge Innovation Program (grants 20036040), Chinese Academy of Sciences (CAS).

Therefore section 2 of this paper will probe the basis for determining NB theoretically, and try to construct an emulation model of HPL test. In section 3 the emulation model is verified further. Section 4 forecasts the Linpack test prevision with system improvement using the verified model. Finally section 5 concludes.

2 Model Introduction

The main problem of HPL is to solve dense linear algebra equations

$$Ax = b. \tag{2}$$

Clearly, LU factorization[4, 5, 6, 7] is the main part of linear equations solution, accounting for O (n^3) operations. The two-dimensional block-cyclic data distribution of dense matrix among processors is confirmed after series of analysis and comparison[8]. But the efficient determination of NB is still hanging.

2.1 Theoretical Determination of Block Size NB

In experiments, we found that the efficiency factor γ, defined as ratio of test time and the amount of operation, varies little with N gemination increase, but does distinct change with NB to a great extent. After our inference verified, it seems proper to choose suitable NB in quality referring to tendency of small-scale matrix efficiency curve. It will undoubtedly reduce the blindness of NB determination, as well as benefit emulation model construction. Detail description is in Reference [9].

2.2 Emulation by Constructing Model

Since computing is the main part in single processor, it is convinced to construct an emulation model for cluster based on computing and communication parts. By $T_{comm} = α + βL$, communication time can be defined. Where α = Communication latency, β = 1/bandwidth, L = Communication package size. By efficiency factor γ, computing time can be emulated utilizing operation amount. Then according to analyzed HPL panel execution flow and execution logic, detailed in reference [9], an estimating emulation model was implemented. Using the law in section 2.1 proper NB is chosen, then emulation model was activated after acquiring the parameters correlative to architecture, such as communication latency, bandwidth and so on.

3 Model Verification

To emulation model, verification is done on an AMD64 node and part of Dawning 4000A, which listed No. 10 in the newest Top500 lists with 11.264 Tflops theoretical peak performance, respectively described in Table 1. Clearly, estimated results shown in Table 2 are really close to real ones, and obviously better than the rough estimate time, denoted as D in table, described in reference [1].

Table 1. Summary of tested architectures

Arch.	Proc. no.	Freq. (Ghz)	Peak Perf. (Gflops)	Bandwidth (Gb/s)	Latency (ns)	Mem. Size per node (GB)
AMD64 single Node (Ⅰ)	1×2	1.6	3.2	0.664	25000	2
Dawning 4000A Nodes (Ⅱ)	16×4	2.4	4.8	2.5	5000	8

Table 2. Comparison of real system record time and estimated one

Tested Arch.	Matrix Dim. N	Proc. Array P×Q	Real Time (s)	Estimate Time (s)	Diff. (%)	D Time (s)	D Diff. (%)
Ⅰ	8000	2×1	70.04	70.25	0.2998	71.88	2.6271
	14140	1×2	354.88	354.13	-0.2113	356.49	0.4537
Ⅱ	57780	2×2	8196.63	8159.42	-0.4540	8123.27	-0.8950
	115760	2×8	16625.72	16552.34	-0.4414	16309.29	-1.9032

Note: The arch. I and II are corresponding to the ones in Table 2.

The above verification further strengthens the credibility of emulation model.

4 Prevision of Potential Test Performance

Referring to system status of Dawning 4000A, rough forecast is attempted on potential performance to be brought by intending system improvement by the above model. It can be beneficial to leverage efforts on architecture optimization.

4.1 When Memory Size Changes by Gemination

The adjustment of memory size is simulated by matrix dimension N. Estimated results in Table 3 indicate that, with memory size increasing by gemination, the increase scale of system efficiency decreases and the difference is only several permillage. Although reference [10] arguments that there is no limitation for the factorization of huge matrix, due to the time completing once row copy of 9000 elements is 1.8 milliseconds, it seems that reducing matrix size to store in row and column simultaneously is more of feasibility than enlarging the matrix size auxiliary by hard disk prefetching. Moreover, in large scale, the difference between Dongarra estimated time and the model estimated is not up to 3%, which assures the model estimation believable to some extent.

Table 3. To estimate performance improving scale with simulated memory change

Mem. Size (G)	Matrix Dim. N	Estimate Time (s)	D Time (s)	FP Op Rate (Gflops)	Gflops Diff (Gflops)	Effi. (%)	Incr. Scale (%)
2560	509120	12609.8	12256.2	6976.8		61.9391	
5120	720000	35164.0	34476.2	7076.3	99.4847	62.8223	0.8832
10240	1018230	98483.1	97134.4	7146.3	70.0316	63.4441	0.6217
20480	1440000	276643.6	273984.8	7195.7	49.3921	63.8826	0.4385

4.2 When Bandwidth Increases by Ten Times

Based on estimated results shown in Table 4, to gigabit per second bandwidth level, it will be only 0.8% performance improvement to HPL with ten times increase of bandwidth. It is obviously inferior to the former ten times improvement to Gb/s acquiring 8.6% increase. The developing potential distinctly diminishes.

Table 4. To estimate performance improving scale with bandwidth change

Bandwidth (Gb/s)	Estimate Time (s)	FP Op Rate (Gflops)	Efficiency (%)	Incr. Scale (%)
0.25	40768.19	6103.6	54.1864	
2.5	35163.99	7076.3	62.8223	8.6359
25	34714.57	7167.9	63.6356	0.8133

4.3 When Latency Decreases by Thousand Times

Clearly, results in Table 5 show that in large scale once the latency reduces to nanosecond level its decrease is of little real meaning to performance improvement.

Table 5. To estimate performance improving scale with latency change

Latency	Estimate Time (s)	D Time (s)	Time Diff. (%)
5ms	53756.93	53050.33	1.3144
5us	35182.56	34494.77	1.9549
5ns	35163.99	34476.21	1.9559

4.4 When Efficiency Factor Varies

The estimated results in Table 6 indicate that, varying +/- 0.1 from current γ value 0.35 causes system floating point operation rate to change by a large scale, and the increased scale increases with further improvement. Unfortunately, γ is acting by complex factors, related to main frequency and number of FPU etc. Limited by its complexity, further insulated analysis on correlated factors for γ should be done.

Table 6. To estimate performance improving scale with efficiency factor change

Effi. factor	Estimate Time (s)	FP Op Rate (Gflops)	Incr. Scale (%)
0.45	45438.57	5476.2	
0.35	35441.02	7021.0	28.2090
0.25	25443.48	9779.8	39.2931

5 Conclusions

Theoretical analysis and actual system verification show that, it is feasible and credible to use the characteristic of efficiency factor to determine NB and further construct emulation model to estimate HPL execution time. The estimated results demonstrate that to expectative system improvements, such as memory size, communication bandwidth etc., HPL performance is not improved as obviously as that brought by little change of γ. It further indicates that updating computation accounts for absolute portion of Linpack test, whose little change will cause obvious performance increase. Yet due to acting by complex factors, work here corresponding to γ is to be done further. The model is to be amended to do more precise forecast.

References

1. Jack J. Dongarra, Piotr Luszczek and Antoine Petitet. The LINPACK Benchmark: Past, Present, and Future, Concurrency and Computation: Practice and Experience 15, 2003
2. A. Petitet, R. C. Whaley, J. Dongarra, A. Cleary. HPL – A Portable Implementation of the High-Performance Linpack Benchmark for Distributed-Memory Computers, http://www.netlib.org/benchmark/hpl/
3. Hans W. Meuer, Erik Strohmaier, Jack J. Dongarra and H.D. Simon. Top500 Supercomputer Sites, 17th edition, November 2 2001. (The report can be downloaded from http://www.netlib.org/benchmark/top500.html)
4. Zhang BL, etc. Theory and method of numeric parallel computing, Beijing: National defense industry press, 1999,7
5. Lin CS. Numeric computing method (Column A), Beijing: Science press, 1998
6. Chen GL. Parallel computing: structure, algorithm, programming (modified version), Beijing: Advanced education press, 2003.8

7. Sun ZZ. Numeric analysis (second edition), Nanjing: South-east university press, 2002.1
8. http://www.cs.utk.edu/~dongarra/WEB-PAGES/SPRING-2000/lect08.pdf
9. Zhang Wenli, Fan Jianping, Chen Mingyu. Efficient Determination of Block Size NB for Parallel Linpack Test. Proceedings of the IASTED International Conference on Parallel and Distributed Computing and Systems (PDCS 2004), MIT, Received.
10. Eddy Caron, Gil Utard. On the performance of parallel factorization of out-of-core matrices, parallel computing, 30 (2004) 357-375

A Novel Fuzzy-PID Dynamic Buffer Tuning Model to Eliminate Overflow and Shorten the End-to-End Roundtrip Time for TCP Channels

Wilfred W. K. Lin[1], Allan K. Y. Wong[1], and Tharam S. Dillon[2]

[1] Department of Computing, Hong Kong Polytechnic University, Hong Kong SAR
{cswklin, csalwong}@comp.polyu.edu.hk
[2] Faculty of Information Technology, University of Technology, Sydney, N.S.W. 2000
tharam@it.uts.edu.au

Abstract. The novel Fuzzy-PID dynamic buffer controller/tuner eliminates overflow at the user/server/application level adaptively. As a result it shortens the end-to-end roundtrip time (RTT) of a client/server TCP interaction due to improved fault tolerance. The Fuzzy-PID, which is independent of what occurs at the system/router level, is formed by combining fuzzy logic with the extant algorithmic model, the *pure* PID (P^2ID) tuner. It eliminates the shortcomings from the P^2ID component but preserves its power. Its operation is independent of the traffic pattern, and this makes it suitable for the Internet, where the traffic pattern switches suddenly, for example, from LRD (long-range dependence) to SRD (short-range dependence) or multi-fractal.

1 Introduction

The proposed novel dynamic buffer controller/tuner, the *Fuzzy-PID*, eliminates overflow at the user/server/application level. It adaptively tunes the buffer size so that it always covers the queue length by the given margin Δ. The fuzzy logic strengthens the power of its component *pure PID* (P^2ID) controller [1]. The result is that the channel roundtrip time (RTT) is shortened due to better fault tolerance, reliability, availability and dependability. Although the deployment data for the extant P^2ID tuner working alone shows that it always eliminates server buffer overflow, it has two distinctive shortcomings: a) it locks up too much unused buffer memory even when remedial control is no longer needed, and b) the queue length can get dangerously close to the buffer length with a sudden influx of requests.

2 Related Work

The "P+D" algorithm was among the first models to deal with user-level overflow [2] by dynamic buffer tuning. It uses queue length (Q) changes for *proportional* (P) control and the rate of Q changes, dQ/dt for *derivative* (D) control. Although it worked perfectly well in simulations, it failed frequently in real-life deployments. The desire to strengthen the "P+D" control led to the P²ID development [3], which adds *integral* (I) control to improve the anticipative power of the "P+D" model. The I control uses the IEPM (*Internet End-to-End Performance Measurement*) technique to sample the trend of Q changes automatically and quickly at runtime as the feedback to auto-tune the control process.

3 The Novel Fuzzy-PID Controller

The novel Fuzzy-PID controller eliminates the P²ID shortcomings and preserves its merits at the same time. It achieves this by dynamic maintenance of the given *safety margin* Δ for the chosen $\{0, \Delta\}^2$ objective function. The fuzzy logic in the controller divides the P²ID control domain into a set of small fuzzy regions. Table 1 shows the fuzzy regions of a Fuzzy-PID design, which is a *FLC (Fuzzy Logic Controller) matrix* (the FLC(4x4) matrix in this case). The fuzzy rules for the fuzzy regions together form the *fuzzy knowledge base* for fuzzy control. The adaptive adjustment of the buffer length (i.e. dynamic buffer tuning) by addition or subtraction depends on the current fuzzy region, which is either manned by an inert "*don't care*" state (marked by X in Table 1) or a unique fuzzy rule. The purpose of the inert states is to shorten the execution time of Fuzzy-PID by requiring no computation.

Table 1. A Fuzzy-PID controller design example; FLC(4x4)

			dQ/dt			
			NL	NS	PS	PL
QOB	0.7 0.8 0.9	ML	-	-	-	-
		L	-	X	X	+
		G	-	X	X	+
		MG	+	+	+	+

4 Experimental Results

The Fuzzy-PID model was verified with simulations in the Aglets environment, (Figure 1), where the driver and the server are aglets (*agile applets*) that collaborate within a single computer. The driver picks a waveform (e.g. Poisson) or a trace (e.g. self-similar [4]) from the table and uses it to generate the inter-arrival times for the simulated merged traffic for the server queue buffer. A trace is a pre-collected set of requests over a TCP channel (e.g. between Hong Kong PolyU and LaTrobe University in Australia). The trace, which embeds a real-life traffic pattern, verifies that the Fuzzy-PID precision and stability is indeed independent of traffic patterns. The traffic characteristic of a chosen waveform or trace, for example, LRD or SRD, are checked and analyzed at the same time. The checking is called "*traffic pattern analysis*" [5] as shown by the box in Figure 1. In the experiments different tools were used to identify the exact waveform/trace character. For example, the *Selfis Tool* [21], which provides the R/S (*rescaled adjusted statistics*) plot, can estimate the *Hurst* (H) effect/value for the results presented here. The usefulness of the H value is as follows: $0.5 < H \leq 1$ indicating LRD, and $0 < H \leq 0.5$ implying SRD.

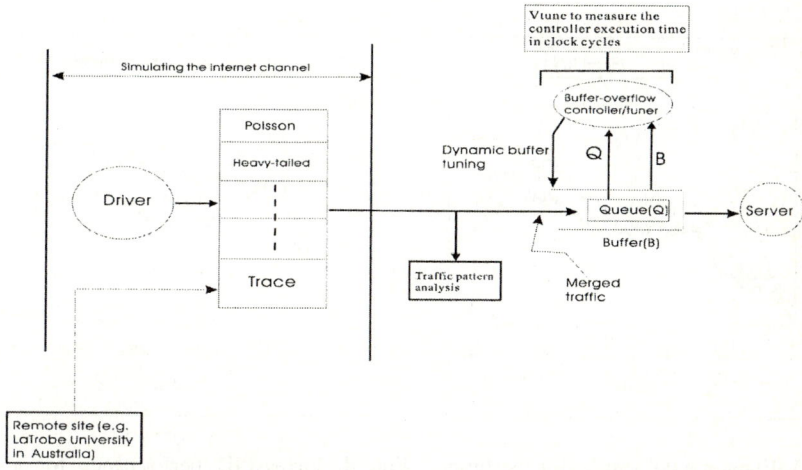

Fig. 1. The setup for the verification experiments

4.1 Case 1 – SRD (Random Traffic)

For the random RTT traffic trace chosen for demonstration the R/S plot of the *Selfis Tool* yields H=0.482, with 99.84% confidence for its SRD character (Figure 2).

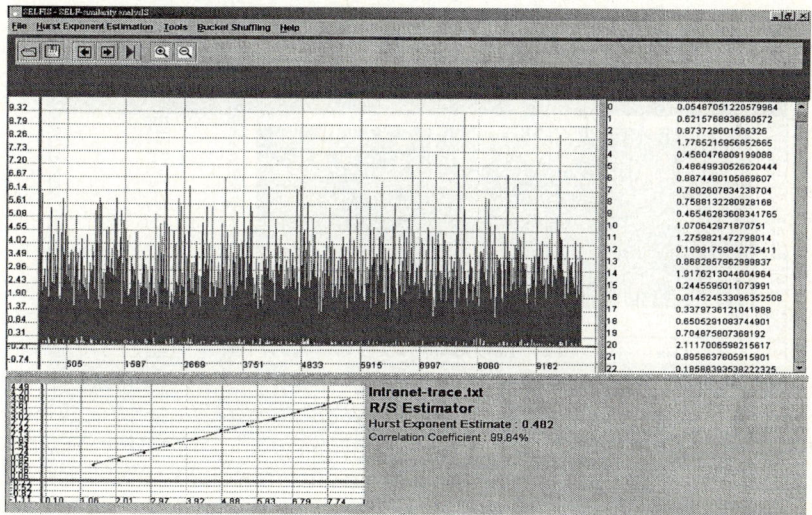

Fig. 2. Trace analysis with the Selfis Tool (R/S Estimator invoked)

Figure 3 shows that both the Fuzzy-PID and P²ID controllers produce no overflow for this trace. The exponential or random nature of the RTT trace is confirmed by calculating and comparing its mean (m) and standard deviation (δ), which are 99 ms and 93 ms respectively. The "$99 \approx 93$" (i.e. $m \approx \delta$) condition indicates that the traffic comes from a Poisson process of the SRD nature.

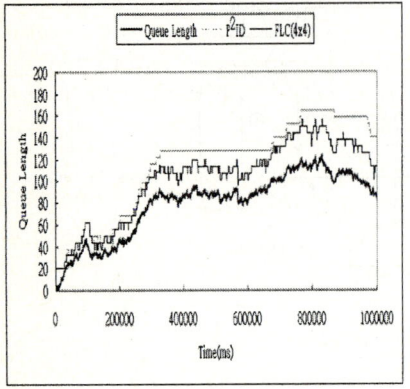 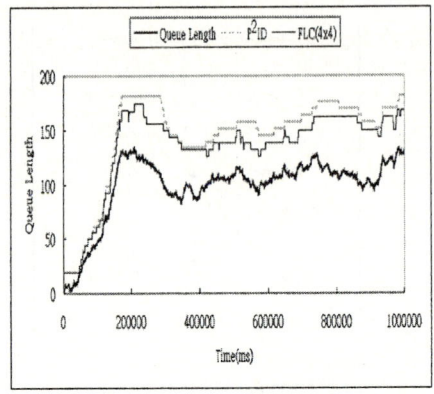

Fig. 3. Experimental results for the Intra-net Traffic

Fig. 4. Fuzzy-PID performance for a self-similar traffic pattern (Figure 4)

4.2 Case 2 – LRD (Self-Similar Traffic)

Self-similar traffic contains bursts that can inundate the server queue buffer. It is important for the Fuzzy-PID to nullify its ill effect by auto-tuning. Figure 4 shows the Fuzzy-PID performance for the self-similar pattern confirmed in Figure 5.

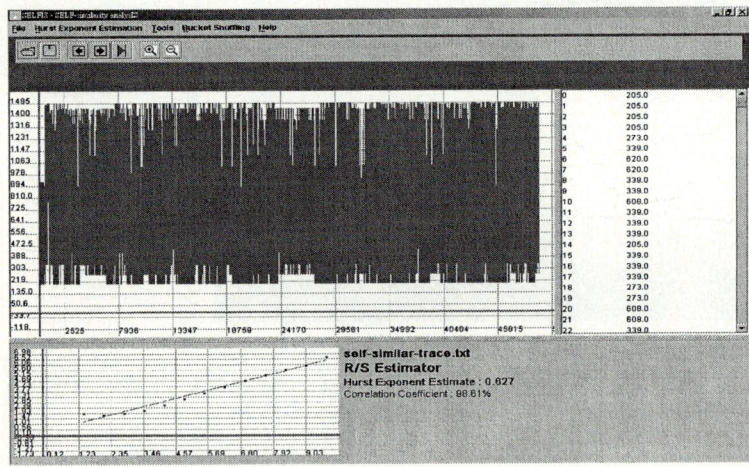

Fig. 5. Trace analysis with the Selfis Tool (R/S Estimator invoked

5 Conclusion

The novel Fuzzy-PID dynamic buffer tuner improves the reliability and the response timeliness of the end-to-end client/server interaction over a TCP channel. It achieves this by eliminating buffer overflow at the server/user level. Together with the AQM mechanism(s) at the system/router level it forms a unified solution for buffer overflow prevention along the path of client/server interaction. This enhances the chance of success for running time-critical applications on the Internet, especially in the soft sense. The Fuzzy-PID is, however, independent of the system operations and traffic patterns. In this way it enhances the client/server interaction reliability, which means better service continuity to satisfy the QoS requirements.

Acknowledgement

The authors thank the Hong Kong PolyU for the HZJ1 and A-PF75 research grants.

References

[1] May T.W. Ip, Wilfred W.K. Lin, Allan K.Y. Wong, Tharam S. Dillon and Dian Hui Wang, An Adaptive Buffer Management Algorithm for Enhancing Dependability and Performance in Mobile-Object-Based Real-time Computing, Proc. of the IEEE ISORC'2001, Magdenburg, Germany, May 2001, 138-144
[2] Allan K.Y. Wong and Tharam S. Dillon, A Fault-Tolerant Data Communication Setup to Improve Reliability and Performance for Internet-Based Distributed Applications, Proc. of the1999 Pacific Rim International Symposium on Dependable Computing (PRDC'99), Dec.1999, Hong Kong (SAR), China, 268-275
[3] Allan K.Y. Wong, Tharam S. Dillon, Wilfred W.K. Lin and T.W. Ip, M^2RT: A Tool Developed for Predicting the Mean Message Response Time for Internet Channels, Journal of Computer Networks, vol. 36, 2001, 557-577
[4] Glen Kramer, Generator of Self-Similar Network Traffic, http://wwwcsif.cs.ucdavis.edu/~kramer/code/trf_gen2.html
[5] T. Karagiannis, M. Faloutsos, M. Molle, A User-friendly Self-similarity Analysis Tool, ACM SIGCOMM Computer Communication Review, 33(3), July 2003, 81-93 (http://www.cs.ucr.edu/~tkarag/Selfis/Selfis.html)
[6] B. Tsybakov and N.D. Georganas, Self-similar Processes in Communications Networks, IEEE Transactions on Information Theory, 44(5), September 1998, 1713-1725
[7] S.I. Resnick, Heavy Tail Modeling and Teletraffic Data, The Annals of Statistics, 25(5), 1997 1805-1869

Communication Using a Reconfigurable and Reliable Transport Layer Protocol

Tan Wang and Ajit Singh

Department of Electrical and Computer Engineering,
University of Waterloo,
Waterloo, Ontario, Canada, N2L 3G1
t7wang@engmail.uwaterloo.ca, asingh@etude.uwaterloo.ca

Abstract. Although TCP is known to be inefficient over networks such as wireless, satellite, and log-fat-pipes, it is still the most widely used transport layer protocol even on these networks. In this paper, we explore an alternative strategy for designing a reliable transport layer protocol that is much more suitable for today's mobile and other types of non-conventional networks. The objective here is to have a single protocol that is compatible with today's communication software and can be easily made to perform better over all types of network. The outcome of the research is a reconfigurable, user-level, reliable transport layer protocol, called RRTP (Reliable and Reconfigurable Transport Protocol) that is TCP-friendly, i.e. it asymptotically converges to fairness as in the case of LIMD (Linear Increase Multiplicative Decrease) algorithms. The protocol is implemented on top of UDP, but it can also easily be incorporated into OS kernels. The paper presents the RRTP algorithm and the key parameters that are necessary for its reconfiguration. We evaluate our protocol using the standard network simulation tool (ns2). Several representative network configurations are used to benchmark the performance of our protocol against TCP in terms of network throughput and congestion loss rate. It is observed that under normal operating conditions, our protocol has a performance advantage of 30% to 700% over TCP in lossy, wireless environments as well as high bandwidth, high latency networks.

1 Motivation

The ubiquitous reliable transport protocol TCP leaves a lot to be desired under certain modern network environments. For instance, according to a previous research work, TCP treats all losses as signs of network congestion [1]. As a result, deploying TCP over wireless network, where wireless losses instead of congestion losses are commonplace, will result in poor performance. In addition to providing unsatisfactory performance in wireless environments, TCP is also ill suited for high bandwidth high latency networks [2]. In this paper, we propose a solution that targets several non-conventional categories of network environments where the performance of TCP is known to be unsatisfactory. At the same time, our solution should provide competitive performance in other environments. Our approach differs from the traditional routes for improving the performance of TCP in several ways:

- We design and implement a reliable transport protocol that would meet or exceed the performance level of TCP under various types of networks.
- Instead of requiring the algorithm to be implemented in the kernel of an operating system, the algorithm can be demonstrated at the user level. At the same time, OS developers can adopt the algorithm later for implementation at the kernel level.
- We suggest the approach of designing a single algorithm that is reliable, robust, and is configurable to provide better performance over different types of networks. For this approach, the research suggests a few key network characteristics that can be used to configure the algorithm.
- The approach can take advantage of an application developer's or an end-user's knowledge of the operating environment and provide better performance. However, it is capable of working well even in the absence of such knowledge.

The new algorithm is called RRTP. For the purpose of evaluating RRTP, we study its behavior under several representative network environments that include: wireless last-hop topology, wireless backbone topology, as well as high bandwidth high latency networks (also known as long-fat-pipes). From these studies, the throughput of RRTP is compared to that of TCP in each scenario. Our simulation results demonstrate that significant improvements can be made to enhance a user's experience with wireless networking through the appropriate usage of parameters for congestion avoidance and loss differentiation. In addition, re-configurability is shown to be of key importance for the superior performance of RRTP. The remainder of the paper details our work so far with the RRTP protocol.

2 The RRTP Approach

This section presents the design approach of the RRTP protocol. First, we focus on the congestion control mechanism of RRTP. According to Chiu and Jain [3], the LIMD (Linear Increase Multiplicative Decrease) approach to congestion control is the only paradigm that will settle down to a state of fairness with an arbitrary starting send rate. The congestion control mechanism of RRTP, like many other TCP variants, follows the basic framework of LIMD approaches but with a significant difference. Instead of taking TCP's approach of flow control window ramping and adjustment, RRTP uses a rate-based algorithm that reacts to incipient congestion and consequently limits the rate of traffic flow below the maximum available bandwidth most of the time. RRTP implements a 4-way handshake connection establishment in order to avoid the DoS (Denial of Service) phenomenon suffered by TCP. During the handshaking process, the nominal value of network RTT (round trip time) is determined. This RTT value refers to the ideal situation in which no network congestions are present.

Once the connection is established, the sender will send out two successive packets for the purpose of probing the network capacity and determining the initial send rate. Let us suppose the send interval of these two packets is X milliseconds. Once the receiver gets both packets, it will advertise to the sender the observed receive interval (Y milliseconds) for the two packets. The sender will calculate the initial send rate based on *max(X, Y)*.

After the initial send rate is determined, the upper layer applications will be able to start using RRTP to transfer information. In the ideal situation where the application user/programmer has an accurate knowledge of the network throughput capability and configures the send rate accordingly, RRTP should be able to instantaneously operate at just below the maximum network capacity. This ensures both minimum wasted bandwidth and stress-free network conditions.

Without user configuration, RRTP will make an educated guess as to the approximate network configuration based on the measured initial send rate and RTT. Each type of network configuration has a pre-defined set of parameter values associated with it. These parameters are: $SendRate_{max}$ and $SendRate_{min}$ which, as the names imply define the upper and lower bounds respectively for the data send rate.

$SendRate_{max}$ serves the purpose of preventing the newly computed send rate from exceeding the maximum network capacity. $SendRate_{min}$ prevents the underutilization of the network that sometimes occurs due to the downward fluctuations of the newly computed send rate.

In protocol design terminology, an epoch refers to a certain interval of packet interchange. In RRTP, we define an epoch to be the interval in which ten packets are sent or received. Since the packet interval time is a key network parameter that we use in RRTP's rate-based congestion control mechanism, we keep two running averages of it: the long-term and the short-term running average. The long-term packet interval average is used for calculating the send/receive rate ratio and adjusting the current send rate. The short-term packet interval average is computed during each epoch. If it significantly deviates from the long-term average, the network would most likely be under stress (congestion due to link failure or additional traffic). At such times, the short-term average is used for the purpose of send rate adjustment instead of the long-term average in order to accurately reflect the network conditions.

Because of the fact that the send rate is only adjusted at the end of each epoch, constant fluctuation of network traffic is minimized. This results in a more stable network connection. At the time of send rate adjustment, the newly adjusted rate is subjected to comparison with two parameters mentioned earlier: *$SendRate_{max}$* and *$SendRate_{min}$*. In other words, the new rate must fall within the range of *$SendRate_{min}$* to *$SendRate_{max}$*. This is done to minimize the chance that an overshoot occurring when RRTP ramps up the send rate during the linear increase phases and the occurrence of unnecessary reduction in the send rate during the multiplicative decrease phases. *$SendRate_{max}$* and *$SendRate_{min}$* are not fixed values. They are re-calculated based on changes of network dynamics as discussed in the previous paragraph.

The send rate adjustment is carried out using the following algorithm: first, we define an additive increase factor α with different initial values based on the type of network RRTP is operating on as well as a multiplicative decrease factor β with an initial value of 0.05. If the send/receive rate ratio is greater than 1.05, RRTP is operating at a level above the maximum network throughput capacity. Our protocol treats such situations as signs of incipient congestion and will carry out the following adjustment: *$SendRate_{new}$* = *$SendRate_{prev}$* $\times max((1-\beta), 0.5)$. The value of β is doubled for every consecutive multiplicative decrease phase until it reaches the upper bound of *$1-\beta > 0$*. Here, we take *$max((1-\beta), 0.5)$* to be the adjustment factor to ensure that the rate reduction factor will never drop below 0.5. In other words, when RRTP first detects signs of incipient congestion, it gently reduces the send rate with a small value of β. If the incipient congestion persists over several epochs, the value of β will be doubled every epoch to more effectively suppress incipient congestions. Now on the other hand, if the send/receive rate ratio is less than 0.95, RRTP is operating well below the maximum network capacity. This results in a linear increase phase in which $SendRate_{new} = SendRate_{prev} + \alpha$. In addition, β is reset to its initial value of 0.05.

With the rate-based congestion avoidance mechanism described above, RRTP is able to avoid several situations for congestion that would be encountered by TCP. However, there are situations that will result in congested network even with RRTP as the end-to-end transport mechanism. Such situations include temporary link failures and sudden surges of new traffic. Under ill-fated network conditions like this, packets may be lost due to severe congestion. RRTP aggressively reduces the send rate (by 50% for each congested epoch) in response to detected congestions. Such efforts are needed to avoid a total network collapse. When the signs of congestion disappear, instead of carrying out the slow start used in TCP, RRTP performs an instantaneous send rate recovery by using the last recorded characteristic send rate as the one for the next send/receive cycle.

For the case in which the user mis-configures the initial send rate, our algorithm is smart enough to detect that. Send rate convergence is still guaranteed in this scenario due to the nature of RRTP's rate control mechanism.

Let us now discuss the issues of reliability and re-configurability. The ability to re-configure to adapt to different network platforms is the key feature that sets RRTP apart from most of the other protocols of its kind. Reconfigurability is built into RRTP by the means of the parameterization of a set of key network parameters. Our experiments indicate that only a small set of parameters is needed to design a re-configurable transport protocol algorithm that would provide a good performance on different types of networks. These parameters are: (1) $SendRate_{nominal}$ that denotes the normal channel capacity, (2) $RoundTripTime_{nominal}$ which defines the normal end to end latency, and (3) $LossRate_{nominal}$ which is the characteristic data loss rate for the channel.

To ensure a reliable transport, the receiver sends two kinds of acknowledgements to the sender: cumulative acknowledgement, and negative acknowledgement. Negative acknowledgements are coupled with timeouts. Our timeout mechanism uses RTT. Cumulative acknowledgements serve as confirmation of received packets during normal network operations. When a cumulative acknowledgement is received by the sender, the sender can safely remove the corresponding acknowledged buffered packets. The cumulative acknowledgement interval is defined to be the period during which 32 packets are received.

Finally, we turn our focus to loss differentiation algorithm of RRTP. Several published research works on the issue of TCP performance enhancement over wireless networks have considered sender-based loss differentiation. RRTP, on the other hand, is based on the intuition that the receiver usually has more accurate and timely knowledge of packet losses. Consequently, the receiver is responsible for figuring out the cause of a particular loss and informing the sender to take the appropriate action.

For wireless last hop networks, RRTP makes two assumptions regarding the path characteristics. The first assumption states that the wireless link has the lowest bandwidth and thus is the bottleneck of the network. Secondly, the wireless base station is assumed to serve strictly as a routing agent between the wired and the wireless network with no additional smart capabilities. As one can, quite easily see, with the big difference in bandwidth between wired LAN (100 Mbps) and cellular wireless (around 19.2 Kbps), packets traveling on the wired network would get congested at the base station while adapting to the lower send rate imposed by the wireless network. As a result, the packets transmitted on the wireless connection tend to be clustered together. If a packet loss occurs due to random wireless transmission errors, the receiver should be able to observe a certain time interval in which the packet is expected but not received. Such an event can be interpreted to be the sign of wireless loss due to transmission errors. Following this reasoning, RRTP can distinguish between wireless losses form congestion losses using the following heuristics: let T_{min} be the minimum observed packet interval for the receiver and $T_{separation}$ be the interval between the time when the last correct packet is received and the time when the lost packet is detected by the receiver. Suppose n packets were lost, the loss is characterized as wireless loss if the following relation holds: $(n + 1) T_{min} < T_{separation} < (n + 1.75) T_{min}$. The number we choose are experimentally determined to cause the lowest misclassification rate between congestion and wireless losses.

For the wireless LAN topology, the assumptions that we made in the previous situation are usually not true. Conventional wired LAN is not much faster than high-speed wireless LAN. As a result, packets don't necessarily travel in close succession on the wireless LAN connection. Consequently, the previous LDA heuristic will not perform as well as in wireless last hop topology. As a result, an alternative approach is used in this case to distinguish between wireless loss and congestion losses.

In order to achieve good accuracy in distinguishing between the two types of packet losses for the wireless LAN topology, RRTP uses the ROTT (Relative One-way Trip Time) measurements as congestion indicators. ROTT is defined to be the time between the moment when the packet is sent and the moment when the packet is received. It is measured at the receiver end. During periods of smooth traffic flow, ROTT measurements will remain relatively stable. When the network starts to become congested, the receiver will detect rising ROTT values. The default behavior of RRTP in this situation is that the receiver will issue an explicit incipient congestion notification to the sender to throttle the send rate. In the event that the rise in ROTT values is coupled with packet losses, the receiver can be confident that the packet losses are caused by congestion. However, if the packet losses are not accompanied by a rise in ROTT value, the receiver will categorize these losses to be due to wireless errors. As it was discussed above, two different LDA schemes are used by RRTP. Depending on the actual wireless network in use, RRTP selects the appropriate LDA to achieve optimum performance.

3 Simulation Results

To evaluate the actual performance of RRTP, we have created various simulation scenarios using the ns2 simulator [4]. Tests were conducted under the simulated environments with RRTP, TCP Reno, TCP New Reno and TCP Vegas. The characteristics of these environments are summarized in Table 1.

Table 1. Testing Environment Specifications

Environment	Bandwidth	One-way Latency	% Loss
High Latency & High BW	100 Mbps	100 ms	1%
CDMA	19.2 Kbps	100 ms	1%
Satellite	256 Kbps	100 ms	1%
LAN	100 Mbps	5 ms	0%
Wireless LAN	11 Mbps	10 ms	1%

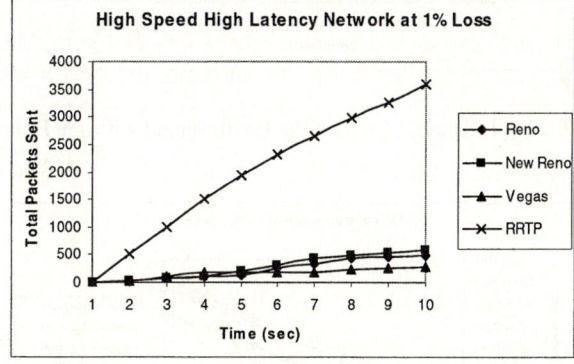

Fig. 1. Protocol Performance for High Speed High Latency Environment with 1% Data Loss Rate

In a high latency high bandwidth topology, a typical protocol that relies on sender-receiver feedbacks will inevitably suffer from the slowness of its response to changing network condition. This is due to the fact that round trip return time is extremely large and consequently, it is difficult to rely on feedbacks to adjust the send rate. Fairness can be severely limited as newly entered traffic will almost always be starved by previously established traffic. However, because of the fact that RRTP is reconfigurable, good estimates of the network conditions can be provided to the application before the transfer starts, allowing a much higher throughput than conventional TCP as demonstrated in Figure 1.

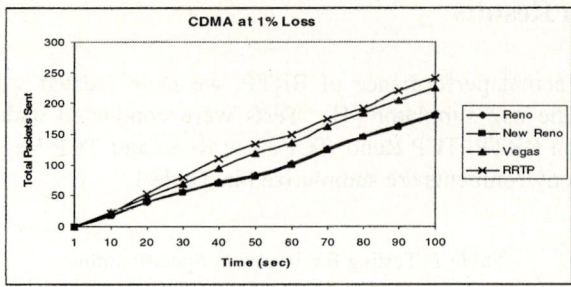

Fig. 2. Protocol Performance for CDMA Environment with 1% Data Loss Rate

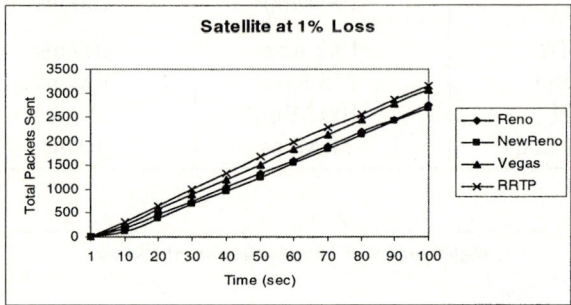

Fig. 3. Protocol Performance for Satellite Environment with 1% Data Loss Rate

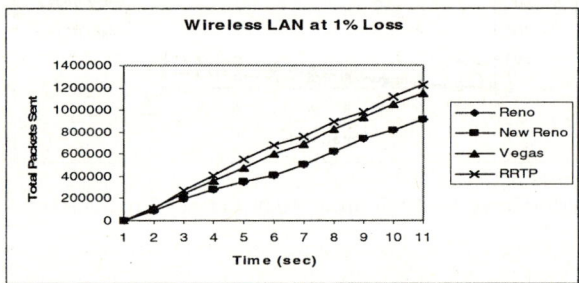

Fig. 4. Protocol Performance for Wireless LAN Environment with 1% Data Loss Rate

Both CDMA and satellite network can be considered to be roughly wireless last hop topologies. As demonstrated in Figure 2 and Figure 3, RRTP performs much better than TCP Reno and TCP New Reno on both types of network platforms. This is expected since when losses are encountered, TCP invokes its congestion control mechanisms right away without making an effort to distinguish among the different types of losses. In this scenario, results are quite similar to the wireless last hop topology. The Spike LDA enables RRTP to differentiate between congestion losses and wireless losses, resulting in a superior performance in term of throughput as shown in Figure 4. In addition to the three network platforms mentioned above in which RRTP demonstrates superior performance, the simulation result shown in Figure 5 also demonstrates that the performance of RRTP on conventional LAN closely matches that of TCP Vegas. This implies that not only could RRTP outperform TCP in certain network configurations, it could also serve as a viable substitute in the more traditional network settings.

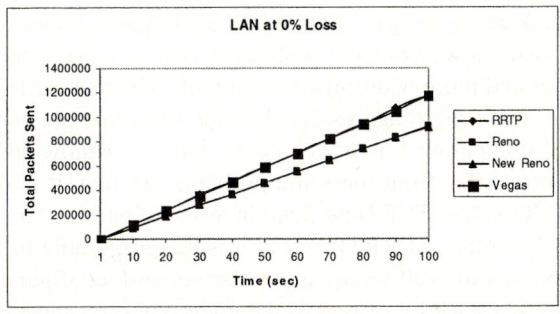

Fig. 5. Protocol Performance for LAN Environment with 0% Data Loss Rate

4 Comparison with Related Works

Many approaches to improve the performance of TCP over wireless have been presented in the data communications literature. The first category of approach uses link-layer retransmissions and thus shields wireless losses from TCP as proposed by DeSimone et al. [5]. Such approaches work well when the latency over the wireless link is small as compared to the coarse grain TCP timer. There are also TCP-aware snoop mechanisms that have a snoop-agent module at the wireless base station as proposed by Balakrishnan et al. [6]. In WWAN environments, snoop does not work well because it exacerbates the problem of large and varying round trip times by suppressing negative ACKs.

In the past, a number of researchers have also proposed end-to-end solutions to improve the performance of TCP in certain cases. Casetti et al. [7] proposed an end-to-end modification of the TCP congestion window algorithm, called TCP Westwood that relies on end-to-end bandwidth estimation to discriminate the cause of packet loss. However, most of their evaluations are based on the wireless link being the last link to the receiver. This algorithm is also highly dependent on the TCP ACKing scheme.

Biaz and Vaidya have looked at two different approaches to the end-to-end loss differentiation for TCP connections. They first looked at a set of "loss predictors" based upon three different analytic approaches to congestion avoidance that explicitly model connection throughput and/or round-trip time (e.g., TCP Vegas) [8]. Their results were negative in that these algorithms, formulated to do loss differentiation, were poor predictors of wireless loss. In subsequent work, they proposed a new algorithm that uses packet inter-arrival time to differentiate losses. Using simulation, they show that it works well in a network where the last hop is wireless and is also the bottleneck link. But they failed to evaluate their algorithm when the wireless link is not the last hop and nor the bottleneck of the network.

One of the fundamental design decisions we made in the making of RRTP is the conscientious effort of congestion avoidance. By promoting congestion avoidance, network throughput can be significantly enhanced as less congestion related situations are encountered during the lifetime of the network connection. This design approach can also be seen in TCP Vegas. However, in the case of TCP Vegas, there is one significant drawback in its design. Lai and Yao [9] have shown in their study that when different traffic flows compete with each other in the same channel, traffic running under older and more widespread version of TCP such as TCP Reno and TCP Tahoe tends to be much more aggressive than the ones that are running under TCP Vegas in terms of competing for the available network bandwidth. RRTP, on the other hand, does not suffer from the same problem. In fact, it is observed to be as aggressive as TCP Reno and TCP New Reno in terms of bandwidth acquisition.

Another major advantage of RRTP is that it is reconfigurable in nature. The user does not have to restrict himself to any particular network configuration for optimum network conditions when RRTP is used as the underlying transport layer protocol. In a way, RRTP tries to be a generic protocol like TCP. The main deviation from TCP's design philosophy is that RRTP takes advantage of user's knowledge of the network. By doing so, RRTP can perform just as well as the various solutions discussed in this section in each individual special cases while still remaining insensitive to the varying network configurations.

The research done by Sinha et al. [10] on WTCP has significant commonality with the present work. WTCP is an end-to-end transport layer protocol that uses a rate-based mechanism for congestion control and the Biaz [8] LDA for differentiating between congestion losses and wireless losses. Although it is able to achieve good results on wireless last hop networks, the authors did not test WTCP on other types of wireless platforms such as wireless backbone network and wireless LAN. In fact, we believe that WTCP will likely perform poorly on the two latter network platforms. The reason is that the Biaz LDA is only optimized for wireless last hop networks. When we tested the Biaz LDA on networks with wireless LAN configuration, we found that the algorithm resulted in a lower throughput than the Spike [11] LDA. RRTP addresses this shortcoming of WTCP by designing a LDA mechanism that is closer to the Spike LDA for better performance on wireless backbone and wireless LAN networks.

Another advantage of RRTP over WTCP is its faster send rate convergence. Since RRTP allows the user to specify the ideal sending rate for the network platform of interest, accurate user inputs could potentially help RRTP to converge to the ideal send rate within the initial connection establishment period. This really translates into the avoidance of many unnecessary overshoots that would otherwise be encountered if WTCP were used as the transport layer protocol. For short-lived connections, RRTP will be able to out perform WTCP by several folds since the user inputs for the initial send rate essentially eliminate the need for network capacity probing phase.

5 Summary

The paper presents a novel reliable and reconfigurable transport protocol, called RRTP, that is able to not only provide better performance on non-conventional networks where TCP's performance is known to be unsatisfactory, it also provides a performance competitive to TCP on conventional networks such as the common LAN environment. The option of user-level implementation facilitates quicker adoption of the protocol on most platforms whereas optional re-configuration would allow tuning of the protocol for better performance under various types of networks. Once the user community gathers enough experience with the protocol, it could be adapted for implementation at the kernel level. The work is continuing on creating an application programmer's interface (API) for RRTP, based on sockets, that is very similar to the socket interface for TCP in Linux and Windows OS environments.

References

1. Balakrishnan, H., Padmanabhan, V., Seshan, S., Katz, R.: A Comparison of Mechanisms for Improving TCP Performance over Wireless Links. IEEE/ACM Transactions on Networking, Vol.5, no. 6, (1997) 756-769
2. Jacobson, V., Braden, R., Borman, D.: TCP Extensions for High performance. RFC 1323, (1992)
3. Chiu, D., Jain, R.: Analysis of the Increase/Decrease Algorithms for Congestion Avoidance in Computer Networks. Journal of Computer Networks and ISDN Systems, vol. 17, no. 1, (1989)
4. ns-2 Network Simulator (version 2). LBL, URL: http://www.isi.edu/nsnam/ns
5. DeSimone, A., Chuah, M.C., Yue, O.: Throughput Performance of Transport Layer Protocols over Wireless LANs. Proceedings of IEEE GLOBECOMM, (1993)
6. Balakrishnan, H., Seshan, S., Amir, E., Katz, R.: Improving TCP/IP Performance over Wireless Networks. Proceedings of ACM MOBICOM, (1995)
7. Casetti, C., Gerla, M., Mascolo, S., Sanadidi, M.Y., Wang, R.: TCPWestwood: Bandwidth Estimation for Enhanced Transport over Wireless Links. Proc. ACM Mobicom 2001 Conference, Rome Italy (2001) 287-297
8. Biaz, S., Vaidya, N.: Distinguishing Congestion Losses from Wireless Transmission Losses: A Negative Result. Proc. 7th Intl. Conf. on Computer Communications and Networks, Lafayette LA (1998)
9. Lai, Y.C., Yao, C.-L.: The Performance Comparison between TCP Reno and TCP Vegas. Proc. of Seventh International Conference on Parallel and Distributed Systems, Iwate, Japan, (2000)
10. Sinha, P., Nandagopal, T., Venkitaraman, N., Sivakumar, R. Bharghavan, V.: WTCP: A Reliable Transport Protocol for Wireless Wide-Area Networks. Wireless Networks 8, (2002) 301-316
11. Tobe, Y., Tamura, Y., Molano, A., Ghosh, S., Tokuda, H.: Achieving Moderate Fairness for UDP Flows by Path-status Classification. Proc. 25th Annual IEEE Conf. on Local Computer Networks (LCN 2000), Tampa FL (2000) 252-261

Minicast: A Multicast-Anycast Protocol for Message Delivery

Shui Yu, Wanlei Zhou, and Justin Rough

School of Information Technology, Deakin University,
Geelong, Victoria, Australia
{syu, wanlei,ruffy}@deakin.edu.au

Abstract. Anycast and multicast are two important Internet services. Combining the two protocols can provide new and practical services. In this paper we propose a new Internet service, Minicast: in the scenario of n replicated or similar servers, deliver a message to at least m members, $1 \leq m \leq n$. Such a service has potential applications in information retrieval, parallel computing, cache queries, etc. The service can provide the same Internet service with an optimal cost, reducing bandwidth consumption, network delay, and so on. We design a multi-core tree based architecture for the Minicast service and present the criteria for calculating the subcores among a subset of Minicast members. Simulation shows that the proposed architecture can even the Minicast traffic, and the Minicast application can save the consumptions of network resource.

1 Introduction

Dramatic development of the Internet has led to many communication paradigms providing all sorts of Internet-based services. Multicast delivers a packet from a source to n members in the multicast group [1] and has applications of data synchronization, Internet meeting, etc. Anycast delivers a packet from a source to the "best" receiver among n replicated servers [2] and has applications in information retrieval; it can reduce the cost for the relative Internet services. Combining the two protocols can provide new practical Internet services. PAMcast [3] is a good example, generalizing both anycast and multicast by delivering packets to m out of the total n group members ($1 \leq m \leq n$). However in PAMcast m is constant, although we can change m from time to time. PAMcast emphasizes message delivery over information retrieval and with a fixed parameter m is not flexible and costs network resources.

Information retrieval is increasingly important given the growth content on the Internet. In this paper, we propose a new packet delivery service, Minicast, which generalizes both anycast and multicast for the purpose of information retrieval. Minicast delivers packets to any m out of n total group members ($1 \leq m \leq n$). Minicast has potentially a wide range of applications, including: parallel information retrieval (searching for a message from a group of similar web sites), parallel cache queries (given several cache replicas a client can query a subset for desired data), parallel grid computing (selection of computers to service a job given a large number of members), and parallel downloading (request m servers send separate portions of the same file).

This paper is organized as follows. Section 2 introduces related work on anycast and the multicast, including combinations. The Minicast architecture is presented in Section 3. A performance evaluation and analysis is in Section 4. Finally in Section 5, conclusions and future work are presented.

2 Related Work

Multicast, recognized as an important facility because of its growing usage in distributed systems, sends packets to all members of a group [1]. A recent survey concluded that there are five classes of multicast routing algorithms [4]: flooding, spanning tree, reverse path forwarding, core-based tree, and solution to the travelling salesman problem. There is also good research on core based multicast algorithms [5–8]. Partridge, Mendez, and Milliken proposed the idea of anycast for the delivery of packets to at least one host in a group, preferably only one host [2]. Initial research focused on network-layer anycast [9–12] but due to limitations development moved to application-layer anycast [13, 14].

Jia et al. proposed integration of multicast and anycast, where multicast provides update consistency and anycast assists multicast requests to reach the "nearest" member in the multicast group [10]. The combination of anycast and multicast offers bi-directional service for distributed data processing systems: multicast provides data synchronization among the multicast group, and anycast finds the "best" server in the anycast group. Furthermore, anycast is a good methodology for both server load balance and network load balance.

3 The Minicast Multi-core Architecture

In this section, we propose a multi-core architecture for the Minicast service. To describe the architecture simply and clearly, we do not address network reliability. Instead, we assume a reliable network with no local failures. The multi-core architecture for the Minicast service is shown in Fig. 1. All Minicast receivers are defined as members of a Minicast group; all the Minicast group members are connected by Minicast routers (routers capable of Minicast routing). Minicast routers may be distributed anywhere in a network. The network is partitioned into N domains by organizations, regions or other metrics. A router is selected in each domain as a local core, and a local Minicast tree rooted on the local core is established of which Minicast members become leaves. The local core holds all information about the local tree, such as number of members and number of hops to each member. All local cores exchange tree information when necessary.

For the whole network, we have N cores for a Minicast group. For performance reasons, we organize the N cores into an anycast group. Anycast addresses are used to ensure client queries are delivered to the "nearest" anycast member (the

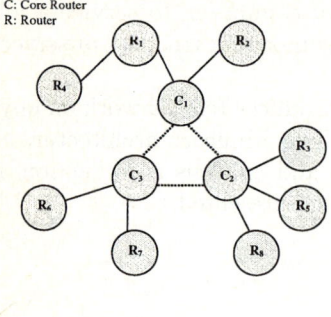
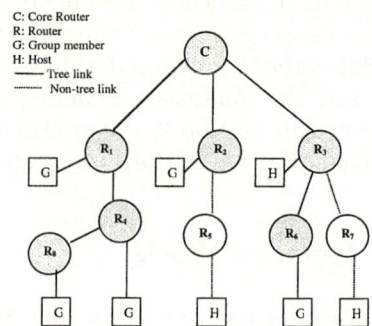

Fig. 1. Multi-core Architecture

Fig. 2. Local Minicast Tree Example

core). Without loss of generality, we suppose that a local tree has k members. If $k \geq m$, where m is the Minicast parameter, the packet is delivered to at least m members and close to k members by the local core setting a suitable *TTL* and multicasting the packet on the local tree. If $k < m$, the local core similarly multicasts the packet to the local tree but simultaneously forwards the Minicast query with parameter $m-k$ to the nearest remote core. The remote core continues the procedure until at least m Minicast members receive the packet.

Fig. 2 shows an example local Minicast tree. All local domain members are included in the tree, and there is a core (members are shaded). Note that a router may be in a local Minicast tree without any directly connected members, such as router R_3. The local core knows the hops to each member, so that it can set the *TTL* when sending packets. We define Minicast radius as the number of hops for delivering a packet to at least m members. For example, to send to at least three members the Minicast radius is set to three. Packets are discarded after three hops, and within that distance, the message is delivered to four members. Local core selection is achieved similar to Gupta and Srimani [5], as follows:

1. Randomly select one node from the member in a given domain, assume the node is the local core, and build the local Minicast tree base on the core.
2. The core calculates the sum of the weights of each subtree, and detects when it is no longer a centroid of the local Minicast tree.
3. The current local core starts a migration towards the current centroid of the local Minicast tree.

4 Performance Analysis

We conducted several simulations of Minicast in an experiment environment consisting of 20 Minicast members, 3 local core and 17 non-core. All local trees containing a similar number of members is called a symmetric Minicast tree, otherwise asymmetric. We conducted traffic concentration simulations for both

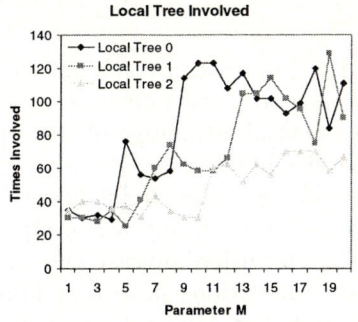 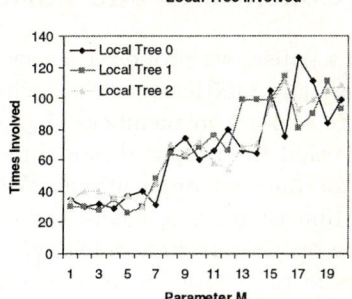

Fig. 3. Asymmetric Tree   **Fig. 4.** Symmetric Tree

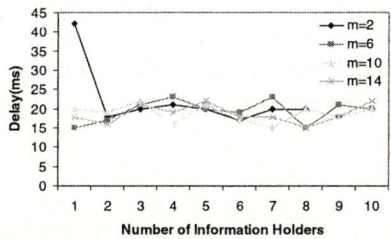

Fig. 5. Minicast Application in Information Caching

symmetric and asymmetric Minicast trees. During each experiment, Minicasts are randomly initiated from the local trees, and count the times a local core is used in packet delivery to measure the distribution of the traffic on the Minicast tree given an increasing m.

For the asymmetric tree (Fig. 3), local trees 0, 1, and 2 have 10, 6, and 4 Minicast members respectively. Based on the analysis of the curves, we can conclude that in general, the packet distribution is even among the local Minicast trees. For the symmetric tree (Fig. 4), the local trees have 7, 6, and 7 members. It is obvious that the packet distribution is even among the three local Minicast trees. The two experiments show that the Minicast solves the traffic concentration problem and has even packet distribution among members.

To simulate the application of Minicast in information retrieval, we examine an information caching application (Fig. 5). Here there are 23 hosts in a Minicast group, some of which hold information (we do not know which). Given that we want to use minimal bandwidth, we measure the network delay as a metric. We did the simulation on the asymmetric Minicast tree scenario with m values of two, six, ten, and four. The simulation shows that the performance (delay) is mostly not sensitive to the value of m or the number of the information holders. Therefore, in practical applications a small m can be used to save bandwidth.

5 Conclusions and Future Work

In this paper, we proposed a new Internet service, Minicast, a combination of anycast and multicast services which given n replicated servers delivers a message to at least m members ($1 \leq m \leq n$). The essential advantage of this service is that the proposed model provides same service in terms of QoS with less cost for Internet applications. We proposed a multi-core tree architecture for the Minicast service, where the cores are organized into an anycast group to ensure queries are first delivered to the "nearest" core using anycast and then delivered to at least m members using multicast. Simulations show that Minicast can handle traffic concentration very well, and that Minicast can reduce network resource consumption while maintaining performance for information caching.

Further issues such as the relationship between the performance and the number of local cores for a Minicast group, the issue of fault-tolerance in the Minicast services, a comparison of the performance of Minicast and related services, should be explored. Further simulations will be completed in the near future to assess the Minicast service.

References

1. Deering, S.E., Cheriton, D.R.: Multicast routing in datagram internetworks and extended lans. ACM Transactions on Computer Systems **8** (1990)
2. Partridge, C., Mendez, T., Milliken, W.: Host anycasting service. Technical Report RFC1546 (1993)
3. Chae, Y., Zegura, E.W., Delalic, H.: Pamcast: Programmable any-multicast for scalable message delivery. In: Proceedings of IEEE OpenArch. (2002)
4. Vincent, R., Luis, C., Rolland, V., Anca, D., Serge, F.: A survey of multicast technologies (2000)
5. Gupta, S.K.S., Srimani, P.K.: Adaptive core selection and migration method for multicast routing in mobile ad hoc networks. IEEE Transactions on Parallel and Distributed Systems **14** (2003)
6. Jia, W., Xu, G., Zhao, W.: Efficient internet multicast routing using anycast path selection. Journal of Network and Systems Management **12** (2002) 417–438
7. Thaler, D., Ravishankar, C.V.: Distributed center location algorithms. IEEE Journal on Selected Areas in Communications **15** (1997) 291–303
8. Yoon, J., Bestavros, A., Matta, I.: Somecast: A paradigm for real-time adaptive reliable multicast. In: IEEE Real-Time Technology and Applications Symposium (RTAS), Washington DC (2000)
9. Basturk, E., Engel, R., Haas, R., Peris, V., Saha, D.: Using network layer anycast for load distribution in the internet. Technical report (1997)
10. Jia, W., Xu, G., Zhao, W.: Integrated fault-tolerant multicast and anycast routing algorithms. IEE Proceedings of Computers and Digital Techniques **147** (2000)
11. Katabi, D., Wroclawski, J.: A framework for scalable global ip-anycast (gia). In: SIGCOMM, Stockholm, Sweden (2000)

12. Xuan, D., Jia, W.: Distributed admission control for anycast flows with qos requirements. In: IEEE International Conference on Distributed Computing Systems. (2001)
13. Bhattacharjee, S., Ammar, M.H., Zegura, E.W., Shah, V., Fei, Z.: Application-layer anycasting. In: IEEE INFOCOM, Kebe, Japan (1997)
14. Fei, Z., Bhattacharjee, S., Zegura, E.W., Ammar, M.H.: A novel server selection technique for improving the response time of a replicated server. In: IEEE INFOCOM. (1998)

Dependable WDM Networks with Edge-Disjoint P-Cycles

Chuan-Ching Sue[1], Yung-Chiao Chen[2], Min-Shao Shieh[1], and Sy-Yen Kuo[2]

[1] Department of Computer Science and Information Engineering,
National Cheng Kung University, Tainan, Taiwan
{suecc, sms}@mail.ncku.edu.tw
[2] Department of Electrical Engineering, National Taiwan University,
Taipei, Taiwan
{ycchen, sykuo}@cc.ee.ntu.edu.tw

Abstract. In this paper, we propose a fault tolerant mechanism on the optical network design with edge-disjoint P-cycles (EDPC). EDPC is a special type of traditional P-cycles, that is, no common edges are allowed to exist between any two P-cycles. Previously published schemes for computing P-cycles are time consuming and do not survive multiple link failures when P-cycles have common edges. Instead of using the complex ILP, a heuristic method based on link state routing which is compatible to the traditional open shortest path first (OSPF) network protocol is proposed to speed up the construction of EDPC. The results show that the EDPC method can tolerate more link failures and improve the restoration efficiency for traditional P-cycles with the decrease of two working units for every two P-cycles.

1 Introduction

Protection and restoration are both important issues in dependable WDM networks. On such topic, there are four approaches as following: loop-back and redundant trees [8], [9], [12], Protection cycles (without chords) [3], P-cycles (with chords) [5], [6], and shared path protection in WDM networks [11], [12]. In this paper, we will focus on the improvement of the P-cycles.

The simplest way to think about P-cycles is that they are like rings, but with support for the protection of chord spans as well as the usual ring spans of the ring itself. A chord span is one that has its end-nodes on the P-cycles, but is not itself part of the P-cycles. The key distinction of P-cycles as opposed to any kind of ring or cycle covers is the protection of chord spans which themselves can each bear two units of working capacity and zero spare capacity.

P-cycles introduced by Grover and Stamatelakis [1-2] can be characterized as pre-configured protection cycles in a mesh network. With the hybrid cycle and mesh approach, the P-cycle concept is able to benefit from the advantages of both worlds. Ring protection mechanisms offer very fast recovery times (about 50-60 ms), but the required spare to working resources ratio is at least 100%, in real networks sometimes more than 200%. For mesh networks, however, the required spare to working resources ratio can typically be in the range of only 50-70% for well-connected physical network graphs. The concept of P-cycle utilizes the benefits of both alternatives: the efficiency of mesh restoration and the recovery speed of ring networks.

In this paper, we assumed WDM node can often be very reliable, e.g., internal redundancy is used and only the link failure is considered. Although one P-cycle provides protection against single link failure, it cannot protect against double simultaneous failures on that P-cycle. If multiple link failures happen in such a way that there is at most one link failure on each P-cycle, the P-cycles can actually protect against such multiple failures. However, without careful consideration, there are many overlapped edges among the P-cycles. Single link failure on the overlapped edge causes multiple associated P-cycles failing to survive one additional link failure. Instead of computing P-cycles with Integer Linear Programming (ILP), we are interested in efficient construction of many edge-disjoint P-cycles (EDPC) that can really protect multiple link failures. The construction scheme is based on the link state routing which is compatible to the traditional open shortest path first (OSPF) network protocol. We then extend the scheme by selective reconfiguration or rerouting when EDPC lost its protection ability due to double link failures in a P-cycle.

In order to explain the difference between the traditional P-cycle and our EDPC, Fig. 1(a) shows only one P-cycle with a long restoration path, Fig. 1(b) shows four P-cycles with shorter restoration paths but the limited fault tolerance for multiple link failures due to shared edges between P-cycles, and Fig. 1(c) compromises the length of restoration path but with the increased ability for multiple link failures.

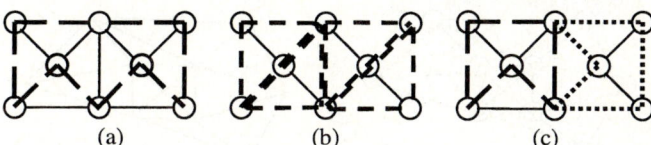

Fig. 1. (a) One P-cycle. (b) Four P-cycles with overlapped edges. (c) Two p-cycles with edge-disjoint edges

The rest of this paper is organized as follows. In Section 2, we give the preliminaries for our edge-disjoint P-cycles scheme. In Section 3, the heuristic method of constructing the edge-disjoint P-cycles is detailed. The time complexity of EDPC is also discussed. Section 4 depicts the performance evaluation and discusses the pros and cons of our EDPC. Finally, we conclude our study in Section 5.

2 Preliminaries

Edge-disjoint P-cycles (EDPC) inherit many advantages of P-Cycle. One is that for chord link failures, the length of protection paths are on average half the length of the individual small protection cycle circumference, whereas in rings, protection paths are essentially the full circumference of the ring. Because EDPC are formed in the spare capacity only, they can be adapted to suit the working path layer at any time, without any impact on working demands. In contrast, rings assert the routing that demands must take, rather than adapting to the routes they want to take. Jointly optimizing the working path routing with EDPC placement should yield even further capacity savings and the efficiency of restoration.

EDPC can not only speed up the recovery time, but also offer more resilience for link failures when the network size is large. Fig. 2(a) shows the allocated spare capacity with dashed lines in the traditional P-cycle and Fig. 2(b) shows the protected working capacity with solid lines using the traditional P-Cycle. The length of the average recovery route is eight hops long and only one link failure can be survived. On the other hand, Fig. 3(a) shows the spare capacity by way of EDPC with three overlapped nodes and none overlapped edge. Fig. 3(b) shows another possible allocation of the spare capacity by way of EDPC with two overlapped nodes and none overlapped edges. Both Fig. 3 (a) and (b) have the same protected working capacity with the traditional P-Cycle in Fig. 2(b). But EDPC can increase the efficiency of failure recovery, i.e., achieving shorter restoration route and surviving two simultaneous link failures. Further, between the edge-disjoint P-cycles, the number of overlapped nodes is proportional to the number of the allocated spare capacity. Fig. 3(b) can support the same number of working capacity but with the fewer spare capacity than Fig. 3(a) because it has only two not three overlapped nodes. EDPC needs two additional spare units at least with one additional overlapped node. Thus, our method would construct the EDPC with the fewer number of overlapped nodes as possible.

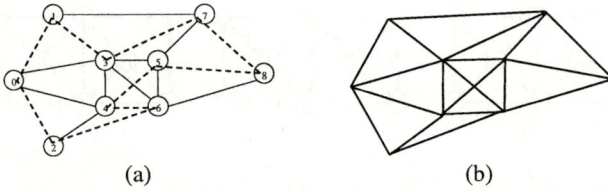

Fig. 2. (a) Spare capacity in the traditional P-cycle. (b) Working capacity in the traditional P-cycle

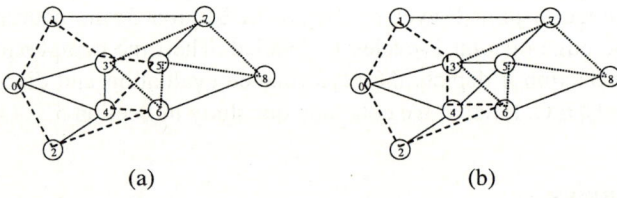

Fig. 3. (a) Spare capacity in the EDPC with 3 overlapped nodes. (b) Spare capacity in the EDPC with 2 overlapped nodes

Take an example that we have a request from node3 to node6 and the result of the shortest path routing is edge 3-6. If the edge 3-6 failures, the traditional P-Cycle in Fig. 2 must be routed four or five hop counts on average from this failure. But EDPC is routed three or four hop counts on average in Fig. 3(a), and two or four hop counts on average in Fig. 3(b). Our design can reduce two hop counts distance on average for recovering from failure.

EDPC provide guaranteed protection against single link failure that is on a P-cycle of EDPC, it can't protect against simultaneous failures to two links on that cycle (for example, (1,3) and (2,6) in Fig. 3). If failures of multiple links happen in such a way that there is at most one failure on each P-cycle of EDPC, then the EDPC can actually protect against such multiple failures (e.g. (1,3) and (7,8)). More P-cycles can protect against more simultaneous failures. The probability of double on-cycle failures should increase with the growth of network scale. Our method should take the proper measure to adjust the node number of each P-cycle in EDPC, and form a better protection partition to protect against more simultaneous link failures.

EDPC allows at least two overlapped nodes when it finds the secondary P-cycle. Because the property of P-cycle is two-connected, one link failure along the cycle does not affect the ability of spare capacity. This property is similar to the short leap share protection (SLSP) [4]. SLSP uses a path-switch LSR (PSL) and a path-merge LSR (PML), which switches over and merges back the affect traffic during a failure, respectively. The design of SLSP has four nodes overlapped with each adjacent p-domain, but EDPC has at least two nodes overlapped for each adjacent P-cycle.

3 Heuristic Method

According the explanation of the above section, we need to propose a method to decide the number of P-cycles in EDPC and the number of nodes in each P-cycle. For simplicity, we assumed that each partition (P-cycle) has the same number of nodes.

In accordance with above assumption, we generate a mathematical formulation to probe how many partitions we should make with different network sizes. With the number of partition increases, the number of overlapped nodes and the number of induced spare edges will grow up. In the same situation, we can endure more links failure simultaneously for EDPC when the partition regions grow in number. We can obtain a better performance tradeoff on spare usage and restoration efficiency. Let the increasing spare edges due to overlapped nodes be represented by formula K, the overhead of spare edges utilization for all partitions by formula X and the effect of hop counts for restoration paths on average by formula Y. Our goal is to minimize the value of objective function for the weighted sum of X and Y to decide the number of partition.

Define:
R: partition number.
K: the increasing spare edges due to overlapped nodes.
X: the overhead of spare resource utilization.
Y: the overhead of failure recovery.
V: the network area's total node number.

Formula K:

When one partition is added and is edge-disjoint with the adjacent partitions, there are at least two overlapped nodes and thus at least two spare edges are increased, e.g. (R,K) = (1,0), (2,2), (3,4), (4,6), (5,8) …etc.

$K = 2*(R-1)$.

Formula X:

We extend the probability of spare edges utilization to total edges for each partition in [6] to consider the effect of increased spare edges. The formula X represents the overhead of the spare edges. The increasing number of partitions can increase the number of spare edges but also increase the number of tolerated link failures. On the other hand, fewer partitions can reduce the number of spare edges but also decrease the number of tolerated link failures. In order to detail the formula, we further define the following variables.

M: the number of spare edges when only one P-cycle is allowed.
N_i: the number of spare edges in each P-cycle i of EDPC.
S: the number of fully connected edges in each P-cycle of EDPC, i.e., a 4-node P-cycle has the same number of edges in K4 for simplicity.
$M = N_i$, $i=1\ldots R$.

We can use $N_i/N_i+2(S-N_i)$ to represent the utilization of spare edges in one of R partitions. And $R*[N_i/N_i+2(S-N_i)]$ is used to represent the relationship between the number of partitions and the utilization of spare edges without considering the effect of overlapping nodes. K/M is to used to consider the effect of overlapped nodes on the usage of spare edges.

Thus $X = R*[N_i/N_i+2(S-N_i)] + K/M$ can simultaneously take the number of partitions and the number of overlapped nodes into account to represent the overhead of spare edges.

Formula Y:

One advantage of EDPC as well as P-cycle is that for chord link failures, the hop counts of protection paths are on average half that of the P-cycle circumference, whereas in rings protection paths are essentially the full circumference of the ring [7]. With the increasing number of partitions, the failure recovery time will also increase in our model.

Thus $Y = 0.5*(V/R)$ is used to represent the overhead of the restoration speed.

As for the objective function, we use the weighted sum of X and Y, i.e., min $[BX+(1-B)Y]$. B is a constant value. This value is set 0.5 because we don't favor X or Y.

We follow this formula to analyze how many partitions we should make with different network sizes (e.g, according to the number of nodes). We depict one of the results of the objective function with different network sizes. For example, when $V = 30$, the minimum value is $R=5$, $N_i=6$, $S=15$, $K=8$, $M=30$, $X=5[6/6+2(15-6)]+(8/30)=1.517$, $Y=0.5(30/5)=3$, and $BX+(1-B)Y=2.259$.

The complete result for the number of partitions is further shown in Fig. 4. It shows the number of partitions can be increased with the increasing network sizes.

After deciding the number of partition number, the heuristic method based on the link-state routing can construct the EDPC. Each link-state routing table of individual node has recorded the weighted value to other nodes, so we can choose a node as the starting node before partitioning with the whole network area. Then we sort these metric values of starting node's routing table and decide the range of each partition and the starting node of each partition according to the number of partitions decided by Fig. 4.

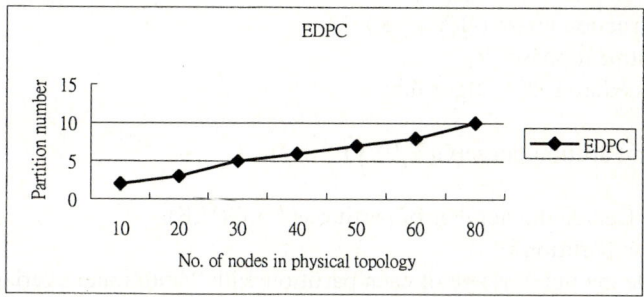

Fig. 4. The results of the proper partition number under various network sizes

For example, we observe the node0's routing table in Fig. 5. At first, we take the node 0 as the starting node of partition I. Second, we allow two nodes to overlap, so the number of partition I is 6 and the number of partition II is 5. This is because we take the ceiling of (9/2) + 1 to obtain the range of partition I. And the last two nodes in the range of partition I are also in the partition II. Third, we enlarge the range of each partition with overlapped nodes to avoid the condition of no disjoint edges between adjacent partitions. The enlargement of the range is an adjustable parameter. We take no more than 1/3 of the original number of nodes in one partition. It is to bound the number of overlapped nodes between each partition. Fourth, each starting node in each partition finds an edge-disjoint P-cycle according to the range and finally an adjustment stage is performed in our heuristic method to try to reduce the number of overlapped nodes to 2.

Node(0)'s routing table	
Next node number	cost
0	-
1	1
2	1
3	1
4	1
5	2
6	2
7	2
9	3

Fig. 5. Link-state routing table for node 0 in Fig. 2

The whole procedure of our EDPC is depicted in the following. The physical topology is represented by G(V, E), where V is the set of nodes and E is the set of edges. T is the link-state routing table of the first starting node. R is the set of partitions' information including each partition range and the starting node.

Procedure EDPC algorithm ()
Begin
   Partition phase (G(V,E),T)
   Construction phase (G(V,E),R)
   Adjustment phase (R)
End Procedure EDPC algorithm.

Function Partition phase(G(V,E),T)
Begin
   Set r = Decide the number of partitions for G(V,E).
   For each partition i
      Decide the initial range of each partition with 2 additional overlapped nodes
      Enlarge the partition range according to degree factor, i.e., adding the nodes with larger degree in partition i-1 and partition i+1 to partition i.
   End for
   Decide the starting node of all the partitions by the resulting partition range through the help of T
End

Define the operation of routing a P-cycle as using the depth-first search (DFS) [10] to find the path in the partition and verify that the ending node of the path has a back edge to the root node to form a cycle.

Function Construction phase (G(V,E),R)
Begin
  For each partition R[i]
     If R[i] is the first partition do
        Route a P-cycle from its range.
     Else
        Route a P-cycle from R[i] and make such P-cycle edge-disjoint with the P-cycle of R[i-1] as possible.
     End If
  End for
End

The final stage is to adjust the overlapped nodes between partition i and partition i-1 to be minimal. Its purpose is to decrease the utilization of spare edges.

Function Adjustment phase (R)
Begin
  For each partition R[i] and R[i+1]
     Fix one larger partition of the partition R[i] and partition R[i-1], and mark the other partition adjustable.

For any three consecutive overlapped nodes in the adjustable partition
   If there is an edge existed between two ending nodes
   Delete the middle node from the member of this adjustable partition.
   End For
   End For
End

In the adjustment phase, let the middle node for any three consecutive overlapped nodes in the adjustable partition be node M, and the other two nodes are nodes S1 and S2. Deleting the middle node M implies that the two edges between the middle node and the ending nodes (ie. (M, S1) and (M, S2)) are automatically released and the edge between two ending nodes (S1, S2) is formed.

The time complexity of our EDPC heuristic method can be decomposed into three components. The most time-consuming component is the time of the construction phase. Although using the depth-first search (DFS) only requires $O(n+m)$ for a physical topology with n nodes and m edges, the P-cycle is not guaranteed to be found in one run. Therefore, routing a P-cycle for a partition with k nodes takes $O(k!)$ time in the worst case. Thus $O(r \times k!)$ time is necessary for r partitions in the worst case. Due to the proper partition, the nodes in each partition k can be thought as a constant. In addition, the time complexity of the partition phase and the adjustment phase is also polynomial. Thus, the time complexity of our EDPC algorithm is polynomial in practice.

4 Evaluation and Discussion

In this section, we compare the ability of tolerating multiple link failures between EDPC and the traditional P-cycles. Three example networks: New Jersey LATA network, APPANET network, and National Network are protected by our EDPC and traditional P-cycles.

For New Jersey LATA network with 11 nodes, two partitions are constructed with no overlapped edges for EDPC and two overlapped edges for traditional P-cycles. For APPANET network with 20 nodes, three partitions are constructed with two overlapped edges for EDPC and three overlapped edges for traditional P-cycles. For National network with 24 nodes, four partitions are constructed with eight overlapped edges for EDPC and two overlapped edges for traditional P-cycles. When the multiple link faults is assumed to be occurred in the different P-cycles, we found that the number of link failures can be tolerated is the same as the number of P-cycles for EDPC with the probability at least 80% while for traditional P-cycles with the probability as low as 43%. The complete result is shown in Table 1.

As the above comparison, our EDPC heuristic method is not guaranteed to generate edge-disjoint P-cycles. The possible results are classified as three cases. For case 1, the EDPC is successfully constructed. For case 2, the EDPC can not be found. For case 3, the EDPC is not complete with some edge-joint P-cycles. Fig. 6(a) shows the possibility of case 2. There exists two edge-disjoint P-cycles but we can not obtain the result from EDPC because the partition phase can not adapt to such unbalanced partitions. Fig. 6(b) shows the edge-joint P-cycles can be constructed when there are shared edges between two partitions. Our algorithm uses the enlargement to reduce the possibility of case 2 and 3, i.e., the worst case of our EDPC is degenerated to traditional P-cycles.

Table 1. The probability of the successful multiple link failure protection

Probability	Tolerating 2 faults	
	EDPCs	P-cycles
New Jersey LATA network	100%	83%
APPANET network	87.31%	81.29%
National network	81.29%	66.59%

Probability	Tolerating 3 faults	
	EDPCs	P-cycles
New Jersey LATA network	0%	0%
APPANET network	81.29%	72.88%
National network	86.68%	53.91%

Probability	Tolerating 4 faults	
	EDPCs	P-cycles
New Jersey LATA network	0%	0%
APPANET network	0%	0%
National network	82.45%	43.39%

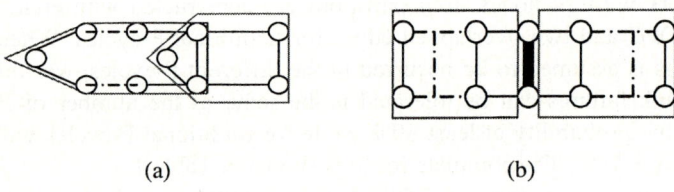

(a)   (b)

Fig. 6. (a) Case 2. (b) Case 3

As for the case 1, we can use table 2 to compare the four performance criteria with three different methods. The four performance criteria would include restoration speed, the number of risk sharing edges, the spare resource utilization, and the protection range.

Table 2. Comparison of four different methods

	Restoration speed	The number of risk sharing edge	Spare resource utilization	Protection range
One protection cycle	Slow	Low	Low	All (high)
Many protection cycle	Fast	High	Medium	Medium
Mesh protection	Medium	High	High	Low
EDPC	Fast	Low ~ medium	Low ~ medium	Medium ~ high

5 Conclusions

This paper proposed an improved protection mechanism for WDM networks without the help of edge-disjoint P-cycles (EDPC) to speed up the restoration time and tolerate more link failures. The proposed heuristic method is based on the local link-state routing table to decide the partition range, construct the EDPC, and adjust the EDPC for tradeoff between the overhead of spare resources and the protection ability. The time complexity of our EDPC is only polynomial instead of NP-hard. Compared with the traditional P-cycles, EDPC can not only tolerate multiple link failures but also shorten the construction time. In the future, we will propose the failure recovery method when the EDPC loses the protection ability due to double link failures in a P-cycle.

References

1. W.D. Grover and D. Stamatelakis: Cycle-oriented distributed preconfiguration: ring-like speed with mesh-like capacity for self-planning network restoration, Proc. IEEE International Conf. Commun., pp. 537-543, June 1998.
2. D. Stamatelakis and W.D. Grover: IP layer restoration and network planning based on virtual protection cycles, IEEE JSAC Special Issue on Protocols and Architectures for Next Generation Optical WDM Networks, vol.18, no.10, pp. 1938 – 1949, October 2000.
3. G. Ellinas, G. Halemariam, and T. Stern: Protection cycle in mesh WDM network, IEEE Journal on Selected Areas in Communications, vol. 18, no. 10, pp. 1924-1937, Oct. 2000.
4. P.-H Ho and H. T. Mouftah: A framework for service-guaranteed shared protection in WDM mesh network, IEEE Communications Magazine, Vol. 40, No. 2, pp. 97–103, 2002,.
5. Guoliang Xue and Ravi Gottapu: Efficient Construction of Virtual P-Cycles Protecting All Cycle-Protectable Working Links, Workshop on High Performance Switching and Routing, pp. 305 – 309, June 2003.
6. A. Sack and W. D. Grover: Hamiltonian p-Cycles for Fiber-Level Protection in Homogeneous and Semi-Homogeneous Optical Networks, IEEE Network, Special Issue on Protection, Restoration, and Disaster Recovery, vol.49-56, 2004

7. W. D. Grover, J. Doucette, M. Clouqueur, D. Leung, and D. Stamatelakis: New Options and Insights for Survivable Transport Networks, IEEE Communications Magazine, vol. 40, no. 1, pp. 34-41, January 2002
8. Muriel Médard, Richard A. Barry, Steven G. Finn, Wenbo He, and Steven S. Lumetta: Generalized Loop-Back Recovery in Optical Mesh Networks, IEEE Trans. Networking, vol. 10, pp.153-164, 2002.
9. Muriel Medard, Steven G. Finn, Richard A. Barry, and Robert G. Gallager: Redundant Trees for Prepalnned Recovery in Arbitrary Vertex-Redundant Graphs, IEEE/ACM Transactions on Networking, vol. 7, no. 5, pp. 641-652, Oct. 1999.
10. R.E. Tarjan: Depth First Search and linear graph algorithm, SIAM Journal on Computing, vol. 1, pp. 146-160, 1972.
11. C.Qiao and D.Xu: Distributed partial information management (DPIM) schemes for survivable network, IEEE INFOCOM, pp.301-311, 2002.
12. L. Sahasrabuddhe, S. Ramamurthy and B. Mukherjee: Fault management in IP-over-WDM network: WDM protection versus IP restoration, IEEE Journal on Selected Areas in Communications, vol. 20 , pp 21 – 33, Jan. 2002.

An Efficient Fault-Tolerant Approach for MPLS Network Systems

Jenn-Wei Lin and Hung-Yu Liu

Department of Computer Science & Information Engineering,
Fun Jen Catholic University, Taipei, Taiwan
`jwin@csie.fju.edu.tw`

Abstract. Multiprotocol label switching (MPLS) has become an attractive technology for the next generation backbone networks. To provide high quality services, fault tolerance should be taken into account in the design of a backbone network. In an MPLS based backbone network, the fault-tolerant issue concerns how to protect traffic in a carried path (label switched path (LSP)) against node and link failures. This paper presents a new efficient fault-tolerant approach for MPLS. When a node or link failure occurs in a working LSP, the traffic of the faulty LSP (the affected traffic) is distributed to be carried by other failure-free working LSPs. To minimize the affections on the failure-free LSPs, the affected traffic distribution is transferred to the minimum cost flow to be solved. Finally, extensive simulations are performed to quantify the effectiveness of the proposed approach over previous approaches.

1 Introduction

With rapid growth of Internet and increase in real-time and multimedia applications, hop-by-hop packet forwarding is insufficient to support current networking demands. The IETF has proposed multiprotocol label switching (MPLS) as a new forwarding technology for meeting the requirement of explosive Internet traffic. To provide high quality services, fault tolerance is also an important issue in addition to fast forwarding. If an Internet service provider (ISP) adopts the MPLS technology to design its backbone network, a fault-tolerant mechanism is necessitated to protect the carried traffic of a label switched path (LSP) against node and link failures. The LSP is a transmission path in the MPLS network. Currently, many papers have addressed the fault-tolerant issue of MPLS [1-7], which are based on the IETF two MPLS recovery models: protection switching and rerouting [8].

In this paper, we propose a new fault-tolerant approach. In the proposed approach, once detecting a failure in a working LSP, other failure-free working LSPs are organized as a recovery path set. The traffic of the faulty LSP (the affected traffic) is distributed to be carried by the failure-free working LSPs in the recovery path set. To minimize the affections on the failure-free LSPs, the minimum cost flow is applied to perform the affected traffic distribution. Unlike the protecting switching, the recovery path of an LSP is not pre-established. The resources in the MPLS network can be all used to created working LSPs to carry traffic. In addition, the proposed approach utilizes failure-free working LSPs to constitute a recovery path set. The routes of failure-free working LSPs and their resources have been set up at their creations. Compared to rerouting, the proposed approach can reduce much recovery time.

The rest of this paper is organized as follows. Section 2 gives background knowledge. Section 3 proposes our approach. Section 4 compares the proposed approach with previous approaches. Concluding remarks are made in Section 5.

2 Background

This section describes the background knowledge of this paper. First, the MPLS network model is given. Then, related work is reviewed.

2.1 Network Model

The network model referred to this paper is shown in Fig. 1, which consists of an MPLS backbone network and two IP based access networks, and an OAM center. In the MPLS backbone network, a number of label switched paths (LSP) are established. Each LSP consists of an ingress label switching router (ingress LSR), one or more intermediate label switched routers (intermediate LSRs), and an egress label switching router (egress LSR). The creation of an LSP is accomplished by the label distribution protocol (LDP) [9], which assigns appropriate labels to the LSRs on the created LSP. By the label assignment, the created LSP is responsible for carrying the packets with a particular forwarding equivalence class (FEC) type. The FEC represents a set of packets with the same traffic requirements.

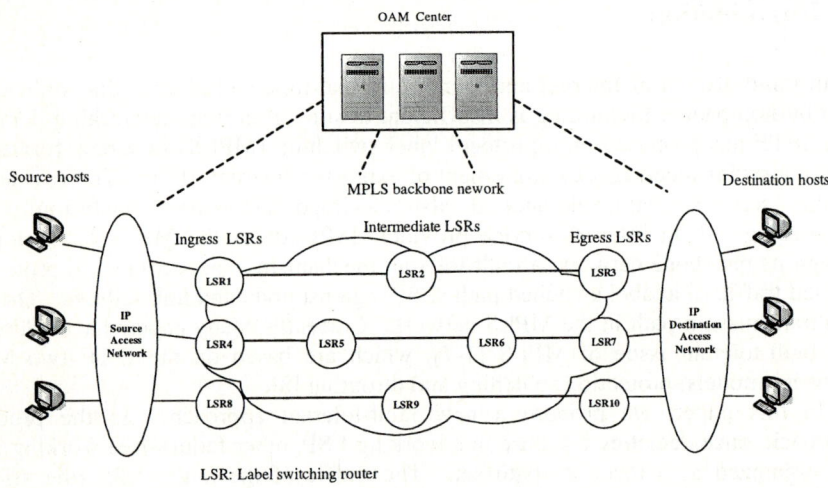

Fig. 1. The MPLS network model

When a source host sends an IP packet to a destination host, the packet is sent through the IP source access network to an entry point (an ingress LSR) of the MPLS backbone network. The ingress LSR determines the FEC of the packet based on the some header fields of the packet. According the FEC, the packet is carried by one corresponding LSP. Then, the ingress LSR inserts a label into the packet and forwards the packet to the next LSR on the carried LSP. Next, each intermediate LSR on the carried LSP use label switching to forward the packet within the MPLS backbone network. Finally, the packet will be forwarded at the exit point (the egress LSR) of the carried LSP. The egress LSR removes the label of the packet and uses the conventional IP forwarding to forward the packet to the destination host via the IP destination access network.

To be an ISP backbone network, an OAM center is also necessitated for managing network components, verifying network performance, and monitoring network status. The IETF has defined the OAM functions for an MPLS network system in [10].

2.2 Related Work

All existing MPLS recovery approaches are based on the IETF two MPLS recovery models (protection switching and rerouting) [8] to pre-establish or dynamically establish a recovery path for a faulty LSP.

The previous approaches of [1-4] belong to the protection switching model. In the approach of [1], each working LSP has a disjoint backup path between its ingress LSR and egress LSR. The backup path is pre-established, and it does not share any intermediate LSRs with the working LSP. When a failure is detected in a working LSP, a failure indication signal (FIS) is sent to the ingress LSR of the faulty LSP. Upon receiving the FIS, the ingress LSR reroutes the incoming packets through the disjoint backup path. However, the approach of [1] has the packet loss problem.

To solve the packet loss problem, the approach of [2] additionally pre-establishes a backward path. The route of the backward backup path is reverse with the working LSP. Although the approach of [2] can solve the packet loss problem, it additionally introduces the packet disorder problem. To overcome the packet disorder problem, the approach of [3] uses tagging and buffering techniques to improve the approach of [2].

Unlike the above protection switching approaches, the approach of [4] pre-establishes several backup paths to protect the traffic of a faulty LSP (the *affected traffic*). In the approach of [4], its fault-tolerant idea is not novel, which is fully based on the local recovery to protect the affected traffic. The main contribution in the approach is to quickly compute the minimum capacity requirement for each backup path.

For the approaches of [5-7], they belong to the rerouting model. In the approach of [5], a pre-qualified recovery mechanism was proposed, as follows. Whenever a working LSP is created, each LSR on the LSP also determines the route of its corresponding recovery path. Unlike traditional rerouting approaches, the approach of [5] does not take time to find the route of the recovery path after a failure since the route has been determined before a failure. However, the approach of [5] incurs the route calculation overhead during the normal time.

In the approach of [6], a technical term: candidate protection merging LSR (candidate PML) is first defined. Then, the concept of the candidate PML is utilized to reroute affected traffic along the least-cost recovery path. The approach of [6] can easily handle multiple simultaneous failures since the route of the recovery path is dynamically found and established. But, the approach incurs a non-trivial recovery time due to dynamically finding the least-cost recovery path.

So far, the above described approaches do not allow failures to occur at the edge LSRs (the ingress or egress LSRs). Now, only the approach of [7] proposed a rerouting solution to tolerate the ingress LSR failure, but the approach cannot be applicable to the intermediate or egress LSR failure.

3 Proposed Approach

This section presents a new fault-tolerant approach for MPLS. Compared to the rerouting based approaches, the proposed approach can reduce much recovery time. In addition, the resource utilization in the proposed approach is more efficient in comparison to the protection switching based approaches.

3.1 Basic Idea

In an MPLS network, the bandwidth capacity of an LSP is usually larger than the bandwidth requirement of its carried traffic. Therefore, when a failure occurs in a working LSP, other failure-free working LSPs may have residual bandwidth capacities. This observation triggers an idea that one or more failure-free working LSPs can substitute the faulty LSP to carry the affected traffic. To achieve the idea, the following problems need to be solved first.

- How to redirect the affected traffic to failure-free LSPs
- How to distribute the affected traffic to failure-free LSPs without degrading the bandwidth requirement
- How to forward the affected traffic along the route of a failure-free LSP

3.2 Problems to be Solved

For the first problem, a switchover mechanism is used to redirect the affected traffic. As shown in Fig. 1, the packets to the MPLS backbone network are via the IP source access network. For a source host, it has a corresponding access router in the IP source access network. In theory, the access router can forward the packets of the source host to any ingress LSR in the MPLS backbone network. After one ingress LSR receives the packets, the packets are then carried by the corresponding LSP. With the assistance of the access router, the redirection of the affected traffic can be done as follows.

Upon detecting a failure in the desired LSP of the source host, the corresponding access router of the source host initiates a switchover mechanism to forward incoming packets to the ingress LSRs of other failure-free LSPs, as shown in Fig. 2. Then, the failure-free LSPs can substitute the faulty LSP to carry the packets of the source host. Here, the access router can be regarded as a load redirector to redirect the affected traffic to other failure-free LSPs. The switchover mechanism in the access router can be implemented by a software module which deletes the routing entry corresponding to the faulty LSP and adds the routing entries corresponding to failure-free LSPs. Based on the added routing entries, the access router can split the affected traffic to multiple failure-free LSPs.

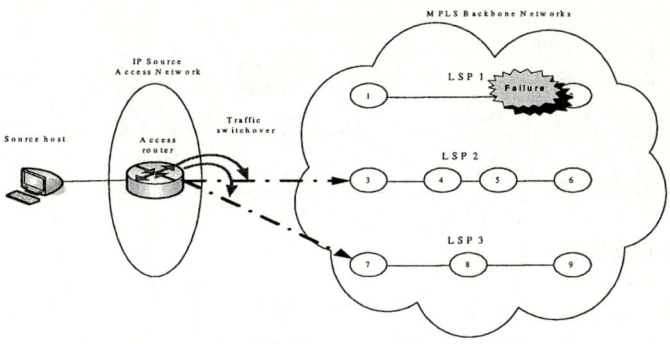

Fig. 2. The redirection of the affected traffic

Due to using multiple failure-free LSPs to carry the affected traffic, the proposed approach has the problem of the affected traffic distribution (the second problem). However, each failure-free LSP has different the residual bandwidth capacity and delivery delay. The two factors will affect the cost of the affected traffic distribution. To obtain the optimal distribution, the affected traffic distribution is transferred to the minimum cost flow [11] to be solved, described as follows.

- Extract all failure-free LSPs to form a simple graph, as shown in Fig. 3. First, the two edge LSR of each failure-free LSP are put in the simple graph as the ingress and egress nodes. Between each ingress-egress node pair, an edge is set to connect them. The cost and capacity in the edge are set to be the delivery delay and residual bandwidth of its corresponding failure-free LSP. Based on the above graph modeling, the available bandwidths in failure-free LSPs and their corresponding delivery delay are mapped in the simple graph. However, from Fig. 1, we can see that the packets from the source host to the destination host are through two IP access networks in addition to the MPLS backbone network. Therefore, the affected traffic distribution should also consider the redirection costs in the two IP access networks. To take this consideration, a source node and a destination node are additionally put in the most left and right positions of the simple graph, respectively. For the source node of the simple graph, it corresponds to the above mentioned load redirector. A number of edges are set from the source node to all the existing ingress nodes in the simple graph. Each of such edges corresponds to one transmission path in the IP source access network. The cost of such each edge is set to the number of transit hops in its corresponding transmission path. The capacity of each such edge is set to infinity. The reason is that a transmission path in an IP access network has no QoS guarantee and it is based on the best effort to forward packets. For the destination node of the simple graph, it corresponds to the access router of the destination host in the IP destination access network. Between the destination node and each egress node, there is also an edge to connect them, which corresponds to a transmission path of the IP destination access network. The transmission path is used to forward the affected traffic from the egress LSR of one failure-free LSP to the destination host. As for each edge of the destination node, its cost and capacity are set based on the same way as the edge of the source node.

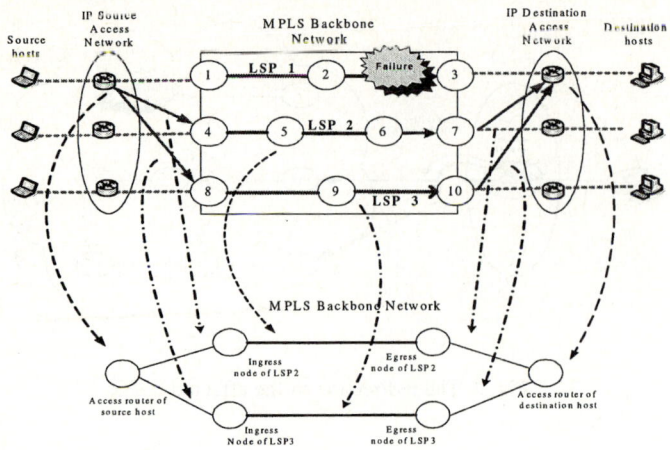

Fig. 3. The modeling of a simple graph

- Transfer the affected traffic distribution to the minimum cost flow. After forming the simple graph, a traffic flow with x units of data rate is supplied to the source node, where x is set to be equal to the bandwidth requirement of the affected traffic. Given the simple graph with an input traffic flow, the affected traffic distribution is transferred to the problem how to send the traffic flow from the source node to the destination node with a minimum cost (the minimum cost flow problem). Based on [11], the minimum cost flow problem can use the following linear equation to represent it:

$$\text{Minimize} \sum_{k=1}^{n} c_k x_k \quad (1)$$

Subject to

$$\sum_{k=1}^{n} x_k = b_{LSP_f} \quad (2)$$

$$0 \leq x_k \leq r_k \text{ for all } k = 1 \cdots n$$

where x_k represents what amount of the affected traffic is redirected to failure-free LSP_k. c_k is the unit cost of a packet carried by failure-free LSP_k, r_k is the residual bandwidth of failure-free LSP_k, b_{LSP_f} is the bandwidth requirement of the affected traffic, n is the number of failure-free LSPs.

Next, the third problem (how to forward the affected traffic along the route of a failure-free LSP) is solved by using the header encapsulation and decapsulation techniques. As mentioned in section 2.1, the packets carried by an LSP have the same FEC type. The FEC type decision of a packet is based on some fields in its header (e.g. source address and/or destination address, or port number). In another words, if two packets are carried by the same LSP, some header fields in them must be same since an LSP associates one type of packet header. This triggers an idea that if a packet is encapsulated with a header type, it can be carried by the LSP associated with the header type. The idea can be further utilized to make the affected traffic be carried by a failure-free LSPs. An example is given in Fig. 4.

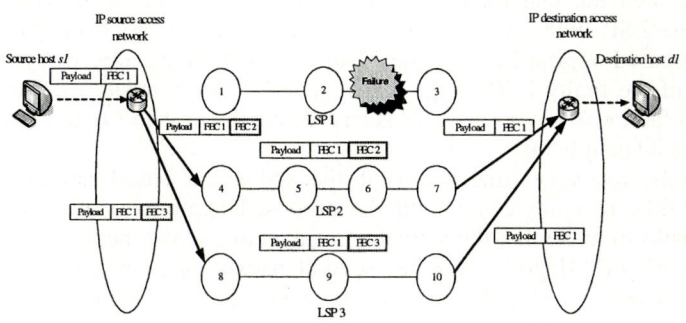

Fig. 4. Header encapsulation and decapsulation

4 Comparison

For making the quantitative comparisons between the previous approaches and the proposed approach, simulations are performed by extending the given MPLS simulation module in the Network Simulator version 2 (NS-2) [12]. In the simulations, there are 18 LSR nodes and 30 links in the MPLS network. The capacity and delay for each link are fixedly set to 20 Mbps and 1 ms, respectively. Three working LSPs are set in the MPLS network. The traffic flows carried by the three working LSP are CBR traffic. The bandwidth requirements for the three traffic flows are 10Mbps, 8Mbps, and 9Mpbs, respectively. Based on the above network model and traffic parameters, each comparison approach perform 4 different simulation runs to make the number of failures in each failure occurrence be 1, 2, 4, and 8, respectively. Note that the failures randomly occur in 18 nodes and 30 links, and the time of a simulation run is set to 200 seconds. For each failure occurrence, if one approach can tolerate the failures successfully, 256 packets are next fed into the recovery path to observe the transmission delay of the 256 packets. Here, the transmission delay is used to represent the recovery quality. If the transmission delay in one approach is smaller than one another, it represents that the recovery path of the former approach is better than that of the later approach.

The simulation results for the fault-tolerant capability, recovery time and recovery quality are illustrated in Fig.5-7, respectively. From Fig. 5-7, we can obviously see that the proposed approach has the best performance in the fault-tolerant capability, packet loss, packet disorder, and recovery qualify. With the fault-tolerant capability, it is quantified as the restoration ratio: $\dfrac{\text{the number of failures tolerated}}{\text{the number of failures occurring}}$. From Fig. 5, we see that the protection switching based approaches have a larger drop in the restoration ratio if the number of simultaneous failures is more than 2. The reason is explained as follows. The protection switching based approaches worry about failure occurrence at the pre-established paths. If the number of simultaneous failures is more than 2, there is a high probability that there are also failures in the pre-established backup paths. For the proposed approach, its fault-tolerant capability is superior to the rerouting based approaches. Unlike the rerouting based approach, the proposed approach uses failure free working LSPs as the recovery path candidates. The proposed approach can tolerate many simultaneous failures as long as there is a failure-free LSP with sufficient residual bandwidth in the MPLS network. The rerouting based approaches use the redundant path being able to bypass the failure segment of the faulty LSP as the recovery path. By simulations, the probability of existing a failure-free working LSP is larger than the probability of existing repairable paths in redundant paths.

With the recovery time, most of the protection based approaches have less recovery time in comparison with the proposed approach since they pre-establish backup paths to make that the activation of a recovery path become easy. Especially, the approach of [4] pre-establishes several backup segments to perform the local recovery; therefore, it has the least recovery time. However, these protection switching based approaches do not consider the packet loss and packet disorder problems. Only the approach of [3] considers the two problems. If the time for handling the two problems is taken into the recovery time, it is not true that the protection switching based approaches are always better than the proposed approach in the recovery time. For example, comparing between the proposed approach and the approach of [3], the proposed approach has a better recovery time.

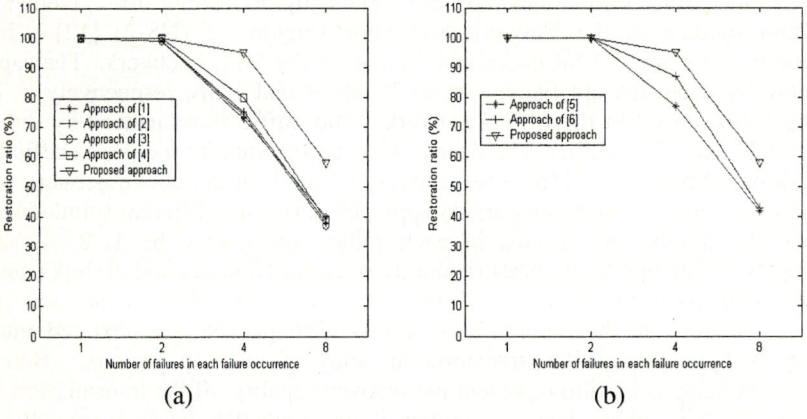

Fig. 5. Fault-tolerant capability comparison (a) protection switching (b) rerouting

As for the recovery quality, the proposed approach has the best performance. In the protection switching based approaches, the pre-establishment of a backup path only considers the bandwidth requirement for the traffic carried on a working LSP. The delivery delay of the backup path is not concerned. For the rerouting based approaches, they dynamically establish a shortest recovery path without considering the bandwidth requirement for the affected traffic. In contrast to the proposed approach, it considers the bandwidth requirement and delivery delay to distribute the affected traffic to multiple failure-free working LSPs. Due to using multiple LSPs to carry the affected traffic, the proposed approach has a better recovery quality.

However, when multiple failures occur simultaneously, few failure-free working LSPs can be used to carry the affected traffic. In such case, the recovery quality of the proposed approach becomes same as other approaches.

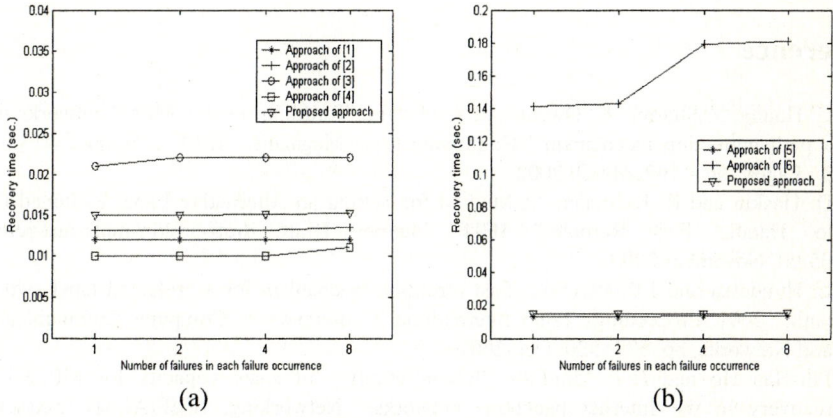

Fig. 6. Recovery time comparison (a) protection switching (b) rerouting

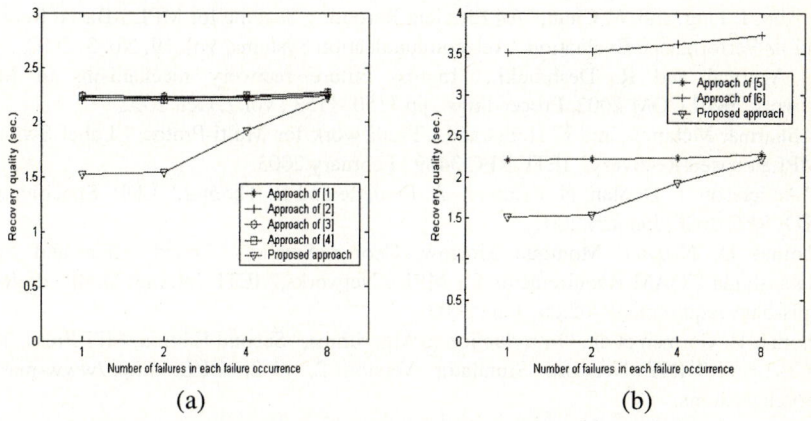

Fig. 7. Recovery quality comparison (a) protection switching (b) rerouting

5 Conclusions

This paper has presented an efficient approach to protecting traffic in an MPLS network. The proposed approach utilizes the available resources in other failure-free LSPs to constitute a virtual LSP to be the backup of the faulty LSP. Based on the virtual LSP, the traffic of the faulty LSP can be distributed to be carried by multiple failure-free LSPs. Extensive simulations were also performed to make the detailed comparisons between the proposed approach and previous approaches.

Acknowledgement

This research was partially supported by the Nation Science Council, Taiwan, under Grant NSC 93-2745-E-030-005-URD.

Reference

1. C. Huang, V. Sharma, K. Owens and S. Makam, "Building reliable MPLS networks using a path protection mechanism," Communications Magazine, IEEE , Volume: 40 , Issue: 3 , Pages:156 – 162, March 2002
2. D. Haskin and R .Krishnan, "A Method for Setting an Alternative Label Switched Paths to Handle Fast Reroute," IETF Internet Draft, draft-haskin-mpls-fast-reroute-05.txt, November 2000
3. L. Hundessa and J.D. Pascual, "Fast rerouting mechanism for a protected label switched path," 2001. Proceedings Tenth International Conference on Computer Communications and Networks, pp. 527 -530.,Oct. 2001
4. Pin-Han Ho and H.T. Mouftah, "Reconfiguration of spare capacity for MPLS-based recovery in the Internet backbone networks," Networking, IEEE/ACM Transactions on, Volume 12, Issue 1,Pages73 - 84, Feb. 2004
5. S. Yoon, H. Lee, D. Choi, Y. Kim, G. Lee and M. Lee, " An efficient recovery mechanism for MPLS-based protection LSP," 2001. Joint 4th IEEE International Conference on ATM (ICATM 2001) and High Speed Intelligent Internet Symposium, pp. 75 -79, April 2001
6. G. Ahn, J. Jang, and W. Chun, "An Efficient Rerouting Scheme for MPLS-Based Recovery and Its Performance Evaluation," Telecommunication Systems, Vol. 19, No. 3, 2002
7. A. Agarwal and R. Deshmukh, "Ingress failure recovery mechanisms in MPLS network," MILCOM 2002. Proceedings , pp.1150 -1153 , vol.2, Oct. 2002
8. V. Sharma, Metanoia and F. Hellstrand, "Framework for Multi-Protocol Label Switching (MPLS)-based Recovery," IETF RFC 3469 , February 2003
9. L. Andersson, P. Doolan, N. Feldman, A. Fredette and B. Thomas," LDP Specification," IETF RFC 3036, January 2001
10. Thomas D. Nadeau, Monique Morrow, George Swallow, David Allan and Satoru Matsushima, "OAM Requirements for MPLS Networks," IETF Internet Draft, draft-ietf-mpls-oam-requirements-02.txt, June 2003
11. Thomas H. Cormen et al., "Introduction to Algorithms," Second Edition, MIT Press, 2001
12. UCB/LBNL/VINT Network Simulator Version 2, ns-2, URL: http://www-mash.cs.berkeley.edu/ns.

A Novel Technique for Detecting DDoS Attacks at Its Early Stage

Bin Xiao[1], Wei Chen[1,2], and Yanxiang He[2]

[1] Department of Computing,
The Hong Kong Polytechnic University,
Hung Hom, Kowloon, Hong Kong
{csbxiao, cswchen}@comp.polyu.edu.hk
[2] Computer School, The State Key Lab of Software Engineering,
Wuhan University, Wuhan 430072, Hubei, China
yxhe@whu.edu.cn

Abstract. Spoofing source IP addresses is always utilized to perform Distributed Denial-of-Service (DDoS) attacks. Most of current detection and prevention methods against DDoS ignore the innocent side, whose IP is utilized as the spoofed IP by the attacker. In this paper, a novel method has been proposed to against the direct DDoS attacks, which consists of two components: the client detector and the server detector. The cooperation of those two components and their interactive behavior lead to an early stage detection of a DDoS attack. From the result of experiments, the approach presented in this paper yields accurate DDoS alarms at early stage. Furthermore, such approach is insensitive to the false suspect alarms with adopted evaluation functions.

1 Introduction

Distributed Denial-of-Service (DDoS) attacks are one of the most serious threats to the internet. As more business and commerce services depend on the internet, DDoS attacks can bring numerous financial loss to these e-business companies. As Moore [1] reported, the majority of attack packets is the TCP type. In the TCP case, SYN flooding is the most common attack behavior [2,3].

Current TCP based DDoS attacks are usually performed by exploiting TCP three-way handshake [4]. The SYN flooding attack is launched by sending numerous SYN request packets towards a victim server. The server reserves lots of half-open connections which will quickly deplete system resource, thus preventing the victim server from accepting legitimate user requests.

Lots of work has been done to detect and prevent the TCP based DDoS attack. Current counteracting methods are mostly deployed at the victim server side, or the attacker side or between them. The information at the innocent host, whose IP is utilized as the spoofed IP, is totally ignored. Detection at the side of a victim server can is more practical but can hardly produce an alarm at the early stage of a DDoS attack because abnormal deviation can only be easily found until the DDoS attack turns to the final stage. Counteracting methods near the side of attack source mainly belong to the prevention mechanism, such as filtering suspicious packets before forwarding them to the internet. Providing an early DDoS alarm near the side of an attacker is a difficult task because the attack signature is not easy to capture at this side. The information at the innocent host side will be used in our approach because the innocent side will receive abnormal TCP control packets for the three-way handshake. These abnormal packets for three-way handshake provide valuable information to give alarms at the early stage.

In order to detect a TCP based DDoS attack at its early stage, in this paper we have made the contributions as follows:

- The client detector performs detection at the side of innocent hosts because this side can provide valuable information. Another benefit of such deployment is to make the detection system itself invulnerable to direct DDoS attacks. To our best knowledge, there is no literal report about detection at the innocent host side.
- A novel cooperative detection system composed of the client detector and the server detector is presented. The cooperation improves the alarm accuracy and shortens the detection time.
- The server detector can actively send queries to the client detector to confirm the existence of a DDoS attack. This active scheme enhances the earlier DDoS detection.

The remainder of the paper is organized as follows: Section 2 will introduce some related works in spoofed DDoS detection. In Section 3 the TCP-based DDoS attack will be discussed. The techniques for cooperative DDoS detection, including the client detector and the server detector, will be presented in section 4. Some experiment results will be given in Section 5 to evaluate the performance of the proposed method. In Section 6 we will conclude our work and discuss future work.

2 The Related Work

Most of current DDoS attack detection and prevention schemes can be classified into three categories according to the location of the detector: at the victim server side, at the attack source side and between them.

Detecting DDoS at the victim server side is more concerned by researchers. In [3] Wang detected the SYN flooding attacks at leaf routers that connect end hosts to the Internet. The key idea of their method is the SYN-FIN pair's behavior should show invariant in normal network traffic and a non-parameter CUSUM method is utilized to accumulate these pairs. In Cheng's work [5], their approach utilizes the TTL(Time-To-Live) value in the IP header to estimate the Hop-Count of each packet. The spoofed packets can be distinguished by Hop-Count deviation from normal ones. A method incorporating SYN cache and cookies method is evaluated in Lemon [6] work. The basic idea is to use cache or cookies to evaluate the security status of a connection before establishing the real connection with a protected server. For those methods that provide detection at the victim server side, the main challenge is to consume the least recourse to record the state of numerous connections and evaluate the safety of each connection.

The detection at the attack source side seems more difficult than at the server side because the signatures at the attack source side are not easy to detect however prevention at the attack source side is more effective. For example the RFC2827 [7] is to filter spoofed packets at each ingress router. Before the router forwards one packet to the destination, it will check whether the packet belongs to its routing domain. If not, this packet is probably a spoofed one and the router will drop it.

Detection and prevention between these two sides mainly include traceback [8–11] and pushback [12]. Tracing back attempts to identify the real location of the attacker. During a DDoS attack, the source IPs are often forged and can not be used to identify the real location of the attack source. Most of the traceback schemes are to mark some packets along their routing paths or send some special packets [11], together with monitored traffic flow. With these special marks, the real routing path can be reconstructed and the true source IP can be located. With the identification of real path of the spoofed packets, pushback techniques can be applied to perform advanced filtering. The pushback is always performed at the last few routers before traffic reaches target.

3 The TCP-Based DDoS Attack

Detecting DDoS attack at its early stage is a challenging problem. For a DDoS attacking packet, its source IP address is usually forged. The normal three-way handshake to build a TCP connection would be changed consequently.

The normal three-way handshake is shown in Figure 1(a). First the client C sends a $Syn(k)$ request to the server S_1. After receiving such request, server S_1 replies with a packet, which contains both the acknowledgement $Ack(k+1)$ and the synchronization request $Syn(j)$(denoted as $Ack(k+1) + Syn(j)$ in the following paper). Then client C sends $Ack(j+1)$ back to finish the building up of the connection. k and j are sequence numbers produced randomly by the server and client during the three-way handshake. During the normal three-way handshake procedure, $Syn(k)$, $Ack(k+1) + Syn(j)$ and $Ack(j+1)$ can be observed at the edge router(R_c in Figure 1(a)) near the client.

If the IP-Spoofed attack happens, the normal authentication process is modified. As Figure 1(b) shown, the innocent host I, whose IP is used as spoofed source IP, is usually not in the same domain with the attacker host A. In fact, to avoid being traced back, the attacker usually uses the IP address belonging to other domains to make a spoofed packet. In Figure 1(b) the edge router R_a in the attacker domain forwards the $Syn(k)$ packet with the spoofed address P_I, the IP address of the innocent host I, to the server S_2. The sever S_2 replies with an $Ack(k+1) + Syn(j)$ packet. This $Ack(k+1) + Syn(j)$ will be sent to the innocent host I because the server S thinks the $Syn(k)$ packet is from I according to the spoofed source IP P_I in it. So the edge router R_I at the innocent host side will receive the $Ack(k+1) + Syn(j)$ packet from victim server S_2. But there is no previous $Syn(k)$ request forwarded by the client detector at R_I. This scenario is different from the normal one. Our approach is proposed on the base of this difference.

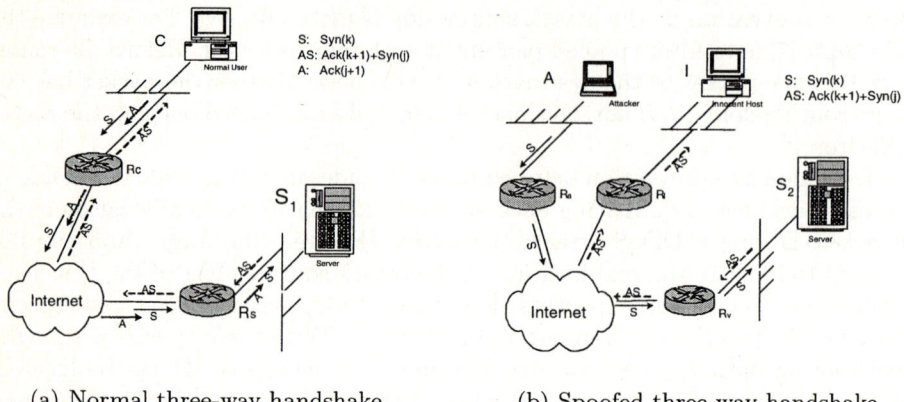

(a) Normal three-way handshake (b) Spoofed three-way handshake

Fig. 1. The process of the TCP three-way handshake

4 Techniques for DDoS Cooperative Detection

In order to detect DDoS at its early stage, the client detector and the server detector are introduced in the presented techniques. The client detector is deployed at the edge router of innocent hosts. For example, the client detector will be installed on the router R_I in Figure 1(b). It checks the TCP control packets flowing through the edge router. When it captures suspicious events, the alarm about the potential DDoS attack will be sent to the side of protected server. The server detector, employed by the protected server, can perform detection not only by passively listening the warning from the client detector, but also by actively sending queries to the client detector to confirm alarms.

4.1 The Client Detector

One of the main tasks for the client detector is to monitor the TCP control packets that comes in and goes out of the domain. Although a TCP connection may hold for several seconds or even for several minutes, most three-way handshake can be finished in a very short period(e.g., less than 1 seconds) at the beginning phrase of the connection. Compared with other stateful defense mechanisms which maintain states for the whole TCP connection, keeping the states only for the three-way handshake will reduce the computing and storage overhead.

Monitoring States for the Three-Way Handshake. To keep the states for a three-way handshake, a record is created in the hash table. There exist three states for each record: 'syn', 'ack' and 'suspicious'. To illustrate the state transition well, R is denoted for the record mapped by the hash function using source and destination IP address values as input. The $Syn(k)$, $Ack(k+1) + Syn(j)$ and $Ack(j+1)$ are denoted for TCP control packets used by the three-way handshake. These packets will trigger the creation of a new record R or the state transition of a existing record R. Two sub-detectors, egress detector(ED) and ingress detector(ID), are designed to process the outgoing and incoming packets individually. When abnormal traffic flowing through the edge router is observed, Suspect Alarm(SA) will be generated and will be sent to the client detector. The final alarm will be decided by evaluating these SAs. The total state transition for R is shown in Figure 2.

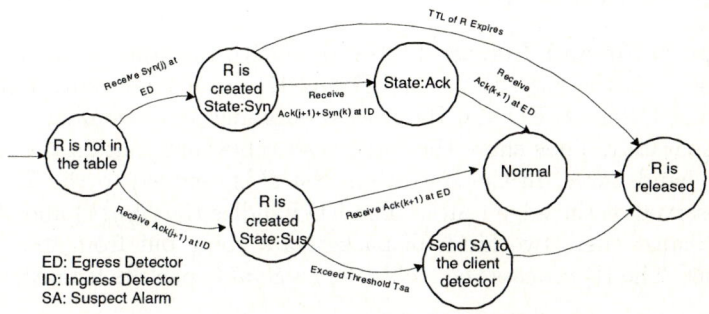

Fig. 2. States transition for R in the hash table

At the first phase of three-way handshake, A $Syn(k)$ packet will be sent from the client C to the server S to initialize a new TCP connection. When this packet is observed in outgoing traffic by ED, a new record R is created and the state of the record is set to 'syn'. This 'syn'-state record will be stored in the table with a Time To Live(TTL). If the TTL expires, this record will be deleted from the table.

When the server S receives the $Syn(k)$ packet from the client C and is ready to start the TCP session, it sends a $Ack(k+1) + Syn(j)$ packet to C. When the $Ack(k+1) + Syn(j)$ packet arrives, ID will check if there exists a corresponding 'syn'-state record previously created. If the matched one is found, the state 'syn' will be modified to 'ack'. Otherwise, a 'suspicious'-state record R will be created. It means there may exist IP spoofing behaviors, or the previous R has been deleted because the TTL of R has expired. It is 'suspicious' because it need to be verified in the third step of three-way handshake. Though the packet is suspicious, the edge router will continue to forward the packet to the destination C. If the $Ack(k+1) + Syn(j)$ packet is a legitimate one, the client C will reply with a $Ack(j+1)$ packet to finish the three-way handshake. If the $Ack(k+1) + Syn(j)$ packet is caused by a spoofed packet, there will be no any response for the handshake.

During the final stage of handshake, when the $Ack(j+1)$ packet arrives the edge router, the ED will search for the corresponding record of this three-way handshake. If it is in the table and the current state is 'ack', it means a normal three-way handshake has been finished. The record will be released from the table. If the current state is 'suspicious', the record will also be deleted because this 'suspicious' state is caused by expiration instead of spoofed packets. As we have mentioned above, if the TTL of a 'syn'-state record expires, this record will be deleted from database and a 'suspicious'-state record will be generated. So this 'suspicious' record can be safely released from the hash table.

On the contrary, in the scenario of the spoofed handshake, no response will be generated because the spoofed IP is unreachable or the client C will send back a RST instead of a $Ack(j+1)$ packet. The 'suspicious'-state record will not be released from the table. If a 'suspicious'-state record stays for a certain time longer than the threshold, T_{time}, a Suspect Alarm(SA) will be generated and sent to the client detector.

Egress Detector and Ingress Detector. With the state transition description in Figure 2, the algorithms for ED and ID are defined individually in the Figure 3 and Figure 4. ED and ID process outgoing and incoming TCP control packets separately. They share the same hash table that contain the records for three-way handshakes. In the algorithm, the P is denoted to the TCP control packet observed at the edge router. The ED handles the $Syn(k)$ and $Ack(j+1)$ packets because these two kinds of packets are going out from the domain to the internet. The ID process the $Ack(k+1) + Syn(j)$ packets from the incoming traffic.

```
VOID Outgoing_Packets_Process (INPUT: P) {
  if P is a Syn(k) packet then
    R=hash(source IP, destination IP)
    Create R
    Set R state = syn
  else if  P is a Ack(j+1) packet then
    R=hash(source IP, destination IP)
    if R is found in the Hash Table AND R state == ack then
      Release R
    else if R is found in the Hash Table AND R state == suspicious then
      Release R
    end if
  end if
}
```

Fig. 3. The algorithm for the Egress Detector

```
VOID Incoming_Packets_Process (INPUT: P) {
if P is Ack(k+1)+Syn(j) packet then
    R=hash(destination IP, source IP)
    if R is found in Hash Table AND R state == syn then
        Set R state=ack
    else if R is not found then
        Create R
        Set R state= suspicious
    end if
end if
}
```

Fig. 4. The Algorithm for Ingress Detector

Detection Scheme for Client Detector. If a 'suspicious' record in the hash table stays for a period of time longer than the threshold T_{sa}, a Suspected Alarm(SA) will be generated and will be stored in the local database. The client detector will analyze the source IP distribution of SAs in database. When SAs with the same source IP P_{victim} are reported in a short period, there probably exists a DDoS attack targeting the host P_{victim}. But if each SA has a different source IP, it is most likely caused by some other reasons. To evaluate the distribution of the source IP of alarms, an expression is presented below:

$$score = \sum_{s \in IPList} (|X_s| - 1)^2$$

Where X_s stands for a subset of the total SA set. All the elements in X_s are SAs that have the same IP value s in a certain period. The score will increase dynamically when the number of SAs with the same source IP increases. On the other hand if each of the SAs has a different source IP, the score will reach minimum.

4.2 The Server Detector

The server detector is deployed at the protect server, such as S_2 in Figure 1(b). On the one hand, the server detector may passively wait for the potential direct DDoS alarm from the client detector. When enough potential DDoS attack alarms come, the server detector will give the confirmed direct DDoS attack alarm to server.

On the other hand it also can perform more active detection by sending queries to client detectors as soon as any suspicious event is found at the protected server. Sometimes the source IPs of spoofed packets is widely distributed. The number of SAs at one client detector is not enough for it to send a potential DDoS alarm to the server detector. In this scenario, the server detector will select several cooperative client detectors to query the number SAs. The client detectors will report the number of SAs with the special source IP, then the server detector will analyze these SAs replied from different client detectors and confirm whether the half-connection is caused by spoofed DDoS packets or by some other reason.

Many other DDoS detection methods have to wait for capturing sufficient DDoS attack evidences before taking further action. This waiting delays detection and prevention against the forthcoming DDoS. In our approach, both the client detector and the server detector do not need to wait passively for special evidences, so this cooperative detection system can give DDoS alarm at the earlier stage.

5 Experiment

To evaluate the cooperative detection system, experiments are designed to test whether the cooperative detection approach can detect DDoS at the early stage. In the experiment, 10 zombies are simulated to perform SYN flooding attacks toward the server. The cooperative detection system include five client detectors and one server detector. The rate of the attack packets rises from 10 packets/sec to 1000 packets/sec in 10 seconds and 100 seconds respectively. The maximum rate is set to 1000 packets/sec because it is enough to shut down some services as Chang reported [13]. In simulation only 1% of $Ack(k+1) + Syn(j)$ packets replied by the victim server are designed to arrive client detectors. These packets will trigger client detectors to generate SAs. The number of SAs generated by the client detectors is shown in the Figure 5.

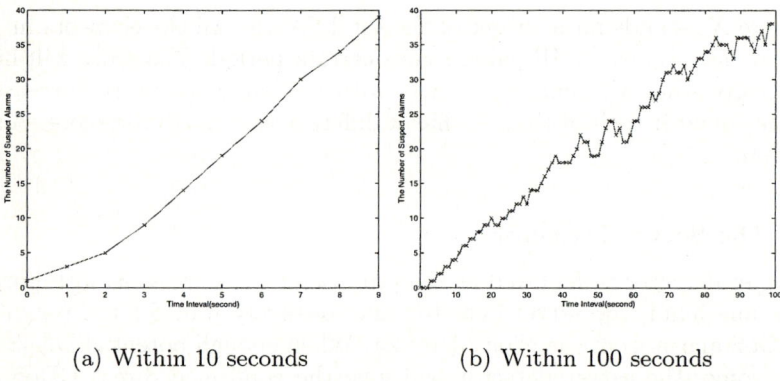

(a) Within 10 seconds   (b) Within 100 seconds

Fig. 5. The number of SA generated by the client detector after receiving 1% of $Ack(k+1) + Syn(j)$ packets from the protected server. The attacking packets rate reaches to 1000 packets/sec within (a)10s (b)100s

Although only 1% spoofed attack packets can be received by client detectors, client detectors still give accurate SAs. The number of SAs is enough for client detectors to give a further potential DDoS alarm at the early stage of the

DDoS attack. The detection results are satisfying even when the DDoS attacker increases the attack packets slowly. From experiment results,the SAs number raises stably in the Figure 5(a) because the DDoS attack is launched in a short time. In the Figure 5(b) the number of SAs seems a little vibrated because the attack packets rise up at a much slower rate.

Considering there exist network errors or latency which may cause false SA at the client detector, the false tolerance ability of the approach is evaluated. To test the sensibility to true SA and tolerance to false one , three different data sets are involved in the second experiment. The first data set contains no false alarms. The number of false alarms contained in the second data set is as many as the true alarms. The number of false alarms in the third one is as many as two times of the true alarms. The experiments result is given in Figure 6.

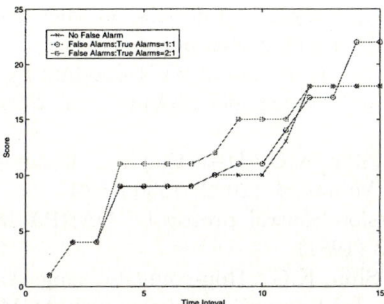

Fig. 6. Insensitive to false alarms

The experiment results show that even the false SAs are more than 50% of total SAs, the score will not be influenced much. The reason is that the score evaluated by the client detector raises rapidly when the SAs have the same IP while reach minimum when the SAs have different IP addresses. So SA evaluation expression is insensitive to false alarms. In the real code running on real machine, more than 50% suspect alarms are expect to be true because false SA caused by the network errors or latency is rather occasional.

6 Conclusion and Future Work

In this paper a novel detection method against the direct DDoS attack is presented. The detection system consists of the client detector and the server detector. The client detector performs detection at the side of innocent hosts and this is quite different from current methods which are often deployed at the sides of the victim server or the attacking source. The server detector is employed by the protected server and can perform both passively and actively DDoS detection. The benefit of this method is to yield accurate alarms at the early stage of DDoS attack. The key idea in the cooperative detection is based on the difference between the normal TCP three-way handshake and the spoofed one. The experiment results are given in section 5 and the proposed approach is effective to detect DDoS at its early stage. The results also show the SA evaluation method is insensitive to the false SAs.

In the future research work we will apply method to real environment to test the memory and computing cost for actual running because the presented method will need more storage and computing cost than some other stateless methods [14] [3]. Some advanced hash algorithms for recording the states of three-way handshake will be researched to compress the storage space and improve the query efficiency.

References

1. Moore, D., Voelker, G., Savage, S.: Inferring internet denial of service activity. In: Proceedings of USENIX Security Symposium. (2001)
2. Chen, Y.: Study on the prevention of SYN flooding by using traffic policing. In: Network Operations and Management Symposium 2000 IEEE/IFIP. (2000) 593–604
3. Wang, H., Zhang, D., Shin, K.G.: Detecting SYN flooding attacks. In: Proceedings of IEEE INFOCOM. Volume 3. (2002) 1530–1539
4. Postel, J.: Transmission control protocol : DARPA internet program protocol specification,RFC 793 (1981)
5. Jin, C., Wang, H.N., Shin, K.G.: Hop-count filtering: An effective defense against spoofed DDoS traffic. In: Proceedings of the 10th ACM conference on Computer and communication security(CCS), ACM Press (2003) 30–41
6. Lemon, J.: Resisting SYN flood DoS attacks with a SYN cache. In: In Proceedings of the BSDCon 2002 Conference. (2002)
7. Ferguson, P., Senie, D.: (Network ingress filtering: Defeating denial of service attacks which employ IP source address spoofing,RFC2827)
8. Song, D.X., Perrig, A.: Advanced and authenticated marking schemes for IP traceback. In: INFOCOM 2001. (2001) 878–886
9. Sung, M., Xu, J.: IP traceback-based intelligent packet filtering: A novel technique for defending against internet DDoS attacks. IEEE Transactions on Parallel and Distributed Systems **14** (2003) 861–872
10. Snoeren, A.C.: Hash-based IP traceback. In: Proceedings of the ACM SIGCOMM Conference, ACM Press (2001) 3–14
11. Bellovin, S.M.: ICMP traceback messages. Technical report (2000)
12. Ioannidis, J., Bellovin, S.M.: Implementing pushback: Router-based defense against DDoS attacks. In: Proceedings of Network and Distributed System Security Symposium, Catamaran Resort Hotel San Diego, California, The Internet Society (2002)
13. Rocky.K.Chang: Defending against flooding-based distributed denial-of-service attacks: a tutorial. Communications Magazine, IEEE **40** (2002) 42–51
14. Yaar, A., Perrig, A., Song, D.: SIFF: A stateless internet flow filter to mitigate DDoS flooding attacks. In: Proceedings. 2004 IEEE Symposium, Security and Privacy. (2004) 130–143

Probabilistic Inference Strategy in Distributed Intrusion Detection Systems

Jianguo Ding[1,2], Shihao Xu[1], Bernd Krämer[2], Yingcai Bai[1],
Hansheng Chen[3], and Jun Zhang[1]

[1] Shanghai Jiao Tong University, Shanghai 200030, P.R. China
Jianguo.Ding@sjtu.edu.cn
[2] FernUniversität Hagen, D-58084 Hagen, Germany
Bernd.Kraemer@FernUni-Hagen.de
[3] East-china Institute of Computer Technology, P.R. China

Abstract. The level of seriousness and sophistication of recent cyber-attacks has risen dramatically over the past decade. This brings great challenges for network protection and the automatic security management. Quick and exact localization of intruder by an efficient intrusion detection system (IDS) will be great helpful to network manager. In this paper, Bayesian networks (BNs) are proposed to model the distributed intrusion detection based on the characteristic of intruders' behaviors. An inference strategy based on BNs are developed, which can be used to track the strongest causes (attack source) and trace the strongest dependency routes among the behavior sequences of intruders. This proposed algorithm can be the foundation for further intelligent decision in distributed intrusion detection.

1 Introduction

Increased complexity, availability of vulnerability information and distributed network services make the whole network society attractive and easy to be attacked. Intruders in cyberspace benefit greatly from the anonymity, speed, and vast amounts of information present in that environment. Tracing of intruder in DIDS (Distributed Intrusion Detection System) will help to stop and prevent the potential attack in the future and provide important evident for prosecution and civil legal proceedings.

There are two categories of intrusion detection techniques: anomaly detection and misuse detection. Anomaly detection uses models of the intended behavior of users and applications, interpreting deviations from this "normal" behavior as a problem [4, 10, 6]. Misuse detection systems essentially define what's wrong. They contain attack descriptions (or "signatures") and match them against the audit data stream, looking for evidence of known attacks [7, 17, 14].

Most current approaches to intrusion detection attempt only to detect and prevent individual attacks. However, it is not the attack but rather the intruder against networks, which must be defended. Through understanding the behaviours of intruders, it is possible to develop a clearer picture of what is occurring. To do this, the information being provided by IDS must be gathered and and be analyzed, so that the activity of individual intruder is made clear. Some researches have been done some research to improve the intrusion detection based on the intruders' behaviour property [8, 13].

Since intruders jump from one computer to another, obfuscating the source of the attack, intruders are able to make themselves difficult to be traced. But the management systems in the network still keep lots of discontinue information which the intruders left such as login record, ports scanning, files modification or deleting, account modification, database searching and so on. Thus it provides possibility to understand more about the intruders.

For a sensitive network, it suffers numerous attacks from different intruders. Under this situation, it is really hard to provide one intrusion detection method to deal with all kinds of intruders. Hence, divide and rule is more effective to defend against intruders. The challenge is how to identify the intruders by their behaviours. Fortunately the activities of individual intruders as they move across networks expose the characteristic of themselves. The existing data that is already being collected by IDS scattered across networks is helpful to understand the intruders' behaviour. Generally, the individual intruder has stable characteristics in behaviours during the attacking and this is recorded by his/her activity sequence. But the individual action is random and not certain. That means the probability is another property for intruders' behaviour.

It is therefore proposed to apply Bayesian networks to model the behaviour sequence and provide efficient methods to locate the intruder in uncertain situations, and eventually to automate part of the IDS.

In this paper Strongest Dependence Route (SDR) algorithm for backward inference in BNs will be developed. The SDR algorithm will allow user to trace the the attack source based on the backward inference. It also provides the dependency sequence of the causes from considered effects in face of the multiple attacks at the same time, such as quick localization of attack sources under DDoS (Distributed Denial of Service) attacks.

BNs model for intrusion detection is discussed in Section 2. The Strongest Dependency Route algorithm for backward inference in Bayesian Networks is presented in Section 3. Section 4 concludes and identifies directions for further research.

2 Bayesian Networks for DIDS

2.1 Model of Bayesian Networks

Bayesian networks (BNs), known as Bayesian belief networks, probabilistic networks or causal networks, are an important knowledge representation in Artificial Intelligence [18, 2]. BNs use DAGs (Directed Acyclic Graphs) with probability labels to represent probabilistic knowledge. BNs can be defined as a triplet (V, L, P), where V is a set of variables (nodes of the DAG), L is the set of causal links among the variables (the directed arcs between nodes of the DAG), P is a set of probability distributions defined by: $P = \{p(v \mid \pi(v)) \mid v \in V\}$; $\pi(v)$ denotes the parents of node v. The DAG is commonly referred to as the dependence structure of a BN.

An important advantage of BNs is the avoidance of building huge JPD (Joint Probability Distribution) tables that include permutations of all the nodes in the network. Rather, for an effect node, only the states of its immediate predecessor need to be considered. Fig. 1 shows a simple example of DIDS. The JPD of the dependency between the activities of intruders are denoted in Fig. 2.

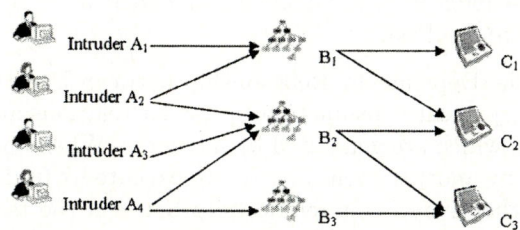

Fig. 1. Example of Bayesian Network for Distributed Intrusion Detection

A_1: login from 64.211.*.12 (proxy X_1)
A_2: login from 135.27.*.88 (proxy X_2)
A_3: login from 202.166.*.31
A_4: login from 211.34.*.101
B_1: display files
B_2: change properties of files
B_3: search logs files
C_1: search the account information
C_2: edit audit logs
C_3: restart machine

$p(A_1) = 82\%$
$p(A_2) = 77\%$
$p(A_3) = 95\%$
$p(A_4) = 65\%$
$p(B_1|A_1 A_2) = 76\%$
$p(B_1|\overline{A_1} A_2) = 58\%$
$p(B_1|A_1 \overline{A_2}) = 29\%$
$p(B_1|\overline{A_1 A_2}) = 0\%$
$p(B_2|A_2 A_3 A_4) = 96\%$
$p(B_2|\overline{A_2} A_3 A_4) = 76\%$
$p(B_2|A_2 \overline{A_3} A_4) = 72\%$
$p(B_2|A_2 A_3 \overline{A_4}) = 66\%$
$p(B_2|\overline{A_2 A_3} A_4) = 59\%$
$p(B_2|A_2 \overline{A_3 A_4}) = 56\%$
$p(B_2|\overline{A_2} A_3 \overline{A_4}) = 17\%$
$p(B_2|\overline{A_2 A_3 A_4}) = 0\%$

$p(B_3|A_4) = 87\%$
$p(B_3|\overline{A_4}) = 0\%$
$p(C_1|B_1) = 90\%$
$p(C_1|\overline{B_1}) = 5\%$
$p(C_2|B_1 B_2) = 88\%$
$p(C_2|\overline{B_1} B_2) = 75\%$
$p(C_2|B_1 \overline{B_2}) = 33\%$
$p(C_2|\overline{B_1 B_2}) = 3\%$
$p(C_3|B_3) = 92\%$
$p(C_3|\overline{B_3}) = 2\%$

Fig. 2. The Probability Distribution for Bayesian Network in Fig. 1

In the example (see Fig. 1), the event A_1, A_2 are the causes for event B_1. The annotation $p(B_1|\overline{A_1}A_2) = 58\%$ (see Fig. 2) can be explained as: B_1 happens in the probability 58% while A_1 does not happen but A_2 happens. Other annotations can be read in the similar way. From the viewpoint of management, some evidences (effect events)are easy to be detected, but the cause events are not obvious to be observed. One important task in DIDS is to infer the hidden factors from the available evidences.

Due to the dense knowledge representation of BNs, BNs can represent large amounts of interconnected and causally linked data as they occur in DIDS. Generally speaking: (1) BNs can represent knowledge in depth by modeling the functionality of the transmission network in terms of cause and effect relationships among the activities of intruders. (2) They have the capability of handling uncertain and incomplete intrusion information due to their grounding in probability theory. (3) They provide a compact and well-defined problem space since they use an exact solution method for any combination of evidences.

2.2 Mapping Distributed Intrusion Detection Systems to BNs

When the event sequences in a DIDS are modelled as a BN, two important processes need to be resolved:

(1) Ascertain the Dependency Relationship between Events.

Dependencies represent consumer and provider relationship between various cooperating events which are generated by intruders. When one event requires a service performed by another event in order to execute its function, or one event is invoked by another event, this relationship between the two events is called a dependency. A dependency graph of a system may be obtained using direct or indirect methods [1]. The notion of dependencies can be applied at various level of granularity. Two models are useful to get the dependency between events which are invocated by intruders in DIDS.

Functional model defines generic service dependencies and establishes the principle constrains to which the other models are bound. A functional dependency is an association between two events, typically captured first at design time, which says that one event requires some services from another.

Time sequence model contains the detailed descriptions of events which one is invoked by another based on time sequence.

(2) Obtain the Measurement of Dependencies.

Single-cause and multi-cause are two kinds of general assumptions to consider the dependencies between events in DIDS. A non-root node may have one or several parents (causal nodes). Single-cause means any of the causes must lead to the effect. While multi-cause means that one effect is generated only when more than one cause happens simultaneity. Intrusion detection information statistics are the main source to get the dependencies between the events in DIDS. The empirical knowledge of experts and experiments are useful to determine the dependency. Some researchers have done useful works to discover the dependencies from the application view in distributed systems [5,9]. The principles can be used as dependency obtaining in DIDS.

3 Backward Inference in Bayesian Networks

3.1 Inference in Bayesian Networks

The most common task in an uncertain reasoning system for DIDS is probabilistic inference which traces the causes from effects. In BNs, one node may have one or several parents (if it is not a root node), and we denote the dependency between parents and their child by a JPD.

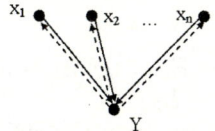

Fig. 3. Basic Model for Backward Inference in Bayesian Networks

Consider the basic model for backward inference in BNs (see Fig. 3), which X be the set of causes of Y, $X = (x_1, x_2, \ldots, x_n)$. Then the following variables are known: $p(x_1), p(x_2), \ldots, p(x_n), p(Y|x_1, x_2, \ldots, x_n) = p(Y|X)$. Here $x_1, x_2, \ldots, x_n$ are mutually independent, so $p(X) = p(x_1, x_2, \ldots, x_n) = \prod_{i=1}^{n} p(x_i)$. By Bayes' theorem, $p(X|Y) = \frac{p(Y|X)p(X)}{p(Y)} = \frac{p(Y|X)p(X)}{\sum_X [p(Y|X)p(X)]} = \frac{p(Y|X)\prod_{i=1}^{n} p(x_i)}{\sum_X [p(Y|X)\prod_{i=1}^{n} p(x_i)]}$,

which computes to $p(x_i|Y) = \sum_{X \setminus x_i} p(X|Y)$.

Hence the individual conditional probability (backward dependency) $p(x_i|Y)$ can be achieved from the JPD $p(Y|X)$, $X = (x_1, x_2, \ldots, x_n)$.

3.2 Strongest Dependency Route Algorithm for Backward Inference

Before we describe the SDR algorithm, the definition of strongest cause is given as follows:

Definition 1. *In a BN, let C be the set of causes, E be the set of effects. For $e_i \in E$, C_i be the set of causes based on effect e_i, iff $p(c_k|e_i) = Max[p(c_j|e_i), c_j \in C_i]$, then c_k is the strongest cause for effect e_i.*

The detailed description of the SDR algorithm is described as follows:

Pruning of the BNs. Generally, multiple effects (symptoms) may be observed at a moment, so $E_k = \{e_1, e_2, \ldots, e_k\}$ is defined as initial effects. In the operation of pruning, in every step only current nodes' parents are integrated and their brother nodes are omitted, because their brother nodes are independent with each other. The pruned graph is composed of the effect nodes E_k and their entire ancestor.

SDR Algorithm. After the pruning operation, a simplified sub-BN is obtained. The SDR algorithm use product calculation to measure the serial strongest dependencies between effect nodes and causal nodes.

Input: $BN = (V, L, P)$; $E_k = \{e_1, e_2, \ldots, e_k\}$, $E_k \subset V$.

Output: T: a spanning tree of the BN, rooted on E_k.

Variables: depend[v]: the strongest dependency between v and all its descendants; $p(v|u)$: the probability can be calculated from JPD of $p(u|\pi(u))$, v is the parent of u; $\varphi(l)$:temporal variable to record the strongest dependency between nodes.

 Initialize the SDR tree T as E_k; // E_k is added as root nodes of T
 Write label 1 on e_i; //$e_i \in E_k$
 While SDR tree T does not yet span the BN
  For each frontier edge l in BN
   Let u be the labelled endpoint of edge l;
   Let v be the unlabelled endpoint of edge l; //v is one parent of u
   Set $\varphi(l) = depend[u] * p(v|u)$;
  Let l be a frontier edge for BN that has the maximum φ-value;
  Add edge l (and vertex v) to tree T;
  $depend[v] = \varphi(l)$;
  Write label $depend[v]$ on vertex v;
 Return SDR tree T and its vertex labels;

The result of the SDR algorithm is a spanning tree T. Every cause node $c_j \in C$ is labeled with $depend[c_j] = p(c_j|M_k, e_i)$, $e_i \in E_k$, M_k is the transition nodes between e_i and c_j.

Proof of the Correctness of SDR Algorithm. When a vertex u is added to spanning tree T, define $d[u] = weight(e_i, u) = -lg(depend[u])$. Because $0 < depend[\delta_j] \leq 1$ so $d[\delta_j] \geq 0$. Note $depend[\delta_j] \neq 0$, or else there is no dependency relationship between δ_j and its offspring.

Proof: suppose to the contrary that at some point the SDR algorithm first attempts to add a vertex u to T for which $d[u] \neq weight(e_i, u)$.

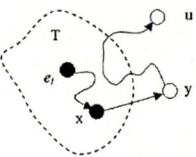

Fig. 4. Proof of SDR Algorithm

See Fig. 4. Consider the situation just prior to the insertion of u and the true strongest dependency route from e_i to u. Because $e_i \in T$, and $u \in V \backslash T$, at some point this route must first take a jump out of T. Let (x, y) be the edge taken by the path, where $x \in T$, and $y \in V \backslash T$. We have computed x, so

$$d[y] \leq d[x] + weight(x, y) \tag{1}$$

Since x was added to T earlier, by hypothesis,

$$d[x] = weight(e_i, x) \qquad (2)$$

Since $< e_i, \ldots, x, y >$ is sub-path of a strongest dependency route, by Eq.(2),

$$weight(e_i, y) = weight(e_i, x) + weight(x, y) = d[x] + weight(x, y) \qquad (3)$$

By Eq. (1) and Eq. (3), we get $d[y] \leq weight(e_i, y)$. Hence $d[y] = weight(e_i, y)$.

Since y appears midway on the route from e_i to u, and all subsequent edges are positive, we have $weight(e_i, y) < weight(e_i, u)$, and thus $d[y] = weight(e_i, y) < weight(e_i, u) \leq d[u]$. Thus y would have been added to T before u, in contradiction to our assumption that u is the next vertex to be added to T. So the algorithm must work. Since the calculation is correct for every effect node. It is also true that for multiple effect nodes in tracing the strongest dependency route.

Analysis of the SDR Algorithm. In SDR algorithm, every link (edge) in BN is only calculated one time, so the size of the links in BN is consistent with the complexity. The number of edges in a complete DAG is $n(n-1)/2 = (n^2 - n)/2$, where n is the size of the nodes of the DAG. Normally a BN is an incomplete DAG. So the calculation time of SDR is less than $(n^2 - n)/2$. The complexity of SDR is $O(n^2)$.

Compared to other inference algorithms in BNs [2, 19], the SDR algorithm belongs into the class of exact inferences and it provides an efficient method to trace the strongest dependency routes from effects to causes and to track the dependency sequences of the causes. It is useful in fault localization, and it is beneficial for performance management. Moreover it can treat multiple connected networks modelled as DAGs.

Researchers have used hidden Markov model for anomaly detection [20], but Markov model is lack of historic remembering, thus can not take good use of the great deal of statistic data in DIDS. Statistic technology is another popular approach in intrusion detection, but BNs model expose the advantages in deal with the the correlation and dependency between multiple variables.

Simulation Measurement in Backward Inference. For the simulation in DBNs, a robust and reliable random number generator plays an important part. Based on the TGFSR (Twisted Generalized Feedback Shift Register) algorithm [15, 16] for the random number generation, we develop a simulator. The simulator can generate the BNs (see Fig. 5)and the data set (see Fig. 6) for inference test. In Fig. 5, the cause set $C = \{A, B, C, D, E\}$, the effect set $E = \{R, T, S\}$. The JPD which describes the Bayesian network is denoted in Fig. 6.

Consider the backward inference in BNs. Based on the SDR algorithm, a spanning tree, which holds the strong dependency routes, is obtained (see Fig. 7).

From the spanning tree, the strongest routes between effects and causes can be obtained by depth-first search. Meanwhile the inference also provides a cause sequence, in which the strongest route between effect nodes and cause nodes can be achieved too.

842   J. Ding et al.

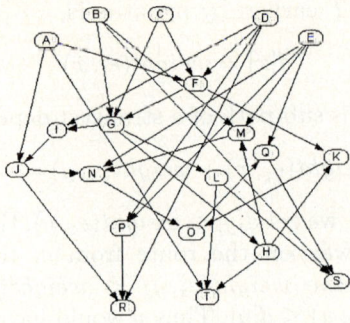

Fig. 5. Simulation of Backward Inference in BNs

p(A=1) = 0.083538	p(I=1|G=1,D=1) = 0.981829	p(N=1|E=1,M=1,J=1) = 0.953636	p(R=1|J=1,D=0,P=1) = 0.680547
p(B=1) = 0.080265	p(I=1|G=1,D=0) = 0.925088	p(N=1|E=1,M=1,J=0) = 0.467355	p(R=1|J=1,D=0,P=0) = 0.528922
p(C=1) = 0.008221	p(I=1|G=0,D=1) = 0.284676	p(N=1|E=1,M=0,J=1) = 0.937806	p(R=1|J=0,D=1,P=1) = 0.256740
p(D=1) = 0.008968	p(I=1|G=0,D=0) = 0.000000	p(N=1|E=1,M=0,J=0) = 0.170258	p(R=1|J=0,D=1,P=0) = 0.053764
p(E=1) = 0.007021	p(J=1|A=1,I=1) = 0.993701	p(N=1|E=0,M=1,J=1) = 0.321906	p(R=1|J=0,D=0,P=1) = 0.242892
p(F=1|A=1,D=1,C=1) = 0.910209	p(J=1|A=1,I=0) = 0.527593	p(N=1|E=0,M=1,J=0) = 0.191096	p(R=1|J=0,D=0,P=0) = 0.000000
p(F=1|A=1,D=1,C=0) = 0.089841	p(J=1|A=0,I=1) = 0.316958	p(N=1|E=0,M=0,J=1) = 0.054857	p(S=1|Q=1,L=1,B=1) = 0.909654
p(F=1|A=1,D=0,C=1) = 0.283011	p(J=1|A=0,I=0) = 0.000000	p(N=1|E=0,M=0,J=0) = 0.000000	p(S=1|Q=1,L=1,B=0) = 0.353975
p(F=1|A=1,D=0,C=0) = 0.028596	p(K=1|F=1,H=1) = 0.986441	p(O=1|N=1,K=1) = 0.933324	p(S=1|Q=1,L=0,B=1) = 0.505480
p(F=1|A=0,D=1,C=1) = 0.866795	p(K=1|F=1,H=0) = 0.928134	p(O=1|N=1,K=0) = 0.498004	p(S=1|Q=1,L=0,B=0) = 0.277857
p(F=1|A=0,D=1,C=0) = 0.051313	p(K=1|F=0,H=1) = 0.866430	p(O=1|N=0,K=1) = 0.509813	p(S=1|Q=0,L=1,B=1) = 0.696711
p(F=1|A=0,D=0,C=1) = 0.186593	p(K=1|F=0,H=0) = 0.000000	p(O=1|N=0,K=0) = 0.000000	p(S=1|Q=0,L=1,B=0) = 0.066841
p(F=1|A=0,D=0,C=0) = 0.000000	p(L=1|G=0) = 0.000000	p(P=1|D=1,E=1) = 0.852289	p(S=1|Q=0,L=0,B=1) = 0.243365
p(G=1|B=1,C=1,F=1) = 0.993147	p(L=1|G=1) = 0.051438	p(P=1|D=1,E=0) = 0.017071	p(S=1|Q=0,L=0,B=0) = 0.000000
p(G=1|B=1,C=1,F=0) = 0.665915	p(M=1|B=1,H=1,E=1) = 0.975844	p(P=1|D=0,E=1) = 0.726409	p(T=1|A=1,L=1,H=1) = 0.934892
p(G=1|B=1,C=0,F=1) = 0.641661	p(M=1|B=1,H=1,E=0) = 0.460523	p(P=1|D=0,E=0) = 0.000000	p(T=1|A=1,L=1,H=0) = 0.866575
p(G=1|B=1,C=0,F=0) = 0.507484	p(M=1|B=1,H=0,E=1) = 0.490122	p(Q=1|E=1,O=1) = 0.835802	p(T=1|A=1,L=0,H=1) = 0.558033
p(G=1|B=0,C=1,F=1) = 0.193340	p(M=1|B=1,H=0,E=0) = 0.091350	p(Q=1|E=1,O=0) = 0.229321	p(T=1|A=1,L=0,H=0) = 0.222810
p(G=1|B=0,C=1,F=0) = 0.141931	p(M=1|B=0,H=1,E=1) = 0.100454	p(Q=1|E=0,O=1) = 0.213323	p(T=1|A=0,L=1,H=1) = 0.623162
p(G=1|B=0,C=0,F=1) = 0.070825	p(M=1|B=0,H=1,E=0) = 0.022430	p(Q=1|E=0,O=0) = 0.000000	p(T=1|A=0,L=1,H=0) = 0.084838
p(G=1|B=0,C=0,F=0) = 0.000000	p(M=1|B=0,H=0,E=1) = 0.098186	p(R=1|J=1,D=1,P=1) = 0.389462	p(T=1|A=0,L=0,H=1) = 0.475918
p(H=1|G=0) = 0.000000	p(M=1|B=0,H=0,E=0) = 0.000000	p(R=1|J=1,D=1,P=0) = 0.708615	p(T=1|A=0,L=0,H=0) = 0.000000
p(H=1|G=1) = 0.130612			

Fig. 6. The JPD of BN in Fig. 6. (0: normal state, 1: abnormal state.)

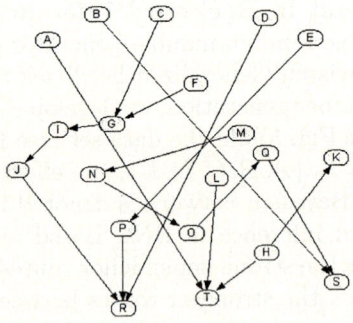

Fig. 7. The Spanning Tree for BN in Fig. 5

During the 10000 tests, the simulation results are presented in Fig. 8. From the simulation result, more than 80% cause nodes can be detected by only checking less than 50% of the whole cause nodes.

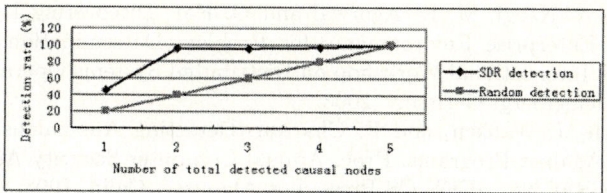

Fig. 8. Comparison between SDR Detection and Random Detection

Comparing this with the random or exhaustive detection, the backward inference in DBNs provides a more efficient approach to catch the causal nodes. When the set of causal nodes is larger, the detection rate is more optimal.

4   Conclusions and Future Work

In DIDS, the unstable, uncertain and incomplete information about the behaviour of intruders should be concerned. Fortunately Bayes' theory provides a scientific foundation to deal with probability of real systems. Hence it is reasonable to use BNs to represent the knowledge of events and model the probabilistic reasoning between them. Bayesian inference is the popular operation for a reasoning system. The SDR algorithm presents an efficient method to trace the causes from effects in backward inference. It is useful for intruders' localization even in face of multiple intrusions.

However, dynamically update from their structures, topologies and their dependency relationship between intrusion events bring more challenges to DIDS. To accommodate sustainable changes and to maintain a healthy DIDS, learning strategies that allows us to modify the cause-effect structure and also the dependencies between the nodes of a Bayesian networks correspondingly should be investigated. Further the proactive intrusion detection strategies based on the dynamic inference and prediction will greatly improve the efficiency in DIDS.

References

1. S. Bagchi, G. Kar, J. L. Hellerstein. Dependency Analysis in Distributed Systems using Fault Injection: Application to Problem Determination in an e-commerce Environment. 12th International Workshop on Distributed Systems: Operations & Management 2001.
2. R. G. Cowell, A. P. Dawid, S. L. Lauritzen, D. J. Spiegelhalter. Probabilistic Networks and Expert Systems. New York: Springer-Verlag, 1999.

3. Eugene Charniak, Robert P. Goldman. A Semantics for Probabilistic Quantifier-Free First-Order Languages, with Particular Application to Story Understanding. Proceedings of IJCAI-89, pp. 1074-1079, Morgan-Kaufmann, 1989.
4. D.E. Denning. An Intrusion Detection Model. IEEE Trans. Software Eng., vol. 13, no. 2, Feb. 1987, pp. 222-232.
5. M. Gupta, A. Neogi, M. K. Agarwal and G. Kar. Discovering Dynamic Dependencies in Enterprise Environments for Problem Determination. Proc. of 14th IEEE/IPIP International Workshop on Distributed Systems Operations and Management. Heidelberg, Germany, 2003.
6. A.K. Ghosh, J. Wanken, and F. Charron. Detecting Anomalous and Unknown Intrusions Against Programs. Proc. Annual Computer Security Application Conference (ACSAC'98), IEEE CS Press, Los Alamitos, Calif., 1998, pp. 259-267.
7. K. Ilgun, R.A. Kemmerer, and P.A. Porras. State Transition Analysis: A Rule-Based Intrusion Detection System. IEEE Trans. Software Eng. vol. 21, no. 3, Mar. 1995, pp. 181-199.
8. K. Julisch and M. Dacier, Mining Intrusion Detection Alarms for Actionable Knowledge, Proc. 8th ACM International Conference on Knowledge Discovery and Data Mining, Edmonton, July 2002.
9. A. Keller, U. Blumenthal, G. Kar. Classification and Computation of Dependencies for Distributed Management. Proceedings of 5th IEEE Symposium on Computers and Communications. Antibes-Juan-les-Pins, France, July 2000.
10. C. Ko, M. Ruschitzka, and K. Levitt. Execution Monitoring of Security-Critical Programs in Distributed Systems: A Specification-Based Approach. Proc. 1997 IEEE Symp. Security and Privacy, 1997, pp. 175-187.
11. S. Klinger , S. Yemini , Y. Yemini , D. Ohsie , S. Stolfo. A coding approach to event correlation. Proceedings of the fourth international symposium on Integrated network management IV, p.266-277, January 1995.
12. I. Katzela and M. Schwarz. Schemes for fault identification in communication networks. IEEE Transactions on Networking. 3(6): 733-764, 1995.
13. W. Lee. A Data Mining Framework for Constructing Features and Models for Intrusion Detection Systems. PhD thesis, Columbia University, June 1999.
14. U. Lindqvist and P.A. Porras. Detecting Computer and Network Misuse with the Production-Based Expert System Toolset. IEEE Symp. Security and Privacy, IEEE CS Press, Los Alamitos, Calif., 1999, pp. 146-161.
15. M. Matsumoto and Y. Kurita. Twisted GFSR generators. ACM Trans. on Modeling and Computer Simulation, 2(1992), pp. 179-194, .
16. M. Matsumoto and Y. Kurita. Twisted GFSR generatos II. ACM Trans. on Modeling and Computer Simulation, 4(1994)pp. 254-266.
17. V. Paxson. Bro: A System for Detecting Network Intruders in Real-Time. Proc. Seventh Usenix Security Symp., Usenix Assoc., Berkeley, Calif., 1998.
18. J. Pearl. Probabilistic Reasoning in Intelligent Systems: Networks of Plausible Inference. Morgan Kaufmann, San Mateo, CA, 1988.
19. J. Pearl. Causality: Models, Reasoning, and Inference. Cambridge, England: Cambridge University Press. New York, NY, ISBN: 0-521-77362-8, 2000.
20. Q. Yin, L. Shen, R. zhang, X. Li, H. Wang. Intrusion Detection Based on Hidden Markov Model. Proc. of 2003 IEEE Conference on Machine Learning and Cybernetics , Volume: 5, pp. 3115 - 3118, 2003.

An Authorization Framework Based on Constrained Delegation*

Gang Yin, Meng Teng, Huai-min Wang, Yan Jia, and Dian-xi Shi

Institute of Network Technology & Information Security,
Department of Computer Science,
National University of Defense Technology, China
jack_nudt@yahoo.com.cn

Abstract. In this paper, we distinguish between authorization problems at management level and request level in open decentralized systems, using delegation for flexible and scalable authorization management. The delegation models in existing approaches are limited within one level or only provide basic delegation schemes, and have no effective control over the propagation scope of delegated privileges. We propose REAL, a Role-based Extensible Authorization Language framework for open decentralized systems. REAL covers delegation models at both two levels and provides more flexible and scalable authorization and delegation policies while capable of restricting the propagation scope of delegations. We formally define the semantics of credentials in REAL by presenting a translation algorithm from credentials to Datalog rules (with negation-as-failure). This translation also shows that the semantics can be computed in polynomial time.

1 Introduction

Many new style applications are emerging in today's Internet or large scale Intranet and playing important role in daily life, such as resource sharing in decentralized systems (coalitions, multi-centric collaborative systems and grid computing), electronic commerce, health care systems, etc. Authorization in these scenarios is significantly different from that in centralized or closed distributed systems. Among numerous new features of these large-scale, open, decentralized systems, there are mainly two kinds of features essentially making authorization challenging, which lead to new authorization problems at management level and request level respectively:

Multi-centric Management. Today's open decentralized systems are often composed of many systems and each system has its own authorization management domain.

---

* This work is supported by the National Grand Fundamental Research 973 Program of China under Grant No.G1999032703; the National High Technology Development 863 Program of China under Grant No.2003AA115210; Foundation of Weapons Research in Advance under Grant No.51415030203KG01, "security technologies in multi-database systems".

Specially, if a domain has a very large scale, its authorization management structure may also be multi-centric.

Cascaded Request. When a user calls on a service to perform some operations, the service may not complete the operations itself, but will call on other services to do so. This will usually result in cascaded request. Cascaded request mainly caused by service integration or service composition in open decentralized systems.

Traditional access control mechanisms usually make authorization decisions based on the identity of the resource requester [13]. We can show that this may be ineffective in above scenarios. First, *multi-centric management* may fall out that the resource owner (authorizer) and the requester are unknown to each other. The authorizer cannot make authorization decision because requester's identity is outside of authorizer's management domain and it is not a trusted identity. Second, *cascaded request* is a chain of calls on authorizer's services. It is still difficult for the authorizer to make authorization decisions because there maybe more than one identity (or privileges) related to the request.

In this paper, we use delegation to solve the two kinds of problems described above. Section 2 studies the usage of delegation in different authorization scenarios and sum-up two kinds of delegation models and corresponding constraint models based on trusted scope. In Section 3, we propose a language framework based on our delegation model and introduce its usage with a serial of examples. We translate credentials of REAL into Datalog rules (with negation-as-failure) in section 4. Then we discuss the implementation issues, future and related work in section 5 and 6, and conclude in section 7.

2 Delegation Models in REAL

Delegation is a promising approach to realize more flexible and scalable authorization management for the distributed systems. We distinguish three kinds of authorization scenarios based on the way they use delegation, and find that the delegation can be mainly classified into two levels, as shown in fig. 1.

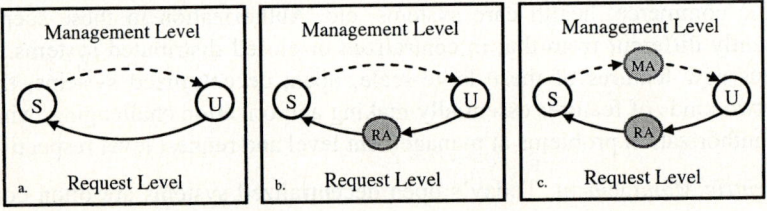

Fig. 1. The Role of Delegation in the Authorization Activities

In fig. 1, S is the server that has services to be protected; U is the user making requests. In centralized multi-user systems, authorization and access request are made directly, and there is no delegation between U and S, as fig. 1.a. The delegation existing in this scenario happens within the operation system context on S, such as the delegation mechanism in hydra [20]. When distributed systems were used, user may firstly logon to workstation, delegating his identity or capabilities to the workstation, and then the workstation accesses the server on behalf of the user. The workstation is called RA (request agency) here, as shown in fig. 1.b. This scenario is the background of some well-known authentication logic [1] and delegation protocols [9, 18], and it originates the form of *cascaded request*. When the small closed distributed systems merged into large-scale, open, decentralized systems such as Internet, the authorization management structure of these merged systems are *multi-centric*. S delegates some of its authority to third party to make authorization decision collaboratively. This third party is called MA (management agency) here, as shown in fig. 1.c. Many trust managements [3, 11, 12, 13] have been proposed to address this kind of delegation. At request level of this scenario, there are more complex and dynamic forms of *cascaded request*. In fact, delegation activities at management level and request level are not unattached, they are interdependent.

REAL distinguish between above two kinds of delegation, delegation of authority and delegation of capability, which belong to delegation activities of management level and request level respectively. This section then describes our delegation models based on the elements of RBAC.

2.1 Delegation at Management Level

Authority is a basic concept for security management. In [2], the authors use the notion of "authority" as a prerequisite for creating and changing management structures as well as for creating and deleting permissions. We define authority as follows,

Definition 1 (Authority). Authority is the privilege to manage authorization of permissions for a specified entity. The idea of delegation of authority is that one entity hands off authorities to another entity to make authorization (within the scope of the delegated authority) on behalf of the former. It will expand the administrative scope of an authorizer and make the security administrative work more efficient.

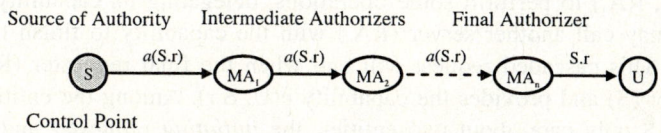

Fig. 2. The control model of delegation of authority

We propose to use administrative roles [15] to express authority. For a role S.r (means the role "r" is defined by S), we use $a(S.r)$ to denote its administrative role. In fig. 2, S wants to share its resources controlled by S.r to the users (e.g., U) in another domain, S may delegate the authority $a(S.r)$ to a MA (MA_1) within or close to this domain. MA_1 may directly authorize the role S.r to U, or may re-delegate the authority to another MA (MA_2) more close to U, and so on. The final MA (MA_n) in the chain will do the actual authorization to U. In the process of delegation of authority, S is the source and the control point of the delegated authority, totally controls the propagation of authorities and permissions. MA_1 to MA_{n-1} are intermediate authorizers, they can pass on the authority and may also set constraints on the delegation process, and the final authorizer MA_n authorizes the role S.r to U.

2.2 Delegation at Request Level

We use delegation of capability to model the delegation activities within the context of a request. In centralized systems, a capability is an unforgettable pair made up of an object identifier and a set of authorized operations (an interface) on that object [7], and capabilities are only given to processes. In distributed role-based authorization systems, capabilities can be modeled as role activations [12].

Definition 2 (Capability). Capability is the privilege to exercise the permission of a specified entity. The idea of delegation of capability is that one entity hands off capabilities to another entity to access resources on behalf of the former. A good model for delegation of capability will cut the complexity of authorization decision making for *cascaded requests*.

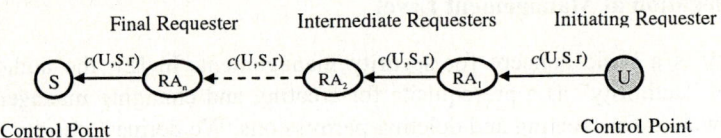

Fig. 3. The control model of delegation of capability

RT [12] uses "U as S.r" to express the capability holding by a session (process), while the subject of the session is U, and its activated role is S.r. We follow this approach, using $c(U, S.r)$ to denote the capability with the same meaning.

In fig. 3, U creates a process with the capability $c(U, S.r)$. The process calls a server (e.g., RA_1) to perform some operations, delegating its capability $c(U, S.r)$ to RA_1. RA_1 may call another server (RA_2) with the capability to finish the operation, and so on. This cascaded request will end when the final requester (RA_n) calls the target server (S) and provides the capability $c(U, S.r)$. Among the entities in delegation chain, S only care about two entities, the *initiating requester* and the *final requester*. S only needs to check whether *final requester* can be authorized with the roles in delegated capabilities (e.g. S.r), which are only related to *initiating requester*. There are two control points in this model. U may control which RAs can exercise his capability, while S may also need to control which RAs are trusted to use the privileges delegated from *initiating requester*.

2.3 Propagation Scope of Delegation

Delegation is a flexible and scalable approach for authorization in decentralized systems. However, abuse or misuse of delegation will cause undesirable propagation of privileges. We propose to use trusted scope to restrict the propagation scope of delegation.

Definition 3 (Trusted Scope). Trusted scope is a collection of entities with certain attributes, which can be trusted by the delegator not to abuse or misuse the privileges (permissions, authorities or capabilities) assigned to them within a specified context. We use role intersections to denote trusted scope. For example, A.r∩B.r' is a role intersection, expressing the collection of entities that are both members of the role A.r and B.r'.

Administrative Scope. We use trusted scope to constrain the administrative scope (*AS*) of delegated authority, as shown in fig. 4.a. An authorizer with the delegated authority can only authorize the permission to entities within the *AS* defined by the preceding entities in the delegation chain. The first *AS* is defined by the control point of the delegated authority (e.g. S), which is represented by the dashed rectangle in fig. 4.a. Any authorizer (from MA_1 to MA_n) in the delegated chain can never authorize the role S.r to a user outside of this combined *AS*.

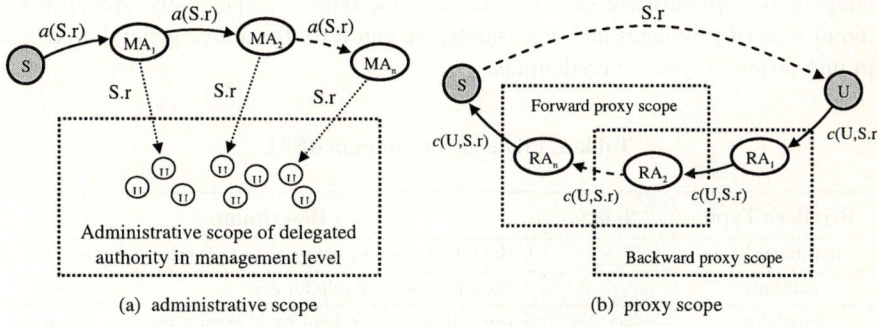

Fig. 4. Propagation scope of delegated privileges

Proxy Scope. We can also use trusted scope to constrain the proxy scope (*PS*) of the delegated capability, as shown in fig. 4.b. The control model at request level is more complex than that at management level (there are two control points in delegation of capability), we further distinguish the *PS* between *Forward PS* (*FPS*) and *Backward PS* (*BPS*) for the delegated capability. Suppose S has authorize (directly or indirectly) the role S.r to U. The *FPS* of the delegated capability is the *PS* specified by requesters (U or RAs). A requester with the delegated capability can only delegate the capability to entities within the *FPS* defined by the preceding requesters in the delegation chain. Fig. 4.b shows the *FPS* set by U, which means the following requesters (from RA_2 to RA_n) must be in the *FPS*. On the other hand, the *BPS* of the delegated capability is the *PS* specified by S. When S makes authorization decision for a request from the direct requester (e.g. RA_n), S may need to constrain the *PS* for the capabilities the requester exercising, which means the preceding requesters must belong to the *BPS* as shown in fig. 4.b (e.g. from RA_{n-2} to RA_1). If the requester uses the capabilities coming from untrusty entities (e.g. from outside of *BPS*), S will not allow the delegation. We call this the authorization of capabilities from S to RA_{n-1}.

3 The REAL Language

In this section, we propose REAL (Role-based Extensible Authorization Language), a language framework for above two delegation models and corresponding constraint models using credentials. The general form of a credential in REAL is:

issuer → subject: < privilege, constraint >

The general meaning of above credential framework is that the *issuer* authorizes (or delegate) the *privilege* to the *subject* while the *constraint* must be satisfied. The credential is signed by the *issuer*. The *issuer* and *subject* are single entities, which are denoted using capitals or words begin with capitals, such as A, B, Alice, etc (in a practical system, entities are public keys). The *subject* can also be the set of entities, which are denoted using distributed roles in this paper. We use different types of structures to express *privilege* and *constraint*. The syntax and brief introductions for *privilege* and *constraint* are shown in table 1 and table 2 respectively. According to different type of privileges and constraints, we can reify the above general credential form into different types of credentials.

Table 1. Privilege Structures in REAL

Privilege Type	Syntax	Description
distributed role	A.r	The role defined by entity A, and r is the role name.
authority	$a(A.r)$	The administrative role for A.r.
capability	$c(D, A.r)$	The role activation held by a session process, which is created by the entity D using role A.r.
abstract capability	$ac(A.r)$	The capability set which are created by the entities using role A.r, where the entities are members of the role A.r.

Table 2. Constraint Structures in REAL

Constraint Type	Syntax	Description
constraint on authorizations	$cau(ts)$	*cau* denotes the prerequisite roles in authorization credentials.
constraint on delegated authorities	$cda(ts)$	*cda* is the *AS* for the delegated authorities.
constraint on authorization of capabilities	$cac(ts)$	*cac* is the *BPS* for the authorized capabilities.
constraint on delegated capabilities	$cdc(ts)$	*cdc* is the *FPS* for the delegated capabilities.
In above constraint structures, *ts* has the type of trusted scope and can be "*", which means all entities in the system. The corresponding constraint structure are also denoted as "*".		

Credentials in REAL can be used to express four kinds of policies: 1) authorization; 2) delegation of authority; 3) authorization of capability; 4) delegation of capability. Each kind includes basic credentials (*subject* is an entity) and scalable credentials (*subject* is a role). Now let's get down to the details.

3.1 Authorization (Type-I Credentials)

The type-I credentials are used to express authorization polices. Credential (I) means that A authorizes B with the role C.r, and B must comply with the constraint *cau*(ts). Credential (I⁺) expresses the authorization to entities who are members of B.r'.

$$(\text{I}) \quad A \rightarrow B: \quad < C.r, cau(ts) >$$
$$(\text{I}^+) \quad A \rightarrow B.r': \quad < C.r, cau(ts) >$$

Example 1. A book store (BS) wants to give a discount (BS.discount) to students of a nearby university (Univ.student). But the students must be the members of BS (BS.member). Suppose Alice is a student of Univ and a member of BS. We can express these policies using following authorization credentials:

(1) BS → Univ.student: < BS.discount, *cau*(BS.member) >
(2) BS → Alice: < BS.member, * >
(3) Univ → Alice: < Univ.student, * >

According to REAL semantics, Alice will be authorized with the role BS.discount by BS, so she will enjoy the discount provided by BS. Example 1 is motivated by the example in [12].

3.2 Delegation of Authority (Type-II Credentials)

The type-II credentials are used to express delegation of authority. Credential (II) means that A delegates the authority a(C.r) to B, and the administrative scope of the delegated authority is constrained by *cda*(ts). Credential (II⁺) expresses the delegation to entities who are members of B.r'.

$$(\text{II}) \quad A \rightarrow B: \quad < a(C.r), cda(ts) >$$
$$(\text{II}^+) \quad A \rightarrow B.r': < a(C.r), cda(ts) >$$

Example 2. Continue with example 1, BS trusts its security administrator (BS.sa) to determine the membership (BS.member). BS also requires that its members must have been registered (BS.registered). Suppose Bob and Carol both are BS's security administrators. Alice has registered in BS and Bob accepts Alice as the member of BS. These policies can be expressed using following credentials:

(4) BS → BS.sa: < *a*(BS.member), *cda*(BS.registered) >
(5) BS → Bob: < BS.sa, * >
(6) BS → Carol: < BS.sa, * >
(7) BS → Alice: < BS.registered, * >
(8) Bob → Alice: < BS.member, * >

According to credentials (4)(5)(7)(8), the semantics for REAL should support that Alice is a member of BS.

3.3 Authorization of Capability (Type-III Credentials)

The type-III credentials are used to express the policies of authorization of capability in request level. Credential (III) means that A allows B to exercise the capability $c(?x, A.r)$, where $?x$ is the entity who is the member of $A.r$. The *BPS* of the delegated capability is constrained by $cac(\text{ts})$. Credential (III$^+$) expresses the authorization to entities who are members of $B.r$'.

$$\begin{aligned}&\text{(III)} \quad &A \to B: \quad &< ac(A.r), cac(\text{ts}) >\\&\text{(III}^+\text{)} \quad &A \to B.r': \quad &< ac(A.r), cac(\text{ts}) >\end{aligned}$$

Example 3. Continue with above examples. BS allows its agencies (BS.agency) to act as the proxies for BS's members (BS.member). VBS is a virtual book store, which is an agency for BS (BS.agency). We can describe the policies using following credentials:

(9)   BS → BS.agency: $< ac(\text{BS.member}), cac(\text{BS.agency}) >$
(10)   BS → VBS: $< \text{BS.agency}, * >$

According to REAL semantics, if a request is initiated by a BS's member, with VBS acting as the proxy for the request, then BS will finally authorize the request.

3.4 Delegation of Capability (Type-IV Credentials)

The type-IV credentials are used to express delegation of capability in request level authorization activities. Credential (IV) means that A delegates the capability $c(C, D.r)$ to B, and the *FPS* of the delegated capability is constrained by $cdc(\text{ts})$. Credential (IV$^+$) expresses the delegation to entities who are members of $B.r$'.

$$\begin{aligned}&\text{(IV)} \quad &A \to B: \quad &< c(C, D.r), cdc(\text{ts}) >\\&\text{(IV}^+\text{)} \quad &A \to B.r': \quad &< c(C, D.r), cdc(\text{ts}) >\end{aligned}$$

Example 4. Continue with above examples, Alice buys a book from VBS, delegating her role BS.member to VBS, but restricting the forward proxy scope of delegated capability within BS.agency. Suppose VBS can not process the request from Alice and forward to another virtual book store (AVBS), AVBS is also an agency for BS and finally submit the request to the BS server successfully. We can describe the delegation process using following credentials:

(11)   BS → AVBS: $< \text{BS.agency}, * >$
(12)   Alice → VBS: $< c(\text{Alice}, \text{BS.member}), cdc(\text{BS.agency}) >$
(13)   VBS → AVBS: $< c(\text{Alice}, \text{BS.member}), * >$

According to REAL semantics, BS will finally authorize the request from AVBS and provide discount for the current bill. This example also introduces a mechanism for the security problem in fault tolerance scenarios.

4 Semantics

We now define a translation from each credential to one or more Datalog rules (with negation-as-failure) [19]. This translation serves both as a definition of the semantics and also as one possible implementation mechanism. We define the translation using the semantic translation algorithm (*STA*) shown in table 3. In the output language, we use 3 special predicates: *auth*, *doa* and *doc*.

The core predicate is *auth(issuer, subject, r)*, which means that *issuer* authorizes *subject* with the role *r*. The predicate *doa(issuer, subject, au, ts)* means that *issuer* delegates (directly or indirectly) the authority *au* to *subject*, while the *AS* of *au* is *ts*. The predicate *doc(issuer, subject, capa, proxy, id)* means that *issuer* delegates (directly or indirectly) the capability *capa* to *subject*, where *proxy* is one of the entities in the delegation chain; *id* is an integer uniquely identifies the predicate, 0 is the default value for *id*. When *issuer* directly delegates *capa* to *subject*, *proxy* takes "Θ" as the value, which is the default value for *proxy*; if there is an entity in the delegation chain does not belong to *BPS* of *capa* (which is specified in type-III credentials), then we use "Ω" to denote that entity. The first 2 arguments of each predicates take an entity as the value.

To simplify the presentation of the translation rules, we add a new predicate *auths_ts*, which takes an entity and trusted scope (role intersection) as arguments. *auths_ts* is used to check whether the entity complies with the trusted scope. For example, *auths_ts*(D, A.r∩B.r') is true if and only if both *auth*(A, D, A.r) and *auth*(B, D, B.r') are true.

We do not present the translation for scalable credentials (such as I$^+$,II$^+$,III$^+$ and IV$^+$) because it is similar to the translation in table 3, we just need to add one *auth* predicate denoting that B authorizes B.r' to a variable and replace B with this variable in each rule. We call the translation algorithm for all credentials *Full-STA* (*FSTA*).

Table 3. Datalog-based Semantics for REAL

Semantic Translation Algorithm (STA)
(I) Translate credential "A → B: < C.r, *cau*(ts) >" to,   1. *auth*(A, C, B.r) :- *auths_ts*(B, ts).
(II) Translate credential "A → B: < *a*(C.r), *cau*(ts) >" to,   1. *doa*(A, B, *a*(C.r), ts).   2. *auth*(A, ?x, C.r) :- *auth*(B, ?x, C.r), *auths_ts*(?x, ts).   3. *auth*(A, ?z, C.r) :- *doa*(B, ?x, *a*(C.r), ?y$_u$), *auth*(?x, ?z, C.r), *auths_ts*(?z, ts), *auths_ts*(?z, ?y$_u$).
(III) Translate credential "A → B: < *ac*(A.r), *cac*(ts) >" to,   1. *auth*(A, B, A.r) :- *doc*(?x, B, *c*(?x, A.r), ?y), ¬ *doc*(?x, B, *c*(?x, A.r), Ω, id).   2. *doc*(?x, B, *c*(?x, A.r), Ω, id) :- *doc*(?x, B, *c*(?x, A.r), ?y), ¬ *auths_ts*(?y, ts).   where id is a global unique identifier generated during translation.
(IV) Translate credential "A → B: < *c*(C, D.r), *cdc*(ts) >" to,   1. *doc*(A, B, *c*(C, D.r)) :- *auth*(D, C, D.r).   2. *doc*(A, ?x, *c*(C, D.r), B) :- *doc*(B, ?x, *c*(C, D.r)), *auths_ts*(?x, ts).   3. *doc*(A, ?x, *c*(C, D.r), ?y) :- *doc*(B, ?x, *c*(C, D.r), ?y), *auths_ts*(?y, ts).

When a server (S) receives a request, which is denoted by a structure R(*creds, op*), R is the identity of the final requester to S, *creds* is the credentials provided by R and *op* are the operations on S which R wants to invoke. Suppose an entity must be authorized with role $S.r_i$ (i=1,...,s) before it can perform the *op* on S. Now we can define the POC (proof of compliance) problem [11] for the current request: given the set of credentials $C \cup creds$, whether the rule set $FSTA(C \cup creds)$ can prove $auth(S, R, S.r_i)$, where C is the credential set held by S and i=1,...,s.

Given a set of REAL credentials C, let $FSTA(C)$ be the Datalog program resulting from the translation. The implication of C, defined as the set of *auth* predicates implied by C, is determined by the minimal model of $FSTA(C)$. In the following, we show that this model can be constructed in polynomial time.

Proposition 1. Given a set C of REAL credentials, computing the implications of C can be done in time $O(N^5)$, where N is the number of credentials in C.

Proof. We follow the analysis approach in [12]. Consider the translation of one credential c; let $FSTA(c)$ be the resulting rule. $FSTA(c)$ has up to 4 variables (*FSTA* will add *one* variable to each rule in table 3) introduced after the translation. When instantiate these variables, only the (at most 2N) entities that appear in the *issuer* and *subject* field of basic credentials in C need to be considered, and only the (at most N) *trusted scope* that appears in the *cau* constraint of type-II credentials need to be considered. So there are O(N) ways to instantiate each variable. For each rule there are at most $O(N^4)$ ways to instantiate it, and there are at most 3N (*FSTA* will translate each credential of type-II or type-III into 3 Datalog rules) rules in $FSTA(C)$. Therefore the size of the ground program is $O(N^5)$. ∎

5 Implementing Issues

We use SICStus Prolog (SICSP) [6] as Datalog inference engine. SICSP supports negation as failure (NSF), for example, the NSF form of a predicate P is "\+P". The compound terms in SICSP make it very convenient to express roles, authorities and capabilities in REAL. We use list structure to express role intersection. For example, we design the predicate *auths_ts* using three SICSP rules:

❷*auths_ts*(_, *). ❶*auths_ts*(_, []). ❷*auths_ts*(X, [r(Y,R)|TS]) :- *auth*(Y, X, r(Y,R)), *auths_ts*(X, TS).

We have implemented TMB (Trust Management Broker), a new-style authorization management middleware based on a CORBA [17] platform, using REAL as its authorization language and policy compliance checking engine. We are now designing the credential distribution and revocation schemes.

6 Related Work and Future Extensions

Delegation is attracting much research interest all along. In this paper, we contribute in two aspects, delegation classification and constraint on delegation.

Delegation Classification. We classify delegation based on two types of privileges being delegated, i.e., authority and capability, and clearly describe the application scenarios for the two kinds of delegations. E. Barka and R. Sandhu focus on the human to human form of delegation [4], classifying delegation by the characteristics such as such as permanence, totality, levels of delegation, etc. Their work is limited to management level and belongs to some kind of delegation of authority. B.S. Firozabadi distinguishes between delegation as creation of new privileges and delegation by proxy [2], which can be regarded as the rudiment of delegation of authority and delegation of capability. But their classification only based on whether the delegatee receives his own privilege or not, and they only realize the first type of delegation.

Constraint on Delegation. It seems a contradiction that using delegation to provide more flexible and scalable authorization management while at the same trying to restrict it. The key point is to find the balance. O. Bandmann proposed constrained delegation model [14], using regular expressions to constrain the shape of delegation trees. The model is essentially centralized because it uses group information to restrict the delegation tree, which is stored in a central authorization server. Furthermore, the author does not give the computational model. N. Li propose to use delegation depth to control the propagation of delegation in DL [13]. Delegation depth is oppugned by some authors [3] and may not reflect the trustworthiness in delegation or authorization (e.g. delegation within one domain for several steps may be safer than one-step delegation across different domains). We use trusted scope to restrict the propagation scope of delegated privileges, which has following merits,

(1) Security administrators can directly control the *AS* of authorities without caring about delegation process. In practical security systems, administrators must be careful to set *AS* for delegated authorities, since *AS* can be strongly affected (or even controlled) by the issuers of the roles in *AS*.

(2) The capabilities coming from its owner to the finial requester through a delegation chain, every entity in the chain must sign a type-IV credential for the authorization checking in the server. The *FPS* and *BPS* of delegated capability allow both user and server to control the use of delegated capabilities. This is very applicable for the delegation policies in distributed object systems [17].

(3) We distinguish between authorization and delegation of authority. Many authorization models do not distinguish this, such as SPKI [3] and KeyNote [10]. In these systems, entities will be authorized with the permission by default if they are delegated with the corresponding authority. This is unreasonable in many scenarios. REAL support constraint on *AS* of delegated authorities, so even if an entity has the authority of some permission, it still can not be authorized with the permission if it is outside of the *AS* of this authority.

Future Extensions 4.4. REAL has been influenced by RT [12]. It borrows the idea of ABAC (attribute-based access control) and the concept of role activations in RT. But REAL can express more flexible delegation policies (credentials of type-II and IV), more scalable authorization policies (type-I credentials) and authorization of capabilities (type-III credentials), as well as the constraints to restrict the propagation of delegated privileges.

We can extend the constraint structure in REAL to support more flexible trusted scope (now we only use role intersection to express trust scope) and different types of constraints, such as time and trust level constraints [16]. We can also extend subject structure to support separation of duty policies. Future research work includes credential retrieval and credential revocation. We are now designing a revocation checking algorithm based on *FSTA* semantics to enforce different revocation policies.

7 Conclusions

We introduce REAL framework, a Role-based Extensible Authorization Language framework for open decentralized systems. REAL supports delegation of authority and delegation of capability, covering authorization activities at management level and request level. We also propose to use trusted scope to prevent undesirable propagation of authorizations in the delegation models. Using simple credential forms, REAL provides more flexible and scalable authorization and delegation policies while capable of restricting the propagation scope of delegations. REAL ensures the privacy of policy storage and policy transmission in authorization process based on public cryptography. REAL can be used to address the policies for proxy mechanisms in Globus [5], all kinds of delegation policies in CORBA CSIv2 specification [17] and the distributed authorization policies for virtual organization in data grid [8].

Acknowledgements

The authors would like to thank Hai-ya Gu, Jian-qiang Hu and Ning-hui Li for their helpful discussions and the anonymous reviewers for their valuable comments.

References

1. Butler Lampson, Martin Abadi, Michael Burrows, and Edward Wobber. Authentication in distributed systems: Theory and practice. ACM Transactions on Computer Systems, 10(4):265–310, November 1992.
2. B. S. Firozabadi, M. Sergot, and O. Bandmann. Using Authority Certificates to Create Management Structures. In Proceeding of Security Protocols, 9th International Workshop, Cambridge, UK, April 2001. Springer Verlag. In press.
3. C. M. Ellison, B. Frantz, B. Lampson, R. Rivest, B. M. Thomas, and T. Ylonen. SPKI Certificate Theory. IETF RFC 2693, 1998.
4. Ezedin Barka and Ravi Sandhu. Framework for Role-Based Delegation Models. In Proceedings of 16th Annual Computer Security Application Conference, New Orleans, LA, December 11-15 2000, pages 168-176
5. Grid Computing - Making the Global Infrastructure a Reality, Edited by Fran Berman, Geoffery Fox, Tony Hey, John Wiley & Sons Ltd, The Atrium, Southern Gate, Chichester, West Sussex PO19 8SQ, England, 2003
6. Intelligent Systems Laboratory, Swedish Institute of Computer Science, SICStus Prolog User's Manual, Release 3.11.1, February 2004

7. Jonathan S. Shapiro, Jonathan M. Smith, David J. Farber, EROS: a fast capability system, 17th ACM Symposium on Operating Systems Principles (SOSP'99).
8. L. Pearlman, C. Kesselman, V. Welch, I. Foster, S. Tuecke, The Community Authorization Service: Status and Future, CHEP03, March 24-28, 2003, La Jolla, California
9. M. Gasser, E. Mcdermott, An architecture for practical delegation in a distributed system. In Proceedings of the IEEE Symposium on Security and Privacy (May 1990), pp. 20-30.
10. M. Blaze, J. Feigenbaum, John Ioannidis, and Angelos D. Keromytis. The KeyNote trust-management system, version 2. IETF RFC 2704, September 1999.
11. M. Blaze, Joan Feigenbaum, John Ioannidis, and Angelos D. Keromytis. The role of trust management in distributed systems. In Secure Internet Programming, volume 1603 of Lecture Notes in Computer Science, pages 185–210. Springer, 1999.
12. Ninghui Li, John C. Mitchell, and William H. Winsborough. Design of a role-based trust management framework. In Proceedings of the 2002 IEEE Symposium on Security and Privacy, pages 114-130. IEEE Computer Society Press, May 2002.
13. Ninghui Li, Benjamin N. Grosof, and Joan Feigenbaum. Delegation logic: A logic-based approach to distributed authorization. ACM Transaction on Information and System Security (TISSEC), February 2003.
14. Olav Bandmann, Mads Damy, Babak Sadighi Firozabadi, Constrained Delegation, Proceedings of the 2002 IEEE Symposium on Security and Privacy (S&P'02)
15. Ravi Sandhu, Venkata Bhamidipati and Qamar Munawer, "The ARBAC97 Model for Role-Based Administration of Roles", ACM Transactions on Information and System Security, Volume 2, Number 1, February 1999, pages 105-135.
16. S. Schwoon, S. Jha, T. Reps, S. Stubblebine, On Generalized Authorization Problems, Proceedings of the 16th IEEE Computer Security Foundations Workshop (CSFW'03)
17. The Common Object Request Broker: Architecture and Specification, Object Management Group, July 2002 Version 3.0
18. Vijay Varadharajan, Philip Allen, Stewart Black. An Analysis of the Proxy Problem in Distributed systems. IEEE Symposium on Research in Security and Privacy. Oakland, CA 1991.
19. Van Gelder, A., Ross, K.A., Schlipf, J.S. The Well-Founded Semantics for General Logic Programs. JACM 38(3): 620~650
20. W. Wulf, E. Cohen, W. Corwin, A. Jones, R. Levin, C. Pierson, and F. Pollack. HYDRA: The kernel of a multiprocessor operating system. Communications of the ACM, 17(6):337–345, June 1974.

A Novel Hierarchical Key Management Scheme Based on Quadratic Residues

Jue-Sam Chou[1], Chu-Hsing Lin[2], and Ting-Ying Lee[2]

[1] Department of Information Management, Nanhua University Chiayi,
622 Taiwan, R.O.C
`jschou@mail.nhu.edu.tw`
[2] Department of Computer Science and Information Engineering,
Tunghai UniversityTaichung, 407 Taiwan, R.O.C
`chlin@mail.thu.edu.tw`

Abstract. In 1997, Lin [1] proposed a dynamic key management scheme using user hierarchical structure. After that, Lee [2] brought to two comments on Lin's method. In 2002, Lin [3] proposed a more efficient hierarchical key management scheme based on Elliptic Curve. Lin's efficient scheme solves the weaknesses appearing in Lee's scheme in [1]. In this paper, we further use Quadratic Residues (Q.R.) theorem to reduce the computing complexity of Lin's method.

1 Introduction

Assume that in a hierarchical key management scheme, there exists n groups, $S_1, S_2, ..., S_n$ with a relationship "$\succ$" in a set S. If $S_j \succ S_i$, it represents group S_j has a higher privilege than group S_i. This means that S_j can derive the group secret key k_i of S_i, but not vice versa.

In 2002, Lin proposed an efficient hierarchical key scheme using Elliptic Curve cryptography (ECC). The newer scheme reduces the computing complexity and solves the two comment problems proposed by Lee [2]. The two comments which will be stated in Section 2 are aimed at the scheme proposed by Lin [1] in 1997.

2 Background

In this section, we will introduce both Lin's schemes, Lee's two comments on Lin's dynamic key management scheme, and in the last, we also describe quadratic residue theorem used in our scheme.

2.1 Lin's Scheme (Proposed in 2002)

In this scheme, CA is responsible for generating all the related parameters. There are three phases in it listed as follows:

2.1.1 Group Key Generation Phase

Step 1: CA must choose an elliptic curve EC over Z_p and a generation point $G \in EC(Z_p)$, then find a large prime q satisfying $q \times G = O$.

Step 2: CA computes its public key $P = K \times G$, where K is the private key selected by CA in advance.

Step 3: Each group S_i selects a random number $k_i \in [1, q-1]$ as his group secret key, then computes $p_i = k_i \times G = (x_i, y_i)$ as its corresponding group public key. We call group secret key as group key for an abbreviation throughout the rest of this article.

Step 4: Each group encrypts this key pair by using CA's public key P and then sends $E_p(k_i)$ to CA.

Step 5: CA decrypts message $E_p(k_i)$ by his private key K. Finally, CA computes the related parameters r_{ji} for each relation $S_j \succ S_i$ according to the following equation.

$r_{ji} = X((k_j \oplus h(x_i \parallel y_i)) \times G) \oplus k_i$, where $h(\ )$ denotes a one-way hash function and $X(\ )$ is defined in equation (1).

$$X(p(x_i, y_i)) = x_i \oplus y_i, \text{ where } p_i(x_i, y_i) \text{ is a point in } EC(Z_p) \qquad (1)$$

2.1.2 Group Key Derivation Phase

For the group relation $S_j \succ S_i$, group S_j requests parameter r_{ji} and p_i from CA and derives k_i with the parameters r_{ji}, k_j and p_i by using the following equation.

$$k_i = X((k_j \oplus h(x_i \parallel y_i)) \times G) \oplus r_{ji} \qquad (2)$$

2.1.3 Group Key Modification Phase

If group S_i wants to modify his group key k_i, S_i will generate a new key $k_i^* \in [1, q-1]$ and send $E_p(k_i^*)$ back to CA. CA will then decrypt the message $E_p(k_i^*)$ and compute a new related parameter r_{ji}^*.

2.2 Lee's Two Comments

In 1997, Lin proposed a scheme based on discrete logarithm, he used the equation $r_{ji} = (Z^{k_j \oplus ID_i} \bmod p) \oplus k_i$ to generate the related parameter r_{ji}. Afterward, Lee proposed two comments on Lin's scheme. We describe the comments as follows:

Comment 1: What happens if the old used group key is obtained by an attacker after it has been changed and discarded.

Assume that a group S_i changes his old group key k_i to a new one k_i^*. If the attacker can obtain the old group key k_i, he can perform the following two steps and obtain the new group key k_i^*.

Step 1: The attacker computes $r_{ji} \oplus k_i$ to get $Z^{k_j \oplus ID_i} \bmod p$.
Step 2: In order to obtain S_i's new group key k_i^*, the attacker computes $(Z^{k_j \oplus ID_i} \bmod p) \oplus r_{ji}^*$, where r_{ji}^* is $(Z^{k_j \oplus ID_i} \bmod p) \oplus k_i^*$.

The reason why this comment can work successfully is that the value $Z^{k_j \oplus ID_i} \bmod p$ is never changed when group S_i modifies its group key k_i, so the attacker can perform a computation like Step 1 to get $Z^{k_j \oplus ID_i} \bmod p$. Then, he can obtain $r_{ji}^* (= (Z^{k_j \oplus ID_i} \bmod p) \oplus k_i^*)$ from CA and compute $(Z^{k_j \oplus ID_i} \bmod p) \oplus r_{ji}^*$ to get the new group key k_i^*.

Comment 2: *A few bits difference between the IDs of any two groups will suffer a sibling attack.*

Assume ID_i of group S_i and ID_l of group S_l have only a few bits difference and they have the same parent S_j, then S_i can easily obtain $(Z^{k_j \oplus ID_l} \bmod p)$ by comparing ID_i with ID_l. Thus, S_i can crack the group key k_l of S_l by computing $(Z^{k_j \oplus ID_l} \bmod p) \oplus r_{jl}$.

2.3 Quadratic Residue Theorem

We use Q.R. theorem as our method's basis; thus, we briefly introduce the feature of Q.R. theorem [5] first. Under the assumption that there exist two large primes p and q such that $n = p \cdot q$. If $x^2 \equiv a \pmod{n}$ has a solution, then a is called a quadratic residue $\bmod n$. The symbol QR_n denotes the set of all quadratic residue number in $[1, n-1]$. It is computationally infeasible to solve x by just knowing a and n, because of the difficulty of factoring n [12].

3 Proposed Scheme

In this section, we propose a new scheme based on quadratic residue theorem. Like in Lin's scheme except for the initial step, there are also three phases in our method described as follows:

3.1 Group Key Generation Phase

Initial Step: In this step, CA pre-generates some parameters for the system construction. CA generates a secret key SK, sends it to each group S_i through a secure channel and chooses two large primes p and q, satisfying $(p+1)|4$ and $(q+1)|4$ respectively, then computes $n = pq$. CA also computes all QR_p and QR_q then encrypts all of QR_p and QR_q using his secret key SK. Finally, CA sends $E_{SK}(QR_p, QR_q)$ to each group.

Step 1: Each group S_i, $i = 1, 2, ..., n$, decrypts the message $E_{SK}(QR_p, QR_q)$, then selects a group key $k_i \in QR_p \cap QR_q$. He encrypts k_i using SK, then sends $E_{SK}(k_i)$ to CA.

Step 2: CA decrypts the encrypted message $E_{SK}(k_i)$ and obtains k_i. Then, CA computes a_i for each group S_i by equation (3).

$$a_i = k_i^2 \bmod n \qquad (3)$$

The relation between k_i and a_i is one-to-one. That is, whenever we are given a $QR_n (= a_i)$, we can uniquely determine k_i. We prove that the key k_i can be determined uniquely when given a_i by contradiction. If there are two keys, k_1 and k_2, satisfying $k_1^2 \equiv a_i \bmod n$ and $k_2^2 \equiv a_i \bmod n$ correspondingly. Subtract both sides of the two equations gives $k_1^2 - k_2^2 \equiv 0 \bmod n$, and using factoring yields $(k_1 + k_2)(k_1 - k_2) \equiv 0 \bmod n$. We can easily see that $(k_1 + k_2) \neq 0 \bmod n$, as required.

Step 3: For each group relation $S_j \succ S_i$, CA computes the related parameters r_{ji} by equation (4).

$$r_{ji} = h(k_j \oplus a_i) \oplus k_i \qquad (4)$$

Step 4: CA maintains all parameters r_{ji} and a_i on each group S_i's requests.

3.2 Group Key Derivation Phase

For the group relation $S_j \succ S_i$, S_j can derive k_i by performing equation (5).

$$k_i = h(k_j \oplus a_i) \oplus r_{ji} \qquad (5)$$

Basically, there are three steps in this phase, we describe it as follows:

Step 1: S_j requests the parameters a_i, r_{ji} from CA.
Step 2: CA sends $E_{SK}(a_i, r_{ji})$ to S_j.
Step 3: S_j derives S_i's group key k_i by equation (5).

3.3 Group Key Modification Phase

Step 1: S_i selects a new group key $k_i^* \in QR_p \cap QR_q$ and sends $E_{SK}(k_i^*)$ to CA.
Step 2: CA will then compute a new Q.R. $a_i^* (= (k_i^*)^2 \bmod n)$ for group S_i.
Step 3: CA computes the related parameter r_{ji}^* by using equation (6).

$$r_{ji}^* (= h(k_j \oplus a_i^*) \oplus k_i^*) \qquad (6)$$

Step 4: CA uses r_{ji}^*, a_i^* to replace r_{ji}, a_i respectively and maintains them.

4 Example

4.1 Group Key Generation Phase

Initial Step: CA selects a secret key SK, sends it to each group S_i through a secure channel and chooses two large primes $p = 19$ and $q = 23$, then computes $n = p \cdot q = 19 \cdot 23 = 437$. CA also computes $QR_{19} = \{1, 4, 5, 6, 7, 9, 11, 16, 17\}$ and $QR_{23} = \{1, 2, 3, 4, 6, 8, 9, 12, 13, 16, 18\}$ then encrypts all of QR_{19} and QR_{23} using his secret key SK and sends $E_{SK}(QR_{19}, QR_{23})$ to each group.

Step 1: Group S_1, decrypts the message $E_{SK}(QR_{19}, QR_{23})$, then selects a group key $k_1 = 9$. He sends $E_{SK}(k_1)$ to CA.
Step 2: CA decrypts the encrypted message $E_{SK}(k_1)$ and obtains k_1. Then, CA computes $a_1 = k_1^2 \bmod 437 = 9^2 \bmod 437 = 81$ for group S_1.
Step 3: We assume that the group key k_2 selected by S_2 is 6 and the relation between S_1 and S_2 is $S_2 \succ S_1$. CA computes the related parameters $r_{21} = h(k_2 \oplus a_1) \oplus k_1 = h(6 \oplus 81) \oplus 9$.
Step 4: CA maintains all parameters r_{21} and a_1 on each group S_2's requests.

4.2 Group Key Derivation Phase

If group S_2 want to derive group S_1's group key k_1, he must do some steps as fallows:

Step 1: S_2 requests the parameters a_1, r_{21} from CA.
Step 2: CA sends $E_{SK}(a_1, r_{21})$ to S_2.
Step 3: S_2 derives S_1's group key $k_1 = r_{21} \oplus h(k_2 \oplus a_1) = 9$.

4.3 Group Key Modification Phase

Step 1: S_1 selects a new group key $k_1^* = 4$ and sends $E_{SK}(k_1^*)$ to CA.
Step 2: CA will then compute a new Q.R. $a_1^* = 16$ for group S_1.
Step 3: CA computes the related parameter $r_{21}^*(= h(k_j \oplus a_i^*) \oplus k_i^*) = h(6 \oplus 16) \oplus 4$.
Step 4: CA uses r_{21}^*, a_1^* to replace r_{21}, a_1 respectively and maintains them.

5 Security of Analysis

In this section, we will propose some attacks on our scheme. After analyzing, we find that each attack can't work successfully.

5.1 Attack 1: Contrary Attacks

We assume that there exists a group relation $S_j \succ S_i$. If S_i wants to obtain k_j, he will perform the calculation as equation (7), but k_j is protected by the one-way hash function.

$$h(k_j \oplus a_i) = r_{ji} \oplus k_i \qquad (7)$$

5.2 Attack 2: Interior Collecting Attacks

If there exists a user U in group S_i which has m parents, namely S_j, S_{j+1},..., S_{j+m}. The user U can collect the related parameters r_{ji}, $r_{j(i+1)}$,..., $r_{j(i+m)}$ of all his parent groups as well as his own. Then, he can get $h(k_{j+v} \oplus a_i), v = 1, 2,..., m$, but he has to face the difficulty in trying to reverse the one-way hash function. Obviously, it is infeasible to find any relationship in these values.

5.3 Attack 3: Exterior Collecting Attacks

If an outside attacker wants to perform a process like Attack 2, it is computationally infeasible for him to crack the group key k_j. Since, he not only has no group key k_i but also need to reverse the one-way hash function.

5.4 Attack 4: Collaborative Attacks

Suppose that a group S_j has two child groups, S_i and S_l, and suppose an user in group S_i and another user in group S_l want to crack S_j's group key k_j collaboratively. Although they can obtain r_{ji} and r_{jl} by equation (8) and equation (9) from CA successfully, they will face the same problem that exists in Attack 2 eventually.

$$r_{ji} = h(k_j \oplus a_i) \oplus k_i \tag{8}$$

$$r_{jl} = h(k_j \oplus a_l) \oplus k_l \tag{9}$$

5.5 Attack 5: Sibling Attacks

Another situation is that when group S_i, with a parent group S_j, wants to crack the group key k_l of S_l which has the same parent as group S_j, and that the attacker S_i only owns r_{ji} computed by equation (8), under this circumstance, S_i will then face a more difficult problem than that exists in Attack4. Thus, he can't crack the group key k_l of S_l.

5.6 Attack 6: Using Lee's Comments on Lin's Scheme

Comment 1: What happens if the old used group key is obtained by an attacker after it has been changed and discarded.

The value a_i in equation $r_{ji} = h(k_j \oplus a_i) \oplus k_i$ will be changed to a_i^* each time when group S_i modifies its group key k_i to a new one k_i^*. So that the value $h(k_j \oplus a_i)$ will be changed as well. Therefore, the weakness mentioned in comment 1 will not exist.

Comment 2: A few bits difference between IDs of two groups will suffer a sibling attack.

For identity of each group doesn't be used in the equation $r_{ji} = h(k_j \oplus a_i) \oplus k_i$ in our method, so that the weakness doesn't exist as well.

6 Time Complexity Comparison

In this section, we discuss the performance of our proposed scheme in worst case. For simplicity, we first define some notations as follows:

T_{MUL} : the time needed by a 1024-bit modular multiplication.
T_H : the time needed by a 160-bit one-way hash function operation.
T_{EC_MUL} : the time needed by the elliptic curve multiplication with 160-bit multiplier.

According to [7], we known that $T_{EC\_MUL} \approx 29 T_{MUL}$.

We list the time needed in the three phases of each method, Lin's and ours in Table 1. We can see that our scheme has lower time complexity than Lin's schemes. Especially in group key derivation phase, our scheme only needs nT_H. For group S_i just need to perform the equation $r_{ji} \oplus h(k_j \oplus a_i)$. In group key generation phase, CA must generate each group's a_i by computing $k_i^2 \bmod n$ and compute the related parameter r_{ji} by performing the equation $r_{ji} = h(k_j \oplus a_i) \oplus k_i$, so the time complexity needed is $n(T_{MUL} + T_H)$. In group key modification phase, CA performs the equation as does in the group key generation phase, so the time complexity needed is also $n(T_{MUL} + T_H)$. Therefore, our scheme is more efficient than Lin's scheme.

Table 1. Time complexity comparisons

	Lin's scheme	Our scheme
Group key generation phase	$(58n+29)T_{MUL}$	$n(T_{MUL}+T_H)$
Group key derivation phase	$(58n+29)T_{MUL}$	nT_H
Group key modification phase	$(58n+29)T_{MUL}$	$n(T_{MUL}+T_H)$

7 Conclusion

In our scheme, we use the group key to compute a_i, a Q.R. modulo of n. The value can be made public, but it is computationally infeasible to compute the group key k_i inversely for the hard factorization problem. Furthermore, the time complexity of computing a_i just needs one multiplication operation. Thus, our scheme is far more efficient than Lin's schemes. For more secure, we can choose two large prime numbers additionally and do one more computation of square in equation (3) by adding some parameters [13].

References

1. C. H. Lin, "Dynamic key management schemes for access control in a hierarchy," *Computer Communications*, 20, 1997, pp.1381-1385.
2. N. Y. Lee and T. Hwang, "Research note Comments on 'dynamic key management schemes for access control in a hierarchy'," *Computer Communications*, 22, 1999, pp.87-89.
3. C. H. Lin and J. H. Lee, "An Efficient Hierarchical Key Management Scheme Based on Elliptic Curves," *Journal of Interdisciplinary Mathematics*, Vol. 5, No. 3, October 2002, pp.293-301.

4. C. C. Chang, S. M. Tsu, "Remote scheme for password authentication based on theory of quadratic residues," *Computer Communications*, 18, 1995, pp.936–942.
5. K. H. Rosen, "Elementary Number Theory and Its Applications," *Addison-Wesley*, Reading, MA (1988).
6. K. J. Tan, H. W. Zhu, "Research note A conference key distribution scheme based on the theory of quadratic residues," *Computer Communications*, 22, 1999, pp.735–738.
7. N. Koblitz, A. Menezes and S. Vanstone, "The State of Elliptic curve Cryptography," *Design, Codes and Cryptography*, 19, 2000, pp.173-193.
8. Draft FIPS 180-2, Secure Hash Standard (SHS), U.S. Doc/NIST, May 30, 2001.
9. N. Koblitz, "A Course in Number Theory and Cryptography," *New York, NY:Spring-Verlag*, Second edition, 1994.
10. A. Menezes, "Elliptic curve Public Key Cryptosystems," *Kluwer Academic Publishers*, 1993.
11. L. Harn, L. Y. Lin, "A Cryptographic Key Generation Scheme for Multi-Level Data Security," *Computer and Security* 9, 1990, pp.539-546.
12. W. Patterson, "Mathematical Cryptology for Computer Scientists and Mathematicians," *Rowman*, 1987.
13. C. I. Fan, and C. L. Lei, "Low Computation Partially Blind Signatures for Electronic Cash," to appear in *IEICE Trans. on Fundamentals of Electronics, Communications and Computer Sciences*, Vol. E81-A, No. 5, 1998.

Soft-Computing-Based Intelligent Multi-constrained Wavelength Assignment Algorithms in IP/DWDM Optical Internet[*]

Xingwei Wang[1], Cong Liu[1], and Min Huang[2]

[1] Computing Center, Northeastern University, Shenyang, 110004, China
wangxw@mail.neu.edu.cn
[2] School of Information Science & Engineering, Northeastern University,
Shenyang, 110004, China

Abstract. Wavelength assignment is one of the most important research areas in IP/DWDM optical Internet. Taking multiple constraints into account, including cost, power, network performance etc., the wavelength assignment is made much fit to the actual network configurations, however, the problem complexity increases correspondingly, leading to the adoption of a layered solution framework. As each layer sub-problem is NP complete, soft-computing algorithms, including Simulated Annealing Algorithm (SAA) and Simulated-annealing-Genetic Algorithm (SGA), and heuristic algorithms are used jointly to design the intelligent multi-constrained wavelength assignment algorithms respectively. Simulation results have shown that these proposed algorithms are feasible and effective.

1 Introduction

IP/DWDM optical Internet is considered to be a promising candidate for the Next Generation Internet (NGI) backbone. Routing and Wavelength Assignment (RWA) is one of the most important means of improving resource utilization in IP/DWDM optical Internet [1]. A lot of static WA algorithms, using integer linear programming and graph-coloring methods etc., and dynamic WA algorithms, using first fit, least used, most used, min-product, least loaded, max-sum strategies etc., have been presented [2-7]. However, most of them solve WA problem under the relatively ideal conditions, often only taking the network topology and wavelength resource into account. In order to make the wavelength resource allocation much fit to the actual network configurations, multiple constraints should be considered, such as cost, power [8], network performance etc. In this paper, the cost considered concludes Wavelength Converter (WC) [9] cost and wavelength amplifier cost.

---

[*] This work is supported by the National Natural Science Foundation of China under Grant No. 60473089, 60003006 (jointly supported by Bell Lab Research China) and No. 70101006; the National High-Tech Research and Development Plan of China under Grant No. 2001AA121064; the Natural Science Foundation of Liaoning Province in China under Grant No. 20032018 and No. 20032019; the Modern Distance Education Engineering Project of China MoE.

From this new viewpoint, an intelligent solution is proposed in this paper, and more practical and better configuration can be obtained under the multi-constraints. Due to the problem's complexity, a four-layered solution framework is adopted, and optimal WA solution can be approached by the cooperation among these four layers. As each layer sub-problem is NP complete [10], two soft-computing algorithms, i.e., Simulated Annealing Algorithm (SAA) [11] and Simulated-annealing-Genetic Algorithm (SGA) [10], and heuristic algorithm are used. Simulation results have shown that the proposed algorithms are feasible and effective.

2 Problem Formulations

2.1 Basic Assumptions

IP/DWDM optical Internet can be modeled as a graph $G(V,E)$, where V is the set of nodes representing wavelength routers or Optical Cross-Connect (OXC) and E is the set of edges representing optical fibers.

Assume that each node is equipped with enough amounts of optical receivers and transmitters. OXC can be classified into Wavelength Selective Cross-Connect (WSXC) and Wavelength Interchanging Cross-Connect (WIXC). In addition, assume that the attenuation of signal power only exists in optical fibers and all other optical equipments are active.

2.2 Related Concepts and Parameters

Node pair set is denoted by $Z = \{(z_1, z_2) | z_1 \in V \land z_2 \in V\}$. Minimum Number of Hop (MNH) of node pair $z \in Z$ equals the minimum number of the spanned edges minus 1, denoted by $MNH(z, G(V,E)) = m(z), \forall z \in Z$. The lightpath set of $z \in Z$ is denoted by $A_{z,e} = \{p | l(p) \leq m(z) + e\}$, where p represents one of the lightpaths of $z \in Z$ with the non-looped edges, $l(p)$ represents the number of the spanned edges along p, and e is called path-extending width, $e = 0,1,2,\ldots$. Each fiber can carry W wavelengths, labeled $1 \sim W$ correspondingly. Demand matrix represents requests arrived at the network, and demand number represents the total number of the arrived requests.

2.3 Solution Framework Descriptions

Under multiple constraints, a four-layered solution framework is designed, composed of WC, Pure RWA (PRWA), EDFA and power layers from top to down (see Fig.1). Correspondingly, all these four layers have their own optimization objectives, and cooperate through them. For WC layer, the goal is to minimize the number of WCs deployed in the network, at the same time approach network performance optimization. For PRWA layer, the goal is to establish lightpaths as much as possible, based on demand matrix and WC layer's result. For EDFA layer, the goal is to minimize the number of EDFAs on links based on PRWA layer's result. For power layer, the goal is to optimize the lightpath establishment further based on the above three layers' results, with power limitations in mind.

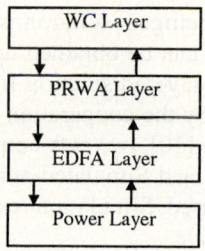

Fig.1. A four-layered solution framework

3 Soft-Computing-Based Multi-constrained WA Algorithms

3.1 SAA-Based Intelligent Algorithm

Intelligent Algorithm for WC, EDFA and Power Layers

Solution Expression. Adopt binary encoding in WC layer, each bit of the binary cluster indicates whether the corresponding node is configured with WC or not, the length of solution S equals the total node number in the network topology. Adopt floating encoding in EDFA layer, the $NUM_{EDFAMAX}$ amount of EDFAs are pre-configured on each link, every pre-configured EDFA has a corresponding item in S, its value representing its position on the link. Adopt floating encoding in power layer, each item represents the initial power for the corresponding connection, valued within $[0, P_{max}]$, P_{max} denotes the maximum power of transmitter, and the length of the solution equals demand number.

Neighborhood Structure. In WC layer, do reverse operation on one bit or two bits of the solution, corresponding to single node or double nodes. Call the former fine-tuning and the latter coarse-tuning. Use fine-tuning at the early stage, and then do coarse-tuning, so that the algorithm efficiency is improved. In EDFA layer, using the formed template as the reference point, position of each EDFA can float forwards or backwards. The floating range is $[0, LEN_{AVE}]$, where LEN_{AVE} is the average distance between adjacent EDFAs on the link. Then, using it as a framework, get a new solution randomly for EDFA deployment. The only difference between power layer and EDFA layer is that the floating range in the former is $[0, P_{max}]$.

Fitness Function. In WC layer, define the following fitness function: $G_1 = a_1 \sum_{0 < i \leq snum} S_i \times Cost_{wc} + b_1 \times G_2$, where a_1 and b_1 are two coefficients, representing cost and network performance weights respectively, $Cost_{wc}$ represents the cost of WC, $snum$ is the length of the solution, S_i is the value of the ith item, and G_2 represents the result in PRWA layer. In EDFA layer, define the following fitness function: $G_3 = a_3 \sum_{0 < j \leq snum} check(j) \times Cost_{EDFA} + b_3 \times G_4$, where a_3 and b_3 are two coefficients, representing cost and power performance weights respectively, $Cost_{EDFA}$ represents the cost of wavelength amplifier, j is the sequence number of item in the solution, and G_4 represents the result in power layer. Define $check(j)$ as follows: if $0 < S_j \leq$ the length of the corresponding link, its value is 1, otherwise 0. If the value of the jth item is within the range of the link length, the cost of the corresponding EDFA should be considered. In power layer, define the following fitness function: $G_4 = Getnum(S)$, where $Getnum(S)$ represents the maximum connection number obtained under the corresponding power distribution strategy with regard to all items of the solution. The procedure of getting $Getnum(S)$ is described as follows:

Step1. Compute path gain for every connection, sort them in descending order and put them into set P; initialize the number of the successfully established connections to be 0.
Step2. If P is not empty, take a connection with maximum path gain out of it, go to Step 3; otherwise, return the number of the successfully established connections, the algorithm ends.
Step3. Add this connection to the network topology.
Step4. Judge whether fiber or EDFA power saturation along this added connection emerged or not. If emerged, remove that connection from the network topology; otherwise, conserve it in the network topology, and increase the number of the successfully established connections by 1. Go to Step 2.

Generating Initial and Feasible Solution. In WC layer, if the scale of the network topology is smaller, initialize all the items of the initial solution to be 0, otherwise initialize them randomly.

In EDFA layer, the maximum number of EDFAs on each link is determined as follows: $NUM_{EDFAMAX} = \lceil LEN_{MAX}/(LEN_{AVE}/\alpha) \rceil + 1$, where LEN_{MAX} represents the maximum link length in the network and α is the signal attenuation ratio in optical fiber. Then, $P_{sigmax} = MIN\{NUM_{max\lambda} \times P_{max}, P_{fbmax}\}$, where P_{sigmax} is the maximum total power of the transmitted signal in optical fiber, P_{fbmax} is the maximum total power of optical fiber and $NUM_{max\lambda}$ is the maximum number of wavelengths carried by optical fiber. Thus, $L \times NUM_{EDFAMAX}$ indicates the length of the solution.

As EDFAs are configured according to the requirement of the longest link, not all the links need such amount of EDFAs. When the value of the item is bigger than the length of its link, it means that the corresponding EDFA has been deployed beyond the link, i.e., that EDFA does not exist actually. The optimization goal is to reduce the number of EDFAs deployed along fibers as much as possible.

EDFAs are configured as inline amplifiers and deployed according to LEN_{AVE} link by link. By this way, an initial solution template is obtained.

In power layer, it is almost the same as that in EDFA layer, and the only difference is that the items' random floating values of the solution are within $[0, P_{max}]$.

Cooling Schedule. The cooling schedules for WC, EDFA and power layers are similar. The initial temperature is determined as follows: $T_{INI} = (T_{max} - T_{min})/c$, $T_{max} = a_{11} \times P_{net} + a_{12} \times Numc \times Cost_{wc} + a_{13}$, $T_{min} = b_{11} \times P_{net} + b_{12}$, where a_{11}, a_{12}, a_{13}, b_{11} and b_{12} are coefficients, valued according to the actual situation, $0 < c < 1$, P_{net}

represents the network performance metric value, denoted by the number of lightpaths with wavelengths successfully assigned in PRWA layer, *Numc* is the total node number in the network, T_{max} and T_{min} represent the maximum and minimum temperature respectively, denoted by the corresponding maximum and minimum cost.

Adopt constant coefficient declining temperature strategy. Here, let *TC* denote the constant coefficient.

Iterative times criterion for each temperature and termination rule are both based on non-improvement rule. Let *MetroMax* be the maximum iterative times under certain temperature and *SAMax* be the maximum times of declining temperature. When temperature declines from T_0 to $T_{\lfloor SAMax/2 \rfloor}$, coarse-tuning is adopted, however, from $T_{\lfloor SAMax/2 \rfloor +1}$ to T_{SAMax}, fine-tuning is adopted.

Heuristic Algorithms for PRWA Layer

Routing Algorithm. The routing criterion is to allocate the least congested lightpath for each request. The algorithm is described as follows:

Step1. According to demand matrix, for $\forall z \in Z$, use extended shortest path algorithm [10] to compute $A_{z,e}$.

Step2. For $\forall z \in Z$, sort lightpaths of $A_{z,e}$ in descending order according to their MNH. Those lightpaths with equal MNH are sorted randomly.

Step3. For $\forall j \in E$, $c_j = 0$, where c_j represents the number of the carried lightpaths by link j.

Step4. Select node pair z from Z.

Step5. $MAX_{load} = \infty$ and $PATH_{load} = \infty$.

Step6. Select one lightpath p from $A_{z,e}$.

Step7. For $\forall j \in p$, $M_c = \text{Maximum}(c_j)$ and $S_c = \sum_{j \in p} c_j$.

Step8. If ($M_c < MAX_{load}$) or ($M_c = MAX_{load}$ and $S_c < PATH_{load}$), $targetP = p$, $MAX_{load} = M_c$, $PATH_{load} = S_c$ and $A_{z,e} = A_{z,e} - \{p\}$.

Step9. If $A_{z,e} \neq \phi$, go to Step6.

Setp10. For $\forall j \in targetP$, $c_j = c_j +1$, $A_{z,e} = \{targetP\}$ and $Z = Z - \{z\}$.

Setp11. If $Z \neq \phi$, go to Step4; otherwise, the algorithm ends.

Then, a ligthpath has been established for every node pair, and all these established ligthpaths have distributed in the network topology as uniformly as possible, avoiding such phenomenon that some lightpaths occupy the same physical link while other physical links are quite idle.

WA Algorithm. WA algorithm is based on the results of WC deployment. Due to no wavelength continuity constraint within paths passing WIXC and due to wavelength continuity constraint within paths only passing WSXC (i.e., passing no WIXC), a method splitting the existing lightpaths for approaching the optimal wavelength assignment is proposed. The so-called splitting is that make lightpath passing WIXCs be split into sub-lightpaths passing no WIXC, then use these sub-lightpaths as the basic units to assign wavelengths. The algorithm is described as follows:

Step1. Select node pair z from Z.
Step2. If OXCs along *targetP* contain x WIXCs, split *targetP* into $x+1$ sub-lightpaths SP_i ($0 < i \le x+1$), replacing *targetP* with all its corresponding SP_i into $A_{z,e}$.
Step3. Sort all SP_i in $A_{z,e}$ in descending order according to their MNH. Those SP_i with equal MNH are sorted randomly.
Step4. For $\forall j \in E$, $W_j = \phi$, where W_j is the set of wavelength sequence number, representing the occupied wavelengths on link j.
Step5. Select one SP_i from $A_{z,e}$.
Step6. Examine W_j in this SP_i, find out the maximum sequence number of the used wavelengths, mark it as w, and $w = w+1$. If $w > W$, there is no wavelength available and the corresponding request is blocked, go to Step8; otherwise, assign wavelength w to this SP_i, $W_j = W_j \cup \{w\}$, and $A_{z,e} = A_{z,e} - \{SP_i\}$.
Step7. If $A_{z,e} \ne \phi$, go to Step5.
Step8. $Z = Z - \{z\}$. If $Z \ne \phi$, go to Step1; otherwise, the algorithm ends.

Using the above algorithm, an initial result can be obtained. However, whether the ligthpath with wavelength assigned can be established actually under the power limitations or not still need to be verified in EDFA layer. Thus, the fitness function in PRWA layer is defined as follows: $G_2 = a_2 \times Num_{con} + b_2 \times G_3$, where a_2 and b_2 are coefficients, whose values can be set according to the actual situation, Num_{con} is the maximum number of the established connections according to demand matrix.

By the way, heuristic algorithms for PRWA layer in the SGA-based solutions are the same as the above ones.

3.2 SGA-Based Intelligent Algorithm

The main differences of the SGA-based intelligent algorithm from the SAA-based one are described as follows.

Traditionally, only the current solution is conserved from the beginning to the end in SAA [11]. In some modified versions of SAA, one variable is added to conserve the best solution up to the present iteration [12], however, some inherent limitations in SAA itself still cannot be well solved. Thus, the population concept and the competition mechanism of Genetic Algorithm (GA) [11] are introduced into SAA to construct SGA. In this paper, based on SAA framework, GA is embedded into the neighborhood structure of SAA. Firstly, generate initial population after a new solution produced by SAA, and then do crossover and mutation operations on it. Use fitness function (refer to section 3.1) to deal with the current population, generate the new population, choose the best chromosome, and return it as the new state for the other SAA operations. Making full use of the available knowledge for search in the solution space, the solution quality is improved further by SGA.

4 Performance Evaluations

A simulation software for the proposed intelligent multi-constrained WA algorithms has been developed, and simulations have been done over some actual network topologies, including NSFNET and CERNET, etc [10].

For SAA-based algorithm, the influence of WC/EDFA cost on network performance is shown in Fig.2. It can be seen that the cost of WC and EDFA has an effect on network performance to some extent. The influence of *SAMax / MetroMax / TC* on network performance and runtime are shown in Fig.3 and Fig.4 respectively. It can be seen that the number of the established connections increases with *SAMax*, *MetroMax* and *TC*, the influence of *SAMax* being much significant. However, increasing tendency of the number of the established connections becomes relatively flat after reaching certain critical point. The runtime increases significantly with *SAMax / MetroMax*, while not obviously with *TC*. Similar evaluations on both EDFA and power layers have also been done and the results are rather satisfied [10].

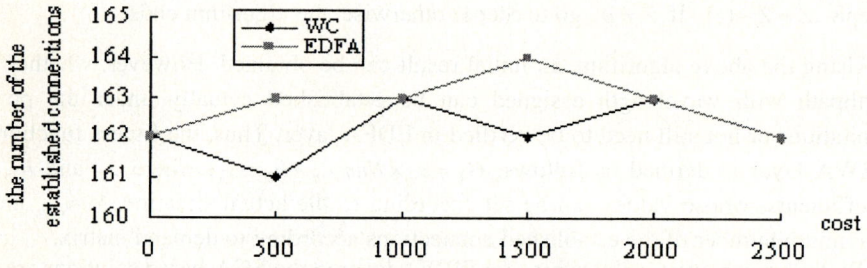

Fig. 2. Influence of WC/EDFA cost on network performance

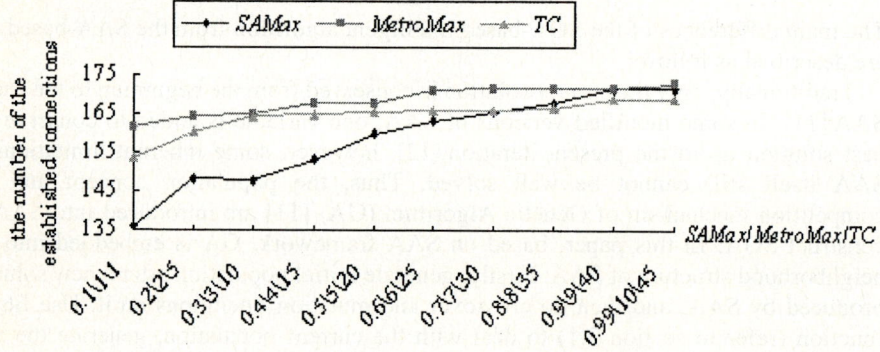

Fig. 3. Influence of *SAMax/MetroMax/TC* on network performance

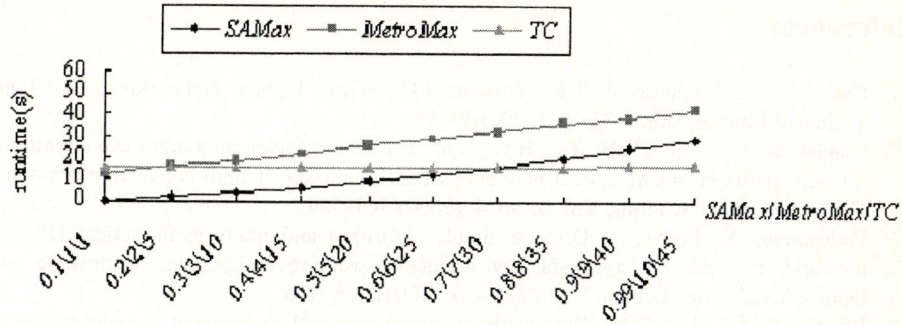

Fig. 4. Influence of *SAMax/MetroMax/TC* on runtime

For SGA-based algorithm, its performance evaluation is similar to the above [10].

Table 1 is the comparisons of the established connection number, deployed WC number, deployed EDFA number and runtime under the same simulation conditions between these two proposed algorithms, using the result of the SGA-based one as the benchmark. It can be seen that the results of the SGA-based one are superior to the SAA-based one in most cases, whereas the former commonly has higher runtime overhead.

Table 1. Performance comparisons between the two proposed algorithms

Established Connection Number (SAA: SGA)	Deployed WC Number (SAA: SGA)	Deployed EDFA Number (SAA: SGA)	Runtime (SAA: SGA)
1.0172:1	1.1429:1	1.0059:1	0.5645:1
0.9701:1	0.7500:1	1.0030:1	0.5703:1
0.9825:1	1.200:1	1.0000:1	0.5699:1
1.0058:1	1.3333:1	1.0029:1	0.5669:1
0.9661:1	0.8750:1	1.0118:1	0.5716:1
0.9181:1	0.8000:1	1.0030:1	0.5533:1
0.9769:1	1.0000:1	1.0088:1	0.5689:1

5 Conclusions

In this paper, WA problem in IP/DWDM optical Internet is solved from a new viewpoint, considering multiple constraints, such as network topology, wavelength, WC cost, EDFA cost and power etc., and helping to guarantee the network performance at the same time. A four-layered solution framework is proposed with soft-computing and heuristic algorithms used jointly to solve it, meeting with user requests as much as possible under these multi-constraints. Simulation results have shown that the proposed algorithms are both feasible and effective.

References

1. Daniel, Y.A., Mohammad, T.F., William, J.G., *et al.*: Optical Networking. Bell Labs Technical Journal, Vol. 3, No. 1. (1998) 39–61
2. Gangxiang, S., Sanjay, K.B., Tee, H.C., *et al.*: Efficient wavelength assignment algorithms for light paths in WDM optical networks with/without wavelength conversion. Photonic Network Communications, Vol. 2, No. 4. (2000) 349–359
3. Podcameni, A., Lopes, J.: Using a simple algorithm and platform in optical DWDM networks for reaching a satisfactory wavelength-routing assignment. Microwave and Optical Technology Letters, Vol. 28, No. 6. (2001) 406–410
4. Bampis, E., Rouskas, G.N.: The scheduling and wavelength assignment problem in optical WDM networks. Journal of Lightwave Technology, Vol. 20, No. 5. (2002) 782–789
5. Ozdaglar, A.E., Bertsekas, D.P.: Routing and wavelength assignment in optical networks. IEEE/ACM Transactions on Networking, Vol. 11, No. 2. (2003) 259–272
6. Kuri, J., Puech, N., Gagnaire, M., *et al.*: Routing and wavelength assignment of scheduled lightpath demands. IEEE Journal on Selected Areas in Communications, Vol. 21, No. 8. (2003) 1231–1240
7. Wang, X.W., Cheng, H., Li, J., Huang, M.: A multicast routing algorithm in IP/DWDM optical Internet. Journal of Northeastern University (Natural Science), Vol. 24, No. 12. (2003) 1165–1168(in Chinese)
8. Ali, M., Ramamurthy, B., Deogun, J.S.: Routing and wavelength assignment with power considerations in optical networks. Computer Networks, Vol. 32, No. 5. (2000) 539–555
9. Chu, X.W., Li, B., Chlamtac, L.: Wavelength converter placement under different RWA algorithms in wavelength-routed all-optical networks. IEEE Transactions on Communications, Vol. 51, No. 4. (2003) 607–617
10. Liu, C.: Research and simulated implementation of multi-constrained wavelength assignment algorithms in IP/DWDM optical Internet. Shenyang: Northeastern University, (2004) (in Chinese)
11. Xing, W.S., Xie, J.X.: Modern optimization computational methods. Beijing: Tsinghua University Press, (1999) (in Chinese)
12. Kang, L.H., Xie, Y., *et al.*: Non-numerical parallel algorithm. Beijing: Science Publishing, (1994) (in Chinese)

Data Transmission Rate Control in Computer Networks Using Neural Predictive Networks*

Yanxiang He[1], Naixue Xiong[1], and Yan Yang[2]

[1] The State Key Lab of Software Engineering, Computer School of Wuhan University,
Wuhan, 430072 P.R. China
yxhe@whu.edu.cn, xiongnaixue@hotmail.com
[2] Department of Computer Science, Central China Normal University, Hubei Wuhan,
430079 P.R. China
Y.Yang@mail.ccnu.edu.cn

Abstract. The main difficulty arising in designing an efficient congestion control scheme lies in the large propagation delay in data transfer which usually leads to a mismatch between the network resources and the amount of admitted traffic. To attack this problem, this paper describes a novel congestion control scheme that is based on a Back Propagation (BP) neural network technique. We consider a general computer communication model with multiple sources and one destination node. The dynamic buffer occupancy of the bottleneck node is predicted and controlled by using a BP neural network. The controlled best-effort traffic of the sources uses the bandwidth, which is left over by the guaranteed traffic. This control mechanism is shown to be able to avoid network congestion efficiently and to optimize the transfer performance both by the theoretic analyzing procedures and by the simulation studies.

1 Introduction

With the rapid development of computer networks, more and more severe congestion problems have occurred. Designing efficient congestion control scheme is, therefore, a crucial issue to alleviate network congestion and to fulfill data transmission effectively. The main difficulty in designing such scheme lies in the large propagation delay in transmission that usually leads to a mismatch between the network resources and the amount of admitted traffic. The crucial issue of the network control is that we should adapt the controllable flows to the changing network environment, so as to achieve the goal of the data transfer and to alleviate network congestion. Congestion is the result of a mismatch between the network resources capacity and the amount of traffic for transmission.

Many control schemes are presented to resolve the network congestion. The paper [1] reviews all kinds of congestion control schemes having been proposed

---

* This research has been supported by National Natural Science Foundation of China under Grant No. 90104005 and by the Key Project of Natural Science Foundation of Hubei Province under Grant No. 2003ABA047.

for computer networks. Among these schemes, the representative one, which is in common use, is the rate-based congestion control (see, e.g., [2-3]). The basic techniques include the Forward Explicit Congestion Notification (FECN) and the Backward Explicit Congestion Notification (BECN) [3-4].

The time delay in data transmission will result in slow transient behavior of buffer occupancy. The responsiveness of the congestion control scheme is crucial to the stability of the whole network system. The non-stability of dynamic network influences the network's performance. To deal with this difficulty, the authors in [5] suggest using the method of fuzzy control to realize the rate-based network congestion control, and the application of heredity algorithm in queue strategy is presented in [6-7]. Furthermore, the recent papers [8-9, 19-22] use a multi-step neural predictive technique to predict the congestion situation in computer networks, but the longer predictive steps has still existed and the effectiveness is greatly limited in existed papers. And yet the responsiveness of the congestion control scheme is crucial to the stability of the whole network system and the relevant performance, this issue is, however, not considered in these works. So this paper aims to improve the predictive scheme. We implement the neural predictive controller at the sources rather than at the switch. This is due to the fact the less prediction horizon usually leads to better accuracy, whereas in the proposed scheme the predictive horizon is linked with the network structure. Under the same circumstance, we use less predictive steps than that in [8-9], this then usually brings forth better performance in terms of predictive accuracy and efficiency.

Our main contribution is the significant development of a multi-step neural network predictive technique for the congestion control. Through simulations of actual trace data from the real-time traffic, we demonstrate that the technique improves the control performance. Compared with the methods discussed in [8-9], this paper introduces a BP neural network, analysis the neural network architectures and evaluates control performance.

The rest of this paper is organized as follows: In section 2, we introduce a noval improved congestion control scheme based on neural networks. In section 3, we describe the predictive control scheme for resource management and in section 4, we use simulation to validate and evaluate the performance of our scheme. Finally, in section 5, we present the conclusions and the future work.

2 Congestion Control Model

The congestion control technique in this paper provides such an approach for the dynamic evaluation of the low priority traffic in the network: the evaluation and distribution functions which compute the rate allocated to each individual source are based on a neural network control strategy, and the functions control the filling level of the low priority traffic buffer.

The paper considers a general model as shown in Figure 1-2 with different connections and with various traffic requirements being mapped into different classes. The rate control algorithm computes the low priority bandwidth $\lambda_L(t)$

left by the sum of the highest priority traffic λ_{H1} and the higher priority traffic λ_{H2}. $x_L(t)$ is the number of λ_L packets waiting at time t in the queue, $x_0(t)$ is queue threshold at time t, usually $x_0(t)$ is a constant [8-9].

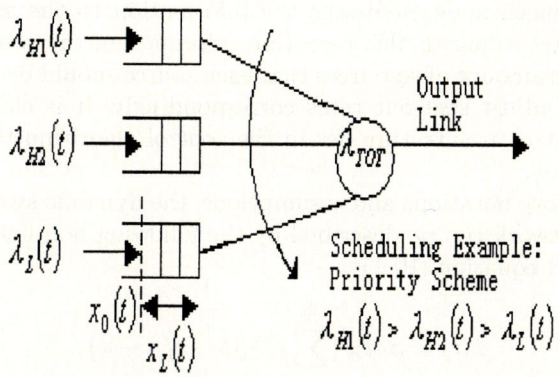

Fig. 1. A simple model of one source, λ_L is controlled

2.1 The Predictive Control Model of a Bottleneck Buffer

It describes the control procedures for multiple sources transmitting data to the buffer of a common bottleneck node. A control algorithm running at the source node evaluates the resource need of each source and distributes the estimated available resources accordingly [8].

In modelling the traffic through these nodes, one has to know the number of source/destination pairs and the rates at which these sources send control packets (CPs) to the network. It's assumed to be N though the number of

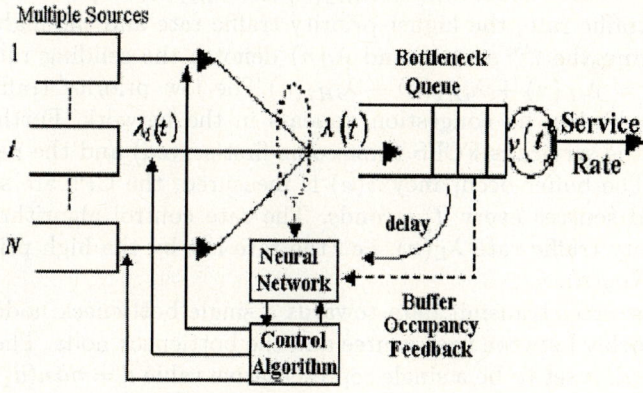

Fig. 2. A model of multiple sources and a bottleneck with controller

active sources denoted by M may vary with time t. The switching node has a finite buffer space K to store the incoming CPs and has an output link to serve them at a constant data rate of v.

The control procedure works in the following manner: each source sends data to the bottleneck node at regular intervals. According to the current loading state, the bottleneck node feedbacks the information to the source along the original route. According to this reception information, the sources can decide the most suitable amount of resources that each source should be available. Thus, the sources can adjust sent-out rates correspondingly. It is clear that the key point of this control architecture lies in the control algorithm that is employed at the source node.

Under the above notations and assumptions, the dynamic system of a switching node in a network can be described by the following non-linear time-variant and time-delayed equation [10-11].

$$\dot{x}(t) = Sat_K\{\sum_{i=1}^{N} e_i \lambda_i(t - \tau'_{1i}) - \nu\}, \qquad (1)$$

where K is the buffer size, $\dot{x}(t)$ is the buffer occupancy at time t, and

$$e_i = \begin{cases} 1, activesource; \\ 0, otherwise. \end{cases}$$

$$Sat_K\{x\} = \begin{cases} K, x > K; \\ x, 0 \leq x \leq K; \\ 0, x < 0. \end{cases}$$

If a feedback control is applied to the above system, we assume the signals get sampled every T seconds. It is reasonable because one can always add a small delay to the input delay so that it is a multiple of T when timing. So we can come to the virtual connection (VC) delay d_i = the input delay τ_{1i} from the ith source node to the switching node + the feedback delay τ_{2i} from the switching node to the ith source node. $\lambda_{iL}(n)$, $\lambda_{iH2}(n)$ and $\lambda_{iH1}(n)$ respectively denote the low priority traffic rate, the higher priority traffic rate and the highest priority traffic rate from the ith source, and $\lambda_i(n)$ denotes the sending rate of source i, i.e., $\lambda_i(n) = \lambda_{iL}(n) + \lambda_{iH1}(n) + \lambda_{iH2}(n)$. The low priority traffic can only be transmitted when no congestion appears in the network. Furthermore, we assume that the service is FCFS (first-come-first-served) and the packet length is constant. The buffer occupancy $x(n)$ is measured, the CPs are sent back to the controlled sources every T seconds. The rate control algorithm computes the low priority traffic rate $\lambda_L(n)$, i.e., the rate left by the high priority traffic $\lambda_{H1}(n)$ and $\lambda_{H2}(n)$.

When N sources transmit data towards a single bottleneck node, there is a control-loop delay between each source and the bottleneck node. The round trip delay (RTD), d, is set to be a single representative value $d = min(d_1, d_2, ..., d_N)$, and the input representative delay,τ_1,is set as $\tau_1 = min(\tau_{11}, \tau_{12}, ..., \tau_{1N})$. So $d = \tau_1 + \tau_2$ (τ_2 is the backward path delay). The best result in system performance is taken for granted the minimum delay [11]. Let $\lambda_i(n) = T \cdot \lambda_i(nT)$ denote the total numbers of data packets flowing into the destination node from the ith VC during the nth interval of T. The component $\mu = T\nu$ denotes the number of packets sent out from the switching destination node during the nth interval of T. The equation can be written into

$$x(n+1) = Sat_K\{x(n) + \Sigma_{i=1}^{N} e_i \lambda_i(n - \tau_{1i}) - \mu\}. \qquad (2)$$

The control algorithm employs the following four steps [8-9]:

(i) Predict the buffer occupancy $\hat{x}(n+1)$ using the multi-step predictive technique.

(ii) Compute the total expected rate of the all sources $\lambda(n)$ at the time n and $\lambda(n) = \Sigma_{i=1}^{N} \lambda_i(n)$. This value varies dynamically with the buffer occupancy.

(iii) Compute the proportion of each source, $\delta_i(n)$, which is the most efficient share of the available resources to be attributed to source number i, $(1 \leq i \leq N$, $\Sigma_{i=1}^{N} \delta_i(n) = 1)$, $\delta_i(n) = \lambda_i(n)/\lambda(n)$.

(iv) Compute the adjusted low priority traffic rate $\lambda_{iL}(n)$. In this section, every source equally shares the available network bottleneck bandwidth, $\lambda_i(n)$ can be expressed as: $\lambda_i(n) = \delta_i(n) \cdot \lambda(n)$. Based on the equation (4), the source i regulates the lowest priority traffic rate $\lambda_{iL}(n)$.

3 The Predictive Control Technique

3.1 The BP Neural Network Architecture

The BP neural network algorithm is introduced into this paper as a predictive mechanism. We assume the number of input neuron is N, and the number of sample study group is M_0. The sample study groups are independent from each other. We further assume the output of the study sample group (teaching assigns) is $R_j^{(k)}$ ($j \in [0, N]$, $k \in [1, M_0]$), and the actual output for output element j in the network is $O_j^{(k)}$. So $E^{(k)}$ is set to be the kth group input goal function. Therefore, we have $E^{(k)} = \Sigma_j (R_j^{(k)} - O_j^{(k)})^2/2$. The total goal function is $J = \Sigma_k E^{(k)}$. If $J \leq \varepsilon_0$, ε_0 is a constant that is small enough and $\varepsilon_0 > 0$, then the algorithm is terminated; Otherwise adjust the weight W between the implicit layer and output layer until it satisfy the expected difference value [12-15].

3.2 Multi-step Neural Predictive Technique

We apply a neural network technique to determine how a BP-based algorithm satisfies its data transfer requirement by adjusting its data transfer rate in a network. As shown in figure 2, the BPNN predictive controller is located at the sources. In order to predict the buffer occupancy efficiently, the neural model for the unknown system above can be expressed as:

$$\hat{x}(n+1) = \hat{f}[x(n), ..., x(n-l+1), \lambda(n-\tau_1-1),$$

$$..., \lambda(n-\tau_1-m-L)], \qquad (3)$$

where $x(n-i)$ $(1 \leq i \leq l-1)$ is the history buffer occupancy and $\lambda(n-j)(\tau_1+1 \leq j \leq \tau_1+m+L)$ is the history sending rate of the source j. L is predictive step, $L = \tau_1+1$, and L, m are constant integers. $\hat{f}[\cdot]$ is the unknown function, which may be expressed by the neural network. The explicit mechanism of BP neural network

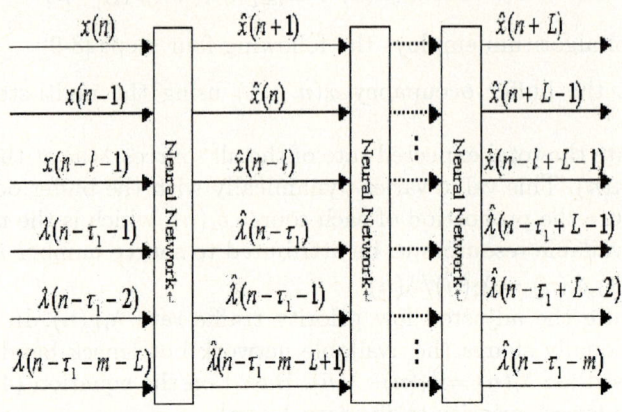

Fig. 3. The Back Propagation (BP) L-step ahead prediction, and $\hat{x}(n+L)$ is the L-step predictions of $x(n)$

L-step ahead prediction is shown in Figure 3, the value of buffer occupancy $x(n)$ and the history value (the past buffer occupancy: $x(n-1), ...x(n-l+1)$; the past source sending rates: $\lambda(n - \tau_1 - 1), ..., \lambda(n - \tau_1 - m - L)$) are used as the known inputs of neural network. Every layer denotes one-step forward predictive, so $\hat{x}(n + L)$ in the output layer is the L-step prediction of $x(n)$. We can compute the expected total rate $\hat{\lambda}(n)$ of the N sources using the following equation:

$$\hat{x}(n+L) = Sat_K\{\hat{x}(n+L-1) + \hat{\lambda}(n) - \mu\}, \qquad (4)$$

Based on the rate $\hat{\lambda}(n)$ the source i adjusts the sending rate $\lambda_{iL}(n) = \hat{\lambda}(n)\delta_i(n) - \lambda_{iH1}(n) - \lambda_{iH2}(n)$, and $\delta_i(n)$ is a factor of share the available resources to source i $(1 \leq i \leq N)$. The specific algorithm is given in the following (Figure 4), At the next instant $n + 1$, we can get new real measured value $x(n + 1)$ and new history measure values: $x(n), ..., x(n-l+2)$; $\lambda(n-\tau_1), ..., \lambda(n-\tau_1-m-L+1)$ which can be used as the next instant inputs of neural network. Then the buffer occupancy $\hat{x}(n+L+1)$ can be predicted.

4 The Simulation Results

To evaluate the performance of the proposed congestion control method based on neural network, we focus upon the following simulation model with eleven

```
Procedure Neural_Network_Prediction(x, λ, x̂, λ̂)
{ At time instant n :
Step 1: Get value of buffer occupancy x(n), and
    the history measurement value by the controller:
    the past buffer occupancy: x(n−1),..., x(n−l+1);
    the past total sending rate: λ(n−τ₁−1),...,λ(n−τ₁−M−1).
    All values above are used as available known input
    of neural network.
Step 2:  Predict L step x̂(n + L).
(1) Predict x̂(n+i), i ∈ [1, L];
    [x̂(n+i−1), x̂(N+i−2),..., x̂(N+i−l)]
    as the neural network input. These predictions are to
    be used to predict the next set.
(2) Train neural network computes the goal function
    J and adjust the weight;
(3) Back propagation for the next L step ahead
    prediction;
(4) Go ahead until L step ahead predictions. Then,
    x̂(n + L) is the L-step prediction.
Step 3:Compute λ̂(n), λ̂, and λ̂ₗ(n).
Step 4: At the next instant n+1, update the history
    value, (n+1) takes the place of n, go to step 1.
Return  (x̂(n+L) ∈ (x̂, û)).  }
```

Fig. 4. Algorithm for on-line control and neural network training at sources

sources and one switch bottleneck node (Figure 5), and assume that the sources always have data to transmit. The congestion controller is used to adjust sending rate over time in sources. The higher priority traffic, i.e., the sum of λ_{iH1} and λ_{iH2} traffic in source i with multiplexing of actual trace data, is acquired from the real time traffic.

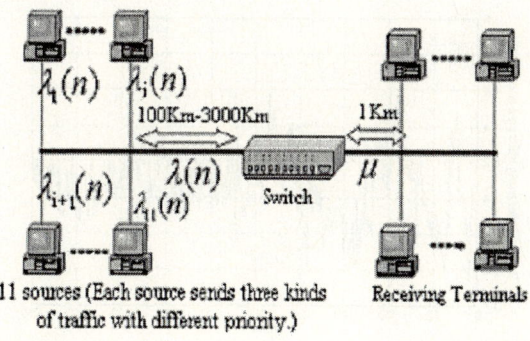

Fig. 5. A simulation model of multiple sources single buffer network

As shown in Figure 6, the maximum sending rate of every source is $\lambda_0 = 15.5 Mbps$. We use a simple resource sharing policy, i.e., the network bottleneck node equally shares the available bandwidth among every source. The sources start to transmit data at time $t = 1 msec$ together. We assume the sending rate of the switch node is $\nu = 155 Mbps$. The sampling time T is $1 msec$ and the congestion threshold is set as $x_0 = 1000 Kb$.

We propose to use a direct multi-step neural predictive architecture with 3 layer neural network, wherein the number of the input data, the input neurons, the hidden neurons and the output neurons are all $(L + m + l)$. There are $l(l = 8)$ terms of buffer occupancy x and $(L + m)$ terms of the total input μ. The prediction horizon is $L = \tau_1 + 1$, and the control horizon is $N = L - \tau_1 + 1 = 2$.

To investigate the performance of this model, we set the distance from sources to switch node to be $300 Km$ with the forward path delay and the feedback path delay being $\tau_{1i} = 3 msec$, $\tau_{2i} = 3 msec$ $(i = 1, 2, ..., 11)$ respectively. Therefore the RTD is $d = 6 msec$. We assume that the RTD is dominant compared to other delays such as processing delays and queuing delay, etc.

For this case, the prediction horizon is $L = 4$, and $m = 4$. Figure 6 shows the rate of higher priority $(\lambda_{H1} + \lambda_{H2})$ traffic. The dynamic of buffer occupancy is shown in Figure 7, where the predictive buffer occupancy and the actual buffer occupancy are described with broken line and real line respectively. The predictive value of the buffer occupancy is acquired beginning from the time $(\tau_1 + L + 9)$. Figure 8 shows the transmitting rate of the lowest priority traffic, which is yielded on the basis of the equation (1) and the predicted buffer occupancy from the time slot 12 to $(500 - \tau_1 - L) = 493$, and Figure 9 shows the total input rates.

From Figure 7, one observes that buffer occupancy is acquired beginning from the time slot $n = 16$ and that the queue size is maintained to be close to the threshold of $1000 Kb$ by the proposed neural networks predictive technique. The average relative error between the predictive buffer occupancy and actual buffer occupancy is 1.5099e-002, which is excellent in terms of accuracy.

Figure 10-12 show the performance that we set the sources $2600 Km$ away from the switch node, and assume the forward delay and the feedback delay

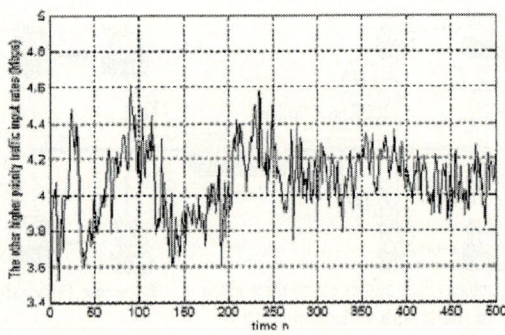

Fig. 6. High priority traffic rate sampled from the real time video traffic

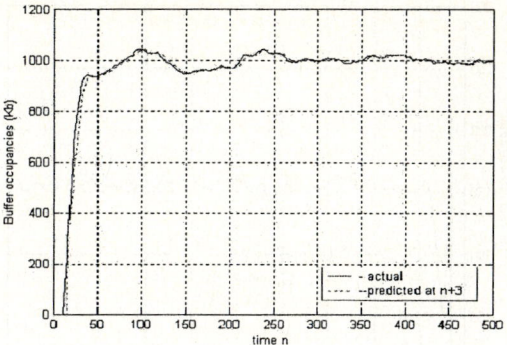

Fig. 7. The buffer occupancy for $L = 4$ step prediction

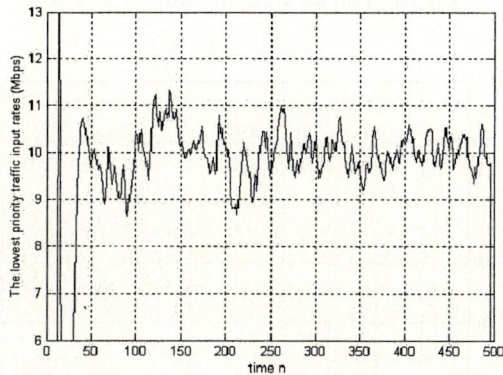

Fig. 8. The lowest priority traffic rate for $L = 4$ step prediction, based on the predictive value in Figure 7

being $\tau_{1i} = 13msec$, $\tau_{2i} = 12msec, (i = 1, 2, ..., 11)$ respectively. Therefore the RTD is $d = 25msec$. We take the prediction horizon $L = 14$ and $m = -6$. Figure 10 shows the buffer occupancy has the value that begins from the time slot at $n = 36$. The neural predictive congestion control technique is also able to maintain the queue size close to the threshold of $1000Kb$, and the average relative error between the predicted buffer occupancy and the actual buffer occupancy is 3.7026e-002. Figure 11 shows the lowest priority traffic rate for $L = 14$ step prediction, and it is yielded on the basis of the original flow equation (1) and the predicted buffer occupancy, and Figure 12 shows total input rate prediction.

The performance of the system is excellent for queue service rates. However, the performance is found to be better in the 4-step prediction than in 14-step prediction case. This is probably due to the fact that the less prediction horizon usually leads to better accuracy, whereas in our scheme the predictive horizon is linked with the forward path delay τ_1.

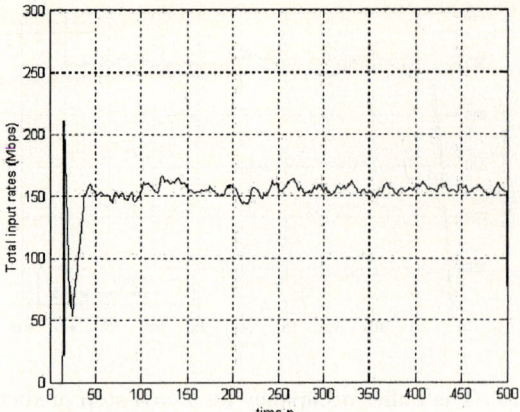

Fig. 9. The total input rates of $L = 4$

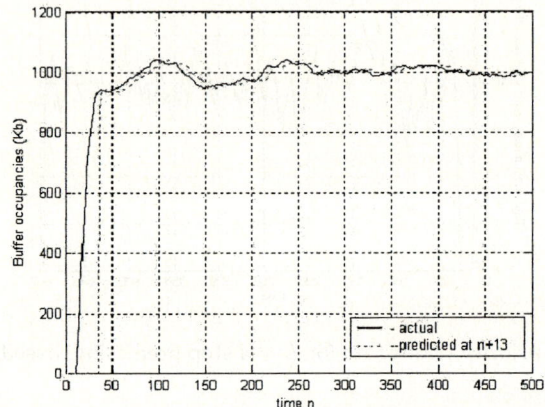

Fig. 10. The buffer occupancy for $L = 14$ step prediction

To compare our algorithm with the conventional approaches like in [8-9], the following remarks can be given.

(i) This paper introduces a new congestion control model based on neural network. The BP network model and algorithm develop the ideas and methods in [8-9].

(ii) The quicker transient response of the source rates is acquired in our mechanism. Under the same circumstance, we use less predictive steps than that in [8-9], because in this paper the neural predictive controller is located at the sources rather than at the switch, this usually brings forth the better performance in terms of prediction accuracy.

(iii) The authors of [14 -17] suggest that only one implicit layer is enough, and it could be randomly mapped into R^m space. With the same number of the

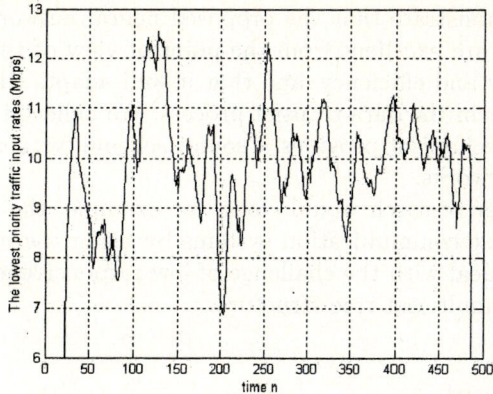

Fig. 11. The lowest priority traffic rate for $L = 14$ step prediction, based on the predictive value in Figure 10

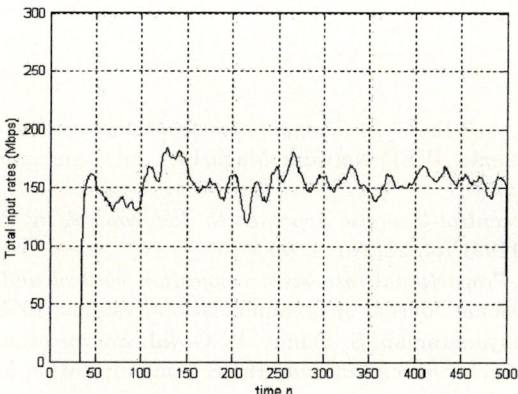

Fig. 12. The total input rates of $L = 14$

implicit layer node, the algorithm will be more efficient if there are less layers. So the implicit layer of BP algorithm in this paper has just one layer and it could improve study efficiency with reasonable study accuracy.

(iv)We have explored the relevant theory on BPNN multi-step predictive architecture and training algorithm, and give relevant simulation analysis.

5 Conclusion

This paper has described a dynamic resource management mechanism for computer communication networks on the basis of an adapting BP neural network control technique. Also we further explored the relevant theoretic foundations as well as the detailed implementation procedure for congestion control. The simulation results demonstrate that the proposed neural network architecture and training algorithm are excellent from the point of view of the system response, predictive accuracy and efficiency, and that it well adapts the data flows to the dynamic conditions in the data transfer process. We believe that the neural network predictive mechanism provides a sound scheme for congestion control in communication networks.

Areas for further research would cover, for example, the issue of congestion control for multicast communication systems by using the neural network predictive method to deal with the challenge of low responsiveness, which is due to the heterogeneous multicast tree structure.

Acknowledgment

This research has been supported by National Natural Science Foundation of China under Grant No. 90104005 and by the Key Project of Natural Science Foundation of Hubei Province under Grant No. 2003ABA047.

References

[1] C. Q. Yang, A. A. S.Reddy,*A taxonomy for congestion control algorithms in packet switching networks*, IEEE Network Magazine, July/August 1995, Vol. 9, No.5, pp.34 - 45.

[2] S. Keshav, *A control-theoretic approach to flow control*, in: Proceedings of ACM SIGCOMM'91 1991 Vol. 21, No. 4, pp.3-15.

[3] D. Cavendish,*Proportional rate-based congestion control under long propagation delay*, International Journal of Communication Systems, 1995, Vol. 8, pp. 79-89.

[4] R. Jain, S. Kalyanaraman, S. Fahmy, R. Goyal. *Source behavior for ATM ABR traffic management: an explanation*, IEEE Communication Magazine, 1996, Vol. 34, No. 11, pp. 50-57.

[5] Rose Qingyang Hu and David W. Petr, *A Predictive Self-Tuning Fuzzy-Logic Feedback Rate Controller*, IEEE/ACM Transactions on Networking, December 2000, Vol. 8, No. 6, pp. 689 - 696.

[6] Giuseppe Ascia, Vincenzo Catania, and Daniela Panno, *An efficient buffer management policy based on an integrated Fuzzy-GA approach*, IEEE INFOCOM 2002, New York, June 23 - 27, 2002, No.107.

[7] G. Ascia, V. Catania, G. Ficili and D. Panno, *A Fuzzy Buffer Management Scheme for ATM and IP Networks*, IEEE INFOCOM 2001, Anchorage, Alaska, April 22-26, 2001, pp.1539-1547.

[8] J. Aweya, D.Y. Montuno, Qi-jun Zhang and L. Orozco-Barbosa, *Multi-step Neural Predictive Techniques for Congestion Control -Part 2: Control Procedures*, International Journal of Parallel and Distributed Systems and Networks, 2000, Vol. 3, No. 3, pp. 139-143.

[9] J. Aweya, D.Y. Montuno, Qi-jun Zhang and L. Orozco-Barbosa, *Multi-step Neural Predictive Techniques for Congestion Control -Part 1: Prediction and Control Models,* International Journal of Parallel and Distributed Systems and Networks, 2000, Vol. 3, No. 1, pp. 1-8.

[10] L. Benmohamed and S. M. Meerkov, *Feedback Control of Congestion in Packet Switching Networks: The Case of Single Congested Node*, IEEE/ACM Transaction on Networking, December, 1993, Vol. 1, No. 6, pp. 693-708.
[11] J. Filipiak, *Modeling and Control of Dynamic Flows in Communication Networks*, Springer Verlag Hardcover, New York, May 1, 1988.
[12] S. Jagannathan, and G. Galan, *A one-layer neural network controller with preprocessed inputs for autonomous underwater vehicles*, IEEE Trans. on Vehicular Technology, Vo. 52, no. 5, Sept. 2003.
[13] D. H. Wang, N. K. Lee and T. S. Dillon, *Extraction and Optimization of Fuzzy Protein Sequence Classification Rules Using GRBF Neural Networks*, Neural Information Processing - Letters and Reviews, Vol.1, No.1, pp. 53-59, 2003.
[14] R. Yu and D. H. Wang, *Further study on structural properties of LTI singular systems under output feedback*, Automatica, Vol.39, pp.685-692, April, 2003.
[15] S. Jagannathan and J. Talluri, *Adaptive Predictive congestion control of High-Speed Networks*, IEEE Transactions on Broadcasting, Vol.48, no.2, pp.129-139, June 2002.
[16] Simon Haykin, *Neural Networks: A Comprehensive Foundation* ,(2nd Edition), Prentice Hall, New York, July 6, 1998.
[17] F. Scarselli and A C Tsoi, *Universal Approximation Using FNN: A Survey of Some Existing Methods and Some New Results*, Neural Networks, 1998, Vol. 11, pp. 15-37.
[18] J. Alan Bivens, Boleslaw K. Szymanski, Mark J. Embrechts, *Network congestion arbitration and source problem prediction using neural networks*, Smart Engineering System Design, vol. 4, N0. 243-252, 2002.
[19] S. Jagannathan, *Control of a class of nonlinear systems using multilayered neural networks*, IEEE Transactions on Neural Networks, Vol.12, No. 5, September 2001.
[20] P. Darbyshire and D.H. Wang, *Learning to Survive: Increased Learning Rates by Communication in a Multi-agent System*, The 16th Australian Joint Conference on Artificial Intelligence (AI'03), Perth, Australia, 3-5 December 2003.
[21] Lin, W. W. K., M. T. W. Ip, et al. , *A Neural Network Based Proactive Buffer Control Approach for Better Reliability and Performance for Object-based Internet Applications*, International Conference on Parallel and Distributed Processing Techniques and Applications (PDPTA 2001), Las Vegas, Nevada, USA, CSREA Press.

Optimal Genetic Query Algorithm for Information Retrieval

Ziqiang Wang and Boqin Feng

Department of Computer Science, Xi'an Jiaotong University, Xi'an 710049,
P.R.C(China)
wzqagent@xinhuanet.com

Abstract. An efficient immune query optimization algorithm for information retrieval is proposed in this paper. The main characteristics of this algorithm are as follows: The genetic individual is a query, each gene corresponds to a weighted term, immune operator is used to avoid degeneracy, local search procedure based on the concept of neighborhood is used to speed up the abilities of finding better query vector. Experimental results show that the proposed algorithm can efficiently improve the performance of the query search.

1 Introduction

With the increasingly widespread use of information networks, there is emerging an environment in which the user can access tremendous amounts of online information. As a consequence, the role of information retrieval (IR) systems is becoming more important.Recently, there has been a growing interest in applying genetic algorithm(GA) to the information retrieval domain with the purpose of optimizing document descriptions and improving query formulation[1-3]. In addition,Immune Algorithm(IA)[4] has also attracted many researchers interests and successfully applied to several NP-hard combinatorial optimization problems, but the use of the algorithm for query optimization, in the context of information retrieval, is a research area where few people explored. Our goal is to build an IA that can find an optimal set of documents which best match the user's need by exploring different regions of the document space simultaneously.

2 Immune Algorithm for Information Retrieval

The proposed system is based on a vector space model in which both documents and queries are represented as vectors. The goal of our IA is to find an optimal set of documents which best match the user's need by exploring different regions of the document space simultaneously.

2.1 Encoding of Query Individual

The genetic individual is a query. Each gene or chromosome corresponds to an indexing term or concept. The locus (the existence or the absence of certain gene) is represented by a real value and defines the importance of the term in the considered query. Each individual representing a query is of the form:

$$Q_u(q_{u1}, q_{u2}, \cdots, q_{uT}) \qquad (1)$$

where T is total number of stemmed terms automatically extracted from the documents and q_{ui} is the weight of the ith term in Q_u. Initially, a term weight is computed as the following formula[5]:

$$q_{ui} = \frac{(1 + log(tf_{ui})) \cdot log(\frac{N}{n_i})}{\sqrt{\sum_{k=1}^{T}((1 + log(tf_{ui})) \cdot log(\frac{N}{n_i}))^2}} \qquad (2)$$

where tf_{ui} is the frequency of term t_i in document d_u, N is the total number of documents, and n_i is the number of documents containing the term t_i.

2.2 Fitness Function

A fitness is assigned to each query in the population. This fitness represents the effectiveness of a query during the retrieving stage. Its definition is as follows:

$$Fitness(Q_u^{(s)}) = \frac{1}{N} \cdot \frac{\sum_{d_j \in D_r^{(s)}} Sim(d_j, Q_u^{(s)})}{\sum_{d_j \in D_{nr}^{(s)}} Sim(d_j, Q_u^{(s)})} \qquad (3)$$

where N is the total number of documents, $D_r^{(s)}$ is the set of relevant documents retrieved at the generation(s) of the IA, d_j is the jth document, $D_{nr}^{(s)}$ is the set of non-relevant documents retrieved at the generation(s) of the IA, and $Sim(d_j, Q_u^{(s)})$ is a similar measure function defined as follows:

$$Sim(d_j, Q_u^{(s)}) = Cos(d_j, Q_u^{(s)}) = \frac{\sum_{i=1}^{T}(q_{ui}^{(s)} \cdot d_{ji})}{\sqrt{\sum_{i=1}^{T} q_{ui}^2} \cdot \sqrt{\sum_{i=1}^{T} d_{ji}^2}} \qquad (4)$$

2.3 Genetic Operator

The selection procedure is based on the well-known tournament selection. The crossover operator does not use a crossing point, our method is to modify the term weights according to their distribution in the relevant and non-relevant documents. Let $Q_u^{(s)}$ and $Q_v^{(s)}$ are two (query) individuals selected for crossover, the reproduction new individual $Q_y^{(s+1)}$ is defined as follows:

$$Q_{yi}^{s+1} = \begin{cases} max(q_{ui}^{(s)}, q_{vi}^{(s)}) : w(t_i, D_r^s) \geq w(t_i, D_{nr}^s) \\ \frac{(q_{ui}^{(s)} + q_{vi}^{(s)})}{2} : \qquad otherwise \end{cases} \qquad (5)$$

where $w(t_i, D_r^s)$ is the weight of term t_i in the set of relevant documents $D_r^{(s)}$, and $w(t_i, D_{nr}^s)$ is the weight of term t_i in the set of non-relevant documents $D_{nr}^{(s)}$. Their definition are as follows:

$$w(t_i, D_r^{(s)}) = \sum_{d_j \in D_r^{(s)}} d_{ji} \qquad (6)$$

$$w(t_i, D_{nr}^{(s)}) = \sum_{d_j \in D_{nr}^{(s)}} d_{ji} \qquad (7)$$

Our mutation operator essentially consists of exploring the terms in the documents in order to adjust the corresponding gene values in the query selected for the mutation. Let $Q_u^{(s)}$ is the selected individual query, the produced new individual query is $Q_u^{(s+1)}$ after mutation. The mutation operator definition is as follows:

$$Q_{ui}^{s+1} = \begin{cases} avg(Q_u^{(s)}) : & random(p) < p_m \\ max(Q_u^{(s)}) - min(Q_u^{(s)}) : & otherwise \end{cases} \qquad (8)$$

where $random(p)$ generates a random number p in the range $[0,1]$, $max(Q_u^{(s)})$ is the maximum term weight of $Q_u^{(s)}$, $min(Q_u^{(s)})$ is the minimum term weight of $Q_u^{(s)}$, and $avg(Q_u^{(s)})$ is the term weight average of $Q_u^{(s)}$.

2.4 Immune Operator

The idea of immunity is mainly realized through two steps based on reasonably selecting vaccines, i.e., a vaccination and an immune selection, of which the former is used for raising fitness and the latter is for preventing the deterioration.

The Vaccination: Given an individual query $Q_u^{(s)}$, a vaccination means modifying the genes on some term weight $q_{ui}^{(s)}$ in accordance with the similarity between the query and documents retrieved so as to gain higher fitness with greater probability. This operation must satisfy the following two conditions. Firstly, if the information on each term weight of an individual query $q_{ui}^{(s+1)}$ is is wrong, i.e., each term weight of it is different from that of the optimal one, then the probability of transforming from $Q_u^{(s)}$ to $Q_u^{(s+1)}$ is 0. Secondly, if the information on each term weight of $Q_u^{(s)}$ is right, i.e., is the optimal one, then the probability of transforming from $Q_u^{(s)}$ to $Q_u^{(s)}$ is 1. Suppose a population is $Q_u^{(s)} = (q_{u1}^{(s)}, q_{u2}^{(s)}, \cdots, q_{un_p}^{(s)})$, then the vaccination on $Q_u^{(s)}$ means the operation carried out on $n_p = \alpha n$ individuals which are selected from $Q_u^{(s)}$ in proportion as α.

The immune selection: This operation is accomplished by the following two steps. The first one is the immune test, i.e., testing the antibodies. If the fitness is smaller than that of the parent, which means serious degeneration must have happened in the process of crossover or mutation, then instead of the individual the parent will participate in the next competition; the second one is the annealing selection, i.e., selecting an individual $q_{ui}^{(s)}$ in the present offspring $Q_u^{(s)} = (q_{u1}^{(s)}, q_{u2}^{(s)}, \cdots, q_{uT}^{(s)})$ to join in the new parents with the probability as follows:

$$p(q_{ui}^{(s)}) = \frac{e^{\frac{q_{ui}^{(s)}}{T_k}}}{\sum_{i=1}^{T} e^{\frac{q_{ui}^{(s)}}{T_k}}} \qquad (9)$$

where T_k is the temperature-controlled series approaching 0, i.e.,

$$T_k = \ln(\frac{T_0}{k} + 1) \qquad (10)$$

2.5 Local Search Procedure

To reinforce the local search abilities of IA, our algorithm adopts a neighborhood-based local search procedure to find a better query vector near the original query vector after applying the immune operator. Let $Q_u^{(s)+}$ and $Q_u^{(s)-}$ be the neighbors of the query vector $Q_u^{(s)}$, their definition is as follows:

$$q_{ui}^{(s)+} = q_{ui}^{(s)} \cdot (1 + \beta) \qquad (11)$$

$$q_{ui}^{(s)-} = q_{ui}^{(s)} \cdot (1 - \beta) \qquad (12)$$

where the value of β decides the ratio of increase or decrease. Each weight in a query vector generates two neighboring vectors. From all neighboring vectors, the vector $Q_u^s(new)$ which has the best fitness function value is selected. If $avg(Q_u^s(new))$ is larger than Q_u, then the Q_u is replaced by $Q_u^s(new)$.

2.6 Retrieved Relevant Documents Merging

At each generation of IA, these retrieved relevant documents by all the individual queries of the query population are merged to a single document list, and presented to user. Our adopted merging methods according to following range formula:

$$Rel^s(d_j) = \sum_{Q_u^{(s)} \in Pop^{(s)}} Fitness(Q_u^{(s)}) \cdot RSV(Q_u^{(s)}, d_j) \qquad (13)$$

where $Pop^{(s)}$ is the population at the generation(s) of the IA, $RSV(Q_u^{(s)}, d_j)$ is the retrieval status value(RSV) of the document d_j for the query $Q_u^{(s)}$ at the generation(s) of the IA.

3 Experimental Results and Comparison

To test the performance of the proposed immune query optimization algorithm, we used the best known TREC collections[6], and evaluated the results of the retrieval via the classical measures of recall and precision. The parameter settings of the immune algorithm are as follows: mutation probability $p_m = 0.02$, vaccination proportion factor $\alpha = 1.5$, the ratio β of increase(decrease) in local search is set

Table 1. Comparison the Number of Relevant Document

Algorithm	Iter-1	Iter-2	Iter-3	Iter-4	Iter-5
IA	107(107)	58(165)	53(218)	55(273)	43(314)
GA	96(96)	64(160)	55(215)	51(266)	32(298)

0.05 , and the number of iterations is fixed at 5.The comparison of the number of relevant document retrieved using IA and GA are shown in Table 1.We can clearly see that IA more effective than GA in retrieving relevant documents. Indeed the cumulative total number of relevant documents using IA through all the iterations is higher than using GA. Therefore, our proposed immune query optimization algorithm efficiently improves the performance of the query search.

4 Conclusions and Future Works

This paper proposes a novel immune query optimization algorithm.Experimental results show that the proposed algorithm can improve the precision of document retrieval compared with genetic algorithm. In future, we plan to combine other efficient heuristics methods to further improve the document retrieval performance.

References

1. Chen,H.:Machine learning for information retrieval:neural networks, symbolic learning and genetic algorithms.Journal of the American Society for Information Science 46(1995)194–216
2. Horng,J.T.,Yeh,C.C.:Applying genetic algorithms to query optimization in document retrieval.Information Processing and Management 36(2000)737–759
3. Boughanem,M.,Chrisment,C.,and Tamine,L.:Genetic approach to query space exploration.Information Retrieval 1(1999)175–192
4. Jiao,L.C.,Wang,L.:A novel genetic algorithm based on immunity.IEEE Transactions on Systems,Man,and Cybernetics-Part A 30(2000)552-561
5. Singhal,A.,Buckley,C.,Mitra,M.:Pivoted document length normalisation.Proc.of the 19th Annual International ACM SIGIR Conference on Research and Development in Information Retrieval,Zurich, Switzerland,ACM Press(1996)21-29
6. Harman,D.K.:Overview of the first text retrieval conference(TREC-1).Proc.of the 1st Text Retrieval Conference,Gaitherburg,USA,(1992)32-59

A Genetic Algorithm for Dynamic Routing and Wavelength Assignment in WDM Networks

Vinh Trong Le[1], Son Hong Ngo[1], Xiaohong Jiang[1], Susumu Horiguchi[1,2], and Minyi Guo[3]

[1] Graduate School of Information Science, Japan Advanced Institute of Science and Technology, Japan
`{vt-le, sonhong, jiang, hori}@jaist.ac.jp`
[2] School of Information Sciences, Tohoku University, Sendai, Japan
[3] School of Computer Science and Engineering, The University of Aizu, Japan
`minyi@u-aizu.ac.jp`

Abstract. In this paper, we study the challenging problem of Dynamic Routing and Wavelength Assignment (DRWA) in WDM (Wavelength-Division-Multiplexing) networks with wavelength continuity constraint, and propose an improved genetic algorithm (GA) for it. By adopting a new fitness function to consider simultaneously the path length and number of free wavelengths in cost estimation of a route, the new genetic RWA algorithm can achieve a good load balance among route candidates and result in a lower blocking probability than both the fixed alternative routing algorithm and the previous GA-based algorithm for DRWA, as verified by an extensive simulation study upon the ns-2 network simulator and some typical network topologies.

1 Introduction

All optical networks using wavelength-division-multiplexing (WDM) technology have been one of key technology to realize the next generation Internet because WDM can provide huge bandwidth capacity effectively. In wavelength-routed optical networks, data are routed in optical channels called lightpaths. To establish a lightpath without wavelength conversion, the same wavelength is required on all the links along the path, which is referred to the *wavelength continuity constraint*. Given a set of connection requests, lightpaths are setup by routing and assigning wavelength to each request. These problems are known as the *routing and wavelength assignment* (RWA) problem [1].

The RWA problem is usually divided in two types: *static RWA* and *dynamic RWA*. In the static RWA problem, the entire set of connections is known in advance, and the problem is to set up lightpaths for these connections so that used network resources such as the number of wavelengths or the number of fibers in the network are minimized. This problem is well known as an NP-complete problem [2]. In the dynamic RWA problem, since connection requests arrive randomly, it is more difficult to solve the problem. Generally, dynamic RWA algorithms aim to minimize the total blocking probability in the entire network. To make the RWA problem more tractable, it is usually decoupled into two sub-problems that are solved separately: the routing problem and the wavelength assignment problem. There are three major routing schemes: fixed routing, fixed-alternate routing and adaptive routing. In the first scheme, a single fixed route is predetermined for each source-destination pair. Whenever a request arrives, its fixed route is attempted for wavelength assignment. Fixed routing scheme is simple for implementation but usually causes a high blocking probability. In the fixed-alternate routing scheme, a set of routes is pre-computed for each source-destination pair. As a connection request arrives, one route is selected from the set of pre-computed routes. Fixed alternate routing always achieves a better performance than that of fixed routing. In the third scheme, the route is computed at the arrival of a request based on the current network state, thus it obtains the best performance. However, it is difficult to implement due to high computational complexity. As for wavelength assignment sub-problem, many algorithms are available now, among them the First-Fit (FF) algorithm is a simple yet efficient one. In the FF algorithm, each wavelength has a number associated with it, and the searching starts from the lowest-numbered wavelength and stops as soon as an available wavelength has been found or all the wavelengths have been searched. A survey of routing and wavelength assignment problem can be found in [1].

Genetic algorithm (GA) is one promising approach to solve dynamic RWA problems in WDM networks, and some GA-based RWA algorithms have been proposed recently. In [3][4], the authors formulated both the static RWA and dynamic RWA problems as a multi-objective optimization problem and solved it using genetic algorithms. In their approach, each gene in an individual represents one of the k-shortest paths between the source and destination nodes. Whenever a new connection arrives, the genetic algorithm will consider all the individuals in order to find the optimal RWA solution for the entire network. This approach can solve well the static RWA problem. However, to find the optimal solution, some lightpaths need to be re-routed because all the existing connection requests in the network are considered together in the optimization. This re-routing scheme is not suitable for a network with highly dynamic traffic because of the high cost for setting up and tearing-down the lightpath. Thus, this approach is not suitable for the dynamic RWA problem. Bisbal et al. [5] proposed a novel GA-based distributed algorithm for dynamic routing and wavelength assignment problem. In this approach, an individual represents a route between a source-destination node pair and a generation is a set of possible routes between the source and the destination. Whenever a new connection arrives, the genetic algorithm will find the best individual based on the current network state. This best individual corresponds to the best route to be setup in term of blocking probability. This algorithm not only obtains low total blocking probability but also employs a very short computation time. In addition, the algorithm can be extended easily to provide fault tolerance capability and fairness among connections.

In this paper, we focus on the dynamic RWA problem and propose an improved GA for it based on the algorithm presented in [5]. We will show through extension simulation by adopting a more general fitness function to consider both the path length (number of hops) and the number of free wavelength in the cost estimation of a route, the new algorithm can achieve a better load balance and thus it results in a lower blocking probability than that of both the fixed alternative routing algorithm and the previous genetic algorithm in [5].

The rest of this paper is organized as follows: Section 2 introduces briefly the main idea of GA algorithms and also the GA-based RWA algorithm proposed in [5]. Section 3 presents our improved GA algorithm for dynamic RWA under the constraint of wavelength continuity. Section 4 presents our simulation and analysis results, and finally, section 5 concludes this paper.

2 Dynamic Routing and Wavelength Assignment Using Genetic Algorithms

In this section, we first present the main idea of genetic algorithm, and then we introduce briefly the GA-based dynamic RWA algorithm proposed in [5].

2.1 Genetic Algorithms (GAs)

GA is a class of search strategies based on the mechanism of biological evolution. The GA is able to reduce search space and also converge to a global good solution of the problem. In a GA application, the first step is to specify the representation of each possible solution as an individual. The next step is to define a population of N individuals with theirs initial values, a fitness value for each individual in the population, the genetic operators such as crossover, mutation, and reproduction. The main steps of the operation in GAs are as follows:

Initialization: In most GAs, the first generation of individuals is initialized with random values.

Determination of Fitness: The effectiveness of an individual in a population is evaluated by the fitness function. This function assigns a cost to each individual in the current population according to its capability to solve the problem. The better the solution solves the problem, the higher its fitness value is.

Crossover: This is a variety-generating feature of GA, where pairs of individuals (parents) mate to produce offspring. Each offspring draws a part from one parent and a part from the other.

Mutation: Implement a change in an individual to a new individual.

Reproduction and Stopping Conditions: GA applies genetic operators on current population. After that, GA selects individuals to generate next generation. This process is called reproduction, which is repeated until a good individual of the problem is found. However, it is not assured that the optimal solution can be found since GA is stochastic searching process. Hence, the reproduction should be stopped after a certain number of generations. More details about genetic algorithms can be found in [6][7].

2.2 The Original GA-Based RWA Algorithm

We introudce briefly here the Genetic Routing and Wavelength Assignment (GRWA) algorithm proposed in [5]. The GRWA is designed for a WDM network without wavelength conversion capability. The GRWA is executed when a lightpath is requested. It works with a population where each individual is a possible route between the requested source-destination node pair. The coding of a route is a vector of integer where each number identifies a node of the route. For example, with the network shown in Fig.1, the coding of two routes from node 0 to node 5 are vector (0, 1, 2, 5) and (0, 2, 4, 5). The main steps of GRWA are as follows:

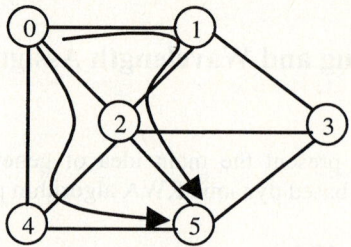

Fig. 1. The two routes from node 0 to node 5 are encoded as (0 1 2 5) and (0 2 4 5)

Initialization

Initialization randomly creates a population with P individuals (P routes between source-destination nodes pair of the request and P is a design parameter). To randomly create a route between a source-destination nodes pair, the source node is firstly set as the beginning of the route. It is marked as visited and also the current node. The next node is randomly selected among the adjacent nodes of current one that are not marked as visited. Then the selected node becomes the current node and is marked as visited. The process is repeated until it reaches the destination node.

Determination of Fitness

The fitness value is evaluated as the inverse of the cost of the route. In case there is least one common free wavelength on all the links of the route, the cost is the number of fiber links it traverses (number of hops). Otherwise, the cost of the route is infinite. According this definition, shorter routes are always preferred as they have higher fitness values.

Crossover Operator

This operator can only be applied to a pair of routes that have at least one node in common, apart from the source and destination nodes. This common node is called crossover point. If there are many common nodes, one of them is chosen randomly. The offspring is generated by interchanging the second halves of its parent as illustrated in Fig. 2.

In the crossover stage, GRWA examines all the possible pairs of routes, beginning with the pairs that include the individual with higher fitness value, until either all combinations is considered or the population size becomes twice of the original size.

Mutation Operator

To do mutation, a node is first randomly selected from the route and the selected node is called mutation point (node). Then, the part from the mutation node to the destination node is randomly generated again. In the mutation stage, the mutation operator is applied to all individuals whose fitness value is below a threshold, which is chosen from the mean fitness value of current generation.

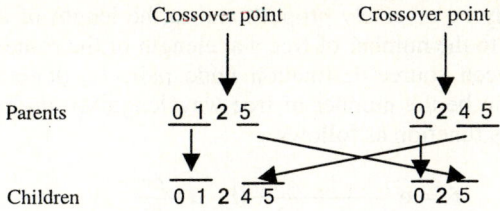

Fig. 2. Example of crossover operation

Reproduction and Stopping Conditions
After applying the genetic operators above, the reproduction stage selects P fittest individuals that have higher value from both parents and children. This process is repeated until the stopping condition is fulfilled and the best individual is selected. Let G denotes the maximum number of generations and S denotes the satisfactory cost value of a route between a node-pair with its initial value being the cost value of the shortest rout between the node-pair, then is pseudo code of GRWA algorithm can be summarized as follows:

```
t = 0;
Initialize randomly P routes & evaluate fitness values;
S = shortest distance between (s, d) nodes;
while ( t < G AND doesn't exist a route that have
length lower or equal S ) do
    Crossover & evaluate fitness value for children;
    Mutation & evaluate fitness value for children;
    Select P fittest individuals;
    S = S + 1;
    t = t + 1;
end while
```

3 The Improved GRWA

In the GRWA algorithm proposed in [5], the fitness value f_i of an individual route i can be computed from the number of hops hi of the route as follows:

$$f_i = \frac{1}{h_i} \quad (1)$$

This fitness function considers only the number of hops of the route. As a result, the GRWA tends to find the shortest available path and take this path for the connection request. It was shown in some previous work [8][9] that this shortest-path based approach usually lead to a case in which one link has to tolerate more load while other links remain unloaded, thus result in an unbalanced link utilization and degraded performance in the network. To increase the utilization and load balancing in the network, it is necessary to distribute the traffic evenly among links.

The above observation motivates us to introduce a new fitness function to improve the GRWA algorithm's performance. To maintain load balancing as much as possible while keeping the GA search towards the shortest available path, the new fitness value should not only be inversely proportional to the length of the route but should also be proportional to the number of free wavelength of the route. Let l_i be the length of the route i between source-destination node pair, l_{min} denote the length of the shortest route, and fw_i be the number of free wavelength on the route i. If $fw_i>0$, we introduce new fitness function as follows:

$$f_i = \alpha \frac{1}{l_i - l_{min} + 1} + (1-\alpha) \frac{fw_i}{W} \qquad (2)$$

Where $\alpha \in [0,1]$ is a design parameter and W is total number of wavelengths on a link. If $fw_i=0$, we just set f_i as zero.

The parameter α should be chosen such that the shorter route has a higher fitness value. Let $d = l_i - l_{min} + 1$, then α should meet the following requirement:

$$\alpha \frac{1}{d} + (1-\alpha).\frac{1}{W} > \alpha \frac{1}{d+1} + (1-\alpha)\frac{W}{W} \qquad (3)$$

which equivalents to:

$$\alpha > \frac{(W-1)(d+1)d}{W + (W-1)(d+1)d} \qquad (4)$$

For a given value of W, the right hand side of (3) increases as d increases. From the definition of d we can see easily that $1 \leq d \leq N-2$, where N is number of the network nodes. Thus, we have:

$$\alpha > \frac{(W-1)(N-1)N}{W + (W-1)(N-1)N} \qquad (5)$$

The performance of genetic RWA algorithm for each value of α will be next discussed in our experimental works.

The complexity of the original GRWA algorithm is $O(G.(P.W.N+P^2.N))$, where N is the number of nodes of the network, W is the number of available wavelength per link, G (maximum number of generations) and P (population size) are two evolution parameters of genetic algorithm [5]. In fact, the complexity to compute the number of free wavelengths on a route (improved GRWA algorithm) is similar to the complexity in the worst case to verify if a route has a free wavelength (original GRWA algorithm). Thus, the costs to compute the fitness value in both algorithms are the same. By consequence, our improvement does not increase the complexity of this genetic algorithm.

4 Simulation Results and Analysis

In this section, we examine the performance of the improved GRWA algorithm by an extensive simulation study upon the *ns*-2 [10] and three typical network topologies (NSF, APRA2 and EON network topologies) as illustrated in Fig. 3.

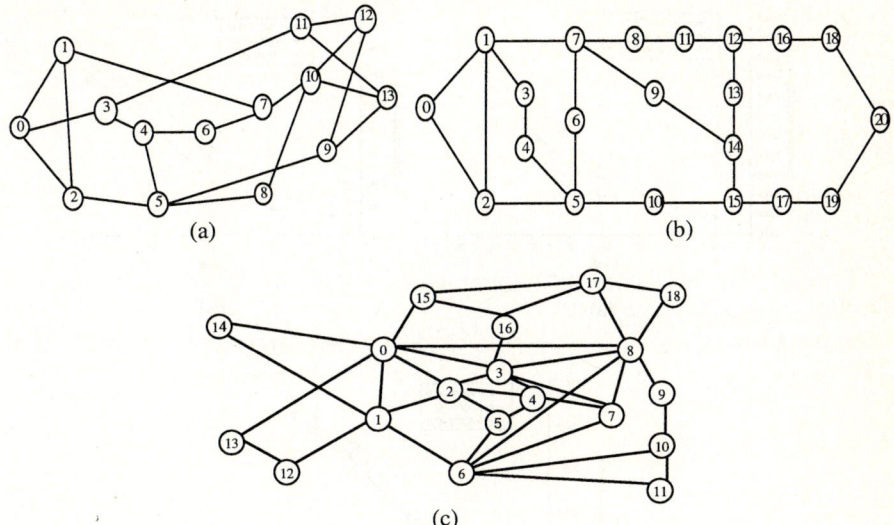

Fig. 3. Network topologies adopted in simulation. (a) NSF network with 14 nodes and 21 links. (b) APRA2 network with 21 nodes and 26 links. (c) EON network with 19 nodes and 35 links

For the performance comparison, we run each experiment with three routing algorithms: FA (Fixed-Alternate routing algorithm), Old-GRWA (the original GRWA algorithm in [5]) and New-GRWA (our improved GRWA algorithm with new fitness function). As explained in many related works [1], Fixed-Alternate Routing is a very promising algorithm for RWA problem. In our experiments, two alternative routes are used ($k=2$).

In our experiments, we use a dynamic traffic model in which the connection requests arrive at the network according to Poisson process with an arrival rate λ (call/seconds). The session holding time is exponentially distributed with mean holding time μ (seconds). The connection requests are distributed randomly on all the network nodes. If there are N sessions over the network, then the total network load is measured by $N*\lambda*\mu$ (Erlands). Thus we can modify N, λ, μ parameters to have different values of workload.

For each experiment, the same values of GA parameters are used in both Old-GRWA and New-GRWA algorithms in order to fairly show the effect of our new fitness function. We take $P=16$ and $G=8$ for the NSF network, $P=32$ and $G=8$ for the ARPA2 network, and $P=16$ and $G=8$ for the EON network. Fig.4 show the sensitiveness of the performance of New-GRWA algorithm to the variations of parameter α.

As shown in Fig.4 and explained in Section 3, the value $\alpha=0.9$ should be used because it satisfies inequation 5. (The case in which selected route is the one that has the biggest number of free wavelengths among the shortest available paths).

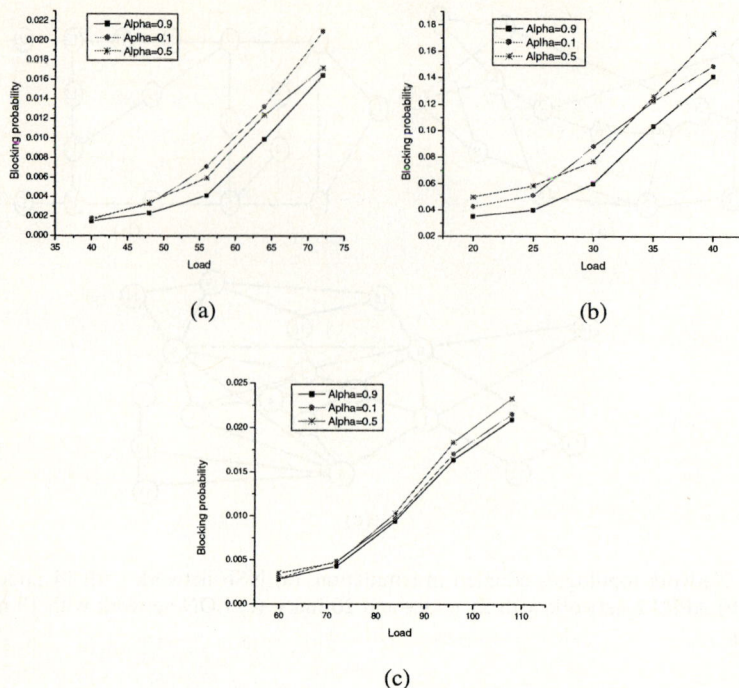

Fig. 4. Blocking probability Versus Network load with different value of α: (a) NSF network, (b) ARPA2 network, (c) EON network

Fig. 5 shows the relation between the blocking probability and the network load over three algorithms. The comparisons in Fig.5 show clearly that the GRWA algorithms always outperform significantly the Fixed Alternate routing algorithm. In all the network topologies, the New-GRWA algorithm is always better than the Old-GRWA algorithm in terms of blocking probability. The running time of both GRWA algorithms are the nearly same because we keep the same genetic parameters. Moreover, the above experiments show that the New-GRWA algorithm is better than the Old-GRWA algorithm in a practical range of blocking probability (0-5%), this is an practical advantage of GA approach with the new fitness function.

We also notice that both of GRWA algorithms perform much better than the FA routing algorithm in case of EON and NSF topologies. However, the difference is not very significant with the ARPA2 topology. The reason is that the NSF and EON network have bigger average node degree than ARPA2 network, so that the GRWA algorithms can be more adaptive because it has more chances to find a good path among more alternate routes (Fig.5 (a), (c)). However, for a large network with a relatively small average node degree (such as the APRA-2 network), there are only a

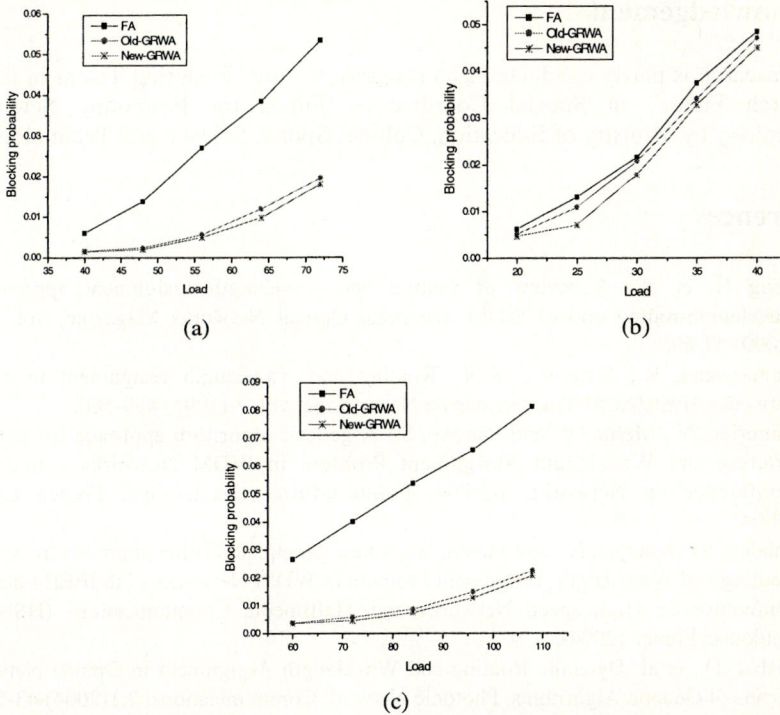

Fig. 5. Blocking probability vs. Network load: (a) NSF network, (b) ARPA2 network, (c) EON network

few alternate routes (about two) between two nodes. Hence the GRWA algorithms cannot be much better than FA algorithm (Fig.5 (b)).

5 Conclusion and Future Works

In this paper, we proposed an improved genetic algorithm for dynamic routing and wavelength assignment in optical networks. By using a new fitness function that considers both the path length and the number of free wavelengths in route evaluation, the genetic RWA algorithm can not only find a good route for a connection request but also keep the load balance among the possible routes. Our simulation shows that the improved RWA algorithm always outperforms the Fixed-Alternate routing and the original genetic RWA algorithms in terms of blocking probability. Moreover, our algorithm does not increase the complexity time in comparison with the original genetic RWA algorithm. In the future, we will extend this algorithm for the dynamic RWA problem in optical networks with wavelength converters.

Acknowledgement

This research is partly conducted as a program for the "Fostering Talent in Emergent Research Fields" in Special Coordination Funds for Promoting Science and Technology by Ministry of Education, Culture, Sports, Science and Technology.

References

1. Zang H. et al.: A review of routing and wavelength assignment approaches for wavelength-routed optical WDM networks. Optical Networks Magazine, vol. 1, no. 1 (2000) 47-60.
2. Ramaswami, R., Sivarajan, K.N.: Routing and wavelength assignment in all-optical networks. IEEE/ACM Transactions on Networking, vol. 3 (1995) 489-500.
3. Banerjee, N., Mehta, V. and Pandey, S.: A genetic algorithm approach for solving the Routing and Wavelength Assignment Problem in WDM Networks - International Conference on Networks, ICN'04, Pointe-a-Pitre, Guadeloupe, French Caribbean (2004).
4. Pandey, S., Banerjee, N., and Mehta, V.: A new genetic algorithm approach for solving the Routing and Wavelength Assignment Problem in WDM Networks - 7th IEEE International conference on High speed Networks and Multimedia Communications (HSNMC'04), Toulouse, France (2004).
5. Bisbal, D., et al: Dynamic Routing and Wavelength Assignment in Optical Networks by Means of Genetic Algorithms, Photonic Network Communications, 7:1(2004) 43-58.
6. D.E. Goldberg: Genetic Algorithms in Search, Optimization, and Machine Learning, Addison-Wesley Publishing Company, Inc., 1997
7. Michalewicz, Z.: Genetic Algorithms + Data Structres = Evolution programs, Springer-Verlag, 1992.
8. Hsu, C.F, Liu, T.L, Huang, N.F: Perfomance of adaptive routing in wavelength-routed networks with wavelength conversion capability, Photonic Network Communications, 5:1(2003), 41-57.
9. Chu, X., Li, B., Zhang, Z.: A dynamic RWA algorithm in a wavelength-routed all-optical network with wavelength converters. IEEE INFOCOM'03 (2003), 1795-1804.
10. The Network Simulator, ns-2. http://www.isi.edu/nsnam/ns/index.html (2003).

Ensuring E-Transaction Through a Lightweight Protocol for Centralized Back-End Database

Paolo Romano, Francesco Quaglia, and Bruno Ciciani

DIS, Università "La Sapienza", Roma, Italy

Abstract. A reasonable end-to-end reliability guarantee for three-tier systems, called e-Transaction (exactly-once Transaction), has been recently proposed. This work presents a lightweight e-Transaction protocol for centralized back-end database. Our protocol does not require coordination among the replicas of the application server and does not rely on any assumption for what concerns the processing order of messages exchanged among processes, as instead required by some existing solution.

1 Introduction

The concept of "e-Transaction" (exactly-once Transaction) has been recently introduced in [5] as a desirable, yet realistic, form of end-to-end reliability guarantee for three-tier systems. In this paper we present an e-Transaction protocol for three-tier systems in which the application servers interact, as in the case of most e-Commerce applications, with a centralized back-end database. Our e-Transaction protocol handles failures (or suspect of failures due to reduced system responsiveness possibly caused by host/network overload) by simply letting the client perform a timeout based retransmission logic of its request to different replicas of the application server. On the other hand, we use some recovery information, locally manipulated at the database side, to guarantee that the corresponding transaction is committed exactly one time. Manipulation of the recovery information does not require coordination among the application server replicas, which do not even need to know each other existence. Our proposal is therefore inherently scalable, and well suited for both local and geographic distribution of the replicas themselves.

Beyond providing the description of the protocol, together with its correctness proof and experimental measures demonstrating its minimal overhead, we present an extended comparative discussion with existing proposals in support of reliability, pointing out the advantages from our solution. The discussion will also outline that the proposal closest to our protocol (both in terms of structure and overhead compared to a baseline approach that does not provide reliability guarantee), namely the one in [4], relies on specific assumptions for what concerns the order of message processing to avoid duplication of transactions at the back-end database. Our solution does not require any of those assumptions, thus being suitable for a wider range of system settings.

2 System Model

The target three tier system we consider, consists of a set of processes, which communicate through message exchange. Processes can fail according to the crash-failure model. If a process does not fail, we say that the process is correct, and we assume there is at least a correct application server process at any time. Communication channels between processes are assumed to be reliable, therefore each message is eventually delivered unless either the sender or the receiver crashes during the transmission. In what follows we present features of each class of processes in the system, i.e. client, application server and database server, together with basics about the recovery information maintained at the back-end database.

Client. A client process does not communicate with the database server, it only interacts with the application servers in the middle-tier. It sends a request to an application server in order to invoke the transactional logic on this server (i.e. the application server is requested to perform some updates on the back-end database) and then waits for the outcome. The client sends the request by invoking the function issue, which takes the request content as a parameter. issue returns only upon the receipt of a positive outcome (commit) for the transaction ([1]).

Application Server. Application server processes have no affinity for clients. Moreover, they are stateless, in the sense that they do not maintain states across requests from clients. A request from a client can only determine changes in the state of the back-end database. Application servers have a primitive compute, which embeds the transactional logic for the interaction with the database. This primitive is used to model the application business logic while abstracting the implementation details, such as SQL statements, needed to perform the data manipulations requested by the client. compute executes the updates on the database inside a transaction that is left uncommitted, therefore the changes applied to data are not made permanent as long as the database does not decide positively on the outcome of the transaction. The result value returned by the primitive compute captures the output of the execution of the transactional logic at the database, which must be communicated to the client. We assume the primitive compute returns, together with the output of the execution of the transactional logic, the identifier assigned by the database server to the corresponding transaction. compute is assumed to be non-blocking, which means it eventually returns unless the application server crashes.

Database Server. The system back-end consists of a database server which is assumed to eventually recover after a crash. Also, there is a time after which

---

[1] For simplicity of presentation, we model with positive outcome also transactions for which the application logic cannot admit update on the database, e.g. like in a bank transfer operation with not enough money on the corresponding account, but that are correctly handled by the database, e.g. with no rollback caused by the concurrency control mechanism.

the database stops crashing and remains up, allowing any legal transaction to be eventually committed ([2]). In practice, this means assuming the database can experience a period of instability during which it can crash and recover, and then experiences a period during which it does not crash, which is long enough to allow a legal transaction to be eventually committed.

The database server has a primitive `decide` which can be used to invoke the commitment of a pending transaction. This primitive is called by the database server (with a transaction identifier as the unique parameter) upon the receipt of a message from the application server that asks for a final decision (commit/rollback) for a pending transaction. `decide` returns commit/rollback depending on the final decision the database takes for the transaction, together with any exception possibly raised during the decision phase. We assume that the `decide` primitive is non-blocking, i.e. it eventually returns unless the database server crashes after the invocation. Also, as in conventional database technology, if the database server crashes while processing a transaction, then, upon recovery, it does not recognize that transaction as an active one. Therefore, if the `decide` primitive is invoked with an identifier associated with an unrecognized transaction, then the return value of this primitive is rollback.

Recovery Information. The database offers an abstraction called "testable transaction" originally presented in [4]. With this abstraction, the database stores recovery information that can be used to determine whether a given transaction has already been committed. Specifically, each transaction is associated with a unique client request identifier, which is stored within the database as a part of the transaction execution itself, together with the result of the transaction. If the identifier is stored within the database, then this means that the corresponding client request originated a transaction that has already been committed. Note that the testable transaction abstraction can be implemented in a transparent way to the client by simply modifying the application server transactional logic. Specifically, as in [4], we assume application servers have an additional primitive `insert` allowing them to ask the database server to write the identifier of a client request within the database together with the result obtained by the execution of the `compute` primitive. Additionally, just like `compute`, the primitive `insert` is non-blocking, i.e. it eventually returns unless the application server crashes.

3 The Protocol

The protocol we present ensures the following two properties synthesizing the e-Transaction problem as introduced in [4, 5]:

Safety. The back-end database does not commit more than one transaction for each client request.

---

[2] We use the term "legal" to refer to a transaction that does not violate any integrity constraint on the database. As an example, the attempt to duplicate a primary key makes a transaction illegal.

Liveness. If a client issues a request, then unless it crashes, it eventually receives a commit outcome for the corresponding transaction, together with the result of the transaction.

It is important to note that, according to the specification of liveness guarantees as proposed in [4, 5], an e-Transaction protocol is not required to ensure liveness in the presence of client crash. This is because the e-Transaction framework deals with thin clients having no ability to maintain recovery information. This reflects a representative aspect of current Web-based systems where access to persistent storage at the client side can be (and usually is) precluded for a variety of reasons. These range from privacy and security issues (e.g. to contrast malicious and/or intrusive Web sites invasively delivering cookies) to constraints on the available hardware (e.g. in case of applications accessible through cell phones).

We present the protocol describing its behavior separately for every class of processes in the system, i.e. client, application server and database server.

Client Behavior. The pseudo-code defining the behavior of the function `issue` used by the client is shown in Figure 1. The client generates an identifier associated with the request, selects one application server and sends a Request message to this server, together with the request identifier. It then waits for the reply. In case it receives commit as the outcome for the corresponding transaction, `issue` simply returns. In any other case, it means that something wrong might have occurred. Specifically: (i) Timeout expiration means the application server and/or the database server might have crashed. (ii) Rollback outcome means instead that the database could not commit the transaction, for example because of decisions of the concurrency control mechanism. In both cases, `issue` reselects an application server (possibly different from the last selected one) and retransmits the Request message to that application server, with the already selected request identifier. Upon successive timeout expirations, the client keeps on retransmitting the Request message (with that same identifier) until it receives the commit outcome.

```
result issue(request_content req){
1.    generate a new id;
2.    select an application server AS;
3.    set outcome=ROLLBACK;
4.    send Request[req,id] to AS;
5.    while (outcome is not COMMIT){
6.        await receive Outcome[outcome,res,id] or TIMEOUT;
7.        if (TIMEOUT or outcome is not COMMIT){
8.            select an application server AS;
9.            send Request[req,id] to AS;
10.       } /* end if */
11.   } /* end while */
12.   return res;
13. }
```

Fig. 1. Client Behavior

```
Application Server:
1.    result res;
2.    transaction_identifier tid;
3.    while(true){
4.         await receive Request[req,id] from client;
5.         [res,tid]=compute(req);
6.         outcome=TestableTransaction(res,id);
7.         send Outcome[outcome,res,id] to client;
8.    } /* end while */

outcome TestableTransaction(result res, request_identifier id){
9.    insert(res,id); /* where id is a primary key */
10.   repeat{
11.        send Decide[tid] to the database server;
12.        await receive Outcome[outcome,exception,tid] or TIMEOUT;
13.   }until(received Outcome[outcome,exception,tid]);
14.   if(exception.type = duplicated_primary_key_exception){
15.        set res=exception.result;
16.        return COMMIT;
17.   } /* end if */
18.   return outcome;
19. }
```

Fig. 2. Application Server Behavior

Application and Database Server Behaviors. The application server behavior is shown in Figure 2. Upon the receipt of a Request message, this server invokes the primitive compute to start a transaction on the back-end database. The transaction identifier assigned by the database server is returned to the application server and maintained by *tid*. The application server then invokes TestableTransaction. Within this function, the application server first executes insert, in order to store the client request identifier within the database, together with the result of the transaction. It then sends a Decide message with that *tid* to the database server and waits for the outcome. This same message is periodically retransmitted in case of subsequent timeout expirations.

We assume the client request identifier to be a primary key for the database, which is the mechanism we adopt to guarantee the safety property. Therefore, any attempt to commit multiple transactions associated with the same client request identifier is rejected by the database itself, which is able to notify the rejection event by rising an exception. This makes the client request for updating data within the database an idempotent operation, i.e. the request can be safely retransmitted multiple times to different application servers. We note that assuming the client request identifier to be a primary key is a viable solution in practice. In case we can modify the database schema, this primary key can be easily added. In case the schema is predetermined and not modifiable (e.g. legacy databases), as suggested in [4] while describing supports for the testable transaction abstraction, an external table can be used.

Upon the receipt of the Outcome message in reply from the database server, the flag *exception* is checked to determine whether the same request identifier was already in the database. In the positive instance, a transaction associated with that same client request has already been committed. As a result, the exception allows the application sever to return an Outcome message with the commit indication to the client together with the already established result. In any other case (i.e. *exception* is not raised), the outcome received by the

```
Database Server:
1.    while(true){
2.        await receive Decide[tid] from an application server;
3.        [outcome,exception]=decide(tid);
4.        send Outcome[outcome,exception,tid] to the application server;
5.    }
```

Fig. 3. Database Server Behavior

database server is sent back to the client. The outcome might be rollback, e.g., due to decisions of the concurrency control mechanism.

The behavior of the database server is shown in Figure 3. For simplicity we only show the relevant operations related to transaction commitment, while skipping the data manipulation associated with the business logic. This server waits for a Decide message from an application server which asks to take a final decision for a transaction associated with a given *tid*, and then attempts to make the transaction updates permanent through the `decide` primitive. The final result (commit/rollback) is then sent back to the application server, together with the *exception*, possibly indicating the attempt to duplicate a primary key (i.e. the identifier of the client request) within the database.

3.1 Proof of Correctness

Theorem 1. - Safety
The back-end database does not commit more than one transaction for each client request.

Proof. **(By Contradiction).** Given the structure of the protocol, it is possible that multiple transactions associated with the same client request are activated by the application servers. Assume, by contradiction, that a generic number $N > 1$ of them are eventually committed. In this case, the database server must have received multiple Decide messages from the application servers for transactions associated with the same client request. By the application server pseudo-code, this server sends the Decide message to the database server (see line 11 in Figure 2) only after it has executed a whole transaction that encompasses both the data manipulation proper of the application business logic through the `compute` primitive (see line 5 in Figure 2), and the storing of the request unique identifier together with the result of the data manipulation through the `insert` primitive (see line 9 in Figure 2). As a consequence, the $N > 1$ transactions associated with the same client request, which are eventually committed, must perform a successful insertion of the unique request identifier within the database. However, this is impossible since the database maintains a primary key constraint on the request identifier, hence no more than one of those N transactions can perform that insertion successfully. Therefore the assumption is contradicted and the claim follows.

Lemma 1. *If a correct application server sends a Decide message to the database server asking for a decision on a transaction, the application server eventually receives an Outcome message for that transaction from the database server.*

Proof. **(By Contradiction).** Assume by contradiction that a correct application server sends a Decide message to the database server and that no Outcome message from the database server is ever received for the corresponding transaction. In this case, the correct application server keeps on retransmitting the Decide message to the database server indefinitely (see lines 10-13 in Figure 2). Hence, a Decide message will be sent by the application server to the database server at time $t' > t$, where t be the time after which the database server stops crashing and remains up. Given that after time t both the correct application server and the database server are always up, for the assumption on the reliability of the communication channels we can claim that the database server will eventually receive the Decide message. Also, the database server will eventually take a decision through the `decide` primitive (since it does not crash anymore) and will send an Outcome message to the application server. Again, since communication channels are assumed to be reliable, the correct application server will eventually receive that Outcome message. Therefore the assumption is contradicted and the claim follows.

Theorem 2. - Liveness
If a client issues a request, then unless it crashes, it eventually receives a commit outcome for the corresponding transaction, together with the result of the transaction.

Proof. **(By Contradiction).** Assume by contradiction that the client issues a request, does not crash and does not eventually receive a commit outcome. In this case, the client keeps on retransmitting the Request message to the application servers indefinitely (see lines 5-11 in Figure 1). As we have assumed that channels are reliable and that at least an application server is correct (i.e. it does not crash), an unbounded amount of Request messages will eventually be delivered to a correct application server. Moreover, if an Outcome message is sent back by a correct application server to the client, this message will eventually be received, since the client does not crash. Then, to show that the previous assumption is wrong, we only need to show that a correct application server receiving an unbounded amount of Request messages will eventually send to the client an Outcome message with a commit indication.

When a correct application server receives a Request message from the client, it calls the primitives `compute` and `insert` (see lines 5 and 9 in Figure 2). These primitives, being non-blocking, eventually return, therefore the application server eventually sends to the database server the Decide message (see line 11 in Figure 2). By Lemma 1, an Outcome message for the transaction is eventually sent back by the database server and is eventually received by the correct application server (recall this message also carries the value of the *exception* flag). There are two possible cases:

A.1 If the Outcome message received from the database server carries a commit indication or the *exception* flag notifies the attempt to duplicate a primary key, then an Outcome message with commit is sent back to the client together with the transaction result. Therefore, the assumption is contradicted.

A.2 If the Outcome message received from the database server carries a rollback indication, with the *exception* flag notifying no attempt to duplicate a primary key, then the transaction was legal but such a reply from the database server implies that the database was unable to commit the required operations (e.g. due to decisions of the concurrency control mechanism). In this case, an Outcome message with a rollback indication is sent to the client.

We note anyway that case A.2 (i.e. the only one that does not contradict the assumption) can't occur indefinitely as we have assumed that there is a time after which the database server remains up and is able to commit any legal transaction. We can therefore assert that we eventually fall in case A.1, which contradicts the assumption. Hence, the claim follows.

3.2 Protocol Overhead

Our protocol is essentially based on logging recovery information (i.e. the client request identifier and the result of the transaction) at the back-end database while processing the transaction associated with the client request. The cost of logging this recovery information is actually the unique overhead we pay as compared to a baseline protocol for the three-tier organization, which is not able to provide any end-to-end reliability guarantee. We argue that this overhead is negligible in practice since it only consists of the cost for a single SQL INSERT statement. To support this claim, we have performed some measurements related to the New-Order and the Payment Transactions specified by the TPC BENCHMARK^{TM} C [10], both reflecting on-line database activity, as typically found in production environments, but exhibiting different profiles for what concerns read/write operations. The measurements have been taken by running the Solid FlowEngine 4.0 DBMS on top of a multi-processor system, equipped with 4 Xeon 2.2 GHz, 4 GB of RAM and 2 SCSI disks in RAID-0 configuration, running Windows 2003 Server. The application logic was implemented in JAVA2 with stored procedure technology. The below table reports the cost of database activities for both the baseline protocol and our proposal. Each reported value, expressed in msec, is the average over a number of samples that ensures confidence interval of 10% around the mean at the 95% confidence level. These experimental data clearly show that the overhead exhibited by our protocol for logging the recovery information is minimal, never exceeding 2%, even for the lighter transaction profile, namely the Payment Transaction.

	Baseline	Our protocol	Overhead
New-Order Transaction	72.2	73.1	+1.21%
Payment Transaction	46.4	47.3	+2.06%

4 Related Work and Discussion

A typical solution for providing reliability consists of encapsulating the processing of the client request within an atomic transaction to be performed by the middle-tier (application) server [6]. This is the approach taken, for example, by Transaction Monitors or Object Transaction Services such as OTS or MTS. However, this solution does not deal with the problem of loss of the outcome due, for example, to middle-tier server crash. The work in [7] tackles the latter issue by encapsulating within the same transaction both processing and the storage of the outcome at the client. This solution imposes the use of a distributed commit protocol, such as two-phase commit (2PC), since the client is required to be part of the transactional system. Therefore, it exhibits higher communication/processing overhead as compared to our protocol.

Several solutions based on the use of persistent queues have also been proposed in literature [1, 2], which are commonly deployed in industrial mission critical applications and supported by standard middleware technology (e.g. JMS in the J2EE architecture, Microsoft MQ and IBM MQ series). However, persistent queues are transactional resources, whose updates must be performed within the same transactional context where the application data are accessed (i.e. the request message must be dequeued within the same distributed transaction that manipulates application data and enqueues the response to the client). This needs coordination among several transactional resources just through a distributed commit protocol (e.g. 2PC). Therefore, compared to our protocol, also in this case the communication/processing overhead is higher. Additionally, as discussed in [3, 5], the use of persistent queues, in combination with classical 2PC as the distributed commit protocol, imposes explicit coordination among the application servers to support fail-over of an application server (i.e. the coordinator of the distributed transaction) suspected to have crashed. This originates additional overhead and reduces scalability. Since our protocol does not use any coordination scheme among the application server replicas, it provides better system performance and scalability, thus being attractive especially in the case of high degree of replication of the application access point, with the replicas possibly distributed on a geographic scale, e.g. like in Application Delivery Networks (ADNs) such as those provided by Sandpiper, Akamai or Edgix ([3]).

Message logging has also been used as a mean to recover from failures in multi-tier systems [8]. A client logs any request sent to the server, which also logs any request received. This allows the server to reply to multiple instances of the same request from a client without producing side effects on the back-end database multiple times. The server also logs read/write operations on the database, in order to deal with recovery of incomplete transaction processing. Differently from our proposal, this solution primarily copes with stateful client/middle-tier applications, e.g. like CAD or work-flow systems.

Frolund and Guerraoui have presented three different e-Transaction protocols [3, 4, 5]. The solutions in [3, 5] are based on an explicit coordination scheme among the replicas of the application server, so they have to pay an additional overhead due to coordination. As a consequence, they are mainly tailored for the case of replicas of the application server hosted by, e.g., a cluster environment, where the cost of coordination can be kept low thanks to low delivery latency of messages among the replicas. Since coordination among the replicas is not required in our protocol, we can avoid that overhead at all, with performance

---

[3] These infrastructures result as a natural evolution of classical Content Delivery Networks (CDNs), where the edge server has not only the functionality to enhance the proximity of contents to clients, but also to enhance the proximity between clients and the application (business) logic, and to increase the application availability.

benefits especially in case of high degree of replication of the application server and distribution of the replicas on a geographical scale (e.g. like in ADNs).

Like our solution, the third protocol by Frolund and Guerraoui [4] relies on the testable transaction abstraction ([4]), and has the advantages of not requiring explicit coordination across the middle-tier and of not using any distributed commit scheme. However, differently from our proposal, it handles failure suspicions through a "termination" phase executed upon timeout expiration at the client side. During this phase, the client sends, on a timeout basis, terminate messages to the application servers in the attempt to discover whether the transaction associated with the last issued request was actually committed. An application server that receives a terminate message from the client tries to rollback the corresponding transaction, in case it were still uncommitted (possibly due to crash of the application server originally taking care of it). At this point the application server determines whether the transaction was already committed by exploiting the testable transaction abstraction. In the positive case, the application server retrieves the transaction result to be sent to the client. Otherwise, a rollback indication is returned to the client in order to allow it to safely send a new request message (with a different identifier) to whichever application server.

Our protocol avoids the termination phase since it makes retransmissions of a same request idempotent operations thanks to the use of a primary key constraint imposed on the recovery information. From the point of view of performance, the avoidance of the termination phase reduces the fail-over latency as compared to [4]. More importantly, avoiding the termination phase makes our protocol a more general solution. In fact, by admission of the same authors, the employment of such a phase limits the usability of their protocol to environments where it can be ensured that a request message is always processed before the corresponding terminate messages. This is due to the fact that, according to the specifications of the standard interface for transaction demarcation, namely XA, when a rollback operation is performed for a transaction with a given identifier, the database system can reuse that identifier for a successive transaction activation (see [9] - state table on page 109). Hence, if a terminate message was processed before the corresponding request message in the protocol in [4], the latter message could possibly give rise to a transaction that gets eventually committed. On the other hand, upon the receipt of a reply to a terminate message, the client might activate a new transaction, with a different identifier, which could eventually get committed, thus leading to multiple updates at the database and violating safety. In order to achieve the required processing order constraint for request and terminate messages, the authors suggest to delay the processing of the terminate messages at the application servers. This expedient might reveal adequate in case the application is deployed over an infrastructure with controlled message delivery latency and relative process speed, e.g. a (virtual) private network or an Intranet. However, if the system can experi-

---

[4] Also this protocol logs some recovery information at the database while processing the transaction through a similar `insert` primitive.

ence periods during which the message delivery latency gets unpredictably long and/or the process speeds diverge, e.g. like in an asynchronous system, simply delaying the processing of a terminate message would not suffice to ensure such an ordering constraint. The latter constraint could be enforced through additional mechanisms (e.g. explicit coordination among the servers), but these would negatively affect both performance and scalability of this protocol. By all means, delaying the processing of terminate messages, even if adequate for specific environments, would further penalize the user perceived system responsiveness during the fail-over phase as compared to our solution. Conversely, our protocol does not rely on any constraint on the processing order of messages exchanged among processes, thus it requires no additional mechanism to enforce such an order and can be straightforwardly adopted in an asynchronous system, e.g. an infrastructure layered on top of public networks over the Internet.

References

1. P. Bernstein, M. Hsu and B. Mann, "Implementing Recoverable Requests Using Queues", *Proc. 19th ACM Conference on the Management of Data*, pp.112-122, 1990.
2. E.A. Brewer, F.T. Chong, L.T. Liu, S.D. Sharma and J.D. Kubiatowicz, "Remote Queues: Exposing Message Queues for Optimization and Atomicity." *Proc. 7th ACM Symposium on Parallel Algorithms and Architectures*, Santa Barbara, CA, pp.42-53, 1995.
3. S. Frolund and R. Guerraoui, "Implementing e-Transactions with Asynchronous Replication", *IEEE Transactions on Parallel and Distributed Systems*, vol.12, no.133-146, pp.2001.
4. S. Frolund and R. Guerraoui, "A Pragmatic Implementation of e-Transactions", *Proc. 19th IEEE Symposium on Reliable Distributed Systems*, pp.186-195. 2000.
5. S. Frolund and R Guerraoui, "e-Transactions: End-to-End Reliability for Three-Tier Architectures", *IEEE Transactions on Software Engineering*, vol.28, no.4, pp. 378-398, 2002.
6. J. Gray and A. Reuter, "Transaction Processing: Concepts and Techniques", Morgan Kaufmann, 1993.
7. M.C. Little and S.K. Shrivastava, "Integrating the Object Transaction Service with the Web", *Proc. 2nd IEEE Workshop on Enterprice Distributed Object Computing*, pp.194-205, 1998.
8. D. Lomet and G. Weikum, "Efficient Transparent Application Recovery in Client-Server Information Systems", *Proc. 27th ACM Conference on the Management of Data*, pp.460-471, 1998.
9. The Open Group, "Distributed Transaction Processing: The XA+ Specification Version 2", 1994.
10. Transaction Processing Performance Council (TPC), "TPC Benchmark^{TM} C, Standard Specification, Revision 5.1", 2002.

Cayley DHTs — A Group-Theoretic Framework for Analyzing DHTs Based on Cayley Graphs

Changtao Qu[1], Wolfgang Nejdl[1], and Matthias Kriesell[2]

[1] L3S and University of Hannover
Expo Plaza 1, D–30539 Hannover, Germany
{qu, nejdl}@l3s.de
[2] Institute of Mathematics (A), University of Hannover
Welfengarten 1, D-30167 Hannover, Germany
kriesell@math.uni-hannover.de

Abstract. Static DHT topologies influence important features of such DHTs such as scalability, communication load balancing, routing efficiency and fault tolerance. While obviously dynamic DHT algorithms which have to approximate these topologies for dynamically changing sets of peers play a very important role for DHT networks, important insights can be gained by clearly focussing on the static DHT topology as well. In this paper we analyze and classify current DHTs in terms of their static topologies based on the Cayley graph group-theoretic model and show that most DHT proposals use Cayley graphs as static DHT topologies, thus taking advantage of several important Cayley graph properties such as vertex/edge symmetry, decomposability and optimal fault tolerance. Using these insights, Cayley DHT design can directly leverage algebraic design methods to generate high-performance DHTs adopting Cayley graph based static DHT topologies, extended with suitable dynamic DHT algorithms.

1 DHTs and Static DHT Topologies

Two important characteristics of distributed hash tables (DHTs) are network degree and network diameter. As DHTs are maintained through dynamic DHT algorithms, high network degree means that joining, leaving and failing nodes affect more other nodes. Based on network degree, we group static DHT topologies into two types: *non–constant degree DHT topologies*, whose network degree increases (logarithmically) with the number of nodes in the network, and *constant (average) degree DHT topologies*, whose network degree stays constant even when the network grows. Consequently, DHTs can be classified into *non-constant degree DHTs* such as HyperCup(hypercubes), Chord (ring graphs), Pastry/Tapestry (Plaxton trees), etc., and *constant degree DHTs* such as Viceroy (butterfly), Cycloid (cube connected cycles), and CAN (tori).

Though this classification is certainly useful, the listed DHT topologies seem to have nothing more in common. Each topology exhibits specific graph properties resulting in specific DHT system features. Consequently, DHTs have so far been analyzed comparing individual systems, without a unified analytical framework which allows further insight into DHT system features and DHT system design.

The unified analytical framework discussed in this paper – *Cayley DHTs* – allows us to compare DHT topologies on a more abstract level and characterizes common features of current DHT designs. In a nutshell, we show that most current static DHT topologies such as hypercubes, ring graphs, butterflies, cube-connected cycles, and d-dimensional tori fall into a generic group-theoretic model, Cayley graphs, and can be analyzed as one class. These Cayley graph based DHTs (hereafter *Cayley DHTs*), including both non-content degree DHTs and constant degree DHTs, intentionally or unintentionally take advantage of several important Cayley graph properties such as vertex/edge symmetry, decomposability, good connectivity and hamiltonicity to achieve DHT design goals such as scalability, communication load balancing, optimal fault tolerance, and routing efficiency. Several non-Cayley DHTs also utilize techniques in their dynamic DHT algorithms that try to imitate desirable Cayley graph properties, again showing the close relationship between Cayley graph properties and desirable DHT system features.

2 Cayley DHTs — A Group-Theoretic Model for Analyzing DHTs

2.1 Groups and Cayley Graphs

Cayley graphs were proposed as a generic group-theoretic model for analyzing symmetric interconnection networks [1]. The most notable feature of Cayley graphs is their universality. Cayley graphs embody almost all symmetric interconnection networks, as every vertex transitive interconnection network can be represented as the quotient of two Cayley graphs [2]. They represent a class of high performance interconnection networks with small degree and diameter, good connectivity, and simple routing algorithms. The following paragraphs give the formal definitions.

A *group* is a pair $\Gamma := (V, \cdot)$ such that V is a (nonempty) set and $\cdot : V \times V \longrightarrow V$ maps each pair (a, b) of elements of V to an element $a \cdot b$ of V with $a \cdot (b \cdot c) = (a \cdot b) \cdot c$ for all $a, b, c \in V$, such that there exists an element $1 \in V$ with the following properties: (i) $1 \cdot a = a$ for all $a \in V$ and (ii) for every $a \in V$, there exists some $b \in V$ with $b \cdot a = 1$.

1 is the unique element having properties (i) and (ii). It is called the *neutral element* of Γ, and $a \cdot 1 = a$ holds for all $a \in V$. b as in (ii) is uniquely determined by a and is called the *inverse* of a, written as $b = a^{-1}$. It is the unique element b for which $a \cdot b = 1$ holds. If $a \cdot b = b \cdot a$ holds for all $a, b \in V$ then Γ is called an *abelian* group. This is usually expressed by *additive notation*, i. e. by writing $\Gamma = (V, +)$, 0 for the neutral element, and $-a$ for the inverse of a. Groups are fundamental objects of mathematics, and the foundation for Cayley graphs.

Let $\Gamma := (V, \cdot)$ be a finite group, 1 its neutral element, and let $S \subseteq V - \{1\}$ be closed under inversion (i. e. $x^{-1} \in S$ for all $x \in S$). The *Cayley graph* $G(\Gamma, S) = (V, E)$ of $(V, \cdot)$ and S is the graph on V where x, y are adjacent if and only if $xy^{-1} \in S$.

We can also define *directed versions* of this concept, which are obtained by omitting the symmetry condition $S^{-1} = S$ to S. The condition $1 \notin S$ keeps Cayley graphs *loopless*. Note that Cayley graph are sometimes called *group graphs*.

2.2 Non-constant Degree Cayley DHTs

HyperCup [3] Though HyperCup itself is not a DHT system, it is a topology for structured P2P networks which could also be used for DHT design, and which represents an important type of Cayley graphs, hypercubes. So far there are no DHTs which use pure hypercubes as static DHT topologies, even though some literature (i.e. [4,5]) argue that Pastry/Tapestry and Chord emulate approximate hypercubes when taking into account the dynamic DHT algorithm design. However, differentiating cleanly between static DHT topologies and dynamic DHT algorithms, it is more appropriate to describe their static topologies as Plaxton trees and ring graphs respectively.

Hypercubes are typical Cayley graphs. For a natural number m, let $(\mathcal{Z}_m, +)$ denote the group of residuals mod m. Consider the group $\Gamma := (\mathcal{Z}_2^d, +)$, where $\mathcal{Z}_2^d$ denotes the set of all $0, 1$-words of length d and $+$ is the componentwise addition mod 2. We want to make a, b adjacent whenever they differ in exactly one digit, i.e. whenever $a - b$ is a word containing precisely one letter 1. So if S is the set of these d words then S is closed under inversion, and $H_2^d := G(\Gamma, S)$ is called the (binary) *d-dimensional (binary) hypercube*.

It is also possible to give a *hierarchical description* of H_2^d by means of the following recursion. Set $H_2^1 = (\{0, 1\}, \{01\})$, and for $d > 1$ define H_2^d recursively by $V(H_2^d) := \{xv : x \in \{0, 1\}, v \in V(H_2^{d-1})\}$ and $E(H_2^d) := \{xvyw : xv, yw \in V(H_2^d)$ and: $(x = y \land vw \in E(H_2^{d-1}))$ or $(x \neq y \land v = w)\}$. Roughly, in every step, we take two joint copies of the previously constructed graphs and add edges between pairs of corresponding vertices.

This concept can be generalized by looking at *cartesian products* of graphs: For graphs G, H, let their *product* $G \times H$ be defined by $V(G \times H) := V(G) \times V(H)$ and $E(G \times H) := \{(w, x)(y, z) : (w = y \in V(G) \land yz \in E(H))$ or $(y = z \in V(H) \land wx \in E(G))\}$. Clearly, $G \times H$ and $H \times G$ are isomorphic (take $(x, y) \mapsto (y, x)$ as an isomorphism). Defining $K_2 := (\{0, 1\}, \{01\})$ to be the complete graph on two vertices, we see that H_2^1 is isomorphic to K_2 and H_2^d is isomorphic to $H_2^{d-1} \times K_2$ for $d > 1$, which is in turn isomorphic to $K_2 \times \cdots \times K_2$ (d factors K_2).

As every finite group is isomorphic to some group of permutations, it is possible to unify the Cayley graph notion once more. Without loss of generality, we can assume that the generating group Γ is a permutation group. This is certainly useful when describing algorithms on general Cayley graphs. For the presentation here, it is, however, more convenient to involve other groups as well, e.g. permutation groups: For Γ, we take the subgroup of the permutation group $S_6 := (\{f : \{1, \ldots, 6\} \longrightarrow \{1, \ldots, 6\} : f$ bijective$\}, \circ)$ generated by $S := \{213456, 124356, 123465\}$ (here $a_1 \cdots a_6$ denotes the permutation f of $\{1, \ldots, 6\}$ with $f(1) = a_1, f(2) = a_2, \ldots, f(6) = a_6$). $f \circ g$ is the permutation defined by $(f \circ g)(x) = f(g(x))$ for all possible x.)

Chord [6] Chord uses a 1-dimensional circular key space, in which the node responsible for a key is the node whose identifier most closely follows the key in the numeric order (the key's *successor*). All nodes in Chord are arranged into a *ring graph*. In a m-bit Chord key space, each Chord node maintains two sets of neighbors: a successor list of k nodes that immediately follow it in the key space, and a finger list of $O(\log N)$ nodes spaced exponentially around the key space. The ith entry of the finger list points to the node that is 2^i away from the current node, or to that node's successor if that node does not exist.

The graphs approximated here are special *circulant graphs*,i. e. Cayley graphs obtained from the cyclic group $(\mathcal{Z}_n, +)$ and an arbitrary (inversion–closed) generator set.The most prominent example is the *cycle* $C_n := G(\mathcal{Z}_n, \{\pm 1\}) = (\mathcal{Z}_n, \{01, 12, 23, \ldots, (n-1)n, n0\})$ of length n. For the topology of the ideal d-bit Chord key space, we simply take the Cayley graph $R_d := G((\mathcal{Z}_{2^d}, +), \{\pm 2^k : k \in \{0, \ldots, d-1\}\})$.

2.3 Constant Degree Cayley DHTs

Cycloid [4] Cycloid is a constant degree DHT emulating a cube connected cycle as its static DHT topology. In Cycloid, each node is specified by a pair of cyclic and cube indices. In order to dynamically maintain connectivity of the DHT topology, the dynamic DHT algorithm of Cycloid forces each node to keep a routing table consisting of 7 entries. Among them, several entries (so-called leaf sets) only make sense for the dynamic DHT algorithm to deal with network connectivity in sparsely populated identifier spaces. A *d-dimensional cube connected cycle graph* is obtained from a d-dimensional cube by replacing each vertex with a cycle of d nodes. It contains $d \cdot 2^d$ nodes of degree d each. Each node is represented by a pair of indices (k, v), where $k \in \mathcal{Z}_d$ is a cyclic index and $v \in \mathcal{Z}_2^d$ is a cube index. A cube connected cycle graph can be viewed as a specific case of Cayley Graph Connected Cycles (CGCC) [7], defined as:

Let $\Gamma = (V, \cdot)$ be a group and $S := \{s_1, \ldots, s_d\} \subseteq V - \{1\}$ closed under inversion with $d \geq 3$. The Cayley graph connected cycles network $CGCC(\Gamma, S) = (V', E')$ is the graph defined by $V' := \mathcal{Z}_d \times V$ and $E' := \{(i, x)(j, y) : (x = y \wedge i = j \pm 1)$ or $(i = j \wedge x = s_i \cdot y)\}$.

$CGCC(\Gamma, S)$ is obtained by replacing each vertex of the Cayley graph $G(\Gamma, S)$ with a cycle of length d and replacing each edge of $G(\Gamma, S)$ with an edge connecting two members of a cycle in a certain way. The edges $(i, x)(j, y)$ with $i = j$ form *cycle connections*, the others form *cayley graph connections*. [8] proves that these graphs are Cayley graphs. Following the definition of CGCC, the n-dimensional cube connected cycle is a graph built from a n-cube replacing each node with a cycle of length n.

Viceroy [9] Viceroy is a constant degree DHT emulating an approximate butterfly graph as its static DHT topology. The dynamic DHT algorithm of Viceroy is rather involved. It works based on a rough estimate of the network size and forces each node to keep a routing table containing 5 to 7 entries [9]. Similar to Cycloids, part of the entries only make sense for the dynamic DHT algorithm to deal with a sparsely populated identifier space (i.e. ring links [9]). For Viceroy we can only guarantee with high probability that the constructed DHT topology is a butterfly graph.

The d-dimensional binary wrapped directed butterfly $\boldsymbol{B}_2^d$ is a graph with vertices $V = V(\boldsymbol{B}_2^d) = \mathcal{Z}_{d-1} \times \mathcal{Z}_2^d$ such that there is an edge from $a = (i, v_1 \cdot v_d) \in V$ to $b = (j, w_1 \cdot w_d) \in V$ if and only if $i \in \{0, \ldots, d-1\}$, $j = i+1$ and $v_k = w_k$ for all $k \in \{0, \ldots, d-1\} - \{i\}$. One can think of i, j as of the *levels* of a and b, respectively, and some level i vertex (i, v) has precisely two neighbors $(i+1, v)$ and $(i+1, v')$, where v' is obtained from v by adding 1 (mod 2)in the ith component of v. The *d-dimensional binary wrapped butterfly* B_2^d is the underlying graph of the digraph $\boldsymbol{B}_2^d$, where there is a (single) edge ab whenever there is an edge (a, b) or an edge (b, a) in $\boldsymbol{B}_2^d$. As we can see, B_2^d is 4-regular for $d \geq 3$.

The advantage of taking the wrapped rather than the unwrapped version of the butterfly is that B_2^d is a Cayley graph, whereas unwrapped ones are not even regular, since for $d \geq 3$ the vertices on the border levels have degree 2 and the others have degree 4. We represent B_2^d as a Cayley graph of the *wreath product* of the groups $(\mathcal{Z}_d, +)$ and $(\mathcal{Z}_2^d, +)$. For (i, v) and (j, w) in V, we define $(i, v) \bullet (j, w) := (i+j, (v_0 + w_{-\ell}, v_1 + w_{-\ell+1}, \ldots, v_{d-1} + w_{-\ell+d-1}))$. Note that $i+j$ and the indices at the components of v and w are to be taken mod d. This operation constitutes a *group* $\Gamma = (V, \bullet)$, with neutral element $(0, 0)$. By taking $S = \{(1,0), (1, (1, 0, \ldots, 00 \cdots 0))\} \subseteq V$ we obtain the representation $B_2^d = G(\Gamma, S)$ of B_2^d as a Cayley graph (for more details see [8]).

CAN [10] CAN is an (adjustable) constant degree DHT using a virtual d-dimensional Cartesian coordinate space to store $(key, value)$–pairs. The topology under this Cartesian coordinate space is a d-dimensional torus. Let $T_{m,n} := C_m \times C_n$ of length m and n be the Cartesian product of two cycles C_m, C_n. The componentwise addition $+$ establishes a group $\Gamma(\mathcal{Z}_m \times \mathcal{Z}_n, +)$ on its vertices, and clearly $T_{m,n} = G(\Gamma, \{(0, \pm 1), (\pm 1, 0)\})$. Hence the torus is a Cayley graph as well. One could consider such a toroidal graph as a rectangular grid, where the points on opposite borders are identified. We can extend this definition easily to higher dimensions: Let $n_1, \ldots, n_d$ be numbers ≥ 2. Componentwise addition $+$ of elements in $V := \mathcal{Z}_{n_1} \times \cdots \times \mathcal{Z}_{n_d}$ establishes a group $\Gamma(V, +)$, and by taking S to be the set $\{(z_1, \ldots, z_d) \in P :$ there is an $i \in \{1, \ldots, d\}$ such that $z_i = \pm 1$ and $z_j = 0$ for all $j \neq i$ in $\{1, \ldots, d\}\}$ and we obtain a d-*dimensional torus* $T_{n_1, \ldots, n_d} = G(\Gamma, S)$. Explicitly, $T_{n_1, \ldots, n_d}$ is a graph on the vertex set V, where $(v_1, \ldots, v_d)$ and $(w_1, \ldots, w_d)$ are adjacent if and only if they differ in exactly one component and the difference in this component is either $+1$ or -1. As the presence of i's with $n_i = 2$ stretches formal arguments slightly (for example, when considering degrees), some authors force $n_i \geq 3$ for all $i \in \{1, \ldots, d\}$. They lose then, however, the possibility to consider the d-dimensional hypercube as a special torus, namely as $T_{2, \ldots, 2}$ (d indices 2).

2.4 Non-cayley DHTs

P-Grid [11] Among non-Cayley DHTs, to the best of our knowledge, only P-Grid [11] still retains most of the advantages of Cayley networks. P-Grid uses prefix based routing, and can be considered as a randomized approximation of hypercube. The routing network has a binary trie abstraction, with peers residing only at the leaf nodes. Each peer is thus responsible for all data items with the prefix corresponding to the peer's path in the trie. For routing, peers need to maintain routing information for the complimentary prefix for each of the intermediate nodes in its path. However, routing choice can be made for any peer belonging to the complimentary paths, and P-Grid exploits these options in order to randomly choose routing peer(s), which in turn provides query-forwarding load-balancing and by choosing more than one routing options, resilience. Additionally, the choices can be made based on proximity considerations, and though routing is randomized, since it is to complimentary key-space partitions, P-Grid routes have the added flexibility to be either bidirectional or unidirectional.

Pastry/Tapestry [12] [13] The static DHT topology emulated by Pastry/ Tapestry are Plaxton trees. However, when taking the dynamic DHT algorithms of Pastry/Tapestry into account, we find that the static DHT topology of Pastry/Tapestry behaves quite similar to an approximation of hypercubes. As analyzed in [5], in the Pastry/Tapestry identifier space, each node on the Plaxton tree differs from its ith neighbor on only the ith bit, dynamic routing is done by correcting a single bit at a time in the left-to-right order. This turns out to be the same routing mechanism adopted by DHTs using hypercubes as static DHT topologies, even though hypercube based DHTs allow bits to be corrected in any order.

3 Cayley Graph Properties and DHTs

Cayley graphs have a specific set of properties which can be closely associated with important DHT system features. The following paragraphs include a discussion of these Cayley DHT properties and provide a good insight into Cayley DHT design.

Symmetry and Load Balancing. The most useful properties of Cayley graphs are *symmetry properties*. Recall that an *automorphism* of some graph G is a bijection $\varphi : V(G) \longrightarrow V(G)$ with $\varphi(x)\varphi(y) \in E(G)$ if and only if $xy \in E(G)$.

A graph G is called *vertex symmetric* or *vertex transitive* if for arbitrary $x, y \in V(G)$ there exists an automorphism φ of G such that $\varphi(x) = y$. As the automorphism $z \mapsto z \cdot x^{-1} \cdot y$ maps x to y, we obtain the following classical observation.

Theorem 1. Every Cayley graph is vertex transitive.

This property results in an important feature of Cayley graphs — routing between two arbitrary vertices can be reduced to the routing from an arbitrary vertex to a special vertex [1]. This feature is significant for Cayley DHTs because it enables an algebraic design approach for the routing algorithm. Suppose that $\Gamma = (V, \circ)$ is a group of permutations, let $S \subseteq V - \{id_V\}$ be closed under inversion and consider the Cayley graph $G = G(\Gamma, S)$. For a path $P = x_0, \ldots, x_\ell$ from x_0 to x_ℓ set $s_i := x_{i-1}x_i^{-1}$ for $i \in \{1, \ldots, \ell\}$. Then the sequence $s_1, \ldots, s_\ell$ in S represents the path P, and it also represents the path from $x_0 x^{-1}$ to id_V. Consequently the routing problem G is equivalent to a certain sorting problem [1]. Taking V to be the set of all permutations of some set and $S \subseteq V$ to be the set of all transpositions will produce a *bubble sort graph* (see [8]).

We can leverage this property to implement optimized routing algorithms for Cayley DHTs through purely algebraic approaches supported by sets of mature algebraic methods. Furthermore, vertex transitivity provides a unified method to evaluate communication load on DHT nodes. In Cayley DHTs, the communication load is uniformly distributed on all vertices without any point of congestion. In contrast, non-Cayley DHTs exhibit congestion points. As communication load balancing is one of the principal design concerns of DHTs, this points out major drawback of non-Cayley DHTs.

In addition to vertex transitivity, Cayley graphs may also have another important property, edge transitivity. A graph G is *edge symmetric* or *edge transitive* if for arbitrary edges wx, yz there exists an automorphism φ such that $\varphi(w)\varphi(x) = yz$. Clearly, every edge transitive graph without isolated vertices is vertex transitive, but the converse is not true. For a discussion of the problem of determining the edge transitiv Cayley graphs we refer to [1] and [8].

Among Cayley DHTs, HyperCup (hypercubes), CAN(d-dimensional torus), and Viceroy (butterfly) are edge transitive, whereas Chord (ring graphs) and Cycloid (cube connected cycles) are not. Non–Cayley DHTs are not edge transitive. Edge transitivity results in a unified method to evaluate communication load on edges. In edge transitive Cayley DHTs communication load is uniformly distributed on all edges without points of congestion. For constant degree Cayley DHTs such as Cycloid, the loss of edge transitivity can be seen as a reasonable tradeoff against the constant degree property. For non-constant degree Cayley DHTs such as Chord, the loss of edge transitivity is disadvantageous, and has to be compensated through the design of the routing algorithm.

Hierarchy, Fault Tolerance, and Proximity. Recall that $<S>_\Gamma$ is the subgroup of $\Gamma = (V, \cdot)$ generated by $S \subseteq V$ i. e. the smallest subgroup of Γ which contains S. Let $\Gamma = (V, S)$ be a group and $S \subseteq V(G) - \{1\}$ such that $S^{-1} = S$. The Cayley graph $G(\Gamma, S)$ is *strongly hierarchical* if S is a *minimal generator* for G, i. e. if $<S>_G = G$ but $<S - \{s, s^{-1}\}>_G$ is a proper subgroup of G for every $s \in S$.

Among Cayley DHTs, HyperCup (hypercubes) and Chord (ring graphs) can be proven to be hierarchical [8]. Hierarchical Cayley graphs "often allow inductive proofs by decomposing (stripping) the graph into smaller members of the same family, thus are scalable in the sense that they recursively consist of copies of smaller Cayley graphs of the same variety" [8]. In DHT design, hierarchy can strongly affect the node organization and aggregation, which is closely associated with two important DHT system features: fault tolerance (i.e. network resilience) and proximity (i.e. network latency). Most hierarchical Cayley DHTs, except for a very particular family, are optimally fault tolerant as their connectivity is equal to their degree [14]. Furthermore, in hierarchical Cayley DHTs, there usually support easy solutions to dynamically organize nodes (or node aggregations) to ensure proximity of DHTs. Hierarchical Cayley graphs have not yet been intensively investigated for DHT design. Two promising hierarchical Cayley graphs not yet utilized in DHT design are star graphs and pancake networks [15], which have smaller network diameter than hypercubes of the same degree.

Connectivity and Fault Tolerance. A graph G is *disconnected* if it contains two vertices x, y such that there is no x, y-path in G. The *connectivity* $\kappa(G)$ of a finite (nonempty) graph is the minimum cardinality of a set X of vertices such that $G - X$ is disconnected or has less than two vertices.

A graph is called *d-regular* if every vertex has degree d. For example, every vertex transitive graph is regular. Clearly, d is an upper bound for the connectivity of a d-regular graph. Let us call a d-regular graph G *optimally fault tolerant* if its connectivity equals d. For example, complete graphs are optimally fault tolerant, so are hypercubes (as one can prove by induction on the dimension, using the recursive characterizations). For edge transitive graphs, we have the following.

Theorem 2. [16, 17, 18] (cf. [19]) *Every connected edge transitive graph is optimally fault tolerant.*

In general, connected Cayley graphs are not optimally fault tolerant; the smallest example showing this is the 5-regular circulant graph $G := G(\mathcal{Z}_8, \{\pm 1, \pm 3, 4\})$, as

$G - \{0, 2, 4, 6\}$ is disconnected. However, the following theorem on connected vertex transitive graphs shows that connectivity and degree can't differ too much.

Theorem 3. [16, 17, 18] (cf. [19]) The connectivity of a connected vertex transitive d-regular graph is at most d and at least $\frac{2}{3}(d+1)$.

In particular, for $d \in \{2, 3, 4\}$, every d-regular connected vertex transitive graph is d-connected, i.e. optimally fault tolerant. For $d = 5$, this statement is wrong even for Cayley graphs as seen in the previous example, but for $d = 6$ it's true "again": Every 6-regular vertex transitive graph is 6-connected. This follows easily from the main result in [16] which implies that *every triangle free connected vertex transitive graph is optimally fault tolerant*. More generally, every vertex transitive graph without four pairwise adjacent vertices is optimally fault tolerant [17]. This gives alternative proofs of the optimal fault tolerance of hypercubes and of d-dimensional tori $T_{n_1,\ldots,n_d}$ with $n_i \geq 4$ for all $i \in \{1,\ldots,d\}$. The graph $G(\mathcal{Z}_8, \{\pm 1, \pm 3, 4\})$ indicates that it might be already a problem to characterize the optimally fault tolerant circulants (solved in [20]).

Edge connectivity is less interesting from the point of view of optimal fault tolerance, as every d-regular vertex transitive graph has edge connectivity equal to d [17, 18] (cf. [19]). Hierarchical Cayley graphs as in Definition 3 and as in [8] or [14] are also known to be optimally fault tolerant unless they belong to a particular family of graphs whose d-regular members still have connectivity $d - 1$. For the technical details, we refer the reader to [8] or [14].

Among Cayley DHTs, HyperCup (hierarchical Cayley graphs), Chord (hierarchical Cayley graphs), Cycloid (3-regular Cayley graphs) and Viceroy (4-regular Cayley graphs) are optimally fault tolerant based on their static DHT topology perspective. CAN (d-regular Cayley graphs) can also be proven optimally fault tolerant based on its dynamic DHT algorithm features such as multiple realities and multiple dimensions [10]. For non-DHTs it is much harder to prove optimal fault tolerance. However, as fault tolerance is one of the principal design concerns of DHTs, most non–Cayley DHTs have included various techniques in their dynamic DHT algorithms to pursue possibly higher fault tolerance, although optimality cannot guaranteed. One possible such technique is to force each node to maintain a successor list in dynamic DHT algorithms.

For DHTs whose static DHT topologies are optimal fault tolerant, it is much easier to also ensure this in the dynamic algorithm design for sparsely populated DHT identifier spaces, or frequently leaving / failing nodes. Possible techniques include the successor list in Chord [6] or the state-machine approach based replication in Viceroy [9].

Hamiltonicity and Cyclic Routing. A path or cycle which visits every vertex in a graph G exactly once is called a *hamiltonian path* or *hamiltonian cycle*, respectively. Hamiltonicity has been received much attention of theorists in this context, as it is still open whether every 2-connected Cayley graph has a hamiltonian path.

The question of *hamiltonian cycles and paths* in Cayley graphs has a long history [21]. All aforementioned topologies of Cayley DHTs such as hypercubes, ring graphs, butterfly, cube-connected cycles, and d-dimensional tori have been proven to be hamiltonian.

Hamiltonicity is important for DHT design because it enables DHTs to embed a ring structure so as to implement ring based routing in dynamic DHT algorithms. Ring based routing, characterized by the particular organization of the DHT identifier space and ensuring the DHT fault tolerance in a dynamic P2P environment by means of maintaining successor/predecessor relationships between nodes, is used by almost all DHT proposals. Gummadi et al. [5] observes that the ring structure "allows the greatest flexibility and hence achieves the best resilience and proximity performance of DHTs". Although in terms of our analytical framework, we do not fully agree with Gummadi et al. on the conclusion that ring graphs are the best static DHT topologies, we agree that an hamiltonian cycle should exist in static DHT topologies in order to ease the dynamic DHT algorithm design. From the static DHT topology perspective, all aforementioned DHTs are hamiltonian except for Pastry/Tapestry (Plaxton trees), which, however, maintain a ring structure through their dynamic DHT algorithm.

4 Discussion and Related Work

Some desirable DHT system features are inconsistent with each other, which means that tradeoffs must be considered when deciding on a static DHT topology. As a general conclusion, we have shown that Cayley DHTs have clear advantages over non–Cayley DHT designs, naturally supporting desirable DHT features such as communication load balancing and fault tolerance.

Cayley DHTs cover both non-constant degree DHTs and constant degree ones, so in each case we can start from Cayley graphs as underlying topology for DHT design. Constant-degree Cayley graphs have the main advantage that their "maintainability" (regarding leaving / failing nodes) is independent of the size of the network. In a dynamic P2P environment, maintainability of nodes might be preferable to other desirable DHT system features such as communication load balancing and fault tolerance, since the loss of other DHT system features can often be compensated through some additional techniques in the dynamic DHT algorithm design, whereas maintainability is almost uniquely determined by the static DHT topology.

When designing constant degree Cayley DHTs, cube connected cycles are an especially promising family of static DHT topologies in terms of our analytical framework, taking into account the simplicity they enable for dynamic DHT algorithm design in comparison to for example butterfly graphs. This conclusion can be extended to a generalized type of constant degree Cayley graphs: Cayley Graph Connected Cycles (CGCC), as we have discussed in Section 2.3. We therefore expect that different variants of CGCC will heavily influence the design mainstream for future constant degree Cayley DHTs.

Looking at non-constant degree Cayley DHTs, the most promising family are hypercubes, as they achieve all desirable DHT system features except for the constant degree property. This conclusion can be extended to k-ary n-cube, which can be regarded as a generalization of the d-dimensional hypercube by taking $k = 2$. Formally, the k-ary d-cubes can be defined as in [22]:

Consider the group $\Gamma := (\mathcal{Z}_k^d, +)$, where $V := \mathcal{Z}_k^d$ denotes the set of all words of length d over the alphabet $\mathcal{Z}_k$ and where $+$ is the componentwise addition mod k. Let S be the set of all $(k-1) \cdot d$ words in V which have exactly one entry ± 1 and all others entries being 0. The graph $H_k^d := G((\mathcal{Z}_k^d, +), S)$ is the k-ary d-cube. By definition, k-ary n-cubes are Cayley graphs. They can be defined recursively as well: Denoting by

C_k the cycle of length k, we see that H_k^1 is isomorphic to C_k and H_k^d is isomorphic to $H_k^{d-1} \times C_k$ for $d > 1$, which is in turn isomorphic to $C_k \times \cdots \times C_k$ (d factors).

Most current Cayley DHTs such as HyperCup, CAN, and Chord use static DHT topologies that are either k-ary d-cubes or isomorphic to k-ary d-cubes such as ring graphs, tori, direct or undirected d-cubes [22]. Even for constant degree Cayley DHTs or non-Cayley DHTs, the static DHT topologies of Cycloid (cube-connected cycles) and Pastry/Tapestry (Plaxton trees) are closely associated with k-ary d-cubes. As we have mentioned, Plaxton trees can be viewed as approximate hypercubes, whereas cube-connected cycles can be viewed as a variant of hypercubes.

Gummadi et al. [5] investigate some commonly used static DHT topologies and explore how these topologies affect static resilience and proximity routing by analyzing the flexibility of different DHTs, i.e. the algorithmic freedom left after the static topologies has been chosen. Manku's [23] analysis starts from static DHT topologies, but then heavily involves dynamic DHT algorithms. His classification for DHT systems (deterministic and randomized) are certainly of value, but cannot serve as an analytical framework for comparing static DHT topologies. Datar [24] provides an in-depth investigation to butterfly graphs and further proposes a new DHT system using multi-butterflies as the static DHT topology. Castro et al. [25] make a comparative study of Pastry, taking Chord and CAN as reference systems.

5 Conclusions

We have discussed DHT topologies in the framework of Cayley Graphs, which is one of the most important group-theoretic models for the design of parallel interconnection networks. Associating Cayley graphs with DHTs enables us to directly leverage the research results for interconnection networks for the DHT design without the need of starting from scratch. Cayley graphs explicitly support an algebraic design approach, which allows us to start with an arbitrary finite group and construct symmetric DHTs using that group as the algebraic model, concisely specifying a DHT topology by providing the appropriate group plus a set of generators. This algebraic design approach also enables us to build new types of structured P2P networks in which data and nodes do not necessarily need to be hashed in order to build content delivery overlay networks, as discussed in [3, 26] for hypercube topologies. Such non-hashed, structured P2P networks allow us to apply semantic Web and database technologies for data organization and query processing and implement expressive distributed information infrastructures which are not implemented easily based on pure DHT designs.

Our analytical framework and its notion of Cayley DHTs provides a unified view of DHTs, which gives us excellent insight for designing and comparing DHT designs. Identifying a DHT design as Cayley DHTs immediately allows us to infer all generic properties for this design, and, through the correspondence of Cayley graph properties to DHT system features, allows us to directly infer the generic DHT features implemented by this design. Furthermore, we can investigate the various tradeoffs between different DHT designs features and use them to guide the design of future DHTs.

Casting and understanding static DHT topologies in a common framework is but the first important step towards principled DHT design. In order to cover all features of a particular design, we also have to explore the general design of dynamic DHT algorithms which can in principle be used to emulate any Cayley graph based static DHT topologies. Such dynamic DHT algorithms need not necessarily be bound to any individual Cayley graph, instead they could be universally applicable to any Cayley graphs, leveraging algebraic design methods in order to build arbitrary Cayley DHTs. Some of these methods and design issues are currently investigated in more detail in our group.

References

1. Akers, S.B., Krishnamurthy, B.: A group-theoretic model for symmetric interconnection networks. IEEE Trans. Comput. **38** (1989) 555–566
2. Sabidussi, G.: Vertex–transitive graphs. Monatsh. Math. **68** (1964) 426–438
3. Schlosser, M., Sintek, M., Decker, S., Nejdl, W.: Hypercup - hypercubes, ontologies and efficient search on p2p networks. In: Intl. Workshop on Agents and Peer-to-Peer Computing, Bologna, Italy (2002)
4. Shen, H., Xu, C., Chen, G.: Cycloid: A constant-degree and lookup-efficient p2p overlay network. In: Intl. Parallel and Distributed Processing Symposium, Santa Fe (2004)
5. Gummadi, K., Gummadi, R., Gribble, S., Ratnasamy, S., Shenker, S., Stoica, I.: The impact of dht routing geometry on resilience and proximity. In: ACM Annual Conference of the Special Interest Group on Data Communication, Karlsruhe, Germany (2003)
6. Stoica, I., Morris, R., Karger, D., Kaashoek, M.F., Balakrishnan, H.: Chord: A scalable peer-to-peer lookup service for internet applications. In: Annual Conference of the ACM Special Interest Group on Data Communications, San Diego, CA, USA (2001)
7. Oehring, S.R., Sarkar, F., Das, S.K., Hohndel, D.H.: Cayley graph connected cycles : A new class of fixed-degree interconnection networks. In: 28th Annual Hawaii Intl. Conference on System Sciences, Hawaii, USA (1995)
8. Heydemann, M.C., Ducourthial, B.: Cayley graphs and interconnection networks. Graph Symmetry, Algebraic Methods and Applications,"NATO ASI C" **497** (1997) 167–226
9. Malkhi, D., Naor, M., Ratajczak, D.: Viceroy: A scalable and dynamic emulation of the butterfly. In: 21st ACM Symposium on Principles of Distributed Computing (PODC 2002), Monterey, California, USA (2002)
10. Ratnasamy, S., Francis, P., Handley, M., Karp, R., Shenker, S.: A scalable content-addressable network. In: Annual Conference of the ACM Special Interest Group on Data Communications, San Diego, CA, USA (2001)
11. Aberer, K., Datta, A., Hauswirth, M.: Ch. 21, "Peer-to-Peer-Systems and Applications". In: P-Grid: Dynamics of self-organization in structured P2P systems. Springer LNCS (2004)
12. Rowstron, A., Druschel, P.: Pastry: Scalable, distributed object location and routing for large-scale peer-to-peer systems. In: IFIP/ACM Intl. Conference on Distributed Systems Platforms (Middleware), Heidelberg, Germany (2001)
13. Zhao, B.Y., Huang, L., Stribling, J., Rhea, S.C., Joseph, A.D., Kubiatowicz, J.D.: Tapestry: A resilient global-scale overlay for service deployment. IEEE Journal on Selected Areas in Communications **22** (2004)
14. Alspach, B.: Cayley graphs with optimal fault tolerance. IEEE Trans. Comput. **41** (1992) 1337–1339
15. BerthomT, P., Ferreira, A., Perennes, S.: Optimal information dissemination in star and pancake networks. IEEE Tran. on Parallel and Distrubuted Systems **7** (1996)
16. Mader, W.: Eine Eigenschaft der Atome endlicher Graphen. Arch. Math. (Basel) **22** (1971) 333–336

17. Mader, W.: Über den Zusammenhang symmetrischer Graphen. Arch. Math. (Basel) **21** (1970) 331–336
18. Watkins, M.E.: Connectivity of transitive graphs. J. Combinatorial Theory **8** (1970) 23–29
19. Boesch, F., Tindell, R.: Connectivity and symmetry in graphs. In: Graphs and applications (Boulder, Colo., 1982). Wiley-Intersci. Publ. Wiley, New York (1985) 53–67
20. Boesch, F., Tindell, R.: Circulants and their connectivities. J. Graph Theory **8** (1984) 487–499
21. Curran, S.J., Gallian, J.A.: Hamiltonian cycles and paths in Cayley graphs and digraphs—a survey. Discrete Math. **156** (1996) 1–18
22. Dally, W.J.: A VLSI Architecture for Concurrent Data Structures. Hingham, MA: Kluwer (1987)
23. Manku, G.S.: Routing networks for distributed hash tables. In: 22nd ACM Symposium on Principles of Distributed Computing (PODC 2003), Boston, USA (2003)
24. Datar, M.: Butterflies and peer-to-peer networks. In: 10th Annual European Symposium. Lecture Notes in Computer Science, Rome, Italy, Springer (2002)
25. Castro, M., Druschel, P., Hu, Y.C., Rowstron, A.: Exploiting network proximity in distributed hash tables. In: Intl. Workshop on Future Directions in Distributed Computing, Bertinoro, Italy (2002)
26. Nejdl, W., Wolpers, M., Siberski, W., Schmitz, C., Schlosser, M., Brunkhorst, I., L÷ser, A.: Super-peer-based routing and clustering strategies for rdf-based peer-to-peer networks. In: 12th Intl. World Wide Web Conference, Budapest, Hungary (2003)

BR-WRR Scheduling Algorithm in PFTS

Dengyuan Xu, Huaxin Zeng, and Chao Xu

School of Computer and Communication Engineering,
Southwest Jiaotong University, Chengdu, Sichuan, China, 610031
xudave@tom.com, xudave@126.com

Abstract. The novel concept of Physical Frame Time-slot Switching (PFTS) over DWDM, proposed by Sichuan Network and Communication technology key laboratory (SC-Netcom Lab), has been around for some time. It differs from existing switching techniques over DWDM by its superior QoS mechanisms embedded in and its capability to simplify Internet into a Single physical-layer User-data transfer Platform Architecture (SUPA).

This paper proposed a Borrow & Return Weighted Round Robin (BR-WRR) algorithm of output scheduling and dispatching in a multiple-priority queue environment in PFTS nodes. In such nodes, there are multi-ports in a DWDM-based PFTS node and each port contains multi-lambdas. Furthermore, multiple queues with different priorities plus burst queue with the highest priority are devised for each output lambda. A Borrow-and-Return mechanism is introduced to improve the orthodox WRR in PFTS, which cannot satisfy continuous transmitting privileged data of a burst to maintain its integrity. Comparison of the results between simulation of BR-WRR and that of WRR is provided in this paper and shows that BR-WRR has better performance with regard to fairness and important QoS parameters such as transit delay, jitters, and non-disordering.

1 A Brief Introduction to PFTS

PFTS (Physical Frame Time-slot Switching)[1,2] is a physical-layer switching technique, which takes the format of Ethernet MAC-frame as that of its physical frame called EPF (Ethernet-like Physical Frame) and the time duration of transmitting the fixed length of EPF (1530 Byte) is defined as the basic time-slot for interleaving user-data.

When entering the PFTS domain, a MAC frame is mapped onto an EPF at the physical layer and the destination-address field of the MAC frame will be replaced by a PFTS switching field. When an EPF exits from the PFTS domain, it will be treated as a MAC frame and the destination-address of MAC frame will be restored. PFTS is connection-oriented and it provides Permanent Virtual Connection (PVC) and Switched Virtual Connection (SVC) services.

Based on PFTS and out-band signaling concept, a Single-layer User-data transfer Platform Architecture Network (SUPANET) [3] was proposed, where user data is transferred in a single-layer platform and Internet protocol stacks are remained in transfer signaling and management information. To further enhance QoS provisioning, a few protocols are introduced at the application layer, such as QoS Negotiation protocol (QoSNP), Admission Control Protocol (ACP), and Traffic Monitoring information Exchange Protocol (TMEP) [4]. Consequently, SUPANET is directly interoperable with Internet and more acceptable for Internet communities compared with complicated No 7 and 2 signaling protocol based protocol stacks [5].

Various QoS support-mechanisms are defined in PFTS, some of which are embedded in the PFTS switching field of an EPF (Ethernet-like Physical Frame) loaded in PFT (Physical Frame Timeslot) as shown in Figure 1.

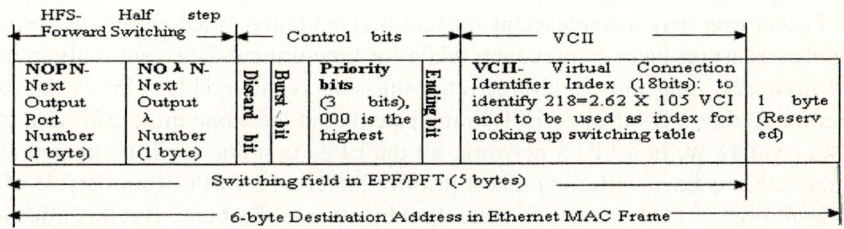

Fig.1. Switching Field of an EPF to loaded in Physical Frame Timeslots

Similar to VPI/VCI in ATM cells, VCII (Virtual Connection Identifier Index) in figure 1 is a local VC identifier for a virtual connection within a lambda between adjacent PFTS nodes while an end-to-end path is uniquely identified by concatenated VCIs. Edge routers in PFTS domain are responsible for assigning a value to the 3-bit priority-field according to QoS negotiation agreement; while in congestion condition, all nodes in PFTS domain may discard an EPF with D-bit set. The E-bit is used to signal the end of a one-way connection and to remind the switch to return resources reserved for the VC to the system. Two special fields in an EPF need further explanation:

A HFS (Half Step Forward Switching)

Destination IP address in Internet, label in MPLS, and VPI-VCI in ATM networks, are used to look up the routing/switching tables to find the right output port. Unlike these techniques, PFTS takes fully the advantage of connection-oriented service by utilizing information about the output lambda and output port of the next node determined during QoS negotiation in signaling plane. Therefore, NOPN and NO N fields carried in an EPF can be used to direct it to the appropriate output queue as soon as it arrives without looking up the switching table. Looking-up function is left for output unit to find the new VCII, NOPN, and NOλN while the input of an EPF (1530 bytes) is still in progressing. With the highest lambda data rate of 80 Gbps up to date, the receiving time for an EPF is 153 ns. Considering that SRAM available with a 333 MHz clock, its access latency can be as low as 6-ns, there should be enough time left for looking-up. This novel technique is called the Half step Forward Switching (HFS).

B   Bursting Phenomenon and the B-Bit in EPF

Bursting has been observed by long-term network monitoring; moreover, traffic/time diagrams thus plotted are very similar disregarding variation in sampling durations (in seconds, milliseconds, or microseconds). This phenomenon has been referred to as "Self-similar nature" of network traffic [6]. Bursts appear as contiguous packet/frame sequences, whose integrity is particularly important for time-critical traffic. An output queues may be piled up with packets/frames when long bursts or simultaneous bursts from different input ports are instantly directed to the same output lambda. And worse still, congestion may occur unless excessive frames/packets can be dropped in time. Packet loss may not important for text transportation since end-to-end retransmission at transport layer can recover, while for time-critical data, especially in high-speed networks, long time delay for retransmission is intolerable. This is the exact reason for having a B-bit in an EPF to indicate that at least one more EPF in a burst follows right away. In a PFTS network, all the EPFs in a burst will be treated as an integral data sequence with the highest priority therefore no EPF from outside of the burst is allowed to be inserted into the output bit-stream of a burst. But this imposes a new problem to be solved as discussed in the following sections.

This paper is devoted to scheduling and dispatching issues in PFTS switching nodes, where there are multi-priority queues plus burst queues for each lambda. Analysis shows that existing scheduling algorithms, such as Round Robin (RR) and Weighted Round Robin (WRR) [7], are not catered for such working environment. A Borrow-and-Return Weighted Round Robin (BR-WRR) dispatching algorithm is proposed in this paper to make a trade-off between fairness among data in different queues and other criteria such as maintaining the integrity of bursts. Results of simulating WRR and emulating BR-WRR are provided in this paper and analysis shows that BR-WRR can make a better trade-off between fairness and maintaining the integrity of bursts.

2   Weighted Round Robin (WRR)

Round Robin (RR) is the simplest and fair scheduling algorithm by taking out one packet/frame each time from different queues by turn. However, RR does not consider the priority of the data to be output. An improved algorithm to RR is called Weighted Round Robin (WRR) [7], with which numbers of packets/frames taken from a queue in an output cycle is proportional to the priority of the queue. Assuming that there are four Queues numbered with 1st through 4th respectively; let us assign Queue1 with the highest priority and 4-frame to be sent out in a scheduling cycle, and assign Queue4 with the lowest priority hence only one frame can be sent out in the same cycle. With such an assignment, output order for individual frames can still be different. For example, we can start from Queue1 and dispatch all the packets within its quota first in a cycle, and then Queue2, Queue3, and Queue4 subsequently. The arrangement can ensure that only when all the frames within its cycle quota in a higher priority queue have been sent out, can frames in the queue of next higher priority be sent out within its quota. Alternatively, an output cycle can be divided into multiple sub-cycles.

It should be emphasized that fairness of RR or WRR is valid to CBR (Constant Bit Rate). In Internet, packets or frames are most likely variable in lengths; it might not be fair with regard to bits sent out per cycle from different queues. Considering that output bits per cycle is more or less associated with throughput (one of important QoS parameters), it would be more meaningful in the context of fixed-length packet switching, where throughput of a user stream in queue can be quantifiably determined to some extent with RR or WRR.

Indeed, these scheduling approaches can be used into PFTS, since EPFs are always 1530 bytes in length. However, further complication arises with WRR in handling bursting phenomenon, which has been observed by long-term traffic monitoring in different networks [6].

A burst in a queue is treated as a continuous output EPF-sequence without interleaving with EPFs from any other services. One could argue that with PQ (Priority Queue) scheduling, the continuity of a burst can be ensured. It is true under the condition that the quota assigned for a given VC during QoS negotiation is enough for transmit a long burst in PQ, but it is unfair for the lower services; otherwise the timeslots has to be stolen from other lower queues to ensure the integrity of a burst. To make the trade-off between fairness and burst integrity, a Borrow & Return WRR algorithm is introduced.

3 Borrow and Return WRR (BR-WRR)

Since burst queues have the highest priority in output process, bursts for time-critical streams, such as CBR (Constant Bit Rate) and RT-VBR (Variable Bit Rate)[8] stream should be transmitted first before EPFs in other queues. However, a VC has an output quota, i.e. the number of EPFs per second agreed between service users and PFTS service providers (i.e. intermediate nodes), and its upper limit is EBR (Extended Burst Rate) [4]. A long burst may easily exhaust its assigned quota for a scheduling cycle. It is better to borrow timeslots from lower priority queues to maintain its integrity. The basic idea of BR-WRR is to reduce unfairness caused by such instant timeslot borrowing through a returning mechanism.

A BR-WRR scheduler is best described with event/state/action transition diagram as shown in figure 2, where:

• States: Representing the status of the scheduler: Idle, WRR-scheduling (WRR-S), BR-WRR–scheduling (BR-S). WRR-S represents the conventional WRR scheduling when there is no Burst to be scheduled in Queue1. In BR-S state, all EPFs of a Burst in Queue1 are transmitted continuously, and during that time it may have to borrow time-slots from those of lower priority Queues when the time-slots needed exceed the quota for that VC, and borrowed time-slots should be registered and be returned in next few cycles. An Idle state is defined for a condition that no EPF is in any queue to

be sent out and in this state the only thing for a scheduler is to poll all the queues in a round robin fashion until an EPF arrives at a queue.
• Events: The most important event is the commencement of scheduling a Burst, which causes re-scheduling in the state of BR-S and the events leading to borrow or return time-slots are also important events.
• Actions: representing all the process at a given state as response to the event.
Figure 2 illustrates the state transition state transition graph of BR-WRR.

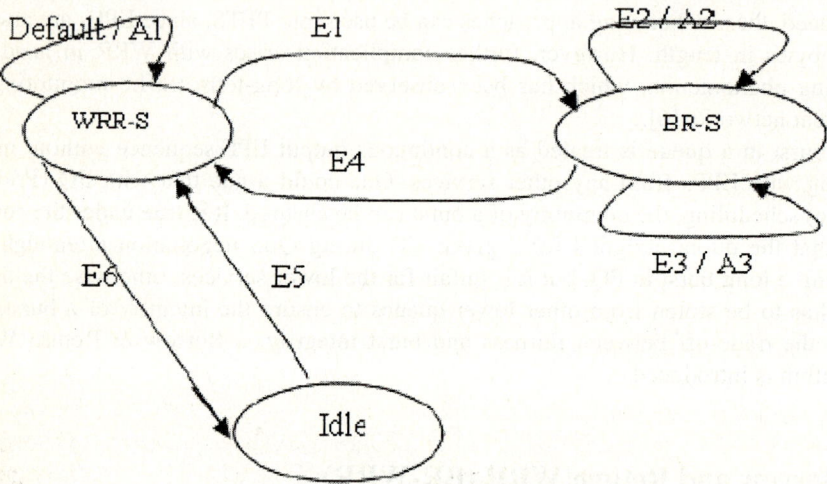

E1: the 1st frame of a Burst in Queue1 starts to be scheduled;
E2: the B-bit of current frame is equal to 1, which means that it belongs to the Burst and is scheduled continuously; and the slots of Queue1 in WRR are used up;
E3: current Burst is scheduled completely and the number of borrowed time slots exceeds upper threshold;
E4: the number of time-slots borrowed is less than upper threshold;
E5: there exists frame in Queues;
E6: all Queues are empty;
A1: schedule according to traditional WRR;
A2: borrow a time-slot from the lower priority Queues;
A3: return a time-slot to the lower priority Queues

Fig. 2. State transition graph of BR-WRR

There are two important issues in application of BR-WRR: Time to start to borrow and time to return borrowed timeslots.
Borrowing time-slot is done when the Burst in high priority queue starts to be scheduled, that is, the high priority Burst queue is polled in WRR and the first frame of the Burst is at the head of the queue. The frames of the same Burst are dispatched continuously until the whole Burst is transmitted completely and once one time-slot

of the lower priority services is borrowed, the variable BT, which means the number of borrowed time-slots, is increased by 1.

Scheduling starts from conventional WRR in the state "WRR-S" as figure 2 and the state "BR-S" is arrived into when high priority Burst starts to be scheduled. Supposing that Burst starts to be transmitted at Queue1 and the first time-slot used by the Burst belongs to its own, the variable BT is equal to 0, next time-slot is owned by Queue 2, which means the Burst borrows one time-slot from Queue 2, and BT is increased by 1. This is done by action "E2" for several times until the Burst is transmitted completely. The value of BT is the total number of timeslots borrowed by the Burst.

It should be noted that Burst can only borrow time-slots from the lower priority queues in order to provide that higher priority service, such as CBR service, doesn't be affected in BR-WRR.

When current Burst is transmitted completely and the number of borrowed time-slots (identified by variable BT) is larger than the upper threshold (identified by variable UT), borrowed timeslots should be returned, which is the action "E3" done; During this time, scheduler polls Queue1 (Burst is in Queue 1), the frames in Queue1 don't be scheduled and polling pointer moves to next in WRR scheduling table, and the variable BT decreases by 1, which means that Queue1 returns one time-slot. This is done several times until the value of BT is less than upper threshold of UT, then BR-WRR comes into the state "WRR-S". It should be noted that if scheduler polls Queue1 and it is empty in the state "WRR-S", Queue 1 is ignored and the value of BT decreases by 1, which means that lower queues borrow Queue1's timeslot. Once there exists a burst in Queue1 being transmitted, Queue1 will take back its loaned timeslots.

The BR-WRR scheduling algorithm can be described as below with a pseudo-programming language:

<variable definition>
  BT_i: the number of time-slots borrowed by queue i from lower priority queues
  UT: the upper threshold of the number of time-slots borrowed by Burst
  Ptr: polling pointer in conventional WRR algorithm
  Ptr_i: the polling pointer of queue i in BR-WRR, which assures the Burst in queue i can only borrow the time-slots from the lower queues
<algorithm>
  Initialization: $BT_i=0$; Ptr points to one timeslot in WRR scheduling table in fig. 2

Borrowing Time-Slots:
  if (BT_i<UT)
  ptri=ptr;
  do /*Transmitting frames of the same Burst continuously */
  if (ptri>i)
  { send one frame of the Burst in queue i;
  BT_i__}_/* queue i borrows one time-slot*/
  else/*The time-slots of higher priority queue can NOT be borrowed by queue i*/
  { transmitting according to conventional WRR_}
  while (B-bit==1)/*until all frames of the Burst are transmitted completely*/

Returning Time-Slots:
/*it is done after all frame of one Burst are scheduled completely */
if (BTi > UT)
do
{ ptr polls by conventional WRR;
if (ptr polls queue i)
{the frame of queue i does NOT be scheduled;
ptr++;
BTi--; /*Queue i returns one time-slot */
} while (BTi> BT) || _ Ci<BT && Queue i is empty_

4 Simulation and Analysis

In order to simulate the behavior of WRR and that of BR-WRR, and to further make a comparison between these two scheduling methods, a simulation configuration is defined as shown in figure 3, which consists of 4 data generators, an EPF classifier, 4 output queues with Queue1 (burst queue) assigned to the highest priority, an EPF-Scheduler..

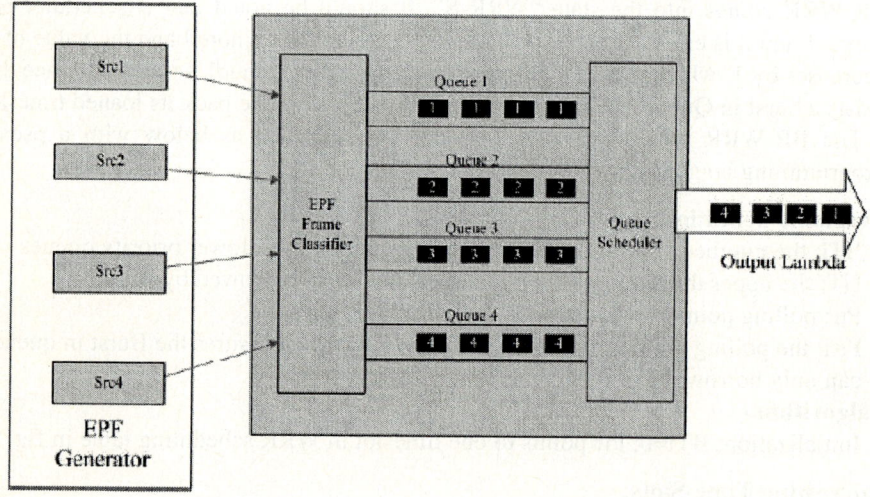

Fig. 3. Simulating model

4.1 EPF Generators (EPF-G)

EPF Generators (EPF-G) are simulation-data sources, which will be fed into input source of individual queues. The EPF Generator Src1 generates class 1 stream which has both Burst and non-Burst, we make the Src1 generates Burst with a probability, such as 1/3, and then the probability of nonBurst is 2/3. And the length of Burst belongs to uniform distribution. Assuming that bursts appear only in class 1 data streams, we could concentrate on how to ensure the integrity of bursts in class 1 and other QoS criteria. Source Src2, Src3 and Src4 generate the non-Bursting data of class 2, class 3 and class 4 separately.

4.2 Average Queue Length of Each Class

Supposing that total generating rate of all classes is equal to the scheduling rate and generating rate of all classes are equal and the length of all output queues is long enough to hold all waiting frames. The average numbers of frames in waiting for different classes of data with WRR and BR-WRR are listed in table 1.

It is considered in conventional WRR that the higher the priority is, the shorter the average queue is, which is right to non-bursting classes, such as the frames in Queue 2 and Queue 3. Due to the priority of class 2 is higher than that of class 3, the average queue length of class 2 is shorter than that of class 3. The case will be changed if class 1 has a burst, For example, even that class 1 has the highest priority, its queue length is longer that of class 2 and class 3. The reason is that once a burst comes into queue 1, several frames have to wait there to be scheduled in conventional WRR even if in majority time Queue1 is empty. While class 2 has no Burst, only one frame arrives into the queue each time, so its average queue length is shorter. But Burst frames in class 1 can be transmitted continuously in BR-WRR, which can reduce the average queue length in class 1 from 5.5 frames to 2.6 frames which can be seen in table 1.

Table 1. A comparison of average queue length of WRR and BR-WRR

Frame Schedule procedure / Type of class	WRR (Frame)	BR-WRR (Frame)
class 1	5.5	2.6
class 2	2.4	2.9
class 3	3.1	3.4
class 4	8.6	8.8

It should be noted that average queue lengths of other lower priority classes are increased in different degrees in BR-WRR. This is because that Burst of class 1 borrows the timeslots from lower priority classes temporarily, which makes more frames of lower priority classes wait in their own queues.

4.3 Queue Delay

Figure 4 is the queue delay of 4 different priority classes in conventional WRR. Although class 1 have the highest priority, its waiting time is longer than that of class 2 and class 3, even longer than that of class 4 sometimes because of the Burst service in class 1. Also it can be seen from figure 4 that the time jitter of class 1 is intense.

However, Figure 5 shows the queuing delay of different priority classes in BR-WRR. It can be seen that the waiting time for class 1 is shorter than those of other classes, even though there exist bursts in class 1. And the jitter in class 1 in figure 5 is shorter than that in figure 4. The reason is that once a burst starts to be forwarded, all frames in this Burst are done with continuously, so its jitter becomes shorter. It should be noted that QoS guarantee of Burst in class 1 makes the waiting time in other classes increase. Table 2 is the average time delay of 4 classes in a PFTS node. It can be seen that the average waiting time in class 1 reduces by 1.04 millisecond (ms) in BR-WRR, while the average delays in class 2, class 3 and class 4 increase 0.55, 0.94 and 1.03 millisecond (ms) respectively, which also shows that BR-WRR reduces the waiting time in class 1 at the cost of increasing the queue time of other classes.

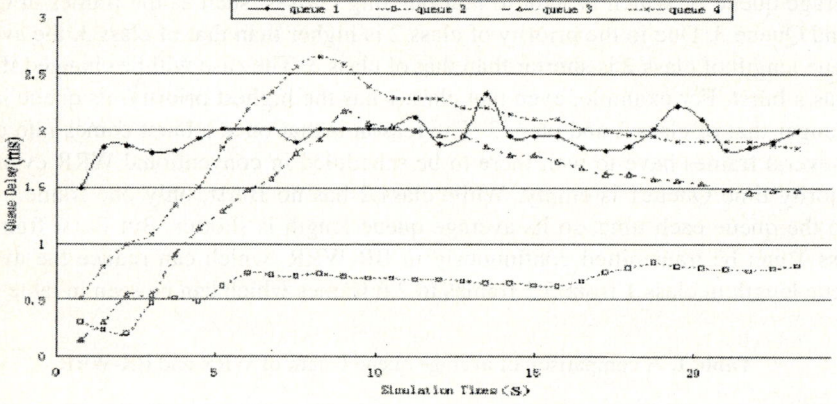

Fig. 4. Queue delay in WRR

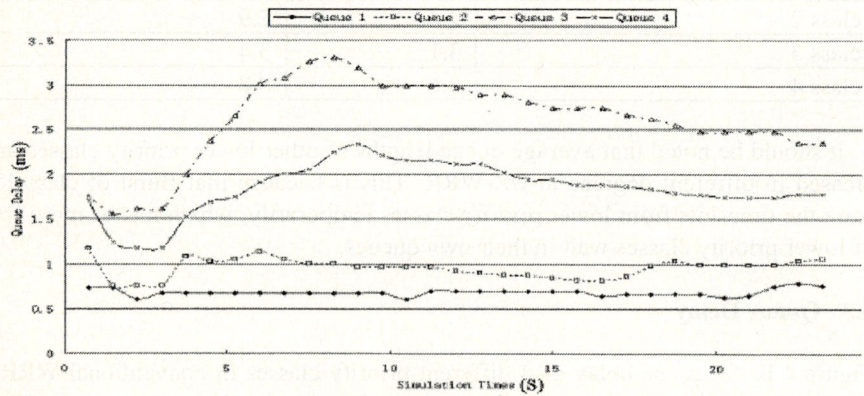

Fig. 5. Queue delay in BR-WRR

Table 2. A comparison of average delay of WRR and BR-WRR

Frame Schedule procedure Type of class	WRR (ms)	BR-WRR (ms)
class 1	1.78	0.74
class 2	0.57	1.12
class 3	0.83	1.77
class 4	1.10	2.13

5 Conclusion

In PFTS, time-critical streams demand a data transfer service with high throughput and low latency, and with tolerable jitters. This paper improves the WRR algorithm in a PFTS environment with a novel concept of "borrow and return". Simulations show that BR-WRR makes a better tradeoff between fairness and maintaining integrity of a burst. Our simulation results have also shown that very long bursts or frequent occurrence bursts in class 1 could cause an increase in waiting time for other classes of data and data losses. Therefore, it is necessary to impose admission control to prevent such conditions. Through our preliminary simulation, we are convinced that BR-WRR is a promising technique in improving the traditional WRR, especially in a switching environment like PFTS nodes.

Acknowledgement

The work presented in this paper has been sponsored by Natural Science Foundation of China (Project ratification No: 60372065) and Ph.D student Creative Foundation of Southwest Jiaotong University. The authors would like to acknowledge the financial support from them.

References

1. Huaxin Zeng, Dengyuan Xu, Jun Dou: On Physical Frame Time-slot Switching over DWDM, Proceedings of PACAT03, IEEE press, Aug (2003) 535-540
2. Huaxin Zeng, Dengyuan Xu, Jun Dou: Promotion of Physical Frame Timeslot Switching (PFTS) over DWDM, to be published in: IEC Annual Report of Communications, Vol. 57 (2004).
3. Huaxin Zeng, Jun Dou, Dengyuan Xu: Single physical layer U-plane Architecture (SUPA) for Next Generation Internet, to be published in: IEC Annual Review of IP Applications and Services (2004)
4. Zeng Hua-xin, Dou Jun, Wang Hai-ying: On the Single Physical Layer User Plane Architecture Network, Journal of Computer Applications (Chinese) vol. 24. June (2004) 1-5
5. ITU, I.120 – Integrated Services Digital Networks (ISDN)

6. An Ge, et al: On Optical Burst Switching and Self-Similar Traffic, IEEE Communication Letters, vol. 4. Mar (2000)
7. D. Stiliadis et al.: Weighted round-robin cell multiplexing in a general-purpose ATM switch chip, IEEE J.Sel. Areas Commun., vol. 9. (1991).
8. "Traffic Management Specification Version 4.0",ATM Forum, Apr (1996)

VIOLIN: Virtual Internetworking on Overlay Infrastructure

Xuxian Jiang and Dongyan Xu

Purdue University, West Lafayette, IN 47907, USA
{jiangx, dxu}@cs.purdue.edu

Abstract. We propose a novel application-level virtual network architecture called VIOLIN (Virtual Internetworking on OverLay INfrastructure). VIOLINs are isolated virtual networks created on top of an overlay infrastructure (e.g., PlanetLab). Entities in a VIOLIN include virtual end-hosts, routers, and switches implemented by software and hosted by physical overlay hosts. Novel features of VIOLIN include: (1) a VIOLIN is a "virtual world" with its own IP address space. All its computation and communications are strictly confined within the VIOLIN. (2) VIOLIN entities can be created, deleted, or migrated on-demand. (3) Value-added network services not widely deployed in the real Internet can be provided in a VIOLIN. We have designed and implemented a prototype of VIOLIN in PlanetLab.

1 Introduction

Current Internet only provides basic network services such as IP unicast. In recent years, overlay networks have emerged as application-level realization of value-added network services, such as anycast, multicast, reliable multicast, and active networking. While highly practical and effective, overlays have the following problems: (1) Application functions and network services are often closely coupled in an overlay, making the development and management of overlays complicated. (2) The development of overlay network services is mainly individual efforts, leading to few standards and reusable protocols. Meanwhile, advanced network services [1][2][3][4][5] have been developed but not widely deployed. (3) It is hard to *isolate* an overlay from the rest of the Internet, making it easy for a compromised overlay node to attack other Internet hosts.

In this paper, we propose a novel virtual network architecture called VIOLIN (Virtual Internetworking on OverLay INfrastructure), motivated by recent advances in virtual machine technologies [6][7]. The idea is to create virtual isolated network environments on top of an overlay infrastructure. A VIOLIN[1] consists of virtual routers, LANs, and end-hosts, all being software entities hosted by overlay hosts. The key difference between VIOLIN and application-level overlay is

---

[1] With a slight abuse of terms, VIOLIN stands for either the virtual network technique or one such virtual network.

that VIOLIN re-introduces *system(OS)-enforced* boundary between applications and network services. As a result, a VIOLIN becomes an "advanced Internet" running value-added network-level protocols for routing, transport, and management.

The novel features of VIOLIN include: (1) Each VIOLIN is a "virtual world" with its own IP address space. All its computation and communications are strictly confined within the VIOLIN. (2) All VIOLIN entities are software-based, leading to high flexibility by allowing on-demand addition/deletion/migration/configuration. (3) Value-added network services not widely deployed in the real Internet can be provided in a VIOLIN. (4) Legacy applications can run in a VIOLIN without modification, while new applications can leverage the advanced network services provided in VIOLIN.

We expect VIOLIN to be a useful complement to application-level overlays. First, VIOLIN can be used to create testbeds for network-level experiments. Such a testbed contains more realistic network entities and topology, and provides researchers with more convenience in experiment setup and configuration. Second, VIOLIN can be used to create a service-oriented (virtual) IP network with advanced network services such as IP multicast and anycast, which will benefit distributed applications such as video conferencing, on-line community, and peer selection.

We have designed and implemented a prototype of VIOLIN in PlanetLab. A number of distributed applications have also been deployed in VIOLIN. The rest of the paper is organized as follows. Section 2 provides an overview of VIOLIN. Section 3 justifies the design of VIOLIN and its benefit to distributed applications. Section 4 describes the implementation and ongoing research problems of VIOLIN. Section 5 presents preliminary performance measurements in PlanetLab. Section 6 compares VIOLIN with related works. Finally, section 7 concludes this paper and outlines our ongoing work.

2   VIOLIN Overview

The concept of VIOLIN is illustrated in Figure 1. The low-level plane is the real IP network; the mid-level plane is an overlay infrastructure such as PlanetLab; and the top-level plane shows one VIOLIN created on the overlay infrastructure. All entities in the VIOLIN are hosted by overlay hosts; and there are three types of entities like in the real network: end-host, LAN, and router.

- A *virtual end-host (vHost)* is a virtual machine running in a physical overlay host. Meanwhile, it is possible that one physical overlay host supports multiple vHosts belonging to different VIOLINs.
- A *virtual LAN (vLAN)* is constructed by creating one *virtual switch (vSwitch,* not shown in Figure 1) that connects multiple vHosts.
- A *virtual router (vRouter)* is also a virtual machine with multiple *virtual NICs (vNICs)*. A vRouter interconnects two or more vLANs.

VIOLIN: Virtual Internetworking on Overlay Infrastructure

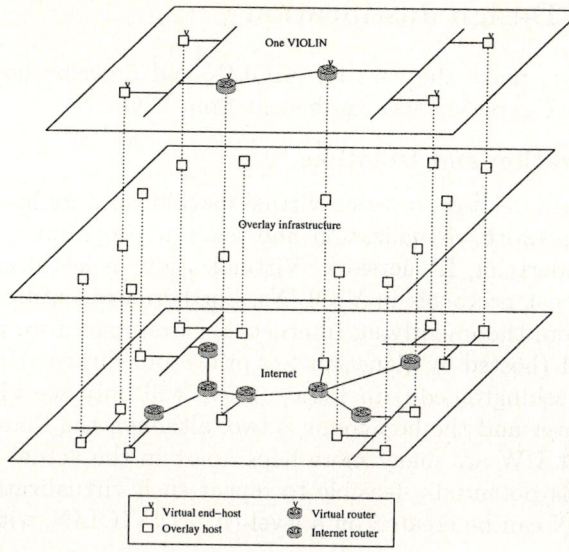

Fig. 1. VIOLIN, overlay infrastructure, and underlying IP network

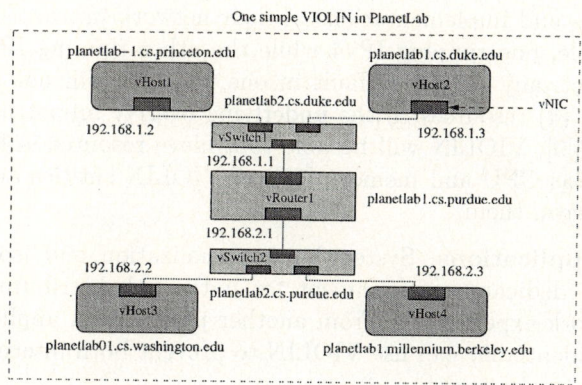

Fig. 2. A VIOLIN in PlanetLab (with names of physical PlanetLab hosts and virtual IP addresses)

Figure 2 shows a simple VIOLIN we create in PlanetLab. Two vLANs are interconnected by one vRouter (vRouter1 hosted by *planetlab1.cs.purdue.edu*): One vLAN comprises vHost1, vHost2, and vSwitch1; while the other one comprises vHost3, vHost4, and vSwitch2. The links between these entities emulate cables in the real world. The IP address space of the VIOLIN is completely independent. Therefore, it can safely overlap the address space of another VIOLIN or the real Internet.

3 VIOLIN Design Justification

In this section, we make the case for VIOLIN and describe how applications (including network experiments) can benefit from VIOLIN.

3.1 Virtualization and Isolation

Analogous with the relation between virtual machine and its host machine, VIOLIN involves network virtualization and leads to isolation between a VIOLIN and the underlying IP network. Virtualization makes it possible to run unmodified Internet protocols in VIOLINs. Furthermore, entities in a VIOLIN are decoupled from the underlying Internet. For example, if we perform *traceroute* from vHost1 (hosted by planetlab-1.cs.princeton.edu) to vHost3 (hosted by planetlab01.cs.washington.edu) in Figure 2, we will only see vRouter1 as the intermediate router and the hop count is two, although the PlanetLab hosts at Princeton and at UW are many more hops apart in the actual Internet. More interestingly, it is potentially feasible to repeat such virtualization *recursively*: a level-n VIOLIN can be created on a level-$(n-1)$ VIOLIN, with level-0 being the real Internet.

Network isolation is with respect to (1) administration: the owner of a VIOLIN has full administrator privilege - but *only* within this VIOLIN; (2) address space and protocol: the IP address spaces of two VIOLINs can safely overlap and the versions and implementations of their network protocols can be different - for example, one running IPv4 while the other running IPv6; (3) attack and fault impact: any attack or fault in one VIOLIN will not affect the rest of the Internet; (4) resources: *if* the underlying overlay infrastructure provides QoS support [8][9], VIOLIN will be able to achieve resource isolation for local resources (such as CPU and memory [10]) of VIOLIN entities and for network bandwidth between them.

Benefit to Applications. System-level virtualization and isolation provide a confined and dedicated environment for untrusted distributed applications and risky network experiments. From another perspective, applications requiring strong confidentiality can use VIOLIN to prevent both internal information leakage and external attacks.

3.2 System-Enforced Layering

Contrary to application-level overlays, VIOLIN enforces strong layering in order to disentangle application functions and network services. In addition, OS-enforced layering provides better protections to network services after the application level software is compromised. We note that layering itself does *not* incur more performance overhead compared with application-level overlays. We also note that layering is between application and network functions, *not* between network protocols. In fact, VIOLIN can be used as a testbed for the *protocol heap* architecture [11].

Benefit to Applications. Application developers will be able to focus on application functions rather than network services, leading to clean design and easy implementation. In addition, legacy applications can run in a VIOLIN without modification and re-compilation.

3.3 Network Service Provisioning

VIOLIN provides a new opportunity to deploy and evaluate advanced network services. There exist a large number of well-designed network protocols that are not yet widely deployed. Examples include IP multicast, scalable reliable multicast [2][4], IP anycast [3], and active networking [1][5]. There are also protocols that are still in the initial stage of incremental deployment (e.g., IPv6). VIOLIN is a platform to make these protocols a (virtual) reality.

Benefit to Applications. VIOLIN allows applications to take full advantage of value-added network services. For example, in a VIOLIN capable of IP multicast, applications such as publish-subscribe, layered media broadcast can be more conveniently developed than in the real Internet. We further envision the emergence of *service-oriented* VIOLINs, each with high-performance vRouters and vSwitches deployed at strategic locations (for example, vRouters close to Internet routing centers, vSwitches close to domain gateways), so that clients can connect to the VIOLIN to access its advanced network services.

3.4 Easy Reconfigurability

Based on all-software virtualization techniques, VIOLIN achieves easy reconfigurability. Different from a physical network, vRouters, vSwitches, and vHosts can be added, removed, or migrated dynamically. Also, vNICs can be dynamically added to or removed from vHosts or vRouters; and the number of ports supported by a vSwitch is no longer a hardware constraint.

Benefit to Applications. The easy reconfigurability and hot vNIC plug-and-play capability of VIOLIN is especially useful to handle the dynamic load and/or membership of distributed applications. Not only can a VIOLIN be created/torn down on-demand for an application, its scale and topology can also be adjusted in a demand-driven fashion. For example, during a multicast session, a new vLAN can be dynamically grafted on a vRouter to accommodate more participants.

4 VIOLIN Implementation

4.1 Virtual Machine

All VIOLIN entities are implemented as virtual machines (VMs) in overlay hosts. We adopt User-Mode Linux (UML) [12] as the VM technology. UML allows most Linux-based applications to run on top of it without any modification. Based on *ptrace* mechanism, UML - the *guest OS* for a virtual machine, performs system call redirection and signal handling to emulate a real OS. More specifically, the guest OS will be notified when an application running in the virtual machine issues a system call, the guest OS will then redirect the system call into its own implementation and nullify the original call. One important feature of UML is that it is completely implemented at user level without requiring host OS kernel modifications.

Unfortunately, the original UML has a serious limitation: both virtual NICs and virtual links of virtual machines are restricted *within* the same physical host. *Inter-host* virtual links, which are essential to VIOLIN, have not been reported in current VM projects [6][7][13]. To break the physical host boundary, we have performed non-trivial extension to UML and introduced transport-based *inter-host tunneling*.

More specifically, we use UDP tunneling in the Internet domain to emulate the physical layer in the VIOLIN domain. For example, to emulate the physical link between a vHost and a vSwitch, the guest OS for the vHost opens a UDP transport connection for the vNIC and obtains a file descriptor - both in the host OS domain. To receive data from the vSwitch, SIGIO signal will be generated by the host OS for the file descriptor whenever data are available. The vSwitch maintains the IP address and UDP port number (in the Internet domain) for the vNIC of the vHost, so that the vSwitch can correctly emulate data link layer frame forwarding. Such virtualization is transparent to the network protocol stack in the guest OS. Finally, inter-host tunneling enables hot plug-and-play of vNICs (Section 3.4); and it does not exhibit MTU effect as in the EtherIP [14] and IP-in-IP [15] approaches.

4.2 Virtual Switch

A vSwitch is created for each vLAN and is responsible for packet forwarding at the (virtual) data link layer. Figure 3 shows a vSwitch which connects multiple vHosts. vSwitch is emulated by a UDP daemon in the host OS domain. The *poll* system call is used to poll the arrival of data and perform data queuing, forwarding, or dropping. More delicate link characteristics may also be implemented in the UDP daemon. The *poll* system call also notifies the UDP daemon of the arrival of a connect request from a new vHost joining the vLAN, so that a new port can be created for the vHost, as shown in Figure 3.

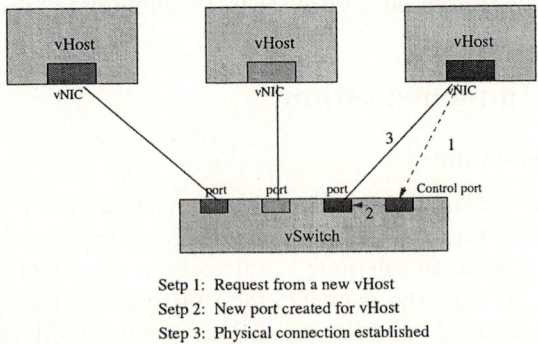

Setp 1: Request from a new vHost
Setp 2: New port created for vHost
Step 3: Physical connection established

Fig. 3. vSwitch and steps of port creation

4.3 Virtual Router

Interestingly, there is no intrinsic difference in implementation between vHost and vRouter, except that the latter has additional packet forwarding capability and user level routines for the configuration of packet processing policies. Linux source tree makes it possible to accommodate versatile and extensible packet processing capabilities.

When a UML is bootstrapped, a recognizable file system will be located and mounted as root file system. Based on UML, the vRouter requires kernel-level support for the capability of packet forwarding, as well as user-level routines, namely *route, iproute2, ifconfig* for the configuration of interface addresses and routing table entries. Beyond the packet forwarding capability, it is also easy to add firewall, NAT, and other value-added services to the UML kernel. In the VIOLIN implementation, we adopt the *zebra* [16] open-source routing package, which provides a comprehensive suite of routing protocol implementations. Recently, to enable active network services, we have also incorporated Click [17] as an optional package for vRouters.

5 VIOLIN Performance

We have implemented a VIOLIN prototype and deployed it in PlanetLab. To evaluate the performance of VM communications in a VIOLIN, we have performed end-to-end throughput and latency measurement between VMs. Figures 4(a) and 4(b) show a set of representative results. Two VMs are hosted by PlanetLab nodes planetlab8.lcs.mit.edu and planetlab6.cs.berkeley.edu, respectively. We measure the TCP throughput and ICMP latency between the VMs, with and without the vSwitches performing UDP payload encryption. As a comparison, we also measured the TCP throughput and ICMP latency between the two PlanetLab hosts. Our measurement results show that VIOLIN introduces an average of 5% degradation in TCP throughput, compared with the TCP throughput

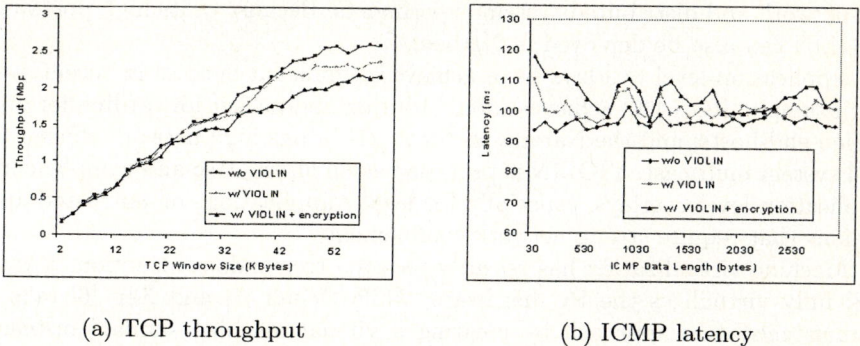

(a) TCP throughput (b) ICMP latency

Fig. 4. TCP throughput and ICMP latency between two VMs in VIOLIN

between the two underlying physical hosts. The addition degradation due to VM traffic encryption is 5% on the average. The degree of ICMP latency degradation (increase) is even less than that of TCP throughput.

To demonstrate VIOLIN's support for advanced network services, we run WaveVideo, a legacy video streaming application, in VIOLIN. WaveVideo requires IP multicast and therefore is not runnable in PlanetLab. However, WaveVideo is able to execute in a VIOLIN with 9 VMs. The VM hosted by planetlab2.cs.wisc.edu is the source of the video multicast session. It streams a short 300-frame video clip using (virtual) IP multicast address 224.0.0.5. The other 8 VMs are all receivers in three different domains: Princeton, Purdue, and Duke. The average peak signal noise ratios (PSNR) of video frames observed by VMs in the three domains are shown in Figure 5.

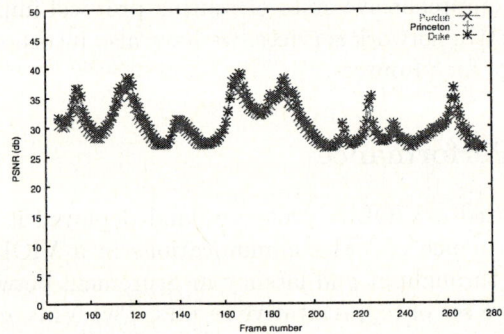

Fig. 5. Video streaming quality in an IP-multicast-enabled VIOLIN

6 Related Work

VIOLIN is made possible by PlanetLab [18], which itself provides resource virtualization capability called slicing. Netbed [19] is another wide-area testbed for network and distributed system experiments. Because of its high portability, VIOLIN can also be deployed in Netbed.

Application-level overlays have achieved significant success in recent years. For example, RON [20] achieves robust routing and packet forwarding for application end-hosts; and the Narada protocol [21] brings high network efficiency to end system multicast. VIOLIN is proposed as an alternative and complement to application-level overlays, especially for legacy applications or untrusted applications that require strong network confinement.

Machine virtualization has recently received tremendous attention. VMware [13] fully virtualizes the PC hardware, while Denali [7] and Xen [6] take the *paravirtualization* approach by creating a virtual machine similar (instead of identical) to the physical machine. Inspired by machine virtualization, VIOLIN is our initial effort toward *network* virtualization.

The X-Bone [15] provides automated deployment and remote monitoring of overlays, and allows network entities (hosts, routers) to participate in multiple overlays simultaneously. By taking the two-layer "IP-in-IP" tunneling approach, X-Bone makes real Internet IPs visible to entities in the overlay domain, leading to a lower degree of isolation and confinement than VIOLIN.

7 Conclusion and Ongoing Work

We present VIOLIN as a novel alternative and useful complement to application-level overlays. Based on all-software virtualization techniques, VIOLIN creates a virtual internetworking environment for the deployment of advanced network services, with no modifications to the Internet infrastructure. The properties of isolation, enforced-layering, and easy reconfigurability make VIOLIN an excellent platform for the execution of high-risk network experiments, legacy applications unaware of overlay APIs, as well as untrusted and potentially malicious applications. Our ongoing work includes:

- *Performance evaluation and comparison.* VIOLIN involves virtualization techniques and is based on the overlay infrastructure. How to evaluate the performance, resilience, and adaptability of VIOLIN, compared with the real Internet and with application-level overlays? Especially, to match the performance of application-level overlays, how much additional computation and communication capacity need to be allocated? Our video multicast application in VIOLIN demonstrates performance comparable to its counterpart in an application-level overlay. However, more in-depth evaluation and measurement are needed before these questions can be answered.
- *Refinement of network virtualization technique.* Our inter-host tunneling implementation is initial and there is plenty of room for refinement and improvement. For example, how to improve the reliability of virtual links? Should we adopt another transport protocol (such as TCP), or integrate error correction (such as FEC) into UDP, or simply let the transport protocols in the VIOLIN domain to achieve reliability? To monitor the status of virtual links, is it possible to leverage the *routing underlay* [22] for better Internet friendliness?
- *Topology planning and optimization.* Our implementation provides mechanisms for dynamic VIOLIN topology setup and adjustment. However, we have not studied the the problem of VIOLIN topology planning and optimization. More specifically, given the overlay infrastructure, where to place the vRouters and vSwitches, in order to achieve Internet bandwidth efficiency and satisfactory application performance? How should a VIOLIN react to the dynamics of Internet condition and application workload using its dynamic reconfigurability (Section 3.4)?

Acknowledgements

We thank the anonymous reviewers for their reviews. This work is supported in part by the National Science Foundation (NSF) under the grant SCI-0438246.

References

1. Calvert, K., Bhattacharjee, S., Zegura, E., Sterbenz, J.: Directions in Active Networks. IEEE Communications Magazine (1998)
2. Kasera, S., Hjalmtysson, G., Towsley, D., Kurose, J.: Scalable Reliable Multicast Using Multiple Multicast Channels. IEEE/ACM Trans. on Networking (2000)
3. Katabi, D., Wroclawski, J.: A Framework for Scalable Global IP-Anycast (GIA). Proc. of ACM SIGCOMM 2000 (2000)
4. Liu, C., Estrin, D., Shenker, S., Zhang, L.: Local Error Recovery in SRM: Comparison of Two Approaches. IEEE/ACM Trans. on Networking (1998)
5. Wetherall, D., Guttag, J., Tennenhouse, D.: ANTS: Network Services without the Red Tape. IEEE Computer **32** (1999)
6. Dragovic, B., Fraser, K., Hand, S., Harris, T., Ho, A., Pratt, I., Warfield, A., Barham, P., Neugebauer, R.: Xen and the Art of Virtualization. Proc. of ACM SOSP 2003 (2003)
7. Whitaker, A., Shaw, M., Gribble, S.D.: Scale and Performance in the Denali Isolation Kernel. Proc. of USENIX OSDI 2002 (2002)
8. Stoica, I., Shenker, S., Zhang, H.: Core-Stateless Fair Queueing: a Scalable Architecture to Approximate Fair Bandwidth Allocations in High-speed Networks. IEEE/ACM Trans. on Networking **11** (2003)
9. Subramanian, L., Stoica, I., Balakrishnan, H., Katz, R.: OverQoS: Offering Internet QoS Using Overlays. Proc. of ACM HotNets-I (2002)
10. Jiang, X., Xu, D.: vBET: a VM-Based Emulation Testbed. Proc. of ACM SIGCOMM 2003 Workshops (MoMeTools) (2003)
11. Braden, R., Faber, T., Handley, M.: From Protocol Stack to Protocol Heap Role-Based Architecture. Proc. of ACM HotNets-I (2002)
12. Dike, J.: User Mode Linux. (http://user-mode-linux.sourceforge.net)
13. : VMware. (http://www.vmware.com)
14. Housley, R., Hollenbeck, S.: EtherIP: Tunneling Ethernet Frames in IP Datagrams. http://www.faqs.org/rfcs/rfc3378.html (2002)
15. Touch, J.: Dynamic Internet Overlay Deployment and Management Using the X-Bone. Proc. of IEEE ICNP 2000 (2000)
16. Ishiguro, K.: Zebra. (http://www.zebra.org/)
17. Kohler, E., Morris, R., Chen, B., Jannotti, J., Kaashoek, M.F.: The Click Modular Router. ACM Trans. on Computer Systems (2000)
18. Peterson, L., Anderson, T., Culler, D., Roscoe, T.: A Blueprint for Introducing Disruptive Technology into the Internet. Proc. of ACM HotNets-I (2002)
19. White, B., Lepreau, J., Stoller, L., Ricci, R., Guruprasad, S., Newbold, M., Hibler, M., Barb, C., Joglekar, A.: An Integrated Experimental Environment for Distributed Systems and Networks. Proc. of USENIX OSDI 2002 (2002)
20. Andersen, D.G., Balakrishnan, H., Kaashoek, M.F., Morris, R.: Resilient Overlay Networks. Proc. of ACM SOSP 2001 (2001)
21. Chu, Y.H., Rao, S.G., Zhang, H.: A Case For End System Multicast. Proc. of ACM SIGMETRICS 2000 (2000)
22. Nakao, A., Peterson, L., Bavier, A.: Routing Underlay for Overlay Networks. Proc. of ACM SIGCOMM 2003 (2003)

Increasing Software-Pipelined Loops in the Itanium-Like Architecture

Wenlong Li[1], Haibo Lin[2], Yu Chen[1], and Zhizhong Tang[1]

[1] Department of Computer Science and Technology, Tsinghua University,
100084 Beijing, P.R. China
{liwenlong00, chenyu00}@mails.tsinghua.edu.cn,
tzz-dcs@tsinghua.edu.cn
[2] Intel China Research Center, 100080 Beijing, P.R. China
jason.h.lin@intel.com

Abstract. The Itanium architecture (IPF) contains features such as register rotation to support efficient software pipelining. One of the drawbacks of software pipelining is its high register requirement, which may lead to failure when registers provided by architecture are insufficient. This paper evaluates the register requirements of software-pipelined loops in Itanium architecture and presents two new methods, which try to reduce static general register requirements in software pipelined loops by either reducing instructions in the loop body or allocating unused rotating registers for variants using static registers. We have implemented our methods in the Open Research Compiler (ORC) targeted for Itanium processors, and experiments show that number of software-pipelined loops in NAS Benchmarks increased. For some benchmarks, the performance is improved by more than 18%.

1 Introduction

The Itanium architecture (IPF) contains many features to enhance parallel execution, such as an explicit parallelism, large register files, predication, and register rotation [1] to support efficient software pipelining. Software pipelining [2] tries to improve the performance of a loop by overlapping the execution of several successive iterations. This improves the utilization of available hardware resources by increasing the instruction level parallelism (ILP).

The drawback of aggressive scheduling techniques [3] such as software pipelining is that they increase register requirements [4]. There have been proposals to perform register allocation for software-pipelined loops [5]. If the number of registers required is larger than available ones provide by architecture, actions must be taken. Introducing spill code [6] or increasing II could reduce register usage. However spill codes also reduce performance and increase schedule complexity, and increasing II reduces parallelism exploited from software pipelining; therefore some compilers just skip software pipelining when available registers are insufficient.

Itanium provides 128 general registers, 128 floating-point registers, and 64 predicate registers. The general registers are partitioned into two subsets. GR0-GR31 are termed the *static general registers*. GR32-GR127 are termed the *stacked general registers*. The stacked registers are made available to a program by allocating a register stack frame consisting of a programmable number of local and output register. The floating-point registers and predicate registers are also partitioned into two subsets respectively. FR0-FR31 (PR0-PR15) are termed the static floating-point (predicate) registers. FR32-FR127 (PR16-PR63) are termed the rotating floating-point (predicate) registers.

2 Register Requirements of Software Pipelining and Software Pipelining Failure in the Itanium Architecture

We have evaluated the register requirements of all the innermost loops in the NAS Benchmarks that are suitable for software pipelining in the Itanium architecture. These loops have been obtained with the ORC compiler. A total of 372 software-pipelinable loops have been used.

Fig. 1 shows the cumulative distribution of the requirements for static general registers for all 372 software pipelinable loops in NAS benchmarks. There are three type of variants requiring static general registers. *Dedicated variants* require special general registers, such as global pointer (gp), memory stack pointer (sp). *Loop invariants* are repeatedly used by a loop in each iteration. *Base post-increment variants* perform both a read and a write in one single operation[1]. Notice that 4 or 5 static general registers are occupied by architecture for special purpose; therefore only 27-28 out of 32 are available for software pipelining. 6.2% of the loops require such many static general registers, and these loops will fail in software pipelining phase.

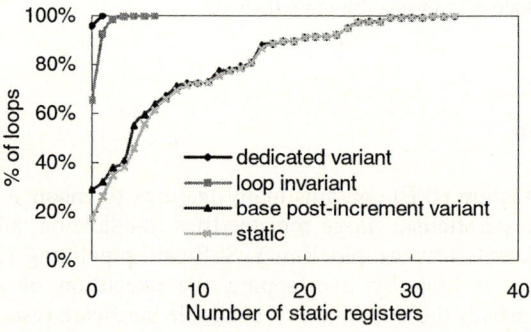

Fig. 1. Cumulative distribution of variants requiring static general registers. Each point (x,y) of the graph represents the percentage y of loops having less than x registers

We also estimated the cumulative distribution of rotating general registers. Experiments shows that 95% of the loop can be scheduled with 40 registers and all loops can be scheduled with 64 registers. This is much less than the maximum number of 96, which Itanium provides. In general, it is indicated that software

---

[1] For example, in memory operation "ld4 r1=[r2],4", r2 is read and then added by 4.

pipelining fails mainly due to insufficient static general registers in the Itanium architecture.

3 Register Sensitive Unrolling (RSU)

Theorem 1. Assume loop L has N_d dedicated variants, N_i loop invariants, and N_b base post-increment variants, then after unrolling L by K times, the number of static registers required by unrolled loop L' is less than or equal to $N_d + N_i + K*N_b$.

Proof. The number of dedicated variants (N_d) and loop invariants (N_i) doesn't change however the loop body is unrolling. Only the number of base post-increment variants, which hold the value of memory address that ld/st instruction refers, increases linearly with the degree of unrolling. After redundant load/store elimination, the number of base post-increment variants in unrolled loop body will be less than or equal to $K*N_b$. Given that 1 variable requires 1 static register, the number of static registers required by L' is less than or equal to $N_d + N_i + K*N_b$.

Register Sensitive Unrolling (RSU) is a new unrolling heuristic considering non-rotating register pressure based on above theorem. Generally speaking, the number of dedicated variants and loop invariants will not increase with loop unrolling, while the number of base post-increment variants often increases linearly with unrolling degree. RSU firstly computes a maximal unrolling degree (K_{max}) allowable for software pipelining according to the number of base post-increment variants, then takes the minimum between K_{max} and original unrolling degree K_{old} as new unrolling degree K_{new}. The detailed formulations are given below:

$$K_{max} = \begin{cases} +\infty & \text{if } N_b = 0 \\ \dfrac{N_a - N_d - N_i}{N_b} & \text{otherwise} \end{cases} \quad (1)$$

$$K_{new} = \begin{cases} K_{old} & \text{if } K_{max} = 0 \\ min(K_{old}, K_{max}) & \text{otherwise} \end{cases} \quad (2)$$

N_a, N_d, N_i and N_b above refer to the number of available static general registers, the number of dedicated variants, the number of loop invariants, and the number of base post-increment variants, respectively.

4 Variable Type Conversion (VTC)

As illustrated in section 2, rotating registers in Itanium are sufficient for software pipelining and base post-increment variants use most of the static general registers. Here we propose VTC method to convert base post-increment variant to loop variant which is allocated on available rotating registers, and thus reduce static general register pressure.

For the base post-increment variant in instruction `ld4 r1=[r2],4`, we convert this instruction to `ld4 r1=[r2]` and adds `r2=r2,4`, which perform the same task of original `ld4 r1=[r2],4`. For the two instructions, variant r2 is a loop variable and then could be allocated a rotating register instead of a static register.

VTC just converts base post-increment variant to loop variant through splitting instruction with the form of REG1=OPCODE [REG2], Imm into two instructions REG1=OPCODE[REG2] and ADDS REG2=REG2, Imm or REG1=OPCODE [REG2], REG3 into REG1=OPCODE[REG2] and ADD REG2=REG2,REG3 instructions, respectively. In Itanium processor, OPCODE could be ld, st, lfetch, and their relative variants. The number of base post-increment variants to be converted to loop variants is determined as follows:

$Num\_Convert = N_d + N_i + N_b - N_a$ where N_a, N_d, N_i and N_b above refer to the number of available static general registers, the number of dedicated variants, the number of loop invariants, and the number of base post-increment variants, respectively.

5 Experimental Results

We have implemented RSU and VTC methods in ORC [7] for Itanium. In this section, we compare the results of applying these two techniques and the original implementation in ORC in the NAS benchmarks. The benchmarks were compiled at the –O3 optimization level without profile feedback and Inter-Procedural Analysis (IPA). The measurements were performed on an HP workstation i20000 equipped with single 733MHz Itanium processor running Redhat Linux 7.2.

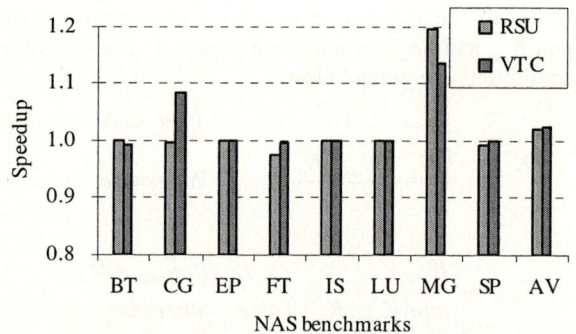

Fig. 2. Performance of RSU and VTC

Fig. 2 shows the percentage improvements of the two methods. RSU provides an overall 2% gain and a peak 19.7% gain in measured performance for all of NAS Benchmarks. VTC shows an overall of 2.6% and a peak 13.7% performance improvement. This result is rather exciting because only 6.2% of loops are optimized.

Note that RSU degrades the performance of FT by 2.3%. Detailed observation shows that only one more loop is software-pipelined after changing unrolling degree from 8 to 7. When performing loop scheduling, the scheduling length is 17 cycles, so each iteration is completed in 17/8=2.125 cycles on average. When performing software pipelining, the kernel contains only 1 stage, and II is 15 cycles. As a result, each iteration is completed in 15/7=2.143 cycles on average. Taking into account the prolog and epilog overhead, the performance of software pipelining is even worse than that of loop scheduling. In such case, software pipelining should be skipped.

6 Conclusions

In this paper we have evaluated the static and rotating register requirements of software-pipelined loops in NAS Benchmarks on Itanium architecture. We also showed that some loops with high static general register requirements fail in software pipelining phase. Then we proposed two new methods for increasing software-pipelined loops on Itanium architecture. One of them reduces the number of instructions in a loop body by limiting the degree of unrolling. The other tries to convert base post-increment variants which occupy static general registers to loop variants. The experimental results that we have obtained with our methods show significant improvements in the execution time of NAS Benchmarks over existing techniques for software pipelining loops.

Acknowledgements

Comments from colleagues at Intel China Research Center (ICRC) and University of Delaware improve the presentation of the paper. We deeply appreciate their support and encouragement.

References

1. Dehnert J. C., Hsu P. Y., Bratt J. P.: Overlapped Loop Support in the Cydra 5. Proceedings of ASPLOS'89. (1989) 26-38
2. Allan V. H., Jones R. B., Lee R. M., Allan S. J.: Software Pipelining. ACM Computing Surveys. **27** (1995) 367-432
3. Huff R. A.: Lifetime-sensitive modulo scheduling. Proceedings of PLDI'93. (1993) 58-267
4. Llosa J.: Reducing the Impact of Register Pressure on Software Pipelining. PhD thesis. Universitat Politècnica de Catalunya (1996)
5. Rau B. R., Lee M., Tirumalai P., Schlansker P.: Register allocation for software pipelined loops. Proceedings of PLDI'92. (1992) 283-299
6. Llosa J, Valero M, Ayguadé E: Heuristics for register-constrained software pipelining. In Proceedings of the MICRO-29 (1996). 250-261
7. Ju R, Sun C, Wu C Y: Open Research Compiler for Itanium Processor Family(IPF). In Proceedings of MICRO-34 (2001)

A Space-Efficient On-Chip Compressed Cache Organization for High Performance Computing*

Keun Soo Yim[a], Jang-Soo Lee[b], Jihong Kim[a], Shin-Dug Kim[c], and Kern Koh[a]

[a] School of Computer Science and Engineering, Seoul National University, Seoul, Korea
[b] IBM Poughkeepsie, NY, USA
[c] Department of Computer Science, Yonsei University, Seoul, Korea
{ksyim, kernkoh}@oslab.snu.ac.kr, jangsoo@us.ibm.com,
jihong@davinci.snu.ac.kr, sdkim@cs.yonsei.ac.kr

Abstract. In order to alleviate the ever-increasing processor-memory performance gap of high-end parallel computers, on-chip compressed caches have been developed that can reduce the cache miss count and off-chip memory traffic by storing and transferring cache lines in a compressed form. However, we observed that their performance gain is often limited due to their use of the coarse-grained compressed cache line management which incurs internally fragmented space. In this paper, we present the fine-grained compressed cache line management which addresses the fragmentation problem, while avoiding an increase in the metadata size such as tag field and VM page table. Based on the SimpleScalar simulator with the SPEC benchmark suite, we show that over an existing compressed cache system the proposed cache organization can reduce the memory traffic by 15%, as it delivers compressed cache lines in a fine-grained way, and the cache miss count by 23%, as it stores up to three compressed cache lines in a physical cache line.

Keywords: Parallel processing, processor-memory performance gap, on-chip compressed cache, fine-grained management, internal fragmentation problem.

1 Introduction

As the performance gap between processor and memory has increased by 28-48% every year, the memory system performance typically dominates the whole computer system performance [1]. In order to improve the memory performance, high-end computers are based on large size on-chip caches with a high off-chip memory bandwidth. Although these are effective in improving the memory performance, they are restricted by physical device limits such as the on-chip area and off-chip pin count.

On-chip compressed cache is an alternative approach of improving the memory performance. Compressed caches, e.g. SCMS [2] and CC [3], store and transfer some cache lines in a compressed form, thereby, reducing both the on-chip cache miss count and off-chip memory traffic without having to face the physical limits. The existing compressed caches manage variable-size compressed cache lines in a coarse-grained manner. Specifically, as exemplified in Figure 1, if a cache line can be com-

---

* "This research was supported by University IT Research center project in korea".

pressed to less than half of the original size, they treat the cache line size as half of the original size, thereby, incurring internally unused space, namely internal fragmentation. Otherwise, they do not handle the cache line in a compressed form. Thus, at most two compressed cache lines can be stored in a physical cache line, and only 1 bit is required to specify the status of a physical cache line whether it embeds two compressed lines or one uncompressed line.

However, their performance gain is often limited by this coarse-grained management. Figure 2 shows a cumulative compression rate distribution of L2 data cache lines on an Alpha machine simulator using the SPEC CPU2000 benchmark suite [7] where the compression rate is defined as the ratio of the compressed data size and the original data size. It shows that over 60% of cache lines are compressed to less than 25% of the original cache line size. This high compression efficiency, mainly due to the frequent value locality [4], strongly suggests that the coarse-grained management is overly conservative.

In order to fully exploit this high compression efficiency, in this paper we present the *Fine-grained Compressed Memory System* (FCMS) based on the four key techniques. First, the FCMS manages compressed cache lines in a fine-grained manner so that it reduces the fragmented space. This implies that the FCMS is more effective in reducing the off-chip memory traffic than the existing compressed caches. Second, based on this fine-grained management, the FCMS stores up to 3 compressed cache lines in a physical cache line in order to further reduce the cache miss count. Unfortunately, this fine-grained management can bring out a large size of metadata. Third, we thus present two additional techniques that limit an increase in the size of both on-chip cache tag address and VM page table without diminishing the obtained performance gain. Fourth, we firstly apply the cooperative parallel decompression technique [5] to on-chip compressed caches in order to reduce the decompression time without harming the compression efficiency significantly.

In order to evaluate the effectiveness of the FCMS, we modified the SimpleScalar [6] simulator and used the SPEC benchmarks. The experimental results show that the FCMS reduces the average execution time by 5% and 12% over the SCMS and a conventional cache system, respectively. In particular, the FCMS reduces the on-chip L2 cache miss count by 23% and 25% and the off-chip memory traffic by 15% and 46% over the SCMS and the conventional system, respectively, in an average case.

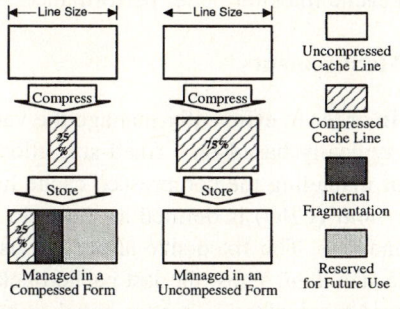
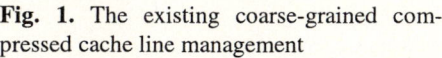

Fig. 1. The existing coarse-grained compressed cache line management

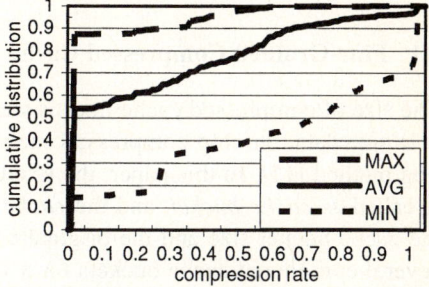

Fig. 2. A cumulative compression rate distribution of on-chip L2 data cache lines

The rest of this paper is organized as follows. In Section 2, we provide the organization of the FCMS. The experimental methodology is described in Section 3, while the evaluation results are given in Section 4. We review the related work in Section 5 and conclude this paper with a summary and a future work in Section 6.

2 Fine-Grained Compressed Memory System (FCMS)

Figure 3 illustrates the overall cache and memory organization of the FCMS, in which both the on-chip unified L2 cache and the main memory are managed in a compressed form. When a data memory page is firstly loaded into the main memory, all L2 cache lines in the page are individually compressed using a hardware compressor. We use the X-RL compression algorithm [8] because of some of its desirable properties, such as the high compression efficiency with small size data and fast (de)compression speed of at least four bytes per cycle. If a compressed cache line stored in the memory is accessed, the off-chip memory bandwidth can be expanded as the line is transferred to the on-chip cache in a compressed form. Moreover, as the line is stored in the on-chip L2 cache in a compressed form, the effective L2 cache capacity is also expanded.

While the line is stored in the L2 cache, the line is concurrently decompressed on the fly using a hardware decompressor in order to deliver the required L1 cache line to the L1 cache. As the remaining L1 cache lines in the decompressed L2 cache line have a high probability of accessing in the near future due to spatial locality, the decompressed L2 line is stored in a decompression buffer which can be accessed in one cycle. The decompression buffer consists of a small fully-associative cache (8 entries [2]) and is managed in the same way as the victim cache does [9]. In this paper, we assume that the decompression buffer can be concurrently accessed with the L1 caches for the fair performance evaluation with the conventional cache systems.

In the FCMS, only data cache lines are managed in a compressed form. As instruction cache lines result in the lower compression efficiency while incurring the larger decompression overhead, the overall cache and memory system performance can be degraded if they are managed in a compressed form. Fortunately, as instruction cache lines are not generally modified at runtime, several off-line code compaction techniques [10] can be used for instruction cache lines for better performance.

2.1 Fine-Grained Compressed Cache Line Management

The size of compressed cache lines is various. In order to efficiently manage the variable size data, on-chip compressed caches are typically based on a fixed-size allocation method [11]. In this paper, the fixed unit of managing the compressed cache line is called as *cache bucket*, and the *cache bucket unit* (*CBU*) is defined as the ratio of the cache bucket size and the original cache line size. The fixed-size allocation uses several consequent cache buckets for a variable size data. Thus, the last cache bucket can incur internal fragmentation, and the average fragmentation size is equal to half of the cache bucket size.

We can notate 1/2 as *CBU* of the coarse-grained management of the SCMS and the CC. As the CC is targeting for embedded processors, we only use the SCMS as a compressed cache which uses the coarse-grained management scheme. This implies that due to the internal fragmentation the coarse-grained management wastes about 25% of compressed cache space in an average case. In order to reduce the fragmented space, we use the fine-grained compressed cache line management in the FCMS. Thus, the variable-size compressed cache lines can be stored in a more fitting cache bucket while reducing the internal fragmentation. Figure 4 visualizes the fraction of internal fragmentation as a function of the compression rate where *CBU* of the FCMS is assumed to be 1/16. In the figure, white area means the compressed data size, red areas mean the internally fragmented space for both the FCMS and the SCMS, blue areas mean the saved space for the FCMS and the fragmented space for the SCMS, and gray area means the saved space for both the FCMS and the SCMS. Thus, the amount of internal fragmentation space can be greatly reduced in the FCMS.

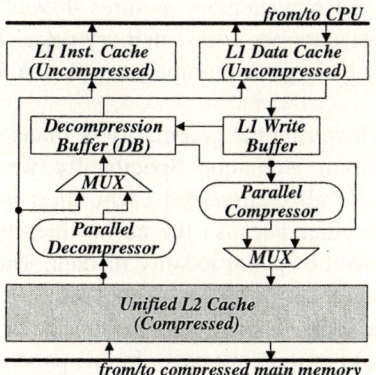 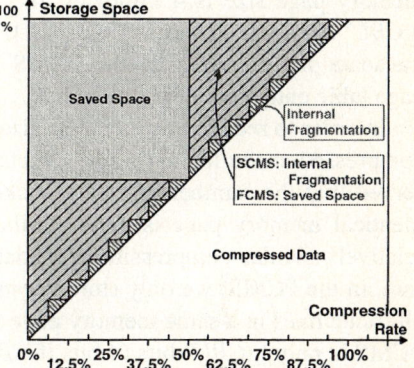

Fig. 3. Overall cache and memory hierarchy organization of the FCMS

Fig. 4. Space efficiency of the fine-grained management of the FCMS (*CBU*=1/16)

In the SCMS, cache lines whose compression rate is higher than or equal to 50% do not provide any space-efficiency although they are managed in a compressed form because of the internal fragmentation. On the other hand, in the FCMS, cache lines that have high compression rate of up to $(CBU^{-1}-1)/CBU^{-1}$ % (93.75% if *CBU* is 1/16) can provide performance benefits. However, the provided benefits of the inefficiently compressed lines are quite small, while the compressed lines bring out the decompression overhead. Thus, we use the selective compression technique [2] where a cache line is managed in a compressed form only if its compression rate is less than a specified threshold value (*THRD*). If the threshold value is set to be a lower value, both the performance benefit and the decompression overhead are reduced because only small fraction of cache lines are compressed. Otherwise if the value is set to be a higher one, conversely both the benefit and the overhead are increased.

In the FCMS, maximum CBU^{-1} (16 if CBU = 1/16) numbers of compressed cache lines can be stored in a physical cache line when all the compressed cache lines can be stored in a bucket whose size is CBU. Actually, in order to do this, the FCMS has to use CBU^{-1} numbers of tag addresses, valid bits, and dirty bits per every physical cache line. However, this requires a large overhead in terms of on-chip area and energy consumption, while typically only a part of the CBU^{-1} tag addresses is used. Thus, the number of tag address (*NTAG*) is another design parameter of the FCMS. In addition, we use LRU as the replacement algorithm of compressed cache lines stored in a physical cache line.

2.2 Metadata Size Reduction Techniques

The fine-grained management at the same time requires a large amount of metadata, $\lg CBU^{-1}$ bits (e.g. 4 bits if CBU is 1/16), to specify the number of cache buckets used for a compressed cache line, while the coarse-grained management where CBU is 1/2 requires only 1 bit for a cache line. When the L2 cache line size is 128 bytes and the memory page size is 4 kilobytes, the fine-grained management requires *4096/128* $\lg CBU^{-1}$* bits (e.g. 128 bits if CBU is 1/16) of metadata per every memory page. This metadata size required by the FCMS is relatively large as compared with the VM page table entry size of about 32 bits.

In order to reduce this metadata size without lowering the granularity of managing compressed cache lines, we use a metadata grouping technique. Specifically, we observed that the number of cache buckets used for all compressed cache lines in an identical memory page is quite similar to each other because the cache lines have relatively similar compression rate mainly due to the spatial locality of data. Therefore, in the FCMS, we only store the maximum number of all cache buckets used for all cache lines in a same memory page as the metadata. Then, the metadata size of the FCMS is only $\lg CBU^{-1}$ bits (4 bits if CBU is 1/16) per memory page. In this paper, we assume that this small size of metadata can be embedded in the VM page table entry, which typically has some unused bits.

Moreover, if the physical memory capacity is 256 megabytes and cache line size is 128 bytes, the size of tag addresses for a cache line in the compressed cache is *NTAG**lg(256M/128) bits (63 bits if *NTAG* is 3). In order to reduce this tag field size, we use a segmented addressing technique. Specifically, a tag address is divided into a tag segment and a tag offset, and all compressed cache lines stored a physical cache line should have an identical tag segment. With this technique, if the tag offset size is *TOFF* bits, the tag field size per a cache line is only lg(256M/128) − *TOFF* + *NTAG* * *TOFF* bits (36 bits if *TOFF* is 7 bits and *NTAG* is 3), while that for a conventional cache is lg(256M/128) = 21 bits. As the FCMS aims to an L2 unified cache where the cache line size is between 512 bits and 2048 bits, we believe that the additional bits used for tag fields in the FCMS are acceptable design overhead.

2.3 FCMS-Based Direct-Mapped Cache Organization

Figure 5 shows the compressed L2 cache organization of the FCMS in a direct-mapped scheme. The same organization technique applies for set-associative caches.

In this organization, the tag address count (*NTAG*) is 3, the tag offset size (*TOFF*) is 4 bits, and the cache bucket unit (*CBU*) is 1/16. Thus, the tag RAM has three valid bits, three dirty bits, a tag-segment address, and three tag-offset addresses per every cache line, and the data RAM is divided into 16 cache buckets.

When a request is generated, the tag segment and the 3 tag offsets of the requested cache set are concurrently compared with that of the generated address. The results are analyzed by using three OR-gates and an AND-gate so as to determine the hit or miss. Because only three gates are additionally used in the critical path of the FCMS as indicated by the bold lines in Figure 5, we assume that the FCMS-based compressed caches do not cause any additional delay as regarded in the access time. Moreover, we can use a three-input OR-gate instead of the three two-input OR-gates to decide the hit or miss, and this reduces the gate delay as regarded in the access time.

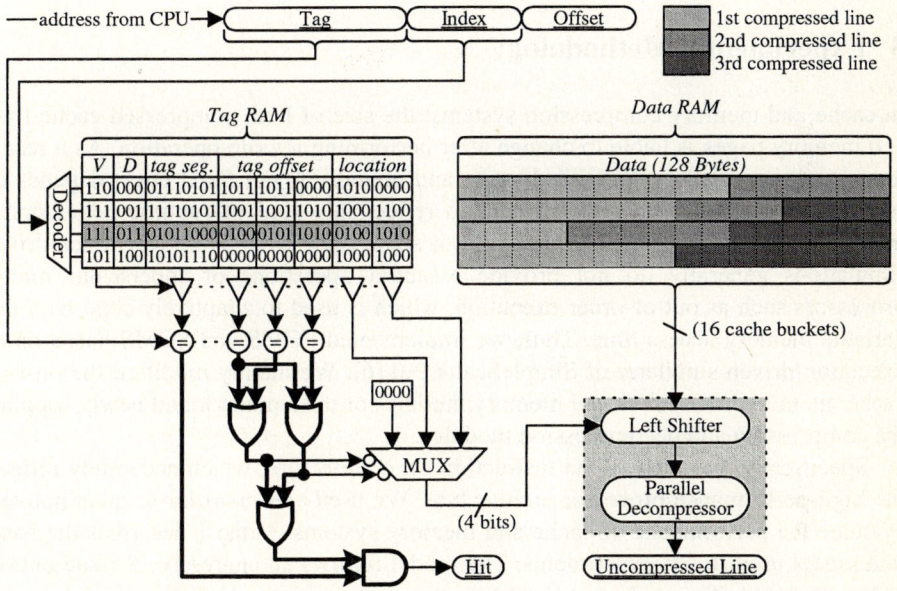

Fig. 5. A fine-grained compressed cache architecture in direct-mapped scheme

We use $\lg CBU^{-1} = 4$ bits to encode the location of a compressed cache line stored in the data RAM. Since the location of the first compressed cache line is always fixed as 0, we use only 8 bits for the location field. The location information is used as an input of the MUX logic, which selects an appropriate one and routes to the left shifter and parallel decompressor logic. The size of the requested compressed cache line is concurrently calculated by subtracting the two adjacent location values.

For example, if the second compressed cache line whose tag area is marked by gray color is accessed, a cache hit occurs and the location data of $0100_{(2)} = 4$ is routed to the left shifter logic, which performs a left shift operation for 4 cache buckets. Also its size is calculated by subtracting its location value from the location value of the third one. The result is $1010_{(2)} - 0100_{(2)} = 6$ cache buckets. Then, the selected cache buckets are routed to the cooperative parallel decompressor logic. Finally, they are delivered to a decompression buffer and an L1 cache in an uncompressed form.

In on-chip compressed caches, the decompression time can seriously degrade the effectiveness of the data compression technique. For example, the X-RL decompressor takes up to $3+LS/4$ cycles for decompressing a cache line, whose size is LS bytes. This means that a smaller compression unit size results in the shorter decompression time. However, it simultaneously incurs the lower compression efficiency. Fortunately, the cooperative parallel decompressor [5] can reduce the decompression time by about 75% while it slightly degrades the compression efficiency. Thus, we use the parallel (de)compressor in the FCMS and evaluate the performance impact.

3 Experimental Methodology

In cache and memory compression systems, the size of both compressed cache lines and memory pages is liable to change after performing a write operation. As a result, the access time of the compressed cache and memory is not fixed but it depends on runtime status. However, it is difficult to reflect this kind of runtime behaviors in trace-driven simulations due to their use of static trace data. Moreover, trace-driven simulations generally do not provide essential operations of superscalar microprocessors such as out-of-order execution, which is used to adaptively cope with this variable memory access time. Thus, we implemented FCMS and SCMS based on an execution-driven simulator of SimpleScalar 3.0 [6]. We mainly modified the on-chip cache, memory bus, and virtual memory modules of the simulator and newly supplied the compression and decompression modules.

Specifically, we used Alpha instruction set architecture, which accurately reflects the high-performance processor architecture. We used *sim-outorder* to quantitatively evaluate the performance of cache and memory systems. Table 1 describes the baseline model used in our experiments. The model follows an aggressive 8-issue out-of-order processor. The cache configuration parameters for the base line model are assumed to be two 32 kilobytes L1 caches and a unified 256 kilobytes L2 cache with four-way associativity and 128 bytes cache line size. We referenced an accurate cache timing model of CACTI for calculating the access time of on-chip caches [12].

We used the SPEC CPU2000 benchmark suite [7] with reference input workload. The benchmark suite is compiled by using the Compaq Alpha compiler with SPEC peak settings. The virtual memory image of this benchmark suite is captured after full execution by using *sim-safe*. For the *sim-outorder* simulations, we used a fast forwarding technique [13] where 1.5 billion instructions are accurately executed after a coarse-grain simulation of 0.5 billion instructions so as to reduce the simulation time without notably compromising the simulation accuracy.

Table 1. Base line model

Parameter	Value
Processor Core	2.4 GHz (6 x 400 MHz), 0.13 Micron, Alpha ISA, 8 fetch/issue/decode/commit, 128-RUU, 128-LSQ.
Branch Predictor	Bimodal 2K, 512-entry 4-way BTB, 8-entry RAS.
TLB (Inst. / Data)	16 / 32 entry, 4KB page size, 4-way, LRU, 72 cycle latency.
L1 Cache (Inst. / Data)	Each 32KB, 1-way, 32B block, LRU, 1 cycle latency, write-back.
L2 Cache (Unified)	256KB, 4-way, 128B blocks, LRU, 7 cycle latency, write-back.
Main Memory	72 cycle latency, 8 bytes bandwidth, 400 MHz bus clock.

4 Performance Evaluations

In this section, we evaluate the performance of the FCMS over a conventional cache system (CS), a conventional system with a decompression buffer (CSDB), and the SCMS. In CSDB, the decompression buffer is only used for prefetching the L2 cache lines to L1 caches.

First, as shown in Figure 6, we measured the average memory cycles spent to transfer a data cache line where the memory latency is excluded. Because L2 cache line size is 128 bytes and memory bus bandwidth is 8 bytes, both CS and CSDB require 8 memory bus cycles to deliver a cache line. On the other hand, in both SCMS and FCMS, the required bus cycles are significantly reduced as they deliver data cache lines in a compressed form. Although they use the same compression algorithm

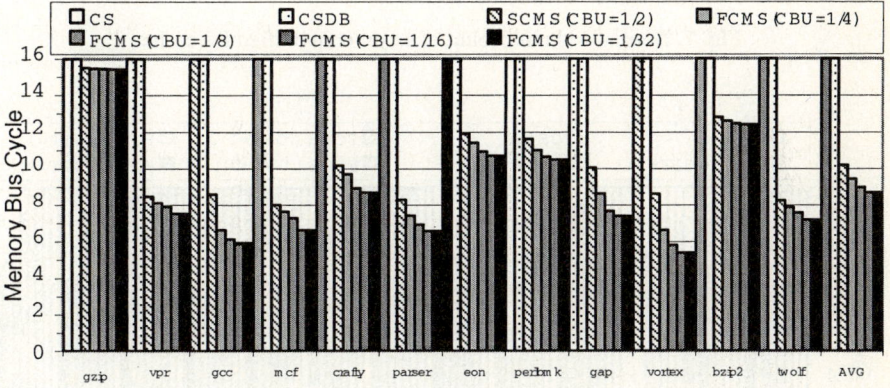

Fig. 6. Average memory bus cycles elapsed to deliver a data cache line

and the same compression threshold value (50%), the FCMS with a finer-grained cache bucket size uses the smaller memory bus cycles. This is because the fact that the fine-grained management more accurately specifies the actual size of compressed cache lines, thereby, reducing the internal fragmentation. As this tendency is stabilized when *CBU* is larger than 1/16, we set 1/16 as *CBU* of the FCMS. The FCMS where *CBU* is 1/16 requires about 8.7 memory bus cycles, while the SCMS requires about 10.2 cycles in an average case. This means that the FCMS reduce the amount of memory traffic by 15% and 46% as compared with the SCMS and CS, respectively.

Second, we measured the miss count of the on-chip unified L2 cache as shown in Figure 7. The miss count of CSDB is slightly higher than that of CS because its decompression buffer filters several L2 cache accesses and consequently disturbs the reference history of the L2 cache, incurring inefficient cache replacements. The miss count of the SCMS is reduced by about 2% as compared with both CS and CSDB in an average case. This is because in the SCMS two compressed cache lines whose tag addresses are same except for the least significant bit can be stored in a physical cache line. Thus, even cache lines are sufficiently compressed as shown

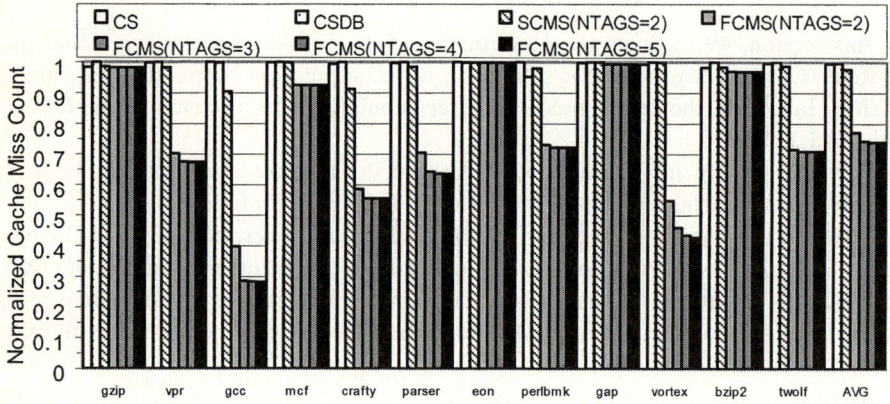

Fig. 7. Normalized miss count of an on-chip unified L2 cache

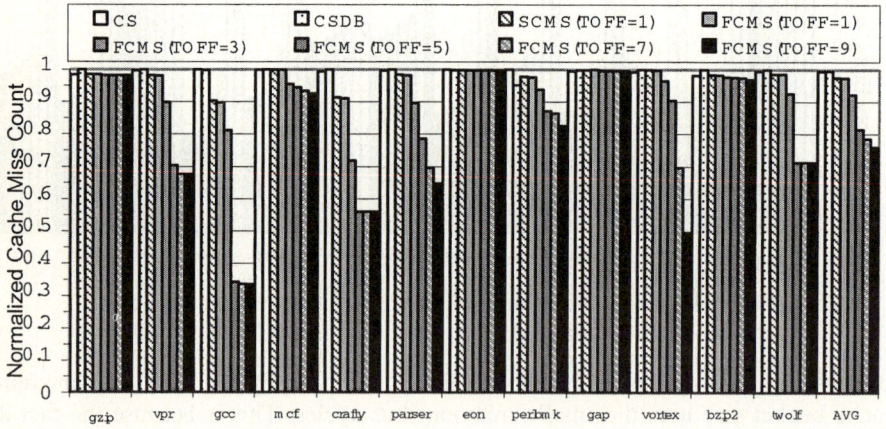

Fig. 8. Normalized miss count of the FCMS as a function of *TOFF* value

in Figure 6, the probability of storing two compressed lines in a physical cache line is quite low. On the other hand, in the FCMS the miss count is reduced by 22%, 25%, 25%, and 26% when the *NTAG* value is 2, 3, 4, and 5, respectively. Thus, we set the *NTAG* to be 3 in the FCMS. For benchmarks *gzip*, *eon*, *gap*, and *bzip2*, the miss count is slightly reduced in both the SCMS and the FCMS because their workload sizes are relatively small, thereby, even if a twice large size cache is used the miss count is seldom reduced.

Third, we measured the cache miss count of the FCMS by changing the *TOFF* value as shown in Figure 8. In the FCMS, the miss count is reduced by 2%, 7%, 18%, 21%, and 23% when the *TOFF* value is 1, 3, 5, 7, and 9, respectively, in an average case as compared with CS and CSDB. The miss count is stabilized when the *TOFF* value is larger than 7 because the accessed cache lines have strong spatial locality, and the address space covered by the tag-offset and cache line size, $2^7 \times 128 = 16$ kilobytes, is sufficient to cope this locality. Thus, we set the *TOFF* of the FCMS to be 7.

Fourth, we evaluated the average memory access time (AMAT) in order to analyze the decompression overhead of the FCMS and the SCMS. We calculated the code AMAT of these two systems in a similar way of calculating AMAT of CS and CSDB [1]. On the other hand, Eq. 1 is used to calculate the data AMAT of the FCMS and the SCMS. In the formula, *A*, *M*, *C*, and *DO* mean the access time, miss rate, fraction of compressed lines, and decompression cycles respectively, while the small symbols of *L1*, *DB*, *L2*, and *MM* mean the L1 data cache, decompression buffer, unified L2 cache, and main memory, respectively.

$$AMAT_{Data} = A_{L1} + M_{L1}\left[A_{DB} + M_{DB}\left\{\begin{matrix}A_{L2} + C_{L2}DO_{L2;avg} + \\ M_{L2}\left(A_{MM} + C_{MM}DO_{MM;avg}\right)\end{matrix}\right\}\right] \quad (1)$$

Based on this, we measured the data AMAT as shown in Figure 9. It shows that the FCMS reduces the data AMAT by 29%, 14%, 16%, and 8% in an average case as compared with CS, CSDB, SCMS, and the SCMS with the cooperative parallel decompressor (SCMS-Parallel). Due to the use of parallel decompressor, the decompression overheads (*L2_DO* and *MM_DO*) are significantly reduced in both the SCMS-Parallel and the FCMS. We used the early restarting technique, which provides the ability of accessing the critical word as early as possible without waiting for the complete decompression of the whole cache line, and the decompression time overlapping technique, which overlaps the transfer time of a compressed cache line from the main memory and its decompression time, in order to lessen DO_{L2} and DO_{MM}, respectively. We observed that most of the decompression times are absorbed by the decompression buffer in both the FCMS and the SCMS since the hit ratio of the decompression buffer in the FCMS is over 40% in an average case. We also observed that the FCMS reduces the code AMAT by about 1% as compared with CS, CSDB, the SCMS, and the SCMS-Parallel as it reduces the miss count of the unified L2 cache.

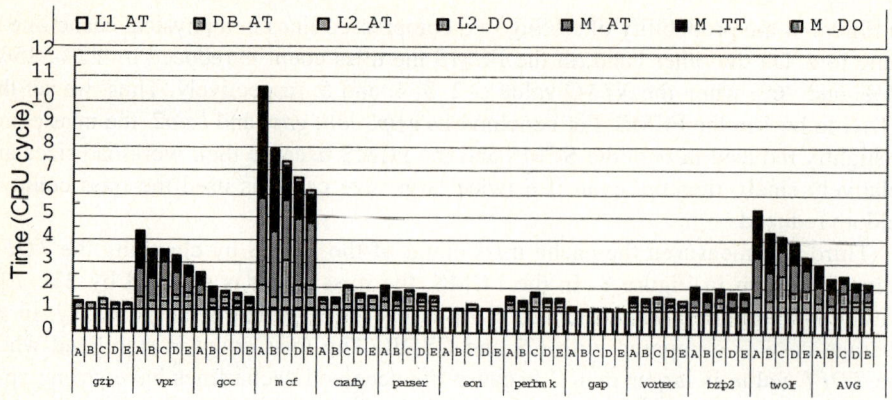

Fig. 9. Average data memory access time (Data AMAT)
* DB: decompression buffer; MM: main memory; AT: access time; DO: decompression overhead; TT: transfer time. (A: CS; B: CSDB; C: SCMS; D: SCMS-Parallel, E: FCMS)

Fifth, we measured the instructions per cycle (IPC) as shown in Figure 10. The average IPC is 1.62, 1.65, 1.72, 1.74, and 1.82 in CS, CSDB, the SCMS, the SCMS-Parallel, and the FCMS, respectively. This implies that the execution time of the FCMS is reduced by 12%, 10%, 6%, and 5% in an average case as compared with CS, CSDB, the SCMS, and the SCMS-Parallel, respectively. In this experiment, the compression threshold of the FCMS is set to be 50%, which is identical configuration to that of the SCMS. If the compression threshold is set to be a higher value, both the performance gain and the decompression overhead are increased. Because of this, we observed that in the FCMS, IPC is slightly influenced by the compression threshold value, and it is maximized when the threshold value is set to be 50%.

Furthermore, when we use 93.75% (=15/16) as the threshold value, IPC of the FCMS is only reduced by less than 1% in an average case. Since a higher threshold value means that a large amount of cache lines is managed in a compressed form and

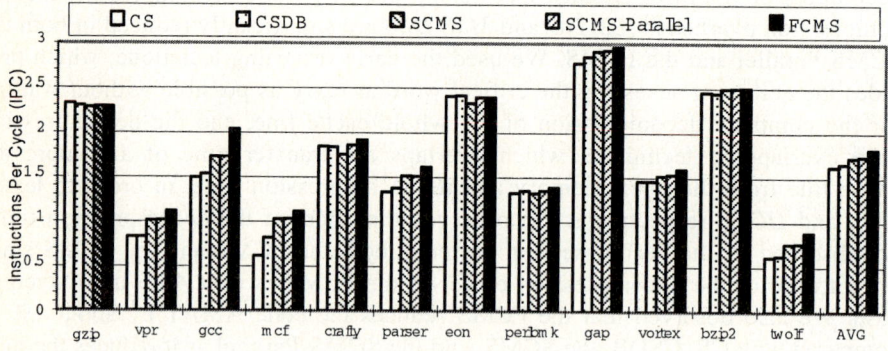

Fig. 10. Instructions per cycle (IPC)

the fine-grained management of the FCMS significantly reduces the fragmented space over the coarse-grained management of the SCMS, the FCMS has a higher potential of expanding the effective main memory capacity than the SCMS. Therefore, we believe that the real improvement in the execution time obtained with the FCMS will be much greater than that we presented in this paper.

5 Related Work

Over the past ten years, several research groups have been studied the on-chip cache and main memory compression systems in order to alleviate the performance gap between processor, memory, and hard disk [14] as well as reduce the energy consumption of memory systems [15] in high-end parallel computers. The existing on-chip compressed caches are typically based on the coarse-grained compressed cache line management because of its simplicity [2, 3]. Although as shown in this paper the coarse-grained management is overly conservative, so far as we know, none has been developed a compressed cache system in this perspective of managing compressed cache lines in a fine-grained manner and storing up to three compressed cache lines in a physical cache line with the appropriate metadata reduction techniques.

6 Conclusion

Recently on-chip compressed caches have been developed to alleviate the processor-memory performance gap in high-end parallel computers. However, we have observed that the performance gain of the existing compressed caches is often limited mainly due to the high compression efficiency of on-chip cache lines. In order to fully exploit the high compression efficiency, in this paper we have presented a novel on-chip compressed cache system based on the four key techniques. First, the proposed system manages the compressed cache lines in a fine-grained manner so that it reduces the fragmented space and consequently reduces the memory traffic over the existing compressed cache systems. Second, based on this, the proposed cache stores up to three compressed cache lines in a physical cache line, thereby, reducing the cache miss count over the existing systems. Third, in order to avoid an increase in the metadata size, the proposed system uses two novel metadata reduction techniques. Fourth, we firstly have applied a parallel (de)compressor to the on-chip cache systems and have shown the performance impact of using this. The execution-driven simulation results have shown that the FCMS reduces the average execution time by 5% and 12% over the an existing compressed cache system and a conventional cache system, respectively. In particular, the FCMS reduces the on-chip L2 cache miss count by 23% and 25% and the off-chip memory traffic by 15% and 46% over the compressed system and the conventional system, respectively, in an average case.

References

1. J. L. Hennessy, D. A. Patterson, and D. Goldberg, *Computer Architecture – A Quantitative Approach*, 3rd Ed., Morgan Kaufmann Publishers, 2002.
2. J. S. Lee, W. K. Hong, and S. D. Kim, "Design and Evaluation of On-Chip Cache Compression Technology," *In Proceedings of the IEEE International Conference on Computer Design*, pp. 184–191, 1999.
3. J. Yang, Y. Zhang, and R. Gupta, "Frequent Value Compression in Data Caches," *In Proceedings of ACM/IEEE International Symposium on Microarchitecture*, pp. 258–265, 2000.
4. Y. Zhang, J. Yang, and R. Gupta, "Frequent Value Locality and Value-centric Data Cache Design," *In Proceedings of the ACM International Conference on Architectural Support for Programming Languages and Operating Systems*, 2000.
5. P. A. Franaszek, J. Robinson, and J. Thomas, "Parallel Compression with Cooperative Dictionary Construction," *In Proceedings of the IEEE Data Compression Conference*, pp. 200–209, 1996.
6. T. Austin, E. Larson, and D. Ernst, "SimpleScalar: an Infrastructure for Computer System Modeling," *IEEE Computer*, Vol. 35, Issue 2, pp. 59–67, 2002.
7. J. L. Henning, "SPEC CPU2000: Measuring CPU Performance in the New Millennium," *IEEE Computer*, Vol. 33, Issue 7, pp. 28–35, 2000.
8. M. Kjelso, M. Gooch, and S. Jones, "Design and Performance of a Main Memory Hardware Data Compressor," *In Proceedings of the 22nd EuroMicro Conference*, IEEE Computer Society Press, pp. 422–430, 1996.
9. N. P. Jouppi, "Improving Direct-Mapped Cache Performance by the Addition of a Small Fully Associative Cache and Prefetch Buffers," *In Proceedings of the ACM/IEEE International Symposium on Computer Architecture*, pp. 364–373, 1990.
10. A. Beszedes, R. Ferenc, T. Gyimothy, A. Dolenc, and K. Karsisto, "Survey of Code-Size Reduction Methods," *ACM Computing Surveys*, Vol. 35, No. 3, pp. 223–267, 2003.
11. A. Silberschatz, P.B. Galvin, and G. Gagne, *Operating System Concepts*, 6th Ed., pp. 285–287, John Wiley & Sons Inc., 2003.
12. P. Shivakumar and N. P. Jouppi, "CACTI 3.0: An Integrated Cache Timing, Power, and Area Model," *Compaq Computer Corporation Western Research Laboratory, Research Report 2001/2*, 2001.
13. I. Gomez, L. Pifiuel, M. Prieto, and F. Tirado, "Analysis of Simulation-adapted Benchmarks SPEC 2000," *ACM Computer Architecture News*, Vol. 30, No. 4, pp. 4–10, 2002.
14. K. S. Yim, J. Kim, and K. Koh, "Performance Analysis of On-Chip Cache and Main Memory Compression Systems for High-End Parallel Computers," *In Proceedings of the International Conference on Parallel and Distributed Processing Techniques and Applications*, pp. 469–475, 2004.
15. L. Benini, D. Bruni, A. Macii, and E. Macii, "Hardware-Assisted Data Compression for Energy Minimization in Systems with Embedded Processors," *In Processing of the IEEE Design, Automation and Test in Europe Conference and Exhibition*, pp. 449–453, 2002.

A Real Time MPEG-4 Parallel Encoder on Software Distributed Shared Memory Systems

Yung-Chang Chiu[1], Ce-Kuen Shieh[1], Jing-Xin Wang[2],
Alvin Wen-Yu Su[2], and Tyng-Yeu Liang[3]

[1] Department of Electrical Engineering National Cheng Kung University,
Tainan, Taiwan, R.O.C.
{qson, shieh}@hpds.ee.ncku.edu.tw
[2] Department of Computer Science and Information Engineering, National Cheng Kung University, Tainan, Taiwan, R.O.C.
m8902005@chu.edu.tw, alvinsu@mail.ncku.edu.tw
[3] Department of Electrical Engineering, National Kaohsiung University of Applied Sciences,
Kaohsiung, Taiwan, R.O.C.
lty@hpds.ee.ncku.edu.tw

Abstract. This paper is dedicated to developing a real-time MEPG-4 parallel encoder on software distributed shared memory systems. Basically, the performance of a MPEG-4 parallel encoder implemented on distributed systems is mainly determined by the latency of data synchronization and disk I/O, and the cost of data computation. For reducing the impact of data synchronization latency, we invent a pipeline algorithm to minimize the number of data synchronization points necessary for video encoding. In addition, we employ a master-slave node structure to overlay computation and I/O in order for alleviating the impact of I/O latency. On the other hand, we propose a two-level partitioning method to minimize the cost of data computation, and overlap the encoding times of two different GOVs. We have implemented the proposed MPEG-4 encoder on a test bed called Teamster. The experimental results show the proposed MPEG-4 encoder has successfully met the requirement of real time through the support of previous techniques via 32 SMP machines, which are equipped with dual 1.5 GHz Itanium II processors per node and connected by Gigabit Ethernet.

1 Introduction

Since the amount of video data is huge, it needs a large storage device. Therefore, we must compress the huge video data to reduce the size of video data before we store them into storage device. At present, the proposed standards of video encoding are MPEG-1, MPEG-2[1] and MPEG-4[2]. Compared to MPEG-1 and MPEG-2, MPEG-4 can support lower bitrate, error-correction and rate control. Consequently, MPEG-4 recently becomes the most popular standard of video encoding. However, the algorithm of MPEG-4 is so complex such that it needs huge data computation for encoding video data. Although the technology of microprocessors has been greatly improved, it is still impossible for one PC and workstation to encode video in real time by using MPEG-4 algorithm. Therefore, how to effectively minimize the encoding time is an important problem for applying the MPEG-4 specification.

For resolving this problem, one possible solution is to develop a parallel MPEG-4 encoder via software technology. There are two common methods to implement a software MPEG-4 parallel encoder. One is to implement the MPEG-4 encoder on shared memory multiprocessors by multithreading and another is to implement the MPEG-4 encoder on a cluster of computers by multi-processes. In general, shared memory multiprocessors are more expensive than PCs or workstations clusters, and the processor number of shared memory multiprocessors machines is bound to the contention of system bus. Therefore, computer clusters are better for implementing parallel MPEG-4 encoders than shared memory multiprocessors machines based on economics and scalability.

On the other hand, the proposed programming interfaces of cluster computing mainly are classified into message passing such as MPI [3] and shared memory such as DSM [4]. MPI is a library specification for message passing, which is proposed as a standard by a broadly based committee of vendors, implementers, and users. When users write their programs with the MPI interface, they must explicitly uses send/receive primitives in the programs for data communication between processes executed on different nodes. In other words, they need to control data distribution and inter-node communication by themselves. Contrast to MPI, distributed shared memory (DSM) is a run time system to emulate a virtual shared memory over a computer network. With the support of DSM systems, users can use shared variables instead of message passing to develop their programs on computer clusters. When a DSM application is executed on a computer cluster, the cluster nodes can communicate with each other though reading/writing shared variables. The DSM systems will automatically maintain the data consistency of shared variables. Consequently, users can put their attention on the development of program algorithm but data communication. Therefore, DSM systems provide an easier programming interface for users to develop applications on computer clusters than MPI.

According to the previous discussion, we devote ourselves to developing a real time MPEG-4 parallel encoder on DSM systems in this paper. To accomplish the goal of real time, i.e., encoding 30 frames per second, the proposed encoder uses a pipeline algorithm for encoding video frame in parallel. With this algorithm, the proposed encoder can effectively minimize the number of data synchronization happening in video encoding such that the communication cost of video encoding can be hugely minimized. In addition, the proposed encoder exploits a two-level data partition method to minimize the cost of data computation, and overlap the encoding times of different video frames. Moreover, it employs a master-and-slave node architecture to overlap disk I/O and computation in order for alleviating the impact of disk I/O latency. We have implemented the proposed MPEG-4 encoder on a test bed called Teamster[5] and have evaluated the performance of the proposed MPEG-4 encoder on a Linux-IA64 cluster which consists of thirty-two Intel SMP machines that are equipped with two Itanium II 1.5GHz processors per node and are connected with Gigabit Ethernet. The experimental results show that the proposed MPEG-4 parallel encoder has successfully accomplished the goal of real time.

The rest of this paper is organized as follow: Section 2 discusses the related work and Section 3 presents the proposed parallel encoder. Section 4 discusses our experimental results. Finally, Section 5 gives the conclusions of this paper and our future work.

2 Related Works

Several software MPEG-4 parallel encoders have been proposed in [6][7][8]. In the study [6], the authors proposed a scheduling policy to achieve load balance via buffer synchronization. Since their work is carried on shared memory multiprocessors, they did not consider the communication cost of video encoding. Therefore, their proposed encoder is not suitable to be implemented on distributed systems. In the paper[7], the authors proposed a dynamic shape-adaptive data partitioning strategy in the spatial domain. They used the information of alpha plane to get the statistical distribution of the contour and standard macroblocks. With the statistical distribution, the data partition method of their proposed encoder can redefine the rectangular sub-alpha plane to avoid the unnecessary computation and achieve load balance. However, this partition method cannot make MPEG-4 encoders accomplish the goal of real time when the size of images is large, i.e., 640x480, and the requirement of frame rate is high, i.e., 30 frames per second. In the paper [8], the authors proposed a GOV-based data partitioning method. The concept of this partition method is to partition video sequence into a set of GOVs(Group Of Video object planes) and then assign different GOVs onto different nodes for parallel execution. In addition, they proposed four scheduling approaches: round robin, adapted batch size round robin, dynamic scheduling, and adapted batch size dynamic scheduling. These scheduling approaches are used for reducing the idle time of CPUs when video data are transferred over network. However, their proposed encoder also cannot achieve the goal of real time as the previous encoder. Compared to the previous work, the MPEG-4 parallel encoder proposed in this paper can achieve the goal of encoding 30 frames per second even when the resolution of image is 640x480. In addition, most of the past work was implemented with MPI while this work is carried out with DSM.

3 Proposed Parallel Encoder

To accomplish the goal of real time, it is necessary to minimize the encoding time of video frames as possible as we can. The MPRG-4 parallel encoder proposed in this paper employs a pipeline-based algorithm, a two-level data partition method and a master-and-slave node architecture to minimize the cost of data synchronization, the cost of data computation and the impact of disk I/O latency. These technological details of the proposed parallel encoder are described in the following subsections.

3.1 Pipeline-Based Algorithm

In DSM systems, the communication cost of maintaining data consistency is dependent on the adopted consistency protocol. According to the release consistency protocol [9], processors have to propagate their updates to shared data for maintaining data consistency only when their local threads arrive at synchronization points such as locks or barriers. Since the process of maintaining data consistency is time consuming and processors keep idle during update-data propagation, it is necessary to minimize the number of synchronization points in the MPEG-4 algorithm in order to minimize the execution cost of data encoding.

Basically, the algorithm of the MPEG-4 sequential encoder can be divided into two phases. One is the intra-frame (I frame) phase and the other is the inter-frame (P frame) phase. In intra-fram, the first step is Loading Video Object Plane (LVOP) that is used to load the data of VOPs (Video Object Planes) from disk to shared memory. Then, the motion estimation (ME) is performed on the current block to find the best matched block within the search window in the previous frame; the block size can be either 16x16 or 8x8. The motion vector is the displacement between the current and the best matched block from the previous frame. Motion compensation (MC) is used to predict the error values between the previous reconstructed frame by motion vectors and the current encoded frame. Discrete Cosine Transform (DCT), which deals with the intra and residual data after motion compensation of VOPs. The DCT is performed on each macroblock. The DC and AC coefficients, generated by DCT, are then quantized by MPEG quantization method. For I-VOP and P-VOP, the intra DC and AC coefficients can be predicted from the corresponding coefficients in the previous neighboring blocks to get the differential DC/AC values. The process of predicted DC/AC is called as ACDC prediction (ACDCP). As same as ACDCP, the motion vector can be predicted from the respect to the neighboring three motion vectors to get the differential motion vector values. After using a scanning method, the quantized transform coefficients are further coded by variable length coding (VLC).

A common method to implement a MPEG-4 parallel encoder is to use a master-slave programming model. In other words, there is a master thread and a set of slave threads created in the program. The master thread is responsible for loading video frames from disk and storing the encoding results into disk. In addition, the master thread has to propagate the video frames to slave threads for parallel encoding and collect the encoding results from the slave threads. By contrast, the main job of slave threads is responsible for data computation in each encoding step except for LVOP and I/O. However, due to the data dependency between the steps of the encoding process, it is necessary to insert three barriers into the algorithms of above mention in order for data synchronization. After inserting the three barriers, the algorithms are transferred to be Figure 1 and Figure 2, respectively. The first barrier is located between LVOP and DCT. This is because each slave thread has to receive macroblocks generated by the master thread in the step of LVOP. The second barrier is located between Quantization and ACDCP since each slave thread need to reference the macroblocks generated by the other slave threads in the step of Quantization to perform ACDCP. The third barrier is located below the step of VLC since the master thread has to collect the VLC results from all of the slave threads and then stores the VLC results into disk. As same as Figure 1, the first barrier in Figure 2 is located between LVOP and ME. This is because the operation of ME needs macroblocks generated in the two steps of LVOP and RVOP (Reconstruct VOP) in order for predicting motion vector. The second barrier is located before the step of ACDCP if macroblocks are in intra-mode or is located before the step of MVP(Motion Vector Prediction) if macroblocks are in inter-mode. This is because ACDCP and MVP need the result of Quantization and ME, respectively. The third barrier is located between VLC and I/O. The reason of adding this barrier is as same as that for the third barrier in Figure 1. Figure 3 is an example of encoding a video sequence. In this example, the process of encoding four frames needs 12 barriers. In other words, the MPEG-4 parallel encoder needs 3N barriers for encoding N frame. To minimize the cost of maintaining data consistency, we invent a pipeline-based algorithm to reduce the number of barriers necessary for encoding video frames in this paper. The flow of the pipeline-based algorithm is shown in Figure 4. The horizontal axis is the number of barriers in the process of video encoding, and the vertical axle is the identifier number of video frames. In this figure, the first frame is I-frame. The second and third frames are P-frame. The fourth frame is the I-frame. Between B1 (Barrier 1) and B2, all of working threads perform the DCT of the first frame, i.e., I-frame, and the I/O thread execute the LVOP of the second frame, i.e., P-frame. Between B2 and B3, the threads perform the ACDCP of the first frame and the ME of the second frame. Between B3 and B4, the threads execute the I/O of the first frame, the ACDCP of the second frame or the MVP of the second frame, and the ME of the third frame. After B3, the threads will complete the process of encoding the first frame, and they will finish the encoding of a frame by an interval of barriers. Consequently, the MPEG-4 parallel encoder needs N+2 barriers for encoding N frame when the pipeline algorithm is adopted. Compared to the previous algorithm, the number of the minimized barriers is theoretically 2N-2.

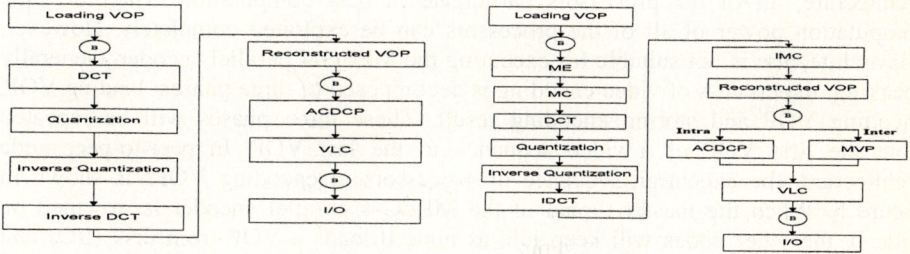

Fig. 1. Intra-frame flowchart with barriers   **Fig. 2.** Inter-frame flowchart with barriers

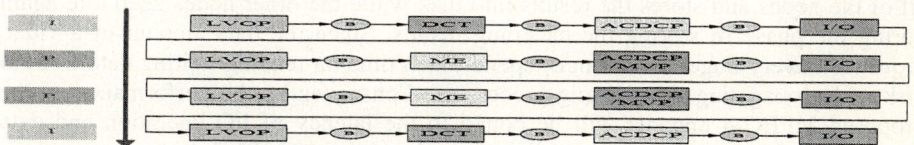

Fig. 3. Flow of video encoding with a common parallel algorithm

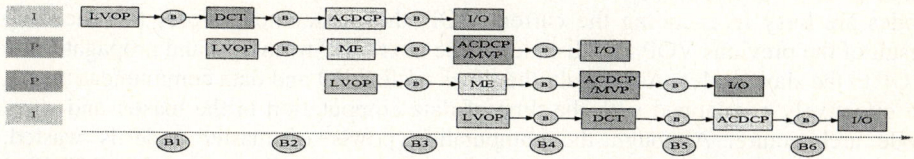

Fig. 4. Flow of video encoding with the pipeline-based algorithm

3.2 Two-Level Data Partition

The video sequence of the MPEG-4 simple profile is composed of a set of GOVs. A GOV is composed of a I-VOP and a set of P-VOPs which follows to the I-VOP. The number of VOPs in a GOV is typically 15. Most of the proposed MPEG-4 parallel encoders usually partition the pending data of the problem into GOVs and distribute different GOVs to different nodes since there is no data dependency between GOVs and this way can make it easy to write programs. Different to the past work, the proposed MPEG-4 parallel encoder employs a two-level partition method which not only partitions GOVs into different clusters but also partition VOPs into the nodes of each cluster. It is obvious that this two-level partition method is better for load balance and parallelism than the previous method since the problem is partitioned into finer granularity. Therefore, our partition method is effective for minimizing the cost of data computation and overlap the encoding times of different frames.

3.3 Master and Slave Node Architecture

Common node architecture to execute parallel programs is peer-to-peer. In this architecture, all of the processors participate in data computation. Therefore, the computation power of all of the processors can be exploited completely. However, this architecture is not suitable for executing the MPEG-4 parallel encoder. Generally speaking, the process of video encoding is decomposed of three phases: loading VOP, encoding VOP and storing encoding result. These three phases will be repeated from the first VOP of a video sequence to the last VOP. In peer-to-peer node architecture, the execution sequence of processors in encoding VOPs is shown in Figure 5. When the master thread of the MPEG-4 parallel encoder is executed on node 0, the other nodes will keep idle as node 0 loads a VOP from disk (I/O) and propagate the VOP, called propagation delay (DP), to them. After the phase of loading VOP, all of the nodes share the workload of encoding the loaded VOP, called data computation (DC) in Figure 5. After all of the execution nodes cooperatively finish the job of encoding the loaded VOP, node 0 collects the encoding results from all of the nodes and stores the results into disk while the other nodes keep idle again during the phase of storing the encoding results. Since the data amount of a video sequence is very huge, node 0 must spend much time on reading/writing data from/to disk and propagating data among processors. Consequently, the performance of the proposed MPEG-4 encoder will be bound to the latency of I/O operation and data propagation. In order to resolve this problem, the proposed MPEG-4 encoder employs a master-and-slave node architecture as shown in Figure 6 to overlap the time of data computation and the time of disk I/O and data propagation. In this architecture, the master thread is located on an individual node called master node and the slave threads are even distributed onto the other nodes called slave nodes. When the slave nodes are busy in encoding the current VOP, the master node stores the encoding result of the previous VOP into disk, loads the next VOP from disk and propagates the VOP to the slave nodes. As a result, the times of disk I/O and data communication can be effectively overlapped with the time of data computation in the master-and-slave node architecture. Although the computation power of master node is wasted, overlapping I/O and computation contributes more to the performance of the MPEG-4 encoder than making use of the computation power of master node in data computation according to our current experience from performance evaluation.

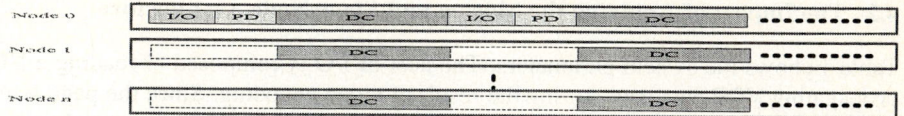

Fig. 5. The execution sequence of processors in peer-to-peer node architecture

Fig. 6. The execution sequence of processors in master-and-slave node architecture

4 Performance

The proposed parallel encoder has been implemented on a DSM system called Teamster, which is built on a cluster of Intel SMP machines, which run Red Hat Linux Enterprise Advance Server V2.1 and are connected with Gigabit Ethernet. There were thirty-two SMP computers used in this performance evaluation, and each computer has dual Itanium II 1.5GHz processors and 1,064 Gbytes memory. We only use a video sequence in this performance evaluation since the complexity of data computation is dependent to the length but the content of video sequence according to the definition of proposed MPEG-4 simple profile. The size of VOP in our experiments is 640x480. Since the motion estimation is time-consumed, we not only executed the work of motion estimation in parallel but also adopt a three-steps search algorithm [10] to improve data computation of motion estimation and keep a visual quality close to that of the full search.

4.1 The Impact of Applying the Pipeline-Based Algorithm

Figure 7 shows the encoding rate for using different algorithms on IA-64 cluster. The horizontal axel is the number of executing nodes, i.e., n, and the number of processors on each executing node, i.e., p. In this experiment, each processor is assigned with a single thread for execution. The vertical axel is the frame rate of encoding. In Figure 7, 3B indicates the algorithm needing three barriers for compressing a VOP, and 1B indicates the algorithm applying a programming model of pipeline and then needing only a barrier for encoding a VOP. Since the number of synchronization points of the 3B algorithm is more than that of 1B, the performance of 3B is worse than that of the 1B algorithm. Therefore, the more barrier needs the encoding algorithm, the more cost spends this algorithm in joining all the slave threads at the same barriers. In addition, the larger is the thread number, the more is the cost. The experimental result shows that the performance of the 1B algorithm but the performance of the 3B algorithm is obviously improved by using 15 nodes for data compressing. The performance of the 1B algorithm is better than that of the 3B algorithm about 18% when 15 nodes are used.

4.2 The Impact of Applying the Master-and-Slave Node Architecture

Table 1 shows the system parameters. The cost of I/O is composed of loading a VOP, propagating VOP data, and storing the result of video encoding. Since the page size of IA-64 cluster is 8192 bytes and the size of a VOP is 640*480, the parallel encoder need to allocate 58 data pages for storing the VOP data. Consequently, the worst case occurs when the content of the current VOP is different from the previous VOP. In the worst case, the cost of propagating the data of a VOP is 51ms, i.e., 58*0.88ms. Since the cost of the VOP propagation is lager than the cost of loading a VOP and storing the result of video encoding, these two costs can be neglected. Table 2 shows the ratio of execution time to propagation cost when the pipeline-based algorithm is applied. The data-propagation cost of a VOP is increased while the execution time of encoding VOP is reduced when the more nodes is used for data encoding. In Figure 7, when the more than 8 nodes are used for video encoding, the latency of propagating VOPs is lager than the time of computation. Therefore, the performance is bound to the data-propagation cost of VOPs. When the master-and-slave node architecture is adopted, the I/O operations included loading VOPs, storing the encoding results of VOPs, and propagating VOPs, can be effectively overlapped with data computation. In Figure 7, the MS indicates the master-and-slave node architecture. Compared with the performance of 1B, the performance of the 1B+MS can be improved about 58% in the case of 15 nodes. This experimental results show that overlapping I/O and computation is very important for enhancing the performance of a MPEG-4 parallel encoder on distributed systems, and applying the master-and-slave node architecture can effectively overlap I/O and computation such that the performance of the proposed MPEG-4 parallel encoder can be obviously improved.

Table 1. System parameters

Page size	Propagation a page	Load a VOP	Store 4K bytes
8192	0.88 (ms)	0.1 (ms)	0.01 (ms)

Table 2. The ratio of execution time to propagation cost with pipeline-based algorithm

	n=2	n=4	n=8	n=15
Propagation cost per VOP(ms)	25	38	44	47
Execution time per VOP (ms)	205	128	90	82
Ratio (%)	12%	29%	48%	57%

4.3 The Impact of Applying the Two-Level Partition Method

In the Figure 7, 2LP indicates the two-level partition method. Since the video sequence can be partitioned into several independent GOVs, different GOVs can be concurrently processed on different clusters. In the experiment, we use two clusters and each cluster is composed of 15 nodes for encoding GOVs. The experimental result shows that the performance of the 1SP+MS+2LP is two times that of 1SP+MS in the cases of the number of nodes is smaller than 8. However, the impact of the two-level partition method is degraded as the node number of each cluster is larger than 8. The reason is that the bitstream center has to collect the results of data compressing from each cluster and store the results into disk such that it becomes a bottleneck of system performance. Compared to the 1SP+MS, the performance of the 1SP+MS+2LP is better about 56%.

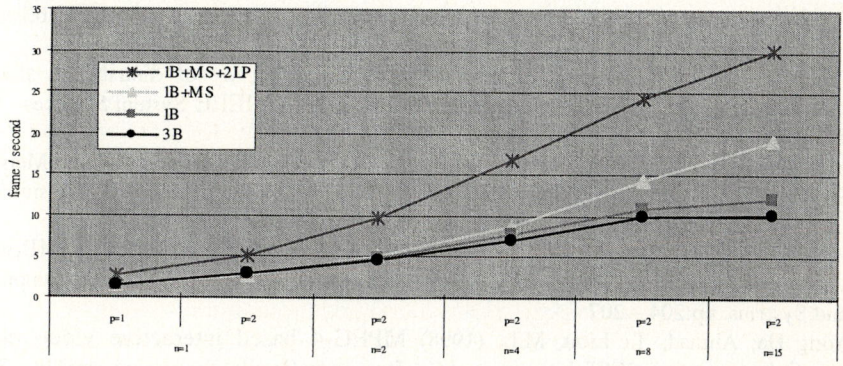

Fig. 7. Encoding rate on IA-64 cluster

5 Conclusions and Future Work

We have successfully developed a real time MEPG-4 parallel encoder on a DSM system in this paper. To meet of the time constrain of a real time encoder, we reduce the number of barriers necessary for encoding VOPs by applying a pipeline-based algorithm in programming. In addition, we apply the master-and-slave node architecture to overlap I/O and computation. In order to further improve the performance of the encoder, we also propose a two-level partitioning method to increase the parallelism of data computation and overlap the time of encoding different GOVs. The experimental results show that the impact of minimizing the number of barriers and overlapping computation and I/O is significant for improving the performance of the MPEP-4 parallel encoder, and the two-level partition method can effectively increase the frame rate of the proposed MPEG-4 encoder.

Recently, a new video encoding specification called H.264 is proposed. This specification is to support a higher encoding rate than MEPG-4. However, the algorithm of H.264 encoder is much more complex than MPEG-4. Therefore, we will challenge in developing a real time H.264 parallel encoder in the future. In addition, we will develop a parallel algorithm to retrieve objects from video sequence, and individually encoding these objects in parallel in order to increase the encoding rate of the proposed MPEG-4 parallel encoder.

Acknowledgements

The authors are grateful to the support received from the NCHC cluster of computers. The web site is http://www.nchc.org.tw/.

References

1. ISO/IEC IS 13818 (1996) Information Technology-Generic Coding of Moving Pictures and Associated Audio Information. (MPEG-2).
2. ISO/IEC (1999) MPEG-4 Overview –(Melbourne Version). JTC1/SC29/WG11 N2995, Oct.
3. Peter S. Pacheco (1997) Parallel Programming with MPI. Morgan Kaufmann Publishers, Inc, San Francisco, California.
4. Protic, J., Tomasevic M. and Milutinovic V. (1995) A survey of distributed shared memory systems. 28th Hawaii International Conference on IEEE System Sciences ,vol: 1 , pp:74 - 84 vol. 1
5. J. B. Chang and C. K. Shieh (2001) Teamster: A Transparent Distributed Shared Memory for Clustered Symmetric Multiprocessors. 1st IEEE/ACM International Symposium on Cluster Computing and the Grid 2001, pp.508-513
6. Hamosfakidis, A.; Paker, Y.; Cosmas, J. (1998) A study of concurrency in MPEG-4 video encoder. Proceedings. IEEE International Conference on Multimedia Computing and Systems, pp:204 – 207
7. Yong He; Ahmad, T.; Liou, M.L. (1998) MPEG-4 based interactive video using parallel processing. IEEE International Conference on Parallel Processing, pp:329 – 336
8. Miguel Ribeiro, Oliver Sinnen, Leonel Sousa (2002) MPEG-4 Natural Video Parallel Implementation on a Cluster. Image and Video Coding, RECPAD2002, pp:3
9. Kourosh Gharachorloo, Daniel Lenoski, James Laudon, Philip ibbons, Anoop Gupta, and John Hennessy (1990) Memory Consistency and Event Ordering in Scaleable Shared-Memory Multiprocessors. 17th Annual International Symposium on Computer Architecture, pp: 15-26.
10. T. Koga, K. Iinurna, A. Hirano, Y. Iijima, and T. Ishiguro (1981) Motion-compensated interframe coding for video conferencing. NTC'81 (IEEE), Orleans, LA, pp. C9.6.1-9.6.5.

A Case of SCMP with TLS

Jianzhuang Lu, Chunyuan Zhang, Zhiying Wang, Yun Cheng, and Dan Wu

School of Computer,
National University of Defense Technology,
Changsha, Hunan, China, 410073
`lujz1977@163.com`

Abstract. As an alternative way of chip design, Single Chip Multi-Processors (SCMP) has been a hot topic in microprocessor architecture research all the while. It achieves higher performance by extracting thread-level parallelism (TLP). Thread-level speculation (TLS) is an important way to simplify TLP extraction. This paper presents a new SCMP architecture called Griffon, which aims at general-purpose applications. It implements thread partition in assembly language. It supports thread-level speculation with simple logics and maintains data dependence using a dual-ring structure. Simulation and synthesis results show that Griffon can achieve ideal speedup, less design complexity and accessorial hardware cost.

1 Introduction

The advancement of semiconductor processing helps us obtain more space when designing a chip. To achieve more parallelism, conventional processors have sought to extract ILP in superscalar or VLIW architectures. But all of these techniques are preferable in a single thread. The parallelism in a single thread is limited for an acceptable cost. The cost manifests in aspects of the die area, the complexity in design and verification, the time of time-to-market and the time of compiling, etc. [1][2]

Alternative ways have been explored, and we take SCMP as an example. It uses relatively simple cores to exploit moderate parallelism in a thread, while executing multiple threads in parallel across multiple processor cores [1]. Several projects have been investigated on the research of SCMP architecture and several products are developed with this architecture. All of them can be classified into two categories. The first is throughout-focus [3]-[6]. Multiple threads running on these systems may come from different processes. Most of them work well in special purpose applications. For there are enough TLP to obtain, they do not support TLS.

The second category is latency-focus [7] [8]. TLS are supported by these systems. The false read may occur if a more speculative thread prematurely read a word before a less speculative one writes it. To avoid the violation of data dependencies, Hydra uses a snooping-base method with special write buffers for each speculative thread on L1 cache [9]. A directory-based MDT is designed to resolve these problems in Illinois' SCMP. The shortcoming of these architectures is the conflicts in maintaining data dependence operations: snooping on bus or accessing the common structure.

In order to make SCMP more balanced in use, it should be able to provide competitive performance when running general-purpose applications. Most of them are sequential programs. Extracting TLP is difficult in these applications, since some inter-thread data dependencies cannot be determined accurately when partitioning threads with software. Conservative strategies, synchronization on all possible data dependencies, will draw back the performance. More aggressive strategies may result in false execution. TLS makes it possible to detect and maintain data dependence in runtime and simplifies the extraction of TLP. This paper presents a new SCMP architecture called Griffon, which aims at general-purpose applications. It implements thread partition in assembly language. A distributed control mechanism is adopted in Griffon for scalable. It achieves thread-level speculation with simple logics. Simulation and synthesis results show that Griffon can achieve ideal speedup as well as less design complexity and accessorial hardware cost.

The remainder of this paper is organized as follows. Section 2 presents the Griffon architecture. Section 3 describes the software support in Griffon. And section 4 is the hardware support. Section 5 gives the evaluations of Griffon. Conclusions are given in section 6.

2 Architecture

2.1 Design Guideline

Several problems must be resolved when running sequential applications on SCMP. The first is to partition a sequential program into multiple threads. Hardware method will increase the complexity of chip design. But for software method, recompiling all sources of an application is not advisable. We prefer a trade-off method. The software inserts thread control instructions in assembly codes based on its analysis of the control and data dependence. Then, it is up to the hardware to check and maintain the dependence and guarantee the correction of the execution.

The second is in execution mode. More threads can be run on a core concurrently. Unfortunately it increases the complexity of the core. So another thread will not be scheduled on a core until the running one finishes and a thread can spawn only one thread in Griffon.

Communications among the cores is the key to achieve the control and data dependence. Griffon uses explicit instructions for register level communications. The data dependence in shared variables and memory disambiguation are resolved using the cache level data path, dual-ring structure in Griffon.

2.2 Architecture

Fig. 1. illustrates the architecture of Griffon. It includes four cores and two-level caches. The L1 caches are private. A core with its instruction cache (L1/I) and data cache (L1/D) compose a process element (PE). The CPU core is designed as pipeline execution. I_BUF is a buffer between L1/I and the core. Thread control unit (TCU) is distributed in each core. The L2 cache is shared by all PEs. A common bus connects L1 caches and L2 cache. Bus AB is an arbiter to determine which one of the L1s can access L2 when conflict occurs.

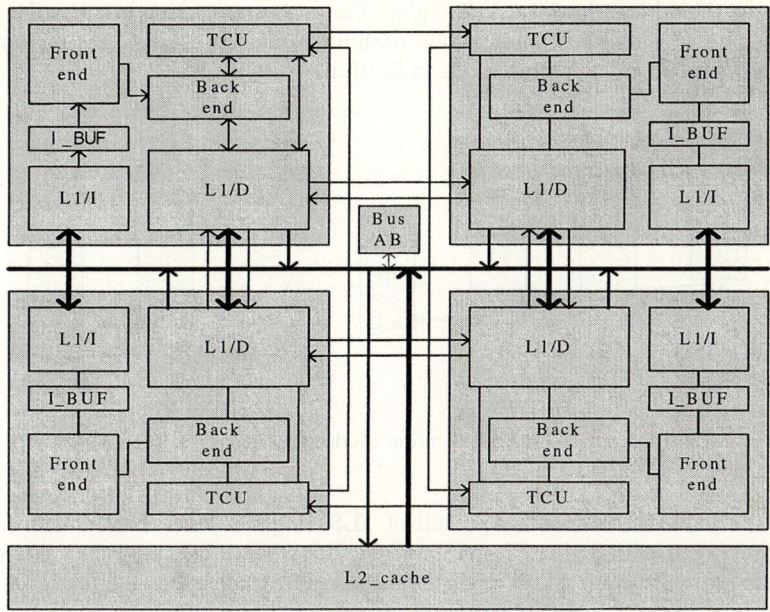

Fig. 1. Architecture of Griffon

This is a typical SCMP structure, similar to Hydra [7] and Illinois' SCMP [8]. The Superthreaded Architecture [10] developed in University of Minnesota adopted a shared L1 data cache. But it introduces private memory buffers in each core. The main differences in each SCMP are the ways to control TLS and the strategies to maintain data dependence. The core in Griffon is very simple. It only has a single- issue pipeline. The ISA is extended for thread control and explicit register level data transfer. Several additional tags are expended in the data cache for data dependence. Both write-through and write back policies are implemented in primary cache, but only the non-speculative thread adopts write-through strategy. Write back is used in L2 cache. The dual-ring structure, two rings in opposite directions among L1/Ds, is introduced to maintain the inter-thread data dependence and cache coherence.

3 Software Support

3.1 Execution Mode

Before running on Griffon, applications will firstly be divided into threads. Thread control instructions and register level transfer instructions are inserted in assembly codes. An application will be loaded on one PE at the beginning. With the Griffon's running, threads are spawned in other PEs. A speculative thread can spawn an even more speculative one. The creation of a thread may occurs before determine whether the thread is in the right execution path. So control speculation arises naturally in thread spawning. Speculation Level (SL) is defined here. At a moment, the non-speculative thread has the lowest SL value; the most speculative one has the highest SL value. This guarantees applications running in a foreseeable speculation mode. Fig.2. shows the ideal execution mode in Griffon.

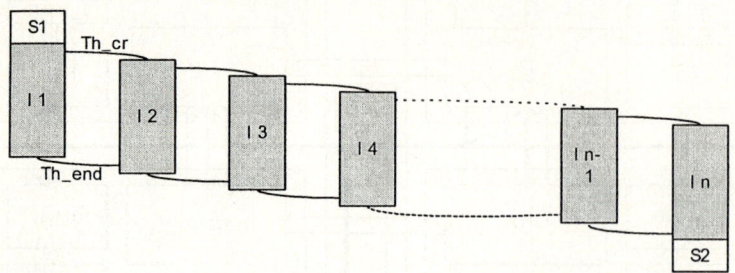

Fig. 2. Execution Mode of Griffon

Data speculation is a inevitable result of TLS. Register level data transfer methods ensure the thread with higher SL get the right data when it uses register value directly from a thread with lower SL. The dual-ring structure propagates the loads and stores information among the PEs when a core accesses its L1/D. The cache controller checks these messages and transfers data from thread with lower SL to higher one if necessary. If a pre-mature load is detected, the cache controller will inform the TCU in the core. When the TCU receives this message it will execute some recover codes or just restart. The information about its operations will be carried to more speculative threads. The L1/D is used to store the speculative version data. Only the thread with lowest SL can store its data to the common L2 cache.

Thread level pipeline is proposed in Superthreaded architecture, but it needs special works in source codes. The SCMP in Illinois partitions threads based on the loops and functions and the synchronizing scoreboard increase the overhead of thread spawning. Wisconsin Multiscalar [11] uses speculation in multiple tasks execution. The ARB makes it difficult to implement. Later, they proposed the speculative versioning cache. The SVC [12] uses write-back primary caches to buffer speculative writes in the primary caches, using a sophisticated coherence scheme.

3.2 Thread Partition

A simple algorithm for thread partition is proposed in this paper, which works on assembly codes. It includes four steps. First, this algorithm constructs a weighted control-flow graph (WCFG) on basic blocks by scanning and profiling the assembly codes. Second, it builds candidates of threads from WCFG. Data dependence between blocks and profiling information are considered in block selection and combination. Thread size is also used as a criterion in this step. Third, it collects blocks not belong to any thread, and inserts them into current thread candidates if possible or treats them as independent threads if not. Thread control instructions and register level data transfer instructions are inserted into threads candidate. In the last step, instruction scheduling is done in threads to make loads advance and stores defer as possible. Fig.3. shows an example of thread partition. The codes are the main loops in WC.

Profiling information is not considered in Hydra and SCMP in Illinois. The Superthreaded architecture inserts special codes in source codes [10], which increase the overhead of thread partition and the size of codes.

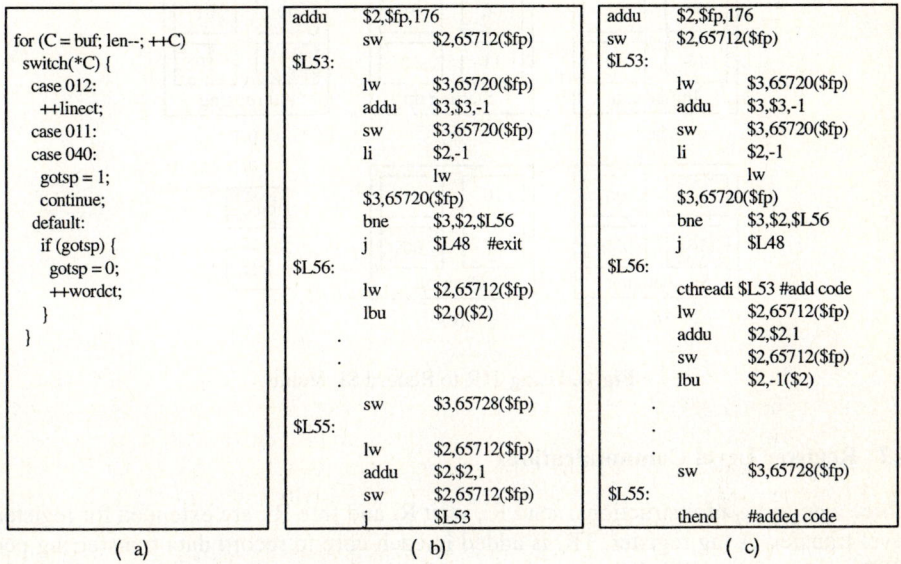

Fig. 3. Thread Partition

4 Hardware Support

4.1 Thread Control Units

Thread control instructions includes thread-creating instructions: cthread Ri and cthreadi Imm; thread-ending instructions: thend, thendez Ri, thendnz Ri. They are processed in two steps. First, they are fetched, decoded and executed in pipeline. The address and the comparing result are computed in execution stage. TCU executes the next step. So, local pipeline is not stalled by these instructions.

TCU is responsible for the following works. 1), It sends information to the successor's TCU when receiving spawning, committing or clearing requests from its pipeline. Clearing is triggered when a mis-speculation is detected. When a committing request is encountered, the TCU of speculative threads stalls the pipeline and waits for becoming non-speculative. 2), It treats the information from the predecessor's TCU. It creates thread on the core, maintains the SL of the thread or clears the thread according to the information. The clearing request is propagated if it is not the thread with highest SL. If the core is busy, the remote spawning request is stalled. 3), It controls thread-restart or recover if a premature load occurs.

Thread ID Register (TIR) in TCU is used to record the SL value of the thread. The core with minimum value in TIR runs non-speculatively and the one with maximum TIR executes the most-speculative thread. TCU changes the value in TIR to cope with the changing of the thread's dynamic SL. A core ID is set in each core to distinguish the core and initial the TIR. Fig.4. shows the way of using TIR to record the SL value.

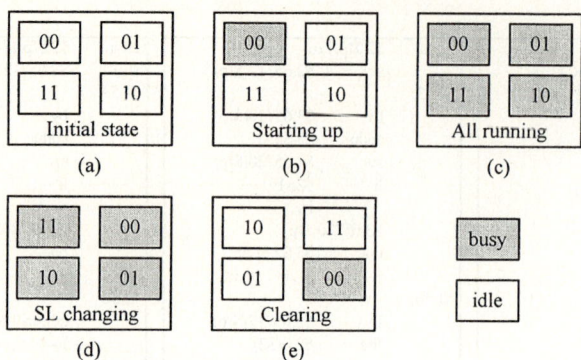

Fig. 4. Using TIR to Record SL Value

4.2 Register Level Communications

Three categories of instructions: send Ri, wait Ri and free Ri, are extended for register level transfer. A tag register, TR, is added in each core to record data transferring per GPR, one bit for a GPR. When a new thread is spawned on a core, the TR is cleared.

When a thread encounters a send Ri instruction, it sends the value of Ri to successor. The successor updates Ri with the value and set the corresponding bit in TR. No data but a message is sent to successor if a free Ri met by a thread. The receiver just set the corresponding bit in TR as answer. When a thread encounters a wait Ri instruction, it stalls the pipeline until the bit in TR for that register is set to 1.

There may be register-level data dependencies in non-adjacent threads. In Griffon, the thread partition algorithm checks these dependencies and changes them into adjacent threads data dependence by inserting proper instructions. This work is completed in step 3.

4.3 Memory Data Dependence

Griffon uses the dual-ring to maintain data dependence. Fig.5. shows it. Part (a) is an extension of cache line's tags. The V and D bits are used as in general cache. The RS bit records if this line is modified by thread with higher SL. The U bit at word level indicates whether the local processor writes that word. The L bit is set to 1 if the thread loads this word speculatively. The S bit indicates if a store message needs to be sent. Part (b) shows the dual-ring structure: the L-ring and the S-ring and the forwarding bus. L-ring is used to propagate load messages from threads with lower SL to higher ones. On the contrary, the store messages are transmitted in opposite direction on the S-ring. The forwarding bus and arbiter (Forward_AB) are used to transfer data from lower SL threads to higher ones if necessary. The massages on these two rings both include the SL of the sponsor, the tag for word i (Wi) in a cache line and cache line address. A restart (R) bit is added in S-ring messages.

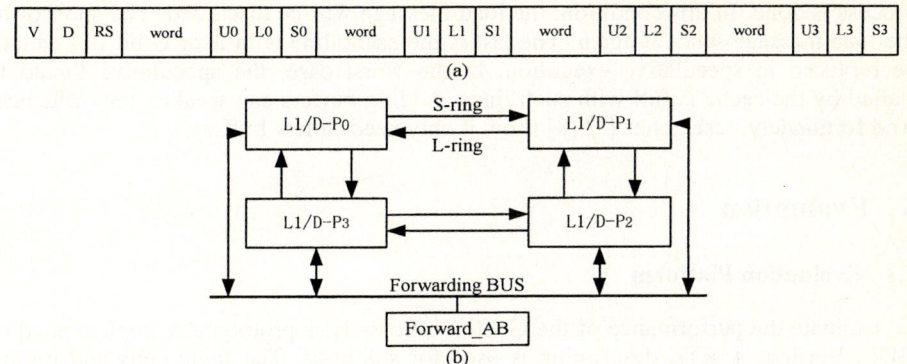

Fig. 5. Dual-ring Structures

For load operations, the non-speculative thread needs not to perform any special actions. Speculative threads may perform unsafe loads. They check the U and L bits of the word firstly. If (U+L) is 0, it must propagate the load address to its predecessor on the L-ring for the word in its cache may be not the new version. If a load miss occurs, both accessing L2 and sending message do concurrently. For all stores, the U bit is set to 1 and the S bit is checked at the same time. If S is 1, no more things should be done. Otherwise, a store message must be sent on the S-ring to successor and the S bit is set to 1. If a store miss occurs, the cache line will be got from L2 before tag checking and message sending.

When a L1/D receives a load message, it checks the U bit of the word the message point to. If the U bit equals to 1, the word will be forwarded on the forwarding bus and the S bit of the word is cleared. Also the Wi bit of the word in message is cleared. If not all Wi bits in message are 0, the speculative threads will propagate the message continuously. The non-speculative thread will send an ACK to the sponsor in that condition. The load message is cleared if all Wi bits are 0 or it reaches the non-speculative thread. When a store message is received, the L bit of the word is checked if the receiver has higher SL. If L bit is 1, the L1/D cache informs the local TCU a premature load occurs. The R bit in message is set to 1. Then the message is propagated. If the receiver has lower SL according to the sponsor, the RS bit is set to 1 to indicate it should be invalided when new thread starts on this PE. Then it propagates the message. The sponsor will clear the message in the end.

The message is propagated to next PE in one cycle. Based on the execution mode in Griffon, more data dependencies occur in adjacent threads. The dual-ring structure reduces the conflicts of messages. The forwarding bus transfers the data directly. So only a few dependencies may be delayed by the propagation. For all loads and stores, the thread needs not send a message only if the L or S bit has been set to 1. This also reduces the message conflicts.

4.4 Issues on Hardware

Some issues must be settled for right execution. First, the message of the dual-ring must be flushed when the TCU changes the TIR value. Because the th.id field in message is invalid in the moment. Second, load and store messages for the same word may be propagated concurrently. This will result in wrong operation if no special process is done. In this condition, the load message will be discarded. The sponsor of the load message sends it again. The last is the cache line with L or U bit is 1 cannot be replaced in speculative execution. In the worst case, the speculative thread is stalled by the cache is full with such lines. Adding buffers can weaken the influence. And fortunately, researches [7] [8] show it only needs a few buffers.

5 Evaluation

5.1 Evaluation Platform

To evaluate the performance of the Griffon efficiently, a prototype is implemented in HDL, Verilog. A RTL description is used for synthesis. The logic cells and timing information can be obtained after synthesis. The ISA of the core is base MIPS R3000. Thread control instructions and register level data transfer instructions are extended. The core only has a single-issue pipeline. Separated private L1 cache is used. Also a common L2 cache and MIU is implemented.

For the model is very simple, relative results are more meaningful than absolute ones. Table 1 shows the logic cells of a general core and the improvement in the Griffon. Table 2 displays the influence on clock rate. From the tables we draw the conclusion that Griffon brings little negative effect on hardware cost and clock rate. The clock rate of L2 drops more deeply is according to appending the bus arbiter.

Table 1. The logic cells of components

	Core	TCU(vs. core)	L1 cache	L2 cache
Single-processor	100%	0	100%	100%
In Griffon	108.2%	6.2%	110.6%	106.5%

Table 2. The clock rates

	Core	TCU(vs. core)	L1 cache	L2 cache
Single-processor	100%	0	100%	100%
In Griffon	96.7%	118.3%	97.8%	89.3%

5.2 Evaluation

The following benchmarks are run on the platform. They are Matrix, Fibonaci, Wc, Compress95, Queens and Eratosthenes' Filter. Matrix has more loop level parallelism. Typical inter-thread data dependencies exist in computing the Fibonacis. Wc and Compress95 are real applications. Queens and Filter have many control dependencies.

Table 3. The Size Increment of Thread Partition

	Matrix	Fibonaci	Wc	Compress95	Queens	Filter
Original codes	100%	100%	100%	100%	100%	100%
After partition	108.8%	111.4%	105.5%	102.5%	101.8%	102.2%

Table 3 gives the size increment after thread partition. The result indicates there is not more increment in static codes. Matrix and Fibonaci increase larger than others. That is because they are written and optimized manually.

Table 4. Thread Control Overhead

	Thread spawning	SL changing	Thread clearing	Accuracy of control speculation
Time cost	3.1 cycles	2.3 cycles	1 cycle	80.5%

Table 4 is the overhead of thread control. The thread size is from 30 to 60 instructions. So the overhead of thread control can be neglected compared with the thread size. But the accuracy of control speculation is not optimistic. More works should be done for it.

Table 5. The Overhead of Dual-ring Structure

	Best case	Worst case	Average
Load message	1	5	2.2
Store message	4	7	4.1
Data forwarding	1	3	1.3

Table 5 shows the overhead of dual-ring structure. The overhead of load message is 2.2 cycles. Data forwarding is 1.3 cycles in average. The read hit in L2 is 4 cycles. So the delay of load message and data forwarding can be overlapped by the L2 read. Store message should travel all the PEs. The average survival period is close to the shortest 4 cycles.

Table 6. Speedup of Griffon

	Matrix	Fibonaci	Wc	Compress95	Queens	Filter
In Griffon	2.87	2.17	2.37	1.89	2.54	1.74

Table 6 is Griffon's speedup over single core. Useful instructions per cycle (UIPC) are used as the criterion. Matrix achieves the highest speedup for its internal properties. Compress95 and filter get a lower speedup for more mis-speculation caused by controls. With the tracing of accessing L2 cache, we know that blocking access-mode leads more conflicts. Multi-ports or non-blocking access-mode in L2 can resolve this problem.

6 Conclusions

This paper presents a new SCMP architecture with TLS, called Griffon. It aims at general-purpose applications. A thread partition algorithm is developed on assembly codes for Griffon. It use a heuristic method based on WCFG, which is constructed with profiling information. Simulation result shows that it did a small increment in static codes size. TCU in Griffon is distributed in each core for scalable. Thread spawning, committing, restarting and clearing are triggered by it efficiently. The thread control instructions did not stall the pipeline themselves. The overhead of thread control can be neglected compared with the thread size. No explicit synchronization is used in Griffon. Explicit register level data transfer and dual-ring structure are designed for register and memory level data dependence respectively. The dual-ring structure avoids the conflicts as in bus snooping method. Most overheads of dual-ring structure are overlapped by L2 accessing.

The evaluation shows that applications could be more benefit from Griffon. The hardware cost and decrement of clock rate is acceptable. Control dependence and data dependence caused by TLS are settled successfully. Griffon achieves the speedup from 1.74 to 2.87 over single core.

References

1. L. Hammond, B.A. Nayfeh, K. Olukotun Stanford University. A Single-Chip Multiprocessor. IEEE Computer Special Issue on "Billion-Transistor Processors", September 1997,pp. 79-85.
2. J. Hennessy. The Future of Systems Research. In IEEE Computer, Vol.32, No. 8, August 1999, pp. 27-33.
3. L. A. Barroso, K. Gharachorloo, R. McNamara, A..Nowatzyk, S..Qadeer,.B. Sano, S. Smith, R.Stets, and B.Verghese. Piranha: A Scalable Architecture Based on Single-Chip Multiprocessing.In Proceedings of the 27th Annual International Symposium on Computer Architecture, June 2000
4. Kyoung Park et al. On-Chip Multiprocessor with Simultaneous multithreading. ETRI Journal, Volume 22, Number 4, December 2000, pp. 13-24.
5. J. M. Tendler, S. Dodson, S. Fields, H. Le, B. Sinharoy. IBM POWER4 System Microarchitecture. Technical White Paper. IBM Server Group October 2001
6. M. Tremblay. MAJC-5200: A VLIW Convergent MPSOC. In Microprocessor Forum, October 1999.
7. L.Hammond,B. A. Hubbert, M. Siu, M.K. Prabhu, M. Chen and K.Olukotun. The Stanford Hydra CMP. IEEE MICRO Magazine, March-April 2000, pp. 71-84.
8. V. Krishnan, J. Torrellas. A Chip-Multiprocessor Architecture with Speculative Threading. IEEE Transactions on Computers,Vol. 48, No. 9, September 1999
9. L.Hammond, M.Willey and K.Olukotun. Data Speculation Support for a Chip Multiprocessor. Proceedings of the Eighth ACM Conference on Architectural Support for Programming Languages and Operating Systems, San Jose, California, October 1998.
10. J. Tsai, J. Huang, C. Amlo, D.J. Lilja, and P.-C. Yew. "The Superthreaded Processor Architecture".To appear in the IEEE Transaction on Computers, Special Issue on Multithreaded Architectures, September, 1999.
11. Gurindar S. Sohi, Scott E. Breach, and T. N. Vijaykumar. Multiscalar processors. In Proceedings of the 22nd Annual International Symposium on Computer Architecture, pages 414–425, June 22–24, 1995.
12. S. Gopal et al. Speculative Versioning Cache. Proc. Fourth Int'l Symp. High-Performance Computer Architecture (HPCA-4), IEEE Computer Society Press, Los Alamitos, Calif., Feb. 1998, pp. 195-205.

SuperPAS: A *Parallel Architectural Skeleton* Model Supporting Extensibility and Skeleton Composition

Mohammad Mursalin Akon, Dhrubajyoti Goswami, and Hon Fung Li

Department of Computer Science, Concordia University,
Montreal, QC, Canada H3G 1M8
{mm_akon, goswami, hfli}@cs.concordia.ca

Abstract. Application of pattern-based approaches to parallel programming is an active area of research today. The main objective of pattern-based approaches to parallel programming is to facilitate the reuse of frequently occurring structures for parallelism whereby a user supplies mostly the application specific code-components and the programming environment generates most of the code for parallelization. Parallel Architectural Skeleton (PAS) is such a pattern-based parallel programming model and environment. The PAS model provides a generic way of describing the architectural/structural aspects of patterns in message-passing parallel computing. Application development using PAS is hierarchical, similar to conventional parallel programming using MPI, however with the added benefit of reusability and high level patterns. Like most other pattern-based parallel programming models, the benefits of PAS were offset by some of its drawbacks such as difficulty in: (1) extending PAS and (2) skeleton composition. *SuperPAS* is an extension of PAS that addresses these issues. *SuperPAS* provides a skeleton description language for the generic PAS. Using SuperPAS, a skeleton developer can extend PAS by adding new skeletons to the repository (i.e., extensibility). SuperPAS also makes the PAS system more flexible by defining composition of skeletons. In this paper, we describe SuperPAS and elaborate its use through examples.

1 Introduction

The concept of design patterns has been used in diverse domains of engineering, ranging from architectural designs in civil engineering [1] to the design of object oriented software [2]. In the area of parallel computing, (parallel) design patterns specify recurring parallel computational problems with similar structural and behavioral components, and their solution strategies. Examples of such recurring parallel patterns are: static and dynamic replication, divide and conquer, data parallel computation with various topologies, compositional framework for control- and data-parallel computation, pipeline, singleton pattern for single-process (single- or multi-threaded) computation, systolic and wavefront computation.

Several parallel programming systems have been built with the intent to facilitate rapid development of parallel applications through the use of design patterns. Some such systems are *CODE* [3], *Frameworks* [4], *Enterprise* [5], *Tracs* [6], *DPnDP* [7], *COPS* [8], *PAS* [9], and *ASSIST* [10]. Most of the works concentrate on the algorithmic/behavioral aspects of patterns, popularly know as *algorithmic skeletons* [11]. Algorithmic skeletons are best expressed using the various functional and logic programming languages [12].

In contrast, Parallel Architectural Skeletons (PAS) [13,9] specify the architectural/structural aspects of patterns. Unlike algorithmic skeletons, architectural skeletons in PAS can be well expressed using existing object-oriented language(s). Consequently, the PAS approach is well suited to the main-stream of parallel application developers where the popular object-oriented languages like C++ and Java are the languages of choice.

Application development using PAS is hierarchical, and is similar to conventional parallel programming using MPI [14] and PVM [15]. However, PAS provides the added benefit of high level patterns and reusability. A developer, depending upon the specific needs of a parallel application, chooses the appropriate skeletons, supplies the required parameters and application-specific code. Architectural skeletons supply most of the code that is necessary for low-level and parallelism-related issues. In other words, architectural skeletons take care of application-independent parallel programming aspects, whereas the developer largely supplies the necessary application code. Consequently, there exists a clear separation between application dependent code and application independent issues (i.e., *separation of concerns*).

Though reusability is a rather useful benefit, however the lack of extensibility and the lack of support for pattern composition are some of the major concerns associated with most of the pattern-based approaches to parallel programming, including PAS. Most existing systems support a limited and fixed set of patterns that are hand-coded into the systems. Generally, there is no provision for adding a new skeleton without understanding the entire system and writing the skeleton from scratch (i.e., lack of extensibility). So if a required parallel computing pattern demanded by an application is not supported, generally one has no alternative but to abandon the idea of using the particular approach altogether (lack of flexibility).

SuperPAS is an extension of PAS and it addresses the drawbacks mentioned previously. It provides a skeleton description language (SDL) for the generic PAS. Using SuperPAS, a skeleton developer can extend PAS by adding new skeletons to the repository (i.e., extensibility). SuperPAS also makes the PAS system more flexible by defining composition of skeletons, where two or more existing skeletons can be composed to create a new skeleton.

The SuperPAS model is targeted for two different groups of users: (1) *skeleton designers* who design new skeletons using the provided SDL and add the new skeletons to the skeleton repository. (2) *Application developers* who use the skeletons already available in the skeleton repository. Unlike a skeleton developer, an application developer need not have any knowledge about SDL and she can directly develop application using C++. On occasions, a skeleton developer and an application developer may be the same person.

This paper describes the SuperPAS model and its use. The paper is organized as follows: section 2 provides a brief introduction to the PAS model. Section 3 describes the SuperPAS model from the perspectives of the two different user groups. The following section illustrates SuperPAS through an example. The concept of skeleton composition is discussed in section 5. Section 6 describes the performance issues of PAS and SuperPAS, and concludes the paper.

2 Preliminaries

A skeleton in PAS encapsulates the structural/architectural attributes of a specific pattern in parallel computing. Each skeleton is parameterized, where the value of a parameter is determined during the application development phase. As an example: a k-dimensional data-parallel mesh skeleton in PAS encapsulates the structural aspects of a data-parallel mesh pattern, together with the associated communication-synchronization primitives. The parameters of the skeleton are: the number of dimensions of the mesh (i.e., k), and the length of each dimension.

During the rest of the discussion, A PAS skeleton with unbound parameters is called an *abstract skeleton*. An abstract skeleton becomes a *concrete skeleton*, when the parameters of the skeleton are bound to actual values during the application development phase. A concrete skeleton is yet to be filled in with application-specific code. A concrete skeleton which is completely filled in with application-specific code is called a *code-complete parallel module* or simply a *module* (the term skeleton is omitted here, because with application code it is no longer a skeleton). As it will be discussed shortly, a parallel application is a hierarchical collection of modules.

Fig. 1 (a) roughly illustrates the various phases of application development using PAS. As is shown in the figure, binding different parameter values (according to needs of the applications) to the same abstract skeleton can result in different concrete skeletons. A concrete skeleton inherits all the properties associated with an abstract skeleton. In object-oriented terminologies, an abstract skeleton can be described as the *generalization* of a particular design pattern. A concrete skeleton is an application-specific *specialization* of a skeleton.

Irrespective of the pattern type, an abstract skeleton, A_m, consists the following set of attributes. Fig. 1 (b) diagrammatically illustrates the attributes of an abstract and a concrete skeleton, where the skeleton is designed for 2-D mesh topology.

- *Representative* represents the module in its action and interactions with other modules. Initially, the representative is empty and is subsequently filled in with application-specific code (refer to the following discussion).
- *Back-end* of an abstract skeleton A_m can formally be represented as $\{A_{m1}, A_{m2}, \ldots, A_{mn}\}$, where each A_{mi} is itself an abstract skeleton. The type of each A_{mi} is determined after the abstract skeleton A_m is concretized. Note

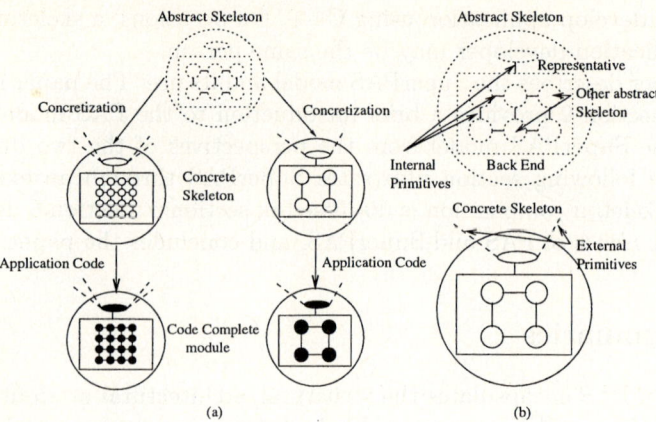

Fig. 1. (a) Abstract skeleton, concrete skeleton and code-complete module, (b) Different components of a skeleton

that collection of concrete skeletons inside another concrete skeleton results in a hierarchy. Consequently, each A_{mi} is called a *child* of A_m, and A_m is called the *parent*. The children of a module are *peers* of one another.
- *Topology* is the logical connectivity between the skeletons inside the back-end. It also includes the connectivity between the children and the representative.
- *Internal primitives* are the pattern-specific communication / synchronization primitives. Interaction among the various modules is performed using these primitives. The internal primitives are the inherent properties of the skeleton and they capture the the parallel computing model of the pattern as well as the topology.

There are pattern-specific parameters associated with some of the previous attributes. For instance: if the topology is a mesh, the number of dimensions, length of each dimensions, and the type of each child in the back-end are the parameters. Fixing these parameters, based on the needs of an application, results in a concrete skeleton. A concrete skeleton C_m becomes a code-complete module when: (i) the representative of C_m is filled in with application-specific code, and (ii) each child of C_m is code-complete. This description obviously indicates that application development using PAS is hierarchical.

In addition, we define the term *external primitives* of a concrete or a code-complete module as the set of communication / synchronization primitives using which the module (i.e., its representative) can interact with its parent and peers. Unlike internal primitives, which are inherent properties of a skeleton, external primitives are adaptable, i.e., a module adapts to the context of its parent by using the internal primitives of its parent as its external primitives. While filling in the representative of a concrete skeleton with application-specific code, the application developer uses the internal and external primitives for interactions with other modules in the hierarchy. Examples of some of these primitives for a mesh structured topology are: *SendToNeighbor(. . .)*, *RecvFromNeighbor(. . .)*,

ScatterPartitions(...), GatherResults(...), etc. Some of these primitives are illustrated in section 4.

Interactions among modules are based on pattern-specific message-passing primitives, which make the PAS model suitable for a network cluster. The high-level abstractions provided by a skeleton hide most of the low-level details which are commonly encountered in any parallel application development. Interested readers can find a comprehensive description of the PAS model with detailed examples in [13,9]. Section 4 illustrates an example of application development using SuperPAS, which is based on the PAS model described here.

3 Introduction to SuperPAS

In this section, we introduce SuperPAS. Subsection 3.1 describes the motivation behind SuperPAS. Subsection 3.2 describes the steps to develop a parallel application using SuperPAS. Finally, in subsection 3.3, SuperPAS model is elaborated.

3.1 Motivation

Like most other pattern-based parallel programming systems, the original PAS system repository of (abstract) skeletons was built by hand-coding and there was no provision for adding new skeletons without writing them from scratch using the associated high-level programming languages, e.g., C++. The problem with this approach is that writing a skeleton from scratch is not easy. It requires in depth knowledge of the implementation of the entire system. This is the reason that PAS and all other similar systems are not extensible or very difficult to extend.

The motivation behind SuperPAS is to make PAS extensible. In a nutshell, SuperPAS is an extension of PAS that includes a skeleton description language (SDL) to describe the generic skeletons in PAS. Using the SDL, a skeleton designer can add new abstract skeletons to the skeleton library with minimum efforts. SuperPAS provides a well-defined way to describe the generic PAS skeletons, along with their primitives and parameters. Moreover, the model of SuperPAS allows a skeleton developer to compose two or more existing skeletons into a new skeleton. More about composite skeleton is discussed in section 5.

3.2 Development Steps in SuperPAS

In SuperPAS, the development process starts with the *skeleton designers* writing the abstract skeletons in SDL and storing them in the repository. When an *application developer* develops a parallel application, she chooses the proper skeletons, concretizes them, and finally fills them with application specific code to create the final parallel application. To concretize a skeleton written in SuperPAS SDL, the developer can directly modify the SDL code or can use the provided tools. Then she uses the SuperPAS tools to generate C++ code for the concretized skeletons. In fact, there is no semantical differences between the generated C++ concrete skeletons and concrete skeletons of original PAS.

The SuperPAS programming environment provides sufficient tools to: (1) verify the SDL syntax of an abstract skeleton, (2) concretize an abstract skeleton, (3) compile a concrete SDL skeleton into C++ code, and (4) manage the skeleton repository.

3.3 The SuperPAS Model for the Skeleton Designer

SuperPAS is a programming environment for creating abstract PAS skeletons. The environment provides an SDL together with other tools which facilitate the skeleton developers to specify each of the components, i.e., topology, parameters, primitives, etc., of a newly designed or composed abstract skeleton in PAS.

In order to be able to specify the (virtual) topology of a newly designed abstract skeleton, SuprePAS provides a collection of multidimensional grids. Each *node* of the grid can be visualized as a virtual processor and hence each grid can be visualized as a virtual processor grid (VPG). SuperPAS also provides the necessary basic communication, synchronization and structural primitives using which the VPG nodes (i.e., virtual processors) can communicate with one another. The choice of the basic primitives is based on research articles [16] and our long experience with parallel programming using PAS, other pattern-based systems and message passing libraries [14,15].

The specification of the topology of a newly designed abstract skeleton starts with mapping rules which maps the individuals or groups of its *children* to the VPG nodes. After the mapping is done, the *internal primitives* of the newly designed skeleton are defined using the primitives of the VPG. Note that both the mapping rules and the primitive definitions are governed by the logical unfolding of the topology onto the grid structure. The example in the next section illustrates these ideas when we design a PAS skeleton with the mesh topology. SuperPAS divides the primitives of a newly designed skeleton into two categories: *private* and *public*. *Private primitives* of a skeleton A_m can only be used by the representative of A_m and are not inherited by its children as external primitives. On the other hand, *public primitives* are available only to the children as external primitives.

The choice of the grid structure is not arbitrary. There are several factors which influenced the selection of a grid: First, a grid is a regular structure, which enables a uniform way to address each node of the grid, which in turn makes it suitable to be implemented on platform like MPI, PVM. Second, processor grid being a very popular structure, a wide collection of suitable communication-synchronization primitives for such a structure can easily be found in the existing literature. Lastly, a regular structure commonly associated with a skeleton can be easily unfolded and mapped into another regular structure. In the existing literature [17, 18], many of such mappings onto processor grid can be found. Moreover, any irregular structure (e.g., an arbitrary graph) also can easily be mapped onto a processor grid.

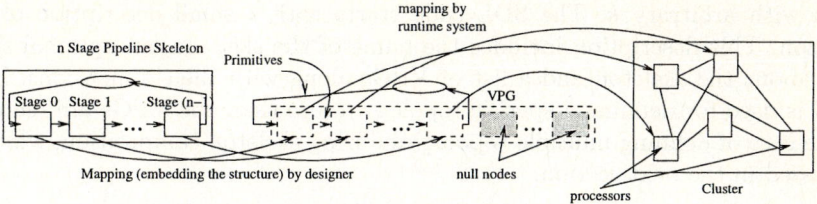

Fig. 2. Mapping of a back-end of a pipeline skeleton into a grid and the grid into physical processors

Note that there may be some VPG nodes which are not mapped onto and they are defined as the *null nodes*. After concretization and at run-time, only non-null nodes are mapped to physical processors. Also note that the VPG and its primitives are completely hidden from the application developer. What the application developer perceives are the skeletons with the required (virtual) topologies and associated set of primitives (designed by skeleton designers).

The SuperPAS approach is shown pictorially in Fig. 2 in the process of designing an abstract pipeline skeleton. In a pipeline skeleton, a child, i.e., a pipeline-stage accepts output of the previous stage as input, computes, and then sends the result to the next stage with the exception of the first and last stages. The unfolding of a pipeline structure into a one dimensional grid structure is simple: the i-th stage of the pipeline is mapped to the i-th node of the grid. The connectivity among the various stages of the pipeline is reflected in the defined primitives for the pipeline, e.g., *SendToNextNode(...)*, *RecvFromPreviousNode(...)*, *ReceiveFromRep(...)*, *SendToRep(...)*, *IsFirstNode()*, *IsLastNode()*, etc. In the figure, the unmapped VPG nodes are shown as *null* nodes. While running the final application program, only the non-null VPG nodes are mapped to the physical processors.

4 Example

In this section, we describe parts of an application, developed using SuperPAS. As an example, let us consider a parallel version of 2-D discrete image convolution application [19]. As will be discussed later, the algorithm uses a 2-D data-parallel mesh skeleton and a singleton skeleton (i.e., a skeleton with an empty back-end and hence no internal primitives). Assuming that both of these skeletons do not exist, the next subsection illustrates the skeleton designer's involvement in designing such skeletons using the SDL. The subsequent subsection illustrates the application developer's involvement in developing the complete parallel application.

4.1 The Abstract Data-Parallel Mesh and Singleton Skeleton

The SDL code for the abstract data-parallel 2-D mesh skeleton is shown in the following. For simplicity, we show a 2-D mesh rather than a generic k-dimensional mesh with arbitrary k. The SDL code starts with a small description of the skeleton. This description includes the name of the skeleton, an optional short note about the skeleton and a list of VPGs along with their dimensions. Each VPG is used for defining a specific topology. More than one VPGs are required in the case of defining multiple topologies during skeleton composition, which is discussed in the next section.

```
# A short description of the skeleton
name "DataParallelSkeleton";
description "A 2-D data parallel mesh skeleton";
# available VPGs and their dimensions
VPG DPGrid(1);   # A 1-dimensional grid

begin DPGrid
    # the parameters
    begin param
        int $HEIGHT, $WIDTH;
    end
    # SDL program variables
    begin var
        int $i;
    end
    # mapping rules
    begin typemap
        $CHILDREN[0] to node {($i)} where
            $i = {(0, $HEIGHT*$WIDTH - 1, 1)};
    end
    # private primitives:
    begin private primitive
        bool ScatterPartitions(MsgVector &mv) {
            assert(mv.size() == HEIGHT * WIDTH);
            ScatterToAllNodes(mv);
        }
        bool GatherResults(MsgVector &mv) { ... }
        ...
end

# public primitives:
begin public primitive
    bool GetPartition(Msg & m) {
        BCastFromRep(m);
    }
    bool SendResult(Msg &m) { ... }
    // is the child located at left edge
    bool IsAtLeftEdge(void) {
        Position p = GetPosition();
        return (p[0] % WIDTH == 0);
    }
    ...
    // send a message to a node located
    // at (dimY, dimX) distance vector away
    bool SendToNeighbor(int dimY, int dimX,
        Msg &m) {
        Position p = GetPosition();
        p[0] += WIDTH * dimY + dimX;
        return ISend(p, m);
    }
    // receive a message from a node
    // at (dimY, dimX) distance vector away
    bool RecvFromNeighbor(int dimY, int dimX,
        Msg &m) { ... }
    ...
end
```

Let us assume that the height and width of the 2-D data parallel mesh are $WIDTH$ and $HEIGHT$ respectively. So there will be $WIDTH \times HEIGHT$ children, where each of them does the same computation, i.e., all of the children are of the same type. To make the problem interesting, here we choose an one dimensional grid structure, rather than a two dimensional structure, to unfold the topology of the skeleton. The unfolding is shown pictorially in Fig. 3. In the SDL code, $DPGrid(1)$ represents the 1-dimensional grid. The corresponding mapping rules and primitives for the associated mesh topology are defined inside the $DPGrid$ code block (i.e., between begin DPGrid and end).

The length of each of the two dimensions (i.e., $HEIGHT$ and $WIDTH$) are the skeleton parameters. In the very beginning of the $DPGrid$ code block, these parameters are declared as $HEIGHT and $WIDTH. SuperPAS SDL has a built-in array ($CHILDREN) to hold the types of the children. In the *typemap* sub-block a mapping rule maps $CHILDREN[0] to node $i, where $i is an SDL program variable. An expression (start, end, inc) is used to iterate from start to end with an increment of inc. So in the mapping rule, $i gets the values of 0, 1, ..., $HEIGHT \times WIDTH - 1$ and hence the rule maps $CHILDREN[0] to all the VPG nodes.

The private primitives for the mesh skeleton: ScatterPartitions(...), GatherResults(...), etc. are implemented by the designer using the available VPG-based primitives. Some examples of the basic VPG primitives, which are

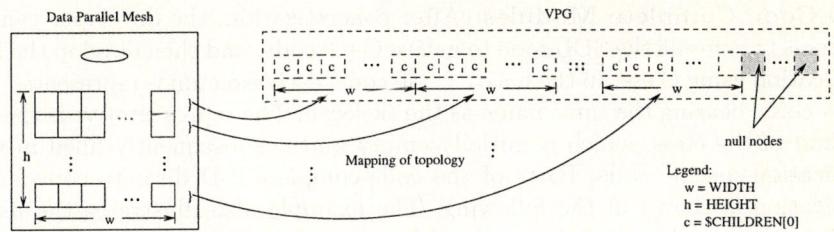

Fig. 3. Mapping a data parallel mesh into a VPG

an integral part of SuperPAS SDL, are: BCastFromRep(...), GetPosition(), ScatterToAllNodes(...), GatherFromAllNodes(...), etc. Similar approach is taken to design the public primitives. Finally, the mapping rule together with the definitions of the primitives complete the description of the topology of the data parallel mesh skeleton.

The singleton skeleton is the simplest skeleton, as it has no child and hence no internal primitive. Consequently, the definition of the singleton skeleton is rather straightforward and is illustrated in the following:

```
name "SingletonSkeleton";
description "A skeleton for sequential computation";
```

4.2 Using the Skeletons

In this subsection, we demonstrate an use of the skeletons designed in the previous subsection. We use the skeletons to develop a parallel 2-D discrete image convolution application.

Parallel 2-D Discrete Image Convolution: A parallel solution to the 2-D discrete image convolution problem can be achieved using the data parallel paradigm. The whole image is partitioned into columns and rows. Discrete convolution of all the partitions can be performed in parallel, where the convolution of individual partition is a sequential operation. In order to compute the convolution of a partition, some additional data from the neighboring partitions are required. Consequently, nearest neighbor communications are involved among partitions.

The Concrete Skeletons: Form the problem definition, it is clear that the parallel 2-D discrete image convolution application can be implemented using the data parallel mesh skeleton. Each child of the mesh is a singleton skeleton and it handles the convolution of an individual partition. In order to develop the final parallel application, the application developer needs to concretize both of the data parallel and the singleton skeletons. To concretize the data parallel skeleton, the $WIDTH and $HEIGHT parameters, as well as the $CHILDREN array (the type of the children) should be bound. Concretization of the singleton skeleton is straight forward as it does not have any parameters. Note that the application developer has the option of using SDL directly, or the associated tools for concretizing a skeleton.

The Code Complete Modules: After concretization, the developer can use the tools to compile the SDL code to native C++ code, and then develop the final application using C++. In the native C++ code, each skeleton is represented by a C++ class, bearing the same name as the skeleton. The representative is the `Rep` method of the class, which is initially empty and is subsequently filled in with application-specific code. Parts of the code-complete 2-D discrete convolution application is shown in the following. The example also illustrates the use of some of the primitives of the mesh skeleton, e.g., `GatherFromChildren(...)`, `SendToNeighbor(...)`, etc.:

```
void TwoDImageConv::Rep(void) {
  // declare variables
  Image Mask, ImgIn, ImgParts, ImgOut;
  ...
  BCastToChildren(Mask);
  ScatterToChildren(ImgParts);
  ...
  // gather the convoluted data from the children
  GatherFromChildren(ImgParts);
  ...
}
```

```
void ConvolutePartition::Rep(void) {
  Msg m; Image ImgIn, ImgOut, Mask;
  ...
  // send image portion to the left neighbor
  if (!IsAtLeftEdge()) {
    SendToNeighbor(0, -1, m);
  }
  ...
  convolute(Mask, ImgIn, ImgOut);
  ...
  SendResult(m);
}
```

5 Composition of Skeletons

SuperPAS defines skeleton composition to build more complex (abstract) skeletons from simpler (abstract) skeletons. In this section, we describe the motivation and the model of skeleton composition.

A large-scale parallel application is a composition of multiple patterns. It will be more desirable to have a single composite skeleton rather than a collection of smaller skeletons, provided that the composite skeleton will be used for developing variations of similar applications. Another reason for having a composite skeleton is performance. Let us consider the example shown in Fig. 4 (a), where a wavefront and a pipeline skeleton are shown. The output of the lower-rightmost child (node 1 in the figure) of the wavefront is send back to the representative; the representative routes it to the representative of the pipeline skeleton, which in turn again routes to the first stage (node 2) of the pipeline. The pseudo code for the representatives are also shown in the figure. One way of composing these two skeletons is shown in Fig. 4 (b). From the figure it is evident that composition reduces the number of routings required from node 1 to node 2.

Let us specify the definition of the skeleton X using SuperPAS SDL as a tuple: $< G_x, A_x >$, where G_x is a set of VPGs and A_x is a set of *aliasing rules* (discussed later). The composition of a skeleton X with another skeleton Y will result in another skeleton $Z = X + Y$. The corresponding default SDL skeleton definition of Z is: $< G_x \cup G_y, A_x \cup A_y >$. Here the $\cup$ operator means a union of two sets. Referring to Fig. 4, number of routings can be further reduced by composing node 1 and node 2 directly into node 3, as shown in Fig. 4 (c). Such types of composition among nodes from two different VPGs are also allowed in SuperPAS and they are expressed as *aliasing rules*. A complete description of these rules is beyond the scope of this paper and hence is omitted.

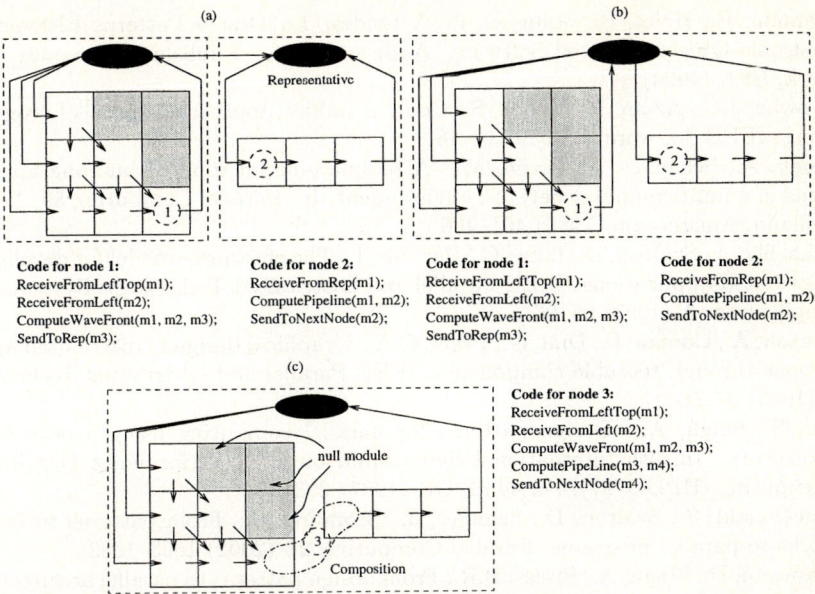

Fig. 4. (a) Two abstract skeletons, (b) Composition of abstract skeletons and (c) Composition of nodes

Note that composition of abstract skeletons to create another abstract skeleton, as described here, is different from hierarchical concretization of skeletons during application development.

6 Conclusion and Future Work

SuperPAS is a on going research work in the domain of pattern-based parallel programming. SuperPAS is targeted to make PAS extensible and more flexible. The performance of the PAS system has already been discussed in detail in [9]. Being a thin layer over MPI, there was no observable performance degradation in using PAS as compared to MPI. SuperPAS is an extension of PAS, which can automatically generate PAS skeletons based on designer's specifications. Obviously, the automatically generated skeletons may not be as optimized as the original hand-coded PAS skeletons. However, the generated code can always be fine tuned for better performance. Another important issue is usability. The performance and usability aspects of SuperPAS are currently being investigated and will be reported in our future works.

References

1. Alexander, C., Ishikawa, S., Silverstein, M.: A Pattern Language: Towns, Buildings, Construction. Oxford University Press, New York, USA (1977)

2. Gamma, E., Helm, R., Johnson, R., Vlissides, J.: Design Patterns Elements of Reusable Object-Oriented Software. Addision-Wesley Publishing Company, New York, USA (1994)
3. Browne, J.C., Azam, M., Sobek, S.: Code: A unified approach to parallel programming. IEEE Software **6** (1989) 10–18
4. Singh, A., Schaeffer, J., Green, M.: A template-based tool for building applications in a multicomputer network environment. In: Parallel Computing 89, North-Holland, Amsterdam (1989) 461–466
5. Schaeffer, J., Szafron, D., Lobe, G., Parsons, I.: The enterprise model for developing distributed applications. IEEE Parallel and Distributed Technology: Systems and Applications **1** (1993) 85–96
6. Bartoli, A., Corsini, P., Dini, G., Prete, C.A.: Graphical design of distributed applications through reusable components. IEEE Parallel and Distributed Technology **3** (1995) 37–50
7. Siu, S., Singh, A.: Design patterns for parallel computing using a network of processors. In: 6th International Symposium on High Performance Distributed Computing (HPDC '97), Portland, OR (1997) 293–304
8. MacDonald, S., Szafron, D., Schaffer, J., Bromling, S.: From patterns to frameworks to parallel programs. Parallel Computing **28** (2002) 1663–1683
9. Goswami, D., Singh, A., Preiss, B.R.: From design patterns to parallel architectural skeletons. Journal of Parallel and Distributed Computing **62** (2002) 669–695
10. Vanneschi, M.: The programming model of assist, an environment for parallel and distributed portable applications. Parallel Computing **28** (2002) 1709–1732
11. Cole, M.: Algorithmic Skeletons: Structured Management of Parallel Computation. MIT Press, Cambridge, Massachusetts (1989)
12. Darlington, J., Field, A.J., Harrison, P.G.: Parallel programming using skeleton functions. In: Lecture Notes in Computer Science. Volume 694., Munich, Germany (1993) 146–160
13. Goswami, D.: Parallel Architectural Skeletons: Re-Usable Building Blocks for Parallel Applications. PhD thesis, University of Waterloo, Canada (2001)
14. MPI: Message passing interface forum (2004) http://www.mpi-forum.org/.
15. PVM: Parallel virtual machine (2004) http://www.csm.ornl.gov/pvm/pvm_home.html.
16. Chan, F., Cao, J., Sun, Y.: High-level abstractions for message passing parallel programming. Parallel Computing **29** (2003) 1589–1621
17. Quinn, M.J.: Parallel computing: Theory and Practice. McGraw-Hill, Inc, New York, NY, USA (1993)
18. Grama, A., Gupta, A., Karypis, G., Kumar, V.: Introduction to Parallel Computing. Addison Wesley (2003)
19. Myler, H.R., Weeks, A.R.: The Pocket Handbook of Image Processing Algorithms In C. Prentice-Hall, Englewood Cliffs, N.J (1993)

Optimizing I/O Server Placement for Parallel I/O on Switch-Based Irregular Networks

Yih-Fang Lin[1,2], Chien-Min Wang[1], and Jan-Jan Wu[1]

[1] Institute of Information Science,
Academia Sinica,
Taipei, Taiwan, R.O.C.
{ice, cmwang, wuj}@iis.sinica.edu.tw
[2] Department of Computer Science and Information Engineering,
National Taiwan University,
Taipei, Taiwan, R.O.C.

Abstract. In this paper, we study I/O server placement for optimizing parallel I/O performance on switch-based clusters, which typically adopt irregular network topologies to allow construction of scalable systems with incremental expansion capability. Finding optimal solution to this problem is computationally intractable. We quantified the number of messages travelling through each network link by a *workload function*, and developed three heuristic algorithms to find good solutions based on the values of the workload function. Our simulation results demonstrate performance advantage of our algorithms over a number of algorithms commonly used in existing parallel systems. In particular, the load-balance-based algorithm is superior to the other algorithms in most cases, with improvement ratio of 10% to 95% in terms of parallel I/O throughput.

1 Introduction

Network contention is one of the major factors that may cause delay in remote data transfers on a distributed-memory system. Network contention occurs when multiple messages want to use the same network links at the same time. Both I/O traffic and inter-processor communication in a processing job can suffer from network contention. Most previous research focused on network contention due to inter-processor communication, whereas our focus is on contention due to parallel I/O traffic. Parallel I/O traffic bears similarity to inter-processor communication, but they differ in two aspects: (1) The size of messages in parallel I/O is much larger, and (2) Since the I/O resources are shared by all the compute nodes in the system, parallel I/O is even more prone to network contention.

Network contention in parallel I/O can be reduced in two steps: (1) Distribute I/O traffic over more network links by choosing proper placement of I/O servers, and (2) Once the locations of I/O servers are known, decide the proper execution sequence of the batch of remote data transfer requests In this paper, we will focus on the problem of I/O server placement.

Although I/O server placement has been extensively studied in multimedia research [4, 5, 12, 13, 15, 16], unfortunately the results from these previous research cannot be applied to parallel I/O, because these works all assume that a client's I/O request can be satisfied entirely by one I/O server, and the goal is to place multiple copies of the server over the network such that each client is within certain distance from at least one copy of the data. Parallel I/O, however, is more complicated in that the data are distributed over multiple I/O servers and each parallel I/O operation involves multiple data transfer requests to multiple I/O servers.

I/O resource placement has also been studied for traditional parallel machines with regular network topologies such as mesh, tori, hypercube, and ring [1, 3, 11, 14]. Switch-based clusters of workstations/PCs, on the other hand, typically adopt irregular topologies to allow the construction of scalable systems with incremental expansion capability. These irregular topologies lack many of the attractive mathematical properties of regular topologies, which makes optimizing resource placement on irregular networks a difficult task.

In this paper, we study I/O server placement on switch-based irregular networks, with the goal to minimize the completion time of a parallel I/O operation. In network-based systems, the link that has the maximum number of messages travelling through it at the same time (also called **workload**) becomes the bottleneck of parallel I/O performance for the whole system. Such link is usually referred to as the "dominating link" or "hot spot". Parallel I/O performance improves with the removal of hot spots. It is known that finding an optimal solution for I/O server placement is computationally intractable. We have developed three heuristic algorithms for this problem. The common goal of these heuristics is to remove "hot spots" in the network. We quantified the number of messages travelling through each link by a *workload function*. The maximum-workload-based heuristic chooses the locations for I/O nodes in order to minimize the maximum value of the workload function. The distance-based heuristic aims to minimize the average distance between the compute nodes and I/O nodes, which is equivalent to minimizing average workload on the links. The load-balance-based heuristic balances the workload on the links based on a recursive traversal of the routing tree for the network.

We conducted extensive simulations to evaluate our algorithms. Our result demonstrates performance advantage of our algorithms over a baseline algorithm, a random selection algorithm, and an even distribution algorithm. In particular, the load-balance-based algorithm is superior to the other algorithms in most cases, with improvement ratio of 10% to 95% in terms of parallel I/O throughput.

The rest of the paper is organized as follows: Section 2 gives an overview of the up-down routing strategy that is commonly used for irregular networks, and defines the workload functions and the I/O-server placement problem. Section 3 presents the three heuristic algorithms we proposed. Section 4 reports our experimental results, and Section 5 concludes.

2 Model

A switch-based system consists of switches and processors. Each switch has a set of ports, which can be used to connect to processors or ports of other switches. The topology of the network can be highly irregular. The connectivity of switches in the network can be represented by a graph $G = (V, E)$, where the set of nodes V represents switches, and the set of edges E represents the bidirectional connection channels among switches.

The up-down routing mechanism [6] first uses a breadth-first search to build a spanning tree T for the switch connection graph $G = (V, E)$. Since T is a spanning tree of G, E is partitioned into two subsets – T and $E - T$. Those edges in T are referred to as *tree edges* and those in $E - T$ as *cross edges* (which provide adaptivity in routing). Since the tree is built with a BFS, the cross edges can only connect switches whose levels in the T differ by at most 1. A tree edge going up the tree, or a cross edge going from a switch with a higher id to a switch with a lower one, are referred to as *up links*. The communication channels going the other direction are *down links*. In up-down routing a message must travel all the up links before it travels any down links.

In this work, we consider the worst case scenerio that generates the heaviest remote data transfer traffic: each compute node read/write data from/to all the I/O nodes. Both read and write operations may exist in the application program and the ratios of read and write operations are known (e.g. by profiling). Under this model, the I/O server placement problem can be stated as follows. Given a graph $G = (V, E)$ which represents an irregular network with $|V|$ switches and $|E|$ links, a network routing function (in this paper we consider up-down routing), the number of processing nodes and the number of I/O nodes in the network, what is the optimal locations for the I/O nodes such that the workload on the dominating link is minimized? The problem is known to be NP-complete by reducing the Partition problem. We developed heuristic algorithms to find good solutions for this problem. The heuristic algorithms are based on *workload functions* defined in the following.

Let N be the set of processing nodes in the system. Workload functions are defined as follows.

- $p(e, i, j)$: the probability of edge e being used by the message with node i as the source and node j as the destination. $p(e, i, j)$ is determined by the routing strategy for the network. For systems that do not support adaptive routing, the value of $p(e, i, j)$ equals to either 0 or 1.
- $w(e) = \sum_{\forall i, j \in N} p(e, i, j)$: the weight of edge e. We give each edge a weight, which is the sum of the probability of every message that travels through this edge.
- $L(e, \mathtt{A}, r_r, r_w) = \sum_{\forall i \in \mathtt{A}} \sum_{\forall j \in \mathtt{N}} \{r_r * p(e, i, j) + r_w * p(e, j, i)\}$: the workload on edge e if all the nodes in set A are assigned to be I/O nodes. r_r and r_w are the ratios of read and write operations respectively. $p(e, i, j)$ and $p(e, j, i)$ are the probabilities for read and write operations respectively that need to go through edge e. Let A, B and C be sets of nodes and $\mathtt{C} = \mathtt{A} + \mathtt{B}$, $\mathtt{A} \cap \mathtt{B} = \phi$, $L(e, \mathtt{C}, r_r, r_w) = L(e, \mathtt{A}, r_r, r_w) + L(e, \mathtt{B}, r_r, r_w)$.

- $M(\mathtt{A}, r_r, r_w) = \max_{\forall e, \forall i \in \mathtt{A}}\{L(e, \{i\}, r_r, r_w)\}$: the maximum load on the links if all the nodes in set A are assigned to be I/O nodes. Let A, B and C be sets of nodes and $\mathtt{C} = \mathtt{A}+\mathtt{B}$, $\mathtt{A} \cap \mathtt{B} = \phi$, $M(\mathtt{C}, r_r, r_w) = \max\{M(\mathtt{A}, r_r, r_w), M(\mathtt{B}, r_r, r_w)\}$.
- $L_s(\mathtt{A}, r_r, r_w) = \sum_{\forall e} L(e, \mathtt{A}, r_r, r_w)$: the sum of the workload of all links if all the nodes in set A are assigned to be I/O nodes. Let A, B and C be sets of nodes and $\mathtt{C} = \mathtt{A}+\mathtt{B}$, $\mathtt{A} \cap \mathtt{B} = \phi$. $L_s(\mathtt{C}, r_r, r_w) = L_s(\mathtt{A}, r_r, r_w) + L_s(\mathtt{B}, r_r, r_w)$.

3 Heuristic Algorithms

In this section, we present the three heuristic strategies: maximum-workload-based, distance-based, and load-balance-based. For clarity of presentation, without loss of generality, in the rest of the paper we will use $M(\mathtt{A})$ to denote $M(\mathtt{A}, r_r, r_w)$ and use $L(e, \mathtt{A})$ to denote $L(e, \mathtt{A}, r_r, r_w)$ when $r_r = r_w = 1$ (i.e. when there are equal numbers of read and write operations).

3.1 Maximum Workload Based Heuristic

The idea of this heuristic is to assign the I/O nodes in the way that minimizes the maximum workload of the dominating link. The selection can be done by enumerating all the possible values of function $(M(\{i\}))$ for all node index i and choose the node index that results in the smallest function value. For multiple I/O nodes, the problem translates to finding a set of nodes A such that $M(\mathtt{A})$ is minimized.

Algorithm Minimize_Maximum_Workload (MM)
Input: m: the number of I/O nodes, **N**: the set of all processing nodes, $[r_r, r_w]$: the ratios of read and write operations.
Output: **A**: the set of nodes assigned to be I/O nodes.
Description
    step 0: $\mathbf{A} \leftarrow \phi$.
    step 1: for $(i = 1, m)$ do {
        Find $p \in \mathbf{N}$ such that $\forall j \in \mathbf{N}$, $M(\mathbf{A}+\{p\}, r_r, r_w) \leq M(\mathbf{A}+\{j\}, r_r, r_w)$.
        $\mathbf{N} \leftarrow \mathbf{N} - \{p\}$, $\mathbf{A} \leftarrow \mathbf{A} + \{p\}$. }
End

3.2 Distance Based Heuristic

The motivation for this heuristic is that the longer the distance between the source and destination of a message, the higher the probability it may contend with other messages in the network. This heuristic attempts to place an I/O node such that each compute node in the network can access the I/O node by traveling the shortest distance. We define the "distance" of multiple compute nodes to an I/O node i to be the sum of the distances of the individual compute nodes to the I/O node. This is equivalent to the sum of the workload of all edges, i.e. $L_s(\{i\})$. Our goal is to select the node id i such that $L_s(\{i\})$ is minimized. For multiple I/O nodes, the problem translates to finding a set of nodes, A, such that $L_s(\mathtt{A})$ is minimized.

Algorithm Shortest_Distance (SD)

Input: m: the number of I/O nodes, $\mathbf{N}$: the set of all processing nodes., r_r, r_w]: the ratios of read and write operations.

Output: $\mathbf{A}$: the set of nodes assigned to be I/O nodes.

Description
  step 0: $\mathbf{A} \leftarrow \phi$.
  step 1: for $(i = 1, m)$ do {
    Find $p \in \mathbf{N}$ such that $\forall j \in \mathbf{N}, L_s(\{p\}, r_r, r_w) \leq L_s(\{j\}, r_r, r_w)$.
    $\mathbf{N} \leftarrow \mathbf{N} - \{p\}, \mathbf{A} \leftarrow \mathbf{A} + \{p\}$ }.
end

3.3 Load-Balance Based Heuristic

This heuristic is motivated by our observation that, on a complete d-ary routing tree T, parallel I/O performance is best when all the subtrees of T have approximately the same number of I/O nodes. This is reasonable because such distribution yields balanced workload on the links that connect the root and the subtrees. The set of links will be referred to as *subtree links* of tree T. Balanced workload on the subtree links avoids hot spots in remote data transfers. This concept can be extended to more general up-down routing trees. Before presenting the LB algorithm, we first define some related variables. Figure 1 gives graphical illustration of these variables. The partial tree T has three subtrees and three subtree links.

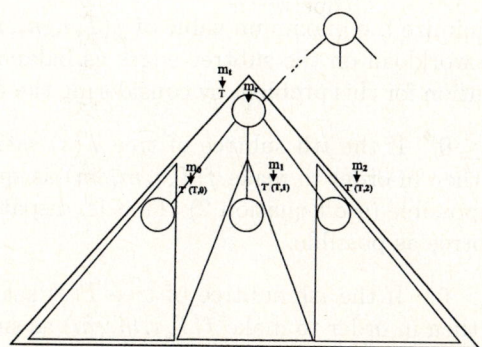

Fig. 1. Subtrees and some related variables

- $T(s)$ the tree whose root is switch s.
- $np(s)$ the number of processing nodes connected to switch s.
- $nm(s)$ the number of I/O nodes connected to switch s.
- $nn(T)$ the number of processing nodes in tree T.
- $ni(T)$ the number of I/O nodes in tree T.
- $rt(T)$ the root(a switch) of tree T and $rt(T(s)) = s$.

- $nt(T)$ the number of subtrees connected to the root of tree T.
- $T'(T, i)$ the ith subtree of tree T.

Let m_i be the number of I/O nodes on the tree $T'(T, i)$, m be the total number of I/O nodes in the network, $\mathbf{N}$ be the set of all processing nodes. We define the *subtree-edge load*, denoted by $f(T, i, m_i, m)$, to represent the workload on the edge, denoted by e, pointing from switch $rt(T)$ to switch $rt(T'(T, i))$. There are two kinds of message traffic tarvelling through edge e, one for writing data to the I/O nodes on the subtree $(T'(T, i))$ and the other for reading data from the I/O nodes outside of subtree $(T'(T, i))$. In the writing case, the workload is $m_i * (|\mathbf{N}| - nn(T'(T, i)))$ and in the reading case, the workload is $nn(T'(T, i)) * (m - m_i)$. The value of $f(T, i, m_i, m)$ for edge e is the sum of these two terms, as shown in Equation 2.

- $f(T, i, m_i, m)$ the *subtree-edge load* between the switch $rt(T)$ and the switch $rt(T'(T, i))$, with m_i I/O nodes dispatched to the subtree $T'(T, i)$ and the total number of I/O nodes in the network is m.
- $q(T, i)$ the load quality factor of the edge connecting the switch $rt(T)$ to the switch $rt(T'(T, i))$.
- $c(T, i, m)$ the constant load quality of the edge connecting the switch $rt(T)$ to the switch $rt(T'(T, i))$.

$$f(T, i, m_i, m) = m_i * (|\mathbf{N}| - nn(T'(T, i))) + nn(T'(T, i)) * (m - m_i) \qquad (1)$$
$$= m_i * q(T, i) + c(T, i, m) \qquad (2)$$

$$q(T, i) = |\mathbf{N}| - 2 * nn(T'(T, i))$$
$$c(T, i, m) = nn(T'(T, i)) * m$$

Our goal is to minimize the maximum value of $f(T, i, m_i, m)$, which is equivalent to making the workload on the subtree edges as balanced as possible. We approximate the solution for this problem by considering the following two cases.

Case 1: $q(T(s), i) \leq 0$. If the ith subtree of tree $T(s)$ satisfies the condition that $q(T(s), i) \leq 0$, then in order to make $f(T, i, m_i, m)$ as small as possible, m_i must be as large as possible (see Equation 2). That is, dispatching as many I/O nodes to the ith subtree as possible.

Case 2: $q(T(s), i) > 0$. If the ith subtree of tree $T(s)$ satisfies the condition that $q(T(s), i) > 0$, then in order to make $f(T, i, m_i, m)$ as small as possible, m_i must be as small as possible (see Equation 2). That is, dispatching fewer I/O nodes to the ith subtree.

Furthermore, in both cases, it is profitable to assign as many I/O nodes as possible to the processing nodes connected to the root switch $rt(T)$, because doing so does not increase the subtree-edge load between any subtree and the root switch, and it also reduces the number of dispatching. Once the number of I/O nodes to be dispatched to each subtree at current tree level is determined, the same process can be applied recursively to the subtrees at the next level until the locations of the I/O nodes at all tree levels are determined. The recursive process is implemented by the algorithm LB.

Algorithm Load_Balance (LB)

Input: T : the up-down routing tree, m : the number of I/O nodes to be dispatched to T.
Output: A the set of the nodes assigned to be I/O nodes.

Description
  step 0: if $nt(T) = 0$ then $mr \leftarrow m$
    else $(mr, \{m_0, m_1, \cdots, m_i, \cdots m_{nt(T)-1}\}) = Dispatch(T, m)$
  step 1: Randomly pick up mr nodes connected to the root switch $rt(T)$ to construct the set **A'**
  step 2: $\forall m_i = 0, \mathbf{A}_i \leftarrow \phi$.
  step 3: $\forall m_i > 0, \mathbf{A}_i \leftarrow LB(T'(T,i), m_i)$.
  step 4: $\mathbf{A} \leftarrow \mathbf{A'} + \mathbf{A}_0 + \mathbf{A}_1 + \cdots + \mathbf{A}_{nt(T)-1}$
end

4 Simulation Experiments and Results

In this section, we present results of simulation experiments to compare the three algorithms we proposed, MM, SD, and LB. We also implemented a baseline algorithm (BL), a random selection algorithm (RAN), and an even distribution algorithm (EVEN) as the basis for comparison. BL always chooses the first m processing nodes for I/O. RAN selects the I/O nodes randomly, and EVEN evenly distributes the I/O nodes to all the switches.

We developed a CSIM-based simulator for our experiments. The simulator can model wormhole routing switches with arbitrary network topologies. Other system parameters, such as communication start-up time, communication link transmission time, and router delay at switch, are all chosen based on the real numbers we obtained from a Myrinet-connected, 16-node Pentium-III PC cluster with an IDE disk hooked to each node. We chose system parameters as follows. Communication start-up time was 5.0 microseconds, link transmission time was 10.5 nanoseconds, and routing delay at switch was 200 nanoseconds. The data rate of I/O operation is 40 MB/sec. The compute nodes issue equal numbers of data transfer requests to a contiguous block of I/O nodes with the start and end of the block being randomly chosen based on the compute node's ids. This type of data transfer patterns commonly occur in many parallel computations.

For all experiments, we assumed a default system configuration of a 160-processor system interconnected by twenty 16-port switches in an irregular topology. Eight ports on a switch are connected to processors and the others are connected to other switches. Links were not allowed between ports of the same switch. A random number generator was used to decide the port and switch or the processing node to which a given switch port should be connected to. We varied the value of IR (the ratio of the number of I/O nodes v.s. the number of switches). The number of I/O nodes is defined as $(NS * IR)/8$, where NS is the number of switches. For each data point in the performance figures, the number was averaged over 150 different network topologies.

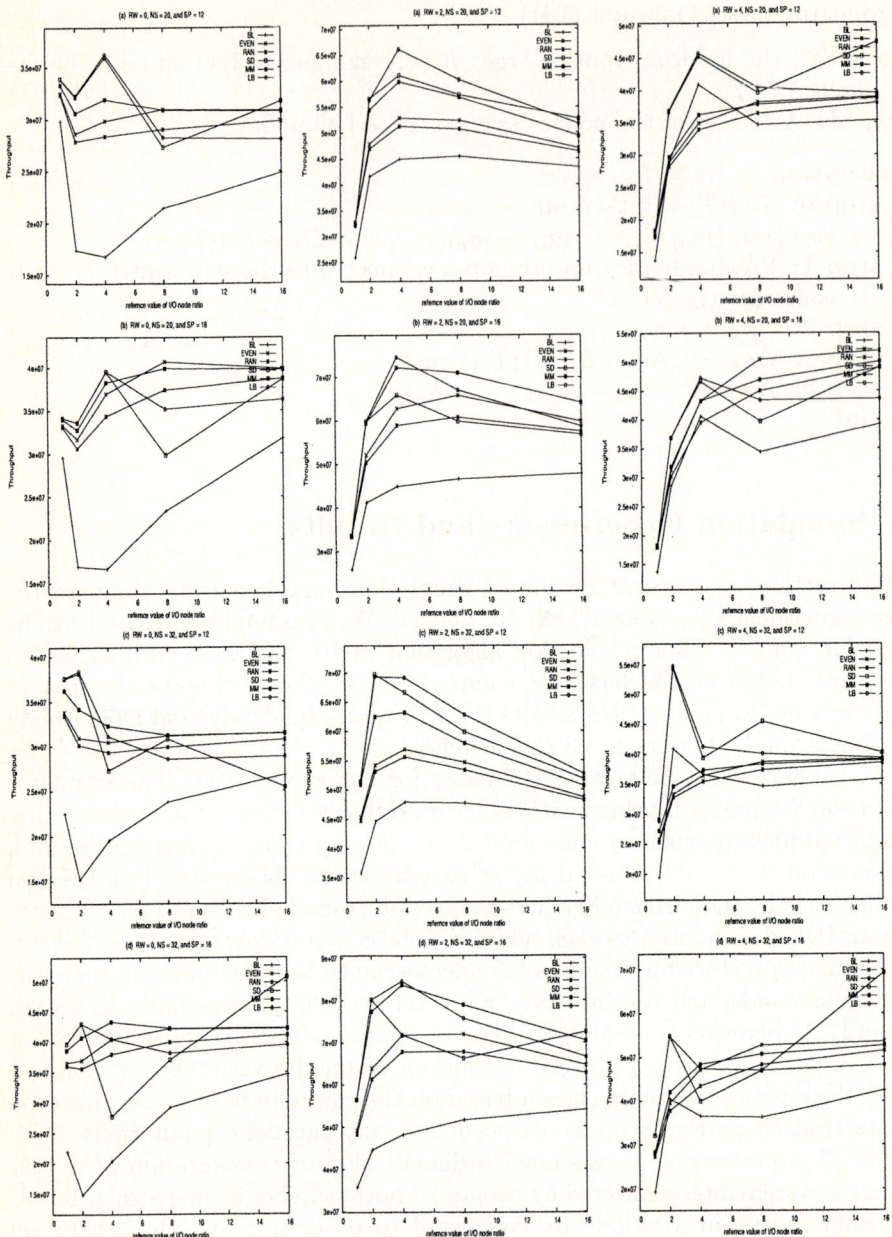

Fig. 2. varied IR, the left column fixed RW = 0 with (a) NS = 20, and SP = 12 (b) NS = 20, and SP = 16 (c) NS = 32, and SP = 12 (d) NS = 32, and SP = 16; the middle column fixed RW = 2 with (a) NS = 20, and SP = 12 (b) NS = 20, and SP = 16 (c) NS = 32, and SP = 12 (d) NS = 32, and SP = 16; the right column fixed RW = 4 with (a) NS = 20, and SP = 12 (b) NS = 20, and SP = 16 (c) NS = 32, and SP = 12 (d) NS = 32, and SP = 16

Figure 2 shows the throughput comparison of these algorithms under RW=0 (all read operations), RW=2 (equal number of read and write operations), and RW=4 (all write operations) respectively. In all cases, the baseline algorithm BL performs the worst, and EVEN and RAN are competitive to each other. Our three algorithms MM, SD and BL are superior to these three basic algorithms until the value of I/O node ratio reaches a threshold value (in this experiment, the threshold value is 8). This is because when the number of I/O nodes increases beyond certain value, data transfer traffic becomes too heavy that it is not possible to avoid contention no matter what locations the I/O nodes are, which makes the effect of optimization less noticeable. In cluster systems, which usually have limited network bandwidth (compared with high-speed networks in traditional parallel computers), the I/O node ratios are likely small.

5 Conclusion

In this paper, we investigated optimization of I/O server placement for switch-based clusters with irregular network topology. We proposed three heuristic algorithms to solve this problem. The maximum-workload-based algorithm (MM) aims to minimize the maximum value of the workload function. The distance-based algorithm (SD) minimizes the average distance between the compute nodes and I/O nodes, which is equivalent to minimizing average workload on the links. The load-balance-based algorithm (LB) balances the workload on the links based on a recursive traversal of the routing tree. Our experimental results show that optimizations of I/O server placement indeed are crucial to parallel I/O performance, and our algorithms, LB in particular, are effective in improving parallel I/O performance.

This work can be extended in several directions. First, in this paper we only consider uniform-length data transfer requests. Currently we are investigating the server placement problem for more general, non-uniform-length data transfers. The other direction is optimizing server placement for heterogeneous systems, in which the compute nodes and the I/O nodes may have different computing power and storage capacity.

Acknowledgement. This work is supported in part by the National Science Council of Taiwan under grant number NSC-93-2213-E-001-027.

References

1. M. Bae and B. Bose. Resource placement in torus-based networks. *IEEE Trans. Computers*, 46(10):1083–1092, October 1997.
2. P. Brezany, T. A Mueck, and E. Schikuta. A software architecture for massively parallel input-output. In *Proc. 3rd International Workshop PARA'96, LNCS* Springer Verlag, 1996.
3. Y. Cho, M. Winslett, M. Subramaniam, Y. Chen, S. W. Kuo, and K. E. Seamons. Exploiting local data in parallel array i/o on a practical network of workstations. In *Proc. fifth Workshop on I/O in Parallel and Distributed Systems (IOPADS)*, 1997.

4. A. Dan and D. Sitaram. An on-line video placement policy based on bandwidth to space ratio. In *ACM SIGMOD International Conf. Management of Data*, pages 376–385, 1995.
5. J. Dukes and J. Jones. Dynamic replication of content in the hammerhead multimedia server. Technical report, Department of Computer Science, Trinity College Dublin, Ireland, 2003.
6. M. D. Schroeder et. al. Autonet: A high-speed, self-configuring local area network using point-to-point links. Technical Report SRC research report 59, DEC, April 1990.
7. M. Harry, J. Rosario, and A. Choudhary. Vipfs: A virtual parallel file system for high performance parallel anddistributed computing. In *Proc. 9th International Parallel Processing Symposium*, 1995.
8. J. Huber, C. L. Elford, D. A. Reed, A. A. Chien, and D. S. Blumenthal. Ppfs: A high performance portable parallel file system. In *Proc. 9th ACM International Conference on Supercomputing*, pages 485–394, 1995.
9. S. Moyer and V. Sunderam. Pious: A scalable parallel i/o system for distributed computing environments. Technical Report Computer Science Report CSTR-940302, Department of Math and Computer Science, Emory University, 1994.
10. Nils Nieuwejaar. *Galley: A New Parallel File System for Scientific Workload*. PhD thesis, Dept. of Computer Science, Dartmouth College, 1996.
11. P. Ramananthan and S. Chalasani. Resource placement with multiple adjacency constraints in k-ary n-cubes. *IEEE Trans. Parallel and Distributed Systems*, 6(5):511–519, May 1995.
12. D. N. Serpanos, L. Georgiadis, and T. Bouloutas. MMPacking: A Load and Storage Balancing Algorithm for Distributed Multimedia Servers. *IEEE Trans. Circuits and Systems for Video Technology*, 8(1):13–17, 1998.
13. S.R. Subramany, B. Narahari, and R. Simha. Placement of storage nodes in a network. In *International Conference on Parallel and Distributed Processing Techniques and Applications*, 1998.
14. N. F. Tseng and G. L. Feng. Resource allocation in cube network systems based on the covering radius. *IEEE Trans. Parallel and Distributed Systems*, 7(4):323–342, April 1996.
15. N. Venkatasubramanian and S. Ramanathan. Load management in distributed video servers. In *Inter. Conf. Distributed Computing Systems*, 1997.
16. Y. Wang, J. Lin, D. Du, and J. Hsieh. Efficient video allocation for video-on-demand services. In *IEEE Multimedia Conference*, 1996.

Designing a High Performance and Fault Tolerant Multistage Interconnection Network with Easy Dynamic Rerouting*

Ching-Wen Chen**, Phui-Si Gan, and Chih-Hung Chang

Department of Computer Science and Information Engineering,
Chaoyang University of Technology,
Wufeng, Taichung County, Taiwan 413, ROC
{chingwen, s9227601, s9227610}@mail.cyut.edu.tw

Abstract. Designing a reliable and high performance multistage interconnection network (MIN) should consider the following issues carefully: (1) fault tolerance guarantee; (2) easy schemes and hardware design of rerouting switches; (3) low rerouting resulting in a low collision ratio. In this paper, we present the High Performance Chained Multistage Interconnection Network (HPCMIN) which has one-fault tolerance, destination tag routing for easy rerouting, one rerouting hop, resulting in a low collision ratio. From our simulation results, the HPCMIN results in a lower collision ratio than other dynamic rerouting networks. The HPCMIN is embedded with the indirect binary n-cube network (the ICube network) which is equivalent to many important MINs. Thus, the design methods used in the HPCMIN can be applied to these MINs so that they have the characteristics of the HPCMIN.

Keywords: Parallel computing, multistage interconnection network (MIN), fault tolerance, collision, performance, destination tag routing.

1 Introduction

Multistage interconnection networks (MINs) are considered as cost-effective ways of providing high-bandwidth communication in multiprocessor systems [1]. To enhance the reliability of MINs, many researchers have investigated fault tolerance issues [2][3][4][5][6][7][8]. In previous work, providing disjoint paths [3][4][5] and using dynamic rerouting [6][7][8][9] were often applied to MINs so that they would have fault tolerance capability. However, when the method of providing multiple disjoint paths is used to tolerate faults, it is necessary to know in advance the location of the faulty element. Then one of the fault-free paths can be taken to deliver message packets. If the location of the faulty element were

---

* This research was supported by the National Science Council NSC-92-2213-E-324-006.
** Corresponding Author. Tel: +886-4-23323000 Ext. 4534 Fax: +886-4-23742375

unknown, it is possible to send multiple packets simultaneously from the source to the destination. However, the former one which chooses a fault-free path cannot solve collision problems and the latter one of sending multiple packets simultaneously arises more collision ratio. Instead of providing disjoint paths, the dynamic rerouting method provides alternative paths to a destination when a packet encounters a faulty or busy element. Thus, this method does not need to know the location of faulty elements before a packet is sent. Consequently, dynamic rerouting can be used to tolerate faults and solve collision problems.

Although, dynamic rerouting can be used to tolerate faults and prevent collisions, some issues still need to be considered carefully in designing networks with high performance and one-fault tolerance. The important issues are: (1) guaranteeing one-fault tolerance; (2) providing easy rerouting schemes and low hardware cost of switches to compute rerouting tags; and (3) taking low rerouting hops to find alternative paths in order to reduce the collision ratio.

The Gamma network [6] and the B-network [7] both provide dynamic rerouting capability to tolerate faults and prevent collision. However, some problems still exist in these networks. For example, the Gamma network provides multiple paths to prevent collisions, but it lacks a mechanism to guarantee one-fault tolerance. Only a single path exists in the Gamma networks when the source and the destination are the same. In addition, although Gamma networks provide multiple paths to prevent collisions, a collision that may occur at the straight output link cannot be prevented. In the B-network, one-fault tolerance cannot be guaranteed, so partial collisions can not be prevented. Besides, the B-network cannot eliminate the possibility of a packet re-countering the same faulty element again after rerouting.

In this paper, we propose a network called the High Performance Chained Multistage Interconnection Network (the HPCMIN). The HPCMIN has the characteristics of dynamic rerouting, one-fault tolerance, destination tag routing and fixed one rerouting hop. The HPCMIN is embedded with the indirect binary n-Cube network (the ICube network) [9], which is equivalent to the multistage cube-type networks and to important multistage networks, for example, the Omega network [10], and the Baseline network [11]. Thus, the destination tag routing function can also be applied to the HPCMIN. In addition, the design features of the HPCMIN can be applied to networks which are equivalent to the ICube network so that these networks have the characteristics of the HPCMIN. With the benefit of destination tag routing, the switches in the HPCMIN do not need to compute the rerouting tag when a packet encounters a faulty or busy element. As a result, the hardware cost and complexity of the switches is lower and simpler than those of distance tag routing. To achieve high throughput (i.e. to reduce the collision ratio), the HPCMIN takes only one rerouting hop to reduce the probability of collisions.

The remainder of this paper is organized as follows. In Section 2, we introduce the topology and destination tag routing in the ICube network. In Section 3, we present the topology and routing/rerouting methods in the HPCMIN. In addition, in the HPCMIN, one-fault tolerance, destination tag routing, and performance improvement resulting from a packet which does not re-encounter the same faulty element are analyzed and presented. In Section 4, we present our experimental results, which show the throughput results of the HPCMIN and other related dynamic networks. Finally, Section 5 concludes our work.

2 Preliminaries

In this section, we present the previous network called indirect binary n-cube network (the ICube network). In Section 2.1, we present the topology of the ICube network. Destination tag routing is discussed in Section 2.2.

2.1 Indirect Binary n-Cube Network (ICube Network)

An ICube network of size $N=2^n$ consists of n stages labeled from 0 to n. Each stage involves N switches [9]. Basically, switches of sizes 1x2 and 2x1 are coupled with the first and last stages respectively. Moreover, each switch located at the intermediate stages is a 2x2 crossbar. Figure 1 shows the ICube network of size 8.

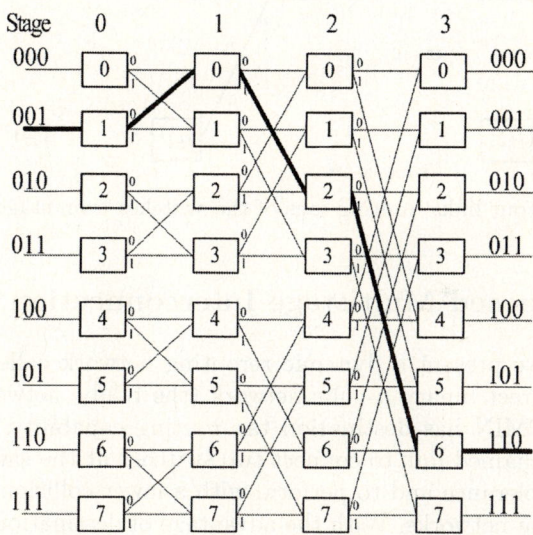

Fig. 1. An ICube network of size N=8. The routing condition is 1 for the source and 6 for the destination

2.2 Destination Tag Routing

In the ICube network, each switch number is given by the n-bit representation, for example, $j_{n-1}j_{n-2} \ldots j_2j_1j_0$, where the bit j_{n-1} is the most significant bit. To make the destination tag routing work, we mark the two output links of a switch: 0 is assigned to the upper link and 1 to the other. We illustrate the general case in Figure 2. After the two output links are marked, the switch can route packets by using their destination routing tag only. When a destination tag T is used to route packets, we reverse the binary representation T from $t_{n-1}t_{n-2}\ldots t_2 t_1 t_0$ to $t_0 t_1 t_2 \ldots t_{n-1}$, and use that as the routing tag. Therefore, the switch sends a packet to the 1 or 0 output link according to the ith bit t_i of the routing tag. That is to say, if the ith bit of the destination tag is 0, then the upper output link is taken. In contrast, if the ith bit of the destination tag is 1, the lower output link is taken. Example 1 illustrates the destination tag routing instance when the source is 1 and the destination is 6. Although the ICube network provides the destination tag routing, yet there is only one path between any source and destination pair.

Example 1: In the ICube network of size $N=8$, the source is 1 and the destination is 6. The routing tag is 011. The routing condition is shown in Figure 1 and is described as follows: 0(stage 1)→ 1(stage 0)→2(stage 2)→6(stage3)

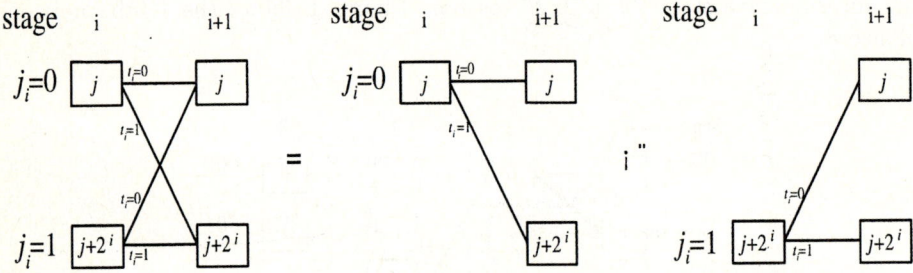

Fig. 2. The output links' routing tags of the switches from stage i to stage $i+1$

3 The Proposed Multistage Interconnection Network

In this section, we present a dynamic rerouting network called the HPCMIN, in which the indirect binary n-cube network (the ICube network) is embedded so that the HPCMIN has destination tag routing capability. In addition, the HPCMIN has a chained link to connect two switches at the same stage to guarantee one-fault tolerance and to perform with a lower collision ratio than other dynamic rerouting networks. With the advantage of destination tag routing, the switches in the HPCMIN do not compute the rerouting tags for rerouting. Thus, hardware cost and complexity can be reduced.

In Section 3.1, the topology of the HPCMIN is proposed. Section 3.2 describes the routing and rerouting situations to guarantee one-fault tolerance.

3.1 Topology of the HPCMIN

A HPCMIN of size $N=2^n$, consists of $n+1$ stages labeled from 0 to n, each stage involving N switches. The HPCMIN is embedded with the ICube network and adds one chained link for each switch. The chained link connects the switch j at stage i to the switch (j-2^{i+1} mod N) at stage i, where $0 \leq i \leq n$-2. At stage n-1, the chained link connects the switch j to the switch (j-2^{n-1} mod N). The chained links are used when a packet encounters a faulty or busy element. Because of destination tag routing, the HPCMIN does not need the hardware for computing rerouting tags, so the architecture of the switches can be size 2x2 and 2x1 crossbars, respectively, at the first and the final stages and be size 3x3 crossbars at other stages. Figure 3 shows the topology of the HPCMIN with size N equal to 8.

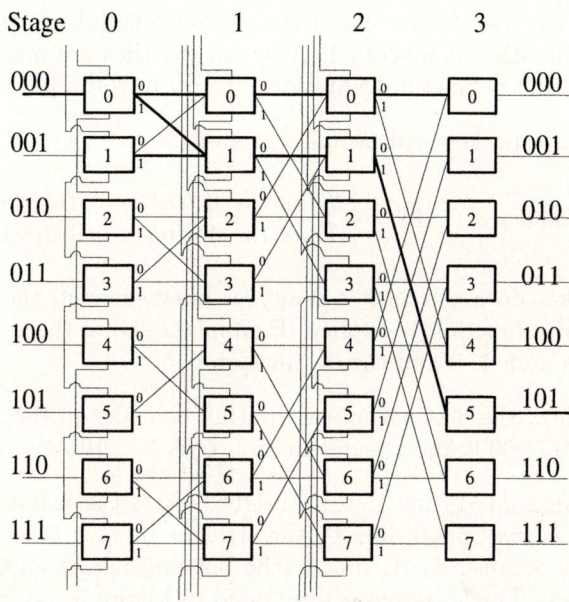

Fig. 3. The NRGIN of size N=8 and the routing condition of S=0 and T=5

3.2 Routing and Rerouting Methods

In this section, we present the routing and rerouting behaviors in the HPCMIN that guarantee one-fault tolerance and prevent collisions. Moreover, the destination tag routing function is introduced in this section. In the final subsection, we discuss the design that prevents a packet from meeting the same faulty element again after rerouting.

Destination Tag Routing in the HPCMIN. To reduce the hardware cost of re-computing the routing tag for the switches, the HPCMIN takes the destination tag as the routing and rerouting tag. If the distance tag routing method is applied and rerouting occurs, the switch must re-compute the routing tag. However, destination tag routing always uses the destination tag to route packets, so the switch does not need to compute the new routing tag. In this section, we introduce the destination tag routing function in the HPCMIN.

The HPCMIN is embedded with the ICube network, so destination tag routing used in the ICube network can be also used in the HPCMIN. In other words, a packet can arrive at a destination via the destination tag routing function, called the Forward function, defined in Definition 1 below. When switch $j = j_0 j_1 j_2 .... j_{n-1}$ at stage i wants to route a packet to the next stage, the switch uses the ith bit of the destination tag $D = d_0 d_1 d_2 .... d_{n-1}$ to route the packet to the next stage. If d_i is 0, the upper output link to the next stage is used. In contrast, if d_i is 1, the other output link to the next stage is used. In order to make destination tag routing work, we must mark the two output links to the next stage 0 and 1. The upper link of the two output links to the next stage is marked 0 and the other is marked 1. After all switches are marked, the switch can use the function defined in Definition 1 below to route packets.

Definition 1: If a packet is at switch j at stage i

$$Forward\ (j, di) = \begin{cases} Take\ the\ 0\ link & if\ d_i=0 \\ Take\ the\ 1\ link & if\ d_i=1 \end{cases}$$

When a packet does not meet a faulty or busy element, the Forward destination tag routing function is applied. Example 2 shows the routing condition when the source node is 0 and the destination is 5.

Example 2: Show the destination tag routing behavior in the HPCMIN when the source S is 0 (=000) and the destination D is 5 (=101).

Solution: Because $d_0=1$, link 1 (the straight link) is taken to switch 1 at stage 1. At stage 1, because $d_1=0$, link 0 (the straight link) is taken to switch 1 at stage 2. Finally, because $d_2=1$, link 1 (the bottom, non-straight link) is taken to the destination. This example is illustrated in Figure 3.

Dynamic Rerouting with a Fault or Busy Element. In this section, we prove that the HPCMIN with the dynamic rerouting method has one-fault tolerance capability. In the HPCMIN, a switch sends packets in the chained link to another switch at the same stage for rerouting. Therefore, we prove that a packet at a switch that is connected by a chained link can also arrive at the destination if the original one can. Before proving this fact, we define the partition concept which means a set of switches at the same stage which can deliver packets to some destination if one of them can deliver packets to the destination. As a result, we prove that the switch connected by a chained link can also route packets to the destination nodes because the switch and the original switch belong to the same partition. Finally, we prove that the HPCMIN has one-fault tolerance with dynamic rerouting.

Definition 2: A partition at stage i is the set of the switches which have the same i least significant bits, where i can be 0, 1, 2, ..., n.

Lemma 1: There are $N/2^i$ partitions at stage i.

Proof: By Definition 2, there are N partitions at stage 0, and each partition contains only one switch. At stage i, each partition has 2^i switches. Thus, there are $N/2_i$ partitions at stage i.

From Definition 2 and Lemma 1, we know that there are $N/2^i$ partitions at stage i, where i can be 0,1,2, ... ,n (=$\log_2 N$). We name the partitions at stage i $P_{i,0}, P_{i,1}, ..., P_{i,N/2^i}$ where i can be 0, 1, 2, ... , n. For example, there are N partitions, $P_{i,0}, P_{i,1}, ..., P_{i,N}$, at stage 0 and there are $N/2$ partitions at stage 1 and so on. In Theorem 1, we prove that the switches in the same partition can deliver packets to the destination if there is a switch in the partition delivering packets to the same destination.

Theorem 1: A packet in the switches belonging to partition $P_{i,j}$ can arrive at some specific destination if one of the switches in the same partition $P_{i,j}$ can deliver a packet to this destination.

Proof: According to the destination tag routing function Forward, when a packet is delivered from stage i to stage $i+1$, the $(i+1)^{th}$ bit of the switch index at stage $i+1$ is the same as that of the destination index. As a result, the routing behavior leads the packets to switch j at stage i and the i least bits of the switch index j are the same as that of the index of the destination node. By Definition 2, the switches in the same partition have the same i least significant bits at stage i. Moreover, the Forward function leads packets to the switch at stage n which has the same n-i most significant bits as the destination. Based on these two facts, the n-bits of the switch at stage n are the same as the destination. Accordingly, when a switch in some partition can lead packets to the destination, the other switches in the partition can also deliver packets to the destination by the Forward destination tag routing function.

From Theorem 1, we know that the other switches can also deliver packets to the destination if one of the switches in the same partition can deliver packets to the destination. In the following, we prove that the HPCMIN has one-fault tolerance.

Theorem 2: The HPCMIN can tolerate one fault located in the middle of the network.

Proof: Once a switch connects a completely faulty element in the middle stages, the switch delivers a packet in the chained link to another switch belonging to the same partition. By Theorem 1, the switch can also deliver the packet to the destination if the packet does not meet the faulty element again. We assume that a packets at switch j at stage i meets a faulty switch located in the middle stage $i+1$. Hence, the packet at stage i is sent to switch $(j$-$2^{i+1})$ mod N of the same stage. Because the output non-straight links of the switch indexed $(j$-$2^{i+1})$ mod N and switch j at stage i are in the same direction (upward or downward), the packet does not meet the same faulty switch at stage $i+1$ again where i can be 0, 1, 2 .... n-2 (middle stages).

Let us assume that the faulty element is a link from stage i to stage $i+1$ where i can be 0, 1, 2 .... n-2 (middle stages). Since the rerouting vertical distance at middle stages is less than N, the packet will not be rerouted to the same switch j at stage i after one rerouting, that is, the packet does not encounter the faulty link again. As a result, the HPCMIN can tolerate one fault in the middle stages using chained links.

4 Experimental Results and Discussion

In this section, we evaluate the performance, under various traffic loads, of four dynamic rerouting networks: Gamma networks, B-networks, and the HPCMIN. In our simulations, we considered two cases, networks without faults and with a faulty switch. However, collisions are considered in our simulation of both cases. A collision means two or more packets at a switch are delivered to the same output. When a collision of two packets occurs at a switch, the switch reroutes one of these two packets to solve the collision situation. With regards to the faulty switch, we assume that the faulty switch is completely faulty and the switches at the previous stage use faulty information to reroute the packets.

Figure 4 and Figure 5 show the performance comparison of these dynamic rerouting networks with one fault and without any fault. In addition, Figure 6 shows the successful rerouting ratio. Although the Gamma network can find an alternative path for rerouting, there is no alternative path when the straight link is taken. Due to the lack of rerouting cases, the Gamma network performs worse than the other networks except the B-network. Although the B-network performs better than the Gamma network in light traffic, the overhead of four rerouting hops in the worst case causes the B-network to have worst throughput in heavy traffic of all the networks. Because of one-fault tolerance guarantee

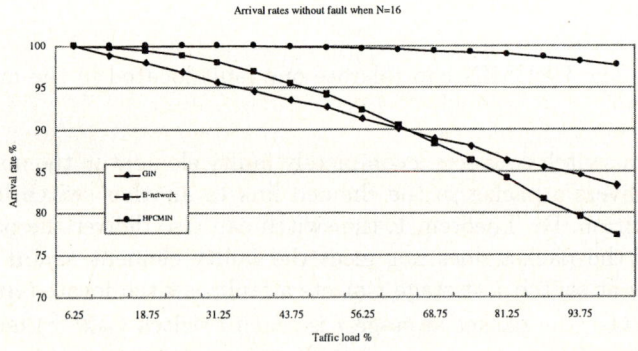

Fig. 4. Arrival rate without fault when the size of networks $N=16$

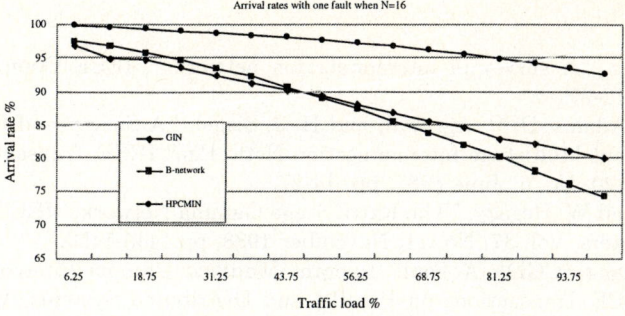

Fig. 5. Arrival rate with one fault when the size of networks $N=16$

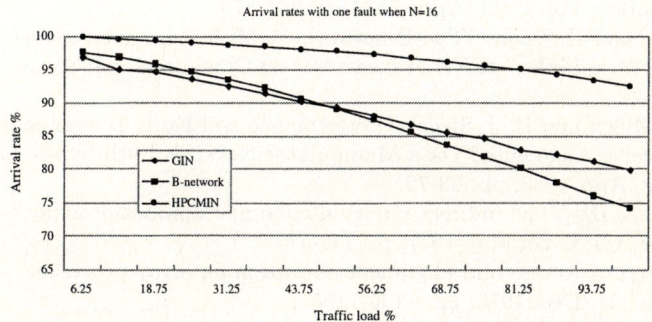

Fig. 6. Rerouting successful rates with one fault when the size of networks $N=16$

and fixed one hop rerouting, HPCMIN has the best throughput of all these networks.

5 Conclusion

In the design of fault tolerant networks, providing disjoint paths and using dynamic rerouting are often used. However, the dynamic rerouting method can guarantee one-fault tolerance and solve collision problems. When the dynamic rerouting method is used, the destination tag routing method can reduce hardware cost and complexity in computing rerouting tags. Based on these reasons, we designed a network called the HPCMIN. The HPCMIN has dynamic rerouting to tolerate faults and prevent collision, destination tag routing, one-fault tolerance and one rerouting hop. Accordingly, the HPCMIN performs better than other dynamic rerouting networks, based on our experimental results. Because of the equivalent property of the ICube network and other important networks, the design features of our work can be applied to those networks so that they have high performance and one-fault tolerance.

References

1. T. Y. Feng, "A survey of interconnection networks", IEEE Computer 14, Dec. 1981, pp. 12-27.
2. G. B. III Adams, D. P. Agrawal, and H. J. Siegel, "A Survey and Comparison of fault-tolerant Multistage Interconnection Networks," IEEE Transactions on Computer Vol. 20, No. 6, June 1987, pp. 14-27.
3. K. Yoon and W. Hegazy, "The Extra Stage Gamma Network," IEEE Transactions on Computers, Vol. 37, No. 11, November 1988, pp.1445-1450.
4. P. J. Chuang "CGIN: A Fault Tolerant Modified Gamma Interconnection Network," IEEE Transactions on Parallel and Distributed Systems. Vol. 7, No. 12, December 1996, pp. 1301-1306.
5. S. W. Seo and T. Y. Feng, "The Composite Banyan Network," IEEE Transactions on Parallel and Distributed Systems, Vol. 6, No. 10, October 1995, pp.1043-1054.
6. D. S. Parker and C. S. Raghavendra, "The Gamma Network," IEEE Transactions on Computers, Vol. C-33, April 1984, pp.367-373.
7. K. Y. Lee and H. Yoon, "The B-network: A Multistage Interconnection Network with Backward Links," IEEE Transactions on Computers vol. 39, no. 7, July 1990, pp. 966-969.
8. R. J. McMillen and H. J. Siegel, "Performance and Fault Tolerance Improvements in the Inverse Augmented Data Manipulator Network," 9th Symp. Computer Architecture, Apr. 1982, pp. 63-72.
9. M. C. Pease III, "The indirect binary n-cube microprocessor array." IEEE Trans. Computer, vol. C-26, May 1977, pp.458-473.
10. D. H. Lawrie, "Access and alignment of data in an array processor", IEEE Trans. Computers 24, Dec. 1975, pp. 1145-1155.
11. C. L. Wu and T. Y. Feng, "On a class of multistage interconnection networks", IEEE Trans. Computer 29, Aug. 1980, pp. 694-702.

Evaluating Performance of BLAST on Intel Xeon and Itanium2 Processors

Ramesh Radhakrishnan[1], Rizwan Ali[1], Garima Kochhar[1], Kalyana Chadalavada[2], Ramesh Rajagopalan[2], Jenwei Hsieh[1], and Onur Celebioglu[1]

[1] Scalable Systems Group, Dell Inc, Round Rock, Texas
http://www.dell.com/hpcc
[2] HPCC Enterprise Solutions Engineering, Dell Inc, Bangalore, India

Abstract. High-performance computing (HPC) has increasingly adopted the use of clustered Intel architecture–based servers. This paper compares the performance characteristics of three Dell PowerEdge (PE) servers that are based on three different Intel processor technologies. They are the PE1750 which is an IA-32 based Xeon system, PE1850 which uses the new 90nm technology Xeon processor at faster frequencies and the PE3250 which is an Itanium2 based system. BLAST (Basic Local Alignment Search Tool), a high performance computing application used in the field of biological research, is used as the workload for this study. The aim is to understand the performance benefits of the different features associated with each processor/platform technology to BLAST and explain the observations using other standard micro-benchmarks like STREAM and LMBench.

1 Introduction

Computing clusters built from standard components using Intel processors are becoming the fastest growing choice for high-performance computing (HPC). Twice yearly, the 500 most powerful computing systems in the world are ranked on the Top 500 Supercomputer Sites Web page [1]. In June 2002 the ranking listed 44 entries using Intel processors; two years later in June 2004, that number reached 287.

Industry standard Intel processors-based servers are changing the landscape of the enterprise server market and in particular high-performance computing. Intel has continuously introduced processors at higher frequencies, larger caches and faster front-side buses. Technological innovations such as hyper-threading, Streaming SIMD (Single Instruction Multiple Data) instructions and NetBurst micro-architecture significantly increased 32-bit Intel processor performance [2]. 64-bit Itanium-2 processors are based on the EPIC (Explicitly Parallel Instruction Computing) architecture and have claimed the top spot in several industry-standard benchmarks [3]. Intel recently introduced a 90nm version of the 32-bit Xeon processors. The key architectural differences are faster processor and front side bus (FSB) frequencies and larger caches. In addition there are several bandwidth-enabling technologies like DDR2 (Double Data Rate) memory and PCI Express available in the new Dell servers supporting the 90nm Xeon processors.

This paper compares the performance characteristics of Dell PowerEdge server models equipped with Intel Xeon (32-bit) processors, the new 90nm Xeon (Nocona) and Itanium-2 (64-bit) processor using an HPC application, BLAST (Basic Local Alignment Search Tool), that is widely used in the field of bioinformatics. The aim is to understand the impact of the different features and technologies that are featured in the Intel processors, as well as the impact of the memory technology on the performance of BLAST. To achieve this end, we use micro-benchmarks like LMbench [5] and Stream [6], in addition to analysis tools that read the processor performance counters. Furthermore we also look at the effect of the processor clock frequency on the performance of BLAST.

Table 1. Servers used in this study

	PE1750 (IA-32)	PE1850 (IA-32)	PE3250 (IA-64)
CPU	Dual Xeon @3.2 GHz (130 nm)	Dual Xeon @3.2 GHz and @3.6 GHz (90nm)	Dual Itanium2 @1.5 GHz
FSB	64-bits @ 533 MHz	64-bits @ 800 MHz	128-bits @ 400 MHz
Cache size	L2: 512 KB L3: 1 MB	L2: 1 MB	L2: 256MB L3: 6 MB
Memory	4 GB DDR-266 MHz	4 GB DDR2-400 MHz	4 GB DDR-266MHz
Bandwidth	4.8 GB/s	6.4 GB/s	6.4 GB/s

Table 1 lists the servers used along with the processor technology in each of these servers. In this paper we will use the Dell model names to avoid ambiguity between the current generation 130nm process based IA-32 Xeon and the newer 90nm Xeon processor. Empirical studies have shown that small-scale symmetric multiprocessing (SMP) systems make excellent platforms for building HPC clusters. Thus in our study all the servers used were two-processor systems. [4]

In the next section we describe the features of the three different processor and server architectures. Section 3 illustrates the results and analysis done on these platforms using BLAST and the micro benchmarks. We conclude in Section 4.

2 Comparison of the Intel Processor Based Systems

We use three servers - Dell PowerEdge 1750, PowerEdge 1850 and PowerEdge 3250 servers in this study. In the following sections we provide an architectural overview of the processor and memory subsystems used in these servers since they have the biggest impact on the performance of BLAST.

2.1 Processor Architecture Overview

The NetBurst™ micro-architecture is the core of Intel's 130nm technology based 32-bit Xeon processor used in the Dell PE1750. The NetBurst architecture uses a 20-stage pipeline which allows higher core frequencies. The 130 nm Xeon processors were introduced at speeds of 1.8 GHz and are currently available at speeds of up to 3.2 GHz. The system bus or FSB scaled from 400MHz in the initial 180nm Xeon offerings to 533 MHz on the current 3.2 GHz 130nm technology based processors.

The Xeon processor is a superscalar processor that combines out-of-order speculative execution with register renaming and branch prediction to improve performance. The processor uses an Execution Trace Cache that stores pre-decoded microoperations. Streaming SIMD (Single Instruction Multiple Data) instructions (SSE2) are used to speedup media types of workload.

The PE1850 is the follow-on to the PE1750 which will use the new Intel Xeon processor fabricated using a 90-nm process technology. The 90nm Xeon processor is an extension of the 130-nm based Xeon processor. However, there are some architectural changes between these two Xeon processors that will have an impact on application performance. The 90nm Xeon processor is being introduced at a frequency of 3.6 GHz coupled with a faster 800 MHz system bus. It uses a longer 31-stage processor pipeline that will facilitate higher frequencies in future versions.

The Dell PE3250 system is based on the Itanium Processor Family (IPF) which uses a 64-bit architecture and implements the Explicitly Parallel Computing (EPIC) architecture. Instructions in groups, called bundles, are issued in parallel, depending on the available resources. The Itanium2 differs from the Xeon processors in the fact that it uses software (compiler) to exploit parallelism, as opposed to complex hardware to detect and exploit instruction parallelism. Software compilers provide the information needed to execute parallel instructions efficiently.

Itanium2 processors are available in frequencies ranging from 1.0 GHz to 1.6 GHz and use varying sizes of L3 caches ranging from 1.5 MB to 6MB. A 128-bit 400 MHz system bus (FSB) is used to connect the processors. The Itanium2 processor has a large number of registers compared to the Xeon processors (128 64-bit GPRs) that are used by the compiler to keep the 6 Integer Functional Units busy.

2.2 Cache and Memory Subsystem Differences

The Dell PE1750 and PE3250 use DDR 266 or PC-2100 memory. The PE3250, however, operates at the speed of DDR 200. The PE1850 uses the new DDR2 memory running at 400MHz (PC2-3200) which has a theoretical bandwidth of 3.2GB/s. DDR2 architecture is also based on the industry-standard DRAM (Dynamic Random Access Memory) technology. DDR2 standard contains several major internal changes that allow improvements in areas such as reliability and power consumption. One of the most important features is its ability to pre-fetch 4-bits of memory at a time compared to 2-bits in DDR [7].

DDR2 transfer speed starts where the current DDR technology ends at 400 MHz. In the future DDR2 will support 533 and 667 mega-transfers/sec (MT/s) to enable memory bandwidths of 4.3GB/s and 5.3GB/s. Currently only DDR2-400 is available, which is the memory technology used in the PE1850 system. The performance advantage of the new DDR2 technology is shown in Section 3, where we measure the bandwidth and latency for the PE1850.

In addition to memory, the cache hierarchy also plays an important part in an application's performance The 130nm Xeon used in the PE1750 and Itanium2 used in the PE3250 come with a Level 3 (L3) cache (optional in the Xeon processor). The 90nm Xeon only has 2 levels of caches. Section 5.2 enumerates the measured cache latencies for the three processors.

3 Performance Evaluation and Analysis

In this section we measure the cache latencies, memory latency and bandwidth on the Dell PowerEdge servers. Results from executing BLAST [8] on the servers are also studied in this section. BLAST was executed on each of the four configurations using a database of about 2 million sequences with about 10 billion total letters. For our study we executed blast against single queries of three different lengths – 94K characters (small), 206K characters (medium) and 510K characters (large). Runs were conducted using both single and dual threads.

3.1 Cache/Memory Latency and Bandwidth results

The cache and memory latencies measured using LMbench is shown in Table 3. The latencies are shown in absolute time (nanoseconds or 10^{-9} seconds), as well as in terms of processor clock cycles (where one clock cycle = 1/CPU frequency).

Table 2. Cache and memory latencies measured using LMbench

Cache/ Memory Levels	PE1750 3.2 GHz (130nm Xeon)		PE1850 3.2 GHz (90nm Xeon)		PE3250 1.5 GHz (Itanium2)	
	Time	cycles	Time	cycles	Time	cycles
L1	0.63ns	2	1.3ns	4	1.34ns	2
L2	5.7ns	18	9.0ns	29	4.02ns	6
L3	8.5ns	27	N/A	-N/A-	13.7ns	21
memory	128ns	410	116ns	371	201ns	302

The L1 data cache access latency is observed to double for the new 90nm Xeon, compared to the 130nm Xeon. This is due to the cache size being doubled. In terms of clock cycles, the cache and memory access times for Itanium2 processor are lower compare to the Xeon processors. However, when comparing absolute memory access times, the PE3250 has the highest value since it runs at a slower speed of 200 MHz.

Figure 1 shows the measured memory bandwidth using the Stream benchmark. The PE3250 and PE1850 show good improvements over the PE1750 due to a wider system bus (128 bits) and faster memory clock speed (200 MHz) respectively.

3.2 BLAST Performance on the Three Intel Processors

In this section we evaluate the performance of BLAST on the four systems using different query sizes, and running single and dual threads. The importance of processor frequency, architecture and memory subsystem design can be inferred from the results obtained on the four configurations that the workload was tested against.

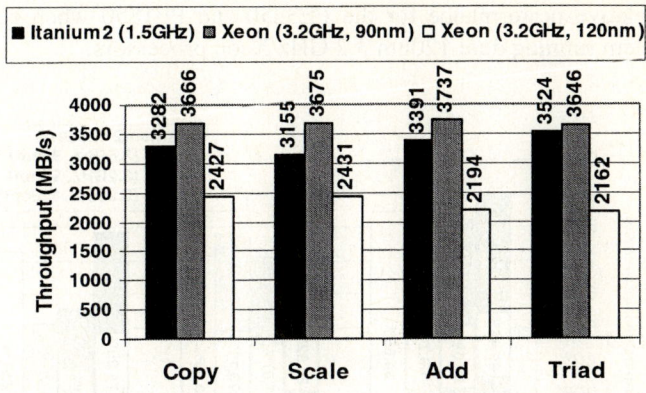

Fig. 1. Sustainable memory bandwidth measured using Stream benchmark

Figure 2 shows run time in seconds to complete the various query sizes on the four Dell systems. Three query sizes – 94k, 206k and 510k were chosen to represent a small, medium and large query size respectively. The database that these queries were matched against remained constant. The medium and large queries are 2.2 and 5.4 times larger than the small query size (94k). When comparing the run times from Figure 1, the query completion times for medium and larger query sizes are 2.6 and 8.3 times longer for the single threaded runs and 2.7 and 8.8 times longer for the dual threaded runs. This shows that as query sizes become larger, the run times increase exponentially as opposed to increasing in a linear fashion.

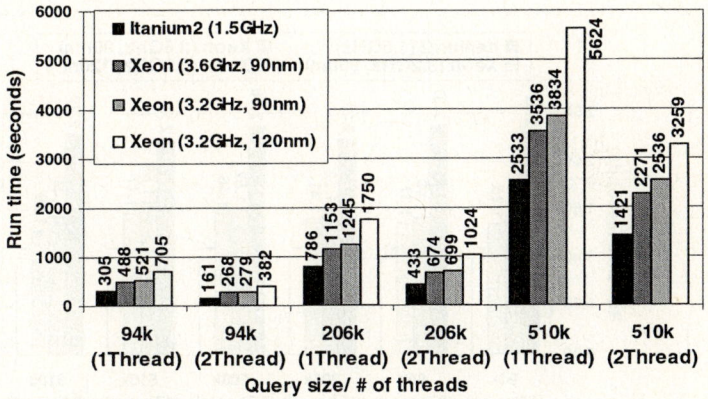

Fig. 2. Completion time in seconds for the four test-beds

Figure 3 illustrates the speedup obtained when running queries using dual threads compared to running in single threaded mode. When going to larger query sizes the speedup from dual threads is lower compared to smaller query sizes. Figure 4 illustrates the relative performance for the PE3250 and PE1850 when compared to the PE1750 system running dual 120nm 3.2 GHz Xeon processors.

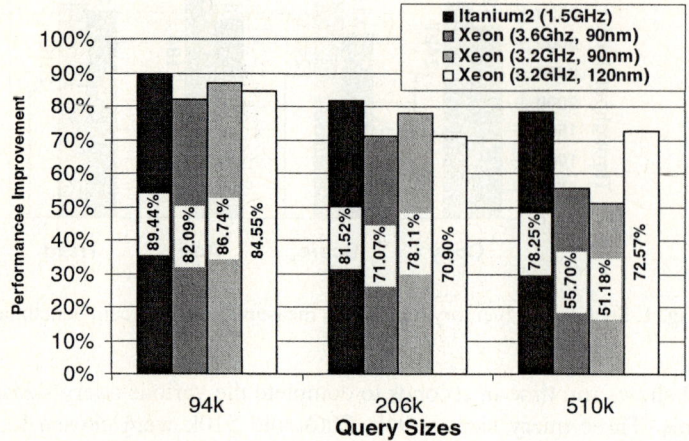

Fig. 3. Percentage improvement for the three query sizes when running two threads

The PE3250 with the Itanium2 processor has the best performance for all query sizes and exhibits speedup ranging from 123% to 137% over the PE1750. It was observed that BLAST scaled more efficiently on the PE3250 with Itanium2 processor. Therefore in Figure 4 the speedup for Itanium2 is slightly higher for dual threaded runs.

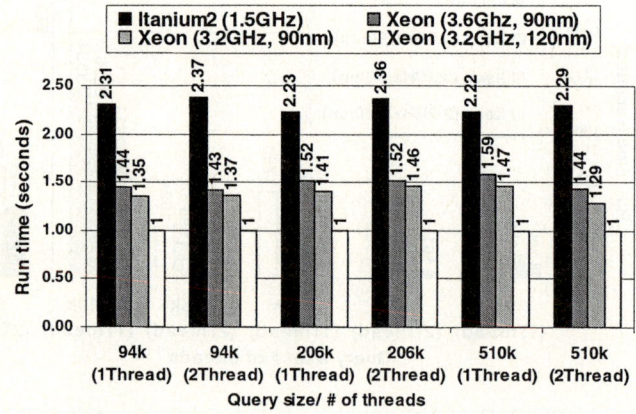

Fig. 4. Performance speedup for PE3250 and PE1850 over the PE1750

Figure 4 also shows that frequency increase from 3.2 to 3.6 GHz (13% frequency scaling) for the 90nm Xeon only resulted in a 6-11% improvement for small and medium query sizes. For the larger query size improvement from frequency scaling was slightly higher at 12-15%. Therefore BLAST is observed to scale well with increasing processor frequency on the Xeon processors, especially on larger query sizes. Figure 4 also shows that the speedup for single and dual threads is in the same ballpark for all systems except for the PE1850 when using a 510k query size. This was observed in Figure 3, which showed that the scaling for dual threads on the Nocona processors were only ~50% compared to 70 and 80% for smaller query sizes.

4 Conclusions

On the PE1750 and PE1850 the difference in performance is mainly from the improved DDR2 technology. The measured cache miss rates were same, and so were other metrics like branch prediction etc. Therefore, the cache miss penalty is higher on the PE1850 due to larger cache access times and branch misprediction penalty is higher due to a longer processor pipeline. In spite of this, the performance was better on the PE1850 by 29% for the 3.2 GHz processor and 44% for the 3.6 GHz processor. The 29% speedup is mainly due to faster memory and 44% is due to faster memory plus the frequency scaling. Larger cache size on the Itanium2 processor, along with ability to execute higher number of integer instructions per cycle help achieve speedups of 130% on the larger query size.

References

1. Top 500 Supercomputer Sites. http://www.top500.org/
2. D. Koufaty, D. T. Marr,"Hyperthreading technology in the netburst microarchitecture", Intel, IEEE Micro, Volume 23, Number 2, March/April 2003, p 56-65
3. H. Sharangpani, K. Arora,"Itanium Processor Microarchitecture", Intel, IEEE Micro, Volume 20, Number 5, September/October 2000, p 24-43
4. J. Hsieh, T. Leng, V. Mashayekhi, and R. Rooholamini, "Impact of Level 2 Cache and Memory Subsystem on the Scalability of Clusters of Small-Scale SMP Servers", Cluster 2000, Chemnitz, Germany, November 2000.
5. LMBench: http://www.bitmover.com/lmbench/
6. The Stream benchmark: http://www.streambench.org/
7. DDR2 Advantages for Dual-Processor Servers, http://www.memforum.org/memorybasics/ddr2/DDR2_Whitepaper.pdf, August 2004
8. BLAST Home Page at NCBI. http://www.ncbi.nih.gov/BLAST/

PEZW-ID: An Algorithm for Distributed Parallel Embedded Zerotree Wavelet Encoder

Zhi-ming Chang, Yan-huang Jiang, Xue-jun Yang, and Xiang-li Qu

School of Computer,
National University of Defense Technology, ChangSha, Hunan, China
cazimi@163.com

Abstract. The image compression algorithm based on EZW can get any compression ratio as specified, which is widely used in the field of image processing. However, with massive computation during wavelet transformation and multiple scans of the transformed wavelet coefficient matrix during encoding processing, much time is consumed. Therefore, both the parallelism of wavelet transformation and zerotree encoding for EZW are necessary, and then we present PEZW-ID: an algorithm for distributed parallel embedded zerotree wavelet encoder. This paper describes the flow of the algorithm, proves the validity, and shows the performance analysis. Finally, with the experimental results under MPP, the parallelism and scalability of the algorithm have been verified.

1 Introduction

Embedded Zerotree Wavelet encoder or EZW encoder algorithm[1, 2], first proposed by Shapiro in 1993, which is based on progressive encoding to compress an image into a bit stream with increasing accuracy. This means that when more bits are added to the stream, the decoded image will contain more details; a property similar to JPEG encoded images. Additionally, EZW encoder can be used in both loss compression and lossless compression. The ratio of lossless compression with EZW can approach 3, which are close to JPEG-LS encoder algorithm[3]. Therefore, it became the milestone of static wavelet-based image compression. Afterwards, zerotree encoder is more and more recognized by many researchers, and many improved algorithms were proposed, the most popular one is set partitioning in hierarchical trees (SPIHT) [4] algorithm, given by A. Said and W. A. Pearlman in 1996. EZW has comprehensive applications in many domains, such as medicinal images database, ISDN, and remote-sensing images.

EZW algorithm, however, needs massive convolutions during wavelet transformation before encoding; furthermore, it needs scan each coefficient in the whole matrix to determine zerotrees and the scan times mainly depends on the maximum coefficient of the transformed matrix. Both of these two sub procedures make EZW encoder algorithm more time-consuming. It is surveyed that the percentage of wavelet transformation and encoding is almost above 95%. Therefore, it is necessary to parallelize EZW algorithm and we present an algorithm for Parallel Embedded Zerotree Wavelet encoder based on Image Division (PEZW-ID) to solve the problem.

The rest of the paper is organized as follows: in section 2 the flow of PEZW- ID is described in detail, and then we prove the correctness and analyze the performance for PEZW-ID. Section 3 gives the experimental results which can show the parallelism. Finally, we draw a conclusion in PEZW-ID algorithm.

2 Description of PEZW-ID Algorithm and Its Analysis

In this section, we present a parallelization approach similar to an SPMD scheme, and suppose that a management node is responsible to global operations and P computing nodes each complete their own local computation.

2.1 Flow of PEZW-ID Algorithm

From the analysis above, we can know that almost all the computation of EZW algorithm focuses on wavelet transformation and zerotree encoding. Therefore, the main idea of PEZW-ID algorithm is as follows:

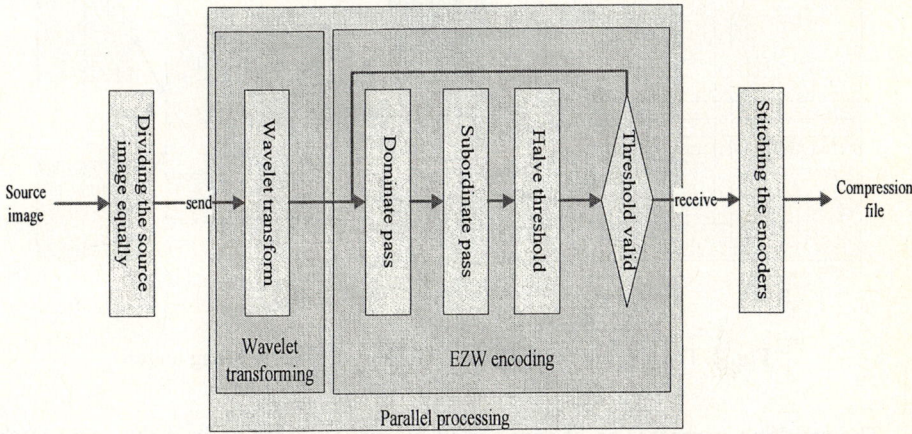

Fig. 1. The main procedures of PEZW-ID algorithm

First, Management node divides the source image into several subimages with equal sizes by the number of computing nodes P and then sends these subimages (broadcast) to all the computing nodes, each subimage should contain some redundancy data of neighbors'. Second, each computing node begins to do wavelet transformation and zerotree encoding, both of which are the same as the serial algorithm. The initial threshold can be determined after wavelet transformation, for each computing node broadcasts its local maximum of wavelet coefficients. Third, these encoders and subband information will be sent (point to point) to management node. Finally, management node would receive all the encoders and produce a compressed file, which is the same as the serial algorithm does. The whole procedure can be referred from Figure 1.

2.2 Validity of PEZW-ID Algorithm

However, the motivation of PEZW-ID algorithm is to improve processing speed and decrease processing time under the guarantee that the compressed stream is the same as the stream generated by the serial algorithm.

After each computing node does wavelet transformation for its own subimage, every transformed wavelet coefficient does has a one-by-one correspondence to the coefficient transformed by serial wavelet transformation. We use Mallet decomposing algorithm [5] to do wavelet transformation. Thus we should add some redundancy image data, which depend on the half-length of the wavelet base in order to ensure the correctness of wavelet coefficients on each computing node. PEZW-ID algorithm divides the source image equally by 4 computing nodes, as is shown in figure 2(A).

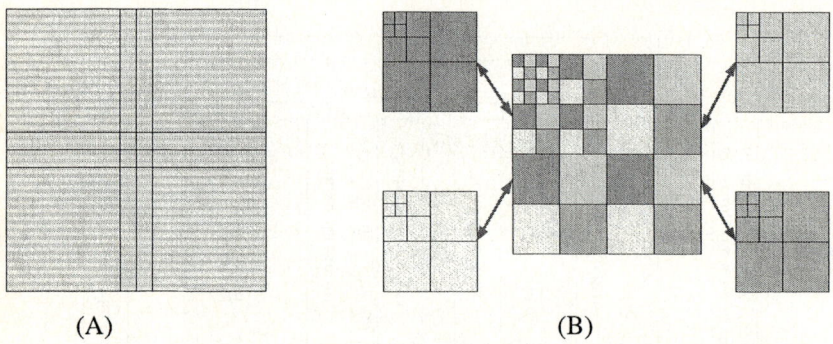

(A)   (B)

Fig. 2. The validity of PEZW-ID algorithm for 4 computing nodes

The resulting compression stream sorted by management node is the same as the one generated by serial algorithm. Because of the correctness of wavelet coefficients in all the computing nodes, the maximum of wavelet coefficients can be determined if each computing node broadcasts its local maximum to all the computing nodes, and the initial threshold will be determined as well. Furthermore, it is known that there exists a kind of parent-child relationship [1] among wavelet coefficients seen in figure 2(B). Therefore, if we add scan order and subband values to encoders, management node can restore the compression encoder of serial algorithm. This processing uses the radix sort algorithm, which can sort all the encoders by priority keywords (subband value <subimage id < scan order).

2.3 The Analysis of PEZW-ID Algorithm

In PEZW-ID algorithm, since both sending image data and encoder data are much fewer (ignore to broadcast threshold) than the massive computation, we can learn that the computation-to-communication is very ideal.

The algorithm has good data locality, since wavelet coefficients have a parent-child relationship, thus a wavelet coefficient on one computing node can't appear in other nodes. Additionally, dividing source image equally can reach the goal of load balancing of wavelet transformation while the same scan times of EZW encoder will lead to similar time consumption. In general, we believe that PEZW-ID achieves high performance because of good load balance and data locality.

3 Experimental Results

The fundamental environment of MPP for implementing PEZW-ID algorithm is as follows: it has 32 homogeneous nodes, in which CPU processes 16.6 hundred millions float operations per a second and has 1GB local memory, and the network topology is based on Fat Tree with transfer speed of 1.2Gb/s; the operating system is UNIX and the parallel communication is based on MPI message library. In the following results, we set one node as management node; the maximum of computing nodes correspondingly is 31.

Table 1 shows the results for lossless compression of Lena's images with different sizes on MPP. Since global operation will have some effects on the performance of the parallel processing, the speedup will increase slowly in contrast to the number of computing nodes. Additionally, larger image will obtain better speedup because the ratio of communication-to-computing is less, as can be seen from figure 3(A).

Table 1. Results of lossless compression of Lena's images

The number of nodes		1	2	4	8	16	31
1024^2 (ms)	Transformation	811	408	212	108	61	37
	Encoding	4573	2291	1150	582	297	153
	Total time	5935	3427	2174	1481	1261	1134
2048^2 (ms)	Transformation	3211	1612	809	407	212	114
	Encoding	17716	8861	4435	2223	1122	584
	Total time	22081	11415	6359	3673	2341	1749
4096^2 (ms)	Transformation	13384	6705	3358	1686	849	432
	Encoding	75183	37603	18812	9415	4713	2362
	Total time	89531	46204	24192	12612	6647	4230
8192^2 (ms)	Transformation	54870	27484	13806	6953	3504	1878
	Encoding	313890	156997	78520	39268	19645	9886
	Total time	371673	189770	97416	51414	27431	16414

PEZW-ID algorithm can be extended to loss compression, but the ratio of loss compression is based on scan times. To implement any ratio compression, the management node should omit some redundant data of the lossless compression encoder, thus the result is similar to the above one. Results of different compression ratio listed in Table 2 come from the processing for Lena's image with size 2048 based on different scan times; also the speedup can be seen in figure 3(B).

Table 2. Results of loss compression of Lena's images

The number of computing nodes	Compression ratio in contrast to lossless compression (ms)				
	100%	96%	70%	40%	12%
1	22081	21018	17196	12390	7885
2	11415	11061	9136	7112	4478
4	6359	6073	5061	3885	2762
8	3673	3512	3070	2566	1923
16	2341	2372	2084	1861	1490
31	1749	1730	1575	1418	1249

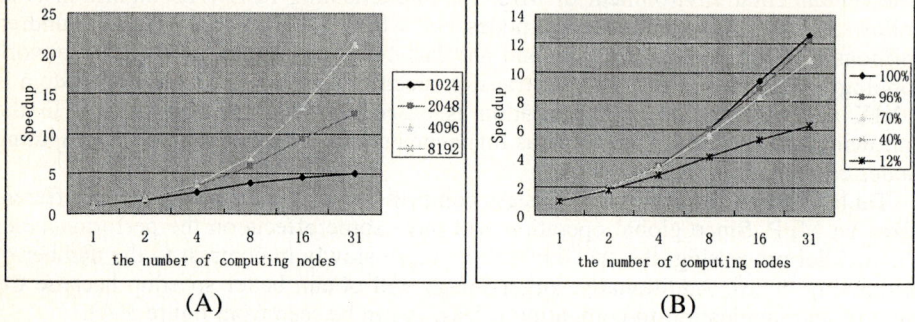

Fig. 3. Speedup of PEZW-ID algorithm

4 Conclusions

This paper targets the problem that much time will be consumed in wavelet transformation and encoding in EZW, and presents PEZW-ID: an algorithm for distributed parallel embedded zerotree wavelet encoder. This algorithm has the some advantages: first, it can reach nearly linear speedup; second, it has the property of good data locality and load balancing.

In the paper, we describe the flow of the algorithm, prove the validity, and show the performance analysis. Finally, the experimental results in MPP have verified the speedup and scalability of the algorithm.

But currently PEZW-ID can't reach any compression ratio effectively, and SPIHT isn't used to optimize the algorithm. All awaits our solutions in the near future.

References

1. Shapiro, J. M.: Embedded Image Coding Using Zerotrees of Wavelet Coefficents. IEEE Transactions on Signal Processing (1993) 3445-3462.

2. Creusere, C. D.: A New Method of Robust Image Compression Based on The Embedded Zerotree Wavelet Algorithm. IEEE Transactions on Image Processing (1997) 1436-1442.
3. Paul G. Howard, Jeffrey Scott Vitter.: Fast and Efficient Lossless Image Compression. IEEE Computer Society/NASA/CESDIS Data Compression Conference (1993) 351-360.
4. Amir Said, William A. Pearlman.: A New, Fast, and Efficient Image Codec Based on Set Parition in Hierarchical Trees. IEEE transactions on circuits and systems for video technology. (1996).
5. Stephane G. Mallat: A Theory for Multiresolution Signal Decomposition: The Wavelet Representation. IEEE Transactions on Pattern Analysis and Machine Intelligence (1989) 674-693.

Enhanced-Star: A New Topology Based on the Star Graph

Hamid Reza Tajozzakerin[1] and Hamid Sarbazi-Azad[1,2]

[1] School of Computer Science, IPM, Tehran, Iran
[2] Computer Engineering Dept., Sharif Univ. of Technology, Tehran, Iran
azad@sharif.edu, {tajozzakerin, azad}@ipm.ir

Abstract. The star graph, though an attractive alternative to the hypercube, has a poor network bandwidth due to a lower number of channels compared to that in an equivalent hypercube. In order to alleviate this drawback, this paper presents a generalization of the star graph topology with a richer connectivity, called the *enhanced-star*. We also examine some topological properties of the proposed network. Some useful functions such as multi-node broadcasting, scattering, total exchange, and group communication, in the enhanced-star are also addressed. We show these operations can be completed faster in the enhanced-star (compared to the star).

1 Introduction

The interconnection network is a crucially important part of any parallel processing or distributed system. A large variety of topologies have been proposed and analyzed. Design features for an efficient interconnection topology include properties such as low degree, regularity, small diameter, high fault tolerance, high connectivity, and the existence of simple and efficient routing algorithms.

One of the most efficient interconnection networks is the well-known hypercube; it has been employed in various commercial multiprocessor machines. Another family of regular graphs called the star graphs has also received much attention as an attractive alternative to the hypercube [1], [6]. The star graph belongs to the class of Cayley graphs [2], is symmetric and strongly hierarchical, and has lower diameter and node degree compared to those of an equivalent hypercube. The star graph of dimension n (or the n-star), denoted as S_n, is regular of degree $n-1$, has $N = n!$ nodes and a diameter of $\lfloor 3(n-1)/2 \rfloor$.

The S_n has $n!(n-1)$ links which is smaller than that in a comparable hypercube. This causes more traffic on the network channels that may dramatically reduce the overall performance of the network. The enhanced-star reduces this effect while preserving many properties of S_n, such as node symmetry and simple minimal routing. It has some better properties discussed in the next sections.

The rest of the paper is organized as follows. Section 2 defines the enhanced-star and introduces a minimal routing algorithm for it. In Section 3, we derive some topological properties of the enhanced-star. In Section 4, some time-optimal basic communication functions are introduced in the enhanced-star. In Section 5, an efficient group communication function in the enhanced-star is proposed. In section 6, compare our topology with other topologies developed base on star previously. Finally, Section 7 concludes this study.

2 The Enhanced-Star: Definition and Properties

Definition 1. Generator gf_i is defined as $gf_i(p_1p_2\cdots p_{i-1}p_ip_{i+1}\cdots p_n)=(p_ip_2\cdots p_{i-1}p_1p_{i+1}\cdots p_n)$ and generator gl_i is defined as $gl_i(p_1p_2\cdots p_{i-1}p_ip_{i+1}\cdots p_n)=(p_1\cdots p_{i-1}p_np_{i+1}\cdots p_{n-1}p_i)$ where $p_1p_2\cdots p_n$ is a permutation of $123\ldots n$ and $1 < i < n$.

Definition 2. The enhanced-star, denoted as ES_n, consists of $n!$ nodes labeled with $n!$ different permutation of n distinct symbols $1, 2, \cdots, n$. Two nodes u and v are connected, in an ES_n, if and only if there exists a generator gf_i or gl_i such that $gl_i(u) = v$ or $gf_i(u) = v$. Note that $gl_i(u) = v$ if $gl_i(v) = u$, and $gf_i(u) = v$ if $gf_i(v) = u$. For example, $gf_3(1234)=3214$ and $gl_3(1234)=1243$.

Alternatively, an ES_n includes $n!$ nodes, each addressed by a permutation of n symbols $1, 2, \cdots, n$. A node $A = (a_1a_2\cdots a_{i-1}a_ia_{i+1}\cdots a_n)$ is connected to node $A' = (a_ia_2\cdots a_{i-1}a_1a_{i+1}\cdots a_n)$ and node $A'' = (a_1a_2\cdots a_{i-1}a_na_{i+1}\cdots a_i)$ in dimension i. Figure 1 shows the 3-dimensional and 4-dimensional enhanced-stars, ES_3 and ES_4.

2.1 Unicast Routing

We need two simple rules to find the path between two nodes. Due to node symmetry property, we may, without lose of generality, show how to route from any node to the identity node $I_k = (123\cdots n)$.

RULE 1. If symbol 1 is the first symbol and n is the last symbol, move one of them to any position not occupied by a correct symbol.
RULE 2. If x is the first symbol (where $x \in \{2, 3, \cdots, n\}$) or x is the last symbol (where $x \in \{1, 2, 3, \cdots, n-1\}$), then move it to its correct position.

Like the algorithm for the star, the above rules ensure a minimal path between any two nodes. As mentioned in [1], any permutation can be viewed as a set of cycles. For example, node 3541762 consists of the cycles $(134), (257)$ and (6). Any symbol already in its correct position (e.g. cycle 6), appears as a 1-cycle. We may consider the above cycle representation for the enhanced-star. Now for a given permutation π,

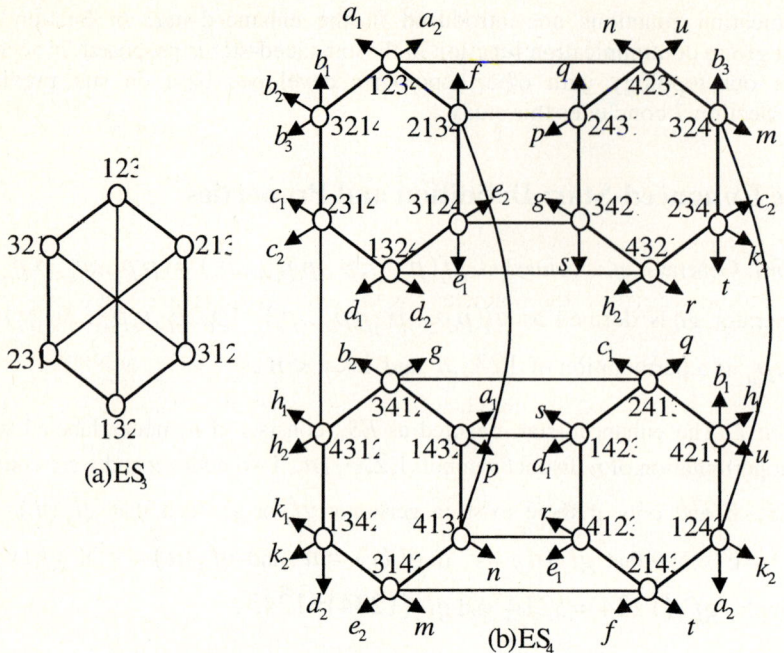

Fig. 1. Some enhanced-stars, (a) 3-D enhanced-star, and (b) 4-D enhanced-star

let c_1, the cycle of length of at least 2, include symbol 1 but not symbol n. Similarly, assume that c_n, the cycle of length of at least 2, includes symbol n but not symbol 1, and $c_{1,n}$ is the cycle including both symbols 1 and n. Let c be the number of cycles including other symbols (all symbols excluding 1 and n) and m be the total number of symbols in these cycles (the number of symbols not in their correct positions). The minimum distance $d(\pi)$ from the identity node I_k can be derived using Theorem 1.

Theorem 1. *The distance $d(\pi)$ from any node π to the identity node I_k, in ES_n, is given by*

$$d(\pi) = c + c_1 + c_n + c_{1,n} + m - \begin{cases} 0, & \pi(1) = 1 \text{ and } \pi(n) = n \\ 4, & c_1 \neq 0 \text{ and } c_n \neq 0 \\ 2, & \text{otherwise} \end{cases} \quad (1)$$

Proof. Assume the two rules mentioned above. In the ES_n, we can swap the last symbol and the first symbol, if required. Thus, it is not necessary to move a symbol, wrongly occupying the last position, to the first position and then to its correct position. That is, we can move such a symbol directly to its correct position, reducing two additional steps when we have disjoint cycles including 1 and n (one cycle includes 1 and another cycle includes n).

Suppose that $c_n \neq 0$. It means that a symbol (not symbol n) has occupied the last position. It can move to its correct position at one step (unlike the star which requires 3 steps). For example, if $\pi = 21453$, the cycle representation is $(12)(345)$. So we have, $c_1=1$, $c_n=1$, $c_{1,n}=0$, $c=0$ and $m=5$ resulting in a distance of $d(\pi)=3$. Note that the distance between node $\pi = 21453$ and the identity node in the star is 5.

2.2 Diameter and Average Distance

In this section, we derive the diameter for the enhanced-star and the average inter-node distance.

Assume n is even and the source node is $A = n \; \overbrace{32}^{pair\,1} \; \overbrace{54}^{pair\,2} \cdots \overbrace{[n-1][n-2]}^{pair\,n/2-1} 1$. It is obvious that the distance between node A and the identity node is the maximum inter-node distance (or diameter) in the network. To move from A to I_k, we require one swap to correct the first and last positions and 3 swaps for every remaining symbol pair. Thus, the distance to the identity node (network diameter) is

$$1 + 3(n-2)/2 = \left\lfloor \frac{3(n-1)}{2} \right\rfloor \quad (2)$$

When n is odd, the node at the maximum distance from the identity node may have an address in the form $A = n \; 1 \; \overbrace{43}^{pair\,1} \; \overbrace{65}^{pair\,2} \cdots \overbrace{[n-1][n-2]}^{pair\,n/2-1} 2$. We require 2 swaps to correct positions of symbols 1, 2, and n, and 3 swaps for every remaining symbol pair. Therefore, the maximum distance can be given by

$$2 + 3(n-3)/2 = \frac{3n-5}{2} \quad (3)$$

The average inter-node distance for any symmetric network is determined by the summation of the distance between given node and all other nodes in the network divided by the total number of destination nodes (network size minus 1). Since the enhanced-star is node symmetric, the average inter-node distance equals the average of distances between the identity node I_k and all other nodes in the network. In [2] the average inter-node distance of the n-dimensional star is given to be

$$n + H_n + \frac{2}{n} - 4 \quad (4)$$

where H_n is the n-th harmonic number given by $H_n = \sum_{i=1}^{n} \frac{1}{i}$.

Lets assume that average inter-node distance in an n-dimensional enhanced-star is given by

$$d = n + H_n + \frac{2}{n} - 4 - x \tag{5}$$

where x accounts for the reduction achieved in the enhanced-star compared to its equivalent star. In what follows we try to calculate x. To this end, two different cases must be considered.

a) when 1 is the first symbol and n is not the last, the distance is reduced by 2 (compared to the equivalent star). The number of address patterns with such property is $1 \times (n-2) \times (n-2)!$. Thus, the distance will be decreased by $2(n-2)(n-2)!$ hops in the total summation of inter-node distances. Hence, an overall reduction of

$$x_1 = \frac{2(n-2)(n-2)!}{n!-1} \approx \frac{2(n-2)}{n(n-1)} \tag{6}$$

may be achieved for the average inter-node distance.

b) when there are two disjoint cycles, in the node address pattern, one including symbol 1 and the other including symbol n (as discussed above), some reduction may be achieved for the average distance inter-node distance. Since one symbol in each cycle is known (1 and n), the remaining n-2 symbols may be included in any of these two cycles. The first cycle may take i, $0 \leq i \leq n-2$, remaining symbols leaving n-2-i symbols for the other cycle. For each case (corresponding to some i), there are $(i+1)!$ permutations in cycle 1 (for symbol 1 and other i symbols) and $(n-i-1)!$ permutations in cycle 2 (for symbol n and other n-i-2 symbols). Therefore, the total number of node address patterns with two cycles (the first cycle with symbol 1 and the second cycle with symbol n) can be given as

$$\sum_{i=0}^{n-2} \left[\binom{n-2}{i} \times (i+1)! \times (n-i-1)! \right] \tag{7}$$

For some of these patterns we might save 0, 1, or 2 swaps (compared to the star). Thus, an upper bound of

$$x_2 = \frac{2 \sum_{i=0}^{n-2} \left[\binom{n-2}{i} \times (i+1)! \times (n-i-1)! \right]}{n!-1} \tag{8}$$

hops is reduced from the average inter-node distance in the enhanced-star (compared to the star).

Therefore, the reduction x, for the average inter-node distance in the n-dimensional enhanced-star, can be given by $x = x_1 + x_2$. To have a fair comparison between the star and the enhanced-star, let us consider the upper bound of the average inter-node diameter in the enhanced-star. To this end, we must use the lower bound for x, i.e. $x = x_1$.

Table 1 shows the average inter-node distance in the enhanced-star (an upper bound calculated above) and the star as a function of network dimensionality, n. As can be seen in this figure, even when considering upper bounds, the enhanced-star has a lower average inter-node distance compared to the star.

Table 1. Average inter-node distance in the enhanced-star and star

	$N=4$	$N=5$	$N=6$	$N=7$	$N=8$	$N=9$	$N=10$
ES	2.3	3.4	4.55	5.67	6.75	7.86	8.9
Star	2.58	3.68	4.79	5.88	6.97	8.05	9.13

3 Group Communication

In this section, we obtain lower bounds on multi-node communication operations including broadcast, scatter, and total exchange [3].

Multi-node broadcasting is the problem where each node wishes to send a message to all other nodes. In [3], lower bounds are determined under two assumptions: *SLA* (single-link availability) and *MLA* (multiple-link availability). Let us consider these assumptions for the enhanced-star. With the first assumption, the results are similar to those in the star as it is independent of the node degree. Therefore, the lower bound would be $n!-1$ steps since there are $n!$ links available at each time step. Under the second assumption, all available $2n!(n-1)$ links can be used to transmit messages at each time step. Thus, similar to the approach taken in [3] for the star, the minimum time required to complete a broadcast is given by $\frac{n!-1}{2(n-1)}$ which is half of that for the star.

Single node scattering is the problem where a specific node wishes to send different messages to any other node. Again, under *SLA* assumption, the minimum time required is $n!-1$ steps (like that in the star), and under *MLA* assumption the minimum time required to complete a single node scatter can given by $\frac{n!-1}{2(n-1)}$ which is again half of that for the star.

Total exchange is the problem where each node wishes to send a distinct message to every other node, i.e. it is equivalent to $n!$ different single node scattering. The minimum required number of message transmissions is $n! \times n! \times d$. With *SLA* condition, because of improvement in the average inter-node distance, the minimum time required is smaller than that in the star. As discussed previously, $n!$ links are available resulting in a minimum time of $n! \times d$ steps. Under *MLA* assumption, all available $2n!(n-1)$ links may be used resulting in a minimum time of $\frac{n! \times d}{2(n-1)}$ steps to complete a total exchange. Table 2 shows the total exchange time in the enhanced-star and the star as a function of network dimensionality, with both *SLA* and *MLA* assumptions.

Some special group communication function. Let us now assume some special type of group communication operation. Let $X_{n-1}(i)$ denote all address patterns of length n with i being the last symbol in the n-dimensional enhanced-star. Given $X_{n-1}(i)$ and $X_{n-1}(j)$, with $i \neq j$, it is required to exchange the contents of the nodes in $X_{n-1}(i)$ with those of the nodes in $X_{n-1}(j)$. Such a group communication function can be effectively done in one communication step in the enhanced-star as shown in Procedure *Copy* below.

```
Procedure COPY (i, j)
    For all vertices *i and *j do in parallel
        Exchange contents along dimension r, where r is
        the position at which symbol j appears in *i or
        the position at which symbol i appears in *j ;
    End For;
End Copy;
```

Note that in Step 2, every node in sub-network $X_{n-1}(i)$ (or $X_{n-1}(j)$) exchange its content along dimension r with its neighbor in sub-network $X_{n-1}(j)$ (or $X_{n-1}(i)$).

Table 2. The total exchange time in the enhanced-star and star with *SLA* and *MLA* assumptions

	N=4	N=5	N=6	N=7	N=8	N=9	N=10
ES under MLA	6.74	40.41	270.53	2031.74	17050.6	158787.43	1628618.8
Star under MLA	20.67	110.5	688.8	4938	40134.86	365201.9	3680800
ES under SLA	40.43	323.25	2705.3	24380.83	238708.4	2540598.9	29315138.06
Star under SLA	62	442	3444	29428	280944	2921615.96	33127199.94

4 Comparison to Other Star-Based Networks

In [7], incomplete star (IS) was introduced to achieve more scalability by filling the wide gap between two consecutive sizes of the star graphs. However, it has some drawbacks such as complex labeling and routing schemes. The IS network has the same diameter as the star although some links and nodes have been removed that can cause more delay and traffic on links and may reduce network performance. Another flavor of the star graph is the star-connected cycles (SCC) as an alternative to the CCC (cube-connected cycles). Compared to the star graph, the SCC has a longer diameter but a desirable fixed node degree. The third network from the family of the star graph is the Macro-star (MS) [9]. It has many desirable properties but longer diameter compared to the star [9].

Figure 2 shows the diameter and the cost of the network for different star-based graph topologies. As can be seen in this figure the diameter of the ES is smaller that that in any other compared network. The cost of the network is, however, worse than others for large network sizes. Note that for practical network sizes, it is comparable to other star-based networks.

The main advantage of the ES is more visible when networks are operating in real conditions. Note that the arrival traffic rate on network channels and average internode distance determine the message latency and overall network performance. Thus, the ES with more channels, and hence, with a lower arrival traffic rate on its network channels, and a lower diameter must look good among other star-based network topologies.

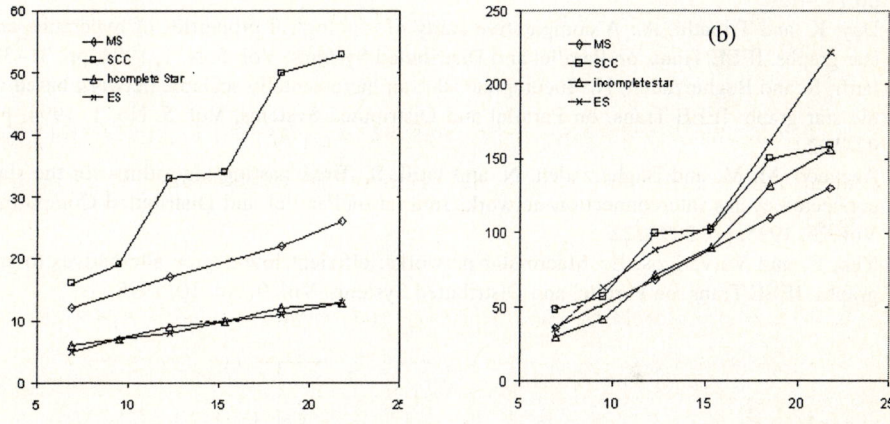

Fig. 2. Comparison of different star-based topologies; (a) Diameter of networks vs. log(number of nodes); (b) Cost of network vs. log(number of nodes)

5 Conclusions

The star graph is an attractive alternative to the hypercube with many desirable properties. The lower number of channels, however, compared to the hypercube may be a drawback when considering network traffic. In this paper, we proposed a new topology, called the enhanced-stars, based on the star with a richer connectivity while preserving most desirable properties of the star. We studied some topological properties and routing functions in the enhanced-star.

Our next objective is to realize accurate and detailed simulation of these networks and comparing their performance empirically under different working conditions. Developing some important algorithms, e.g. FFT, on the proposed network can also be a challenging research line for future work.

References

1. Akers, S.B. and Harel, D. and Krishnamurthy, B. The star graph: an attractive alternative to the hypercube. Proceedings of the International Conference on Parallel Processing, St. Charles, IL, 1987, pp. 393–400.
2. Akers, S.B. and Krishnamurthy, B.: A group-theoretic model for symmetric interconnection networks. IEEE Transaction on Computers, Vol. 38, No. 4, 1989, pp. 555–566.
3. Fragopoulou, P. and Akl, S.G. Optimal communication algorithm on star graphs using spanning tree constructions. Journal of Parallel and Distributed Computing, Vol. 24, 1995, pp. 55–71.
4. Akl, S.G. and Qiu, K.: A novel routing scheme on the star and pancake networks and its applications. Parallel Computing, Vol. 19, No. 1, 1993, pp.95–101.
5. Latifi, S.: On fault diameter of star graphs. Information Processing Letters, Vol. 46, 1993, pp.143–150.
6. Day, K. and Tripathi, A.: A comparative study of topological properties of hypercube and star graphs. IEEE Trans. on Parallel and Distributed Systems, Vol. 5, N. 1, 1994, pp. 31–38.
7. latifi, S. and Bagherzadeh, N.: Incomplete star: an incrementally scalable network based on the star graph. IEEE Trans. on Parallel and Distributed Systems, Vol. 5, No. 1, 1994, pp. 97–102.
8. Azevedo, M. M. and Bagherzadeh, N. and latifi, S.: Broadcasting algorithms for the star-connected cycles interconnection network. Journal of Parallel and Distributed Computing, Vol. 25, 1995, pp. 209–222.
9. Yeh, C. and Varvarigos, E.: Macro-star networks: efficient low-degree alternatives to star graphs. IEEE Trans. on Parallel and Distributed Systems, Vol. 9, No. 10, 1998.

An RFID-Based Distributed Control System for Mass Customization Manufacturing

Michael R. Liu[1], Q. L. Zhang[2], Lionel M. Ni[1], and Mitchell M. Tseng[2]

[1] Department of Computer Science, Hong Kong University of Science and Technology,
Clearwater Bay, Kowloon, Hong Kong,
{lrcomp, ni}@cs.ust.hk

[2] Department of Industrial Engineering & Engineering Management, Hong Kong University of Science and Technology, Clearwater Bay, Kowloon, Hong Kong,
{qlzhang,tseng}@ust.hk

Abstract. Mass customization production means to produce customized products to meet individual customer's need with the efficiency of mass production. It introduces challenges such as drastic increase of varieties, very small batch size and random arrival of orders, thus brings to the manufacturing control system requirements of flexibility and responsiveness which make mass customization production a fertile ground for intelligent agents. With RFID (Radio Frequency Identification) integration, an applicable agent-based heterogeneous coordination mechanism is developed to fulfill the requirements. In this paper, we propose a distributed system framework including a number of intelligent agents to collaborate in a virtual market-like environment. The proposed price mechanism in our system has the advantage of improving the production efficiency and total profit that the manufacturer will receive from a certain amount of jobs by utilizing a mechanism of both resource competition and job competition. Based on the simulation results, we compare the total profit, average delay and average waiting time of our agent-based price mechanism with those based on the widely exploit FIFO (First In First Out) and EDD (Earliest Due Date) scheduling mechanisms to show that when resources are with large queue length, the agent-based price mechanism significantly outperforms the other two.

1 Introduction

The increasing diversity of the customer requirements and the attraction of the mass production efficiency shift the major manufacturing mode from mass production to mass customization. Unlike mass production in which finished products need to be stocked in inventory and wait to serve customer's demands, mass customization considers fulfilling individual customer needs while maintaining near mass production efficiency [12]. Unique information is provided by each customer so that the product can be tailored to his or her requirements [15]. This mode of manufacturing requires the production system to be very flexible and its control system adaptive to the rapid changing customer demands.

While centralized planning and control systems are no longer suited to handle the increasing system complexity in the high variety and rapid changing manufacturing environments [2], distributed collaborative control and scheduling approaches have been proposed. Early works, appeared from 1970s, started to introduce the auction based distributed control mechanisms in the manufacturing applications [7] [9] [10] [11]. Recently, multi-agent systems (MASs) for resolving manufacturing control problems have drawn wide interest in many literatures [3] [4] [12] [14]. The MAS incorporates various types of market-like protocols to facilitate negotiations among agents with different functionalities. It provides more flexibility and quicker reactions to the control systems in dynamic changing environment such as mass customization manufacturing environment.

Although a lot of research has been carried on in the area of collaborative control and agent-based auction systems, very few have considered the implementation of such systems in a real time basis. With the emerging of real time information technologies such as RFID technology, this application opportunity has been enabled. Object (part, component, sub-assembly, etc.) in the manufacturing environment is a attached with an RFID tag which can carry the information such as its identification, attribute values and production status. The data can be read by an RFID reader and be forwarded to a subsystem such as PC, robot control system which will use the received data to decide the correct operation to be performed at the position to the object, without human intervention. The RFID reader can not only read and forward data, but also write to the tag which makes possible the documentation of any state changes of the object and therefore keeping track of the system's status and predicting the future. Enabled by RFID technology, the control system could become more dynamic and flexible in tackling instant changes in the manufacturing systems.

In this paper, we bring further the idea of using collaborative intelligent agents to a more agent-suited stage with the help of RFID technology. An agent-based job shop scheduling system is proposed. The agents will access, manage, and utilize the information carried with RFID tags, and intelligently anticipate, adapt and actively seek ways to manage the manufacturing procedure. A two-stage control mechanism is proposed in this system. In the first stage, job agents make decisions about the job routings based on the job status (carried in RFID tags) and a set of bidding processes with resource agents. In the second stage, the resource agents manage the queues and make dispatching decisions in a collaborative manner based on job *priority* and *emergency* factors.

This paper is organized as follows. Section 2 gives a brief introduction to RFID technology and explains how it can be utilized in a decentralized control system. Section 3 explains in detail our agent-based collaborative control mechanism. Section 4 shows the simulation results and performance evaluation. Section 5 concludes the paper and presents future work.

2 RFID Technology

RFID (Radio Frequency Identification) is a means of storing and retrieving data through magnetic or electromagnetic field. An RFID system is made up of two components: RFID tag and RFID reader.

An RFID tag is a data-carrying device and normally consists of a coupling element and an electronic microchip. A tag is categorized as either passive or active. A passive tag does not possess its own voltage supply (battery). It absorbs power from the RF field of the reader and reflects RF signal to the reader after adding information by modulating the received RF signal. An active tag possesses its own power supply. Thus it can maintain data in RAM, a temporary working memory for microprocessor. Active tags usually have a bigger read range than passive tags and are suited to more applications. However, active tags have limited operational lifetime due to power constraint and are more expensive.

An RFID reader can read and write data received from RFID tags. It operates on a defined radio frequency according to a certain protocol. A reader typically contains a high frequency module (transmitter and receiver), a control unit, and a coupling element to the transponder. In addition, many readers are fitted with an additional interface (e.g., RS 232 and RS 485) to interconnect with another system such as PC and robot control system.

As shown in Figure 1, the power required to activate the tag is supplied to the tag through the coupling unit (contactless) as is the timing pulse and data [5].

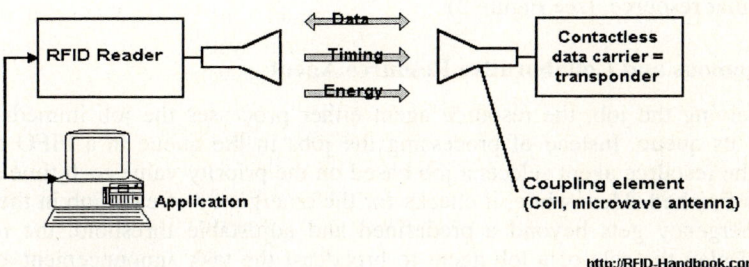

Fig. 1. The reader and the tag are the main components of every RFID system

The characteristic of being contactless, which is achieved by using magnetic or electromagnetic fields for data exchange and power supply instead of galvanic contacts, gives RFID a broad range of applications from secure internet payment systems to industrial automation and access control.

3 Agent-Base Collaborative Control System

In this section, we will present detailed introduction to the proposed collaborative control system for the mass customization manufacturing. We consider a job shop environment with m flexible machines. Incoming customer orders will come in a random manner with detailed customer requirements such as *product type* and *due date*. Different *product type* requires unique set of operations and setups and therefore can create different routings in the system. Each flexible machine can perform a set of operations and its working efficiency defers from other machines. The ultimate goal of the control system is to maximize the total profit which includes the basic job price and the penalty cost if any delay occurs.

3.1 Distributed Control Mechanism

The control system is designed to coordinate a set of autonomous workstation that are individually controlled by a local computer running agent programs and connected by a local area network (see Figure 2). Similar to the mechanism proposed by [12], the system schedule and control are integrated as an auction-based bidding process with a price mechanism that rewards product similarity and responds to customer's needs. However, instead of making decision according to its own interests and market price fact as [12], the mechanism proposed in this paper will coordinate the autonomous workstation to act toward the optimal goal of the whole system.

A. Auction Based Bidding Process

The job agents broadcast the task announcements with their payments to resource agents. Each resource agent collects the task announcements, calculates the profit rate with which tasks are ranked, and constructs the bids which contains the information of the resource cost. Each resource agent then sends its bid to the requesting job agent. The job agent then collects all the bids and selects the resource with minimal cost. The job agent then informs the select resource agent and assigns the operation to the particular resource. (see Figure 3)

B. Autonomous and Collaborative Resource Agent

After receiving the job, the resource agent either processes the job immediately or puts it in its queue. Instead of processing the jobs in the queue in a FIFO or EDD manner, the resource agent selects a job based on the priority value each time a previous job is finished. Meanwhile, it checks for the emergency of each job in the queue. If the emergency gets beyond a predefined and adjustable threshold, the resource agent will play the role of a job agent to broadcast the task announcement and start another round of bidding and select one resource agent to handle the job and assign the operation to the selected resource agent. Notice that the selected resource agent could be the originating resource agent, because the originating resource agent will also take part in the bidding as shown in Figure 4.

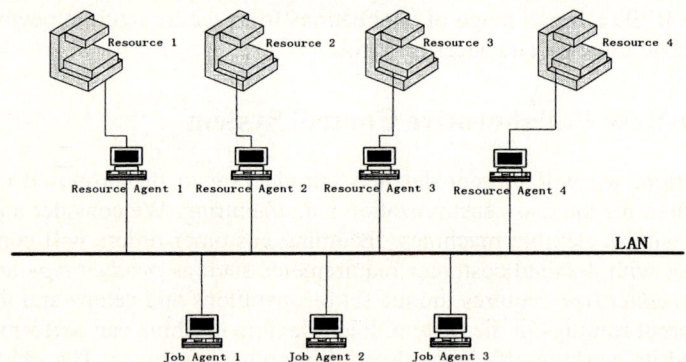

Fig. 2. The layout of the computer controlled autonomous workstations

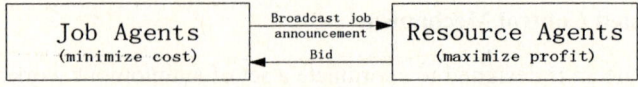

Fig. 3. The market-oriented price mechanism

C. Price Mechanism

The bidding process and the agent activities presented above are the steps that should be performed one after another in the negotiation procedure. The price mechanism actually defines the negotiation principles and the information transmitted between job agent and resource agent. We modify the market price mechanism proposed in [12], which is based on market-oriented programming for distributed computation, in order to make the autonomous agents cooperate towards the optimization of the whole system (see Figure 5).

3.2 Parameter Definition:

- c_i: Operation i completion time
- ct: Current time
- d_i: Due date of operation i which is calculated from the due date of the job. $d_i = ct + \sum(pt_i/(\sum pt_j))$
- Opportunity_Cost: Represents the cost of losing the particular slack for other job agents due to the assignment of resource for one job agent's operation i
- pt_i: Standard processing time of operation i
- st_i: Standard setup time of operation i
- PFAIndex: Represents the product family consideration in setting up the manufacturing system. We set PFAIndex = 0, if the two consecutive jobs, i and j, are in the same product family. Thus the setup cost is eliminated for operation j.
- qt_i: The arrival time that entity i enters the latest queue
- s: setup cost

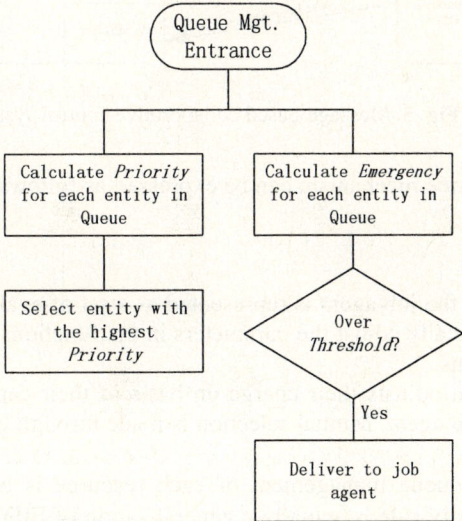

Fig. 4. Autonomous queue management flowchart

3.3 Functions

1. Job_Price = Basic_Price + Penalty_Price
2. Penalty_Price = $penalty \times e^{c_i - ct}$

$$\text{Priority} = \begin{cases} (\text{Basic\_Price} + penalty \times e^{d_i - ct} - \text{PFAIndex} \times s)/pt_i, \\ \quad \text{if } ct > d_i \\ k \times (\text{Basic\_Price} + penalty \times e^{d_i - ct - offset} - \text{PFAIndex} \times s)/pt_i, \\ \quad \text{if } ct < d_i \end{cases}$$

3. Emergency = Penalty/Basic_Price = $penalty \times e^{qt_i - ct}$ /Basic_Price
4. Emergency Threshold = 0.5 (*Justifiable*)
5. Resource_Cost = Opportunity_Cost + Penalty_Cost
6. Penalty_Cost = $penalty \times e^{-(d_i - ct - \sum pt_j - pt_i - st_i)}$ ($\sum pt_j$ stands for total processing time of entities in queue)

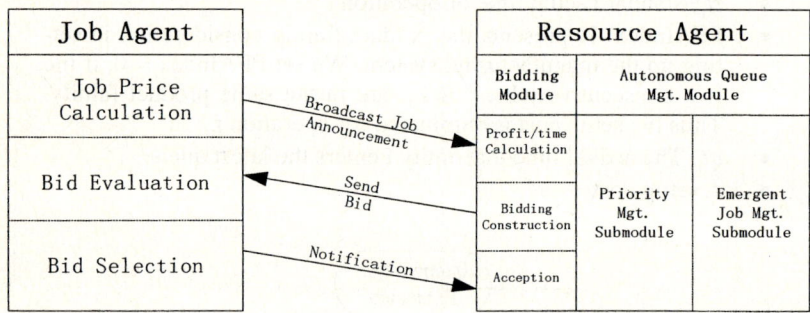

Fig. 5. Message-based collaborative control system

Specifically, the price mechanism can be expressed as follows:

3.4 Explanation

In the market model, the job agent is represented as a set of evaluation functions. The job agent can dynamically adjust the parameters in the functions based on the internal and external conditions.

The resource agent adjusts their charge on basis of their capability, queue length and the requesting job agent. Mutual selection is made through bi-directional communication.

The autonomous queue management of each resource is based on priority and emergency. The priority rule is actually a generalization of FIFO and EDD. If we restrict that (1) all the entities in the queue have the same due date and (2) different resources have the same processing time for all types of entities, the priority rule becomes EDD. If we add to the previous two restrictions that the incoming entities are of the same type, the priority rule becomes FIFO. Besides, it takes into consideration the effect of product family architecture, which can be utilized to reduce the complexity of the mass customization manufacturing. We distinguish the calculation of priority on two different conditions by giving the second function an adjustable weight k and an adjustable offset in order to balance the manufacturing polarization. Coefficient k and offset can be adjusted according to the simulation result. Figure 6 shows the curve of the priority function. Offset stands for the parallel movement of the right half curve to the left of d_i. The half curves on both sides of d_i are necessary. If there only exists the left half curve, all resources will follow the trend of trying to pick up the jobs whose due date are very close to the current time, the closer the due date of the job is to the current time, the higher possibility that the job will be picked up, leaving the overdue ones not concerned. Similarly, if there only exists the right half curve, all resources will be busy with the overdue ones, thus the new jobs in the queue tend to be ignored until they are overdue. Therefore, we should neither merely encourage the resource to try to finish the jobs as early as possible, nor should we merely try not to delay the finishing of jobs, what we want is the balance, the aggregation of the finishing time of each job to the point d_i.

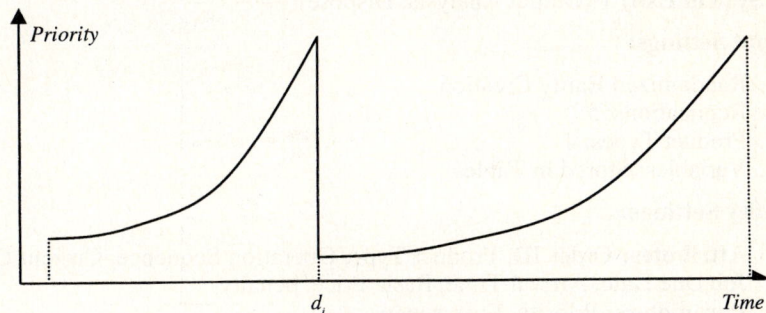

Fig. 6. Priority Curve

The emergency rule is used to settle the potential problem that one single functional resource has many entities in its queue in which some entities with low priority can also be processed by other resources, which could be idle at the busy time of the owner resource.

The decision rules of job agents are consistent with those of resource agents by the function of Resource_Cost and the calculation of the final profit of manufacturing a certain amount of products of different types. In other words, the goal of job agents is to minimize the cost and the goal of resource agents is to maximize the profit that can be achieved with no confliction.

The resource agent will dynamically raise its charge (Resource_Cost) when it has a large queue length, so that it tends to diminish the demands and drive jobs to other resources and lower its charge when the queue length is small in order to increase the demand and attract more jobs. Thus, the global optimization is achieved with the distributed collaboration.

4 Simulation Results and Performance Analysis

"Computer simulation refers to a collection of methods for studying models of real-world systems by numerical evaluation using software designed to imitate the system's operations or characteristics." "It is a process of designing and creating a computerized model of a real or proposed system for the purpose of conducting numerical experiments to give us a better understanding of the behavior of that system for a given set of conditions" [13].

The software we used for simulation is Arena 5.0 and we model the collaborative control system as the message-based system as shown in Figure 5.

4.1 Simulation Setup

System Configuration:

1. **System Entrance:** 1 (Create Input Entities; Initialize Tag for Each Entity; Job Agent)
2. **Workstation:** 3 (Queue up Incoming Entities; Modify Entity Priority and Select Item; Check for Emergency; Resource Agent (Manage It's Own Queue); Job Agent (Auction for the Next Operation))
3. **System Exit:** 1 (Output Analysis; Dispose)

Input Settings:

1. Randomized Entity Creation
2. Replications: 5
3. Product Types: 4
4. Variables: Stored in Tables

Entity Settings:

1. **Attributes:** Order ID, Product Type, Operation Sequence, Current Operation, Job Due Date, Arrival Time, Basic Price, penalty
2. **Parameters:** Priority, Emergency

Process Settings:

1. **Attributes:** Processing Time, Charge (Cost)

Workstation Setting:

1. **Attributes:** ID, Capability
2. **Parameters:** Queue (*Change with Time*)

In a physical system with RFID, the entity attributes will be carried in tags and encoded in EPC. Figure 7 shows a picture description of the EPC. The attributes of processes and workstations will be stored in the corresponding PCs.

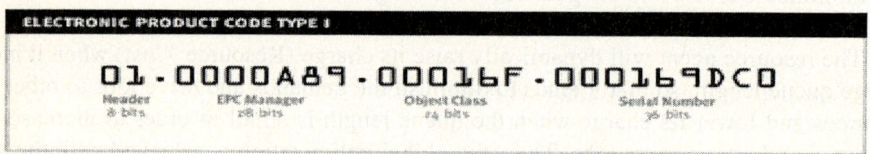

Fig. 7. A picture description of the EPC

4.2 Simulation Results and Analysis

The following performance metrics are used in our system. The total profit is the ultimate goal of the manufacturers. The average delay of different products reflects the manufacturing efficiency. The average waiting time in queue of all finished products reflects the utilization of different resources. We made comparison on the metrics with the commonly used manufacturing system for the present in below output performance measures.

Figure 8 shows the comparison of total profit between three systems with different queue manage mechanism, namely EDD, FIFO and Agent-based price mechanism. We made five replications in order to counteract the input randomicity. As shown in Figure 8, in the first two replications, the three systems perform similarly and our agent-based collaborative control does not significantly outperform the other two. That is because at the beginning stage, the queues of resources have not grown to the extent of releasing the power of agent-based collaborative control. As the simulation goes, the queue length is increasing and the advantage of our agent-based control system is becoming obvious.

Figure 9 shows the average delay of the completion time in terms of different products. Agent-based control system has the smallest delay, that is, the best performance in all four types of products. Lateness leads to penalty, while earliness does not bring any profit. A small delay means a great manufacturing efficiency.

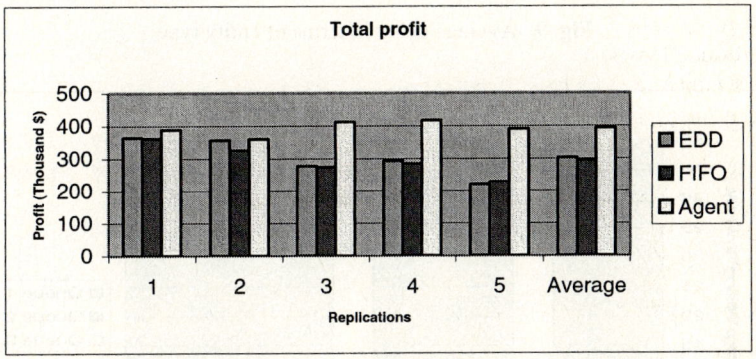

Fig. 8. Total profit of each replication and the average

Figure 10 shows the average waiting time in the queue under different queue management mechanism. The greatly imbalanced distribution of products to resources incurs a huge waiting time in some queues and little in the others. The distribution with intelligent collaborative decision will greatly balance the strong bias, thus, increasing the utilization of resources.

5 Conclusion and Future Work

The dynamic changes of global markets pushed the advent of strategy of mass customization. The fallible monitoring of the production processes and the lack of flexibility and responsiveness of centralized control makes it no longer suited for the increasing complexity and speed in the manufacturing industry. "While RFID makes possible the interlink of object and data and creates an information flow between the individual processing workstations, the pressure can be taken off the centralize control system." [5]. In this paper, we have presented an agent-based decentralized collaborative control system with RFID integration. The market-like model and the auction-based bidding process used in our system mimic the resource competing and decision making procedure as in a real world market environment, hence, cope with dynamically changing demands.

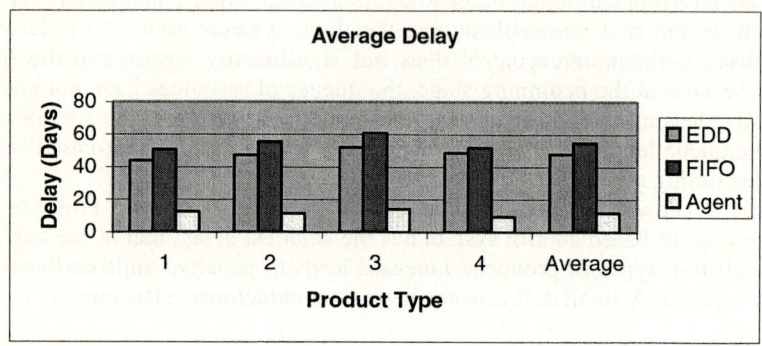

Fig. 9. Average delay in terms of entity type

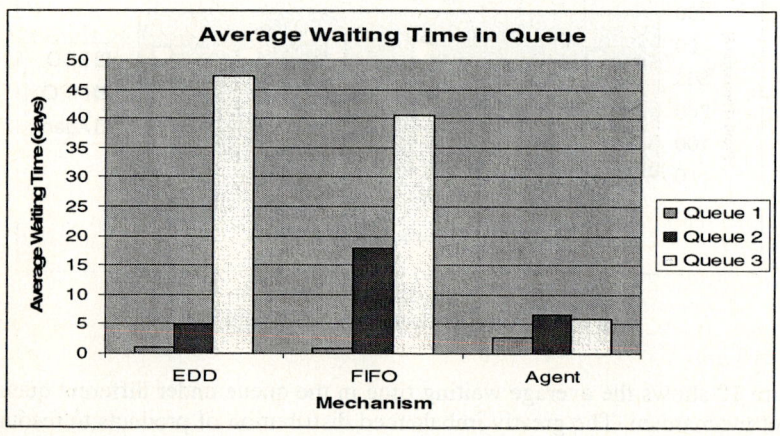

Fig. 10. Average waiting time in queue for different queue management mechanism

References

1. Babayan, Astghik and He, David, Solving the n-job 3-stage Flexible Flowshop Scheduling Problem Using an Agent-based Approach INT. J. PROD. RES., 2004, VOL. 42, NO. 4, 777–799

2. Baker, A.D. and Merchant, M. E., 1993, Automatic factories how will they be controlled? IEEM Potentials, 12:15–20
3. Chen, Y. Y. , Fu, L. C. and Chen, Y. C., 1998, Multi-agent Based Dynamic Scheduling for a Flexible Assembly System. Proceedings of the 1998 IEEE International Conference on Robotics & Automation.
4. Darbha, S. and Agraval, D. P., 1998, Optimal scheduling algorithm for distributed-memory machines. IEEE Transactions on Parallel and Distributed Systems, 9(1).
5. Finkenzeller Klaus RFID-Handbook, 2nd edition (April 2003) "Fundamentals and Applications in Contactless Smart Cards and Identification" Wiley & Sons LTD ISBN: 0-470-84402-7
6. Kotha, S., 1996, Mass-customization: a strategy for knowledge creation and organizational learning. Int. J. of Technology Management, 11(7-8): 846–858
7. Maley, J. G. 1988, Managing the Flow of Intelligent Parts, Robotics and Computer Integrated Manufacturing, 4(3/4):525–530.
8. Nwana Hyacinth S. Software Agents: An Overview Knowledge Engineering Review, Vol. 11, No 3, pp.1–40, Sept 1996. © Cambridge University Press, 1996
9. Parunak, H. V. D. 1987, Manufacturing Experience with the Contract Net, Distributed Artificial Intelligence, Michael M. Huhns (eds), Pitman, London, pp. 285–310.
10. Shaw, M. J. 1988, Dynamic Scheduling in Cellular Manufacturing Systems: a framework for networked decision making, Journal of Manufacturing Systems, 7(2):83–94.
11. Smith, R. G. and Davis, R.., 1981, Framework for co-operation in distributed problem solving, IEEE Transaction on System, Man and Cybernetics, SMC-11(1):61–70.
12. Tseng, M. M. , Lei, M. , Su, C., 1997, A Collaborative Control System for Mass Customization Manufacturing. Annals of the CIRP, Vol. 46 (1).
13. Kelton, W. David, Sadowski, Randall P., Sadowski; Deborah A. Simulation with Area; pg 7
14. Zhou, Z. D. , Wang, H. H. , Chen, Y. P. , Ai, W. , Ong, S. K. , Fuh, J. Y. H. and Nee, A. Y. C., 2003, A Multi-Agent-Based Agile Scheduling Model for a Virtual Manufacturing Environment. International Journal of Advanced Manufacturing Technology, 21:980–984.
15. Zipkin Paul, MIT Sloan Management Review; Spring 2001; 42, 3; ABI/INFORM Global pg. 81

Event Chain Clocks for Performance Debugging in Parallel and Distributed Systems*

Hongliang Yu, Jian Liu, Weimin Zheng, and Meiming Shen

Department of Computer Science and Technology,
Tsinghua University, Beijing, 100084, China
{hlyu, liujian98, zwm-dcs, smm-dcs}@tsinghua.edu.cn

Abstract. In this paper, the Event Chain Clock synchronization algorithm is presented. This algorithm can maintain a global physical clock that reflects both the partial order and the elapsed time of all events occurred. This algorithm, which repeats some basic operations, has good astringency, and is suitable for parallel program performance debugging.

1 Introduction

Distributed computing is the simultaneous use of multiple computer resources to solve a computational problem. Distributed computing can be implemented in parallel and distributed systems. Performance is always the thing we should mention in distributed computing, it can suffer from load balance, communication overhead, and synchronization loss, etc.

A distributed system may consist of a group of servers or workstations that communicate with each other by sending and receiving messages. There doesn't exists a global clock in these systems. Many tools, such as fault-tolerant tools and debugging tools[1], require that synchronized clocks[2] be available to have approximately the same view of time. So, we can order distributed events and accurately measure the duration time of two events.

Physical clock is the actual time of a processor or computer. When a process needs to record the timestamp of an event, physical clock can be used. But in distributed systems, there is no global unique physical clock. Every node has its own local physical clock. So, logical clock[3][4] was developed.

Logical clock doesn't indicate the real time ticks. It is just a number which reflects the order of corresponding events[5][6]. Events in a distributed system use this number as their timestamps. Events can be ordered by these timestamps correctly. Other clock synchronization algorithms[7][8] were also developed for various clock usage.

In this paper, we present a clock synchronization algorithm. This algorithm can maintain a global physical clock that reflects both the partial order and the

---

* Supported by the National High-Tech Research and Development Plan of China under Grant No.2002AA1Z2103 and Grant No.2003AA104021.

elapsed time of all events occurred. This algorithm, which repeats some basic operations, has good astringency, and is suitable for parallel program performance debugging.

2 Event Chain Clocks

There is no global unique physical clock in distributed systems. But it is quite possible to find a physical clock which have some characteristics of a global unique physical clock. As we know, in parallel debugging, we always care about the elapsed time in some events. So, if we could judge both the event durations and the event orders according to this clock, things will be going well.

Definition 1. *Event Chain Clock is a simulated physical clock which reflect the partial order relations among events correctly.*

As a physical clock, Event Chain Clock reflect both the execution time length of an event, and the partial order relations of all events, it is very useful for parallel program performance debugging.

3 Clock Synchronization Algorithm

3.1 Definitions

Considering a message passing event, there should be one message sending event a from process i and one message receiving event b from process j (assume $i \neq j$). We define function t_i to represent the local clock of process i. There is some drift between the global time T and the local time t_i, we can express it as: $t_i(e) = \tau_i + \alpha_i T(e)$ where τ_i is the initial drift between T and t_i, and α_i is the drift rate between T and t_i.

For most modern hardware clocks the rate α_i should be a number that is very closer to 1. Furthermore, in parallel debugging, we only care about the clock ticks of a period of time, but the execution time of a process is always not long enough, so, the effect of α_i in that small period is always small enough, we will ignore it and assume α_i is equal to 1. Then, the time delay in this message passing should be counted as below:

$$d_{ij} = t_j(b) - t_i(a) = (\tau_j - \tau_i) + (T(b) - T(a))$$

If $d_{ij} > 0$, this message passing event will have the same partial order with the clocks, that is: $a \rightarrow b \Leftrightarrow t_i(a) < t_i(b)$. If this is true for all events, we get the Event Chain Clock.

Definition 2. M_{ij} *is the set composed of all message passing events between process numbered i and j. We assume: $md_{ij} = min(t_j(b) - t_i(a), (a, b \in M_{ij}))$. We calculate md_{ij} for each two different processes in the system, and get a matrix noted as D: $D[i,j] = md_{ij}$. We define D as the minimum transmit-delay matrix.*

Furthermore, if the system contains n processes, D will be a $n \times n$ dimension matrix, the diagonal of D can be counted as a large natural number. Obviously, if we can find a minimum transmit-delay matrix with each element larger than zero, we can get the Event Chain Clock.

Supposing that we want to speed the local clock of process k with δ(that is: $t_k = t_k + \delta$), we will have a new minimum transmit-delay matrix D':

$$D'(i,k) = D(i,k) + \delta;$$
$$D'(k,i) = D(k,i) - \delta;$$
$$i \in [1,n]; \quad \text{n is the process number in the system}$$

That's the basic adjustment operation which will be used below, we must know δ for adjustment.

Definition 3. *We define H_k as an operation on minimum transmit-delay matrix D explained as below. We note $D' = H_k(D), d_{ik} = D(i,k), d'_{ik} = D'(i,k)$.*

$$d'_{ik} = d_{ik} + X_k;$$
$$d'_{ki} = d_{ki} - X_k;$$
$$X_k = \begin{cases} |\min(d_{ik})|, & \min(d_{ik}) < 0 \\ 0 \end{cases} \quad i \neq k$$

Furthermore, we have some labels below:

$$H(D) = H_n(H_{n-1}((H_2(H_1(D))))), \text{n is the process number in the system;}$$
$$H^{(2)}(D) = H(H(D));$$
$$H^{(p)}(D) = H(H^{(p-1)}(D));$$
$$H^{(p,q)}(D) = H_q(H_{q-1}((H_2(H_1^{(p)}(D))))$$

We can see that $H^{(p,q)}(D)$ actually operates $p \times (n+q)$ times on matrix D. Elements in $H^{(p,q)}(D)$ is noted as $d_{ik}^{(p,q)}$.

3.2 The Synchronization Algorithm

Our clock synchronization algorithm is shown as below:
1. Get the minimum transmit-delay matrix D
2. Calculate the matrix D', where $D' = H(D)$
3. Repeat procedure 2 until all element in new matrix is not negative, still note the new matrix as D.
4. Find a row k in D, where $d_{kj} > 0, \forall j[1,n]$, and there exist a j, $d_{jk} = 0$
5. Let $\delta = min(d_{kj}, j = 1, 2, \ldots, n)/2$, adjust the matrix D using δ:

$$d_{kj} = d_{kj} - \delta, \forall j \in [1,n]$$
$$d_{jk} = d_{jk} + \delta, \forall j \in [1,n]$$

6. Repeat procedure 4 to 5, until all elements in the matrix is positive.
7. Adjust all local clocks according to the matrix D.

The whole synchronization is based on some simple operations. We have proved the astringency of this algorithm, but we can not show it here for the space limit.

3.3 Sample Result

To simple the problem, here we take a parallel program that has three processes as the example of our clock synchronization algorithm.

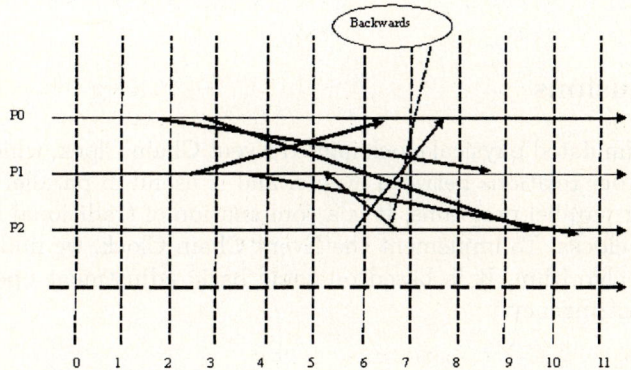

Fig. 1. Space-Time Diagram before Adjustment

First, after executing the parallel program, we can easily obtain the corresponding minimum transmit-delay matrix D. Technically, we add another row of zero at the bottom of the matrix. And the corresponding space-time diagram is showed in Figure 1.

N	6.508	6.804
3.926	N	6.853
1.559	-1.650	N
0	0	0

After doing the $H(D)$ operation, we get a new matrix:

N	8.158	6.804
2.276	N	5.203
1.559	0.000	N
0	1.650	0

We can find that there is still one zero element in the Matrix (neglecting the last row). So, do the second step of adjustment operation. Here $\delta =$

$min(6.804, 5.203)/2 = 2.602$, add 2.602 to every element in the third row, and decrease 2.602 from every element in the third column, then we get the final matrix:

N	8.158	4.202
2.276	N	2.602
4.161	2.602	N
0	1.650	-2.602

The last row of this matrix tells us that the timestamps in process 0 maintain unchanged; every timestamp in process 1 will be added by 1.650; and every timestamp in process 2 will be decreased by 2.602.

4   Conclusions

We define a simulated physical clock named Event Chain Clock, which can reflect the partial order relations between events, and is useful in parallel performance debugging for parallel programs. It is a combination of traditional logical clocks and physical clocks. To implement the Event Chain Clock, we find a clock synchronization algorithm. It is based on some basic adjustment operations, and has good constringency.

References

1. J. Chassin de Kergommeaux, B. de Oliveira Stein , "Flexible performance visualization of parallel and distributed applications", Future Generation Computer Systems, Volume: 19 Issue: 5, July 2003
2. Johannessen S., Time synchronization in a local area network, IEEE Control Systems Magazine, Vol. 24, Issue 2, Apr 2004, 62-69
3. Leslie Lamport. Time, clocks and the ordering of events in distributed systems. Communications of the ACM, 21(7):558-565, 1978.
4. Laurence Duchien, Gerard Florin, Lionel Seinturier. Partial order relations in distributed object environments, ACM SIGOPS Operating Systems Review (October 2000), Volume 34, Issue 4
5. David J. Taylor and Michael H. Coffin, Integrating Real-Time and Partial-Order Information in Event-Data Displays, CASCON'94
6. Hofmann R., Hilgers U., Theory and Tool for Estimating Global Time in Parallel and Distributed Systems, Proceeding of the Sixth Euromicro Workshop on Parallel and Distributed Processing PDP98, Los Alamitos: IEEE Computer Society, 1998; 173-179.
7. F. Cristian. Integrating External and Internal Clock Synchronization, Real-Time Systems, Volume 12, Number 2, 123-171 (1997)
8. Ted Herman, Phase Clocks for Transient Fault Repair, IEEE Transactions on Parallel and Distributed Systems, Vol. 11, No. 10, Oct 2000, 1048-1057

Author Index

Ahn, Seongjin 725
Akon, Mohammad Mursalin 985
Ali, Rizwan 1017
Arenaz, Manuel 4
Au, Puion 59

Bacigalupo, David 635
Bagarinao, E. 290
Bai, Yingcai 835
Baker, Mark 604
Beaumont, A.J. 85
Bernard, Thibault 146
Berten, Vandy 367
Bui, Alain 146

Cai, Wentong 274, 625
Cao, Jiannong 75, 340, 568
Cao, Xiaolin 757
Cao, Zhenfu 715
Cecchet, Emmanuel 115
Celebioglu, Onur 1017
Chadalavada, Kalyana 1017
Chan, Alvin T.S. 529
Chan, Edward 169
Chan, Keith C.C. 568
Chang, Chih-Hung 1007
Chang, Ruay-Shiung 584
Chang, Zhi-ming 735, 1024
Che, Yonggang 91
Chen, Chi-Hsiu 268
Chen, Ching-Wen 1007
Chen, Daoxu 75
Chen, Guihai 357
Chen, Hansheng 835
Chen, Hao 330
Chen, Huaping 489
Chen, Huo-wang 433
Chen, Jianer 698
Chen, Kefei 705
Chen, Mingyu 777
Chen, Ningjing 451
Chen, Po-Hung 584
Chen, Wei 825
Chen, Xiaolin 75

Chen, Xin 489
Chen, Yingwen 534
Chen, Yu 947
Chen, YueQuan 357
Chen, Yung-Chiao 804
Cheng, Shudong 544
Cheng, Yun 975
Cheung, L. 136
Chiu, Yung-Chang 965
Chou, C.-F. 136
Chou, Jue-Sam 858
Chuang, Siu-Nam 529
Chung, Jinwook 725
Ciciani, Bruno 903

Dai, Han 75
Deng, Dafu 330
De Rose, César 392
Dillon, Tharam S. 410, 783
Ding, Jianguo 835
Doallo, Ramón 4
Dongarra, Jack 1
Donghua, Liu 484
Dou, Wenhua 178

Fan, Jianping 777
Francis, Lau 280
Feng, Boqin 888
Feng, Yuhong 274
Ferreto, Tiago 392
Flauzac, Olivier 146
Fu, Haohuan 59
Fu, Jung-Sheng 105
Fu, Yingjie 59

Gan, Phui-Si 1007
Gao, Bo 352
Gao, Qing 156
Ge, He 484
Ge, Jinming 666
Goh, Jen Ye 54
Golubchik, L. 136
Gong, Zhenghu 463
Goossens, Joël 367

Goscinski, Andrzej M.   648
Goswami, Dhrubajyoti   126, 985
Guidec, Frédéric   44
Guo, Minyi   893
Guo, XiaoFeng   357

Haddad, Ibrahim   217
Hagihara, Kenichi   245
Han, Bo   519
Han, Song   169
Harwood, Aaron   233
He, Jing   746
He, Ligang   635
He, Yanxiang   825, 875
Hickernell, Fred J.   257
Hommel, Günter   64
Horiguchi, Susumu   893
Hsiao, Hung-Chang   604
Hsieh, Jenwei   1017
Hsu, Ching-Hsien   268
Huang, Joshua Zhexue   499
Huang, Min   866
Huang, Tao   451
Huang, Weili   314

Ino, Fumihiko   245
Inoguchi, Yasushi   578

Jansson, Carl Gustaf   509
Jarvis, Stephen A.   635
Jayaputera, James   49
Jeong, Hong   263
Jeong, Young-Sik   382
Jia, Weijia   519
Jia, Yan   845
Jiang, Xiaohong   893
Jiang, Xuxian   937
Jiang, Yan-huang   735, 1024
Jin, Beihong   451
Jin, Hai   330
Johnson, David B.   3
Jonsson, Martin   509

Kaneko, Keiichi   556
Karunasekera, Shanika   233
Kilander, Fredrik   509
Kim, Hyuncheol   725
Kim, Jihong   952
Kim, Jongkyung   725
Kim, Shin-Dug   952

King, Chung-Ta   604
Knoke, Michael   64
Kobayashi, Hiroaki   16
Kochhar, Garima   1017
Koh, Kern   952
Krämer, Bernd   835
Kriesell, Matthias   914
Kühling, Felix   64
Kuo, Sy-Yen   804
Kwon, Ohyoung   469
Kwong, Kin Wah   319

Lai, Andy S.Y.   85
Lam, Kam-Yiu   188
Lau, Francis C.M.   280, 688
Le, Vinh Trong   893
Leangsuksun, Chokchai   217
Lee, Bu-Sung   625
Lee, Jang-Soo   952
Lee, Sungyoung   677
Lee, Ting-Ying   858
Li, Chunjiang   594
Li, Haiyan   28
Li, Hon F.   126, 985
Li, Huaizhong   656
Li, Juan   446
Li, Li   28
Li, Minglu   474, 499
Li, Mingshu   446
Li, Shanping   156
Li, Shiqun   705
Li, Wei   509
Li, Wenjie   550
Li, Wenlong   947
Li, Xiangxue   705
Li, Xiaomei   91
Li, Xiaoming   615
Lian, Chiu Kuo   268
Liang, BiYu   188
Liang, Tyng-Yeu   965
Liao, Xiaofei   330
Libby, Richard   217
Lin, Chen   314
Lin, Chu-Hsing   858
Lin, Haibo   947
Lin, Jenn-Wei   815
Lin, Wilfred W.K.   783
Lin, Yih-Fang   997
Liu, Bin   550
Liu, Cong   866

Liu, Hui   340, 474
Liu, Hung-Yu   815
Liu, Jian   1050
Liu, Kwong-Ip   257
Liu, Michael R.   1039
Liu, Min   730
Liu, Tong   217
Liu, Yudan   217
Lou, Wei   223
Lu, Jianzhuang   975
Lu, Xicheng   399
Lu, Yonggang   746
Luo, Han   421
Luo, Jun   568
Luo, Junzhou   372
Luo, Zongwei   499

Ma, Fanyuan   544
Ma, Huiye   352
Ma, Teng   372
Matsui, Manabu   245
Matsuo, K.   290
Mo, Zeyao   757

Nakai, T.   290
Nejdl, Wolfgang   914
Ngo, Son Hong   893
Nguyen, Ngoc Chi   677
Ni, Linoel M.   2, 1039
Northfleet, Caio   392
Nudd, Graham R.   635

Pan, Yi   340
Pang, Henry C.W.   188
Park, Hyoungwoo   469
Park, Kumrye   469
Park, Sungchan   263
Park, Sungyong   469
Peng, Gang   156
Pontelli, Enrico   746
Premaratne, Malin   233

Qu, Changtao   914
Qu, Xiang-li   1024
Quaglia, Francesco   903

Radhakrishnan, Ramesh   1017
Rajagopalan, Ramesh   1017
Romano, Paolo   903
Rough, Justin   798

Roussain, Hervé   44

Sarbazi-Azad, Hamid   1030
Sarmenta, L.   290
Scott, Stephen L.   217
Shan, Jiulong   489
Shen, Hong   578
Shen, Ji   519
Shen, Meiming   1050
Shi, Dian-xi   845
Shieh, Ce-Kuen   965
Shieh, Min-Shao   804
Shu, Jiwu   200, 421, 765
Singh, Ajit   788
Son, Sang H.   188
Spooner, Daniel P.   635
Su, Alvin Wen-Yu   965
Su, Jinshu   399
Sue, Chuan-Ching   804
Sun, Guangzhong   489
Sundararajan, Elankovan   233

Tajozzakerin, Hamid Reza   1030
Takizawa, Hiroyki   16
Tanaka, Y.   290
Tang, Feilong   499
Tang, Junjun   544
Tang, Zhizhong   947
Taniar, David   49, 54
Teng, Meng   845
Touriño, Juan   4
Tsang, Danny H.K.   319
Tseng, Mitchell M.   1039
Tsujita, Yuichi   34

Vo, Nhat Minh Dinh   677

Wang, Bing   200
Wang, Chen   387
Wang, Chien-Min   997
Wang, Chih-Min   584
Wang, Cho-Li   280, 499
Wang, Dongsheng   212
Wang, Guojun   568, 698
Wang, Huai-min   845
Wang, Ji   433
Wang, Jing-Xin   965
Wang, Lechun   463
Wang, Lin   126
Wang, Qing   446

Wang, Tan   788
Wang, Tianqi   280
Wang, Xingwei   866
Wang, Xuehui   178
Wang, Zhenghua   91
Wang, Zhiying   975
Wang, Ziqiang   888
Wei, Zunce   126
Wen, Mei   28
Wong, Adam K.L.   648
Wong, Allan K.Y.   410, 783
Wong, Peter Y.H.   529
Wu, Dan   975
Wu, Jan-Jan   997
Wu, Jie   223
Wu, Nan   28
Wu, Richard S.L.   410
Wu, Zhanchun   446

Xavier, Percival   625
Xiao, Bin   825
Xiao, Jitian   656
Xiao, Nong   594
Xiao, Xiaolin   698
Xiong, Naixue   875
Xu, Chao   926
Xu, Cheng-Zhong   382
Xu, Dengyuan   926
Xu, Dongyan   937
Xu, Jian   156
Xu, Jianliang   156
Xu, Ming   534
Xu, Shihao   835
Xue, Qingshui   715
Xue, Wei   765

Yan, Jiong   433
Yang, Bo   451
Yang, Laurence T.   91
Yang, Xue-jun   735, 1024
Yang, Xuejun   594

Yang, Yan   136, 875
Yang, Yuhang   352
Yang, Zhe   730
Ye, Xinfeng   303
Yim, Keun Soo   952
Yin, Gang   845
Yu, Chang Wu   268
Yu, Hongliang   1050
Yu, Kun-Ming   268
Yu, Shui   798
Yuen, Man-Ching   519
Yuzhong, Sun   484

Zeng, Huaxin   926
Zeng, QingKai   357
Zhang, Chao   330
Zhang, Chunyuan   28, 975
Zhang, Dalu   314, 730
Zhang, Jinyu   615
Zhang, Jun   835
Zhang, Lei   178
Zhang, Liang   544
Zhang, Longjun   705
Zhang, Q.L.   1039
Zhang, Wenbo   451
Zhang, Wenli   777
Zhang, Youhui   212
Zhang, Yuanyuan   578
Zhang, Zhijiao   534
Zheng, Jing   399
Zheng, Weimin   200, 212, 765, 1050
Zhiwei, Xu   484
Zhou, Bing Bing   387
Zhou, Jingyang   75
Zhou, Jipeng   688
Zhou, Wanlei   798
Zhu, Peidong   463
Zhu, Ye   372
Zimmermann, Armin   64
Zomaya, Albert Y.   387
Zou, Futai   544